BELIZE
HONDURAS
Tegucigalpa
NICARAGUA
Managua
San José
COSTA RICA
PANAMA
Panama
Lago de Nicaragua

ST. LUCIA
ST. VINCENT AND THE GRENADINES
GRENADA
BARBADOS
TRINIDAD AND TOBAGO

Caribbean Sea

Barranquilla
Maracaibo
Caracas
San Cristóbal
VENEZUELA
Orinoco
Ciudad Guayana
Georgetown
GUYANA
Paramaribo
Cayenne
SURINAME
French Guiana
(FRANCE)

Medellín
Bogotá
Cali
COLOMBIA
Mitú

*Malpelo I.
(COLOMBIA)*

Negro
Macapá

Equator

Quito
ECUADOR
Guayaquil
Iquitos

Amazon
Manaus
Santarém

Amazon
Belém
São Luís
Teresina
Fortaleza

Trujillo

PERU

Lima
Ica
Cusco
Arequipa
Arica

Pôrto Velho
Rio Branco

BRAZIL

Natal
Recife

Marañon
Ucayali
Tocantins

BOLIVIA
Lago Titicaca
La Paz
Cochabamba
Santa Cruz
Sucre
Trinidad

Cuiabá
Goiânia
Brasília

São Francisco

Juàzeiro
Aracajú
Salvador

PACIFIC OCEAN

*Isla San Félix
(CHILE)*
*Isla San Ambrosio
(CHILE)*

Antofagasta

CHILE

Xingu

PARAGUAY
Asunción

Belo Horizonte
Vitória

São Paulo
Rio de Janeiro
Curitiba
Tropic of Capricorn

San Miguel
de Tucumán
Resistencia

Paraná

Florianópolis

*Isla Juan Fernández
(CHILE)*

Córdoba

Valparaíso
Santiago
Mendoza
Rosario
Paraná
Salto
URUGUAY

Pôrto Alegre

Concepción
ARGENTINA
Buenos Aires
Montevideo

Bahía Blanca
Mar del Plata

Puerto Montt

Rawson

ATLANTIC OCEAN

Comodoro Rivadavia

ATLANTIC OCEAN

Estrecho de Magallanes
Stanley
Falkland Islands (UK)

Punta Arenas
Ushuaia

South Georgia (UK)

SOUTH AMERICA

0   250   500 Miles

0   250   500 Kilometers

CARIBBEAN inset

*British Virgin Is.
(UK)*
Anguilla (UK)
St. Martin (FR and NETH)
St. Barthélemy
ANTIGUA AND BARBUDA
St. John's
Guadeloupe
(FR)
Basse-Terre
Virgin Is. (US)
Rico
Netherlands
Antilles (NETH)
Montserrat
(UK)
ST. KITTS
AND NEVIS
Basseterre
DOMINICA
Roseau
Martinique
(FR)
Fort-de-France
ST. LUCIA
Castries
Bridgetown
Kingston
BARBADOS
ST. VINCENT AND
THE GRENADINES
GRENADA
St. George's
Caribbean Sea

Port-of-Spain
TRINIDAD
AND
TOBAGO

CARIBBEAN

50   100 Miles

50   100 Kilometers

WORLDMARK ENCYCLOPEDIA OF THE NATIONS

Volume 3

WORLDMARK
ENCYCLOPEDIA OF THE NATIONS

AMERICAS

Formerly published by Worldmark Press, Ltd.

An International Thomson Publishing Company

NEW YORK • LONDON • BONN • BOSTON • DETROIT • MADRID
MELBOURNE • MEXICO CITY • PARIS • SINGAPORE • TOKYO
TORONTO • WASHINGTON • ALBANY NY • BELMONT CA • CINCINNATI OH

Gale Research Inc. Staff

Jane Hoehner, *Developmental Editor*
Allison McNeill, Rebecca Nelson, and Kelle S. Sisung, *Contributing Developmental Editors*
Marie Ellavich, Jolen Gedridge, and Camille Killens, *Associate Developmental Editors*
Lawrence W. Baker, *Senior Developmental Editor*

Mary Beth Trimper, *Production Director*
Evi Seoud, *Assistant Production Manager*
Mary Kelley, *Production Associate*

Cynthia Baldwin, *Product Design Manager*
Barbara J. Yarrow, *Graphic Services Supervisor*
Todd Nessell, *Macintosh Artist*

Library of Congress Cataloging-in-Publication Data

Worldmark encyclopedia of the nations. — 8th ed.
    5 v.
    Includes bibliographical references and index.
    Contents: v. 1. United Nations — v. 2. Africa — v. 3. Americas —
v. 4. Asia & Oceania — v. 5. Europe.
    ISBN 0-8103-9878-8 (set). — ISBN 0-8103-9893-1 (v. 1). — ISBN
0-8103-9880-x (v. 2)
    1. Geography—Encyclopedias. 2. History—Encyclopedias.
3. Economics—Encyclopedias. 4. Political science—Encyclopedias.
5. United Nations—Encyclopedias.
G63.W67   1995
903—dc20   94–38556
                                                           CIP

This book is printed on acid-free paper that meets the minimum requirements of American National Standard for Information Sciences—Permanence Paper for Printed Library Materials, ANSI Z39.48-1984. ∞™

ISBN 0-8103-9881-8

Printed in the United States of America by Gale Research Inc.
Published simultaneously in the United Kingdom
by Gale Research International Limited
(An affiliated company of Gale Research Inc.)

10 9 8 7 6 5 4 3 2 1

The trademark **ITP** is used under license.

CONTENTS

For Conversion Tables, Abbreviations and Acronyms, Glossaries, World Tables, Notes to the Eighth Edition, and other supplementary materials, see Volume 1.

Antigua and Barbuda 1

Argentina 7

Bahamas 25

Barbados 31

Belize 37

Bolivia 43

Brazil 55

Canada 73

Chile 93

Colombia 107

Costa Rica 121

Cuba 131

Dominica 141

Dominican Republic 147

Ecuador 157

El Salvador 169

French American Dependencies 179

Grenada 181

Guatemala 187

Guyana 197

Haiti 205

Honduras 215

Jamaica 225

Mexico 235

Netherlands American Dependencies 251

Nicaragua 253

Panama 263

Paraguay 273

Peru 283

St. Kitts and Nevis 299

St. Lucia 305

St. Vincent and the Grenadines 311

Suriname 317

Trinidad and Tobago 323

Turks and Caicos Islands 331

United Kingdom American Dependencies 335

United States of America......................................339

Uruguay 375

Venezuela 385

Index to Countries..399

GUIDE TO COUNTRY ARTICLES

All information contained within a country article is uniformly keyed by means of small superior numerals to the left of the subject headings. A heading such as "Population," for example, carries the same key numeral (6) in every article. Thus, to find information about the population of Albania, consult the table of contents for the page number where the Albania article begins and look for section 6 thereunder. Introductory matter for each nation includes coat of arms, capital, flag (descriptions given from hoist to fly or from top to bottom), anthem, monetary unit, weights and measures, holidays, and time zone.

FLAG COLOR SYMBOLS

| Yellow | Red | Green | Blue | Orange | Brown | White | Black |

SECTION HEADINGS IN NUMERICAL ORDER

1 Location, size, and extent
2 Topography
3 Climate
4 Flora and fauna
5 Environment
6 Population
7 Migration
8 Ethnic groups
9 Languages
10 Religions
11 Transportation
12 History
13 Government
14 Political parties
15 Local government
16 Judicial system
17 Armed forces
18 International cooperation
19 Economy
20 Income
21 Labor
22 Agriculture
23 Animal husbandry
24 Fishing
25 Forestry
26 Mining
27 Energy and power
28 Industry
29 Science and technology
30 Domestic trade
31 Foreign trade
32 Balance of payments
33 Banking and securities
34 Insurance
35 Public finance
36 Taxation
37 Customs and duties
38 Foreign investment
39 Economic development
40 Social development
41 Health
42 Housing
43 Education
44 Libraries and museums
45 Media
46 Organizations
47 Tourism, travel, and recreation
48 Famous persons
49 Dependencies
50 Bibliography

SECTION HEADINGS IN ALPHABETICAL ORDER

Agriculture 22
Animal husbandry 23
Armed forces 17
Balance of payments 32
Banking and securities 33
Bibliography 50
Climate 3
Customs and duties 37
Dependencies 49
Domestic trade 30
Economic development 39
Economy 19
Education 43
Energy and power 27
Environment 5
Ethnic groups 8
Famous persons 48
Fishing 24
Flora and fauna 4
Foreign investment 38
Foreign trade 31
Forestry 25
Government 13
Health 41
History 12
Housing 42

Income 20
Industry 28
Insurance 34
International cooperation 18
Judical system 16
Labor 21
Languages 9
Libraries and museums 44
Local government 15
Location, size, and extent 1
Media 45
Migration 7
Mining 26
Organizations 46
Political parties 14
Population 6
Public finance 35
Religions 10
Science and technology 29
Social development 40
Taxation 36
Topography 2
Tourism, travel, and recreation 47
Transportation 11

FREQUENTLY USED ABBREVIATIONS AND ACRONYMS

ad—Anno Domini
am—before noon
b.—born
bc—Before Christ
c—Celsius
c.—circa (about)
cm—centimeter(s)
Co.—company
Corp.—corporation
cu ft—cubic foot, feet
cu m—cubic meter(s)
d.—died
e—east
e—evening
e.g.—exempli gratia (for example)
ed.—edition, editor
est.—estimated
et al.—et alii (and others)

etc.—et cetera (and so on)
f—Fahrenheit
fl.—flourished
FRG—Federal Republic of Germany
ft—foot, feet
ft^3—cubic foot, feet
GATT—General Agreement on Tariffs and Trade
GDP—gross domestic products
gm—gram
GMT—Greenwich Mean Time
GNP—gross national product
GRT—gross registered tons
ha—hectares
i.e.—id est (that is)
in—inch(es)
kg—kilogram(s)
km—kilometer(s)

kw—kilowatt(s)
kwh—kilowatt-hour(s)
lb—pound(s)
m—meter(s); morning
m^3—cubic meter(s)
mi—mile(s)
Mt.—mount
Mw—megawatt(s)
n—north
n.d.—no date
NA—not available
oz—ounce(s)

pm—after noon
r.—reigned
rev. ed.—revised edition
s—south
sq—square
St.—saint
UK—United Kingdom
UN—United Nations
US—United States
USSR—Union of Soviet Socialist Republics
w—west

A fiscal split year is indicated by a stroke (e.g. 1987/88).
For acronyms of UN agencies and ther intergovernmental organizations, as well as other abbreviations used in text, see the United Nations volume.
A dollar sign ($) stands for us$ unless otherwise indicated.
Note that 1 billion = 1,000 million = 10^9.

ANTIGUA AND BARBUDA

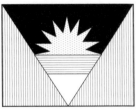

CAPITAL: St. John's.

FLAG: Centered on a red ground is an inverted triangle divided horizontally into three bands of black, light blue, and white, the black stripe bearing a symbol of the rising sun in yellow.

ANTHEM: Begins "Fair Antigua and Barbuda, I salute thee."

MONETARY UNIT: The East Caribbean dollar (EC$) is a paper currency of 100 cents, pegged to the US dollar. There are coins of 1, 2, 5, 10, 25 cents and 1 dollar, and notes of 5, 10, 20, and 100 dollars. EC$1 = US$0.3704 (or US$1 = EC$2.70).

WEIGHTS AND MEASURES: Imperial measures are used, but the metric system is being introduced.

HOLIDAYS: New Year's Day, 1 January; Labor Day, 1st Monday in May; CARICOM Day, 3 July; State Day, 1 November; Christmas, 25 December; Boxing Day, 26 December. Movable holidays include Good Friday, Easter Monday, and Whitmonday.

TIME: 8 AM = noon GMT.

¹LOCATION, SIZE, AND EXTENT

The state of Antigua and Barbuda, part of the Leeward Islands chain in the eastern Caribbean, is approximately 420 km (261 mi) SE of the US Commonwealth of Puerto Rico and 180 km (110 mi) N of the French overseas department of Guadeloupe. The total land area of 440 sq km (170 sq mi) includes Antigua (280 sq km/108 sq mi); Barbuda, 40 km (25 mi) to the N (161 sq km/62 sq mi); and uninhabited Redonda, 40 km (25 mi) to the SW (1.3 sq km/5 sq mi). This total area comprises slightly less than 2.5 times the size of Washington, D.C. The total coastline is 153 km (95 mi). Antigua and Barbuda's capital city, St. John's, is located on the northwestern edge of the island of Antigua.

²TOPOGRAPHY

Partly volcanic and partly coral in origin, Antigua has deeply indented shores lined by reefs and shoals; there are many natural harbors and beaches. Boggy Peak (405 m/1,329 ft), in southwestern Antigua, is the nation's highest point. Antigua's northeastern coastline is dotted by numerous tiny islets; the central area is a fertile plain. Barbuda, a coral island with a large harbor on the west side, rises to only 44 m (144 ft) at its highest point. Redonda is a low-lying rocky islet.

³CLIMATE

Temperatures average 24°C (75°F) in January and 29°C (84°F) in July, with cooling tradewinds from the east and northeast. Rainfall averages 117 cm (46 in) per year; September through November is the wettest period. The islands have been subject to periodic droughts and to autumn hurricanes.

⁴FLORA AND FAUNA

Most of the vegetation is scrub, but there is luxuriant tropical growth where fresh water is available. Many varieties of fruits, flowers, and vegetables are grown. Palmetto and seaside mangrove are indigenous, and about 1,600 hectares (4,000 acres) of red cedar, white cedar, mahogany, whitewood, and acacia forests have been planted. Barbuda is heavily wooded, with an abundance of deer, wild pigs, guinea fowl, pigeons, and wild ducks.

⁵ENVIRONMENT

Water management is a principal environmental concern on Antigua. Antigua and Barbuda have a shortage of water due to limited rainfall and drought. The existing water supply is threatened by pollution from distilleries, food processing facilities, and other industrial operations. The nation's energy demands, combined with agricultural development, have created the problem of deforestation. The loss of the forests contributes to erosion of the land. Rainfall, which is concentrated in a short season, quickly runs off. The nation's main city, St. John's, has developed a problem with waste disposal. Construction of a desalination plant in 1970 relieved some of the water shortage. The government of Antigua and Barbuda has created a Historical, Conservation, and Environmental Commission to address the nation's environmental problems.

⁶POPULATION

The population in May 1991 was 65,962 (the 1970 census figure was 65,525), of which about 98% resided on Antigua. The overall population density was 149 per sq km (386 per sq mi) in 1991, but the density on Antigua was 231 per sq km (597 per sq mi). St. John's, the capital, had an estimated population of 36,000 in 1986; Codrington is Barbuda's only settlement.

⁷MIGRATION

The UK has been the historic destination of Antiguan emigrants, but in recent years St. Martin, Barbados, the US Virgin Islands, and the US mainland have been the principal recipients of the outflow. The primary motive for emigration is the search for work.

⁸ETHNIC GROUPS

Antiguans are almost entirely of African descent. There are small numbers of persons of European, Arab, and Asian Indian ancestry.

⁹LANGUAGES

English is the official and commercial language. An English patois is in common use.

10RELIGIONS

St. John's, as capital, serves as the episcopal seat of both the Anglican and Roman Catholic churches. The Church of England claims almost 45% of the population, and other Protestant groups, including Baptist, Methodist, Pentecostal, Seventh-day Adventist, Moravian, and Nazarene, some 42%. Roman Catholics are in the minority at 8.7%.

11TRANSPORTATION

There are about 240 km (149 mi) of highways. Total motor vehicle registration was 17,000 in 1992. The railway consists of 78 km (48 mi) of narrow-gauge track, used mainly to haul sugarcane. The islands have no natural deepwater harbors; a deepwater facility was constructed at St. John's in 1968. The merchant fleet in 1991 consisted of 105 ships with a cargo capacity of 392,000 GRT. Vere Cornwall Bird International Airport, 7 km (4 mi) northeast of St. John's, accommodates the largest jet aircraft; Coolidge Airport, also on Antigua, handles freight. There is also a landing strip at Codrington. The islands had 80 commercially licensed pilots in 1992.

12HISTORY

The first inhabitants of Antigua and Barbuda were the Siboney, whose settlements date to 2400 BC. Arawak and Carib Indians inhabited the islands at the time of Christopher Columbus' second voyage in 1493. Columbus named Antigua after the church of Santa Maria de la Antigua, in Sevilla, Spain. Early settlements were founded in 1520 by the Spanish, in 1629 by the French, and in 1632 by the British. Antigua formally became a British colony in 1667, under the Treaty of Breda.

In 1674, Sir Christopher Codrington established the first large sugar estate in Antigua. He leased Barbuda to raise slaves and supplies for this enterprise. In 1834 slavery was abolished, but this was a mere technicality, since no support was provided for the new freemen. In 1860, Barbuda was formally annexed to the island of Antigua. The islands were governed under the Federation of the Leeward Islands from 1871 to 1956, and from 1958 to 1962 belonged to the Federation of the West Indies.

Antigua became an associated state with full internal self-government as of 27 February 1967. Opposition to complete independence came from the residents of Barbuda, who sought constitutional guarantees for autonomy in land, finances, and local conciliar powers. With these issues still not fully resolved, Antigua and Barbuda became an independent state within the Commonwealth of Nations on 1 November 1981. In May 1987, the prime ministers of the members of the Organization of Eastern Caribbean States (OECS) agreed on a merger proposal, creating a single nation out of their seven island states. The accord was subject to a national referendum in each of the states.

13GOVERNMENT

Universal adult suffrage on the islands dates from 1951, and ministerial government from 1956. The bicameral legislature was established in its present form in 1967, and formal independence was granted in November 1981. Under the constitution, the British monarch, as head of state, is represented in Antigua and Barbuda by a local governor-general who is appointed on the advice of the prime minister. The bicameral legislature consists of a 17-member House of Representatives, elected from single-member constituencies for up to five years by universal adult suffrage at age 18; and a 17-member Senate, appointed by the governor-general, of whom 11 (including at least 1 inhabitant of Barbuda) are named on the advice of the prime minister, 4 on the advice of the leader of the Opposition, 1 at the governor-general's discretion, and 1 on the advice of the Barbuda council. The prime minister, who must have the support of a majority of the House, is appointed by the governor-general, as is the cabinet.

14POLITICAL PARTIES

The Antigua Labour Party (ALP) has held power since 1946, except for a period from 1971 to 1976, when the Progressive Labour Movement (PLM), led at the time by George H. Walter, held a parliamentary majority. In the election of April 1987, the ALP won 16 of 17 seats in the House of Representatives, and an independent won the Barbuda seat. Other political groups include the Antigua Caribbean Liberation Movement, the United People's Movement, and the National Democratic Party, which was founded in 1985.

The ALP grew out of a trade union movement in the 1940s. Under colonial supervision and tutelage, the party became the dominant force in national politics. The party's preeminence continued after independence with electoral victories in 1984 and 1989.

15 LOCAL GOVERNMENT

The island of Antigua is divided into 6 parishes and 2 dependencies, Barbuda and Redonda. Local government affairs are conducted by 29 community councils, each with 9 members, 5 elected and 4 appointed.

16JUDICIAL SYSTEM

The legal system is based on English common law and local statutory law and is administered by the Eastern Caribbean Supreme Court, based in St. Lucia, which also provides a High Court and Court of Appeal. Final appeals may be made to the Queen's Privy council in the United Kingdom. A court of summary jurisdiction on Antigua, which sits without a jury, deals with civil cases involving sums of up to EC$1500; three magistrates' courts deal with summary offenses and civil cases of not more than EC$500 in value. The Industrial Court, for arbitration and settlement of trade disputes, was reintroduced in 1976. There is currently a plan to establish a regional supreme court with the six other Caribbean Community (CARICOM) member states (Montserrat, Anguilla, Dominica, Grenada, St. Kitts and Nevis, St. Lucia, and St. Vincent and the Grenadines), but a number of legal and constitutional obstacles will have to be overcome before this can be accomplished.

17ARMED FORCES

There is a Royal Antigua and Barbuda Defense Force of some 90 men that forms a part of the Eastern Caribbean Regional Security System. The Royal Antigua Police Force numbers about 250.

18INTERNATIONAL COOPERATION

Antigua and Barbuda joined the UN on 11 November 1981. It belongs to ECLAC, FAO, ICAO, ILO, IMF, UNESCO, and WHO. It is also a member of the Commonwealth of Nations, CARICOM, CDB, G-77, IDB, OECS, and OAS. The nation is a signatory of the Law of the Sea. Two areas totaling about 365 hectares (900 acres) are leased to the US for use as tracking stations for space vehicles and missiles. In May 1987, the prime ministers of Antigua and Barbuda and the six other OECS member-states voted to create a single nation; national referendums on the issue were to be held. The referendums were later defeated and the seven nations remained separate.

19ECONOMY

Throughout most of the islands' history, sugar and cotton production were by far the principal economic endeavors. Since the 1960s, however, tourism has been the main industry, accounting indirectly for some 60% of GDP, while the sugar industry has been insignificant and the annual cotton output has declined to a small fraction of the 1 million pounds formerly produced each year. Recent efforts to offset the economy's dependence on tourism have focused on rehabilitation of the sugar industry on a smaller scale to meet domestic needs, along with diversification of fruit and vegetable production. The West Indies Oil Refinery, which had closed in 1975, was reopened in 1982 as a government enterprise to process Venezuelan and Mexican oil for reexport to eastern Caribbean neighbors.

Agriculture accounts for only 5% of GDP. Sweet potatoes, tomatoes, coconuts, carrots, pineapples, pumpkins, cucumbers, and mangoes account for most agricultural production. On the other hand, tourism has continued to grow to the extent of becoming the largest tourism industry in the Windward and Leeward Islands, with about 480,000 visitors annually. It contributes approximately 60% to GDP and employs 50% of the labor force. Air service improved recently, with several direct European flights opening up. Also, there has been a steady expansion in hotels and tourist sites since 1990. In spite of this growth, the country hasn't reduced both its substantial foreign debt and large trade deficit. In 1992, the country registered its first economic growth in three years. GDP grew by 2.8%.

20INCOME

In 1992, Antigua and Barbuda's GNP was US$395 million at current prices, or $4,870 per capita. For the period 1985–92 the average inflation rate was 5.9%, resulting in a real growth rate in per capita GNP of 1.1%.

In 1992, the GDP was $433 million in current US dollars. It is estimated that in 1986, agriculture, hunting, forestry, and fishing contributed 4% to GDP; mining and quarrying, 2%; manufacturing, 3%; electricity, gas, and water, 3%; construction, 8%; wholesale and retail trade, 21%; transport, storage, and communication, 14%; finance, insurance, real estate, and business services, 13%; community, social, and personal services, 7%; and other sources, 25%.

21LABOR

The total labor force in 1991 was estimated at 20,500. About 82% of the employed labor force worked in occupations connected with tourism or other services; 7% in industry, and 11% in agriculture, hunting, forestry, and fishing.

Unemployment in 1991 officially stood at 5%, although actual unemployment levels among Antiguans are likely to be higher due to high numbers of work permits issued to foreigners. There are four trade unions with a total of about 20,500 members; the labor movement is well organized but politically divided. In April 1992, a general strike was called for political reasons rather than for an industrial dispute.

22AGRICULTURE

Some 30% of land on Antigua is under crops or potentially arable, with 18% in use. Sea-island cotton is a profitable export crop. A modest amount of sugar is harvested each year, and there are plans for production of ethanol from sugarcane. Vegetables, including beans, carrots, cabbage, cucumbers, plantains, squash, tomatoes, and yams, are grown mostly on small family plots for local markets. Over the past 30 years, agriculture's contribution to the GDP has fallen from over 40% to 12% (1991). The decline in the sugar industry left 60% of the country's 66,000 acres under government control, and the Ministry of Agriculture is encouraging self-sufficiency in certain foods in order to curtail the need to import food, which accounts for about 25% by value of all imports. Crops suffer from droughts and insect pests, and cotton and sugar plantings suffer from soil depletion and the unwillingness of the population to work in the fields. Agricultural exports, including mangoes, bananas, coconuts, and pineapples, were estimated at US$1.5 million in 1992. In 1992, an estimated 4,000 tons of sugar were produced from about 8,000 hectares (19,800 acres) harvested.

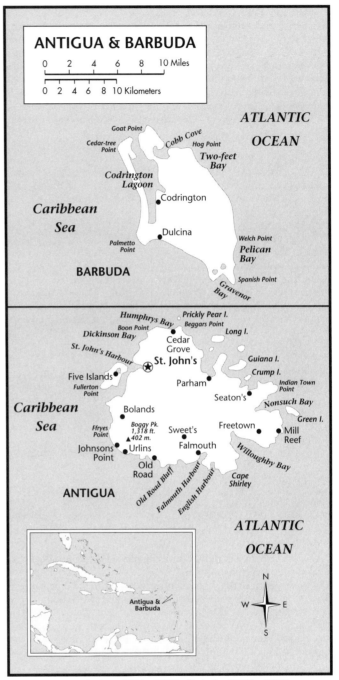

LOCATION: Antigua: 17°9′N; 61°49′w. Barbuda: 17°41′N; 61°48′w. **TOTAL COASTLINE:** 153 km (95 mi). **TERRITORIAL SEA LIMIT:** 12 mi.

23ANIMAL HUSBANDRY

Livestock estimates in 1992 counted 16,000 head of cattle, 13,000 sheep, and 12,000 goats; there were some 4,000 hogs in the same year. Most livestock is owned by individual households. Milk production in 1992 was an estimated 6,000 tons. The government has sought to increase grazing space and to improve stock, breeding Nelthropp cattle and Black Belly sheep. There is a growing poultry industry. In 1992, the European Development Bank provided US$5 million to the government to help develop the livestock industry.

24FISHING

Most fishing is for local consumption, although there is a growing export of the lobster catch to the US and of some fish to

Guadeloupe and Martinique. Antiguans annually consume more fish per capita (46 kg/101.4 lb) per year live weight than any other nation or territory in the Caribbean. The main fishing waters are near shore or between Antigua and Barbuda. There are shrimp and lobster farms operating, and the Smithsonian Institution has a Caribbean king crab farming facility for the local market. The government has encouraged modern fishing methods and supported mechanization and the building of new boats. Fish landings in 1991 were 2,300 tons; the lobster catch, 200 tons. Barbuda annually exports over 100 tons of lobster.

25FORESTRY
About 11% of the land is forested, mainly by plantings of red cedar, mahogany, white cedar, and acacia. A reforestation program was begun in 1963, linked with efforts to improve soil and water conservation.

26MINING
Little of the islands' mineral resources, which include limestone, building stone, clay, and barite, were exploited until recently. Some limestone and volcanic stone has been extracted for construction purposes, and the manufacture of bricks and tiles from local clay has begun on a small scale. Barbuda produces a small amount of salt, and phosphate has been collected from Redonda in the past.

27ENERGY AND POWER
Electric power produced in 1991 totaled 95 million kwh, based on a capacity of 26,000 kw. The Antigua Public Utilities Authority, run by the Ministry of Public Works and Communications, operates generating stations at Cassada Gardens and Crabbes Peninsula. Gas is now produced and refined locally. Offshore oil exploration took place during the early 1980s.

As part of the government's energy conservation program, incentives are offered for the manufacture and use of solar-energy units, and there are import surcharges on automobiles with engine capacities exceeding 2,000 cc. Under study as alternatives to fossil fuels are wind power, surplus bagasse from the sugar refinery, and fast-growing tree species.

28INDUSTRY
Industrial activity has shifted from the processing of local agricultural produce to consumer and export industries using imported raw materials. Industrial products include rum, refined petroleum, pottery, paints, garments, furniture, and electrical components. The Arawak Hustler Motor Car Co. assembles English-made parts for reexport. The government encourages investment in manufacturing establishments, and most industries have some government participation. The manufacturing sector has not changed much, consisting mainly of light, export-oriented assembly industries and accounting for 15% of GDP. It is expected to grow further with the opening of tax-free zones and the Antigua Brewery.

29SCIENCE AND TECHNOLOGY
Technological services for the fishing industry, such as the introduction of depth finders and hydraulic gear, are provided by the government. An extramural department of the University of the West Indies offers technical courses, as does Antigua State College. The University of Health Sciences, founded in 1982, has 46 students.

30DOMESTIC TRADE
General business is usually conducted from 8:30 AM to 4 PM, Monday–Saturday, except for Thursday afternoon, when many shops close. Banks are open from 8 AM to 12 noon five days a week, and on Friday additionally from 3 to 5 PM. St. John's is the main commercial center.

31FOREIGN TRADE
In 1991, exports totaled $345.7 million. Primary exports included chemicals, manufacturing goods and materials, food and live animals, machinery and transport equipment. Imports reached $403.1 million, including mineral fuel lubricants and related materials, food, live animals, machinery and transport equipment. The country exported mostly to the US, UK, Canada, Trinidad and Tobago, and Barbados. Its major providers were the US, UK, CARICOM, Canada, and the former Yugoslavia.

32BALANCE OF PAYMENTS
Inward capital movements, of which the largest part is foreign investment (especially for tourism-related construction), compensated for the imbalance in trade and services in the 1980s, but Antigua and Barbuda now maintains a consistently large trade deficit from heavy import dependence, in spite of growing tourism earnings.

In 1992 merchandise exports totaled us$16.67 million and imports us$304.30 million. The merchandise trade balance was us$–287.63 million. The following table summarizes Antigua and Barbuda's balance of payments for 1991 and 1992 (in millions of US dollars):

	1991	1992
CURRENT ACCOUNT		
Goods, services, and income	–70.07	–72.64
Unrequited transfers	18.20	13.17
TOTALS	–51.87	–59.48
CAPITAL ACCOUNT		
Direct investment	35.60	23.36
Other long-term capital	–21.60	–28.23
Other short-term capital	–24.80	41.07
Exceptional Financing	43.60	—
Reserves	–5.04	–5.56
TOTALS	27.76	30.64
Errors and omissions	24.12	28.84
Total change in reserves	–5.04	–17.98

33BANKING AND SECURITIES
There were eight commercial banks in 1994, five of which were foreign. Their liabilities and assets at that time stood at EC$113.4 million. The Antigua and Barbuda Development Bank, wholly owned by the government, began operations in 1975. Currency is issued by the Eastern Caribbean Central Bank.

34INSURANCE
There are several life insurance companies on the islands.

35PUBLIC FINANCE
Since the abolition of the income tax on residents, government revenues have been derived mainly from indirect taxes, principally customs and excise duties and consumption taxes. A major source of revenue is the EC$4.1 million paid by the US for its two bases. Current receipts and expenditures in 1991 totaled us$106 million and us$110 million, respectively; capital expenditures amounted to us$13 million that year. The 1992 deficit was estimated at us$5.3 million, and the external debt in 1992 was us$338 million.

36TAXATION
The income tax, introduced in 1924, was abolished for residents (based on a six-month stay) at the end of 1976. However, a 25% flat tax is levied on an individual's income from trade or business. A property tax is levied, as well as taxes on life and general insurance premiums, property transfers, share transfers, and hotel

stays. The company income tax is 40%, with double-taxation relief provisions for foreign-owned enterprises.

³⁷CUSTOM AND DUTIES
Antigua and Barbuda adheres to the common external tariff schedule of CARICOM; rates (which range from 5–70%) are generally ad valorem, based on c.i.f. value, and a wide range of goods is permitted duty-free entry. Additional special rates are applied for tobacco, cement, petroleum products, vans and trucks, and certain types of timber.

³⁸FOREIGN INVESTMENT
The government's efforts to improve the local investment climate have met with some success. Tax holiday periods of 10–15 years are specified under the Fiscal Incentives Act of 1975. Tax rebates of 25–50% are offered to export-oriented industries. Import duty exemptions for machinery, equipment, spare parts, and raw materials may also be granted to approved companies. The offshore financial sector has grown primarily due to instability in Argentina and Brazil. The offshore banking system alone holds close to us$5 billion in deposits, 80% of which are denominated in foreign currency. It offers tax haven facilities to international business companies, trusts, banks and insurance companies. In addition, the country has no capital gains or personal income tax.

³⁹ECONOMIC DEVELOPMENT
The government participates actively in the national economy. Various government ministries and the Antigua Employers' Federation have created job-training programs and special concessions to stimulate industrial expansion and diversification. Plans in the late 1980s called for the redevelopment of St. John's harbor area to provide shopping and entertainment facilities for cruise ship visitors. Agricultural policy has focused on revitalization of the sugar industry, expansion of cotton production, and vegetable and poultry production for export. The government, which has acquired unused sugar estates, now owns some 60% of all agricultural land. Its agricultural development plan includes a land tenure policy to help farmers buy land or obtain long-term leases. For the development of certain public utilities and services, Antigua and Barbuda has been reliant on funds from the UK, the US, and the CDB.

In mid-1992, the European Development Fund provided the government with a us$5 million grant to further develop the island's livestock industry. In addition, the government plans to continue fomenting foreign investment in order to generate more jobs.The 80% foreign owned Antigua Brewery, a us$6.3 million company, employs about 65,000 workers.

⁴⁰SOCIAL DEVELOPMENT
Effective from 1973, a social security fund provides for compulsory coverage of all persons between the ages of 16 and 60 years who are engaged in insurable employment. Contributions to the fund amount to 8% of wages (5% from employers and 3% from workers). It provides old age, disability, and survivor benefits. A medical insurance scheme is financed by a 5% levy on earnings, shared equally between employers and employees and includes maternity benefits. The government operates day-care centers for children under 5 years of age. The seven-member Board of Guardians meets each month to interview destitute persons who have applied for relief.

Although there are no legal restrictions on women's roles in society, traditions tend to limit their activities outside the home. The Directorate of Women's Affairs provides programs in areas including health, crafts, and business skills to help women advance in the public and private sectors. The total fertility rate in 1991 was 1.7, and 53% of married women practiced contraception between 1985 and 1990.

⁴¹HEALTH
Four institutions are maintained for the care of the sick and aged: Holberton Hospital, with 226 beds; the Fiennes Institute for the aged, with 150 beds; the Pearns Leper Home, with 40 beds; and the Mental Hospital, with 200 beds. In addition, 4 health centers and 16 dispensaries receive about 50,000 outpatient visits a year. Health personnel in 1991 included 59 physicians, 13 dentists, 13 pharmacists, 179 nurses, and numerous midwives. The infant mortality rate in 1992 was 20 per 1,000 live births, with 1,100 births that year. The average life expectancy in 1992 was 74 years. Between 1991 and 1992, 87% of the population was immunized against measles.

⁴²HOUSING
The Central Housing and Planning Authority advises on suitable sites, rehabilitates houses in the event of disaster, develops new housing tracts, and redevelops blighted areas. Between 1988 and 1990, 100% of the urban and rural population had access to sanitation services.

⁴³EDUCATION
Education for children between the ages of 5 and 16 years is compulsory. Primary education begins at the age of five years and normally lasts for six years. Secondary education lasts for five years (comprising of two cycles; i.e., three years and two years). The government administers the majority of the schools. In 1987–88, there were 43 primary schools and 15 secondary schools. There were 9,097 students enrolled at the primary schools and 4,413 students at the secondary schools. There currently are two colleges. The University of Health Sciences, Antigua, was founded in 1982. It had, in the 1990s, an enrollment of 46 students and 16 teachers. The University of the West Indies School of Continuing Studies (Antigua and Barbuda) was found in 1949 and offers adult education courses, secretarial skills training programs, summer courses for children and special programs for women.

The University of the West Indies has campuses in Barbados, Trinidad and Jamaica and it maintains extramural departments in several other islands, including Antigua. Those interested in higher education also enroll at schools in the UK, the US, Europe and Canada.

⁴⁴LIBRARIES AND MUSEUMS
Two public libraries are on the islands. A school library service has existed since the mid-1960s. There is an archaeological museum near St. John's.

⁴⁵MEDIA
The islands' automatic telephone system, operated by the Antigua Public Utilities Authority, has approximately 7,000 telephones. International telephone and telex services are supplied by Cable and Wireless (West Indies), Ltd. Eight broadcasting stations—four AM, two FM, and two television—were received in 1992 by some 27,000 radios and 23,000 television sets. There are also two shortwave stations, one coaxial submarine cable, and one Atlantic Ocean INTELSAT earth station.

The *Antigua & Barbuda Herald*, with a circulation of about 2,500, is published biweekly. The *Workers' Voice*, the official organ of the ALP and the Antigua Trades and Labour Union, appears weekly and has a circulation of 1,200. The *Outlet*, published weekly by the Antigua Caribbean Liberation Movement, has a circulation of 5,000. *The Nation*, with a circulation of about 2,000, appears weekly, published by the government.

⁴⁶ORGANIZATIONS
Four employers' organizations represent workers' interests in Antigua and Barbuda. The Antigua Chamber of Commerce has its headquarters in St. John's. Many missionary and charitable

organizations have operations on the islands, including Planned Parenthood, the Inter-American Foundation, the American Bible Society, and the People-to-People Health Foundation.

[47]TOURISM, TRAVEL, AND RECREATION

Tourism is the main source of revenue in Antigua and Barbuda. Antigua's plethora of beaches—said to number as many as 365—and its charter yachting and deep-sea fishing facilities have created the largest tourist industry in the Windward and Leeward Islands. The international regatta and Summer Carnival are popular annual events. Cricket is the national pastime; local matches are played Thursday afternoons, Saturdays, and Sundays. A wide range of hotels and restaurants served 196,571 tourists in 1991, 128,771 from the Americas and 64,133 from Europe. Cruise ships brought another 281,000 visitors to the islands during the same year and receipts from tourism climbed to US$313 million. A valid passport is required by all visitors except nationals of the US, Canada and the UK. Visas are not required for short stays.

[48]FAMOUS ANTIGUANS AND BARBUDANS

The first successful colonizer of Antigua was Sir Thomas Warner (d.1649). Vere Cornwall Bird, Sr. (b.1910), has been prime minister since 1981. (Isaac) Vivian Alexander ("Viv") Richards (b.1952) is a famous cricketer.

[49]DEPENDENCIES

Antigua and Barbuda has no territories or colonies.

[50]BIBLIOGRAPHY

Coram, Robert. *Caribbean Time Bomb: The United States' Complicity in the Corruption of Antigua.* New York: Morrow, 1993.

Gooding, Earl. *The West Indies at the Crossroads.* New York: Schenkman, 1981.

Henry, Paget. *Peripheral Capitalism and Underdevelopment in Antigua.* New Brunswick, N.J.: Transaction Books, 1985.

Russell, Richard J., and William G. McIntire. *Barbuda Reconnaissance.* Baton Rouge: Louisiana State University Press, 1966.

The State of Antigua and Barbuda: Information Memorandum, May 1982. St. John's, 1982.

Thome, James A. *Emancipation in the West Indies: A Six Months' Tour in Antigua.* New York: Abolitionists' Press, 1838.

ARGENTINA

Argentine Republic
República Argentina

CAPITAL: Buenos Aires.

FLAG: The national flag consists of a white horizontal stripe between two light blue horizontal stripes; centered in the white band is a radiant yellow sun with a human face.

ANTHEM: *Himno Nacional*, beginning "Oíd, mortales, el grito sagrado Libertad" ("Hear, O mortals, the sacred cry of Liberty").

MONETARY UNIT: The peso (A$) is a paper currency of 100 centavos. There are coins of 1, 5, 10, 25 and 50 centavos, and notes of 1, 2, 5, 10, 20, 50, and 100 pesos.

WEIGHTS AND MEASURES: The metric system is the legal standard.

HOLIDAYS: New Year's Day, 1 January; Labor Day, 1 May; Anniversary of the 1810 Revolution, 25 May; Occupation of the Islas Malvinas, 10 June; Flag Day, 20 June; Independence Day, 9 July; Anniversary of San Martín, 17 August; Columbus Day, 12 October; Immaculate Conception, 8 December; Christmas, 25 December. Movable religious holidays include Carnival (two days in February or March) and Good Friday.

TIME: 9 AM = noon GMT.

¹LOCATION, SIZE, AND EXTENT
Shaped like a wedge with its point in the south, Argentina, the second-largest country in South America, dominates the southern part of the continent. Argentina is slightly less than three-tenths the size of the United States with a total area of 2,766,890 sq km (1,068,302 sq mi); the length is about 3,650 km (2,270 mi) N–S and the width, 1,430 km (890 mi) E–W. To the N Argentina is bounded by Bolivia; to the NE by Paraguay; to the E by Brazil, Uruguay, and the Atlantic Ocean; and to the S and W by Chile, with a total boundary length of 14,654 km (8,106 mi).

Argentina lays claim to a section of Antarctica of about 1,235,000 sq km (477,000 sq mi). Both Argentina and the UK claim the Falkland Islands (Islas Malvinas), with the UK exercising effective occupancy. In 1978, Argentina almost went to war over three Chilean-held islands in the Beagle Channel. The case was referred to papal mediation; on 29 November 1984, the two countries signed a treaty that confirmed Chile's sovereignty over the three islands.

Argentina's capital city, Buenos Aires, is located along the eastern edge of the country on the Atlantic coast.

²TOPOGRAPHY
Except for the mountainous western area, Argentina is for the most part a lowland country. It is divided into four topographical regions: the Andean region, Patagonia, the subtropical plain of the north, and the pampas. The Andean region, almost 30% of the country, runs from the high plateau of the Bolivian border southward into western Argentina. Patagonia comprises all the area from the Río Negro to the southern extremity of the continent, or about 777,000 sq km (300,000 sq mi). Rising from a narrow coastal plain, it extends westward in a series of plateaus. In most places, the altitude range is 90–490 m (300–1,600 ft), although it may rise to 1,500 m (5,000 ft). Patagonia is a semi-arid, sparsely populated region. It includes the barren island of Tierra del Fuego, part of which belongs to Chile. A portion of the Gran Chaco, covering the area between the Andean piedmont

and the Paraná River, consists of an immense lowland plain, rain forests, and swampland, little of which is habitable. The most characteristic feature of Argentine topography, however, is the huge expanse of lush, well-watered level plains known as the pampas. Stretching from the east coast estuary, Río de la Plata, the pampas spread in a semicircle from the Buenos Aires area to the foothills of the Andes, to the Chaco, and to Patagonia, forming the heartland of Argentina, the source of its greatest wealth, and the home of 80% of its people.

The major Argentine rivers, which originate in the Andean west or the forested north, flow eastward into the Atlantic Ocean. The Paraná, Uruguay, Paraguay, and Alto Paraná rivers all flow into the Río de la Plata, which reaches a maximum width at its mouth of 222 km (138 mi), between Uruguay and Argentina. The highest peaks in Argentina are Mt. Aconcagua (6,960 m/22,835 ft), also the highest mountain in South America; Mt. Tupungato (6,800 m/22,310 ft); and Mt. Mercedario (6,770 m/22,211 ft). There is a region of snow-fed lakes in the foothills of the Andes in western Patagonia. Many small lakes, some of which are brackish, are found in the Buenos Aires, La Pampa, and Córdoba provinces.

³CLIMATE
Argentina's climate is generally temperate, but there are great variations, from the extreme heat of the northern Chaco region, through the pleasant mild climate of the central pampas, to the subantarctic cold of the glacial regions of southern Patagonia. The highest temperature, 49°C (120°F), was recorded in the extreme north, and the lowest, –16°C (3°F), in the southern tip of the country. Rainfall diminishes from east to west. Rainfall at Buenos Aires averages 94 cm (37 in) annually, and the mean annual temperature is 16°C (61°F). Light snowfalls occur occasionally in Buenos Aires. Throughout Argentina, January is the warmest month and June and July are the coldest. North of the Río Negro, the winter months (May–August) are the driest period of the year. The wide variations of climate are due to the

great range in altitude and the vast extent of the country. In the torrid zone of the extreme north, for example, the Chaco area has a mean annual temperature of about 23°C (73°F) and a rainfall of about 76 cm (30 in), whereas Puna de Atacama has a temperature average of 14°C (57°F) and a rainfall of about 5 cm (2 in). The pampas, despite their immensity, have an almost uniform climate, with much sunshine and adequate precipitation. The coldest winters occur not in Tierra del Fuego, which is warmed by ocean currents, but in Santa Cruz Province, where the July average is 0°C (32°F).

⁴FLORA AND FAUNA

More than 10% of the world's flora varieties are found in Argentina. The magnificent grasslands have figured prominently in the development of Argentina's world-famous cattle industry. Evergreen beeches and Paraná pine are common. From yerba maté comes the national drink immortalized in gaucho literature, while the shade-providing ombú is a national symbol.

Many tropical animals thrive in the forests and marshes of northern Argentina; among them are the capybara, coypu, puma, and various wildcats. In the grasslands and deserts are the guanaco, rhea, and many types of rodents. The cavy, viscacha, tuco tuco, armadillo, pichiciago, otter, weasel, nutria, opossum, various types of fox, and hog-nosed skunk are common. The ostrich, crested screamer, tinamou, and ovenbird are a few of the many species of birds. Caimans, frogs, lizards, snakes, and turtles are present in great numbers. The dorado, a fine game fish, is found in larger streams, and the pejerrey, corvina, palameta, pacu, and zurubi abound in the rivers.

Spanish cattle on the pampas multiplied to such an extent that the role of wild cattle herds in Argentine history was the same as that of the buffalo herds in the US West. Argentina is richly endowed with fossil remains of dinosaurs and other creatures.

⁵ENVIRONMENT

The principal environmental responsibilities are vested in the Ministry of Public Health and the Environment; the Subsecretariat of Environmental Planning in the Ministry of Transportation and Public Works; and the Subsecretariat of Renewable Natural Resources and Ecology within the Secretariat of State for Agriculture and Livestock. In 1977, the Metropolitan Area Ecological Belt State Enterprise was created to lay out a 150-km (93-mi) greenbelt around Buenos Aires, with controls on emission and effluents as well as on building density. The project was well under way in 1987.

In 1994, the major environmental issues in Argentina are pollution and the loss of agricultural lands. Most of the air and water pollution is in Argentina's cities because that is where most of the people live. The soil is threatened by erosion, salinization and deforestation. Air pollution is also a problem due to chemical agents from industrial sources. Argentina contributes .5% of the world's gas emissions which damage the ozone layer. The water supply is threatened by uncontrolled dumping of pesticides, hydrocarbons and heavy metals. Argentina has a water supply of 166.5 cu mi. Twenty-seven percent of all city dwellers and 83% of people living in rural areas do not have pure water.

As of 1987, endangered species in Argentina included the ruddy-headed goose, Argentinian pampas deer, South Andean huemul, Puna rhea, tundra peregrine falcon, black-fronted piping guan, glaucous macaw, spectacled caiman, and the broadnosed caiman. By 1991, the list included the Lear's macaw, the guayaquil great green macaw, and the American crocodile. Of the 255 species of mammals in Argentina, 23 are considered endangered. Fifty-three of the 927 species of birds are threatened. Four species of a total of 204 reptiles are also threatened. Of the nation's 9,000 plant species, 159 are endangered as of 1994.

⁶POPULATION

Argentina's census population was 27,862,771 in 1980 and 32,615,528 in 1991, an increase of 17%. The projected population for the year 2000 was 36,238,000, assuming a crude birthrate of 19.8 per 1,000 population, a crude death rate of 8.6, and a net natural increase of 11.2 during 1995–2000. Average population density was 11.7 per sq km (30.3 per sq mi) in 1991. In that year, 87% of the population lived in urban communities of 2,000 or more; more than one-third of all Argentines live in or around Buenos Aires. The population of Buenos Aires as of the 1991 census was 2,965,403. The estimated metropolitan-area populations in 1992 were: Buenos Aires, 11,662,050; Córdoba, 1,179,420; Rosario, 1,157,372; Mendoza, 801,920; La Plata, 676,128; and San Miguel de Tucumán, 642,473.

⁷MIGRATION

Migration to Argentina has been heavy in the past, especially from Spain and Italy. Under the rule of Juan Domingo Perón, immigration was restricted to white persons, exceptions being made for relatives of nonwhites (Japanese and others) already resident. More recently, immigrants from across the border in Paraguay have numbered at least 600,000; Bolivia, 500,000; Chile, 400,000; Uruguay, 150,000; and Brazil, 100,000. Some 300,000 illegal aliens were granted amnesty in 1992. Foreigners, on application, may become Argentine citizens after two years' residence. A total of 16,738 were naturalized in 1991, of which 13,770 were from other American countries. In 1992, Argentina's refugee population was estimated at 11,500. Few Argentines emigrated until the 1970s, when a "brain drain" of professionals and technicians began to develop. In the mid-1980s, some 10,000 of the estimated 60,000 to 80,000 political exiles returned home.

Of much greater significance to Argentina has been the tendency for workers in rural areas to throng to the cities. This had particular political and economic overtones during the Perón regime of 1946–55. Perón's encouragement of workers to move to Buenos Aires and surrounding industrial areas drained rural areas of so many persons that agriculture and livestock raising, the base of Argentina's wealth, suffered severely. Moreover, the inability of the economy to absorb all of the new urban masses led to a host of economic and social problems that still besiege the nation. Both the federal government and provincial governments have since vainly entreated aged workers to return to rural areas.

⁸ETHNIC GROUPS

Argentina's population is overwhelmingly European in origin (principally from Spain and Italy); there is little mixture of indigenous peoples. An estimated 97% of the people are of European extraction, and some 3% are Amerindian or of mixed lineage. The pure Amerindian population has been increasing slightly through immigration from Bolivia and Paraguay.

⁹LANGUAGES

The national language of Argentina is Spanish. Argentine Spanish has diverged in many ways from Castilian, showing the effects of the vast influx of foreigners into Buenos Aires, as well as of Spaniards from Andalucía, Galicia, and the Basque provinces. First- and second-generation Italians have added their touch to the language, and French settlers have contributed many Gallicisms.

The outstanding phonetic feature of Argentine Spanish is the yeísmo, in which the *ll* and *y* are pronounced like the *z* in *azure*. The meaning of many Castilian words also has been modified. The Porteños, as the inhabitants of Buenos Aires are called, rely heavily upon a variety of intonations to express shades of meaning.

English has become increasingly popular as a second language, especially in metropolitan areas and in the business and professional community. There are pockets of Italian and German immigrants speaking their native languages. Some Amerindian languages are still spoken, including a version of Tehuelche in the pampas and Patagonia, Guaraní in Misiones Province, and Quechua in some parts of the Jujuy and Salta provinces.

10 RELIGIONS

In 1993, an estimated 92% of the population was Roman Catholic and 2% Protestant; the remainder belonged to other religions or indicated no preference. Supervision of religious bodies is the responsibility of the Ministry of Foreign Affairs and Worship. Argentina retains national patronage, a form of the old Spanish royal patronage, over the Roman Catholic Church. Under this system, bishops are appointed by the president of the republic from a panel of three submitted by the Senate; papal bulls and decrees must be proclaimed by the president and sometimes must be incorporated into an act of the Congress. The 215,000 Jews (as of 1990) constitute the third-largest Jewish concentration in the Americas; their numbers have declined in recent years, however, in part because of alleged anti-Semitism in the military.

Relations between church and state have been relatively free of the conflict that has taken place in other Latin American countries. During the early part of the Perón era, there was no great opposition by the Church, but several of Perón's later actions, especially the legalizing of divorce in 1954, drew strong criticism. In 1955, Perónists pillaged and burned many noted churches of Buenos Aires, and the Church's opposition hardened. After the revolution in 1956, divorce was again made illegal. (It was legalized again in 1987.) The Church made no formal effort to halt the government's program of using terror against left-wing terrorists during the late 1970s. In the early 1980s, however, Church leaders launched a program of national reconciliation, holding meetings with military leaders, politicians, labor and business representatives, and human rights activists. A Roman Catholic revival, especially among the young, was bolstered by the visit of Pope John Paul II in June 1982, the first visit by any pope to Argentina.

11 TRANSPORTATION

Argentina has the largest railway system in South America, with 34,172 km (21,234 mi) of track as of 1991. Although railroads link all the provinces, the three provinces of Buenos Aires, Córdoba, and Santa Fe contain about one-half the total track and are the destinations of about two-thirds of all goods carried. The seven major railroads and all other lines belong to the state and are administered by Argentine Railways. Until 1947, when Perón bought them at a price exceeding their real value, the railroads were mainly under the control of British interests. Since then, they have been in decline and have regularly run up large deficits. Gauges are often incompatible, forcing virtually all interregional freight traffic to pass through Buenos Aires. The railroads' share of merchandise transported has declined gradually since 1946, dropping, for example, from 12,557 million ton-km in 1973 to 9,504 million ton-km in 1985 and 9 million ton-km in 1991.

A five-year railroad modernization and rationalization plan was initiated by the military government in 1976, but the general decline of the railway system was not halted, and the number of passengers carried dropped from 445 million in 1976 to 300 million in 1991. The subway system in Buenos Aires, completely state owned since 1978, consists of five lines totaling 36 km (22 mi). The number of passengers in 1985 was 182.8 million.

The continued deterioration of the railroads has resulted in a sharply increased demand for road transportation, which the present highways cannot handle. By 1991, the nation had 208,350 km (129,469 mi) of roads, of which 47,550 km (29,549 mi) were paved. In late 1969, a tunnel under the Río Paraná was opened, connecting Santa Fe with the nation's eastern region. The road system is still far from adequate, especially in view of

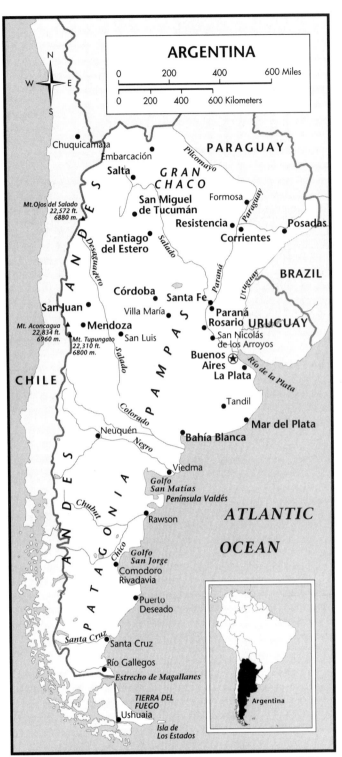

LOCATION: 21°47′ to 55°16′s; 53°38′ to 73°35′w. **BOUNDARY LENGTHS:** Bolivia, 832 km (517 mi); Paraguay, 1,880 km (1,168 mi); Brazil, 1,224 km (761 mi); Uruguay, 579 km (360 mi); Atlantic coastline, 4,989 km (3,100 mi); Chile, 5,150 km (3,200 mi). **TERRITORIAL SEA LIMIT:** 200 mi.

Argentina's rapidly increasing automotive industry. A total of 138,958 vehicles were produced in 1991, and in 1992 the total number registered reached 5,970,775 (including 4,417,882 passenger cars).

The main river system of Argentina consists of the Río de la Plata and its tributaries, the Paraná, Uruguay, Paraguay, and Alto Paraná rivers. There is a total of 10,950 km (6,800 mi) of navigable waterways, offering vast possibilities for efficient water transportation. The river system reaches Paraguay, northeastern Argentina, and regions of Brazil and Uruguay. The La Plata estuary, with its approaches and navigation channels, is the basis of the entire river system. The La Plata ports (Buenos Aires and La Plata) account for more than half of all maritime cargo, including more than two-thirds of all cargo transported on the river system. The Paraná is easily navigable up to Rosario, but the 171-km (106-mi) stretch between Rosario and Santa Fe has considerably less depth and is less suitable for oceangoing vessels. Upriver from Santa Fe, the Paraná rapidly loses depth and is navigable only by small ships.

The port of Buenos Aires handles about four-fifths of the country's imports and exports, and it is the focus of river traffic on the La Plata system. Other major ports are Rosario, Quequén, Bahía Blanca, Campana, and San Nicolás. Most port storage facilities are owned and operated by the government. In 1961, the State Merchant Fleet and the Argentine Overseas Navigation Fleet were merged to form the Argentine Maritime Lines. This state company carries approximately one-half of all Argentine overseas freight. In 1991, the merchant marine consisted of 108 vessels with a GRT of 1.3 million. Tankers accounted for about 38% of total tonnage, and bulk carriers for 27%.

Buenos Aires is the most important air terminal in South America. Ezeiza Airport, about 40 km (25 mi) from Buenos Aires, is one of the largest in the world and serviced 2,609,000 passengers in 1991. Argentina has a total of 137 airports and paved landing fields, 10 of which handle international traffic. The government line is Aerolíneas Argentinas; in 1992 there were four other major Argentine airlines and many foreign lines operating in the country. In 1992, the total scheduled civil aviation services flew 9,995 million passenger-km (6,211 million passenger-mi) and 156 million freight ton-km (96.9 freight ton-mi) in 1992.

12HISTORY

Before the Spaniards arrived, about 20 Amerindian groups comprising some 300,000 people lived in the region now called Argentina. They were mainly nomadic hunter-gatherers, although the most developed group, the Guaraní, practiced slash-and-burn agriculture.

Spaniards arrived in Argentina in 1516. They called the region "La Plata" (literally, "silver") under the mistaken impression that it was rich in silver. Colonists from Chile, Peru and Asuncion created the first permanent settlements in Argentina, including Buenos Aires in 1580. In 1776, Rio de la Plata became a viceroyalty, with Buenos Aires as the main port and administrative center.

During the colonial period, there was little interest in Argentina. The region had no mineral wealth, and Spaniards overlooked the fertile soil and temperate climate of the region. As a result, Buenos Aires had a population of only about 25,000 at the time of the viceroy's arrival. Even the creation of the viceroyalty was more the result of fears of Portuguese expansion than any genuine interest in the area.

In May 1810, following the example set by Spanish cities after the capture of King Ferdinand VII by the French, Buenos Aires held an open town meeting (cabildo abierto). A junta was elected, which deposed the viceroy and declared itself in authority. On 9 July 1816, a congress of provincial delegates in San Miguel de Tucumán signed a declaration of independence, and in 1817, Gen. José de San Martín led an army across the Andes to liberate Chile and Peru.

There followed a period of provincial strife between Buenos Aires and the regional strongmen (caudillos). It lasted until Juan Manuel de Rosas became governor of Buenos Aires Province. He imposed order and centralism from 1835 until 1852, when he was defeated by the forces of Gen. Justo José de Urquiza. A new

constitution was adopted in 1853, and Urquiza was elected president in 1854. Urquiza attempted to shift power away from Buenos Aires, so much so that Buenos Aires seceded. Urquiza gave way to Bartolomé Mitre in 1862, who finally consolidated the nation with Buenos Aires at its center. The new Federal Republic of the Argentine joined Uruguay and Brazil against Paraguay in the War of the Triple Alliance from 1865 to 1870. The administrations of Domingo Faustino Sarmiento (1868–74) and his successor, Nicolás Avellaneda (1874–80), promoted railroad construction, immigration, and education. It was a period of rapid economic development, and the fertile pampas began to shift from pastoral to agricultural production. In 1880, tension between Buenos Aires and the other regions of Argentina was resolved by the creation of a neutral Federal District as capital.

Conflict in Argentina shifted to the social arena. The National Party, under the leadership of Gen. Julio Roca (who served two terms as president, 1880–86 and 1898–1904) and supported by the military and landowners, dominated the nation. To combat this powerful coalition, a middle-class party called the Radical Civic Union (Unión Cívica Radical) was formed. The Radicals stressed democratic practices and attempted to expand the political system beyond its elite-restricted boundaries. The Radicals' efforts came to fruition in 1916, when Hipólito Yrigoyen was elected president for a six-year term. After six years out of office, Yrigoyen was again elected president in 1928, but because of an economic crisis caused by the world depression, he was overthrown by a military coup in 1930.

For the next thirteen years, Argentina was ruled by the old Conservative oligarchy. The military-landowner alliance brought both economic recovery and political corruption, as well as the exacerbation of social tensions. Particularly divisive was the matter of Argentina's foreign relations. While opening Argentina to trade with Europe improved the economic picture, many felt that the leadership had sold out to foreign interests. Argentina's careful neutrality toward the Axis powers masked considerable Fascist sympathies, further dividing the nation.

Another military coup in 1943 brought to power an even more axis-sympathetic group, but also launched a new era in Argentine politics. Argentina had undergone an industrial expansion, accelerated by the war. This expansion led to the formation of a large blue-collar workforce, which in 1943 came under the direction of the military head of the Labor Department, Col. Juan Domingo Perón. Perón used his new constituency to build a power base that allowed him in 1946 to be elected president, while his supporters won majorities in both houses of Congress.

Perón then inaugurated sweeping political, economic, and social changes. His ideology was an unusual blend of populism, authoritarianism, industrialism and nationalism. His strong personal appeal was buttressed by the charm of his wife Eva ("Evita"), who captivated the masses with her work on behalf of the poor. Peronist rhetoric stressed the rights of the "descamisados" (literally, "shirtless"), the poor of Argentina.

Perón sought to establish a foreign policy that allied Argentina with neither the West nor East, while acting as protector of weaker Latin American nations against US and British "imperialists." He called this approach "justicialismo", a word which has no meaning in Spanish, much less English. After reelection in 1951, Perón became increasingly dictatorial and erratic, especially after the death of Evita a year later. Economic hardship led to reversals in policy that favored the old oligarchy. Newspapers were shut down and harassed. Perón legalized divorce and prostitution, and began to incite violence against churches. Finally, a disillusioned military group took over in September 1955.

For the next twenty years, Argentina felt the shadow of Perón. From exile in Spain, Perón held a separate veto power. Under the military's watchful eye, a succession of governments attempted unsuccessfully to create a new political order.

The first of these efforts came from Gen. Pedro Eugenio Aramburu, who repressed Perón's followers and declared their party illegal. After two years of provisional government, elections were held. Rival factions of the old Radical Civic Union competed in a contest won by Arturo Frondizi of the more left-leaning UCRI. With the initial support of the peronistas, Frondizi attempted to balance that support with the military, which grew nervous at the mention of Perónism. Frondizi curbed inflation through an austerity program and increased Argentina's petroleum production by extending concessions to foreign companies. These economic measures helped increase political tensions, and in the elections of 1961 and 1962, Perónist candidates, running under the banner of the Justicialist Front (Frente Justicialista), won sweeping victories. A military junta removed Frondizi from the presidency in March 1962 and annulled the elections, thus denying governorships to the supporters of Perón. Divisions among the military leaders kept the nation in a state of tension until mid-1964, when new elections were held. Dr. Arturo Illía of the rightist UCRP won the presidency. Illía's administration was beset by rising government debt, inflation, labor unrest, and political agitation, but was most seriously threatened by the military. The chief of the armed forces, Lt. Gen. Juan Carlos Onganía resigned in November 1965, after Illía appointed a Perónist sympathizer as war minister.

In June 1966, following election victories by the Perónist faction, the military leaders installed Onganía as President. Onganía dissolved the nation's legislative bodies and suspended the constitution. Onganía announced a revolutionary program to restore economic prosperity and social stability, saying that only after this restoration would the democratic system be reestablished. Inflation was cut by means of rigid wage controls, and by the end of 1969, the economy was growing at a rate of 7% annually. These successful economic policies were overshadowed, however, by growing political tension. With the help of the military, strict controls were imposed on the press and all means of mass communication. Students led in denouncing these repressive policies, and in the early months of 1969, violence erupted in Córdoba and Rosario.

Dissatisfaction mounted early in 1970, and acts of terrorism increased. Several groups were active, some of which claimed to be Perónist, others Marxist, still others claiming to be both. The most serious incident was the kidnapping and killing of former President Aramburu by a Perónist group. Although President Onganía stiffened in response to the disorder, it was becoming clear that Argentina would never be stabilized without the participation of the Perónists. For his part, Perón encouraged these groups from abroad.

In June 1970, a junta of high-ranking military officers removed Onganía, and began to move toward democratic reform. Under two ensuing military governments, preparations were made for elections that would include the Perónists, now organized as the Justicialist Liberation Front (FREJULI). In general elections held in March 1973, the winner was Dr. Héctor J. Cámpora, whose unofficial slogan was "Cámpora to the presidency; Perón to power." Cámpora was elected president with 49% of the vote, while FREJULI won a congressional majority and 11 of the 22 provincial governorships. However, Cámpora, who assumed office in May 1973, was no better able than his predecessors to cope with a rising tide of terrorism, much of it from extreme Perónist factions. After a consultation with Perón in Madrid, Cámpora announced his resignation, effective in July.

Perón, who had returned to Argentina in June 1973, ran for the presidency and took 61.9% of the vote in a special election in September. His running mate was his third wife, María Estela ("Isabel") Martínez de Perón, a former exotic dancer. There was no magic left in the elderly Perón. He cracked down on the very terrorist groups he had encouraged, but the economy sagged. When he died in July 1974, his widow succeeded to the presidency.

Isabel had none of Evita's appeal, and her administration plunged Argentina more deeply into chaos. The first year of Isabel Perón's regime was marked by political instability, runaway inflation, and a renewal of guerrilla violence. In September 1975, Perón vacated her office for 34 days, ostensibly because of ill health. During her absence, the military strengthened its position. In March 1976, she was arrested in a bloodless coup, and a military junta consisting of the commanders of the army, navy, and air force took over. The leading member of the junta was Army Commander Lt. Gen. Jorge Rafael Videla, who became president.

The junta dissolved Congress, suspended political and trade union activity, and mounted a concerted campaign against leftist guerrillas. For seven years, the military attempted to "purify" Argentina by removing all vestiges of leftism, Perónism, trade unionism and any other political positions deemed divisive. Meanwhile, they attempted a complete liberalization of the economy, including the privatization of banking and industry. However, the military was never able to solve the problem of inflation, which remained in triple digits for most of this period.

In March 1981, Gen. Roberto Viola succeeded Videla as president, and in December, Lt. Gen. Leopoldo Galtieri took over. Troubled by economic woes and lacking any political support from the general populace, the military turned to foreign affairs in an attempt to gain support. In April 1982, Argentina invaded the Falkland Islands, claiming sovereignty over them, but in the ensuing war with the UK, Argentina's armed forces were routed, surrendering in June. The defeat led to Galtieri's resignation, and a new junta was formed in July under Maj. Gen. Reynaldo Benito Antonio Bignone. Liberalization measures during the remainder of 1982 led to strikes and antigovernment demonstrations, including a one-day general strike in December in which 90% of the work force reportedly took part. In addition to demands for a return to civilian rule, more and more Argentines began to insist on a clarification of the fate of at least 6,000 and perhaps as many as 15,000 persons who had "disappeared" during 1975–79. A government report issued in April 1983 claimed that those who had "disappeared" were casualties in the war against leftist violence, but practically all political groups, as well as the Vatican, termed the report unsatisfactory and continued to press for a full accounting.

In elections for a civilian president held in October 1983, the upset winner was a human-rights activist from the UCRP, Dr. Raúl Alfonsín. After taking office in December, Alfonsín called for a new inquiry into the "disappearances" and ordered the prosecution of former junta members. In December 1985, five were convicted, including Lt. Gen. Videla. The legacy of the "dirty war" between 1976 and 1983 preoccupied the Alfonsín government. The president saw the need to close the 50-year cycle of military intervention and political instability by building a stable democracy. However, the political reality of Argentina could not be changed by wishes. The human rights trials of leading military officers irked the military, and in April 1987, an abortive military uprising spread to a number of bases. Although Alfonsín refused to yield to the rebels, he soon afterward retreated from his position, getting approval from Congress for a law that would limit the trials to a few superior officers, thereby accepting the defense of "taking orders" for the lower-ranking officers.

The Alfonsín administration also acted to halt rampant inflation with the "Austral Plan" of mid-1985, which froze wages and prices and created a new unit of currency, the austral, to replace the beleaguered peso. The initial success of the plan was weakened by a resurgence of inflation and labor intransigence over wage demands.

With the failure of the Alfonsín administration to stabilize the economy or bring military leaders to justice, Argentines sought change from an old source: the Perónists. In May 1989, Carlos Saul Menem, running under the Justicialist banner, was elected with 47% of the popular vote. Menem was to have taken office in December, but the Alfonsín government was in such dire straits that the president resigned in July and Menem was immediately installed.

Menem, a former soccer player, proved to be anything but a classic Perónist. Abandoning his party's traditional support of state enterprises, he has cut government spending and generally liberalized the Argentine economy. As in the rest of Latin America, the pressures of international economics are somewhat responsible, but Menem has not shied away from taking responsibility for these actions. At times, he has met strong resistance from his own party, and from some sectors of the Radical Party. He has met with significant success in the economic sphere, so much so that he speaks of trying to continue past his constitutional term limit in 1995.

13 GOVERNMENT
Argentina's government is ruled by its 1853 constitution, although that document has been suspended many times. The basic structure is federal and republican. In 1949, the Perón government adopted a new constitution, but the subsequent military government expunged that document. Some modifications in the original constitution were subsequently made by a constituent assembly that met in October 1957. In July 1962, a system of proportional representation was adopted.

The constitution provides for a federal union of provinces that retain all powers not specifically delegated to the federal government by the constitution. There is a separation of powers among the executive, legislative, and judicial branches, but the president is powerful within this arrangement. The president can draw up and introduce his own bills in Congress, appoint cabinet members and other officials without the consent of the Senate, and possesses broad powers to declare a state of siege and suspend the constitution. The president is commander-in-chief of the army, navy, and air force and appoints all major civil, military, naval, and judicial offices, with the approval of the Senate in certain cases. The president is also responsible, with the cabinet, for the acts of the executive branch and has the right of patronage (control over appointments) in regard to bishoprics. The president and vice-president are elected by an electoral college for six-year terms. They must be Roman Catholics, and either they or their parents must be native-born citizens. Voting is compulsory for all citizens 18 to 70 years of age.

The constitution calls for a National Congress consisting of a Senate and a Chamber of Deputies. The 46 senators are indirectly elected by local legislatures. The term of office is set at nine years, with staggered elections every three years for one-third of the membership. The Chamber of Deputies is the result of direct elections for 254 seats. Seats are allocated to each province in proportion to its population. The deputies' term of office is four years, with one-half of the membership being elected every two years. The Chamber of Deputies is authorized to receive the budget and initiate fiscal legislation and has the exclusive right to impeach officials before the Senate.

The most recent suspensions of the constitution were between 1966 and 1973, and then again from 1976 until 1983. During the most recent suspension, a military junta performed the executive, legislative, and judicial functions. Since the resumption of civilian government in 1983, there has been an uneasy relationship between the military and the government. The controversial trials of military leaders led to serious questions about the credibility of the judiciary. This controversy did not abate after the rather mild sentences these trials produced.

14 POLITICAL PARTIES
Political party activity in Argentina has been sporadic, given the frequency of military takeovers and the many years during which

parties have been banned. Still, several parties reformed in the 1980s and continue to be active in the 1990s.

Traditionally, the alignment of Argentine political parties has been along socioeconomic and religious lines. The landowners, the high clergy, and the more conservative lower class supporters have formed an alliance that defends the Church and the status quo. On the other side have been the advocates of change: merchants and professionals who resent the preeminence of the aristocracy and who tend also to be anticlerical. This second group has supported separation of church and state and decentralization. However, in modern times, new parties have emerged to represent the working class, small farmers, and intellectuals.

During the first half of the 20th century, the Radical Party in Argentina was either the governing party or the chief opposition. The Radicals were committed to the expansion of Argentine politics to the middle and lower classes, and a transformation of the nation's economic and social life. This party was as close to a mass-based party as Argentina had ever had. The core was middle class, but the party was also supported by upper- and lower-class elements. Only radical by the standards of Argentine politics, it occupied a middle ground between the Conservatives and the Socialist left. However, with a heterogeneous membership, tensions and schisms were frequent. The party split into the Radical Intransigent Civic Union (Unión Cívica Radical Intransigente—UCRI), which formed the major support for Arturo Frondizi in 1958, and the People's Radical Civic Union (Unión Cívica Radical del Pueblo—UCRP), led by Ricardo Balbín. The UCRP was somewhat more nationalistic and doctrinaire than the UCRI, but shifting policies made the differences difficult to define. Balbín's party survived into the 1980s as the Radical Civic Union (Unión Cívica Radical—UCR). After the military stepped down in 1983, that party was one of the few viable political entities in Argentina, and emerged victorious in the 1983 elections. However, with the failure of the Alfonsín administration, the UCR finds itself again in its old role as loyal opposition. Currently, the UCR holds 85 seats in the Chamber of Deputies and 14 seats in the Senate.

The Conservatives dominated Argentine politics from about 1874 to 1916 and again from 1932 to 1945, when they were known as the National Democrats. This era of Argentine politics was known as the "Concordancia." The Conservatives were the chief spokesmen for the landed interests, from whom they drew their main support. During the Perón regime, the right lost most of its influence. In 1958, conservative parties banded together to form the National Federation of Parties of the Center (Federación Nacional de Partidos del Centro). Years of military rule in the name of conservatism yielded no mass-based conservative parties, mainly because the military professed a disdain for partisan politics. Currently, there are several small right-wing parties, the largest of which, the Union of the Democratic Center (UCD), holds 10 seats in the Chamber of Deputies.

Although leftism in Argentina has a long tradition, it was dealt a serious blow during the 1976–1983 military governments. Those governments were committed to the extermination of all leftist influences. This meant the jailing and "disappearance" of leaders of the socialist and communist movements. In addition, Perónism preempted much of the ideological appeal of these parties, as well as their traditional working-class constituencies. The earliest leftist party was the Communist Party, founded in 1918 by Juan B. Justo, who split from Yrigoyen and the Radicals. The Communists were never terribly revolutionary, but concentrated instead on the trade union movement.

In the 1970s, Argentine leftism was thrown into confusion by the appearance of several substantial "urban guerrilla" movements. The Trotskyist People's Revolutionary Army (Ejército Revolucionario del Pueblo—ERP), the Montoneros, and the Perónist Armed Forces (FAP), among others, became major players on the Argentine political scene, if only because of the dramatic impact of their actions. Their presence may well have hastened the return of Perón in 1974, but their persistence became a major justification for the military repression that followed. Refusing to make any distinction between a leftist and a terrorist, the government decimated the Argentine left.

Perónism, which defies political classification, is still very much alive in Argentina, in name if not in anything else. Perónism went underground for nearly two decades after the coup of 1955. Operating under the names Popular Union Party, Populist Party and Laborite Party, a variety of Perónist organizations put up candidates wherever possible. The movement was alternately wooed, tolerated, or repressed, depending on the degree to which the military was involved. In 1973, elections were held in which the Perónists were allowed to field a candidate, Hector J. Cámpora, representing a coalition of various Perónist factions and other smaller parties. This coalition, the Justice Liberation Front (Frente Justicialista de Liberación—FREJULI), took 49% of the vote. Under Cámpora's successors, Juan Perón and Isabel Perón, FREJULI remained the governing coalition until the March 1976 coup, after which political activity was suspended until 1980. A "reform" movement led to infighting that crippled the party in the 1983 elections.

In 1989, Carlos Menem's victory was accompanied by solid legislative majorities in both houses of the legislature. The Justicialist Party (JP) holds 122 seats in the Chamber of Deputies and 27 seats in the Senate.

Argentina's party politics have been contentious and vicious over the years, with various sides coalescing in order to defeat rivals. One notable exception was the formation in July 1981 of the Multipartidaria, an alliance among Argentina's five leading parties—FREJULI, the UCR, the Democratic Christian Federation (Federación Demócrata Cristiana), the Movement for Integration and Development (Movimiento de Integración y Desarrollo—MID), and the Intransigent Party (Partido Intransigente). Claiming the support of about 80% of the voters, this opposition alliance began to negotiate with the military concerning a return to constitutional government, and in July 1982, political parties were formally permitted to resume their activities.

15LOCAL GOVERNMENT

Argentina is a federation of 22 provinces, the federal capital of Buenos Aires, and the territories of Tierra del Fuego, the Antarctic, and the South Atlantic Islands. During the 19th century there was a bitter struggle between Buenos Aires and the interior provinces, and there has long been an element of tension regarding the division of powers between the central government and provincial bodies. The federal government retains control over such matters as the regulation of commerce, customs collections, currency, civil or commercial codes, or the appointment of foreign agents. The provincial governors are elected every four years.

The constitutional "national intervention" and "state of siege" powers of the president have been invoked frequently. The first of these powers was designed to "guarantee the republican form of government in the provinces." Since the adoption of the 1853 constitution, the federal government has intervened over 200 times, mostly by presidential decree. Under this authority, provincial and municipal offices may be declared vacant, appointments annulled, and local elections supervised. Between 1966 and 1973, all local legislatures were dissolved and provincial governors were appointed by the new president. A restoration of provincial and municipal government followed the return to constitutional government in 1973. After the March 1976 coup, the federal government again intervened to remove all provincial governors and impose direct military rule over all municipalities. Since 1983, representative local government has been in force again.

16JUDICIAL SYSTEM

Justice is administered by both federal and provincial courts. The former deal only with cases of a national character or those to which different provinces or inhabitants of different provinces are parties. The Supreme Court, which supervises and regulates all other federal courts, is composed of five members nominated by the president and confirmed by the Senate. Other federal courts include nine appellate courts, with three judges for each; single-judge district courts, at least one for each province; and one-judge territorial courts. The federal courts may not decide political questions. Judges of the lower courts are appointed by the president.

Provincial courts include supreme courts, appellate courts, courts of first instance, and minor courts of justices of the peace (alcaldes) and of the market judges. Members of provincial courts are appointed by the provincial governors. Trial by jury was authorized by the 1853 constitution for criminal cases, but its establishment was left to the discretion of Congress, resulting in sporadic use.

A 1991 law provides a fund for compensating prisoners who were illegally detained during the 1976-1983 military dictatorship.

In 1992, a system of oral public trials was instituted in order to speed up the judicial process while improving the protection of procedural rights of criminal defendants.

In practice, there is not a truly independent judiciary. The courts lack power to enforce orders against the executive and federal judges who actively pursue charges of police or military corruption. In 1989, President Menem, in a court-packing maneuver, expanded the number of Supreme Court justices from fine to nine.

17ARMED FORCES

The Argentine armed forces are being reduced and reorganized, and numbered 65,000 in 1993. They included 12,000 conscripts of whom 8,000 served in the army of 35,000. The other services are the navy (which has air and marine units) of 20,000, a national gendarmerie of 17,000, a coast guard of 13,000 and an air force of 10,000. The gendarmerie and coast guard are not considered military, but have military capabilities.

The new army will have 6 regional divisions of 10 mixed brigades with armored, mechanized infantry, light infantry, and artillery battalions. The navy has 4 submarines, 1 aircraft carrier, 7 frigates, 6 missile-equipped destroyers, and 13 coastal combatants. The air force had an estimated 10,000 personnel and 174 combat aircraft. Defense spending has fallen from an estimated $4 billion to $700 million in the last decade or 1.5% of domestic product. Much of the reduction comes from abolishing internal security forces and unwanted reserves and paramilitary units.

Since the 1930 revolution headed by Gen. José Félix Uriburu, the armed forces have frequently intervened in political affairs; from that year through 1983, 14 of the 18 presidents were military officers. The military remained one of the most potent political forces in Argentina until it suffered the humiliating defeat in the Falklands War in 1982.

To increase efficiency and reduce military influence, the position of commander in chief of each service has been replaced by that of a chief of staff. Control of the coast guard has been removed from the navy and placed with the Defense Ministry, which has also taken control of the gendarmerie.

The reformed Argentine armed forces import only $10 million and export $5 million in arms, an import reduction from $300 million in 1988.

18INTERNATIONAL COOPERATION

Argentina is a charter member of the UN, having joined on 24 October 1945; it belongs to ECLA and all the nonregional specialized agencies. It is also a member of the OAS and of many other inter-American and intergovernmental organizations, such as G-77, IDB, LAIA, and PAHO. In 1962, Argentina signed an agreement of cooperation and exchange of information with EURATOM. It led in the attempt to form a Latin American common market, and in 1967, it joined other Latin American nations in claiming an offshore territory extending 200 mi. Argentina is a signatory to GATT and the Law of the Sea.

[19]ECONOMY

Argentina has one of the most highly developed economies and most advantageous natural resource bases of Latin America, but political instability and conflicts between various sectors of the economy have delayed the realization of its potential. In 1960, manufacturing industries contributed 29% to the GDP, and by 1974, the share from manufacturing had increased to 32.2%; however, partly as a result of the recession, the share from manufacturing (including mining) fell to 24.3% in 1986. The country has to a large degree overcome its dependence on imported machinery and finished products, but in their place there has grown a great demand for parts and raw materials that are assembled or finished within the country. Basic industries, such as iron and steel, petroleum and petrochemicals, aluminum, plastics, and electrical equipment are being established, but these will continue to require extensive raw material imports for some time, or even permanently, because of the absence of certain minerals, such as bauxite.

The GNP has risen slowly since 1960. For the period 1961–68, the gain averaged only 3.1% per year, or only 1.5% per capita. This economic stagnation was due partly to the tremendous government bureaucracy and partly to political and social unrest, which caused serious stoppages. The annual increase rose to 4.4% in 1970 but declined during the next two years. The economy was reactivated in 1973 and further accelerated in 1974, when increases of 7.6% were registered for agriculture, 7.5% for commerce, and 22.3% for construction. Prior to the early 1960s, Argentina suffered serious deficits in trade balance, but with increased exports, favorable trade balances were achieved, continuing throughout the 1970s except for 1975. The trade balance became negative in 1980 but returned to the positive side from 1981 through 1986.

There was widespread inflation in the early 1960s, but rigid controls initiated in 1967 decreased the rate of growth. From 1973, however, a combination of political instability, massive budget deficits, and union wage demands pushed the consumer price index (1970 = 100) to 872.3 in June 1975, with an index of 828.7 for food. At the time of the March 1976 coup, the annual inflation rate was 444%; it decreased in the late 1970s but rose again to 209% in 1982. The peso had so depreciated by 1982—a Sunday newspaper cost $18,000 pesos and bus fare $6,000 pesos—that the government decided to redenominate the currency in 1983, with 10,000 old pesos equaling one new peso. Also in 1983, Argentina received a stabilization loan of US$2.15 billion from the IMF to compensate for the effects of inflation and recession. As a condition for the loan, the government agreed to reduce the inflation rate to 165% in 1983. By autumn, however, inflation was running at an annual rate of over 900%; the inflation rate for the whole of 1983 was 434%, the highest in the world. Thereafter, it rose without interruption until it reached some 1,200% in mid-1985. At that time, the government introduced the Austral Plan—a bold attempt to halt inflation by freezing wages and prices, revaluing (and redenominating) the currency, and resolving to finance public spending with real assets only (not by printing money); under this plan the annual rate for 1985 was cut to 385%. By the end of 1986, the rate had been cut further to 82%; by early 1987, however, inflation had begun to surge again, and it was expected to exceed 100% by the end of the year.

In July 1989, President Carlos Menem of the Justicialista Party took office at a time when the economy was entangled in a hyper-inflationary spiral. In 1991, President Menem unveiled an innovative stabilization/reform program which has so far been implemented successfully. The cornerstone of the stabilization/reform plan was to link the peso to the dollar at a fixed rate of 0.99 pesos per dollar, and requiring congressional approval for devaluation. In 1992, economic policy continued to focus on structural adjustment involving an strategy of cleaning up the fiscal accounts. The reforms took on a four-pronged approach: fiscal revenues were strengthened through a broadening of the Value Added Tax and an enhancement in revenue enforcement; administrative reforms included a substantial cut in public payroll and revamping of national fiscal accounting; the use of the Central Bank's rediscount window to finance deficits of provincial governments was curtailed; and public enterprises were privatized. Both revenue enhancements and expenditure cutbacks have sharply reduced the relative size of the deficit. A key priority for the plan was compliance with the provisions of the March 1992 Extended Fund Facility with the IMF. The plan's success facilitated an important rapprochement with the international financial community by making Argentina eligible for a Brady Plan debt reduction agreement. This process was successfully concluded on 7 April 1993 with the signing of an accord that offered creditors two payment options which reduced medium and long-term debt by some $2.23 billion and save about $3 billion in interest payments over a 30 year period. During 1991–92 GDP growth averaged a very strong 8.8% per annum.

The country's reform program is deepening, although the rate of economic growth has slowed. The market has embraced the credibility of a stabilization program anchored on a fixed exchange rate and involving fiscal surplus, deregulation, privatization, and an open economy. Based on the successful implementation of the current reform program, the Argentinean economy should continue to post healthy growth in 1994. In last October's congressional elections, the President's party won an impressive victory which would allow Mr. Menem to seek another term in office. The Government is expected to hold the course in terms of fiscal and monetary policy through 1994.

[20]INCOME

In 1992, Argentina's GNP was $200.28 million at current prices, or $6,050 per capita. For the period 1985–92 the average inflation rate was 495.7%, resulting in a real growth rate in per capita GNP of 0.5%.

In 1992 the GDP was $228.78 million in current US dollars. It is estimated that in 1991, agriculture, hunting, forestry, and fishing contributed 7% to GDP; mining and quarrying, 2%; manufacturing, 24%; electricity, gas, and water, 2%; construction, 5%; wholesale and retail trade, 15%; transport, storage, and communication, 5%; finance, insurance, real estate, and business services, 15%; community, social, and personal services, 25%.

[21]LABOR

According to the 1970 census, the economically active population of Argentina was 9,011,450; by 1990, it was estimated at 12,305,000. According to 1985 estimates, the labor force was divided as follows: 21% in industry, 53% in services, and 13% in agriculture. Unemployment began to rise again in the 1970s, exceeding 7% by 1972; for the rest of the decade, however, Argentina's unemployment rate was among the lowest in Latin America (2.2% in 1978). In the wake of the 1981 recession, unemployment and underemployment rose to about 14%. The unemployment rate was 5.3% in October 1991.

Organized labor has probably had a greater effect on the modern history of Argentina than any other group. Before Perón, unions were poorly organized, few in number, and lacking in political power. Perón used the labor movement as his chief vehicle in achieving and holding dictatorial power. He built the General Confederation of Labor (Confederación General del Trabajo—CGT) from a disjointed membership of about 250,000 into a highly centralized organization of 6 million workers encompassing every aspect of the Argentine economy. With strong backing from the government, union organization proceeded at a spectacular rate, giving Perón control over landowners and management. In return for their solid support, labor unions won tremendous benefits in the form of higher wages and improved working conditions. These advances, however, were built on a political rather than a sound economic foundation, and the fortunes of the CGT waxed and waned with those of Perón and his followers. In 1991, 28% of laborers were unionized.

Following the 1966 coup, Gen. Onganía attempted to reach an understanding with various factions of the CGT, including the Perónists. Early in 1970, Onganía was successful in negotiating a series of agreements with confederation leaders, but inflationary pressures created a growing dissatisfaction with the economic policies of the government. When Gen. Levingston assumed office in mid-1970, he promised the CGT a stronger voice in labor affairs. The CGT regained much of its former influence under the Cámpora and two Perón governments, pushing for numerous inflationary wage increases. In 1973, an estimated 4 million workers were organized into 500 trade unions, 45 federations, and 3 confederations. The CGT, composed of 94 affiliated organizations, had an estimated membership of 2.5 million. All trade union activity was banned after the March 1976 coup, but the ban was rescinded in 1981, and the right to belong to a union was reestablished in September 1982. In December 1982, the unions organized a one-day general strike to protest inflation and human-rights abuses, with about 90% of all workers participating. Another successful one-day general strike, called by the CGT and enlisting about 85% participation, took place in March 1983. The government, which had sought to appease the unions by announcing a 12% wage increase, declared the strike legal.

Shortly after taking office in 1983, the Alfonsín administration attempted to legislate democratic practices into the labor movement. While the bill was defeated in the Senate and the government was forced to water it down, elections were held in most unions in 1984, and a reform movement, known as the Committee of 25, began in the CGT. In 1987, another pragmatic group of reformist union leaders, the Group of 15, called for negotiations rather than the one-day general strikes that the CGT's traditional leadership had been using to bring pressure on the government. The CGT split into two factions in 1989, over the question of whether to support Menem's economic policies. In March 1992, having seen their influence and power diminish, the CGT was reunified.

Through deficit reduction, tight fiscal policy, privatization of inefficient industries, the encouragement of domestic recapitalization and foreign investment, both inflation and unemployment have been reduced under the Menem administration. In late 1991, a deregulation decree and "flexibilization laws" were passed, affecting national union organizations. Under the decree, wage negotiation at the shop-floor level was encouraged, as was linkage of wages to productivity. Also under the deregulation, centralization of the vast union social welfare funds was to occur in the Labor Ministry. The "flexibilization laws" allowed businesses to cut union and pension contributions up to 50% for temporary and training positions, and set ceilings on severance and work-related accident payments. Argentine law specifies that all workers are entitled to a minimum wage (about $98 per month in 1991); family allowances for child care, educational expenses, and special occasions; a minimum paid vacation of 14 days annually; and an annual bonus equal to one-twelfth of an employee's total remuneration for the previous 12-month period. The legal workweek for men and women 18 years and over is 48 hours, with a maximum of 8 hours a day. Overtime rates are ordinarily determined through collective bargaining. A minimum of one month's notice is required for dismissal; the employer is obligated to pay full wages during that period and to grant the employee two paid hours a day to seek another job. Provincial laws and regulations provide additional protection for workers.

[22]AGRICULTURE

Agriculture and agro-industry in Argentina focus on the production of cereal, oil grains and seeds, sugar, fruit, wine, tea, tobacco, and cotton. Argentina is one of the greatest food-producing and food-exporting countries of the world, with an estimated 27,200,000 hectares (67,210,000 acres) of arable and permanent cropland. Agriculture and animal husbandry have traditionally supplied the nation with 70–95% of its export earnings, and the landowners have alternated the two activities in accordance with prices on the world market. As of 1992, agriculture made up 70% of the GDP, contributing to the growth of the GDP by 1% from the previous year. Agricultural products also account for 70% of exports by value. One of the most important factors in Argentine agriculture is the advanced degree of mechanization; in 1991, an estimated 202,000 tractors and 48,800 harvester-threshers were in use. In 1989, agriculture's contribution to GDP growth was down by 0.4%, but from 1990 to 1992, the contribution to GDP growth was positive, ranging from 0.7% to 1.4%.

The principal agricultural region consists of the humid pampas, one of the world's greatest reaches of arable land. Argentine agriculture is virtually coextensive with this region, although efforts have been made to spread it into other areas. Citrus fruit, tobacco, cotton, and sugarcane are cultivated outside the pampas.

Wheat is the leading crop. Argentina normally accounts for about 64% of all wheat produced in South America and was the world's fourth-leading wheat exporter, following the US, France, and Australia. The area sown in 1991/92 was estimated at 4.5 million hectares (11.1 million acres), and production at 9 million tons (down from 10.9 million tons in the previous year). Argentina is the third-largest corn-growing country of Latin America, following Brazil and Mexico. The area sown in 1991/92 was 2.4 million hectares (5.9 million acres), and production was 10.5 million tons. Barley is favored as the grain of greatest yield and resistance to disease; types for feed and beer are grown in the pampas areas having soil unfavorable or a climate too rigorous for wheat. Harvests once amounted to 659,000 tons per season in the early 1970s, but by 1990/91 production had fallen to 303,000 tons; in 1991/92 production reportedly climbed back to 565,000 tons.

Rice is a major crop, with a 1992 production of 753,000 tons on plantings of 148,000 hectares (365,700 acres). Argentina is one of the world's biggest producers of flaxseed (linseed); production in 1990/91 was 480,000 tons. Most of the crop is exported in the form of linseed oil. The province of Tucumán dominates the sugar-raising industry, which dates from 1646; sugarcane production in 1990/91 was 12.5 million tons. To control overproduction, the government formed the National Sugar Co. in 1970 and forbade the construction of new sugar mills through the end of the decade.

Cotton growing dates from 1909 and is concentrated in Chaco Province. In 1992 the production of cotton fiber was 253,000 tons. Sunflower seed oil is a major industrial plant product; 2.5 million hectares (6.1 million acres) of sunflowers were harvested in 1991/92. Tobacco is raised in several northern provinces, especially Misiones; production in 1991 was 94,443 tons. Soybean production, only 78,000 tons in 1971/72, increased to 7.1 million tons by 1985/86, and to an estimated 10.3 million tons in 1991/92.

Fruit growing has developed rapidly since the 1940s. Estimates for 1991/92 fruit production (in tons) were apples, 1,100,000;

oranges, 750,000; lemons and limes, 570,000; peaches and nectarines, 250,000; and grapefruit, 180,000. The output of bananas was 400,000 tons in 1974, 10 times the 1961–65 average; it fell to 144,000 tons in 1978 and rebounded to 280,000 tons in 1992.

The province of Mendoza is the center for the nation's vineyards. In 1991, grape production was 2 million tons. Argentina is one of the world's leading producers of wine, accounting for an estimated 1.15 million tons in 1992, or 4% of the world's total production.

The following table shows 1985/86 and 1990/91 crop production (in thousands of tons):

Barley	118	303
Corn	12,400	12,600
Cotton	330	324
Oats	400	434
Rice	450	299
Rye	94	96
Soybeans	7,100	11,500
Sugarcane	—	12,500
Tobacco	62	68
Wheat	8,700	10,900

[23]ANIMAL HUSBANDRY

Argentina is one of the world's preeminent producers of cattle and sheep, possessing approximately 4% of the entire world's stock of the former and 2% of the latter. Livestock and meat exports play an essential part in the nation's international trade. Annual meat exports (including meat extracts) were 598,900 tons in 1978, but fell to 394,900 tons in 1981 and 125,290 tons in 1992. Because of extremely favorable natural conditions, Argentina, with more than 50 million head of cattle in 1992, is one of the world's leading cattle-raising countries.

Cattle were introduced into Argentina by Pedro de Mendoza in 1536, and these cattle, together with those brought by other explorers, quickly became wild and began to multiply on the lush grasses of the pampas. There was no attempt to control the vast herds; when the inhabitants wanted meat and hides, they would merely kill the animals at random and take the desired parts. The most important single advance was the invention of refrigeration, which enabled ships to transport meat without spoilage. The policy followed by foreign-owned meat-packing firms of purchasing cattle by quality rather than weight led to the introduction of new breeds and selective cross-breeding, which have brought the cattle industry to its present advanced state.

Argentine pastures covered an estimated 142.1 million hectares (351.1 million acres) in 1991 and are most productive in the provinces of Buenos Aires, Santa Fe, Córdoba, Entre Ríos, and Corrientes. The most important beef-producing breeds are Shorthorn, introduced in 1823; Hereford, 1858; Aberdeen Angus, 1879; and in recent years, zebu and Charolais.

The dairy industry has shown steady development. In 1992, the following quantities were produced: milk, 6,700,000 tons; cheese, 280,000 tons; and butter, 42,000 tons. The most important dairy breeds are Holstein-Friesian, Jersey, and Holando Argentino. Egg production was 327,600 tons in 1992. The number of poultry in 1992 reached 58 million.

In sheep raising, Argentina ranks thirteenth in the world, with an estimated 23.7 million animals in 1992. Before World War II, Argentina accounted for 14% of the world's wool production, but in the 1970s, its production declined; the wool clip (greasy basis) was 128,000 tons in 1992. In 1992, production of mutton and lamb was 85,000 tons. Patagonia has approximately 40% of all the sheep in Argentina.

Total meat production was 3.54 million tons in 1992, of which 2.6 million tons consisted of beef.

In 1992, Argentina had 3.3 million horses, placing it among the top five countries in the world. Argentine horses, especially favored as polo ponies and racehorses, have won many international prizes. Other livestock in 1992 included 4.7 million pigs and 3.3 million goats. In 1990, Argentina accounted for 4.8% of the world's exports of leather, valued at over US$417.6 million. Argentina is South America's largest producer of honey, with an output of 45,000 tons in 1992.

[24]FISHING

In a country that is among the world's leaders in meat production, fishing has not been able to develop as an industry of any significance. In recent years, the government has tried with some success to induce the public to eat more fish in order to export more beef, one of the country's largest earners of foreign exchange. Since 1970, the government has offered fiscal incentives to encourage the modernization of the fishing industry. The catch has increased from 475,043 tons in 1982 to 559,777 tons in 1987 to 640,636 tons in 1991.

The most favored saltwater fish are the pejerrey, a kind of mackerel; the dorado, resembling salmon but of a golden color; and the zurubí, an immense yellow and black spotted catfish. There is a limited whaling industry.

Argentina established a 200-mi territorial sea limit in December 1966. In 1982, the government moved to protect Argentina's coastal waters from foreign exploitation, declaring that only 16 foreign vessels would be allowed in Argentine waters at any one time.

[25]FORESTRY

Argentina's forests, estimated at some 59.1 million hectares (146 million acres), or about 21% of the total area, constitute one of its greatest underexploited natural resources. Of the 570 species of trees sold in international commerce, Argentina possesses 370, but of these it exploits only about a dozen species. A major factor in the industry's lack of development is the great distance of most forests from the markets and the resultant high cost of transportation. In the Río Paraná Delta, the woods currently exploited are softwoods, such as the elm and willow, used in the cellulose and container industries; in the Gran Chaco, white quebracho, used as a fuel and in the refining of coal, and red quebracho, from which tannin is extracted; in Misiones Province, several varieties, including cedar for furniture manufacturing; in the Salta-Tucumán region, cedar and oak; and in Patagonia, araucania, pine, cypress, larch, and oak.

The most important tree is the red quebracho, which contains 21% tannin, the extract used for tanning. Argentina possesses four-fifths of the world's supply of this wood. Many quebracho trees now being used are from 200 to 500 years old, and trees younger than 75 years are of little commercial use. Since the trees are not being replaced, it is estimated that the quebracho forests will be largely exhausted by the end of the current century.

Production of roundwood was 10.8 million cu m in 1991.

[26]MINING

Until relatively recently, Argentina did not produce minerals in significant volume, but the future of the mining industry appears secure. The mountainous northwest, especially the province of Jujuy, is rich in a variety of minerals. New privatization laws were introduced in 1989 to give investors greater guarantees, reduce bureaucratic delays, and attract foreign capital. In 1991, mineral production and trade only nominally contributed to the GDP and total exports. Mineral exports, excluding hydrocarbons, dropped to an estimated US$67 million (60.5% in metal, 39.5% in industrial minerals), which was down 14% from 1990.

Output of iron ore was 239,400 tons in 1970; it peaked at 1.04 million tons in 1988 but by 1991 had declined to 980,000 tons. The largest deposits of iron ore are in Río Negro Province.

In 1991, the nation's production of important metal concentrates included zinc, 39,253 tons; lead, 23,697 tons; tin, 230 tons; copper, 408 tons; silver, 56,359 kg; and gold, 1,478 kg. Asphaltite, fluorspar, copper, mica, manganese, gold, silver, and antimony are found mainly in the northwest. There are deposits of varying amounts of salt, sodium sulfate, sulfur, and borax. There are also deposits of lithium, beryllium, and columbium. In 1991, Argentina produced 250,000 tons of boron, ranking 4th in the world after the US, Turkey, and Russia.

[27]ENERGY AND POWER

Despite a shortage of energy resources, production of electric power has steadily increased since 1958, after more than a decade of neglect. Installed generating capacity was estimated at 17.4 million kw in 1991, about 40% of which was hydroelectric; in that year, electrical energy production totaled 54.05 million kwh. The government places great emphasis on the development of hydroelectric projects and nuclear power, even though installed capacity exceeds projected demand. The Yac003retá-Aripe project on the Paraná will add 2,700 Mw to Argentina's generating capacity when completed.

Argentina was the first country in Latin America to install nuclear-powered electric generating plants. The Atucha power station in Buenos Aires Province has a capacity of 370 Mw; Embalse (600 Mw) in Córdoba Province started up in 1983, but was closed for five weeks for maintenance in late 1990. Nuclear energy accounted for 14% of total electricity production in 1991. Construction of a second reactor at Atucha began in 1980 and was expected to be complete and operating by 1995, after a decade of delays and extra costs. Argentina is rich in uranium; production of concentrate was 60 tons in 1991.

The modern petroleum industry dates from 1907; after 1940, it became necessary to supplement domestic production with large-scale imports of foreign fuels. In 1958, ownership of all crude oil and natural gas was taken over by the state, and petroleum was then placed under the control of the state oil corporation, Yacimientos Petrolíferos Fiscales (YPF). Production of crude oil fell from 25.6 million tons in 1982 to 22.3 million tons in 1987 but rose to 27.9 million tons in 1992. In 1978, foreign-owned companies were allowed to drill for oil, after decades of policy changes on the role of foreign companies. In August 1985, the Alfonsín government announced more liberal rules on foreign-company participation; in 1987, YPF's influence was reduced. Between 1990 and 1992, under the Menem administration, YPF leased or sold 56 of its secondary fields to the private sector. YPF also has entered joint ventures on four major fields, including its largest producing field. Since complete deregulation took place at the beginning of 1991, YPF no longer has a monopoly; US, British, Spanish, and Italian investors have become active in the petroleum market. Large deposits have been found in the San Jorge Gulf near Comodoro Rivadavia. In recent years though, production rates have exceeded the rate at which depleted reserves have been replaced by new discoveries. Proven reserves as of 1 January 1992 were estimated at only 1.57 billion barrels, down from the previous estimate of 2.3 billion barrels.

In conjunction with petroleum extraction, the significant natural gas industry has rapidly expanded. Production in 1992 was 23.3 billion cu m, compared with 5.3 billion cu m in 1969 and 9.8 billion cu m in 1982; about 80% was produced by YPF and about 20% by contractors. In 1992, known reserves were estimated at 579.05 billion cu m. Natural gas is imported from Bolivia (2.38 billion cu m in 1990). There is a network of over 9,900 km (6,150 mi) of gas pipelines.

A major coal deposit in Santa Cruz Province is estimated to contain 552 million tons of coal, nearly 80% of the nation's total. Production as a whole was reported at 291,546 tons in 1991, down from 505,000 tons in 1988. The state coal entity, Yacimientos

Carboníferos Fiscales (YCF), had planned to increase production to 650,000 tons by 1990; lack of investment and operating problems in the Rio Turbio mine led to the reduction in coal output.

[28]INDUSTRY

Argentina's principal industrial enterprises are heavily concentrated in and around the city of Buenos Aires. The plants are close to both the many raw materials imported by ship and the vast productive area of the pampas. The major industries in Buenos Aires are meat packing, food processing, machinery manufacturing and assembly, flour milling, tanning and leather goods manufacturing, oil refining, oilseed milling, and textile, chemical, pharmaceutical, and cement manufacturing. Other industrial areas include Rosario, with important steel-producing plants and oil refineries, tractor and meat-packing plants, and chemical and tanning industries; Córdoba, the center of the automobile and rolling-stock industries; Santa Fe, with zinc- and copper-smelting plants, flour mills, and dairy industry; San Miguel de Tucumán, with sugar refineries; Mendoza and Neuquén, with wineries and fruit-processing plants; the Chaco region, with cotton gins and sawmills; and Santa Cruz, Salta, Tierra del Fuego, Chubut, and Bahía Blanca, with oil fields and refineries. During 1960–70, the average annual growth rate of industry was 5.9%; during 1970–80, it was only 1.8%. In the early 1980s, industrial production went into recession, declining by 16% in 1981 and by 4.7% in 1982. The sharp cutback in imports due to the foreign debt crisis spurred local manufacturing to growth of 10.8% and 42%, in 1983 and 1984, respectively; 1985 brought a sharp plunge of 10.5%, but 1986 saw a growth of 12.8%, aided by the "Austral Plan."

In workers employed, value of production, and importance in foreign trade, packing and processing of foodstuffs is the oldest and most important industry. Beginning in the last part of the 19th century, the great frigoríficos, or meat-packing plants, were founded to prepare beef for export to Europe. In recent times, the Argentine government has entered directly into the meat-processing enterprises, which for many years were under British ownership. The textile industry was also developed quite early, making use of wool from the vast herds of sheep and the cotton from Chaco Province. In addition to these traditional products, a variety of synthetic fibers are now produced.

Portland cement is the country's leading construction material, with 5.55 million tons produced in 1986. A major chemical industry produces sulfuric, nitric, and other acids and pharmaceuticals. The petrochemical industry is related to the increasing production of oil and has received special benefits from the government. The most important center of this industry is San Lorenzo on the Río Paraná. Output of petroleum fuels reached 21.3 million tons in 1985, continuing the trend toward slow decline from 23.8 million tons in 1981. More important was the fact that in 1985, exports of petroleum fuels exceeded imports (US$690 million compared with US$440 million) for the first time.

In 1961, a giant integrated steel mill began production at San Nicolás. The output of crude steel totaled 3.24 million tons in 1986. Dependent on steel is the automobile industry, which experienced sustained growth during the 1960s and early 1970s. Production rose from 33,000 units in 1959 to 293,742 in 1973 but declined to 194,687 (including commercial vehicles) in 1978. In 1980, the total was 288,917; in 1986, 170,000. Tractors, motorcycles, and bicycles also are manufactured, primarily in the city of Córdoba. Argentina also produces electric appliances, communications equipment—including radios and television sets—motors, watches, and numerous other items.

Industry continues to restructure to become competitive after decades of protection. A firm's profit center was its treasury. Nowadays, the focus has shifted to more productive activities.

Capacity utilization rates have increased substantially in the past three years and companies are now focusing on modernization and expansion of their plants to meet both domestic and foreign demand. New technologies are being adopted, work forces pared, and management is focusing on just what its clients want. Overall industrial output reached record levels by mid 1993, capping a three-year climb from the trough of the hyperinflationary 1989–90. Output from the industrial sector is expected to increase by 9 percent in 1993 and 11 percent in 1994. Industrial production continues to demonstrate strong growth, particularly in the auto and energy sectors. The automobile sector provided the engine for growth-creating ripple effects throughout an extensive network of suppliers. Demand for consumer durable goods climbed as well, rising 11 percent in the first eight months of 1993. On the contrary, non-durable goods—in particular petroleum products—fell substantially.

The availability of credit as well as continuing strong demand for final products was felt throughout Argentina's basic industries. Output of cement, trucks, machinery, plastics, petrochemicals and other chemicals all rose, while production of basic metal goods held flat or rose off a low base.

29 SCIENCE AND TECHNOLOGY

Argentina has five scientific academies: an academy of agronomy and veterinary science (founded in 1909); an academy of exact, physical, and natural sciences (1874); an academy of medicine (1822); and the National Academies of Sciences of Córdoba (1869) and Buenos Aires (1935). In 1986 there were 30 scientific and technological research institutes. Research and development in 1986 amounted to 2.1% of budgeted government expenditures. Argentina has 48 universities and colleges offering training in basic and applied sciences.

In 1992, the public budget allocated to science and technology was $466 million. In 1988, 6,241 technicians and 11,088 scientists and engineers were engaged in research and development.

30 DOMESTIC TRADE

Many leading mercantile firms have their head offices in Buenos Aires and branches or agents in the other large cities. Department stores, retail shops, and specialty shops in Buenos Aires are on a par with similar establishments in most world capitals. The number of supermarkets and self-service shops is increasing, although most retail trade is still conducted in small specialty shops and grocery stores or in public markets supervised by the municipal governments. Stores are usually open from 9 AM to 7 PM, Monday–Friday, and close midday on Saturday; banks are generally open on weekdays from 10 AM to 4 PM. Domestic demand absorbs most of the nation's industrial production. Buenos Aires is the center for the nation's advertising agencies, including branches of several US firms.

31 FOREIGN TRADE

The main imports are raw materials, fuels, intermediate goods, and capital goods. Many industrial products imported prior to 1960 are now produced in Argentina. Argentina has removed virtually all non-tariff barriers to trade and reduced the highest tariff rate to 20 percent, except for automobiles, auto parts, and consumer electronics. The only non-tariff barrier is the tariff/quota system applicable to auto and auto parts imports. The Argentina/ Brazil auto agreement establishes preferential market access treatment for both countries. In view of the liberalization of trade policies and the real appreciation of the peso, the current account deficit has deteriorated sharply since 1990. This has been more than offset by a large influx of foreign capital that was enticed by the Government's new economic program. A surge in imports during 1991–92 shifted the trade balance from a large surplus to a deficit position. The strong increase in imports is explained by

several factors: first, the dynamic growth of the domestic economy which is resulting in greater import demand; second, the reduction of import tariffs and elimination of non-tariff barriers which has released pent-up demand for imports; and third, the real appreciation of the peso which has made imports much less expensive since the local currency cost of these goods has risen by much less than the accumulated inflation since the beginning of the Convertibility Plan.

	1991	1992	1993
Exports ($mm)	$11,978	$11,965	$12,619
growth (%)	–3.04	–0.11	5.47
Imports ($mm)	$ 7,400	$13,649	$14,877
growth (%)	98.6	84.5	9.0

Growth of imports of consumer goods has tapered off quite a bit since 1991–92, while growth in the import of capital goods continues to climb. Even more important is the incipient growth of exports, despite the overvaluation of the peso. Exports went up by 5.47% reaching US$12.6 billion in 1993 while imports grew 9.0% totalling US$14.88 billion. The export value of many non-traditional goods—including transport goods, prepared foods, metal goods, and machinery—is increasing, as firms prove competitive in world markets.

The creation of NAFTA is viewed as an extremely positive development and presents Argentina with the possibility of acceding to NAFTA as either a member of MERCOSUR or alone. The government remains fully committed to seeing the creation of MERCOSUR (a common market incorporating Argentina, Brazil, Uruguay, and Paraguay) through its completion (on January 1, 1995). The trade imbalance accelerated during the first months of 1994. During the first semester of 1994, exports to the US grew by 36.7%, to MERCOSUR grew by 9.8% and to the European Union by 5%. Imports showed a growing share of capital goods mainly from the US. On the other hand, MERCOSUR is relatively more important as the origin of imported goods, vehicles and consumption goods.

32 BALANCE OF PAYMENTS

Until 1952, Argentina's foreign-payments position was excellent, owing mainly to its large exports of basic commodities. In that year, however, because of widespread crop failures and unfavorable terms of trade, export value decreased sharply while imports remained high. The Argentine deficit was met by foreign credits, with dollar shortfalls partially covered by large credits from the Export-Import Bank, the IMF, and US banks. From 1961 to 1967, with the slow expansion of exports, large fiscal deficits continued. Most of the foreign exchange earned was used to service the external debt. Between 1963 and 1968, the nation enjoyed a trade surplus of US$2.05 billion, but 56% of it was absorbed by external debt burden. The strict economic controls enacted by the Onganía government in 1967 helped curb the inflationary trend and thus stabilized the nation's economy.

After a decline during the early 1970s, Argentina registered a surplus of US$863 million in 1973, raising its international reserve holdings to US$1.32 million as of 31 December 1973. Between 1975 and 1979, the annual payments surplus increased from US$921 million to US$4.43 billion, but since 1981, the current account has been in deficit (despite a persistent surplus in the trade account) because of heavy debt-servicing costs.

Since 1991, the decline in interest rates internationally and the lack of attractive alternatives for foreign direct investment have helped to generate a massive inflow of foreign capital, much of which was actually owned by Argentines but held abroad. A debt restructuring plan in early 1993 permitted the investment of reserves by eliminating the threat of seizure.

In 1992, merchandise exports totaled $12,235 million and imports $13,623 million. The merchandise trade balance was $–1,388 million.

The following table summarizes Argentina's balance of payments for 1991 and 1992 (in millions of US dollars):

	1991	1992
CURRENT ACCOUNT		
Goods, services, and income	−2,833	−8,329
Unrequited transfers	29	−32
TOTALS	−2,804	−8,361
CAPITAL ACCOUNT		
Direct investment	2,439	4,179
Portfolio investment	−195	−1,161
Other long-term capital	−117	−978
Other short-term capital	190	8,602
Exceptional financing	3,436	2,153
Other liabilities	22	−21
Reserves	−2,630	−4,550
TOTALS	3,145	8,224
Errors and omissions	−341	137
Total change in reserves	−1,977	−4,250

33 BANKING AND SECURITIES

In 1935, the Central Bank of the Argentine Republic was established as a central reserve bank, having the sole right of note issue, with all capital held by the state. On 31 December 1993, total reserves of the Central Bank were A$14,989 million. The bank acts as the fiscal agent of the state. Its board of directors is appointed by and responsible to the government. The bank administers banking laws, regulates the volume of credit and interest rates, supervises the securities market, and applies government laws and decrees regarding banking and foreign exchange. Legislation in August 1973 increased its control over the commercial banking system.

The National Mortgage Bank, founded in 1886, is the most important institution for housing credit. Other institutions include the National Development Bank, the National Bank for Savings and Insurance, and the Cooperative Credit Bank. Money supply, as measured by M2, totaled A$45,803 million at the end of 1993.

In 1993 there were 23 government-owned provincial banks, 5 government-owned municipal banks, 31 private commercial banks, and 30 foreign-owned banks. As of 1993, foreign assets of commercial banks totaled A$8.02 million pesos.

The Buenos Aires Stock Exchange is one of the 23 markets that form the Buenos Aires Commercial Exchange, which has over 12,000 members and is often confused with the Stock Exchange. The Commercial Exchange, founded in 1854, established the Stock Exchange, which the government subsequently separated from it. The Commercial Exchange now includes a grain market, a foreign currency exchange, a general produce exchange, and the securities exchange. There are also stock exchanges in the cities of Córdoba, San Juan, Rosario, Mendoza, and Mar del Plata, although more than 90% of stock transactions are conducted on the Buenos Aires exchange.

34 INSURANCE

In 1984, about 200 insurance companies were in operation in Argentina. Although various legal restrictions have been placed on foreign insurance companies, many retain offices in Buenos Aires. Between 1989 and 1991, insurance sales nearly doubled, rising from US$1.5 billion to US$3.1 billion, and then climbing to US$3.25 billion in 1992. Factors cited in this dramatic growth are increased economic stability, government deregulation, greater competition among private companies, and growth in the automotive industry. Automotive insurance accounted for 67% of the new policies issued from September 1990 to September 1991.

35 PUBLIC FINANCE

Since 1970, Argentina's budget picture has steadily worsened. By the late 1970s, deficit spending annually ranged from 10–14% of GDP, and topped 15% in the early 1980s, when public expenditures consumed some 40% of GDP. By the late 1980s, hyperinflation and depletion of reserves necessitated a public finance reform. Since 1991, the government has considerably narrowed the deficit gap through structural reform efforts. Stricter controls on public spending and more efficient tax collection methods resulted in an overall public sector accounts surplus of 1% of GDP in 1992, compared to a deficit equivalent to 21.7% of GDP in 1989. Fiscal income from the privatization of public enterprises added $6 billion in revenue during the first three quarters of 1992.

The following table shows actual revenues and expenditures for 1988 and 1989 in millions of pesos.

	1988	1989
REVENUE AND GRANTS		
Tax revenue	9,411.0	286.3
Non-tax revenue	834.5	27.8
Capital revenue	30.9	5.3
Grants	—	—
TOTALS	10,285.4	319.4
EXPENDITURES & LENDING MINUS REPAYMENTS		
General public service	735.8	25.4
Defense	1,030.7	30.4
Public order and safety	619.6	16.9
Education	1,119.8	30.3
Health	244.0	9.2
Social security and welfare	4,864.0	119.4
Housing and community amenities	43.5	1.0
Recreation, cultural, and religious affairs	106.9	2.5
Economic affairs and services	2,467.7	48.9
Other expenditures	893.3	22.6
Adjustments	−114.3	−1.1
Lending minus repayments	361.5	26.2
TOTALS	12,372.5	331.8
Deficit/Surplus	−2,087.1	−12.4

In 1988, Argentina's total public debt stood at A$16,442.2 million pesos, of which A$13,869.8 million pesos was financed abroad. In April 1993, Argentina rescheduled US$28 billion in debt principal and US$8 billion in arrears with its 750 commercial bank creditors.

36 TAXATION

In 1993, the principal national taxes included personal income tax (ranging between 6% and 30%), wealth tax (1%), value-added tax (18%), and excise taxes on tobacco, alcohol, soft drinks, perfumes, jewelry, precious stones, automobile tires, insurance policies, gasoline, lubricating oils, and other items. There is no inheritance tax. Corporate taxes are levied at 30% for domestic and foreign companies. Provincial and municipal governments impose various taxes.

37 CUSTOMS AND DUTIES

The Perón regime abolished in large measure the traditional system of exports and imports. Through the use of multiple exchange rates and through control of Argentine agricultural exports by the Argentine Institute for the Promotion of Exchange, Perón was able to obtain goods from producers at low prices and sell them abroad at great increases, employing the difference to promote the development of industry. In 1959, this cumbersome system of import permits and multiple exchange rates was abolished.

The government has since employed surcharges on imports to promote the growth of Argentine industries, and special import benefits are allowed to industries and regions regarded as significant contributors to the national economy. The petrochemical, cellulose, and steel industries have shared in these benefits, which include exemption from customs duties and exchange premiums on imports of machinery, spare parts, and raw materials. In 1993, the maximum tariff rate was 22% for finished products destined for consumption. Duties of 11% were applied to raw materials and medicines, and a 35% duty was applied to many electronic items and small appliances.

38FOREIGN INVESTMENT

During the 19th century, Argentina offered a favorable climate for foreign investment, and the basic development of the nation's transportation system and shipping facilities was financed with British capital. This system placed ownership of extensive properties in foreign hands, arousing the resentment of Argentine nationalists, who advocated a policy of reducing dependence on outside interests. The organization of a national petroleum agency, YPF, in 1922 was one of the first important steps in implementing that policy. The high point in the drive for nationalization came during the Perón era, when railroads were purchased from foreign owners and numerous state-owned enterprises were established. These measures led to substantially reduced foreign investment (only US$8 million between 1953 and 1955).

In 1958, President Frondizi negotiated contracts with a number of foreign companies, allowing them to join YPF in the exploitation of Argentine petroleum. In 1958, he promoted a bill designed to attract foreign capital under close government supervision. As a result, foreign companies invested, between 1959 and 1961, over US$387 million, of which more than half came from the US.

Between 1961 and 1966, direct foreign investment declined, with the question of foreign ownership constantly entering the political picture; in 1963, it amounted to US$78 million; in 1966, US$28 million. After the military coup, President Onganía declared that his government would renew an "open door" policy and would provide legal guarantees to investors. The net capital inflow continued through the late 1960s.

In the 1970s, government policies toward investors underwent a significant reappraisal. Restrictions on direct foreign investment, as specified in a foreign investment law enacted in 1973, required specific congressional approval for formation of companies with foreign capital that exceeded 50% of the total. Profit remittances and capital repatriation were limited, and new foreign investments were prohibited in several major areas, including national defense, banking, mass media, agriculture, forestry, and fishing. This restrictive legislation was superseded by a new foreign investment law enacted in 1976 and substantially amended in 1981. Incentives are provided for investments in mining, shipbuilding, iron and steel, petrochemicals, forest industries, silo construction, wine, and maritime fishing. Industrial investments within 60 km (37 mi) of the city of Buenos Aires and within the city limits of Córdoba and Rosario are not eligible for benefits.

Capital inflows remain strong. The privatization has generated an enormous source of US dollars with the sale of state owned enterprises to local and foreign investors.

Argentina in 1991 signed a Bilateral Investment Treaty with the US. A key provision for US investors is the right to seek, after 6 months, international arbitration of investment disputes. The treaty provides for national treatment in virtually all sectors of the Argentine economy. Those exceptions sought by Argentina have, in some instances, been unilaterally extended to foreign investors by subsequent Argentine laws, such as the Mining Law, which permits mining within 50 miles of Argentina's frontiers.

Foreign investors in Argentina face new limitations on their activities, except for some minor points, such as constraints on foreign banks opening new branches.

Principal foreign investors include: the US, Italy, Spain, France, Brazil, Chile, Germany, Switzerland, and the Netherlands.

39ECONOMIC DEVELOPMENT

Argentine economic policy has undergone several cycles of change since the 1940s. During World War II, the demand for Argentine beef and wheat boosted the country's exchange reserves to their highest point in history. Under the Perón regime, however, declining terms of trade and increasing state benefits and subsidies, as well as Perón's attempt to industrialize Argentina at the expense of the agrarian sector, disrupted the nation's economic system. Although inherently a wealthy country, Argentina, with a crushing foreign debt and a shattered economy, was nearly bankrupt.

When Perón fell, steps were taken to fund foreign obligations with long-term provisions for Argentine repayment and to create a climate favorable to private investment. Complicated multiple exchange rates were abolished, and massive financial assistance was extended by the IMF and other foreign agencies. Staggering deficits in the railroads and other state enterprises were a constant problem. The government sought to turn over some of these to private hands, and it also encouraged livestock raising and agricultural production, the chief earners of foreign exchange.

The government abolished many of the state subsidies and at the same time tried to hold wages steady. The austerity program fell hardest on the workers, who saw wages increase sluggishly while prices skyrocketed. They sought political solutions to the economic problems through crippling strikes, which in turn robbed the government of the increased production on which it was relying for a solution to the economic crisis. Between 1960 and 1966, the problems continued, with the government fluctuating between economic nationalism and liberal policies designed to seek foreign investment. Inflation, unemployment, and commercial failures reached new highs. Economic strife formed the backdrop for the military coup of 1966 and the suspension of the constitutional government. Despite widespread opposition, steps were taken in the late 1960s to turn over some state enterprises to private owners; other measures sought to put state-owned businesses on a paying basis.

The 1970s brought a resurgence of economic and political instability. The return to constitutional government—and especially the return of the Perónists to power—brought a period of increased labor influence, extraordinary wage demands, accelerating inflation, and huge government deficits, largely financed through short-term borrowing. The government's Three-Year Plan for Reconstruction and Liberation, announced in December 1973 during Juan Perón's presidency, called for more equitable distribution of income, elimination of unemployment and underemployment, better regional distribution of wealth, and extension of government housing, health, welfare, and education programs and services. More specifically, the plan projected annual growth rates of 7.5% for GDP, 6.5% for agriculture, 10% for industry, 14.8% for construction, and 11% for public utilities. Exports were to grow by 19.6% annually and imports by 13.1% annually. Wage earners were to receive 47.7% of the national income in 1977, compared with 42.5% in 1973, and urban unemployment was to decline to 2.5% in 1977, chiefly through labor-intensive projects in the social sector.

Perón's death in July 1974 and the subsequent political instability aborted this program and led to an economic crisis. In 1978, a medium-term economic adjustment plan, based on free-market principles, was announced. It included regular devaluations of the peso, cuts in public investment, and return of some state enterprises to private ownership; but instead of improving the nation's economic performance, the new policies led to triple-digit inflation and increasing unemployment. In the fall of 1982, the government began to negotiate with the IMF for a standby loan and committed itself to an austerity program, consisting of cuts in government spending, higher interest rates on bank loans to the private sector, and continuing regular devaluations. Additional financial controls, including a temporary ban on the issuance of new import licenses, were imposed the following autumn.

The "Austral Plan," launched in June 1985, was the Alfonsín government's attempt to break out of the stagflation that has characterized the economy since 1982. The combination of a wage-price freeze, a new currency pegged to the dollar, and a commitment to austerity in public spending was initially successful in curbing inflation, although somewhat at the expense of development. Since then, the government has attempted to manage price and wage increases and has offered several public corporations for sale.

Multilateral assistance to Argentina totaled us$6,332.7 million between 1962 and 1986, of which 51% came from the IDB and 41% from the IBRD. In December 1987, Argentina and Italy signed a us$5 billion pact to promote economic development in Argentina during 1988–92.

Based on the successful implementation of the current reform program, the Argentinean economy should continue to post healthy growth in 1994, but somewhat below the rapid pace of 1991–93. The President's party won an impressive 42.3% of the seats in the House, opening the way for Constitutional reform which would allow President Menem to seek another term in office. GDP is likely to post a healthy 5.0% growth in 1994. The industrial sector's performance is excellent, in particular the food processing, construction, and automotive industries. Demand for consumer durable goods is strong as a result of ample credit availability. Construction activity has been boosted by infrastructure projects associated with the privatization.

The government is expected to hold course in terms of fiscal and monetary policy through 1994. Nevertheless, slower economic growth may dampen growth in fiscal revenues, making it difficult for the government to meet its targets with the IMF. The current account deficit is expected to worsen as a result of continued import growth although the pace of import growth will subside. Exports should post moderate growth although the real appreciation of the currency will be further penalizing exporters despite government efforts to grant subsidies.

The key to the success of the government's economic program will be the ability to control inflation which in turn hinges on the stability of the currency. There is a growing consensus that a fixed exchange rate is not sustainable.

The credibility of the government's program is increased every time another structural inefficiency—one of the fabled "Argentine costs"—is reduced or eliminated. The privatization of many state owned enterprises has led to dramatic increases in efficiency. For instance, gas, electricity, water, and telephone services are showing noticeable improvement.

40 SOCIAL DEVELOPMENT

The election of Hipólito Yrigoyen as president of Argentina in 1916 initiated a series of profound changes in the nation's social structure. Although the urban middle class furnished the principal strength for his Radical Party, Yrigoyen also drew widespread support from the urban working class. The Radical-controlled legislature enacted a series of economic and social measures, including a measure to establish retirement funds. These social benefits were partly offset by continued policies of laissez-faire on the part of the president. Labor became hostile after a series of strikes was crushed by the army in 1919, an event thenceforth called the tragic week (semana trágica). Despite these differences between Radical leadership and labor, limited social welfare mea-

sures were continued until 1930, when Yrigoyen was expelled from office. The Conservative regime in power for the next 13 years took little cognizance of demands for social benefits.

Juan Perón was associated with important elements of the Conservative government while still a colonel in 1943, and he saw that by an alliance with labor he could have a powerful political base. Taking control of the relatively insignificant Department of Labor and Social Welfare, he began to extend benefits to workers and thereby win them to his political movement. Organized labor played a key role in his rise to power as president in 1946. Perón did not forget this, and in the following years he took aggressive steps to enact far-reaching social and economic measures. The 44-hour workweek that had been enacted in 1933 was for the first time put into effect. New provisions established salary increases, paid holidays, sick leave, job tenure, and many other benefits. Perón's wife Eva joined him in extending legislation to working women and to the poorer classes, called the shirt-less ones (descamisados). By 1945, a National Social Security Institute administered social insurance programs and the pension system. In the early 1950s, these measures continued and were extended also to the rural sector. The entire system was highly centralized under state control, with Perón exercising dictatorial powers to accomplish his goals.

Financial problems, including the decline of exports and a growing national debt, caused Perón to change his social policies radically after 1953, placing greater emphasis upon production than upon benefits. Many advantages that the working class had enjoyed were drastically curtailed. The failure of the Argentine welfare system to live up to Perón's promises helped to bring about his overthrow in 1955. During the 1960s, the pension funds were often diverted for other purposes, and there was a general breakdown in the system. By 1970, many of the persons eligible for welfare payments received none at all, and Francisco Manrique, secretary of social welfare under the Levingston administration, charged former government authorities with misappropriating millions of pesos.

Most of the social legislation enacted during the Perón years has remained on the statute books. In 1991, workers paid 10% of their wages into a pension fund, and this amount was supplemented by an 11% contribution from the employer. The employees also paid 3% (supplemented by 4.5% from employers) to a social health fund. Since Argentina has a relatively low rate of population growth (about 1.6% annually during 1980–85) and an abundant supply of fertile land, the government has felt no pressure to promote birth control. The fertility rate for 1985–1990 was 3.2. Abortion is permitted on broad health grounds.

Although guaranteed equality under the Constitution, women are fighting for equal advancement and pay in the labor force. The National Women's Council was created in 1991 to coordinate women's policies.

41 HEALTH

In the field of health and medical care, Argentina compares favorably with other Latin American countries. National health policy is determined by the Department of Public Health, an agency of the Ministry of Social Welfare. Nutritional requirements are comfortably met and, from 1988 to 1991, 65% of the population had access to safe water and 69% had adequate sanitation. Health and medical services for workers are provided by clinics of unions, and employers are usually required to provide free medical and pharmaceutical care for injured workers. From 1985 to 1992, 71% of the population had access to health services. Argentina's total health expenditures for 1990 were us$4.44 million.

There were 3 doctors per 1,000 people in 1992, and 150,000 hospital beds (4.8 beds per 1,000). In 1991, there were 21,900 dentists.

In 1992, the infant mortality rate was 22 per 1,000 live births, with 673,000 births that year. Between 1980 and 1993, 74% of married women (ages 15–49) used contraception.

Of the major infectious diseases, smallpox, malaria, and diphtheria have been virtually eliminated, and poliomyelitis has been greatly reduced. The incidence of tuberculosis in 1990 was 50 per 100,000 people, and from 1990 to 1992, one-year-old children were immunized against the following diseases: tuberculosis (99%); diphtheria, pertussis, and tetanus (78%); polio (83%); and measles (89%). Life expectancy averaged 71 years (75 for females and 68 years for males). The death rate in 1993 was 8.6 per 1,000 people (about the same rate as in 1960).

⁴²HOUSING

Housing in Argentina reflects the Italian and Spanish ethnic backgrounds of the population. Except for marginal rural dwellings and urban shanty towns, concrete, mortar, and brick are favored as the principal construction materials. Wood is generally considered less durable and feared as a fire hazard. In the late 1960s, the Ministry of Social Welfare initiated a program to eliminate the extensive shantytowns in Buenos Aires and other cities. In 1968, the government provided $50 billion (in old pesos) to finance the construction of 26,000 dwellings, and private capital totaling $200 billion (in old pesos) was set aside for an additional 100,000 units. The goal for 1969 was nearly double the previous year's investment, with major attention given to low-cost housing. Various international agencies, including the IBRD and the IDB, cooperated in providing housing funds, as did the US through AID programs. In the 1980s, however, the national housing program languished due to a fiscal policy of austerity, while sources of credit for would-be home buyers practically vanished. The total number of dwellings in 1992 was 9.22 million. As of 1993, there was a housing deficit of roughly 2.5 million houses in Argentina.

A new public credit agency, the National Housing Fund, was established in 1973; the fund is financed by a payroll tax on employers of 5% (as of 1986) on gross earnings.

⁴³EDUCATION

Argentina has one of the highest literacy rates in Latin America, estimated at 95.3% (95.5% for males and 95.1% for females in 1990). Education is free, secular, and compulsory for all children at the primary level. Secondary education lasts for four to six years depending on the type of course. The general certificate of education lasts for five years; commercial, four or five years, and technical or agricultural lasts for six years. The Ministry of Education supervises the National Council on Technical Education and the National Administration of Middle and Higher Education.

In 1991 there were 21,703 primary schools with 275,162 teachers and 4,874,306 students enrolled. Since 1966, the national universities have been under the supervision of the Ministry of Education. Traditionally, university students have played an active role in campus policy, based in part on the concept of university autonomy established in the Córdoba reform movement of 1918. Student organizations have also been outspoken in national politics, denouncing the policies of the military government in the late 1970s and early 1980s.

Argentina has 46 officially accredited universities with a total of 570,000 students (1988). The largest is the University of Buenos Aires with an enrollment of 187,000 in 1988.

Private, foreign, and religious schools are permitted, but they must conform to a nationally prescribed pattern of teaching in the Spanish language.

⁴⁴LIBRARIES AND MUSEUMS

The National Library was founded in 1810 and has occupied its present site in Buenos Aires since 1902; in 1989 it had about 1.95 million volumes. There are thousands of public and school libraries and innumerable private libraries. The libraries of the University of Buenos Aires have combined holdings of over 2 million volumes.

The National Museum of Fine Arts in Buenos Aires contains modern Argentine, American, and European works, as well as paintings attributed to old masters, paintings of the conquest of Mexico executed 300–400 years ago, and wooden carvings from the Argentine interior. Also in Buenos Aires are the National Historical Museum; the Isaac Fernández Blanco Museum of Hispanic-American Art, which contains an interesting and valuable collection of colonial art; the Mitre Museum and Library, containing the manuscripts, documents, printed works, and household objects of Gen. Bartolomé Mitre, which constitute a unique record of Argentine political development; the Natural Science Museum; and the Municipal Museum. There are several important historical museums in the provinces, including the Colonial and Historical Museum at Luján and the Natural History Museum of the University of La Plata, which is world-famous for its important collections of the skeletons of extinct pre-Pliocene reptiles (for which the Argentine pampas form one of the richest burial grounds).

⁴⁵MEDIA

In 1991 there were 3,921,629 telephones in the nationalized system, over half of which were in Buenos Aires and its environs. Internal telegraph facilities and some international circuits to nearby countries are wholly government operated.

In 1992 there were 171 AM radio stations and 231 TV stations. The number of radios was estimated at 22.3 million in 1991, and there were 7.2 million television sets.

Buenos Aires is one of the principal editorial centers of the Spanish-speaking world, with more than 50 publishing houses in the 1980s. Numerous literary magazines and reviews, as well as books, are published. Press coverage in Argentina is one of the most thorough in the hemisphere. Three news agencies (Noticias Argentinas, TELAM, and Diarios y Noticias) were operating in 1991, and the major international news services were also represented. Two of the great dailies of Buenos Aires, *La Nación* and *La Prensa,* enjoy international reputations, and *La Prensa* is probably the most famous newspaper in Latin America. Throughout the early days of the Perón regime, *La Prensa* battled the dictatorship, but it was finally taken over forcibly by Perón and given to the CGT, the dictator's central labor organization. The provisional government of Gen. Eduardo Lonardi returned *La Prensa* to its rightful owner, Alberto Gainza Paz, and it resumed publication in February 1956. In 1969, the Onganía government imposed siege regulations on the press, and in August of that year, two weekly papers were closed down. After the 1976 coup, no formal censorship was introduced, but some journalists were arrested for "subversive" articles. With the restoration of democratic government, harassment of the media stopped.

The largest dailies, with their estimated circulation figures in 1991, are listed in the following table:

	CIRCULATION
BUENOS AIRES	
Clarín	557,700
La Nación	217,000
La Prensa	115,000
Diario Popular	120,000
PROVINCES	
La Voz del Interior (Córdoba)	85,000
La Gazeta (Tucumán)	80,000
La Capital (Rosario)	80,000
El Día (La Plata)	57,000
Los Andes (Mendoza)	50,000
El Litoral (Santa Fe)	38,000

⁴⁶ORGANIZATIONS

Argentine organizations fall into the following main categories: agricultural, business and industrial, and social and cultural. The Argentine Rural Society, established in 1866, with a membership predominantly of owners of large ranches (estancias), occupies itself mainly with the improvement of agricultural and livestock production. The Argentine Association of Cooperatives and the Argentine Agrarian Federation also represent rural interests.

Industrialists and business leaders participate in the Argentine Industrial Union, which originated in 1887 and was reestablished in 1977 through the merger of the Argentine Industrial Confederation and the General Confederation of Industry. The leading chambers of commerce in 1993 were the Argentine Chamber of Commerce; the Chamber of Commerce, Industry, and Production of the Argentine Republic; the Chamber of Foreign Trade of the Federation of Trade and Industry; and the Chamber of Exporters of the Argentine Republic.

Social organizations are found in almost every community of any size. The Athletic and Fencing Club in Parque Palermo, a suburb of Buenos Aires, has extensive recreational facilities. The Argentine capital also sponsors numerous clubs in the delta region. At the other social extreme is the exclusive Jockey Club of Buenos Aires, with a wealthy membership. There are several yacht clubs. The Automobile Club operates a chain of service and rest stations throughout the country, giving travel information and selling gasoline at a slight discount. Many intellectuals belong to the Argentine Writers' Society.

⁴⁷TOURISM, TRAVEL, AND RECREATION

The government promotes tourist trade through the National Tourist Bureau, with headquarters in Buenos Aires. In 1991, 2,870,346 foreign tourists visited Argentina, 34% from Uruguay, 20% from Chile, and 10% from Brazil. Receipts from tourism totaled US$2.3 billion. As of 1990, there were 108,812 hotel rooms with 264,804 beds. Foreigners visiting Argentina must have a valid passport. No visa is required of citizens of Western Hemisphere countries (except Cuba and the US), Western European countries (except Portugal), and Japan.

Mar del Plata, on the South Atlantic, about 400 km (250 mi) from Buenos Aires, is the most popular ocean resort. The delta of the Río Paraná, forming a series of inland waterways, is a center for pleasure boats and launches. Córdoba, with its fine colonial cathedral, and nearby Alta Gracia attract many visitors. San Carlos de Bariloche, at the entrance to Nahuel Huapi National Park in the Andean lake region of western Patagonia, has become famous as a summer and winter resort, with some of the best skiing in the Southern Hemisphere. The Iguazú Falls, in the province of Misiones, on the border of Argentina and Brazil, is a major tourist attraction. Mendoza, situated in a fertile oasis below the towering Andes, offers such historical attractions as the Cerro de la Gloria, with its monument to San Martín, and the Historical Museum, with its collection on San Martín.

The most popular sport is football (soccer); in 1978, the year Argentina hosted the World Cup competition, the Argentine national team won the championship. Tennis, rugby, basketball, and golf are also played. Opportunities for gambling include a weekly lottery, football pools, horse racing at the Palermo and San Isidro tracks (in Buenos Aires), and the casino at Mar del Plata, whose profits go to the Ministry of Social Welfare.

⁴⁸FAMOUS ARGENTINES

The most famous Argentine is José de San Martín (1778–1850), known as the Protector of the South, who was principally responsible for freeing southern South America from the Spanish yoke.

The tyrannical dictator Juan Manuel de Rosas (1793–1877) ruled Argentina from 1829 to 1852. The political tactics and the pen of the statesman and essayist Domingo Faustino Sarmiento

(1811–88) did much to undermine him. While in exile, Sarmiento wrote some of his best works, including *Facundo,* the story of a rival caudillo. The most literary of Argentina's statesmen was Gen. Bartolomé Mitre (1821–1906), who was president from 1862 to 1868. Mitre, the founder and owner of the newspaper *La Nación,* wrote several important historical works and biographies. The most famous Argentine political figures of modern times have been Juan Domingo Perón Sosa (1895–1974) and his second wife, Eva Duarte de Perón (1919–52), known as "Evita." Perón's third wife, María Estela ("Isabel") Martínez de Perón, was vice-president during 1973–74 and, after her husband's death, president from 1974 to 1976.

José Hernández (1834–86), one of the first Argentine literary figures to use the uncultured language of the gaucho in his writings, is the author of *Martín Fierro,* considered the greatest of gaucho poems. Ricardo Güiraldes (1886–1927) kept the "gauchesco" spirit alive in his novel *Don Segundo Sombra,* a spiritual study of an Argentine gaucho. A less romantic view of these hardy horsemen of the pampas appears in the writings of Benito Lynch (1885–1951). The works of the poet Leopoldo Lugones (1874–1938) form a panorama of all Argentine life and landscape. José Mármol (1817–71) gave a good description of life in Buenos Aires under the tyrant Rosas in his novel *Amalia,* and Enrique Rodríguez Larreta (1875–1961) wrote the first Latin American novel to win international fame, *La gloria de Don Ramiro,* a reconstruction of Spanish life during the reign of Philip II. The leading contemporary writer of Argentina is Jorge Luis Borges (1899–1986), best known for his essays and collections of tales such as *Historia universal de la infamia.* Other world-famous writers are Julio Cortázar (1914–84) and Adolfo Bioy Casares (b.1914). Outstanding in the visual arts are the sculptor Rogelio Irurtia (1879–1950) and the painters Miguel Carlos Victorica (1884–1955) and Emilio Pettoruti (1892–1971). Argentina's foremost composers are Alberto Williams (1862–1952), founder of the Buenos Aires Conservatory; Juan José Castro (1895–1968); Juan Carlos Paz (1901–72); and Alberto Ginastera (1916–83).

The most famous Argentine scientist, Bernardo Alberto Houssay (1887–1971), was awarded the 1947 Nobel Prize in medicine for his work on diabetes; French-born Luis Federico Leloir (1906–87) won the Nobel Prize for chemistry in 1970. Notable philosophers include Alejandro Korn (1860–1936), whose work marked a reaction against positivism, and Francisco Romero (1891–1962). Carlos Saavedra Lamas (1878–1959), an authority on international law, received the Nobel Prize for peace in 1936. Adolfo Pérez Esquivel (b.1931), a sculptor and professor of architecture, received the Nobel Peace Prize in 1980 for his work in the Argentine human-rights movement.

⁴⁹DEPENDENCIES

Argentina continues to claim the Falkland Islands (Islas Malvinas), held by the UK, and a sector of Antarctica as dependencies.

⁵⁰BIBLIOGRAPHY

Alexander, Robert J. *Juan Domingo Perón.* Boulder, Colo.: Westview, 1979.

American University. *Argentina: A Country Study.* Washington, D.C.: Government Printing Office, 3d ed. 1985.

Argentina in Pictures. Minneapolis: Lerner, 1988.

Barnes, John. *Evita—First Lady: A Biography of Eva Perón.* New York: Grove, 1978.

Biggins, Alan. *Argentina.* Oxford, England; Santa Barbara, Calif.: Clio Press, 1991.

Calvert, Susan. *Argentina: Political Culture and Instability.* Pittsburgh, Pa.: University of Pittsburgh Press, 1989.

Corradi, Juan E. *The Fitful Republic: Economy, Society, and Politics in Argentina.* Boulder, Colo.: Westview, 1985.

Crassweller, Robert D. *Perón and the Enigma of Argentina.* New York: Norton, 1987.

Crawley, Eduardo. *A House Divided: Argentina, 1880–1980.* New York: St. Martin's, 1984.

Fernández, Julio A. *The Political Elite in Argentina.* New York: New York University Press, 1970.

Ferns, Henry S. *The Argentine Republic, 1516–1971.* New York: Barnes & Noble, 1973.

Foster, David W. and Virginia R. *Research Guide in Argentine Literature.* Metuchen, N.J.: Scarecrow, 1970.

Gillespie, Richard. *Soldiers of Perón: Argentina's Monteneros.* New York: Oxford University Press, 1982.

Goldwert, Marvin. *Democracy, Militarism and Nationalism in Argentina, 1930–1965: An Interpretation.* Austin: University of Texas Press, 1972.

Halperin-Donghi, Tulio. *Politics, Economics and Society in Argentina in the Revolutionary Period.* New York: Cambridge University Press, 1975.

Hastings, Max, and Simon Jenkins. *The Battle for the Falklands.* New York: Norton, 1983.

Horvath, Laszlo. *A Half Century of Perónism, 1943–1993: an International Bibliography.* Stanford, Calif.: Hoover Institution, Stanford University, 1993.

Kirkpatrick, Jeane. *Leader and Vanguard in Mass Society: A Study of Perónist Argentina.* Cambridge, Mass.: MIT Press, 1971.

Mallon, Richard, and Juan Sourrouille. *Economic Policymaking in a Conflict Society: The Argentine Case.* Cambridge, Mass.: Harvard University Press, 1975.

Martínez de Hoz, José Alfredo. *La Agricultura y la Ganadería en el Período 1930–1960.* Buenos Aires: Sudamericana, 1967.

Martínez Estrada, Ezequiel. *X-Ray of the Pampa.* Austin: University of Texas Press, 1975.

Masiello, Francine. *Between Civilization and Barbarism: Women, Nation, and Literary Culture in Modern Argentina.* Lincoln: University of Nebraska Press, 1992.

Palacio, Ernesto. *Historia de la Argentina, 1515–1938.* Buenos Aires: Alpe, 1954.

Peterson, Harold F. *Argentina and the United States, 1810–1960.* Albany: State University of New York, 1964.

Potash, Robert A. *The Army and Politics in Argentina 1928–1945: Yrigoyen to Perón.* Stanford, Calif.: Stanford University Press, 1969.

Potash, Robert A. *The Army and Politics in Argentina, 1945–1962: Perón to Frondizi.* Stanford, Calif.: Stanford University Press, 1980.

Randall, Laura. *An Economic History of Argentina.* New York: Columbia University Press, 1977.

Ratliff, William E. *Changing Course: the Capitalist Revolution in Argentina.* Stanford, CA: Hoover Institution, Stanford University, 1990.

Rock, David. *Authoritarian Argentina: the Nationalist Movement, its History, and its Impact.* Berkeley: University of California Press, 1993.

Scobie, James R. *Argentina: A City and a Nation.* New York: Oxford University Press, 1971.

Solberg, Carl E. *Oil and Nationalism in Argentina: A History.* Stanford, Calif.: Stanford University Press, 1979.

Sunday Times of London Insight Team. *War in the Falklands: The Full Story.* New York: Harper & Row, 1982.

Tella, Guido di and Carlos Rodriguez Braun (eds.). *Argentina, 1946–83: the Economic Ministers Speak.* New York: St. Martin's Press, 1990.

Tella, Guido di, Carlos Rodriguez Braun, and Rudiger Dornbusch (eds.). *The Political Economy of Argentina, 1946–83.* Pittsburgh, Pa.: University of Pittsburgh Press, 1989.

Timerman, Jacobo. *Prisoner Without a Name, Cell Without a Number.* New York: Knopf, 1981.

Waisman, Carlos H., and Monica Peralta-Ramos (eds.). *From Military Rule to Liberal Democracy in Argentina.* Boulder, Colo.: Westview, 1986.

Windhauser, Anselm. *Geología Argentina.* Buenos Aires: Talleres S.A., Casa J. Peuser, 1931.

Wright, Ione S., and Lisa M. Nekhom. *Historical Dictionary of Argentina.* Metuchen, N.J.: Scarecrow, 1978.

BAHAMAS

Commonwealth of the Bahamas

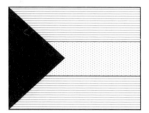

CAPITAL: Nassau.

FLAG: Three horizontal stripes of blue, gold, and blue, with a black triangle at the hoist.

ANTHEM: *March on Bahamaland.*

MONETARY UNIT: The Bahamas dollar (B$) of 100 cents has been in use since May 1966; as of June 1972, the Bahamas dollar ceased to be part of the sterling area and was set on a par with the US dollar. There are coins of 1, 5, 10, 15, 25, and 50 cents, and 1, 2, and 5 dollars, and notes of 50 cents and 1, 3, 5, 10, 20, 50, and 100 dollars. B$1=US$1 (or US$1=B$1).

WEIGHTS AND MEASURES: Imperial weights and measures are in use.

HOLIDAYS: New Year's Day, 1 January; Labor Day, 1st Friday in June; Independence Day, 10 July; Emancipation Day, 1st Monday in August; Discovery Day, 12 October; Christmas Day, 25 December; Boxing Day, 26 December. Movable religious holidays include Good Friday, Easter Monday, and Whitmonday.

TIME: 7 AM = noon GMT.

¹LOCATION, SIZE, AND EXTENT

The Commonwealth of the Bahamas occupies a 13,940-sq km (5,382-sq mi) archipelago which extends 950 km (590 mi) SE–NW and 298 km (185 mi) NE–SW between southeast Florida and northern Hispaniola. Comparatively, the area occupied by the Bahamas is slightly larger than the state of Connecticut. There are nearly 700 islands, of which about 30 are inhabited. New Providence, 207 sq km (80 sq mi), although not the largest, is by far the most populous and most densely populated island. The total coastline is 3,542 km (2,201 mi).

The Bahamas occupy a strategic location adjacent to the US and Cuba.

The Bahamas' capital city, Nassau, is located on Nassau Island in the center of the island group.

²TOPOGRAPHY

The Bahamas were formed as surface outcroppings of two oceanic banks, the Grand Bahama Bank and the Little Bahama Bank. The islands are for the most part low and flat, rising to a peak elevation of about 63 m (206 ft), on Cat Island. The terrain is broken by lakes and mangrove swamps, and the shorelines are marked by coral reefs.

³CLIMATE

The climate is pleasantly subtropical, with an average winter temperature of 23°C (73°F) and an average summer temperature of 27°C (81°F). Rainfall averages 127 cm (50 in), and there are occasional hurricanes.

⁴FLORA AND FAUNA

Because of a favorable combination of soil and climate conditions, the islands abound in such tropical flora as bougainvillea, jasmine, oleander, orchid, and yellow elder. Native trees include the black olive, casuarina, cascarilla, cork tree, manchineel, pimento, and seven species of palm. There are 218 species and subspecies of birds, including flamingos, hummingbirds, and other small birds and waterfowl.

⁵ENVIRONMENT

Among the government's priorities in environmental protection are monitoring industrial operations, providing potable water and regular garbage collection throughout the country, maintenance and beautification of public parks and beaches, and the removal of abandoned vehicles. Other significant environmental issues are the impact of tourism on the environment, waste disposal, and water pollution. The principal environmental agency is the Department of Environmental Health Services. A rookery on Great Inagua affords protection to some 30,000 flamingos as well as to the roseate spoonbill. Land clearing for agricultural purposes is a significant environmental problem because it threatens the habitat of the nation's wildlife.

As of 1987, endangered species included Kirtland's warbler, Bachman's warbler, green sea turtle, hawksbill turtle, Allen Cays rock iguana, and Watling Island ground iguana. The Caribbean monk seal and American crocodile are extinct. Of 17 species of mammals, 2 are endangered. Four species of birds are also threatened. Four species of reptiles in a total of 204 are threatened. One amphibian of 124 species is also considered endangered as of 1994.

⁶POPULATION

The population in 1990 was 255,095, an increase of 22% over the 1980 census figure of 209,505. Some two-thirds of the people reside on the island of New Providence (171,542 as of the 1990 census), the site of Nassau, the largest city. The population of Freeport on Grand Bahama Island (41,035) grew from a few hundred in 1960 to 24,423 in 1990. Population density in 1990 averaged 18.4 persons per sq km (47.7 per sq mi), but the density on New Providence in 1990 was 829 per sq km (2,146 per sq mi).

⁷MIGRATION

Emigration to the UK, considerable in the past, has fallen off since the mid-1960s. Some Bahamians migrate to the US in search of employment. Illegal immigrants from Haiti, many of whom take jobs considered menial by Bahamians, numbered an estimated 30,000–40,000 in 1993. There is interisland migration, chiefly to New Providence and Grand Bahama islands.

8ETHNIC GROUPS
Descendants of slaves brought to the Western Hemisphere from Africa make up about 86% of the legal population. About 8% of the total is of mixed origin; the remainder is white, largely of British origin.

9LANGUAGES
English is the spoken and official language of the Bahamas. Haitian immigrants speak French or a Creole patois.

10RELIGIONS
The population is overwhelmingly Christian, with Baptists comprising 32%. Roman Catholics (an estimated 19% in 1992) and Anglicans (20%) also have sizable memberships, and Methodists (6%) and the Church of God of Andersonville, Indiana (5.7%) are represented. More traditional practices related to witchcraft and known to scholars as voodoo or obeah continue to be observed in some areas.

11TRANSPORTATION
The larger islands have modern road networks. In 1991 there were about 2,400 km (1,500 mi) of highways, of which 56% were paved and 44% were gravel. There were 70,000 passenger cars and 15,000 commercial vehicles that year. About 60% of all vehicles are on New Providence. There are no railways.

The Bahamas established a shipping register in 1976. By 1 January 1992, the archipelago nation had a merchant fleet of 756 ships with a volume capacity of 18.2 million GRT and a deadweight capacity of 30.5 million tons (7th in the world). Nassau is a major port of call for cruise ships, which visit Freeport as well. There are international airports at Nassau and Freeport, with frequent connections to the US, Canada, and the UK. In 1990, approximately 1.23 million passengers embarked or disembarked through Freeport International Airport. Some 54 airstrips link the major islands. Bahamas Air, a state enterprise, is the national airline.

12HISTORY
Christopher Columbus is believed to have made his first landfall on the island now called San Salvador (formerly Watlings Island) on 12 October 1492, but the Spanish made no permanent settlement there. Spanish traders captured the native Lucayan Indians and sold them as slaves. The first permanent Erupoean settlement was established in 1647 by the Eleutherian Adventurers, a group of religious refugees. They and subsequent settlers imported blacks as slaves during the 17th century. The islands were also used as bases for pirates, including the notorious Blackbeard.

The British established a crown colony to govern the islands in 1717. The first royal governor, Captain Woodes Rogers, himself an ex-pirate, drove away the privateers, leaving the slave trade as the main economic enterprise on the islands.

After the end of slavery in 1838, the Bahamas served only as a source of sponges and occasionally as a strategic location. During the US Civil War, Confederate blockade runners operated from the islands. After World War I, prohibition rum-runners used the islands as a base. During World War II, the US used the islands for naval bases.

Like other former British colonies, the Bahamas achieved independence in stages. After self-government was established in 1964, full independence was granted on 10 July 1973. The country's first prime minister was Lynden O. Pindling, leader of the Progressive Liberal Party. Pindling ruled for nearly twenty years, during which the Bahamas benefitted from tourism and foreign investment. By the early 1980s, the islands had also become a major center for the drug trade, with 90% of all the cocaine entering the US reportedly passing through the Bahamas.

In August 1992, the Bahamas had its first transfer of political power, when Hubert Ingraham became prime minister.

13GOVERNMENT
Under the constitution of 10 July 1973, the Bahamas adheres to a republican form of government, formally headed by the British sovereign, who is represented by a governor-general. Executive authority is vested in a prime minister and a cabinet. The bicameral legislature consists of a 16-member Senate, appointed by the governor-general (9 on the advice of the prime minister, 4 on the advice of the opposition leader, and 3 at the governor's discretion), and an elected 49-member House of Assembly. The prime minister is the leader of the majority party in the House. The normal span of the elected legislature is five years, but, as in the U.K., elections can be called at any time. Suffrage is universal at age 18.

14POLITICAL PARTIES
The Progressive Liberal Party (PLP), a leader in the pro-independence movement, emerged as the Bahamas' majority party in the early 1970s. The Free Progressive Liberal Party, a splinter group formed in 1970, merged with another opposition group, the United Bahamian Party, to form the Free National Movement (FNM). After years of loyal opposition, the FNM took power in 1992, winning 32 seats, to 17 for the PLP. Two other parties, the Vanguard Nationalist and Socialist Party (VNSP) and the People's Democratic Force (PDF) are without any representation in the House of Assembly.

15LOCAL GOVERNMENT
There are 21 administrative districts, consisting of various islands and groups of islands. Each is headed by a commissioner responsible to the national minister of local government.

16JUDICIAL SYSTEM
British common law forms the basis of the Bahamas' judicial system. The highest court is the Court of Appeal, consisting of three judges. The Supreme Court is composed of a chief justice and two puisne justices. High court appointments are made by the governor-general. Ultimate appeals go to the Privy Council of the UK. Lower courts include three magistrates' courts on New Providence and one on Freeport. For other islands, commissioners decide minor criminal and civil cases.

The judiciary is independent. Judges are appointed by the executive branch with the advice of a judicial panel.

Long pretrial detentions are not uncommon in cases involving narcotics. In 1993, new magistrate's courts were established in order to work toward a reduction of backlogs requiring long pretrial detentions. A new Supreme Court was established in Freeport (in addition to the Supreme Court in Nassau).

Police abuse of suspects has been a serious problem. In 1993, a coroner's court was established to investigate cases in which criminal suspects die while in police custody.

Criminal defendants have the right to an attorney, but government appointed counsel is provided only in capital cases. There is also a right to be brought before a magistrate within 48 hours, a right to bail, a presumption of innocence and a right to appeal.

17ARMED FORCES
The Royal Bahamas Defense Force of 700 sailors is responsible for external security, manning 15 patrol boats and other small craft. The police of 1,700 provide internal security. Defense expenditures are $65 million (1990) or 2.7 percent of gross domestic product.

¹⁸INTERNATIONAL COOPERATION
The Bahamas joined the UN on 18 September 1973 and belongs to ECLA and all the nonregional specialized agencies except IAEA, IDA, IFAD, IFC, and UNIDO. It is a member of the Commonwealth of Nations, G-77, IDB, OAS, and PAHO. The Bahamas signed the Law of the Sea and is a de facto adherent of GATT.

¹⁹ECONOMY
Tourism, the mainstay of the economy, directly or indirectly involves most of the population. Because of the absence of direct taxation, the Bahamas has also become a financial haven for the activities of a large number of banking and trust companies, mutual funds, investment firms, and offshore sales and insurance companies. Local firms produce a small array of exports, including salt, cement, timber, pharmaceutical, and petroleum products refined on Grand Bahama and reexported. To broaden the economic base, the government is attempting to diversify agriculture and industry.

The agricultural sector, which presently accounts for 5% of GDP and employs 5% of the labor force, expanded during 1992 as the Government's agricultural development program gained momentum. Export-oriented citrus and vegetable production in Abaco grew as export demand rose. On Eleutherea, most agricultural and fishing infrastructure was seriously damaged by Hurricane Andrew. Similarly, tourism, which continues to be the Bahamas' engine of growth, contracted for the second consecutive year in 1992. As a result, the country's economy remained in recession in this year. GDP declined by 2%, following a 4% drop in the previous year. Unemployment reached almost 15% of the labor force and domestic demand contracted as both consumption and investment declined.

²⁰INCOME
In 1992, Bahamas' GNP was $3,161 million at current prices, or $12,020 per capita. For the period 1985–92 the average inflation rate was 5.5%, resulting in a real growth rate in per capita GNP of –1.2%.

In 1992, the GDP was $2.6 million in current US dollars. It is estimated that in 1991 agriculture, hunting, forestry, and fishing contributed 3% to GDP; mining, quarrying, and manufacturing, 4%; electricity, gas, and water, 3%; construction, 3%; wholesale and retail trade, 27%; transport, storage, and communication, 6%; finance, insurance, real estate, and business services, 21%; community, social, and personal services, 9%; and other sources 23%.

²¹LABOR
The total number of workers was estimated at 127,500 in 1991, the overwhelming majority employed in tourism or tourist-related activities. A 1970 Industrial Relations Act, as amended in 1979, requires the registration of trade unions and provides guidelines for labor contracts and job actions. The leading union federation is the Commonwealth of the Bahamas Trade Union Congress, which groups 10 of the country's 44 registered unions, and has a combined membership of 12,000. About 30,000 workers (25% of the labor force) were unionized in 1991. Unemployment in 1992 stood at about 20,000, or 14.8%.

Under the Fair Labor Standards Act of 1970, an employee working more than 8.5 hours a day or 48 hours per week is entitled to overtime pay. A minimum paid vacation of one week is also guaranteed, though a minimum of two weeks is standard. Wage councils may establish minimum rates on an industry-by-industry basis.

²²AGRICULTURE
Agriculture is carried out on small plots throughout most of the islands. Only about 1% of the land area is cultivated. The nature

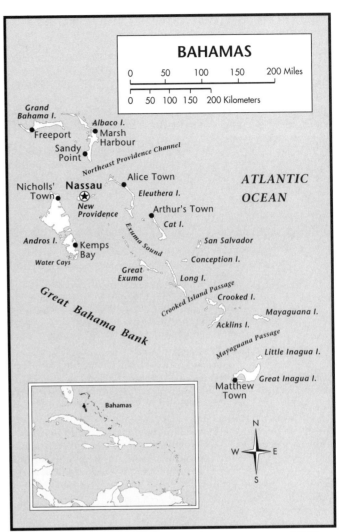

LOCATION: 20°50' to 27°25'N; 72°37' to 82°32'W. **TOTAL COASTLINE:** 3,542 km (2,201 mi) **TERRITORIAL SEA LIMIT:** 3 mi.

of the terrain limits the scope of farming, which is mainly a household industry. The main crops are vegetables: onions, okra, and tomatoes, the last two raised mainly for export. Inadequate production has necessitated the import of some 80% of the islands' food supply. Among steps the government has taken to expand and improve agriculture is the reserving of 182,000 hectares (450,000 acres) exclusively for farming, 8,000 hectares (20,000 acres) of which were converted to fruit farming. Export-oriented orange, grapefruit, and cucumber production in Abaco grew in 1992 as export demand expanded. However, on Eleuthera, most agricultural and fishing infrastructure was seriously damaged by Hurricane Andrew. In 1992, about $34.1 million in agricultural products was exported by the Bahamas. The government is seeking to produce over $100 million annually in food to reduce imports.

²³ANIMAL HUSBANDRY
Except for poultry and egg production, the livestock industry is relatively insignificant. In 1992, the livestock population included 5,000 head of cattle, 40,000 sheep, 19,000 goats, 20,000 hogs, and 2,000,000 poultry. About 3,000 tons of cow's milk, 1,000 tons of goat's milk, and 420,000 tons of eggs were produced in 1992. Poultry meat production in 1992 (7,000 tons) accounted for nearly 90% of domestic meat production. In December 1991, the government banned foreign chicken, in

order to protect local poultry producers from cheaper imports, mainly from the US.

²⁴FISHING
The 1981 catch amounted to 9,202 tons, over 80% of which was spiny lobsters (crayfish). Crayfish and conch exports are commercially important. There is excellent sport fishing for wahoo, dolphin, and tuna in Bahamian waters. In 1991, fisheries exports totaled $51 million. Since the Bahamas imports 80% of its food, the government is interested in expanding the role of domestic commercial fishing. Aquaculture and mariculture development are planned to grow into a $150 million annual business by the government during the 1990s, with the anticipation of 15,000 new jobs created.

²⁵FORESTRY
Caribbean pine and cascarilla bark are the major forestry products, but there is no commercial forestry industry.

²⁶MINING
Salt, with about 1 million tons produced annually from salt beds on Great Inagua and Long Island, and aragonite, a component in glass manufacture, are the only commercially important mineral products.

²⁷ENERGY AND POWER
Most electricity is produced at thermal plants owned by the Bahamas Electricity Corp. Net installed capacity in 1991 was 401,000 kw; production totaled 965 million kwh. As of 1991, a 28,000 kw upgrade was underway at the Clifton Power Plant on the west end of New Providence Island. Two new gas turbines are scheduled to be built at the Blue Hill Power Station. In 1986, a U.S. company began drilling for offshore oil in the Great Bahama Bank area. In 1991, a petroleum exploration permit for the southern Bahamas was being negotiated between British Petroleum Exploration Operating Company and the government. Fuels account for about 50% of all imports by value.

²⁸INDUSTRY
Large-scale oil refining began in 1967 with the installation of a large refinery on Grand Bahama with a daily capacity of 500,000 barrels. The facility was constructed with an investment of $300 million from US oil firms. In 1983, 2.93 million tons of petroleum products were produced, down from 9.53 million tons in 1979. Cement and rum production are also important, and enterprises producing pharmaceuticals and steel pipe have been developed. During the 1970s, the New Providence Industrial Estate was built by the government as a center for new light industry. Heineken, Europe's largest beer company, constructed a brewery in Nassau in 1985 with an annual production capacity of 1.2 million cases of beer.

Due to an increase in export demand, manufacturing output in the free-trade zone of Freeport increased substantially in 1992, particularly of pharmaceutical products. In contrast, rum output declined by almost 45% because the only rum producer in the Bahamas transferred part of its production to another Caribbean country. Still, manufacturing continues to account for only a minimum portion of the GDP. After tourism, financial services is the most important industry, accounting for 12% of GDP. There were 404 banks and trust companies licensed in the Bahamas in 1992, including six major retail banks and several mortgage institutions.

²⁹SCIENCE AND TECHNOLOGY
Agricultural research facilities include the Bahamas Agricultural Research Center, Central Agricultural Station, and the Food Technology Complex.

³⁰DOMESTIC TRADE
Nassau is the principal distribution and import center. Shopping hours are from 9 AM to 5 PM, except Sunday. Banks are open from 9:30 AM to 3 PM, Monday–Thursday, and from 9:30 AM to 5 PM on Friday. The press is the principal advertising medium.

Domestic financing through commercial bank loans and the issuance of government securities continued to increase in 1991. The National Insurance Board, a Bahamian pension fund, purchased $317.5 million in treasury bills and government-registered stock through August 1991.

³¹FOREIGN TRADE
The US is the Bahamas' major trading partner. US firms exported over $800 million worth of goods and services to the Bahamas in 1990.

Canada and the European Community also trade frequently with the Bahamas since several agreements have been signed with these countries. These agreements include the Caribbean Basin Initiative (CBI), the Caribbean-Canada Agreement (CARIBCAN), and the Lomé Convention.

Total exports, totaling $310.2 million in 1992, included pharmaceuticals, shellfish, salt, cement, rum, aragonite, cascarilla bark, tomatoes, and citrus. Total imports totaled $1,156.4 million, primarily composed of foodstuffs, meat, motor vehicles, oil, animal feed, petroleum products, clothing, machinery and appliances. Bahamas' major export-trading partners were the US, Canada, and the UK. Its major import-trading partners included the US, Japan, Canada, and the UK.

³²BALANCE OF PAYMENTS
Income from tourism is a vital offsetting factor in the country's balance-of-payments position. The balance of payments fell by $53 million in 1991 and increased the deficit to $238 million. Merchandise exports, however, increased to about $306 million in 1991. Foreign reserves at the end of 1991 were estimated at $170 million, as banks imported funds to boost reserves. In 1992, merchandise exports totaled $310.2 million and imports $–1,156.4 million. The merchandise trade balance was $–846.2 million.

The following table summarizes the Bahamas' balance of payments for 1991 and 1992 (in millions of US dollars):

	1991	1992
CURRENT ACCOUNT		
Goods, services, and income	–200.7	–126.4
Unrequited transfers	20.5	13.4
TOTALS	–180.2	–113.0
CAPITAL ACCOUNT		
Direct investment	—	7.4
Portfolio investment	—	—
Other long-term capital	165.0	89.8
Other short-term capital	11.9	–49.5
Exceptional Financing	—	—
Other liabilities	—	
Reserves	–10.9	–27.5
TOTALS	166	20.2
Errors and omissions	14.2	92.8
Total change in reserves	–23.1	26.0

³³BANKING AND SECURITIES
The Bahamas Central Bank, established in 1973, is the central issuing and regulatory authority. Funds for local development are made available through the Bahamas Development Bank.

Low taxation and lenient regulations have encouraged the establishment of 391 financial institutions in the country. As of 1993 there were 274 domestic and foreign institutions, allowed to deal with both residents and nonresidents; the remainder were offshore banks, dealing exclusively with nonresidents. Many of the loans of domestic banks are denominated in foreign currency. As of 1993 foreign assets were B$222.7 million. Demand deposits were at B$241.2. At the end of December 1993 the money supply, as measured by M2, stood at B$1,610 million.

There is no securities exchange in the Bahamas, but trading in both foreign securities and currencies is permitted under the authority of the Central Bank.

34INSURANCE
The establishment of a large number of insurance firms in the Bahamas has been encouraged by a 1970 law that permits companies to conduct part or all of their business out of the country, while benefiting from local tax advantages. The government is encouraging the formation of "captive" insurance companies, created to insure or reinsure the risks of offshore companies.

35PUBLIC FINANCE
The 1992 recurrent budget totalled $627.5 million, an increase of $27.5 million (4.5%) over 1991. The budget reflected continuing government priorities on education (17%), health (15%), police (11%), and tourism (7%). In 1991 the government was forced to implement emergency tax and borrowing measures needed to cover $96 million in revenue shortfall. The national debt in 1991 reached $1.15 billion, up from $915.5 million in 1990. Foreign currency debt soared to $406.3 million from $267.7 million, as airport, harbor, and electrification projects required funding.

36TAXATION
The absence of direct taxation has enabled the Bahamas to attract a substantial number of financial enterprises in search of tax-shelter advantages. The country has no income taxes, capital gains taxes, or profit taxes, and residents are free from succession, inheritance, gift, or estate taxes. The only indirect taxation is a real property tax, ranging from 0.75% to 1% for owner-occupied property and 0.5% to 1.5% on commercial property.

37CUSTOMS AND DUTIES
Import duties make up approximately 70% of government revenues. Based on c.i.f. value, import levies range from 1% to 200% and average 35%. Most duties are applied ad valorem; preferential rates apply to imports from Commonwealth countries. Exemptions are available for many commodities. In 1990, customs duties were eliminated on a number of luxury goods to encourage tourism.

38FOREIGN INVESTMENT
The absence of corporate or personal income taxes acts as a direct inducement to foreign capital. In addition, specific investment incentives include the Industries Encouragement Act, providing total exemption from import duties and taxes for development of approved industries; the Hawksbill Creek Act, which provides for tax-free development of the Freeport area; and the Hotel Encouragement Act, which exempts large new hotels from all taxes for 20 years.

The government actively seeks foreign investment in every sector of the economy. The Bahamas is recognized as an international offshore financial center where the government imposes no tax on income, capital gains or distribution to shareholders, inheritance, probate dividends, sales, corporations or royalties. The absence of most forms of direct taxation together with the benefits of profit repatriation are two of the primary incentives to investing in the Bahamas. The Bahamas has signed an OPIC agreement, and a number of OPIC-financed projects have been approved. Canada also has an investment insurance program in the Bahamas. The country has also become a signatory to the Multilateral Investment Guarantee Agency (MIGA) of the World Bank. In addition, a new International Business Companies Act went into effect on 15 January 1990, simplifying the requirements for incorporation and reducing the cost of forming an offshore company. This Act was later revised and the new Companies Act went into effect in 1992.

39ECONOMIC DEVELOPMENT
The promotion of tourism and financial activity by foreign firms continues as a basic tenet of the Bahamas government. Since the late 1960s, increased emphasis has been focused on development of local industry, with the liberal tax structure remaining the key incentive. In 1976, the government began a series of measures to foster greater participation by Bahamians in the economy. The new rulings included increased work-permit fees for foreigners and sharp rises in property-transfer taxes and business licensing fees for non-Bahamians. Since late 1979, government permission has been required for the sale of land to non-Bahamians. The Bahamas Development Bank helps provide financing for Bahamian entrepreneurs. The government is attempting to diversify the economy and attract new industry, as well as to conserve and develop the country's 800,000 acres of forest to build a lumber industry.

The outlook for the Bahamas continues to be closely linked to the performance of the tourism sector, which directly affects consumption and investment decisions. In the near future, this sector is expected to remain sluggish. Bahamas must improve the quality of its traditional tourist product and lower its costs. This should be complemented by the development of more upscale facilities in the Family Islands and the diversification of the economy into agriculture, fishing and services. In addition, structural reforms will have to play a major role in shaping the future pattern of development in the Bahamas. The government has started with privatization activities that do not require parliamentary approval. In early 1993, janitorial and messenger services were in the process of being sold, and other economic activities such as an asphalt plant and tender boat services are likely to follow suit in the near future.

The long-term perspectives for the Bahamas depend critically on the manner and speed by which it is transformed into a more competitive and diversified economy.

40SOCIAL DEVELOPMENT
Principal welfare legislation is contained in the National Insurance Act of 1972, as amended. Workers' compensation and retirement, maternity, survivors', and funeral benefits are provided for, with specific terms set out by the National Insurance Board. Contributions are shared by employers and employees; rates are progressive, depending on weekly wages earned.

Bahamian women are well represented in business, the professions, and government. However, activists are calling for improved child care availability to encourage more women to enter the labor force. As of 1993, the Constitution did not allow foreign-born husbands of Bahamian women to become citizens.

41HEALTH
The government operates the 478-bed Princess Margaret Hospital in Nassau and 4 other hospitals. In addition, 85 clinics and satellite clinics are maintained throughout the islands, with emergency air links to Nassau.

In 1985, there were 228 physicians, and 120 midwives. In 1990, there were 52 pharmacists and 682 nurses. In 1991, the fertility rate was 2.1. In 1992, the infant mortality rate was 24 per 1,000 live births, with 5,200 births that same year. In 1993, the birth rate was 19.3 per 1,000 people, and the general mortality rate was 5.2 per 1,000. Average life expectancy in 1992 was 72 years. From 1991 to 1992, 93% of one-year-old children were immunized against measles. The country is free from tropical diseases.

[42]HOUSING
Overcrowding is a problem on New Providence, and adequate low-cost housing is in short supply. In 1980, 70% of housing units were detached houses, 14% were apartments, and 12% were single attached dwellings. Over 50% of all homes were stone, concrete and/or brick and 32% were wood. Housing starts in the Bahamas in 1985 totaled 1,956, valued at more than B$109 million; 1,409 houses were completed. The Bahamas Housing Authority was established by the government in 1983, with a mandate to develop housing for low-income people.

[43]EDUCATION
Education is under the jurisdiction of the Ministry of Education and Culture and is free in all government maintained schools. English is the official language. Government expenditure on education (1988) was 20.7% of the total budget. Secondary education begins at age 11 and consists of two cycles each of three years duration. Education is compulsory for children aged 5–14. In 1990, 101 primary schools enrolled 25,452 pupils; 37 junior/senior high schools enrolled 23,502 students; 86 all-age schools, 10,739; and 5 special schools, 268. Postsecondary training is provided by the government through the College of the Bahamas, with an enrollment of about 2,050 students. Over 96% of Bahamians are reported literate.

There are several schools of continuing education which offer academic and vocational courses. The government-maintained Princess Margaret Hospital and the College of the Bahamas each offer a nursing degree through the School of Nursing. The College of Bahamas (founded in 1974) provides a two-year/three-year program which leads to an associate degree. It also offers a Bachelor of Arts degree in education. Since 1960, the Bahamas has been affiliated with the University of the West Indies. Other important institutions of higher learning are the Bahamas Hotel Training College and the Industrial Training College.

[44]LIBRARIES AND MUSEUMS
The Nassau Public Library is the largest of four public libraries on New Providence, with some 80,000 volumes. The Ranfurly Out Island Library, a private institution, distributes free book packages to school libraries throughout the country. The library of the College of the Bahamas in Nassau maintains a collection of 60,000 volumes. The Bahamia Museum is in Nassau, as is the Public Records Office, which contains historical documents dating from 1700.

[45]MEDIA
All telephone, telegraph, and teletype service is provided by the Bahamas Telecommunications Corp. In 1990, 132,522 fully digital telephones were in service, with automatic equipment in use on the major islands. A submarine cable connects New Providence with Florida, and direct dialing to the US has been available since 1971. There are three AM and two FM stations, and one television station. In 1991 there were an estimated 140,000 radios and 56,000 television sets on the islands.Two daily newspapers are published in Nassau. The *Tribune* (circulation 14,000

in 1991) appears evenings; the Nassau *Guardian* (17,000), mornings. The daily Freeport *News* has a circulation of 6,000.

[46]ORGANIZATIONS
Commercial associations include the Bahamas Chamber of Commerce and service groups such as Kiwanis, Rotary, and Lions clubs. Employers' groups include the Bahamas Employers' Confederation and the Hotel Employers' Association. International amateur sports activities are coordinated by the Bahamas Olympic Association.

[47]TOURISM, TRAVEL, AND RECREATION
Tourism in the Bahamas is recovering from a decline which began in 1990, due to the US recession and competition from other Caribbean nations. In 1991, 1.43 million tourists visited the islands. Some 1.17 million of those spending at least one night on land were from the US, 90,120 from Canada, and 112,04 from Europe; cruise ships brought an additional 2,020,000 visitors to the Bahamas. In the same year, tourists spent a total of $1.22 billion in the islands, and there were 13,165 hotel rooms and 26,332 beds with a 56.3 percent occupancy rate.

Visitors are attracted to the Bahamas' excellent climate, beaches, and recreational and resort facilities. Water sports (including excellent deep-sea fishing) are the favorite pastimes. Gambling is legal for non-Bahamians.

Passports are not required for tourists from the UK, Canada, and US for stays of less than three weeks. Passports but not visas are required of most visitors from Western Europe, Commonwealth countries, and Latin America.

[48]FAMOUS BAHAMIANS
Lynden Oscar Pindling (b.1930), a lawyer and leader of the PLP, became the Bahamas' first prime minister following independence in 1973.

[49]DEPENDENCIES
The Bahamas has no territories or colonies.

[50]BIBLIOGRAPHY
Boultbee, Paul G. *The Bahamas.* Oxford, England; Santa Barbara, Calif.: Clio Press, 1989.
Craton, Michael. *A History of the Bahamas,* 3d ed. Waterloo, Ontario: San Salvador Press, 1986.
———.*Islanders in the Stream: A History of the Bahamian People.* Athens: University of Georgia Press, 1992.
Evans, F. C., and N. Young. *The Bahamas.* New York: Cambridge University Press, 1977.
Hughes, Colin A. *Race and Politics in the Bahamas.* New York: St. Martin's, 1981.
Keegan, William F. *The People Who Discovered Columbus: The Prehistory of the Bahamas.* Gainesville: University Press of Florida, 1992.
Lewis, James A. *The Final Campaign of the American Revolution: Rise and Fall of the Spanish Bahamas.* Columbia, S.C.: University of South Carolina Press, 1991.

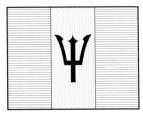

BARBADOS

CAPITAL: Bridgetown.

FLAG: The national flag has three equal vertical bands of ultramarine blue, gold, and ultramarine blue and displays a broken trident in black on the center stripe.

ANTHEM: *National Anthem of Barbados,* beginning "In plenty and in time of need, when this fair land was young. . . ."

MONETARY UNIT: Officially introduced on 3 December 1973, the Barbados dollar (BDS$) of 100 cents is a paper currency officially pegged to the US dollar. There are coins of 1, 5, 10, and 25 cents and 1 dollar, and notes of 1, 2, 5, 10, 20, 50, and 100 dollars. BDS$1 = US$0.4972 (or US$1 = BDS$2.0113)

WEIGHTS AND MEASURES: The metric system is used.

HOLIDAYS: New Year's Day, 1 January; Errol Barrow Day, 23 January; May Day, 1 May; Kadooment Day, 1st Monday in August; CARICOM Day, 1 August; UN Day, 1st Monday in October; Independence Day, 30 November; Christmas Day, 25 December; Boxing Day, 26 December. Movable religious holidays are Good Friday, Easter Monday, and Whitmonday.

TIME: 8 AM = noon GMT.

¹LOCATION, SIZE, AND EXTENT

Situated about 320 km (200 mi) NNE of Trinidad and about 160 km (100 mi) ESE of St. Lucia, Barbados is the most easterly of the Caribbean islands. The island is 34 km (21 mi) long N–S and 23 km (14 mi) wide E–W, with an area of 430 sq km (166 sq mi) and a total coastline of 97 km (60 mi). Comparatively, Barbados occupies slightly less than 2.5 times the area of Washington, D.C.

The capital city of Barbados, Bridgetown, is located on the country's southwestern coast.

²TOPOGRAPHY

The coast is almost entirely encircled with coral reefs. The only natural harbor is Carlisle Bay on the southwest coast. The land rises to 336 m (1,102 ft) at Mt. Hillaby in the parish of St. Andrew. In most other areas, the land falls in a series of terraces to a coastal strip or wide flat area.

³CLIMATE

The tropical climate is tempered by an almost constant sea breeze from the northeast in the winter and early spring, and from the southeast during the rest of the year. Temperatures range from 24° to 29°C (75–84°F). Annual rainfall ranges from about 100 cm (40 in) in some coastal districts to 230 cm (90 in) in the central ridge area. There is a wet season from June to December, but rain falls periodically throughout the year.

⁴FLORA AND FAUNA

Palms, casuarina, mahogany, and almond trees are found on the island, but no large forest areas exist, most of the level ground having been turned over to sugarcane. The wide variety of flowers and shrubs includes wild roses, carnations, lilies, and several cacti. Natural wildlife is restricted to a few mammals and birds; finches, blackbirds, and moustache birds are common.

⁵ENVIRONMENT

Principal environmental agencies are the Ministry of Housing, Lands, and Environment, established in 1978, and the Barbados Water Authority (1980). Soil erosion, particularly in the northeast, and coastal pollution from oil slicks are among the most significant environmental problems. The government of Barbados created a marine reserve to protect its coastline in 1980. As of 1994, the most pressing environmental problems result from the uncontrolled handling of solid wastes which contaminated the water supply. Barbados is also the recipient of air and water pollution from other countries in the area. Despite its pollution problems, 100% of Barbados' urban and rural populations have safe water.

In 1987, the Barbados yellow warbler, Eskimo curlew, tundra peregrine falcon, and Orinoco crocodile were endangered species. In addition, one plant species is considered endangered as of 1994.

⁶POPULATION

The population in 1990 was 257,082, an increase of 14.7% over the 1980 census figure of 224,228. The projected population for the year 2000 was 268,000. In 1990, the population density was 598 per sq km (1,549 per sq mi), the highest of all American nations. The birth rate in 1992 was 16.2 per 1,000, and the death rate was 9.1 per 1,000. Bridgetown, the capital, had a population of 6,720 in 1990; with its suburbs, it had a population of about 110,000, or over 40% of the island's total. Other major towns are Holetown, Speightstown, and Oistins.

⁷MIGRATION

To meet the problem of overpopulation, the government encourages emigration. Most emigrants now resettle in the Caribbean region or along the eastern US coast.

⁸ETHNIC GROUPS

About 90% of all Barbadians (colloquially called Bajans) are the descendants of former African slaves. Some 5% are mulattos and another 5% are whites.

[9]LANGUAGES

English, the official language, is spoken universally, with local pronunciations.

[10]RELIGIONS

About 40% of the population is Anglican. Of the remainder, Methodists (7.1%), Pentecostals (8%), and Roman Catholics (4%) predominate. There are more than 70 smaller Christian sects and other religious groups as well.

[11]TRANSPORTATION

The highway system had a total length of 1,570 km (976 mi) in 1991, of which about 94% was paved; there were 40,951 passenger cars and 6,724 commercial vehicles registered in 1992. Grantley Adams International Airport is situated 18 km (11 mi) southeast of Bridgetown. Barbados is served by 14 international and 1 local airlines and had 17 civil aircraft and 77 commercial pilots registered in 1992. There is a deepwater harbor at Bridgetown, with berthing facilities for cruise ships and freighters.

[12]HISTORY

Barbados originally supported a considerable population of Arawak Indians, but invading Caribs decimated that population. By the time the British landed, near the site of present-day Holetown in 1625, the island was uninhabited. Almost 2,000 English settlers landed in 1627–28. Soon afterward, the island developed the sugar-based economy, supported by a slave population. Slavery was abolished in 1834, and the last slaves were freed in 1838.

During the following 100 years, the economic fortunes of Barbados fluctuated with alternating booms and slumps in the sugar trade. In 1876, the abortive efforts of the British to bring Barbados into confederation with the Windward Islands resulted in the "confederation riots."

In the 1930s, the dominance of plantation owners and merchants was challenged by a labor movement. Riots in 1937 resulted in the dispatch of a British Royal Commission to the West Indies and the gradual introduction of social and political reforms, culminating in the granting of universal adult suffrage in 1950. In 1958, Barbados became a member of the West Indies Federation, which was dissolved in 1962. The island was proclaimed an independent republic on 30 November 1966. Political stability has been maintained since that time. Barbados helped form CARICOM in 1973, the same year the nation began issuing its own currency. The country was a staging area in October 1983 for the US-led invasion of Grenada, in which Barbadian troops took part.

[13]GOVERNMENT

The constitution of Barbados, which came into effect on 30 November 1966, provides for a crown-appointed governor-general (who in turn appoints an advisory Privy Council) and for independent executive, legislative, and judicial bodies. The bicameral legislature consists of a Senate and a House of Assembly. The Senate, appointed by the governor-general, has 21 members: 12 from the majority party, 2 from the opposition, and 7 of the governor-general's choice. The 28-member House of Assembly is elected at intervals of five years or less. The voting population is universal, with a minimum age of 18. The governor-general appoints as prime minister that member of the House of Assembly best able to command a majority. The prime minister's cabinet is drawn from elected members of the House of Assembly.

[14]POLITICAL PARTIES

The leading political groups grew out of the labor movement of the 1930s. The BLP was established in 1938 by Sir Grantley Adams. The Democratic Labour Party (DLP) split from the BLP in 1955. The National Democratic Party (NDP) was formed in 1989 by dissident members of the DLP. The parties reflect more personal than ideological differences.

Errol W. Barrow, the DLP leader, was prime minister from independence until 1976. The BLP succeeded under J.M.G. ("Tom") Adams, the son of Grantley Adams. In 1981, the BLP retained its majority by 17-10, and Adams continued in that office until his death in 1985. On 28 May 1986, Barrow and the DLP won 24 House of Assembly seats to three for the BLP. After Barrow's death on 2 June 1987, Deputy Prime Minister Erskine Sandiford, minister of education and leader of the House of Assembly, assumed the prime ministership.

Despite the resignation of finance minister Dr. Richie Haynes in 1989, and his subsequent formation of the National Democratic Party, the DLP under Sandiford continued in power, retaining 18 of the now 28 seats in the House of Assembly. The BLP won the remaining 10 seats, leaving the dissident NDP without any representation.

[15]LOCAL GOVERNMENT

All local governments, including those on the district and municipal levels, were abolished on 1 September 1969; their functions were subsumed by the national government. The country is divided into 11 parishes and the city of Bridgetown for administrative and electoral purposes.

[16]JUDICIAL SYSTEM

The Barbados legal system is founded in British common law. The Supreme Court of Judicature sits as a high court and court of appeal; vested by the constitution with unlimited jurisdiction, it consists of a chief justice and three puisne judges, appointed by the governor-general on the recommendation of the prime minister after consultation with the leader of the opposition party. Magistrate courts have both civil and criminal jurisdiction. Final appeals are brought to the Committee of Her Majesty's Privy Council in the UK.

The courts enforce respect for civil rights and assure a number of due process protections in criminal proceedings including a right of detainees to be brought before a judge within 72 hours of arrest.

[17]ARMED FORCES

The Barbados Defense Force was established in 1979, consisting of the former Barbados Regiment and Barbados Coast Guard. This force and the Royal Barbados Police Force number about 1,000. The defense budget is US$10 million (1989).

[18]INTERNATIONAL COOPERATION

Barbados became a member of the UN on 9 December 1966 and belongs to ECLAC and all the nonregional specialized agencies except IAEA and IDA. It is a member of the Commonwealth of Nations, G-77, IDB, OAS, and PAHO. On 1 August 1973, Barbados joined Guyana, Jamaica, and Trinidad and Tobago as an initial signatory to CARICOM; it had been a member of CARIFTA, CARICOM's forerunner, since 1968. The country also belongs to the Caribbean Development Bank, established in 1970. Barbados is a signatory to GATT and the Law of the Sea.

[19]ECONOMY

The economy has traditionally been dependent on the production of sugar, rum, and molasses. In recent years, however, tourism and manufacturing have surpassed sugar in economic importance.

As of 1994, Barbados was in the grip of an economic crisis which began in mid-1990. A severe downturn led to a contraction of GDP of about 3.3% in 1990 and a loss of foreign reserves of over US$50 million. The economic recession deepened in 1992 as a consequence of austere government stabilization policies and the impact on tourism of sluggish economic growth in the US and competition from other Caribbean countries. Real GDP fell by 4.1%. Reversals in the key agricultural and manufacturing sectors have contributed strongly to the recession, with real output in agriculture declining by 1.1%, and manufacturing by 9.2%. Unemployment for 1992 was 23%, up from 17.1% in 1991.

[20]INCOME

In 1992, Barbados' GNP was US$1,693 million at current prices, or US$6,530 per capita. For the period 1985–92 the average inflation rate was 4.3%, resulting in a real growth rate in per capita GNP of 0.6%.

In 1992 the GDP was $1,740 million in current US dollars. It is estimated that in 1991 agriculture, hunting, forestry, and fishing contributed 5% to GDP; mining and quarrying, 1%; manufacturing, 7%; electricity, gas, and water, 3%; construction, 5%; wholesale and retail trade, 26%; transport, storage, and communication, 7%; finance, insurance, real estate, and business services, 13%; community, social, and personal services, 3%; and other sources, 30%.

[21]LABOR

Total civilian employment as of 1992 was 96,100. In 1992, services accounted for about 39% of the labor force; commerce and tourism, 14.3%; manufacturing, 10%; construction, 7.3%; agriculture, 6%; and other sectors, 23.4%. In 1991, the tourism industry accounted for 9% of all employment. Most employees work a 40-hour week. A minimum of a three-week paid holiday each year (four weeks for those employed at least five years) is required by law. There is one major union, the Barbados Workers' Union, with some 22,000 members (65% of all union membership), and several smaller specialized ones. Major anti-government demonstrations and industrial actions occurred in 1991 (by teachers, doctors, civil servants and other public workers), due to the severe cutbacks in personnel in the private sector, the planned privatization of state-run enterprises, and wage cuts and layoffs within the public sector. Unemployment, traditionally high, was reported at 23% in 1992.

[22]AGRICULTURE

About 33,000 hectares (81,500 acres), or 76.7% of the total land area, are classified as arable. At one time, nearly all arable land was devoted to sugarcane, but the percentage devoted to ground crops for local consumption has been increasing. In 1992, 600,000 tons of sugar were produced, down 44% from 1979–81 levels. In 1990, sugar exports amounted to US$29 million, or 23.7% of total exports. Major food crops are yams, sweet potatoes, corn, eddo, cassava, and several varieties of beans. Some cotton is also grown.

[23]ANIMAL HUSBANDRY

The island must import large quantities of meat and dairy products. Most livestock is owned by individual households. Estimates for 1992 showed 21,000 head of cattle, 56,000 sheep, 45,000 hogs, 34,000 goats, and 1,600 tons of hen eggs.

[24]FISHING

The fishing industry employs about 2,000 persons, and the fleet consists of more than 500 powered boats. The catch in 1991 was 2,697 metric tons. Flying fish, dolphinfish, tuna, turbot, kingfish, and swordfish are among the main species caught. A new fisheries terminal complex opened at Oistins in 1983.

[25]FORESTRY

Fewer than 20 hectares (50 acres) of original forests have survived the 300 years of sugar cultivation. In 1992, imports of forest products totaled US$15.1 million.

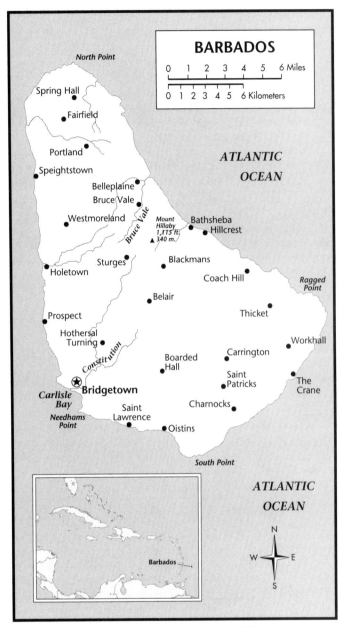

LOCATION: 13°2′ to 13°20′N; 59°25′ to 59°39′W.
TOTAL COASTLINE: 101 km (63 mi). TERRITORIAL SEA LIMIT: 12 mi.

[26]MINING

Deposits of limestone and coral are quarried to meet local construction needs. Clay, sand, and carbonaceous deposits provide limited yields. Cement production amounted to 200,000 tons in 1991.

[27]ENERGY AND POWER

Supply and distribution of electricity are undertaken by a private firm, Barbados Light and Power, under government concession. Installed capacity in 1991 was 152,000 kw; production that year totaled 527 million kwh. Oil accounted for more than 70% of energy usage in 1991; bagasse burning, 20%; solar water heating, less than 1%; natural gas satisfied the remaining energy demand. The world oil crisis of the mid-1970s initiated an active search for commercial deposits of oil and natural gas. Limited pockets of natural gas were discovered, and oil was found in St. Philip Parish. Crude oil production in 1991 was 470,000 barrels, meeting 30% of domestic requirements; natural gas production was 35 million cu m. Reserves of crude oil were estimated at 3.5 million barrels at the end of 1992.

28INDUSTRY

In 1984 there were 300 industrial firms on the island. Traditionally, sugar production and related enterprises were Barbados' primary industry, but light industry, especially the manufacture of electrical components, has become more important. By 1983 sugar production had reached a production low of 86,000 tons, but by 1986 production had risen by 29.1% as a result of the government's policy of modernizing harvesting and refining methods. Important by-products of sugarcane are molasses and rum. Other important food manufactures are yams, sweet potatoes, poultry, and milk. Items manufactured for export include soap, glycerine, pharmaceuticals, furniture, household appliances, plastic products, fabricated metal products, and cotton garments. The Barbados Industrial Development Corp. encourages diversification through the development of small industrial enterprises.

As a result of fiscal and monetary austerity, industrial output in the early 1990s declined by 9.2%, as did construction (17.1%). Sugar subsector GDP was down by 18% due to a very late harvest caused by labor conflicts. However, nonsugar output soared as land was transferred from sugar cane to other agricultural uses, mainly yams and other food crops. As a result, value added by the agricultural sector as a whole fell by 1.1%. Although tourism has not been performing well either, it has become Barbados' most important industry, leaving behind both sugar production and the offshore business sector. In 1992, the tourism sector experienced a modest recovery, with cruise ship arrivals increasing by 7.4%.

After tourism, the offshore business sector is now the second largest earner of foreign exchange, contributing close to US$100 million in foreign revenues in 1991. Recently Government of Barbados has been promoting data processing, mainly for US companies as another "offshore" industry.

29SCIENCE AND TECHNOLOGY

Barbadian learned societies include the Barbados Astronomical Society and the Barbados Pharmaceutical Society. The Bellairs Research Institute, associated with McGill University in Montreal, is a center for the study of the tropical environment. The University of the West Indies has faculties in medicine, the natural and social sciences, as well as two research institutes. Barbados Community College, founded in 1968, offers training in science and technology.

30DOMESTIC TRADE

There are fish, fruit, and vegetable markets and stores for all types of shopping. General business is conducted on weekdays from 8 AM to 4 PM. Most shops are also open Saturdays from 8 AM to noon. Banks are open from 9 AM to 3 PM, Monday–Thursday, and 3 to 5 PM each Friday.

31FOREIGN TRADE

Merchandise exports in 1992 totaled US$196 million, reflecting a marginal increase over 1991. Merchandise imports totaled US$525 million, reflecting a considerable decrease as the government's stabilization program decreased local demand. Its primary exports included clothing, electronic components, building cement, chemicals, rum, furniture, machinery transport equipment and sugar. Barbados' main imports were composed of foodstuffs, other non-durables, fuels and lubricants, building materials, chemicals, textiles, and machinery.

The country exported mostly to the US, Canada, CARICOM, and the UK. Barbados imported mainly from the US, the UK, CARICOM, and Japan.

32BALANCE OF PAYMENTS

The consistently adverse trade balance is substantially alleviated by foreign currency remittances from various emigrants and by tourist expenditures. In 1992 merchandise exports totaled US$158 million and imports US$464.7 million. The merchandise trade balance was $–306.7 million. The following table summarizes Barbados' balance of payments for 1991 and 1992 (in millions of US dollars):

	1991	1992
CURRENT ACCOUNT		
Goods, services, and income	–62.9	96.9
Unrequited transfers	33,0	40.3
TOTALS	–29.9	137.1
CAPITAL ACCOUNT		
Direct investment	–6.1	–13.5
Portfolio investment	–8.8	–4.1
Other long-term capital	–0.9	–28.4
Other short-term capital	2.8	–81.4
Exceptional financing	—	—
Other liabilities	—	
Reserves	39.1	–28.3
TOTALS	39.1	–128.7
Errors and omissions	–9.2	–8.5
Total change in reserves	33.7	–1.7

33BANKING AND SECURITIES

The bank of issue is the Central Bank of Barbados. Commercial banks include Barbados National Bank and branches of several UK, and Canadian banks. Public institutions include the Barbados Development Bank and the Sugar Industry Agricultural Bank. Barbados has begun to seek offshore banking business, but with very limited success. There is no stock exchange in Barbados, although the Central Bank has established the Barbados Securities Marketing Corp. in anticipation of the future development of a Securities Exchange.

34INSURANCE

In 1985 there were nine local and approximately 30 foreign insurance companies in Barbados. A full range of life and non-life insurance is available.

35PUBLIC FINANCE

Revenues are derived mostly from import duties, internal consumption taxes, and income tax. Public sector deficits grew during the 1980s as the economy weakened. The international recession of 1990/91 magnified problems of debt service and debt management. By the end of 1990, the national debt was US$928.3 million, 9.5% higher than 1989. By 1991, the fiscal deficit had become unsustainable; in February 1992, the government began a stabilization program in fiscal policies with assistance from the IMF.

36TAXATION

Tax on individual income as of mid-1993 ranged from 25% to 40%. The corporate income tax rate was 40% but companies that improve the national trade position receive special tax rates and exemptions. Other taxes were levied on insurance premiums, property transfers, land value, bank assets, and rental income.

37CUSTOMS AND DUTIES

Most imports, except those from other CARICOM members, are subject to import duties that include a customs duty, a consumption tax, and a stamp tax.

38FOREIGN INVESTMENT

Various investment incentives, including exemption from customs duties, tax reductions and exemptions, and training grants are available to both domestic and foreign investors. Foreign ownership of Barbadian enterprises or participation in joint ventures must be approved by the central bank.

The government favors productive foreign investment, with an emphasis on tourism and manufacturing because of their employment and foreign exchange generating potential. Recently, Barbados has concluded a Bilateral Double Taxation Treaty and has a Tax Information Exchange Agreement with the US, as well as an extensive tax treaty network with other developed countries. Foreign investors are eligible for all available tax incentives including full exemption from all income and withholding taxes for investors in offshore industries.

39ECONOMIC DEVELOPMENT

The government has been making efforts to relieve the island's economic dependence on sugar by establishing small manufacturing industries and encouraging tourism. Government planning has been in operation since 1951. In 1957, the Barbados Development Board (now the Barbados Industrial Development Corp.) was established. The development plan for 1979–83 called for expansion of the construction sector. During 1962–86, multilateral aid to Barbados totaled US$123.1 million, of which 24% was from the IDB and 57% from the IBRD.

The stabilization program implemented in 1992 slightly improved the country's public finances and its external accounts. Unfortunately, this was at the expense of real growth and unemployment. Prospects for the Barbadian economy in the near future will depend greatly on the government's ability to stimulate foreign investment as well as the country's capacity to earn foreign exchange, which implies improving the competitiveness of exports and tourism. The tourism industry is expected to continue growing at a moderate rate. Sugar output is presumed to decrease further, while output of other crops and livestock is expected to grow as some land is transferred from sugar cane to other agricultural uses. The performance of commerce and most other services will reflect that of tourism, while manufacturing and construction will depend largely on the government's economic policies.

40SOCIAL DEVELOPMENT

A national social security system began operations in 1937, providing old age and survivors' pensions, sickness, disability, and maternity benefits, and (under a January 1971 extension) employment injury benefits. People between the ages of 16 and 65 are covered. The first stage of a national health service, the National Drug Plan, began in 1980, and in 1981 unemployment insurance was introduced. All government hospitals and clinics maintain public wards for free medical treatments. The US contributes foodstuffs to a school lunch project. Barbados has an active population planning program, and the Family Planning Association receives government support. The fertility rate was 1.7 in 1991.

Although women are well-represented in all aspects of national life, women's rights advocates cite domestic violence as a serious problem. The 1992 Domestic Violence law requires a police response to violence against women and children.

41HEALTH

In 1986, government health facilities included 11 hospitals and 8 public health centers, with a total of more than 2,100 beds. In 1990, there were 294 doctors, 836 nurses, and 33 dentists. In 1990, 100% of the population had access to health care services. Life expectancy during 1992 was 76 years. The infant mortality rate was 10 per 1,000 live births in 1992, with 4,100 births that year.

The 1993 birth rate was 1.58 per 1,000 people, and the general mortality rate was 9.1 per 1,000 people. Between 1991 and 1992, 90% of one-year-old children were vaccinated against measles.

42HOUSING

The Barbados Housing Authority is empowered to acquire land, construct housing projects, and redevelop overcrowded areas. In 1980, 90% of all housing consisted of detached homes and 6% of apartments. Over 70% of all homes were owner occupied, 16% were rented privately, 5% were occupied rent free, another 5% were rented from the government, and 3% were occupied by squatters or otherwise specified.

43EDUCATION

The official language in Barbados is English. Primary education begins at the age of five. Secondary education begins at the age of 11 and lasts for six years. The state provides tuition for approximately 86% of the eligible children for primary and secondary education, the latter offering certificate courses equivalent to the general certificate of education in both ordinary and advanced level curricula.

Education is compulsory for children between the ages of 5 and 16. The education program in Barbados is administered by the Ministry of Education and is free in all government-run schools. In 1991, children in 106 primary schools numbered 26,662. Scholarships are awarded for study in the UK and in Caribbean institutions. The Barbados branch of the University of the West Indies opened at Cave Hill in 1963. It has 266 teachers and 2,264 students. The government pays the fees of all Barbadian students at the Cave Hill Campus of the University of West Indies. The Barbados Community College was established in 1968. Barbados' adult literacy rate in 1992 was estimated at 99%.

There is an advanced education for adults at the Extramural Center of the University of West Indies. There are also special schools including two residential institutions for disabled persons.

44LIBRARIES AND MUSEUMS

A free library is maintained by the government in Bridgetown. There are seven branches, with bookmobile stops throughout the island. By 1986, the system had 174,923 volumes. The library of the Barbados branch of the University of the West Indies has 137,000 volumes. The Barbados Museum and Historical Society is a general museum with collections showing the geology, history, natural history, marine life, and plantation home furnishings of the island, as well as Arawak artifacts.

45MEDIA

Automatic telephone service is provided by a private firm, the Barbados Telephone Co. Ltd.; there were 110,600 telephones in 1991. A wireless telephone service provides overseas communications, and a telex cable connects Barbados with the UK. The Congor Bay Earth Station, opened in 1972, links Barbados with the global satellite communications system. Barbados has a government-controlled television and radio broadcasting service and a commercial rediffusion service which broadcasts over a cable network. Altogether, there are three AM and two FM stations and two television stations (of which one is pay television). An estimated 226,000 radios and 68,000 television sets were in use in 1991.

There are two daily newspapers in Bridgetown, the *Advocate-News* (circulation, 32,000 in 1991) and the *Daily Nation* (77,000), as well as some periodicals, including a monthly magazine, the *New Bajan*.

46ORGANIZATIONS

Barbados has a chamber of commerce in St. Michael. Branches of other international organizations include the Lions Club, Rotary Club, 4-H Clubs, Boy Scouts, Girl Scouts, YMCA, and YWCA.

[47]TOURISM, TRAVEL, AND RECREATION

Barbados, with its fine beaches, sea bathing, and pleasant climate, has long been a popular holiday resort. In 1991, 394,222 tourists visited Barbados, of whom 119,069 were from the US, 46,286 from Canada, and 153,954 from Europe; tourist spending was an estimated us$453 million. There were 6,650 hotel rooms with 14,537 beds and a 50.5 percent occupancy rate. Cricket is the national sport, followed by surfing, sailing, and other marine pastimes.

A valid passport is required of all visitors except nationals of the US and Canada traveling directly to Barbados with other acceptable identification.

[48]FAMOUS BARBADIANS

Sir Grantley Adams (1898–1971) was premier of the Federation of the West Indies (1958–62). His son, John Michael Geoffrey Manningham "Tom" Adams (1931–85) was prime minister from 1976 until his death; succeeding Errol Walton Barrow (1920–87) who again assumed the office of prime minister until his death. Erskine Sandiford (b.1938) succeeded Barrow. Barbados-born Edwin Barclay (1882–1955) was president of Liberia from 1930 to 1944. George Lamming (b.1927) is a well-known West Indian novelist. Sir Garfield Sobers (b.1936) has gained renown as the "world's greatest cricketer."

[49]DEPENDENCIES

Barbados has no territories or colonies.

[50]BIBLIOGRAPHY

Beckles, Hilary. A History of Barbados: From Amerindian Settlement to Nation-State. New York: Cambridge University Press, 1990.

———. White Servitude and Black Slavery in Barbados, 1627–1715. Knoxville: University of Tennessee Press, 1989.

Handler, Jerome S. The Unappropriated People: Freedom in the Slave Society of Barbados. Baltimore: Johns Hopkins University Press, 1974.

Harlow, V. T. A History of Barbados, 1625–1685. London: Oxford University Press, 1972.

Hoyos, F. A. Grantley Adams and the Social Revolution. London: Macmillan, 1974.

Hunte, George. Barbados. London: Batsford, 1974.

Potter, Robert B. Barbados. Oxford, England; Santa Barbara, Calif.: Clio Press, 1987.

Puckrein, Gary A. Little England: Plantation Society and Anglo-Barbadian Politics. New York: New York University Press, 1986.

Tree, Ronald. A History of Barbados. 2d ed. Woodstock, N.Y.: Beekman, 1979.

BELIZE

CAPITAL: Belmopan.

FLAG: The national flag consists of the Belize coat of arms superimposed on a white disk and centered on a blue rectangular field with a narrow red stripe at the top and the bottom.

ANTHEM: *Land of the Free.*

MONETARY UNIT: The Belize dollar (B$), formerly tied to the UK pound sterling and now pegged to the US dollar, is a paper currency of 100 cents. There are coins of 1, 5, 10, 25, 50 cents and 1 dollar, and notes of 1, 1, 5, 10, 20, 50, and 100 dollars. B$1=US$0.50 (or US$1=B$2.00).

WEIGHTS AND MEASURES: Imperial weights and measures are used. The exception is the measuring of petroleum products, for which the US gallon is standard.

HOLIDAYS: New Year's Day, 1 January; Baron Bliss Day, 9 March; Labor Day, 1 May; Commonwealth Day, 24 May; National Day, 10 September; Independence Day, 21 September; Columbus Day, 12 October; Garifuna Day, 19 November; Christmas, 25 December; Boxing Day, 26 December. Movable holidays are Good Friday and Easter Monday.

TIME: 6 AM = noon GMT.

¹LOCATION, SIZE, AND EXTENT
Belize (formerly British Honduras), on the Caribbean coast of Central America, has an area of 22,960 sq km (8,865 sq mi), extending 280 km (174 mi) NNE–SSW and 109 km (68 mi) WNW–ESE. Comparatively, the area occupied by Belize is slightly larger than the state of Massachusetts. Bounded on the N by Mexico, on the E by the Caribbean Sea, and on the S and W by Guatemala, Belize has a total boundary length of 902 km (560 mi).

The capital city of Belize, Belmopan, is located in the center of the country.

²TOPOGRAPHY
The country north of Belmopan is mostly level land interrupted only by the Manatee Hills. To the south the land rises sharply toward a mountainous interior from a flat and swampy coastline heavily indented by many lagoons. The Maya and the Cockscomb mountains (which reach a high point of 1,122 m/3,681 ft at Victoria Peak, in the Cockscombs) form the backbone of the country, which is drained by 17 rivers. The coastal waters are sheltered by a line of reefs, beyond which there are numerous islands and cays, notably Ambergris Cay, the Turneffe Islands, Columbus Reef, and Glover's Reef.

³CLIMATE
The climate is subtropical and humid, tempered by predominant northeast trade winds that keep temperatures between 16° and 32°C (61–90°F) in the coastal region; inland temperatures are slightly higher. The seasons are marked more by differences of humidity than of temperature. Annual rainfall averages vary from 127 cm (50 in) in the north to more than 380 cm (150 in) in the south. There is a dry season from February to May and another dry spell in August. Hurricanes occur from July to October.

⁴FLORA AND FAUNA
Most of the forest cover consists of mixed hardwoods—mainly mahogany, cedar, and sapodilla (the source of chicle). In the flat regions there are extensive tracts of pine. The coastal land and the cays are covered with mangrove. Indigenous fauna include armadillo, opossum, deer, and monkeys; common reptiles include iguana and snakes.

⁵ENVIRONMENT
Because of its low population density, Belize has suffered less than its neighbors from such problems as soil erosion and pollution. However, substantial deforestation has occurred, and water quality remains a problem. Water pollution in Belize is a problem because of the seepage of sewage along with industrial and agricultural chemicals into the water supply. It is estimated that 47% of the country's rural population does not have access to pure water. Pollutants also threaten Belize's coral reefs. Removal of coral, picking orchids in forest reserves, spear fishing, and overnight camping in any public area (including forest reserves) are prohibited. As of 1987, endangered species in Belize included the tundra peregrine falcon, hawksbill, green sea, and leatherback turtles, American crocodile, and Morelet's crocodile. By 1991, the iguana, Larpy eagle, spoonbill, wood stork, and several types of hawks and parrots were endangered in a total of 121 species of mammals. Of 504 bird species, four are threatened. Thirty-six of 3,240 plant species were also endangered as of 1994.

⁶POPULATION
As of 1991, the total population was 189,392, with 59,220 living in Belize City. The population density was 8.2 persons per sq km (21.4 per sq mi), the lowest in Central America. The annual population growth rate was almost 3% in the 1980s.

⁷MIGRATION
Because of its high emigration rate, Belize encourages immigration. A colony of Mennonites near the Guatemalan border numbered about 6,000 by 1986, and by the end of 1992 Belize had some 20,000 Central American refugees and perhaps as many nonrefugee Central American illegal aliens. Families who wish to

farm are each given a 20-hectare (50-acre) plot of land. The government has also sought to attract Jamaican immigrants. As many as 65,000 Belizeans were living in the US by mid-1988.

[8]ETHNIC GROUPS

Some 29.8% of the population was Creole (of African descent) in 1991, while 43.6% was Mestizo (mixed White and Maya). Another 14.6% was Maya and 6.6% Garifuna (Carib). There were also people of European, Chinese, Asian Indian, and Syrian-Lebanese ancestry.

[9]LANGUAGES

The official language is English. At least 80% of the people can speak standard English and/or a Creole patois. Spanish is spoken by approximately 60% of the population; for one-third to one-half it is the first language. Although English is the language of instruction, other languages spoken include Garifuna, Mayan and other Amerindian languages, and, in the Mennonite colony, Low German.

[10]RELIGIONS

In 1993, 62% of the inhabitants were Roman Catholic with 30% members of various Protestant denominations, including Anglicans (12%), Methodists (6%), and Maronites (4%). Afro-American spiritists, Hindus, Baha'is, Muslims, and Jews are also represented by small communities.

[11]TRANSPORTATION

In 1991 Belize had 2,575 km (1,600 mi) of roads, most of them paved or gravel-surfaced. The UK has proposed to donate $38 million to upgrade the roads from Guatemala to Belize's ports. That year there were about 4,800 motor vehicles. Belize City is the main port; in the late 1970s, deepwater facilities were constructed through financing from the Caribbean Development Bank (CDB). Seven shipping lines provide regular services to North America, the Caribbean, and Europe. International airports at Belize City and Punta Gorda handle services to the US and Central America. Maya Airways provides domestic service, and there are various international air carriers. The P. S.W. Goldson Airport at Belize City handled 272,000 passengers in 1991.

[12]HISTORY

Numerous ruins indicate that the area now called Belize was once heavily populated by Maya Indians, whose civilization collapsed around AD 900. Columbus sailed along the coast in 1502, but did not land. The first permanent settlement was established in 1638 by shipwrecked English seamen. Later immigrants included African slaves and British sailors and soldiers.

In its early colonial history the area was a virtual backwater, used only for logging and as a pirate base. A power struggle between England and Spain ensued over possession of the area, with the British prevailing by the 19th century. In 1862 the British organized the area as the colony of British Honduras. For the next century, forestry continued as the main enterprise until eventually supplanted by sugar.

On 1 January 1964, a constitution was promulgated, providing for self-government, although the UK maintained the defense force. That force remained in place partly because of a border dispute with Guatemala, going back to an 1859 treaty. The Guatemalan government pressed territorial claims over the southern quarter of the area. A settlement guaranteeing the country's independence by 1970 seemed to resolve the dispute, but rioting in British Honduras in May 1968 led to the repudiation of the agreement by both the UK and Guatemala.

The country dropped the appearance, if not the reality, of colonial dependence in 1973, adopting Belize as the official country name. The border dispute continued unabated until 1977, when

Guatemala and the UK began new negotiations on Belize. The UK, Guatemala, and Belize reached agreement on a solution in March 1981, but disagreement soon followed. Finally, the UK decided to take matters into its own hands and granted Belize independence as of 21 September 1981. Guatemala refused to recognize the new nation, severed diplomatic relations with the UK, and declared the date of independence a national day of mourning. In December 1986, the UK and Guatemala resumed full diplomatic ties, but the 1800-member British garrison remained in Belize.

[13]GOVERNMENT

The independence constitution of 21 September 1981 (based on that of 1 January 1964) vests governmental authority in a governor-general appointed by the UK monarch, a cabinet headed by a prime minister, and a bicameral National Assembly. The cabinet ministers are appointed by the governor-general on the advice of the prime minister. The National Assembly consists of a 28-member House of Representatives elected by universal adult suffrage, and a Senate of eight members appointed by the governor-general (five on the advice of the prime minister, two on the advice of the opposition, and one on the recommendation of the Belize Advisory Council). Parliamentary elections must be held at intervals of no longer than five years. The voting age is 18.

[14]POLITICAL PARTIES

The two major parties in Belize are the current majority People's United Party (PUP) and the United Democratic Party (UDP). The PUP, led by George C. Price, holds 15 of the 28 seats in the House of Representatives. The UDP, a merger of three smaller opposition parties, is led by Manuel Esquivel.

Price has dominated Belize's politics since becoming the country's premier in 1964. At independence in 1981, Price became prime minister and ruled for three years. The UDP coalition took 21 House seats in 1984 and ruled until 1989, when the current government was formed.

[15]LOCAL GOVERNMENT

Belize is divided into six administrative districts: Corozal, Orange Walk, Belize City, El Cayo, Stann Creek, and Toledo. Except for Belize City, which has an elected city council of nine members, each is administered by a seven-member elected town board. Local government at the village level is through village councils.

[16]JUDICIAL SYSTEM

The independent judiciary is appointed by the crown. The law of Belize is the common law of England, augmented by local legislation. There is a Supreme Court, presided over by a chief justice. Appeals are to the court of appeal, established in 1968, and, ultimately, to the UK Privy Council. Six summary jurisdiction courts (criminal) and six district courts (civil) are presided over by magistrates.

The judiciary has protected individual rights and fundamental freedoms. Detainees must be brought before a judge within 72 hours of arrest. Bail is liberally afforded. A jury trial is required in capital cases.

[17]ARMED FORCES

The Belize Defense Force, established in 1978 consists of 1,160 men in five regular and three reserve companies and a training organization. As of 1993, 1,500 British soldiers and airmen served in Belize on a rotational basis. The defense budget is around US$9 million.

[18]INTERNATIONAL COOPERATION

Belize was admitted to the UN on 25 September 1981, four days after independence; during the late 1970s, the UN General

Assembly had repeatedly supported Belize's rights of self-determination and territorial integrity, in contradistinction to Guatemala's land claims. Belize participates in ECLAC, IBRD, IDA, IFAD, IFC, ILO, IMF, ITU, UPU, UNESCO, and WMO, and is a signatory to the Law of the Sea and their GATT. A member of the Commonwealth of Nations, Belize also belongs to CARICOM and G-77.

19ECONOMY

The economy is dependent on agriculture and fishing. Sugar, citrus products, fish, and garments make up the bulk of the country's exports. Until a recent depletion, the country's main export had been forest products, especially mahogany. The country continues to import most of its consumer goods, including much of its food and all of its petroleum requirements. The country's chronic economic problems have been aggravated by destruction from hurricanes, notably in 1961, 1974, and 1978. A five-year plan for 1985–89 aimed to promote private investment in industry and tourism.

Belize started out the decade of the 1990s with signs of economic recovery. The country's economic growth accelerated during 1992, recording a healthy GDP growth rate of 8.1%. The progress of the economy was due to strong performance of the agricultural sector, excluding sugar cane production. Agricultural production accounts for 20% of employment. Major agricultural exports still include sugar, citrus, bananas, and seafood. Citrus and bananas experienced a significant increase in output in 1992. Citrus export revenues more than doubled that of the previous year by reaching US$27.4 million. The banana industry recorded a record harvest in 1992, largely as the result of improvements in technology. Exports of bananas, mostly to the UK, increased by 36% in 1992. Production of nontraditional export crops as well as production of seafood registered an increase in 1992.

Belize's economic growth is due in part to the improvement of technology and physical infrastructure. The US Agency for International Development (USAID), the World Bank, the UK, the EC, the Caribbean Development Bank (CDB), and Canada are providing assistance to Belize in the reconstruction and pavement of several portions of major highways. Rural electrification is progressing and urban electric power is becoming more dependable.

20INCOME

In 1992, Belize's GNP was US$442 million at current prices, or US$2,210 per capita. For the period 1985–92 the average inflation rate was 3.2%, resulting in a real growth rate in per capita GNP of 6.3%.

In 1992 the GDP was $456 million in current US dollars. It is estimated that in 1991 agriculture, hunting, forestry, and fishing contributed 16% to GDP; mining and quarrying, 1%; manufacturing, 12%; electricity, gas, and water, 2%; construction, 7%; wholesale and retail trade, 16%; transport, storage, and communication, 10%; finance, insurance, real estate, and business services, 9%; community, social, and personal services, 6%; and other sources, 21%.

21LABOR

The labor force in 1992 was estimated at 60,000, or about 25% of the total population. About 30% of the labor force is employed in agriculture, forestry, and fishing. Although industrial employment tends to be seasonal, Belize City workers rarely accept employment in agriculture.

Labor legislation covers minimum wages, work hours, employment of young persons, and workers' safety and compensation. The National Trades Union Congress of Belize is the major union federation, and the United General Workers' Union is the leading trade union. In 1992, there were 13 independent

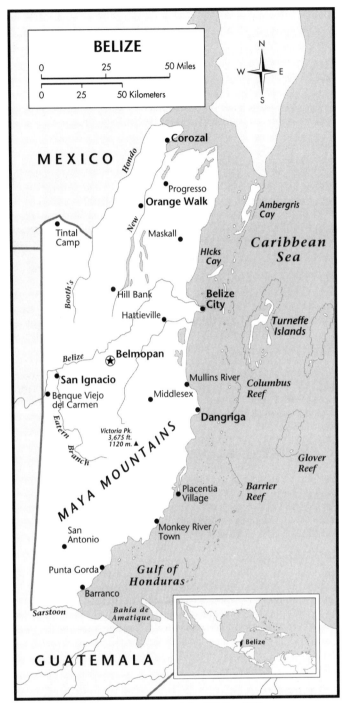

LOCATION: 15°53′ to 18°31′N; 87°16′ to 89°8′W.
BOUNDARY LENGTHS: Mexico, 251 km (156 mi); Caribbean coastline, 475 km (295 mi); Guatemala, 269 km (167 mi). TERRITORIAL SEA LIMIT: 3 mi.

unions, comprising 9.8% of the labor force, which represented a cross-section of white-collar, blue-collar, and professional workers (including most civil service employees).

22AGRICULTURE

Only 2.5% (57,000 hectares/141,000 acres) of total land area is used for the production of seasonal and permanent crops. Most Mayans still practice the traditional slash-and-burn method of farming, under which at any one time some 80% of the land is left idle. More efficient agricultural colonies have been established by Mennonite immigrants. Sugar and citrus are the leading agricultural exports. Sugarcane production, centered in the north-ern lowland around Corozal and the town of Orange Walk, totaled 984,000 tons in 1992. Citrus production is concentrated in the Stann Creek valley; the 1992 output included 70 million lb of oranges and 40 million lb of grapefruit. The US-based Hershey Foods Corp. has invested B$4 million in cacao cultivation in El Cayo.

Because agriculture is not sufficiently diversified, the country relies heavily on food imports. By establishing a marketing board to encourage production of rice, beans, and corn, the government hopes eventually to become self-sufficient in these crops. Rice paddy production, which averaged 9,000 tons annually during 1979–81, fell to a reported 4,000 tons in 1990 but climbed to 5,000 tons by 1992. Corn production, which had been hovering at 18,000 tons per year, rose to a reported 22,000 tons in 1992. Dry bean production was 2,000 tons. The agricultural sector grew by 16% in 1992, despite a modest decline in sugar cane pro-duction, which represents about 40% of value added by crops. Citrus output (exported in concentrate form), expanded by 170% as new acreage planted in the late 1980s came into production and weather conditions were very favorable. In 1985, a consor-tium that included Coca-Cola paid B$12 million for 383,000 hectares (946,400 acres) northwest of Belmopan for a citrus farming project. Citrus thus displaced bananas as the country's second most important crop, although banana production went up by 32% following a decline in 1991. Banana production was aided by recent privatization and restructuring in the production and marketing areas, which has acted as a catalyst to improve technology (success in combating sigatoka disease) and infra-structure. Papaya exports total over 10,000 lb monthly and mango, peanut, pineapple, and winter vegetable exports are also on the rise.

23ANIMAL HUSBANDRY
Mennonite farms account for much of Belize's dairy and poultry output. In 1992, the nation had an estimated 26,000 hogs, 4,000 sheep, and 1,000,000 chickens. Cattle suited for breeding or cross-breeding with local cattle are Red Poll, Jamaica Black, Hereford, and Brahman (zebu). Some 4,000 tons of poultry meat and 7,000 tons of milk were produced in 1992.

24FISHING
Fishing resources and development are good. In 1991, the total catch was 1,639 tons. Lobster is the leading product; of the US$5.6 million in export earnings derived from fishing in 1991, 94% was accounted for by lobster sales. In 1992, shrimp produc-tion increased by 75%, as three new shrimp farms came into pro-duction. Marine exports (including shrimp, lobster, fish, and conch) rose by 22.1% from 1991 (817,000 tons) to 1992 (998,000 tons).

25FORESTRY
Although Belize is still rich in forest resources, the accessible stands of commercial timber have been depleted. Reforestation and natural regeneration in the pine forest (mainly in Cayo, Stann Cree, and Toledo Districts) and artificial regeneration of fast-growing tropical hardwood species are creating a resur-gence in forestry. Timber cutting is usually done during the short dry season. The principal varieties of trees cut are mahogany, pine, cedar, and rosewood. Exports of forest produce (including chicle) in 1965 amounted to one-third of total exports; in 1991, however, the export value was B$5.9 million, or about 3% of the total.

26MINING
The Belize, Sibun, and Monkey Rivers, as well as North and South Stann Creeks, were the sites of clay, gravel, and sand operations.

In 1991, clay production amounted to 2 million tons; dolomite, 100,000 tons; limestone, 300,000 tons; sand and gravel, 200,000 tons; marl, 1,000 tons; and gold, 5 kg.

27ENERGY AND POWER
Electric power supplied by ten diesel-powered generators is inade-quate. A central authority, the Belize Electricity Board, supplies and operates the national power system. In 1991, total capacity of the Board's generators was 25,000 kw; an additional 9,000 kw is supplied by private industries and individuals. A 22,000 kw hydroelectric power station on the Macal River has been pro-posed by two US companies. The $50 million station is scheduled to be complete in 1994 and would be privately owned until 2034, after which the plant would be transferred to the government. Imports of fuels and energy accounted for 13% of total imports in 1990.

28INDUSTRY
The industrial sector is small but has been expanding. Industrial products in 1985 included 3,670 tons of fertilizer, 2,734,000 gar-ments, 2,528,000 lb of wheat flour, 102,000 tons of sugar, 28,100 tons of molasses, 600,000 gallons of beer, and 74,000 cig-arettes. The Development Finance Corp. promotes private capital investment in industry.

Aside from the processing of sugar and citrus, the manufactur-ing sector in Belize continues to be quite small. The garment industry revenues increased by 5.7% in 1992. Similarly, the tour-ism industry is expanding at a controlled rate. According to gov-ernment policy, there must be no large-scale investment in expansion of tourism in order to avoid social and environmental pressures. Still, tourism earnings rose by 21% in 1992, accumu-lating US$44 million in revenues.

29SCIENCE AND TECHNOLOGY
University College of Belize and Wesley College offer some scien-tific and technical training, but Belizean students must go abroad for advanced study.

30DOMESTIC TRADE
Except for warehouses and shops in Belize City, open markets still predominate in Belize. Normal business hours in Belize cities are 8 AM to noon and 1 to 6 PM, Monday through Friday, and 8 AM to 12:30 PM on Saturday. Banks are open from 8 AM to 1 PM, Monday through Thursday, and from 8 AM to noon and 3 to 6 PM on Fridays.

31FOREIGN TRADE
In 1992, Belize's merchandise imports reached $243 million, while exports totaled $141 million. Belize's most important imports were composed of machinery, food, fuels, chemicals, and other manufactured goods. The country's major exports included sugar, garments, citrus concentrate, seafood, vegeta-bles, lumber, and bananas. Belize's major export partners were the US, the UK, Mexico, CARICOM, the EC, and Canada. Its major providers included the US, the UK, Mexico, Japan, and the Netherlands.

32BALANCE OF PAYMENTS
The visible trade deficit is counterbalanced by overseas aid, Brit-ish military expenditures, foreign remittances from expatriates, and receipts from tourism. Belize has had a balance of payments surplus since 1985. Foreign reserves stand at US$80 million and business has avoided foreign exchange constraints. In 1992 mer-chandise exports totaled US$140.6 million and imports US$244.5 million. The merchandise trade balance was $–103.9 million.

The following table summarizes Belize's balance of payments for 1991 and 1992 (in millions of US dollars):

	1991	1992
CURRENT ACCOUNT		
Goods, services, and income	−54.6	−66.5
Unrequited transfers	27.9	30.0
TOTALS	−26.7	−36.5
CAPITAL ACCOUNT		
Direct investment	12.8	18.1
Portfolio investment	—	0.2
Other long-term capital	9.8	10.3
Other short-term capital	−7.6	−2.1
Exceptional Financing	—	—
Other liabilities	—	—
Reserves	20.2	−1.3
TOTALS	35.2	25.2
Errors and omissions	−8.5	11.3
Total change in reserves	16.3	0.1

³³BANKING AND SECURITIES

The bank of issue is the Central Bank of Belize. Four foreign banks conduct commercial banking. As of 1993, reserves of commercial banks amounted to B$44.0 million; demand deposits were B$82 million. The money supply at that date, as measured by M2, amounted to B$423.0 million. The Banking Ordinance has been amended to authorize offshore banking.

There is no securities exchange in Belize

³⁴INSURANCE

No recent information is available.

³⁵PUBLIC FINANCE

About half of Belize's recurrent expenditures are financed by customs duties; nearly all capital spending is funded by foreign loans and grants.

Since an IMF standby stabilization program was implemented in 1985, fiscal responsibility has improved. The government typically budgets over 50% of projected spending to capital development, and raises 60% of current revenues from trade taxes. Despite high public sector savings, increased expenditures for public investment in 1991/92 adversely affected the fiscal accounts. The following table shows actual revenues and expenditures for 1992 and 1993 in thousands of US dollars.

	1992	1993
REVENUE AND GRANTS		
Tax revenue	206,628	215,906
Non-tax revenue	35,806	41,521
Capital revenue	2,780	4,000
Grants	9,848	13,781
TOTAL	255,062	275,208
EXPENDITURES & LENDING MINUS REPAYMENTS		
General public service	53,119	50,043
Defense	10,991	11,069
Public order and safety	22,263	24,069
Education	53,843	60,315
Health	21,120	35,993
Social security and welfare	12,697	12,286
Housing and community amenities	19,446	45,435
Recreation, cultural, and religious affairs	7,848	16,071
Economic affairs and services	104,534	80,820
Other expenditures	19,918	18,619
Lending minus repayments	−32,898	−6,110
TOTAL	287,881	348,610
Deficit/Surplus	−32,819	−73,402

³⁶TAXATION

Income tax is levied on companies and individuals. Corporate taxes are set at a fixed rate of 35% of the chargeable income; personal income tax is based on a graduated scale from 5% to 45%. A company granted a development concession has a tax holiday of 5–15 years.

³⁷CUSTOMS AND DUTIES

Customs duties are generally ad valorem. Belize uses the CARICOM Common External Tariff, which ranges from 5% to 45%. There is also a stamp tax (normally 12%) on certain goods.

³⁸FOREIGN INVESTMENT

Investments in public projects are coordinated through the Development Finance Corp. and the Office of Economic Development. Private investment is actively encouraged through 10-year development concessions in sectors such as tourism, agricultural export diversification, fertilizers, and clothing manufacture. An Aliens Landholding Ordinance governs real estate investment through licensing procedures.

Although Belize offers a number of fiscal and other incentives for foreign investment, these concessions must first be negotiated and evaluated by the government. However, under its designation as a Caribbean Basin Initiative (CBI) beneficiary, investor interest has increased markedly. CBI opportunities are concentrated primarily in agribusiness and projects of national interest. In 1991, Belize, as part of CARICOM, signed a trade and investment agreement with the US, whose major investments include tourism, textiles, fuel distribution, livestock, citrus, tropical fruits, cocoa, and aquaculture.

³⁹ECONOMIC DEVELOPMENT

The government has opted to concentrate on developing agriculture, livestock, forestry, fishing, and tourism as foreign currency earners. The main sources of bilateral aid are the US and the UK; of multilateral aid, the UN and CDB. Multilateral and bilateral aid during 1983–86 totaled US$68.3 million.

Belize extended its hemispheric ties during 1992 by joining the Organization of American States, the Inter-American Development Bank and several subregional organizations, which will increase its access to development financing and external technical cooperation and strengthen its ties to Central America and Mexico. The future growth of Belize's primary sector depends largely on continued preferential access arrangements offered by its major trading partners. Improvements in service quality and infrastructure facilities are the major requirements for the tourism markets, while production diversification and marketing infrastructure are crucial to agricultural export development. The new government of Prime Minister Manuel Esquivel, which took office in July 1993, has indicated a strong commitment to these matters.

⁴⁰SOCIAL DEVELOPMENT

Workers' compensation covers agricultural workers. A social security scheme is in effect. Employed persons aged 14-64 are eligible to make contributions for old age, disability, survivor, and health benefits, for all of which the employer contributed amounts ranging from B$1.63 to B$7.80 per week.

Women have access to education and are active in all areas of national life, but face domestic violence and certain types of discrimination in the business sector. The Women's Bureau of the Ministry of Labor and Social Services develops programs to improve the status of women.

In 1991, the contraception prevalence rate was 47%, and the fertility rate was 4.5.

⁴¹HEALTH

Belize is relatively free of endemic diseases; since 1976, however, there has been an increase in the incidence of reported malaria cases. The government maintains a hospital in Belize City and rural health centers throughout the country. As of 1988, there were 85 physicians and 12 dentists. Ten hospitals provided a total of 583 beds that year.

Life expectancy was 69 years in 1992, and the infant mortality rate in 1992 was 41 per 1,000 live births. There were 7,200 births in 1992, with 72% of one-year-old children having been vaccinated against measles between 1991 and 1992.

[42]HOUSING

Housing is inadequate, and the situation has been aggravated by hurricane devastation. The government has put aside small sums for low-cost housing programs. As of 1980, 80% of dwellings were detached houses and 16% were apartments. Slightly under 60% were owner occupied, 26% were rented privately, 11% were occupied rent free, and 2% were rented from the government. In the same year, 70% of houses were wood, 12% were concrete, 7% adobe, and 3% were wood and concrete.

[43]EDUCATION

The adult literacy rate presently exceeds 93%. Primary education is free and compulsory for children between the ages of 6 and 14. In 1991, there were 46,023 pupils in 236 primary schools, 8,557 in 32 secondary schools and 1,191 in eight postsecondary schools. Most schools are run by the churches, others by the government.

Belize's first university, the University College of Belize, was opened in 1986. The University of the West Indies maintains a School for Continuing Education (SCE) in Belize. There are also several colleges providing specialized training such as the Belize Technical College; the Belize Teachers' College; Belize College of Agriculture; and the Belize Vocational Training Center. There are also two special schools maintained by the government for children with mental and physical disabilities.

[44]LIBRARIES AND MUSEUMS

The National Library Service maintains a central library in the Bliss Institute, a children's library, and a branch library in Belize City, as well as branches in all main towns and villages for a total of over 40 service points. The total book stock is approximately 130,000 volumes. The remains of the ancient Maya civilization—the best known are at Xunantunich—are being excavated by the government. The Department of Archaeology in Belmopan houses artifacts thus far uncovered.

[45]MEDIA

Belize is connected by radiotelegraph and telephone with Jamaica, Guatemala, Mexico, and the US. This service, along with cable and telex services, is operated by Cable and Wireless Ltd. An automatic telephone network, covering the entire country, is operated by the Belize Telecommunications, which was fully privatized in early 1992. In 1993, Belize had 21,000 telephones. The Belize National Radio Network, a government station in Belize City,

transmits in English and Spanish. Altogether, there are 6 AM and 5 FM radio stations and 1 television station. In 1991, there were 112,000 radios and 32,000 television sets.

There is no daily newspaper. The five weeklies in 1991 represented a spectrum of government and opposition views.

[46]ORGANIZATIONS

Cooperatives have been actively encouraged. There are 78 agricultural cooperatives, 18 fishing, 7 bee keeper, 5 housing, 3 transport, 3 consumer, 1 arts and crafts, 1 women's, and 5 cooperative societies in the country. The Belize Chamber of Commerce and Industry has its headquarters in Belize City.

[47]TOURISM, TRAVEL, AND RECREATION

Belize is attracting growing numbers of tourists to its Mayan ruins, its barrier reef (the longest in the Western Hemisphere), and its beaches, forests, and wildlife. Tourist arrivals totaled 222,779 in 1991, 26,016 from Europe and 84,033 from the Americas. In 1991, tourist expenditures totaled US$95 million. There were 2,922 hotel rooms and 4,990 beds. Tourists from the US and UK who arrive directly from their own countries do not need passports or visas; visitors from Commonwealth and most Western European countries need passports but not visas. In May, 1992, Belize was the site of the First World Congress on Tourism and the Environment.

[48]FAMOUS BELIZEANS

George C. Price (b.1919), leader of the PUP, became the country's first premier in 1964. Manuel Esquivel (b.1940), leader of the UDP, became prime minister in December 1984.

[49]DEPENDENCIES

Belize has no territories or colonies.

[50]BIBLIOGRAPHY

Bolland, O. Nigel. *Belize: A New Nation in Central America.* Boulder, Colo.: Westview, 1986.
———. *The Formation of a Colonial Society: Belize, from Conquest to Crown Colony.* Baltimore: Johns Hopkins University Press, 1977.
Fernandez, Julio A. *Belize: Case Study for Democracy in Central America Fernandez.* Brookfield, Vt.: Avebury, 1989.
Grant, Cedric Hilburn. *The Making of Modern Belize: Politics, Society, and British Colonialism in Central America.* New York: Cambridge University Press, 1976.
Merrill, Tim, ed. *Guyana and Belize: Country Studies.* 2nd ed. Washington, D.C.: Government Printing Office, 1993.
Setzkorn, William David. *Formerly British Honduras: A Profile of the New Nation of Belize.* Rev. ed. Athens, Oh.: Ohio University Press, 1981.
Wright, Ronald. *Time among the Maya: Travels in Belize, Guatemala, and Mexico.* New York: Weidenfeld & Nicolson, 1989.

BOLIVIA

Republic of Bolivia
República de Bolivia

CAPITALS: Sucre (legal and judicial capital); La Paz (administrative capital).

FLAG: The flag is a horizontal tricolor of red, gold, and green stripes, representing the animal, mineral, and vegetable kingdoms, respectively, with the coat of arms centered on the yellow band.

ANTHEM: *Himno Nacional,* beginning "Bolivianos, el hado propicio coronó nuestros volos anhelos" ("Bolivians, propitious fate crowned our outcries of yearning").

MONETARY UNIT: The boliviano (B) was introduced on 1 January 1987, replacing the peso at a rate of P1,000,000 = B1. There are coins of 2, 5, 10, 20, and 50 cents and 1 boliviano and notes of 2, 5, 10, 20, 50, 100, and 200 bolivianos. B1 = $0.2186 (or $1 = B4.5750).

WEIGHTS AND MEASURES: The metric system is the legal standard, but some Spanish weights are still used in retail trade.

HOLIDAYS: New Year's Day, 1 January; Labor Day, 1 May; National Festival, 5–7 August; Columbus Day, 12 October; All Saints' Day, 1 November; Christmas, 25 December. Movable holidays include Carnival, Ash Wednesday, Holy Thursday, Good Friday, Holy Saturday, and Corpus Christi.

TIME: 8 AM = noon GMT.

¹LOCATION, SIZE, AND EXTENT

Situated in South America just N of the Tropic of Capricorn, Bolivia has a total area of 1,098,580 sq km (424,164 sq mi), extending about 1,530 km (950 mi) N–S and 1,450 km (900 mi) E–W. Comparatively, the area occupied by Bolivia is slightly less than three times the size of the state of Montana. Completely landlocked, Bolivia is bounded on the N and NE by Brazil, on the SE by Paraguay, on the S by Argentina, on the SW by Chile, and on the W by Peru, with a total boundary length of 6,743 km (4,190 mi).

The capital city of Bolivia, La Paz, is located in the west-central part of the country.

²TOPOGRAPHY

Bolivia has three geographic zones: the Andean highlands in the west, running north to south; the moist slopes and valleys on the eastern side of the Andes, called the Yungas and Valles; and the eastern tropical lowland plains, or Oriente. In Bolivia, the Andes, divided into two chains, attain their greatest width, about 640 km (400 mi), and constitute about one-third of the country. Between the Cordillera Occidental, forming the border with Chile and cutting Bolivia off from the Pacific, and the complex knots of the Cordillera Oriental lies a broad sedimentary plateau about 4,000 m (13,000 ft) above sea level, called the Altiplano, which contains about 28% of Bolivia's land area and more than half of its population. In the north of this plateau, astride the border with Peru, lies Lake Titicaca, 222 km (138 mi) long and 113 km (70 mi) wide; with its surface at an altitude of more than 3,660 m (12,000 ft), it is the highest navigable lake in the world. The lake is drained to the south by the 320-km (200-mi) Desaguadero River, which empties into shallow, salty Lake Poopó. Farther south are arid salt flats. The Cordillera Oriental has high habitable basins and valleys collectively referred to as the Puna. Bolivia's most majestic mountains are in the northern part of the Cordillera Oriental around Lake Titicaca, where the mountain sector is capped with snow; the highest of these is Ancohuma

(6,550 m/21,489 ft). Illimani and Illampu, both rising more than 6,400 m (21,000 ft), overlook the city of La Paz, which is protected from cold winds by its position in the spectacular gorge formed by the headwaters of the La Paz River. The three important valleys of this region, Cochabamba, Sucre, and Tarija, are from 1,830 to 3,050 m (6,000 to 10,000 ft) in altitude.

Bolivia's important rivers descend across the Yungas and Valles into the low tropical plains of the Oriente, which comprises three-fifths of the land but only about one-fifth of the population. The Guaporé, the Mamoré, the Beni, and the Madre de Dios rivers cross the often-flooded northern savanna and tropical forests, all converging in the northeast to form the Madeira, which flows into Brazil. The plains become drier in the southeast, forming Bolivia's scrub-covered Chaco. Crossing the Chaco to the southeast, the Pilcomayo River leaves Bolivia to form the border between Paraguay and Argentina.

³CLIMATE

Although Bolivia lies entirely in the tropics, extreme differences in altitude and rainfall give it a great variety in climate. The mean annual temperature of La Paz, at 3,697 m (12,130 ft), is about 8°C (46°F); that of Trinidad, in the eastern lowlands, is 26°C (79°F). In the western highlands, cold winds blow all year round; at night the temperature often drops below freezing, but the sun is intense and the air brilliant during the day. The rainy season lasts from December to February, but during most of the year the high Altiplano plateau is parched and inhospitable. Around Lake Titicaca, rainfall is adequate, but there is less than 5 cm (2 in) a year in the extreme southwest. The fertile valleys in the Cordillera Oriental have a warmer, semiarid Mediterranean climate.

The Yungas and Valles have a semitropical, moist climate that gradually becomes hotter as one descends from the eastern slopes of the Andes to the tropical eastern lowlands. Rainfall is heavy in the northeast, and floods are common in March and April. The lowland plain becomes drier to the south, until it reaches drought conditions near the Argentine border.

4FLORA AND FAUNA

Bolivia shares much of the wide variety of flora and fauna found in the four countries surrounding it. Because of the wide range in altitude, Bolivia has plants representative of every climatic zone, from arctic growth high in the sierra to tropical forests in the Amazon basin. On the high plateau above 3,050 m (10,000 ft) grows a coarse bunch grass called ichu, used for pasture, thatching, and weaving mats. A reed called totora, which grows around Lake Titicaca, is used for making small fishing boats (balsas). The low bushlike tola and the resinous mosslike yareta are both used for fuel. The Lake Titicaca region is believed to be the original home of the potato.

In the tropical forest, the quinine-producing quina tree grows, as does the Pará rubber tree. There are more than 2,000 species of hardwoods. Aromatic shrubs are common, as are vanilla, sarsaparilla, and saffron plants. Useful native plants include palms, sweet potatoes, manioc, peanuts, and an astonishing variety of fruits. The Chaco is covered with a prickly scrub collectively called monte; tannin-producing quebracho trees also abound there.

On the Altiplano, the most important animal is the llama, one of the most efficient carrier animals known; alpaca and guanaco and several varieties of cavy (guinea pig) are found there, too. Lake Titicaca has several varieties of edible fish. In the tropical Amazon region are the puma, coati, tapir, armadillo, sloth, peccary, capiguara (river hog), and ant bear, as well as several kinds of monkeys. Birdlife is rich and varied. Reptiles and an enormous variety of insects are found below 3,050 m (10,000 ft).

5ENVIRONMENT

As of 1994, the chief environmental problem in the densely populated Altiplano was soil erosion, resulting from poor cultivation methods (including slash-and-burn agriculture) and overgrazing. Erosion currently affects 30% of the land in Bolivia. Salinity and alkolinization are also a significant problem. Inadequate sanitation and solid-waste disposal, as well as effluents from mining activities, contribute to the Altiplano's declining water quality, which poses a threat both to fish life and to human health. Bolivians have 72.0 cubic miles of water supply, but 24% of the city dwellers and 70% of all rural people do not have pure water. The main sources of water pollution are fertilizers, pesticides, and mining. Most environmental legislation dates from the 1970s, when Bolivia enacted the Health Code of 1978 (which contains provisions governing water quality), the National General Forest Act of 1974, and the Law of Wildlife, National Parks, Hunting, and Fishing (Decree Law No. 12,301) of 1975.

In July 1987, the Bolivian government became the first government in history to agree to protect a part of its environment in return for a reduction of its foreign debt, when Conservation International, a US nonprofit group, purchased $650,000 of the debt in return for Bolivia setting aside 1.5 million hectares (3.7 million acres) of tropical lowlands in three conservation areas. The Department of Science and Technology, within the Ministry of Planning and Coordination, plans and coordinates all governmental and intergovernmental activities related to the environment.

As of 1987, endangered species in Bolivia included the puna rhea, South American river turtle, broad-nosed caiman, spectacled caiman, black caiman, jaguar, jaguarundi, margay, ocelot, emperor tamarin, and giant anteater. The llama and the alpaca are also threatened with extinction. As of 1994, out of a total of 267, 21 mammals were considered endangered. There were 34 threatened species of birds in a total of 1,177 species. Four species of reptiles out of 180 and one species of fish were also considered endangered. Of 15–18,000 plant species, 39 were in danger of extinction.

6POPULATION

The total population as of the last national census in 1992 was 6,420,792. The projected population for the year 2000 was 9,038,000, assuming a crude birthrate of 32.1 per 1,000 population, a crude death rate of 8.4, and a net natural increase of 23.7 during 1995–2000. Annual population growth between 1985 and 1990 averaged 2.46%. The population density in 1992 was 5.8 persons per sq km (15.1 per sq mi). About 46% of the population was rural in 1995; three-fourths of the total population lives on the Altiplano or in the western mountain valleys; the southeastern lowlands are sparsely populated. In 1988, La Paz, the administrative capital, had an estimated population of 1,049,800; Sucre, the legal and judicial capital, had 95,635. Other important cities, all departmental capitals, are Santa Cruz, with an estimated 1988 population of 615,122; Cochabamba, 377,259; Oruro, 195,239; and Potosí, 114,092.

7MIGRATION

Aside from Spaniards during the colonial period, European immigration has been insignificant. Small numbers of Italians, Poles, and Germans have settled mainly in the vicinity of La Paz and Cochabamba, and some Jewish refugees from Nazi Germany arrived in the 1930s. After World War II, about 1,000 Japanese settled in colonies around Santa Cruz and became successful in truck farming, and several hundred Okinawan families established themselves as rice growers in the same area.

Since the 1950s, migration to neighboring countries has increased: 30,000 left Bolivia in 1950–55; 40,000 left in 1980–85. About 675,000 Bolivians were estimated to reside outside the country in the late 1980s, in search of employment and better economic opportunities. Since the emigrants tend to have basic training or technical skills, a drain of important human resources is occurring. A number of Bolivian braceros (contract agricultural laborers) go to northwestern Argentina to work in rice and sugar harvests. In the 1970s, Brazilian settlers, drawn by improved railroad and highway links, migrated to northeastern Bolivia in growing numbers; these immigrants had a substantial influence on the region, since they continued to speak Portuguese and to use Brazilian currency as their medium of exchange. Within the country, migration is swelling the sparsely populated lowlands, particularly in Santa Cruz and its environs. High unemployment among agricultural laborers and miners has caused significant migration to the cities.

8ETHNIC GROUPS

Estimates of the make-up of the Amerindian population are Quechua, about 30% and Aymará, 25%. Cholos (Bolivians of mixed white and Amerindian lineage) make up another 25 to 30%, and those of wholly European background account for virtually all of the remainder. One reason for the uncertainty of these estimates is that although the distinction between Amerindian, cholo, and white was at one time racial, it has gradually become at least partially sociocultural: Amerindians become cholos when they abandon their native costumes, learn to speak Spanish, and acquire a skill or trade. Not all those classified as whites are without some Amerindian mixture.

The rapidly disappearing Amerindians who populate the tropical plains in the southeast, the Chiriguanos, are believed to be a Guaraní tribe that moved west from Paraguay before the Spanish conquest. The Mojenos, Chiquitanos, and Sirionós inhabit the forest-grassland border in the far east. In all, Amerindians number about 100,000.

9LANGUAGES

Spanish, Quechua, and Aymará are all official languages. About 40% of Bolivians speak Spanish as a mother tongue. As spoken by educated Bolivians, it differs less from Castilian than do the

LOCATION: 9°40′ to 22°53′s; 57°29′ to 69°35′w. **BOUNDARY LENGTHS:** Brazil, 3,125 km (1,942 mi); Paraguay, 756 km (470 mi); Argentina, 742 km (461 mi); Chile, 861 km (535 mi); Peru, 1,048 km (651 mi).

dialects of many regions in Spain itself. Approximately 37% of the people still speak Quechua, and 24% speak Aymará, although an increasing number of Amerindians also speak Spanish.

¹⁰RELIGIONS

The 1961 constitution abolished state support of the Roman Catholic Church, thus formally separating church and state. All religions are accepted; only civil marriage is legal. In 1993, an estimated 92.1% of the population was Roman Catholic; most Amerindians incorporate their own religious symbolism into the Christian context. An active Protestant minority was established

in 1985 at 335,000; there were also about 65,900 tribal religionists, 2,800 Buddhists, and 600 Jews.

¹¹TRANSPORTATION

Transportation in Bolivia has been seriously impeded both by the geographic configuration of the country and by the concentration of population and mineral wealth in the mountain regions. Railroads and highways twist along the Andean Range, and are often blocked by mudslides during the rainy season. The shortage of transportation facilities is one of the most serious barriers to economic development. Railroads are single-track meter gauge, totaling 3,675 km (2,284 mi) in 1991. All of the trackage is government owned. A major portion of the railway system services the Altiplano, the western mountainous region, providing vital international connections with Pacific coast ports. The remaining track connects the eastern city of Santa Cruz with Brazil and Argentina. An important route to Puerto Suárez eventually reaches the Brazilian port of Santos, while the line to Argentina via Villazón continues on to Buenos Aires. The two systems are administered by the government-owned National Railway Co.; two smaller lines (157 km/98 mi) are run by the Mining Corp. of Bolivia and by the Pulacayo mining enterprise.

In 1991, of a total of 38,836 km (24,133 mi) of roads, less than 4% were paved. The Cochabamba-Santa Cruz highway, completed in 1963, was a major achievement in connecting lowland and highland Bolivia. In 1992 there were 335,000 motor vehicles, of which 270,000 were passenger cars, and 65,000 were commercial vehicles.

Airlines are particularly important in view of Bolivia's topography and the underdevelopment of other means of transportation. The hub of air traffic is El Alto airport near La Paz, the world's highest commercial airport; the other international airport is at Santa Cruz. Lloyd Aéreo Boliviano (LAB), with 50% government capital, services most of the country. In 1992, LAB carried 1,214,400 passengers. Military Air Transport, operated by the air force, provides some civilian freight and passenger service, and numerous air taxi companies are also in service.

Little use has been made of Bolivia's 14,000 km (8,700 mi) of navigable waterways. There are no regular riverboat services. Bolivia has free port privileges at Antofagasta and Arica (Chile), at Mollendo (Peru), and at Santos (Brazil).

12HISTORY

By about AD 600, Amerindians (believed to belong to the Aymará-speaking Colla tribe) were settled around the southern end of Lake Titicaca. As they came into contact with coastal tribes, the highly developed classic Tiahuanaco civilization emerged, reaching its peak about AD 900. Lake Titicaca became a place of worship and a great commercial center. Then cultural and political disintegration set in, and by 1300, the Quechua-speaking Incas had conquered the region and had colonized villages in most of what is now Bolivia.

The demise of the Inca empire began in 1527 with the death of the Inca Emperor Huayna Capac. His two sons, Huáscar and Atahualpa, fought a civil war over succession. Francisco Pizarro, taking advantage of the civil war raging between the two heirs, led the Spanish conquest of the Inca Empire in 1532–33. In 1539, Pedro de Anzures established La Plata, subsequently called Charcas and Chuquisaca and now known as Sucre, Bolivia's legal and judicial capital.

The Spaniards did not become interested in the land called Alto Peru, or Upper Peru, until the discovery in 1545 of the fabulously rich silver mine called the Cerro Rico (Rich Hill) de Potosí. Three years later, La Paz was founded on the main silver transport route between Potosí and the coast. In 1559, the audiencia (region under a royal court) of Charcas was established in Upper Peru under the viceroyalty of Lima. The mines continued to produce vast amounts of wealth for the Spanish Empire, and for years the city of Potosí was the largest city in the Western Hemisphere. In 1776, the audiencia was appended to the viceroyalty of La Plata (Buenos Aires).

The independence of Upper Peru came from the revolt of the small, native-born Spanish ruling class. In 1809, a year after Napoleon's invasion of Spain, the Spanish authorities in Chuquisaca (Sucre) were temporarily overthrown, and the local elite proclaimed independence. The movement was quickly put down by Spanish arms. The young government in Buenos Aires showed some interest in the region, having included delegates from Upper Peru when independence was declared at the Congress of Tucumán in 1816. However, independence came from Peru, after Simón Bolívar's victory at the battle of Ayacucho in December 1824. Bolívar then sent his young general, Antonio José de Sucre, to free Upper Peru. On 6 August 1825 a congress at Chuquisaca formally proclaimed the independence of the Republic of Bolívar, a name soon changed to Bolivia. Sucre was chosen as the first president in 1826, and Chuquisaca was renamed Sucre in his honor.

Sucre was driven out of office after only two years. He was succeeded by Gen. Andrés de Santa Cruz, a man with imperial ambitions. In 1836, Santa Cruz conquered Peru and formed the Peruvian-Bolivian Confederation. In 1839, Chilean forces defeated and dissolved the confederation and ended the life term of Santa Cruz.

A period of instability followed, with civilians and army officers succeeding one another, usually by force of arms. The almost constant civil war retarded Bolivia's economic organization and helped bring about the loss of a large part of its land. The first of these losses came after the War of the Pacific (1879–84), pitting Chile against Bolivia and Peru. Chile's superior military force routed the Bolivians and seized what was then the Bolivian port of Antofagasta. The postwar settlement took away Bolivia's only coastal territory, as well as the nitrate-rich coastal area around it. Bolivia was forever after a landlocked country, with only rights of access to the Pacific under a 1904 treaty. Another territorial loss came in 1903 with the cession to Brazil of the Acre region, rich in natural rubber, in exchange for an indemnity and other minor concessions.

The economy was aided in the late 19th century by a silver boom. When prices collapsed, silver production gave way to tin mining. The dominance of mining in Bolivia's economy conditioned the political system. A few wealthy mine and plantation owners, allied with various foreign interests, competed for power. Indians, excluded from the system, found their lot unchanged after almost 400 years.

This arrangement began to unravel with yet another loss of Bolivian territory. In 1932, Bolivia warred with Paraguay over the Chaco, the lowland area believed at the time to be rich in oil. Despite their numerical superiority, the Bolivians were defeated by 1935, and Paraguay controlled about three-fourths of the disputed territory. The formal settlement in 1938 gave most of this land to Paraguay, although Bolivia was promised a corridor to the Paraguay River.

The Chaco war pointed out the weaknesses in Bolivia's political and social structure. Bolivia's loss was in part due to the poor morale of its soldiers, an army of conscripted Indians with no loyalty to the elite officer corps. In 1936 Bolivia's rigid caste system cracked, and Col. David Toro came to power with labor support and a vaguely socialist/nationalist platform. The government expropriated Standard Oil of New Jersey's Bolivian properties in 1937. Toro's government attempted social reform, and its efforts to control mining and banking led to fierce opposition. The tension continued after Toro was forced out of office by Col. Germán Busch. Busch challenged Bolivia's three large tin-mining interests, owned by Patiño, Aramayo and Hochschild. With strong labor backing, Busch arranged for the constitution of 1938, a document guaranteeing the right of labor to organize, universal education, and nationalized subsoil rights. The very next year, Busch died in what was officially ruled a suicide.

World War II brought further strains to Bolivia. As world demand skyrocketed, the tin market boomed, but working conditions remained miserable and wages remained low. In 1942, protests by tin workers against the "tin barons" and their American financiers was met with force by the government of Gen. Enrique Peñaranda, resulting in the "Catavi massacre." Wishing to retain the strategic materials in mid-war, the United States commissioned a US-Bolivian commission to study working conditions. This report confirmed the workers' grievances, but was completely ignored by Peñaranda. In December of 1943, a coalition of the army and the Nationalist Revolutionary Movement (Movimiento Nacionalista Revolutionario—MNR), which had gained considerable support among the mine workers, engineered a coup, ousting Peñaranda and putting Maj. Gualberto Villaroel into power. The tin market collapsed at the war's end, weakening the government's power base. In 1946, Villarroel was overthrown and hanged, along with others, by a mob of workers, soldiers, and students, and a conservative government was installed.

In 1951, the MNR's candidate, Víctor Paz Estenssoro, a former associate of Villarroel, apparently won the presidential election, but a military junta stepped in, denying the legality of the vote. Paz, representing the left wing of the MNR, became president in 1952 as a result of a party-led uprising. For the next twelve years, Bolivian politics would be dominated by the MNR.

The leadership of the MNR was shared by four men: Paz, Juan Lechín Oquendo, leftist head of the miners' union, Hernán Siles Zuazo, close ally of Paz, and the right-wing Walter Guevara Arze. A pact among the four was to allow them to take turns in the presidency over the next sixteen years. The Paz government made dramatic moves in an attempt to transform Bolivian society. The tin holdings of the three dominant family interests were expropriated, and a comprehensive land reform program was begun, along with wide-scale welfare and literacy programs. Industry was encouraged, the search for oil deposits was accelerated, and a new policy gave Amerindians the right to vote and sought to integrate the Amerindian community more fully into the national economy. The right to vote, previously restricted to literate Bolivian males (who constituted less than ten per cent of the population), was made universal for all Bolivians over 21.

In 1956, as expected, Hernán Siles succeeded to the presidency. But Siles only governed under Paz's watchful eye, and in 1960, Paz challenged the candidacy of Guevara Arze. Guevara went into exile, and Paz again assumed the presidency, with Lechín as his vice-president. Paz became increasingly dictatorial, and the splits within the MNR worsened. Paz conspired to give himself yet another presidential term, complete with rigged elections in June 1964. Siles, now leading the right wing of the MNR, and Lechín, now leading the leftist opposition, conducted a hunger strike protesting Paz's authoritarian designs. Finally, the military defected when it became clear that Paz was without any allies. The military coup occurred in November 1964, with the junta selecting as president Paz's vice-president René Barrientos Ortuño.

Barrientos moved quickly to consolidate his new government, removing Paz's old supporters and sending Lechín into exile. In the following year, a military faction forced Barrientos to allow Gen. Alfredo Ovando Candia to become his "co-president." This odd arrangement was resolved in 1966 with new elections. Barrientos and his newly formed Popular Christian Movement won a resounding victory.

In 1967, an active guerrilla movement with pro-Castro tendencies emerged in southeastern Bolivia. The Bolivian authorities imprisoned the French intellectual Jules Régis Debray, who revealed that the famous comrade of Fidel Castro, Ernesto "Che" Guevara, was leading the guerrilla movement. Later in the year, the Bolivian army apprehended and killed Guevara.

Barrientos died in a helicopter crash in April 1969, and a civilian, Vice-President Adolfo Siles Salinas, became president. Siles was overthrown in September by Barrientos' former rival, Gen. Ovando, who presented himself as a presidential candidate for 1970 but then canceled the election. In October 1970, President Ovando was overthrown by rightist elements of the military, but the next day a leftist faction succeeded in making Gen. Juan José Torres Gonzales the new president.

The Torres regime was marked by increasing political instability. Backed by students and the Bolivian Labor Council, Torres expelled the US Peace Corps, permitted the expropriation of both US and privately owned Bolivian properties, sanctioned the seizure of land by landless peasants, established a labor-dominated People's Assembly, and declared his support for the reestablishment of diplomatic relations with Cuba. In a bloody three-day revolution in August 1971, the Torres government was ousted by a coalition of the armed forces and political leaders from the MNR and the Bolivian Socialist Falange (Falange Socialista Boliviana—FSB), together with other middle-class groups. The leader of the coup was Hugo Banzer Suárez, who was installed as president later in the month. Banzer consolidated his support with the founding of the Nationalist Popular Front, which became the political framework of the new government. Ex-president Paz returned from exile to head the MNR.

The first threats to the Banzer government came from the left. There were reports late in 1971 of renewed activity by the Guevarist National Liberation Army. The government launched a vigorous antiguerrilla campaign and claimed nearly complete success. In 1973, however, Banzer's coalition began to splinter. In 1974, when the MNR threatened to withdraw from the coalition, Paz went into exile again. After two coup attempts had been crushed in the fall of 1973 and two others in the summer of 1974, Banzer formed a new all-military cabinet. In November 1974, the MNR, the FSB, and other political parties were abolished, and trade union meetings were declared illegal.

In response to industrial and political unrest, Banzer announced the restoration of political parties in 1977 and of unions in 1978. He promised to hold new elections in July 1978. Paz again returned from exile to run. The election results were annulled, however, and a new military government came to power in a bloodless coup. Another election took place in July 1979, but because no candidate received a majority and the Congress could not decide whom to select from among the three main candidates, an interim president was named. Another coup followed in November, but constitutional government was restored only two weeks later, in the wake of popular resistance. New presidential and congressional elections in June 1980 again failed to produce a majority winner, and in July there was another coup, staged by Gen. Luis García Meza, who promptly suspended the Congress, banned most political parties and all union activity, and established strict censorship in order to remove the "Marxist cancer" from Bolivia. Paz again went into exile. During the García regime there were frequent reports of arbitrary arrests, use of torture, and other human rights violations. In August 1981, García, who was suspected, along with other top officials in the government, of involvement with the cocaine trade, was deposed in a coup—the 190th in Bolivian history. He went into exile in Argentina in October 1982, and in May 1983, he was ordered arrested on charges of "corruption and economic crimes"—specifically, the fraudulent use of government funds in agricultural, construction, and oil refinery deals. Meanwhile, under two more military governments, political and union rights were gradually restored.

In October 1982, amid a worsening economic situation and increasing labor unrest, the Congress elected Hernán Siles Zuazo to the presidency. Siles, returning to office 22 years after the end of his previous presidency, could still count on electoral support, and had received a plurality of votes in the 1979 and 1980 elections. His shaky coalition faced continued economic problems, including food shortages and rampant inflation, and a right-wing threat from paramilitary groups whose activities were reportedly financed by cocaine smuggling. In November 1983, the Bolivian government announced an austerity program that included a 60% devaluation of the peso and hefty food price increases. By mid-1985, Siles had so mismanaged the economy and the political situation that labor unrest and social tension forced him to call national elections and to agree to relinquish power a full year before the expiration of his term. Banzer won a plurality of the popular vote, but the MNR won more seats in the congressional elections, resulting in a fourth term of office for the 77-year-old Paz. In a departure from the norm, the MNR and Banzer's party agreed to cooperate, allowing a comprehensive economic reform package to pass through the legislature.

Faced with runaway inflation, which reached an annual rate of 14,000% in August 1985, the government abandoned controlled exchange rates, abolished price controls, liberalized external trade, and instituted more restrictive monetary and wage policies. The result was sharply lower inflation and interest rates, and a more stable economy, although the shocks of this liberalization were felt through government layoffs and falling consumer buying power.

More importantly, Paz was able to forge a fundamental consensus among competing political parties in support of a continuing democracy. In 1989, despite a hotly contested presidential race, power passed from the MNR to the left-wing movement of the Revolutionary Left (MIR.) The peaceful transfer of power from one party to another was a milestone in itself. An equally hopeful sign was the fact that the MIR leader, Jaime Paz Zamora, was able to hold together a coalition with the right-wing Democratic National Alliance (ADN) to serve a full four-year presidential term. In a country like Bolivia, this is no mean feat.

The elections of 1993 brought the MNR back to power, with Gonzalo Sánchez de Lozada assuming the presidency. Sánchez chose as his running mate Victor Hugo Cárdenas, an advocate of Bolivia's Aymara-speaking Amerindians. While some saw the move as a cynical ploy, others expressed hope that Bolivia's long-suffering native population might be brought into the political system.

13 GOVERNMENT

Constitutionally, Bolivia is a centralist republic. The constitution of 3 February 1967 provides for a representative democracy, with its government divided into an executive branch, a bicameral legislature (a Congress consisting of a Chamber of Deputies and a Senate), and the judiciary.

Bolivia has had a spotty constitutional history. The current constitution is the result of a series of actions begun by the military junta that took control in November 1964. The junta replaced the 1961 constitution with the 1945 constitution, as amended in 1947. At the same time, it retained those sections of the 1961 constitution that dealt with universal suffrage, nationalization of the tin mines, land reform, and compulsory education. The 1967 constitution was further amended to circumscribe the power of militia forces. In practice, the constitution has not been rigorously observed. Coups and states of siege have been frequent. Congress was dissolved by the armed forces from 1969–79 and again between 1980 and 1982.

Under the constitution, the president and the vice-president are elected by direct popular vote for a four-year term, and cannot serve consecutive terms. If no candidate receives a majority in a presidential election, the Congress chooses among the three leading candidates. However, between 1966 and 1978, no presidential elections were held. The president's powers are considerable, and presidential authority often extends beyond the constitution. The president has the prerogative to declare a state of siege and may then rule by decree for 90 days. The Congress consists of 27 senators (3 from each department) and 130 deputies. Members of both houses are elected for four-year terms. Bolivia utilizes a form of proportional representation to ensure minority representation in the Chamber of Deputies and an incomplete-list system for the Senate. The regular session of Congress lasts for 90 days.

Universal suffrage, with no literacy or property qualifications, was decreed in 1952 for married persons at 18 years and single persons at 21. The constitution includes a bill of rights, which guarantees the right to express ideas freely, petition the government, and obtain a release under a writ of habeas corpus in case of illegal detention.

14 POLITICAL PARTIES

Bolivia's proportional representation system has encouraged the formation of several political parties. Numerous parties and coalitions have formed and dissolved over the years, usually tied to the personalities of the various leaders.

The Nationalist Revolutionary Movement (Movimiento Nacionalista Revolucionario—MNR) was founded by Víctor Paz Estenssoro, Hernán Siles Zuazo and others in 1941. Although militant originally, the years have moderated the party's stance. The MNR came to power in 1952, with the help of the Revolutionary Workers Party, the carabineros (national police), and the miners' and peasants' militias. In the subsequent years, the MNR began to rely increasingly on foreign aid, especially from the US, and became increasingly autocratic and corrupt. Finally, quarreling among the party leadership weakened the party, and by 1964 the MNR's monopoly on power had dissolved. In November 1964, Paz was sent into exile in Peru.

The MNR was then eclipsed by the charisma of President René Barrientos and his Popular Christian Movement. The MNR returned as part of the Nationalist Popular Front, organized by Hugo Banzer Suárez. Banzer then outlawed the MNR in November 1974. In the late 1970s, the MNR reappeared, along with a dissident MNRI (the MNR "of the left") headed by Hernán Siles.

With the restoration of Bolivian democracy, the MNRI won the presidency under Siles along with a coalition of leftist parties that included the and the Communist Party of Bolivia. Movement of the Revolutionary Left (Movimiento de la Izquierda Revolucionaria—MIR, headed by Jaime Paz Zamora. As part of the Siles government, the MIR was an active partner. Paz Zamora was the vice-president, and several MIR officials were in the cabinet. In January of 1983, six MIR ministers resigned, which in retrospect was not a bad strategic move. The MIR won the presidency in 1989 after an extremely close election, and only after seeking support from the right-wing Democratic Nationalist Alliance (Alianza Democrática Nacionalista—ADN).

The ADN is closely tied to former President Hugo Banzer Suarez. Banzer, a former military officer, came to power in an alliance with the MNR, but eventually ruled as a military dictator. This right-wing party was denied power in 1985 by the MNR/MIR coalition. However, the ADN was instrumental in bringing Paz Zamora to power, and held half the ministerial positions in that government.

The United Left (Izquierda Unida—IU) is a coalition of leftist parties. The center-right Christian Democratic Party (PDC) and far-right "Conscience of the Fatherland" (CONDEPA) complete the list of parties with significant electoral support.

15 LOCAL GOVERNMENT

Bolivia is essentially a unitary system, with a highly centralized national government. Bolivia's nine departments—La Paz, Cochabamba, Chuquisaca, Potosí, Oruro, Santa Cruz, Tarija, El Beni, and Pando—are administered by prefects appointed by the president for four-year terms. The departments are subdivided into 94 provinces, each headed by a subprefect recommended by the prefect, appointed by the president and responsible to him through the minister of the interior. The provinces are further divided into about 1,000 cantons, each of which is under the jurisdiction of a magistrate (corregidor). There are no local legislatures. Important towns and cities have more self-government. Each has a popularly elected council of from 5 to 12 members, but municipal tax ordinances must be approved by the Senate. Mayors (alcaldes) are also elected. The Amerindian communities, although they are not formal administrative units, are recognized by law.

¹⁶JUDICIAL SYSTEM
The Bolivian judiciary usually defers to the political direction of the nation's executive. Judicial power is exercised by the Supreme Court, the superior district courts in each department (courts of second instance), and the local courts (courts of first instance). The Supreme Court, which sits at Sucre, is divided into four chambers: two deal with civil cases, one with criminal cases, and one with administrative, mining, and social cases. The 12 Supreme Court judges, called ministros, are chosen for 10-year terms by a two-thirds vote of the Chamber of Deputies from a list of three names submitted for each vacancy by the Senate. They may be reelected indefinitely.

Most cases that reach the Supreme Court are appellate; its area of original jurisdiction is limited mainly to decisions on the constitutionality of laws and to disputes involving diplomats or important government officials. Each district court judge is elected by the Senate for six years from a list of three submitted by the Supreme Court.

The district courts usually hear appeals from the courts of first instance. Judges of the courts of first instance (tribunales and juzgados) are chosen by the Supreme Court from a list submitted by the district courts. There is also a separate national labor court and an agrarian court, dealing with agrarian reform cases.

¹⁷ARMED FORCES
Since the 1985 elections, relative political stability and civilian control has reduced the power and cost of the military. In theory there is universal compulsory military training, but only 20% of the registered youths (19,000 in 1993) are called for service. As of 1993, armed strength totaled 31,500 men (army, 23,000; a navy for lake and river patrol, 4,500; and air force, 4,000), and paramilitary police of 16,200. Defense expenditures in 1993 were $80 million or 1.6% of the gross domestic product.

¹⁸INTERNATIONAL COOPERATION
Bolivia is a charter member of the UN, having joined on 14 November 1945, and participates in ECLAC and all the nonregional specialized agencies except IMO and WIPO. It is a signatory of the Law of the Sea. As a member of the International Tin Council, Bolivia retains part of its tin exports to help stabilize prices. Bolivia is also a member of many inter-American and other intergovernmental organizations, notably the Andean Pact, G-77, IDB, LAIA, OAS, and PAHO.

¹⁹ECONOMY
Bolivia is one of the Western Hemisphere's poorest countries, despite an abundance of mineral resources. Its economy has always been dependent on mineral exports, principally of tin, but these have gradually declined since World War II. Little of the nation's great agricultural and forest potential has been developed; agriculture remains little above the subsistence level, and Bolivia must import large quantities of food. The economic system is basically state capitalism; about 80% of the GDP is produced by government enterprises.

Bolivian economic production grew at a steady annual rate of 5.8% over the period 1960–70 but declined to 4.5% during 1970–72. During 1973–77, the average annual growth rate of the GDP was 6.3%, but in the 1980s, growth has been negative, from a low of –8.7% in 1982 to –3.7% in 1986. Inflation, very low by Latin American standards in the 1960s, when the average annual rate was 3.5%, began to increase during the 1970s, averaging more than 22% yearly, and reached 14,000% in August 1985, before declining to 10.5% at the end of July 1987.

In August 1985, President Paz implemented a drastic antiinflationary program; he floated the peso, froze public-sector wages, cut public spending, eliminated controls on bank interest rates, authorized banks to make foreign-currency loans and offer foreign-currency accounts, initiated a comprehensive tax reform, eliminated price controls, established a uniform 20% tariff and removed tariff exemptions, eliminated virtually all import and export restrictions, and modified labor laws to permit greater flexibility in hiring and firing. The immediate result was a jump in capital repatriation and retention as a result of the rise in interest rates. The government also restructured several public-sector institutions, including the Central Bank and COMIBOL, the national mining corporation; the bank closed several branches and reduced staff by 70%, and COMIBOL closed numerous mines and dismissed nearly 20,000 workers.

As of 1994, Bolivia was in its second decade of democratic rule and its tenth consecutive year of economic expansion. Market reforms are firmly in place, investment is growing steadily and inflation is under control. Real GDP grew by 4.2% in 1993, up from 3.4% in 1992. Growth was led by construction, manufacturing, and services. Inflation was substantially reduced from the incredible 14,000% in 1985 to only 10.7% in 1992. The country's non-traditional exports, mainly beef, cotton, soybeans, coffee, and shelled chestnuts declined in 1992. The decrease is blamed on a decline in world prices for food products, but also on domestic policy factors such as excessive bureaucracy.

²⁰INCOME
In 1992, Bolivia's GNP was $5,084 million at current prices, or $680 per capita. For the period 1985–92 the average inflation rate was 25.7%, resulting in a real growth rate in per capita GNP of 1.0%.

In 1992 the GDP was $5,270 million in current US dollars. It is estimated that in 1988 agriculture, hunting, forestry, and fishing contributed 18% to GDP; mining and quarrying, 7%; manufacturing, 16%; electricity, gas, and water, 1%; construction, 3%; wholesale and retail trade, 14%; transport, storage, and communication, 11%; finance, insurance, real estate, and business services, 8%; community, social, and personal services, 5%; and other sources, 16%.

²¹LABOR
The economically active population was 2.4 million in 1991. Of 2.2 million salaried employees in 1991, 27.2% engaged in industry; and 36.4% each in agriculture and services. The state mining company, COMIBOL, employed 35,000 miners in 1985, but pared that amount to 7,000 recently because of a sharp decline in world mineral prices. Urban unemployment was officially estimated at 7% in 1991, but the overall rate was hypothesized to be somewhat higher. As of February 1992, the monthly minimum wage was 135 bolivianos (worth $36 at that time). In 1991, approximately 20% of the work force was employed in the informal sector, and therefore not covered by the minimum wage laws.

Virtually the entire nonagricultural labor force, as well as part of the peasantry, is unionized. The Central Bolivian Workers' Organization (Central Obrera Boliviana—COB) is the central labor federation, to which nearly all unions belong. The ministries of mines and petroleum, rural affairs, labor, and transportation are nominated by the COB and appointed by the president. Since 1985, the COB has lost influence as the government has pursued policies to stabilize the economic chaos and hyperinflation.

Workers' militias were dissolved by order of the military junta after they had taken control of the mines in May 1965; there was also a purge of pro-Communist trade union leaders. In November 1974, union meetings were declared illegal, but union activity was permitted again in 1978. After the July 1980 coup, trade unions were again suspended, but they were permitted to resume activity in May 1982. Throughout 1991–1992, COB policies had been ineffective; stoppages and demonstrations against privatization were weakly supported or virtually ignored by many workers.

22AGRICULTURE

An estimated 2.1% of Bolivia's land area was under cultivation in 1991; another 24% was permanent pasture. Agricultural development has been impeded by extremely low productivity, poor distribution of the population in relation to productive land, and a lack of transportation facilities. Prior to 1953, about 93% of all privately owned land was controlled by only 6.3% of the landowners. The agrarian reform decree of August 1953 was aimed at giving ownership of land to those working it and abolishing the large landholdings (latifundios). By 1980, 30.15 million hectares (74.5 million acres) had been distributed to 591,310 families.

Except around Lake Titicaca, about two-thirds of the cultivated land on the Altiplano lies fallow each year. Dry agriculture is the rule, and the most important crops are potatoes, corn, barley, quinoa (a milletlike grain), habas (broad beans), wheat, alfalfa, and oca (a tuber). The potato is the main staple; dehydrated and frozen to form chuño or tunta, it keeps indefinitely. The Yungas and Valles contain about 40% of the cultivated land. The eastern slopes, however, are too steep to permit the use of machinery, and erosion is a serious problem despite the practice of terracing. The most lucrative crop in the Yungas is coca, which is chewed by the local population and from which cocaine is extracted. It has been estimated that from 1982 through 1986, about 71,000 hectares (175,000 acres), twice as much land as in 1980, were planted with coca and that some 100,000 people were engaged in its cultivation. Coffee, cacao, bananas, yucca, and aji (a widely used chili pepper) are also important. In the fertile irrigated valleys, the important crops are corn, wheat, barley, vegetables, alfalfa, and oats. The Tarija area is famous for grapes, olives, and fruit. The region east of Santa Cruz de la Sierra, where most of the nation's unused fertile lands lie, is considered the "promised land" of Bolivian agriculture. Lowland rice production is increasing rapidly and already satisfies domestic need. The sugar grown there is used mostly for alcohol, but in the 1960s, the mills increased their refining capacity, thus meeting internal consumption requirements. In the tropical forests of the northeast, the Indians practice slash-and-burn agriculture.

In the late 1970s, sugar, cotton, and coffee were the most important export crops, but cotton exports plummeted from 9,729 tons in 1979 to 900 tons in 1984 and are no longer significant to the international market. Area harvested and production in 1992 for selected crops are shown in the following table:

	AREA (HECTARES)	PRODUCTION (TONS)
Potatoes	110,000	671,000
Corn	240,000	358,000
Rice	100,000	195,000
Wheat	92,000	79,000
Habas	13,000	13,000
Onions	5,000	27,000
Tomatoes	4,000	45,000

Droughts and freezing weather in the west during 1992 caused harvests to fall for basic crops like quinoa, potatoes, barley, and garden vegetables. In 1991, agriculture contributed 2.1% to the growth of the GDP.

23ANIMAL HUSBANDRY

In 1992 there were an estimated 5.7 million head of cattle, 7.3 million sheep, 1.4 million goats, 2.2 million hogs, 6,340,000 donkeys, and 323,000 horses. Poultry numbered 28 million in 1992.

The main cattle-raising department is El Beni, in the tropical northeast, which has about 30% of the nation's cattle. Cochabamba is the leading dairy center, and improved herds there supply a powdered-milk factory. The Amerindians of the high plateau depend on the llama because it can carry loads at any altitude and provides leather, meat, and dung fuel. Leading animal product exports are hides, alpaca and vicuña wool, and chinchilla

fur. Breeding of alpacas and llamas is by and large left to chance; disease is rampant, and production is low, considering the relatively large numbers of animals.

24FISHING

Fishing is a minor activity in Bolivia. A few varieties of fish are caught in Lake Titicaca by centuries-old methods and sent to La Paz. The catch was estimated at 5,637 tons in 1991. Bolivia has some of the world's largest rainbow trout, and Bolivian lakes are well stocked for sport fishing.

25FORESTRY

Bolivia is potentially one of the world's most important forestry nations. More than half of the total area is held as public land by the state, and more than 40 million hectares (100 million acres) of forest and woodland are maintained as reserves or for immediate exploitation. Trees are mostly evergreens and deciduous hardwoods, with the richest forests on the Andes' eastern slope along the tributaries of the Amazon. More than 2,000 species of tropical hardwoods of excellent quality, such as mahogany, jacaranda, rosewood, palo de balsa, quina, ironwood, colo, and cedar, abound in this area. Sawmills are few, however, and the almost total lack of transportation facilities has made exploitation expensive. Most of the sawmills are in the eastern department of Santa Cruz. Roundwood production in 1991 was only 1,632,000 cu m, up from 1,412,000 cu m in 1986 and 1,369,000 cu m in 1981. Exports of wood and wood products accounted for only $29.8 million in 1992, down from $45.3 million in 1991. Bolivia is one of South America's leading rubber exporters.

26MINING

Bolivia is the fourth largest tin-producing nation, after Brazil, Malaysia, and Indonesia. Mineral exports usually constitute about 70% of the nation's exports, when natural gas is included. Tin, which used to account for over half the total mineral exports (54.2% in 1968), accounted for only 38.5% by 1973, largely because of the increasing importance of other mineral exports, including crude petroleum. In 1985, tin represented 29% of all export earnings; the collapse of the international tin market in October 1985 was a major blow to the economy. By 1991, the entire mining sector was only contributing about 9% ($540 million) of the GDP. Siglo XX, at Llallague (south of Oruro), was once the world's largest single tin mine, but the mine closed in 1986. Total mine production in Bolivia was 16,830 tons in 1991; smelter production was 14,663 tons.

For two centuries following the discovery of silver at Cerro Rico de Potosí in 1545, the area that is now Bolivia was the world's largest silver producer. Tin production began about 1870 and had surpassed silver in value by the beginning of the 20th century. Before 1952, 80% of the nation's mineral production was in the hands of three mining companies: the Patiño, Hochschild, and Aramayo interests. In that year, they were expropriated and turned over to the Mining Corp. of Bolivia (Corporación Minera de Bolivia—COMIBOL), a government enterprise. The overwhelming problems confronting COMIBOL included falling tin prices, rising production costs, and a shortage of technical personnel. By 1961, one pound of Bolivian tin, worth about $1.00 on the world market, cost $1.30 to produce. Another basic problem facing economic planners was the near exhaustion of the richest tin deposits. To revive the industry, COMIBOL has been installing new equipment to improve recovery from low-grade ores, and the Bolivian government has encouraged the export of processed metals, which are more profitable than raw ores. Despite these efforts, in the late 1970s, COMIBOL ran annual deficits of more than $30 million, and in August 1981 the company was reportedly close to bankruptcy. As part of the government's 1985 economic reform program, COMIBOL, which had employed 40% of the nation's miners, dismissed almost 20,000 workers and closed several mines. COMIBOL is now focusing on the attraction of private firms to operate its mines under joint-venture or operating contracts. New mining codes enacted on 3 April 1991 allow foreign firms to operate with fewer restrictions, and also replace royalties with a 30% tax on profits.

Silver, zinc, tungsten, bismuth, lead, copper, gold, asbestos, and other metals are also exported. Bolivia was the world's second largest producer of antimony in 1991, with 7,287 tons mined; the nation's first antimony smelter began operations in 1975. Iron deposits have been discovered, and large salt deposits are found near Lake Poopó. Mineral output in 1991 was as follows (in tons): zinc, 129,778; tin, 16,830; lead, 20,810; antimony, 7,287; tungsten, 1,065; arsenic, 463; tantalum, 3,735 kg; silver, 375,702 kg; gold, 3,500 kg; and rough amethyst, 31,893 kg.

[27]ENERGY AND POWER

Installed capacity at Bolivian electric power plants rose from 267,000 kw in 1970 to 605,000 kw in 1991. Total electric power output in 1991 was 2,150 million kwh, of which 59% was hydroelectric; much of the total was produced by the Canadian-owned Bolivian Power Co., which supplies the cities of La Paz and Oruro. Another large producer is COMIBOL, which operates two plants on its own and also buys power from the government-run National Electricity Co. Electricity consumption grew by 7% in 1992, more than triple the population growth.

Petroleum production peaked in 1974 but then began to decline because of well depletion; in 1980, Bolivia became a net importer of oil. In 1991, the total average daily production of crude oil increased 5.6% to 22,174 barrels from 20,928 barrels in 1990. Production of natural gas gradually increased from the mid-1980s until the early 1990s; the estimate for 1991 was 5,432,000 cu m, up from 4,565,000 cu m in 1987. In 1991, natural gas sold to Argentina accounted for 40.1% of production and brought $230.1 million in foreign revenue. In February 1993, Bolivia and Brazil signed a contract for natural gas sales, but financing for the necessary $2 billion pipeline to Brazil is still pending.

[28]INDUSTRY

Industrial development has been severely restricted by political instability, the small domestic market, the uncertain supply of raw materials, and the lack of technically trained labor. Domestic industry supplies less than one-fourth of the processed food and manufactured goods consumed. Manufacturing output has declined by over 40% since 1980; in 1986, it represented only 10% of the GDP. Over one-half of output is in nondurable consumer goods—food, beverages, tobacco, and coffee. Handicrafts and hydrocarbons account for much of the remainder.

The 1985 economic stabilization program has been a mixed blessing to industry. The easing of foreign-exchange restrictions, the new uniform 20% tariff, and the significant reduction in duties on nonessential consumer goods have improved the availability of raw materials and semi-manufactured goods, thereby stimulating industrial growth. Persistently high interest rates, however, running in early 1987 at 36–46%, have deferred expansion; also, public-sector wage increases have not kept pace with inflation, so that demand has dampened.

In 1992, growth in the construction industry was a remarkable 15.3%, sustained both by the larger number of public works projects and by private investment. The manufacturing sector grew by 4.3%, with the largest gains occurring in agriculture-based industries despite the problems resulting from the precarious state of agriculture. The mining and hydrocarbon sector contracted because of the decline in mining output and stagnation in the production of petroleum and natural gas. The drastic reduction of COMIBOL's production resulted from the closing of several mines and frequent labor disputes. The slump in the

hydrocarbons subsector was because of the depletion of a number of wells, lack of investment in exploring for new deposits, and the torrential rains that damaged the infrastructure of the state-owned company.

[29]SCIENCE AND TECHNOLOGY

The Bolivian National Academy of Sciences was founded in 1960. There are some 14 scientific and technological research institutes and learned societies, notably the Bolivian Geological Service and Bolivian Petroleum Institute, both in La Paz. Bolivia has eight universities offering courses in basic and applied sciences.

[30]DOMESTIC TRADE

La Paz is the chief marketing center. Oruro is second to La Paz as a market for imported goods and is the main distributing center for mining supplies. Cochabamba distributes its agricultural production to La Paz and the mining districts. Buyers visit agricultural settlements by truck, purchase food for resale at city markets, and sell manufactured goods. Outside the main cities, most buying and selling is carried on at weekly markets and village fairs. In less accessible areas, barter is still common.

Advertising is not highly developed. Newspapers, radio stations, and movie theaters are the main advertising media. Regular retail store hours are weekdays, 9 AM to noon and 2 to 6 PM. Bank hours are 9 AM to noon and 2 to 4:30 PM.

[31]FOREIGN TRADE

Bolivia depends primarily on its mineral exports, especially tin and natural gas. Tin exports, however, have been gradually decreasing since 1946. In 1992, Bolivia's total exports decreased by 20% to $608 million, primarily due to a decrease in the contract price for gas to Argentina and a drop in international prices of Bolivia's principal mineral exports. The country has signed a 20-year contract to sell gas to Brazil, but pipeline financing is still not determined. In 1993, export earnings increased to $630 million while imports reached $1,131 million (no major change from previous years. Bolivia's major trading partners include the US, Brazil, and Japan.

[32]BALANCE OF PAYMENTS

Unlike many nations, Bolivia has no large earnings from tourism or shipping to compensate for trade deficits. After World War II, falling exports and rising imports led to depletion of the nation's gold and foreign currency reserves. By 1969, in part because of increased US aid, the unfavorable balance had been considerably reduced; five years later, thanks to import restrictions and a sharp rise in export earnings, Bolivia had a favorable payments balance of $72.5 million. In the late 1970s, Bolivia's international financial position again began to worsen, and by the end of 1986, the country had accumulated $3.7 billion in foreign debt (perhaps $100 million of it in the private sector), an amount virtually equal to the GDP. In 1992, the trade deficit was $561 million, more than double the restated 1991 deficit of $234 million. The widening deficit was caused by a simultaneous 18% increase in imports and a 20% decrease in exports in 1992 over 1991, primarily due to a decrease in the contract price for natural gas sales to Argentina.

In 1992 merchandise exports totaled $608.4 million and imports $1,040.8 million. The merchandise trade balance was $–432.4 million. The following table summarizes Bolivia's balance of payments for 1991 and 1992 (in millions of US dollars):

	1991	1992
CURRENT ACCOUNT		
Goods, services, and income	–445.1	–776.5
Unrequited transfers	183.0	243.2
TOTALS	–262.1	–533.3

	1991	1992
CAPITAL ACCOUNT		
Direct investment	50.0	91.1
Portfolio investment	—	—
Other long-term capital	56.3	248.3
Other short-term capital	8.3	44.5
Exceptional financing	169.5	158.1
Other liabilities	−52.9	−6.5
Reserves	−22.4	−26.5
TOTALS	208.8	499.0
Errors and omissions	53.3	34.3
Total change in reserves	−20.7	90.7

33BANKING AND SECURITIES

The Central Bank of Bolivia, established in 1928 and reorganized in 1945, is the sole bank of issue and operates as a commercial bank. There are five state-owned development banks, including the Mining Bank and the Agricultural Bank. The Bolivian Development Corp. channels credits from the IDB into industrial expansion projects.

In addition to the state institutions, there are private Brazilian, Argentine, Peruvian, US, Germany, and domestic banks. Private banks, which had been under strict control since 1953, were largely deregulated in mid-1985. In 1986 there were 13 private commercial banks and 6 foreign banks. The commercial bank demand deposit base totaled $1,466 million as of the end of 1992. The money supply, as measured by M2, reached в10,242 million at the same time.

There is no stock exchange in Bolivia. Bolivian securities are traded on the New York, London, and Santiago exchanges.

34INSURANCE

In the mid-1980s there were 15 domestic insurance companies, 3 US-owned companies and 1 company with mixed capital.

35PUBLIC FINANCE

Many of the expenditures and revenues of autonomous agencies (government development, mining, petroleum corporations, and the universities) do not appear in the central budget. Since April 1992, comprehensive privatization has helped decrease the need for public sector expenses.

The following table shows actual revenues and expenditures for 1991 and 1992 in millions of bolivianos.

	1991	1992
REVENUE AND GRANTS		
Tax revenue	1,763.3	2,244.2
Non-tax revenue	1,077.0	1,095.7
Capital revenue	1.4	37.8
Grants	297.4	732.2
TOTAL	3,139.1	4,110.1
EXPENDITURES & LENDING MINUS REPAYMENTS		
Defense	375.1	697.2
Public order and safety	249.2	267.8
Education	600.8	742.6
Health	105.9	365.4
Social security and welfare	599.2	560.6
Housing and community amenities	6.3	6.3
Recreation, cultural, and religious affairs	5.1	3.9
Economic affairs and services	543.6	717.9
Other expenditures	314.3	663.6
Adjustments	—	—
Lending minus repayments	−66.5	111.2
TOTAL	659.8	1,211.3
Deficit/Surplus	−422.5	−660.1

Bolivia has refinanced much of its debt from commercial banks and reduced its commercial debt from $683 million in 1987 to $1.7 million in May 1993. Public debt at the end of 1992 stood at $3.5 billion, a decrease of $1.1 billion from 1987.

36TAXATION

The Bolivian revenue system contains an unusually large number of taxes, leading to complexity and confusion that make the system difficult to enforce. In May 1986, the tax structure was radically revamped, with the intention of raising tax revenues from 1% to 12% of the GDP. Income taxes are extremely low, with a flat individual rate of 13% and a corporate rate of 3% of taxable net worth. Both local and foreign corporations can receive tax holidays, exemptions, and other benefits if they invest in new companies or in production of nontraditional exports. There are taxes on the value of particular assets, such as motor vehicles, boats, and airplanes, which can be deducted from income tax, and taxes of 30–50% on alcohol, perfume, jewelry, tobacco, and other items classed as luxuries. A value-added tax of 13%, which can be deducted from income tax, was in effect as of 1 March 1992.

37CUSTOMS AND DUTIES

Export and import duties have traditionally been an important source of government revenue, but in mid-1985, as part of a drive to stimulate the economy, the Paz government established a uniform 20% duty on all imports, eliminated tariff exemptions, removed import restrictions except for those related to health and state security, and eliminated all export controls except those on dangerous substances, endangered species, and cultural treasures. In 1990, Bolivia lowered duties on capital goods from 10% to 5% and on non-capital goods from 17% to 10%. There is a 13% value-added tax on all imports.

38FOREIGN INVESTMENT

The Patiño, Hochschild, and Aramayo mining groups, expropriated in 1952, accounted for nearly all the foreign capital in mining at that time. In 1955, Bolivia issued the Petroleum Code, safeguarding foreign investment in the exploitation of petroleum, and US oil companies began large-scale exploration and development. By 1969, foreign oil companies had spent an estimated $90 million in petroleum exploration. Although the investment law of December 1971 grants substantial benefits to foreign investors, political instability, inadequate infrastructure, and Bolivia's poor debt-repayment record have tended to hold foreign investments down.

The US has been Bolivia's major foreign investor and trading partner. Most foreign investment in Bolivia is in the hydrocarbons, mining, and electricity generation sectors, and the bulk of that is US-owned.

INTI RAYMI, an 85% US-owned gold and silver mining company, is an excellent case study for how to do business in Bolivia. In January 1993, INTI RAYMI brought into operation a $163 million carbon-in-leach plant that will lower its costs of production significantly and will triple Bolivia's official gold output.

In addition, the government has a four-year $2 billion road construction plan underway for 1993-1996. Many of the projects already have been promised funding by the Spanish, German, and Japanese governments.

39ECONOMIC DEVELOPMENT

President Gonzalo Sanchez de Lozada, elected in 1993, is entering a crucial period when his ability to turn political promises in to practice will be tested. The flagship of the government's economic policy is its privatization scheme, which it is calling the capitalization program in order to deflect popular opposition to privatization. It seems that the IMF will agree to a new structural adjustment facility if, as expected, the government makes a credible case that the public-sector deficit, which was twice as large as the target in the 1993 election year, can be cut to about 3% in 1994.

The economy is forecast to grow by about 4.5% by the end of 1994 and, depending upon the success of government policies, by a similar amount in 1995. Agricultural investment in the eastern lowland will drive private-sector growth in the region. Mining should benefit from slightly higher metal prices, some fresh investment, and a larger volume of output. Exports are expected to grow by 11%, led by a recovery in agriculture and growing mining exports. Imports will also rise, but more slowly, at a projected 5% in both 1994 and 1995. As a result the current-account deficit will stabilize in 1994 and fall slightly in 1995.

40SOCIAL DEVELOPMENT

Social security coverage is compulsory for both salaried employees and rural workers. Those covered by the program receive medical, hospital, dental, and pharmaceutical care for themselves and their families. Old age pensions begin at age 55 for men and at 50 for women. A worker's family is eligible for survivors' benefits equal to 40% of the insured person's pension. Maternity benefits cover female workers and workers' wives. Family allowances include cash payments for marriage, birth, and burial and monthly subsidies for each unmarried child.

After the census of 1976, the government decided to promote a higher rate of population growth; among the steps taken was the closing down of family planning clinics and the encouragement of immigration. Since 1985, the Paz government has taken a more liberal attitude toward family planning. Bolivia continues to allow abortion on broad health grounds. The fertility rate in 1991 was 4.7, and the contraceptive prevalence rate was 30%.

Although guaranteed equal protection under the Constitution, women by and large do not enjoy the same social status as men due to limited political power and social traditions. In most cases, women earn less than men for doing similar work.

41HEALTH

Health conditions have been notably poor, owing to poor hygiene and an insufficient number of doctors and hospitals, especially in rural areas. The most common disorders are acute respiratory diseases, tuberculosis, malaria, hepatitis, and Chagas' disease. In 1990, 335 per 100,000 people were diagnosed with tuberculosis. Malnutrition is a serious and growing problem, with 18% of children under 5 considered malnourished in 1990. From 1988 to 1991, only 52% of the population had access to safe water, and 26% had adequate sanitation. In 1991, there were 2,868 physicians and 333 dentists. From 1985 to 1990, there were 1.3 hospital beds per 1,000. From 1985 to 1992, 63% of the population had access to health care. The country's total health care expenditures for 1990 were $181,000,000.

There were 261,000 births in 1992 (with a total country population of 7.7 million in 1993). Approximately 30% of married women (ages 15–49) were using contraception from 1980 to 1993. From 1990 to 1992, one-year-old children were immunized at the following rates: tuberculosis, 86%; diphtheria, pertussis, and tetanus, 77%; polio, 84%; and measles, 80%. The infant mortality rate has declined from 117 per 1,000 live births in 1985 to 80 per 1,000 in 1992.

Life expectancy in 1992 was estimated at 61 years. The overall death rate in 1993 was 9.4 per 1,000 people.

42HOUSING

As of 1988, 67% of all housing units were detached private dwellings, 25% were detached rooms for rent with common facilities, 5% were huts, and 2% were apartments. Although the government intended to provide adequate drinking water systems for all places of 2,000 or more inhabitants and to alleviate the sewage system shortage, water systems remained inadequate in the 1980s. As of 1988 only about 50% of the population had access to piped indoor water, and about 26% lived in dwellings with adequate sanitary facilities. Owners occupied 70% of all dwellings; 13% were rented and about 15% were occupied rent free.

43EDUCATION

In 1990, Bolivia's estimated adult literacy rate was 77.5% (males, 84.7%; and females, 70.7%). Primary education, which lasts for eight years, is compulsory and free of charge. Secondary education lasts for another four years. Educational expenditure declined from 3.7% of the GDP in 1980 to an estimated 1.6% in 1985. In 1990, it rose to 18%. In 1990, there were 1,278,775 students enrolled at the primary level with 51,763 teachers; there were also 219,232 secondary students with 12,434 teachers.

Bolivia has eight state universities, one in each departmental capital except Cobija; there are also two private universities. The University of San Andrés (founded in 1930) in La Paz is Bolivia's largest university; the University of San Francisco Xavier in Sucre, dating from 1624, is one of the oldest universities in Latin America.

44LIBRARIES AND MUSEUMS

The number of public libraries in Bolivia is small. The National Library in Sucre (150,000 volumes) also serves as a public library for that city. Another important public library is the Mariscal Andrés de Santa Cruz Municipal Library (80,000 volumes) in La Paz. The most important university libraries are those of the University of San Andrés in La Paz (121,000 volumes) and the University of San Simón in Cochabamba (56,000 volumes). The National Museum of Archaeology in La Paz is the most prominent museum.

45MEDIA

The government supervises all broadcasting and communications. Bolivia had about 194,180 telephones in 1991. The telegraph system is owned by the Ministry of Communications; remote parts of the country are connected by wireless. There were 129 AM radio stations in 1992 and 43 television stations. In 1991, Bolivia had an estimated 4,590,000 radio receivers and 755,000 television sets. A government-owned television station broadcasts from La Paz.

Although freedom of the press is constitutionally guaranteed, newspapers have often been closed or their freedom impaired. In 1991 there were 14 daily newspapers. The most important La Paz daily newspapers, with their estimated circulations in 1991, are *El Diario,* 45,000; *Hoy,* 20,000; and *Última Hora,* 20,000. Important provincial dailies are *Los Tiempos* (Cochabamba), with 15,000 circulation; and *El Mundo* (Santa Cruz), with 18,000.

46ORGANIZATIONS

Learned societies include the Institute of Bolivian Sociology, the Society of Geographic and Historical Studies, the Center of Philosophical Studies, the Bolivian Language Academy, the National Academy of Fine Arts, the Archaeological Society of Bolivia, and the Tiahuanaco Institute of Anthropology, Ethnology, and Prehistory. There is a national chamber of commerce, with headquarters in La Paz, as well as departmental chambers of commerce.

47TOURISM, TRAVEL, AND RECREATION

A valid passport and a visa (or tourist card) are required for entry. Typhoid-paratyphoid inoculation is advisable, as are precautions against yellow fever and malaria before visiting the lowlands. In 1991 there were 220,902 tourist arrivals in hotels and other establishments, 16% from Peru, 12% from the United States, and 11%

from Argentina. There were 11,013 rooms in hotels and other facilities, with 19,627 beds and a 27% occupancy rate, and tourism receipts totaled US$90 million.

The dry season (May–November) is the best time to visit Bolivia. La Paz and Sucre have many colonial churches and buildings; and there are Inca ruins on the islands of Lake Titicaca, which also offers opportunities for fishing and sailing. The world's highest ski run is located at Chacaltaya, and mountain climbing and hiking are available on the country's "coldilleras" and other peaks.

⁴⁸FAMOUS BOLIVIANS

Pedro Domingo Murillo (1757–1810) was the precursor and first martyr of Bolivian independence. Andrés de Santa Cruz (1792–1865), who considered himself the "Napoleon of the Andes," dominated the early years of the independent nation. The most infamous of the 19th-century Bolivian dictators was Mariano Melgarejo (1818–71). Ismael Móntes (1861–1933), who was president of Bolivia from 1904 to 1909 and from 1913 to 1917, is identified in Bolivian history as the "great president." Simón Patiño (1861–1947), the richest of the "big three" tin barons, began his career as a loan collector and acquired his first mine by chance; he later became one of the world's wealthiest men. Víctor Paz Estenssoro (b.1907), architect of the national revolution of 1952 and founder of the MNR, served as president during 1952–56 and was reelected in 1960 and 1964; he was deposed shortly thereafter by a military junta but returned to office in 1985. Hernán Siles Zuazo (b.1914), also connected with the MNR and later founder of the MNRI, was president in 1956–60 and again in 1982–85. Juan Lechín Oquendo (b.1914?), a leader of the 1952 uprising, led the powerful Bolivian Workers' Federation from its formation in 1952 until 1987.

Bolivia's outstanding literary figure is Gabriel René-Moreno (1836–1909), a historian, sociologist, and literary critic. The highly original poet and philosopher Franz Tamayo (1879–1956), although belonging to the landed aristocracy, was a champion of the downtrodden Amerindian. Tamayo was elected president in 1935, but an army revolt prevented him from taking power. Alcides Argüedas (1879–1946) achieved fame throughout Latin America with his historical works on Bolivia and his novels *Wata wara* and *Raza de bronce*, concerned with the plight of the Indian; his critical sociological study *Pueblo enfermo* provoked an enduring controversy. The archaeologist and anthropologist Arturo Posnansky (1874–1946), born in Austria, did pioneering work in studying the civilization that once flourished at Lake Titicaca. Jaime Laredo (b.1941) is a world-famous violinist.

⁴⁹DEPENDENCIES

Bolivia has no territories or colonies.

⁵⁰BIBLIOGRAPHY

Barton, Robert. *A Short History of Bolivia*. Detroit: Ethridge, 1969.

Blair, David Nelson. *The Land and People of Bolivia*. New York: J.B. Lippincott, 1990.

Crandon–Malamud, Libbet. *From the Fat of Our Souls: Social Change, Political Process, and Medical Pluralism in Bolivia*. Berkeley: University of California Press, 1991.

Dunkerly, James. *Rebellion in the Veins: Political Struggle in Bolivia, 1952–82*. New York: Schocken, 1984.

Fifer, J. Valerie. *Bolivia: Land, Location and Politics since 1825*. New York: Cambridge University Press, 1972.

Gallo, Carmenza. *Taxes and State Power: Political Instability in Bolivia, 1900–1950*. Philadelphia: Temple University Press, 1991.

Heath, Dwight B. *Historical Dictionary of Bolivia*. Metuchen, N.J.: Scarecrow, 1972.

Hudson, Rex A. and Dennis M. Hanratty, *Bolivia, a Country Study*. 3rd ed. Washington, D.C.: Government Printing Office, 1991.

James, Daniel (ed.). *The Complete Bolivian Diaries of Che Guevara and Other Captured Documents*. New York: Stein & Day, 1968.

Klein, Herbert S. *Bolivia: the Evolution of a Multi–ethnic Society*. 2nd ed. New York: Oxford University Press, 1992.

Malloy, James M and Eduardo Gamarra. *Revolution and Reaction: Bolivia, 1964–1985*. New Brunswick, N.J.: Transaction Books, 1988.

Morales, Waltraud Q. *Bolivia: Land of Struggle*. Boulder, Colo.: Westview Press, 1992.

Sanabria, Harry. *The Coca Boom and Rural Social Change in Bolivia*. Ann Arbor: University of Michigan Press, 1993.

Yeager, Gertrude Matyoka. *Bolivia*. Oxford, England; Santa Barbara, Calif.: Clio Press, 1988.

BRAZIL

Federative Republic of Brazil
República Federativa do Brasil

CAPITAL: Brasília.

FLAG: The national flag consists of a green field upon which is imposed a large yellow diamond twice as wide as it is high. Centered within the diamond is a blue globe showing constellations of the southern skies dominated by the Southern Cross. Encircling the globe is a white banner bearing the words *Ordem e Progresso.*

ANTHEM: *Hino Nacional Brasileiro,* beginning "Ouviram do Ipiranga" ("Listen to the cry of Ipiranga").

MONETARY UNIT: On 3 March 1990, the cruzeiro (CR$), a paper currency of 100 centavos, replaced the cruzado (CZ$) at the rate of CR$1:CZ$1,000 (the cruzado had replaced an older cruzeiro at the same rate in 1986). There are currently no coins in circulation, with notes of 50, 100, 200, 500, 1,000, 5,000, 10,000, 50,000, 100,000, and 500,000 cruzeiros. CR$1 = US$0.0011 (or US$1 = CR$913.345).

WEIGHTS AND MEASURES: The metric system is the legal standard, but some local units are also used.

HOLIDAYS: New Year's Day, 1 January; Tiradentes, 21 April; Labor Day, 1 May; Independence Day, 7 September; Our Lady of Aparecida (Patroness of Brazil), 12 October; All Souls' Day, 2 November; Proclamation of the Republic, 15 November; Christmas, 25 December. Movable holidays include the pre-Lenten carnival, usually in February, Good Friday, and Corpus Christi.

TIME: At noon GMT, the time in Fernando de Noronha is 10 AM; Rio de Janeiro, 9 AM; Manaus, 8 AM; Rio Branco, 7 AM.

¹LOCATION, SIZE, AND EXTENT

Situated in east-central South America, Brazil is the largest country in Latin America and the fourth-largest in the world in coterminous area, ranking after Russia, Canada, and China (the US is larger with Alaska, Hawaii, and the dependencies included). Occupying nearly half of the South American continent, it covers an area of 8,511,965 sq km (3,286,488 sq mi), extending 4,320 km (2,684 mi) N–S and 4,328 km (2,689 mi) E–W. Contiguous with all continental South American countries except Ecuador and Chile, Brazil is bounded on the N by Venezuela, Guyana, Suriname, and French Guiana, on the NE, E, and SE by the Atlantic Ocean, on the S by Uruguay, on the SW by Argentina and Paraguay, on the W by Bolivia and Peru, and on the NW by Colombia, with a total boundary length of 22,182 km (13,783 mi). Brazil is divided into 23 states, 3 federal territories, and a Federal District. The Federal District, including the capital of Brasília, inaugurated on 21 April 1960, is surrounded on three sides by the state of Goiás and on the fourth by Minas Gerais.

Brazil's capital city, Brasília, is located in the southeastern part of the country.

²TOPOGRAPHY

The northern part of Brazil is dominated by the basin of the Amazon River and its many tributaries, which occupies two-fifths of the country. The Amazon Basin itself occupies 7,049,975 sq km (2,722,000 sq mi), or about 40% of South America's total area. The Amazon River (Rio Amazonas) is, at 6,280 km (3,900 mi), the world's second-longest river after the Nile, although the Amazon ranks first in volume of water carried; rising in the Peruvian Andes, the Amazon eventually empties into the Atlantic Ocean at an average rate of 200,000 cu m (7 million cu ft) per second. The Amazon lowlands east of the Andes constitute the world's largest tropical rain forest. In the northernmost part of the Amazon Basin lies a series of mountain ranges, known as the Guiana Highlands, where Brazil's highest mountain, Pico da Neblina (3,014 m/9,888 ft), is located. South of the Amazon Basin is a large plateau called the Brazilian Highlands, ranging in elevation from 300 to 910 m (1,000 to 3,000 ft) above sea level. From the city of Salvador (Bahia) southward to Pôrto Alegre, the highlands meet the Atlantic Ocean in a steep, wall-like slope, the Great Escarpment, which in southeastern Brazil is surmounted by mountain ranges with elevations from 2,100 to 2,400 m (7,000 to 8,000 ft) above sea level.

The Atlantic coast of Brazil has no real coastal plain, but there are stretches of lowlands along the northeast coast, and there are many baylike indentations, where Brazil's principal cities are located. Along the southwest border is a small portion of the upper Paraguay lowlands. The Paraná, Paraguay, and Uruguay rivers flow through southern Brazil; the São Francisco flows 2,915 km (1,811 mi) through northeastern and central Brazil; and the Tocantins (2,699 km/1,677 mi) empties into the Pará and from there into the Atlantic Ocean at an estuary south of the Amazon proper.

³CLIMATE

Brazil is a tropical country but extends well into the temperate zone. The Amazon Basin has a typically hot, tropical climate, with annual rainfall exceeding 250 cm (100 in) in some areas; the Brazilian Highlands, which include roughly half of the total area, are subtropical. The narrow coastal lowland area ranges from tropical in the north to temperate in the south. The cool upland plains of the south have a temperate climate and an occasional snowfall. The coolest period is from May to September, and the hottest is from December to March. October to May is the rainy season. Rainfall is excessive in the lowlands and in the upper Amazon Basin, along the northern coast, at certain points on the east coast,

55

and in the southern interior, while there are periodic droughts in the northeast. The average high temperature in Rio de Janeiro in February is 29°C (84°F); the average low in July is 17°C (63°F).

⁴FLORA AND FAUNA
About one-fourth of the world's known plant species are found in Brazil. The Amazon Basin, the world's largest tropical rain forest, includes tall Brazil nut trees, brazilwood, myriad palms, kapok-bearing ceiba trees enlaced with vines and creepers, rosewood, orchids, and water lilies, and is the home of the wild rubber tree.

South of the vast Amazonian forest is a mixture of semideciduous forest (mata) and scrub forests. The characteristic flora of the northeast interior is the carnauba wax-yielding palm in the states of Ceará and Piauí, and to the east big areas of thorn scrub, the result of generally poor soils and periodic devastating droughts. Along the humid coast are many mango, cajú, guava, coconut, and jack-fruit trees, as well as large sugar and cotton plantations, the latter indigenous. Within the savanna, sparse forests, and "campos cerrados" (enclosed fields) of badly deforested, populous Minas Gerais, are various woody shrubs, lianas, and epiphytes, the staghorn fern, and an abundance of herbs, especially grasses. Brazil has many fair to good pasturage grasses, on which millions of beef cattle graze, not always of high grade, and some dairy cattle in the favored southern states.

In the southern states are exotic flowers, such as papagaias; flowering trees, such as the quaresma, which blossoms during Lent; and the popular ipê tree with its yellow petals, planted on some São Paulo streets. In the southernmost part of the Brazilian plateau forests, where temperate climate prevails, is found a mixture of araucarias (umbrella pines) and broadleaf species. The pampas of Rio Grande do Sul are extensive grasslands. Maté, of economic importance as a beverage, is made from the roasted, powdered leaves of a tree harvested extensively in the southern states.

The Amazon rain forest is host to a great variety of tropical fauna, including hundreds of types of macaws, toucans, parrots, and other brightly colored birds; brilliant butterflies; many species of small monkeys; anacondas, boas, and other large tropical snakes; crocodiles and alligators; and such distinctive animals as the Brazilian "tiger" (onca), armadillo, sloth, and tapir. The rivers in that region abound with turtles and exotic tropical fish, and the infamous "cannibal fish" (piranha) is common; in all, more than 2,000 fish species have been identified.

⁵ENVIRONMENT
As of 1986, it was estimated that the forests of the Amazon were being cleared for colonization, pasturage, timber development, and other commercial purposes at a rate of up to 20 million hectares (50 million acres) a year. As of 1994, this figure has decreased, although no exact figures are available. A 20-year US-Brazilian project, initiated by the World Wildlife Fund, in Washington, D.C., and the National Institute for Research on Amazonia, in Manaus, has been studying the Amazon forest since 1978 in order to recommend appropriate measures for its protection. A Brazilian law requiring that developers leave 50% of each Amazon land parcel untouched is erratically enforced. Federal agencies with environmental responsibilities include the National Environment Council of the Ministry of the Interior, the Brazilian Institute of Forest Development, and the Ministry of Planning. Other environmental problems in Brazil include water pollution and land damage. The pollution of rivers near urban industrial centers is from mercury, toxic industrial wastes, and untreated waste. Brazil lacks fertile soil for agriculture. The existing soils are threatened by erosion from the clearing of the forests.

The damage to the rain forest environment is reflected in the number of endangered species which inhabit the region. Between 1900 and 1950, 60 species of birds and mammals became extinct. Forty species of birds and 318 plant species are currently endangered. The list of endangered species as of 1991 included the Lear's macaw, the guayaquil great green macaw, and the American crocodile. As of 1987, endangered species in Brazil included two species of marmoset (buffy-headed and white-eared); three species of tamarin (golden lion, golden-headed lion, and golden-rumped lion); the black saki; the woolly spider monkey; and the maned sloth. Among endangered amphibians and reptiles were the Anegada ground iguana, three species of turtle (South American river, leatherback, and green sea) and three species of caiman (spectacled, broadnosed, and black). Endangered bird species included Pernambuco solitary tinamou, tundra peregrine falcon, black-fronted piping guan, red-billed curassow, red-tailed parrot, two species of macaw (glaucous and Lear's), southeastern rufous-vented ground cuckoo, two species of hermit (hook-billed and Klabin Farm long-tailed), black barbthroat, black-hooded antwren, fringe-backed fire-eye, and cherry-throated tanager. Also on the endangered list was Harris' mimic swallowtail butterfly. In December 1987, the Brazilian senate enacted a law to ban whale hunting within the country's territorial waters.

⁶POPULATION
Brazil is the most populous country in Latin America and the fifth most populous in the world. According to the 1991 census, the total population was 146,917,459, an increase of 23% since 1980. A population of 172,777,000 was projected by the UN for the year 2000, assuming a crude birthrate of 20.8 per 1,000 population, a death rate of 7.2, and a net natural increase of 13.6 during 1995–2000. In 1991, 75% of the population was urban and 25% rural. Population density in 1991 was 17 per sq km (45 per sq mi). The population is concentrated in the Atlantic coastal region, with the states of Rio de Janeiro, São Paulo, and Minas Gerais containing approximately 41% of the total; the states of Bahia, Rio Grande do Sul, Pernambuco, and Ceará have about 23%, and the remaining units about 36%. Major cities and their 1991 populations are as follows: São Paulo, 9,626,894; Rio de Janeiro, 5,473,909; Belo Horizonte, 2,017,127; Salvador, 2,072,058; Fortaleza, 1,765,794; Brasília, 1,598,415; Recife, 1,296,995; Curitiba, 1,313,094; Porto Alegre, 1,263,239; Belém, 1,244,688; and Manaus, 1,010,544.

⁷MIGRATION
Between 1821 and 1945, approximately 5.2 million European immigrants entered Brazil, most of them settling in the south. Brazil has the largest expatriate Japanese colony in the world, numbering more than one million. In recent years, because of the increasing prosperity of Europe and Japan, there has been less desire to migrate to underdeveloped rural Brazil or its inflation-harassed industrial cities. Moreover, immigration is controlled by laws limiting the annual entry of persons of any national group to 2% of the total number of that nationality that had entered in the preceding 50 years.

⁸ETHNIC GROUPS
The indigenous inhabitants were Indians, chiefly of Tupi-Guaraní stock, and other small groups in the Amazon Basin and the lowlands of the Paraguay and Paraná rivers. The Portuguese settlers had few taboos against race mixture, and centuries of large-scale intermarriage have produced a tolerant and distinctly Brazilian culture. Within the Brazilian nationality are blended the various aboriginal Indian cultures; the Portuguese heritage, with its diverse strains; the traditions of millions of persons of African descent; and European elements resulting from sizable immigration since 1888 from Italy, Spain, Germany, and Poland. The influx of Japanese and some Arabs during the 20th century has contributed to the complex Brazilian melting pot.

According to the 1991 census, 55% of Brazil's population was white, 39% mixed, and 5% black; the remaining 1% consisted of

LOCATION: 5°16′19″N to 33°45′9″S; 34°45′54″ to 73°59′32″W. **BOUNDARY LENGTHS:** Venezuela, 1,495 km (929 mi); Guyana, 1,606 km (998 mi); Suriname, 593 km (368 mi); French Guiana, 655 km (407 mi); Atlantic coastline, 7,408 km (4,603 mi); Uruguay, 1,003 km (623 mi); Argentina, 1,263 km (785 mi); Paraguay, 1,339 km (832 mi); Bolivia, 3,126 km (1,942 mi); Peru, 2,995 km (1,861 mi); Colombia, 1,644 km (1,022 mi). **TERRITORIAL SEA LIMIT:** 12 mi.

Japanese, Indians, and other groups. The Indian population was estimated at 250,000 in 1993.

⁹LANGUAGES

The language of Brazil is Portuguese, which is spoken by virtually all inhabitants except some isolated Indian groups. Substantial variations in pronunciation and word meaning, however, distinguish it

from the language as it is spoken in Portugal. German, Italian, and Japanese are also spoken in immigrant communities. A large percentage of the educated have learned either French or English.

¹⁰RELIGIONS

Brazil is the largest Catholic nation in the world, with an estimated 88% (in 1993) of the population professing the Roman

Catholic faith. The primate is the archbishop of Salvador; Rio de Janeiro and São Paulo are also archdioceses. In 1989, Protestants constituted upwards of 6% of the population. Two other large religious groups are the Spiritists (about 2%), followers of the writings of Allan Kardec; and the Afro-American Spiritists (about 2%), whose rites evolved from the animism of black Africans. It was estimated in 1975 that about 30% of all Roman Catholics were affiliated in some degree with organized spiritism. Brazil's Japanese and Okinawan population still included some 395,000 Buddhists in 1985, but the younger generation has tended to acculturate by embracing Christianity. In 1985 there were approximately 194,000 Orthodox Christians and, in 1990 some 100,000 Jews. Brazilians are notably tolerant of and receptive to new religious cults.

11TRANSPORTATION

Roads are the primary carriers of freight and passenger traffic. Brazil's road system totaled 1,448,000 km (899,787 mi) in 1991. The total of paved roads increased from 35,496 km (22,056 mi) in 1967 to 48,000 km (29,827 mi) in 1991. Motor vehicles registered as of 1992 included 12,974,991 passenger cars, and 1,371,127 commercial vehicles. Although the bulk of highway traffic is concentrated in the southern and central regions, important roads have been constructed to link the northeastern and northern areas with the industrialized south. Roads of all types have been built with federal aid, the most important being the network of more than 14,000 km (8,700 mi) of paved roads south of Brasília; aid is also supplied for their maintenance. In September 1970, construction began on the 5,000-km (3,100-mi) Trans-Amazon Highway, possibly the most ambitious overland road project undertaken in this century, linking Brazil's Atlantic coast with the Peruvian border; when completed, a 4,138-km (2,571-mi) north–south section will link Santarém, on the Amazon River, with Cuiabá. The project has had a profound effect on the Amazon Basin, among the world's last great wildernesses. A recent World Bank study, however, has shown that 28% of the country's existing highways were in bad condition, up from only 10% in 1979. Lack of proper road maintenance possibly adds 10–15% to total transportation costs in Brazil.

Brazil's railway system has been declining since 1945, when emphasis shifted to highway construction. The total extension of railway track was 28,828 km (17,914 mi) in 1991, as compared with 31,848 km (19,789 mi) in 1970. In 1985, the government announced plans to invest $422 million to modernize and expand the railway network. Most of the railway system belongs to the Federal Railroad Corp., with a majority government interest; there are also seven private lines.

Coastal shipping links widely separated parts of the country. Of the 36 deep-water ports, Santos and Rio de Janeiro are the most important, followed by Paranaguá, Recife, Vitória, Tubarão, Maceió, and Ilhéus. Bolivia and Paraguay have been given free ports at Santos. There are 50,000 km (31,070 mi) of navigable inland waterways. In 1991, the merchant shipping fleet, which included 259 vessels of over 1,000 tons, had a total GRT of 5,645,000.

Air transportation is highly developed. In 1992, Brazilian airlines performed 28,447 million passenger-kilometers (17,677 million passenger-miles) of service. Of the 48 principal airports, 21 are international; of these, Rio de Janeiro's Galeão international airport and Guarulhos International Airport at São Paulo are by far the most active. The main international airline is Empresa de Viação Aérea Rio Grandense (VARIG). Other Brazilian airlines are Transbrasil Linhas Aéreas, Cruzeiro do Sul, (associated with VARIG since 1983), and Viação Aérea São Paulo (VASP), which handles only domestic traffic and is run by the state of São Paulo. All except VASP are privately owned.

12HISTORY

The original inhabitants of Brazil were hunter-gatherers, except in the lower Amazon, where sedentary agriculture developed. There are no reliable population estimates from pre-European times, but probably there were no more than 1 million.

After the European discovery of the New World, Spain and Portugal became immediate rivals for the vast new lands. Portugal's claim was established by a papal bull of Pope Alexander VI (1493) and by the Treaty of Tordesillas (1494), which awarded to Portugal all territory 370 leagues west of the Cape Verde Islands. On Easter Sunday in 1500, the Portuguese admiral Pedro Álvares Cabral formally claimed the land for the Portuguese crown. Cabral's ship returned to Portugal with a cargo of red dyewood, which had been gathered along the shore, and from the name of the wood, pau-brasil, the new land acquired the name Brazil.

In 1532, the first Portuguese colonists arrived, bringing cattle, seed, and the first slaves from Africa. In 1549, the Portuguese governor-general, Tomé da Souza, founded the city of São Salvador, and established the first Portuguese government in the New World. The same year marked the arrival of the missionary Society of Jesus (the Jesuits) to begin their work among the Indians.

Other Europeans began to move in on the Portuguese colony. In 1555, the French established a settlement in the Bay of Rio de Janeiro. In 1624, the Dutch attacked Bahia and began to extend throughout northeastern Brazil. Under the Dutch, who remained until ousted in 1654, the area flourished economically. Colonists planted sugarcane, and during the 17th century, the large sugar plantations of northeastern Brazil were the world's major source of sugar.

In 1640, Portugal appointed a viceroy for Brazil, with his seat first in Bahia and after 1763 in Rio de Janeiro. The discovery of gold in 1693 and of diamonds about 1720 opened up new lands for colonization in what are now the states of São Paulo, Minas Gerais, Paraná, Goiás, and Mato Grosso. From their base in São Paulo, Brazilian pioneers (Bandeirantes) pushed inland, along with their herds of cattle and pigs, in search of Indian slaves and mineral riches. By the 1790s, when the primitive surface gold and diamond mines were largely exhausted, the Brazilian plateau became thinly populated.

Brazil's first attempt at independence came in 1789 in the mining state of Minas Gerais. A plot, known as the Miners' Conspiracy (Conjuração Mineira) was led by Joaquim José da Silva Xavier, a healer known as Tiradentes ("tooth-puller"). The plot was betrayed and crushed, and Tiradentes was captured and eventually executed, but Tiradentes remains a national hero. In 1807, the invading armies of Napoleon forced the Portuguese royal family and 15,000 Portuguese subjects to flee to Brazil. Rio de Janeiro became the seat of the Portuguese royal family until 1821, when King John (João) VI returned home, leaving his son Pedro to rule Brazil as regent. Meanwhile, Portugal's monopolistic trade practices, the suppression of domestic industry, and oppressive taxation had brought about a strong movement for independence, which Pedro supported.

Pedro proclaimed Brazil's independence on 7 September 1822, and later that year was crowned Emperor Pedro I. In 1831, a military revolt forced him to abdicate. The throne passed to his 5-year-old son, Pedro. In 1840, Pedro was crowned Emperor. Under Pedro II, Brazil enjoyed half a century of peaceful progress. New frontiers were opened, many immigrants arrived from Europe, railroads were built, and the gathering of rubber in the Amazon Basin stimulated the growth of cities, such as Belém and Manaus. The abolition of slavery in 1888 brought about an economic crisis that disrupted the Brazilian Empire. In 1889, a bloodless revolution deposed Pedro II and established the Republic of the United States of Brazil. A new constitution modeled after the US federal constitution, was promulgated by the Brazilian government in 1891. At first, the republic was ruled by military regimes, but by 1894 constitutional stability was achieved.

Meanwhile, empty areas of good soil were settled in the southern plateau by over 2.5 million Italian, Portuguese, German, Polish, and Levantine immigrants. The rapid spread of coffee cultivation in the state of São Paulo transformed Brazil into the world's largest coffee-producing country. By the end of the 19th century, coffee had become the nation's principal source of wealth. Brazil soon entered a period of economic and political turmoil. Malayan and Indonesian rubber plantations had overwhelmed the Brazilian rubber market, while coffee revenues were reduced by falling world prices of coffee. Regionalism and military rivalries contributed to instability, and by 1930, the nation was in a state of unrest. In that year, a military coup with widespread civilian support placed into power Getúlio Vargas, the governor of Rio Grande do Sul.

Vargas' ideology was a blend of populism and corporatism. He sought reforms for Brazil's middle and lower classes, but discouraged dissent and was often repressive. Between 1930 and 1937 Vargas brought a minimum wage and social security to Brazil, but also crushed a leftist uprising in 1935. Vargas formalized his system in 1937, calling it the New State ("Estado Novo"). For eight years, Vargas attempted to industrialize Brazil, while organizing both workers and their employers into state-run syndicates. Vargas was nationalist in foreign policy, although he encouraged foreign investment. He exploited the US-German rivalry over Latin America to get large amounts of aid until joining the allies in 1942.

Conservative elements of the military, convinced that Vargas was a dangerous force, removed him from office, and promulgated a new constitution in 1946. The "Second Republic" was initiated with the presidency of Eurico Dutra. Vargas was returned to the presidency in the election of 1950, and did not attempt to rejuvinate the New State. He did continue to press for industrialization under state control, establishing a National Development Bank and a state petroleum company. Eventually he ran afoul of the military, who demanded his resignation. He committed suicide in August 1954, a few months before his term of office was due to expire.

He was succeeded from 1955 to 1961 by Juscelino Kubitschek de Oliveira. Kubitschek embarked on an ambitious program of development, spending huge amounts of money and attracting large foreign investments in Brazil. Kubitschek's most ambitious program was the building of a new federal capital, Brasília, in the highlands of central Brazil. Inflation and a burdensome national debt proved to be his undoing, and in January 1961 Jânio da Silva Quadros was inaugurated after a campaign promising an end to corruption and economic stability. The situation proved too difficult for Quadros, and he resigned after only seven months. João Goulart, who had been vice-president under both Kubitschek and Quadros, became president only after the conservative Congress combined with the military to reduce his powers and institute an unwieldy form of parliamentary government. In January 1963, in a national plebiscite, Brazil chose to restore presidential powers. But Goulart was caught between pressures from the left, demanding the acceleration of social programs, and the right, increasingly alarmed by trends toward populism.

On 1 April 1964 the military deposed Goulart, and arrested 40,000 people, including 80 members of Congress. In the same month, Congress appointed Humberto de Alencar Castelo Branco to the presidency, and in July it approved a constitutional amendment extending Castelo Branco's term of office to March 1967. National elections were postponed, and Brazil entered an era of military supremacy.

In March 1967 Arthur da Costa e Silva, a former army marshal, took office under a new constitution. That constitution was suspended in December 1968, and military hard-liners took the upper hand. Costa e Silva suffered a stroke in September 1969 and died in December. Gen. Emilio Garrastazú Médici, former head of the secret police was chosen to replace him. In March 1974, Gen. Ernesto Geisel, a high official in the Castelo Branco government, became president.

The military governments of the previous ten years had brought Brazil rapid economic expansion, but there was a dramatic reversal during the oil crisis of 1973–74. Opposition began to mount, encouraged by religious and trade union leaders. President Geisel gradually instituted some degree of political liberalization (abertura), but the military split on the wisdom of this policy.

During the late 1970s, continuing economic difficulties led to labor unrest and numerous strikes, including a strike of 300,000 metalworkers in metropolitan São Paulo in April and May of 1980 that ended only after troops in tanks and trucks occupied the region. Meanwhile, Gen. João Baptista de Oliveira Figueiredo became president in March 1979. That August, Figueiredo continued Geisel's policy of liberalization by signing a political amnesty law that allowed many political exiles to return home. Also in 1979, censorship of the press and the controlled two-party system were abolished. In November 1982, Brazil had its first democratic elections since 1964. Opposition parties won the governorships of 10 populous states and a majority in the lower house of Congress, but the ruling party remained in control of the upper house and the electoral college, which was to choose the next president. Moreover, the military retained broad powers to intervene in political affairs under national security laws.

The 1985 election was indirect, yet the opposition managed to turn the campaign in 1984 into a reflection of popular choice and capture the presidency. The ruling party chose São Paulo governor Paulo Maluf, who proved unable to distance himself from the unpopularity of the military-controlled regime. The opposition capitalized on the groundswell of hostility and coalesced behind the paternal figure of Tancredo Neves, a senator from Minas Gerais who had held office under Vargas and who campaigned as if the ballot were direct. The election went against the government, and in January 1985, the electoral college duly chose Neves as Brazil's first civilian president in a generation. In March, however, just before his inauguration, Neves fell gravely ill, and he died in April without having been formally sworn in. Brazilians feared another military strike, but Vice-President José Sarney was allowed to take office as president. Sarney, who represented a small center-right party allied with Neves' party, consolidated his position after an impressive showing in regional and legislative elections in November 1986.

A new constitution, passed in 1988, was followed by elections a year later. Brazil's first direct presidential elections in 29 years resulted in the victory of Fernando Collor de Mello. Collor received 53% of the vote in the runoff elections. Collor took office in March 1990, and launched an ambitious liberalization program, which attempted to stabilize prices and deregulate the economy. Collor was in the process of renegotiating Brazil's huge debt with foreign creditors and the IMF when massive corruption was revealed inside the Collor administration. Allegations implicated Collor himself, who was forced to resign in December 1992. Itamar Franco took over, promising to continue Collor's programs, but long-standing structural problems continued.

¹³GOVERNMENT

The Federative Republic of Brazil is a constitutional republic composed of 26 states and the Federal District. This district surrounds the federal capital, Brasília. The constitution of October 1988 establishes a strong presidential system.

From 1969 to 1985, the president was elected for a five-year term, later extended to a six-year term, by an electoral college. In

1985, the constitution was amended to allow for direct popular election. Between 1964 and 1978, presidents were preselected by the military. The president is the head of the armed forces and is in charge of the bureaucracy, assisted in that task by a cabinet of ministers. He also appoints justices to the Supreme Federal Tribunal, the highest court in Brazil.

The Congress consists of the Senate and the Chamber of Deputies. The Senate has 81 members, three for each state plus the Federal District. Senators serve for eight-year terms, with half the members retiring every four years. The 503 deputies (as of 1994) are elected for four-year terms by a system of proportional representation in the states, territories, and Federal District. The constitution stipulates that Congress meet every year from 15 March to 15 December. In practice, from 1964 to 1985, the military used the office of the president to dominate the Congress and the state legislatures, suspending them from time to time. The Supreme Federal Tribunal heads the judicial branch. That body includes 11 Justices, appointed by the president and confirmed by the Senate.

Voting is compulsory for most men and employed women between the ages of 18 and 65 and optional for persons over 65, unemployed women, and officers of the armed forces. Illiterates were permitted to vote in 1985. Enlisted servicemen may not vote.

14POLITICAL PARTIES

During the last days of the Brazilian Empire, a group of positivists advocating abolition of the monarchy organized the Republican Party (Partido Republicano—PR) along military lines. After the fall of the empire in 1889, the government was controlled by PR-supported military regimes and opposed locally by the established Conservative and Liberal parties. An opposition group, the Civilian Party (Partido Civilista), organized by Ruy Barbosa, overcame the military regime but was soon absorbed into the Conservative and Liberal groups from Minas Gerais and São Paulo, which instituted a system of alternating the presidency between the two states.

Getúlio Vargas was responsible for the success of three successive parties, one of which survives to this day. In 1930, Vargas formed the Liberal Alliance Movement (Aliança Nacional Liberal—ANL). After Vargas resigned the presidency in 1945 his supporters formed the Social Democratic Party (Partido Social Democrático—PSD). Eurico Dutra, who succeeded Vargas, ran under this party. In 1950, Vargas was elected under the banner of the Brazilian Labor Party (Partido Trabalhista Brasileiro—PTB). Finally, Vargas inspired the National Democratic Union (União Democrática Nacional—UDN) to put up candidates against him. The UDN won the presidency in 1961 for Quadros.

The PSD continued on without Vargas, but formed a coalition with the PTB in 1955. The PSD candidate, Kubitschek, became president, while the PTB's leader, Goulart, became vice-president. In the 1958 congressional elections, however, the PTB broke with the PSD. The PTB survives as a small party, having lost many of its members to other laborite parties.

After the military takeover in 1964, parties disappeared. In 1966, the military allowed the formation of two official parties: the Alliance for National Renewal (Aliança Renovadora Nacional—ARENA) and the Brazilian Democratic Movement (Movimento Democrático Brasileiro—MDB). ARENA was created as the ruling party, with the MDB playing the role of "loyal opposition." ARENA began with two-thirds majorities in both houses of Congress and increased its majorities in the elections of 1970, while also maintaining control of nearly all state legislatures. ARENA scored further gains in the 1972 municipal elections. However, beginning in November 1974, the MDB began to score legislative gains. Moreover, in the 1974 election the MDB was able to raise issues of social justice and civil liberties.

In November 1979, in accordance with the government's liberalization policy, Congress passed a law abolishing ARENA and the MDB and permitting the formation of new parties. Over the next decade, a number of groups emerged. The government created the Social Democratic Party (Partido Democrático Social—PDS) to replace ARENA. It is a conservative party, a strong advocate of foreign investment, and is strong among upper middle-class voters. The Democratic Workers' Party (Partido Democrático Trabalhista—PDT) is headed by Leonel Brizola, a frequent critic of the military regime and leader of a similar party before 1964. The Party of the Brazilian Democratic Movement (Partido de Movimento Democrático Brasileiro—PMDB) was a moderate successor to the MDB. Dissidents from the PDS soon joined, including Neves. Because of its origins as an umbrella opposition party, there was never a clear ideology, and it continues to change its approach as new dissidents are attracted. The Workers' Party (Partido dos Trabalhadores—PT) was led by Luis Inacio da Silva, also known as "Lula," the popular leader of the metalworkers' union. In 1989, Lula placed second in the presidential race, running under a coalition of laborite parties called the Popular Front. Lula lost the run-off election to Collor, receiving 47% of the vote. The Brazilian Workers' Party (Partido Trabalhista Brasileiro—PTB) is a populist party, which suffers from having no leader even remotely approaching the popularity of Vargas, who was the party's candidate in 1950. The party has a working class appeal, but is conservative on a variety of economic issues.

In 1985, the Liberal Front Party (PFL) was organized by dissident PDS members. It formed the National Alliance with the PMDB, an alliance that won the 1985 elections. After the death of Neves, José Sarney of the PFL became president. Although the PFL lost the 1989 presidential elections, it soon allied with President Collor, although the scandal of 1992 did little to help their fortunes.

The Communists had been banned since 1957, but were allowed to organize after 1985. The Brazilian Communist Party (Partido Comunista Brasileiro) was founded in 1922, but spent most of its life underground. It is now a Euro-Communist party firmly committed to conventional politics. After a purge of a leftist faction in 1984, it suppported the Sarney administration, and seeks alliances with other parties. The Communist Party of Brazil (Partido Comunista do Brasil) is a more radical faction, Maoist in its origins but now expressing solidarity with any socialists who resist reforms.

15LOCAL GOVERNMENT

Brazil is a highly centralized system, in which local units have very little authority. Each of Brazil's 23 states has its own constitution and popularly elected legislature and governor. The states are divided into nearly 4,000 municipalities, which are, in turn, divided into districts. Each municipality has its own legislative assembly, elected by proportional representation, and its own elected executive authority. In practice, the state and municipal legislative bodies are ineffectual before the executive branches. Municipal authorities are responsible for the construction and maintenance of roads, the creation and upkeep of public parks and museums, and for the program of primary education. As districts increase in population, they, in turn, become municipalities. The large municipalities are important political units and may rival the state in political power. The largest city in each municipality serves as the capital, and usually the largest city in the largest municipality serves as the state capital. The Federal District government in Brasília is appointed by the president with Senate approval.

In 1960, after Brasília had become the new capital, the former Federal District, comprising Rio de Janeiro and the 1,165 sq km (450 sq mi) surrounding it, became the state of Guanabara. Eventually this state was amalgamated into the state of Rio de Janeiro. From 1979 on, a few previously unincorporated territories became states.

16JUDICIAL SYSTEM

The legal system is based on continental European principles. Although the jury system has been used in criminal cases for more than 100 years, there is a general tendency away from the use of juries. The Supreme Federal Court is composed of 11 justices, chosen by the president with Senate approval, who serve until age 70. It has final jurisdiction, especially in cases involving constitutional precepts and the acts of state and local authorities. The Federal Appeals Court deals with cases involving the federal government. Immediately below it are federal courts located in the state capitals and in the Federal District, as well as military and labor courts. Codes of criminal, civil, and commercial law are enacted by Congress, but in order to preserve the jurisdiction of state courts, the federal courts will not accept original jurisdiction solely because a law of Congress is involved. Electoral tribunals deal with registration of political parties, supervision of voting, infractions of electoral laws, and related matters.

Each state and municipality has its own judicial system. Justices of the peace and magistrates deal with commercial and other civil cases of the first instance. Decisions from state or municipal courts may be appealed to the federal courts and on up to the Supreme Federal Court.

The judiciary is independent from the executive and legislative branches. Judges are appointed for life and may not accept other employment.

17ARMED FORCES

The Brazilian armed forces had a total strength of 296,700 (128,500 conscripts) in 1993. Military service for a minimum of one year is compulsory. Draftees are inducted at the age of 18, but in the event of mobilization, all male citizens up to 45 years of age are liable for service.

The army had 196,000 personnel in 1993; the navy, 50,000 (including 15,000 marines and 700 naval airmen); air force personnel totaled 50,700, with 307 combat aircraft; the army, largely armed with Brazilian weapons, has the equivalent of about sixty mechanized, motorized, cavalry, jungle, mountain, artillery, engineers, airborne, coast defense and border patrol brigades. The navy has one carrier and 18 destroyers and frigates, supplemented by 36 patrol and mine warfare craft. Naval aviation is an ASW helicopter force.

In 1992 Brazil spent $2.12 billion on defense, which was about 1% of gross domestic product. Brazil imported us$20 million in arms and exported us$70 million (1991).

18INTERNATIONAL COOPERATION

Brazil is a charter member of the UN, having joined on 24 October 1945; it belongs to ECLAC and all the nonregional specialized agencies. The nation contributed a battalion of troops to the UNEF in the Gaza area after the Suez crisis of 1956 and also sent troops to the Congo (now Zaire) in the early 1960s. A signatory of the GATT and the Law of the Sea, Brazil is also a member of G-77, IDB, LAIA, OAS, PAHO, and other inter-American organizations. Brazil participated in the Alliance for Progress program, under which it received US funds for development and stabilization.

19ECONOMY

The history of the Brazilian economy before World War II is characterized by six principal cycles, each centered on one particular commodity: brazilwood, livestock, sugar, gold, rubber, and, most recently, coffee. At the height of each cycle, Brazil led the world in production of that commodity. Even during the postwar era, variations in price and market conditions for coffee have largely determined the degree of national prosperity.

Attempts to diversify the economy through rapid industrialization have made Brazil one of the two leading industrial nations of South America, but spiraling inflation has offset many of the economic advances. The rising inflationary trend of the mid-1960s was due mainly to public budgetary deficits resulting from losses incurred by the government-owned railroads and shipping lines and by official subsidy expenses for imports, such as wheat and petroleum. At the same time, wages increased at a higher rate than productivity; expansion of credit to private enterprises also lagged. The pace of further industrial expansion was determined largely by the availability of foreign exchange, derived chiefly from the sale of coffee, to buy the necessary equipment and raw materials, especially wheat and crude oil.

After the period of the "economic miracle" (1968–73), during which GDP growth averaged more than 11% annually, the economy cooled to an annual growth rate of 6% between 1974 and 1980, mainly because of increased costs of imported oil. Throughout this period, industrial growth rates outstripped those for the economy as a whole, and industrial products claimed an increasingly large share of exports. In the early and mid-1980s, agriculture accounted for about 40% of Brazil's annual foreign exchange earnings, as compared with 80% a decade earlier. Industrial production accounted for about 60% of export value in the early and mid-1980s.

Brazil's chronic economic problem is inflation. Between 1973 and 1980, the wage index for laborers increased from 420 to 7,212 (1966 = 100). Indeed, inflation has been so chronic that in the late 1960s, the government created "monetary correction," whereby fixed payments were indexed to past inflation. Thus, interest rates, pension payments, mortgage payments, and so forth, kept pace with rising prices, but inflation fed on itself. Even as economic growth surged in the mid-1980s, and annual exports soared past us$12 billion, triple-digit inflation persisted. Finally, in February 1986, as the projected inflation rate for the year approached 500%, the government imposed a package of sweeping economic reforms, the "Cruzado Plan," which created a new currency (the cruzado), eliminated "monetary correction," and froze wages and prices. While inflation plunged to near-zero initially, by mid-1987, it had surged beyond 100%, fueled by increased customer spending due to the price freeze. The government then imposed an austerity program and began negotiations with the IMF for a rescheduling of the staggering foreign debt.

The Brazilian economy was hit by a deep recession and record inflation in 1990. GDP fell by an unprecedented 4%, while inflation hit an all-time high of 2,938%. In March 1990, upon assuming office, President Collor announced sweeping economic reforms designed to stop inflation and integrate Brazil into the developed world economy. In addition, the Collor Plan imposed a price freeze, as well as a freeze on bank deposits, resulting in a precipitated capital flight. Trade barriers were significantly reduced but the attempt to reduce Brazil's large fiscal deficit resulted in the continual resurgence of inflation and a lack of confidence in the government's economic policies. By mid-1990 the monthly inflation rate was around 10% and by the end of the year it was in the 20% range.

The Collor Government introduced on 31 January 1991 another package, "Collor II," attempting to reduce inflation. The package included wage and price controls and eliminated the overnight market. The economy experienced a lackluster recovery with GDP growth of 1.2%. However, the failure to reduce the structural fiscal deficit, intermittent tightening and loosening of monetary policy, the unfreezing of prices and wages by the third quarter, and the unfreezing of remaining blocked accounts undermined the efforts to reduce inflation.

Under IMF guidance, monetary policy continued to tighten liquidity in 1992. The failure of the governments's stabilization efforts produced a new inflationary spiral with monthly inflation rates in the mid 20% range. High real interest rates combined with the acceleration of inflation and the political uncertainty over the outcome of the impeachment proceedings produced another recession with GDP decline of 1.5% for 1992.

GDP growth for 1993 is estimated at 4.5%. Most of the growth occurred in the first quarter and was concentrated in industrial production. The economy expanded strongly in the first half of the year, led by strong domestic demand fueled by government spending and higher real wages. GDP is expected to grow by 3.8 percent in 1994. Rising taxes, interest rates and inflation dull the outlook for the first quarter of 1994 and only after the political negotiation of the 1994 federal budget will there be any basis for predicting growth of the economy and the possibility of controlling inflation. The degree of uncertainty is considerable, though the consensus is that the economy should continue to grow in 1994, but a lower rate than in 1993.

[20]INCOME

In 1992, Brazil's GNP was US$425,412 million at current prices, or US$2,770 per capita. For the period 1985–92 the average inflation rate was 731.3%, resulting in a real growth rate in per capita GNP of –0.7%.

In 1992 the GDP was $395,289 million in current US dollars. It is estimated that in 1989 agriculture, hunting, forestry, and fishing contributed 8% to GDP; mining and quarrying, 1%; manufacturing, 27%; electricity, gas, and water, 2%; construction, 8%; wholesale and retail trade, 7%; transport, storage, and communication, 5%; finance, insurance, real estate, and business services, 33%; and community, social, and personal services, 12%.

[21]LABOR

A 1990 estimate reported the economically active population of Brazil at 64,467,981, or 43.8% of the total population. In 1990, 27% were engaged in agriculture, 42% in services, and 27% in industry.

Generally, there is little unemployment among skilled workers, but there is persistent structural unemployment and high underemployment in rural areas, especially in the northeast. The economic recovery since mid-1984 has cut unemployment dramatically: in 1985, for the first time in over a decade, the economy generated enough jobs to absorb the entrants into the labor market, as well as many of those who had been unemployed during the recession. Official statistics give unemployment rates for large cities only; in May 1991, unemployment was about 5.7% in the six largest urban areas. In order to cope with an annual inflation rate that has exceeded 2,000%, many Brazilians with fixed income employment have multiple jobs.

Several major schemes have been under way since the 1960s to expand Brazil's industrial work force. The National Industrial Apprenticeship Service provides training for industrial workers; similar training for workers in commerce is provided by the National Commercial Apprenticeship Service. The Center for Technical Education was founded in 1967 in accordance with a plan to improve production in the north and northeast regions. Another national service agency, established in 1976, trains rural workers. Between 1950 and 1990, the labor force increased from 17 million to more than 64 million. The economically active female population increased from about 18% of the labor force in 1960 to over 35% by 1990.

Major labor union federations include the National Confederation of Industrial Workers, the National Confederation of Commercial Workers, the National Confederation of Ground Transport Workers, and the National Confederation of Bank Workers. Union financing depends largely on the compulsory checkoff (contribuição sindical) administered by the government, which applies to nearly all workers, except government and domestic employees, irrespective of union affiliation, and amounts to one day's wages per year from each employed worker. Employees who belong to a union must also pay union dues. The

union structure in Brazil was created after 1930 by the Getúlio Vargas regime along the "corporative" lines of the syndical organizations of fascist Italy and Portugal. Employee syndicates are paralleled by employer organizations under federal control. The right to strike was outlawed in 1937, but was restored by the 1946 constitution, subject to strict government regulation. Strikes were further limited in the 1964 constitution to disputes in which employers have already rejected arbitration by competent authorities. The right to strike was fully restored in 1984. Strikes in 1992 included those by port workers and stevedores, airport workers, public and private bus drivers and fare collectors, public and private teachers, and federal and state government employees. Strike activity in the private sector was low in 1993, with workers concerned mostly with job preservation.

The many benefits provided by government legislation have the effect of making workers look to government rather than unions for economic advancement. Comprehensive labor legislation is contained in the consolidated labor laws of 1943 and later amendments. The basic working conditions include provisions for an eight-hour day, a six-day week, a minimum wage, an annual bonus, paid civil and religious holidays, an annual paid vacation, compensation for overtime work, profit sharing, equal pay for equal work regardless of sex, a family allowance, and severance pay. Special protection is provided for women and minors, as are special benefits for pregnant women. The minimum wage, adjusted annually, varies from region to region. Overtime and night pay must be at least 20% higher than normal hourly wages. These laws generally apply to industrial and commercial workers and, through the Rural Workers' Law, to agricultural workers, but they exclude government employees and domestic workers, who are covered by other regulations. Workers' compensation, provided under the national social security program, was extended to farm workers in 1975. Many labor organizations, which had supported the constitutional revision process begun in 1988, altered their views in 1993. Worries about employment and salaries during the 1992 recession contributed to such opposition.

Enforcement of labor laws is the responsibility of the Ministry of Labor and Social Security through regional delegates of the ministry located in each state. Enforcement, however, is lax. Labor courts are separate from the regular judiciary, and consist of local conciliation and judicial boards, regional labor councils, and a central tribunal. Worker and employer representatives are appointed to each of the labor courts by the president of Brazil, from a list of names submitted by workers and employers. A national mediation service is housed in the Ministry of Labor.

[22]AGRICULTURE

Despite advances in large-scale manufacturing in recent years, 26% of all Brazilians still lived in the countryside in 1985 (as compared with 44% in 1970). In 1992, the total arable and permanent crop area comprised 61.3 million hectares (152 million acres), and the number of farms was about 5 million. Although agriculture's share of exports has declined relative to industrial goods, the value has continued to increase, so that Brazil in 1977 became the world's second-largest exporter of agricultural products. Except for grain (particularly wheat), of which some 2.8 million tons had to be imported in 1990, Brazil is virtually self-sufficient in food. The growth rates for agriculture as a whole averaged 3.0% during 1980–85, 2.3% during 1985–90, and stood at 2.6% in 1991. In 1992, real value added of agriculture rose 6%, as producers benefitted from good weather and increased access to subsidized credit. In 1991, agriculture accounted for 10.8% of the total GDP. Export crops are significant—in addition to the traditional exports of coffee and cocoa, Brazil is also a major exporter of soybeans and orange juice. In 1990, Brazil ranked fifth in the export of cocoa beans (after Côte D'Ivoire, Ghana, Malaysia, and Nigeria) at 118,126 tons, or

7.1% of the world's cocoa bean exports. Almost 15% of all cocoa bean production came from Brazil in 1992. That same year, nearly 22% of the world's coffee came from Brazil.

The Land Statute Law of 1964 was designed to modify the agrarian structure and increase agricultural output in selected regions over a 20-year period. The law empowered the federal government to expropriate unused or underutilized land by offering indemnification in bonds in the case of large properties and cash payment for smallholdings. In redistributing expropriated lands, priority is given to those who work the land under tenancy, sharecropping, or ordinary labor agreements. Responsibility for implementing the law is divided between the Brazilian Agrarian Reform Institute and the National Institute of Agricultural Development. In October 1984, a law was passed to facilitate the distribution of 43.1 million hectares (106.5 million acres) of state-owned land and nonproductive private estates to 1.4 million peasant families, primarily in the impoverished northeast, through 1989. The formation of cooperatives was to be encouraged.

Further agricultural reforms have been carried out under the Carta de Brasília of 1967. The Carta included an incentive program for the construction of storage facilities, to permit farmers to hold products off the market in expectation of better prices. Agricultural research in Brazil is conducted by the Agriculture and Cattle Raising Institute of Research. The expansion of power, transportation, and communications systems during the 1970s further contributed to agricultural development.

Coffee, until 1974 preeminent among export earners, has been declining in importance since the early 1960s, while soybeans, sugarcane, cotton, wheat, and citrus fruits have shown dramatic increases. Sugarcane production, in which Brazil ranked first in the world in 1992, is grown not only for refined sugar but also as a source of alcohol for fuel. Agricultural production in 1991 and 1992 (in millions of tons) is shown in the following table:

	1991	1992
Sugarcane	260.8	270.6
Oranges	18.9	19.6
Cassava (manioc)	24.5	22.6
Soybeans	14.9	19.1
Rice	9.4	9.9
Coffee	1.5	1.3
Wheat	2.9	2.8
Cotton	0.68	0.62
Cashews	0.17	0.10

23 ANIMAL HUSBANDRY

Brazil is a leading livestock-producing country, and 185,500,000 hectares (458,370,000 acres—more than one-fifth of the total national area) were devoted to open pasture in 1991. Since World War I, cattle production has become one of the country's major sources of wealth. Hereford and polled angus are raised in the southern states of Rio Grande do Sul and Santa Catarina, and Dutch and Jersey cattle supply dairy products in the uplands of Minas Gerais, Rio de Janeiro, São Paulo, and Paraná. The humped zebu was first introduced in Minas Gerais, where intense cross-breeding produced the Hindu-Brazil breed that is now most common throughout Brazil because it resists tick fever and heat. There were an estimated 153 million head of cattle in 1992, as compared with an annual average of 116.6 million during 1979–81 and 59.8 million during 1961–65.

Hog raising, marked by an improvement in breeds, has doubled since 1935, making Brazil the world's third-largest producer. In 1992 there were an estimated 33 million hogs. Berkshires and Poland Chinas have been introduced in quantity, and, since vegetable oils are increasingly replacing lard, the emphasis is on production of pork, ham, and sausages. Brazil is not a major sheep country, since most of the area is too tropical. The bulk of Brazil's 19.5 million sheep are in Rio Grande do Sul. Of the other domestic

animals raised commercially in 1992, there were some 12 million goats, 6.2 million horses, 1.3 million asses, and some 570 million chickens.

The government encourages production and seeks more efficient methods of conservation and distribution of meat products. Meat production was estimated at 7.18 million tons in 1992, including 3 million tons of beef and veal. The dairy industry is most highly developed in the vicinity of large cities. Estimated output of dairy products in 1992 included 15.5 million tons of fresh cow's milk and 1,455,000 tons of eggs. In 1992, livestock production expanded by 5.3%.

In 1985, financing for the rural sector was turned over in large measure to the commercial banks, with government banks being responsible for the balance. Previously, financing had been through central government credits.

24 FISHING

Although Brazil has a seacoast of some 7,400 km (4,600 mi) and excellent fishing grounds off the South Atlantic coast, the nation has never fully utilized its commercial potential. Traditionally, fishing has been carried on by small groups of individual fishermen using primitive techniques and equipment and seldom venturing out of sight of land. Lack of storage facilities, canneries, and adequate methods of distribution have limited the supply and led to the importation of dried fish. Swordfish is caught in large quantities off the coast of Paraíba and Rio Grande do Norte, and shrimp is caught and dried along the coasts of Maranhão, Ceará, and Bahia. The fish resources of the Amazon River are not exploited, except for the commercial processing of the pirarucú and an aquatic mammal, the sea cow. The annual fish catch is so modest that there has traditionally been a scandalous scarcity during Holy Week, about the only time when Brazilians eat much fish. The total catch in 1991 was 800,000 tons. Small quantities of lobster are exported.

A fisheries development agency was established in the early 1970s to exploit Brazil's coastal potential. The discovery of large quantities of tuna off the coast of Rio Grande do Sul has interested foreign fishing companies, and Japanese and US concerns have obtained the right to fish in Brazilian waters and to establish storage and canning facilities. Normally, foreign fishing rights are reserved to the Portuguese.

25 FORESTRY

Nearly three-fifths of South America's forests and woodlands are in Brazil. Sylvan areas in Brazil are nearly three-quarters as large as the forests of all African nations combined. Brazil's forests and woodlands cover 58% of the total area and are among the richest in the world, yielding timber, oil-bearing fruits, gums, resins, waxes, essential oils, cellulose, fibers, nuts, maté, and other products. In the rainforest, as many as 3,000 different species per sq mi (2.6 sq km) may coexist. However, only a limited percentage of forestland is being exploited, in part because of a lack of adequate transportation. Brazil is one of the leading producers of tropical hardwood products. Brazilian timber is of fine quality, ranging from wood as light as cork to the wood of the Brazilian pepper tree, with a density one and one-half times that of water. By 1991, rapid deforestation during the previous thirty years in the Amazon (from migration, road building, mining, and tax incentives) had caused the rainforest to shrink by an estimated 8.5% since colonial times. The hardwood trees of the Amazon rain forest are of excellent quality, but because of a thriving domestic furniture industry, they are used mainly locally. The Paraná pine (a conifer related to the genus Araucaria) is in greatest demand. It grows in the southern states in stands that comprise about 420 million trees. A Brazilian ban on log exports has focused exports on value-added products (mostly lumber, plywood, hardboard, and veneers). Production of paper and woodpulp has expanded considerably since 1975; exports of paper intensified between 1981 and 1991, from 337,000 tons to 840,000 tons.

Production of roundwood in 1991 was estimated at 264.6 million cu m; sawn wood production was an estimated 17.1 million cu m. Brazil's production of rubber in 1991 was 304,082 tons; the natural rubber industry, once a world leader, was dealt a strong blow by the development of cheaper synthetics. Forest products like rubber, Brazil nuts, cashews, waxes, and fibers now come from plantations and no longer from wild forest trees as in earlier days.

26MINING

The discovery of gold in Minas Gerais ("general mines") in 1693 made Brazil the world's leading gold producer, but rapid exploitation under the Portuguese colonial system exhausted the mines in less than a century. The dissipation of the nation's gold wealth for the benefit of an overseas power instilled in modern Brazilians an acutely protective attitude toward their mineral reserves. Thus, although there is a vast potential of mineral wealth, resources have hardly been developed. Lack of capital has restricted development by domestic firms, and Brazilian mining laws, as well as adverse geographic conditions, have discouraged foreign capital. In 1991, the total value of minerals produced was about US$10 billion, or 3% of GDP. Nevertheless, Brazil is the world leader in the production of bauxite, columbium, gemstones, gold, iron ore, kaolin, manganese, tantalum, and tin. Within Latin America, Brazil is a major producer of aluminum, cement, and ferroalloys.

In 1965, the government introduced a 10-year master plan for the evaluation of Brazil's mineral resources. The government also took steps to encourage private capital investment in the development of phosphate reserves, the exploitation of oil-bearing shale, and the expansion of iron ore output. Through the Company for Mineral Resources Research, created in 1969, the government has sought to expand existing mineral industries and to establish new ones. The Amazon region has been a particular focus of the company's efforts. The Carajás mining project in the Amazonian state of Pará, begun in the early 1980s, was expected to lead to the mining there of manganese, copper, tin, silver, gold, nickel, molybdenum, bismuth, and zinc. During 1982–85, mining was Brazil's fastest growing sector, averaging some 15% annually. The investment levels of the 1980s have considerably decreased; in 1990–91 investment in mining fell from US$1 billion to US$400 million.

Brazil's iron ore reserves are known to total 19.2 billion tons. In 1942, the government-controlled Rio Doce Valley Co. began mining operations in the high-grade deposits of Minas Gerais and increased its activities with the aid of loans from the US Export-Import Bank. Iron ore production for 1974 was 59.4 million tons, more than double the 1969 total; by 1991, output had increased to 150 million tons. New deposits in the state of Pará were discovered during the 1970s. In 1981, Brazil became the world's leading exporter of iron ore; 113,469,000 tons were exported in 1990, mostly to Japan and Germany. Output of manganese (mined in Minas Gerais, Mato Grosso, and the territory of Amapá) in 1991 was 2.5 million tons. The absence of good coking coal is a handicap to industrial plans, and coal production, mainly from Santa Catarina, has been hurt by cutbacks in state investment. Production for 1991 was estimated at 7.2 million tons, only two-thirds of consumption (10.9 million tons). There are important deposits of beryllium, chrome, graphite, magnesite, mica, quartz crystal, thorium, titanium, and zirconium; copper, lead, nickel, tin, and zinc deposits occur in several states. Brazil has become a major exporter of tin in recent years. Major deposits of high-quality bauxite have been discovered in the Amazon region; output has risen rapidly, from 6.5 million tons in 1987 to 10.3 million tons in 1991. Large phosphate deposits exist in Minas Gerais.

Diamonds and other precious and semiprecious stones are mined in Minas Gerais, Goiás, and Bahia; in 1991, production of diamonds amounted to 1,500 carats (6th in the world). Gold deposits found at Serra Pelada in 1980 raised gold production to 103,000 kg tons by 1989, making Brazil the world's sixth-largest producer; production averaged 90,380 kg during 1987–91, for a relative ranking of 7th in the world. Recent discoveries of platinum indicate that Brazil may have half the world's reserves. There are large supplies of good building stone, as well as sizable deposits of nonmetallic minerals, such as limestone, fluorite, dolomite, and other materials indispensable to the chemical industry.

27ENERGY AND POWER

Brazil's hydroelectric potential is thought to be one of the world's largest. Great strides are being taken to increase production to keep pace with Brazil's expanding industries. Total installed capacity increased from 4.8 million kw in 1960 to 8.5 million kw in 1968 and to 54.1 million kw in 1991; of that total, hydroelectricity accounted for 86%; conventional thermal plants for 13%; and nuclear plants for 1%. Production for 1991 was 234.3 billion kwh, of which hydropower contributed 93%. Construction of the Itaipu Dam on the Paraná River took place between 1975 and 1982; this joint Brazilian-Paraguayan project, the world's largest hydroelectric plant, attained its full capacity of 12.6 million kw in 1986, at a cost of $15 billion. About 70% of Brazil's population is served by Itaipu, which generates about 75 billion kwh per year. Each of Brazil's nine turbines (Paraguay controls the other nine) at Itaipu generates 700,000 kw, which can be transmitted up to 1,000 km (620 mi) away. Brazil regularly purchases a large portion of Paraguay's half of its Itaipu electricity production.

Industrial consumption of electricity has increased rapidly in recent years, especially in highly industrialized areas, such as São Paulo and Minas Gerais, where some 80% of the energy produced is utilized, and where industries are offered generous incentives to convert from fuel oil to electricity. Higher electricity prices since 1993 have permitted electric utilities to invest and improve their financial positions.

Brazil has sharply scaled back its ambitious nuclear power program since 1983. The first nuclear power plant, built with US help near Rio de Janeiro and known as Angra I, began operations in May 1985. Nuclear generation of electricity only accounted for about 1% of production in 1991. Estimates of uranium reserves were put at 163,000 tons in 1991, the fifth largest in the world.

The government-owned Petróleo Brasileiros (Petrobrás), established in 1953, has had a monopoly over the exploration and development of petroleum reserves; the 1988 constitution guaranteed the maintenance of state monopolies in the petroleum and electricity sectors, despite rampant privatization. Foreign participation and investment formerly were forbidden by Brazilian law; however, in the mid-1970s, exploration was opened to foreign companies through risk contracts, and the petrochemicals industry was opened to foreign participation. Production has traditionally come from the state of Bahia; total production, which amounted to 7.8 million tons in 1968, increased only marginally to 9 million tons (or about 186,000 barrels per day) in 1980. By 1992, production totaled 33 million tons, or 685,000 barrels per day. Natural gas production was 6.2 billion cu m in 1991. In 1975, the government initiated a program to develop alcohol from sugarcane as an energy source. By 1985, 83% of the country's cars and light vehicles were powered by alcohol, although slumping oil prices and persistent financial problems within the alcohol industry prompted the government to freeze production at the 1985 level of 11.1 billion liters. By 1991, with over 13 million alcohol-powered vehicles on the road and an acute alcohol shortage, the production of gasoline-powered vehicles had shifted to 70% from the almost entirely alcohol-powered production of 1989.

²⁸INDUSTRY

Industry accounted for 27.7% of the GDP in 1984, and industrial exports accounted for 57% of all export earnings in 1985. Major industries include iron and steel, automobiles, petroleum processing, chemicals, and cement; technologically based industries have been the most dynamic in recent years. Peak industrial growth was achieved in 1973, when the manufacturing sector grew by 15.8%; growth rates averaging about 7% were posted during 1978–80, rising to 8.3% in 1985 and 11.3% in 1986.

In 1969, 3.7 million tons of crude steel were produced; by 1974, production had nearly doubled to 7.3 million tons, and in 1980, the output reached 15.3 million tons; production in 1981 dipped to 13.2 million tons, but had recovered to 20.5 million tons by 1985. Domestic production exceeds domestic needs. Vast reserves of accessible, high-grade ore, plus a rapidly expanding domestic demand for these products, favor continued expansion of the steel industry. The major negative factor is lack of domestic soft coal. Production of pig iron was 18.9 million tons in 1985.

The Brazilian automotive industry is the major producer of vehicles in all Latin America and the sixth-largest producer in the world, providing direct employment for more than 200,000 factory workers and more than 1 million jobs in related fields. Production, which started with 5,097 units in 1956, rose to 286,944 in 1969, 858,479 in 1974, and 1,057,257 in 1986. Domestic sales are high, given the country's large and rapidly increasing population, although prices are high, retail financing difficult, and highways deficient. The 1981 recession hit automobile production severely; by 1985, however, production could not keep pace with demand. Volkswagen of Brazil accounts for more than 50% of passenger car production. A major growth industry of the 1970s was shipbuilding, absorbing an investment of about US$1 billion annually; during the mid-1980s, however, shipbuilding declined, and in 1986, the industry was operating at 50% of capacity.

Because of increased domestic refining capacity, imports of petroleum products declined in the late 1970s; by 1979, Brazil was a net exporter of petroleum derivatives. Petroleum products in 1985 included diesel oils, 21,750,000 cu m; gasoline, 10,757,000 cu m; and fuel oil, 12,201,000 cu m. Brazil's developing petrochemical industry emphasizes the production of synthetic rubber. Petrobrás' synthetic rubber plant started production of styrenebutadiene synthetic rubber in 1962. Brazil's natural rubber production, which at one time supplied the world, cannot now meet the nation's own needs. Production of synthetic rubber in 1985 was 265,015 tons (as compared with 59,000 tons in 1968). Other chemical products included 11.1 billion liters of distilled alcohol in 1985, much of it for the nation's burgeoning alcohol-fuel fleet. There are over 500 pharmaceutical laboratories and plants in Brazil, the majority in São Paulo. Over 80% of the industry is foreign-owned. Increased construction demands boosted Brazil's cement production to 6.4 million tons in 1967; output rose to 13.4 million tons in 1973 and 20.6 million tons in 1985. Brazil's electrical equipment industry manufactures television sets, transistor radios, refrigerators, air conditioners, and many other appliances.

According to the Brazilian Statistical Institute (IBGE), industrial production rose 107% in the first nine months of 1993, following declines of 5.3% and 0.3% in the same periods of 1991 and 1992. Manufacturing has grown rapidly, led by durable goods which are up 42%. Manufacturing output has been weakening since the second semester of 1993, however.

The growth of the industrial sector decelerated during the second semester of 1993 as a result of high inflation, declining purchasing power, high interest rates, layoffs, and labor stoppages.

Motor vehicle production, Brazil's industrial backbone, increased 29 percent in comparison to 1992. Output for 1993 has been projected at 1.36 million units compared to 1.07 million

units last year, which surpasses the 1980 all time high of 1.1 million units.

Brazil is responsible for approximately 3% of world textile production, with total sales averaging $19 billion a year. The country has the largest textile industry in South America in terms of installed capacity and output, with nearly half of the spindles and looms in operation on the continent.

The Brazilian pulp and paper sector consists of more than 220 companies, which together employ approximately 80,000 people in industrial operations, as well as another 57,000 in forestry work and operations. The pulp and paper sector is almost fully privately-owned.

²⁹SCIENCE AND TECHNOLOGY

The National Council of Scientific and Technological Development, created in 1951 as the National Research Council, formulates and coordinates Brazil's scientific and technological policies. The Brazilian Academy of Sciences was founded in 1916; by 1993 there were 26 learned societies and 53 research institutes covering virtually every area of scientific and technological endeavor. Among the most important scientific institutions are the Oswaldo Cruz Foundation for biological research in Rio de Janeiro and the Butantan Institute in São Paulo which produces serums for the bites of venomous snakes, a field in which Brazil leads the world. Total expenditures on research and development in 1985 amounted to CR$5.4 billion, with 52,863 scientists, engineers, and technicians engaged in research and development.

Brazil entered the space age in 1973, with the launching of the SONDA II rocket as part of a program to determine electron density in the low ionosphere, a queston of practical importance for aircraft navigation. Under the government's Amazon development program, Humboldt City, a scientific and technological center, has been established in Mato Grosso. Atomic research is conducted at the Energetics and Nuclear Research Institute of São Paulo; other research reactors are located at Belo Horizonte and Rio de Janeiro. The Nuclear Energy Center for Agriculture was established in 1966. A total of US$550 million was allocated to the nuclear energy program under the 1975–79 development plan, but the program languished in the 1980s.

³⁰DOMESTIC TRADE

Rio de Janeiro and São Paulo are the principal distribution centers; the largest numbers of importers, sales agents, and distributors are located in these cities, having branch offices in other areas. Other major commercial centers are Recife and Pôrto Alegre, in the northeast and the south; Belém, which serves as a distribution center for the whole Amazon River Valley; and Salvador, which is the main distribution center for Bahia and the neighboring states.

The Brazilian commercial code permits the exercise of trade by all persons who make trade their habitual occupation and register with the appropriate government body. Goods are sold in department stores, in specialty shops, by street vendors, and in supermarkets in the larger cities, but most commercial establishments have fewer than six employees. There are a number of consumer cooperatives that are generally sponsored by ministries, trade unions, and social security institutes. Producer cooperatives are found mostly in agriculture and fishing. Credit is extended to higher-income customers on open accounts and to lower-income groups on installment payment plans.

Business hours are from 9 AM to 6:30 PM, Monday through Friday, and from 9 AM to 1 PM on Saturday. Banks transact business from 10 AM to 4:30 PM, Monday through Friday. In many smaller cities and towns, stores are closed for over an hour at lunchtime.

Since 1947, the advertising sector (in all the various media) has increased its expenditures many times over. Most advertising agencies maintain headquarters in São Paulo; some major agencies have branch offices in other large cities. Advertisements are presented on television and on all radio stations, with the exception of the special broadcasting system of the Ministry of Education. Newspapers, magazines, periodicals, motion pictures, billboards, posters, and electric signs are used for advertising. Mobile advertising units equipped with loudspeaker systems are common in the larger cities.

³¹FOREIGN TRADE

Brazil's long-favorable foreign-trade balance deteriorated substantially between 1958 and 1974 as a result of industrial expansion, which necessitated increased imports of industrial capital goods and petroleum. The government's policy of granting incentives to exporters and liberalizing trade restrictions, begun in the late 1960s, partially alleviated the trend. During 1975–76 and again from 1978 to 1982, the foreign-trade balance was in deficit. Beginning in 1983, Brazil recorded trade surpluses: $5.1 billion in 1983, $11.8 billion in 1984, and $11.3 billion in 1985. This achievement was the result of policies that restricted imports and offered substantial incentives to exporters.

Between 1963 and 1981, exports expanded at an average annual rate of 17%, but they grew by only 9.1% between 1982 and 1985. Coffee has long been Brazil's dominant export, but the proportion of its export earnings declined from 41.3% in 1968 to 21.7% in 1973 and to 10.2% in 1985. The main export item after coffee in 1985 was soybeans and soy products, which accounted for 9.9% of total export value. As a result of an ambitious energy development program, Brazil's reliance on imported oil has dropped from 70% of its needs in 1980 to 45% in 1985. Fuels and lubricants, chiefly imported petroleum, accounted for 46.9% of Brazil's import value in 1985.

The notable accomplishment of former President Fernado Collor de Mello to open the Brazilian market remains the cornerstone of Brazil's economic and trade policies. After almost 30 years of import substitution, which initially brought high growth and short term industrialization in the 1960–70s, this policy was finally recognized in the late 1980s as the principal culprit of Brazil's economic woes, particularly high inflation and industrial decline. Over the last three years, most of Brazil's non-tariff barriers to trade, such as import quotas, import prohibitions, restrictive import licensing, and local content requirements, which for many years were the hallmark of Brazil's restrictive trade regime, were eliminated or drastically reduced. Import duties were reduced from an average of about 50% in the late-1980s to 14.2% and a maximum of 35%. While the overall level and pervasiveness of non-tariff barriers have been drastically reduced, some import duties remain high in comparison with other countries. While the depression/inflation problems of recent years have reduced purchasing power by about 50% among the working and lower middle classes and further skewed the already highly uneven distribution of income, the Brazilian market remains enormously attractive to US businesses.

Although trade barriers continue to recede with the present government of President Itamar Franco who assumed office on 2 October 1992, trade liberalization has lost some of its momentum, and there are some serious concerns regarding automobile, telecommunications, and informatics sectors.

Trade Balance (US$ billion)	1992	1993	1994
Exports	36.2	38.0	42.9
Imports	20.5	24.0	28.8
Balance	15.7	14.0	14.0

³²BALANCE OF PAYMENTS

After a decline in the mid-1960s, Brazil's reserve holdings grew spectacularly between 1969 and 1973, reaching US$6,536 million

(US$6,150 million in foreign exchange) by 31 March 1974. The prime reason was a steadily rising inflow of long-term capital investment, coupled with trade balances that were favorable or only minimally unfavorable. In 1974, however, a decline in the value of coffee exports and a doubling of import costs (partly attributable to increased oil prices) more than offset a further rise in capital investment, resulting in Brazil's first payments deficit in nearly a decade. Between 1976 and 1978, Brazil had a positive balance of payments, but large deficits were registered in 1979 and 1980. A surplus was achieved in 1981, in part because of Brazil's excellent trade showing. The surpluses in 1984 and 1985 were sufficient to pay all interest on the foreign debt.

In 1992 merchandise exports totaled $36,103 million and imports reached $20,578 million. The merchandise trade balance was $15,525 million. The following table summarizes Brazil's balance of payments for 1991 and 1992 (in millions of US dollars):

	1991	1992
CURRENT ACCOUNT		
Goods, services, and income	–2,964	4,219
Unrequited transfers	1,556	2,056
TOTALS	–1,408	6,275
CAPITAL ACCOUNT		
Direct investment	–42	1,308
Portfolio investment	3,808	14,466
Other long-term capital	–1,723	–4,760
Other short-term capital	–1,188	1,581
Exceptional financing	4,023	–2,670
Other liabilities	–739	—
Reserves	–197	–15,068
TOTALS	556	–5,143
Errors and omissions	852	–1,132
Total change in reserves	–209	–15,001

During the 1970s and early 1980s, Brazil increasingly came to rely on international borrowing to meet its financing needs. The foreign debt grew rapidly after 1974 as the government pressed for continued economic growth without regard for balance of payments pressures generated by the oil shocks of 1973–74 and 1978–79 and without increasing domestic savings or improving the tax base. The huge trade surpluses of 1984 and 1985 halted the upward trend. However, the 1986 surge in consumer spending drained reserves to such an extent that by early 1987, the government was forced to suspend payments on $68 billion of the estimated $108 billion debt, the highest of any developing nation. An agreement was reached in April 1991 on 1989/90 arrears. In 1992, Brazil and the advisory committee representing foreign commercial banks agreed to a debt and debt service reduction for $44 billion.

³³BANKING AND SECURITIES

A banking reform enacted in December 1964 provided for the establishment of the Central Bank of the Republic of Brazil (changed in 1967 to the Central Bank of Brazil), with powers to regulate the banking system and the stock market. The Central Bank serves as the financial agent of the federal government and functions as a depository for the reserves of private banks. The reform also created the National Monetary Council, which formulates monetary policies for the Central Bank.

There once were about 340 commercial banks in Brazil, with hundreds of branch offices; however, banking reforms have reduced that number. In 1993 there were 29 commercial banks, 10 state banks, 7 development banks, and 9 foreign banks. The largest banks, which together account for about 50% of all deposits, are the Bank of Brazil (56% owned by the government), the Banco do Estado de São Paulo, the Banco Brasileiro de Descontos, the Banco de Credito Real de Minas Gerais, the Banco Real, the Banco Mercantil de São Paulo, and the Banco Nacional de Minas Gerais. Important non-commercial banks include the Federal Savings Bank, with several hundred branches; the National Bank for Economic and Social Development (BNDE); and the Northeast Bank of Brazil.

Commercial banks are the main source of short-term credit. The Agricultural and Industrial Credit Department of the Bank of Brazil provides both short-term and medium-term capital to private industrial firms. Long-term credit is provided by finance companies, which were first introduced in 1946 with the establishment of the Brazilian Company of Investments. Since then, their number has increased rapidly.

In the late 1970s, interest rates were freed from controls; some domestic rates subsequently reached 250-350%, figures that reflect, in large part, Brazil's high rate of inflation. The money supply grew from CR$462,655 million in 1978 to CR$102,413,000 million at the end of 1985. Lending by commercial banks increased from CR$1,947 billion in 1979 to CR$269,806 billion in 1985. Currency in circulation totaled CR$290.7 billion at the end of 1980 and cz$125.1 billion in July 1987. In February 1986, Brazil created a new unit of currency, the cruzado, worth 1,000 cruzerios, and devalued it by 25%. It was devalued by a further 7.5% later in the year and then by 50% in early 1987. The attempt to "fix" the exchange rate, as part of the "Cruzado Plan," was a reversal of the policy of frequent adjustments, which had been followed for a decade and which was seen as inflationary.

Of the 12 functioning stock exchanges (bôlsas) in Brazil in 1979, those in Rio de Janeiro and São Paulo accounted for about 95% of all trading. All stock exchanges are subject to loose supervision by the state governments, except the Rio de Janeiro exchange, which is under federal government jurisdiction. Daily transactions are published in the leading newspapers, and periodic transactions and trends are published in the *Bôlsa de Fundos Públicos*. A system of national registration introduced in 1970 permits trading of a company's shares on any of the 12 exchanges; the Securities and Exchange Commission is the principal regulatory agency.

Brazilian tax laws provide incentives for investment in stocks and bonds. Up to 24% of individual tax liability may be invested annually in share certificates from authorized financial institutions. Furthermore, individuals may deduct from gross income up to 42% of capital subscription in companies or agencies involved in developing the northeastern or Amazon regions, up to 25% of amounts invested in open capital companies, up to 20% of investments in approved forestation or reforestation projects, up to 13.5% of investments in mutual funds, and up to 4% of savings under the National Housing System. Earnings from all the above investments are also excluded from gross income, up to certain limits.

34INSURANCE

The operations of insurance companies in Brazil are supervised by the Superintendency of Private Insurance, the National Private Insurance Council, and the Institute of Reinsurance of Brazil. In 1986 there were 24 major national insurance companies. Life insurance is sold entirely by domestic companies, and the principal non-life branches, such as fire and transport accident insurance, are extensions of the social insurance system organized in 1954. Life insurance in force was CR$400 billion in 1985. Insurance premiums paid in 1987 totaled us$1,522, corresponding to us$1.7 per capita for life insurance and us$9.1 per capita for nonlife insurance. Total premiums were 0.86% of the GDP.

35PUBLIC FINANCE

The Brazilian fiscal year coincides with the calendar year. The budget, prepared under the supervision of the Ministry of Planning and Economic Coordination, represents the government's plans for financing administrative operations and capital expenditures. Budgetary deficits increased considerably in the 1960s, rising from CR$102.5 million in 1961 to CR$775 million in 1965. The government's objective—to hold total expenditures fairly constant while raising the share of capital outlays—was achieved by a policy of increasing receipts through tax rate changes and by cutting all non-essential expenditures. Although the federal budget deficit was reduced in real terms during the late 1960s, fiscal problems continued to be a major source of inflationary pressure, with revenues hovering between 14.5% and 16.5% of GNP and total government expenditures in the range of 17.5% to 19%. Government revenue increased considerably, and each year the real deficit was reduced below the previous year's level both absolutely and relatively. Public expenditure was stepped up, particularly transfer payments; transfers of capital to decentralized agencies increased, although direct investment by the central government fell off. Thus, more capital was invested in basic infrastructural projects.

Increases in both revenues and expenditures were rapid during the 1970s, but the pattern of decreasing deficits continued. The budgets for 1973 and 1974 actually showed a surplus, although the realized surplus of CR$3,882 million in 1974 fell far short of the budgeted surplus of CR$11,471 million. There was a budget deficit of CR$3,248 million in 1975, but surpluses were recorded annually during 1976–80. One of the principal causes of Brazil's financial instability in the 1980s has been the rate at which public spending has exceeded revenues. Following another stabilization program in 1990, a budget surplus of 1.4% of GDP was recorded, but deteriorated to a deficit of 1.7% of GDP by 1992. At the beginning of 1994, the public deficit was estimated at $22 billion. It is still uncertain whether or not proposed tax increases and budget cuts will be able to eliminate the public deficit in the near future.

The following table shows actual revenues and expenditures for 1990 and 1991 in millions of cruzeiros.

	1990	1991
REVENUE AND GRANTS		
Tax revenue	6,044.9	26,310.0
Non-tax revenue	1,186.7	2,347.6
Capital revenue	2,751.5	8,308.0
Grants	22.6	109.7
TOTALS	10,000.2	42,879.3
EXPENDITURES & LENDING MINUS REPAYMENTS		
General public service	1,465.1	6,518.8
Defense	391.1	1,231.2
Public order and safety	125.3	1,007.2
Education	347.8	1,543.7
Health	742.4	2,844.5
Social security and welfare	2,799.1	14,172.4
Housing and community amenities	21.6	270
Recreation, cultural, and religious affairs	11.0	39.4
Economic affairs and services	356.4	3,819.7
Other expenditures	8.585.5	14,400.5
Lending minus repayments	772.0	3,208.0
TOTALS	11,847.4	44,390.3
Deficit/Surplus	−1,847.2	−1,511.0

As of 1993, Brazil's external debt totalled approximately us$120 billion, of which us$42 billion was medium-term debt owed by the government.

36TAXATION

Brazil's tax structure has been modified repeatedly in recent years, and traditional tax evasion has come under strong attack. A series of income tax reforms, during the 1960s and 1970s closed many loopholes and expanded the roster of taxpayers, mainly through wider use of withholding taxes. The welter of indirect taxes, both federal and state, was simplified, eliminating the old stamp taxes and systematizing the two major sales levies. In the late 1960s, the government took steps to ease the tax burdens on corporations and shareholders in order to encourage equity investment and offset the erosion of asset value by inflation.

The basic income tax rate on corporations and other legal entities in 1993 was 30%; profits are taxed at approximately 50.5% and capital gains at 25%. Firms may effectively reduce income tax liability by up to 24% by investing part of the tax due in government-approved incentive projects or by purchasing quotas in funds that invest in such projects. Depreciation allowances are customarily 4% for buildings, 10% for machinery and equipment, and 20% for vehicles, on a straight-line basis. Corporate income earned overseas is not taxed.

There is a progressive personal income tax with a maximum rate of 25%. Numerous exclusions from ordinary taxable income include profits on sales of shares, profits from certain real estate sales, and dividends and interest on stocks and bonds up to certain limits. Under a 1975 revision, foreigners transferring residence from abroad need not pay income tax on returns from their overseas holdings for the first five years of residence in Brazil.

Turnover, sales, and excise taxes underwent a general reorganization and simplification in 1967. The sales and consignment tax, formerly levied with cumulative effect by state governments at rates varying for different goods and in different states, has been replaced by two value-added taxes. The sales tax varied between 9 and 17% on both intrastate and interstate transactions in 1993, while the federal excise tax, payable on all goods or products imported to or produced in Brazil, ranged up to 20% for most products (but up to 365% for some luxury items). Other taxes include a financial operations levy; taxes on the production and distribution of minerals, fuels, and electric power; a real estate transfer tax; and municipal service and urban real estate taxes.

37CUSTOMS AND DUTIES

Between the late 1980's and July 1993, the government reduced import duties incrementally to encourage trade. Tariffs are assessed on an ad valorem basis, which ranged from zero to 40% of the c.i.f. price in January 1993, plus specific duties in some cases. All imports and exports are controlled by DECEX, the Foreign Trade Department of the Ministry of Economy. Imports of goods on Brazil's national list from LAIA members receive tariff preferences. Imports are grouped into three categories: those that do not need import certificates, those that do need certificates, and prohibited imports. In order to encourage exports, in early 1987, the government waived the industrial products tax for companies on an amount equivalent to 10% of the increase in the companies' exports in the preceding year. Payment procedures for exporters have been speeded up as well.

Certain sectors, including petroleum products, arms and ammunition, and wheat, require departmental or ministerial approval for imports. Importers must pay state and federal value-added taxes at ports, but these may be recovered for goods to be manufactured or sold in Brazil. The free-trade zone in Manaus and some energy development projects are exempted from import duties.

38FOREIGN INVESTMENT

Brazil's great potential in natural resources and its economic growth and industrial expansion have played a major role in attracting foreign capital. Between 1974 and 1978, foreign investment represented 6%–7.3% of the GDP. Since 1975, industry has received about 75% of all foreign investment. Total cumulative

foreign investment in Brazil reached US$23.7 billion by June 1985. The largest investors were the US (32%), the FRG (13%), Japan (9.1%), Switzerland (9%), and the UK (5%).

According to a report prepared for the US Senate Subcommittee on Multinational Corporations in 1975, multinational enterprises controlled all of Brazil's automobile production, about 94% of the pharmaceutical industry, 82% of rubber production, and substantial shares of other industries. Multinationals held controlling interest in 59 of Brazil's 100 largest manufacturing enterprises and accounted for about half of manufacturing sales.

Brazilian law gives the same protection and guarantees to foreign capital investments that it gives to investments made by Brazilian nationals. Special incentives are offered for investments in mining, fishing, tourism, shipbuilding, and reforestation and for projects undertaken in the northeast and Amazon regions. The registration of foreign capital and of reinvestment, profit remittance, interest, royalties, and payments, for technical assistance, as well as repatriation of foreign capital, are regulated by Brazil's Foreign Capital and Profit Remittance Law of 1962, as amended. There is no limitation on the repatriation of capital; reinvestment of profits is considered an increase of the original capital for the purposes of the law.

Exploration and drilling for oil constitute a federal monopoly, but foreign companies are permitted to search for oil under risk contract with Petrobrás. Newspapers, television, and radio are closed to foreign ownership, and hydroelectric energy, coastal shipping, financial institutions, and rural real estate are subjected to restrictions.

Although Brazil welcomes foreign direct investment, many of its regulations are often restrictive and discriminatory. For example: foreign investment is prohibited in several sectors, including petroleum production and refining, public utilities, media, real estate, shipping, and various strategic industries. Brazil also imposes certain performance requirements.

Many of the problems and issues that relate to foreign investment stem from the government's interest in protecting Brazil's existing industries and controlling unemployment. Sectorial accords emerge with disturbing elements of restrictive equity participation, local content requirements, and incentives which are linked to export performance targets.

The Brazilian Congress was expected at the end of 1993 to make a full scale review of the 1988 Constitution. Among the issues to be considered are possible modifications to the Constitution affecting foreign investment, including revisions of the state monopolies in the petroleum and telecommunications sectors, as well as a lifting of the restriction on foreign participation in the mining sector. The Constitution also imposes prohibitions on foreign capital participation in land, river, coastal, maritime, and internal air transportation as well as foreign ownership of television, radio, and print media. In addition, foreigners can not own land in certain coastal zones and other areas on national security grounds.

Prohibitions on remittances for royalty and technical service payments between related parties were removed under the 1992 tax code. In addition, the base tax rate on profits and royalty remittances has been reduced from 25% to 15%.

The informatics industry has been protected for 17 years. In spite of drastic reductions in the protection rate, significant barriers which discriminate against foreign products exist. Local computer firms are also seeking to maintain a tax advantage over imported products.

39ECONOMIC DEVELOPMENT

Economic policy since the late 1960s has had three prime objectives: control of inflation, gradual improvement of the welfare of the poorest sector, and a high economic growth rate. Generally, under the stewardship of finance ministers Roberto de Oliveira

Campos and, subsequently, Antonio Delfim Netto (who became minister of planning in 1979), Brazilian policy has sought to prevent inflation from eroding economic growth by a process of "monetary correction"—that is, by the legal revaluation of "fixed" assets, such as real estate, debits in arrears, and the face value of bonds, to reflect inflation. This technique, which requires extensive government control over the economy, is intended to prevent inflation from distorting the relative values of various types of holdings. It is also inflationary, however.

The central stress of the 1975–79 development plan was on economic growth. Economic infrastructure (energy, transportation, and communications) received top priority, with a 25% share of the total investment. A special development plan, known as Palomazonia (the Program for Agriculture, Cattle Raising, and Agrominerals for the Amazon), concentrated on expansion of agriculture, forestry, mineral exploitation, and hydroelectric power in the region. The third national plan (1980–85) placed the greatest emphasis on agricultural development, energy, and social policies. Its main aim was improvement of the public welfare through continued economic growth and more equitable income distribution.

The policy of the new civilian government, couched in the First National Plan of the New Republic (1986–89), sought to maintain high levels of economic growth, introduce a wide range of basic institutional and fiscal reform in the public sector, and reduce poverty significantly. When inflation continued to mount, however, the "Cruzado Plan" was introduced; it froze wages and prices for a year, and introduced a new unit of currency, the cruzado. While inflation did drop dramatically, the ensuing consumer spending boom, caused by the desire to take advantage of the price freeze, rekindled inflation. By 1987, Brazil had reverted to orthodox austerity and "monetary correction" in an attempt to bring the economy under control.

Since the early 1960s, the government has offered special incentives to agricultural and industrial enterprises that further the development of the northeastern and northern regions of the country. The development of these areas is under the supervision of the Development Superintendency of the Northeast, whose activities cover the states of Maranhão, Piauí, Ceará, Rio Grande do Norte, Paraíba, Pernambuco, Alagoas, Sergipe, Bahia, the zone of Minas Gerais located within the "Drought Polygon," and the territory of Fernando de Noronha. The Superintendency for the Development of Amazonia, the Superintendency for the Development of the West-Central Region, and the Superintendency for the Development of the Southern Region are the other regional development agencies.

The indecisiveness of President Itamar Franco in terms of a comprehensive economic package to combat the ever growing inflationary problem has significantly eroded his credibility. The nationalistic bend of his policies raised serious concerns about the sustainability of market reforms initiated by his predecessor. With presidential elections slated for November 1994, the scenario raises the possibility of military intervention. The imminence of an economic crisis and the political uncertainties regarding upcoming elections are hindering investment decisions. Among the major candidates are Lula, a populist against market economy reforms, and Mr. Cardoso, the former finance minister and the architect of a new Brazilian currency.

40SOCIAL DEVELOPMENT

The Organic Social Security Law of Brazil, passed during the Vargas reform years of the 1930s, covered only some 4 million urban workers by the 1960s, including metallurgical, textile, and other industrial workers and commercial, bank, and store clerks. In 1976, social security laws were consolidated, and in the following year, the National System of Social Security and Welfare was established. Benefits include modest insurance against

accidents; old age, invalids', and survivors' pensions; funeral insurance; and medical, dental, and hospital coverage. Maternity benefits were introduced under the new Constitution in 1988. However, as a reaction against the maternity leave law, some employers require prospective female employees to present sterilization certificates or try to avoid hiring women of childbearing age. The social security system is financed by contributions from the employer, the worker, and the government. The maximum rates in 1991 were 8–10% of the monthly salary for employees and 13% of payrolls for employers. Under the compulsory checkoff (contribuição sindical), the federal government deducts one day's pay per year from the wages of most urban workers. The law specifies that these funds may be spent for a wide variety of health, educational, and welfare services, as well as for trade union activities. In addition, a special fund established in 1971 is available for employees' retirement, marriage, home purchase, or other special occasions; the fund is financed through a turnover tax on employers.

The government has a long-standing pronatalist attitude, but it is generally supportive of family planning. The private Brazilian Civil Society for Family Welfare provides family planning services. The fertility rate for 1985–90 was 3.2, and the contraceptive prevalence rate was 66%. Abortion is permitted for limited medical reasons only.

41HEALTH

A health and welfare program, Prevsaúde, introduced by the government in 1981, was intended to double health services by 1987. In 1993, however, Brazil's national health care system came to an end, chiefly due to extensive fraudulent activity by hospitals, physicians, and state and municipal health secretariats. The new Brazilian Minister of Health planned to implement a new system and supported legislation to increase funds for the public health sector.

Health and sanitary conditions vary widely from region to region. The large cities have competent physicians, generally with advanced training abroad, but there is a dearth of doctors, hospitals, and nurses in most towns in the interior. Between 1988 and 1992, there were 1.46 physicians per 1,000 people. Between 1985 and 1990, there were 3.5 beds per 1,000 people (a slight decline since 1988).

Between 1988 and 1991, 87% of the population had access to safe drinking water. In 1990, there were 56 cases of tuberculosis per 100,000 inhabitants. Between 1990 and 1992, children one-year-old and under were immunized as follows: tuberculosis (87%); diphtheria, pertussis, and tetanus (69%); polio (62%); and measles (93%).

The infant mortality rate is high, but it has declined over the last 10 years. In 1992, the infant mortality was 54 per 1,000 live births, with 3,626,000 births that same year. Between 1980 and 1993, data shows that 66% of married women (between the ages of 15 and 49 years) used contraception. The 1993 death rate was 7 per 1,000 people, while the average life expectancy in 1992 was 66 years.

42HOUSING

Despite major urban developments, both the housing supply and living conditions in Brazil remain inadequate. Large, sprawling slums are endemic in the large cities, while most rural dwellers live without amenities such as piped water and electricity. In 1992, there were 32,705,000 residences. As of 1984, 82% of all housing units were detached houses of brick, stone, wood or concrete, 8% were apartments, 8% were rural dwellings of wood or clay, and 2% were semi-private units called "quartes". Roughly 63% of all dwellings were owner occupied, 22% were rented, and 14% were issued by employers. Some 30% had sewage facilities and 17% had septic tanks; 80% of all homes had electric lighting.

In 1964, the federal government enacted the National Housing Act and suspended rent controls, with the stipulation that rents could be brought in line with private market levels. The law provided for the establishment of the National Housing Bank (Banco Nacional de Habitação, or BNH), whose main purposes are to stimulate savings to finance home construction through lending institutions, to coordinate the activities of both the public and private sectors, and to introduce financial incentives. The BNH can raise funds through bond issues and may also receive deposits from governmental agencies, public cooperatives, and mixed companies.

Under the National Sanitation Plan, announced in 1971, the government aimed, with the help of us$772 million provided by the IBRD, to provide 90% of urban dwellings with safe water and 65% with sewerage facilities by the end of the 1980s.

43EDUCATION
Education in colonial times was carried on first by the Jesuits and then by a few royal schools. Brazil's public school system, always weak, was under the Ministry of Justice and Interior until 1930, when the Ministry of Education and Health was created by Vargas. Responsibility for public education, as defined by the 1946 constitution and the 1961 directives and standards for national education, is divided between the federal, state, and municipal governments. Public elementary and secondary instruction is almost exclusively a function of the municipalities and states, while higher education is the responsibility of the federal Ministry of Education. Public education is free at all levels and non-profit private schools also receive public funding. The federal government has been active, however, on all three levels through the Federal Council of Education, established in 1961 to coordinate the implementation of the 1961 directives and to advise the Ministry of Education.

The 1961 directives required the federal government to contribute at least 12% of its tax revenues to education, and state and municipal governments were required to contribute a minimum of 20% of their tax revenues for this purpose. The first National Plan of Education, formulated in 1962, called for the extension of compulsory elementary education to five and, eventually, six years and, by 1980, eight years of schooling was required. The 1988 Brazilian constitution allocates 25% of state and local tax revenues to education. In 1964 there were 10 million students attending school at all levels. Since then, great strides have been made in the Brazilian system of education. In 1990 there were 37.6 million students; 3.9 million in preschool; 28.2 million at the elementary level; 3.8 million at the secondary level; and 1.7 million are at the university.

In 1990 there were 93 universities, including the Federal University of Rio de Janeiro (founded 1920) and the universities of Minas Gerais (1927), São Paulo (1934), Rio Grande do Sul (1934), Bahia (1946), Recife (1946), Paraná (1946), and Brasília (1961). The federal government maintains at least one federal university in each state. Entrance to a college or university is through an examination called "vestibular"; students may either earn a "bacharel" degree or, with an additional year spent in teacher training, obtain a "licenciado" degree.

Adult education campaigns have functioned sporadically since 1933, backed by the federal government with some assistance from social, fraternal, Catholic, Protestant, professional, and commercial organizations. Although millions of Brazilians have received literacy training, adult illiteracy in 1990 was still 18.9% (males: 17.5% and females: 20.2%), according to estimates by UNESCO.

44LIBRARIES AND MUSEUMS
Of Brazil's many hundreds of public libraries, the largest municipal library system is that of São Paulo, with over one million volumes. The National Library in Rio de Janeiro (founded in 1810) houses more than eight million documents and many rare manuscripts. Included in the collection is 60,000 volumes brought to Brazil by the Royal Family of Portugal in 1808. Various government ministries maintain separate libraries in Brasília. The largest academic library is the São Paulo University Integrated Library System with over 1.3 million volumes.

Brazil has nearly 200 museums. The National Museum (founded in 1818), one of the most important scientific establishments in South America, is especially known for its Brazilian ethnographic collections. Other important museums in Rio de Janeiro include the National Museum of Fine Arts, the Museum of Modern Art, the Historical Museum, the Museum of Mineralogy, the Museum of the Bank of Brazil, the Getúlio Vargas Gallery, and the Indian Museum. The most important museums in São Paulo are the Paulista Museum and the São Paulo Museum of Art. The Goeldi Museum in Belém is famous for exhibits covering every phase of Amazonian life and history.

45MEDIA
The principal telegraph network is operated by the Brazilian Postal and Telegraph Administration, in which the government holds part ownership. National trunk routes and international connections are also operated directly by another mixed corporation, the Brazilian Telecommunications Corp. (EMBRATEL), which inaugurated an earth satellite station in 1969 linking the Brazilian network with member countries of INTELSAT. EMBRATEL has rapidly modernized and extended the domestic telecommunications system with the introduction of microwave networks, including long-distance direct dialing, throughout much of the country. In the Amazon region, the company relies on a tropodiffusion system because of the area's large empty spaces. In 1991 there were 14,059,524 telephones in use, as compared with 2.4 million in 1973. In 1992, Brazil had 1,223 AM radio stations and 112 commercial television stations; some 58,500,000 radios and 31,400,000 television sets were in use in 1991.

Freedom of the press is guaranteed under the constitution of 1967, and no license is required for the publication of books, newspapers, and periodicals. However, under the newspaper code of 19 September 1972, newspapers were forbidden to publish "speculative" articles on politics or unfavorable reports on the economy. The interests of Brazilian journalists are defended by the Inter-American Press Association and the influential Brazilian Press Association. Newspaper staffs have traditionally enjoyed privileged status and exemption from income taxes; nevertheless, in the early 1970s, reporters were sometimes subject to arrest and alleged brutalities at the hands of the government. In 1979, practically all restrictions on the press were lifted.

The following table shows Brazil's leading dailies, with their political orientations and estimated circulations in 1986:

	ORIENTATION	CIRCULATION
RIO DE JANEIRO		
O Globo	Conservative	539,000
O Dia	Labor	425,000
Jornal do Brasil	Conservative	216,000
SAO PAULO		
O Estado de São Paulo	Conservative	424,000
Folha de São Paulo	Independent	426,000
Jornal da Tarde	Independent	150,000
Gazeta Mercantie	NA	75,000
Folha da Tarde	Conservative	59,000
BELO HORIZONTE		
Estado de Minas	Independent	147,000
ORTO ALEGRE		
Zero Hora	NA	228,000

	ORIENTATION	CIRCULATION
RECIFE		
Diário de Pernambuco	Independent	85,000
BRASÍLIA		
Correio Brasiliense	NA	75,000
Jornalde Brasília	NA	35,000

The largest Brazilian-owned magazine, which competes with the Portuguese-language edition of the *Reader's Digest*, called *Seleções*, is the popular illustrated *Manchete* of Rio de Janeiro (1991 circulation 130,000).

46ORGANIZATIONS

Owners of large farms and plantations, particularly coffee plantation owners, usually belong to one or more agricultural associations. The largest of the national agricultural organizations is the National Confederation of Agriculture, but, in general, local member groups of the federation, such as the Paraná Coffee Producers' Association, the Association of Coffee Farmers, and the Brazilian Rural Society, are more powerful than the national organization. Agricultural societies are organized for the primary purpose of promoting favorable legislation toward agriculture in the Congress, and they have become important political units over the years. Other agricultural groups include cattlemen's associations, dairymen's associations, and rice growers' and grain producers' organizations, usually organized on a statewide basis. These are strongest in São Paulo, Minas Gerais, Paraná, Goiás, and Rio Grande do Sul. Chambers of commerce function in every state.

Most international service, social, and fraternal organizations are represented in Brazil. The Rotary International is well organized in the industrial cities of the south, the Boy Scouts of Brazil are active nationwide, and there are Lions clubs and societies of Freemasons. The Catholic Church and the growing number of Protestants maintain various organizations. Catholic Action and the Catholic hierarchy have actively addressed themselves to combating misery and disease, especially in the big-city slums and in the northeast.

47TOURISM, TRAVEL, AND RECREATION

Rio de Janeiro is one of the leading tourist meccas in South America. Notable sights include Sugar Loaf Mountain, with its cable car; the Corcovado, with its statue of Christ the Redeemer; Copacabana Beach, with its mosaic sidewalks; and the Botanical Gardens. Large numbers of visitors are also drawn to the churches of Bahia; the historic city of Ouro Preto in Minas Gerais; and the colorful Amazon Valley cities of Belém and Manaus. Brazil is also famous for its colorful celebrations of Carnival, especially in Rio de Janeiro; neighborhood samba groups rehearse all year for this occasion. Brazil's African heritage can best be savored at the Carnival in Salvador. Eco-tourism attracts growing numbers of visitors to the world's largest rain forest in the North, the Iguacu Falls in the South, and the Mato Grosso wetlands in the Central West region. The United Nations Conference on Environment and Development—the Earth Summit—was held in Rio in June 1992.

No visa is required for tourists from most Western European or Latin American countries entering Brazil, but a valid passport, a tourist card issued by a transportation company, and a round-trip ticket or transportation to another country are required. Citizens of the US and most other countries must secure a visa in advance. A total of 1,352,000 tourists visited Brazil in 1991, 736,000 from South America and 163,000 from the US. Estimated tourism revenues were US$1.5 billion. As of 1990, there were 137,000 hotel rooms and 274,158 beds with an occupancy rate of 42.8%.

Soccer is by far the most popular sport. Brazil hosted the World Cup competition in 1950, and Brazilian teams won the championship in 1958, 1962, 1970, and 1994. The Maracanã soccer stadium in Rio de Janeiro seats more than 180,000 spectators. Other favorite recreations include water sports, basketball, tennis, and boxing.

48FAMOUS BRAZILIANS

Joaquim José da Silva Xavier, also known as Tiradentes (d.1792), led an unsuccessful uprising in 1789 against Portuguese colonial rule. The patriarch of Brazilian independence was José Bonifacio de Andrada e Silva (1763–1838), a geologist, writer, and statesman. Pedro I (Antonio Pedro de Alcántara Bourbon, 1798–1834), of the Portuguese royal house of Bragança, declared Brazil independent and had himself crowned emperor in 1822; he became King Pedro IV of Portugal in 1826 but gave up the throne to his daughter, Maria da Gloria. His Brazilian-born son, Pedro II (Pedro de Alcántara, 1825–91), emperor from 1840 to 1889, consolidated national unity and won respect as a diplomat, statesman, and patron of the arts and sciences. Other famous Brazilians during the imperial period include the Brazilian national hero, Luís Alves de Lima e Silva, Duque de Caxias (1803–80), a patron of the Brazilian army; and Joaquim Marques Lisboa, Marques de Tamandaré (1807–97), a naval hero, soldier, and statesman. In the field of international politics, Joaquim Nabuco (1849–1910) won distinction as a diplomat, journalist, and champion of the abolition of slavery; José Maria de Silva Paranhos, Barão de Rio Branco (1847–1912), was a famous minister of foreign affairs, who represented Brazil at many international conferences; and Ruy Barbosa (1849–1923) was a lawyer, diplomat, statesman, and jurist. A leader of industrial and economic development was Irineu Evangelista de Souza, Barão de Mauá (1813–89). Brazilian aviation pioneer Alberto Santos Dumont (1873–1932) is called the father of flight for his invention of a gasoline-powered airship in 1901. Oswaldo Cruz (1872–1917) founded the Brazilian Public Health Service and helped eradicate yellow fever in Rio de Janeiro. Marshal Cândido Rondon (1865–1957), an explorer of Amazonia, organized the Brazilian Indian Bureau. Dr. Vital Brasil (1865–1950) developed São Paulo's snake-bite serum institute at Butantã.

Joaquim Maria Machado de Assis (1839–1908), author of *Memórias Póstumas de Braz Cubas* and other novels and poems, is generally considered the greatest Brazilian literary figure. The poet Euclides da Cunha (1866–1909) wrote *Os Sertões* (1902), one of the foremost works by a Brazilian. Other literary figures include Antônio Gonçalves Dias (1824–64), a romantic poet who idealized the Brazilian Indian; Castro Alves (1847–71), who influenced the abolition of slavery; and contemporary writers such as Gilberto de Mello Freyre (1900–1987), José Lins do Rego (1900–1959), Erico Verissimo (1905–1975), and Jorge Amado (b.1912).

Aleijadinho (Antônio Francisco Lisboa, 1739–1814) was an 18th-century church architect and carver of soapstone religious statues in Minas Gerais. Contemporary artists include the painter Emiliano di Cavalcanti (1897–1976); the painter and muralist Cândido Portinari (1903–62), considered the greatest artist Brazil has produced; and the sculptor Bruno Georgio (b.1905). Lúcio Costa (b. France, 1902–85), regarded as the founder of modern Brazilian architecture, designed the new capital city of Brasília, and Oscar Niemeyer (b.1907) designed most of the government buildings. Robert Burle Marx (b.1909) has originated an unusual form of landscaping to complement modern architectural form. Another Brazilian architect of note, Alfonso Eduardo Reidy (b. France, 1909), designed the Museum of Modern Art in Rio de Janeiro.

The greatest figure in Brazilian music is the composer and educator Heitor Villa-Lobos (1887–1959), who wrote prolifically in many styles and forms. Other musicians include the composers Carlos Gomes (1836–96), Oscar Lorenzo Fernandez (1897–1948), Francisco Mignone (1897–1987), and Carmargo Guarnieri (b.1907); the concert pianist Guiomar Novaës Pinto (1895–1979); the operatic soprano Bidu Sayao (Balduina de Oliveira Sayao, b.1902); and the folklorist and soprano Elsie Houston (1900–1943). One of the best-known Brazilians is soccer star Edson Arantes do Nascimento (b.1940), better known as Pelé.

Other noted figures are Getúlio Vargas (1883–1954), president-dictator in the period 1930–45, who increased the power of the central government; Francisco de Assis Chateaubriand Bandeira de Melo (1891–1966), a publisher, diplomat, and art collector; Oswaldo Aranha (1894–1960), president of the UN General Assembly during 1947–49; and Marcelino Candau (1911–83), director-general of WHO during 1953–73. Gen. Ernesto Geisel (b.1907) and his presidential successor, Gen. João Baptista de Oliveira Figueiredo (b.1918), guided Brazil through a period of political liberalization.

49DEPENDENCIES

Brazil has no colonies, but three national territories are contiguous with or incorporated within the national domain. The constitution provides for the creation of a territory from part of an established state, the incorporation of a territory into an established state, or the organization of a territory into a new state if that territory can demonstrate its ability to meet the requirements of statehood.

The territories are Amapá, on the French Guiana border; Roraima (formerly Rio Branco), on the Venezuelan border; and the Fernando de Noronha islands, off the northeastern coast. Territorial governors are appointed by the president of the republic, and each territory has one representation in the federal Chamber of Deputies. Territories have no representation in the federal Senate. Trindade and the Ilhas Martin Vaz, small islands in the Atlantic, also belong to Brazil.

50BIBLIOGRAPHY

Baaklini, Abdo I. *The Brazilian Legislature and Political System.* Westport, Conn.: Greenwood Press, 1992.

Bacha, Edmar L. and Herbert S. (eds.). *Social Change in Brazil, 1945–1985: the Incomplete Transition.* Albuquerque: University of New Mexico Press, 1989.

Becker, Bertha K. *Brazil: A New Regional Power in the World-economy.* New York: Cambridge University Press, 1992.

Burns, E. Bradford. *A History of Brazil.* 3rd ed. New York: Columbia University Press, 1993.

Chacel, Julian, et al. *Brazil's Economic and Political Future.* Boulder, Colo.: Westview, 1985.

Erickson, Kenneth P. *The Brazilian Corporative State and Working Class Politics.* Berkeley: University of California Press, 1978.

Fritsch, Winston. *External Constraints on Economic Policy in Brazil, 1889–1930.* Pittsburgh, Pa.: University of Pittsburgh Press, 1988.

Graham, Richard (ed.). *Brazil and the World System.* Austin: University of Texas Press, 1991.

Hartness, Ann. *Brazil in Reference Books, 1965–1989: An Annotated Bibliography.* Metuchen, N.J.: Scarecrow Press, 1991.

Haverstock, Nathan A. *Brazil in Pictures.* Minneapolis: Lerner, 1987.

Kinzo, Maria D'Alva G. (ed.). *Brazil, the Challenges of the 1990s.* New York: St. Martin's Press, 1993.

Levine, Robert M. *Historical Dictionary of Brazil.* Metuchen, N.J.: Scarecrow, 1979.

———. *The Vargas Regime.* New York: Columbia University Press, 1970.

Mahar, Dennis J. *Government Policies and Deforestation in Brazil's Amazon Region.* Washington, D.C.: World Bank, 1989.

Parkin, Vincent. *Chronic Inflation in an Industrialising Economy: the Brazilian Experience.* New York: Cambridge University Press, 1991.

Payne, Leigh A. *Brazilian Industrialists and Democratic Change.* Baltimore: Johns Hopkins University Press, 1993.

Policymaking in a Newly Industrialized Nation: Foreign and Domestic Policy Issues in Brazil. Austin: University of Texas at Austin, 1988.

Poppino, Rollie E. *Brazil: The Land and People.* New York: Oxford University Press, 1973.

Roett, Riordan. *Brazil: Politics in a Patrimonial Society.* Westport, Conn.: Praeger, 1992.

Selcher, Wayne A. *Brazil's Multilateral Relations Between First and Third Worlds.* Boulder, Colo.: Westview Press, 1978.

Smith, Peter Seaborn. *Oil and Politics in Modern Brazil.* Toronto: Macmillan of Canada, 1976.

CANADA

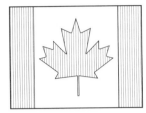

CAPITAL: Ottawa.

FLAG: The national flag, adopted in 1964, consists of a red maple leaf on a white field, flanked by a red vertical field on each end.

ANTHEM: Since 1 July 1980, *O Canada* has been the official anthem.

MONETARY UNIT: The Canadian dollar (c$) is a paper currency of 100 cents. There are coins of 1, 5, 10, 25, and 50 cents and 1 dollar, and notes of 2, 5, 10, 20, 50, 100, and 1,000 Canadian dollars. Silver coins of 5 and 10 dollars, commemorating the Olympics, were issued during 1973–76. c$1 = us$0.7227 (or us$1 = c$1.3837). US currency is usually accepted, especially in major cities and along the border.

WEIGHTS AND MEASURES: The metric system is the legal standard.

HOLIDAYS: New Year's Day, 1 January; Good Friday; Easter Monday; Victoria Day, the Monday preceding 25 May; Canada Day, 1 July; Labor Day, 1st Monday in September; Thanksgiving Day, 2d Monday in October; Remembrance Day, 11 November; Christmas Day, 25 December; Boxing Day, 26 December. Other holidays are observed in some provinces.

TIME: Newfoundland, 8:30 AM = noon GMT; New Brunswick, Nova Scotia, Prince Edward Island, and Quebec, 8 AM = noon GMT; Ontario east of 90° and western Quebec, 7 AM = noon GMT; western Ontario and Manitoba, 6 AM = noon GMT; Alberta and Saskatchewan, 5 AM = noon GMT; British Columbia and Yukon, 4 AM = noon GMT.

¹LOCATION, SIZE, AND EXTENT

Canada consists of all of the North American continent north of the US except Alaska and the small French islands of St. Pierre and Miquelon. Its total land area of 9,976,140 sq km (3,851,808 sq mi) makes it the second-largest country in the world (slightly larger than China and the United States), extending 5,187 km (3,223 mi) E–W from Cape Spear, Newfoundland, to Mt. St. Elias in the Yukon Territory and 4,627 km (2,875 mi) N–S from Cape Columbia on Ellesmere Island to Pelee Island in Lake Erie. Canada is bounded on the N by the Arctic Ocean, on the E by Kennedy Channel, Nares Strait, Baffin Bay, Davis Strait, and the Atlantic Ocean, on the S by the US, and on the W by the Pacific Ocean and the state of Alaska. The coastal waters of Canada also include the Hudson Strait and Hudson Bay. The country's total boundary length is 252,684 km (157,602 mi). Canada's capital city, Ottawa, is located in the southeastern part of the country.

²TOPOGRAPHY

Canada's topography is dominated by the Canadian Shield, an ice-scoured area of Precambrian rocks surrounding Hudson Bay and covering half the country. This vast region, with its store of forests, waterpower, and mineral resources, is being increasingly developed. East of the Shield is the Maritime area, separated from the rest of Canada by low mountain ranges pierced by plains and river valleys, and including the island of Newfoundland and Prince Edward Island. South and southeast of the Shield are the Great Lakes–St. Lawrence lowlands, a fertile plain in the triangle bounded by the St. Lawrence River, Lake Ontario, and Georgian Bay. West of the Shield are the farmlands and ranching areas of the great central plains, some 1,300 km (800 mi) wide along the US border and tapering to about 160 km (100 mi) at the mouth of the Mackenzie River. Toward the north of this section is a series of rich mining areas, and still farther north is the

Mackenzie lowland, traversed by many lakes and rivers. The westernmost region of Canada, extending from western Alberta to the Pacific Ocean, includes the Rocky Mountains, a plateau region, the coastal mountain range, and an inner sea passage separating the outer island groups from the fjord-lined coast. Mt. Logan, the highest peak in Canada, in the St. Elias Range near the Alaska border, is 5,951 m (19,524 ft) high. The Arctic islands constitute a large group extending north of the Canadian mainland to within 885 km (550 mi) of the North Pole. They vary greatly in size and topography, with mountains, plateaus, fjords, and low coastal plains.

The central Canadian Shield area is drained by the Nelson-Saskatchewan, Churchill, Severn, and Albany rivers flowing into Hudson Bay. The 4,241-km (2,635-mi) Mackenzie River—with its tributaries and three large lakes (Great Bear, Great Slave, and Athabasca)—drains an area of almost 2.6 million sq km (1 million sq mi) into the Arctic Ocean. The Columbia, Fraser, and Yukon rivers are the principal drainage systems of British Columbia and the Yukon Territory. The Great Lakes drain into the broad St. Lawrence River, which flows into the Gulf of St. Lawrence. Other rivers flow laterally from the interior into Hudson Bay or the Atlantic or Pacific ocean.

³CLIMATE

Most of northern Canada has subarctic or arctic climates, with long cold winters lasting 8 to 11 months, short sunny summers, and little precipitation. In contrast, the populated south has a variety of climatological landscapes. The greatest temperature range is in the Northwest Territories, where the average temperature at Fort Good Hope ranges from –31°C (–24°F) in January to 16°C (61°F) in July.

Cool summers and mild winters prevail only along the Pacific coast of British Columbia. There the mean temperatures range from about 4°C (39°F) in January to 16°C (61°F) in July, the least range in the country. On the prairies there are extreme differences in temperature between day and night and summer and winter. In Ontario and Quebec, especially near the Great Lakes and along the St. Lawrence River, the climate is less severe than in western Canada. This region has abundant precipitation that is highly uniform from season to season. The growing season is short, even in the south. Much of the interior plains area does not get enough rain for diversified crops.

East of the Rockies across the flat prairie lies the meeting ground for air from the Arctic, Pacific, and American interior. The mixing of air masses leads to a turbulent atmosphere and the emergence of cyclonic storms, producing most of the rain and snow in the country. The northwest and the prairies, having fewer or weaker storms, are the driest areas, although the prairies are the site of some heavy blizzards and dramatic thunderstorms. The windward mountain slopes are exceptionally wet; the protected slopes are very dry. Thus, the west coast gets about 150–300 cm (60–120 in) of rain annually; the central prairie area, less than 50 cm (20 in); the flat area east of Winnipeg, 50–100 cm (20–40 in); and the Maritime provinces, 115–150 cm (45–60 in). The annual average number of days of precipitation ranges from 252 along coastal British Columbia to 100 in the interior of the province.

4FLORA AND FAUNA

A great range of plant and animal life characterizes the vast area of Canada, with its varied geographic and climatic zones. The flora of the Great Lakes–St. Lawrence region resembles that of the adjacent US section, with white pine, hemlock, sugar and red maples, yellow birch, and beech trees. Coniferous trees—particularly red spruce—predominate in the Maritime region, black spruce in the eastern Laurentian zone, white spruce in the western. In the east are also found the balsam fir, white cedar, tamarack, white birch, and aspen, with jack pine in the drier areas. From the prairie grassland to the Arctic tundra there are aspen, bur oak, balm of Gilead, cottonwood, balsam poplar, white birch, and other deciduous trees. Conifers dominate the northern section. Many types of grasses grow on the interior plains. The wet area along the west coast is famous for its tall, hard conifers: western hemlock and red cedar, Douglas fir, Sitka spruce, and western white pine. Subalpine forests cover the Rocky Mountain area, and here are found such conifers as alpine fir, Engelmann spruce, lodgepole pine and aspen, and mountain hemlock. The great Arctic region is covered with low-growing grasses, mosses, and bushes.

The fauna of the Great Lakes–St. Lawrence region includes deer, black bear, opossum, gray and red squirrel, otter, beaver, and skunk; birds include eastern bluebird, red-winged blackbird, robin, wood thrush, woodpecker, oriole, bobolink, crow, hawk, bittern, heron, black duck, and loon. In the boreal forest area there are moose, caribou, black bear, lynx, timber wolf, marten, beaver, porcupine, snowshoe rabbit, red squirrel, and chipmunk. Typical mammals of the Rocky Mountain area are grizzly bear, mountain goat, moose, wapiti, cougar, and alpine flying squirrel. In the plains are rabbits, gophers, prairie birds, and waterfowl. Abundant on the west coast are deer, Cascade mountain goat, red squirrel, mountain beaver, various species of mice, and Puget striped skunk; common birds include northern Pigmy-owl, band-tailed pigeon, black swift, northern flicker, crow, rufous-sided towhee, and black brant. Over the stretches of the Arctic are the musk ox and reindeer, polar bear, caribou, white and blue fox, arctic hare, and lemming, as well as the snowy owl, ptarmigan, snow bunting, arctic tern, and other birds. Walrus, seals, and whales inhabit Canada's coastal waters.

5ENVIRONMENT

Canada's principal environmental agency is the Department of the Environment, established in 1971 and reorganized in 1979.

Responsibilities of this department, also known as Environment Canada, include air and water pollution control, land-use planning, and wildlife preservation. Responsibility for maritime resources was vested in the Department of Fisheries and Oceans under the 1979 reorganization. Among Canada's most pressing environmental problems in the mid-1980s was acid rain, dilute solutions of nitric and sulfuric acids created when pollutants mix with water vapor in the atmosphere. Acid rain poses a threat to natural resources in an area of eastern Canada of about 2.6 million sq km (1 million sq mi). Canadian sources estimate that about 14,000 lakes in eastern Canada are acidified and another 300,000 lakes will remain in danger if adequate emission reductions are not implemented. As of 1994, acid rain has affected 150,000 lakes in total throughout Canada. Waterfowl populations have already been depleted. About half the acid rain comes from emissions from Canadian smokestacks, but Canada blamed US industry for 75% of the Ontario pollution, as the two nations sought to negotiate a solution to the problem.

Canada is trying to protect its plant resources through the national Green Plan. Out of a total of 3,220 plant species nationwide, 13 are endangered. Canada's rivers suffer from the presence of toxic pollutants from agricultural and industrial sources. As of 1994, 50% of Canada's coastal shellfish areas are closed because of the dangerous levels of pollutants. In 1994, Canada ranks 12th in the world for hydrocarbon emissions with a total of 2,486.1 metric tons.

Canada has more than 90 bird sanctuaries and 44 National Wildlife Areas, including reserves in the western Arctic to protect waterfowl nesting grounds. In May 1986, Canada and the US signed an agreement to restore the breeding habitat of mallard and pintail ducks in the midcontinental regions of both countries. The project, which would span 15 years and cost c$1.5 billion, would protect and improve 1,200,000 hectares (3,000,000 acres) of duck habitat. It was hoped that the project would reverse the decline in waterfowl populations, and raise the average annual fall migration to 100 million birds—the level of the 1970s. The project also called for the protection of waterfowl habitats in the lower Mississippi River and Gulf Coast region, and the black duck habitat in eastern Canada and the East Coast of the US.

The annual Newfoundland seal hunt, producing seals for pelts and meat, drew the ire of environmentalists chiefly because of the practice of clubbing baby seals to death (adult seals are shot). Approval by the European Parliament of a voluntary boycott on sealskin imports undercut the market, and the Newfoundland seal catch dropped from about 1,400 in 1981/82 to 360 in 1982/83. In 1987, Canada banned the offshore hunting of baby seals, as well as blueback hooded seals.

As of 1987, endangered species in Canada included the Vancouver Island marmot, eastern puma, wood bison, sea otter, right whale, St. Lawrence beluga, Acadian whitefish, mountain plover, piping plover, spotted owl, leatherback turtle, cucumber tree, Furbish's lousewort, Eskimo curlew, Kirtlands warbler, American peregrine falcon, whooping crane, and the southern bald eagle. In 1991, the brown bear, the black-footed ferret, the gray wolf, the eastern timber wolf, the right whale and the California Condor were also endangered. In a total of 197 mammals, five are on the endangered species list. The list also includes six species of birds which are endangered as of 1994.

6POPULATION

The total population according to the census of 1991 was 27,296,859, an increase of 12.1% since the 1981 census. In late 1993, the population was estimated at 28,500,000. Of the 1991 population, 13,656,370 (51%) were female. The population growth rate declined from 12.9% for 1971–81 to 12.1% for 1981–91. The highest growth rates for 1981–91 were in the Northwest Territories (26%), British Columbia (19.6%), and

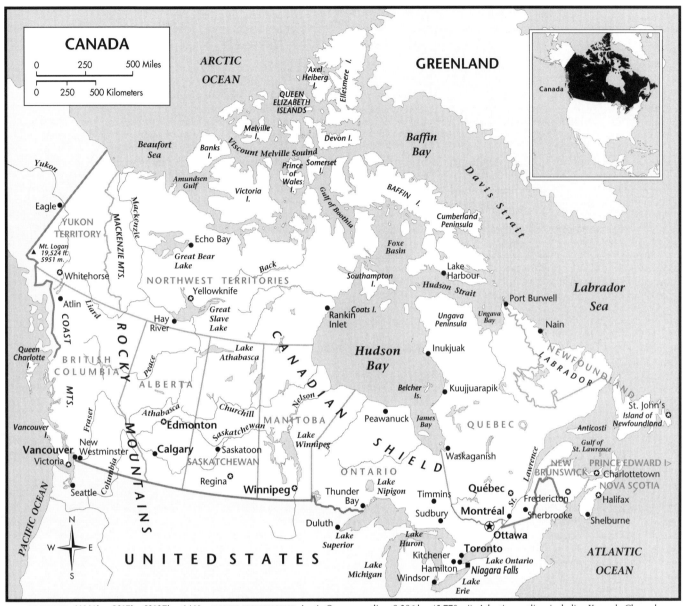

CANADA

ARCTIC OCEAN

GREENLAND

LOCATION: 41°41′ to 83°7′N; 52°37′ to 141°W. **BOUNDARY LENGTHS:** Arctic Ocean coastline, 9,286 km (5,770 mi); Atlantic coastline, including Kennedy Channel, Nares Strait, Baffin Bay, Davis Strait, 9,833 km (6,110 mi); US, 6,416 km (3,987 mi); Pacific coastline, 2,543 km (1,580 mi); Alaska, 2,477 km (1,539 mi); Hudson Strait and Hudson Bay shoreline, 7,081 km (4,400 mi). **TERRITORIAL SEA LIMIT:** 12 mi.

Ontario, 16.9%. In 1991, Canada's crude birthrate was 14.9 per 1,000 population, the death rate was 7.2, and the net natural increase 7.7. A population of 30,425,000 was forecast for the year 2000.

The average population density in 1991 was 3 per sq km (7.7 per sq mi). The population is unevenly distributed, ranging from 0.02 per sq km (0.045 per sq mi) in the Northwest Territories to 22.8 per sq km (59 per sq mi) on Prince Edward Island. Nearly two-thirds of the people live within 160 km (100 mi) of the US boundary. All except the Maritime provinces have large areas virtually uninhabited. The population movement has long been from rural to urban areas.

Some 76.6% of all Canadians lived in urban communities at the time of the 1991 census, and nearly 3% lived on farms. Ontario, with 81.8% of its population classed as urban, was the most urbanized province, followed by British Columbia (80.4%), Alberta (79.8%), and Quebec (77.6%). Only New Brunswick (47.7%), Prince Edward Island (39.9%) and the Northwest Territories (36.7%) have less than half their population in urban areas.

The Toronto metropolitan area had a population of 3,893,046 in 1991; Montreal, 3,127,242. Other large metropolitan areas are Vancouver, 1,602,502; Ottawa-Hull (Ottawa is the federal capital), 920,857; Edmonton, 839,924; Calgary, 754,033; Winnipeg, 652,354; Quebec, 645,550; Hamilton, 599,760; St. Catherines-Niagara, 364,552; London, 381,522; Kitchener, 356,421; Halifax, 320,521; Victoria, 287,897; Windsor, 262,075; Oshawa, 240,104; and Saskatoon, 210,023.

[7]MIGRATION

Canadians of French origin are descendants of about 10,000 settlers who arrived in the 17th century and in the first half of the 18th century. Later in the 18th century, thousands of British settlers came to Canada from New England and other colonies to the south. Up to about 1850, about 500,000 persons left the British Isles for Canada; between 1846 and 1854, an additional 500,000 arrived, mainly from Ireland. Thereafter, there was a continuing influx. The peak year was 1913, when 400,870 immigrants arrived. In the decade 1921–30 there were 1,230,202 immigrants; in 1931–40, only 158,562; in 1946–65, 2,504,120. Many reemigrated, mainly to the US; by 1950, Canadian-born persons formed the second-largest group of foreign-born US inhabitants. Between 1951 and 1956, however, the excess of immigration over emigration was almost 600,000. After a lull in the early 1960s, immigration reached a peak of 222,876 in 1967. The total declined to 121,900 in 1971 but rose again to 218,465 in 1974; in that year immigration controls were tightened, and between 1975 and 1985, the number of immigrants per year averaged 118,656, and between 1986 and 1993, 193,881, rising from 99,219 in 1986 to 252,042 in 1993. Of the 1993 total of 252,042, immigrants from Asia numbered 134,532; Europe accounted for 50,050; Africa, 19,033; the Caribbean, 19,028; the US, 6,565; and South America, 11,327. Emigration is mainly to the US. In 1990 there were 871,000 Canadian-born people living in the US.

Canada is a major source of asylum for persecuted refugees. At the end of 1992 it was harboring 568,200 such persons.

Interprovincial migration is generally from east to west. During 1990–91 British Columbia gained 37,620 more people from other provinces than it lost, and Alberta 7,502, while Ontario lost 22,301 more than in gained, Saskatchewan 9,941, and Quebec, 7,690.

8ETHNIC GROUPS

According to the 1991 census, 83.5% of the population was Canadian-born, a decrease of 0.4% from 1981. In general, the percentage of population born outside Canada increases as one goes westward from Newfoundland to British Columbia. Persons of whole or partial British (including Irish) origin made up 44.6% of the total population in 1991; those of whole or partial French origin (centered mainly in Quebec, where they constitute 80% of the population), 31.1%. Other European groups (of single origin) included Germans (3.3%), Italians (2.8%), Ukrainians (1.5%), Dutch (1.3%), and Poles (1%). Nearly 28.9% of the total population claimed multiple ethnic origin.

Amerindians, commonly known as Indians, numbered 365,375 (1.4%) in 1991 and formed the sixth-largest ethnic group. As of fiscal 1992/93 there were 604 Indian bands living on 2,364 reserves comprising a total of 2,750,000 hectares (6,792,500 acres). These Indians were classified into ten major ethnolinguistic groups; the métis (75,150), of mixed European and Indian extraction, were recognized as an aboriginal people in the Constitution Act of 1982. Most of the 30,090 Inuit (Eskimos) live in the Northwest Territories, with smaller numbers in northern Quebec and northern Newfoundland (Labrador). Since 1959, Inuit cooperatives have been formed to finance fishing, fish processing, retail, housing, and tourist enterprises, and to promote the graphic arts.

9LANGUAGES

English and French are the official languages of Canada and have equal status and equal rights and privileges as to their use in all governmental institutions. The federal constitution also gives the English and French minorities the right to publicly funded education in their own language at the primary and secondary levels, wherever the number of children warrants it.

The constitution provides for bilingualism in the legislature and courts of Quebec, New Brunswick, and Manitoba. Although there are no similarly entrenched constitutional rights in Ontario and Saskatchewan, these provinces have made English and French the official languages of the courts. In 1984, the Northwest Territories Council adopted an ordinance providing for the use of aboriginal languages and establishing English and French as official languages.

English was proclaimed the sole official language of Manitoba in 1890, and French was made the official language of Quebec in

1974. However, the 1890 Manitoba legislation was declared unconstitutional in 1979, as was a Quebec law passed in 1977 declaring French to be the sole language of the legislature and the courts.

Although Canada is frequently referred to as a bilingual country, in 1991 only 16.3% of the people were able to speak both English and French. In Quebec, 82.3% of the people spoke French as a native language in 1991; in the other provinces, most of the people spoke only English, although there were sizable proportions of people able to speak French in New Brunswick (42%) and parts of Ontario (11.9%) and Manitoba (9.3%). That year, 16,169,880 Canadians (60%) reported that their only mother tongue was English, and 6,502,865 (24%) that it was French. In 1991, the number of persons reporting a single mother tongue other than English or French was 3,991,045 (15%). Native speakers of Italian numbered 510,980; German, 466,240; Chinese, 498,845; Ukrainian, 187,015; Portuguese, 212,090; and Polish, 189,815. There were 73,870 native speakers of Cree, the most common Indian language; there are at least 58 different Indian languages and dialects, in 10 major language groups. About 1% of the population reported that they had more than one mother tongue.

10RELIGIONS

In 1990, the principal religious denominations and their adherents in Canada were the Roman Catholic Church, 11,582,350; United Church of Canada, 2,013,258; Anglican Church of Canada, 848,256; Presbyterian Church in Canada, 245,883; Lutherans, 78,566; and Baptists, 201,218. Also represented were Greek Orthodox, Russian Orthodox, Greek Catholic, Mennonite, Pentecostal, and other groups. The estimated Jewish population in 1990 was 310,000

Freedom of religion has been specifically protected since the enactment of the Canadian Bill of Rights in 1960, now incorporated into the constitution.

11TRANSPORTATION

With such a vast land area, and with most production inland, all forms of transportation are vital. Since 1945, with the rapid growth of road, air, and pipeline services, the trend has been away from railways for carrying both freight and passengers. But because they can supply all-weather transportation in large volume over continental distances, railways are still important. The federal government, through the Canadian Transport Commission, has allowed a few rate rises and has insisted on a slow curtailment of services; nevertheless the companies have traditionally operated at a deficit or very low margin of profit because of competition and rising costs. There were 93,544 km (58,134 mi) of railways in 1991. Two great continental systems operate about 90% of the railway facilities, the government-owned Canadian National Railways (CNR), with trackage of 51,745 km (32,153 mi), and the privately owned Canadian Pacific Ltd. (CP), with 34,016 km (21,137 mi). They compete in some areas but cooperate where duplication of service is not profitable. In addition to their railway operations, CNR and CP maintain steamships and ferries, nationwide telegraph services, highway transport services, and hotel chains.

The populated sections are generally well supplied with roads and highways, but because of difficult winter weather conditions, road maintenance is a recurring and expensive task and puts a tremendous strain on road-building facilities. There are about 884,272 km (549,487 mi) of roads, including 250,023 km (155,364 mi) of paved highway. The 7,820-km (4,860-mi) paved Trans-Canada Highway, a c$500-million project financed jointly by the federal and provincial governments, was completed in 1962. Canada ranks next to the US in per capita use of motor transport, with one passenger car for every 2 persons. Motor vehicles in use in 1991 totaled 16,805,096, including 13,061,084 passenger cars, 3,679,804 trucks and 64,208 buses. Plans to construct a bridge from Prince Edward Island to the mainland were underway in 1992.

Bounded by water except for the Alaskan and southern land boundaries with the US, and with many inland lakes and rivers that serve as traffic arteries, Canada makes much use of water transport in domestic as well as foreign commerce. The major part of Canada's relatively small merchant fleet—480,000 GRT in 1991—consists of tankers. That year there were 68 ships of more than 1,000 GRT; most overseas commerce is carried by foreign ships. Montreal is Canada's largest port and the world's largest grain port. Others among the many well-equipped ports are Toronto, Hamilton, Port Arthur, and Fort William on the Great Lakes, and Vancouver on the Pacific Coast. The Montreal and lake ports are closed by ice from December to April, during which time Halifax on the Atlantic and Saint John on the Bay of Fundy are the only Atlantic Ocean traffic terminals.

The St. Lawrence Seaway and Power Project, constructed jointly by Canada and the US, and its many canals provide an 8-m (27-ft) navigation channel from Montreal to Lake Superior. The Athabasca and Slave rivers and the Mackenzie, into which they flow, provide an inland, seasonal water transportation system from the end of the railway in Alberta to the Arctic Ocean. The Yukon River is usually open from mid-May to mid-October. All Canadian inland waterways are open on equal terms to the shipping of all nations.

International air service is provided by government-owned Air Canada and Canadian Airlines. Regional service is provided by some 570 smaller carriers. Air transport is the chief medium in the northern regions for passengers and freight. Canada has 1,400 airports, with 1,155 of them usable, including 443 with permanent runways. The Lester Pearson airport in Toronto is by far the busiest, with 17,278,000 passengers handled in 1991.

[12]HISTORY

The first inhabitants of what is now Canada were the ancient ancestors of the Inuit. Exactly where they originated or when they arrived is uncertain, but they probably crossed from eastern Siberia to Alaska, Canada, and Greenland between 15,000 and 10,000 BC. Their descendants, the Dorset people, who inhabited the central Canadian Arctic region from about 700 BC to AD 1300, were primarily hunters of walrus and seal. The shorter-lived Thule culture, which may have assimilated the Dorset, lasted from about 1200 to the first arrival of the Europeans. Although most Inuit lived near the coast, some followed the caribou herds to the interior and developed a culture based on hunting and inland fishing.

Although the Norse had occupied a settlement at L'Anse aux Meadows in Newfoundland by AD 1000, the first fully documented arrival by Europeans was in 1497 by the Italian-born John Cabot, who led an English expedition to the shore of a "new found land" (Newfoundland) and claimed the area in the name of Henry VII. In 1534, the French, under Jacques Cartier, planted a cross on the tip of the Gaspé Peninsula; the following year, his expedition discovered and ascended the St. Lawrence River. By 1604, Pierre du Guast, Sieur de Monts, along with Samuel de Champlain had founded the first permanent French colony, Port Royal (now Annapolis Royal, Nova Scotia). Four years later, Champlain established the town of Quebec. The great St. Lawrence waterway led Étienne Brulé and others after him to the Great Lakes and the rivers flowing south through the center of the North American continent. Missionaries and fur traders soon arrived, and an enormous French territory was established. Between 1608 and 1756, about 10,000 French settlers arrived in Canada. In the hope of protecting French settlers and the fur trade, Champlain supported the Huron Indians against their enemies, the Iroquois. When the Iroquois demolished the Hurons, the French colony was almost destroyed.

In the 17th century, England pressed its claim (by virtue of Cabot's expedition) to the rich fur-trading colony, and during the frequent skirmishing between New France and New England the English conquered Quebec (1629). Restored to France in 1632, Quebec, together with the rest of New France, was placed under the absolute control of a chartered commercial organization, the Company of One Hundred Associates, with the twofold purpose of exploiting the fur trade and establishing settlements. In 1663, New France became a royal province of the French crown. Thereafter, three important officials—the royal governor, the intendant, and the bishop—competed in exercising control of the government. Under the seigneurial system, which had been founded in 1598, large land grants were made to seigneurs, who made other grants to settlers. The actual farmers owed some quasi-feudal dues and could sell the property only by paying a large duty to the seigneur.

The movement of exploration, discovery, commercial exploitation, and missionary enterprise, which had begun with the coming of Champlain, was extended by such men as Jacques Marquette, Louis Jolliet, and Robert Cavelier, Sieur de la Salle, reaching its climax in the last three decades of the 17th century. At that time, French trade and empire stretched north to the shores of Hudson Bay, west to the head of the Great Lakes, and south to the Gulf of Mexico. Meanwhile, a British enterprise, the Hudson's Bay Company, founded in 1670, began to compete for the fur trade.

The European wars between England and France were paralleled in North America by a series of French and Indian wars. The imperial contest ended after British troops commanded by James Wolfe defeated Marquis Louis Joseph de Montcalm on the Plains of Abraham, bringing about the fall of Quebec in 1759. The French army surrendered at Montreal in 1760, and the Treaty of Paris in 1763 established British rule over what had been New France. The Quebec Act of 1774 established English criminal law but secured seigneurial tenure, a modified oath of office allowing Roman Catholics to serve in the conciliar governments, and the right of the Roman Catholic Church to collect tithes.

These concessions, which reflected the sympathy of the British ruling class for the French upper classes, instituted the separateness of French-speaking Canada that has become a distinctive feature of the country. It also secured the loyalty of the French clergy and aristocracy to the British crown during the American Revolution. Although the poorer French settlers (habitants) sympathized with the Revolutionists, efforts to take Canada by arms for the revolutionary cause failed in the Quebec campaign. Some 40,000 Loyalists from the colonies in revolt fled northward to eastern Canada and did much to change the political character of their new country. The Constitutional Act of 1791 divided Lower Canada (now southern Quebec) from Upper Canada (now southern Ontario) and provided for elected assemblies with limited powers, the first organs of self-government in the territory.

In the 1780s, the newly organized North West Company began to challenge the Hudson's Bay Company's fur-trade monopoly. The period was one of expansion, marked by Alexander Mackenzie's journey to the Arctic Ocean in 1789 and his overland voyage to the Pacific Ocean in 1793. British mariners secured for Britain a firm hold on what is now British Columbia.

The War of 1812, in which US forces attempting to invade Canada were repulsed by Canadian and British soldiers, did not change either the general situation or the US-Canadian boundary. After amalgamating the North West Company in 1821, the Hudson's Bay Company held undisputed sway over most of the north and west. Eastern border problems with the US were resolved by the Webster-Ashburton Treaty in 1842; in the west, however, US expansionists sought to fix the border at 54°40′N. In 1846, the border was resolved at 49°N, and since then, except for minor disputes, the long border has been a line of peace.

The continuing influx of immigrants stimulated demands for political reforms. In Nova Scotia and New Brunswick the reformers had some early success, but in the two Canadas it was not until groups led by Louis Joseph Papineau in Lower Canada and William Lyon Mackenzie in Upper Canada had conducted separate futile rebellions in 1837–38 that the British government acted. John George Lambton, Earl of Durham, was sent to Canada as governor-general in 1838; he resigned later that year, but in 1839 submitted a report to the crown in which he recommended the granting of some forms of self-government. He also advised the immediate union of the two Canadas for the express purpose of Anglicizing the French Canadians. Union of the two provinces was approved in 1840, but responsible government was not achieved until 1849, after strenuous efforts by leaders in the various provinces. There was, however, no single unified nation—only a string of provinces in the east and the Hudson's Bay Company domain in the west and north.

The movement for Canadian confederation—political union of the colonies—was spurred in the 1860s by the need for common defense and the desire for a common government to sponsor railroads and other transportation. John Alexander Macdonald and George Brown, rival political leaders, agreed in 1864 to unite Upper Canada and Lower Canada under a common dominion government. Already the Maritime provinces were seeking union among themselves; their Charlottetown Conference in 1864 was broadened to admit delegates from the Canadas. After two more conferences, in 1864 and 1866, the dominion government was established under the British North America Act of 1867. The dominion was a confederation of Nova Scotia, New Brunswick, and the two provinces of Canada. There had been much opposition, and Nova Scotia and New Brunswick were brought to accept the union only through the efforts of Sir Charles Tupper and Sir Samuel Leonard Tilley and by the fear and indignation roused by the invasion of Canada by Fenians (militant Irish nationalists) from the US in 1866. Since the name Canada was chosen for the entire country, Lower Canada and Upper Canada became the provinces of Quebec and Ontario, respectively.

In 1869, the Hudson's Bay Company relinquished its territorial rights to Rupert's Land and the Northwest Territories. In 1870, the province of Manitoba was established and admitted to the confederation, and the Northwest Territories were transferred to the federal government. In 1871, British Columbia, on the Pacific shore, joined the confederation, largely on the promise of a transcontinental railroad. Prince Edward Island did not join until 1873. Pushing through the Canadian Pacific (CP) Railway was a main achievement of Macdonald's Conservative administration. The CP was given large grants of land in return for its promise to aid in settling these lands, a policy that is still being carried on. Objection in the west to being taken over by the east led to two métis rebellions, headed by Louis Riel, in 1869–70 and 1885, but the west was opened to settlement nonetheless.

Under the long administration (1896–1911) of the Liberal Party under Sir Wilfrid Laurier, immigration to the prairie provinces was greatly accelerated. The prairie agricultural empire bloomed. Large-scale development of mines and of hydroelectric resources helped spur the growth of industry and urbanization. Alberta and Saskatchewan were made provinces in 1905. In 1921, Manitoba, Ontario, and Quebec were greatly enlarged to take in all territory west of Hudson Bay and south of 60°N and all territory east of Ungava Bay. In February 1931, Norway formally recognized the Canadian title to the Sverdrup group of Arctic islands (now the Queen Elizabeth Islands); Canada thus held sovereignty in the whole Arctic sector north of the Canadian mainland. Newfoundland remained apart from the confederation until after World War II; it became Canada's tenth province in March 1949.

Canadian contributions of manpower and resources were immensely helpful to the Allies when Canada joined the British side in World War I; more than 600,000 Canadians served in Europe, and over 60,000 were killed. The war contributions of Canada and other dominions helped bring about the declaration of equality of the members of the British Commonwealth in the Statute of Westminster of 1931. The wartime struggle over military conscription, however, deepened the cleavage between French Canadians and other Canadians. After the war, the development of air transportation and roads helped weld Canada together, and the nation had sufficient strength to withstand the depression that began in 1929 and the droughts that brought ruin to wheat fields. The farmers developed huge cooperatives, especially in Nova Scotia and the prairie provinces, and also took up radical political doctrines, notably through the Social Credit and the Socialistic Cooperative Commonwealth Federation parties.

Canada was again vitally important in World War II, under the premiership of William Lyon Mackenzie King. More than 1 million Canadians took part in the Allied war effort, and over 32,000 were killed. The nation emerged from the war with enhanced prestige, actively concerned with world affairs and fully committed to the Atlantic alliance.

Domestically, a far-reaching postwar development was the resurgence in the 1960s of French Canadian separatism, symbolized by a series of cultural agreements between France and Quebec. In 1970, terrorist acts by the Quebec Liberation Front led to the banning of that organization and to the federal government's first invocation in peacetime of emergency powers under the War Measures Act. The emergency measures, imposed on 16 October, were not lifted until 30 April 1971. Although administrative reforms—including the establishment of French as Quebec's official language in 1974—helped meet the demands of cultural nationalists, separatism continued to be an important force in Canadian politics. In the 1976 provincial elections, the separatist Parti Québécois came to power in Quebec, and its leader, Premier René Lévesque, proposed that Quebec become politically independent from Canada, in a relationship termed sovereignty-association. In a referendum on 20 May 1980, in which 82% of those eligible voted, the proposal was defeated, 59.5% to 40.5%. Meanwhile, other provinces had their own grievances, especially over oil revenues. Alberta objected to federal control over oil pricing and to reduction of the provincial share of oil revenues as a result of the new National Energy Program announced in late 1980; the failure of Newfoundland and the federal government to agree on development and revenue sharing hindered the exploitation of the vast Hibernia offshore oil and gas field in the early 1980s.

Since 1927, when discussions first began on the question of rescinding the British North America Act, disagreements between the provinces and the federal government over constitutional amendment procedures had stood in the way of Canada's reclaiming from the UK authority over its own constitution. In 1980, Liberal Prime Minister Pierre Elliott Trudeau made "patriation" of the constitution a principal priority of his administration. Initially he faced considerable opposition from eight of the ten provincial premiers, but a compromise on amending procedures and a charter of rights eventually proved acceptable to all but Quebec. The Constitution Act, passed in December 1981 and proclaimed by Queen Elizabeth II on 17 April 1982, thus replaced the British North America Act as the basic document of Canadian government. In 1987, Quebec was to sign the new constitution, after winning the inclusion of a clause acknowledging that Quebec is a "distinct society." The Meech Lake Accord of 1987, however, failed to compel Quebec into signing the constitution, and Quebec's status has been in limbo ever since. New Brunswick and Manitoba failed to ratify the Accord because of the perceived preferential status Quebec would have received. The

Charlottetown Accord also proposed recognizing Quebec as a "distinct society" in addition to acknowledging aboriginals' inherent right to self-government and converting the senate into an elected and more effective legislative body. On 26 October 1992, however, the majority of Canadians chose not to support the Charlottetown Accord in a national referendum.

Canada joined with the US and Mexico to negotiate the North American Free Trade Agreement (NAFTA), which was built upon the US–Canada Free Trade Agreement (FTA). The three nations came to an agreement in August 1992, and signed the text on 17 December 1992. NAFTA seeks to create a single market of 370 million people with a combined GNP exceeding $6 trillion and was implemented in 1994.

13 GOVERNMENT

Canada is a federation of 10 provinces and 2 northern territories. Under the British North America Act of 1867, which united the 4 original provinces of Quebec, Ontario, Nova Scotia, and New Brunswick into one dominion under the name of Canada, the federation was provided with a powerful central government, which, besides its areas of exclusive authority, held residual authority in matters beyond the powers of local or private concern specifically assigned to the provincial legislatures. The British North America Act—which effectively served, together with a series of subsequent British statutes, as Canada's constitution—could be amended only by the British Parliament. In 1982, the British North America Act was superseded by the Constitution Act (or Canada Act), the principal innovations of which are the Charter of Rights and Freedoms and the provision for amendment. For passage, an amendment requires approval by the federal Parliament and the legislative assemblies of at least two-thirds of the provinces, which must hold an aggregate of at least half the population of all the provinces. However, when an amendment derogates from provincial rights, it will not apply in any province in which the legislative assembly dissented by majority vote. When such an amendment deals with education or other cultural matters, the federal government must pay compensation to any dissenting province, to make up for the funds that would have been transferred had the province accepted the amendment.

Under the Constitution Act, the British sovereign remains sovereign of Canada and head of state; for the most part, the personal participation of Queen Elizabeth II in the function of the crown for Canada is reserved to such occasions as a royal visit. The queen's personal representative in the federal government is the governor-general, appointed by the crown on the advice of the prime minister of Canada; the governor-general is usually appointed for a term of five years. Active executive authority resides in the cabinet, or ministry, headed by the prime minister.

The federal Parliament is made up of the House of Commons and the Senate. A new House of Commons, with 295 members as of 1993, is elected at least once every five years by all Canadian citizens and British subjects 18 years of age or older who have resided in Canada for at least 12 months prior to polling day. Representation by provinces and territories is based on population, ranging from 1 for the Yukon Territory to 99 for Ontario.

The leader of the party that wins the largest number of seats in a newly elected House of Commons is asked to form the government. The governor-in-council (cabinet), responsible for determining all important government policies and for securing the passage of legislation, financial measures, and administrative provisions, is chosen by the prime minister.

The 104 members of the Senate, or upper house, are appointed for life, or until age 75, by the governor-general on the nomination of the prime minister, with equality of representation for regional divisions. There are equal proportions of senators from the Maritime provinces, Ontario, Quebec, and the western provinces. In

October 1992, Canadian voters declined a constitutional amendment that would have made the Senate an elected body.

14 POLITICAL PARTIES

Throughout most of the 20th century, national unity has been the primary aim of every Canadian government: leaders of both the English-speaking majority and the French-speaking minority have cooperated to develop a united Canada with a great destiny to which differences arising from national origin were subordinate. In the 1970s, this unity was challenged by a growing demand for French Canadian autonomy. Despite cultural division, national unity has remained a basic factor in Canadian foreign policy. Two elements have contributed to the growth of Canadian nationalism—deliberate government policy and reaction against overidentification with either the UK or the US.

Continuity of policy characterizes party relationships. The Liberal Party (LP), which held office from 1935 to 1957 and again (except for part of 1979) from 1968 to 1984, is nationwide in its representation but has its main strength in Quebec. It traditionally emphasizes trade and cultural relationships with the US, while its principal rival, the Progressive Conservative Party (PC), which held power from 1957 to 1968, from May to December 1979, and again since the end of 1984, stresses Canada's relationships with the UK. In economic policy, the Liberals generally champion free trade, while the Conservatives favor a degree of protection; but practical political considerations have modified this distinction.

The Cooperative Commonwealth Federation (CCF) was a farmer-labor party with its main strength in Saskatchewan. Its foreign policy was much like that of the British Labour Party, but with an admixture of traditional Canadian prairie radicalism. It merged with the Canadian Labour Congress to form the New Democratic Party (NDP) in 1961. The Social Credit Party (SCP) has headed governments in Alberta and British Columbia but has not done well nationally. In June 1962, the group collapsed into independent factions, leaving only five representatives in the Commons. In September, the Quebec wing of the party united to form the Ralliement des Créditistes, which after the 1965 elections became the new focal point of French Canadian interests.

After 22 years of uninterrupted rule, the Liberals were defeated by the PC in the 1957 elections. This was widely interpreted as a vote of protest against individual Liberal ministers and high taxes, as a reflection of concern over US economic penetration, and as a demonstration of widespread feeling that it was "time for a change." In the general election on 31 March 1958, the PC was returned to power with an unprecedented majority, taking 208 of the 265 seats. The LP was reduced to 49 seats, the smallest number in its history. In the election of June 1962, the PC lost 92 seats. The following February, the PC government lost a vote of confidence, the major issue being defense policy and the refusal of the prime minister to accept nuclear weapons from the US. In the election of April 1963, the resurgent Liberals gained an additional 29 seats for a total of 129 (4 short of a parliamentary majority). With some support from the SCP, Liberal leader Lester B. Pearson formed a new government.

In April 1968, the new Liberal Party leader, Pierre Elliott Trudeau, was elected prime minister in a colorful campaign emphasizing personality more than specific issues. In the June general election, which he called for almost immediately, the LP took 155 seats and the PC 72; the SCP lost all 5 of its seats. In the general elections of 30 October 1972, the Liberals lost their parliamentary majority, winning only 109 seats to the PC's 107. The NDP increased its representation from 22 seats to 31, and the Créditistes, who had resumed calling themselves the SCP in 1971, won 15 seats. When the NDP decided to support the continuance of Liberal rule, Prime Minister Trudeau formed a new cabinet. The Liberal-NDP alliance collapsed on 8 May 1974, when, for the first time in Canadian history, the government received a vote of no confidence on a budget bill. Elections were called, and the campaign was fought largely on the issue of inflation, with the PC calling for a system of wage and price controls. In the elections of 8 July 1974, the Liberals regained their majority.

In the general elections of 22 May 1979, the Liberals lost to the PC, taking 114 seats of the now 282-seat Parliament to the PC's 136, and were unable to form a government in any province. However, on 13 December 1979, the government of Prime Minister Joe Clark was defeated by a Liberal and NDP coalition on a vote of no confidence on a budget bill that called for an increase of 18 cents a gallon in the excise tax on gasoline. Trudeau, who in November had announced his planned retirement, decided to continue as Liberal leader, and again became prime minister after elections on 18 February 1980 gave the Liberals 147 seats. Four years later, on 29 February 1984, Trudeau again announced his impending retirement, and his party chose John Turner as successor. Brian Mulroney became prime minister following a landslide PC victory in the September 1984 elections, which gave the PC 211 seats, the Liberals 40 (their lowest number ever), the NDP 30, and an independent 1. However, the Liberals regained strength over the next year, and in 1985 won the Quebec general election and, in a coalition with the NDP, ended 42 years of PC government in Ontario.

In 1993, the PC fell from power, primarily due to one of the worst Canadian recessions in nearly 60 years and the failure of the PC government to implement constitutional reforms. Brian Mulroney resigned, and was succeeded by Kim Campbell. Liberals soundly defeated the PC in the October 1993 election, with 177 of the 295 seats (up from only 80 in 1988). The PC retained only 2 of their 157 seats. The Liberal party named Jean Chretien as the new Prime Minister

15LOCAL GOVERNMENT

Canada is made up of 10 provinces and 2 territories. Each province has a premier and a legislature. They function like those of the central government. However, the provincial parliaments are unicameral. In each province, the sovereign is represented by a lieutenant-governor appointed by the governor-general. The provinces are empowered to regulate their own affairs and dispose of their own revenues. Civil and property rights, civil law, education, health, labor conditions, licenses, management and sale of public land, municipal government, and direct provincial taxation are within the jurisdiction of the provinces. Although the federal government still exercises considerable authority over the Yukon Territory and the Northwest Territories, both now have elected legislative bodies, and in the Yukon, the powers of the federal commissioner have been greatly reduced.

Each province is divided into municipalities, the number and structure of which vary from province to province. In Prince Edward Island, Nova Scotia, New Brunswick, Ontario, and Quebec the first order of municipalities consists of counties, which are further subdivided into cities, towns, villages, and townships, although there are minor variations. In Newfoundland and the four western provinces there are no counties; municipalities are either rural or urban, the latter being made up of cities, towns, and villages, but again with minor variations. Municipalities are usually administered by an elected council headed by a mayor, overseer, reeve, or warden. Local governments are incorporated by the provinces, and their powers and responsibilities are specifically set forth in provincial laws.

16JUDICIAL SYSTEM

The civil law follows English common law everywhere except in Quebec, where it follows the Napoleonic Code. The main body of criminal law is derived from English sources; most criminal statutes, being federal, are uniform throughout the country. Police

magistrates and justices of the peace are appointed by the provincial governments. Civil and criminal courts exist on county, district, and superior levels; all judges of the superior, federal, tax, district, and county courts are appointed for life (but not beyond age 75) by the governor-in-council (the cabinet) and are paid by the federal Parliament. The Supreme Court in Ottawa has appellate, civil, and criminal jurisdiction throughout Canada; its chief justice and eight associate justices (at least three of whom must come from Quebec) are appointed by the governor-general. The Federal Court of Canada (formerly the Exchequer Court) hears cases having to do with taxation, claims involving the federal government, copyrights, and admiralty law. The death penalty in Canada was abolished in 1976; that decision was upheld in a vote by the House of Commons in June 1987.

The judiciary is independent of the legislative and executive branches. The Canadian Charter of Rights and Freedoms, part of the 1982 revised Constitution, guarantees a number of individual fundamental rights.

Criminal defendants are afforded a wide range of procedural due process protections including a presumption of innocence, a right to counsel, public trial and appeal.

17ARMED FORCES

All service in the armed forces is voluntary. The armed forces were unified in 1968 as functional commands that have land, sea, and air missions, but not separate service headquarters; total forces as of 1993 were 84,000; there were 29,700 reserves. The Land Forces had 22,000 troops in 1993; the Maritime Forces had 17,000 and 3 submarines, 18 frigates or destroyers and 220 other vessels, deployed in both the Atlantic and Pacific; and the air force had a strength of 22,400, with 8 fighter squadrons (some 300 aircraft). There were 22,600 personnel in the Canadian forces training system, communications command, Canadian forces in Europe, and 15 other unified armed forces organizations. In 1993, 576 members of the Canadian armed forces were deployed in Cyprus, 216 in Syria and Israel, and 24 in the Sinai, in Angola, 208 in Cambodia, 1200 in Croatia, 50 in Iraq, and 50 elsewhere in Africa on peacekeeping operations. Defense expenditures in 1992 totaled $10.4 billion or perhaps 1.5 percent of gross domestic product.

The Royal Canadian Mounted Police (RCMP) is a civil force maintained by the federal government, originally to police federal territories. It is the sole police force in the Yukon Territory and the Northwest Territories. However, all the provinces except Ontario and Quebec, which have their own police forces, have entered into contracts with the RCMP to enforce provincial laws (under the direction of the provincial authorities). In some municipalities the RCMP also enforces municipal laws. Other urban centers maintain their own police forces or (as in Ontario and Quebec) hire the services of the provincial police under contract.

18INTERNATIONAL COOPERATION

A Commonwealth nation, Canada became a charter member of the UN on 9 November 1945 and participates in ECE, ECLAC, and all the nonregional specialized agencies. A Canadian, Lester B. Pearson, served as president of the General Assembly in 1952/53. Maj. Gen. E. L. M. Burns of Canada was chief of staff of the UN Truce Supervision Organization in the Middle East from August 1954 to November 1956, when UNEF was established, and he served as UNEF commander for the next three years. Canada has also contributed to UN peacekeeping efforts in Kashmir, the Congo (now Zaire), West Irian (now Irian Jaya, Indonesia), Cyprus, and the Middle East.

Canada is a member of NATO; as part of its contribution to the collective defense of the alliance, Canada has land and air units in the Federal Republic of Germany (FRG). Canada cooperates with the US in North American defense through the North

American Air Defense Command (NORAD). Canada also belongs to many inter-American, inter-Commonwealth, and other intergovernmental organizations, including the Asian Development Bank, IDB, OECD, and PAHO. It is a permanent observer at the OAS. Canada is a signatory to GATT and the Law of the Sea.

[19]ECONOMY

The Canadian economy is the seventh largest among the western industrialized nations. The post-war period has seen a steady shift from the production of goods toward increased emphasis on services. In 1993, 7 out of 10 employed Canadians worked in the services industry, compared to only 5 out of 10 in 1960. Canada is a world leader in the production and export of asbestos, nickel, silver, zinc, uranium, cadmium, cobalt, magnesium, gypsum, molybdenum, potash, aluminum, gold, iron ore, copper, fluorspar, and lead. Although no longer the foremost sector of the economy, agriculture is of major importance to the economy as a whole and still is basic in many areas; Canada accounts for approximately 20% of the world wheat trade. Canada is also the world's leading producer of newsprint and ranks among the leaders in other forestry products. Canada has changed from a country producing and exporting mainly primary products to one that is increasingly producing and exporting manufactured goods. In the 1980s, machinery and equipment joined automotive products among the country's leading exports; at the same time, the importance of natural resource products declined (partly reflecting the 1986 collapse of oil prices). Canada was hard hit by the recession of the early 1980s, with interest rates, unemployment, and inflation all running higher than in the US. The effects of the recession on minerals and manufacturing were especially severe. By the end of 1982, all mining operations in the Yukon were closed, and throughout the country more than 70,000 of 115,000 miners were unemployed. The economy recovered during the mid-1980s, and Canada's economic growth rate was among the highest of OECD countries during 1984–86. However, differences in prosperity among the provinces increased during the 1980s, with the central provinces relatively robust, the western provinces suffering declines in growth because of lower prices for oil and other natural resources, and the Atlantic provinces depressed. By the second quarter of 1990, the economy had begun to contract, affected by the recession and the central bank's tight monetary policy. Recovery began in the second half of 1991, although the early 1990s were marked by continuing unemployment and restrained domestic spending.

[20]INCOME

In 1992, Canada's GNP was $565,787 million at current prices, or $20,320 per capita. For the period 1985–92 the average inflation rate was 3.7%, resulting in a real growth rate in per capita GNP of 0.3%.

In 1992 the GDP was $537,100 million in current US dollars. It is estimated that in 1989 agriculture, hunting, forestry, and fishing contributed 2% to GDP; mining and quarrying, 3%; manufacturing, 17%; electricity, gas, and water, 3%; construction, 6%; wholesale and retail trade, 13%; transport, storage, and communication, 6%; finance, insurance, real estate, and business services, 18%; community, social, and personal services, 5%; and other sources, 26%.

[21]LABOR

Employment decreased 0.8% in 1992 to 12,240,000, down 2.6% (331,000) from a peak in 1990. There were 13,797,000 in the total civilian labor force in 1992. Of those in civilian employment in 1992, 27.7% were in industry; 3.5% in agriculture; and 73% in services. The rate of participation by men in the employed labor force fell from 78.4% in 1981 to 73.8% in 1992, whereas that of women rose from 57.6% to 55.1% during the same period.

Cold weather and consumer buying habits cause some regular seasonal unemployment, but new techniques and materials are making winter construction work more practicable, and both government and many industrial establishments plan as much work as possible during the winter months. Unemployment ranged between 7.5% and 11.8% in the period 1982–1992, with the lowest rate occurring in 1989 and the highest in 1983. In 1992, it stood at 11.3%. Payments of c$19.3 billion went to the unemployed in 1992, a record high, and up 9.1% from 1991.

At the beginning of 1992, labor organizations active in Canada reported a total membership of 4,089,000. Of this number, 57.8% were represented by unions affiliated solely with the Canadian Labor Congress (CLC) and 6.2% and 5.2% with the Confederation of National Trade Unions and the Canadian Federation of Labor, respectively. All together there were 235 national unions in Canada, and 62 international unions. The public sector is the most highly organized, with three of the four largest unions in Canada. The largest unions, with 1992 membership totals, are the Canadian Union of Public Employees, 406,000; National Union of Public and General Employees, 307,500; United Food and Commercial Workers International Union, 180,000; and the Public Service Alliance of Canada, 165,000.

Federal and provincial laws set minimum standards for hours of work, wages, and other conditions of employment. Minimum wage rates for experienced adult workers ranged from c$4.75 in Newfoundland and Prince Edward Island to c$6.50 in Northwest Territories in 1992. Many collective agreements are negotiated every other year and are concluded without work stoppage, although sometimes with the assistance of government conciliation services. Collectively bargained group health insurance, pension, and unemployment benefit plans supplement payments under the Unemployment Insurance Act, which was reformed in 1990 in order to be funded entirely by employer/employee taxes on insurable earnings. Since 1951, the federal government and the provinces have enacted laws prohibiting discrimination in employment because of race, color, religion, or national origin, as well as laws requiring equal pay for men and women. Safety and health regulations and workers' disability compensation have been established by federal, provincial, and municipal legislation.

Weekly earnings for the industrial aggregate averaged c$520.16 in December 1992, up from c$501.08 in 1991.

[22]AGRICULTURE

Until the beginning of the 1900s, agriculture was the predominant occupation, and farmers and their families made up the majority of the population. Since then, however, the farm population has been shrinking both relatively and absolutely. Even in Saskatchewan, the province with the highest proportion of farm population, farm families account for no more than 25% of the total population. For Canada as a whole, agriculture engaged only 3.4% of the economically active population in 1992. Farm production continues to increase, as have the size of holdings, crop quantity, quality and variety, and cash income. Canada is still one of the major food-exporting countries of the world; agriculture engages about 466,000 people (3.4% of the economically active population) and generates about $23 billion in cash farm receipts (about 2.4% of GDP).

Of Canada's total land area, about 5% is classified as arable land; another 3% is considered as permanent pasture land. More than 90% of the cultivated area is in the three prairie provinces. The trend is toward fewer and larger farms and increased mechanization and specialization. Of total farm cash receipts of c$20,380 million in 1986, Ontario and Saskatchewan together received 47%. Sale of field crops provide more than 50% of farm cash income in the prairie region, but less than 10% elsewhere in Canada.

The following table shows the estimated extent and output of principal field crops in 1992:

AREA	YIELD (HECTARES)	(1,000TONS)
Wheat	13,830,000,000	29,870,000
Barley	3,792,000	10,919,000
Corn	750,000	4,531,000
Oats	1,238,000	2,823,000
Rapeseed	2,904,000	3,689,000
Potatoes	124,000	3,529,000

Formerly, Canada imported only such items as could not be grown domestically—coffee, tea, cane sugar, spices, and citrus fruits—while exporting large surpluses of wheat, barley, and livestock. However, food imports have risen sharply in recent years. Nevertheless, Canada remains a net food exporter; in 1992, Canada's agricultural exports were 8.1% of total exports. During 1989–90, Canada was the second largest exporter of wheat through the International Wheat Agreement (after the US); over 9.9 million tons were exported, mostly to Japan, Pakistan, and Iran.

Federal and provincial departments of agriculture provide guidance and aid to farmers in almost every field of operation. Activities include research and experimentation, protection of animals and crops, irrigation and reclamation, and price stability and farm credit measures. The government can stabilize the price of any agricultural product (except wheat, for which separate provision is made) by outright purchase or by supporting the market with guarantees or deficiency payments.

The departments of agriculture apply fundamental scientific research to soil management and crop and animal production, promote agricultural production, and enact financial measures to ensure greater stability of the farm economy. Long-term and short-term mortgages are made available; other loans are granted for equipping, improving, and developing farms. Various federal acts assist the marketing of produce. Governments, working with product organizations, also set limits on the production of milk, eggs, tobacco, and chicken and turkey meat. Price supports may be given to any designated natural or processed product but are mandatory for cattle, sheep, hogs, dairy products, wheat, oats, and barley. Farmers who have suffered severe crop losses through drought may obtain compensation, and prairie farmers who cannot deliver all their grain to market are given temporary financial assistance. The rail freight rates paid by western farmers to ship their grain to eastern markets, basically unchanged since 1897, were scheduled to increase fivefold between 1983 and 1991. The increase, partially subsidized by the federal government, would pay for improvements in the western rail system.

23 ANIMAL HUSBANDRY
Canada traditionally exports livestock products, producing more than the domestic market can use. Animal production (livestock, dairy products, and eggs) now brings in about half of total farm cash income. Stock raising is the foundation of agricultural economy in the foothills of the Rockies, across northern Alberta and Saskatchewan and southern Manitoba, on the interior plateaus of British Columbia, in the Georgian Bay district of Ontario, in Prince Edward Island, and in western Nova Scotia. One of the great ranching sections is located in the Palliser Triangle of southern Saskatchewan and Alberta.

Livestock on farms in 1992 numbered 13,002,000 head of cattle; 10,395,000 pigs and hogs; 914,000 sheep and lambs; and 118,000,000 hens and chickens. In 1992, livestock slaughtered included 3,339,000 head of cattle, and calves, 15,285,000 hogs, and 538,000 sheep. Chicken and turkey production totaled 733,000 tons. Milk production in 1992 was 7.3 million tons; butter production amounted to about 100,000 tons, and cheese production to 291,139 tons. Most dairy products are consumed within Canada. In 1992, 317,060 tons of eggs were produced.

The wild fur catch, which was important in Canada's early history, is now limited to the northern parts of the provinces, the Northwest Territories, and the Yukon.

24 FISHING
With a coastline of nearly 29,000 km (18,000 mi) and a lake-and-river system containing more than half the world's fresh water, Canada ranks among the world's major fish producers and in 1991 was the world's third-leading exporter of fresh, chilled, and frozen fish by value (after the US and Norway), the exports of which were valued at $1,091,845,000. Exports of dried, salted and smoked fish in 1991 amounted to $338.5 million, more than any other nation except Norway.

Two of the world's great fishing grounds are located off Canada. One lies along the Atlantic coast of the Maritime provinces, and in this region the Grand Banks of Newfoundland constitute the largest area. More than 1 billion lb of cod, haddock, halibut, pollock, and other fish are caught every year along the Atlantic in deep-sea and shore operations. Most of the cod and about a third of the total catch is dried and salted for export to Mediterranean and Latin American countries; another third is sold fresh; and the rest is canned. Vast numbers of lobsters and herring are caught in the Gulf of St. Lawrence and the Bay of Fundy. The other great fishing region includes the bays, inlets, river mouths, and fjords of British Columbia. Salmon, the specialty of the Pacific fisheries, is canned for export and constitutes the most valuable item of Canadian fish production. Also exported are fresh halibut and canned and processed herring. Other important export items are whitefish, lake trout, pickerel, and other freshwater fish caught in the Great Lakes and some of the larger inland lakes. Feed and fertilizer are important by-products.

Export sales in 1991 amounted to $2.17 billion. The US imported more than $1.45 billion of Canada's fish product exports in 1992.

The government protects and develops the resources of both ocean and inland waters and helps expand the domestic market for fish. It extends loans to fishermen for the purchase of fishing craft. Canadian-US action has helped restore Pacific salmon runs and halibut stocks and the Great Lakes fisheries, but pollution represents a threat to freshwater sport fishing, especially in Ontario.

25 FORESTRY
In 1991, forests covered 360 million hectares (890 million acres) or 39% of Canada's total land area. Of this area, some 96,000 sq km (37,000 sq mi) went for uses other than timber production, including parks, game refuges, water conservation areas, and nature preserves. In 1991, 62,000 persons were engaged in forestry and logging, which typically accounts for 0.5% of GDP. About 92% of productive forestland is crown-owned, 83% is provincial crown forestland (most of which is in Quebec, British Columbia, and Ontario), and 10% is under federal jurisdiction. The remaining 7% is privately owned. The crown forests are leased to private individuals or companies. Canada ranks as the third largest producer of coniferous wood products (after the US and Russia), and is the leading supplier of softwood products to world markets. Chief forest products in eastern Canada are pulp and paper manufactures, especially newsprint, three-fourths of which goes to the US. In the west, the chief product is sawn timber. Somewhat less than half the lumber production is exported. The manufacture of pulp and paper has been a leading industry for many years. In 1991, an estimated 52 million cu m of sawn wood was cut. In addition, 52,040,000 cu m of pulp wood and 8,977,000 tons of newsprint were produced. Exports of forest products in 1992 were valued at over $16.9 billion, or 12.6% of total exports.

26MINING

With such a large annual forestry output, conservation and reforestation are stressed. Both government and industry promote improvements in management practices and in the use of forest products. New manufacturing methods permit the use of inferior classes of wood.

Mining has been conducted in Canada since the 17th century, but the remarkably rapid development of mineral exploitation dates from the end of World War I. Currently being commercially produced are some 52 minerals. Petroleum has been found in the midwest; iron ore deposits in Labrador, Quebec, and Ontario; uranium in Ontario and Saskatchewan. Canada is the world's largest producer of mine zinc and uranium and is among the leaders in silver, nickel, aluminum (from imported bauxite), potash, gold, copper, lead, salt, sulfur, and nitrogen in ammonia. Yet the country has only just begun to fully develop many of its most important mineral resources, and resources developed earlier continue to display great growth potential. Beginning in 1981, large new deposits of gold ore were discovered at Hemlo, Ontario, north of Lake Superior; by 1991, more than 50% of Ontario's gold production came from the three mines in the Helmo district.

Output totals for principal Canadian metals in 1991 are shown in the following table:

	OUTPUT (METRIC TONS)	VALUE (1,000,000 US$)
Iron ore	35,961,000	982
Copper	797,603,000	1,834
Zinc	1,148,189	1,179
Gold	178,712*	2,056
Uranium	9,124	412
Nickel	196,868	1,596
Silver	1,338*	162
Molybdenum	11,333	61
Lead	278,141	178
Platinum group	11,532*	124
Cobalt	2,158	54

*Kilograms

The aggregate value of mineral production decreased by 16% to us$13,062 million in 1991. Of that total, the value of metals produced was $9,101 million, down 16.5% from 1989; nonmetal production was worth $1.965 million, down 9.7% from 1990; and construction materials produced were worth $1.996 million, droping by 18.2% from 1990. Of the total value in 1991, Alberta accounted for 46.4%, Ontario 14.5%, British Columbia 10.9%, Saskatchewan 8.6%, and Quebec 8.2%, Manitoba 3.3%, Northwest Territories 2.3%, New Brunswick 2.3%, and the remainder form the otehr provinces and territories.

What are believed to be the world's largest deposits of asbestos are located in the eastern townships of Quebec, approximately 130 km (80 mi) east of Montreal. Asbestos production in 1991 amounted to 670,000 tons, and was valued at us$240 million. Other nonmetallic mineral production included salt (11,585,000 tons, worth us$226 million), sand and gravel (200,497,000 tons, worth us$551 million), peat (762,116 tons, valued at us$80 million), and potash (7,012,000 tons, with a value of us$802 million). An area extending from central Saskatchewan southeast into Manitoba is probably the largest and richest reserve of potash in the world, and at the present rate of consumption it could probably supply all the world's needs for 1,000 years. Steady production began in 1962 and in 1991 known national reserves amounted to 14 billion tons.

27ENERGY AND POWER

Canada is abundantly endowed with fossil fuels and hydroelectric resources. Coal production reached 71.1 million metric tons in

1991, with a value of $1.66 billion. Although Canada accounted for less than 2% of world coal production in 1991, it exported about 50% of its production, making it the fourth largest coal-exporting nation. The increase in total output since 1970, especially the increased output from Alberta and British Columbia, is almost entirely due to the growth of the Japanese and South Korean export markets. In eastern Canada, however, domestic coal must be augmented by US coal imports.

Petroleum production in quantity began in 1947 with the discovery of oil 29 km (18 mi) south of Edmonton. Output of oil in 1992 was 98.2 million tons, at a rate of 2,065,000 barrels per day. Petroleum is now the largest single contributor to mineral output ($9,269 million, or 54% in 1991). Natural gas production rose to 116,600 million cu m in 1992, third in the world after Russia and the US. In 1992, proved crude oil reserves were estimated at 7.6 billion barrels, and natural gas reserves at more than 95.7 trillion cu ft (2% of the world's total). Crude oil pipelines totaled 23,564 km (14,642 mi) in length in 1991. There are two major oil pipeline systems, both originating at Edmonton, one extending east to Toronto, and the other southwest to Vancouver and the state of Washington, in the US. Natural gas pipelines extended 74,980 km (46,592 mi) in 1991 and are constantly increasing. As of 1993, the Iroquois pipeline serving the eastern US from Alberta was complete. As of 1992, the Canadian government had an 8.5% stake in the massive c$5.2 billion Hibernia oil project off the coast of Newfoundland. Athabasca, Peace River, and other bitumen and heavy oil deposits in Alberta amounted to 2.5 trillion barrels of oil in place, about 40% of the world's known bitumen in 1991.

Canada ranks sixth in the production of electric power in the world and first in the production of hydroelectricity. In 1991, Canada's total net installed capacity reached 104.6 million kw. The marked trend toward the development of thermal stations, which became apparent in the 1950s, is due in part to the fact that most of the hydroelectric sites within economic transmission distance of load centers have already been developed. When the Churchill Falls project reached completion in 1974, the capacity of the plant was 5,225 Mw, making it, at the time, the largest single generating plant of any type in the world. It has since been surpassed by Hydro-Quebec's 5,328-Mw generator, the first completed station of the massive James Bay project. Total electric power generation in Canada in 1991 was 507,913 million kwh, 61% of it hydroelectric, 23% conventional thermal, and 16% nuclear. Net exports of electricity decreased 28% in quantity between 1988 and 1991, amounting to 24,522,000,000 kwh in 1991.

Low-cost electricity generated from waterfalls and fast-flowing rivers has been a major factor in the industrialization of Quebec, Ontario, and British Columbia, most significantly in the establishment of metal-smelting industries. In other areas, hydroelectric power is not as abundant, but all provinces have turbine installations. Canada's hydroelectric power totaled 284.1 billion kwh in 1992, more than any other nation.

Atomic Energy of Canada Ltd. is responsible for research into reactor design and the application of nuclear power in the electric power field. In 1962, commercial electric power was first generated in Canada by a nuclear reaction when the Nuclear Power Demonstration Station at Rolphton, Ontario, became operative. Canada's first full-scale nuclear power station, completed in 1956 at Douglas Point on Lake Huron, produced its first power early in 1967. Nuclear energy consumption in Canada is projected to increase by about 19% from 1991 to 1995; approximately 8% from 1995 to 2000; and about 24% from 2000 to 2010.

28INDUSTRY

Manufacturing accounted for 17% of GDP in 1992 and employed 15% of the labor force. The leading industrial sectors are foods and beverages, transport equipment, petroleum and coal products, paper and paper products, primary metals, chemicals, fabricated metals, electrical products, and wood products. The value of industrial production in 1990 amounted to c$91.3 billion. Between 1985 and 1990 the added value of manufacturing rose by 10.4% in real terms, less than the strong overall growth rate, and the sector was strongly affected by Canada's recession. In 1990, real value added fell by approximately 4%, compared with an average 1% increase in all sectors.

Of the total manufacturing output, about half is concentrated in Ontario, which not only is the center of Canadian industry but also has the greatest industrial diversification. Some important industries operate there exclusively.

Quebec ranks second in manufacturing production, accounting for more than 25% of the value of Canadian manufactured goods. British Columbia ranks third. Manufacturing is also the leading industry in Manitoba, New Brunswick, Nova Scotia, and Newfoundland.

29SCIENCE AND TECHNOLOGY

Canadian science and technology depends on a base of about 600,000 scientists, engineers, and technologists, of whom nearly 15% are engaged in scientific research and development. In 1991, research and development expenditures were estimated at about c$9.7 billion, or about 1.4% of GDP. Slightly more than 50% of this expenditure was carried out by the business sector, much of it by the telecommunication equipment, aircraft, and business machines industries. Researchers in higher education were responsible for a little under 25% of total research and development activity; the federal government accounted for another 20%.

The Ministry of State for Science and Technology, established in 1971, is the chief federal policymaking body. In 1986, the National Advisory Board for Science and Technology, chaired by the prime minister, was created, and merged with the ministry. In the following year, a National Science and Technology Policy (NSTP) was approved by ministers of the federal, provincial, and territorial governments. The NSTP has emphasized a strong push linking national research to national needs.

30DOMESTIC TRADE

Between 1986 and 1989, wholesale trade grew 27% to c$298.8 billion. Wholesalers' and manufacturers' sales branches are the most prominent wholesale and distribution agencies. Wholesaling is particularly prominent in foodstuffs, lumber and building supplies, hardware, coal, clothing, dry goods, automotive equipment, and machinery. In producer goods generally, however, direct relations are often maintained by resident or traveling agents.

Retail trade in 1991 was c$181,2 billion. Large-volume outlets, including department stores, large mail-order houses, and chain stores, often buy direct from the manufacturer. A wide variety of local and imported goods is available in all major towns and cities. Vast indoor shopping complexes have been developed in the larger cities, including Eaton Centre in Toronto with over 300 stores and the West Edmonton Mall in Alberta.

Because of Canada's size and its regional economic differences, distribution is essentially regional. Toronto and Montreal dominate merchandising, are the headquarters of much of Canada's trade and financial apparatus, and do by far the greatest share of import business. Winnipeg is the business center for grain and agricultural implements. Vancouver is the center of the growing British Columbia market.

In 1989, c$9,186 million was spent on advertising, nearly ten times the amount spent a decade earlier. There is considerable advertising overflow from the US. In 1989, there were 8,202 Canadian advertising agencies. Shopping hours are 9:30 AM to 6 PM, Monday through Saturday; many stores stay open to 9 PM on

Thursday and Friday nights. Normal banking hours are from 10 AM to 3:30 PM, Monday through Friday.

31FOREIGN TRADE

Canada's exports are highly diversified; the principal export groups are industrial goods, forestry products, mineral resources (with crude petroleum and natural gas increasingly important), and agricultural commodities. Imports are heavily concentrated in the industrial sector, including machinery, transport equipment, basic manufactures, and consumer goods. Trade balances are almost invariably favorable: in 1990, exports were at c$120,521 million and imports c$146,057 million.

The principal exports in 1990 (in millions of Canadian dollars) were as follows:

Wheat	3,383.0
Other farm and fish products	9,942.4
Crude petroleum and natural gas	8,878.2
Other energy products	5,479.7
Lumber	6,566.9
Motor vehicles and parts	33,875.6
Other manufactured goods	34,904.0
Pulp and paper	14,271.5
Chemicals and fertilizers	7,105.5
Metals and minerals	19,170.6
Other exports	2,479.5
TOTAL	146,056.9

The principal imports in 1990 (in millions of Canadian dollars) were as follows:

Construction materials	2,511.2
Industrial materials	24,513.5
Motor vehicles and parts	30,490.9
Machinery and equipment	42,501.9
Food	9,119.3
Other consumer goods	15,844.0
Chemicals and chemical products	3,283.0
Crude petroleum	5,381.1
Other energy products	2,686.5
Other imports	2,210.5
TOTAL	135,258.9

The US is by far Canada's leading trade partner, with the latter exchanging raw materials such as crude petroleum and processed items such as paper for US machinery, transportation and communications equipment, and agricultural items such as citrus fruits. In 1992, the US accounted for 77% of Canada's exports and 69% of imports, and total Canadian-US trade was the largest between any two countries in the world in 1986. In October 1987, Canada and the US reached an agreement that would eliminate tariffs and nontariff trade barriers between the two countries by 1 January 1989. In 1992, the United States, Canada, and Mexico signed the North American Free Trade Agreement (NAFTA), which was ratified by all three countries the following year. The US-Canadian FTA will be superseded by this trilateral agreement.

The principal trade partners in 1990 (in millions of Canadian dollars) were as follows:

	EXPORTS	IMPORTS
US	110,282	92,892
EC (excluding UK)	8,304	9,931
Japan	7,638	8,223
UK	3,461	4,935
Other OECD countries	3,488	4,950
All other countries	12,883	14,328
TOTALS	146,057	135,259

³²BALANCE OF PAYMENTS

Canada's merchandise balances, although fluctuating, showed consistent surpluses between 1961 and 1992, except for 1975. These, however, have been offset by persistent deficits from other transactions. Sources of these deficits include Canada's indebtedness to other countries, travel of Canadians abroad, payments for freight and shipping, personal remittances, migrants' transfers, official contributions, and other Canadian government expenditures abroad. The deficit on the current account was expected to widen to c$30 billion in 1992, or 4.5% of GDP. Despite strong merchandise export growth and a trade account surplus reaching 1% of GDP by 1994, large deficits in services trade (especially tourism) and investment income (primarily interest) may prevent much improvement in the current account through 1994.

In 1992 merchandise exports totaled $133,303 million and imports $125,120 million. The merchandise trade balance was $8,183 million.

The following table summarizes Canada's balance of payments for 1991 and 1992 (in millions of US dollars):

	1991	1992
CURRENT ACCOUNT		
Goods, services, and income	–25,293	–23,274
Unrequited transfers	–49	262
TOTALS	–25,343	–23,012
CAPITAL ACCOUNT		
Direct investment	–233	2,015
Portfolio investment	15,144	8,248
Other long-term capital	987	308
Other short-term capital	10,903	3,218
Exceptional financing	—	—
Other liabilities	—	—
Reserves	2,486	5,807
TOTALS	29,287	19,596
Errors and omissions	–3,944	3,416
Total change in reserves	2,231	5,843

³³BANKING AND SECURITIES

The Bank of Canada, which began operations in 1935, is a government-owned institution that regulates the total volume of currency and credit through changes in the cash reserves of 7 domestic (at the end of 1991) chartered banks and 55 foreign bank subsidiaries. The Bank of Canada also acts as the government's fiscal agent, manages the public debt, and has the sole right to issue paper money for circulation in Canada. It is empowered to buy and sell securities on the open market, to fix minimum rates at which it will make advances, and to buy and sell bullion and foreign exchange. The Bank rate rose to a high of 21.24% in August 1981, before declining to 7.05% in March 1987. The bank rate was 4.11% in 1993. Total assets of the Bank of Canada were c$28.5 billion at the end of 1993.

The Federal Business Development Bank, established as the Industrial Development Bank in 1944 as a subsidiary of the Bank of Canada, has operated as a separate entity since 1975. It does not engage in the business of deposit banking but supplements the activities of the chartered banks and other agencies by supplying medium- and long-range capital for small enterprises.

The 7 domestic chartered banks are commercial and savings banks combined, and they offer a complete range of banking services. Canada's banks were reorganized in 1992 under the Banking Act. Every 10 years the banks' charters are subject to renewal and the Banking Act is revised to keep abreast of changing trends, a practice unique to Canada. The banks were reorganized into Schedule I and II banks. The Schedule I banks are banks whose ownership is public. No one shareholder in Schedule I banks controls more than 10% of the shares. Schedule II banks are subsidiaries of foreign or domestic banks that are held privately or semi-privately. There are 7 Schedule I and 55 Schedule II banks in Canada. As of 1991, they had 7,583 branch bank offices, an average of 1 office for every 3,600 Canadians. The Schedule I and II banks' personal savings and foreign currency deposits were c$301.33 in 1993.

The Toronto Stock Exchange was founded in 1852 and incorporated in 1878; the Standard Stock and Mining Exchange, incorporated in 1908, merged with it in 1934. Its members have branch offices in principal Canadian cities and in some US financial centers.

The Montreal Stock Exchange was incorporated in 1874. In 1974, it merged with the Canadian Stock Exchange, which was organized in 1926 as the Montreal Curb Market. Other securities exchanges are the Winnipeg Stock Exchange, founded in 1903; the Vancouver Stock Exchange, founded in 1907; and the Alberta Stock Exchange (formerly Calgary Stock Exchange), founded in 1913.

³⁴INSURANCE

Of the billions of dollars worth of coverage that Canadians buy every year, most is either life and health insurance or property and casualty insurance. In 1993, Canada had about 900 insurance companies, of which 395 were federally registered. Canadians lead the world in purchases of life and health insurance, which, in 1991, totaled nearly c$1,122 billion with federally registered companies. In 1992, property and casualty insurers wrote c$14.6 billion worth of premiums. About half were for automotive insurance. Life insurance in force in 1992 totaled c$1,317 billion.

³⁵PUBLIC FINANCE

By far the largest item of expenditure of the federal government is for social services, including universal pension plans, old age security, veterans benefits, unemployment insurance, family and youth allowances, and assistance to disabled, handicapped, unemployed, and other needy persons.

Through the early 1970s, federal budgets remained relatively in balance, fluctuating between small surpluses and small deficits. Since then, however, the budget has been in continuous and growing deficit. The federal government made some progress in slowing down the growth of the public debt after 1984, reducing the annual federal deficit from c$38.3 billion in fiscal 1985 to c$28.1 billion in fiscal 1991. Government options to reduce the deficit are constrained by the high level of non-discretionary spending in the federal budget. Statutory social transfers to individuals and to provincial and local governments accounted for 40% of the 1990/91 budget, while public debt service payments accounted for and additional 28% of spending. The federal budget deficit in 1991/92 was c$34.6 billion. The deterioration was largely due to weak revenues, as the recession ended up longer and deeper than the government anticipated. Aggregate provincial deficits were a record c$22.2 billion in 1991/92.

Sources of provincial revenue include various licenses, permits, fines, penalties, sales taxes, and royalties, augmented by federal subsidies, health grants, and other payments. Federal grants and surpluses and federal payments to the provinces under the federal-provincial tax-sharing arrangements constitute a major revenue source of the provinces. Corporation and personal income taxes provide a considerable portion of the revenue of Quebec. The largest provincial expenditures are for highways, health and social welfare, education, natural resources, and primary industries. Real property taxes account for more than two-thirds of revenue for municipalities and other local authorities. Almost one-third of their expenditures goes to support local schools.

The following table shows actual revenues and expenditures for 1988 and 1989 in millions of dollars.

	1988	1989
REVENUE AND GRANTS		
Tax revenue	105,467	117,112
Non-tax revenue	13,844	15,425
Capital revenue	—	—
Grants	45	–94
TOTAL	119,356	132,631
EXPENDITURES & LENDING MINUS REPAYMENTS		
General public service	13,303	14,802
Defense	10,417	11,055
Public order and safety	—	—
Education	4,250	4,344
Health	7,682	7,783
Social security and welfare	47,397	52,328
Housing and community amenities	1,860	1,914
Recreation, cultural, and religious affairs	949	901
Economic affairs and services	15,212	16,748
Other expenditures	36,915	41,441
Adjustments	–4,730	–2,354
Lending minus repayments	–391	237
TOTALS	132,864	149,199
Deficit/Surplus	–13,508	–16,588

In 1989 Canada's total public debt stood at c$295,800 million, of which c$62,082 million was financed abroad. Gross external debt was c$329.7 billion in 1990, with debt service payments of c$32,923 million.

36 TAXATION

The federal government levies direct and indirect taxes, of which the individual and corporation income taxes yield the largest return. Excise taxes (including a general sales tax), excise duties, and customs duties also produce a substantial revenue. Federal inheritance taxes were eliminated as of 1 January 1972. The federal goods and services tax (GST) went into effect on January 1, 1991. It is a 7% value-added tax on most goods and services.

For personal income tax purposes, exemptions in 1993 were a basic personal exemption of c$1,098; additional exemptions of c$915 for married persons, c$592 for persons 65 or older, and c$720 for the disabled; and c$269 for dependents who are older or infirm. The federal rate structure for 1993 ranged from 17% on the first c$29,590 of taxable income to 29% on taxable income in excess of c$59,180. The general tax rate payable by corporations on their taxable incomes was 38%. The "typical" provincial rate of corporate income tax in 1993 was 15.5%.

37 CUSTOMS AND DUTIES

Customs duties, once the chief source of revenue, have declined in importance as a revenue source. The tariff, however, still is an important instrument of economic policy. There is a wide range of duties, progressing from free rates on raw materials to higher duties as goods become more highly processed. Producer goods, including machinery of a kind not made in Canada, are subject to lower rates or are admitted free. Imports from the UK, most Commonwealth countries, and some crown colonies receive a tariff preference on a basis of reciprocity. Imports from nonmembers of GATT that have not negotiated a trade agreement with Canada are subject to the general or highest duty category.

In October 1987, Canada and the US reached agreement to establish a free trade area between the two countries, with all tariffs being eliminated within 10 years. Canadian commercial policy is generally opposed to the use of quantitative restrictions except as permitted by GATT, or for sanitary reasons, or in emergencies, or to allocate scarce supplies, or to meet balance-of-payments problems. Canada does not adhere to a general system of import licensing but does require permits for a limited number of products, such as electric power, petroleum, and natural gas and by-products. There are no free ports, but bonded facilities are operated at many ports. Except in grain, for which storage facilities are extremely large, customs warehousing is not extensive.

The United States, Canada, and Mexico signed the North American Free Trade Agreement (NAFTA) in December 1992. Approved by the legislatures of all three countries in 1993, NAFTA will replace the existing free trade agreement between Canada and the United States but retain many of its major provisions and obligations.

38 FOREIGN INVESTMENT

At the end of 1992, foreign direct investment in Canada, which has increased steadily since the early 1950s, amounted to c$137 billion. In addition, non-residents held c$231 billion worth of Canadian bonds. Since the 1970s, foreign investment in bonds has grown substantially, while growth in direct foreign investment has abated. Canada's external assets reached c$239 billion in 1992, 41% in the form of direct investment abroad. The United States receives roughly 58% of Canada's exported capital, more than any other nation. The Foreign Investment Review Agency, empowered to regulate foreign investments in Canada, was disbanded by the Conservative government that took office in 1984. To encourage foreign investment, an agency called Investment Canada was subsequently established. The investment review process has been simplified, and except for certain investments and acquisitions, screening has been abolished.

39 ECONOMIC DEVELOPMENT

Basically, Canada has a free-enterprise economy. However, the government has intervened in times of economic crisis and to accomplish specific social or economic goals. For example, in October 1963, the Canadian government announced a plan, involving tariff rebates, designed to induce US automobile companies to increase the export of vehicles and parts from their plants in Canada; subsequently, US companies markedly increased the scale of their Canadian operations. To dampen speculative buying of the Canadian dollar, the government permitted the dollar to float in the foreign exchange markets as of 31 May 1970; the government's intent was also to make imports cheaper in terms of Canadian dollars, and thereby to dampen domestic inflation. Another attempt at economic intervention, the Canada Anti-Inflation Act, became effective on 16 December 1975. This legislation established an Anti-Inflation Board and an Anti-Inflation Appeal Tribunal to monitor wage and price guidelines, which are mandatory for key sectors of the economy. The act was part of a government program to limit the growth of public expenditures and public service employment, to allow the money supply to increase at a rate consistent with moderate real growth, and to establish new agencies and policies to deal with energy, food, and housing.

A recurrent problem for Canada has been the dominant position of US corporations and investors. Attempts to limit US influence have included tightened tax policies, the Foreign Investment Review Act, and, in 1980, the National Energy Program (NEP), which aimed at reducing foreign ownership of Canada's oil and gas industry, principally through assisting Canadian companies to take over foreign holdings. One beneficiary of the NEP was the government-owned Petro-Canada, created in the mid-1970s; by the end of 1985, Petro-Canada had become the country's second-largest oil company, ranked by assets. However, much of the NEP was eliminated in the mid-1980s by the Conservative government, which sought to encourage foreign investment and to privatize government-owned enterprises. Between 1984 and 1991, the government sold or dissolved over 20 federal corporations, deregulated much of the energy, transportation, and financial sectors, and removed many controls on foreign investment.

Canadian official development assistance during 1990/91 amounted to c$3 billion. Of bilateral aid, c$541.8 million went to African countries and c$368.9 million to Asia, Europe, and Oceania. In 1990–91, Canada contributed c$382.3 million in food aid, the largest per capita amount of any nation in the world.

40SOCIAL DEVELOPMENT

Welfare needs are met by federal, provincial, and municipal governments as well as by voluntary agencies. Community chests, federated funds, and welfare councils throughout the country, affiliated with the Canadian Welfare Council, provide cooperative planning, financing, and consultative services.

Federal programs include family allowances, old age security, and earning-related disability and survivors' pensions. In general, all children under 16 (and youths aged 16 or 17 who attend school full time), regardless of means, are eligible for small monthly allowances (the average was c$33.93 per child in 1991). Persons aged 65 and over receive monthly pensions, supplemented in some provinces on a means-test basis. Under federal-provincial programs, monthly allowances are paid to needy persons aged 65 to 69 and to needy persons aged 18 or over who are blind or totally and permanently disabled. The federal government reimburses each participating province for half the cost of unemployment insurance payments.

The provinces provide allowances to needy mothers and their dependent children, widows, and mothers whose husbands have deserted them, are disabled, or are in mental hospitals. Some provinces provide for divorced, separated, and unmarried mothers, and for mothers whose husbands are in penal institutions. Most provinces reimburse municipalities for part of the costs of aid to transients and to needy residents on the basis of a means test. Municipalities, provinces, and voluntary agencies finance child welfare services. Homes for the aged are generally maintained by municipalities and voluntary organizations.

Since 1941, a contributory scheme of unemployment insurance and a nationwide free employment service have been in operation. Workers contribute 2.25% of their earnings, and employers contribute at 1.4 times that rate. In every province, employers are also required to contribute to a workers' compensation insurance fund. Workers are indemnified for accidents or occupational diseases and receive medical aid during the period they do not work. In industries, burial expenses in case of accidental death are included.

Liberalization of divorce and abortion laws since the 1960s, coupled with the increased participation of women in the labor force, have brought significant changes to Canadian family life. In 1988, there were 7 marriages per 1,000 population, down from 8 in 1980. The median age at first marriage, having fallen for both sexes between 1951 and 1972, thereafter rose to 24 for women and 26 for men by 1988. Since 1968, the crude divorce rate (per 100,000 population) has risen by more than 400%, reaching 244 in 1985, when there were a total of 61,980 divorces. The fertility rate for 1985–90 was 1.7, up 2.4% from the previous 5-year period.

Women participate fully in the Canadian labor force, including business and the professions, although government reports show that their average earnings are still less than those of men. In July 1993, the Canadian Panel on Violence Against Women published a major study on this issue, including a controversial survey indicating that over 50% of Canadian woman had experienced a rape or attempted rape.

41HEALTH

Canada adopted a national health insurance scheme in 1971. Administered regionally, each province runs a public insurance plan with the government contributing about 40% of the cost

(mostly from taxes). Government regulations ensure that private insurers can only offer particular types of health care provision. Drug prices are low. Most hospitals and doctors operate privately. Hospitals are paid by allocated budgets and doctors receive fees per treatment. The system offers considerable choice, but there is little competition and the government has used rationing measures to limit health care expenditures. Access to health care and cost containment are good, but there are strains on the budget and an aging population increases the strain. Major health planning is carried on by provincial governments, most of which offer substantially free care for patients suffering from tuberculosis (8 cases per 100,000 people reported in 1990), poliomyelitis, and venereal diseases and for certain cancer victims. They also assume responsibility for mental health treatment. Municipalities are responsible for sanitation; communicable disease control; child, maternal, and school health care; public health nursing; health education; and vital statistics. In some cases they supply hospital care and medical service to the poor. The federal government provides consultant and specialist services to the provinces, assists in the financing of provincial programs, provides services to veterans and Indians, exercises control over the standard and distribution of food and drugs, maintains quarantine measures, and is responsible for carrying out certain international health obligations. The federal Department of National Health and Welfare provides financial assistance for provincial health and hospital services through the National Health Program, and for provincial hospital insurance programs through the Hospital Insurance and Diagnostic Services Act of 1957, under which the federal government shares the provinces' costs (since 1977, by means of tax transfers and cash payments). By 1973, this program had been established in all provinces and territories, covering more than 99% of the total population of Canada. Federal and provincial governments contribute toward construction costs of new hospitals. Total health care expenditures for 1990 were $51,594 million. The Canadian death rate of 7.7 per 1,000 population in 1993, the maternal death rate (1990–91) of 5 per 100,000 live births, and the infant mortality rate (1992) of 7 per 1,000 live births are among the lowest in the world. Diseases of the heart and arteries account for more than 40% of all deaths, and cancer accounts for just under one-third; the proportion of deaths from causes related to old age is rising. Accidents are the leading cause of death in childhood and among young adult males, and rank high for other population groups. From 1985 to 1990, the leading causes of death per 100,000 people were: (1) communicable diseases, maternal/perinatal causes (39); (2) noncommunicable diseases (395); and (3) injuries (48). In 1992, life expectancy at birth was estimated at 77 years. Canada had 391,000 births in 1992 (14 per 1,000 people), and a birth rate in 1993 of 14.2 per 1,000. Between 1980 and 1993, 73% of married women (ages 15 to 49) were using contraception. And, between 1990 and 1992, children up to one year of age were immunized as follows: tuberculosis (85%); diphtheria, pertussis, and tetanus (85%); polio (70%); and measles (85%). As of 1986 there were 1,048 hospitals of all types, with 170,721 beds. From 1985 to 1990, there were 16.1 beds per 1,000 people. In the far eastern part of Canada, hospital beds available in 1993 were as follows: New Brunswick (5,025); Newfoundland (2,900); Nova Scotia (5,357); and Prince Edward Island (675). From 1988 to 1992, Canada had 2.2 doctors per 1,000 people and a nurse to doctor ratio of 4.7. In 1990, there was one doctor for every 450 people. In 1991, there were 60,559 physicians, 14,621 dentists, and 262,288 nurses. In 1990, there were 22,121 pharmacists.

42HOUSING

There were slightly more than 10 million occupied private dwellings in Canada in 1991. With the economy recovering only slowly from recession, housing starts were estimated at just over

156,000 during 1991, a 14% drop from 1990. Single detached homes are the predominant type of housing accommodation, although their relative numbers have gradually declined in favor of multiple dwellings. In 1951, 66.7% of all dwellings were single detached; by 1991, the proportion had dropped to 57%, while 32% were categorized as low density, 9% high rise, and 2% mobile dwellings. As of the 1991 Census, almost half of Canada's homes had been built since 1970.

In 1992, the federal government introduced the programs to promote home ownership: the First Home Insurance initiative and the Buyers' Home Plan.

43EDUCATION

Virtually the entire adult population is literate. The age limits of compulsory school attendance are roughly from age 6 to age 15. Primary schools lasts for eight years and secondary or high school another three to five years. In 1990, primary schools numbered 12,220. There were 154,698 teachers and 2,371,558 students in primary schools. The same year, secondary schools had 164,125 teachers and 2,292,735 students.

Each province is responsible for its own system of education. While the systems differ in some details, the general plan is the same for all provinces except Quebec, which has two parallel systems: one mainly for Roman Catholics and speakers of French, the other primarily for non-Catholics and speakers of English. Quebec, Newfoundland, Alberta, Saskatchewan, and, to a lesser extent, Ontario provide for public support of church-affiliated schools. Primary and secondary education is generally free, although nominal fees are charged for secondary education in some schools or provinces. Public elementary and secondary schools are administered by the provinces and Yukon Territory.

During 1992 there were 69 degree-granting colleges and universities in Canada. There are provincial universities in all provinces; other degree-granting institutions are private or connected with a religious denomination. In 1991, full-time enrollment in all higher level institutions, colleges and universities was 1,942,814. The federal government operates one military college with degree-granting powers conferred by the province of Ontario. Since 1977, the federal government has contributed to postsecondary education by cash payments and tax transfers, independently of provincial program costs. In 1985/86, the federal government was responsible for nearly 60% of the c$9.4 billion spent on postsecondary education.

Canadian higher education began with the founding of the Collège des Jésuites in Quebec City in 1635. The Séminaire de Québec, another Jesuit institution, established in 1663, became Laval University in 1852. Other early institutions on the French collegiate model were the Collège St. Boniface in Manitoba (1827), the University of Ottawa (1848), and St. Joseph's University in New Brunswick (1864). Although many French institutions survive—most notably the University of Montreal, which separated itself from Laval in 1920—most university-level instruction is conducted in English on the Scottish, British, or US model. The first English-language college in Canada was King's College in Windsor, Nova Scotia (1789). Two private universities on the Scottish model are Dalhousie University in Halifax (1818) and McGill University in Montreal (1821). The first state-supported institution, founded in 1827 on the principles of Anglicanism and loyalty to the British crown, was King's College at York in Upper Canada, which became the University of Toronto, the largest and one of the most distinguished of Canadian institutions. Universities in each of the four western provinces—Manitoba, Saskatchewan, Alberta, and British Columbia—founded in the late 19th century, represent a Canadian adaptation of the US state land-grant universities.

Canada also has numerous community colleges, teachers' colleges, technical institutes, nursing schools, and art schools. Adult education is sponsored by universities, colleges, school boards, government departments, and voluntary associations, each of which has some other primary function. The Canadian Broadcasting Corp., the National Film Board, and many museums, art galleries, and libraries engage in adult education as part of their work. Instructors are represented by the Canadian Association of University Teachers, and students by the Canadian Federation of Students.

44LIBRARIES AND MUSEUMS

Municipal public libraries serve the large cities and many small towns and rural areas, and regional units supply library service to scattered population areas. Traveling libraries, operated by provincial governments or university extension departments, also provide mail services for more isolated individuals and communities. Although public libraries are organized and financed by municipalities, in most provinces the provincial government supervises library services and makes grants to the municipal units. Special libraries of various kinds and at various levels serve limited groups.

In 1988, 158.4 million volumes were borrowed from 6,157 municipal, regional, and provincial libraries, which have a stock of about 60 million volumes. The National Library at Ottawa (7,200,000 volumes) publishes *Canadiana*, a monthly catalog of books and pamphlets relating to Canada, and maintains the National Union Catalogue of books in all major Canadian libraries.

There were about 1,900 museums, art galleries, and related institutions in Canada in 1990. The National Arts Center is located in Ottawa, as are Canada's four national museums: the National Gallery of Canada (including the Canadian Museum of Contemporary Photography), the Canadian Museum of Civilization (including the Canadian War Museum), the National Museum of Natural Sciences, and the National Museum of Science and Technology (including the National Aviation Museum and the Agriculture Museum). The major museums and art galleries, located in the principal cities, provide valuable educational services to adults and children; many supply traveling exhibitions for their surrounding areas or regions. The National Gallery conducts extension work throughout the country and sends many exhibitions on tour.

45MEDIA

There were 20,126,490 telephones in Canada in 1991. The 10 public and private companies in Telecom Canada provide a major share of the nation's telecommunications services, including all long-distance service, and link regional networks across Canada. Telegraph services are operated by the two transcontinental railroads and by the federal government to outlying districts. All external telecommunication services are operated by the Canadian Overseas Telecommunication Corp., a crown agency. The Post Office became a crown corporation in 1981.

The Broadcasting Act of 1968 entrusted the Canadian Radio-Television Commission with the regulation and supervision of all aspects of the broadcasting system. The publicly owned Canadian Broadcasting Corp. (CBC) provides the national broadcasting service in Canada. Its radio and television facilities extend from the Atlantic Ocean to the Pacific and north to the Arctic Circle. The CBC has broadcasting stations in the principal cities and operates both English- and French-language national networks. Privately owned local stations form part of the networks and provide alternative programs. As of 1991, there were 900 AM broadcasting stations, 800 FM stations, and 2,039 television stations. In the same year there were 27,776,000 radios and 17,252,000 television sets. Radio Canada International, the CBC's shortwave service, broadcasts in 12 languages to Europe, Africa, Latin America, Asia, the Middle East, the South Pacific, and the US.

The Canadian communication satellites play an increasingly significant role in efforts to bring radio and television services to the more remote parts of the country, particularly in the north. Beginning in late 1980, a new television network began broadcasting programs in the Inuit language via satellite, offering viewers the opportunity to "talk back" through their television sets to people in other communities. As of May 1987, radio and television services reached 99% of Canadian homes. A new broadcasting policy announced in April 1983 increased the quantity of Canadian content required in programming, established a fund to assist private and independent producers, and relaxed licensing requirements for the use of satellite earth-stations for radio and television reception.

In 1991 there were 107 daily newspapers. Although some newspapers in Montreal, Quebec, Toronto, Winnipeg, and Vancouver have more than local influence, most circulate only on a regional basis and have a limited number of readers. Rural areas are served by some 1,100 monthly and weekly publications. There are many consumer magazines, but only *Maclean's* is truly national. Three large news-gathering organizations are the Canadian Press, a cooperatively owned and operated venture, the British United Press, and United Press International of Canada.

Canada's leading newspapers (with their 1991 daily circulations) include the following:

	LANGUAGE	CIRCULATION
Toronto Star (all day)	English	494,681
Globe and Mail (Toronto, m)	English	330,000
Le Journal de Montréal (m)	French	281,686
Toronto Sun (m)	English	252,895
Vancouver Sun (e)	English	193,749
La Presse (Montreal, m)	French	186,590
Ottawa Citizen (all day)	English	173,091
Province (Vancouver, m)	English	172,387
Edmonton Journal (m)	English	159,159
Gazette (Montreal, m)	English	158,493
Winnipeg Free Press (e)	English	151,928
Calgary Herald (m)	English	121,282
Hamilton Spectator (e)	English	111,461
Le Soleil (Quebec, e)	French	96,799

46ORGANIZATIONS

Cooperatives are very important in Canadian agriculture and fishing, and also provide housing, medical insurance, transportation, and other services. The Canadian Red Cross Society, affiliated with the International Red Cross Society, has branches in all 10 provinces. Almost every city has a chamber of commerce, affiliated with the national Canadian Chamber of Commerce. Among organizations active in education are the Canadian Association for Adult Education, the Canadian Association of University Teachers, and the Industrial Foundation on Education, a research organization aiming to promote aid to education by business. The Canada Council is the official national agency for promotion of the arts, humanities, and the social sciences. The Royal Canadian Academy of Arts is the oldest arts organization with national prestige. The Canada Arts Council, a federation of professional cultural organizations, includes the Royal Architectural Institute of Canada, the Canadian Authors' Association, the Canadian Music Council, the Sculptors' Society of Canada, and similar societies. Many voluntary societies are active in the field of health.

47TOURISM, TRAVEL, AND RECREATION

One of Canada's principal attractions for tourists is its extraordinary geographic variety: from the polar ice cap to the mountains, fjords, and rain forests of the west coast, from the lakes, forests, and ranchlands of the interior to the rugged shores and fine beaches of the east, Canada offers a remarkable range of scenic wonders. The excavation of L'Anse aux Meadows in Newfoundland, with its Norse artifacts and reconstructed dwellings, has been designated a world heritage site by UNESCO, as have Nahanni National Park in the Northwest Territories and Dinosaur Park in Alberta's Red Deer Badlands. Among the most spectacular parks are the Kluane National Park in the Yukon and the Banff (with Lake Louise) and Jasper national parks in the mountains of Alberta. Other attractions include the Cabot Trail in Nova Scotia; the Bay of Fundy, between New Brunswick and Nova Scotia; and the Laurentians and the Gaspé Peninsula in Quebec.

The arts and crafts of the Dene Indians and the Inuit may be seen in cooperative workshops in Inuvik in the Northwest Territories; and of the North West Coast Indians, at the reconstructed Indian village Ksan in British Columbia. Quebec City is the only walled city in North America; picturesque old fishing villages are to be found in the Atlantic provinces. Fishing and hunting attract many sportsmen to Canada, and ice hockey attracts many sports fans, particularly to the Forum in Montreal. Major league baseball teams play in Montreal and Toronto. In 1992, the Toronto major league baseball team, the Blue Jays, became the first non-American team to both play in and win the World Series. Toronto again won the World Series in 1993 on a dramatic 9th inning home run by outfielder Joe Carter. It was only the second time the World Series had ended on a home run.

One of the world's foremost summer theatrical events is the Shakespeare Festival at Stratford, Ontario. Toronto is known for its many theaters, the tower, and a fine zoo; Montreal, the second-largest French-speaking city in the world (after Paris), is famous for its fine French cuisine, its nightlife, its vast underground shopping and entertainment network, and its excellent subway system.

Montreal in 1967 hosted a major world trade exhibition, EXPO 67; the Summer Olympics took place in that same city in 1976. A world's fair, EXPO 86, was held in Vancouver in 1986, and Calgary was the site of the 1988 Winter Olympics.

In 1991, Canada was the world's tenth most popular tourist destination. In that year, 14,988,600 tourists arrived from abroad, 80% of them from the US and 11% from Europe. Gross receipts from tourism were US$5.5 billion.

Citizens of the US do not need passports but should carry documents attesting to their citizenship, such as birth certificates or voter registration cards. Alien residents should carry their green cards. Nationals of other countries must have valid passports and may require visitor visas; they should check with the nearest Canadian embassy, consulate, or high commission. In 1991, a 7% Goods and Services Tax went into effect; however, it is refundable to foreign tourists. Montreal was scheduled to open a government-run casino on the Notre-Dame in the Palais de la Civilisation in early 1994.

48FAMOUS CANADIANS

Political Figures

Because of their exploits in establishing and developing early Canada, then known as New France, a number of eminent Frenchmen are prominent in Canadian history, among them the explorers Jacques Cartier (1491–1557), Samuel de Champlain (1567?–1635), Étienne Brulé (1592?–1633), Jacques Marquette (1637–75), Robert Cavelier, Sieur de la Salle (1643–87), and Louis Jolliet (1645–1700); François Xavier de Laval de Montigny (1623–1708), first and greatest bishop of Quebec; Jean Baptiste Talon (1625?–94), first and greatest intendant, who re-created the colony on a sound economic basis; and Louis de Buade, Comte de Palluau et de Frontenac (c.1622–98), greatest of the French royal governors. Great explorers of a later period include Pierre Gaultier de Varennes, Sieur de la Vérendrye (1695–1749), Sir Alexander

Mackenzie (1764–1820), David Thompson (1770–1857), Simon Fraser (1776–1871), Joseph E. Bernier (1852–1934), and Joseph Burr Tyrrell (1858–1957). Louis Riel (1844–85), of Indian and French-Irish ancestry, led the métis in rebellion in 1869–70 and 1885, when he was captured and hanged for treason.

Fathers of confederation and other important 19th-century political figures include Louis Joseph Papineau (1786–1871) and William Lyon Mackenzie (1795–1861); Sir John Alexander Macdonald (1815–91), first prime minister of the confederation; George Brown (1818–80), Sir Samuel Leonard Tilley (1818–96), and Sir Charles Tupper (1821–1915). The greatest political leader at the turn of the century was Sir Wilfrid Laurier (1841–1919), prime minister from 1896 to 1911. The outstanding national leader of the first half of the 20th century was William Lyon Mackenzie King (1874–1950), Liberal prime minister for over 21 years (1921–26, 1926–30, 1935–48), who retired with a record of the longest service as prime minister in Commonwealth history. Charles Vincent Massey (1887–1967), governor-general from 1952 to 1959, was the first Canadian to represent the British crown in Canada. Lester Bowles Pearson (1897–1972), prime minister and Canada's longtime UN representative, won the Nobel Prize for peace in 1957. Pierre Elliott Trudeau (b.1919) served as prime minister from 1968 to 1979 and again from 1980 to 1984, when he was succeeded by Brian Mulroney (b.1939). The best-known French-Canadian separatist was René Lévesque (1922–87), leader of the Parti Québécois, who became premier of Quebec in 1976.

Artists

Highly regarded Canadian painters include James Edward Hervey MacDonald (1873–1932), Thomas John ("Tom") Thomson (1877–1917), Frederick Horsman Varley (1881–1969), and Lawren Stewart Harris (1885–1970) of the Group of Seven; James Wilson Morrice (1864–1924); and Emily Carr (1871–1945). Paul-Emile Borduas (1905–60) and Jean-Paul Riopelle (b.1923) both were part of the Montreal School; however, after settling abroad, they probably became better known in France and the US than in their native country. Two other artists of distinction are James W. G. MacDonald (1897–1960) and Harold Barling Town (b.1924). The portrait photographer Yousuf Karsh (b.Armenia-in-Turkey, 1908) is a longtime Canadian resident.

Musicians

Well-known Canadian musicians include the composer Healey Willan (1880–1968); the conductor Sir Ernest Campbell MacMillan (1893–1973); the pianist Glenn Gould (1932–82); the singers Edward Johnson (1878–1959), Jon Vickers (b.1926), and Maureen Forrester (b.1931); the bandleader Guy Lombardo (1902–77); and, among recent popular singers and songwriters, Gordon Lightfoot (b.1938), Paul Anka (b.1941), Joni Mitchell (b.1943), and Neil Young (b.1945).

Actors

Canadian-born actors who are known for their association with Hollywood include Marie Dressler (Leila Koerber, 1869–1934), Walter Huston (Houghston, 1884–1950), Mary Pickford (Gladys Mary Smith, 1893–1979), Raymond Hart Massey (1896–1983), Walter Pidgeon (1897–1984), Norma Shearer (1904–83), Lorne Greene (1915–87), Raymond Burr (b.1917), William Shatner (b.1931), and Donald Sutherland (b.1935). Stage personalities include Beatrice Lillie (b.1894), Hume Cronyn (b.1911), and Christopher Plummer (b.1929).

Sports

Notable in the world of sports are ice-hockey stars Maurice ("Rocket") Richard (b.1921), Gordon ("Gordie") Howe (b.1928), Robert Marvin ("Bobby") Hull, Jr. (b.1939), Robert ("Bobby") Orr (b.1948), and Wayne Gretzky (b.1961).

Authors

Thomas Chandler Haliburton (1796–1865), author of *Sam Slick,* was the first Canadian writer to attain more than a local reputation. Sir Charles George Douglas Roberts (1860–1943) and Bliss Carman (1861–1929) were widely read poets and short-story writers. Archibald Lampman (1861–99) wrote sensitive poems about nature. Narrative poems about the northwest frontier by Robert William Service (1874–1958) achieved mass popularity, as did the backwoods novels of Ralph Connor (Charles William Gordon, 1860–1937). The animal stories and bird drawings of Ernest Evan Thompson Seton (b.UK, 1860–1946) are still highly regarded. Stephen Butler Leacock (1869–1944), economist and essayist, is regarded as Canada's leading humorist. The *Anne of Green Gables* novels of Lucy Maud Montgomery (1874–1942) have been popular with girls of several generations. Mazo de la Roche (1885–1961) achieved fame for her romantic Jalna novels about an Ontario family. Well-known contemporary novelists are Morley Edward Callaghan (b.1903), Hugh MacLennan (b.1907), Farley McGill Mowat (b.1921), Alice Munro (b.1921), Margaret Lawrence (b.1926), Mordecai Richler (b.1931), and Marian Passmore Engel (b.1933). The novels and plays of Robertson Davies (b.1913), newspaper editor, actor, music critic, and university administrator, crackle with wit. Lorne Albert Pierce (1890–1961) was a prominent editor and literary critic. Herbert Marshall McLuhan (1911–80) was a communications theorist and cultural critic. Herman Northrop Frye (b.1912) is a well-known literary critic, and Margaret Atwood (b.1939) is a noted novelist and poet. The British newspaper publisher William Maxwell Aitken, 1st Baron Beaverbrook (1879–1964), was born in Canada.

The *Histoire du Canada* (1845) of François Xavier Garneau (1809–66) stimulated a great interest in French Canada's heritage. Joseph Octave Crémazie (1827–79) was the first notable French Canadian poet. The poems of Louis Honoré Fréchette (1839–1908) were crowned by the French Academy. Louis Hémon (1880–1913), a French journalist who came to Canada in 1910 and spent only 18 months there, wrote the classic French Canadian novel *Maria Chapdelaine* (1914). Authors of realistic novels dealing with social and economic problems of French Canada include Claude-Henri Grignon (1894–1976), author of *Un Homme et son péché* (1933); Jean-Charles Harvey (1892–1967), author of *Les Démi-civilisées* (1934); Ringuet (Dr. Philippe Panneton, 1895–1960), author of *Trente Arpentes* (1938); Germaine Grignon Guevremont (1900–1968); Roger Lemelin (b.1919), author of *Au pied de la pente douce* (1944); and Gabrielle Roy (Carbotte, b.1909-1993). Gratien Gélinas (b.1909) is an actor, director, and dramatic satirist. Abbé Félix Antoine Savard (1896–1982) wrote a poetic novel of pioneer life, *Menaud, maître-drayeur.*

Scientists and Inventors

Among the famous Canadian scientists and inventors are Sir Sanford Fleming (1827–1915), inventor of standard time; Sir William Osler (1849–1919), the father of psychosomatic medicine; and Sir Charles Saunders (1867–1937), who developed the Marquis wheat strain, which revolutionized wheat growing in northern latitudes. The codiscoverers of insulin, Sir Frederick Grant Banting (1891–1941) and John James Richard Macleod (1876–1935), were awarded the Nobel Prize for medicine in 1923. George Brock Chisholm (1896–1971) was an eminent psychiatrist and former head of WHO. Gerhard Herzberg (b.Germany, 1904) won the 1971 Nobel Prize for chemistry for his work on molecular spectroscopy. Marius Barbeau (1883–1969), anthropologist and folklorist, was an authority on totem poles and Canadian folk music.

⁴⁹DEPENDENCIES

Northwest Territories

The Northwest Territories constitute all of Canada north of 60°N except the Yukon and the northernmost parts of Quebec and Newfoundland. Also included are certain islands in Hudson Bay, James Bay, and Ungava Bay, plus the vast islands to the north. Total land area is 3,293,020 sq km (1,271,438 sq mi). Most of the people who live in the territories are Inuit or Indians. The 1991 estimated population was 54,317. The Mackenzie River and its tributaries, the Athabasca and Slave, provide an inland transportation route of about 2,700 km (1,700 mi). There is some traffic on Lakes Athabasca, Great Bear, and Great Slave. Most of the settlements in Mackenzie are linked by scheduled air service.

The territory is governed by a commissioner, for whom policy is determined by the governor-general-in-council or the Ministry of Indian Affairs and Northern Development, and by a 22-member elected territorial council. The territorial capital since 1967 has been Yellowknife. Following the approval by voters in April 1982 of a proposal to divide the territory, the federal government agreed in principle to the division pending settlement of native land claims and other issues. The Inuit of the east call their proposed eastern jurisdiction Nunavat; Indians of the west—where opinion on the proposal was sharply split—call their proposed division Deneden. Land claims of Dene Indians and the Inuit overlap.

Northwest Territories' mineral resources include rich deposits of gold, silver, lead, tungsten, and zinc. The Northwest Territories contain some of Canada's total mineral resources. It was the Yukon gold rush in 1897 that triggered a large migration of people northward. In one year 30,000 people from the lower parts of Canada were in the Northwest and Yukon Territories looking for gold.

Today, Canada leads in the production of oil, natural gas and coal. Canada's mining industry contributes US$20 billion annually to the economy and employs 145,000 people. In 1989 Canada's oil and gas industry was valued at US$19 billion with 55% of the revenue coming from oil production. Whitefish and trout are caught in Great Slave Lake; the 100,000 or more lakes of the territories provide an "angler's last frontier" in North America for sport fishermen. Fur production is a sizable industry in the Northwest Territories. A handicrafts industry supplies Inuit-made sculpture and prints to the Canadian Handicrafts Guild.

Yukon Territory

The Yukon Territory, located north of British Columbia and east of Alaska, has a land area of 478,970 sq km (184,931 sq mi) and had an estimated population in 1990 of 29,708, of whom some one-fifth were of Indian origin. The principal town is Whitehorse, the capital. An all-weather roadway connects the territory with Alaska and British Columbia, and a railroad connects Whitehorse with ocean shipping at Skagway, Alaska. Air service is available to and from Edmonton, Vancouver, and Fairbanks, Alaska. There are local telephone services in the three chief towns. The territory was separately constituted in June 1898. Since 1978, the Yukon has had a legislative assembly, consisting of 16 elected members. In late 1982, the federal government gave its consent for the Yukon cabinet to call itself the Executive Council and officially to take over some powers hitherto reserved by the federally appointed commissioner, as representative of the governor-general-in-council, or by the minister of Indian affairs and northern development. The Yukon government has recently pressed for provincial status.

Mineral resources include rich deposits of gold, silver, lead, tungsten, and zinc. Mining and tourism are Yukon's principal industries.

⁵⁰BIBLIOGRAPHY

Baillargeon, Samuel. *Littérature canadienne-française*. Montreal: Fides, 1957.

Bell, David V. J. *The Roots of Disunity: A Study of Canadian Political Culture*. Rev. ed. Toronto: Oxford University Press, 1992.

Brebner, John Bartlett. *Canada: A Modern History*. Ann Arbor: University of Michigan Press, 1970.

Bumsted, J. M. *The Peoples of Canada*. New York: Oxford University Press, 1992.

Canada in Pictures. Minneapolis: Lerner, 1993.

The Canadian Encyclopedia. 2d ed. Edmonton: Hurtig Publishers, 1988.

Canadian Institute of International Affairs. *Canada in World Affairs*. Toronto: Oxford University Press, 1952-date.

Clement, Wallace and Glen Williams (eds.). *The New Canadian Political Economy*. Kingston: McGill-Queen's University Press, 1989.

Creighton, Donald. *Canada's First Century, 1867–1967*. New York: St. Martin's, 1970.

Dawson, Robert MacGregor, *Dawson's the Government of Canada*. 6th ed. Toronto: University of Toronto Press, 1987.

Dickey, John Sloan (ed.). *Canada and the American Presence*. New York: New York University Press, 1975.

Easterbrook, W. T., and Hugh G.J. Aitken. *Canadian Economic History*. Toronto: University of Toronto Press, 1988.

Fox, Annette B., et al. (eds.). *Canada and the United States: Transnational and Transgovernmental Relations*. New York: Columbia University Press, 1976.

Halpenny, Frances, et al. (eds.). *Dictionary of Canadian Biography*. 7 vols. to date. Toronto: University of Toronto Press, 1972-date.

Ingles, Ernest B. *Canada*. Santa Barbara, Calif.: Clio, 1990.

Jenness, Diamond. *The Indians of Canada*. Ottawa: National Museum of Canada, 1958.

Joy, Richard J. *Canada's Official Languages: The Progress of Bilingualism*. Toronto: University of Toronto Press, 1992.

Lanctot, Gustave. *A History of Canada*. Cambridge, Mass.: Harvard University Press, 1963.

Lipset, Seymour Martin. *Continental Divide: The Values and Institutions of the United States and Canada*. New York: Routledge, 1990.

Lower, Arthur Reginald Marsden. *Canadians in the Making: A Social History of Canada*. Toronto: Longmans, Green, 1958.

Malcolm, A. H. *The Canadians*. New York: Times Books, 1985.

Martin, Chester Bailey. *Foundations of Canadian Nationhood*. Toronto: University of Toronto Press, 1955.

Pearson, Lester B. *Mike: The Memoirs of the Right Honorable Lester B. Pearson*. 2 vols. New York: Quadrangle Books, 1972–73.

Putnam, Donald Fulton, and Donald P. Kerr. *Regional Geography of Canada*. Toronto: Dent, 1958.

Russell, Peter H. *Constitutional Odyssey: Can the Canadians Become a Sovereign People?* Toronto: University of Toronto Press, 1992.

Simeon, Richard, and Ian Robinson. *State, Society, and the Development of Canadian Federalism*. Toronto: University of Toronto Press, 1990.

Stewart, Gordon T. *The American Response to Canada since 1776*. East Lansing, Mich.: Michigan State University Press, 1992.

Story, Norah. *Oxford Companion to Canadian History and Literature*. Toronto: Oxford University Press, 1967.

Wade, Mason. *The French Canadians, 1760–1945*. Toronto: Macmillan, 1955.

Webber, Jeremy H. A. *Reimagining Canada: Language, Culture,*

Community and the Canadian Constitution. Montreal: McGill-Queen's University Press, 1994.

Woodcock, George. *The Canadians.* Cambridge, Mass.: Harvard University Press, 1980.

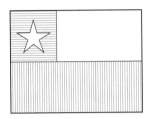

CHILE

Republic of Chile
República de Chile

CAPITAL: Santiago.

FLAG: The flag, adopted in 1817, consists of a lower half of red and an upper section of white, with a blue square in the upper left corner containing a five-pointed white star.

ANTHEM: *Canción Nacional (National Song)* beginning "Dulce Patria, recibe los votos".

MONETARY UNIT: The new peso (P) of 100 centavos replaced the escudo as the nation's monetary unit in October 1975. There are coins of 1, 5, 10, 50, 100, and 500 pesos, and notes of 500, 1,000, 5,000 and 10,000 pesos. P1 = US$0.0023 (or US$1 = P426.89).

WEIGHTS AND MEASURES: The metric system is the legal standard, but local measures also are used.

HOLIDAYS: New Year's Day, 1 January; Labor Day, 1 May; Navy Day (Battle of Iquique), 21 May; Assumption, 15 August; Independence Day, 18 September; Army Day, 19 September; Columbus Day, 12 October; All Saints' Day, 1 November; Immaculate Conception, 8 December; Christmas, 25 December. Movable religious holidays include Good Friday and Holy Saturday.

TIME: 8 AM = noon GMT.

¹LOCATION, SIZE, AND EXTENT

Situated along the southwestern coast of South America, Chile has an area of 756,950 sq km (292,260 sq mi). Comparatively, the area occupied by Chile is slightly smaller than the state of Montana. A long string of land pressed between the Pacific and the towering Andes, Chile is 4,270 km (2,653 mi) long N–S; it is 356 km (221 mi) wide at its broadest point (just north of Antofagasta) and 64 km (40 mi) wide at its narrowest point, with an average width of 175 km (109 mi) E–W. It is bordered on the N by Peru, on the NE by Bolivia, on the E by Argentina, on the S by the Drake Passage, and on the W by the Pacific Ocean. At the far SE, at the end of the Strait of Magellan (Estrecho de Magallanes), it has an opening to the Atlantic Ocean. Chile's boundary length is 12,606 km (7,833 mi).

Included in the national territory are the Juan Fernández Islands, Easter Island, and other Pacific islands. A dispute with Argentina over three small islands in the Beagle Channel almost led to war between the two countries in 1978, but papal intervention prevented hostilities. The issue was resolved peacefully by a treaty signed in the Vatican on 29 November 1984 and ratified on 2 May 1985, granting Chile sovereignty over the three islands, giving Argentina rights to waters east of the Strait of Magellan, and dividing the territorial waters south of Cape Horn between the two countries. There is another outstanding boundary problem with Bolivia over its claim for an opening to the sea. Chile also claims the Antarctic Peninsula and other areas of Antarctica, comprising 1,250,000 sq km (482,500 sq mi).

Chile's capital city, Santiago, is located in the center of the country.

²TOPOGRAPHY

Chile is divided into three general topographic regions: the lofty Andean cordillera on the east; the low coastal mountains of the west; and the fertile central valley between. The Andes, occupying from one-third to the entire width of the country, stretch from the Puna de Atacama in the north, a high plateau with peaks averaging 4,600 m (15,000 ft), to middle Chile, where, on the border

with Argentina, rises the highest peak in the Western Hemisphere, Aconcagua (7,265 m/23,834 ft), and then, diminishing in height, run south into the Chilean lake country, with its snow-capped volcanoes and several passes. The coastal range, verging from 300 to 2,100 m (1,000 to 7,000 ft) in height, rises from the sea along most of the coast. In the extreme north, the coastal mountains join with the Andean spurs to form a series of plateaus separated by deep gorgelike valleys. In the south, the valleys and the coastal range plunge into the sea and form a western archipelago; fjords reach into the range at about 42°S. The central valley, an irregular alluvial plain 965 km (600 mi) long, 73 km (45 mi) wide at its maximum, and up to 1,200 m (4,000 ft) high, begins below the arid Atacama Desert of the north and ends at Puerto Montt in the south. Fertile between the Aconcagua and Bío-Bío rivers, this valley is the center of agriculture and of population. Although some 30 rivers rise in the Andes and descend to the Pacific, cascades and great waterfalls severely limit navigation; the ocean itself facilitates transportation between the different regions of this narrow country.

The northern side of the Strait of Magellan, part of Patagonia (a region shared by Chile and Argentina), and part of the island of western Tierra del Fuego (divided between Chile and Argentina) is low, glaciated, morainal country.

³CLIMATE

Climatic zones range from the subtropical deserts in the north to the temperate rain forests of Aisén and the tundras of Magallanes in the extreme south. The cold Humboldt Current, traveling northward from the Antarctic, affects the climate of the coastal regions of central and northern Chile. Generally, however, Chile is divided into three climatic regions: (1) The north, which contains the Atacama Desert, one of the driest regions in the world, is characterized by hot and arid weather in the lowlands and occasional summer showers in the Andean highlands. (2) The middle, extending about 1,450 km (900 mi) from 30° to 43°S, has a Mediterranean climate, with mild, wet winters, averaging 11°C (52°F), and long, dry summers, averaging 18°C (64°F). (3) The south, a

93

region of mountains and fjords, has high winds and heavy rains. Annual rainfall ranges from no recorded precipitation in some parts of the north to 50–100 cm (20–40 in) around Concepción, in south-central Chile, to more than 500 cm (200 in) in some southern regions. South of the Bío-Bío River, rains occur all year round. The Andean highlands, even in the tropical north, are cold and snowy.

⁴FLORA AND FAUNA

Chile's botanical zones conform to the topographic and climatic regions. The northernmost coastal and central region is largely barren of vegetation, approaching the most closely an absolute desert in the world. On the slopes of the Andes, besides the scattered tola desert brush, grasses are found. The central valley is characterized by several species of cactus, the hard espinos, the Chilean pine, and the copihue, a red bell-shaped flower, Chile's national flower. In southern Chile, south of the Bío-Bío River, the heavy precipitation has produced dense forests of laurels, magnolias, and various species of conifers and beeches, which become smaller and more stunted to the south. The cold temperatures and winds of the extreme south preclude heavy forestation. Grassland is found in Atlantic Chile (in Patagonia). The Chilean flora is distinct from that of Argentina, indicating that the Andean barrier existed during its formation. Chilean species include the monkey-puzzle tree and the pinelike araucaria, also found in Australia. True pines have been introduced from the Northern Hemisphere.

Chile's geographical isolation also has restricted the immigration of faunal life, and only a few of the many distinctive Latin American animals are found. Among the larger mammals are the puma or cougar, the llamalike guanaco, the Andean wolf, and the foxlike chilla. In the forest region, several types of marsupials and a small deer known as the pudu are found.

There are many species of small birds, but most of the larger common Latin American types are absent. Few freshwater fish are native, but North American trout have been successfully introduced into the Andean lakes. Owing to the vicinity of the Humboldt Current, ocean waters abound with fish and other forms of marine life, which, in turn, support a rich variety of waterfowl, including different penguins. Whales are abundant, and some six species of seals are found in the area.

⁵ENVIRONMENT

The principal responsibility for environmental matters is vested in the programs department of the environment in the Ministry of Health and in the National Planning Office, as well as in the ecological advisory office in the Ministry of National Welfare and the department of the environment in the Ministry of Foreign Affairs.

As of 1987, endangered species in Chile included the South Andean huemul, tundra peregrine falcon, puna rhea, Chilean woodstar, ruddy-headed goose, and green sea turtle.

Chile's main environmental problems are deforestation, soil erosion, and the pollution of its air, water, and land. The clearing of the nation's forests for commercial purposes contributes to the erosion of the soil. Air and water pollution is especially acute in the urban centers where the population has doubled in the last 30 years. Industry and transportation are the main sources of air pollution. Chile contributes 0.1% of the global total for gas emissions. Water pollution results from the same sources. As of 1994, Chile has 112.3 cubic miles of water. One hundred percent of its urban dwellers have pure water. Seventy-nine percent of its rural population does not.

In 1994, there were nine species of mammals in a total of 90 that were considered endangered. Of 393 bird species, 18 were threatened with extinction. One type of fresh-water fish and 284 plant species in a total of 5,500 were also endangered.

⁶POPULATION

Chile's 1992 population was 13,231,803, compared with 11,329,736 at the time of the 1982 census. The population for the year 2000 was projected at 15,272,000, assuming a crude birth rate of 20.9 per 1,000 population, a crude death rate of 6.5, and a net natural increase of 14.4 during 1995–2000. As of 1992, 34% of the population was under 18 years of age. The annual rate of increase during 1982–92 was 1.55%, compared with 2.03% during 1970–82. It was estimated that 85% of the population was urban in 1993 and that the population density was 17.5 persons per sq km (45 per sq mi) in that year. However, over 80% of the people live in the central region, between La Serena and Concepción, although this part covers little more than a quarter of the area. Santiago, the largest city, had 233,060 inhabitants in 1992. Other large cities, with their estimated 1985 populations, include Viña del Mar, 302,765; Valparaíso, 276,737; Talcahuano, 246,556; Concepción, 329,304; Antofagasta, 226,749; Temuco, 240,880; and Rancagua, 187,134.

⁷MIGRATION

After the Spanish conquest, there were three main waves of immigration: Germans during 1800–50; Spaniards, Italians, Swiss, Yugoslavs, Syrians, Jordanians, and Lebanese around 1900; and Spaniards and European Jews during the 1930s and 1940s. Since World War II, permanent immigration has been minimal.

In the years immediately preceding and after the Allende victory in 1970, about 10,000 political refugees (largely Brazilians, Bolivians, and Argentines) came to Chile. After the military coup of 1973, however, the bulk of them were expelled. The 1970s also witnessed two successive waves of Chilean emigration when, as a reaction to the Allende victory and, later, as a result of the military coup, several hundred thousand Chileans departed the country for political and economic reasons. Many of them later returned. In 1990 a National Office of Refugees was established to facilitate the reincorporation of returning exiles into Chilean society. In its first three years this office assisted more than 13,000 of the 26,000 exiles who returned in this period.

There is a seasonal pattern of trans-Andean immigration to Argentina by Chilean agricultural workers, and for many years the presence of several thousand Chilean settlers in the Argentine part of Patagonia created a minority problem.

⁸ETHNIC GROUPS

Ethnically, the Chilean population is estimated at nearly 75% mestizo (mixed white and Amerindian), almost 20% white, and about 5% Amerindian. Mixtures between the conquering Spaniards, largely Andalusians and Basques, and the Mapuches (Araucanians) produced the principal Chilean racial type. An indigenous population of perhaps as many as 600,000 pure Mapuches live mainly in Temuco and in the forest region south of the Bío-Bío River. Remnants of other small tribal groups are found in isolated oases within the northern desert or live a nomadic existence on the archipelagos and islands of the extreme southern coast. A small minority of Germans and their descendants live in the Valdivia-Puerto Montt area.

⁹LANGUAGES

Spanish is the national language. A sizable segment of Mapuche (Araucanian) Amerindians use Spanish in addition to their native tongue. The only other language of any importance is German, spoken mainly in the Valdivia region.

¹⁰RELIGIONS

The Roman Catholic Church was disestablished by the constitution of 1925, but Roman Catholicism remains the principal religion, with about 89% of the population being at least nominally Catholic in 1993. The remaining population includes several

Protestant denominations, totaling about 11% of the population; an estimated 15,000 Jews (1990); and about 95,000 Amerindians who still practice an indigenous religion involving shamanism.

11 TRANSPORTATION

In 1991, Chile had 8,613 km (5,352 mi) of railways, the fourth largest rail system in Latin America. Rail lines in the desert area are used mainly for mineral transport. By 1988, cargo transportation by rail had reached nearly 6 million tons annually. There are five international railroads from Chile: a line to Tacna, Peru; two to La Paz, Bolivia; and two to Argentina. In 1975, the first section of a new subway was opened in Santiago; the second section was opened in 1980.

There were 79,025 km (49,106 mi) of roads in 1991, of which only 12% were paved. The Pan American Highway, extending 3,460 km (2,150 mi) from the Peruvian border to Puerto Montt, is Chile's principal road artery. In 1992 there were about 1,007,713 passenger cars and 206,790 commercial trucks, buses, and taxis. The Carretera Austral Presidente Augusto Pinochet, a highway, is under construction in the south; when complete it will link Cochrane in Coyhaique with Puerto Montt.

Chile has some 20 ports, 10 of which are used principally for coastal shipping. Valparaíso, the principal port for Santiago, is by far the most important. Arica, Iquique, Tocopilla, Antofagasta, Coquimbo, San Antonio, Talcahuano, and Punta Arenas are other important ports. In 1991, the Chilean merchant marine had 38 vessels over 1,000 tons and a GRT of 470,000.

Air transportation has become increasingly important. Chile's largest airline is the state-owned National Airlines of Chile (LAN-Chile), which provides both domestic and international service, and flew 922,700 passengers in 1992. LAN-Chile's only significant domestic competitor is Copper Airlines (LADECO), a privately owned company, which handled 760,500 passengers in 1992. Santiago has the principal international airport, which serviced 1,948,000 arriving and departing passengers in 1991.

12 HISTORY

Before the Spanish conquest, central and southern Chile were sparsely inhabited by Araucanian Amerindians, who came under the influence of the Incas in the early 15th century. The conquistador Pedro de Valdivia founded Santiago in 1541, and brought Chile north of the Bío-Bío River under Spanish rule. The Araucanians resisted Spanish rule and killed Valdivia in battle. Continued Amerindian resistance effectively barred Spanish settlement south of the Bío-Bío. Although subject to the viceroyalty of Peru, Chile enjoyed the status of captaincy-general and was largely administered from Santiago.

Chile had one of Latin America's first independence movements. A cabildo abierto ("town meeting") declared independence in 1810 in response to the French usurpation of the Spanish crown. Rival independence leaders Bernardo O'Higgins and José Miguel Carrera fought each other, then were overcome by Spanish troops. Eventually, Gen. José de San Martín, with O'Higgins as his chief ally, defeated the Spanish in 1817, and in 1818 Chile formally proclaimed independence. O'Higgins ruled from 1818 to 1823, during which time he built a navy and consolidated the Chilean government under his dictatorial regime. However, his anti-clerical and anti-nobility policies proved to be his undoing.

The next few years saw the growth of two political parties, the Conservative and the Liberal. While both were narrow elite factions, they differed in that Liberals favored a parliamentary, secular, federal system, while Conservatives wanted a traditional, religious, centralized system. The two groups fought bitterly, plunging Chile into civil strife until 1830, when Conservative Diego José Victor Portales assumed control of the political system.

Portales ruled as behind-the-scenes dictator from 1830 until his assassination in 1837. He launched a successful three-year

LOCATION: 17°31′ to 56°33′s; 66°25′ to 80°47′w. **BOUNDARY LENGTHS:** Peru, 169 km (105 mi); Bolivia, 861 km (535 mi); Argentina, 5,308 km (3,298 mi); coastline, 5,338 km (3,317 mi). **TERRITORIAL SEA LIMIT:** 12 mi.

war with Peru (1836–39), which destroyed a threatening Bolivian-Peruvian confederation. He also initiated a Conservative rule which was to last until 1861. During that period, Chile's territory expanded with new claims to Patagonia and the island of Tierra del Fuego, and in 1847, the founding of Punta Arenas on the Strait of Magellan.

Between 1861 and 1891, the Conservatives were forced to share power with the Liberals, who had won several legislative victories. A wave of liberal reforms curtailed the power of the Roman Catholic Church and the presidential office. At the same time, both parties suffered a series of splits and realignments. But most notable during this period was Chile's greatest military achievement. In the War of the Pacific (1879–83), Chile again fought Peru and Bolivia, this time over possession of the Atacama Desert and its nitrate deposits. After victories on land and sea, Chilean forces entered Lima in 1881. By a treaty signed in 1883, Peru yielded Tarapacá, while Bolivia surrendered Antofagasta. The disposition of the other contested areas, Tacna and Arica, was not finally settled until 1929, when, with US mediation, Tacna went to Peru and Arica to Chile.

In 1891, Jorge Montt, a naval officer, led a revolt that resulted in eight months of civil war. The triumph of Montt marked the beginning of a 30-year period of stable parliamentary rule. Bolstered by nitrate revenues, Chile's national treasury grew, especially during World War I. But at the same time the seeds of revolt were sown. Miners, farm workers and factory workers, sharing none of this prosperity, began to agitate for change. After the war ended, there was a recession, and the country was on the verge of civil war. In 1920, a coalition of middle and working class groups elected Arturo Alessandri Palma president. Alessandri, the son of an Italian immigrant, found himself in between the left's demands for change and the right's intransigence. He was deposed in a coup in 1924 but recalled in a countercoup in the following year. His second administration lasted only six months, but he left the legacy of a new constitution passed on 18 October 1925. The new system created a strong, directly elected executive to replace the previous parliamentary system.

The military strongman Gen. Carlos Ibáñez del Campo ruled Chile from behind the scenes until 1927, then served formally as president until 1931. US banks loaned large sums to Chilean industry, and efforts were made to salvage the foundering nitrate trade and boost the copper sector. World depression struck, however, bringing an end to foreign loans and a catastrophic drop in world copper prices. A general strike caused Ibáñez to flee in 1931. After two years marked by short-lived juntas and presidencies and a 100-day "socialist republic," Alessandri was again elected.

Chile pulled out of the depression by 1938, but popular demand for social legislation remained unsatisfied. The 1938 election was narrowly won by Radical Party member Pedro Aguirre Cerda, running under the banner of a catch-all coalition called the "Popular Front." His ambitious "new deal" program was never enacted, as Aguirre found himself in the crossfire of Chilean politics. His coalition dissolved formally in January 1941, and Aguirre died in November. In 1942, the Radicals won election easily over former dictator Ibáñez.

Juan Antonio Ríos governed moderately amid political conflict aroused by World War II. Ríos at first cooperated with Argentina in toning down the US-sponsored anti-Axis program but later led his country into a pro-Allied position, entering the war on the side of the US in 1944. After World War II, Chile went into an inflationary cycle, and riots and strikes broke out throughout the country. Ríos died in 1946, and a special election brought to power a coalition of Communists and former Popular Front supporters under Gabriel González Videla. González's coalition soon broke down, as the Communists organized demonstrations and strikes. Within months, González fired the three Communists he had appointed to cabinet positions. He then broke off relations with the Soviet Union, and outlawed the Communist party. Strikes and violence grew, and Chile, an example of stability by Latin American standards for so long, seethed with tensions. Chile's pursuit of industrialization, which had started with the Aguirre and Ríos administrations, had led to increasing social

problems as the cities bulged with unemployable rural workers. As the cost of living soared, the radicalism of the workers intensified.

The 1952 election brought the seventy-five-year-old Carlos Ibáñez del Campo back to power. The ex-dictator, who had been plotting to return to power for years, defeated González Videla by exploiting a split among the Radicals and the disaffected Communists. Despite his reputation as an authoritarian and his connection with Argentina's Perón, Ibáñez ruled democratically until 1958.

By 1958, the cost of living had soared and Chile's trade balance had moved from a large surplus to a deficit. Evidence of a general discontent could be seen in the 1958 presidential election. A narrow victory was won by Jorge Alessandri Rodríguez (a son of President Arturo Alessandri Palma), who received support from both Liberals and Conservatives. The Socialist Salvador Allende Gossens, supported by his own party and the newly legalized Communist Party, won 29% of the vote (compared with only 5% in 1952), and Eduardo Frei Montalva, candidate of the new Christian Democratic Party (Partido Demócrata Cristiano—PDC), ran third with 20% of the vote.

Aware of popular pressure for reform, Alessandri drew up a 10-year development plan, initiated in 1959 with construction projects, tax reforms, and a token start at agrarian reform. A devastating earthquake and tidal wave in 1960 cut drastically into Alessandri's programs, and his government was unable to regain momentum. In 1964, the traditional parties of the right and center lost strength to a wave of reform sentiment that shifted public attention to a choice between the socialist Allende and the moderate reformer Frei. In September 1964, Frei was elected by an absolute majority, and congressional elections in March 1965 gave the PDC a majority in the Chamber of Deputies and a plurality in the Senate.

The Frei government implemented numerous social and structural reforms. These included educational reform, land reform and a scheme to create a majority Chilean interest in Chile's copper mines. Frei became a cornerstone of the Alliance for Progress, a harsh critic of communism, and a leading exponent of Christian democracy. However, the reforms did not deliver as hoped, and overall economic growth was sporadic. The Frei administration was not able to control the endemic inflation that has plagued Chile for more than 80 years.

In the 1970 presidential election there were three contenders: Jorge Alessandri, PDC candidate Radomiro Tomic, and the Socialist Senator Salvador Allende. Allende, who was supported by Popular Unity, a leftist coalition that included the Communist Party, received 36.5% of the total vote. Alessandri followed with 35.2%, and Tomic with 28%, with 0.3% of the ballots left blank as a protest. Since no candidate received a majority of the popular vote, Congress was required by the constitution to select the president from the two leading candidates. The PDC supported Allende in exchange for a promise of full constitutional guarantees. The victory was unique in that for the first time in the Western Hemisphere, a Marxist candidate took office by means of a free election. Dr. Allende, inaugurated on 3 November 1970, called for a socialist economy, a new, leftist constitution, and full diplomatic and trade relations with Cuba, China, and other Communist countries. It was later revealed through US congressional investigations and independent journalistic inquiries that the US, with the help of the International Telephone and Telegraph Corp. (ITT), had secretly worked to thwart the election and confirmation of Allende.

The first full year of rule by Allende witnessed a rise in economic prosperity and employment, as well as an improvement in the standard of living of the poorer elements of the population. Allende expropriated US copper interests, and turned large rural landholdings into peasant communes. By 1972, however, the economy began to lag, and the situation was aggravated by middle- and upper-class resentment over the government's seizures of industrial and agricultural property. In June 1973, against a backdrop of strikes and street brawls beginning in the previous year, an abortive coup attempt was staged by a rightist army contingent. Throughout this period, the US Central Intelligence Agency had secretly supported the 1972 and 1973 strikes and disturbances, especially the truckers' strike, which had caused nationwide shortages of food and consumer goods.

On 11 September 1973, the Allende government was violently overthrown. Allende himself died—officially reported as a suicide. A four-man junta headed by Gen. Augusto Pinochet Ugarte seized power, dissolved Congress, banned all political activities, and declared that Marxism would be eradicated in Chile. At least 2,500 and possibly as many as 10,000 people were killed during and immediately after the coup. The military declared a state of siege, and assumed dictatorial powers.

During its 16 years in power, the military attempted to eradicate not only Marxism, but all vestiges of leftism, trade unionism, reformism, and, for that matter, any other deviation from the official military line. High on their list of priorities was the privatization of the Chilean economy, which had gradually become more dependent on the state over three decades, a movement that had accelerated dramatically under Allende. This included the attracting of foreign investment, virtually untouched by government regulations or requirements.

This powerful dose of economic liberalization was administered within a continuously authoritarian political system. After the original state of siege was lifted in 1978, Chile continued under a "state of emergency" until another state of siege was declared from November 1984 to June 1985. A third state of siege was in effect from September 1986 to January 1987. At each denial of democracy, the Pinochet government insisted that it was not yet done with the task of "redeeming" Chile, and that full political rights could not be restored until then. A constitution which outlawed the advocacy of Marxism, and gave Pinochet eight more years of rule, was passed by 67% of voters in 1980.

Although forced to operate clandestinely, an opposition nevertheless emerged. A collection of political factions found common cause with the Roman Catholic Church, forming a group called the Civic Union. The Church had become increasingly critical of the Pinochet regime, despite the latter's insistence that Catholicism was the cornerstone of the new Chile. When Pope John Paul II visited Chile in 1987, he brought accusations of torture and other human rights abuses. Finally, in 1988, Pinochet called for a plebiscite to determine whether he should become president for another eight years. In February 1988, 16 political parties came together to form the Coalition for the "No." In October 1988, Pinochet was soundly defeated, and in 1989 new elections were held. Christian Democrat Patricio Aylwin, running as the candidate of a 17-party Concert of Parties for Democracy (Concertación de Partidos por la Democracia) received 55.2% of the vote, and assumed office in 1990. Democracy had returned to Chile.

Although Aylwin had some difficult times, he completed his term, as Chilean economic performance improved. There was some trouble with Pinochet, who resisted Aylwin and his efforts to place the military firmly under civilian control. However, in the elections of December 1993 voters gave the Christian Democratic Party candidate, Eduardo Frei Ruiz-Tagle, an impressive 58% of the vote, leaving no doubt that the Chilean electorate had confidence in their civilian leadership, and in the Frei family name.

13GOVERNMENT

After the restoration of democracy in 1990, Chile still felt the legacy of the Pinochet years. The Constitution of 1980 is still in effect, even though it was created with a different Chile in mind.

In 1989, a series of amendments originally planned in 1980 went into effect, restoring the bicameral National Congress. The 1980 constitution, as amended, is the third constitution Chile has used, the first two being the original of 1833, and the 1925 amendments. The constitution had been prepared by the Council of State, established in 1975 and including two former presidents, Gabriel González Videla and Jorge Alessandri.

The constitution provides for a strong executive with a four-year term. The president has the authority to proclaim a state of emergency for up to 20 days. He can dissolve Congress and call for new elections once per term, and has the power to introduce legislation. There is a bicameral National Congress, consisting of a 120-member Chamber of Deputies and a 46-member Senate. The Senate includes eight appointed members, as well as all ex-presidents, who have life membership. The constitution also provides for an independent judiciary, headed by a 17-member Supreme Court.

The constitution guarantees the active participation of the armed forces in government, establishes limitations on the right to strike and on freedom of information and expression, and institutionalizes a free-market economy.

14POLITICAL PARTIES

Except for an initial period of political disorder, independent Chile's first century of political life was dominated by the aristocratic Liberal and Conservative parties. Segments of the two parties split, shifted, entered into new alliances, regrouped, and took on new names. Since electoral law permitted the registration of parties with relatively small popular bases, coalitions were usually formed to elect presidents and control the Congress. Many cabinets had a fleeting existence. After 1860, the Radicals emerged from the Liberal party, and over the next six decades, they increased their following with the rise of the middle class. In the meantime, the Liberals became conservative, and moved close to that party. Although the Conservatives and Liberals disagreed over the status of the Roman Catholic Church and over the matter of relative congressional and presidential powers, they were united in opposing the Radicals.

Designation of Chilean parties as being of the right, center, or left has been a function of shifting national political climates. Parties and party alliances have tended to appear and disappear over time. The trend during the 1950s and 1960s was to fewer and larger parties. Before the 1973 military coup there were five major parties in Chile: the Christian Democratic Party, founded in 1957, the Socialist Party, founded in 1931, the Communist Party, founded in 1921 (and outlawed during 1948-58), the National Party, formed in 1966 by members of the Liberal and Conservative parties, and the old Radical Party, which saw its strength greatly diminished after 1964. The ruling Allende coalition of Popular Unity consisted of Socialists, Communists, and several smaller leftist parties. The most radical political group, the Revolutionary Movement of the Left, was not a coalition member.

In September 1973, all the Allende coalition parties were abolished. The other parties were initially suspended and then banned in March 1977. The 1980 constitution explicitly prohibits the formation of "totalitarian" (read "communist") parties.

The reemergence of political parties in the aftermath of Pinochet has been dramatic. The Concert of Parties for Democracy includes four major parties: the Christian Democrats (PDC), the Party for Democracy (PPD), the Radical Party (PR), and the Socialist Party (PS). The opposition from the right comes from the Independent Democratic Union (UDI) and the National Renewal (RN). Bolstered by the eight Senate appointees (who were originally Pinochet holdovers after the 1989 elections, these parties held sway in the Senate until 1993. The provision against totalitarian parties has not been enforced to proscribe leftist partisan activity. The Chilean Communist Party (PCCh) and the Allende Leftist Democratic Movement (MIDA) remain active, although they lack any representation in Congress.

15LOCAL GOVERNMENT

After the military government came to power in September 1973, local authorities yielded power to the armed forces, and the nation was divided into military districts. The traditional 25 provinces, as well as all municipalities, were placed under military control.

An administrative reform in 1974, confirmed by the constitution of 1981, divided the country into 12 regions and the metropolitan area of Santiago. The regions were subdivided into 40 provinces. Provinces and regions are administered by provincial and regional governors. Municipalities, headed by mayors, are the smallest unit of local government.

The Aylwin administration pledged to decentralize the military's hierarchical scheme. In 1991, a constitutional amendment was passed granting significant autonomy to local areas. In 1992, municipal authorities were elected nationwide, establishing an electoral base for mayors.

16JUDICIAL SYSTEM

The Chilean civil code of 1857, although modified and amended, remained in use until 1973. Although not eliminated by the military in the wake of the 1973 coup, the judicial system had almost all of its major powers removed, with the military code of justice in force as the effective law of the land. In 1975, the junta began to restore some of the traditional powers exercised by the 13-member Supreme Court.

The 1981 constitution, which came into full effect in 1989, provides for an independent judiciary. The Supreme Court, whose members are appointed by the president, would have authority over appellate and lower courts but would not exercise jurisdiction over the seven-member Constitutional Court and the five-member Electoral Court, which was to supervise all elections.

Although independent in theory, the judiciary remains subject to criticism for inefficiency and lack of independence. As of 1993, 10 of the 17 Supreme Court judges are Pinochet appointees. The lower courts are also dominated by appointees of the former military regime. Reforms passed in 1991 (the "cumplido" laws), however, transferred some of the jurisdiction of the military tribunals to the civilian courts.

17ARMED FORCES

Chilean males between the ages of 14 and 49 are eligible for military service, but only about 15,000 are called up from a pool of over 100,000 each year. The period of service for conscripts is two years. After initial service, Chilean men serve in the reserve (estimated at 45,000 active in 1993). The army had 54,000 (half conscripts) personnel in 1993, the navy 25,000, and the air force 12,000 regulars and 800 conscripts. The army is organized as six divisions and one brigade, all with mixed mobile forces and foot infantry armed with European and American heavy weapons and Chilean-made small arms. The navy, which includes aviation and marines, has 10 destroyers and frigates and four submarines. The air force has a mix of British, American, and French combat aircraft, which number 106, and about the same number of support aircraft. In addition, there is a paramilitary national police force, the carabiñeros, of about 27,000. Defense expenditures were $1 billion in 1991 or about 2% of gross domestic product.

18INTERNATIONAL COOPERATION

Chile is a charter member of the UN, having joined on 24 October 1945, and participates in all the nonregional specialized agencies. Chile supports the Pan American system and in the 1960s began to play a leading role in the OAS. Chile is also a member of G-77, LAIA, and PAHO, among other intergovernmental organizations. Chile was among those countries that shared in the formation of the IDB. The headquarters of ECLAC and the Latin American office of the FAO are located in Santiago. Chile is also a signatory to GATT and the Law of the Sea.

On 25 May 1969, Chile signed the Andean Pact, which now includes Bolivia, Colombia, Ecuador, Peru, and Venezuela. During Allende's presidency, Chile established close ties with Cuba, China, and a majority of other Communist states. Following the 1973 coup, the military government broke off ties with Cuba, the USSR, and other Communist countries except China and Romania. In the late 1970s and early 1980s, relations with Western Europe (particularly France, Italy, Portugal, and the Scandinavian countries) were strained or nonexistent.

19ECONOMY

The Chilean economy is strongly oriented toward commerce and industry, although minerals, chiefly copper and iron ore, provide most of the country's foreign exchange earnings. Chile's leading industries are engaged in the processing of local raw materials; they include metallurgy, petroleum products, textiles—both wool and synthetics—and paper products. Chilean agriculture, dwarfed in value by mining and manufacturing, supports less than one-sixth of the population. Arable land is limited, and livestock raising is the dominant rural enterprise.

The economy suffered profound economic disruptions during the Allende period. Legal nationalization of industries and expropriation of large agricultural holdings by the government were accompanied by illegal seizures of property by its militant allies. The chaotic situation was exacerbated by acts of economic sabotage perpetrated by the opposition, by covert destabilization by agents of the US, and by denial of commercial credit by foreign banks and corporations. By the time of the military coup in late 1973, the nation's manufacturing and farm production had fallen by about 10% from 1972 levels, and inflation had soared to 350%.

After the coup, the military government attempted to revitalize the economy by emulating the principles of a free marketplace. Subsidies were removed and tariffs were lowered to increase competition. A policy of privatization of industries and utilities was instituted, including the return of companies nationalized under Allende to their previous owners, the sale of government-owned companies to individuals and conglomerates, and the sale of percentages of companies to employees and the public on the stock exchange. The GDP fell by 12% in 1975, but Chile's economic performance began to improve thereafter. The average annual rate of increase in GDP between 1977 and 1981 was 7.8%, and the inflation rate dropped from 174% in 1976 to 9.7% in 1981. In 1982, however, a severe economic slump (caused by the worldwide recession, low copper prices, and an overvalued peso) led to an inflation rate of 20.7% and a drop in the GDP of 14.1% in real terms.

Beginning in 1984, the economy grew moderately, constrained largely by persistently depressed world copper prices. Inflation proved stubborn: the consumer price index rose in 1985 by 30.5%. An economic adjustment program introduced in 1985 aimed at strengthening exports other than copper. Inflation was down to 26% in 1985 and 17.4% in 1986; the GDP growth rate was 5.7% in 1986 and 6% in the first half of 1987.

The Chilean economy in 1993 completed a decade of strong and sustained expansion. Although the growth of real GDP was below that of the year before, the 5.6% rate estimated by the economic authorities represents a very important achievement in the context of a global economy which generally continued to show very little dynamism. Due to weak signs of recovery in the world economy, national growth was mainly sustained by the non-tradable sector. The most dynamic activities were construction, trade and services, while agriculture and fishing showed the slowest growth. Although annual inflation of 12.2% met authorities previous estimates, price changes during the year provoked some thought for the need for additional efforts in controlling the inflationary inertia and achieving greater stability. The employment rate kept falling and towards the end of 1993 was approaching 4.4%.

20 INCOME

In 1992, Chile's GNP was $37,064 million at current prices, or $2,730 per capita. For the period 1985–92 the average inflation rate was 19.7%, resulting in a real growth rate in per capita GNP of 6.1%.

In 1992 the GDP was $41,203 million in current US dollars. It is estimated that in 1985 agriculture, hunting, forestry, and fishing contributed 7% to GDP; mining and quarrying, 10%; manufacturing, 22%; electricity, gas, and water, 3%; construction, 4%; wholesale and retail trade, 16%; transport, storage, and communication, 5%; finance, insurance, real estate, and business services, 15%; community, social, and personal services, 6%; and other sources, 11%.

21 LABOR

In 1991/92, Chile's civilian labor force numbered 4,790,000, or about 36% of the population. Of the employed work force in 1992, 48.6% was engaged in professional and white-collar occupations, 18% in agriculture, 17% in manufacturing, 7.5% in transportation, communications, and utilities, 7.1% in construction, and 1.8% in the mining sector. Unemployment and underemployment have plagued successive governments during recent decades, reaching nearly 30% (unofficially) by the end of 1982. It was officially 4.73% in 1990, and 4.79% in 1991.

Chile's unions, dating from the turn of the century, developed slowly until the 1930s, at which time labor established itself as a political force. Communists, who had played an important role in Chilean labor affairs, were forbidden by law to hold union offices in 1948. Although they were not legally permitted to reenter the labor movement until 1958, they remained a strong influence. In 1953, the government, fearing Communist influence in the unions, formed a confederation of labor unions. About 90% of Chilean unions joined this confederation, the Central Union of Chilean Workers (Central Única de Trabajadores de Chile—CUTCH). By the late 1960s, however, Socialists and Communists had risen to the highest offices in CUTCH and formed a key base of support for Allende's election in 1970. Union membership dropped from about 30% of the labor force in 1975 to about 14.7% in 1991.

Mining has been the most highly organized sector. Only in recent decades were efforts made to organize agricultural workers. A rural organization law of 1967 doubled the number of rural labor unions and allowed for the organization of 100,000 workers. Many previous provisions, including the right to strike and engage in collective bargaining, were ended with the overthrow of the Allende government in 1973. CUTCH was dissolved, and workers who had formed self-rule and self-defense communities, called cordones industriales, in industrial sections of Santiago bore the brunt of repression by the military.

With the Aylwin administration, a renewed CUTCH works closely with the government and has accepted a market-oriented economic model while encouraging greater social equity. Recently, the Aylwin government has focused on labor code reforms, professional development and training, freedom of association and collective bargaining, and social welfare program adjustments to increase equity in Chilean society.

As of 1992, CUTCH signed an agreement with the Confederation of Production and Commerce, the largest business association, and the Ministers of Labor, Finance, and Economy to increase the minimum wage rate by 17.1%. This tripartite cooperation marked a significant departure from the antagonistic labor relations of the previous two decades, and showed the government's effort to resolve problems affecting workers' interests.

22 AGRICULTURE

Of the total land area of 74.8 million hectares (184 million acres), arable land was estimated in 1991 at 4.4 million hectares (10.8 million acres) and pastures at 13.5 million hectares (33.4 million acres). Until 1940, Chile was substantially self-sufficient in most basic foodstuffs. Since World War II, serious food deficits have developed, adding to the nation's external payments burden.

Agricultural production of major crops in the 1990/91 and 1991/92 growing seasons (in thousands of tons) was as follows:

	1990/91	1991/92
Sugar beets	2,600	2,200
Wheat	1,590	1,400
Potatoes	844	1,023
Corn	830	790
Oats	207	212
Rice	98	78
Beans	117	91
Barley	107	110
Rapeseed	58	62

Food imports valued at $634.2 million were required to supplement the nation's production in 1992. Agriculture was one of the sectors most adversely affected by the recession of 1982, but it quickly recovered and in 1985 grew by 5.6%. In 1992, farm output rose 3.6% (up from a growth rate of 1.8% in 1991). Poor results in the traditional agricultural sector inhibit a more rapid expansion in agriculture. One of the areas of most rapid growth is in fresh fruit, with the production of grapes rising by 235% between 1981 and 1985. The fruit harvest in 1992 (in tons) included grapes, 1,190,000; apples, 830,000; peaches and nectarines, 223,000; pears, 180,000; oranges, 105,000; and lemons and limes, 92,000.

The traditional land system, inherited from colonial times, has retarded maximum use. Chile's first agrarian reform law, passed in 1962 and supplemented by a constitutional reform in 1963, enabled the government to expropriate and subdivide abandoned or poorly cultivated land and compensate the landowner in installments. Another agrarian reform law was passed in 1967 to clarify expropriation and settlement procedures and to permit an increased turnover rate. By the end of the Frei administration in November 1970, some 1,400 agricultural estates, representing 3.4 million hectares (8.4 million acres), had been confiscated and converted to asentamientos (agricultural communities).

The pace of expropriation was accelerated by the Allende government, which by 1972 had doubled the previous administration's figure for land acquisitions. By taking over virtually all of the land subject to redistribution under the 1967 reform act, the Allende government effectively transformed the Chilean land tenure system. In addition, agricultural laborers, often led by militants to the left of the Allende government, illegally seized some 2,000 farms. Following the 1973 coup, the military regime returned almost all farms in the last category to their original owners; the expropriated land was redistributed to 45,000 smallholders. In 1978, the land reform law was replaced by new legislation that removed restrictions on the size of holdings.

23 ANIMAL HUSBANDRY

Stock raising is the principal agricultural activity in most rural areas. In 1992 there were an estimated 6.6 million sheep, 3.4 million head of cattle, 1.3 million hogs, 600,000 million goats, and 568,000 horses, mules, llamas, and alpacas. The extreme south of Chile is noted for sheep production, while cattle are raised in the central regions. Meat products must be imported from Argentina to fulfill domestic demand. In 1992, 195,000 tons of beef and veal, 138,000 tons of pork, and 14,000 tons of mutton and lamb were produced. In north-central Chile, the hills afford pasturage during the rainy season, and fodder or irrigated pasture provides feed during the dry months. In the south-central regions, natural pasturage is available throughout the year.

The dairy industry is small; milk production totaled 1.5 million tons in 1992. Production of raw wool in 1992 was an estimated 16,400 tons.

24FISHING

With 1,016 species of fish within Chilean waters, its commercial fisheries have long been important. The low temperatures and Antarctic current supply the purest and most oxygenated marine waters in the world. Since 1959, their growth has been rapid, largely owing to the development of a fish-meal industry, centered around Iquique. Anchovies are predominant along the northern coast, whiting and mackerel in the central waters, and shellfish in the south.

Leading fish and seafood caught commercially are Spanish sardines and yellow jacks, as well as anchovies, whiting, eels, sea snails, mackerel, and mussels. Tuna fishing has increased, as have catches of clams and lobsters. The total fish catch soared from 340,000 tons in 1960 to 1,237,000 tons in 1976 and 6,002,967 tons in 1991. Chile is ranked fourth in the world in total landings of fish and is the world's principal exporter of fish meal. In 1990, Chile contributed 2% to the world's exports of fresh, chilled or frozen fish, valued at $266.9 million. In 1992, fisheries output increased by 13.6% over 1991, but accounted for less than 1% of GDP.

25FORESTRY

Chile has extensive forests, estimated in the early 1990s at some 8.8 million hectares (21.7 million acres), or 12% of the total land area. Logging operations are concentrated in the areas near the Bío-Bío River. Softwoods include alerce, araucaria, and manio; hardwoods include alamo, laural, lenga, and olivillo. The establishment of radiata pine and eucalyptus plantations, largely as a result of government assistance, has helped Chile to become an important supplier of paper and wood products to overseas markets. Chile is a major source of hardwood in the temperate zone. Native forests are as yet under-utilized and could become an important factor in Chile's growing competitiveness. Most wood products from Chile are exported as logs, chips, and lumber. Government incentives also resulted in an increase of forestry product exports from $36.4 million in 1973 to $468 million in 1980; by 1991, forestry exports rose to $836 million. Production from 1987 to 1991 increased by nearly 60% as a result of maturation of trees planted in the 1970s. During the first half of 1993, sales abroad of Chilean forestry products increased by 1% over the same period in 1992, with bleached pulp accounting for nearly one-quarter of exports. The major markets for Chilean wood are Japan, South Korea, the US, Taiwan, Belgium, Argentina, and Germany. The Chilean-German Technology Transfer Center in Concepción assists in contributing to the technological development of forestry in the Bío-Bío region.

26MINING

In 1991, the mining industry contributed $4.4 billion (48.5%) to export value, and employed 1.7% (76,843) of the labor force. Traditionally, most of Chile's foreign exchange has come from copper alone (75% in 1970, 80% in 1973), but a policy of export diversification reduced copper's share of the total to 19% in 1991.

During 1987–91, about 18% of the world's copper was produced in Chile, but Chile was estimated to have 22% of the world's copper reserves. In 1991, copper output exceeded 1.8

million tons, the highest in the world. Copper earned an estimated $917 million in 1969, $1.1 billion in 1973, and $3.62 billion in 1991, even though copper prices fell by over 50% during the early 1980s.

Legislation passed in 1966 initiated a "Chileanization" policy for the copper industry, which provided for government ownership of a controlling share of the sector; US management of the large mines was permitted to continue. In 1967, the government signed agreements with the three US-owned companies that produced most of Chile's copper. The agreements provided for an increase of Chilean participation, expanded investment to double output by the early 1970s, and a stable tax and exchange rate. The Chilean government was to acquire an equity position in the mines. A law passed unanimously by the Chilean Congress in 1971 provided for the nationalization of the copper holdings of the Kennecott and Anaconda corporations. Copper production grew significantly in the years that immediately followed, despite the emigration—for economic and political reasons—of a large number of trained specialists. The junta subsequently compensated US interests for their expropriated holdings. Foreign investment in copper has since resumed, but by 1991, about 62% of Chile's copper was produced by the state-run Chilean National Copper Corp. (Corporación del Cobre de Chile—CODELCO). The large Escondida Mine, opened in late 1990, produced more than 316,000 tons of copper concentrates in its first year of operation.

Chile leads the world in natural nitrate production, although world production of synthetic nitrates has sharply reduced Chile's share of total nitrate output. Exports were 351,230 tons in 1990. The Salar (salty marsh) de Atacama, which holds significant nitrate reserves, also contains 58% of the world reserves of lithium; in 1991, production of lithium carbonate reached 8,575 tons (2d largest after the US). Chile has been an exporter of iron ore since the 1950s, but production, after doubling between the mid-1960s and mid-1970s, fell from 10.3 million tons in 1974 to 6.5 million tons in 1985, but rose back to 8.4 million tons by 1991.

Production totals for 1991 included gold, 28,879 kg; zinc, 30,998 tons (58% more than in 1987); molybdenum, 14,434 tons; and silver, 676,339 kg. Besides copper, Chile's primary export minerals are gold, iodine, iron ore, lithium carbonate, molybdenum, silver, and zinc. Chile is one of the world's ten leading producers of gold, silver, and molybdenum, as well as iodine, rhenium, lithium, and selenium.

In 1991, Chile also produced 2.5 million tons of coal. Chile's coal reserves have been boosted by recent discoveries in the Bío-Bío area. Reserves are now believed adequate to supply Chile's needs for 100 years. A number of petroleum-fired electric generators have recently been converted to coal.

27ENERGY AND POWER

Electric power generation reached 19.9 billion kwh in 1991, up from 6.9 billion kwh in 1968; 51% of the power production was hydroelectric. Within South America, Chile is exceeded only by Brazil in its hydroelectric power potential (estimated in 1985 at 18 billion kwh annually), much of it located in the heavily populated central part of the country between La Serena and Puerto Montt. The quick descent of Andes-born rivers, together with the narrowness of the country, makes production and transportation of electricity comparatively inexpensive. The Cólbun-Machicura power plant, with a 490,000-kw generating capacity, opened in 1985 on the Maule River.

The state lays claim to all petroleum deposits, and a government agency, the National Petroleum Co. (Empresa Nacional del Petróleo—ENAP), manages oil fields in Region XII. ENAP's oil production only meets 15% of Chile's needs and reserves are decreasing. Production, which began in 1945, is concentrated around the Strait of Magellan, both onshore and offshore; the crude petroleum is transported by sea to the refinery at Concón, north of Valparaíso. A second refinery was completed near Concepción in 1965. In 1991, production totaled 6.5 million barrels, a decline of 10% from 1990. The output of natural gas, a by-product of petroleum extraction, grew in the 1970s; 1991 production totaled 4,067 million cu m, down from 4,353 million cu m in 1987. In 1992, Chile imported 40 million barrels of oil (valued at $800 million), primarily from Nigeria, Gabon, Venezuela, and Ecuador.

28INDUSTRY

Chile ranks among the most highly industrialized Latin American countries. Since the 1940s, manufacturing has contributed a larger share of GDP than has agriculture. Output of selected industrial products in 1985 included assembled vehicles, 8,145; iron pellets, 3,604,700 tons; cement, 1,429,000 tons; wheat flour, 1,310,500 tons; steel ingots, 654,500 tons; paper, 369,400 tons; plate glass, 2,977,000 sq m. Wine production in 1985 was estimated at 450,000,000 liters, nearly all of it consumed domestically.

The basic industrial pattern, established in 1914, included food processing, beverage production, sugar refining, cotton and woolen mills, a hosiery mill, a match factory, an iron foundry, and a cement factory. During the next decade, industrial production rose about 85%, but from 1949 to 1958 the level of output was virtually stationary. With the establishment of the Huachipato steel mill in 1950, the groundwork was laid for the development of heavy industry. Chile's first copper refinery was inaugurated in November 1966. The major industrial region is the Santiago-Valparaíso area. Concepción is in the center of a growing industrial complex.

During 1970–73, 464 domestic and foreign-owned plants and facilities were nationalized by the Allende government. These included the copper installations of the Anaconda and Kennecott corporations and other companies owned by US interests. By 1982, the military government had returned most expropriated installations to their original owners. The free-market policies of the junta, together with a worldwide recession, resulted in a 25.5% drop in manufacturing output in 1975. After the mid-1970s, Chilean industry moved away from concentration on import substitution to become more export-oriented. Over 20% of the 1985 value of industrial production comes from exports. Key sectors include textiles, automobiles, chemicals, rubber products, steel, cement, and consumer goods.

The industrial sector grew at an average rate of 7% between 1976 and 1982. After 1981, however, it was squeezed by the effects of the fixed exchange rate. After that rate was abandoned in 1982, industry started a slow recovery. Industrial production grew by 9.8% in 1984 but by only 1.2% in 1985.

In 1992, construction, which accounted for roughly 50% of gross fixed capital formation, grew strongly as a result of private sector modernization of plants and equipment and public sector investments in housing and infrastructure. Increased construction activity accompanied by a higher production of exports and consumer durables resulted in a 12.2% rise in manufacturing. On the other hand, farm output and mining rose 3.6% and 1.1%, respectively. Nevertheless, the growth rate in farm output doubled from a year ago. Increased output of copper helped offset lower production of iron, coal, and petroleum.

29SCIENCE AND TECHNOLOGY

There are two major scientific academies, both founded in 1964 in Santiago: the Academy of Sciences, which promotes research in the pure and applied sciences; and the Academy of Medicine, which promotes research and disseminates information in the health sciences. In addition, there are dozens of learned societies in the fields of medicine; the natural, biological, and physical sciences; mathematics and statistics; and technology. The

government agency responsible for planning science and technology policy is the National Commission for Scientific and Technological Research. Expenditures on research and development reached P23 billion in 1988. In 1988 there were over 7,000 scientists, engineers, and technicians engaged in research and development.

Chile has 22 institutes conducting research in agriculture, medicine, natural sciences, and technology, and 25 colleges and universities offering degrees in basic and applied sciences.

30DOMESTIC TRADE

The best market for manufactured and imported goods is heavily concentrated in central Chile, particularly in Santiago, Valparaíso, and Viña del Mar. Valparaíso, which serves as the shipping outlet for Santiago, is Chile's chief port. Concepción provides direct access to the markets of southern Chile, and Antofagasta to those in northern Chile.

The predominant elements in the pattern of retail merchandising are the independent merchants. They sell their wares in small specialized stores, in municipally owned markets, or in free markets (ferias libres). There is a growing number of chain groceries and supermarkets. Stores are owned primarily by Chileans, although foreign interests are represented in retail merchandising.

The usual retail business hours are from 10:30 AM to 7:30 PM, with half a day on Saturday. Normal banking hours are from 9 AM to 2 PM, Monday through Friday. It is common practice for stores and factories to close for about 15 days sometime between 1 December and 1 April for summer vacation. In 1986, Chile received $3,565.1 million in multilateral development aid, of which $1,986.6 million came from IDB and $1,343.8 million from IBRD. US nonmilitary loans and grants totaled $1,183.0 million during the same period, nearly all of it prior to 1974.

In looking at the prospects for the Chilean economy in the next couple of years, great importance is given to the capacity of developed countries to react to the recessionary picture of recent years with greater energy. This factor will be decisive in reactivating the exporting business. The signs are ambiguous at the moment, with the US recovering more decisively than the EC and Japan. On the other hand, unemployment levels lead one to believe that future growth will have to be based mainly on productivity increases, a subject to which the whole productive structure, especially the export sector, should address itself with decisive priority. The country should also redouble its stabilization efforts by controlling public and private spending better, a theme which should also be related to strengthening the real exchange rate.

31FOREIGN TRADE

In 1993, Chile's exports were estimated at $10.476 billion while imports came to $10.9 billion. In 1992, copper accounted for 38.9% of exports; fresh fruit, 9.4%; and meat, 12%. Other export items (39.7%) include forestry products and value-added natural resources, such as fish meal (valued at $540 million in 1992, up 17% from 1991), wood chips (valued at $530 million in 1992, 71% higher than in 1991), tomato sauce, preserved foods, wine, and furniture. Fuel and energy accounted for 11% of imports in 1992, which grew by 26% from the previous year to total $9.24 billion. Substantial increases in imports of capital goods (up 37% to $2.79 billion) and consumer goods (valued at $1.9 billion) caused the trade surplus to fall from $1,576 million in 1991 to just $749 million in 1992.

Since 1991, Japan has been Chile's largest export market. In 1992, exports to Japan amounted to about $1.7 billion. Japan, the US, and the EC account for about 70% of Chilean exports, and are also the leading suppliers for Chilean imports. Other important trading partners are Taiwan, China, South Korea, Peru, Hong Kong, and Mexico. Chile has trade agreements with

Mexico (1991), Argentina (1991), and Bolivia and Venezuela (1993) which grant tariff preference under 11% for certain products. The US announced plans to undertake a comprehensive free trade agreement with Chile in mid-1992.

³²BALANCE OF PAYMENTS

Between 1982 and 1984, the combination of world recession, slumping copper prices, rising foreign interest rates, and an unexpected rise in imports prompted the Pinochet regime to impose domestic austerity measures in order to meet IMF fiscal and monetary targets. The unrest that followed forced the government to request a 90-day moratorium on some debt repayments and seek rescheduling of $3.4 billion due in 1983–84. Even after the restructuring was in place, Chile's debt servicing in 1984 amounted to $2.6 billion, or 57% of total export receipts. Chile accumulated $524 million in foreign exchange reserves during the first three quarters of 1991. In a departure from previous years, the surplus was partially financed through the current account, which featured a 28% increase in the trade surplus and a 15% decline in the services deficit. The most dramatic change, however, came in the short-term capital account, which went from a $622 million surplus in 1990 to a $781 million deficit in 1991. Some of this outflow was believed to be associated with Chilean investment in the booming Argentine stock market.

In 1992 merchandise exports totaled $9,986 million and imports $9,238 million. The merchandise trade balance was $748 million. The following table summarizes Chile's balance of payments for 1991 and 1992 (in millions of US dollars):

	1991	1992
CURRENT ACCOUNT		
Goods, services, and income	–197	–1,014
Unrequited transfers	339	431
TOTALS	142	–583
CAPITAL ACCOUNT		
Direct investment	563	737
Portfolio investment	93	–14
Other long-term capital	—	—
Other short-term capital	–220	1,668
Exceptional financing	13	—
Other liabilities	8	47
Reserves	–1,246	–2,548
TOTALS	-424	396
Errors and omissions	282	187
Total change in reserves	–1,340	–1,947

³³BANKING AND SECURITIES

During the Allende period, almost all private banks were taken over by the government, mainly through the purchase of stock. The military government reversed its predecessor's policy, and now the financial market is essentially private. In 1985, Chile had, in addition to the State Bank, 18 domestic banks and 7 finance societies (which have less capital than banks and cannot perform foreign trade operations). There were also 12 foreign banks in the country.

As of 1992, State Bank foreign assets stood at P5,238.5 billion. Commercial bank reserves amounted to P528.6 billion; time, savings and foreign currency deposits, P4,596.7 billion; and demand deposits, P456.7 billion. The total money supply, as measured by M2, was P5,838 billion. Following government intervention in a number of financial institutions in 1983, the Central Bank introduced three major measures: the issue of $1.5 billion in emergency loans; a provision by which banks could sell their risky portfolios to the Central Bank for 10 years with an obligation to use their profits to buy them back; and the "popular capitalism" program,

announced in April 1985, which allowed, among other things, a new share issue for banks in which there had been intervention.

Chile has two stock exchanges, located in Santiago and Valparaíso. They are both under the control of the Superintendency of Commercial Exchanges. Securities trading has been traditionally inhibited by the Chilean investors' preference for real estate investment. Chronic inflation also has had a disruptive effect upon stock exchange transactions. There is free sale of securities, the largest groups of which are in mining, banking, textile, agricultural, metallurgical, and insurance stocks. All corporations with more than 100 shareholders must register with a stock exchange.

³⁴INSURANCE

Private insurance companies are government-supervised. In 1986, Chile had 23 general insurance, 16 life insurance, and 3 reinsurance companies. Premiums paid in 1990 totaled US$56.8 per capita, or 3% of the GDP. In 1992, P28.5 billion of life insurance was in force.

³⁵PUBLIC FINANCE

Chile experienced budget deficits from the early 1960s through the mid-1970s. Expenditures grew steadily with the expansion of public-sector participation in social welfare and economic activities and with increasing government investment in development projects; the resulting deficits were covered by Central Bank loans and foreign borrowing. Budgetary surpluses were recorded from 1975 through 1981, after which the pattern reverted to deficits. The nonfinancial public sector deficit fell from 2.4% of GDP in 1985 a surplus of 4% of GDP in 1990 and 3.2% in 1992, due to rising copper prices and conservative fiscal policies.

The following table shows actual revenues and expenditures for 1991 and 1992 in billions of pesos:

	1991	1992
REVENUE AND GRANTS		
Tax revenue	2,200.49	2,869.83
Non-tax revenue	342.28	419.32
Capital revenue	24.08	18.63
Grants	—	—
TOTAL	2,702.82	3,495.07
EXPENDITURES & LENDING MINUS REPAYMENTS		
General public service	101.66	131.13
Defense	254.36	304.12
Public order and safety	118.18	158.65
Education	315.66	418.55
Health	260.23	351.11
Social security and welfare	852.03	1,052.29
Housing and community amenities	138.82	176.11
Economic affairs and services	315.06	472.22
Other expenditures	278.11	240.89
Lending minus repayments	–23.35	1.17
TOTAL	2,518.67	3,153.15
Deficit/Surplus	184.15	341.92

From 1985 to 1993, Chile reduced its external debt by $11.3 billion through debt-equity conversions. At the end of 1992, the total debt outstanding was equivalent to 50% of GDP; total foreign debt amounted to $18.9 billion. Public debt as a proportion of total debt has fallen from 86% in 1987 to 69% in 1990 to 54% in 1992.

³⁶TAXATION

Prior to 1920, government revenue was derived largely from export and import taxes, but since then, a more varied tax base has been achieved. Taxes on mining production, consistently a dominant revenue source, have ranged from 20% to 50%. Indirect taxes accounted for about 74% of government revenues in 1985, and direct taxes for 16%.

During the Allende period, the Congress, which was strongly influenced by opposition parties, resisted government efforts to introduce redistributive income tax policies. At the end of 1974, the military government eliminated the capital gains tax and established a 15% taxation rate for income from real estate, investments, and commercial activities, which has since been reduced to 10%. Personal income is taxed at rates ranging from 5% for the lowest wage and salary bracket to a maximum of 50%. Corporate income is subject to a basic flat tax of 10% and an additional tax on dividends remitted to non-residents abroad, bringing the total effective tax rate to 35%. Other principal levies include a value-added tax, housing tax, inheritance and gift tax, various stamp taxes, and assessments on real estate, entertainment, gasoline, alcoholic beverages, and tobacco.

37 CUSTOMS AND DUTIES
The Frei administration (1964–70) imposed extensive import duties and taxes to discourage monetary outflow for nonessential goods and to protect local industries. Imports were curbed by the requirement of advance deposits according to six categories, ranging from 10% for essential goods up to 10,000% for nonessential goods and goods competitive with domestic production. The Allende government reimposed a high tariff schedule meant to discourage the expenditure of scarce foreign reserves on luxury items. When the military government came to power at the end of 1973, it adopted a policy of liberalizing trade and returning to an open market. Tariffs were removed on foreign goods to force competition on low-efficiency, high-priced local industries. By 1979, tariffs had been reduced to 10%, the only exception to this uniform rate being some car models. In late 1982, the government introduced legislation permitting higher tariffs on some 20 products, including milk, canned fish products, synthetic fibers, and leather footwear. Most imported goods are taxed at a flat rate of 11% and subject to an 18% value-added tax. In 1977, free-trade zones were established in Iquique and Punta Arenas.

38 FOREIGN INVESTMENT
Through the Decree Law 600 of 1974 (and its subsequent modifications) and the Chilean Foreign Investment Committee, Chile seeks to encourage foreign direct investment. However, broadcasting, fishing, shipping, and hydrocarbon production usually require majority national control. In 1993, total foreign direct investment was estimated at $1.4 billion, of which the US accounted for 11% (down from 44% in 1991). Other principal investors include Japan, Finland, Australia, Canada, the Cayman Islands, the UK, the Netherlands, Singapore, and Hong Kong. As of mid-1993, mining accounted for 43.3% of materialized foreign investment; industry, 33.1%; services (including investment funds), 16%; transportation, 3.4%; construction, 1.8%; forestry, 1.2%; aquaculture, 0.1%; and other sectors, 1.1%.

39 ECONOMIC DEVELOPMENT
Chile has established two free trade zones: the Free Zone of Iquique (ZOFRI) in the northern tip (Region I), and the Free Zone of Punta Arenas (PARANEZON) in the southern tip (Region XII). Through its concentration on value-added exports and increased foreign direct investment, Chile has become one of Latin America's most developed nations. Economic growth has averaged over 5% annually since 1985, and was 10.4% in 1992, one of the highest rates in the world and Chile's highest since 1965. Under the Alywin administration, the population living in poverty has dropped by 800,000 to 4.5 million, and real incomes of the poorest workers have increased by 20%.

Chile's debt management has been very effective. The government negotiated a favorable rescheduling with its creditor banks of its 1991–94 debt maturities. The government also kept inflation limited to 10–12% in 1993 (down from 21.4% in 1989). Social expenditures, especially those aimed at improving human capital, have risen since 1991 to 15% of GNP (through increased surtaxes). For example, the Program for Youth Labor Training focuses on the high levels of poverty and unemployment among youth; the program trains 100,000 and includes experience in the private sector.

40 SOCIAL DEVELOPMENT
Prior to the 1973 coup, Chile had built one of the most comprehensive social welfare systems in the world, with over 50 separate agencies participating in programs. Benefits for the insured included medical care for employees; prenatal and postnatal care for their wives; milk for babies; medical care for dependents up to 2 years of age; some dental care; life insurance; old age, disability, and death benefits; and funeral expenses. The basic welfare legislation was enacted in 1925, and in 1952, the social insurance system was integrated with the National Health Service under a unified administration. In that year, Chile established Latin America's first school of social work, located in Santiago.

Following the military's accession to power in 1973, many of the welfare benefits were suspended, and regulations lapsed. From 1974 to 1981, the junta remodeled the welfare system along the lines of private enterprise. Social security covers all laborers, other employees, and their dependents. The system includes pensions, family allowances, medical care, sickness benefits, and unemployment compensation. Pensions are financed exclusively by workers, whose contributions go into individual accounts with Administrators of Pension Funds, which are private social security institutions. Workers must contribute 10% of their taxable income to pension funds monthly, up to a certain limit, 7% of their remuneration for medical care, and 3.25–3.6% for disability or life insurance. Employers contribute only a 2% tax on remuneration plus a flat rate of 90% of payroll and an additional 3.4–6.8% according to industry and risk for accident insurance.

Between 1966 and 1979, Chilean leaders supported family planning; alarmed by the declining birth rate, however, the government adopted a pronatalist policy in 1978. Accessibility to contraceptives is restricted, and the 1981 constitution makes abortion illegal. The fertility rate for 1985–90 was 2.7%, and the contraceptive prevalence rate was 60%.

Following the transition to democracy, the chief human rights issues involved redress for abuses committed during the military rule. A 1989 law removed many restrictions on women, although some legal distinctions exist, and divorce is illegal. The director of the National Women's Service estimated that one in four women had been subject to domestic violence in 1992.

41 HEALTH
As of 1993, Chile's budget for the public health sector increased for the fourth consecutive year. Chile planned to build and/or expand hospitals in San Felipe (central), Valdivia (south), Iquique (north), and Chillan (south).

The 1993 birth rate was 22.5 per 1,000 people. In 1992, there were 309,000 births (23 per 1,000 people). Approximately 43% of married women (ages 15 to 49) used contraception according to a study made between 1980 and 1993.

Chile has made considerable progress in raising health standards: the crude death rate per 1,000 was reduced from 12 in 1960 to 6 in 1993, and the infant mortality rate declined from 147 per 1,000 live births in 1948 to 15 in 1992. There were 67 reported cases of tuberculosis per 100,000 people in 1990. Average life expectancy in 1992 was 72 years.

In 1991, there were 14,203 physicians; in 1989, there were 5,200 dentists; and, in 1988, there were 3,355 nurses. In 1990, there were approximately 32,931 hospital beds (3.3 beds per 1,000 people).

An estimated 15% of Chileans in the early 1980s, including 10% of children under the age of 5, fell below the minimum nutritional requirements established by the FAO. Protein deficiency among the general population has induced an abnormally high rate of congenital mental handicap. In 1990, however, only 2% of children under 5 years of age were considered malnourished, and between 1988 and 1991, 86% of the population had access to safe water. From 1985 to 1992, 97% of the population had access to health care services.

42HOUSING
The 1970 housing census enumerated 1,860,111 housing units. From 1981 through 1985, the number of new units built was 201,244. In 1991, the total housing stock numbered 3,261,000. The number of new dwellings completed jumped from 88,000 in 1991 to 106,000 in 1992.

The Allende government expanded the housing program. In 1971, 6.5% of the national budget was expended on public housing, mainly for the poor, and the state built 76,079 new housing units. The military government, on the other hand, stressed the role of the private sector in the housing market. In 1974, the number of new units built by the public sector was 3,297, as against 17,084 units built privately; the corresponding figures for 1984 were 276 and 46,493. Construction was one of the hardest-hit sectors in the recession of 1982, but international loans for highway building and maintenance helped offset the losses in housing.

43EDUCATION
Since 1920, primary education has been free, compulsory, and nonsectarian. The adult illiteracy rate, estimated at 50% in 1920, had been reduced to 6.6% by 1990, per UNESCO estimates (males: 6.5% and females: 6.8%).

Chile's present educational system stems from a 1965 reform program that called for curriculum modernization (with new texts for all grade levels), teacher training, and professional educational planning and management. There are both state-run and private schools; all state schools provide free education. An eight-year primary and four-year secondary program, with increased emphasis on vocational instruction at the secondary level, was introduced. In 1991, enrollment in secondary and professional schools was 699,455 (general: 436,892 and vocational: 262,563).

The University of Chile (founded as Universidad Real de San Felipe in 1738) and the University of Santiago de Chile (founded as Universidad Técnica del Estado in 1949) are national universities with branches in other cities. There are numerous institutions which provide vocational and technical education. There are also several Roman Catholic universities. Upon the accession of the military government in 1973, civilian authorities in all levels of the educational system were superseded by the military, which subsequently retained its control over the universities. Higher educational enrollment was approximately 286,962 in 1991, compared with 119,008 in 1980.

44LIBRARIES AND MUSEUMS
Chile's principal libraries and museums are in Santiago. The three most notable libraries are the National Library (3,500,000 volumes in 1989), the central library of the University of Chile (41 libraries with an aggregate of over 1,000,000 volumes), and the Library of Congress (800,000). Other significant collections include the Severín Library in Valparaíso (101,000) and the library of the University at Concepción (340,000).

Chile's most outstanding museums are the National Museum of Fine Arts, the National Museum of Natural History, and the National Museum of History, all in Santiago, and the Natural History Museum in Valparaíso.

45MEDIA
An extensive telegraph service, about three-fourths of which is state-owned, links all the principal cities and towns. Some 90% of Chile's telephone service is provided by Chilean Telephone, formerly a subsidiary of ITT. A total of 866,663 telephones were in use in 1991, the majority located in Santiago. International links are supplied by worldwide radiotelephone service and by international telegraph companies. In 1993, Chile also used two Atlantic Ocean satellite and three domestic satellite stations. There were about 150 AM and 153 FM radio stations and 131 television stations as of 1992. In 1991, Chile had 2.8 million television sets and 4.6 million radios.

Many of Chile's newspapers and periodicals were closed for political reasons in the aftermath of the 1973 military coup. The lifting of the second state of siege in mid-1985 brought a significant improvement in the area of the freedom of the press. Opposition magazines resumed publication, and editors were no longer required to submit copy to government censors prior to publication; radio and television programs featuring political debates reappeared in the last half of 1985. The largest of some 48 newspapers (1991) are in the Santiago-Valparaíso area, where the most important magazines are also published. Among the best-known magazines are *Caras, Analisis,* and *Qué Pasa?.* The newspaper *El Mercurio* (founded in 1827) claims to be the oldest newspaper in the Spanish-speaking world. The *El Mercurio* chain includes *La Segunda* and *Las Últimas Noticias* of Santiago, *El Mercurio* of Valparaíso, and *El Mercurio* of Antofagasta.

The names and approximate 1991 circulation figures of the leading daily newspapers of Santiago, Valparaíso, and Concepción were as follows:

SANTIAGO	CIRCULATION
La Tercera de la Hora	200,000
La Cuarta	160,000
Las Últimas Noticias	140,000
El Mercurio	88,500
La Nación	50,000
La Segunda	28,000
El Diario Oficial	15,000
VALPARAÍSO	
La Estrella	30,000
El Mercurio	20,000
CONCEPCIÓN	
El Sur	41,000

46ORGANIZATIONS
The members of many workers' organizations have formed consumer cooperatives. Producer cooperatives also are common, particularly in the dairy industry. The National Society of Agriculture has been politically very influential, and the minister of agriculture has been frequently drawn from its ranks. Representative of the many industrial, commercial, and professional organizations are the National Mining Society, Society of Industrial Development, Commercial Union Society, National Press Association, Medical Society, Chilean Medical Association, Agronomers' Society, Geographical Society of Chile, and Scientific Society.

Among fraternal organizations, the Masonic Order is prominent. The National Council of Sports is the overall confederation of athletic associations. The Chilean Chamber of Commerce, with its headquarters in Santiago, is the central organ for all chambers of commerce and most trade associations. Rotary and Lions clubs are also active among the business community.

Among the more politically potent organizations are the professional middle-class guilds (gremios), which were instrumental in bringing down the Allende government. Social development corporations, comprising mainly businesspeople, have been organized regionally to deal with various welfare problems. The Confederation for Production and Commerce is an official organization representing the country's industrialists and traders.

47 TOURISM, TRAVEL, AND RECREATION

Tourists need a valid passport and visa or tourist card, with exceptions based on reciprocity agreements. Tourist attractions include the Andean lakes of south-central Chile and the famed seaside resort of Viña del Mar, with casinos rivaling those of Monaco. Also popular is Robinson Crusoe Island, in the Pacific. Another Pacific dependency, Easter Island (Isla de Pascua), with its fascinating monolithic sculptures, is a major attraction. The giant Christ of the Andes statue, which commemorates the peaceful settlement of the Chilean-Argentinian border dispute in 1902, is located on the crest of the Andes overlooking the trans-Andean railway tunnel. Santiago is noted for its colonial architecture, as well as the largest library in South America. Popular national parks include Parque Nacional Lanca in the North, the Nahuelbuta Park near Temuco, and Terres del Paine in the far South. Chilean ski resorts, notably Portillo, near Santiago, have become increasingly popular.

The most popular sport in Chile is soccer. Other pastimes include skiing, horseracing, tennis, fishing in the Pacific for marlin and swordfish, and some of the world's best trout fishing in the Lake District.

In 1991, 1,349,000 tourist visits were reported, 67% from Argentina, 7% from Peru, and 7% from European countries. Tourism receipts totaled US$700 million. There were 25,817 total hotel rooms with 62,340 beds.

48 FAMOUS CHILEANS

Chile's first national hero was the conquistador Pedro de Valdivia (1500?–53), who founded Santiago in 1541. The Indian leader Lautaro (1525–57), another national hero, served Valdivia as stable boy and then escaped to lead his people to victory against the Spanish. His exploits are celebrated in the great epic poem *La Araucana* by Alonso de Ercilla y Zúñiga (1533?–96), a Spanish soldier. Bernardo O'Higgins (1778–1842), a leader of the fight for independence, was the son of the Irish soldier of fortune Ambrosio O'Higgins (1720?–1801), who had been viceroy of Peru. Diego Portales (1793–1837) helped build a strong central government. Admiral Arturo Prat (1848–79) is Chile's most revered naval hero because of his exploits during the War of the Pacific. Arturo Alessandri Palma (1868–1950), who became president in 1921, initiated modern sociopolitical reform. Salvador Allende Gossens (1908–73), the Western Hemisphere's first freely elected Marxist head of state, served three years as Chile's president (1970–73), initiating a broad range of socialist reforms and dying in the throes of a violent military coup in September 1973. The coup's leader was Gen. Augusto Pinochet Ugarte (b.1915), a former commander-in-chief of the army. Outstanding church figures have been Crescente Errázuriz (1839–1931), archbishop of Santiago, and his successor, José Cardinal Caro (1866–1958). Benjamin A. Cohen (1896–1960) was an undersecretary of the United Nations.

Three distinguished historians, Miguel Luis Amunátegui (1828–88), Diego Barros Arana (1830–1907), and Benjamin Vicuña Mackenna (1831–86), brightened the intellectual life of the second half of the 19th century. José Toribio Medina (1852–1930) gained an international reputation with works ranging from history and literary criticism to archaeology and etymology. Important contemporary historians include Francisco Antonio Encina (1874–1965), Ricardo Donoso (b.1896), and Arturo

Torres Rioseco (b.1897), who is also a literary critic. Benjamín Subercaseaux (1902–73) was a popular historian as well as a novelist.

The first indigenous literary movement was that of the "generation of 1842." One of its leaders was the positivist writer José Victorino Lastarria (1817–88). The novelist and diplomat Alberto Blest Gana (1830–1920) wrote panoramic novels about Chilean society in the tradition of Balzac. Twentieth-century writers include novelist Eduardo Barrios (1884–1963), an explorer of the abnormal psyche; Joaquín Edwards Bello (1887–1968), an author of realistic novels of urban life; the symbolic novelist, poet, and essayist Pedro Prado (1886–1952); and José Donoso (b.1925), who is perhaps the best-known contemporary novelist.

Poets of note include Gabriela Mistral (Lucila Godoy Alcayaga, 1889–1957), who won the Nobel Prize in 1945; Pablo Neruda (Neftalí Ricardo Reyes, 1904–73), the nation's greatest poet, who was awarded a Stalin Prize as well as the Nobel Prize (1971); and the poet-diplomat Armando Uribe Arce (b.1933).

The nation's first native-born composer was Manuel Robles (1780–1837); Silvia Soublette de Váldes (b.1923) is a leading composer, singer, and conductor; and Gustavo Becerra (b.1925) is a composer and teacher. Claudio Arrau (1903-91) is one of the world's leading concert pianists. Well-known painters are Roberto Matta (b.1911) and Nemesio Atúnez (b.1918), while sculptors include Lily Garafulic (b.1914) and Marta Colvin (b.1917).

49 DEPENDENCIES

Easter Island

About 3,700 km (2,300 mi) w of Chile, at 27°3'–12's and 109°14'–25'w, is Easter Island (Isla de Pascua or Rapa Nui), a volcanic island roughly 24 km (15 mi) long by 16 km (10 mi) wide. Easter Island is inhabited by a mostly Polynesian-speaking population and a few hundred people from the mainland. Easter Island's population totaled 2,000 in 1993. The people raise bananas, potatoes, sugarcane, taro roots, and yams. The island is famous for its massive monolithic stone heads of unknown origin, carved from tufa, a soft volcanic stone. The cryptic sculptures have attracted increasing numbers of visitors to the island from both mainland Chile and around the world. In 1975, the government engaged Spanish consultants to undertake major tourist development on the island. The number of tourist arrivals increased from 2,705 in 1984 to 4,163 in 1987. In 1986, about one-third of the island was a national park.

Easter Island was discovered by Edward Davis, an English buccaneer, in the late 1680s and was named on Easter Day 1722 by Roggeveen, a Dutch navigator. Claimed by Spain in 1770, the island was taken over by Chile in 1888 and is now administered as part of Valparaíso Province.

Diego Ramírez Islands

About 100 km (60 mi) sw of Cape Horn, at 56°30's and 68°43'w, lies the small, uninhabited Diego Ramírez archipelago.

Juan Fernández Islands

Some 580 km (360 mi) w of Valparaíso, at 33°36' to 48's and 78°45' to 80°47'w, is a group of rugged volcanic, wooded islands belonging to Chile. The two principal islands, about 160 km (100 mi) apart e–w, are Robinson Crusoe, formerly Más a Tierra (93 sq km/36 sq mi), and Alejandro Selkirk, previously Más Afuera (85 sq km/33 sq mi); the smaller island of Santa Clara (or Goat Island) is off the southwest coast of Robinson Crusoe. The chief occupation is lobster fishing. Discovered by Juan Fernández around 1563, the islands achieved fame in 1719, when Daniel Defoe wrote *Robinson Crusoe*, generally acknowledged to have been inspired by the experiences of Alexander Selkirk, a Scottish sailor who quarreled with his captain and was set ashore at his own request on Más a Tierra, where he lived alone until he was rescued (1704–09). The islands are administered by Valparaíso Province.

Sala-y-Gómez Island
About 3,380 km (2,100 mi) w of Chile and some 400 km (250 mi) ENE of Easter Island, at 26°28′s and 105°28′w, lies arid, volcanic Sala-y-Gómez Island. Almost 1,200 m (4,000 ft) long and about 150 m (500 ft) wide, this uninhabited island belongs to and is administered by Valparaíso Province.

San Ambrosio Island
Volcanic San Ambrosio Island, uninhabited, lies 965 km (600 mi) w of Chile, at 26°21′s and 79°54′w, rising to 479 m (1,570 ft).

San Félix Island
Situated 19 km (12 mi) ESE of San Ambrosio Island, at 26°17′s and 80°7′w, is small, uninhabited San Félix Island (about 8 sq km/3 sq mi). Of volcanic origin, the island rises to about 180 m (600 ft). The islet of González is at its southeastern tip. San Félix, along with San Ambrosio, was discovered in 1574.

Chilean Antarctic Territory
Chile claims the section of Antarctica lying between 53°w and 90°w, the Antarctic (or O'Higgins) Peninsula, parts of which are also claimed by Argentina and the UK.

[50]BIBLIOGRAPHY
Allende, Salvador. *Chile's Road to Socialism.* Baltimore: Penguin, 1973.

Bizzarro, Salvatore. *Historical Dictionary of Chile,* 2d ed. Metuchen, N.J.: Scarecrow Press, 1987.

Blakemore, Harold. *Chile.* Oxford, England; Santa Barbara, Calif.: Clio Press, 1988.

Chile Since Independence. New York: Cambridge University Press, 1992.

Davis, Nathaniel. *The Last Two Years of Salvador Allende.* Ithaca, N.Y.: Cornell University Press, 1985.

Debray, Régis. *Conversations with Allende.* London: NLB, 1971.

Drake, W. and Ivan Jaksic. *The Struggle for Democracy in Chile, 1982–1990.* Lincoln: University of Nebraska Press, 1991.

Haverstock, Nathan A. *Chile in Pictures.* Minneapolis: Lerner Publications Co., 1988.

Hojman, D. E. *Chile: The Political Economy of Development and Democracy in the 1990s.* Pittsburgh, Pa.: University of Pittsburgh Press, 1993.

Kaufman, Robert R. *The Politics of Land Reform in Chile: 1950–1970.* Cambridge, Mass.: Harvard University Press, 1972.

Loveman, Brian. *Chile: The Legacy of Hispanic Capitalism,* 2nd ed., New York: Oxford University Press, 1988.

Moran, Theodore H. *Multinational Corporations and the Politics of Dependence—Copper in Chile.* Princeton, N.J.: Princeton University Press, 1975.

Petras, James F. *Democracy and Poverty in Chile.* Boulder, Colo.: Westview Press, 1994.

Sater, William F. *Chile and the United States: Empires in Conflict.* Athens: University of Georgia Press, 1990.

Spooner, Mary Helen. *Soldiers in a Narrow Land: The Pinochet Regime in Chile.* Berkeley: University of California Press, 1994.

Stewart-Gambino, Hannah W. *The Church and Politics in the Chilean Countryside.* Boulder, Colo.: Westview Press, 1992.

Valenzuela, J. Samuel and Arturo. *Military Rule in Chile: Dictatorship and Oppositions.* Baltimore: Johns Hopkins University Press, 1986.

Zeitlin, Maurice. *The Civil Wars in Chile; or, The Bourgeois Revolutions That Never Were.* Princeton, N.J.: Princeton University Press, 1984.

COLOMBIA

Republic of Colombia
República de Colombia

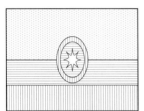

CAPITAL: Bogotá.

FLAG: The national flag consists of three horizontal stripes; the yellow upper stripe is twice as wide as each of the other two, which are blue and red.

ANTHEM: *Himno Nacional,* beginning "O gloria inmarcesible, júbilo inmortal" ("O unwithering glory, immortal joy").

MONETARY UNIT: The Colombian peso (c$) of 100 centavos is a paper currency. There are coins of 10, 20, and 50 centavos and of 1, 2, 5, 10, 20, and 50 pesos, and notes of 100, 200, 500, 1,000, 2,000, 5,000 and 10,000 pesos. Commemorative gold coins of various denominations also have been minted. c$1 = us$0.0038 (or us$1 = c$264.1).

WEIGHTS AND MEASURES: The metric system is the official standard, but Spanish units such as the botella, vara, fonegada, arroba, and quintal are also used.

HOLIDAYS: New Year's Day, 1 January; Epiphany, 6 January; St. Joseph's Day, 19 March; Labor Day, 1 May; Day of St. Peter and St. Paul, 29 June; Independence Day, 20 July; Battle of Boyacá, 7 August; Assumption, 15 August; Columbus Day, 12 October; All Saints' Day, 1 November; Independence of Cartagena, 11 November; Immaculate Conception, 8 December; Christmas, 25 December. Movable religious holidays include Holy Thursday, Good Friday, Holy Saturday, Ascension, Sacred Heart, and Corpus Christi. In addition there are six official commemorative days.

TIME: 7 AM = noon GMT.

¹LOCATION, SIZE, AND EXTENT

Colombia is the only South American country with both Caribbean and Pacific coastlines. The fourth-largest country in South America, it has a total area of 1,138,910 sq km (439,736 sq mi), including insular possessions, and extends 1,700 km (1,060 mi) NNW–SSE and 1,210 km (750 mi) NNE–SSW. Comparatively, the area occupied by Colombia is slightly less than three times the size of the state of Montana. Bordered on the N by the Caribbean Sea, on the NE by Venezuela, on the SE by Brazil, on the SW by Peru and Ecuador, on the W by the Pacific Ocean, and on the NW by Panama, the Colombian mainland is located entirely within the tropics. Its total boundary length is 10,616 km (6,596 mi). Also held by the Republic of Colombia (though claimed by Nicaragua) are the archipelago of San Andrés and Providencia in the Caribbean Sea, about 190 km (120 mi) off the coast of Nicaragua, and the islands of Malpelo, Gorgona, and Gorgonilla in the Pacific Ocean. Colombia also holds the uninhabited Caribbean islands of Quita Sueño Bank, Roncador Cay, and Serrana Bank, to which the US renounced all rights under the Treaty of Quita Sueño, ratified by the US Senate in July 1981; Nicaragua also disputes this claim. Colombia has a dispute with Venezuela over maritime rights in the Gulf of Venezuela. Negotiations have been going on unsuccessfully since 1970, and in August 1987, a Colombian naval vessel entered the disputed region in an apparent attempt to make Venezuela more responsive.

Colombia's capital city, Bogotá is located in the center of the country.

²TOPOGRAPHY

The Andes Mountains divide just north of Colombia's southern border with Ecuador into three separate chains, or cordilleras, known as the Cordillera Occidental (western), the Cordillera Central, and the Cordillera Oriental (eastern). The western and central cordilleras run roughly parallel with the Pacific coast, extending northward as far as the Caribbean coastal lowlands. They are alike in geological structure, both being composed of massive crystalline rocks. The Cordillera Central is the highest range of the Colombian Andes, with several volcanic cones whose snow-covered peaks rise to about 5,500 m (18,000 ft), notably Huila (5,750 m/18,865 ft). The third chain, the Cordillera Oriental, runs northeastward, bifurcating into an eastern branch, the Sierra de los Andes, which slopes down to Venezuela, and a second branch, the Sierra de Perijá, which continues northward to terminate on the border between Venezuela and Colombia just south of the Guajira Peninsula. This range is composed of folded stratified rocks over a crystalline core. On the margin of the Caribbean stands the Sierra Nevada de Santa Marta, an isolated block of mountains composed of a triangular massif of granite, whose highest elevation is Cristóbal Colón (5,797 m/19,020 ft), the tallest peak in Colombia. West of the Cordillera Occidental but not geologically a part of the Andean chain is the low Serranía de Baudó, which skirts the Pacific and extends into the Isthmus of Panama.

Separating the three principal Andean ranges are Colombia's two major rivers, the Cauca (1,014 km/630 mi), which flows northward between the western and central cordilleras, and the Magdalena (1,553 km/965 mi), which divides the central and eastern cordilleras. After emerging from the mountains, the two rivers become one and descend through marshy lowlands to the Caribbean. The area south and east of the Andean ranges is largely composed of river plains divided among the effluents of the Orinoco and Amazon rivers. Open plains immediately adjoin the mountains, but as the distance from the cordillera increases, the plains give way to largely uninhabited and unexplored jungle.

The Pacific coastal area is also characterized by jungle vegetation. Principal rivers on the Pacific coast include the Baudó, San Juan, and Patía.

³CLIMATE

Colombia's climatic variations are determined by altitude, and seasons are periods of lesser or greater rainfall, with little or no temperature change. The country may be divided vertically into four regions. The hot country, or tierra caliente, is the tropical zone, reaching from sea level to roughly 1,100 m (3,500 ft), where the mean annual temperature is 24°C to 27°C (75–81°F); at sea level, temperatures have a mean maximum of 38°C (100°F) and a minimum of 18°C (64°F). Between 1,100 m (3,500 ft) and 2,000 m (6,500 ft) is the temperate zone, or tierra templada, where the average year-round temperature is about 18°C (64°F). Between 2,000 m (6,500 ft) and 3,000 m (10,000 ft) is the cold country, or tierra fría, with temperatures averaging a little over 13°C (55°F). Above the 3,000-m (10,000-ft) level the temperature varies from 13°C to –17°C (55°F to 1°F), according to altitude. The annual mean temperature at the capital, Bogotá (altitude 2,598 m/8,525 ft), is 14°C (57°F). Rainfall is heaviest on the west coast and in the Andean area; rainy and dry seasons, or "winter" and "summer," generally alternate in three-month cycles, as in Bogotá, where precipitation occurs most heavily and consistently during the periods of April to June and October to December. Northern areas have only one long rainy season, from May through October.

⁴FLORA AND FAUNA

More than 45,000 species of plants have been identified in Colombia, and it is predicted that when the region has been thoroughly explored the number may be doubled. At the highest (3,000–4,600 m/10,000–15,000 ft) and coldest level of mountain meadows, called páramos, the soil supports grasses, small herbaceous plants, and dense masses of low bushes. In the intermontane basins some vegetables, European-introduced grains, and corn are found, along with the bushes, trees, and meadow grasses indigenous to the region. The temperate areas support extensive and luxuriant forests, ferns, mosses, trees of the laurel family, Spanish cedars, vegetables, and grain crops. The tropical zone may be divided into four main groups according to the amount of rainfall received: desertlike areas supporting arid plants, deciduous forests, rain forests, and grass plains. Palm trees of various species abound in the tropics, and there are many edible fruits and vegetables.

Animal life is abundant, especially in the tropical area. Among carnivorous species are the puma, a variety of smaller cats, raccoons, and mustelids. Herbivores include the tapir, peccary, deer, and large tropical rodents. Sloths, anteaters, opossums, and several types of monkeys are also found, as well as some 1,665 species and subspecies of South American and migratory birds.

⁵ENVIRONMENT

Columbia's main environmental problems are soil erosion, deforestation and the preservation of its wildlife. Soil erosion has resulted from the loss of vegetation and heavy rainfall. Deforestation has resulted from the commercial exploitation of the country's forests which cover 43% of the country. Without a substantial reforestation effort, Colombia's timber resources could be exhausted by the year 2000. Approximately 908,000 hectares (2,244,000 acres) of natural forest were lost annually in the 1970s to farming, erosion, and the lumber industry, but only 5,000 hectares (12,000 acres) were reforested each year; between 1981 and 1985, 820,000 hectares (2,260,000 acres) were lost each year, and 8,000 hectares (20,000 acres) were reforested.

The Columbian government has initiated several programs to protect the environment. The National Environmental Education Plan for 1991-1994 introduces environmental issues in the elementary schools. In 1973, the government created the National Resources and Environment Code. By 1959, the Amazon forests, the Andean area and the Pacific coast were protected. The main environmental agency is the Institute for Development of Renewable Natural Resources and the Environment (INDERENA), established in 1969. Among other activities, it has undertaken extensive projects in the training of personnel in conservation, fishing, and forestry. The Colombian Sanitary Code, in force since January 1982, establishes pollution control standards.

As of 1987, endangered species in Colombia included the tundra peregrine falcon, Cauca guan, gorgeted wood-quail, red siskin, pinche, five species of turtle (green sea, hawksbill, olive ridley, leatherback, and arrau), two species of alligator (spectacled caiman and black caiman), and two of crocodile (American and Orinoco). By 1994, 25 of Columbia's 388 species of mammals were endangered. Sixty-nine in a total of 1,665 bird species were also threatened, along with 10 reptile species in a total of 383. As of 1994, 327 of the 45,000 plant species in Columbia are endangered.

⁶POPULATION

Colombia is the third most populous country in Latin America, following Brazil and Mexico. The 1985 census set the population total at 27,837,932, and the estimate for 1992 was 33,391,536. The population in the year 2000 was projected at 37,822,000, assuming a crude birth rate of 22 per 1,000 population and a crude death rate of 5.8 during 1995–2000. A rapid transfer of population to urban centers has taken place since the 1950s; by 1995, 72.7% of the people were estimated to live in cities. The population density was an estimated 29.2 per sq km (75.7 per sq mi) in 1992, with about 95% of the population residing in the mountainous western half of the country. In 1992, the estimated population of the capital, Santa Fé de Bogotá, was 4,921,264; the populations of other major cities were as follows: Medellín, 1,581,364; Cali, 1,624,401; Barranquilla, 1,018,763; Bucaramanga, 349,403; Cartagena, 688,306; and Cúcuta, 450,318.

⁷MIGRATION

Despite government inducements, such as the granting of agricultural land in the eastern plains, immigration has been insignificant, partly because of the turbulence and violence of the 1940s and 1950s. Emigration is small but significant, since many of those who leave the country are scientists, technicians, and doctors. Between 1951 and 1985, some 218,724 Colombians settled in the US, at an annual rate that surpassed 10,000 by 1981 and rose to 11,802 in 1985. In 1990 there were more persons in the US of Colombian birth—304,000—than of any other South American nationality. About 300,000 Colombians were internally displaced in 1993 by the violence endemic to the country.

⁸ETHNIC GROUPS

The predominant racial strain in Colombia is the mestizo (mixed white and Amerindian), constituting about 50% of the total population in the late 1980s. An estimated 25% of the inhabitants are of unmixed white ancestry, 20% are mulatto (black–white) or zambo (black–Indian), 4% are black, and 1.5% are pure Amerindian. Blacks and mulattoes are concentrated in the coastal regions and tropical valleys. Pure Amerindians are rapidly disappearing; the remaining few live mainly in inaccessible and barren regions. The principal Amerindian culture of Colombia during the pre-Columbian period was that of the Chibcha, whose descendants are today chiefly concentrated in the departments of Cundinamarca, Boyacá, Santander, and Norte de Santander. The Motilones, one of the few surviving Amerindian groups untouched by civilization in South America, inhabit the region west of Lake Maracaibo and the Venezuelan border; they are famous for their lethal

COLOMBIA

0   50   100   150   200   250 Miles

0   50   100   150   200   250 Kilometers

Caribbean Sea

PANAMA

Gulfo de Panamá

PACIFIC OCEAN

NETHERLANDS ANTILLES

Uribia

Gulfo de Venezuela

Santa Marta

Barranquilla

Cartagena

Cristóbal Colón Pk. 18,947 ft. ▲ 5775 m.

Valledupar

Lago de Maracaibo

Sincelejo

Montería

Turbo

Guapá

Juradó

Medellín

La Fria

Cúcuta

Bucaramanga

Barrancabermeja

Arauca

Elorza

VENEZUELA

LLANOS

Puerto Carreño

Atrato

Cauca

CORDILLERA CENTRAL

CORDILLERA ORIENTAL

Magdalena

Tunja

Yopal

Ambalema

⊛ Bogotá

Meta

Villavicencio

Puerto Inírida

Ibagué

Buenaventura

Palmira

Cali

Neiva

Popayán

Guaviare

ANDES MTS

San José del Guaviare

Guayabero

Tumaco

Pasto

Tres Esquinas

Vaupés

Mitú

Apaporis

Uaupés

BRAZIL

Monclar

ECUADOR

Caquetá

Rocafuerte

Napo

El Encanto

Putumayo

Amazon

Leticia

PERU

NICARAGUA

Providencia

San Andrés

Caribbean Sea

COSTA RICA

PANAMA

COLOMBIA

PACIFIC OCEAN

Malpelo

N
W   E
S

Colombia

LOCATION: 12°27'46"N to 4°13'30"S; 66°50'54" to 79°1'23"W. BOUNDARY LENGTHS: Caribbean Sea, 1,600 km (994 mi); Venezuela, 2,219 km (1,379 mi); Brazil, 1,645 km (1,022 mi); Peru, 1,626 km (1,010 mi); Ecuador, 586 km (364 mi); Pacific Ocean, 1,300 km (808 mi); Panama, 266 km (165 mi). TERRITORIAL SEA LIMIT: 12 mi.

weapon, the black palm bow and arrow. Small, diverse Amerindian groups also inhabit the eastern extremities of the Colombian plains region, the south, and the western coastal jungles.

⁹LANGUAGES

The official language, Spanish, is spoken by all but a few Amerindian tribes. Spanish as spoken and written by educated Colombians is generally considered the closest to Castilian in Latin America.

¹⁰RELIGIONS

A 1973 concordat with the Vatican declares Roman Catholicism to be an essential element of the common welfare and of the integral development of the national community.

The vast majority of the country is Roman Catholic (93.1%). Religious freedom is constitutionally guaranteed to the small Protestant population and 6,500 Jews.

¹¹TRANSPORTATION

Transportation lacks integration, owing to the mountainous terrain. For this reason, air transportation has become the most important means of travel for most passengers. Despite the development of roads and railways, river travel has remained the chief mode of transportation for cargo since the trip up the Magdalena River in 1536 by the Spanish conqueror Gonzalo Jiménez de Quesada. Inland waterways navigable by riverboats totaled 14,300 km (8,886 mi) in 1991. The Magdalena, the fourth-largest river in South America, is navigable for 950 km (590 mi); it carries almost all of Colombia's river traffic.

The railroads, which were nationalized in 1954 and deregulated in 1989, had a length of 3,563 km (2,214 mi) in 1991.

In 1991 there were about 75,450 km (46,885 mi) of roads, of which only about 9,350 km (5,810 mi) were paved. Many roads are plagued by landslides and washouts. The Barco government has announced a program to build highways as a means of countering insurgency (by facilitating the shipment of produce to markets). The principal region affected is the entire eastern edge of the Andes, from the Caribbean to the Ecuadorean border. The 2,800-km (1,700-mi) Caribbean Trunk Highway, completed in 1974, links the Atlantic ports of Cartagena, Barranquilla, and Santa Marta with the Pan-American Highway (south of Panama) and the Venezuelan highway system. Early in 1991, the government proposed a 15-project program, with an estimated cost of $500 million. In 1991 there were 798,606 passenger cars, and 670,000 commercial motor vehicles.

Owing to inadequate land transport, air service is essential and well developed. A flight from Bogotá to Medellín takes only half an hour, while a truck requires 24 hours over a winding mountain road. Colombia's airline, Avianca, is the second-oldest commercial airline in the world and one of the largest in Latin America. Avianca handles about two-thirds of the domestic and international movement of passengers. Most of the country's air transportation is handled by the six principal airports at Bogotá, Barranquilla, Medellín, Cali, Cartagena, and San Andreas. In 1991, these airports serviced about 9,735,000 passengers.

Colombia's merchant marine is dominated by the Grand Colombian Merchant Fleet (Flota), a stock corporation owned by the Colombian Coffee Federation. In 1991, merchant marine companies had an aggregate of 34 vessels, comprising 300,000 GRT. The nation's chief ports on the Caribbean are Barranquilla, Cartagena, and Santa Marta. Buenaventura is the only important Pacific port.

¹²HISTORY

Archaeological studies indicate that Colombia was inhabited by various Amerindian groups as early as 11,000 BC. Prominent among the pre-Columbian cultures were the highland Chibchas, a sedentary agricultural people located in the eastern chain of the Andes.

The first Spanish settlement, Santa Marta on the Caribbean coast, dates from 1525. In 1536, Gonzalo Jiménez de Quesada and a company of 900 men traveled up the Magdalena River in search of the legendary land of El Dorado. They entered the heart of Chibcha territory in 1538, conquered the inhabitants, and established Bogotá. As a colony, Colombia, then called New Granada, was ruled from Lima, Peru, until it was made a viceroyalty. The viceroyalty of New Granada, consolidated in 1740, incorporated modern Colombia, Panama, Venezuela, and Ecuador. The area became Spain's chief source of gold and was exploited for emeralds and tobacco.

In the late 1700s, a separatist movement developed, stemming from arbitrary taxation and the political and commercial restrictions placed on American-born colonists. Among the Bogotá revolutionaries was Antonio Nariño, who had been jailed for printing a translation of the French Assembly's Declaration of the Rights of Man. Independence, declared on 20 July 1810, was not assured until 7 August 1819, when the Battle of Boyacá was won by Simón Bolívar's troops. After this decisive victory, Bolívar was tumultuously acclaimed "Liberator" and given money and men to overthrow the viceroyalty completely.

After 1819 Bolívar's Republic of Gran Colombia included Colombia, Venezuela, Ecuador and Panama. Venezuela and Ecuador seceded, but Panama remained part of Colombia. In 1831 the country became the State of New Granada. Political and financial order was attained under Francisco de Paula Santander, Bolívar's vice-president, who took office in 1832. During Santander's four-year term and in the subsequent decade there was intense disagreement over the relative amount of power to be granted to the central and state governments and over the amount of power and freedom to be given the Roman Catholic Church. Characterized by Bolívar as Colombia's "man of laws," Santander directed the course of the nation toward democracy and sound, orderly government.

By 1845, the supporters of strong central government had organized and become known as the Conservatives, while the federalists had assumed the Liberal label. The respective doctrines of the two parties throughout their history have differed on two basic points: the importance of the central governing body and the relationship that should exist between church and state. Conservatism has characteristically stood for highly centralized government and the perpetuation of traditional class and clerical privileges, and it long opposed the extension of voting rights. The Liberals have stressed states' rights, universal suffrage, and the complete separation of church and state. The periods during which the Liberals were in power (1849-57, 1861-80) were characterized by frequent insurrections and civil wars and by a policy of government decentralization and strong anticlericalism.

As effective ruler of the nation for nearly 15 years (1880-94), the Conservative Rafael Núñez, a poet and intellectual, restored centralized government and the power of the church. During his tenure as president, the republican constitution of 1886 was adopted, under which the State of New Granada formally became the Republic of Colombia. A civil war known as the War of a Thousand Days (1899-1902) resulted in more than 100,000 deaths, and the national feeling of demoralization and humiliation was intensified by the loss of Panama in 1903. After refusing to ratify without amendments a treaty leasing a zone across the Isthmus of Panama to the US, Colombia lost the territory by virtue of a US-supported revolt that created the Republic of Panama. Colombia did not recognize Panama's independence until 1914, in exchange for rights in the Canal Zone and an indemnity from the US.

Conservative presidents held power between 1909 and 1930, and Liberals from 1930 to 1942. During World War II, which

Colombia entered on the side of the Allies, social and political divisions within the country intensified.

The postwar period was marked by growing social unrest and riots in the capital and in the countryside. Politics became much more violent, especially after the assassination of Jorge Eliécer Gaitán, the leftist Liberal mayor of Bogotá. This extended and bloody period of rural disorder (La Violencia) claimed 150,000 to 200,000 lives between 1947 and 1958. Sporadic guerrilla fighting between Liberals and Conservatives, few of whom understood any ideological implications of their loyalty, ripped Colombia apart. La Violencia convinced Colombia's elite that there was a need to bring the rivalry between Liberal and Conservative under control.

The political system in the 1950s had become irrelevant in the midst of the violence. Three years of Conservative government were followed by a populist military government under Gen. Gustavo Rojas Pinilla. Rojas ruled as an absolute dictator, but could not quell the violence still raging in the field. Overthrown largely by a coalition of Conservatives and Liberals that used newsprint as its weapon, Rojas gave up power in May 1957 to a military junta, which promised and provided free elections.

When the fall of Rojas was imminent, Liberal and Conservative leaders met to discuss Colombia's future. Determined to end the violence and initiate a democratic system, the parties entered into a pact establishing a coalition government between the two parties for 16 years. This arrangement, called the National Front, was ratified by a plebiscite in December 1957. Under the terms of this agreement, a free election would be held in 1958. The parties would then alternate in power for four year terms until 1974. Thus, Liberals and Conservatives would take turns in the presidency. Parties were also guaranteed equal numbers of posts in the cabinet and in the national and departmental legislatures.

In 1958, the first election under the National Front was won by Liberal Alberto Lleras Camargo. As provided in the agreement, he was succeeded in 1962 by a Conservative, National Front candidate Guillermo León Valencia. In May 1966, Colombia held another peaceful election, won by Carlos Lleras Restrepo, a Liberal economist. Although lacking the necessary two-thirds majority required under the Colombian constitution to pass legislation, Lleras came to power with the firm support of the press and other important public sectors. His regime occupied itself with increasing public revenues, improving public administration, securing external financial assistance to supplement domestic savings, and preparing new overall and sectoral development plans. In April 1970, Conservative Party leader Misael Pastrana Borrero, a former cabinet minister, was elected president, narrowly defeating former President Rojas. The election results were disputed but later upheld.

In August 1974, with the inauguration of the Liberal Alfonso López Michelsen as president, Colombia returned to a two-party system for presidential and congressional elections. As provided by a constitutional amendment of 1968, President López shared cabinet posts and other positions with the Conservative Party. In 1978, another Liberal candidate, Julio César Turbay Ayala, won the presidency, but because his margin of victory was slim (49.5% against the Conservatives' 46.6%), he continued the tradition of giving a number of cabinet posts to the opposition. In June 1982, just before leaving office, Turbay lifted the state of siege that had been in force intermittently since 1948.

Because of a split in the Liberal Party, the Conservatives won the 1982 elections, and a former senator and ambassador to Spain, Belisario Betancur Cuartas, was sworn in as president in August. He continued the tradition of including opposition party members in his cabinet. But Betancur's most immediate problem was political violence, including numerous kidnappings and political murders since the late 1970s by both left- and right-wing organizations. In 1983, it was estimated that some 6,000 leftist guerrillas were active in Colombia. There were at least four guerrilla groups in the field, of which M-19 was the best known. The Betancur government pursued a policy of negotiation with the guerrillas. He offered amnesty and political recognition in exchange for the cessation of activity and the joining of the electoral process. Betancur's last year in office was marred by a seizure by M-19 of the Palace of Justice. Troops stormed the building, and it was completely destroyed by fire, with over 90 people killed.

The 1986 election went resoundingly to the Liberals under the long-time politician Virgilio Barco Vargas, who campaigned on a platform of extensive economic and social reform, focusing on poverty and unemployment. Barco won a significant majority, but the Conservative Party broke from the traditions started by the National Front, refusing cabinet and other government posts offered. President Barco's rhetoric was not matched by policies of any substance, and the economy continued to stagnate. Barco made no progress with drug traffickers, who arranged for the murder of his attorney-general. However, he was able to initiate a plan aimed at bringing guerrilla groups into the political system.

The election of 1990 brought another Liberal, César Gaviria Trujillo, to the presidency. In that election, three candidates were assassinated. Gaviria continued Barco's outreach to the various leftist guerrilla groups, and in 1991 the notorious M-19 group demobilized and became a political party. The other groups chose to remain active. Gaviria responded to their intransigence in November 1992 by announcing new counterinsurgency measures and a hard-line policy against both guerrillas and drug traffickers.

Gaviria also decided to create a new constitution, which occurred on 5 July 1991. It included a number of reforms aimed at increasing the democratization of Colombia's elite-controlled political system.

In the 1994 elections, Colombians continued their preference for Liberal candidates, with Ernesto Samper Pizano winning a run-off election against Conservative TV newscaster Andrés Pastrana. In the general election, only 18,499 votes separated the two candidates. The campaign was again marked by widespread political violence.

13GOVERNMENT

Colombia is a unitary republic, organized democratically under the constitution of August 1886, substantially amended in 1910, 1936, 1945, 1957, 1959, 1968, and 1979, and superseded by the constitution of 1991.

The Congress consists of a Chamber of Representatives and a Senate. Members of both are elected directly for four-year terms. Colombian congressional representation is determined by the size of the population. Some seats are reserved for blacks, Indians and former guerrillas. In 1990 there were 161 representatives and 102 senators. Under the 1968 constitutional reform, laws may be passed by a simple majority rather than by the previously required two-thirds vote. The chief executive is granted the initiative in fiscal policies and the power to declare a state of emergency during times of economic and social stress. Under such a declaration, the president may rule by decree for a period of not longer than 90 days in any one year.

The president is elected directly for a four-year term and may not succeed himself until one term has passed. For many years, Colombia used an officer called the designado ("designate") who served as a sort of vice-president. Every two years, Congress elects a designado, who exercises the executive function in the president's absence. The cabinet has 13 members. The Council of State is a consultative body with jurisdiction over administrative conflicts. A comptroller general is elected by the Chamber of Representatives. There is universal suffrage for those 18 years of age and over. Women have had the right to vote since 1954.

[14]POLITICAL PARTIES

For many years, the Colombian constitution allowed only two political parties, the Liberal and the Conservative, to participate in the national government. These two parties consistently dominated Colombian politics. Recent changes allow for more parties, and several have emerged.

The Liberal Party (Partido Liberal—PL) continues to support religious toleration and a positive response to the social and economic demands of the masses. The Liberals theoretically support separation of church and state, though in practice a strong church is accepted. Federalism, while important in theory, has been abandoned in practice by Liberal leaders. Liberals have been far more successful electorally than the Conservatives, having won all but one post-National Front elections, and continue to enjoy majority support in both houses.

The policy of the Conservative Party (Partido Conservador Social—PCS) has been characterized by close cooperation with the Roman Catholic Church, a lack of tolerance for non-Roman Catholic religious beliefs, maintenance of class privileges, and highly centralized government, with local authority strictly subservient to national rule. Before universal suffrage, the Conservatives sought to allow only heads of families to vote. The Conservatives currently suffer from a serious split between Misael Pastrana Borrero, the current PCS leader, and Alvaro Gomez Hurtado. Gomez, the son of a former conservative president, formed his own party, the National Salvation Movement (MSN) and placed second in the 1990 elections with 24% of the vote.

Despite the spread of suffrage and the rise of industrialization and a middle class, both parties continue to be dominated by a wealthy oligarchy. Both are controlled at the national level by a convention and a directorate, and congressional discipline is strong. Since the National Front agreement of 1958, the two parties have become increasingly similar ideologically.

Congressional and presidential elections from 1958 through 1982 primarily constituted votes of confidence in the National Front. Perhaps as a means of protest, 60% of eligible voters abstained from the presidential election in 1978, and 80% of the electorate abstained from the municipal and local elections of March 1980. In 1982 and 1986, however, Colombian voters turned out in record numbers, with 55% of the electorate participating in the presidential ballot in 1982 and 57% in 1986.

Parties of the left are all faithful to the memory of Jorge Eliécer Gaitán. The Colombian Communist Party (Partido Communista de Colombia—PCC) is a traditional, Moscow-oriented party. They coalesced with the guerrilla Colombian Revolutionary Armed Forces (Fuerzas Armadas Revolucionarias de Colombia—FARC) to form the UP. After taking only 4.1% of the 1986 legislative vote, they dropped to 1% in 1990.

There is considerable independent party activity in Colombia, and it has been increasing. Traditionally, the third force in Colombian politics has been provided by former dictator Rojas Pinilla. His National Popular Alliance (Alianza Nacional Popular—ANAPO) was a strong party movement, although it is no longer in existence. In 1978, ANAPO coalesced with other groups and won five seats in congress. ANAPO never got over the election of 1970, when Pastrana was elected in what ANAPO claimed were fraudulent elections. The election of 19 April 1970 gave rise to the spectacular rebel group M-19, which stood for the April 19th Movement. After over two decades of actions against the government, M-19 demobilized in 1991 and fielded candidates in the 1990 elections. M-19 candidate Antonio Navarro Wolf received a rather impressive 13% of the vote, outpolling the Conservative candidate, but fell to 3.8% in 1994.

Several militant leftist groups remain outside the system. The National Liberation Army (Ejército de Liberación Nacional—ELN) refuses virtually all government offers to join the regular political system. The FARC is a longstanding group, notable for its close ties with peasants and its involvement in coca trade. Although officially a member of the Patriotic Union, the FARC also refuses to demobilize. The People's Liberation Army (EPL) began to demobilize in 1993, but a dissident faction refused orders to lay down arms, and returned to the field.

[15]LOCAL GOVERNMENT

Colombia is divided into 23 departments (states), 4 intendencies, 5 administrative territories (commissariats), and the Bogotá federal district. However, under the constitution of 1991 the intendencies and commissariats are to become full-fledged departments, thereby giving them a greater degree of autonomy. Departments control their own finances, as well as administration, within the limits set by the constitution. Governors of departments, once appointed by the president, are now elected. Each departmental assembly meets yearly for a session of two months. Assembly members are elected by universal suffrage, one for each 40,000 inhabitants.

The departments have the power to establish municipal districts and to review the acts of the municipal governments to determine their constitutionality. Each municipality has a popularly elected municipal council. Another reform from the 1991 constitution is the direct election of mayors, where previously mayors were chosen by the president and were directly under the control of their respective governors.

[16]JUDICIAL SYSTEM

The Supreme Court in Bogotá is composed of 24 magistrates selected for lifetime terms by justices already in office. The Supreme Court reviews state and municipality laws, frames bills to be submitted to Congress, and proposes reforms. It acts as an advisory board to the government and can veto decrees. It has original jurisdiction in impeachment trials and constitutional interpretation and appellate jurisdiction in ordinary judicial matters. The court is divided into four chambers—civil cassation, criminal cassation, labor cassation, and constitutional procedure.

There is a superior court of three or more judges in each of the judicial districts and a number of municipal courts. A judge of minors in the capital of each department has jurisdiction throughout the department. There are also special labor courts. In criminal cases, the judge chooses a five-member jury; jury duty is obligatory. There is no capital punishment; the maximum penalty for crimes is 20 years in prison. Although the right of habeas corpus is guaranteed by the constitution, suspects in security cases have been detained incommunicado for 10 days or longer.

The 1991 Constitution extensively revises the judicial system. It establishes an independent prosecution system and a national people's defender office to investigate human rights cases. Traditional courts on Indian reservations are validated. A Constitutional Court reviews the constitutionality of proposed legislation.

The judiciary is independent in theory from the executive and legislative branches. In practice, however, it is overburdened and subject to intimidation in cases involving narcotics or offenses by paramilitary units. In 1991, the government set up five regional jurisdictions to handle narcotics, terrorism and police corruption cases in which anonymous judges and prosecutors handle the major trials of narcotics terrorists.

The procedures in the new regional courts have raised concerns over defendant's rights. In 1993, the constitutional court invalidated provisions of the regional court law under which detainees could be held without right to bail.

Criminal defendants have a right to an attorney, although indigents have difficulties in obtaining effective counsel. In some cases, inability to post bail and backlogs in processing result in the accused serving the applicable sentence for the crime charged before the case goes to trial.

[17]ARMED FORCES

Colombia's total armed forces in 1993 amounted to 139,000 personnel. All adult males (18–30) are obliged by the constitution to serve if called, for one or two years, then retain reserve service obligations until age 45. Most conscripts serve in the army (120,000, 38,000 conscripts), but all services have token conscript contingents.

The army is organized in four divisions of twelve infantry brigades, stationed on a regional basis. Special mobile forces, counterinsurgency forces, rangers, commandoes, and mechanized guard forces number an additional seventeen battalions, armed with American and European weapons. The army controls 10,000 of 117,000 reservists.

Air force personnel numbered 7,000 in 1993; combat aircraft numbered 68, armed helicopters 51. The navy had 12,000 personnel, including 6,000 marines. Naval ships included 5 frigates, 2 submarines, and 26 patrol craft and gunboats. There is also a 85,000-member national police force.

Colombia's defense forces are frequently occupied in opposing rural violence, often stemming from militant guerrilla groups and drug lords' armies. Guerrilla forces number perhaps 2,000. The military supplies peasants with arms and instruction in self-defense tactics. Defense expenditures in 1991 amounted to around $1 billion, or 2.5 percent of gross domestic product. Colombia imported $60 million in arms from the United States.

[18]INTERNATIONAL COOPERATION

Colombia is a charter member of the UN, having joined on 5 November 1945, and participates in ECLAC and all the nonregional specialized agencies. It was the only Latin American country to participate actively in the Korean War and the Suez Canal crisis. In Korea, a regiment of 1,000, three times relieved, was continuously in action. Colombia has been active in the OAS since its formation at Bogotá in 1948.

The Andean Pact, signed by five countries in May 1969, has opened up new marketing possibilities for Colombian products, in addition to those already existing in the US and Europe. Colombia also participates in G-77, the IDB, and PAHO, and it is a signatory to GATT and the Law of the Sea.

[19]ECONOMY

Despite gradual expansion of the manufacturing sector, Colombia's economy remains basically agricultural, coffee and sugar being the largest contributors to the GNP. Coffee is by far the most important crop: its share of total exports ranged from about 40% to 65% of the annual total between 1964 and 1986, depending on crop yields and international commodity prices. Colombia has been trying to reduce its dependence on coffee exports, because of widely fluctuating world market conditions that have prevailed since the mid-1950s, and to encourage other exports, especially sugar, bananas, rice, potatoes, and cotton. Mining has been the area of the most impressive growth in recent years; Colombia became a net exporter of petroleum in 1986, and coal has become a major export as well. The Trade Development Bureau (PROEXPO) was established in 1967 to increase the volume of nontraditional exports and to provide a flexible exchange rate and special tax incentives.

During the 1970s, Colombia's economy struggled with an inflationary spiral that had risen from a rate of 15.4% in 1972 to 25% in 1974; inflation was 25.6% in 1981 and about 25% in 1982. Since 1983, however, the economy has improved significantly, and growth rates above the world and hemispheric averages have been recorded—5% in 1986, for example. Inflation was kept close to 20% annually through the 1980s. The government had been hoping that at the end of 1994 inflation would be down to 19% from almost 23% in 1993. However, the results for the first quarter make the target look overly optimistic. By the end of 1993, unemployment had increased to 10.5%; up from 9.8% in the previous year..

Colombia is expected to register continuous economic growth over the decade of the 1990s. In 1993 GDP increased by 4.5%, up from 3.6% at YE 1992. Growth was strongest in mining, oil, and construction. New oil output from the recently discovered Cusiana field will play an important role in GDP growth in the coming years.

The agricultural sector, in contrast, has been performing poorly. Factors in this performance include a drought and the financial difficulties of the Agrarian Credit Bank, which is responsible for disbursing more than half of the funds for agricultural credits.

[20]INCOME

In 1992, Colombia's GNP was $44,555 million at current prices, or $1,290 per capita. For the period 1985–92 the average inflation rate was 26.1%, resulting in a real growth rate in per capita GNP of 2.4%.

In 1992 the GDP was $48,583 million in current US dollars. It is estimated that in 1991 agriculture, hunting, forestry, and fishing contributed 16% to GDP; mining and quarrying, 9%; manufacturing, 20%; electricity, gas, and water, 3%; construction, 5%; wholesale and retail trade, 14%; transport, storage, and communication, 10%; finance, insurance, real estate, and business services, 11%; community, social, and personal services, 5%; and other sources, 7%.

[21]LABOR

In 1970, 50% of the 5,100,000-member labor force worked in agriculture; by 1990, 30% of the labor force of about 10,750,000 held jobs in agriculture and 24% in industry. An average of 350,000 people joined the labor force annually during the 1970s and early 1980s, creating a demand for new positions that could not be met. To alleviate the problem, the government has undertaken an extensive program in which export production and foreign investments figure prominently.

The largest confederation, the Unitary Workers Central (CUT), represents 57% of the organized labor movement. In April 1992, two smaller confederations merged, the Colombian Confederation of Democratic Workers and the General Confederation of Workers, representing 12% and 3% of organized labor, respectively.

The 1991 constitution guarantees the right to strike except in public utilities, equal rights for women workers, workers' participation in management, universal education, expanded social insurance coverage, and the incorporation of ratified international labor conventions into the labor code. The basic source of Colombian labor legislation is the Substantive Labor Code of 1950, as amended. Colombia's basic labor law covers all inhabitants, regardless of nationality, and guarantees individual and collective rights. Maximum daily and weekly hours are 8 and 48, respectively, except in agriculture, stock breeding, and forestry, in which they are 9 and 54, respectively. There are overtime rates, annual vacations, sick benefits, and a compulsory twice-yearly bonus. Social security is compulsory, with the employer paying one-half and the employee and the government one-quarter each. Firms located more than a mile from a school must maintain primary schools for the children of the workers.

Urban unemployment is a major problem. The average unemployment rate was 9.6% in 1991, and 11% by March 1992.

[22]AGRICULTURE

Agriculture, despite an endemic problem of poor productivity, remains the most important segment of the Colombian economy. In 1991, agricultural goods and livestock accounted for 17.4% of the GDP.

Only about 5% of Colombia's land area is cultivated, most of it in elevated regions of the temperate zone. The small area of cultivation is due in part to the rugged Andean terrain and in part to lack of irrigation. In 1991, irrigated districts covered only about 9.6% of all arable land. The flat, fertile valleys are generally devoted to livestock, limiting cultivation to the slopes, an uneconomical practice that is gradually being changed. Hand cultivation, especially by machete and hoe, predominates, but mechanization is making headway on the larger farms. Fertilizer is expensive and not sufficiently used. The small size of farms is another constraint on agricultural growth: in the mid-1980s, farms of less than 20 hectares (50 acres) accounted for 80% of all agricultural land.

Coffee, by far the most important crop, is grown mainly on the Andean slopes at altitudes of 1,300–1,800 m (4,200–6,000 ft). Colombia, the world's second-largest coffee grower, contributes 13–16% of the total world production each year. In 1991, coffee-growing farms, many under 6 hectares (15 acres), accounted for about 1,020,000 hectares (2.5 million acres), or about two-thirds of the land under permanent crops. In 1992, despite a severe drought, the amount of coffee exported increased by over 30% as a result of a successful strategy for expanding the external markets. In 1992, coffee exports were valued at $1.26 billion, or 17.1% of the total value of exports.

Sugar, also important, is grown chiefly in the Cauca Valley, with its center at Cali. Many varieties of bananas are grown; bananas for export are produced in the departments of Magdalena and Antioquia. Corn, yucca, plantains, and, in high altitudes, potatoes have been traditional food staples since before the Spanish conquest. Beans, rice, and wheat, introduced in the 19th century, are also important in the diet. Other export crops include fresh-cut flowers, cotton, and tobacco. Cocoa is produced in limited amounts for domestic consumption. Colombia produces much of its domestic food requirements, but it has to import wheat, barley, fats, oils, and cocoa.

Since 1940, the government has taken an increasing part in the control, organization, and encouragement of agriculture. Through the Colombian Institute of Agrarian Reform (Instituto Colombiano de la Reforma Agraria—INCORA), farmers are given financial support, technical aid, and social assistance for better housing, education, and health facilities. INCORA acquires land for equitable distribution to farmers and helps to develop potentially valuable but uncultivated land and to increase agricultural productivity. The production-oriented Rural Development Program, begun in 1976, gave technical assistance and credit to about 30,000 small landholders. Farmers have also benefited from the us$150-million rural electrification program, introduced in 1981, and from a program to extend irrigation and drainage systems, initiated in the early 1980s. The agricultural annual growth rate averaged 4.4% during 1985–90, and stood at 5.5% in 1991.

Agricultural production (in thousands of tons) for major crops in 1990 and 1992 was as follows:

	1990	1992
Sugarcane	27,791	28,930
Plantains	2,425	2,745
Potatoes	2,313	2,131
Rice	2,117	1,735
Cassava	1,939	1,836
Bananas	1,600	1,900
Corn	1,213	1,056
Coffee	845	1,050
Sorghum	777	752
Cotton	119	117
Palm oil	252	304
Tobacco	33	27

[23]ANIMAL HUSBANDRY

Occupying about 40 million hectares (99 million acres) of pasture in 1991, livestock farming, especially cattle breeding has long been an important Colombian industry. The Ministry of Agriculture maintains experimental stations in Antioquia and Bolívar departments to improve breeds, but the quality of livestock is still low. Cattle are driven to market by truck. This practice often entails crossing high mountains, with much wastage; accordingly, there has been a movement to construct slaughterhouses and meat-packing plants near the ranges. Dairy farming, not important in the past, expanded in the 1970s, especially near the big cities. Colombian sheep produce about one-third of the wool used by the country's textile industry. The government maintains an experimental station for sheep in Cundinamarca and for goats in Norte de Santander. In 1992 there were 24.7 million cattle, 2.5 million sheep, and 2.6 million pigs. The production of beef and veal increased from a 1969–71 average of 422,000 tons a year to 678,000 tons in 1995.

[24]FISHING

Colombia has an abundance of fish in its Caribbean and Pacific coastal waters and in its innumerable rivers. Lake Tota in Boyacá and Lake La Cocha in Nariño abound in trout, as do the artificial reservoirs of Neusa and Sisga in Cundinamarca. About half of the annual catch consists of freshwater fish. Tarpon are caught in the delta waters of the Magdalena, and sailfish, broadbills, and tuna in the Caribbean. The 1991 fish catch was 108,708 tons. In 1991, aquaculture production increased, with frozen shrimp exports of $36.7 million and rock lobster exports of $6.5 million. A cholera outbreak in 1992, however, adversely affected fishing and aquaculture output.

[25]FORESTRY

Colombia's forested area is some 50 million hectares (123.6 million acres), or nearly 50% of the total area. Although much timberland is inaccessible or of limited value, the nation is self-sufficient in lumber. The soft tropical woods that predominate are also suitable for plywood production, for paper pulp, and for furniture manufacture; the wood pulp, paper, and paperboard output in 1991 was 690,000 tons. Roundwood production was 19,702,000 cu m in 1991.

[26]MINING

The new Constitution of 1990 replaced the 1866 Constitution; under the new article 332, the state retains the rights to all surface and subsurface nonrenewable and natural resources. The government grants concessions for exploration and production. In 1989, the new mining code became law. The mining act was written to encourage mineral exploration and development by expediting the processing of claims, by improving the security of mineral occupancy and tenure, and by providing financial aid to small- and medium-scale miners. In 1991, the total value of Colombian mineral exports amounted to about $2.9 billion, or 38% of total exports.

Colombia is the 3d largest coal producer in the Western Hemisphere (after the US and Canada); proven reserves are sufficient for the next millenium at curent extraction rates. Colombia began a modest export program in the mid-1970s; exports are projected to reach 49 million tons per year by 2001. It is also one of the few countries that is self-sufficient in coking coal for its steel industry. In the early 1980s, Carbones de Colombia (CARBOCOL), a state agency, began to expand coal-mining operations in conjunction with Spain, Romania, Brazil, and the International Resources Corp. (INTERCOR), an Exxon affiliate.

Colombia is one of the world's largest producers of precious metals and stones; in 1990, production of emeralds surpassed 3 million carats. The production and sale of emeralds is a joint private-government venture administered by Mineralco, the government-administered mining agency. Colombia is South America's only producer of platinum, and it ranks second in the region in gold production. Colombia has an estimated 2% of the world's nickel production, which amounted to 20,590 tons in 1991.

A large copper deposit, with reserves estimated at 625 million tons, was discovered at Pantanos, Antioquia, in 1973. The country's substantial copper, iron, nickel, and lead reserves are of major importance to the future development of the economy. Colombia also has a variety of quarried resources, such as limestone, sand, gravel, marble, gypsum, clay, and feldspar, as well as minor deposits of sulfur, asbestos, bauxite, mercury, zinc, and dolomite, among others. Salt production is a government monopoly; the production of marine salt in 1991 was 482,000 tons, and of rock salt, 219,000 tons.

The production of principal minerals (excluding petroleum and natural gas) in 1990 and 1991 (preliminary) included the following:

	1990	1991
Coal (tons)	20,400,000	20,200,000
Gold (kg)	29,352	34,844
Iron ore (tons)	628,000	685,000
Silver (kg)	6,591	8,036
Platinum (kg)	1,316	1,603

[27] ENERGY AND POWER

Colombia's mountainous terrain and network of rivers offer one of the highest potentials in the world for the generation of hydroelectric power, estimated at 118,100 Mw in 1985. These resources remain largely undeveloped, despite intensive government efforts. Net installed capacity increased from 9,414,000 kw in 1988 to 9,724,000 kw in 1991. The expansion of electrical capacity in Colombia has benefited significantly from financial support by the IBRD. In 1985, six new hydroelectric projects were under way.

Colombia's recoverable petroleum reserves were estimated at 1.9 billion barrels in 1992. Production of oil has rapidly risen from 7.4 million tons in 1982 to 19.6 million tons in 1987 to 22.2 million tons in 1992. In 1976, Colombia was forced to start importing oil. It was thus a source of considerable pride to Colombians when petroleum exports recommenced late in 1986. Guerrillas have made the export pipeline a target, however, and outflow has been disrupted occasionally. Between 1986 and 1991, there were 195 attacks on oil facilities, costing $800 million. British Petroleum's discovery of large oil reserves (at Cusiana) recently came at a time when declining reserves had increased doubts as to whether or not Colombia could remain an oil exporter through the 1990s. Empresa Colombiana de Petróleo (ECOPETROL), a state agency, has a 20% royalty share and 50% concession rights.

The production and consumption of natural gas have grown since the mid-1970s. In 1992, production totaled 5.3 billion cu m.

[28] INDUSTRY

Industrialization was not significant until 1930, but progress has been sizable since then. The National Association of Manufacturers (Asociación Nacional de Industriales—ANDI) represents firms engaged in some 40 different branches of manufacturing. ANDI was founded in 1944 to assist both large and small businesses. Since 1940, the Industrial Development Institute (Instituto de Fomento Industrial—IFI), a government-operated finance corporation, has been investing in enterprises that otherwise might not be undertaken because of high risk or lack of capital. It provides direct financing for construction, acquisition of essential machinery and equipment installation services, and working capital. Most of the industrial activity in the early 1980s was concentrated in and around Bogotá, Medellín, and Cali.

Manufacturing accounted for 15% of GDP in 1950 and 21.7% in 1985. Colombia is almost self-sufficient in consumer products, which represent about half of total industrial production. The 1970s witnessed a shift in industrial development policies from import substitution to expansion of exports. While the pace of industrial growth declined in the mid-1970s, it increased in the 1980s—by 8% in 1984, 2.5% in 1985, and 7% in 1986.

High growth rates were recorded in many export-oriented industries, including coal and oil derivatives, chemicals, porcelain, and glass. The textile industry, with the exception of the clothing sector, reached annual growth rates between 4% and 8%, as did the paper, furniture, and rubber industries and the food-, beverage, and tobacco-processing sectors. The cement industry slumped, however, and output declined in the production of machinery, transportation equipment, and raw materials. Assembly of motor vehicles fell by 10%; 33,208 passenger cars and 5,878 other vehicles were assembled in 1985. The printing and plastics industries recorded no growth. Production in 1985 included 5,314,963 tons of cement; 1,367,352 tons of refined sugar; 274,416 tons of steel; and 64,870 tons of petroleum products.

The country's industrial output grew by 4.1% in 1993. This was rather less than 1992's growth of 5.6%, but if coffee milling is excluded, the performance was better than that of the previous year, with output up 6% from 3.4%. Coffee milling was adversely affected by the decline in the harvest, which caused a 16.1% fall in the industrial product. Other sectors which showed a decline were leather and leather products (8.3%), paper and paper products (7.1%) and clothing (5.8%). Increases were recorded in transport and transport materials (42.5%), furniture (16.2%), glass and glass products (13.7%), porcelain and china (13.6%), metallic products (13.5%) and other manufacturing industries (13%). The construction sector was one of the most dynamic, with a growth of nearly 10%. The mining and hydrocarbons sector rebounded because of larger petroleum output made possible by the opening of the Magdalena Medio oil pipeline to the Atlantic in late 1991 and the expansion of the country's coal production capability.

[29] SCIENCE AND TECHNOLOGY

In 1982, total expenditure for research and development amounted to 2,754 billion pesos. In that same year, the nation had 2,107 scientists and engineers. The government of President Turbay Ayala (1978–82) emphasized research in farming, in order to help raise agricultural production. In 1993, Colombia had 2 scientific academies (the Academy of Exact, Physical, and Natural Sciences and the National Academy of Medicine, both in Bogotá), 26 scientific learned societies, and 11 scientific research institutes. Colombia has dozens of universities offering degrees in basic and applied sciences.

[30] DOMESTIC TRADE

There are four primary marketing areas: the Caribbean coast region, the Antioquia region, the Cauca Valley region, and the Bogotá region. Firms desiring distribution of their products to all important national markets generally appoint agents in the leading city of each of the four regions (Barranquilla, Medellín, Cali, and Bogotá, respectively). Most small purchases are made for cash, but many stores offer installment credit facilities. Domestic trade lagged during the recession of the early 1980s.

Small, individually owned retail establishments predominate, although chain stores are increasing. Variety stores on the pattern of those in the US are becoming popular, and food supermarkets are increasing in larger cities. Local farmers' markets, however, are still more generally patronized even in the cities, and in rural areas they are often the only trading centers.

Business hours vary largely with climatic conditions; however, the usual workday is from 8 AM to noon and from 2 to 6 PM.

Most businesses close on Saturday afternoons, Sundays, and on state or religious holidays. Banking hours are from 9 AM to 3 PM.

The principal advertising media are newspapers, magazines, radio, and television; in motion picture theaters, it is also customary to display advertisements on the screen between features. There are several advertising and public relations firms.

31FOREIGN TRADE

Colombia trades in traditional commodities, such as coffee, sugar, and bananas, and in an increasing number of nontraditional products, such as metals, chemicals, and pharmaceuticals. Among the newer export items, fresh-cut flowers have been outstandingly successful, with the value of exports increasing from us$300,000 in 1968 to us$141,400,000 in 1985. Government efforts to increase nontraditional exports were increased in 1967 with the establishment of the Export Promotion Bureau (PROEXPO) and with the inauguration of a flexible exchange rate and special tax incentives. In cooperation with other government agencies, the Colombian Institute of Foreign Trade regulates import and export licensing, establishes tax requirements, arranges international contracts, seeks means to increase and diversify exports, and assures the marketing of necessary quotas of national production abroad. A combination of such export promotion efforts and a sharp curtailment of imports have improved the trade balance since 1981, when it registered a deficit of us$3,090.4 million. The 1985 deficit was down to us$925.7 million; exports rose 79.8% between 1981 and 1985, while imports dropped 76.9%. A surplus of us$134 million was achieved in 1986, due to a 68% rise in the value of coffee exports.

As of 1985, 35.2% of Colombia's exports went to the US, and 29.8% to the EEC (more than two-fifths of this amount went to the Federal Republic of Germany, or FRG). The US is by far the leading source of imports, supplying 34.4% of Colombia's total in 1985. The EEC accounted for 18.6%, and Japan 10%.

Colombia has several free-trade zones, the largest of which is Barranquilla, on the Caribbean. Other free-trade zones providing benefits for importers and exporters, as well as for manufacturers located within the zone boundaries, are the Buenaventura Harbor, Cúcuta, Palmaseca (near Cali's international airport), and the Caribbean port of Santa Marta. The illicit trade in marijuana and cocaine, especially to the US, is known to be substantial, but there are no reliable estimates of its volume or its value.

There was little movement in exports in 1993. In contrast, imports soared in response to faster growth in the economy, the strength of the peso and the dismantling of trade barriers over the previous three years, known as the "Apertura program." Exports, excluding temporary exports, re-exports, and commercial samples, totaled $7,100 million, which is only 3.1% more than the comparable figure for 1992. The increase was small due to the fact that coffee export earnings fell by 9.5% as a reduction in the volume sold combined with weak prices. Major export-trading partners include the US, Germany, Venezuela, Japan, UK, and the Netherlands.

On the other hand, imports jumped 51.1% to total $9,000 million (again the figure excluded temporary imports, re-imports, commercial samples and so on). Consumer goods accounted for 75% of the growth of imports. Major import-trading partners included the US, Japan, Germany, Brazil, Venezuela, and Italy.

32BALANCE OF PAYMENTS

When President Carlos Lleras Restrepo took office in August 1966, the economy of Colombia was unstable; inflation was spiraling, and there was a lack of centralized economic planning. Lleras embarked on an austerity program that included trade and exchange controls, tight credit policies, tax reforms, a balanced budget, and the determination of priorities in the field of public investment. The cutback in imports had repercussions in the industrial sector, but controls were then loosened and business activity stepped up rapidly. For the first time in over a decade, Colombia succeeded in building up a national account surplus, showing positive net international reserves of about us$35.2 million by 30 June 1968. Emphasis on export expansion, import substitution, and continuation of foreign assistance led to further progress, and by the end of 1974, gold and foreign exchange reserves totaled us$467 million. Several of the next seven years were boom years for coffee exports, and net reserves expanded to us$5.6 billion by 30 June 1981. Reserves plunged sharply in the ensuing years but in 1985 began to expand again, reaching $2.1 billion by the end of the year. The merchandise trade balance surplus in 1991 was $2.8 billion, led by petroleum, coffee, coal, agricultural products, and non-traditional exports. By the end of 1991, the ratio of total debt to GDP had fallen to 35.7%, and the debt service to exports ratio had dropped to 41.4%. By 1992, gross international reserves stood at over $7.7 billion, or nine months' worth of goods and services imports.

In 1992 merchandise exports totaled $7,263 million and imports $6,030 million. The merchandise trade balance was $1,233 million. The following table summarizes Colombia's balance of payments for 1991 and 1992 (in millions of US dollars):

	1991	1992
CURRENT ACCOUNT		
Goods, services, and income	651	−822
Unrequited transfers	1,698	1,734
TOTALS	2,349	912
CAPITAL ACCOUNT		
Direct investment	433	740
Portfolio investment	81	60
Other long-term capital	−371	−463
Other short-term capital	−930	−54
Other liabilities	2	−117
Reserves	−1,836	−1,092
TOTALS	−2,621	−926
Errors and omissions	271	14
Total change in reserves	−1,891	−1,225

33BANKING AND SECURITIES

In 1993, Colombia had 27 commercial banks. Under a 1975 law, banks must be at least 51% Colombian-owned. The government supervises the banking system by means of a special governmental body, the Superintendency of Banks. In late 1982, in the wake of a scandal that led to the liquidation of a commercial bank and a finance company, the government moved to reform the banking sector by placing limits on the equity any individual (or his family) could hold in a financial institution and on the credit any lending institution could extend to any individual or entity. Several more crises in the ensuing years shook public confidence in the financial system, but tight government control over the sector has brought it back to a state of partial recovery. Beginning in 1989 the government began to privatize the banks.

The Bank of Bogotá, founded in 1879, was the first Colombian credit establishment. The Bank of the Republic was established in 1923 as the semiofficial central bank. This bank is the sole note-issuing authority. The notes must be covered by a reserve in gold or foreign exchange of 25% of their value. The Bank of the Republic also operates the mint and the salt and emerald monopolies for the government. It rediscounts and also makes loans to official and semiofficial institutions. In 1963, the Monetary Board was set up to assume from the bank the responsibility for setting required reserve rates for managing general monetary policy; this board, which formulates monetary, credit, and exchange policy, is thus the most influential financial agency in Colombia. In 1993, the commercial banks had reserves amounting to c$6,597.2 billion.

The Bank of the Republic and the commercial banks supply mainly short-term loans. A key role in industrial development is played by investment corporations, which generally make one- to three-year loans to manufacturing, agricultural, and mining companies. Savings and loan corporations provide financing for construction and housing. Commercial finance companies invest private funds; they issue promissory notes or fixed-term investment securities.

The Bogotá Stock Exchange, organized in 1928, is the largest official stock exchange in the country. The Medellín Exchange was established in 1961.

34INSURANCE

The government regulates the insurance industry through the Insurance Section of the Superintendency of Banks; all insurers, both domestic and foreign, must have its authorization to operate, and all insurance companies must have at least 51% Colombian capital. A social insurance system, established by law in 1946, is gradually expanding but is not yet in operation all over the country. In 1987, there were 24 life and 34 non-life insurance companies in operation. Total premiums paid in 1990 amounted to us$19.4 per capita, or 1.8% of the GDP.

35PUBLIC FINANCE

Considerable sums are spent to stimulate the development of industry, and higher than normal military expenditures have been necessitated by the continuing and disruptive guerrilla activity. The inflationary conditions that have prevailed since 1961 have also stimulated government expenditures. For political reasons, the national government has been unable to raise tax revenues sufficient to cover sharply expanding investment outlays. Loans from external financial agencies (including the IMF, IDB, and IBRD) have been substantial but insufficient to permit a buildup in the level of public investment operations. The recession of the early 1980s brought another round of deficits as spending increased far more rapidly than revenues. By the early 1990s, reforms in the public sector had greatly improved the efficiency of public expenditures; the deficit of the nonfinancial sector fell from 0.6% of GDP in 1990 to 0.2% in 1991.

Disbursed and outstanding debt in 1991 totaled $17.5 billion, or 49% of GDP. In 1991, external public debt amounted to $16,703 million, with foreign debt equivalent to 39.9% of GDP (down from 43.3% in 1988). Much of Colombia's foreign debt has been accumulated by financing infrastructural rather than industrial projects, the latter being more common among Latin American nations during the 1970s and 1980s.

36TAXATION

In addition to income taxes, Colombia levies a 14% value-added sales tax, and municipal and property taxes. Many stamp taxes are imposed on legal documents. Personal income tax rates vary from 10% to 30%. There is a flat tax on corporate income of 30% with a special 25% surcharge for calendar years 1993–97.

The 1960 tax law began the process of modernizing Colombia's tax structure. Recognizing that Colombia's fiscal problems stemmed from insufficient revenues rather than excessive spending, the Lleras Restrepo administration (1966–70) introduced legislation to curb tax evasion and instituted a system of withholding income tax at the source. After President López took office in August 1974, a new series of fiscal measures was put into effect, including a revision of income, property, and sales taxes, removal of tax exemptions for government companies, and a reduction in tax incentives for exporters. The principal objectives of the tax reform were more progressivity in tax assessment scales, promotion of economic stability, and a reduction in the

incidence of tax evasion. A clampdown on tax avoidance and evasion was also the motivation behind the reform measures of December 1982, which, like those under López, came amid a budgetary crisis. As part of an emergency package, a special amnesty was declared so that taxpayers who had omitted assets or declared nonexisting liabilities were allowed to make corrections in their tax returns and pay the tax due without additional penalty. The measure was reportedly targeted at those who had profited from the drug trade.

Indirect taxes include a tax on exported goods and services, a 5% surcharge on imports, a gasoline tax, and local taxes on vehicles and property.

37CUSTOMS AND DUTIES

Colombian import tariffs have been significantly reduced and simplified. There are several levels of duties: primary goods and intermediate and capital goods not already produced in Colombia; goods in the previous category that are already produced in the country, and finished consumer goods. Colombia maintains free trade agreements with Venezuela and Ecuador and was negotiating an agreement with Mexico as of 1993. There are also international agreements for preferential duty rates with the Caribbean Common Market and the Central American Common Market.

38FOREIGN INVESTMENT

Although the government has sought to encourage investment since the early 1970s—especially in manufacturing—petroleum and petroleum-derivative industries remain the most heavily dominated by foreign capital; investment in this sector reached us$408 million by 1980 and more than us$600 million by the end of 1981. There is a high degree of cooperation between local and foreign investors in joint ventures, and Colombian industry has welcomed participation by foreign capital in almost every important field of private enterprise. Indeed, the government's policy has been to seek foreign capital for economic development, especially in the form of joint enterprises, from international lending agencies as well as from private sources. To aid Colombian economic development, the government permits tax concessions and tariff protection for certain products that will either reduce import requirements or increase exports. In the 1980s, new foreign investment was still encouraged in manufacturing, mining, and oil and gas, but it was banned in public services, domestic transportation (the government does, however, seek foreign public investment for the Bogotá Metro), publicity, and mass media, and it was restricted in domestic marketing, tourism, and banking. Due in large part to the drop in oil prices, foreign investment declined in 1986. New investment in oil and gas exploration was us$62.4 million (compared with us$169.1 million in 1985). Direct foreign investment in sectors other than mining was us$54.9 million in 1986. The foreign private debt in 1985 was us$3,600 million.

The Andean Group's Common Code on Foreign Investment, adopted in 1973, anticipated the nationalization of industries operating in the Andean Group at a rate of 51% to 80% by 1988. Companies operating solely in Colombia are excluded from this category.

The government has been promoting foreign investment in recent years. The former Gaviria Administration implemented an aggressive economic liberalization program. Foreign companies were put on an equal legal footing with local ones. As a result, total foreign investment increased in 1992 to us$4,300 million up from us$4,000 million in 1991. Colombia's major foreign investors include the US, Switzerland, and the UK. In order to continue boosting the economy, the government has negotiated several free trade agreements with several countries. A free trade pact among Colombia, Venezuela and Central America (except Costa Rica) came into operation on 1 July 1993. In addition, a preliminary agreement was reached among Colombia, Venezuela, and Mexico (G-3) on establishing a free trade zone from the beginning of 1994. The trade pact with Venezuela is expected to support the continuation of a trade boom.

In 1993, Colombia, Venezuela, Ecuador and Bolivia achieved a customs union, with free trade between the four countries under the auspices of the Andean Pact. Effective January 1, 1994 Pact members implement a common external tariff (CET) based on a four tier tariff range.

39ECONOMIC DEVELOPMENT

Colombia encourages private enterprise and limits state intervention to activities necessary to the national economy that are not sufficiently developed by private capital. On this principle, the national government and municipalities own and operate most telephone and telegraph facilities, power plants, and railroads. The government has long operated certain revenue-producing monopolies, such as salt mining, emerald production, and distillation of liquor.

The government helps establish new industries through the IFI, whose long-term policy has been to raise its capitalization to meet the demands of economic growth. In the 1970s, IFI took major responsibility in the financing of several mining projects, notably in nickel ore and coal.

In 1974, President López outlined to Congress a long-range development plan with a major objective of achieving maximum growth while raising the living standards of the poorer half of the population. Efforts were to be concentrated in four main areas: exports, agriculture, regional development, and industry. An economic program published in 1980, during the administration of Turbay, listed investment in energy, economic decentralization, regional autonomy, improvements in communications and transportation, mining development, and social improvement as its principal aims. A new national development plan for the years 1981–84 provided for acceleration of public works. In October 1982 and again in December 1982, during the Betancur presidency, economic emergencies were declared so that decrees to revise banking and fiscal policies could be issued without the need of congressional approval. The Barco administration has formulated an economic program similar to that proposed in 1974 by López, whose objectives are to attack unemployment and poverty while encouraging growth.

US nonmilitary loans and grants to Colombia during 1946–86 totaled US$1,405.8 million. During the same period, Colombia also received US$7,247.9 million from multilateral sources, including US$5,524.3 million from the IBRD.

Real economic growth is expected to keep pace with the increase of 1992-93. The greatest impetus is expected to come from private investment attracted by low real interest rates, spurring continuation of the reconversion and industrial modernization program. Trade within the region will help to bolster the economy, aided by the trade agreement with Venezuela—a lucrative market and net importer of food products—as well as the agreement signed with Ecuador in November 1992.

In the medium term, the outlook for the growth of the economy is favorable, even though nontraditional exports are growing at a much slower pace than in recent years and the prices and volume of coffee sold remain depressed. Production gains will be even more pronounced if, in addition, the government is successful in its campaign to halt terrorist activities.

40SOCIAL DEVELOPMENT

According to legislation enacted in 1971, social security coverage extends to salaried and nonsalaried people alike, including small business people, and other self-employed workers in both rural and urban areas. The government, employers, and employees pay for the program, which includes disability, old age, and death benefits and coverage of nonoccupational illnesses, maternity, and job-related accidents.

The Colombian Institute of Social Security administers other programs, including severance pay, pensions, vacation benefits, group life insurance, job training, transportation and clothing subsidies, educational benefits for families, and a scholarship fund. The Family Welfare Institute coordinates an estimated 1,000 public and private charities involved in caring for children and destitute families.

The government supports family planning; the Maternal and Child Health Program of the Ministry of Health and the private PROFAMILIA program provide services to urban and rural populations. The fertility rate fell from 6.03 in 1967/68 to 2.9 for 1985–90. The contraception prevalence rate in 1990 was 66%.

In 1975, measures were enacted to end discrimination against women under the law. However, there is still discrimination against women, especially in rural areas. They earn 30–40% less than men for doing similar work, and occupy few of the top positions in government. Urban Colombia, and especially Bogotá, has acquired a reputation for street crime: pickpockets and thieves are a common problem. Drug trafficking flourishes on a large scale, despite government efforts to suppress cocaine smuggling and to eradicate the coca and marijuana crops.

41HEALTH

Health standards have improved greatly since the 1950s, but malaria is still prevalent up to 1,100 m (3,500 ft) in altitude, and many Colombians suffer from intestinal parasites. Malnutrition, formerly a very serious problem, with nutritional goiter, anemia, scurvy, and pellagra frequent, had become less severe by the early 1980s, when the per capita calorie supply was estimated at 102% of requirements. In 1990, 12% of children under 5 years old were considered to be malnourished, and there were 67 reported cases of tuberculosis per 100,000 people that year. From 1988 to 1991, 86% of the population had access to safe water.

Colombia's population is expected to rise to 39 million by the end of the century, overtaking Argentina as the third most heavily populated country of the region (after Brazil and Mexico). Health care provisions (doctors and beds) do not compare favorably with other countries of the region.

From 1988 to 1992, there were 87 physicians for every 100,000 people, with a nurse to doctor ratio of .6. According to a study made between 1985 and 1992, 60% of the population (34 million population in 1993) had access to health care services. Total health care expenditures in 1990 were US$1,604 million.

Average life expectancy in 1992 was 69 years. The infant mortality rate decreased from 99.8 per 1,000 live births in 1960 to 17 in 1992. There were 809,000 births in 1992 (24 per 1,000), with 66% of married women (ages 15 to 49) using contraception (studied between 1980 and 1993). Between 1990 and 1992, children up to one year of age were immunized against the following: tuberculosis (86%); diphtheria, pertussis, and tetanus (77%); polio (84%); and measles (74%). Colombia's death rate for 1993 was 6 per 1,000 inhabitants, and between 1986 and 1992, there were approximately 22,000 civil war-related deaths.

42HOUSING

Colombia's housing shortage is largely a result of the rapid growth of the urban population. With the annual urban population growth rate at over 5%, the housing deficit was estimated to be around 800,000 units in the early 1980s and is expanding annually. Total housing units numbered 6,906,000 in 1992. As of 1985, 86% of all housing units were detached houses, ranches, and huts, 13% were apartments, and the remainder were mobile units, natural shelters, and non-residential housing. Roughly 67% were owner occupied and 24% were rented. Three-fourths of all dwellings were made of bricks, adobe, mud or stone, 14% had external walls of wattle or daub, 7% were wood, and 3% were mostly cane. Public resources are channeled into urban housing through the Central Mortgage Bank, whose mortgages, because of interest rates, down-payment requirements, and repayment terms, have usually been accessible only to upper-middle-income groups; and the savings and loan corporations, whose interest rates are pegged to inflation through daily monetary correction factors known as units of constant purchasing power value (UPACs). In addition, through AID, substantial private funds from US investors have been used for various housing programs. A us$2.6-billion government program for the rehabilitation and construction of new housing in the 1987–90 period was announced by the Barco government, with 131,000 units to be constructed and 258,000 units to be improved.

Rural housing problems in Colombia are dealt with primarily by the Credit Bank for Agriculture, Industry, and Mining, and by the Colombian Institute of Agrarian Reform.

43EDUCATION

Education is free and compulsory for five years in Colombia. Illiteracy is declining, having dropped from an estimated 90% at the end of the 19th century to an estimated 13.3% in 1990 (males: 12.5% and females: 14.1%). The constitution provides that public education shall not conflict with the doctrines of the Roman Catholic Church; courses in the Roman Catholic religion are compulsory, and the Church is in virtual control of the public schools. Private schools have freedom of instruction, and there are a number of Protestant schools, principally in Bogotá. The national government supports secondary as well as university education and maintains a number of primary schools throughout the country.

By law, Colombia must spend at least 10% of its annual budget on education. In 1987, the education budget was c$161,457 million, or 16.6% of the total. Financing and supervision of public education is the joint responsibility of the Ministry of Education, the departments, and the municipalities. Secondary and technical education and universities are administered by the Ministry of Education. The central government also pays teachers' salaries.

In 1985 there were 4,039,533 students enrolled in primary schools and 1,934,032 in secondary schools. Although schooling is compulsory for children in the 7–11 age group, dropout rates are high at the primary level, particularly in rural areas, where the students frequently live at considerable distances from their schools. Almost all secondary schools are in the larger cities; thus, little educational opportunity is open to rural children, except those reached by educational radio and television broadcasting. The government has established two basic programs for improving secondary education—the integration of practical training into high-school academic curricula and the training of agricultural experts at the secondary level—so that students who do not go on to college (relatively few do) are prepared to receive further technical training or to earn a living. The National Apprenticeship Service offers technical and vocational training in fields that contribute to national development; the program is financed by compulsory contributions from private enterprises and employees.

As of 1989 there were 474,787 students enrolled in higher education in Colombia. The National University in Bogotá, founded in 1572, is one of the oldest in the Western Hemisphere. Other important universities include the Universidad Javeriana (founded 1622), which is operated by the Jesuits; the Universidad de los Andes, a private institution based on a US model; and the Universidad Libre, a private university with active liberal leanings. In 1964, the Colombian Overseas Technical Specialization Institute (ICETEX) was formed to coordinate scholarship and fellowship funds for Colombians wishing to study abroad.

Graduates of foreign schools who return to Colombia reimburse ICETEX from their subsequent earnings.

44LIBRARIES AND MUSEUMS

The National Library, founded in 1777 in Bogotá, has over 800,000 volumes. It also acts as a public library and maintains a small circulating collection and a children's room. The cities have municipal libraries and the towns have village libraries, which are under the control of the Colombian Institute of Culture. There are two valuable private libraries in Bogotá, established by Dr. Luis Agusto Cuervo and Dr. Antonio Gómez Restrepo, respectively; each contains about 50,000 volumes. The Bank of the Republic maintains the Luis Angel Arango Library, an important cultural center holding some 400,000 volumes. The library of the National University in Bogotá has approximately 183,000 volumes.

The most notable of Bogotá's museums are the National Museum, which concentrates on history and art since the Spanish conquest, the founding of Bogotá, and the colonial period; the National Archaeological Museum, which exhibits indigenous ceramics, stone carvings, gold objects, and textiles; the Museum of Colonial Art, formerly part of the National University, which specializes in art of the 16th, 17th, and 18th centuries; and the Gold Museum, located in the Bank of the Republic.

45MEDIA

Telephone and telegraph networks link all provincial capitals and connect these centers with surrounding rural areas. Each local system is independent; most are municipally owned, but a few are in private hands. Long-distance service is provided by the national government and is based on an agreement among local and departmental systems. In the early 1980s, the National Telecommunications Enterprise (TELECOM) had one of the largest automatic telephone service networks in Latin America. Its high-capacity microwave system connected the 40 largest cities in the country. In 1991, over 2,500,000 telephones were in service.

Most of the nation's 440 radio stations were privately owned in 1991. In that year there were about 5,800,000 radio receivers and about 3,800,000 television sets. All television is owned and operated by the national government, but there are some commercial programs. Color television was introduced in 1979. Colombia launched its own communications satellite in 1985. In 1992 there were 33 television stations in Colombia.

Almost every town publishes at least one daily newspaper. The press varies from the irregular, hand-printed newspapers of the small towns of the interior to such national dailies as El Tiempo, one of the most influential newspapers of the Spanish-speaking world. Although censorship has been exercised occasionally in times of national emergency, freedom of the press has generally been respected.

The country's principal newspapers and their 1991 daily circulations are as follows:

	ORIENTATION	CIRCULATION
BOGOTÁ		
El Tiempo (m)	Liberal	220,000
El Espectador (m)	Liberal	160,000
El Espacio (e)	Liberal	92,000
El Nuevo Siglo (m)	Conservative	70,000
CALI		
El País (m)	Conservative	66,000
Occidente (m)	Conservative	55,000
MEDELLÍN		
El Colombiano (m)	Conservative	110,000
BARRANQUILLA		
El Heraldo (m)	Liberal	75,000

120   Colombia

	ORIENTATION	CIRCULATION
BUCARAMANGA		
El Vanguardia liberal (m)	Liberal	42,000
MANIZALES		
La Patria (m)	Conservative	33,000

46 ORGANIZATIONS

The National Federation of Coffee Growers, organized in 1927, is a semiofficial organization partly supported by tax revenue. The organization carries great weight as the representative of Colombia's leading industry, and its influence is felt in many spheres. Other trade associations include the National Association of Manufacturers, the People's Association of Small Industrialists of Colombia, and chambers of commerce in the larger cities. The Bank Association is an association of both national and foreign banks in Colombia. The Colombian Livestock Association, the National Federation of Cotton Growers, and the National Association of Sugar Growers serve their respective industries. Most farmers belong to the Agricultural Society of Colombia.

Learned societies include the Academy of History, the Colombian Academy of Language, the Colombian Academy of Exact, Physical, and Natural Sciences, the Academy of Medicine, the Colombian Academy of Jurisprudence, the Colombian Geographical Society, the Colombian Institute of Anthropology, and a number of regional bodies. The National Association of Colombian Writers and Artists includes most of the country's writers, painters, sculptors, and composers. Journalists have national and local organizations.

47 TOURISM, TRAVEL, AND RECREATION

The tourist industry in Colombia developed greatly in the late 1970s but declined in the 1980s. Spurred by the government's economic liberalization program, earnings from tourism rose from US$270 million in 1990 to US$410 million in 1991. In that year, the country had 43,072 hotel rooms. Both the government and private groups have been active in developing hotels and other tourist facilities. Colombia has mountains, jungles, modern and colonial cities, and resorts on both the Pacific and the Caribbean, all of which the Colombian Government Tourist Office (CORTURISMO) has aggressively promoted. The number of tourists entering Colombia reached 857,000 in 1991, with 798,000 from the Americas and 53,691 from Europe. Soccer is the most popular sport, followed by basketball, baseball, boxing, and cockfighting; there are also facilities for golf, tennis, and horseback riding, and bullrings in the major cities.

Tourists from Western countries generally require no visa, but all visitors need a valid passport and a return ticket for entry. No vaccinations are required.

48 FAMOUS COLOMBIANS

Outstanding political and military figures in Colombian history include Francisco de Paula Santander (1792–1840), who served as a general in the war of independence and was the first president of independent Colombia, and José María Córdoba (1800?–1830), a brilliant young soldier of the war of independence, who was made a general at 22 by Simón Bolívar.

Colombia, famous for its literary figures, has produced three outstanding novelists widely read outside the country: Jorge Isaacs (1837–95), whose most famous work, *María*, is a novel in the Romantic tradition; José Eustacio Rivera (1880–1929), whose outstanding novel, *La Vorágine (The Vortex)*, written after World War I, is a drama of social rebellion; and Gabriel García

Márquez (b.1928), a Nobel Prize winner in 1982, who is best known for *Cien años de soledad (One Hundred Years of Solitude)*. Colombia has had a number of noteworthy poets. The 19th-century Romantic school included Julio Arboleda (1814–92), José Eusebio Caro (1817–53), Gregorio Guitiérrez Gonzales (1826–72), and Rafael Pombo (1834–1912). Caro, who was influenced by the English poets, is generally rated as the most important Colombian Romantic. José Asunción Silva (1865–96) is regarded as the father of Latin American symbolism; his *Nocturnos* are among the finest poems in the Spanish language. Guillermo Valencia (1873–1945), the author of *Anarkos*, was a polished poet of the classical school, and León de Greiff (b.1895) is a well-known modern poet. Miguel Antonio Caro (1843–1909) and Rufino José Cuervo (1844–1911) were philologists and humanists of great erudition who influenced scholars and students in the 19th century. The Instituto Caro y Cuervo in Bogotá is devoted to the study and publication of their works. Well-known literary critics include Baldomero Sanín-Cano (1861–1957) and Antonio Gómez-Restrepo (1868–1951).

Colombia's most notable painter was Gregorio Vázquez Arce y Ceballos (1638–1711), whose drawing and coloring have been compared to the work of the Spanish painter Murillo. Francisco José de Caldas (1770–1816) was a brilliant botanist who discovered a system for determining altitude by the variation in the boiling point of water and began the scientific literature of the country. Guillermo Uribe-Holguín (1880–1971) and José Rozo Contreras (b.1894) are noted composers. The works of the contemporary historian Germán Arciniegas (b.1900) are well known to the English-speaking world through translation.

49 DEPENDENCIES

The archipelago of San Andrés and Providencia, administered as an intendancy, is located 729 km (453 mi) from the Caribbean coast northwest of Cartagena and about 190 km (118 mi) off the Nicaraguan coast. Grouped roughly around the intersection of longitude 82°W and latitude 12°N, the archipelago consists of the islands of San Andrés and Providencia and 13 small keys. The population is mostly black. Both English and Spanish are spoken. The principal towns are San Andrés, San Luis, and Loma Alta on the island of San Andrés and Old Town on Providencia.

50 BIBLIOGRAPHY

Bushnell, David. *The Making of Modern Colombia: a Nation in Spite of Itself.* Berkeley: Univ. of California Press, 1993.
Cohen, Alvin and Frank R. Gunter. *The Colombian Economy: Issues of Trade and Development.* Boulder, Colo.: Westview Press, 1992.
Colombia in Pictures. Minneapolis: Lerner, 1987.
Davis, Robert H. *Colombia.* Oxford, England; Santa Barbara, Calif.: Clio Press, 1990.
———. *Historical Dictionary of Colombia,* 2d ed. Metuchen, N.J.: Scarecrow Press, 1993.
Hartlyn, Jonathan. *The Politics of Coalition Rule in Colombia.* New York: Cambridge University Press, 1988.
McFarlane, Anthony. *Colombia Before Independence: Economy, Society, and Politics under Bourbon Rule.* New York, NY, USA: Cambridge University Press, 1993.
Pearce, Jenny. *Colombia: the Drugs War.* New York: Gloucester Press, 1990.
Randall, Stephen J. *Colombia and the United States: Hegemony and Interdependence.* Athens: University of Georgia Press, 1992.

COSTA RICA

Republic of Costa Rica
República de Costa Rica

CAPITAL: San José.

FLAG: The national flag consists of five horizontal stripes of blue, white, red, white, and blue, the center stripe being wider than the others.

ANTHEM: *Himno Nacional,* beginning "Noble patria, tu hermosa bandera" ("Noble native land, your beautiful flag").

MONETARY UNIT: The colón (c) is a paper currency of 100 céntimos. There are coins of 1, 5, 10, 25, and 50 céntimos and of 1, 2, 5, 10, and 20 colones, and notes of 5, 10, 20, 50, 100, 500, and 1,000 colones. c1 = $0.0065 (or $1 = c153.97).

WEIGHTS AND MEASURES: The metric system is the legal standard, but local measures also are used.

HOLIDAYS: New Year's Day and Solemnity of Mary, 1 January; Day of St. Joseph (Costa Rica's patron saint), 19 March; Anniversary of the Battle of Rivas, 11 April; Labor Day, 1 May; Day of St. Peter and St. Paul, 29 June; Anniversary of the Annexation of Guanacaste, 25 July; Feast of Our Lady of the Angels, 2 August; Assumption (Mother's Day), 15 August; Independence Day, 15 September; Columbus Day, 12 October; Immaculate Conception, 8 December; Abolition of Armed Forces Day, 1 December; Christmas, 25 December. Movable religious holidays include Holy Thursday, Good Friday, Holy Saturday, and Corpus Christi.

TIME: 6 AM = noon GMT.

¹LOCATION, SIZE, AND EXTENT
The third-smallest country in Central America, Costa Rica has an area of 51,100 sq km (19,730 sq mi), including some small islands. Comparatively, the area occupied by Costa Rica is slightly smaller than the state of Virginia. Its length is 464 km (288 mi) N–S, and its width is 274 km (170 mi) E–W. Costa Rica is bordered on the N by Nicaragua, on the E by the Caribbean Sea, on the SE by Panama, and on the SW and W by the Pacific Ocean; the total boundary length is 1,929 km (1,199 mi). Costa Rica's capital city, San Jose, is located in the center of the country.

²TOPOGRAPHY
Costa Rica has three main topographic regions. The central highlands, extending from northwest to southeast, reach elevations of more than 3,660 m (12,000 ft) south of San José; the highest point in the country is Chirripó Grande (3,810 m/12,500 ft). Four volcanoes, two of them active, rise near the capital city; one of these volcanoes, Irazú (3,432 m/11,260 ft), erupted destructively during 1963–65. Nestled in the highlands is the Meseta Central, with an elevation of 900–1,200 m (3,000–4,000 ft), covering some 2,000 sq km (770 sq mi) of fairly level, fertile terrain. Half of the population, the centers of culture and government, four of the six main cities, and the bulk of the coffee industry are found on the plateau. The Atlantic coastal plain, on the Caribbean side of the highlands, comprises about 30% of Costa Rica's territory and is low, swampy, hot, excessively rainy, and heavily forested. The Pacific slope, some 40% of the country's area, resembles the Caribbean lowlands, but to the northwest is a dry area producing cattle and grain. Fifteen small rivers drain Costa Rica.

³CLIMATE
Costa Rica has only two seasons: the wet season, from May to November, and the dry season, from December to April. There are three climatic zones. The torrid zone, which includes the coastal and northern plains to an altitude of 900 m (3,000 ft), is characterized by heavy rains, almost continuous on the Atlantic watershed, and by a temperature range of 25–38°C (77–100°F). The temperate zone, including the central valleys and plateaus, has altitudes ranging from 900 to 1,800 m (3,000 to 6,000 ft), with regular rains from April through November and a temperature range of 15–25°C (59–77°F). The cold zone, comprising areas higher than 1,800 m (6,000 ft), has a temperature range of 5–15°C (41–59°F) and is less rainy but more windy than the temperate regions. The average annual rainfall for the country is more than 250 cm (100 in).

⁴FLORA AND FAUNA
Costa Rica supports varied flora and fauna. From the coast to an altitude of about 900 m (3,000 ft) are tropical forests and savannas; oaks and chaparrals are found between 2,070 and 3,050 m (6,800 and 10,000 ft); and subandean and subalpine flora characterizes the highest mountains. The dense tropical forests contain rich stands of ebony, balsa, mahogany, oak, laurel, campana, and cedar. Plant life is abundant, and the country has more than 1,000 species of orchids.

Most of the wild mammals common to South and Central America, such as jaguar, deer, puma, and varieties of monkeys, are found in Costa Rica. There are over 790 species of birds and 130 species of snakes and frogs; fish and insects are plentiful.

⁵ENVIRONMENT
Nearly all of Costa Rica was once covered by forests, but deforestation for agricultural purposes has reduced virgin forest to only 25% of the total area. Most of the wood is wasted by burning or rotting, and there is little incentive for conservation or reforestation. The result has been soil erosion and the loss of soil fertility. Another serious problem, according to the UN, has been contam-

ination of the soil by fertilizers and pesticides used in growing important cash crops, such as bananas, sugarcane, and coffee. Costa Rica's use of pesticides is greater than all the other countries in Central America added together. Under the General Health Law of 1973, the Ministry of Health has broad powers to enforce pollution controls, and the Division of Environmental Health has attempted to set standards for air and water quality. However, trained personnel and equipment are lacking. As of 1994, Costa Rica has 22.8 cu mi of water with 89% of the total used for farming activity. All of the nation's urban dwellers have safe water. Six percent of the rural population does not. The coastal waters are threatened by agricultural chemicals. Costa Rica contributes 0.1% of the world's total gas emissions. It has the highest per capita emissions rate in North America.

Costa Rica's national park system is among the most extensive and well developed in Latin America. As of 1986, the system, covering nearly 4% of the total land area, included 12 parks, six nature reserves, four recreation areas, the Guayabo National Monument archaeological site in the Turrialba region, and the International Peace Park established jointly by Costa Rica and Panama on their common border. Although Costa Rica protects some 80 animal species by law, the IU recognized only nine species as being endangered in 1987: the red-backed squirrel monkey, tundra peregrine falcon, spectacled caiman, American crocodile, four species of sea turtle (green sea, hawksbill, olive ridley, and leatherback), and golden toad. Of 203 mammal species in Costa Rica, 10 are threatened. Fourteen species of birds in a total of 796 are also threatened. Two types of reptiles in a total of 218, and 418 plant species in a total of 8,000, are considered endangered as of 1994.

⁶POPULATION

Costa Rica's population according to the 1984 census was 2,416,809. The 1994 estimate was 3,099,000. The UN projected population for the year 2000 was 3,798,000, assuming a crude birth rate of 24 per 1,000 population, a crude death rate of 3.8, and a net natural increase of 20.2 during 1995–2000. Population density was estimated at 61 persons per sq km (158 per sq mi) in 1992; 50.3% of the population was rural and 49.7% urban, according to an estimate for 1995.

The principal city is San José, the capital, which had an estimated population of 296,625 in 1991. Other large cities (with 1991 estimated populations) are Limón, 67,784; Alajuela, 158,276; Puntarenas, 92,360; Cartago, 108,958; and Heredia, 67,387.

⁷MIGRATION

Large numbers of Nicaraguans migrate seasonally to Costa Rica seeking employment opportunities. By 1986, approximately 200,000 refugees from Guatemala, El Salvador, and Nicaragua were in Costa Rica, constituting a heavy burden on the country's educational and health facilities. This number had fallen to 113,500 by the end of 1992.

⁸ETHNIC GROUPS

The population is fairly homogeneous, primarily of European (mainly Spanish) descent, with a small mestizo (mixed white and Amerindian) minority (about 7%), especially in Guanacaste Province. The remainder are blacks (3%), East Asians (2%), and Amerindians (1%). The blacks for the most part are of Jamaican origin or descent, and some mulattoes live mainly in the Limón port area. Most of the Amerindians reside on isolated reservations.

⁹LANGUAGES

Spanish is the national language, but English is also spoken among the middle class. Descendants of the Jamaican blacks speak an English dialect.

¹⁰RELIGIONS

Roman Catholicism, the predominant religion, is the official religion of the state, but the constitution guarantees religious freedom. Some 91.6% of the population in 1993 was Roman Catholic. San José is an archbishopric; there are four other bishops in the country. Approximately 40,000 Protestants and 2,000 Jews lived in Costa Rica in 1990. Small numbers of Baha'is, Buddhists, and Chinese folk-religionists are also present.

¹¹TRANSPORTATION

San José is linked to both coasts by railroad and by highway. The Inter-American Highway, 687 km (427 mi) long, connects Costa Rica with Nicaragua and Panama. Another major highway runs from San José to the Caribbean coast beyond Limón. As of 1991 there were 7,030 km (4,368 mi) of paved roads. Additionally, there were another 7,010 km (4,356 mi) of gravel roads and 1,360 km (845 mi) of dirt roads. Motor vehicle registrations in 1991 included 168,814 passenger automobiles and 95,066 commercial vehicles.

Of the three railroad lines, the Atlantic Railway between Limón and San José and the government-owned Pacific Electric Railway between San José and Puntarenas are interconnected; in 1974, the government nationalized the Northern Railway, taking ownership away from the British. A railway line operated by the Banana Company of Costa Rica (246 km/153 mi) is used mainly for company business. Some 27% of the total 950 km (590 mi) of railway was electrified as of 1991.

Principal ports are Limón on the Caribbean Sea and Puntarenas, Caldera, and Golfito on the Pacific. As of 1991, Costa Rica had only one merchant vessel of more than 1,000 GRT. River traffic is not significant.

Líneas Aéreas Costarricenses, S.A. (LACSA), the national airline, provides domestic and international services centered at Juan Santamaria International Airport near San José. In 1992, LACSA carried 534,500 passengers.

¹²HISTORY

There were about 25,000 Amerindians in the region when Columbus landed in 1502. He named the area Costa Rica ("rich coast"), probably because he saw gold ornaments on some of the Indians. Costa Rica was organized as a province by the Spanish in 1540 and was eventually placed under the provincial administration in Guatemala. Cartago, the colonial capital, was founded in 1563.

When independence came to Central America in 1821, Costa Rica had fewer than 70,000 inhabitants. In the following year, it was absorbed into the short-lived Mexican Empire proclaimed by Agustín de Iturbide. Following the collapse of Iturbide's rule, Costa Rica became a member of the United Provinces of Central America in 1823. At the same time, the provincial capital of Costa Rica became San José. The United Provinces fell apart in 1838, and in 1848, the Republic of Costa Rica was established. The new state was threatened by William Walker, a US military adventurer who invaded Central America in 1855, but his troops were repelled in 1857, and in 1860 Walker was captured and executed. In 1871, General Tomás Guardia, a dictator, introduced the constitution that, though frequently modified, remained Costa Rica's basic law until 1949.

In the late 19th and early 20th centuries, there was a series of boundary disputes with Panama and Nicaragua, in the course of which Costa Rica annexed Guanacaste Province from Nicaragua. In World Wars I and II, Costa Rica was a US ally, but not a military participant.

Meanwhile, the success of coffee cultivation, introduced in the early 1800s, had encouraged rapid population growth, progress in education, and the beginnings of modern economic development, through the construction of a coast-to-coast railroad from

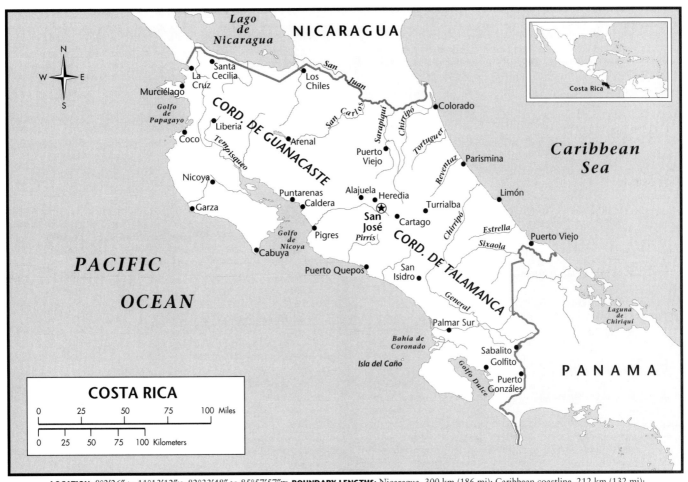

LOCATION: 8°2′26″ to 11°13′12″N; 82°33′48″ to 85°57′57″w. **BOUNDARY LENGTHS:** Nicaragua, 300 km (186 mi); Caribbean coastline, 212 km (132 mi); Panama, 363 km (226 mi); Pacific coastline, 1,016 km (631 mi). **TERRITORIAL SEA LIMIT:** 12 mi.

Limón on the Caribbean through San José to the Pacific. Banana cultivation was started in 1871, and the United Fruit Co. (now United Brands) made Costa Rica a major producer of bananas.

A period of relative political stability that began with the turn of the century ended after World War II, when President Teodoro Picado Michalski annulled the 1948 elections in order to impose Rafael Ángel Calderón Guardia as president rather than the legally elected Otilio Ulate Blanco. José Figueres Ferrer led a civilian uprising, installed his own junta for 18 months, and restored democratic government, turning over the presidency to Ulate. In 1949, a new constitution, based on the constitution of 1871, reinstated free elections and banned a standing army.

Figueres was elected president by an overwhelming majority in 1952 (with women voting for the first time), and under his leadership, Costa Rica was one of the most democratic and prosperous countries in Latin America. Figueres was strongly opposed to all dictatorships, and Costa Rica proceeded to sever diplomatic relations with several Latin American countries. Border skirmishes with Nicaragua in 1955 were resolved through OAS mediation.

In 1958, the candidate of the opposition National Unification Party, Mario Echandi Jiménez, was elected by a bare majority vote. He was unable to enact his program of minimizing the government's role in social and economic matters because the legislature was dominated by other parties. In 1962, Francisco J. Orlich Bolmarich, a candidate of Figueres's National Liberation Party (Partido de Liberación Nacional—PLN), won the elections and continued Figueres's progressive program. Costa Rica joined the Central American Common Market (CACM) in 1963 and has benefited from Central American economic integration, especially

through increasing industrialization. Figueres served as president again from 1970 to 1974 and was succeeded by the PLN candidate, Daniel Oduber Quirós. A conservative, Rodrigo Carazo Odio, was elected president in 1978, and Luis Alberto Monge Álvarez of the PLN became president in 1982.

During the 1980s, Costa Ricans were confronted with a severe economic crisis and with increasing political violence in the region, including sporadic terrorist activities in San José. The Monge government introduced an austerity program to help economic recovery and also tried to avoid being drawn into the war in neighboring Nicaragua between the insurgents known as "contras" and the Sandinista government. Nevertheless, the government resisted pressure from the US to support the contras or to accept US aid toward the building of a military establishment.

In February 1986, the PLN won the presidential election, with Oscar Arias Sánchez defeating Rafael Ángel Calderón Fournier. An important factor in the PLN victory was the popularity of the incumbent, President Monge. In August 1987, a peace plan for Central America outlined by Arias was signed in Guatemala by Nicaragua, Guatemala, Costa Rica, El Salvador, and Honduras. Among the provisions were free elections in all countries, a guarantee of basic democratic freedoms in Nicaragua, a cease-fire by both Sandinistas and contras, an end to outside aid to the contras, amnesty for the contras, repatriation or resettlement of refugees from all countries, and an eventual reduction in the armed forces of all countries. Arias won the Nobel Prize for peace later in that year.

In 1990, amid continued economic troubles, Calderón Fournier was elected, and his Social Christian Unity Party (PUSC) won a paper-thin majority in the Legislative Assembly. Although he campaigned on a platform of liberalization and the reduction of Costa Rica's social welfare system, little change was accomplished. The economy nevertheless rebounded, with peace settlements easing some of Costa Rica's demographic problems.

On 6 February 1994, Costa Ricans returned the PLN to power, as José María Figueres Olsen, son of former president Figueres, was elected president of Costa Rica.

13GOVERNMENT

For the most part, Costa Rica has held to a tradition of orderly, democratic rule. The nation is a republic organized under the constitution of 1949, based on the constitution of 1871. A president, two vice-presidents, and a unicameral congress (the Legislative Assembly) of 62 members (in 1987), apportioned by provinces, are all directly elected for four-year terms. The cabinet (composed of 15 members in 1986) and the president form the Council of Government. For each three assembly deputies, one substitute deputy (suplente) is elected to obviate any subsequent need for by-elections. After each population census, the Supreme Electoral Tribunal proportions the number of deputies for each province. Suffrage is universal and obligatory for all persons of 18 years or more.

The constitution gives legal status to "autonomous institutions," which, in effect, constitute a fourth branch of government. Autonomous institutions include organizations to regulate elections, the state banking system, the state insurance monopoly, a railroad, a housing corporation, the tourist-promotion corporation, an electricity company, and other enterprises that are expected to be free of political pressures. New autonomous institutions can be created by a two-thirds vote of the Legislative Assembly. Public administration has been improved by the institution of a career civil service based on the merit system.

The constitution bars all high government officials from running for the legislature or the presidency while already in office. The president, cabinet ministers, and all government employees are forbidden to interfere with or to participate in election campaigns or to hold party office. The constitution guarantees equality before the law, as well as freedom of speech, assembly, press, and organization. In addition, it guarantees foreigners the same rights as Costa Rican citizens. However, foreigners may not participate in political affairs, nor may members of the clergy.

14POLITICAL PARTIES

The largest political grouping is the PLN, a reformist party that has been the nation's leading party since its formation in 1948. The other major party is the more conservative Social Christian Unity Party (Partido Unidad Social Cristiana—PUSC), which held the presidency during 1978–82, and from 1990–1994. The PUSC has ties to Christian Democratic parties in the Western Hemisphere and Europe. There are also several minor parties.

15LOCAL GOVERNMENT

Costa Rica is divided into seven provinces, which are further subdivided into cantons and districts. The governor of each province is appointed by the president and is responsible to the minister of government. There are no provincial assemblies. The chief city of each canton elects a council (municipalidad), which possesses legislative powers and cooperates with a presiding officer appointed by the national executive. A police agent appointed by the national government oversees each district.

16JUDICIAL SYSTEM

The judiciary consists of justices of the peace, lower courts, labor courts, a court of cassation, two civil courts of appeal, two penal courts of appeal, and the Supreme Court, the highest court in the land. The Supreme Court is composed of 17 justices chosen for eight-year terms by the Legislative Assembly. Justices are automatically reelected for an additional eight-year term unless the Legislative Assembly votes to the contrary by a two-thirds majority. The Assembly also names 25 alternates from a list of 50 names submitted by the Supreme Court, and vacancies on the court are then filled by lot from the list of alternates. Relatives of incumbent justices are ineligible for election. The Supreme Court, by a two-thirds majority, can declare legislative and executive acts unconstitutional. Justices of lower courts are appointed by the Supreme Court, but justices of the peace are appointed by the minister of government acting for the president. Capital punishment has been abolished. The judiciary is independent of the legislative and executive branches and assures fair public trials.

17ARMED FORCES

The 1949 constitution prohibits armed forces. A 4,300-member (in 1993) Civil Guard and a 3,200-member Rural Assistance Guard perform security and police functions. An antiterrorist battalion serves as the presidential guard. Small coast guard and air units (18 planes and helicopters) patrol the coast for revenue purposes. Security expenditures were $60 million in 1992 with no significant arms imports.

18INTERNATIONAL COOPERATION

Costa Rica is a charter member of the UN, having joined on 2 November 1945, and participates in ECLAC and all the nonregional specialized agencies. It also belongs to CACM, G-77, IDB, and PAHO. Costa Rica signed the Rio Pact and Law of the Sea.

Costa Rica is the home of the Inter-American Institute of Agricultural Science, a specialized agency under the OAS, founded in 1944 for the advancement of agriculture, with its main campus at Turrialba. There is also a Graduate School of Public Administration for Central America, organized by the UN and financed by Central American governments. San José is the seat of the Inter-American Human Rights Court.

In November 1974, Costa Rica participated in an unsuccessful move to repeal the OAS sanctions against Cuba. Costa Rica resumed trade relations with Cuba in January 1975, and in July it participated in the successful action on the resolution to lift OAS sanctions. In May 1981, however, Costa Rica severed diplomatic relations with Cuba to protest the latter's support of leftist subversion in Central America. In August 1987, a peace plan for Central America proposed by Costa Rica was signed by Costa Rica, El Salvador, Guatemala, Honduras, and Nicaragua.

19ECONOMY

Traditionally, the economy of Costa Rica, like that of all other countries in Central America, has been based on the production of tropical agricultural commodities for export. There is some forestry but very little mining, although steps have been taken to exploit bauxite, sulfur, and petroleum. Since about 1961 there has been a significant expansion of manufacturing activity, but most industrial plants remain small, concentrating on simple consumer goods to displace more expensive imports. Government efforts to promote diversification of agricultural production have resulted in notable expansion of cattle and dairy farming, now second in value only to coffee among agricultural sectors.

Through the 1960s, prospects for economic expansion were promising, particularly in view of progress toward economic integration in Central America. However, in the 1970s, increases in import prices for raw materials and finished goods caused an inflationary spiral, to which certain internal factors, such as credit expansion and wage increases, also contributed significantly. Chiefly because of a decline in coffee prices in 1978 and the doubling of oil import costs in 1979, the economic growth rate fell sharply from 8.9% in 1977 to –8.8% in 1982. Positive growth rates averaging 3.8% resumed in 1983 and continued through 1985, largely due to government-imposed austerity in return for IMF standby agreements negotiated in 1982 and 1985. Unemployment declined from an official high of 9.1% in 1982 to 6.7% in 1986. The annual inflation rate dropped as well, from 90% to 11.8% in the same period. The foreign debt, rescheduled in 1983 and 1985, remains high—$3.67 billion, or $1,800 per capita, at the end of 1985.

Although per capita income has been slipping in the past two years, as of 1994 Costa Ricans still enjoyed the highest per capita income in Central America. By the end of the last decade, Costa Rica's fiscal deficit was escalating. As a consequence, the country entered a structural adjustment period, eliminating government subsidies, reducing government expenditures, increasing taxation and refinancing the external debt. As a result of these reforms real GDP grew by 7.3% in 1992. In 1993, however, economic growth was slower, declining to a still healthy 4.5%.

Agriculture continues to be the backbone of the economy. By the end of 1993, it accounted for 26% of total employment, 18% of GDP, and roughly half of its exports. The main crops are coffee, bananas and sugar. In recent years, cattle breeding and the production of tropical fruits and vegetables for export have gained in importance. In 1992, agricultural output grew by 3%, despite low coffee prices that year. Coffee prices have been rising steadily in 1994, which will contribute positively to total agricultural earnings.

20INCOME

In 1992, Costa Rica's GNP was $6,261 million at current prices, or $2,000 per capita. For the period 1985–92 the average inflation rate was 18.1%, resulting in a real growth rate in per capita GNP of 2.6%.

In 1992 the GDP was $6,530 million in current US dollars. It is estimated that in 1990 agriculture, hunting, forestry, and fishing contributed 16% to GDP; mining, quarrying, and manufacturing, 19%; electricity, gas, and water, 3%; construction, 3%; wholesale and retail trade, 20%; transport, storage, and communication, 5%; finance, insurance, real estate, and business services, 13%; community, social, and personal services, 6%; and other sources, 15%.

21LABOR

In 1992, the labor force amounted to 1,087,000, or 34.1% of the total population, a 2% increase over the labor force of 1991. About half the total work force is concentrated in San José and Alajuela provinces. The structure of civilian employment in 1992 was as follows: industry, 26.9%; agriculture, 24.1%; and services, 49%. Public sector employment grew from 159,794 in 1991 to 169,371 in 1992, mostly in central government. Unemployment was 4.1% in July 1992.

A labor code governs labor-management relations. Minimum wages are set for two-year periods by a mixed labor-management commission. Strikes in the public services are prohibited. Conciliation systems and labor courts function effectively.

Unionization is about 15% of the total work force, and includes some 420 unions and 156,000 workers. In recent years, Solidarismo, a Costa Rican alternative to traditional trade unions, has grown popular and probably now outnumbers union membership, comprising 18–20% of the total work force. Solidarismo promotes cooperative labor/management relations by offering workers practical benefits (like credit unions), in exchange for which workers renounce their striking and collective bargaining rights. Solidaristas are established with mutual contributions from the employer and the workers, so that the fund serves as savings plan, benefits, and severance pay.

22AGRICULTURE

In 1991, about 10.3% (529,000 hectares/1,307,000 acres) of the total land area was used for crop production, with 45.6% in use as permanent pasture. Nearly half of all farms average less than

10 hectares (25 acres) in size. Over 259,000 persons, or 23.8% of the economically active population, were employed in farming in 1992.

Corn and sugar crops are usually sufficient to meet domestic needs, but beans and rice must be imported from time to time. Agriculture accounted for about 18% of the GDP in 1992. The principal cash crops are coffee, bananas, cocoa, and sugar. Coffee and bananas together accounted for 40% of exports in 1992, with values of $203 million and $485 million, respectively. Notwithstanding low coffee prices, the agricultural sector grew by 3% in 1992.

Over 85% of coffee properties belong to Costa Ricans. The banana industry has been producing more than 1 million tons of bananas annually since the 1970s. The principal owner of Costa Rica's banana plantations is United Brands. Corn, rice, potatoes, beans, sisal, cotton, citrus fruits, pita (used to make hats, baskets, and mats), yucca, vegetables, pineapples and other fruits, tobacco, abaca (hemp), and vegetable oils (especially African and coconut palms) are produced primarily for domestic consumption. Estimated crop production in 1992 (in tons) was sugarcane, 2,840,000; bananas, 1,633,000; rice, 209,000; coffee, 168,000; corn, 40,000; dry beans, 36,000; and cocoa, 2,000. In 1992, the value of agricultural output increased by 3% over that in 1991.

23ANIMAL HUSBANDRY

About 46% of Costa Rica's total land area was devoted to livestock raising in the early 1990s as the result of a major conversion of land to pasturage during the 1970s. In the past, Costa Rica had to import meat, but recent improvements in animal husbandry have made the country self-sufficient, provided a surplus for export, and made cattle exports the third most important source of export earnings. Exports of beef were worth $44.2 million in 1992. In that year there were an estimated 1,707,000 head of cattle, 225,000 hogs, 114,000 horses, and 4,000,000 chickens.

24FISHING

Fish abound in Costa Rican waters, particularly in the Pacific Ocean, where 89% of the annual harvest is caught. Tuna, herring, and shrimp are the most valuable commercial fish; they are caught, processed, and shipped abroad by US firms. A small native fishing industry contributes to the domestic food supply and exports shark, mollusks, and live lobsters. Pearl fishing, once an important industry on the Pacific coast, has declined. In 1991, the total volume of fish landed was an estimated 17,905 tons.

25FORESTRY

Costa Rica's forestland has declined rapidly from about 75% of the total land area in 1940 to 32% in 1991. About 18% of the area still forested is lightly exploited, while 82% is virgin forest. Varieties of commercial woods include laurel, cedar, oak, quina, espavel, campana, cristobal, pochote, maca wood, cedro macho, cedar, and caoba (mahogany). In the Golfo Dulce rain forest of the southern Pacific coast, 135 families of trees embracing some 1,315 species in 661 genera have been identified. Forest products include rubber, chicle, ipecac, roots, medicinal plants, seeds, and other plant products. Although lumber exports have declined, overall timber output increased to 4,201,000 cu m of roundwood cut in 1991, of which about half was used for fuel. Forest product exports in 1992 totaled $15,934,000, only one-fifth of forestry imports. Costa Rica imported some $77,960,000 in paper and paperboard in 1992.

26MINING

Mineral production in 1991 included iron ore, 65,000 tons; cement, 700,000 tons; common clay, 399,000 tons; limestone, 1,300,000 tons; and marine salt, 50,000 tons. Diatomite, lime, pumice, silver, sandstone, and sand and gravel were also mined in 1991. Gold-mining activity plunged in the early 1980s but recovered in 1985 due to the Central Bank's establishment of a support price. The government encourages the mining industry by allowing duty-free importation of machinery and tools.

27ENERGY AND POWER

Costa Rica depends upon imports for an overwhelming proportion of its petroleum needs; this dependency has contributed greatly to the inflationary trends of recent years. Because of a scarcity of fuels, Costa Rica depended on hydroelectric energy for about 85% of its light and power in 1991. Installed capacity at the end of 1991 was 909 Mw, more than double the 1973 total. Production of hydroelectricity totaled 3,647 million kwh in 1991, a 4% increase from 1990. As of 1991, construction of the 55-Mw geothermal station at Miravalle continued. Preliminary work for the 90-Mw Toro hydroelectric complex is now underway.

Most of the nation's electricity is generated by the Instituto Costarricense de Electricidad (ICE), an autonomous government corporation, which owns and operates the central transmission network. In the early 1960s, the IBRD helped finance construction of a 30,000-kw hydroelectric plant on the Río Macho, about 50 km (30 mi) from San José, the installation of 9,000 kw of diesel generating capacity at Colima and Limón, and the extension of the central transmission network. These facilities provided a 40% increase in generating capacity in Costa Rica's central zone. In 1991, Costa Rica's hydroelectric potential was estimated at 25,450,000 kw.

28INDUSTRY

Costa Rica is one of the most industrialized countries in Central America, although industries are predominantly small-scale and primarily involve assembling or finishing imported semifinished components. Of the few larger-scale manufacturing enterprises, the majority are in chemical fertilizers, textiles, coffee and cocoa processing, chemicals, and plastics. Growth was a modest 2% in 1985 (compared with 8.4% in 1984). Food, beverages, tobacco, and wood and furniture showed gains; fuel oils, textiles, and leather and shoes showed declines, largely because of shrinking demand within Central America.

In 1992, manufacturing advanced by 10.5% with strong growth in nontraditional export products, petroleum refining, fertilizers and pesticides, communication equipment, bakery products, and clothing. With the exception of construction, all the productive sectors had substantial positive growth.

29SCIENCE AND TECHNOLOGY

The principal scientific policymaking body in Costa Rica is the General Directorate of Geology, Mining, and Petroleum (founded in 1951). The chief scientific research station is the Alajuela Institute. Several institutes specialize in tropical sciences, including the Organization for Tropical Studies, the Tropical Science Center, and the Tropical Agronomy Center (in Turrialba), as well as medicine, nuclear energy, technology, geology, agriculture, and meteorology. Costa Rica has five colleges and universities offering degrees in basic and applied sciences. In 1986, research and development expenditures totaled 612 million colones. In 1988, 1,528 scientists and engineers were engaged in research and development.

30DOMESTIC TRADE

San José is the commercial center, and most importers, exporters, and manufacturers' agents operate there. San José's importing and exporting firms handle more than 75% of the country's total trade. Most trading is carried on by small merchants in the public markets. Cooperative societies produce, buy, and sell food, clothing, machinery, and other items. There are no formal commercial

credit companies as such, but installment buying at high interest rates is becoming widespread.

Shops are open on weekdays from 8:30 to 11:30 AM and from 2 to 6 PM and on Saturdays in the mornings only. Normal banking hours are 8 to 11 AM and 1:30 to 3 PM, Monday through Friday, and 8 to 11 AM on Saturday. Advertising agencies in Costa Rica, all located in San José, offer advertising services through newspapers, radio, television, and direct mail.

31FOREIGN TRADE

Among Costa Rica's major exports are coffee, bananas, sugar, cocoa, and cattle and meat products, all commodities vulnerable to fluctuations in world market prices. Exports declined by 4.3% from 1984 to 1985 (after growing 15.3% during the previous year), to $963 million. The value of banana exports declined from $251 million to $212 million; beef exports rose to $53 million; and sugar exports, hampered by a 50% cut in US imports, declined to $10 million, 69% below the 1984 total. Coffee exports rose to $310 million as Costa Rica's quota increased and the world price strengthened. Nontraditional industrial exports showed strength in the mid-1980s as the government strove to encourage their growth. Notable examples were flowers and ornamental plants, 32% (sales of these rose some 50% in 1985, from the $12 million recorded in 1984); shellfish, which constituted 30% of this sector; and wood products, 9%.

Imports in the 1980s consisted mainly of raw materials for industry and mining, followed by consumer goods and capital goods (mainly machinery and equipment) for industry, mining, and transportation. The import bill has risen steadily during the 1980s as the economy has recovered, moving from $1,043 million in 1982 to $1,261 million in 1985.

The US is by far Costa Rica's most important trading partner, accounting for 48% of exports and 42% of imports in 1986. The US buys mainly bananas, as well as some coffee and sugar, and sells Costa Rica raw materials and consumer and capital goods. Costa Rica also purchases these items from Western Europe, notably from the Federal Republic of Germany (FRG), the Netherlands, and the UK, and supplies mainly coffee and cocoa in return. After Costa Rica joined the CACM in 1962, the nation's trade within the group increased greatly; other CACM nations supplied about 13% of Costa Rica's imports and bought about 22% of its exports in 1980. Because of regional conflict, however, the CACM has virtually collapsed; El Salvador and Guatemala alone remain significant trading partners.

The country's exports have been following a moderate increase in the last years. In 1993, Costa Rica's total exports were estimated at $1,966.4 million, up by 7.56% in 1992. Exports grew by 5% in the first four months of 1994, compared with the same period in 1993. However, exports to the United States dropped by 4%. Analysts are concerned about the impact of NAFTA on the country's exports to the US. Nevertheless, the US remained the country's main export destination in the first four months of 1994. Costa Rica's main exports include bananas, coffee, beef, sugar, ornamental plants, melons, wood products, pineapples, fresh seafood, textiles and footwear. After the US, Costa Rica exports mainly to Germany, the Netherlands, the UK, Guatemala, and Japan.

At years end 1993, total imports had increased by 11.2% to $2,737 million from $1,828 million in 1992. Principal import items included petroleum, machinery and transportation equipment, consumer goods, food, paper products (carton) and chemical. Costa Rica's major suppliers were the US, Japan, Germany, and Guatemala.

32BALANCE OF PAYMENTS

Costa Rica has traditionally experienced balance-of-payments difficulties because of the vulnerability of its main sources of exchange earnings to fluctuations in world markets. The nation's payments problems in the late 1970s were aggravated by domestic inflationary policies and rising trade imbalances, despite increases in foreign capital receipts. The deficit on capital accounts declined in the mid-1980s due to increases in capital investment from foreign loans and credits and favorable renegotiation of the foreign debt. The trade deficit fell from $577.5 million in 1990 to $260.8 million in 1991, because of a 10% increase in exports and an 8.5% decrease in imports. Exports of bananas, beef, and coffee increased, while import restrictions were also intensified.

In 1992 merchandise exports totaled $1,714.3 million and imports $2,211.9 million. The merchandise trade balance was $–497.6 million.

The following table summarizes Costa Rica's balance of payments for 1991 and 1992 (in millions of US dollars):

	1991	1992
CURRENT ACCOUNT		
Goods, services, and income	–216.8	–531.8
Unrequited transfers	141.6	170.4
TOTALS	–75.2	–361.4
CAPITAL ACCOUNT		
Direct investment	172.8	217.3
Portfolio investment	–13.0	–16.9
Other Long-term capital	–80.0	–66.3
Other short-term capital	75.2	139.7
Exceptional Financing	185.6	117.5
Other liabilities	7.1	–6.6
Reserves	–348.4	–141.1
TOTALS	–24.7	233.6
Errors and omissions	99.9	127.8
Total change in reserves	—	—

33BANKING AND SECURITIES

The Central Bank (Banco Central de Costa Rica), an autonomous governmental body established in 1950, issues currency, holds the nation's gold reserves, formulates general banking policy, and regulates commercial banks. The Central Bank held reserves of $190.3 billion at the end of 1992. Five commercial state banks are operated as autonomous government corporations: the Banco Nacional de Costa Rica, Banco de Costa Rica, Banco Anglo Costarricense, Banco Crédito Agrícola de Cartago, and Banco Popular y de Desarrollo Comunal. There are also branches of the Bank of America and several other private banks. In 1992 commercial banks held demand deposits totaling c6.1 billion. The money, as measured by M2, was c371 billion. Interest rates are kept high to encourage domestic savings and attract capital.

Costa Rican residents can own and deal in gold, own foreign securities and foreign currencies, maintain foreign bank balances, import and export national bank notes, and import goods from abroad, but they cannot export without a license and must repatriate export earnings. Costa Ricans traditionally put their savings into real property rather than securities, but on several occasions during the 1960s and 1970s, the government successfully floated bond issues within the country. Stock sales and foreign currency transactions are handled by the Bolsa Nacional de Valores in San José.

34INSURANCE

Only the government's National Insurance Institute (Instituto Nacional de Seguros), founded in 1924, may write insurance in Costa Rica. It handles all types of insurance, the most important being life, fire, automobile, and workers' compensation. One of the more popular features of life insurance policies is that the holder may borrow up to the full face value of the policy after paying premiums for only two years. Life insurance in force in 1991 was c178.6 billion.

35PUBLIC FINANCE

The central government budget is passed upon by the Legislative Assembly. Municipal budgets are of minor importance, and local government funds are mainly grants from the national government. The financial range of the public sector extends to a large number of publicly owned entities. In 1990, a new government helped decrease the budget deficit from 7% of GDP to 3.5% of GDP; by 1992, the consolidated public sector deficit was 1.1% of GDP.

The following table shows actual revenues and expenditures for 1990 and 1991 in billions of colones.

	1990	1991
REVENUE AND GRANTS		
Tax revenue	102.89	139.78
Non-tax revenue	17.36	22.68
Capital revenue	0.47	0.06
Grants	—	—
TOTAL	120.72	162.52
EXPENDITURES & LENDING MINUS REPAYMENTS		
General public service	12.45	—
Public order and safety	9.16	10.73
Education	25.42	32.75
Health	35.17	54.73
Social security and welfare	18.47	21.45
Housing and community amenities	1.43	1.29
Recreation, cultural, and religious affairs	1.76	1.07
Economic affairs and services	12.68	13.85
Other expenditures	16.21	—
TOTAL	137.13	171.72
Deficit/Surplus	–16.41	–9.20

In 1983 Costa Rica's total public debt stood at c35.36 billion, of which c15.20 billion was financed abroad. The ratio of external debt to GDP is expected to fall from 63% in 1992 to 43% in 1996.

36TAXATION

Indirect taxes, such as import duties, contribute about three-quarters of government revenues. Consumption and other indirect taxes include taxes on coffee processing, cattle slaughter, liquor, cigarettes, sugar, and transportation fares.

Both individuals and businesses residing in Costa Rica are subject to income tax only on income derived from sources within the country. The corporate tax rate is 30%; incentives are available for new industries and for engaging in the export of nontraditional products. Personal income taxes range from 10%–15% for employed persons and 10%–25% for self-employed persons. There is a 12% value-added tax and consumption taxes ranging from 5%–20%, except for luxury items, which can go as high as 70%.

37CUSTOMS AND DUTIES

Initially, the import tariff was primarily for revenue-raising purposes, but in 1954 it was increased to protect Costa Rican industry. In 1962, tariffs were raised so high that they virtually prohibited foreign competition in certain fields. Import duties usually include a specific duty on the gross weight in kilograms and ad valorem duties of varying percentages of the c.i.f. value of the imported goods. Since Costa Rica's entry into GATT, tariffs have been lowered. As of October 1992, the maximum rate for most products was 27%, and this was due to be lowered to 20% by April 1993. Exemptions include textiles, clothing, and footwear, which are scheduled to be reduced by 1995.

38 FOREIGN INVESTMENT

Foreign investment, which is welcomed in Costa Rica, is concentrated in agriculture (chiefly banana interests), railways, tobacco, communications, airlines, government bonds, and real estate. The US, Costa Rica's major foreign investor, has interests chiefly in agriculture, petroleum refining and distribution, utilities, cement, and fertilizers. The continued high level of trade with the US has been conducive to private foreign investment, especially in export industries. Investment incentives include constitutional equal-treatment guarantees, tax incentives, and free-trade zones.

In 1990, Costa Rica signed a framework Agreement with the US, which establishes the context within which future negotiations regarding bilateral trade and investment issues will take place. The country has also signed a Tax Information Exchange Agreement with the US, which qualifies it to receive 936 tax-free funds from Puerto Rico at lower than market rates of interest.

39 ECONOMIC DEVELOPMENT

Costa Rica's development objectives include the promotion of small industry; the expansion of transportation, communications, and power facilities; the strengthening of agriculture; and the betterment of public health and housing. The government has stimulated cooperative enterprises, and it regulates agricultural prices through the National Production Council.

The Industrial Protection and Development Law of 1959 established incentives by way of liberal exemptions from taxes and import duties. In keeping with its provisions, 855 enterprises were established or expanded between 1960 and 1973. The total investment exceeded c1 billion, of which textile production accounted for c198 million, food production for c166 million, nonmetallic mineral refining for c165 million, the chemical industry for c142 million, and paper manufacturing for c110 million.

In May 1974, a National Planning Law was promulgated. Its objectives were to accelerate growth of national output, to promote a more equitable distribution of income and of social services, and to encourage increased citizen participation in the solution of economic and social problems. A National Development Plan for 1974–78 projected current receipts of the public sector to increase by 11.8% annually at constant prices. Since then, expenditures for development have been allocated in the annual government budgets but have been limited by the scarcity of public funds available.

Between 1953 and 1986, Costa Rica received $818.5 million in multilateral development assistance, of which 60% came from the IBRD and 37% from the IDB. US nonmilitary loans and grants during 1946–86 totaled $1,099.3 million. The elections of President Monge and his successor, President Arias, coupled with the heightened political and military tensions elsewhere in Central America, have brought improved relations with the US, which provided $766.8 million in aid between 1983 and 1986. Costa Rica ranks second only to El Salvador among Latin American nations as a recipient of direct US economic assistance.

Stricter fiscal and monetary policies resulted in lower inflation and smaller budget deficits in the 1990's. On the other hand, investment has been hampered by high interest rates and tight credit. The country has encouraged private investment and strengthened the private sector. As a result, tourism and exports experienced significant development. The export structure is increasingly diversifying away from coffee and bananas. The most promising sectors are garments and light manufacturing. Free zone operations are flourishing.

40 SOCIAL DEVELOPMENT

The national Social Security Fund administers social insurance and pension programs. Most of the population is covered by these social services, which have been extended throughout the country and include workers' compensation and family assistance benefits. The social security program is compulsory for all employees under 65 years of age. The program, however, suffered as a result of the fiscal crisis of the 1980s and was overtaxed by a large influx of refugees.

Costa Rica's birth rate and rate of net natural increase have declined steadily in recent decades, in part because of an active family planning program; the fertility rate for 1985–90 was 3.9. New recipients of contraceptive devices numbered 183,600 between 1970 and 1976, when an estimated 46% of married women were practicing birth control; by 1985-90, the proportion had risen to an estimated 65%. Abortion is permitted on broad health grounds.

A 1990 law reinforces existing provisions barring sex discrimination and improves women's property rights. Women and men are usually paid equally for equal work. The Women's Council and the Women's Delegation, two government agencies established recently, act as advisors and advocates for women who have suffered abuse or harassment.

41 HEALTH

Health standards have steadily improved in Costa Rica. The crude death rate per 1,000 population decreased from 22.1 in the early 1930s to 3.7 in 1993. The infant mortality rate, 166.7 per 1,000 live births in 1927, was 62.3 in 1968, and 14 in 1992. During 1992, life expectancy at birth was an average of 76 years. The decreases in mortality rates were attributed to improvements in sanitary and medical facilities under the national health program administered by the Ministry of Health. Between 1985 and 1992, it was estimated that 80% of the population had access to health care services.

Hospitals are located in the principal cities, and about 95% of the hospital beds are in urban areas. In 1990, Costa Rica had 1 physician per 1,030 people.

Health services for the rural population are generally inadequate, and the refugee problem has severely taxed urban services. However, there are sanitary units and dispensaries to care for the health needs of the poor. During the 1980s, the greatest health problem was protein-calorie malnutrition, particularly among infants and children. Diseases of the circulatory system are the leading cause of death. From 1988 to 1991, 93% of the population had access to safe water, and from 1990 to 1992, Costa Rica immunized children up to one year old as follows: tuberculosis (92%); diphtheria, pertussis, and tetanus (90%); polio (90%); and measles (84%).

42 HOUSING

Sources for housing mortgages include private funds, the Central Bank, the Social Security Fund, and the national banking system. The National Institute of Housing and Urban Affairs, established in 1954, administers a national low-cost housing program. As of 1980, 30% of all housing had no electricity. Between 1980 and 1985, there were 1.5 persons per habitable room.

43 EDUCATION

Costa Rica has one of the most literate populations in Latin America; in 1990, adult literacy was 92.8% (males: 92.6%; females: 95.1%) (UNESCO). Nearly one-fifth of the government's expenditure is on education. Primary education lasts for six years followed by three years of secondary education. This again is followed by a two-year highly specialized course. Primary and secondary education is free, and primary-school attendance is compulsory. In 1984, enrollment in primary schools was

353,958, with 12,223 teachers; in secondary schools, 148,032 students, with 9,152 teachers.

The country has five universities, including an open university. The University of Costa Rica (founded in 1843) is supported by the government and enrolls about 28,000 students. The Open University (1977), in San José, operates 28 regional centers for all students who apply. There are also the Autonomous University of Central America (1976), in San José, and the National Autonomous University of Heredia (1973).

44LIBRARIES AND MUSEUMS

The National Library at San José, founded in 1888, is a reference library containing about 175,000 volumes. Other important libraries in San José are the National Archives, with 5,000 volumes, and the library of the University of Costa Rica, which contains about 375,000 volumes. The Inter-American Institute of Agricultural Sciences, in Turrialba, has a library of over 30,000 volumes.

The National Museum of Costa Rica in San José, founded in 1887, is a general museum with collections of pre-Columbian, colonial, republican, and religious art, a herbarium, and bird displays. The Museum of Costa Rican Art was founded at San José in 1977. There are several other art museums in the capital as well.

45MEDIA

Costa Rican telephone, telegraph, and radio systems are owned and operated by both governmental and private firms. In 1993 there were over 300,000 telephones in use. There are 71 AM and no FM radio stations and 18 TV stations in Costa Rica. In 1991 there were about 800,000 radios and 435,000 television sets.

Freedom of the press is guaranteed by the constitution and observed in practice. About 120 newspapers, bulletins, and periodicals are issued. There are four daily newspapers, all published in San José. These papers (with 1991 daily circulations) are *La Nación*, an independent morning paper (97,000); *La Prensa Libre*, an independent evening paper (63,000 each edition); *La República*, an independent morning journal (66,000); and *Extra*, also an independent morning paper (95,000).

46ORGANIZATIONS

The government stimulates the formation of cooperatives, but the movement is limited. Consumer cooperatives purchase, sell, and distribute goods among the membership. The cooperative credit societies procure loans for agriculture, stock raising, and industrial development, and cooperative housing associations provide low-cost housing facilities.

There are chambers of commerce and of industry in San José, in addition to some 50 employers' and industrial organizations. Church-sponsored groups have a sizable membership. An unusual organization, combining features found in credit unions, company unions, and building and loan societies, is the Movimiento Solidarista, which advocates harmony between employers and workers.

47TOURISM, TRAVEL, AND RECREATION

Visitors to Costa Rica must have passports. US citizens may travel on a tourist card available at consulates and transportation companies. In 1991, Costa Rica received 504,649 tourists, of whom 425,505 were from the Americas and 69,087 from Europe. Revenues from tourism reached US $331 million. There were 7,196 hotel rooms with a 67.9 percent occupancy rate. Popular tourist sights in San José are the National Museum, National Theater, and the Central Bank's gold exhibition. Other attractions include the Irazú and Poás volcanoes, brief jungle excursions, and the Pacific beaches. Popular recreations are bird-watching, mountain climbing, swimming, water skiing, and deep-sea fishing.

Football (soccer) is the national sport, and there are matches every Sunday morning in San José from May through October. Horseback riding is widely available.

48FAMOUS COSTA RICANS

José María Castro was Costa Rica's first president (1847–49, 1866–68). Juan Rafael Mora Porras, the second president of the republic (1849–59), successfully defended the country against the invasion of US military adventurer William Walker. General Tomás Guardia (1832–82) led a revolt against the government in 1870, became a dictator, and in 1871 introduced the constitution that remained in force until 1949. José Figueres Ferrer (b.1906), president during 1953–58 and 1970–74, is regarded as the father of the present constitution. Oscar Arias Sánchez (b.1940), president since 1986, won the Nobel Prize for peace in 1987 for his plan to bring peace to Central America. Ricardo Fernández Guardia (1867–1950) is regarded as Costa Rica's greatest historian. Joaquín García Monge (1881–1958) founded the literary review *Repertorio Americano*.

49DEPENDENCIES

Cocos Island—26 sq km (10 sq mi), about 480 km (300 mi) off the Pacific coast, at 5°32′N and 87°2′W—is under Costa Rican sovereignty. It is mainly jungle, with a maximum elevation of 850 m (2,788 ft). There is no permanent population, but the island is popular with transient treasure hunters. Cocos Island has two harbors and is of strategic importance because of its position along the western approach to the Panama Canal.

50BIBLIOGRAPHY

American University. *Area Handbook for Costa Rica*. Washington, D.C.: Government Printing Office, 1970.

Ameringer, Charles. *Don Pepe: A Political Biography of José Figueres of Costa Rica*. Albuquerque: University of New Mexico Press, 1978.

Barry, Tom. *Costa Rica: a Country Guide*. Albuquerque, NM: Inter-Hemisphere Education Resource Center, 1990.

Bell, John P. *Crisis in Costa Rica: The 1948 Revolution*. Austin: University of Texas Press, 1971.

Bension, Alberto et al. *Costa Rica and Uruguay Studies*. New York: Oxford University Press, 1993.

Costa Rica in Pictures. Minneapolis: Lerner, 1987.

Creedman, Theodore S. *Historical Dictionary of Costa Rica, 2nd ed*. Metuchen, N.J.: Scarecrow Press, 1991.

Denton, Charles F. *Patterns of Costa Rican Politics*. Boston: Allyn & Bacon, 1971.

Dorn, Marilyn April. *The Administrative Partitioning of Costa Rica: Politics and Planners in the 1970s*. Chicago: University of Chicago, 1989.

Fernández Guardia, Ricardo. *Costa Rica en el siglo XIX*. San José: Gutenberg, 1929.

Gudmundson, Lowell. *Costa Rica Before Coffee: Society and Economy on the Eve of the Export Boom*. Baton Rouge: Louisiana State University Press, 1986.

Herrick, Bruce, and Barclay Hudson. *Urban Poverty and Economic Development: A Case Study of Costa Rica*. New York: St. Martin's, 1980.

Jones, Chester L. *Costa Rica and Civilization in the Caribbean*. New York: Gordon, 1976 (orig. 1935).

Monge Alfaro, Carlos. *Historia de Costa Rica*. San José: Trejos, 1960.

Rolbein, Seth. *Noble Costa Rica*. New York: St. Martin's Press, 1988.

Rovinski, Samuel. *Cultural Policy in Costa Rica*. New York: UNESCO, 1978.

Seligson, Mitchell A. *Peasants of Costa Rica and the Development of Agrarian Capitalism*. Madison: University of Wisconsin

Press, 1980.
Stansifer, Charles L. *Costa Rica*. Oxford, England; Santa Barbara, Calif.: Clio Press, 1991.
Williams, Philip J. *The Catholic Church and Politics in Nicaragua* *and Costa Rica*. Pittsburgh, PA: University of Pittsburgh Press, 1989.
Winson, Anthony. *Coffee and Democracy in Modern Costa Rica*. New York: St. Martin's Press, 1989.

CUBA

Republic of Cuba
República de Cuba

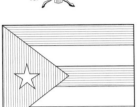

CAPITAL: Havana (La Habana).

FLAG: The flag consists of five alternating blue and white horizontal stripes penetrated from the hoist side by a red triangle containing a white five-pointed star.

ANTHEM: *Himno de Bayamo (Hymn of Bayamo),* beginning "Al combate corred bayameses" ("March to the battle, people of Bayamo").

MONETARY UNIT: The Cuban peso (c$) of 100 centavos is a paper currency with one exchange rate. There are coins of 1, 2, 3, 5, 20, 40, and 100 centavos and notes of 1, 3, 5, 10, 20, 50, and 100 pesos. The value of the peso was set at exactly US$1 until 1971, when the dollar was devalued against gold but the peso was not. c$1 = US$1.3637 (or US$1 = c$0.7333).

WEIGHTS AND MEASURES: The metric system is the legal standard, but older Spanish units and the imperial system are still employed. The standard unit of land measure is the caballería (13.4 hectares/133.1 acres).

HOLIDAYS: Day of the Revolution, Liberation Day, 1 January; Labor Day, 1 May; Anniversary of the Revolution, 25–27 July; Proclamation of Yara, 10 October. Celebration of religious holidays falling during the workweek was prohibited by a 1972 law.

TIME: 7 AM = noon GMT.

¹LOCATION, SIZE, AND EXTENT

The Republic of Cuba consists of one large island and several small ones situated on the northern rim of the Caribbean Sea, about 160 km (100 mi) south of Florida. With an area of 110,860 sq km (42,803 sq mi), it extends 1,223 km (760 mi) E–W and averages about 89 km (55 mi) N–S. Cuba is the largest country in the Caribbean, accounting for more than one-half of West Indian land area. Comparatively, the area occupied by Cuba is slightly smaller than the state of Pennsylvania. It is separated from Florida by the Straits of Florida, from the Bahamas and Jamaica by various channels, from Haiti by the Windward Passage, and from Mexico by the Yucatán Channel and the Gulf of Mexico. Cuba's total coastline is 3,764 km (2,339 mi). The largest offshore island, the Isle of Youth (Isla de la Juventud), formerly known as the Isle of Pines (Isla de Pinos), lies southwest of the main island and has an area of 2,200 sq km (849 sq mi); the other islands have a combined area of 3,715 sq km (1,434 sq mi).

Cuba's capital city, Havana, is located on its northeastern coast.

²TOPOGRAPHY

Cuba's spectacular natural beauty has earned it the name Pearl of the Antilles. The coastline is marked by bays, reefs, keys, and islets. Along the southern coast are long stretches of lowlands and swamps, including the great Zapata Swamp (Ciénaga de Zapata). Slightly more than half the island consists of flat or rolling terrain, and the remainder is hilly or mountainous, with mountains covering about a quarter of its total area. In general, eastern Cuba is dominated by the Sierra Maestra, culminating in Pico Real del Turquino (1,974 m/6,476 ft); around Camagüey are rolling plains and low mountains; central Cuba contains the Trinidad (Escambray) Mountains in addition to flat or rolling land; and the west is dominated by the Sierra de los Órganos. The largest river, the Cauto, flows westward for 370 km (230 mi) north of the Sierra Maestra but is little used for commercial navigation purposes.

³CLIMATE

Except in the mountains, the climate of Cuba is semitropical or temperate. The average minimum temperature is 21°C (70°F), the average maximum 27°C (81°F). The mean temperature at Havana is about 25°C (77°F). The trade winds and sea breezes make coastal areas more habitable than temperature alone would indicate. Cuba has a rainy season from November to April. The mountain areas have an average precipitation of more than 180 cm (70 in); most of the lowland area has from 90 to 140 cm (35–55 in) annually; and the area around Guantánamo Bay has less than 65 cm (26 in). Droughts are common. Cuba's eastern coast is often hit by hurricanes from August to October, resulting in great economic loss.

⁴FLORA AND FAUNA

Cuba has a flora of striking richness, with the total number of native flowering species estimated at nearly 8,000. The mountainous areas are covered by tropical forest, but Cuba is essentially a palm-studded grassland. The royal palm, reaching heights of 15 to 23 m (50–75 ft), is the national tree. Pines like those in the southeastern US grow on the slopes of the Sierra de los Órganos and on the Isla de Juventud (Isle of Youth). The lower coastal areas, especially in the south, have mangrove swamps. There is a small area around Guantánamo Bay where desert plants grow.

Only small animals inhabit Cuba. These include tropical bats, rodents, birds, and many species of reptiles and insects.

⁵ENVIRONMENT

The Cuban government has formed several agencies to protect the environment. Among them are the National Parks Service, the National Commission of Environmental Protection and Rational Use of Natural Resources (1977), the National Environmental Education Program, the Academy of Sciences of Cuba, and the National Commission for the Protection of the Environment and for Conservation of Natural Resources. As of 1994, Cuba's most pressing environmental problems were deforestation and the preservation of its wildlife. The government has sponsored a successful reforestation program aimed at replacing forests that had gradually decreased to a total of 14% of the land area by 1959. Another major environmental problem is the pollution of Havana Bay. The nation has 8.3 cubic miles of water with a total of 89% used for agricultural purposes. One hundred percent of Cuba's city dwellers and 91% of its rural people have pure water. Cuba contributes 0.1% of the world's total gas emissions. As of 1987, endangered species in Cuba included the Cuban solenodon, four species of hutia (dwarf, Cabera's, large-eared, and little earth), two species of crocodile (American and Cuban), and the Cuban tree boa. The ivory-billed woodpecker has become extinct. In 1994, 11 mammal species in a total of 39 were considered threatened. Fifteen bird species in a total of 286 are also threatened. Four types of reptiles of 100 are endangered along with 860 plant species out of a total of 8,000.

6POPULATION

According to the 1991 census, the population was 9,723,605. The estimated population in December 1992 was 10,870,000; and the UN projection for the year 2000 was 11,504,000. In 1992, the crude birthrate was 14.5 per 1,000 people; the death rate was 6.9; and the net natural increase was 7.6. The estimated population density in 1992 was 98 per sq km (254 per sq mi). About 76% of the population was urban and 24% rural in 1995. Havana, the capital, had an estimated population of 2,096,054 at the end of 1989. Other important cities, with their estimated populations at the end of 1989, are Santiago de Cuba (405,354), Camagüey (283,008), Holguín (228,053), Santa Clara (194,354), and Guantánamo (200,381).

7MIGRATION

Before independence, there was a large migration from Spain; the 1899 census reported 129,000 Spanish-born persons living in Cuba. The 1953 census reported about 150,000 persons of foreign birth, of whom 74,000 were Spaniards. From 1959 through 1978, Cuba's net loss from migration, according to official estimates, was 582,742; US figures indicate that during the same period a total of 669,151 Cubans arrived in the US. During the 1960s, the emigrants were predominantly of the higher and middle classes, but in the 1970s urban blue-collar workers and other less educated and less wealthy Cubans came to predominate. The flow of emigrants declined in the late 1970s, but beginning in April 1980, Cubans were allowed to depart from Mariel harbor; by the end of September, when the harbor was closed, some 125,000 Cubans in small boats (the "freedom flotilla") had landed in the US. Of that number, 2,746 were classified as "excludable aliens" and were being held in prisons or mental institutions. According to an agreement of December 1984, Cuba agreed to accept the 2,746 back; repatriation began in February 1985, but in May, Cuba suspended the agreement. Some Cubans have combined to make it to the US in small boats and rafts. A total of 3,656 did so in 1993 and 3,864 in the first half of 1994. By the mid-1980s, well over 500,000 Cuban exiles were living in the Miami area. In 1990 there were 751,000 Cuban-born persons in the US. Large numbers have also settled in Puerto Rico, Spain, and Mexico.

8ETHNIC GROUPS

According to the 1981 census, whites (primarily of Spanish descent) made up about 66% of the total; blacks, 12%; and the mixed group (mulattos), 22%. Virtually the entire population is native-born Cuban.

9LANGUAGES

Spanish is the national language of Cuba.

10RELIGIONS

The Roman Catholic Church has never been as influential in Cuba as in other Latin American countries. In the 1950s, approximately 85% of all Cubans were nominally Roman Catholic, but the Church itself conceded that only about 10% were active members.

From the early-1980s to 1993, Roman Catholics represented about 40% of the population. Other religions in Cuba include Santería (a group of cults that meld African beliefs with Christian dogma), several Protestant churches, including Baptist, Methodist, and Presbyterian, and a very small Jewish population. Since the Castro revolution, the number of people who attend churches has diminished, and it is estimated that half of the population has no religion. A resurgence of religious practice beginning in 1992 has raised church-state relations to a new level of interest.

11TRANSPORTATION

In 1991, Cuba had about 21,000 km (13,050 mi) of roads, of which 9,000 km (5,600 mi) were paved. The first-class Central Highway extends for 1,223 km (760 mi) from Pinar del Río to Guantánamo, connecting all major cities. An extensive truck and bus network transports passengers and freight.

Nationalized railways connect the east and west extremities of the island by 5,295 km (3,290 mi) of standard-gauge track. In addition, large sugar estates have 9,630 km (5,984 mi) of lines of various gauges.

Cuba first began to develop a merchant marine under the revolutionary government. The USSR had supplied oceangoing vessels and fishing boats and, in the mid-1960s, built a huge fishing port in Havana Bay to service Cuban and Soviet vessels. By 1991, the Cuban merchant fleet had 83 vessels totaling 585,000 GRT. Cuba's major ports—Havana, Cienfuegos, Mariel, Santiago de Cuba, Nuevitas, and Matanzas—are serviced mainly by ships of the former Soviet republics, with ships from Spain, the UK, and Eastern Europe making up the bulk of the remainder.

There are daily flights between Havana and the major Cuban cities, and weekly flights to Spain, Mexico, Moscow, Prague, and Jamaica. Cubana Airlines is the national air carrier. The number of air passengers increased from 140,000 in 1960 to 1,000,000 in 1985, but fell to 830,600 in 1991 and to 672,200 in 1992. Between 1975 and 1980, airports at Havana and Camagüey were renovated, and new airports were built at Bayamo, Manzanillo, and Las Tunas. In 1991, Jose Marti airport at Havana serviced 1,199,000 embarking or disembarking passengers.

12HISTORY

Cuba was originally inhabited by about 50,000 Ciboney and Taíno, agricultural Amerindians related to the Arawak peoples, who died from disease and maltreatment soon after the Spanish arrived. Christopher Columbus made the European discovery of Cuba in 1492 on his first voyage to the Americas. The African slave trade began about 1523 as the Amerindian population declined, and grew thereafter, especially with the development of coffee and sugar on the island. During the early colonial years, Cuba served primarily as an embarkation point for such explorers as Hernán Cortés and Hernando de Soto. As treasure began to flow out of Mexico, Havana became a last port of call and a target for French and English pirates. In 1762, the English captured Havana, holding Cuba for almost a year. It was ceded to Spain in exchange for Florida territory in the Treaty of Paris (1763). Spanish rule was harsh, and intermittent rebellions over the next century all ended in failure.

Cuba's first important independence movement came in 1868, when Carlos Manuel de Céspedes, a wealthy planter, freed his slaves and called for a revolution against Spain. For the next 10 years, guerrillas (mambises), mainly in eastern Cuba, fought in vain against the Spanish colonial government and army. Although

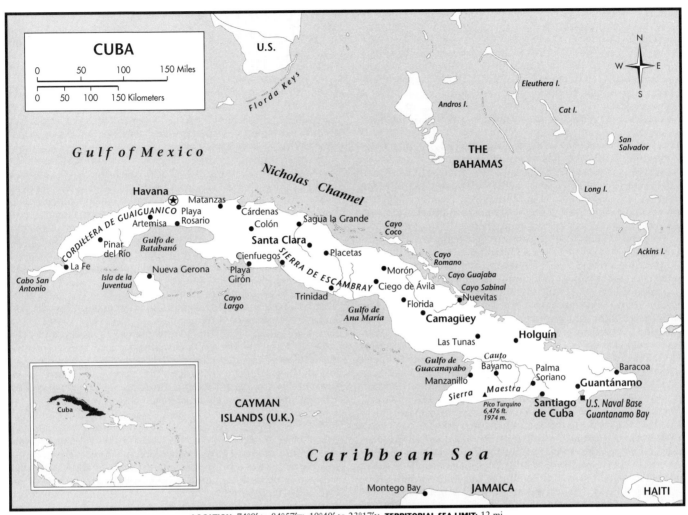

CUBA

0   50   100   150 Miles

0   50   100   150 Kilometers

U.S.

Florida Keys

Gulf of Mexico

Nicholas Channel

Eleuthera I.

Andros I.

Cat I.

San Salvador

THE BAHAMAS

Long I.

Havana ★
Matanzas
Playa
Rosario
Cárdenas
Colón
Sagua la Grande
Cayo Coco
Ackins I.

CORDILLERA DE GUAIGUANICO
Artemisa
Pinar del Río
Gulfo de Batabanó
Santa Clara
Cienfuegos
Placetas
Cayo Romano

La Fe
Nueva Gerona
Playa Girón
SIERRA DE ESCAMBRAY
Morón
Cayo Guajaba

Cabo San Antonio
Isla de la Juventud
Cayo Largo
Trinidad
Ciego de Ávila
Cayo Sabinal
Nuevitas

Gulfo de Ana María
Florida
Camagüey

Las Tunas
Holguín
Baracoa

Gulfo de Guacanayabo
Cauto
Bayamo
Palma Soriano
Guantánamo

Manzanillo
Sierra
Maestra
Santiago de Cuba
U.S. Naval Base Guantanamo Bay

Pico Turquino 6,476 ft. 1974 m.

CAYMAN ISLANDS (U.K.)

Cuba

Caribbean Sea

Montego Bay
JAMAICA
HAITI

LOCATION: 74°8′ to 84°57′w; 19°49′ to 23°17′n. **TERRITORIAL SEA LIMIT:** 12 mi.

eventually subdued, Céspedes is nevertheless viewed as the father of Cuban independence. A second hero was added in the 1890s when poet and journalist José Martí founded the Cuban Revolutionary Party during exile in the US. The call to arms (Grito de Baire) on 24 February 1895 initiated a new war. After landing with a group of recruits gathered throughout the region, Martí was killed at Dos Ríos, in eastern Cuba. The Spanish had the insurrection under control within a year.

In the end, the Cubans had to rely on the US to defeat the Spanish. Anti-Spanish sentiment, fueled by US newspapers, erupted after the battleship *Maine* mysteriously blew up in Havana harbor on 15 February 1898. The US declared war on Spain on 25 April, and in a few months, the Spanish-American War was over. The Treaty of Paris (10 December 1898), established Cuban independence. During the interim period 1899–1902, the US army occupied Cuba. It instituted a program that brought about the eradication of yellow fever, but it was more fundamentally concerned with the establishment of US political and commercial dominance over the island.

On 21 February 1901, a constitution was adopted, and Cuba was nominally a free nation. But the US insisted that Cuba include in its constitution the Platt Amendment, which gave the US the right to intervene in Cuban affairs and maintain a naval base at Guantánamo.

For the next 30 years, Cuba lived through a succession of governments, constitutional and otherwise, all under the watchful eye of the US. American companies owned or controlled about half of Cuba's cultivated land, its utilities and mines, and other natural resources. The US Marines intervened in 1906–9, in 1912, and again in 1920. The period culminated in the brutal dictatorship of Gerardo Machado y Morales (1925–33).

Cuba entered another unstable phase in 1933. A nationalist uprising chased Machado from office. After the US attempted to install a regime, a "sergeants' revolt" headed by 32-year-old Fulgencio Batista y Zaldívar assumed power and named Ramón Grau San Martín provisional president. Grau, a physician and university professor noted for his nationalist zeal, was never recognized by the US, and lasted only four months. From 1934 until 1940, Batista ruled through a series of puppet presidents. During these years, Batista made two major contributions to Cuba. In 1934, President Franklin D. Roosevelt allowed Cuba to abrogate the Platt Amendment, although the US retained its naval base at Guantánamo Bay (which is scheduled to revert to Cuban authority at the end of the century). Batista also allowed the drafting of a new constitution, passed in 1940 under which he became president. In 1944, Batista permitted Grau San Martín, now his political enemy, to take office. The eight years of rule by Grau and his ally, Carlos Prío Socarrás, were ineffective and corrupt, and in 1952, a reform party was expected to win election.

That election was subverted, however, on 10 March 1952, when Batista seized power in a military coup. During the seven years of Batista's second administration, he used increasingly savage suppressive measures to keep himself in office. Under the Batista regime, the US dominated the economy, social services suffered, poverty and illiteracy were widespread, and the bureau-cracy was flagrantly corrupt. Cuba was looking for a redeemer, and Fidel Castro came on the scene.

Castro's insurrection began inauspiciously on 26 July 1953 with an abortive raid on the Moncada Army Barracks in Santiago. Captured, jailed, and then exiled, Castro collected supporters in Mexico, and in 1956 landed in Cuba. Routed by Batista's troops, Castro escaped into the Sierra Maestra with a mere dozen support-ers. The force never grew to more than a few thousand, but clever use of guerrilla tactics evened the score with Batista's poorly trained army. Moreover, there was almost no popular support for Batista, and in 1958 the US ended its military aid to the falling government. On 1 January 1959, the Batista regime collapsed, and Batista and many of his supporters fled the country. Castro's 26th of July Movement took control of the government, and began to rule by decree. The revolutionary government confiscated property that had been dishonestly acquired, instituted large-scale land reforms, and sought to solve Cuba's desperate financial and eco-nomic problems by means of a bold revolutionary program.

After June 1960, Cuban-US relations deteriorated at an accel-erated pace. Largely in retaliation for the nationalization of about $2 billion in US-owned property in Cuba, the US severed diplo-matic relations with the Castro government. Tensions increased when the revolutionary regime nationalized US oil refinery com-panies for refusing to process Soviet crude oil. The US response was to eliminate Cuba's sugar quota. In April 1961, a group of 1,500 Cuban exiles—financed, trained, organized, and equipped by the CIA—invaded Cuba at the Bay of Pigs on the southern coast. The brigade was defeated within 72 hours, and the 1,200 surviving invaders were captured. They were eventually released after US officials and private sources arranged for a ransom of $50 million in food and medical supplies.

Meanwhile, Communist influence in the Cuban government was growing. Castro declared Cuba to be a Socialist country in late 1960, and the following year declared himself to be a Marx-ist-Leninist and a part of the Socialist world. All major means of production, distribution, communication, and services were nationalized. Soviet-style planning was introduced in 1962, and Cuba's trade and other relations turned from West to East. In October 1962, US planes photographed Soviet long-range-missile installations in Cuba. The US blockaded Cuba until the USSR agreed to withdraw the missiles, in exchange for a US government pledge to launch no more offensive operations against the island.

However, the US did continue its attempt, through the OAS and other international forums, to isolate Cuba politically and economically from Latin America and the rest of the non-Com-munist world. All Latin American governments were encouraged (some use stronger terms) to break off diplomatic relations with Cuba. Castro responded with an attempt to destabilize certain Central and South American governments. Inspired by the Sierra Maestra campaign, guerrilla movements became active through-out the region, often with Cuban support. However, by 1967, when Ché Guevara was killed in Bolivia, these movements had collapsed. The US was only slightly more successful in its cam-paign of isolation. The OAS suspended Cuba in 1962, but in July 1975 passed the "freedom of action" resolution allowing coun-tries to deal with Cuba as they pleased.

During the Carter administration, there were moves to nor-malize relations with Cuba. In 1977, the US and Cuba resumed diplomatic contacts (but not full relations) and concluded fishing and maritime rights agreements. However, the advent of the Reagan administration brought increased tensions between the two countries. Citing Cuban involvement in Angola, Ethiopia, Nicaragua, and Grenada, the US took up a more intransigent stance toward Cuba.

Domestically, Castro's administration has had its successes and failures. A strong social welfare system, including free health care

and subsidized housing, was implemented in the 1960s and 1970s. However, an attempt to produce 10 million metric tons of sugar by 1970 seriously crippled the island's economy. Other misman-aged projects have led to economic stagnation or chaos. Cubans live frugally under a highly controlled system of rationing.

Cuba was dealt a serious blow in the late 1980s with the col-lapse of the Soviet Union. This meant a cutoff in economic and military aid, on which Cuba had come to rely over the years. Since 1960, the former USSR had been Cuba's most important trading partner and provided the major market for Cuban sugar. The few consumer goods the former USSR had supplied in the past were no longer available. An additional setback came with the Nicaraguan elections of 1990, in which the Sandinistas lost the presidency. Despite various movements for reform in current and former Communist regimes, Castro remains committed to the original goals of his revolution. He continues to blame Cuba's economic woes on the US embargo, rather than any shortcomings in his own programs. Castro has even gone so far as to accuse the US of waging "bacteriological warfare" against Cuba, blaming the CIA for outbreaks of African swine fever, sugarcane rust, and tobacco mold.

About three-quarters of a million Cubans have fled Cuba since Castro came to power. Most have settled in southern Florida, and many still have hope of returning to a Castro-free Cuba. There have been sporadic attempts to reunite families broken up by the emigration, but political circumstances often curtail these pro-grams. For example, in February 1985 the repatriation of 2,746 "undesirables" from the US began, but after Radio Martí (spon-sored by Voice of America) began broadcasting in Spanish in May 1985, Cuba abrogated the agreement.

13GOVERNMENT

After he became premier on 16 February 1959, Fidel Castro was the effective source of governmental power. The juridical basis for this power rested on the Fundamental Law of the Revolution, which was promulgated on 8 February 1959 and was based on Cuba's 1940 constitution. To regularize government functions, a 10-member Executive Committee, with Castro as premier, was formed on 24 November 1972.

A new constitution, first published on 10 April 1975, then approved by the first congress of the Cuban Communist party in December, and ratified by a 97.7% vote in a special referendum in February 1976, establishes the National Assembly of People's Power as the supreme state organ. The deputies, elected by municipal assemblies, serve five-year terms. The National Assem-bly elects the Council of State, which includes the president, who is both head of state and head of government. There are six vice-presidents in the Council of State, and 23 other members.

In December 1986, the third National Assembly, with 510 del-egates, reelected Castro as president of the state council. He remains the key figure in domestic and foreign policy making. The constitution recognizes the Communist party as the "highest leading force of the society and of the state," which effectively outlaws other political parties.

Suffrage is universal for citizens age 16 and over, excluding those who have applied for permanent emigration.

14POLITICAL PARTIES

Fidel Castro came to power through a coalition group known as the 26th of July Movement. Along with it, in 1959, the Student Revolutionary Directorate (Directorio Revolucionario Estudian-til) and the Communist Party (Partido Socialista Popular—PSP) were permitted to function.

Castro's relationship with the PSP was at first uneasy. The PSP condemned his early attempts at insurrection as "putschism," and did not support the 26th of July Movement until it had reached its final stages in 1958. After June 1959, Castro began to refer to anti-Communists as counterrevolutionaries, and used the PSP as an organizational base and as a link to the USSR. In December 1961, Castro declared his complete allegiance to Marxism-Leninism.

By 1962, the 26th of July Movement, the Student Revolutionary Directorate, and the PSP had merged into the Integrated Revolutionary Organization (Organización Revolucionaria Integrada), which, in turn, gave way to the United Party of the Socialist Revolution (Partido Unido de la Revolución Socialista) and, in 1965, to the Cuban Communist Party (Partido Comunista Cubano—PCC).

On 17 December 1975, the PCC convened its first congress, which ratified a 13-member Politburo; Fidel Castro was reelected first secretary of the PCC. The second congress of the PCC took place in December 1980. The third congress, in February and November-December 1986, witnessed a massive personnel change when one-third of the 225-member Central Committee and 10 of 24 Politburo members were replaced, with Fidel Castro reelected first secretary. PCC membership was about 450,000 in 1986. The Young Communist League and the José Martí Pioneer Organization for children up to 15 years of age are mass political organizations closely affiliated with the PCC.

15LOCAL GOVERNMENT

The country is divided into 14 provinces and 169 municipalities. The Isla de la Juventud is a special municipality. The 1976 constitution provides for a system of municipal assemblies to be elected for 2½-year terms by direct universal suffrage at age 16. Municipal assemblies choose delegates to provincial assemblies and deputies to the National Assembly.

16JUDICIAL SYSTEM

The 1976 constitution establishes the People's Supreme Court, consisting of a president, vice-president, and other judges, as the highest judicial tribunal. All members of the court are elected by the National Assembly, as are the attorney general and deputy attorneys general. Through its Governing Council, the court proposes laws, issues regulations, and makes decisions that must be implemented by the people's courts, whose judges are elected by the municipal assemblies. There are also seven regional courts of appeal, as well as district courts with civil and criminal jurisdiction.

Although the constitution provides for an independent judiciary, the courts are subordinate to the National Assembly and the Council of State.

17ARMED FORCES

The Castro government has maintained conscription since 1963, and it now requires two years of military service from men between the ages of 16 and 50. Two-year service is voluntary for women. Total armed strength in 1993 was estimated at 175,000. The army had 145,000 personnel, organized into 23 divisions of varying levels of manning and readiness. Only one armored division and an airborne brigade are always combat-ready. It is equipped with 1,600 Russian tanks and 2,000 other combat vehicles. Russian equipment dates to 1960s models. Army reserves numbered 135,000.

The navy has 13,500 personnel and over 34 combat vessels, including 3 submarines. The air force has 17,000 personnel and 162 combat aircraft and 85 armed helicopters. Paramilitary forces included 15,000 state security troops, 4,000 border guards, 50,000 civil defense forces, the 100,000-member youth labor army, and the 1,300,000-member territorial militia. Except for small military training missions, Cuba has withdrawn its troops from Africa and Central America. Cuban arms imports were estimated to amount to an average of $1.5 billion a year in the 1980s until Russia cut its aid to around $500 million in 1991.

In 1993, an estimated 4,300 Russian troops remained in Cuba. Cuba's 1991 budget for defense and internal security was $1.2 billion, or about 6% of gross domestic product.

The US maintains a naval base at Guantánamo Bay in southeastern Cuba, under a 1934 leasing treaty. The US government considers the base to be of some strategic and training significance in the Caribbean and has refused to give it up, despite demands by the Castro regime that it do so. About 3,000 military personnel are stationed at Guantánamo.

18INTERNATIONAL COOPERATION

Cuba is a member of the UN, having joined on 24 October 1945, and belongs to ECLAC and all the nonregional specialized agencies except the IBRD, IDA, IFC, and IMF. The nation participates in G-77 and PAHO, and became a full member of the CMEA on 11 July 1972. According to Western estimates, there were 6,000–8,000 Soviet civilian advisers in Cuba in the mid-1980s.

Cuba's charter membership in the OAS was suspended at the second Punta del Este meeting, in February 1962, through US initiative. The isolation of Cuba from the inter-American community was made almost complete when, at Caracas, on 26 July 1964, the OAS voted 15–4 for mandatory termination of all trade with the Castro government. However, relations with Latin America and the US improved markedly during the early 1970s, and at the San José, Costa Rica, meeting of the OAS on 29 July 1975, each member country was left free to determine its own stance toward Cuba. By the end of the year, most OAS members had restored relations with Cuba, which rejoined as a nonparticipating member.

Since the 1960s, Cuba has been very active in the Nonaligned Movement, and held its chairmanship between 1979 and 1983. The sixth summit conference of nonaligned countries took place in Havana on 3–9 September 1979; at the conference, President Castro denounced the US and described Cuba as a friend of the USSR. Cuba is a signatory of GATT and of the Law of the Sea.

19ECONOMY

The world's fourth-leading cane sugar producer, Cuba is primarily an agricultural nation; sugar is the leading earner of foreign exchange, representing about 7–8% of world output. After 1959, the revolutionary government, following policies espoused by Ernesto "Che" Guevara, attempted to liberalize the sugar economy in order to achieve agricultural diversification and industrialization. When this policy proved disastrous to the sugar crop, in 1962 Castro reversed the Guevara program and announced a goal of a 10-million-ton crop by 1970. Despite a severe drought in 1968–69, Cuba did achieve a record 7.6-million-ton output of refined sugar in 1970. Efforts to diversify foreign trade during the early 1970s were aided by record high prices for sugar. Commercial agreements with Argentina, Canada, Spain, France, the UK, Italy, and the FRG indicated Cuba's keen desire to move away from nearly exclusive reliance on the Socialist countries for both imports and exports. Nevertheless, in 1986, trade with the USSR and other CMEA members still made up the bulk of Cuba's foreign commerce, and Soviet aid remained essential to the economy.

Between 1971 and 1975, the Cuban economy grew by about 10% annually. During 1976–80, the gross social product (which does not include housing, education, public administration, defense, and other services) increased by only 4% a year, but during 1981–85, the average annual growth was 7.3%. Economic growth in 1986 was estimated at 1.7%, and growth in 1987 was projected at 1.5–2%; the declining rate of growth was the result of a depressed world sugar market, a falling dollar, and prolonged drought.

Since the collapse of the Soviet Union and its control of the Eastern bloc countries, which resulted in a cutoff of substantial Soviet assistance and the near demise of traditional markets for

Cuban exports, the Cuban economy slowed significantly in the period 1990–91, recording –1.5% and –25% respectively. Real GDP recovered in 1993–94 to a still depressed –14% and –10% respectively.

As measures for economic recovery, the Castro government has restricted public expenditure as well as restructured the economy. It has begun trading with market economies and developed new sources of foreign currency. The government has put special emphasis on the promotion of foreign investment and the development of sugar and tourism. In 1992, tourism surpassed sugar as Cuba's principal source of foreign exchange leaving revenues of $382.4 million.

[20]INCOME
In 1992, Cuba's GNP was estimated to be $14.9 billion in current US dollars, or $1,370 per capita. In 1992, agriculture contributed approximately 11% to GNP.

[21]LABOR
All Cuban workers belong to a trade union, under the central control of the Confederation of Cuban Workers (CTC), which is affiliated with the Communist-oriented World Federation of Trade Unions. The CTC's membership was 2,649,000 in the mid-1980s. Independent unions are explicitly prohibited. Those who attempt to engage in independent union activities face government persecution and harassment. Strikes are not legally permitted, and none were known to have occurred in 1992. Unemployment, which the Castro government estimated at 33% of the work force in 1958, was practically abolished, but a limited unemployment scheme, in which redundant workers received 70% of their former wages until rehired, was introduced in the early 1980s to rationalize certain sectors of the economy. Underemployment is a chronic problem, and has been exacerbated by the idling of thousands of industrial workers whose jobs rely on foreign imports. Labor has been shifted to agriculture to compensate for fuel and machinery shortages affecting food and production.

The minimum wage is supplemented by social security consisting of free medical care and education and subsidized housing and food; a worker must still earn significantly more than minimum wage to support a family. The eight-hour workday, a weekly rest period, an annual paid vacation of one month, and workers' compensation are guaranteed by the constitution. The workweek is 44 hours, with shorter workdays for hazardous occupations.

In June 1990, the state labor force of 3,600,000 was distributed as follows: industry, 22%; agriculture, forestry, and fishing, 20%; education and culture, 13.3%; trade, 11%; services, and government, 16.7%; construction, 10%; transportation and communications, 7%. Women constituted 38% of the work force, up from 27% in 1975.

[22]AGRICULTURE
In 1991, the state owned 3,330,000 hectares (8,228,000 acres) of cultivated land, and 2,970,000 hectares (7,339,000 acres) of pasture. About 20% of employment is in the agricultural sector. An agrarian reform law of June 1959 made the government proprietor of all land in Cuba, created the National Institute of Agrarian Reform (INRA) as administrator, and set a general limit of 30 caballerías (400 hectares/990 acres) of farmland to be held by any one owner. A second agrarian reform, of October 1963, expropriated medium-size private holdings; there remained about 170,000 small private farms, with average holdings of over 16 hectares (40 acres). By 1985 there were 1,378 farm cooperatives. Almost a third of cultivated land is irrigated.

Sugarcane, Cuba's most vital crop and its largest export, is grown throughout the island, but mainly in the eastern half. The government regulates sugar production and prices. Sugar output reached 7.6 million tons in 1970, but that fell short of the 10 million tons projected. Subsequent targets were lowered, and the output was 7.9 million tons in 1979, 6.7 million in 1980 (when crop disease reduced production), 8 million in 1985, and 5.8 million in 1992. Cuba has pioneered the introduction of mechanical cane harvesters, and by 1992 there were 7,370 harvester-threshers (up from 5,717 in the early 1980s). Cuba and Russia signed several finance and investment accords in 1992 and 1993 whereby Russia will supply fuel, spare parts, fertilizer and herbicide in exchange for Cuba's 1993/94 sugar harvest; Russia will import a minimum of 2 million tons of Cuban sugar. The sugar industry also has diversified into exporting molasses, ethyl alcohol, rum and liquor, bagasse chipboard, torula yeast, dextran, and furfural. Tobacco, the second most important crop, is grown on small farms requiring intensive cultivation. In the late 1970s, the average annual production was about 35,000 tons, but crop disease in 1979 resulted in a drop in production to 8,200 tons in 1980; production rebounded to 44,000 tons by 1992. Other crops in 1992 included (in tons) oranges, 570,000; lemons and limes, 60,000; grapefruit, 315,000; rice, 308,000; bananas, 200,000; potatoes, 245,000; sweet potatoes, 240,000; and coffee, 25,000. Other Cuban products with export potential include: mangoes, pineapples, ginger, papayas, and seeds.

[23]ANIMAL HUSBANDRY
In the state sector, milk production in 1992 amounted to 856,000 tons (1,100,000 in 1990) and egg production reached 88,000 tons (11,040 tons in 1990). Livestock in 1992 included an estimated 4.7 million head of cattle, 1.8 million hogs, and 625,000 horses. The populations of most livestock species have declined about 4% since 1990, as a result of input shortages from the worsening economy. In 1992, honey production was an estimated 9,500 tons, higher than any other Caribbean nation.

[24]FISHING
The territorial waters of Cuba support more than 500 varieties of edible fish. The catch in 1991 was 165,236 tons, compared with 244,673 tons in 1986. Tuna, lobster, and shellfish are the main species caught. The Cuban Fishing Fleet, a government enterprise, supervises the industry.

The former USSR aided in the construction of a fishing port in Havana to service the Cuban and former-Soviet fleets as well as to process and store fish. Seafood exports are an important source of foreign exchange; in 1991, fish and fish products exports amounted to $129,612,000 with shellfish accounting for 97% of the total.

[25]FORESTRY
Much of the natural forest cover was removed in colonial times, and cutting between the end of World War I and the late 1950s reduced Cuba's woodland to about 14% of the total area and led to soil erosion. Between 1959 and 1985, about 1.8 billion seedlings were planted, including eucalyptus, pine, majagua, mahogany, cedar, and casuarina. By 1991, state forests covered 2,765,000 hectares (6,832,000 acres), or about 25% of the total area. Sawn wood production in 1991 was 130,000 cu m.

[26]MINING
In 1991, Cuba was the world's sixth-largest nickel producer, with 35,000 tons. Nickel deposits and plants are located in eastern Cuba. The largest plant, at Moa, has a capacity of producing 24,000 tons per year. Copper is extracted around Pinar del Río, the Matahambre mine being by far the largest producer; copper production in 1991 was 3,000 tons (metal content), a substantial decline from prerevolutionary days. Chromite production was 50,000 tons. In 1991, Cuba produced 200,000 tons of salt from

seawater. High quality marble is also produced; in 1991, about 70% of production was bartered or sold to Italy, Mexico, and Spain.

27 ENERGY AND POWER
Cuba has no coal, and its hydroelectric potential is slight. Bagasse (sugarcane waste) has traditionally supplied most of the sugar industry's fuel. In 1991, Cuba's net installed capacity was 3,988,000 kw; total production that year was 16,255 million kwh. A 1,650-Mw nuclear plant, built with Soviet assistance in Cienfuegos, was to have been operational by 1993, supplying 15% of Cuba's annual electrical output. On 6 September 1992, however, President Castro announced that work on the nuclear plant had been indefinitely stopped, since Russia was demanding hard currency payment.

Crude oil was first produced in Cuba in 1942. The domestic petroleum industry was nationalized in July 1960, and all subsoil deposits were also nationalized; domestic and foreign holdings (mostly US) were expropriated or confiscated after foreign-owned refineries refused to process oil from Eastern Europe. Crude oil production dropped from 52,000 tons in 1959 to 10,000 tons in 1961, but rose to 104,000 tons in 1968 and to 868,000 tons by 1985. As of 1991, production totaled less than 700,000 tons. In 1991, Union de Petroleo de Cuba (a state-owned company) granted two French companies a six-year oil exploration contract in an approximately 1,800 sq km (700 sq mi) concession off Cuba's north coast. Natural gas production in 1991 amounted to 34 million cu m.

28 INDUSTRY
All Cuban industrial production has been nationalized; the last private firms were expropriated in March 1968. Between 1981 and 1985, industrial production grew by 8.8% annually.

Cuba had 156 sugar mills in 1985, and about 10% of Soviet exports to Cuba consisted of machinery for the sugar industry. Other food-processing plants produce cheese, butter, yogurt, ice cream, wheat flour, pasta, preserved fruits and vegetables, alcoholic beverages, and soft drinks. Light industry comprises textiles, shoes, soap, toothpaste, and corrugated cardboard boxes. Production of selected items in 1985 included wheat flour, 441,600 tons; soft drinks, 2.7 million hectoliters; and textiles, 205 million sq m. Cuban factories also turned out 12.4 million pairs of shoes in 1985.

With technical and financial help from the USSR and Eastern Europe, Cuba has made significant progress in heavy industry. The country's cement plants had an output of 3.2 million tons in 1985 (as compared with 2.63 million in 1979). Average annual fertilizer production increased from 420,500 tons during 1959–66 to 1,159,700 tons in 1985. In the same year, 606 sugarcane combines were produced, and steel production was 412,900 tons. Other products in 1985 included 2,393 buses, 94,065 television sets, 25,900 refrigerators, and 236,300 radios.

The sugar industry continues to be one of the principal sources of foreign exchange, but has diminished considerably in importance. In 1993, the Cuban government announced plans to create smaller, more autonomous cooperative farms in sugarcane and other agricultural sectors, in an effort to boost productivity. The sugar industry produces final molasses, ethyl alcohol, rum and liquor, bagasse clipboard torula yeast, dextran, and furfural.

Cuban officials say they are on target to reach 1.2 million tons of domestic crude production by the end of 1994, following a record 1.1 million tons in 1993, most of which was used in electricity generation.

29 SCIENCE AND TECHNOLOGY
In 1989, 22 million pesos were invested in science and technology with 8,830 technicians and 12,052 scientists and engineers engaged in research and development. Emphasis fell on agricultural research, new technologies for the recovery of nickel and cobalt, and a joint Soviet-Cuban space-flight research program. The Academy of Sciences of Cuba, founded in 1962, is Cuba's principal scientific institution. Cuba has 20 learned societies and institutes conducting scientific research, mainly in agriculture. The Ministry of Agriculture has research centers throughout Cuba. Cuba has 10 colleges and universities offering degrees in basic and applied sciences.

30 DOMESTIC TRADE
Havana is Cuba's commercial center. Provincial capitals are marketing and distribution centers of lesser importance. Camagüey is a cattle and sugar center, Santa Clara lies in the tobacco belt, and Santiago is a major seaport and mining city. Holguín has been transformed into a major agricultural and industrial center.

By May 1960, the National Institute of Agrarian Reform was operating about 2,000 people's stores (tiendas del pueblo), and by the end of 1962 all retail and wholesale businesses dealing in consumer essentials had been nationalized. In 1984 there were 27,301 retail establishments in Cuba.

Because of the US-organized trade boycott and the inability of Soviet and domestic production to meet Cuban demands, rationing was applied to many consumer goods in the 1960s and 1970s, but by the mid-1980s, rationing had been reduced and accounted for about 25% of individual consumption. Allocation of major consumer items after 1971 was by the "just class" principle, with the "best workers" receiving priority. The availability of basic consumer items increased noticeably after 1980, when the smallholder' free market (mercado libre campesino) was introduced. Under this system, small-scale private producers and cooperatives could sell their surplus commodities directly to consumers once their quotas had been filled. However, the peasant markets were abolished in May 1986, allegedly because they led to widespread speculation and profiteering.

31 FOREIGN TRADE
Cuba has established or reestablished trade relations with many countries of Latin America, the Caribbean, Africa, Asia, and Europe. Sugar remains the predominant export (74% of the 1985 total), followed by nickel and cobalt extracts, citrus fruits, and tobacco manufactures.

Imports in 1985—chiefly machinery and transport equipment, petroleum, and foodstuffs—totaled c$7,906 million, widening the trade deficit to c$1,923 from c$578 in 1980. Cuba has been heavily dependent on the USSR (75% of exports and 67% of imports in 1985), which buys mainly sugar and sells manufactured goods, chiefly capital equipment. Other Communist countries accounted for 14% of exports and 17% of imports in 1985. Trade with the non-Communist world increased during the 1970s, and by the late 1970s, a modest indirect trade with the US had developed. Japan, Spain, Argentina, and France were Cuba's leading non-Communist trade partners in 1984.

The sudden rupture of trade with the former Soviet Union and the Eastern bloc nations, after 30 years has caused severe trauma to the Cuban economy since 1990. However, there remains a clear political will on the part of the former Soviet Union to maintain economic relations with Cuba, with a certain degree of preference. Cuba has diversified its trading partners in recent years. The country now trades more with Spain, Italy, France, the Netherlands, Canada, and several Latin American nations. Trade with China has increase as well, amounting to $420 million in 1990.

Total exports totaled $3.6 billion in 1991, primarily composed of sugar, citrus, shellfish, nickel, coffee, tobacco, and medical products. Cuban products' main destinations were the former USSR (63%), China (6%), Canada (4%), and Japan (4%).

Total imports amounted to $3.7 billion, mainly composed of petroleum, capital goods, industrial raw materials, and food. Cuba's major suppliers were composed of the former USSR (47%), Spain (8%), China (6%), Argentina (5%), Italy (4%), Mexico (3%).

³²BALANCE OF PAYMENTS

Since the US stopped trading with Cuba in 1960, Cuba's dollar reserves have dropped to virtually nothing, and most trade is conducted through barter agreements. By 1991, Cuba's debt to the former USSR was estimated at US$24 billion. With the subsequent demise of the Soviet Union, Cuba has focused on trading with market-oriented countries in order to increase foreign currency reserves, notably by promoting tourism, sugar exports, and foreign investment. In 1992, tourism surpassed sugar as Cuba's primary source of foreign exchange earnings, at $382 million.

³³BANKING AND SECURITIES

All banks in Cuba were nationalized in 1960. The National Bank of Cuba, established in 1948, was restructured in 1967, and in 1993 it had 167 branches and agencies throughout the country. The National Bank began to pay a maximum 2% interest on personal savings in 1983. Commercial banks include Banco Financiero International, SA (1984). Savings banks include Banco Popular dul Ahorro. There are no securities exchanges.

³⁴INSURANCE

All insurance enterprises were nationalized by January 1964. In 1986, there were 2 non-life insurance companies in operation.

³⁵PUBLIC FINANCE

Under the Economic Management System, developed during the 1970s and approved by the PCC Congress in 1975, state committees for statistics and finances have been established, and formal state budgets, abandoned in 1967, have been reintroduced. State revenues come from the nationalized enterprises, income tax, social security contributions, and foreign aid. Cuba's debt to Western nations in 1991 was estimated at $6.5 billion, while debt to the former USSR was estimated at $24 billion.

³⁶TAXATION

A 1962 tax code instituted a sharply progressive income tax as well as a surface transport tax, property transfer tax, documents tax, consumer goods tax, and a tax on capital invested abroad.

³⁷CUSTOMS AND DUTIES

Cuba has a two-column tariff, with one set of rates for imports from countries with which it has trade agreements and members of GATT and a second set of rates for other countries. In December 1990, Cuba introduced new tariff regulations to standardize product descriptions.

³⁸FOREIGN INVESTMENT

In February 1960, Fidel Castro announced that foreign investment in Cuba would be accepted only if delivered to the government to be used as it saw fit. The enterprises in which this capital would be invested were to be "national enterprises," so that Cuba would not be dependent on foreigners. Any new foreign investments were to be controlled by the Central Planning Board. From mid-1960, US holdings in Cuba were systematically seized, partly for political reasons and partly because US corporations refused to accept Cuba's terms of nationalization. Some of the investments of other foreign nationals were left operating under stringent governmental regulation.

Between 1960 and the early 1970s, foreign investment activities were restricted to limited technical and economic assistance

from East European countries and the USSR, with which Cuba concluded over 40 cooperation agreements between 1963 and 1983. Limited investments from the non-Communist world were sought, with some success, in the mid-1970s. In 1982, in a further effort to attract investors from Western Europe, Canada, and Japan, Cuba passed its first foreign investment law, permitting foreign companies to form joint ventures with the Cuban government but to own no more than 49% of the stock. In 1985, however, direct investment in Cuba by OECD countries totaled only US$200,000.

In recent years, the Cuban government has seen the necessity to open its recessed economy to foreign investment, either via joint ventures or other forms of association. In 1992, Cuba further intensified its efforts to attract foreign investment in several key areas of its economy, including sugar, tourism, textiles, tobacco, pharmaceuticals, nickel, and shipping. As of June 1994, there were already 146 joint ventures in force. Of these, 30 were in tourism, 20 in mining, 10 in light industry, 3 in agriculture, and the rest spread over engineering, construction, lubricants, and telecommunications. In addition, many foreign contracts have been signed for oil drilling.

Principal foreign investors include Spain with 32 ventures, France with 12, Italy and the Netherlands with 11 each, and 4 Latin American countries that share a further 49.

³⁹ECONOMIC DEVELOPMENT

Until 1959, the Cuban government followed a policy of free enterprise; government ownership was largely limited to local utilities. When the Castro government came to power in January 1959, it proceeded to create a centrally planned economy. By means of nationalization and expropriation, all producer industries, mines, refineries, communications, and export-import concerns were brought under government control by 1968.

Planning in the 1960s vacillated on the question of whether Cuba should concentrate on the production of sugar, on industrialization, or on a balance between the two. After 1963, an emphasis on sugar became predominant. But the effort that went into the 1970 harvest diverted enormous resources from other sectors of the economy. At the same time, there was growing absenteeism and low productivity in the labor force, attributed to the policy of eliminating material incentives. Under the Economic Management System, pay was once again tied to production through the introduction of a system akin to piecework.

The 1975–80 development plan, approved at the PCC Congress in December 1975, set specific production goals for Cuban industry and protected an overall economic growth rate of 6% annually; it was announced in 1980 that the actual growth rate was 4%. The 1981–85 plan introduced new incentive schemes and gave more freedom to market forces; it also eased restrictive hiring regulations. One of the major aims of the plan was to increase industry's share of the gross social product to 50%, but industry accounted for only 45.3% in 1985. The 1986–90 plan envisioned a 5% annual growth and aimed particularly at an increase in exports. In December 1986, 28 austerity measures were approved by the National Assembly, including increases in transport and electricity prices and rationing of kerosene.

According to unofficial estimates, nonmilitary aid from the USSR during 1961–84 totaled about US$37.6 billion; Soviet assistance in 1986 was estimated at US$4 billion.

The Cuban economy is undergoing a recession mainly as a result of the collapse of the Communist bloc, which constituted the country's major economic support. It is presently in the process of restructuring economic policies and opening to market economies. It has many plans to continue refurbishing key sectors of the economy by allowing a great amount of foreign investment. It will construct 3,000-4,000 hotel rooms per year over the next two years as means for further developing the tourist industry.

Under several finance and investment accords signed by Cuba and Russia in 1992 and 1993, Russia will supply fuel, tires and spare parts for mechanical harvesters and other vehicles, fertilizers, and herbicides for Cuba's 1993/94 sugar harvest. In addition, Russia will import a minimum of 2 million tons of Cuban sugar. Russia also agreed to extend a $350-million credit to Cuba to complete and further develop a number of oil, energy, and nickel mining projects which had previously been backed by the Soviet Union.

40 SOCIAL DEVELOPMENT

Although there was no unified social insurance program during the Batista regime, laws were enacted setting up more than 50 retirement and disability pension plans in various industries, trades, and professions. A single system of social security covering almost all workers and protecting them against the risks of old age, disability, and death was enacted in 1963. Children's homes and rehabilitation centers were established in 1959 to eradicate begging on the streets. Day-care programs were begun in 1961; by 1985 there were 844 nurseries and kindergartens, enrolling 100,600 children.

Family-planning services are integrated within the general health care system. Since 1964, free hospital abortions have been available on request to women 18 years of age or older during the first 10 weeks of pregnancy. The ratio of 849 abortions per 1,000 live births in 1984 was one of the highest in the world. Contraception is widespread, with the intrauterine device (IUD) and the pill being the most popular methods among women. In 1991, the contraception prevalence rate was 70%, and the fertility rate was 1.9%.

The Family Code of 1975 proscribes all sex discrimination. Although women constituted 37.3% of the labor force in 1985, they were still concentrated in traditional jobs. The Maternity Law provides 18 weeks of maternity leave.

41 HEALTH

Sanitation is generally good, and health conditions greatly improved after the 1959 revolution. From 1988 to 1991, 98% of inhabitants had access to safe water, and 92% had adequate sanitation. From 1985 to 1992, 98% of the country had access to health care services. However, with the dissolution of the Soviet Union, Cuba no longer receives the same support and has fallen behind in many of its social services.

Infant mortality declined from more than 60 per 1,000 live births before 1959 to 10 in 1992. Studies show that between 1980 and 1993, 70% of married women (ages 15 to 49) used contraception. The government claims to have eradicated malaria, diphtheria, poliomyelitis, tuberculosis, and tetanus. Between 1990 and 1992, children up to one year of age were immunized as follows: tuberculosis (98%); diphtheria, pertussis, and tetanus (91%); polio (93%); and measles (98%).

Life expectancy was an average of 76 years for women and men in 1992, and the overall death rate was 6.7 per 1,000 people in 1993. Between 1985 and 1990, major causes of death were recorded as follows: communicable diseases and maternal/ perinatal causes (73 per 100,000); noncommunicable diseases (472 per 100,000); and injuries (82 per 100,000). There were 10 reported cases of tuberculosis per 100,000 in 1990.

Between 1985 and 1990, there were 3.75 doctors per 1,000 people, with a nurse to doctor ratio of 1.7. From 1985 to 1990, there were 5.0 hospital beds per 1,000 population. Medical services are now more widely distributed in rural as well as urban areas. All doctors are obliged to work for the rural medical service in needy areas for two years after graduation. All health services are provided free of charge.

42 HOUSING

Cuban housing has not kept pace with the population increase. All large cities have slum problems, despite the construction of 200,000 housing units between 1959 and 1975, 83,000 units between 1976 and 1980, and 335,000 units between 1981 and 1985. As of 1981, 67% of housing units were detached houses, 15% were apartments, 13% were palm huts called hohios, and 5% were cuarterias, housing units in buildings composed of a number of detached rooms where occupants share some or all facilities. In the same year, 61% of dwellings were concrete and brick, 33% were solid wood, and 4% were constructed with palm planks. Water was piped indoors to 53% of homes and outside to 21%; 49% had private bath facilities.

43 EDUCATION

Education has been a high priority of the Castro government. In 1959 there were at least 1 million illiterates, and many more were only semiliterate. An extensive literacy campaign was inaugurated in 1961, when 100,000 teachers went out into the countryside. In 1990, UNESCO reported the illiteracy rate of persons aged 15 years and over as 6.0% (males: 5% and females: 7%). Education is free and compulsory for five years (6–11 years of age). Secondary education lasts for another five years. An innovation of the Castro government was the addition of agricultural and technical programs to the secondary-school curriculum; the work-study principle is now integral to Cuban secondary education. Students in urban secondary schools must spend at least seven weeks annually in rural labor. The first junior high schools, based on the work-study concept, were introduced in 1968. Catholic parochial schools were nationalized in 1961.

Cuba has five universities: the University of Havana (founded 1728), Oriente University at Santiago de Cuba (1947), the University of Las Villas at Santa Clara (1952), University of Camagüey (1974), and the University of Pinar Del-Rio. Workers' improvement courses (superación obrera), to raise adults to the sixth-grade level, and technical training schools (mínimo técnico), to develop unskilled workers' potentials and retrain other workers for new jobs, were instituted after 1961. Today, special worker-farmer schools prepare workers and peasants for enrollment at the universities and for skilled positions in industrial and agricultural enterprises.

44 LIBRARIES AND MUSEUMS

The José Martí National Library in Havana, founded in 1901, had a collection of over 2.4 million volumes in 1989. Besides acting as the National Library, it provides lending, reference, and children's services to the public. Other sizable collections in Havana are found at the Havana University Library (691,000 volumes), the Library of the Institute of Literature and Linguistics (97,000) and the José Antonio Echevarría Library of the House of the Americas (120,000).

Although libraries of private institutions disappeared in the 1960s and many collections were transferred to the National Library, the number of special and research libraries increased, especially with the creation of many departments of the Academy of Sciences. A national library network was established by the Department of Libraries of the National Cultural Council. In 1989 there were 332 libraries with an aggregate collection of over 4.3 million volumes supervised by the Ministry of Culture. The 3,860 school libraries contained over 15.4 million volumes in 1987.

Cuba had 226 museums of all types in 1990. The National Museum of Fine Arts in Havana contains classical and modern art from around the world as well as Cuban art from the colonial period to the present day. The Colonial Municipal Museum and the Felipe Poey Museum in Havana, the Bacardi Municipal Museum in Santiago, the Oscar Rojas Museum in Cárdenas, and the Ignacio Agramonte Museum in Camagüey are also noteworthy.

[45]MEDIA

All telephone service is free; about 95% of the telephones are automatic. In 1993 there were 321,054 telephones in use. In 1992 there were 150 AM and 5 FM radio broadcasting stations and 3,695,000 radios in use. Fifty-eight television stations operate throughout the country; in 1992 there were about 1,746,000 television sets in use.

Like the radio and television stations, the press is controlled and owned by the government. Cuba's major newspapers, published in Havana, include *Granma* (with an estimated 1991 circulation of 420,000) and *Juventud Rebelde* (150,000). *Granma*, the official organ of the Communist party, was established in 1965; it also publishes weekly editions in Spanish, English, and French. *Juventud Rebelde* is the publication of the Union of Young Communists.

Magazines published in Havana include *Bohemia* (weekly, 200,000, general articles and news) and *Mujeres* (monthly, 250,000, women's-interest news). Prensa Latina, the Cuban wire service, covers international affairs and distributes its coverage throughout Latin America.

[46]ORGANIZATIONS

Most of the leading mass organizations in Cuba were founded shortly after the revolution. The Committees for the Defense of the Revolution were founded on 28 September 1960 to combat counterrevolutionary activities. The Federation of Cuban Women was established 23 August 1960. The National Association of Small Farmers, the leading peasants' organization, was established 17 May 1961; in 1989 it had 167,461 members, both private farmers and members of cooperatives. The Confederation of Cuban Workers, the principal trade union federation, antedates the revolution. Founded in 1939, it had a total membership of 3,060,838 workers in 1990.

[47]TOURISM, TRAVEL, AND RECREATION

Before 1959, tourism, especially from the US, was a major source of revenue. Foreign tourism declined in the 1960s, and Cuba's ornate and expensive hotels were used mainly by visiting delegations of workers and students. Renewed emphasis on international tourism characterized the 1976–80 development plan, under which 25 new hotels were opened. Today, the Cuban government actively promotes tourism as a means of offsetting the financial decline brought on by the collapse of the Soviet bloc. In 1991, an estimated 424,000 tourists visited Cuba, 207,000 from the Americas and 171,000 from Europe. Direct revenue from tourism was us$300 million, compared with us$145 million in 1987. In 1992, tourism became the country's main source of foreign exchange.

Among Cuba's attractions are fine beaches; magnificent coral reefs, especially around the Isle of Youth; and historic sites in Old Havana (where some buildings date from the 17th century), Trinidad, and Santiago de Cuba. Passports and visas are required for nationals of countries that do not have visa-free agreements with Cuba. In June 1992, Cuba was admitted to the Caribbean Tourism Organization.

[48]FAMOUS CUBANS

José Martí (1853–95), poet, journalist, and patriot, was the moving spirit behind the revolution that liberated Cuba from Spain. Antonio Maceo (1848–96), the mulatto general known as the "Titan of Bronze," became famous both as a guerrilla fighter and as an uncompromising advocate of independence. Carlos J. Finlay (1833–1915) gained lasting recognition for his theory regarding the transmission of yellow fever.

Cuban literature is most famous for its poetry and essays. The influential Afro-Cuban tradition has been explored by Cuban scholars, most notably by Fernando Ortiz (1881–1916), jurist and ethnographer. Another leading writer was José Antonio Saco (1797–1879), author of a six-volume history of slavery. Ernesto Lecuona (1896–1963) was a composer of popular music, and Juan José Sicre (b.1898) is Cuba's outstanding sculptor.

The major heroes of the revolution against Batista are Fidel Castro Ruz (b.1926); his brother, Gen. Raúl Castro Ruz (b.1931); Argentine-born Ernesto "Che" Guevara (1928–67), who was killed while engaged in revolutionary activities in Bolivia; and Camilo Cienfuegos (d.1959). Cubans notable in literature include poet Nicolás Guillén (b.1902) and playwright and novelist Alejo Carpentier y Valmont (1904–80). Alicia Alonso (b.1921), a noted ballerina, founded the National Ballet of Cuba.

[49]DEPENDENCIES

Cuba has no territories or colonies.

[50]BIBLIOGRAPHY

Balfour, Sebastian. *Castro*. New York: Longman, 1990.

Brundenius, Claes. *Revolutionary Cuba: The Challenge of Economic Growth with Equity*. Boulder, Colo.: Westview Press, 1984.

Cardoso, Eliana A. *Cuba After Communism*. Cambridge, Mass.: MIT Press, 1992.

Chayes, Abram. *The Cuban Missile Crisis*. New York: Oxford University Press, 1974.

Cuba After the Cold War. Pittsburgh: University of Pittsburgh Press, 1993.

Cuba and the Future. Westport, Conn.: Greenwood Press, 1994.

Cuba: a Short History. New York: Cambridge University Press, 1993.

Foner, Philip S. *The Spanish-Cuban-American War and the Birth of American Imperialism, 1895–1902*. 2 vols. New York: Monthly Review Press, 1972.

Halebsky, Sandor, and John M. Kirk (eds.). *Cuba: Twenty-Five Years of Revolution, 1959–1984*. New York: Praeger, 1985.

Haverstock, Nathan A. *Cuba in Pictures*. Minneapolis: Lerner Publications Company, 1987.

Mesa-Lago, Carmelo. *The Economy of Socialist Cuba: A Two-Decade Appraisal*. Albuquerque: University of New Mexico Press, 1981.

Newsom, David D. *The Soviet Brigade in Cuba: a Study in Political Diplomacy*. Bloomington: Indiana University Press, 1987.

Perez, Louis A. *Cuba and the United States: Ties of Singular Intimacy*. Athens: University of Georgia Press, 1990.

———. *Cuba: Between Reform and Revolution*. New York: Oxford University Press, 1988.

Preeg, Ernest H. *Cuba and the New Caribbean Economic Order*. Washington, D.C.: Center for Strategic and International Studies, 1993.

Rudolph, James D. (ed.). *Cuba, a Country Study*, 3d ed. Washington, D.C.: Government Printing Office, 1985.

Ruffin, Patricia. *Capitalism and Socialism in Cuba: a Study of Dependency, Development, and Underdevelopment*. New York: St. Martin's Press, 1990.

Schroeder, Susan. *Cuba: A Handbook of Historical Statistics*. Boston: G.K. Hall, 1982.

Suchlicki, Jaime. *Historical Dictionary of Cuba*. Metuchen, N.J.: Scarecrow Press, 1988.

Thomas, Hugh. *The Cuban Revolution: 25 Years Later*. Boulder, Colo.: Westview, 1984.

DOMINICA

Commonwealth of Dominica
Dominica

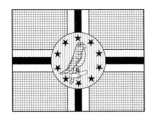

CAPITAL: Roseau.

FLAG: On a green background appears a cross composed of yellow, black, and white stripes; in the center is a red disk with 10 yellow-bordered green stars surrounding a parrot.

ANTHEM: *Isle of Beauty, Isle of Splendor.*

MONETARY UNIT: The East Caribbean dollar (EC$) of 100 cents is the national currency. There are coins of 1, 2, 5, 10, and 25, and 1 dollar, and notes of 5, 10, 20, and 100 East Caribbean dollars. EC$1 = US0.3704 (US$1 = EC$2.70).

WEIGHTS AND MEASURES: The metric system is being introduced, but imperial measures remain in common use.

HOLIDAYS: New Year's Day, 1 January; Labor Day, 1 May; Caricom Day, 2 July; Bank Holiday, 1st Monday in August; National Days, 3–4 November; Christmas, 25 December; Boxing Day, 26 December. Movable religious holidays include Carnival, Good Friday, Easter Monday, and Whitmonday.

TIME: 8 AM = noon GMT.

¹LOCATION, SIZE, AND EXTENT

Although usually classified as one of the Windward Islands, Dominica, located between Guadeloupe to the N and Martinique to the S, marks the midpoint of the Lesser Antilles. To the E lies the Atlantic Ocean, to the W the Caribbean Sea. The island has an area of 750 sq km (290 sq mi) and is 47 km (29 mi) long by 26 km (16 mi) wide, with a coastline of 148 km (92 mi). Comparatively, the area occupied by Dominica is slightly more than four times the size of Washington, D.C.

Dominica's capital city, Roseau, is located on the southwest coast of the island.

²TOPOGRAPHY

The most rugged island of the Lesser Antilles, Dominica is a mass of peaks, ridges, and ravines. Several mountains are over 1,200 m (4,000 ft), of which the highest is Morne Diablotins, with an altitude of 1,447 m (4,747 ft). The whole land mass is of recent volcanic formation, and the mountain peaks are cones of volcanoes with lava craters and small lakes of boiling water. The coastal rim of the island is a thin strip limited by the mountainsides, which extend directly down to the shore.

³CLIMATE

The climate of Dominica is mildly tropical; in the winter months the temperature averages 25°C (77°F); in the summer, 28°C (82°F). The spring months are the driest; the heaviest rains fall during late summer. The average yearly rainfall ranges from about 200 cm (80 in) on the drier Caribbean coast to 635 cm (250 in) in mountainous inland areas. Destructive hurricanes coming in from the Atlantic Ocean can be expected during the late summer months. Dominica was devastated by hurricanes in 1979 and 1980; the cost of restoring agricultural production was estimated at $68.5 million, more than double the 1978 GNP.

⁴FLORA AND FAUNA

Since few plantations could be established on Dominica's rugged terrain, the island is still covered with forests, some of which have never been cut except by the destructive winds of a hurricane. On one 4-hectare (10-acre) plot in the rain forest as many as 60 species of trees may be identified. Some of the most common are chataignier, gommier, carapite, breadfruit, white cedar, and laurier.

There are no large wild animals, but the agouti and manicou can be found. Some 135 species of birds inhabit Dominica, the coastal waters of which abound in fish.

⁵ENVIRONMENT

As a member of the Organization of Eastern Caribbean States (OECS) formed in 1981, Dominica shares environmental problems common to the area with neighboring island countries. Among the nation's environmental concerns, shortages in the supply of water is one of the most significant. The lack of water is complicated by pollution from chemicals used in farming and untreated sewage. The nation's forests are endangered by the expansion of farming activities.

Pollution of the nation's coastal waters threatens the tourist trade in the area. Two extensive areas have been set aside as nature reserves. The southern reserve, which constitutes Morne Trois Pitons National Park, covers an area of 6,500 hectares (16,100 acres). In it are the nesting places of the red-necked and imperial parrots, both endangered species of Dominica; the tundra peregrine falcon and the green sea and hawksbill turtles were also classified as endangered in 1987. Hurricanes are the most destructive environmental force.

⁶POPULATION

In 1991 the population was 71,183, down from 73,795 in 1981. The capital of the island, Roseau, had 15,853 inhabitants. The population density of Dominica in 1991 was 95 persons per sq km (245 per sq mi), one of the lowest in the West Indies.

⁷MIGRATION

There are no restrictions on foreign travel, emigration, or repatriation.

8ETHNIC GROUPS

Up-to-date statistics on racial or ethnic origins are not available. In 1981, the vast majority of Dominicans (91%) were descendants of African slaves brought to the island in the 17th and 18th centuries. Some 6% of the population was of mixed descent, and a small minority of about 0.5% was of European origin. Dominica is the only island of the Caribbean on which descendants of the Carib Indian population still make up a community of significant size. Isolation and the establishment of a 1,500-hectare (3,700-acre) reserve have enabled the Caribs, who constitute about 3,000 people, to preserve their identity.

9LANGUAGES

English is the official language of Dominica. Nearly all Dominicans also speak a French patois, based on a mixture of African and French grammar and consisting mostly of French words, with some English and Spanish borrowings. Some islanders speak French as their first language.

10RELIGIONS

About 80% of the population is Roman Catholic. There are Anglican and Protestant minorities, as well as Baha'is and some Rastafarians.

11TRANSPORTATION

A paved road circles the northern two-thirds of the island, connecting the two main towns, Roseau and Portsmouth, with Melville Hall Airport in the northeast. Much of the road system—about 750 km (450 mi) in 1991—was severely damaged by the 1979 hurricane; reconstruction has averaged about 16 km (10 mi) a year. There are 10,200 motor vehicles on the island. A deepwater harbor has been completed near Roseau on Woodbridge Bay, and both Roseau and Portsmouth also receive ships. Dominica Air Transport and other small airlines connect the main airport with Martinique, Guadeloupe, Antigua, and Barbados. There is a 760-m (2,500-ft) airstrip at Canefield, about 5 km (3 mi) north of Roseau.

12HISTORY

Dominica was the first island sighted in the New World by Christopher Columbus on his second voyage, Sunday (*dies dominica*), 3 November 1493. At that time, the island was inhabited by Carib Indians whose ancestors originally had come from the Orinoco Basin in South America and, during the 14th century, had driven out the indigenous Arawaks. The Caribs resisted conquest, and the Spaniards soon lost interest in the island, which had no apparent mineral wealth.

In 1635, France claimed Dominica, and French missionaries visited the island seven years later, but strong Indian resistance to further contact prevented either the French or the English from settling there. In 1660, England and France declared Dominica a neutral island and left it to the Caribs. Within 30 years, however, Europeans were beginning to settle on the island, and in 1727 the French took formal possession. Under the Treaty of Paris in 1763, France ceded the island to Great Britain, which then developed fortifications for its defense. Coffee plantations were established during the French period of colonization, and sugar production was introduced later by the British, but the large slave plantations that characterized other West Indian islands never developed on Dominica. When Great Britain abolished slavery in the West Indies in 1834, 14,175 Dominican slaves were given their freedom.

In the years following emancipation, blacks first gained considerable political power, then lost it as the British government, acceding to the wishes of Dominican planters, diluted the strength of the Legislative Assembly and, in 1896, reduced Dominica to a crown colony. Dominica was governed as part of the Leeward Islands from 1871 until 1939, and in 1940 was transferred to the Windward Islands administration. From 1958 to 1962, the island

was a member of the Federation of the West Indies. Dominica became an associated state of the Commonwealth of Nations in 1967 and on 3 November 1978 became an independent republic.

In its first years of independence, Dominica had several problems. Some were brought about by destructive hurricanes, especially Hurricane David in 1979, but others were attributable to the corrupt and tyrannical administration of Premier Patrick John. John was ousted in June 1979, and after a year of interim rule, Mary Eugenia Charles became prime minister in July 1980.

Charles, the first female prime minister in the Caribbean, has ruled Dominica ever since. Her Dominica Freedom Party has been given parliamentary majorities in 1985 and 1990, partly because of an improved economic picture. Charles fully supported and sent a token force to participate in the US-led intervention of the island of Grenada in October 1983.

13GOVERNMENT

Under the independence constitution of 3 November 1978, Dominica has a unicameral parliament, the House of Assembly, with 21 members elected by universal adult suffrage (at age 18) and 9 appointed members (5 named on the advice of the prime minister, 4 the advice of the leader of the opposition). The life of the Assembly is 5 years. Parliament elects a president as head of state, who in turn appoints the prime minister and cabinet from the majority party in the assembly.

14POLITICAL PARTIES

The majority party in Dominica is the Dominican Freedom Party (DFP), which holds 11 of the 21 elective seats in parliament. It is led by Mary Eugenia Charles, the first female prime minister in the Caribbean. The United Workers Party (UWP), headed by Edison James, holds 6 seats, with the remaining 4 held by the traditional opposition Democratic Labour Party (DLP), led by Rosie Douglas. The Dominica Liberation Movement (DLM) is a small organization operating on the left.

15LOCAL GOVERNMENT

In contrast to other English-speaking islands in the Caribbean, local government in Dominica is well developed. There are 25 village councils, made up of both elected and appointed members. Both Roseau and Portsmouth have town councils. There are also 10 parishes, which are administrative divisions for the national government.

16JUDICIAL SYSTEM

Dominica's judicial process derives from English common law and statutory acts of the House of Assembly. The courts of first instance are the four magistrates' courts; at the second level is the Court of Summary Jurisdiction. The highest court is the Eastern Caribbean Supreme Court, based in St. Lucia. In exceptional cases, appeals may be carried to the Judicial Committee of the UK Privy Council. Patrick John, who had been convicted in lower courts of planning to overthrow the constitutional government, made such an appeal, and the Privy Council dismissed the charges in 1982. (In 1985, however, he was retried, convicted, and sentenced to 12 years' imprisonment for treason.)

17ARMED FORCES

A police force of 300 is in charge of law and order. Dominica, along with Antigua and Barbuda, Grenada, St. Kitts and Nevis, St. Lucia, and St. Vincent and the Grenadines, is a member of the Regional Security System, established in 1985. Defense from foreign attack would come from the US or UK.

18INTERNATIONAL COOPERATION

Dominica became a member of the UN on 18 December 1978 and belongs to ECLAC and all the nonregional specialized agencies except IAEA, ICAO, ITU, and WIPO. Dominica also belongs to the Commonwealth of Nations, as well as the Caribbean Development Bank, CARICOM, G-77, OAS, and the Organization of Eastern Caribbean States (OECS). In May 1987, the prime ministers of the member-states of the OECS agreed to create a single nation of the seven island states. The accord was subject to a national referendum in each of the states. The referendums were defeated by the OECS countries and they remained separate. The nation is a de facto adherent to GATT and a signatory of the Law of the Sea.

19ECONOMY

Virtually the entire economy is based on agriculture, fishing, and forestry. Bananas are the main crop, and coconuts, from which copra is extracted for export, are also important. Tourism is a potential source of income but is relatively undeveloped; still lacking are a jet airport, a good system of roads, and luxury hotels. The government is promoting industrial development involving agroprocessing and light industry.

In 1991, agriculture accounted for over 33% of GDP, employing approximately 40-50% of the labor force. In 1992, real GDP increased by 2.6%; unemployment reached 15%; and the inflation rate was 5.3%.

20INCOME

In 1992, Dominica's GNP was $181 million at current prices, or $2,520 per capita. For the period 1985–92 the average inflation rate was 5.0%, resulting in a real growth rate in per capita GNP of 5.1%.

In 1992, the GDP was $8.4 million in current US dollars. It is estimated that in 1991, agriculture, hunting, forestry, and fishing contributed 21% to GDP; mining and quarrying, 1%; manufacturing, 6%; electricity, gas, and water, 3%; construction, 6%; wholesale and retail trade, 11%; transport, storage, and communication, 14%; finance, insurance, real estate, and business services, 12%; community, social, and personal services, 1%; and other sources, 25%.

21LABOR

The labor force in 1991 was estimated at 30,600. About 47% of the labor force is employed in agriculture, fishing, and forestry. Among the labor unions operating in Dominica are the Dominica Trade Union, established in 1945; the Dominica Amalgamated Workers' Union, which originated as a banana workers' union; and the Waterfront and Allied Workers' Union. Other unions represent civil servants and teachers. Only slightly over 10% of the work force is organized. Unemployment was officially 11% in 1991.

22AGRICULTURE

About 22.6% of the total land area is arable. Agricultural production was on the decline even before the 1979 hurricane disaster. The main crop of Dominica is bananas, output of which had fallen to 29,700 tons in 1978. As a result of Hurricane David, production hit a low of 15,700 tons in 1979. Agriculture suffered a further blow from Hurricane Allen in August 1980. However, after outside financial support began to rehabilitate the sector, production rose to 27,800 tons in 1981 and 70,000 tons by 1992.

Agriculture accounts for over one-third of GDP, and employs 40-50% of the labor force. Agricultural exports amounted to $35 million in 1992. Most crops are produced on small farms, the 9,000 owners of which are banded together in about 10 cooperatives; there are also several large farms that produce mostly bananas for export. Coconuts and citrus fruits are grown in commercial quantities. Fruits and vegetables are produced mostly for local consumption.

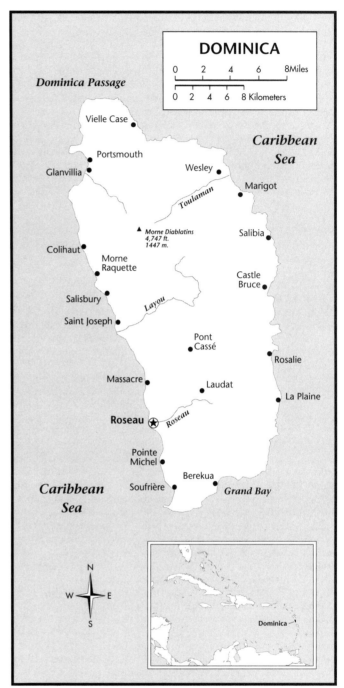

LOCATION: 15°N; 16°W. TOTAL COASTLINE: 148 km (92 mi). TERRITORIAL SEA LIMIT: 12 mi.

23ANIMAL HUSBANDRY

The island does not produce sufficient meat, poultry, or eggs for local consumption. In 1992 there were an estimated 9,000 head of cattle, 5,000 hogs, 10,000 goats, and 8,000 sheep. In 1992, production of meat totaled 1,000 tons; milk, 5,000 tons; eggs, 158 tons; and cow hides, 40 tons.

24FISHING

Before Hurricane David, some 2,000 persons earned a living fishing in coastal waters, producing about 1,000 tons of fish a year and meeting only about one-third of the local demand. The hurricane destroyed almost all of the island's 470 fishing boats; afterward, only about a dozen vessels could be reconstructed for use.

Since then, the catch has averaged about 650 tons annually. In 1991, the catch was only 590 tons.

25 FORESTRY

Dominica has the potential for a lumber industry. Some 31,000 hectares (76,600 acres) are classified as forest and woodland, representing 41% of the total land area. In 1962, Canadian experts produced a study indicating that over a 40-year period the island could produce a yearly output of 800,000 cu ft of lumber. Before Hurricane David, annual output had reached about 265,000 cu ft. There are some 280 hectares (700 acres) of government land allocated to commercial forestry and about 38 hectares (94 acres) of forestland in private hands. Commercially valuable woods include mahogany, blue and red mahoe, and teak.

26 MINING

Commercial mining is limited. Pumice is the major commodity extracted from the island for export. Limestone, volcanic ash, sand and gravel, and clay deposits are also found.

27 ENERGY AND POWER

Of the numerous short streams on Dominica, at least 30 might be sources of hydroelectric power on a small scale; about 60% of the power supply already comes from hydroelectric sources. In 1991, the Trafalgar Hydroelectric Power Station came onstream, making the island nearly self-sufficient in energy. The island also has thermal power potential in the volcanic Boiling Lake, the temperature of which has been recorded in the range of 82–92°C (180–198°F). In 1991, 31 million kwh of electricity were produced.

28 INDUSTRY

Dominica has only light industry, most of it connected with the processing of agricultural products. Industrial establishments include a plant for processing coconuts into oil and copra for export, 4 plants to process limes and other citrus fruits, two bottling plants, two distilleries, four small apparel plants, and four small furniture factories. Soap and toiletries are exported; Dominica exports water to its Caribbean neighbors. Home industries produce some leather work, ceramics, and straw products.

In recent years, the small manufacturing sector has been expanding at a modest pace, including electronic assembly, rum, candles and paints. The Trafalgar Hydro Electric Power Station is now operational, making the island virtually energy self-sufficient.

The tourism industry is also growing slowly. The government has many plans for an expansion of this sector, including a new international airport, redevelopment of the Roseau bay front, and refurbishing and expansion of the existing hotels.

29 SCIENCE AND TECHNOLOGY

Late in 1980, Dominica created a Council for Science and Technology, under the Ministry of Education. However, without a budget and a clear directive of purpose, the council had accomplished little by 1986. The Caribbean Agricultural Research Development Institute has been active in more than a half-dozen projects.

30 DOMESTIC TRADE

Local produce markets exist in all the small villages and towns. Commercial activity is concentrated in the morning hours, since tropical rains impede afternoon travel. Banks are open from 8 AM to 1 PM, Monday–Friday, with additional hours from 3 to 5 PM on Friday. There is no commercial activity on Saturday afternoon and Sunday.

31 FOREIGN TRADE

Bananas, which normally account for about 70% of export earnings, in 1979 made up only 47.8% because of hurricane damage. In 1984, banana exports were 44.6% of the total; soap and toiletries accounted for 22.1%; galvanized sheets, 9.8%; coconut oil, 6.5%; and all other products, 17%. Food and capital goods, including material to rebuild roads, bridges, and public buildings, are the major imports. In 1985, exports amounted to EC$76.8 million and imports to EC$149.4 million.

In 1991, banana exports accounted for 58.9% of total exports; soaps accounted for 21%; pre-fab houses accounted for 5%, bay oil accounted for 25%; plantains and oranges accounted for 1%. Total exports reached $66 million. Total imports, including foodstuffs and consumer goods, reached $110 million. Dominica's major trading partners were CARICOM, the US, Canada, and the UK.

32 BALANCE OF PAYMENTS

Before 1980, the increasingly unfavorable balance of trade produced a growing current-account deficit and boosted the foreign debt. The 1979, 1980, and 1989 hurricanes added to the debt burden.

In 1992, merchandise exports totaled $54.57 million and imports $97.51 million. The merchandise trade balance was $–42.94 million.

The following table summarizes Dominica's balance of payments for 1991 and 1992 (in millions of US dollars):

	1991	1992
CURRENT ACCOUNT		
Goods services and income	–42.64	–42.67
Unrequited transfers	21.17	16.40
TOTALS	–21.47	–26.27
CAPITAL ACCOUNT		
Direct investment	10.66	11.44
Portfolio investment	—	—
Other long-term capital	12.13	10.29
Other short-term capital	2.95	1.59
Exceptional financing	—	—
Other liabilities	—	—
Reserves	–4.22	–3.36
TOTALS	21.51	19.95
Errors and omissions	–0.04	6.32
Total change in reserves	–4.23	–3.53

33 BANKING AND SECURITIES

The principal national banks are the National Commercial and Development Bank of Dominica and the Dominica Agricultural Industrial and Development (AID) Bank. Private commercial banks include Barclays (UK), the Royal Bank of Canada, and the Banque Française Commerciale. Demand deposits totaled EC$47.87 million at the end of January 1993. The money supply, as measured by M2, totaled EC$320.37 as of 1993. Dominica has no stock market

34 INSURANCE

Representatives of British, Canadian, and US insurance companies do business in Dominica.

35 PUBLIC FINANCE

Operating revenues come mostly from customs duties, excise taxes, and other taxes and fees for government services. The leading areas of expenditure are education, health, public services, housing, and defense. In 1991/92, current receipts totaled $12.4 million, while current expenditures came to $50.2 million.

Capital receipts and expenditures in 1991/92 amounted to $13.5 million and $21.7 million, respectively. The external public debt as of the end of 1991 was $84.8 million.

36TAXATION

Taxes levied by the Dominican government include a progressive personal income tax ranging from 0% to 40% in 1991; a business income tax of 35%; social security taxes; a 3% gross receipts tax on retail sales; and taxes on land transfers and land-value appreciation. There is no capital gains tax except the land-value appreciation tax.

37CUSTOMS AND DUTIES

Specific import duties apply to food, and ad valorem duties to other items. The government levies export duties on the principal agricultural products; the charge is heavy on rum and cigarettes but lighter on bananas and coconuts. Under a 1992 Caribbean Community agreement, Dominica will eliminate import licensing. Dominica has adopted CARICOM's Common External Tariff, which ranges from 5–45%.

38FOREIGN INVESTMENT

The amount of foreign investment in Dominica is limited, in large part because the island lacks the infrastructure to support an industrialization program. Investment has, however, been increasing under the Charles government, particularly in agriculture. The marketing of banana production is monopolized by a European multinational corporation. There is some foreign investment in the forestry industry. Tax holidays and import-duty exemptions are offered as investment incentives. The area near the Canefield airstrip is being developed as an industrial estate and export processing zone. The government agency for industrial development has had some success in attracting capital investment funds for the island.

Other investment incentives include repatriation of profits, alien's landholding license fee, factory building/industrial estates, residence/work permits.

39ECONOMIC DEVELOPMENT

Dominica seeks to foster private enterprise, but the Charles government has been willing to intervene in the economy where it perceives a need. In 1986, for example, it created an export-import agency and announced a land-reform program, both to stimulate agriculture. Under the latter, the government purchased 800 hectares (2,000 acres) of land in prime growing areas and then guaranteed a minimum holding with security of tenure, as well as services and equipment, to small farmers and landless farm workers.

As of 1994, the government continues to encourage agricultural expansion by implementing a program of diversification. This program is aimed at improving the marketing of production and providing income guarantees for farmers diversifying into new crops. There is still serious concern at the implications of the restructuring of formerly protected European markets, despite the EC's implementation of favorable banana regime on 1 July 1993.

40SOCIAL DEVELOPMENT

Until the early 1980s, a high rate of unemployment, a markedly high rate of emigration, and very limited resources hindered the development of social service programs in Dominica. In 1976, the government established the Social Security Scheme, which covers all workers from 16 to 60 years of age. Under this plan, both workers and employers contribute specific amounts to a government fund, which provides pensions for workers reaching retirement age, compensation for workers who become incapacitated, and survivor benefits. There are also sickness and maternity benefits. The social security system is supported by worker contributions of 3% of earnings and a 6.75% employer payroll tax. The government supports family planning; the privately run Dominica Planned Parenthood Association is an affiliate of the International Planned Parenthood Federation. Apart from the constitution, there is no specific legislation in force to protect women from sex discrimination. A hotline manned by volunteers is available for battered women, and the Welfare Department often helps them find temporary quarters, although there were no battered women's shelters as of 1993.

41HEALTH

In 1991 there were 38 physicians, and in 1990 there were 4 dentists in the country. The one general hospital on Dominica is in Roseau; two cottage hospitals of very limited resources are located on the Atlantic coast to the north and south of the capital. There are 12 health centers scattered across the island. Serious tropical diseases such as yaws and malaria have been eradicated, but owing to the high humidity and rainy conditions, tuberculosis and other respiratory diseases continue to be a problem. Malnutrition and intestinal parasites afflict particularly those in the early years of life. In 1984, the death rate was 5.2 per 1,000 population. In 1992, the infant mortality rate was 18 per 1,000 live births. Average life expectancy in 1992 was 73 years. Total 1992 population was 72,000 with 1,500 births that year. From 1991 to 1992, 98% of the country's children were immunized against measles.

42HOUSING

Hurricane David destroyed the homes of over four-fifths of the population. Under an emergency housing program, construction supplies were brought into the island, and shelters were built for most of the population. In 1981, 80% of all dwellings were detached houses and 13% were apartments. Over 66% were owner occupied, 22% were rented privately, 9% were occupied rent free, and under 1% were rented from the government. Some 62% of housing units were wooden, 20% were concrete, and 14% were wood and concrete.

43EDUCATION

Much of Dominica's school system had to be rebuilt after the 1979 hurricane. In 1991 there were 65 primary schools, with 605 teachers and 12,120 students enrolled. In the general secondary schools, there were 5,983 students enrolled. Transportation to secondary schools is a problem for students in rural areas. Higher educational facilities include a teacher training institute, a technical college, a nursing school, and a local center maintained by the University of the West Indies. In higher level institutions, there were 40 teaching staff and 658 students in 1991.

44LIBRARIES AND MUSEUMS

A library system maintained by the government includes branches in five of the leading villages; there is also a mobile library unit for rural areas. Before Hurricane David, the main government library contained one of the best collections of Caribbeana in the Lesser Antilles. Those libraries damaged by the hurricane are rebuilding.

45MEDIA

In 1991 there were 42,000 radios and 5,000 television sets. Two weekly newspapers are published in Roseau, the *New Chronicle* (circulation in 1991, 3,000) and the *Official Gazette* (600). Dominica had 14,613 digital telephone lines in operation in 1993. There are 3 AM and 2 FM radio stations and one (cable) television station.

46ORGANIZATIONS

There are many cooperatives and credit unions in Dominica. A chamber of commerce and a small Rotary Club also function.

[47]TOURISM, TRAVEL, AND RECREATION

Dominica's tourist industry is expanding gradually with government support. However, development of tourism has been slow compared with that on neighboring islands, but Dominica expects to benefit from the growth of eco-tourism. Its principal attraction is the rugged natural beauty of its volcanic peaks, forests, lakes, and waterfalls, and over 365 rivers. Day trips to Dominica from Barbados, Guadeloupe, and Martinique have gained increasing popularity. In 1991, 46,312 tourists arrived in Dominica, 26,357 from Caribbean countries, followed by 6,898 from the US and 4,520 from the UK. Revenues from tourism reached $28 million. Hotel rooms numbered 586,000. Cricket is the national sport.

[48]FAMOUS DOMINICANS

Maria Eugenia Charles (b.1919), cofounder of the Dominica Freedom Party, became prime minister in 1980.

[49]DEPENDENCIES

Dominica has no territories or colonies.

[50]BIBLIOGRAPHY

Baker, Patrick L. *Centring the Periphery: Chaos, Order, and the Ethnohistory of Dominica.* Montreal: McGill-Queen's University Press, 1994.

Caribbean and Central American Databook, 1987. Washington, D.C.: Caribbean/Central American Action, 1986.

Investing in Dominica. Washington. D.C.: Caribbean/Central American Action, 1986.

Myers, Robert A. *Dominica.* Santa Barbara, Calif.: Clio Press, 1987.

Trouillot, Michel-Rolph. *Peasants and Capital: Dominica in the World Economy.* Baltimore: Johns Hopkins University Press, 1988.

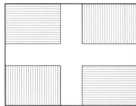

DOMINICAN REPUBLIC

Dominican Republic
Republica Dominicana

CAPITAL: Santo Domingo.

FLAG: The national flag, adopted in 1844, consists of a white cross superimposed on a field of four rectangles, the upper left and lower right in blue, the upper right and lower left in red.

ANTHEM: *Himno Nacional,* beginning "Quisqueyanos valientes, alcemos nuestro canto" ("Valiant Dominicans, let us raise our song").

MONETARY UNIT: The Dominican peso (RD$) of 100 centavos is a paper currency. There are coins of 1, 5, 10, 25, and 50 centavos and 1 peso, and notes of 1, 5, 10, 20, 50, 100, 500, and 1,000 pesos. RD$1 = US$0.0766 (US$1 = RD$13.05).

WEIGHTS AND MEASURES: The metric system is the legal standard, but US and Spanish weights are widely used in commercial transactions.

HOLIDAYS: New Year's Day, 1 January; Epiphany, 6 January; Altagracia Day, 21 January; Duarte Day, 26 January; Independence Day, 27 February; Labor Day, 1 May; Restoration of Independence, 16 August; Day of Our Lady of Las Mercedes, 24 September; All Saints' Day, 1 November; Christmas, 25 December. Movable religious holidays include Good Friday and Corpus Christi.

TIME: Eastern Daylight Savings Time is maintained throughout the year; 8 AM = noon GMT.

¹LOCATION, SIZE, AND EXTENT

The Dominican Republic occupies the eastern two-thirds of the island of Hispaniola (Española) and includes the islands of Beata, Catalina, Saona, Alto Velo, and Catalinita in the Caribbean Sea, and several islets in the Atlantic Ocean. It has an area of 48,730 sq km (18,815 sq mi), with a length of 386 km (240 mi) E–W, extending from Cape Engaño to the Haitian border, and a width of 261 km (162 mi) N–S, extending from Cape Isabela to Cape Beata. Comparatively, the area occupied by the Dominican Republic is slightly more than twice the size of the state of New Hampshire. Bounded on the N by the Atlantic Ocean, on the E by the Mona Passage (which separates it from Puerto Rico), on the S by the Caribbean Sea, and on the W by Haiti, the Dominican Republic has a total boundary length of 1,563 km (971 mi).

The Dominican Republic's capital city, Santo Domingo, is located on its southern coast.

²TOPOGRAPHY

The Dominican Republic is generally mountainous, with deserts in the extreme western regions. The principal mountain range, the Central Cordillera, running from east to west and extensively pine-forested, bisects the republic. Between the Central Cordillera and the Northern Cordillera (and their associated plains) lies the famous Cibao (La Vega Real) Valley (225 km/140 mi long, with an average width of 23 km/14 mi), noted for the excellent quality of its soil. Fertile valleys also abound in the central and eastern areas. The country contains both the highest mountain in the West Indies, Mt. Duarte (3,175 m/10,417 ft), and the lowest-lying lake, Lake Enriquillo (40 m/131 ft below sea level). The Yaque del Norte, the Yaque del Sur, and the Yuna are the principal rivers.

³CLIMATE

Climate and rainfall vary with region and altitude. Generally, however, average minimum and maximum temperatures range

from 18° to 29°C (64° to 84°F) in the winter and from 23° to 35°C (73° to 95°F) in the summer. The coastal plain has an annual mean temperature of 26°C (79°F), while in the Central Cordillera the climate is temperate and the mean is 20°C (68°F). Rainfall varies from an annual average of 135 cm (53 in) in the eastern regions, with an extreme of 208 cm (82 in) in the northeast, to a mean of 43 cm (17 in) in the western areas. The rainy season generally extends from May to November and the dry season from December to April. The nation lies within the hurricane belt, and tropical storms constitute a major weather hazard.

⁴FLORA AND FAUNA

Plants and animal life vary by region. Dense rain forests are common in the wetter areas; scrub woodland thrives along the drier slopes; and savanna vegetation is found on the open plains. Dominican mahogany and highly resinous pine trees grow in the high mountains. The rare hutia (a small rodent) and herds of wild boar are found in the mountainous areas. Ducks, doves, and several varieties of pigeons are seasonal visitors. Lake Enriquillo is the natural habitat of large flocks of flamingos. Spanish mackerel, mullet, bonito, and yellowtail snapper are found in the surrounding waters.

⁵ENVIRONMENT

The main agencies responsible for environmental protection are the Department of National Parks and the Department of State for Agriculture. The Dominican Republic has environmental problems in the areas of deforestation, water supply, and soil erosion. UN sources report that, as of 1993, the nation was losing 20,000 hectares per year of its forest lands largely due to commercial interests. The problem of deforestation is linked to the nation's decreasing water supply since forests help to stop the loss of surface water. The felling of trees was prohibited in 1967 to remedy the ill effects of indiscriminate cutting by commercial producers and farmers and the destruction by fire of large stands of timber. However, many farmers continue to cut trees surreptitiously to make more land available for cultivation. Soil erosion results from a combination of rainfall and the use of land in mountainous areas. As of 1994, the Dominican Republic has an inadequate water supply. The country has 4.8 cubic miles of water with 89% used for farming. Eighteen percent of all urban dwellers and 55% of the rural people have pure water. Water pollution results from the effects of mining along with industrial and agricultural sources.

As of 1987, endangered species in the Dominican Republic included the tundra peregrine falcon, Haitian solenodon, three species of sea turtle (green sea, hawksbill, and leatherback), and American crocodile. In 1994, one of the country's mammal species and five bird species are considered endangered. Four types of reptiles are also threatened. As of 1993, 62 plant species in a total of 1,600 were threatened with extinction.

6POPULATION

According to the 1981 census, the population was 5,647,977. The population in mid-1991 was estimated at 7,313,000. For the year 2000, a population of 8,621,000 was projected, assuming a crude birthrate of 25.2 per 1,000 population, a crude death rate of 5.8, and a net natural increase of 19.4 during 1995–2000. The estimated population density in 1991 was 150 per sq km (389 per sq mi). In 1995, an estimated 65% of the population was urban. Santo Domingo, the capital and largest city, had an estimated population of 2,203,000 in 1990. Other important cities are Santiago de los Caballeros, La Romana, San Pedro de Macorís, San Francisco de Macorís, and Concepción de la Vega.

7MIGRATION

Four overseas immigration movements have taken place in recent decades: approximately 5,000 refugees from the Spanish Civil War in the late 1930s; a Jewish refugee group, which arrived in 1940; a continuous flow of Japanese, mainly farmers, since 1950; and 600 Hungarian refugees invited by the government in 1957. These are dwarfed, however, by the influx of Haitians, some seasonal, others permanent. The number of Haitians living in the Dominican Republic has been estimated at 500,000 to 1,000,000.

Emigration became significant for the first time during the 1960s, when 93,300 Dominicans legally entered the US; during 1971–80, the figure rose to 148,100, and during 1981–85, it was 104,800. In 1990 there were 357,000 Dominican-born people living in the US, mostly along the eastern seaboard. This census total may have been an undercount, for estimates of the Dominican population in the US ranged as high as 1,000,000, including 200,000 in Puerto Rico.

8ETHNIC GROUPS

Ethnic divisions were estimated in 1986 at 16% white, 11% black, and 73% mulatto. Descendants of early Spanish settlers and of black slaves from West Africa constitute the two main racial strains.

9LANGUAGES

Spanish is the official language. Some English is spoken in the capital, and a Creole dialect is used along the Haitian border.

10RELIGIONS

Religious freedom is protected by law. Roman Catholicism is the state religion, in keeping with the Concordat of 1884; the Dominican archbishop is referred to as the Primate of the Indies. In 1993, professing Roman Catholics represented an estimated 91.2% of the population. Protestants, including Baptists and Seventh-day Adventists, accounted for another 1%. Prior to and during World War II, some Jewish refugees settled in the Republic, but by 1990 there were only about 100 Jews left in the country.

Followers of spiritism and voodoo numbered about 60,000 in 1984.

11TRANSPORTATION

The national highway system is the dominant means of inland public transportation. Three main highways emanate from Santo Domingo: the Carretera Sánchez (connecting with Elias Piña on the Haitian border), the Carretera Mella (to Higüey in the extreme southeast), and the Carretera Duarte (to Monte Cristi on the northwest coast). In 1991 there were 5,800 km (3,600 mi) of paved roads and 5,600 km (3,500 mi) of unpaved roads. In 1991, 139,069 passenger cars, and 102,969 commercial vehicles were licensed. During the same year, 1,655 km (1,028 mi) of railways in four different gauges were in service.

The Santo Domingo, Andrés, and Haina harbors, all in the Santo Domingo area, handle the vast majority of imports. Other large ports include Puerto Plata in the northwest; La Romana, Boca Chica, and San Pedro de Macorís in the southeast; and Barahona in the southwest. The Dominican merchant fleet had one ship of 2,000 GRT in 1991.

Dominicana de Aviación provides international service from Las Americas International Airport at Punta Caucedo, 29 km (18 mi) east of Santo Domingo. Cargo and mail service to the US mainland, Puerto Rico, and the US Virgin Islands is provided by Aerolíneas Argo. Alas del Caribe provides domestic passenger service, as does Aerovías Quisqueyanas. There are also international airports at Puerto Plata and La Romana.

12HISTORY

The eastern part of the island of Hispaniola was originally known as Quisqueya, meaning "mother of all lands." It was first settled by the nomadic and warlike Carib Amerindians and later by the agricultural and peace-loving Arawaks. Christopher Columbus made the European discovery of the island and claimed it for Spain in 1492. Santo Domingo, the oldest city in the New World, was founded four years later by Bartholomew Columbus, the explorer's brother. By 1517, Hispaniola had become the springboard for Spanish conquest of the Caribbean and of the American mainland. As with other Caribbean islands, the Amerindian population dwindled, and were replaced by African slaves.

The importance of Hispaniola waned during the 16th and 17th centuries. In 1697, by the Treaty of Ryswick, Spain was forced to recognize French dominion over the western third of the island, an area now known as Haiti. In 1795, under the Treaty of Basel, Spain ceded to France the eastern two-thirds of the island, which by then had been renamed Santo Domingo. The island then came under the rule of the rebellious ex-slave Toussaint L'Ouverture. After Haiti received independence in 1804, the French retained the rest of the island until 1809. After a brief attempt at independence, the Dominicans fell under the control of Spain, which regained the eastern section of the island under the Treaty of Paris (1814).

In 1821, the Dominicans, led by José Múñez de Cáceres, proclaimed their independence. The Dominicans sought to become part of Simón Bolívar's newly-independent Republic of Gran Colombia, but in 1822, the Haitians conquered the entire island. For 22 years, the Haitians ruled with an iron fist. A civil war in 1843 gave the Dominicans the opportunity to try again for independence, and, under Juan Pablo Duarte, they established the Dominican Republic as an independent state.

Between 1844 and 1916, the new republic alternated among personalist leaders, who sought foreign protection against Haiti. The most prominent among these were Pedro Santana and Buenaventura Baez, who dominated Dominican politics until 1882. Santana restored the Dominican Republic to the Spanish Empire during 1861–65. Baez in 1869 negotiated a treaty providing for US annexation, but the US Senate refused to ratify it.

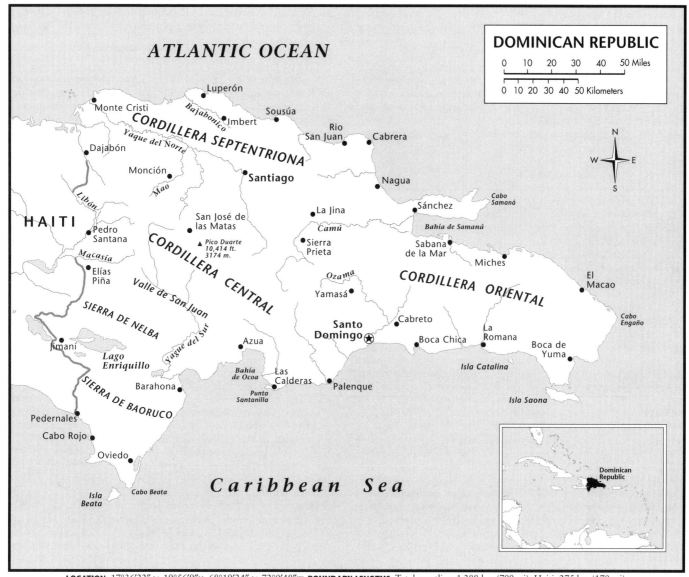

LOCATION: 17°36′22″ to 19°56′9″N; 68°19′24″ to 72°0′48″W. **BOUNDARY LENGTHS:** Total coastline, 1,288 km (799 mi); Haiti, 275 km (170 mi). **TERRITORIAL SEA LIMIT:** 6 mi.

After a 17-year dictatorship, the Dominican Republic entered a turbulent period characterized by general political instability and increasing debt to US interests. In 1905, US President Theodore Roosevelt appointed an American receiver of Dominican customs, and a subsequent treaty provided for repayment of the debt. This first application of the "Roosevelt Corollary" to the Monroe Doctrine was followed in 1916 by the US establishment of a military government under Marine Capt. H. S. Knapp. It ruled the Dominican Republic until 12 July 1924, when sovereignty was restored. US customs control continued until 1941.

In 1930, Rafael Leonidas Trujillo Molina was elected president. For the next 31 years, he ruled the Dominican Republic either directly or indirectly. Trujillo had himself reelected in 1934, 1940, and 1947, then arranged for his brother to become president in 1952 and 1957, and then installed Joaquín Balaguer in 1960. Under Trujillo, the Dominican Republic achieved some economic progress, removing its foreign debt. But Trujillo brutally suppressed fundamental human rights. Only one party was allowed, the press was totally controlled, and constant purges weeded out all but his most servile supporters. His most dubious achievements were a result of his own megalomania. In the capital

city of Santo Domingo, which he renamed Ciudad Trujillo (Trujillo City), there were 1,870 monuments to Trujillo, who gave himself the title "The Benefactor of the Fatherland." He also amassed a personal fortune estimated at $900–$1,500 million.

Trujillo was assassinated on 30 May 1961. In July 1961, the Balaguer cabinet resigned as street rioting broke out. The opposition agreed to an interim coalition government under Balaguer that September, with Rafael Trujillo, Jr. as head of the armed forces. In November, the US sent warships just outside Dominican waters to prevent the armed forces chief and his two uncles from a staging a rumored coup. They went into exile immediately.

After rule by an interim Council of State, Juan Bosch of the Dominican Revolutionary Party was elected president in 1962. He assumed office in February 1963 but remained in power for only seven months, during which time very little of the promised social and economic reform could be accomplished. The military, which overthrew Bosch in September 1963, proceeded to install a three-man civilian junta, called the Triumvirate, which was in turn overthrown by the supporters of Juan Bosch in April 1965. With anarchy threatening—and, according to US allegations, Communists deeply involved in the pro-Bosch insurrection—the

US sent 23,000 troops into the Dominican Republic, ostensibly to protect the lives of US citizens. Within weeks, the OAS had set up an Inter-America Peace Force in the country. This controversial set of arrangements eventually brought order to the country. After the 1966 elections, all US and OAS troops left the island.

In the general elections of June 1966, Balaguer, returned from exile, campaigned vigorously throughout the country, while Bosch, his chief opponent, remained home, apparently fearful of an attempt on his life. Balaguer won with 57.2% of the vote. Four years later, Balaguer ran essentially without opposition, as most parties withdrew from the campaign in response to rising political violence.

Events in 1974 followed a similar pattern. Two days before the election, an opposition coalition announced their withdrawal from the contest. The opposition's principal candidate, Silvestre Antonio Guzmán Fernández of the PRD (Dominican Revolutionary Party), called for general abstention from the election, charging Balaguer with fraudulent practices. Only one candidate—Adm. Luis Homero Lajara Burgos—a former chief of police under Trujillo—remained in the race. The results gave Balaguer an overwhelming majority, with 924,779 votes to Lajara Burgos's 105,320.

Following the 1974 elections, the opposition coalition, an amalgam of widely divergent political elements, decided to disperse in anticipation of the 1978 campaign. Despite their lack of representation in the legislature and in municipal councils as a result of the election boycott, the major opposition parties, aided by a large degree of press freedom, remained active and vocal. In the 1978 elections, the main presidential candidates were Balaguer and the PRD's Guzmán. Guzmán won with a 158,000-vote plurality, and his party gained a majority in the Chamber of Deputies. Numerous other parties, including the newly-legalized Dominican Communist Party, participated in the elections. A right-wing military attempt to prevent Guzmán from assuming office was foiled, partly because of US government pressure.

During Guzmán's term, political prisoners were freed, press censorship was practically abolished, and political parties engaged in open activity. At the same time, however, there were mounting economic difficulties, aggravated by two hurricanes in 1979, which together left 1,300 people dead, 500 missing, and 100,000 homeless. In May 1982, a left-wing PRD senator, Salvador Jorge Blanco, was elected president. On 4 July, six weeks before his term was due to expire, Guzmán committed suicide, after several close associates were accused of fraud. Power was transferred peacefully to Vice-President Jacobo Majluta Azar, and, in August 1982, to Blanco.

Whatever plans President Blanco may have had were soon overtaken by the country's burgeoning foreign debt. Blanco turned to the IMF for assistance. The resultant restrictive economic policy raised production costs and reduced industrial output. A cut in the US sugar quota, as well as a generally low world price for the commodity, was a further blow to the economy.

In presidential elections held 16 May 1986, former president Joaquín Balaguer was returned to office with 857,942 votes (41.6%). Jacob Majlota Azar of the PRD took 814,716 votes (39.5%), and Juan Bosch of the PLD (Dominican Liberation Party) won 379,269 votes (18.4%). Balaguer embarked on an ambitious program of public works that created employment for nearly 100,000 people. But by 1988, inflation was on the rise, and the peso had become unstable. Nationwide strikes in 1989 suggested that the country was headed for further crisis.

In 1990, Balaguer stood for re-election, and won a narrow, hotly-contested victory amid claims of fraud by the opposition. Officially, Balaguer received 35.7% of the vote to Juan Bosch's 34.4%. Balaguer was nevertheless inaugurated, but immediately created controversy by suggesting a set of IMF-style liberal reforms. This brought national strikes in August and November of 1990, and demands for resignation. In the Chamber of Deputies, where Bosch's PLD held a plurality of seats, Balaguer faced serious opposition.

13 GOVERNMENT

The constitution of 28 November 1966 established a unitary republic consisting of 26 provinces and a single National District. The government is effectively controlled by the chief executive, a president directly elected for a four-year term and eligible for reelection. Cabinet ministers (called secretaries of state) are appointed by the president, who must be at least 30 years of age.

The National Congress consists of a Senate, composed of 30 members, and a 120-member Chamber of Deputies, apportioned on the basis of population. Members of the two houses are elected for four-year terms and must be at least 25 years of age. Bills for legislative action may be introduced by the president, by the National Congress, or by the Supreme Court. Voting is by universal suffrage of citizens 18 years or older, although younger citizens, if married, may also vote. Presidential and congressional elections are held simultaneously.

14 POLITICAL PARTIES

The Dominican Republic's 4 major parties and more than 10 minor parties mask the underlying reality of the political system, which is that two men, Joaquín Balaguer Ricardo and Juan Bosch Gavino, have dominated the system.

Joaquín Balaguer, a former president under Trujillo, has been elected to the presidency five times. He founded the Social Christian Reform Party (Partido Reformista Social Christiano—PRSC) while living in exile in New York in 1963. The party is tied to the Christian Democratic political movement, and relies principally on peasant and middle-class support. Currently, it holds 16 of 30 Senate seats and 41 of 120 seats in the Chamber of Deputies.

Juan Bosch, who held the presidency for seven months in 1963 before the Dominican crisis, remains a major voice in Dominican politics. Bosch founded the Dominican Revolutionary Party (Partido Revolucionario Dominicano—PRD) in 1939. After withdrawing from the PRD, Bosch created his own party, the Dominican Liberation Party (Partido de la Liberación Dominicana—PLD) in 1973. Largely subject to Bosch's ideological whims, the party gained 12 seats in the Senate and 44 seats in the Chamber of Deputies.

The PRD continues on without Bosch, and with far more success than Bosch was ever able to achieve. The PRD won the presidential elections of 1978 and 1982, although it was unable to achieve a majority in either house of Congress in 1982. Now headed by Jose Francisco Peña Gomez, the PRD has an association with the Socialist International, and has a "Eurosocialist" ideological thrust of moderate economic and social change. The PRD draws support from landless peasants and urban workers. Pena Gomez received 23% of the vote in 1990 and won 33 seats in the Chamber of Deputies, but won only 2 Senate seats.

The Independent Revolutionary Party (PRI) is the vehicle for Jacobo Majluta Azar, who served as president briefly in 1982. A former PRD member, Majluta received only 7% of the vote in 1990, and the party holds only two seats in the Chamber of Deputies and none in the Senate.

15 LOCAL GOVERNMENT

The Dominican Republic is divided into 29 provinces and a National District, encompassing Santo Domingo. The provinces are further subdivided into municipal districts. The president appoints the provincial governors. Municipal districts must have at least 5,000 inhabitants and produce sufficient revenue to finance their own administrative agencies. The municipalities and the National District are governed by mayors and municipal councils of at least five members, elected by popular vote.

¹⁶JUDICIAL SYSTEM

The judicial system is headed by a Supreme Court with nine judges, which rules on constitutional questions and serves in the last instance on appeals. Supreme Court judges are elected by the Senate and cannot be removed from office. The Supreme Court appoints the judges of the lower courts and of the special courts. All judges are required to hold a law degree.

There are 26 provincial courts, as well as one court of first instance in the National District. The judicial system also includes one judge or court of justice for each of the country's 72 municipal districts, three courts of appeal, a court of accounts, and a land tribunal. There are also justices of the peace. The death penalty was abolished in 1924. Although the constitution provides for an independent judiciary, in practice the executive branch as well as public and private entities exert pressures on the courts. The constitution guarantees public trials, and indigent defendants have a right to a court-appointed attorney at state expense.

¹⁷ARMED FORCES

The army of 15,000 is organized into 4 infantry brigades and 4 independent battalions. The air force has 4,200 personnel and is equipped with 10 combat aircraft of US and UK origin. The navy of 3,000 mans 18 coast defense ships and 2 auxiliaries. The national police numbers 15,000, including 1,000 special operations officers. In 1991, the armed forces were allocated $35 million or about 1% of GDP.

¹⁸INTERNATIONAL COOPERATION

The Dominican Republic is a charter member of the UN, having joined on 24 October 1945, and participates in ECLAC and all the nonregional specialized agencies except WIPO. It also belongs to G-77, IDB, OAS, and PAHO, as well as the International Bauxite Association and the International Coffee Organization. The Dominican Republic is a signatory to GATT and the Law of the Sea. The nation's application to join CARICOM was blocked by Grenada in July 1983.

¹⁹ECONOMY

The economy is based primarily on agriculture, which accounts for more than 16% of the GDP, nearly half of the labor force, and a major share of total exports. By the mid-1980s, however, tourism had overtaken sugar as the country's main source of foreign exchange. Mining produces the significant nonagricultural export items. The introduction of light and medium industries since the early 1950s has increased employment and reduced some imports. The government encourages investment in industry, seeking to diversify the economy and reduce the nation's dependence on sugar production.

Between 1968 and 1974, the average annual growth of the GDP was 10.5%. The improved political climate stimulated public and private investment, both domestic and foreign, and the development of tourism. A sugar boom also contributed to rapid economic growth. During the second half of the 1970s, the growth rate slowed, in part because of rising oil prices and a weakening of the sugar market; damage from two 1979 hurricanes cost an estimated $1 billion to repair. The GDP declined by 2.2% in 1985, reflecting low world prices for the country's exports, declining US sugar quotas, and IMF-imposed austerity. Unemployment soared to 26% (officially), and inflation reached 37.5% the same year. The economy recovered somewhat in 1986 and 1987, however, thanks in part to President Balaguer's capital spending program and a boom in foreign investment.

As of 1994, agriculture and mining are still the mainstay of the economy, despite intensive efforts to develop tourism and light manufacturing in the free trade zones. The Dominican Republic has benefitted immensely from the Caribbean Basin Initiative (CBI). It has been undergoing a dynamic and fast-paced modern-

ization. The development of tourism, free trade zones, and nontraditional agricultural exports received a strong boost from the CBI. With more world-class hotels than any other Caribbean nation, the country was estimated to have attracted 2 million visitors and to have generated revenues of approximately $1.3 billion by 1993. Agriculture achieved a moderate growth of 5.2% in 1992, aided by lower prices for seeds and fertilizer. However, the traditional crops of sugar, coffee, and tobacco registered a dismal performance, due to lower prices for those products in the world markets.

Light manufacturing is contributing an ever more important share of national output, exports, and employment.

Overall economic performance was positive in 1992; real GDP increased by 7.7%, reversing the negative trend in the two previous year. The high inflation prevailing in the late 1980s and early 1990s was finally tamed in 1993, as a result of a dramatic austerity program that brought the annual inflation rate from 53.9% in 1992 to only 4.6% in 1993.

²⁰INCOME

In 1992, the Dominican Republic's GNP was $7,611 million at current prices, or $1,040 per capita. For the period 1985–92 the average inflation rate was 34.4%, resulting in a real growth rate in per capita GNP of 0.3%.

In 1992, the GDP was $7,729 million in current US dollars. It is estimated that in 1991, agriculture, hunting, forestry, and fishing contributed 15% to GDP; mining and quarrying, 3%; manufacturing, 16%; electricity, gas, and water, 2%; construction, 7%; wholesale and retail trade, 15%; transport, storage, and communication, 9%; finance, insurance, real estate, and business services, 13%; community, social, and personal services, 10%; and other sources, 10%.

²¹LABOR

The labor force in 1992 consisted of about 3.24 million persons. In 1989, 18% were engaged in agriculture, 49% in industry, and 33% in services. Unemployment is a persistent problem, affecting an estimated 30% of the work force in 1992. Underemployment was estimated at another 20%.

In 1992, only 10–15% of the Dominican work force was unionized. The central union amalgamation is the National Confederation of Dominican Workers, with a combined membership of over 100,000; the Unitary Workers Confederation also had some 100,000 members. The new 1992 labor code grants employees in non-essential public services the right to strike and eliminates the prohibition against general, political, or sympathy strikes. The code also specifies steps for union registration, entering into collective bargaining pacts, and calling strikes.

²²AGRICULTURE

With almost 30% of the total land area suitable for crop production and almost 40% of the labor force engaged in farming, agriculture remains the primary occupation, accounting for 10% of GDP. Value of agricultural output grew at an average annual rate of 7.1% during 1968–73, but since 1975 the sector has been hampered by droughts (1975, 1977, and 1979), hurricanes (in 1979 and 1980), and slumping world prices and quota allocations for sugar (since 1985). While growth rates of 2–5% were recorded in the early 1980s, 1984–86 saw declines of 2–3%. The fertile Cibao Valley is the main agricultural center. In 1992, arable land totaled 1,000,000 hectares (2,471,000 acres); with land under permanent crops at 446,000 hectares (1,103,000 acres).

After Cuba, the Dominican Republic is the second-largest Caribbean producer of sugarcane, the nation's most important commercial crop. The State Sugar Council (Consejo Estatal de Azúcar—CEA) operates 12 sugar mills and accounts for more than half of total production. Other large producers are the privately owned Vicini, with three mills, and Gulf & Western, whose largest mill is at La Romana. In 1992, sugar cane production was 678 million tons, up from 638.5 million tons in 1990. Output of sugar has declined annually since 1982, and land is being taken out of sugar production and switched to food crops—in 1985, for example, 12,500 hectares (30,900 acres) belonging to the CEA were thus transformed.

Another leading cash crop is coffee. Part of the crop was destroyed by hurricanes in 1979 and 1980, and 1979/80 production was only 670,000 bags (40,200 tons). Although production has averaged at about 57,000–59,000 tons annually since the early 1980s, the acreage harvested has declined from 157,000 hectares (388,000 acres) in the early 1980s to 103,000 hectares (255,000 acres) in 1990, indicating a greater yield per acre. Cocoa and tobacco are also grown for export. In 1992, production of cocoa was 46,000 tons and of tobacco, 23,000 tons. Production of other crops in that year (in thousands of tons) included rice paddy, 514; tomatoes, 145; beans, 47; and cotton, 2.

Under a land reform program initiated in 1962, a total of 178,602 hectares (441,333 acres) had been distributed to 36,480 farmers by the end of 1977. The government encourages fuller use of the nation's arable land through extensive land-clearing and irrigation projects and diversification of crops. Some mechanization has taken place on the large plantations, but primitive techniques are generally used. In 1973, the first stage of the Integrated Agricultural and Livestock Development Plan was initiated, calling for an investment of RD$38.1 million, to be financed by the IDB. The plan was designed to provide credit and technical aid to 45,000 small farmers, improve side roads, and study the country's water resources. The second stage of the plan, in the early 1980s, included extension of farm credits, reforestation, manpower training for irrigation projects, and reorganization of the Dominican Agrarian Institute. In 1992, the agricultural sector grew by 5% due in part to increased financing from the Agricultural Bank and the elimination of taxes on inputs, farm equipment, and machinery, so that outputs of rice, root crops, bananas, and plantains grew. However, external factors, such as falling international market prices, coupled with heavy rainfalls and poor plant health caused production levels of sugar, coffee, tobacco, and cotton to fall from their 1991 levels.

23ANIMAL HUSBANDRY

In 1992, Dominican livestock included 560,000 goats and 122,000 sheep. The hog population was decimated by African swine fever, decreasing from 400,000 in 1978 to 20,000 in 1979; by 1992, however, it had risen to 750,000. In 1992, 44,000 tons of beef and about 350,000 tons of milk were produced.

24FISHING

Although the waters surrounding the Dominican Republic abound with fish, the fishing industry is comparatively undeveloped, and fish for local consumption are imported. In 1991, the total marine catch was 16,098 tons, up from 13,169 tons in 1984. Marlin, barracuda, kingfish, mackerel, tuna, sailfish, and tarpon are found in the Monte Cristi Bank and Samaná Bay, which also supports bonito, snapper, and American grouper.

25FORESTRY

About 13% of the total land area consisted of forests and woodlands in 1991. Virtually all the timber cut is for land clearing and fuel.

26MINING

The Dominican Republic has a variety of mineral resources, and mining was long an important asset to the economy. Production, however, has stagnated since a slump began in the mid-1980s.

The Aluminum Co. of America (Alcoa) mined bauxite, traditionally the principal product of the mining sector, between 1959 and 1983, when it turned its concession over to the state; it ceased operations in 1985. Production of bauxite in 1991 dropped to 7,000 tons, down 92% from the previous year. In 1990, Alcoa agreed to continue purchasing bauxite. In 1991, mining operations were suspended at the largest mine by presidential decree, in response to increasing fears of deforestation, even though reforestation of mined areas was in progress. Ferronickel is another important mining product. About 29,062 tons were mined in 1991.

A new gold and silver mine and a processing plant were opened at Pueblo Viejo, on the north coast, in 1975. The plant, initially owned by New York and Honduras Rosario Mining, was nationalized in October 1979. By 1980, the mine had become the largest gold producer in the Western Hemisphere. Production was declining by the mid-1980s, so mining of the sulfide zone of the gold ore body has commenced; such mining, however, requires more extensive processing facilities than currently exist. Production of gold was 3,160 kg in 1991 (down from 7,651 kg in 1987) and of silver, 21,954 kg (down from 39,595 kg in 1988).

The country is one of the few sources of amber in the Western Hemisphere. Salt Mountain, a 16-km (10-mi) block of almost solid salt about 37 km (23 mi) west of Barahona, is the largest known salt deposit in the world. Extraction of salt and gypsum (of which there are large deposits near Salt Mountain) is carried on by the state, which was negotiating in 1991 with the US firm CIE for expansion of the rock salt mining industry, potentially selling its salt in North America for roadway snow and ice removal. There is also some iron mining, and substantial lignite deposits were found in the early 1980s.

27ENERGY AND POWER

With no coal and little petroleum, the country has depended upon imported diesel oil for its electrical energy, although a substantial development of hydroelectric potential took place during the 1970s and early 1980s. Net installed public capacity was 1,447,000 kw in 1991, when 5.3 billion kwh were produced. The Dominican Electric Corp. (Corporación Dominicana de Electricidad), purchased for us$15 million from US interests in 1972, is responsible for all public production, sale, and distribution of energy. Private sources, such as Falconbridge Dominicana, sell excess electricity into the grid.

Steam thermal units supply about 72% of the nation's energy needs; hydroelectric power is used mainly for irrigation. The Tavera Dam, with a capacity of 60,000 kw, was completed in 1972. The Sabana Yegua hydroelectric and irrigation complex, begun in 1974, opened in 1980. The Habo I coal-fired plant at Haina was inaugurated in 1984.

Under the terms of an agreement with Venezuela and Mexico, the Dominican Republic buys one-third of its oil at discount prices. A petroleum deposit in Haina had an estimated potential output of 30,000 barrels per day in 1991.

28INDUSTRY

Manufacturing, limited largely to the processing of agricultural and mineral commodities, accounted for 17.1% of the GDP in 1985. The industrial sector experienced an average annual growth rate of 8.3% during 1970–80, which slowed to 2% for 1982–83 and registered declines of 3% and 4.1% in 1984 and 1985, respectively.

Including the processing of sugar, food processing represents more than half of total industrial production. Smaller plants, directed principally toward the local market, produce flour, textiles, powdered and condensed milk, ceramics, aluminum furniture and fittings, concrete blocks, earthenware pipes and tiles, air conditioners, barbed wire, and other products. The Dominican

Petroleum Refinery, 50% owned by Shell and 50% state-owned, was opened at Haina in early 1973, with a capacity of 32,000 barrels per day. Industrial production in 1985 included 997,050 tons of cement, 466,426 tons of wheat flour and derivatives, 89,119 tons of refined sugar, 71,965 tons of husked coffee, 3,925 million cigarettes, and 103.8 million liters of beer.

Many industrial firms are government-owned or have government capital investment. Most of the industrial enterprises that were confiscated from the Trujillo family were turned over to the Dominican Corporation of Estate Enterprises. The Dominican Development Foundation was established in 1962 to act as a development bank and to aid diversification and the general development of Dominican industry.

The assembly (*maquiladora*) sector continued to grow through the mid-1980s: 21 new firms opened in 1985, generating 5,000 new jobs. The government, assisted by funding from AID, expanded the industrial parks created for these industries. There were five in 1985: Puerto Plata, San Pedro de Macorís, La Romana, Santiago de los Caballeros, and Baní; a sixth was under construction at Itabo.

In 1992, construction was the star performer with an expansion of 24.7% following a dismal fall of 11.3% the year before. Much of this growth was fueled by government spending on infrastructure improvements. With the budget in surplus, the government found some margin to increase expenditures on public works. However, private construction remained severely hampered by high interest rates.

Manufacturing, which had been falling marginally in 1991, soared by 12.3% in 1993. The most important contribution to growth came from improved electricity supply which reduced blackouts and buoyed production.

On the other hand, the mining sector contracted 19% in 1992, with only a modest recovery in 1993. Declining international prices led to a cutback in ferronickel production, while the difficulties of the state gold mining company—caused by heavy borrowing, rising production costs, and the lack of funds for new investments—contributed to the poor showing of the mining sector.

29SCIENCE AND TECHNOLOGY

The Dominican Institute of Industrial Technology is part of the Central Bank. The Dominican Medical Association and the Dominican Sugar Institute have their headquarters in Santo Domingo. Nine colleges and universities offer degrees in engineering, basic sciences, medicine, and agriculture.

30DOMESTIC TRADE

Santo Domingo is the principal port and commercial center, while Santiago de los Caballeros is the market and distribution center for the Cibao Valley. Most importing and exporting firms are located in Santo Domingo. Importers ordinarily represent numerous foreign manufacturers.

Department stores and supermarkets are increasing in number, but most retail stores are specialty shops. Groceries, meat, and fish are sold in most cities through a large central market, by neighborhood stores, and by street vendors. Retail credit is granted by larger stores and automobile dealers.

Usual business hours are from 8:30 AM to noon and from 2:30 to 6:30 PM, Monday through Friday, and from 8 AM to noon on Saturday. Government offices are open weekdays from 7:30 AM to 2:30 PM. Banks are open from 8:30 AM to 4:00 PM, Monday through Friday. Advertising media include television, radio, newspapers, magazines, and outdoor displays.

31FOREIGN TRADE

The country's trade balance is traditionally in deficit, although a favorable market for Dominican sugar and other commodities

produced a surplus in 1973. The country remains subject to fluctuating export revenues, since the government's efforts to free exports from the vicissitudes of the commodities market have as yet had little impact.

The years 1980–82 saw drastic declines in exports (–9.5% in 1981), caused mainly by the civil war and low world commodity prices, but 1983–86 was marked by modest growth (0.8%–2.3%). The average annual rate of inflation during 1970–80 was 11.3%; the rate jumped to 22.3% in 1985 and 32.0% in 1986.

The avowed aim of the leftist guerrillas had been to destroy as much infrastructure as possible. Between 1979 and 1986, the war was responsible for some $1.2 billion in accumulated losses. The economy was further weakened by the earthquake in October 1986, which caused an estimated $2 billion in damage.

32BALANCE OF PAYMENTS

For more than 20 years prior to 1961, the Dominican Republic had no internal or external debt. In 1961, however, about US$70 million was taken out of the country by the Trujillo family and others. Political instability induced further net outflows of private capital throughout the 1960s. The nation's current accounts position worsened during the 1970s, as trade deficits grew. Declines in export earnings in the 1980s have brought shortages of foreign exchange. Large increases in imports and a sharp drop in commodity exports pushed the merchandise trade deficit to over $1.6 billion by the end of 1992. Rising imports have been attributed to the lifting of trade restrictions and to increases in aggregate domestic demand caused by salary increases and the expansion of the money supply. Export earnings have dropped due to declining world commodity prices for nickel, sugar, coffee, and cocoa, along with serious production problems at the state-owned gold mine.

In 1992 merchandise exports totaled $566.1 million and imports $2,178.1 million. The merchandise trade balance was $–1,612.0 million.

The following table summarizes the Dominican Republic's balance of payments for 1991 and 1992 (in millions of US dollars):

	1991	1992
CURRENT ACCOUNT		
Goods, services, and income	–444.5	–824.5
Unrequited transfers	386.5	431.8
TOTALS	–58.0	–392.7
CAPITAL ACCOUNT		
Direct investment	145.0	179.0
Portfolio investment	—	—
Other long-term capital	–44.6	–51.5
Other short-term capital	–28.1	–85.2
Exceptional financing	39.4	–10.9
Other liabilities	—	—
Reserves	–341.5	–26.2
TOTALS	–229.8	5.2
Errors and omissions	287.8	387.5
Total change in reserves	–340.0	–46.0

33BANKING AND SECURITIES

The Central Bank of the Dominican Republic (Banco Central de la República Dominicana) is the sole bank of issue. As of 31 December 1993, it had gold assets of $6.9 million and total capital reserves valued at $63.4 million. As of 1993 there were 10 commercial banks, 12 development banks, and 3 foreign banks. The state-owned Banco de Reservas, established in 1941, is the largest commercial bank in the country and acts as the fiscal agent and depository for the government. The Agricultural and the Industrial Credit Bank promotes the development of agriculture as well as industry by granting medium-term and long-term credit. The National Housing Bank (Banco Nacional de Vivienda) is a primary investor in low-cost housing.

As of 31 December 1993, commercial bank foreign assets amounted to RD$1,984 million. Demand deposits totaled RD$6,588 million; time deposits, RD$17,226 million. The total money supply, as measured by M2, at that time stood at RD$30,775 million.

There is no securities exchange in the Dominican Republic.

34INSURANCE
Insurance firms are government-supervised and are required to furnish a bond to guarantee their obligations. Net premiums on life insurance are taxable, and the law requires that a certain proportion of the premiums collected must be invested locally. In 1986, there were 42 life, 47 non-life, and 37 composite insurance companies in operation.

35PUBLIC FINANCE
Between 1968 and 1975, dependence on foreign loans and grants to finance the budget was substantially decreased, and by 1975 tax revenues amounted to about 12% of the GNP. In the late 1970s, this trend was reversed, with rising expenditures and increased assistance from abroad. From 1989 to 1992, the consolidated public sector deficit improved from 6.2% of GDP to a surplus of 1.4% of GDP, due to increased petroleum revenues and import taxes.

The following table shows actual revenues and expenditures for 1989 and 1990 (in millions of pesos):

	1989	1990
REVENUE AND GRANTS		
Tax revenue	5,497.5	6,535.1
Non-tax revenue	573.4	726.3
Capital revenue	41.4	72.9
Grants	67.7	50.0
TOTAL	6,180.0	7,384.3
EXPENDITURES & LENDING MINUS REPAYMENTS		
General public service	553.7	1,175.9
Defense	283.5	340.5
Public order and safety	—	264.4
Education	587.0	715.0
Health	695.3	985.9
Social security and welfare	281.0	329.2
Housing and community amenities	1,211.9	1,094.1
Recreation, cultural, and religious affairs	75.6	79.2
Economic affairs and services	2,262.7	2,565.4
Other expenditures	323.2	—
Adjustments	—	—
Lending minus repayments	—	7.6
TOTAL	6,161.3	7,036.7
Deficit/Surplus	18.7	347.6

In 1991, expenditures exceeded revenues by RD$246.8 million. In 1992, the total external debt was RD$4.5 billion, and total debt service amounted to RD$316 million.

36TAXATION
Taxes accounted for 63% of current government revenues in 1986. A personal income tax was reintroduced in 1962, replacing a head tax that had been imposed in 1950. Personal income tax rates in 1992 ranged upward from 15% to 30%, which was slated for reduction to 25% by 1995. As of June 1992, business income was taxed at a flat rate of 30%, scheduled for reduction to 25% by 1995.

37CUSTOMS AND DUTIES
The customs tariff is primarily a revenue-raising instrument, although it is occasionally used to protect local industry. In the 1960s and early 1970s, import duties constituted the greatest single source of government revenue (accounting for 40–45% of the annual total during 1964–70), but by 1986 the share of import duties in government income had fallen below 30%. Export duties provided only about 5% of government revenues in 1986, down from 8.6% in 1973. As of 1993, a government decree had simplified the tariff schedule to 6 categories, with 7 tariff rates, ranging from 5% to 35%. Quantitative import restrictions were replaced by tariffs, and a temporary surcharge to be phased out over 3 years was imposed.

There are no free ports, but 19 industrial free zones have been established at locations including La Romana, San Pedro de Macorís, Santiago de los Caballeros, Baní, and Puerto Plata.

38FOREIGN INVESTMENT
Total registered direct foreign investment in the Dominican Republic totaled US$785.6 million in 1992, of which the US accounted for US$744 million in direct investment. Other principal foreign investors include Spain, Canada, Panama, and South Korea.

As of July 1993, there were 26 industrial free zones with 431 firms. Total cumulative investment in these zones amounted to US$631 million in 1991; new investment in 1992 came to RD$253 million.

39ECONOMIC DEVELOPMENT
Caribbean Basin Initiative and Generalized System of Preferences have provided enormous benefits to the economy. As of mid-1993, the Dominican Republic operated over two dozen industrial free zones, which engaged 431 businesses and over 142,000 workers. The government allows the import of raw materials, equipment, and goods for re-export duty-free. Tax exempt status for up to 20 years is available. Duty-free access to the EC market was granted by joining the Lomé Convention in 1989, which also provides inexpensive financing for economic development projects.

Since 1991, the government has made considerable progress in catching up on its debt arrears with the IMF, World Bank, and IDB. To help keep the economy stable, the government sought to extend its standby agreement with the IMF in December 1993.

40SOCIAL DEVELOPMENT
Social assistance is administered by the Department of Public Health and Welfare, which operates benefit programs dealing with sickness, disability, survivors' and workers' compensation, occupational disease, maternity, incapacity, and old age. The social insurance program is funded, in principle, by compulsory contributions from employees (2.5% of earnings), employers (7% of payroll), and the government (2.5% of employee earnings).

Since 1968, the Dominican government has been firmly committed to reducing population growth through family planning. The fertility rate remained high at 4.2 for 1985–90, however. By 1991, an estimated 56% of married women of childbearing age were using contraception. Abortion is illegal. An Institute for the Development of Women was created in 1975, but women still have lower economic and social status than men. They are often paid less for similar work and occupy few top leadership positions. Divorce is easily obtainable by either spouse.

41HEALTH
In 1992, average life expectancy was 67 years; the infant mortality rate was 42 per 1,000 live births; and overall mortality was 6 per 1,000 people. There were 214,000 births in the same year (a rate of 29 per 1,000 people). The country's total population was

7.6 million in 1993. Between 1980 and 1993, a study indicated that 56% of married women (ages 15 to 49) used contraception.

Modern aqueducts, drainage systems, and garbage disposal plants have been constructed in the principal cities. The National Water Supply and Sewerage Institute was established in 1962, and the National Rural Water Service was formed in 1964. From 1988 to 1991, an estimated 67% of the population had access to safe water, compared with 37% in 1970, and 87% had adequate sanitation. In 1990, however, approximately 13% of children under 5 years of age were considered to be malnourished, and there were 110 cases of tuberculosis per 100,000 people that same year.

From 1988 to 1992, there were 1.08 doctors per 1,000 people, with a nurse to doctor ratio of 0.7. Between 1985 and 1990, there were 2.0 hospital beds per 1,000 inhabitants. Between 1985 and 1992, statistics indicate that 80% of the population had access to health care services.

Between 1990 and 1992, the country immunized children up to 1 year of age against tuberculosis (48%); diphtheria, pertussis, and tetanus (48%); polio (63%); and measles (75%). Major causes of death between 1985 and 1990 were: communicable diseases and maternal/perinatal causes (206 per 100,000); noncommunicable diseases (443 per 100,000); and injuries (88 per 100,000). Total health care expenditures in 1990 were $263 million.

42HOUSING
Rapid population growth and migration to urban areas have combined to create an increasingly serious housing shortage. The National Housing Institute and the National Housing Bank, both established in 1962, have made modest efforts to ease the problem, between them constructing nearly 10,000 units during 1966–72. During 1975–78, construction activity slowed down, but the 1979 hurricanes prompted a construction boom—not, however, to create new housing but to replace units that had been destroyed by the storms. The Guzmán government promoted the building of low-cost housing at a rate of about 6,000 units a year. President Jorge Blanco pledged in 1982 that 25,000 low-cost houses would be built annually during his administration, and President Balaguer, after he returned to office in 1986, also built low-cost housing, though at a far slower pace than announced. As of 1981, 82% of housing units were detached houses, 13% were single rooms, and 4% were apartments. Roughly 65% obtained water through a public system, while 20% used wells. Dwellings were most commonly constructed of concrete (31%), wood (31.3%), or palm (31.1%).

43EDUCATION
The 1990 estimated adult literacy rate was 83.3% (males, 84.8% and females, 81.8%). The foremost educational objective in recent years has been the enrollment of the entire population in the 7–14 age range. Eight years of education is compulsory. Primary education lasts for eight years followed by four years of secondary education.

In 1989 there were 4,854 primary schools with 21,850 teachers and 1,032,055 students enrolled. Teacher training has been expanded dramatically with aid from UNESCO, UNICEF, and other agencies.

The state-run Autonomous University of Santo Domingo, founded in 1538 and the oldest in the hemisphere, has suffered from a lack of resources. There are four private universities, one technological institute, four colleges, and seven schools of art and music. In 1985, about 123,748 students were enrolled in the university and higher level institutions.

44LIBRARIES AND MUSEUMS
There are about 130 libraries located throughout the country, of which the library of the University of Santo Domingo is the most

important, with over 104,000 volumes. The National Library, founded in Santo Domingo in 1970, has a collection of over 150,000 volumes. In general, public library collections are few and small, mostly containing only a few hundred books. An exception is the public library in Santo Domingo, with holdings of about 35,000 volumes.

The leading museums are in and around Santo Domingo. The Museum of Dominican Man (formerly the National Museum) houses 19,000 pre-Columbian, colonial, and contemporary exhibits relating to the country's history. The Gallery of Modern Art promotes the music, painting, sculpture, and poetry of both Dominican and foreign artists and writers. A new complex at the Plaza of Culture houses several museums, the National Library, and the National Theater. The Alcázar de Colón is the 16th-century home of the Columbus family; the reputed tomb of Christopher Columbus is in the Cathedral of Santa María la Menor. The zoological and botanical garden in Santo Domingo is unique because of its natural setting and grottoes.

45MEDIA
The Dominican Telephone Co. operates the domestic telephone system, and the government controls domestic telegraph service. The larger cities have automatic telephone exchanges, and about 90% of the telephones are on an automatic dial system. In 1993 there were some 350,000 telephones, most of which were in the capital. International cable service is provided by All America Cables and RCA. In 1992 there were 130 AM and 40 FM radio stations and 18 television stations, of which the government-owned Radio-Televisión Dominicana is the most important. In 1992 there were approximately 1,250,000 radios and 615,000 television receivers.

The daily newspapers of the Dominican Republic are rated by the Inter-American Press Association as among the freest in Latin America. The leading daily is *El Caribe,* an independent morning daily, published in Santo Domingo (circulation 75,000 in 1991). Two other papers of importance, also published in the capital, are *El Nacional* (circulation 45,000), and the *Listín Diario* (about 60,000). Of the four dailies published outside the capital, *La Información* of Santiago (circulation 10,000) is the best known.

46ORGANIZATIONS
Consumer associations, mainly for low-income groups, deal in basic foods such as rice, plantains, potatoes, and beans. The Confederation of Employers of the Dominican Republic and the National Council of Businessmen are the principal employers' organizations. There are chambers of commerce in Santo Domingo and other large towns.

47TOURISM, TRAVEL, AND RECREATION
Although the Dominican Republic offers fine beaches and historical sites as well as good hotel facilities, it had no organized tourist industry to speak of until 1967, and received no more than 45,000 visitors per year. Increased political stability made the country more attractive to tourists, and by 1973 the number of foreign visitors had grown to 182,036. Today the Dominican Republic has more hotel rooms than any other Caribbean nation, and tourism is a mainstay of its economy. Receipts from tourism amounted to $877 million in 1991. In that year, 1.32 million tourists arrived, and hotel rooms numbered 22,555. The Ministry of Tourism has conducted a vigorous campaign to promote such resort centers as La Romana, Puerto Plata, Samaná, and Playa Grande. Visitors require valid passports. Short-stay visitors may not require visas or may require only tourist cards, depending on nationality.

Baseball is the national sport. Other popular pastimes include basketball, boxing, tennis, golf, hunting, fishing, and scuba diving. The Juan Pablo Duarte Olympic Center is one of the best-equipped sports facilities in the Caribbean.

⁴⁸FAMOUS DOMINICANS

Juan Pablo Duarte (1813–76), national hero of the Dominican Republic, was the leader of the famous "La Trinitaria," along with Francisco del Rosario Sánchez (1817–61) and Ramón Matías Mella (1816–64), which proclaimed and won independence from Haiti in 1844. Emiliano Tejera (1841–1923) and Fernando Arturo de Merino (1833–1906), first archbishop of the Dominican Republic, were noted statesmen.

Rafael Leonidas Trujillo Molina (1891–1961) was the dominant figure in the political life of the country from 1930 until his assassination on 30 May 1961. He served four times as president and was commander-in-chief of the armed forces. Since his appointment as president in 1960, Joaquín Balaguer (b.1909) has continued to be a prominent political figure in the Dominican Republic. Juan Bosch (b.1909), founder of the leftist PRD and later of the PLD, who served for seven months as president in 1963, is another important politician.

Juan Bautista Alfonseca (1810–75), the father of Dominican music, was the first composer to make use of Dominican folklore. José Reyes (1835–1905), musician and soldier, wrote the music for the national anthem. José de Jesús Ravelo (1876–1954) composed the oratorio *La Muerte de Cristo*, which has been performed yearly since 7 April 1939 at the Basílica de Santa María la Menor on Good Friday. Other prominent Dominican musicians are Juan Francisco García (1892–1974), Luis Emilio Mena (1895–1964), and Enrique de Marchena (b.1908). Juan Marichal (b.1938) achieved fame in the US as a baseball pitcher.

⁴⁹DEPENDENCIES

The Dominican Republic has no territories or colonies.

⁵⁰BIBLIOGRAPHY

Atkins, G. Pope. *Arms and Politics in the Dominican Republic.* Boulder, Colo.: Westview, 1981.

Black, Jan K. *The Dominican Republic: Politics and Development in an Unsovereign State.* Winchester, Mass.: Allen Unwin, 1986.

Calder, Bruce J. *The Impact of Intervention: The Dominican Republic During the U.S. Occupation of 1916–1924.* Austin: University of Texas Press, 1984.

Crasweller, Robert D. *Trujillo: The Life and Times of a Caribbean Dictator.* New York: Macmillan, 1966.

Dominican Republic and Haiti: Country Studies. 2d ed. Washington, D.C.: Library of Congress, 1991.

Gleijeses, Piero. *The Dominican Crisis: The 1965 Constitutional Revolt and American Intervention.* Baltimore: Johns Hopkins University Press, 1979.

Gordon, Raul (ed.). *Image of the Dominican Republic: The Dominican Miracle.* New York: Gordon, 1978.

Grasmuck, Sherri and Patricia R. Pessar. *Between Two Islands: Dominican International Migration.* Berkeley: University of California Press, 1991.

Haverstock, Nathan A. *Dominican Republic in Pictures.* Minneapolis: Lerner Publications Co., 1988.

Landau, Luis. *Dominican Republic: Its Main Economic Development Problems.* Washington, D.C.: IBRD, 1978.

Lowenthal, Abraham F. *The Dominican Intervention.* Cambridge, Mass.: Harvard University Press, 1972.

Munro, D. G. *Intervention and Dollar Diplomacy in the Caribbean, 1900–1921.* Princeton, N.J.: Princeton University Press, 1964.

Palmer, Bruce. *Intervention in the Caribbean: the Dominican Crisis of 1965.* Lexington, Ky.: University Press of Kentucky, 1989.

Plant, Roger. *Sugar and Modern Slavery: Haitian Migrant Labor and the Dominican Republic.* Totowa, N.J.: Biblio Dist., 1986.

Schoenhals, Kai P. *Dominican Republic.* Santa Barbara, Calif.: Clio Press, 1990.

Vedevato, Claudio. *Politics, Foreign Trade and Development: A Study of the Dominican Republic.* New York: St. Martin's, 1986.

Wiarda, Howard J. and Michael J. Kryzanek. *The Dominican Republic, a Caribbean Crucible.* 2nd ed. Boulder, Colo.: Westview Press, 1992.

ECUADOR

Republic of Ecuador
República del Ecuador

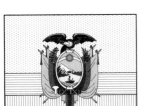

CAPITAL: Quito.

FLAG: The flag consists of three horizontal stripes, the yellow uppermost stripe being equal to the combined widths of the blue center stripe and the red lower stripe; coat of arms superimposed at center of the flag.

ANTHEM: *Salve, O Patria (Hail, O Fatherland)*.

MONETARY UNIT: The sucre (s/) of 100 centavos is a nonconvertible paper currency. There are coins of 10, 20, and 50 centavos and 1 sucre, and notes of 50, 100, 500, 1,500, 5,000, and 10,000 sucres. s/1 = $0.0042 (or $1 = s/238.75).

WEIGHTS AND MEASURES: The metric system is the legal standard, but local and old Spanish units are also used.

HOLIDAYS: New Year's Day, 1 January; Epiphany, 6 January; Labor Day, 1 May; Battle of Pichincha, 24 May; Simón Bolívar's Birthday, 24 July; Quito's Independence Day, 10 August; Guayaquil's Independence Day, 9 October; Columbus Day, 12 October; All Saints' Day, 1 November; All Souls' Day, 2 November; Cuenca's Independence Day, 3 November; Foundation of Quito, 6 December; Christmas Day, 25 December. Movable holidays include Carnival and Holy Week.

TIME: Mainland, 7 AM = noon GMT; Galápagos Islands, 6 AM = noon GMT.

¹LOCATION, SIZE, AND EXTENT

The fourth-smallest country in South America, Ecuador is located on the west coast of the continent and is crossed by the Equator (the country gets its name from the Spanish word for "Equator"). It has a length of 714 km (444 mi) N-S and a width of 658 km (409 mi) E-W. Ecuador borders Colombia on the N, Peru on the E and S (it has lost about two-thirds of the territory it once claimed to these countries), and the Pacific Ocean on the W, with a total boundary length of 4,247 km (2,639 mi).

The Galápagos Islands, a province of Ecuador with an area totaling 8,010 sq km (3,093 sq mi), are approximately 1,130 km (700 mi) off the coast on the Equator at 89° to 92°W. The total area of the republic and its territory is estimated at 283,560 sq km (109,483 sq mi). Comparatively, the area occupied by Ecuador is slightly smaller than the state of Nevada. Ecuador also claims about 200,000 sq km (77,000 sq mi) of land awarded to Peru under the 1942 Protocol of Rio de Janeiro. Armed hostilities flared along a still undemarcated stretch of the border in January 1981, but by 20 February, a 14-km (9-mi) demilitarized zone had been arranged along the disputed line. Official maps of Ecuador show the entire region as Ecuadoran territory. Ecuador's capital city, Quito, is located in the north central part of the country.

²TOPOGRAPHY

Ecuador is characterized by three distinct regions: the coast; the highlands, or Sierra; and the eastern interior jungles, or Oriente. The coast, except for a hilly area west of Guayaquil, is a low alluvial plain from 32 to 185 km (20 to 115 mi) wide, comprising about one-quarter of the national territory. It extends from sea level to the base of the Cordillera Real of the Andes, at an elevation of about 460 m (1,500 ft). The Guayas in the southwest and the Esmeraldas in the northwest form the principal river systems

and serve as important arteries of transportation in their respective regions.

The highlands constitute another fourth of the country. This region is formed by two parallel ranges of the Andes, from 110 to 290 km (70 to 180 mi) wide, and the intervening narrow central plateau, nearly 640 km (400 mi) long. This inter-Andean plateau is divided into 10 basins at altitudes from 2,400 to 2,900 m (7,800 to 9,500 ft), some draining east and some west. The Andes are studded with massive snow-capped volcanoes, the highest of which are Chimborazo, 6,267 m (20,561 ft); Cotopaxi, 5,948 m (19,514 ft), the world's third-highest active volcano; Cayambe, 5,790 m (18,996 ft); Antisana, 5,705 m (18,717 ft); Altar, 5,320 m (17,454 ft); Iliniza, 5,266 m (17,277 ft); Sangay, 5,230 m (17,159 ft); and Tungurahua, 5,016 m (16,457 ft).

The Oriente, forming part of the upper Amazon Basin, begins at the base of the Andes at about 1,200 m (4,000 ft). The land at first drops quickly and is segmented by rushing torrents escaping from the cold highlands. At about 260 m (850 ft), the jungle becomes almost level, and the streams suddenly widen into sluggish, meandering rivers as they begin their journey down the Amazon system to the Atlantic.

³CLIMATE

The climate varies with the region. Most of the coast consists of wet, tropical forest, increasingly humid toward the north. The cold Humboldt Current, which flows northward along the coast of Peru and then heads out into the Pacific off the coast of central Ecuador, limits the rainfall on a strip of the coast extending from as far north as the Bay of Caráquez and widening to include most of the coastal lowlands south of Guayaquil. In the Guayaquil area there are two seasons: a hot rainy period, lasting from January to May; and a cooler dry season, during the rest of the year, when sea breezes modify the equatorial heat. The tropical jungles of the Oriente, east of the Andes, are more humid than the coast; there, temperatures are high, and rain falls all year round.

The climate of the central plateau is governed mainly by the altitude. The capital, Quito, at 2,850 m (9,350 ft), has perpetual spring, with an average temperature of 13°C (55°F) and about 127 cm (50 in) of rainfall annually. The highlands are cut by numerous deep valleys, which bring subtropical climates to within a few miles of the more temperate areas. Cold and wind increase as the slopes surrounding the central plateau ascend to form the páramo, or highland meadow. The higher areas rise to peaks with perpetual snow above 5,200 m (17,000 ft). Exceptionally heavy rains in late 1982 caused extensive flooding, affecting about 4.5 million people and causing widespread crop and property damage.

4FLORA AND FAUNA

The arid savanna strip along about half of Ecuador's coast, with occasional low shrubs and isolated ceiba trees, contrasts sharply with the northern coast and the inner portion of the southern coast. In these humid regions, the typical dense growth of the tropical jungle abounds, extending as wet mossy forests up the Andean slopes to over 2,400 m (8,000 ft) in some places. Beyond the moisture barrier formed by the Western Cordillera, the high mountain slopes above 3,000 m (10,000 ft) are covered with wiry páramo grass and, in the northern province of Carchi, with a mulleinlike plant, the fraylejón (espeletia).

The highland valleys, at an altitude between 2,400 and 3,000 m (8,000 and 10,000 ft), support most of the temperate-zone plants; potatoes and corn, for example, have been raised there for thousands of years. There are few native trees in the highlands; eucalyptus was introduced in the 1860s and has been widely planted. The Oriente has little that is unique to tropical flora except for the delicious naranjilla, a small green orange used in making a conserve.

Ecuadoran jungles support the usual smaller mammals, reptiles, and birds. In the highlands, the condor and a few other species of birds are found. There is relatively little wild game because of the density of the population and the intensive use of the land. The Amerindians still make some use of the llama in southern Ecuador. Throughout the highlands, Amerindians and some mestizos raise cavies (guinea pigs) in their homes as an important source of meat.

5ENVIRONMENT

As of 1994, Ecuador's major environmental problems were erosion in the highland areas; deforestation, especially in the Oriente; and water pollution. The Ecuadoran Institute of Water Resources estimated that the amount of arid land increased by 31.5% between 1954 and 1979, when 7.5% of the coastal lowland and Sierra were classified as arid. Between 1981 and 1985, 340,000 hectares (840,000 acres) of land were deforested annually.

As of 1994, it was estimated that, at current deforestation rates, coastal forests will be completely eliminated within 15 years and the Amazon forests will be gone within 40 years. Traditional farming practices have been blamed for most of these problems, but oil development has also played a role in the clearing of forests. A program for reforestation and maintenance of existing forests was initiated in 1979, but only 4,000 hectares (10,000 acres) were reforested annually during the early 1980s. Ecuador's principal environmental agency is the Ministry of Natural Resources and Energy. Land erosion is accelerated by deforestation.

Flooding and desertification are related problems which have damaged or eliminated valuable soil, particularly in the western coastal region. As of 1993, 10,000 hectares of land were affected. Water pollution is a problem due to the influx of domestic, industrial, and agricultural contaminants.

Ecuador's cities produce 1.2 million tons of solid waste per year. The nation has 75.3 cubic miles of water with 90% used for agricultural purposes. Thirty-seven percent of all urban dwellers and 56% of the rural population do not have pure water.

The expansion of Ecuador's population centers threatens its wildlife. Twenty-one species of mammals in a total of 280 are currently endangered. Sixty-four of 1,447 bird species and eight of 345 reptile species are also endangered. Of the 10,000–20,000 plant species in Ecuador, 256 are threatened with extinction. An extensive wildlife protection system was established in 1970. As of 1987, endangered species on the Ecuadoran mainland included the tundra peregrine falcon, yellow-tailed woolly monkey, five species of turtle (green sea, hawksbill, olive ridley, leatherback, and South American river), and three species of crocodile (spectacled caiman, black caiman, and American).

Endangered species in the Galápagos Islands included the dark-rumped petrel, Galápagos dark-rumped petrel (a subspecies), black petrel, African ass, two species of turtle (green sea and hawksbill), and the Galápagos giant tortoise and 11 of its subspecies. A subspecies of Galápagos giant tortoise is extinct, and another may be.

Ecuador's natural attractions could lead to increased tourism, benefitting the economy. However, environmental problems and further endangerment of native plants and animals could be exacerbated without careful management of the areas attractive to tourists.

6POPULATION

In 1990, the total population was 9,648,189, of whom 44.6% lived in rural areas; the national population density was 35.5 per sq km (92 per sq mi). The average annual rate of increase between 1985 and 1990 was 2.5%. The projected population for the year 2000 was 13,090,000, assuming a crude birthrate of 26.9 per 1,000 population, a crude death rate of 6.5, and a net annual increase of 20.4 during 1995–2000. Guayaquil, the major port, had 1,508,444 inhabitants in 1990, and Quito, the capital, 1,100,847. Other large cities were Cuenca (194,981), Machala (144,197), and Portoviejo (132,937). Of the total population, 39% were less than 15 years old.

7MIGRATION

Ecuador has had no large-scale immigration since the colonial period, and emigrants have generally outnumbered newcomers. There was an influx of European refugees in the late 1930s. In 1959, a modest attempt was made to colonize the northern coastal province of Esmeraldas with Italian families.

Within Ecuador, the largest migration is from rural areas to the cities, as urban employment opportunities widen. There is also a growing movement from the overpopulated highlands to the virgin lands of the Oriente and the coast.

8ETHNIC GROUPS

The population of Ecuador is about 40% mestizo (mixed Amerindian and white). About 40% are Amerindian, 10–15% white, and 5% black. There are only a few groups of unassimilated Amerindians on the coast, notably the Colorados and Cayapas. The blacks live mainly in the northern coastal province of Esmeraldas. The Amazon Basin is inhabited by many primitive tribes, including the Jívaros, once famous for their shrunken-head war trophies, and the Záparos, Aucas, Secoyas, and Cofanes. In the early 1980s, the tribes were organized in the Federación Shuar, which seeks to preserve their cultural identities.

9LANGUAGES

The official language of Ecuador is Spanish, spoken by about 93% of the population. The Spanish of the coastal areas is similar to that of the other lowland areas of Latin America, maintaining something of the Andalusian characteristics, especially the dropping or slurring of the consonants represented by *s* and *d*. In the isolated highlands, a more precise Castilian pronunciation is found, but many words and even some of the singsong intonations of Quechua, the Amerindian language, have crept into the Spanish.

Nearly 7% of the total population speak only Quichua, a dialect of the Quechua language, while another 6% speak it in addition to Spanish. Quechua was imposed on the Amerindians of Ecuador by the conquering Incas in the 15th century, supplanting a number of unrelated languages. Remnants of these forgotten languages are retained in many modern place names. There has been little detailed study of the languages of the jungle tribes of the Oriente.

10RELIGIONS

Roman Catholicism was introduced by the Spaniards with the conquest in 1540; in 1993, an estimated 93% of the population was Roman Catholic. Protestants, including Anglicans, Baptists, and Methodists, make up less than 2% of the population. Animistic religions survive among the jungle Amerindians of the Oriente. Freedom of worship is guaranteed by the constitution.

11TRANSPORTATION

The topography and climate of Ecuador have greatly hindered the development of adequate means of land transportation. In 1991 there were 28,000 km (17,399 mi) of highways, including 3,600 km (2,237 mi) of asphalted roads. The Pan American Highway (1,076 km/669 mi in Ecuador) extends the length of the highlands from Tulcán on the Colombian border to Loja in the south and on to Peru. In 1970, the five-nation Bolivarian Highway was undertaken, as were east-west routes linking the Oriente with the Sierra, and Guayaquil with its hinterland. The most important lateral route connecting the highlands and the coast runs from Latacunga, crossing a pass in the Cordillera Real over 3,650 m (12,000 ft) high, to Quevedo in the lowlands. In 1991 there were 265,000 passenger cars and 60,000 commercial vehicles.

Modern port facilities to serve Guayaquil were opened in 1963 on an estuary 10 km (6 mi) from the Guayas River. The Guayas River basin is important for transportation in the coastal provinces. Other international ports are Esmeraldas, Puerto Bolívar, and Manta; La Libertad and Balao can accommodate oil tankers. In 1991, Ecuador's merchant marine consisted of 49 ships with a gross registered tonnage of 348,000; in addition, Ecuador shared with Colombia an interest in the Grancolombia Merchant Fleet.

Railways, all government owned, are of decreasing importance because of their poor condition and competition from highways. The three railroad networks total 965 km (600 mi); the most important line runs between Guayaquil and Quito. Floods in 1983 damaged much of the system, and by 1986 service had been restored on only some of the sections.

Ecuador's rugged topography has hastened the growth of air travel. There are some 43 airports; those of Guayaquil and Quito are international. In 1992, total scheduled airline traffic amounted to 1,261 million passenger-kilometers (784 million passenger-miles). The government-run Ecuatoriana de Aviación provides service between Ecuador and the rest of Latin America and carried some 252,300 international and domestic passengers in 1991.

12HISTORY

Archaeological explorations indicate that the coastal regions of present-day Ecuador supported corn-cultivating communities as

LOCATION: 1°26′30″N to 5°1′s; 75°11′44″ to 81°1′w. **BOUNDARY LENGTHS:** Colombia, 538 km (334 mi); Peru, 1,316 km (818 mi); Pacific coastline, 848 km (527 mi). **TERRITORIAL SEA LIMIT:** 200 mi.

early as 4500 BC. In the first few centuries AD, the population was divided into dozens of small isolated tribes. By AD 1000, the highland groups had formed a loose federation, the Kingdom of Quito, but they were absorbed into the Inca Empire in the late 15th century. Atahualpa, son of the conquering Inca Huayna Capac and a Quito princess, later became emperor, but by then, the Spanish forces under Francisco Pizarro were gaining a foothold on the coast.

Pizarro's pilot, Bartolomé Ruiz, the first European to see the Ecuadorian coast, arrived in 1526 on a scouting expedition. The actual conquest reached Ecuador in 1531. Except for a few emeralds, from which their first landing place took its name (the city and province of Esmeraldas), the Spanish found those shores valuable only as a stopping place on their way to the riches of the Incas in Peru. Sebastian de Belacázar, a lieutenant of Pizarro, extended Spanish dominion northward from Peru after the conquest of the Incas. He found the northern capital of the Inca Empire left in ashes by the retreating Amerindians, and on that site in 1534, he founded the city of San Francisco de Quito, later to become the capital of the republic.

The Spanish governed the region as the Audencia of Quito, part of the Viceroyalty of Peru. Quito, in the cool highlands, was soon steeped in culture and rich in ornately decorated churches and monasteries. Guayaquil, the principal seaport, grew slowly because of its unhealthy tropical climate, and would not become a major city until much later. The Spanish colonial period was a time of ruthless exploitation of the Amerindians and bickering and bloodshed among the Spanish in the struggle for power and riches.

The early stirrings of Ecuadorian independence were spread, in part, through the writings of the 18th-century satirist Francisco Javier Eugenio de Santa Cruz y Espejo. Abortive revolts against Spanish rule came in 1809 and in 1811. The decisive struggle began on 9 October 1820, with the proclamation of an independent Guayaquil. Finally, on 24 May 1822, with the Battle of Pinchincha, the Spanish were defeated. This victory unified the liberation movements of the continent. Simón Bolívar and José de San Martín met in Guayaquil in 1822 to consider the future of newly freed areas. Liberated Ecuador became part of Bolívar's dream, the Republic of Gran Colombia, consisting of modern Ecuador, Colombia, Venezuela, and Panama. In 1830, when this union collapsed, the traditional name Quito was dropped in favor of La República del Ecuador, "The Republic of the Equator."

Republic of Ecuador
The Republic's first president was Juan José Flores, one of Bolívar's aides. The 15-year period of Flores's domination was noted for iron-handed conservative rule. In 1832, he occupied the Galápagos Islands in a comic-opera invasion witnessed only by the giant tortoises native to the islands. Then, from 1845–60, Ecuador went through 11 presidents and juntas. The nation was split between pro-clerical Conservatives and the more secular Liberals, and regional strongmen vied for power.

From 1860 to 1875, Ecuador was ruled by the fervently religious Conservative Gabriel García Moreno, Ecuador's first great statesman. He sought peace and consolidation for his torn country through a rigid, theocratic government. His administration granted special privileges to the Roman Catholic Church, even dedicating the Republic to "The Sacred Heart of Jesus" by act of congress in 1873. Beyond his religious zeal, García Moreno was also known for developing roads and public education, beginning the Guayaquil-Quito railway, and putting Ecuador on a firm financial footing. However, his relentless conservatism caused bitter strife, culminating in the dictator's assassination in 1875. In the ensuing period of confusion, the Conservatives were not able to carry on the program of García, nor could the opposition take command until the emergence of Gen. Eloy Alfaro, who ushered in the Radical Liberal era with the revolution of 1895. He and the succeeding Liberal presidents were able to counteract much of García's program. Church and state were carefully separated, and liberty of thought, worship, and the press was established. The Guayaquil-Quito railway was completed, uniting the coast and the highlands commercially.

The Liberal era continued until 1944, with numerous interludes of violence and crisis. The economy rose and fell with world prices on such commodities as cocoa. Territory was lost to Brazil in 1904, Colombia in 1916, and finally Peru in 1942. The border dispute with Peru, originating in the colonial period, came to a climax when Peru invaded Ecuador's southern and Oriente (Amazon Basin) provinces. The Rio de Janeiro Protocol awarded to Peru the greater part of the Amazon Basin territory claimed by Ecuador. Ecuadorans never forgot that treaty, and have been trying to rescind it ever since.

In 1944 José María Velasco Ibarra came to power as a nationalist denouncing the Rio agreement. Velasco, who had served as president during 1934–35, ruled for three years until he was sent into exile. After three ineffective presidents in less than one year, Galo Plaza Lasso (1948–52) was elected to the presidency. Plaza,

later chief of the OAS, ruled for four years. In 1952, Velasco Ibarra returned to office for four years, and was again elected in 1960. In his inaugural address, Velasco formally renounced the Treaty of 1942, and embarked on an economic program of "growth through inflation."

By 1961, with Ecuadorian currency in a slump and consumers heavily taxed, the air force revolted and sent Velasco into exile, thus ending Ecuador's unprecedented streak of elected governments. Vice President Carlos Julio Arosemena Monroy assumed the presidency on 7 November 1961. Arosemena lasted less than two years, and in July 1963, he was arrested by the military for "drunkenness" (a charge that could have been substantiated throughout his presidency) and sent into exile.

Military governments
A four-man military junta headed by Capt. Ramón Castro Jijón took over and ruled until March 1966. Elections were scheduled and held in October 1966 for a constitutional assembly. Otto Arosemena Gómez, cousin of Arosemena Monroy, became provisional president. In 1968, new elections were held for the presidency, won yet again by Velasco. On 22 June 1970, following a fiscal crisis, Velasco suspended the 1967 constitution and assumed dictatorial power. He dissolved Congress, reorganized the Supreme Court, and proceeded to rule by executive decree.

In June 1971, Velasco promised new presidential and congressional elections, which were scheduled for the following June. However, on 15 February 1972, Velasco was overthrown in a bloodless coup after he refused demands by senior army officers to postpone the elections. On the following day, Gen. Guillermo Rodríguez Lara was installed as head of a new military government. Velasco, deported to Panama, was granted asylum by Venezuela.

Return to elected government
General Rodríguez lasted for four years, and was ousted on 12 January 1976. A three-member Supreme Council assumed power, promising a return to civilian government within two years. Presidential elections took place in July 1978, but because none of the candidates received the required majority, a runoff election was held in April 1979. The winner was Jaime Roldós Aguilera, a populist running under the banner of the Concentration of Popular Forces. Christian Democrat Osvaldo Hurtado was made vice-president. Both were inaugurated on 10 August 1979, the day Ecuador's current constitution went into effect. Roldós was killed in a plane crash on 24 May 1981, whereupon Hurtado became president until 1984.

Hurtado's term was marked by modest gains in the economy, but by 1984, a flagging economy, caused in part by widespread flooding, led to calls for change. The 1984 election was won by León Febres Cordero Rivadeneira, a conservative Social Christian who advocated a free-enterprise economic policy. Febres formed a coalition government and pressed his platform of reducing state intervention in the economy and making it more responsive to market forces.

Economic recovery falters
Just as it appeared that Febres's fiscal policies were about to bring widespread benefits to the populace, Ecuador was dealt two staggering blows: the 1986 plunge in world oil prices cut revenues by 30%, and a devastating earthquake in March 1987 cut off oil exports for four months and caused more than $1 billion in damage. The government had already suspended repayment of the $11 billion foreign debt in January 1987, but after the earthquake repayment was further postponed. Febres's fortunes took a turn for the worse with the decline in oil prices. His coalition

partners were defeated in elections held on 1 June 1986, and the leftist parties opposed to him gained control of parliament.

In presidential elections held 31 January 1988, Rodrigo Borja Cevallos of the Democratic Left (ID) Party and Abdalá Bucaram Ortiz of the Roldista Party (honoring ex-President Roldós) won the most votes in a field of 10 candidates. Borja won the runoff election, and took office along with a strong contingent in congress. The government made improvements in Ecuador's human rights record, and reached an accord with the terrorist group Alfaro Vive, Carajo (literally, "Alfaro [Eloy] lives, damn it!"). However, economic troubles, particularly inflation, continued, and the ID lost half its congressional seats in mid-term elections in 1990. In 1992, voters elected a conservative government, headed by President Sixto Durán-Ballén of the Republican Unity Party (PUR) and Vice President Alberto Dahik of the Conservative Party (CP).

From its stormy past, Ecuador appears to be in the process of developing a genuine democracy.

13GOVERNMENT

Since 1860, Ecuador has had 17 different constitutions. The most recent constitution came into force on 10 August 1979, in preparation for a return from military to civilian rule. Under this document (as amended in mid-1983), the unicameral Chamber of Representatives consists of 77 provincial and national members. The provincial representatives serve two-year terms and the national representatives four-year terms. The chamber meets in full session for only two months a year, leaving the rest of its business to four permanent committees. The president and vice-president are elected for four-year terms, and the president cannot be reelected. As is traditional in Ecuador, the president initiates the budget and appoints the cabinet, as well as provincial governors, many administrative employees, and diplomatic representatives. The chamber may, however, remove cabinet ministers (it forced the finance minister out of office late in 1986). The president also controls the armed forces and can declare a state of siege.

Voting is compulsory for literate people aged 18 to 65, and optional for illiterates.

14POLITICAL PARTIES

The constitution of 1979 guarantees the right of democratic activities of political parties. To retain its legal standing, a party must field candidates in at least half the nation's provinces (including two of the three most populous ones) and must win at least 1.5% of the total votes cast in the last national election. This procedure, coupled with a proportional representation system, guarantees a proliferation of Ecuadorian parties. There are currently 18 parties active in Ecuadorian politics.

Two major parties played dominant roles prior to the 1960s. The Conservative Party (Partido Conservador—PC), which held sway during the first half of the republic's history, was the political representative for the Roman Catholic Church, and its support came from the large landowners of the highlands. The principal opposition, the Radical Liberal Party (Partido Liberal Radical—PLR), which rose to power in the revolution of 1895, was supported by businessmen and the newer city elite. It sought scrupulous separation of church and state, especially in public education, and called for the development of industry and the attraction of foreign capital.

Modern parties on the right include the Social Christian Party (PSC), the Republican Unity Party (PUR) from which President Durán-Ballén hails, and the Conservative Party (CE) of Vice President Dahik. On the left are the Democratic Left (ID), linked to the Social Democratic Movement, the Popular Democracy Party (DP) of former President Hurtado and the traditional Radical

Liberal Party (PLR).

Ecuador's populist tradition has given rise to many parties, organized along highly personalist lines, such as the Roldista Party (PRE), headed by Abdalá Bucaram, the Assad Bucaram Party (PAB), headed by Avicena Bucaram, and the Concentration of Popular Forces (CFP).

The far left in Ecuador has been beset by factionalism and governmental intrusion. In the 1920s, the original Socialist Party of Ecuador split into the Socialist Party and the Communist Party. Further splits occurred with the advent of the Cuban revolution. Currently, the Popular Democratic Movement (MPD), the Ecuadorian Socialist Party (PSE), and other parties vie for the non-Communist vote. Communists are divided between the Communist Party of Ecuador (PCE), which is identified as a pro-North Korean faction, and the Communist Party of Ecuador/Marxist-Leninist (PCMLE), which is identified as Maoist.

15LOCAL GOVERNMENT

The three levels of local government—province, canton, and parish—are controlled by the central government in a fundamentally unitary system. Ecuador has 20 continental provinces, plus the insular Galápagos Islands. The provincial governors, who are appointed by the president, are responsible to the interior ministry. Each province is divided into cantons, which in urban areas are administratively subordinate to the municipality with which they coincide. A municipal council is popularly elected and in turn elects its officers. In the larger towns, a mayor is popularly elected. The municipality is unique in that it lies somewhat outside the unitary pattern and is less subject to national control than are the other units of local government. The rural canton, of little importance in the sparsely populated Oriente and northern coast, is significant in the more developed regions of the highlands and the coastal provinces of Guayas and Manabí. The highest official of the canton, the political chief, is appointed by the president on the recommendation of the provincial governor.

16JUDICIAL SYSTEM

Traditionally, the judicial function has been carried out by five levels of tribunals. The parochial judge, the political lieutenant appointed by the president to supervise the affairs of the parish, handles only minor civil cases. Cantonal courts, at least one in each canton, try minor civil and criminal actions. Provincial courts handle all but a few of the criminal cases and the more serious civil and commercial suits. Superior courts handle appeals from the lower courts and have other administrative duties in the district; they may try original cases only if these relate to the affairs of their district. The Supreme Court has 31 justices and three alternates chosen by the National Chamber of Representatives for six-year periods.

Although citizens are afforded a wide range of freedoms and individual rights, there remain some shortcomings in the functioning of the judicial system which is susceptible to political pressure. Police officers are tried only in closed session before police courts so that convictions for abuse or other violations are rare. Despite laws restricting arbitrary arrest and detention, such violations continue to occur in practice. Modernization of the court system began in 1993.

17ARMED FORCES

The armed forces of 57,500 are designed for internal security more than external defense. In 1991, Ecuador spent $261 million on defense or around 4% of GDP. Since 1933, a conscription law (now 1-year) has supported a sizable army, now 50,000. In recent decades the army has built roads and bridges and, through the

Military Geographic Institute, has mapped uncharted areas of the country.

In 1993, the navy of 4,500 (including 1,500 marines) mans 2 submarine and 14 small surface combatants. The air force of 3,500 (with 85 combat aircraft) also provides air defense and transport services. The National Civil Police numbers 5,800 and a coast guard 200. Ecuador's arms imports dropped from $100 million (1986) to $10 million (1991).

18 INTERNATIONAL COOPERATION

Ecuador is a charter member of the UN, having joined on 21 October 1945. It belongs to ECLAC and all the nonregional specialized agencies except WIPO. Ecuador is also active in the OAS and its various commissions, as well as in G-77, IDB, LAIA, and PAHO. Ecuador received full membership in OPEC in 1973. As of September 1987, the nation, which claims a 200-mi territorial sea limit, had declined to sign the Law of the Sea.

Under the US-backed Alliance for Progress program, Ecuador financed its private industry, improved and expanded the Quito sewerage system, promoted housing development, increased production of foodstuffs and raw materials, moved toward eradication of foot-and-mouth disease, implemented a mapping program, purchased selected Australian sheep, and implemented the land settlement program. In the early 1970s and again in the early 1980s, relations with the US were strained because of Ecuadoran seizures of US tuna-fishing boats. In mid-1983, the US lifted its prohibition on the importation of Ecuadoran tuna, and the tension between the two countries over the tuna fishing eased.

In 1965, Ecuador and Colombia established a permanent Colombian-Ecuadoran economic integration commission. In May of that year, Ecuador signed the Andean Pact, thereby becoming a member of the Andean Common Market.

19 ECONOMY

Ecuador's average annual GDP growth rate exceeded 9% in the 1970s, due largely to an oil-driven economy. As oil prices fell in the early 1980s, debt began to increase. Furthermore, a major earthquake in 1987 interrupted oil production and exports. In 1990, GDP amounted to $10.9 billion, of which services accounted for 41.7%; manufacturing, 22.8%; petroleum and mining, 15.1%; agriculture and fishing, 13.2%; construction, 3.7%; and other sectors, 3.5%.

Ecuador's economy grew by 3.5% in 1992, mainly from increased petroleum production and expansionary fiscal policy from the first half of the year. The new administration of President Ballen, elected in July 1992, raised petroleum derivatives taxes and electricity tariffs, while cutting public expenditures and freezing public sector employment. As a result, the 1992 public sector deficit fell from 7% to 2.8%. The inflation rate was 60% at the end of 1992; the Ballen administration had set a target of 30% for the next year, but the inflation rate was estimated at 40% in 1993. In 1993, the GDP was estimated at $12.5 billion (up 3.7%).

20 INCOME

In 1992, Ecuador's GNP was $11,843 million at current prices, or $1,070 per capita. For the period 1985–92 the average inflation rate was 49.6%, resulting in a real growth rate in per capita GNP of 0.6%.

In 1992, the GDP was $12,681 million in current US dollars. It is estimated that in 1991 agriculture, hunting, forestry, and fishing contributed 15% to GDP; mining and quarrying, 11%; manufacturing, 21%; electricity, gas, and water, 0%; construction, 4%; wholesale and retail trade, 23%; transport, storage, and communication, 9%; finance, insurance, real estate, and business services,

7%; community, social, and personal services, 4%; and other sources, 7%.

21 LABOR

In 1991, the civilian labor force numbered 4.4 million. In 1990, 30.8% were engaged in agriculture, 24.9% in services, 11.6% in manufacturing and mining, 14.2% in commerce, 5.9% in construction, 3.9% in transportation, and 8.7% in other activities. The unemployment rate officially stood at 8% in 1991, but 61% of the urban labor force was underemployed.

In 1991, only 13.5% of the economically active population was affiliated with labor unions. The largest concentration of unions has been in the Guayas and Pichincha provinces. The dominant groups are the Ecuadoran Confederation of Class Organizations (Confederación Ecuatoriana de Organizaciones Clasistas—CEDOC), the Ecuadoran Confederation of Workers (Confederación de Trabajadores del Ecuador—CTE), and the Ecuadoran Confederation of Free Trade Union Organizations (Confederación Ecuatoriana de Organizaciones Sindicales Libres—CEOSL). These three confederations form the United Front of Workers (Frente Unido de Trabajadores—FUT).

The labor code provides for a 40-hour workweek. Special permits are necessary for women and minors and for workers in certain hazardous industries. There are minimum wages, two-week paid vacations, overtime pay, social security, and severance pay. The minimum monthly wage was about us$40 in 1991. Fringe benefits include three bonuses and paid vacation of 15 days a year, plus a seniority bonus of one day for each year of service in excess of five years, up to a 30-day maximum. Benefits typically can double the basic salary.

22 AGRICULTURE

Although Ecuador's main economic activity has long been agriculture, less than 10% of the land is arable or under permanent crops, and another 18% is permanent pasture. Throughout the 1970s, agricultural development was neglected because of the emphasis on oil exploitation, and the sector showed negative rates of growth, declining by 5.4% in 1978, 2.8% in 1979, and 2% in 1980. During 1985–90, however, agriculture (along with fishing and forestry) showed an average annual increase of 4.8%; in 1991, the growth rate jumped to 6.8%.

The land census of 1974, which covered a much larger area than the previous census of 1954, showed that during the intervening period the number of agricultural units had grown by 172,810. This increase was the result of agrarian reform and colonization, initiated in 1964. Ceilings imposed on the maximum size of holdings ranged from 800 hectares (2,000 acres) of arable land plus 1,000 hectares (2,500 acres) for pasture in the Sierra region to 2,500 hectares (6,200 acres) plus 1,000 hectares (2,500 acres) of pasture in the coastal region. The 1975 plans of the Ecuadoran Land Reform and Colonization Institute called for the redistribution of 80,000 hectares (198,000 acres) of land in the coastal region.

Traditionally, agricultural products have included bananas, coffee, tea, rice, sugar, beans, corn, potatoes, and tropical fruit. Exported products of more recent prominence include roses and carnations, strawberries, melons, asparagas, heart of palm, and tomatoes. The major crops of the highlands are corn, barley, wheat, kidney beans, potatoes, horsebeans, peas, and soybeans, all for domestic consumption. Agriculture on the coast is largely oriented toward the export market. In 1992, agricultural production expanded by 6%, despite the modest growth of banana production and a declining coffee output. Increased acreage and improved yields, as well as the government's price-support program, have caused rapid growth in agriculture. Most of the cacao crop is produced on plantations of 60 hectares (150 acres) or larger, but the more important banana and coffee crops are grown mainly on small landholdings by independent farmers. Banana exports rose from less than 5% of total exports after World War II to 62.2% in 1958; in 1974, they were 10.8%, and in 1992, 21.5% ($647 million). Cacao became a valuable export in the mid-1970s, but low prices and declining harvests had a negative impact on revenues until the early 1980s, when they began to rise again. In 1978 and 1979, coffee brought high export earnings ($281.2 million and $263.1 million, respectively), but with lower world prices, coffee export earnings fell to $106 million in 1981, despite an increase in production. Revenues rose steadily during the 1980s, reaching $290 million in 1986, but fell to $129 million by 1990 and to only an estimated $75 million in 1992.

Principal commodities in 1992 (in tons) were sugar, 6,500,000; bananas, 3,600,000; corn, 500,000; potatoes, 375,000; cocoa beans, 78,000; and coffee, 95,000. The production of paddy rice reached 981,000 tons in 1992.

The agricultural sector of the economy, as of 1994, presents potential for further development and growth. Crops for domestic consumption, particularly rice, barley, maize, African palm, and potatoes, continue to show growth due to increased area planted and improved yields. Other segments likely to experience growth are non-traditional agricultural products such as flowers, fresh fruit, and vegetables, and processed foods.

23 ANIMAL HUSBANDRY

The dairy industry is located in the most fertile valleys of the highland plateau from Ibarra to Riobamba, where irrigation is available. Cattle production increased by more than 3% a year between 1990 and 1992; there were 4,665,000 head of cattle in 1992. Nearly all the sheep (1,511,000 in 1992) are in the highlands; most are raised by Amerindians and are pastured at over 2,700 m (9,000 ft). The wool is of poor quality. Hogs and goats, found throughout the country, are frequently diseased and poorly fed; in 1992 there were an estimated 2,434,000 hogs and 314,000 goats. The use of bananas as hog feed has made hog farming more attractive economically.

24 FISHING

In the waters around the Galápagos Islands, Ecuador has some of the world's richest fishing grounds, particularly for tuna. In the past, these waters were exploited mainly by foreign companies, but in recent years, Ecuadoran enterprises have participated more fully. Shifts in ocean currents can cause great variance in the annual catch. Ecuador is a leading producer of canned tuna. Shrimp farming occupies some 110,000 hectares (over 270,000 acres). Ecuador produces more shrimp than any other nation in the Americas, and exports more than 35,000 tons annually, mostly to the US. Rainbow trout aquaculture is being developed in the Andean highlands. The total catch in 1991 was 383,600 tons, down from 1,003,380 tons in 1986. Exports of fish and fish products in 1991 totaled $587.6 million, or 20% of total exports in 1991.

Ecuador proclaimed sovereignty over its coastal waters to a limit of 200 km. In 1952, along with Peru and Chile, Ecuador signed the Declaration of Santiago (joined later by Colombia) to enforce these rights.

25 FORESTRY

One of Ecuador's vast untapped resources is its forestland. Forests, half of which are government owned, cover 10,600,000 hectares (26,192,000 acres), or 38% of the total mainland area. The tropical jungles contain more than 2,240 known species of trees. Some of the denuded highlands have been planted with eucalyptus trees, which prevent soil erosion and provide both fuel and rough lumber. About 87% of the wood cut was burned as fuel. As of 1991, deforestation was a serious problem.

Ecuador is the world's largest producer and exporter of balsa. Several varieties of hardwoods, including species of mahogany, are used in cabinetmaking. Other forest products having some importance are the fiber for Panama hats (toquilla palm), vegetable ivory (tagua palm), kapok (ceiba tree), quinine (cinchona bark), and rubber. Wood exports of various kinds totaled $26.4 million in 1991.

Forest products represent a segment of Ecuador's economy that has potential for growth.

26 MINING

Aside from petroleum, Ecuador's known mineral resources include gold, silver, copper, bismuth, tin, lead, zinc , clays, kaolin, limestone, marble, and sulfur. A mining law of 1991 offers incentives for exploration, allows foreign individuals and companies access to mining rights, attempts to curtail corruption and unnecessary bureaucracy in the mining sector, and requires environmental impact statements to be filed before any mining activity starts. In 1991, production of nonfuel minerals contributed about 1% to the GDP, even though that sector grew by 60%, especially in the areas of industrial minerals, from the previous year. The total value of mineral output was $1.2 billion in 1991, down from $1.4 billion in 1990.

27 ENERGY AND POWER

In recent years, an increasingly important percentage of Ecuador's national income has come from the petroleum industry. Initially, this industry was slow in developing, and production actually declined from 2,849,000 barrels of crude oil in 1965 to a low of 1,354,000 barrels in 1971. Starting in the 1970s, however, output increased dramatically, from 28,579,000 barrels in 1972 to 77,052,000 barrels in 1981, and to 109,400,000 barrels in 1991. By February 1993, production had risen to 330,000 barrels per day. The value of petroleum exports was $1,110 million in 1992, accounting for 50% of total exports.

Extensive oil fields have been found in the Oriente; proved oil reserves were estimated at 1.6 billion barrels in 1992. At the rate of current extraction, known reserves will be depleted by around 2005. A 1969 agreement between the government and a Texaco-Gulf oil company consortium provided for the construction of a 480-km (300-mi) pipeline with a capacity of 250,000 barrels a day; the pipeline, which carries petroleum across the Andes, was partially destroyed in the March 1987 earthquake and was unusable for five months. Most oil comes from the Oriente Province (the Amazon jungle region) and is transported via the Trans-Ecuador pipeline to the tanker-loading port of Esmeraldas.

In the early 1970s, most of the oil industry was managed by foreign companies, which were granted concessions by the government; in 1993, Petroecuador, the state oil agency, controlled all production, exploration, and marketing . Production of natural gas is relatively small, totaling only 180 million cu m in 1991.

Installed capacity rose from 93,000 kw in 1958 to 357,000 kw in 1972 and 2,240,000 kw in 1991. Total production of electric power in that year amounted to 6.9 billion kwh, 72% of it hydroelectric. In order to conserve oil, energy requirements were increasingly being covered by hydroelectricity, by large projects, and by small-scale decentralized stations. In 1987, a 156 Mw station at Agoyan was completed. As of 1992, recently opened plants capable of producing another 600,000 kwh per year satisfied all national electricity demand, which permitted Ecuador to export energy. Electricity prices are the lowest in the Andean region.

28 INDUSTRY

In 1992, manufacturing output continued the recovery which began in 1991, particularly in tobacco products, textiles, wood and paper, chemicals, and metallic minerals. The expansion was fueled by the greater availability of intermediate goods and by strong domestic demand combined with demand from Colombia.

Ecuador enjoyed a period of prosperity in the 1970s, due in large part to a boom in oil. In the mid-1980s, the government launched an effort to stabilize and improve the competitiveness of the economy by liberalizing trade, reducing subsidies, eliminating price controls, and increasing exchange and interest rate flexibility. These efforts were undermined by the decline in international oil prices, an earthquake in 1987, and political difficulties.

The most promising sectors, outside of oil, are linked to agriculture and natural resources. In the agricultural sector, growth is likely to come from processed foods and nontraditional agricultural products, such as flowers and fresh tropical fruits (mango, babaco, and passion fruit) and vegetables (asparagus and heart of palm).

The vast mineral resources of Ecuador also present an opportunity for industrial development. Forestry also presents a promising sector.

29 SCIENCE AND TECHNOLOGY

In 1993, Ecuador had 17 scientific learned societies and research institutes, most notably the General Directorate of Hydrocarbons, the Institute of Nuclear Sciences, the Ecuadoran Institute of Natural Sciences, and the National Institute of Agricultural Research. Ecuador has 21 colleges and universities offering degrees in basic and applied sciences.

30 DOMESTIC TRADE

In the mountains, businesses are generally open from 8 AM to 6 PM, with a two-hour midday break; lunch hours are longer along the coast during the hottest months (December–April). Banks are usually open from 9 AM to 1:30 PM and from 2 to 6 PM during the week. Urban and suburban factories typically operated from 7 AM to 4 PM. Rural domestic trade among the *camposinos* (indegenous peoples) is limited by frequent deep levels of poverty and underdevelopment. The (mostly urban) middle class is relatively moderate in size, and suffered major declines in purchasing power during the 1980s, when economic growth faltered. Unemployment and underemployment have stimulated the growth of informal domestic economic activity.

31 FOREIGN TRADE

Ecuador's trade balance is generally positive, due overwhelmingly to its oil exports. In 1992, exports exceeded $3 billion, of which oil contributed 44%; bananas, 22%; and manufactured products, 8%. Imports that year totaled almost $2.5 billion, of which food accounted for 10%; fuel and energy, 4%; and capital goods, 38%. The US accounts for about 44% of Ecuador's foreign trade, and the Latin American Free Trade Association for most of the remainder. Under the Andean Pact, tariffs range from 0–35% for imported goods, with the exception of automobiles, which have a 40% tariff. Most non-tariff fees on imports have been eliminated.

32 BALANCE OF PAYMENTS

Ecuador's balance of payments showed repeated deficits on current accounts until the vast increases in petroleum exports in the 1970s; only in 1974 did the net balance finally register a surplus of $26.7 million. International reserves rose steadily from $57.3 million in 1967 to $64.7 million in 1971, increased dramatically (with the rise in oil exports) from $143.4 million in 1972 to $1,013 million by December 1980, and then dropped (as the oil market softened) to $304 million at the end of 1982, but were up to $500 million by 1987. In 1992, currency appreciation caused a narrowing in the trade surplus, which pushed down international reserves. In 1992, merchandise exports totaled $2,851 million and imports $2,207 million. The merchandise trade balance was $644 million.The following table summarizes Ecuador's balance of payments for 1990 and 1991 (in millions of US dollars):

	1990	1991
CURRENT ACCOUNT		
Goods, services, and income	−273.0	−577.0
Unrequited transfers	107.0	110.0
TOTALS	−166.0	−467.0
CAPITAL ACCOUNT		
Direct investment	82.0	85.0
Portfolio investment	—	—
Other long-term capital	−731.0	−598.0
Other short-term capital	−121.0	−65.0
Exceptional financing	1,064.0	1,004.0
Other liabilities	24.0	22.0
Reserves	−276.1	−163.7
TOTALS	41.9	284.3
Errors and omissions	124.1	182.7
Total change in reserves	−252.0	−167.9

33 BANKING AND SECURITIES

The Central Bank, founded as a private bank in 1927, was declared an organ of the state in 1948. It is owned by the national government and by private banks, which are required to invest at least 5% of their capital and reserves in it. The Central Bank issues and stabilizes currency, holds and manages foreign-exchange reserves, issues import and export permits, and regulates international transactions. Its foreign assets reserves in 1992 were s/3,037.1 million. The money supply, as measured by M1, in 1992 totaled s/1,860.2 million. The government-owned National Development Bank was founded in 1928 to provide credit for agricultural and industrial development.

Many of the 26 private domestic banks (in 1993) have branches in the largest cities and in most provincial capitals. There were 4 foreign banks in Ecuador in 1993. In that year, the financial system also included 3 finance corporations (financieras), which serve mainly corporate customers.

Trading in securities is relatively minor. Ecuador has 2 stock exchanges, one in Quito and the other in Guayaquil. Purchase and sale of government and some private securities are functions of the National Financial Corp., which, along with several hundred business people, owns the two stock exchanges. This autonomous agency, founded in 1964, deals in mortgage bonds issued by banks for agricultural and industrial development and in general seeks to mobilize funds for technical assistance to industry.

34 INSURANCE

The principal branches of insurance are fire, marine, and vehicular, with some growth in aviation and aircraft. Other branches, including life, are still undeveloped. In the mid-1980s, there were some 30 insurance companies in Ecuador.

Insurance is closely governed by legal provisions determining necessary reserves and security funds and requiring a percentage of investments in government securities. The Superintendency of Banks periodically examines insurance operations and authorizes the formation of new companies. In accordance with the General Insurance Law of 1965, all insurance companies must apply to the Bank Superintendency for authority to operate.

35 PUBLIC FINANCE

The central government, 20 provincial and some 100 municipal governments, and a large number of decentralized autonomous agencies constitute the public sector of the Ecuadoran economy.

Central government budgets in the late 1970s and early 1980s grew expansively from one year to the next, with oil being the key source of income.

In early 1992, the public sector deficit was estimated at 7% of GDP. A stabilization program begun in 1992 involved cutting public expenditures, freezing public employment, and the proposal for the elimination of many public agencies. By the end of 1992, the public sector deficit had been pared down to 2.8% of GDP.

The following table shows actual revenues and expenditures for 1989 and 1990 in billions of sucres.

	1989	1990
REVENUE AND GRANTS		
Tax revenue	819	1,337
Non-tax revenue	13	23
TOTAL	832	1,360
EXPENDITURES & LENDING MINUS REPAYMENTS		
General public service	62	85
Defense	102	156
Public order and safety	33	51
Education	155	219
Health	84	133
Social security and welfare	16	23
Housing and community amenities	3	7
Recreation, cultural, and religious affairs	2	5
Economic affairs and services	91	142
Other expenditures	164	316
Adjustments	23	68
Lending minus repayments	—	—
TOTAL	735	1,205
Deficit/Surplus	97	155

Ecuador's external debts are per capita one of the largest in Latin America. Medium- and long-term debt (including arrears) amounted to $12.3 billion in 1992.

36TAXATION

Taxation provides the government with most of its income. Levies include the income tax, 10% value-added tax, stamp tax, real estate tax, transfer tax, municipal taxes, and inheritance and gift taxes.

The basic tax on corporate income after the 1990 tax reform was 25% on undistributed profits and 36% on profits paid to foreign stockholders residing abroad. The basic personal income tax rate in 1993 ranged from 10% to 25%. Additional rates and surcharges are applicable to certain classes of income. A stamp tax is levied on almost all commercial and legal documents. Estate, inheritance, and gift taxes vary according to the amount involved and the closeness of family relationship between donor and recipient. Municipal real estate taxes range from 0.3% to 2%.

37CUSTOMS AND DUTIES

Import duties range from 5% to 17%, with the highest rate applied to consumer goods, imports of clothing competing with local products, leather goods, and wood products. Importers also pay a 1% service charge, a 2% special tax, and a 10% value-added tax.

Comparatively few trade barriers remain following measures taken in the 1990s to liberalize trade.

38FOREIGN INVESTMENT

Total foreign direct investment in Ecuador amounted to $100 million in 1993, of which the US accounted for $19 million; Panama, $8 million, and the Bahamas, $7 million. These figures do not include investments made in the petroleum sector, which are estimated at $200 million.

The government welcomes foreign investment and has substantially decreased regulatory barriers in recent years. As a member of the Andean Pact, Ecuador's foreign investment policy is governed largely by the parameters of Andean Pact Decisions 291 and 292 of May 1991, which provide for equal treatment of foreign and domestic investors, and unrestricted remittance of profits overseas. As of 8 January 1993, foreign investment is allowed without previous authorization in virtually all sectors of the economy that are open to domestic private investment, with remittance permitted of 100% of profits. The new regulations nullify the prior limitation of 49% of equity for foreign investment in the financial sector. Foreign investment in mining, fishing, and media require government approval, while investment in telecommunications, electricity, and petroleum are closed to foreigners. New capital markets legislation passed in 1993 should strengthen securities markets, thus attracting foreign portfolio investment.

39ECONOMIC DEVELOPMENT

In 1992/93, a major macroeconomic adjustment program was introduced, featuring a sizable currency devaluation (35%) and substantial increases in domestic fuel and electricity prices. Free trade agreements with Colombia, Venezuela, and Bolivia were signed, and new investment regulations were adopted to open up the economy to foreign investment and eliminate previous bureaucratic impediments. In April 1993, Ecuador qualified for Andean Trade Preference Act benefits, and has officially submitted an application to join the GATT.

High debt service obligations and external payments arrears to commercial banks impair Ecuador's economic growth. Ecuador sought a new IMF stand-by agreement in 1993, even though it rescheduled 1991 and 1992 interest and amortization payments to the Paris Club in 1992. As of September 1993, the Inter-American Development Bank (IBD) had projected lending to Ecuador to develop the following sectors: energy (irrigation and hydroelectric projects); urban and rural road construction; sanitation (potable water and sewerage); rural electrification; agriculture; and ecological projects for the coast. Foreign economic assistance from the US in 1992 amounted to $27.6 million in grants.

40SOCIAL DEVELOPMENT

The Social Welfare and Labor Ministry is active in a number of programs dedicated to the public welfare, participating in such cooperative efforts as the Andean Indian Mission of the UN.

Ecuador's social security program is administered through the Ecuadorian Social Security Institute. This program, based on the Compulsory Social Insurance Law of 1988, is operated through three agencies: the Social Security Fund, the Pension Fund, and the Social Insurance Medical Department. All workers in industry and commerce are covered. The benefits, for those who fulfill the requirements, consist of health and maternity insurance with free medical service, accident and occupational disease insurance, old age pensions, survivors' pensions, funeral expenses, and unemployment insurance. In 1991, workers' contributions to the social security system ranged from 6% to 11.35% of earnings, depending on the type of employment; employers contributed 8.85–10.85%.

In the mid-1970s, the government created the National Population Council within the Department of Public Health. The private Profamilia Association, affiliated with the International Planned Parenthood Federation, offers family-planning assistance. The 1978 constitution gives the government the responsibility of facilitating responsible parenthood; about 90,700 persons availed themselves of public family-planning services in 1991. In 1989, 53% of couples of reproductive age were practicing contraception. Ecuador's fertility rate in 1991 was 3.8.

Despite equal legal status, women have fewer educational and employment opportunities than men. There are fewer women in the professions, and salary discrimination is common.

41 HEALTH

Malnutrition and infant mortality are the country's two basic health problems. In 1990, 38% of children under 5 years old were considered malnourished. Food production has risen slightly, while the population has been increasing by 3% a year, and is expected to rise to 13.3 million by 2000.

Other health problems are largely being controlled. Yellow fever was eliminated by the efforts of the Rockefeller Foundation. Malaria, which as recently as 1942 caused about one-fourth of the deaths in Ecuador, has almost disappeared. Two antituberculosis organizations have helped reduce the mortality from that disease, which earlier was responsible for one-fifth of the nation's deaths. Between 1988 and 1991, still only 55% of the population had access to safe water, and 48% had adequate sanitation.

Health facilities are largely concentrated in the towns and are both too expensive and too distant to be used by most of the highland Amerindian population. Hospitals are operated by agencies of the national government, the municipalities, and private organizations or persons. In 1990 there were 15,737 physicians, 229 pharmacists, 4,847 dentists, and 5,538 nurses. The number of doctors per population has increased (from 1988 to 1992, there were 1.04 doctors per 1,000, with a nurse to doctor ratio of 0.3). From 1985 to 1992, it was estimated that 88% of the population had access to health care services. The country spent a total of $441 million in 1990 on health care.

In 1992, there were 20,000 births, and the 1993 birth rate was 29.7 per 1,000 people. Studies show that from 1980 to 1993, 53% of married women (ages 15 to 49) used contraception. The infant mortality rate in 1992 was 47 per 1,000 live births. Life expectancy at birth in 1992 was estimated at 66 years. Between 1990 and 1992, Ecuador immunized large numbers of children up to 1 year old: tuberculosis (99%); diphtheria, pertussis, and tetanus (83%); polio (83%); and measles (66%). The overall death rate in 1993 was 6.9 per 1,000 people. From 1985 to 1990, major causes of death were cited as follows: communicable diseases and maternal/perinatal causes (210 per 100,000); noncommunicable diseases (448 per 100,000); and injuries (166 per 100,000). There were 166 reported cases of tuberculosis per 100,000 people in 1990.

42 HOUSING

The 1974 housing census counted 2,747,800 housing units in the country. Almost all rural homes and many city dwellings on the coast are made with split bamboo siding and a palm thatch or corrugated iron roof. A housing development bank, Banco de la Vivienda, was established in 1961. In the late 1970s, the Institute of Sanitary Works initiated a program of well digging and latrine construction in rural areas. In 1990, 64% of all housing units were private houses with sanitary facilities; 12% were mediagreas, substandard one-story dwellings; 8% were apartments; 8% were cuartes, or semi-private units, and 7% were ranchos or covachas, dwellings of wood, stone, or brick covered with palm leaves, straw or other vegetation. Owners occupied 68%; 23% were rented; 6% were occupied rent free; and 3% were rented in exchange for services. Half of all dwellings had private toilet facilities and 40% had private baths.

43 EDUCATION

Literacy for persons 15 years of age and over rose from 56% in 1950 to 87.3% in 1990 (males, 90.4% and females, 83.9%). A five-year plan (1963–68) was implemented to teach 450,000 adults to read and write. A distinctive feature of the program was

its goal of "functional literacy," whereby adults not only learned how to read and write but also received further school training to develop their capacities and skills. The Ministry of Education is the principal authority for all educational programs except higher education, which is supervised by a national technical council. Education receives almost 30% of the national budget.

Although education is free and compulsory for six years, 13% of school-age boys and 15% of girls were not attending school in the late 1970s. At the primary level, in 1988 there were 62,451 teachers and 1,827,920 students enrolled. Secondary level had 771,928 students in 1987.

The Central University of Ecuador dates from 1594. There are three Catholic universities, in Quito, Guayaquil, and Cuenca. The National Polytechnic School in Quito offers degrees in industrial science and mechanical engineering, while the Polytechnic School of the Littoral in Guayaquil provides training in naval and petroleum engineering and in the natural sciences. At the universities and all higher level institutions, there were 206,541 students enrolled and 12,856 teaching staff in 1990.

44 LIBRARIES AND MUSEUMS

The oldest and most important library in Ecuador is the Central University Library in Quito. It was founded in 1586 and has 170,000 volumes. Other important collections are maintained at the Cuenca University Library (more than 62,000 volumes) and at the National Library in Quito (over 70,000 volumes). Libraries are also maintained by the other universities; the municipalities of Quito, Guayaquil, Cuenca, and Riobamba; the Ecuadoran Culture House; the Central Bank of Ecuador (100,000 volumes); and other organizations.

A number of museums in Ecuador preserve and display paintings, sculpture, coins, records, and artifacts of historic and scientific interest. One of the best collections in Ecuador of Inca and pre-Inca objects is found in a privately owned museum near Cuenca. Quito has the National Museum and Archives, the Archaeological Museum and Art Galleries of the Central Bank of Ecuador, the Jijón and Caamaño Archaeological and Historical Museum, and the museum of the Ecuadoran Culture House, as well as other smaller collections in schools and government agencies.

45 MEDIA

In 1991 there were 691,460 telephones in use; telephone service is operated by the government. Between the main cities and towns there are radiotelephone links. Quito is connected by telegraph with Colombia and Peru, and there are telephone and cable connections with all parts of the world. Ecuador had more than 300 radio stations in 1991. In addition to the numerous local stations, there was one central government network, Radio Nacional del of Ecuador. Also in 1991, the country had 33 television stations, most of them in color. There were 3,420,000 radio receivers and 910,000 television sets in the same year.

In 1991, Ecuador had 26 daily newspapers. The leading newspapers, with their political tendencies and estimated daily circulations in 1991, were as follows:

	ORIENTATION	CIRCULATION
QUITO		
El Comercio (m)	Independent	140,000
Últimas Noticias (e)	Independent	80,000
Hoy (a)	Independent	70,000
GUAYAQUIL		
El Universo (m)	Independent	190,000
La Razón (e)	Independent	40,000
CUENCA		
El Mercurio (m)	Conservative	13,000

The government requires all mass periodicals to participate in literacy and adult education campaigns. There is no censorship of newspapers or of radio and television stations.

46 ORGANIZATIONS

The outstanding contemporary learned society is the Ecuadoran Academy, founded in 1875, the second academy in Spanish America. It is a correspondent of the Royal Spanish Academy. The Ecuadoran Culture House prints works of contemporary Ecuadoran writers, encourages the investigation of scientific and social problems, and conducts discussions on cultural matters.

Fraternal organizations and service clubs in Ecuador include the Masons, Rotary clubs, and Lions clubs. The International Red Cross is also active. Social and charitable organizations are sponsored by the Roman Catholic Church. In addition, there are about seven chambers of commerce and industry and at least 10 employers' organizations.

47 TOURISM, TRAVEL, AND RECREATION

Tourists may enter Ecuador with tourist cards in place of visas if they are citizens of other Western Hemisphere countries. Visitors need a yellow-fever inoculation certificate if arriving from infected areas. Tourist facilities on the coast include modern resort hotels and fine beaches. Ecuador's highlands, reached by air or by the spectacular railroad or highway, are rich in natural beauty. Quito, the second-highest capital in the world, has modern hotels and transportation. Its churches and monasteries, with their delicately carved doors and altars, and an abundance of exquisite paintings and sculptures, make Quito, in the words of a 1979 UNESCO citation, a "cultural patrimony of mankind."

An important part of Ecuador's cultural life is the feria, or market day, which takes place weekly in many towns. The town of Otavala, about 56 km (35 mi) north of Quito, is well known for its colorful Saturday fairs. The Galápagos Islands, world-famous for their unusual wildlife, have become a popular site for ecotourism.

In 1991, there were 364,585 visitor arrivals in Ecuador, 35% from Colombia, 18% from the US, and 12% from Peru. Tourism receipts totaled $189 million. There were 29,464 hotel rooms with 58,629 beds.

Tourism has the potential to experience significant growth in Ecuador, given the natural attractions and architectural and historical sights. Further development of ecotourism must be carefully managed to avoid negative impact on Ecuador's environment.

48 FAMOUS ECUADORANS

Ecuadorans claim Atahualpa (1500?–33), the last emperor of the Incas, as the first renowned figure in their country's history; during the civil war between him and his half-brother, Huáscar, his administration of the Inca empire was based in what is now Ecuador. The 16th-century Amerindian general Rumiñahui is remembered for his heroic resistance to Spanish conquest. During the colonial period, Quito produced notable artists and sculptors. Among these were Miguel de Santiago (d.1673) and the Amerindians Manuel Chile (Caspicara) and Pampite. Francisco Javier Eugenio de Santa Cruz y Espejo (1747–95), the national hero of Ecuador, inspired much of the independence movement through his political writings. Espejo advocated complete emancipation from Spain, autonomous government for each colony, and nationalization of the clergy. Although he did not live to take part in the War of Independence (he died in prison for his political activities), he was an important figure in its philosophical development.

Vicente Rocafuerte (1783–1847), an early president, made significant contributions to the development of the republic. Another president, Gabriel García Moreno (1821–75), was the first to achieve national consolidation; he also contributed to the literary development of the nation. Juan Montalvo (1823–89) bitterly and brilliantly opposed conservatism in his essays and other works. Eloy Alfaro (1841–1912), another outstanding president, was noted for the honesty of his administration.

Among Ecuador's literary figures were Numa Pompilio Llona (1832–1907), a poet-philosopher, and Juan de León Mera (1832–94), a poet and novelist. Outstanding Ecuadorans of the 20th century include the poets Gonzalo Escudero (1903–71), Jorge Carrera Andrade (1903–78), César Dávila Andrade (1918–67), and Benjamín Carrión (1897–1979); the novelist Jorge Icaza (1906–78); the painter Oswaldo Guayasamin Calero (b.1919); José María Velasco Ibarra (1893–1979), who served five times as president of his country; and Galo Plaza Lasso (1906–87), a former president of Ecuador and of the OAS.

49 DEPENDENCIES

Ecuador's only territory, administered as a province since 1973, is the Archipelago of Columbus (Archipiélago de Colón), more commonly known as the Galápagos Islands, after the Spanish name for the large land tortoise found there. The six largest islands of the group (with their earlier names in parentheses) are Isabela (Albemarle), Santa Cruz (Indefatigable), Santiago (San Salvador or James), Fernandina (Narborough), Floreana (Santa María or Charles), and San Cristóbal (Chatham). Lying on the equator, this cluster of 60-odd islands is scattered over nearly 60,000 sq km (23,000 sq mi) of ocean and has a total land area of 8,010 sq km (3,093 sq mi). The center of the group lies at about 90°w, some 1,130 km (700 mi) from the coast of Ecuador and about 1,600 km (1,000 mi) southwest of Panama.

Most of the islands are small and barren. The largest, Isabela, which is 121 km (75 mi) long and makes up half the land area of the group, has the highest volcano (now only slightly active), reaching 1,689 m (5,541 ft). The climate of these tropical islands is modified by the cold Humboldt Current, which keeps the mean annual temperature as low as 21°C (70°F). Desertlike low-lying areas contrast with mist-shrouded heights at 240 m (800 ft) and higher elevations that have considerable rainfall.

Charles Darwin visited the islands in 1835 during his voyage on the *Beagle*, and his observations there made an important contribution to the development of his theories of evolution and natural selection. The unique forms of plant and animal life found on the various islands include 15 species of giant tortoise, considered to be the longest-lived creatures on earth, with a life span of about 150 years and a maximum weight of more than 225 kg (500 lb). In the mid-1980s, their number was estimated at 10,000. The Galápagos have 85 species of birds. In 1959, Ecuador declared the Galápagos a national park to prevent the extinction of the wildlife. The islands have since become one of the world's most noted focal points for naturalist studies and observations. About 21,000 visitors come to the islands each year.

Early Amerindian navigators, traveling on balsa rafts, frequently went to the Galápagos for the excellent fishing, but there are no evidences of any permanent settlement. Bishop Tomás de Berlanga of Panama landed at the Galápagos in 1535; he was the first of a series of Spaniards to visit the islands. In 1832, the first president of Ecuador, Juan José Flores, declared the islands a national territory. Several were used from time to time for penal colonies, but the practice was discontinued in 1959. The administrative seat is San Cristóbal. Only four of the islands are inhabited, and the estimated population numbers 6,000.

50 BIBLIOGRAPHY

Conaghan, Catherine M. *Restructuring Domination: Industrialists and the State in Ecuador*. Pittsburgh, Pa.: University of Pittsburgh Press, 1988.

Corkill, David. *Ecuador*. Santa Barbara, Calif.: Clio Press, 1989.

Cueva, Augustín. *The Process of Political Domination in Ecuador.* New Brunswick, N.J.: Transaction Books, 1981.

Darwin, Charles. *Voyage of the Beagle.* Garden City, N.Y.: Natural History Press, 1962 (orig. 1840).

Ecuador in Pictures. Minneapolis: Lerner Publications Co., 1987.

Fitch, John S. *The Military Coup d'Etat as a Political Process: Ecuador, 1948–1966.* Baltimore: Johns Hopkins University Press, 1977.

Hanratty, Dennis M. (ed.). *Ecuador, a Country Study.* 3rd ed. Washington, D.C.: Library of Congress, 1991.

Hassaureck, Friedrich. *Four Years Among the Ecuadorians.* Carbondale: Southern Illinois University Press, 1967.

Hurtado, Osvaldo. *Political Power in Ecuador.* Albuquerque: University of New Mexico Press, 1981.

Isaacs, Anita. *Military Rule and Transition in Ecuador, 1972–92.* Pittsburgh, Pa.: University of Pittsburgh Press, 1993.

Martz, John D. *Ecuador: Conflicting Political Culture and the Quest for Progress.* Boston: Allyn and Bacon, 1972.

Martz, John D. *Regime, Politics and Petroleum: Ecuador's Nationalistic Struggle.* New Brunswick, N.J.: Transaction, 1986.

Salomon, Frank. *Native Lords of Quito in the Age of the Incas.* New York: Cambridge University Press, 1986.

Schodt, David W. *Ecuador: An Andean Enigma.* Boulder, Colo.: Westview Press, 1987.

Spindler, Frank MacDonald. *Nineteenth Century Ecuador: A Historical Introduction.* Fairfax, Va.: George Mason University Press, 1987.

Van Aken, Mark J. *King of the Night: Juan Jose Flores and Ecuador, 1824–1864.* Berkeley: University of California Press, 1989.

Whitten, Norman E., Jr. *Sicuanga Runa: The Other Side of Development in Amazonian Ecuador.* Urbana: University of Illinois Press, 1985.

EL SALVADOR

Republic of El Salvador
República de El Salvador

CAPITAL: San Salvador.

FLAG: The national flag consists of a white stripe between two horizontal blue stripes. The national coat of arms is centered in the white band.

ANTHEM: *Saludemos la Patria Orgullosos (Let us Proudly Hail the Fatherland).*

MONETARY UNIT: The colón (C), often called the peso, is a paper currency of 100 centavos. There are coins of 1, 2, 3, 5, 10, 25, and 50 centavos and 1 colón and notes of 1, 2, 5, 10, 25, 50, and 100 colónes. C1 = $0.1147 (or $1 = C8.720).

WEIGHTS AND MEASURES: The metric system is the legal standard, but some old Spanish measures also are used.

HOLIDAYS: New Year's Day, 1 January; Labor Day, 1 May; Independence Day, 15 September; Columbus Day, 12 October; All Souls' Day, 2 November; First Call for Independence, 5 November; Christmas, 25 December. Movable religious holidays include Good Friday, Holy Saturday, Easter Monday, and Corpus Christi; there is a movable secular holiday, the Festival (1st week in August).

TIME: 6 AM = noon GMT.

¹LOCATION, SIZE, AND EXTENT

El Salvador, the smallest Central American country, has an area of 21,040 sq km (8,124 sq mi), extending 269 km (167 mi) WNW-ESE and 106 km (66 mi) NNE-SSW. Comparatively, the area occupied by El Salvador is slightly smaller than the state of Massachusetts. Bounded on the N and E by Honduras, on the S and SW by the Pacific Ocean, and on the NW by Guatemala, El Salvador has a total boundary length of 852 km (529 mi). It is the only Central American country without a Caribbean coastline.

El Salvador's capital city, San Salvador, is located in the west central part of the country.

²TOPOGRAPHY

El Salvador is a land of mountains and once-fertile upland plains. It is divided into three general topographic regions: (1) the hot, narrow Pacific coastal belt, 260 km (160 mi) long and 16–24 km (10–15 mi) wide; (2) the central plateau, at an altitude of about 610 m (2,000 ft), crossing from east to west, between two mountain ranges; and (3) the northern lowlands, formed by the wide Lempa River Valley, bounded by a high mountain range ascending to the Honduran border.

The central plateau, north of the Pacific coastal belt, is an area of valleys endowed with rich volcanic soil. This is the agricultural, industrial, and population center of the country; the capital, San Salvador (682 m/2,237 ft above sea level), is in this region. Almost surrounded by active volcanoes—Santa Ana (2,381 m/7,812 ft), San Vicente (2,173 m/7,129 ft), San Miguel (2,132 m/6,995 ft), San Salvador (1,967 m/6,453 ft), and Izalco (1,965 m/6,447 ft)—the region is a zone of recurrent earthquakes and volcanic activity; Izalco is known as the Lighthouse of the Pacific. El Salvador has several lakes, the largest being Ilopango, Güija, and Coatepeque. The Lempa, the most important of some 150 rivers, rises in Guatemala and runs south into El Salvador, eventually reaching the Pacific.

A major earthquake occurred in the San Salvador area in October 1986, leaving 1,500 dead and 329,165 homeless.

³CLIMATE

Located in the tropical zone, El Salvador has two distinct seasons: the dry season, from November to April, when light rains occur, and the wet season, from May to November, when the temporales, or heavy rains, fall. The coastal plain receives the heaviest rainfall. Some interior areas are relatively dry most of the year, necessitating irrigation and a selection of crops suited for arid land cultivation. The average annual rainfall is 182 cm (72 in). Temperatures vary with altitude, from the hot coastal lowlands to the semitropical central plateau; in general, the climate is warm, with an annual average maximum of 32°C (90°F) and an average minimum of 18°C (64°F). The average temperature at San Salvador is 22°C (72°F) in January and 23°C (73°F) in July.

⁴FLORA AND FAUNA

Indigenous trees include the mangrove, rubber, dogwood, mahogany, cedar, and walnut; pine and oak are found in the northern mountainous region. Varieties of tropical fruit, numerous medicinal plants, and balsam, a medicinal gum, grow in the country. Native fauna (greatly reduced in recent decades) includes varieties of monkey, jaguar, coyote, tapir, and armadillo, along with several kinds of parrots and various migratory birds. Fish, both freshwater and saltwater, turtles, iguanas, crocodiles, and alligators abound. Both venomous and nonvenomous snakes, the latter including the boa constrictor, are common in El Salvador.

⁵ENVIRONMENT

Because of heavy cutting, the forest resources of El Salvador had been reduced to about 2% of the total area by 1980. Forty-five percent of the wood taken from the forests is used for fuel. Peasant farmers burn the small trees and other growth on the hillsides to plant corn and beans, thus hastening the erosion of the topsoil. Seventy-five percent of the land area in El Salvador is threatened by erosion and desertification at a rate of 20 tons per hectare per year. The government enacted forestry conservation measures in 1973, but they have had little effect on the rate of deforestation. Among the environmental consequences of forest depletion, in addition to loss of soil fertility, are diminution of groundwater resources and drastic loss of native flora and fauna. Pollution is widespread and waste disposal is lax. El Salvador's cities produce 0.5 million tons of solid waste per year. Water quality is deteriorating. By 1993, 90% of El Salvador's rivers were polluted. Eighty-five percent of the people living in rural areas have impure water. Eighty-nine percent of the nation's 4.5 cu mi of water is used for agricultural purposes. There is no comprehensive national law controlling environmental protection, and the legislation that is on the books is poorly enforced. The National Environmental Protection Committee, established by decree in 1974, has had little impact.

The pollution of the environment in El Salvador is a serious threat to the survival of its plants and wildlife. As of 1994, six mammal species in a total of 129 were threatened. Two types of birds in a total of 432 and one type of reptile in 92 are also threatened. Of 2,500 plant species, 26 are considered endangered. As of 1987, endangered species in El Salvador included the tundra peregrine falcon, four species of turtle (green sea, hawksbill, leatherback, and olive ridley), American crocodile, ocelot, and spectacled caiman. The jaguar, giant anteater, and Central American tapir are extinct.

⁶POPULATION

As of the 1992 census, the population was 5,047,925. A population of 6,425,000 is projected by the UN for the year 2000, assuming a crude birthrate of 31.5 per 1,000 population, a crude death rate of 6.5, and a net natural increase of 25 during 1995–2000. El Salvador, with an estimated density in 1992 of 240 persons per sq km (621 per sq mi), is the most densely populated country in Central America. In 1995 an estimated 41% of the population was under 15 years of age. Approximately 53% of the population was estimated in 1995 to live in rural areas, and 47% in cities. San Salvador had a population in 1992 of 1,522,126, including suburbs; other major cities are Santa Ana (202,337) and San Miguel (182,817).

⁷MIGRATION

Until the early 1980s, emigration and immigration were negligible in El Salvador, except for migration of Salvadorans seeking economic advantages in Honduras, a trend that inflamed tensions between the two nations and was an underlying cause of their 1969 war. At that time, 300,000 Salvadorans were estimated to have settled in Honduras; following the war, as many as 130,000 Salvadorans may have returned from Honduras.

The civil war in El Salvador has resulted in the emigration of hundreds of thousands of Salvadorans. In mid-1992 there were 4,200 refugees in Mexico, 2,400 in Guatemala, 5,600 in Nicaragua, 5,600 in Costa Rica, 8,800 in Belize, and 400 in Panama. This was only the registered number. The number of foreign-born Salvadorans in the US swelled from 94,000 in 1980 to 473,000 in 1990, indicating the magnitude of the exodus.

While the US granted temporary asylum to thousands of Salvadoran refugees under the Carter administration, the Reagan administration began returning them to El Salvador in 1981. Of the 1 million Salvadorans estimated to be in the US in 1988, an estimated 550,000 had entered the country illegally, about 500,000 of them since 1979. Their remittances to their families in El Salvador are an important component of the stagnant economy, and their absence from the Salvadoran labor force has kept local unemployment lower than it would otherwise have been. Migration from rural to urban areas also is heavy. An estimated 550,000 people were displaced from their homes by warfare between 1979 and 1992.

⁸ETHNIC GROUPS

The population of El Salvador is racially and culturally homogeneous, with about 89% mestizo (mixed white and Amerindian), 10% Amerindian (mainly the Pipil tribes), and 1% white.

⁹LANGUAGES

The official language of the country is Spanish. A few Amerindians continue to speak Nahuatl.

¹⁰RELIGIONS

The constitution of 1962 guarantees religious freedom and exempts churches from property taxes. About 94% of the population is Roman Catholic; San Salvador is an archbishopric. Church officials and clergy have been active in the movement for human rights and social justice in El Salvador and have consequently been the targets of right-wing death squads and government security forces. In March 1980, Archbishop Oscar Arnulfo Romero y Galdames was killed by a sniper as he said mass at a hospital chapel in San Salvador. US aid was temporarily suspended after the murder of three US nuns and a social worker the following December.

About 3% of the population is Protestant; there are many active Protestant missions throughout the country, including Baptist, Lutheran, Church of Jesus Christ of Latter-day Saints, and Seventh-day Adventist denominations. In addition, there are about 7,800 Amerindian tribal religionists.

¹¹TRANSPORTATION

As of 1991 there were 1,500 km (930 mi) of paved highways and a total of 10,000 km (6,200 mi) in the road network. The Pan American Highway links El Salvador with Guatemala in the northwest and Honduras in the southeast. The Cuscatlán Bridge, where the Pan American Highway crosses the Lempa River, was the main thoroughfare for travel between eastern and western El Salvador at the time of its destruction by guerrillas in January 1984. In 1991, registered motor vehicles numbered about 160,000, half of which were passenger cars.

El Salvador had 602 km (374 mi) of public-service railroad tracks in 1991. The major domestic rail lines are operated by the National Railway of El Salvador, which was formed in 1975 by a merger of the Railways of El Salvador and the US-owned International Railways of Central America. There are rail connections to Puerto Barrios, Guatemala, which, under an agreement with the government of Guatemala, provides El Salvador with a port on the Caribbean. Another line runs west from La Unión/Cutuco to the El Salvador Guatemala border, providing a rail link to Mexico. The entire rail system is narrow gauge and single track.

The country's three ports—La Unión/Cutuco, La Libertad, and Acajutla—are located on the Pacific coast. Inland waterway traffic is negligible.

El Salvador is linked with the entire Western Hemisphere, Europe, and the Far East through international air services provided by Transportes Aéreos Centroamericanos (TACA) and other transportation companies. The principal international airport is near San Salvador, and serviced 539,000 international passengers in 1991.

¹²HISTORY

About 3000 BC, nomadic Nahuatl Amerindians, originally from present-day Mexico, migrated to Central America. The Pipil Amerindians were living in the region now known as El Salvador at the time of the Spanish conquest. They were an agricultural people, with a civilization similar to that of the Aztecs, except that the Pipil had abolished human sacrifice. In addition to the Pipil, two smaller groups, the Pocomans and the Lenca, lived in the area.

In 1524, Pedro de Alvarado, a Spanish conquistador sent by Hernán Cortés from Mexico, invaded El Salvador. After being

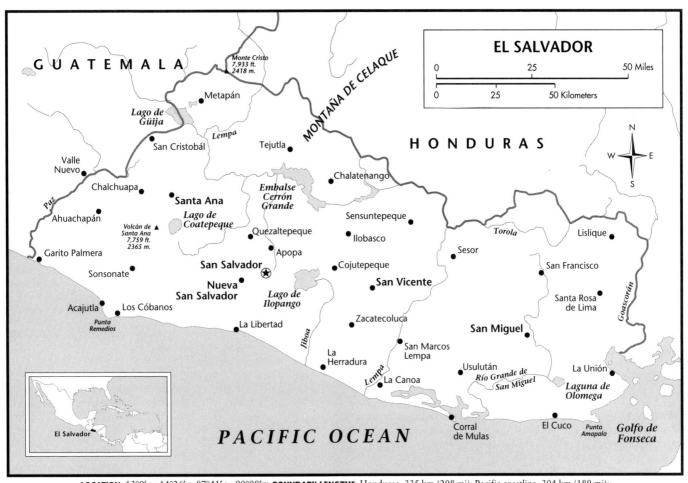

LOCATION: 13°9′ to 14°26′N; 87°41′ to 90°08′W. **BOUNDARY LENGTHS:** Honduras, 335 km (208 mi); Pacific coastline, 304 km (189 mi); Guatemala, 203 km (126 mi). **TERRITORIAL SEA LIMIT:** 200 mi.

forced to retreat by Amerindian resistance, he returned in 1525, defeated the Pipil and founded Sonsonate and San Salvador. During Spanish colonial period, San Salvador was one of six administrative regions under the captaincy-general of Guatemala. Spanish settlement consisted of a few cattle ranchers and some farmers.

The first call for independence from Spain, still commemorated as a national holiday, was made on 5 November 1811 by Father José Matías Delgado from the church of La Merced in San Salvador. However, independence did not come until 15 September 1821, when Guatemala led a movement for the independence of all of Central America. In 1822 the Mexican Emperor Agustín de Iturbide annexed all of Central America, but a year later he was deposed. El Salvador joined the United Provinces of Central America, which dissolved in 1838-39.

The republic was formally proclaimed on 25 January 1859. Turbulence, political instability, and frequent presidential changes characterized Salvadoran history during the second half of the 19th century. This period also saw the growth of coffee as El Salvador's leading product. The Salvadoran elite, known as the "14 families" or simply "the 14," created large coffee plantations, often on the land of displaced Indians. There followed a period of relative stability during 1900–30, but the seizure of power in 1931 by Gen. Maximiliano Hernández Martínez brought a period of constant military rule for almost 50 years. Hernández ruled for 13 years. Known for his mysticism, he also presided over the brutal suppression of a peasant uprising in 1932, with about 30,000 people killed. Few of the governments after Hernández even tried to reduce the gap between the

landowners and the landless classes, and those who did were doomed to failure.

Landless Salvadorans found land available in neighboring Honduras. During the 1960s the influx of Salvadorans increased, provoking counter-measures from the Honduran government. Tensions rose between the two nations, and on 14 July 1969, they went to war for four days. The immediate occasion of the conflict was the Central American soccer championship (which El Salvador won over Honduras on a disputed referee's call), leading US journalists to dub it the "Soccer War." A total of 3,000–4,000 people were killed on both sides. After an OAS-sponsored cease-fire, the two sides worked out a peace settlement, which was signed on 30 October 1980. A leftover border dispute was settled by the International Court of Justice (ICJ) in 1992.

In 1972, the military candidate for president was opposed by José Napoleón Duarte of the Christian Democratic Party (PDC). Duarte was denied election by fraud and sent into exile. Pressure for reform now came from the armed resistance from several leftist factions. The right unleashed "death squads" whose function was to intimidate and eliminate any who attempted to introduce change to the country.

By the late 1970s, the situation had become a civil war. The guerrillas had consolidated under the Farabundo Martí National Liberation Front (FMLN), and right-wing violence escalated. The military had not distinguished itself in its response to the violence, having engaged in widespread repression against suspected rebels. In 1979, a coup brought to power a set of reformist officers, who found common cause with such civilian leaders as Duarte. The junta liberalized the political system, setting legislative elections for 1982. The junta nationalized banks and the coffee export trade, while launching an ambitious and controversial land-reform program. Attacked by both left and right, the junta was unable either to suppress left-wing guerrillas or to control its own security forces, which began their own vigilante campaigns even as the angry landowners hired "death squads" to suppress opposition among peasants, students, clergy, and other groups.

In December 1980, José Napoleón Duarte of the Christian Democratic Party was installed by the junta as president—El Salvador's first civilian head of state since 1931—and a state of siege was proclaimed. Human rights groups could scarcely keep up with the number of violations occurring. In 1980, Archbishop Romero was assassinated while celebrating mass. In January 1981, the guerrillas launched their unsuccessful "final offensive," and throughout the rest of the year the bloodshed continued unabated. It is estimated that at least 62,000 people died between October 1979 and April 1987, most of whom were civilian non-combatants murdered by death squads and government security forces.

The junta headed by Duarte oversaw the drafting of a new constitution, adopted on 20 December 1983 by a constituent assembly that had been elected in March 1982. An interim president, Álvaro Alfredo Magaña, served through the March 1984 presidential election. In 1984, Duarte defeated Roberto D'Aubisson Arrieta of the National Republican Alliance in a runoff election in May. Duarte thus became the first constitutionally elected president in over 50 years.

Duarte's regime was a difficult balancing act between right and left, in the midst of attempts to bring about reform. The Christian Democrats won a majority in the National Assembly in the March 1985 legislative elections, but Duarte's overriding problem was the ongoing civil war. Eventually, economic dislocation brought on by Duarte's austerity measures and charges of corruption led to the election in 1989 of ARENA's (National Republican Alliance) Alfredo Félix Christiani Burkard. As Christiani came to office, many feared the worst, for ARENA had been tied to former candidate D'Aubisson, an extreme anti-communist who had frequently been accused of plotting against the government and supporting the death squads.

Christiani allayed these fears, and then some others. He called for direct dialogue between the government and the guerrillas. Although the process was rocky, and punctuated by military escalations, on 31 December 1991 the government and the FMLN signed an agreement, the Chapultepec Accord, which supplied the long hoped-for political solution to the civil war. A cease-fire took effect 1 February 1992, and held until 15 December 1992, when the FMLN officially laid down its arms.

The accord calls for reforms throughout the military, including the purge of officers linked to human rights abuses, and a reduction by 50% of the force. While the former has been a slow process, the latter goal was achieved ahead of schedule. In addition, paramilitary groups are to be banned, and land reforms are to be initiated.

New elections were held in 1994, and voters returned ARENA to power, with Armando Calerón Sol winning the presidential run-off election.

13GOVERNMENT
The constitution adopted on 20 November 1983 defines El Salvador as a republic. The constitution vests executive power in the president, who is to be elected by direct popular vote for a term of five years. The president, who must be native-born, over 30 years of age, the offspring of native-born parents, and a layperson, is not eligible for immediate reelection. The president is commander-in-chief of the armed forces, enforces the laws, formulates an annual budget, draws up international treaties and conventions (which must be ratified by the National Assembly),

appoints diplomatic and consular officials, and supervises the police. Every two years, the National Assembly elects three substitutes (designados), who can, in order of designation, assume the presidency when the president and vice-president are not available.

Legislative power is exercised by a unicameral National Assembly composed of 84 deputies apportioned among the various departments according to population. Deputies are elected for a three-year term and must be at least 25 years of age. The Assembly levies taxes, contracts loans and arranges for their payment, regulates the money supply, approves the executive budget, ratifies treaties and conventions, declares war, and suspends or reestablishes constitutional guarantees in national emergencies. The deputies, the president's ministers, and the Supreme Court all may propose legislation. The Assembly approves legislation and is technically empowered to override a presidential veto by a two-thirds vote.

Universal male and female suffrage was inaugurated in 1950. However, voting in El Salvador has been a source of controversy. During the 1980s, the government made voting compulsory, while the guerrillas insisted the citizens should not collaborate with the system. Thus, the Salvadorans were posed with a dilemma: vote, and face the wrath of the guerrillas, or refuse to vote, and immediately become suspected of leftist sympathy. At times, voting was not secret. Current practices suggest a more secretive and voluntary system is developing.

14POLITICAL PARTIES
The leading party of the right in El Salvador is the National Republican Alliance (Alianza Republicana Nacionalista—ARENA), organized in 1982 by the redoubtable Roberto D'Aubuisson. Although ARENA controlled the National Assembly until 1985, it could not win the presidency until it moderated its stance and Duarte's government had faltered. The National Conciliation Party (Partido de Conciliación Nacional-PCN) was founded by the military in 1961. With the PCN as its vehicle, the military controlled Salvadoran politics until 1979. Today, it is allied with ARENA. Together, the two parties control a majority of seats in the legislature.

The moderate Christian Democratic Party (Partido Demócrata Cristiano—PDC) was formed in 1960. For three decades, it was associated with its leader and founder, José Napoleón Duarte. Damaged by many splits over the years, and suffering after Duarte's unsuccessful presidency from 1984–1989, the PDC now makes up the largest opposition party, holding 26 of 84 seats in the National Assembly. The current secretary-general of the party, Fidel Chavez Mena, ran unsuccessfully for the presidency in 1989, receiving 36.6% of the vote.

The Democratic Convergence (CD) is a coalition of three leftist parties, dominated by Dr. Ruben Zamora and his Popular Social Christian Movement (MPSC). Zamora, a former member of the FMLN's political directorate, returned from exile in 1987. The CD holds 8 legislative seats.

The Chapultepec Accords introduced a new force into El Salvador's electoral system: the Farabundo Martí National Liberation Front (Frente Farabundo Martí de Liberación Nacional—FMLN). Named for an insurgent leader of the 1930s, the FMLN is a collection of five factions, each with its own version of Marxism or revolutionary ideology. As part of the Accords, the FMLN is now a legal party after having ceased its military operations.

15LOCAL GOVERNMENT
El Salvador is divided into 14 departments, each with a governor and alternate governor appointed by the executive power through the corresponding ministry. The country's 262 municipalities (cities, towns, and villages) are administered by mayors and municipal councils elected by popular vote. Traditionally independent in

their local functions, municipalities may be limited in their activities by the departmental governor.

16 JUDICIAL SYSTEM

The court system includes justices of the peace, courts of the first instance, intermediate level appellate courts, and the Supreme Court made up of 13 justices selected by the National Assembly. Other than justices of the peace, judges are appointed to renewable three-year terms. An 11-member National Council of the Judiciary, appointed by the National Assembly, is an independent body charged with screening judicial candidates for nomination.

According to the constitution, the Supreme Court is the court of last appeal; it passes on writs of habeas corpus, constitutionality of the laws, jurisdiction and administration of lower courts, and appointment of justices below the appellate level. There are also special courts, appointed by the National Assembly, and military tribunals, selected by the Supreme Court. Proposed reforms in 1993 may lead to more independence of the judiciary once constitutional amendments pave the way for the legislative proposals.

17 ARMED FORCES

The armed forces have a prescribed minimum strength of 3,000 men, set by the National Assembly. All Salvadoran males between 18 and 30 years of age are eligible for military service. In 1993, 43,700 personnel were in service, 40,000 of whom were in the army, 1,300 in the navy and marine corps, and 2,400 in the air force. National police and village militia forces number an additional 30,000. In 1991, $220 million, or about 3.6% of the gross domestic product was budgeted for national defense.

US military assistance during the civil way (1979–92) reached over $1 billion. In 1992 the US spent $200 million in El Salvador. As a precondition for authorizing military aid for El Salvador, the US Congress during 1982 and 1983 required the president of the US to certify every six months that progress in human, political, and economic rights had been made there. There are five battalions or their equivalents, assigned for territorial defense or mobile counterguerrilla operations, mostly foot infantry. The air force has 30 light aircraft and 25 armed helicopters. The navy is for coastal and riverine interdiction operations.

The United Nations maintains a civilian military police force of 1,000 to monitor the peace agreement of 1992.

18 INTERNATIONAL COOPERATION

El Salvador is a founding member of the UN, having joined on 24 October 1945, and participates in ECLAC and all the nonregional specialized agencies except UNIDO. It is also a member of the CACM, of which it was a principal advocate, as well as G-77, IDB, OAS, and PAHO. As of 1987, El Salvador was a signatory of the Law of the Sea, and in 1991 of GATT. El Salvador signed a peace plan for Central America outlined by Costa Rican President Oscar Arias Sánchez. Many of its principles, such as amnesty for rebels and repatriation or resettlement of refugees, applied to El Salvador.

19 ECONOMY

The economy is recovering from the civil strife of the 1980s. In 1992, GDP grew by 4.6% (from 1991) to over $6.4 billion, or $1,133 per capita. In 1990, GDP was just under $5.4 billion, of which commerce accounted for 34.8%; manufacturing, 18.6%; agriculture, 11.2%; personal services, 10.4%; government services, 7.6%; transportation and communications, 4.6%; construction, 2.6%; and other sectors, 10.2%

Agriculture is the foundation of El Salvador's economy, providing about two-thirds of the nation's exports and employing one-third of its labor force. Favorable weather caused agriculture to grow by 6.7% in 1992.

In the 1970s, El Salvador was the most industrialized nation in Central America, although a dozen years of civil war have eroded this position. The industrial sector is based in San Salvador and has been oriented largely to the domestic and Central American markets.

The Cristiani administration, which came into power in mid-1989, began a comprehensive economic reform plan oriented toward a free market economy. A market-based currency exchange rate was adopted. Price controls on over 200 products were eliminated, as were the sugar, coffee, and cotton marketing monopolies. The nationalized banking system was largely privatized.

After falling to 11.2% in 1991, inflation rose to around 20% in 1992, due to increased credit availability and public services prices. Although unemployment was officially registered at 7.5% in 1991, underemployment (part-time workers and those paid below minimum wage) may be as high as 49%.

20 INCOME

In 1992, El Salvador's GNP was $6,283 million at current prices, or $1,170 per capita. For the period 1985–92 the average inflation rate was 17.7%, resulting in a real growth rate in per capita GNP of 0.9%.

In 1992 the GDP was $6,443 million in current US dollars. It is estimated that in 1991 agriculture, hunting, forestry, and fishing contributed 10% to GDP; mining and quarrying, 1%; manufacturing, 19%; electricity, gas, and water, 2%; construction, 3%; wholesale and retail trade, 35%; transport, storage, and communication, 5%; finance, insurance, real estate, and business services, 8%; community, social, and personal services, 10%; and other sources, 7%.

21 LABOR

The civilian labor force numbered 2.6 million in 1991, of which 604,000 were engaged in agriculture, 680,000 in services, and 227,000 in industry. Unemployment in 1991 was estimated at 10%, down from 40–50% in 1985 as the result of a 3.5% economic growth rate in 1991.

The first official recognition of labor came in 1946, when a minister of labor and social welfare was added to the cabinet. In 1950, labor was permitted to organize. By 1972, about 4% of all wage and salary earners were unionized, and the proportion increased to about 15% (400,000 workers) in 1992. In 1991, eight labor leaders, all affiliated with the National Union of Workers and Peasants (UNOC) were elected to the national assembly. Strikes were frequent under the Duarte administration, in part as a response to the recession and austerity measures.

The workday is fixed at eight hours, and the workweek at 44 hours. Because of severe rural underemployment, many workers receive far less than the official minimum agricultural wage, which in 1991 was equivalent to $3.18 per day. Compulsory social security provisions cover enterprises employing five or more persons but exclude agricultural, domestic, and government workers. In 1992, there were 44 strikes and lockouts, involving some 124,200 workers, and resulting in 736,230 lost working days.

22 AGRICULTURE

In 1991, arable land comprised 733,000 hectares (1,811,000 acres); land under irrigation, 118,649 hectares (293,062 acres); and permanent pastures, 610,000 hectares (1,507,000 acres).

Coffee, El Salvador's major crop (30% of total agricultural output), is grown principally in the west and northwest at elevations of 460 to 1,520 m (1,500 to 5,000 ft). Primarily as a result of the civil war, coffee production declined in the 1980s. In June 1993, the Ministry of Economy certified the first shipment of organic coffee; the agrarian reform cooperative that produced the coffee had not used chemicals or pesticides for over four years.

Production in 1992 amounted to 147,000 tons (down from 156,000 tons in 1990). Exports of coffee in 1992 amounted to $150 million (25.6% of total exports), down $110 million from 1990.

Cotton production has declined sharply: earnings slumped from $45.2 million in 1979 to $2 million in 1992; production fell from 57,000 tons annually during 1979–81 to 4 tons in 1992. Reduced investment, in response to the civil war and the threat of agrarian reform, was responsible for the change. Cotton is grown primarily in south-central El Salvador.

Sugar production fell between 1979 and 1981 but later recovered; the total in 1991–93 was a record 324,257 tons, which brought $44.7 million in export earnings. Sugar is the most important agricultural product after coffee. Investment has increased, as has the area under cultivation (in contrast to the other two major export crops). The world price has, however, been in decline for several years, and this decline has cut export earnings; the government secured a $30.4 million loan from Venezuela to divert some production to gasohol. Agricultural production export values have declined in recent years; the 1992 value of $230.7 million only amounted to 58% of the 1988 value.

There were record harvests in 1992 of sugar, beans, and corn. Production amounts in 1992 included 706,000 tons of corn, 214,000 tons of sorghum, 72,000 tons of rice, and 62,000 tons of dry beans.

Government statistics in 1975 showed that 2% of the farms occupied 56.5% of the cultivable land, while 91.4% of the farms occupied only 21.9% of the land. An agrarian reform program was initiated in March 1980. The first phase, successfully completed the same year, resulted in the reallocation of 12% of total arable land (in 330 estates, all larger than 500 hectares/1,240 acres) to 30,000 peasants in cooperatives. The second phase, affecting medium-sized farms of 100–500 hectares (250–1,240 acres), was scheduled for 1981 but was effectively suspended. The third phase provided for the division of farms not worked directly by their owners into parcels of up to 7 hectares (17 acres), to be transferred to the tenant farmers or sharecroppers who worked on them. The government began implementing this phase, but after the March 1982 elections, the landowners tried to dismantle the entire program and evicted thousands of farmers. The government then mounted a countercampaign to restore the land to the peasantry. Throughout 1982, the struggle continued until, in early 1983, the National Assembly voted to extend the third phase to the end of the year, despite fierce opposition from ARENA. In June 1984 it revoked the program. By November 1984, 12,000 definitive titles had been issued to small tenant farmers.

23ANIMAL HUSBANDRY
Cattle and hogs are the predominant livestock in El Salvador. Cattle are of the "criollo" type and are used for production of both meat and milk. In 1992 there were 1,276,000 head of cattle. The hog population rose from 390,000 in 1950 to 560,000 in 1979 but then declined to 310,000 in 1992. Other livestock included 95,000 horses, 23,000 mules, 15,000 goats, and 4,000,000 chickens. In 1992, milk production was 350,000 tons. A total of 47,305 tons of eggs were produced during the same year.

24FISHING
The fishing industry, which centers on shrimp, has undergone significant development since it first gained commercial importance in 1957. The best coastal fishing grounds are off the southeastern sector. Scaled fish include freshwater robalo, sea bass, mullet, mackerel, swordfish, and redmouth; a tuna industry has been operating since 1963. The total fish catch was 11,341 tons in 1991.

25FORESTRY
Forests and woodlands covered 104,000 hectares (256,000 acres) in the early 1990s, representing only 5% of the total land area. Virgin forests once covered 90% of El Salvador. Almost all of the lumber used in building and in other Salvadoran industries must be imported, mainly coming from neighboring Guatemala, Honduras, and Nicaragua. In 1992, El Salvador imported $21.8 million in forest products. Forest products include dye woods and lumber, such as mahogany, walnut, and cedar, for furniture and cabinet work. El Salvador is the world's main source of balsam, a medicinal gum; between Acajutla and La Libertad in the southwest is the so-called Balsam Coast, which supports a species of balsam tree unique to El Salvador.

26MINING
Geologically, about 90% of the country is of volcanic origin and is less well endowed with mineral resources than any other Central American country. Although gold, silver, copper, lead, zinc, and sands that contain titanium and ilmenite are found in El Salvador, there is little commercial mineral production; gold and silver mining were halted in the 1980s, due to political unrest. In 1991, industrial minerals, especially limestone mined for domestic cement plants, were the primary commodities of the Salvadoran mineral industry.

27ENERGY AND POWER
El Salvador has no exploitable fossil fuels. Electric power production is controlled by the government's autonomous agency, the Executive Hydroelectric Commission of Río Lempa (CEL), created in 1945 to plan and develop the country's electric power facilities. The total installed power capacity rose from 171,000 kw in 1965 to 669,000 kw in 1991. Production increased from 582 million kwh in 1965 to 2,300 million kwh in 1991. Of that total, hydroelectric sources supplied nearly 60%; petroleum plants contributed 27%; and the remainder was supplied by geothermal resources. The electric power supply system has been a major target for economic sabotage by insurgents, particularly in the east. In 1991, guerrilla sabotage targeted power generation facilities and distribution stations, affecting 50% of generating capacity.

Geothermal energy sources have been explored to determine the possibility of tapping volcanically produced underground steam for power. At the end of 1975, CEL opened a 30,000-kw geothermal plant at Ahuachapán, at a cost of $25 million. This and subsequent projects were expected to reduce Salvadoran dependency on imported fuels, a goal frustrated by the civil war.

28INDUSTRY
The leading industrial region is the department of San Salvador. Other industrial centers are La Libertad, Santa Ana, San Miguel, Usulután, and San Vicente. The vast majority of enterprises are small scale, employing no more than 10 persons. Several new industries were introduced in the 1960s, most of them relying on imports of crude materials. In 1963, an oil refinery at Acajutla began processing Venezuelan crude oil; most of the output is consumed locally. During the early 1970s, the greatest increase in value of manufacturing occurred in chemicals and textiles. There are coffee-processing plants, sugar mills, bakeries, and plants making petroleum products, vegetable oils, fats, confectionery, dairy products, soaps, candles, matches, and organic fertilizers.

The manufacturing sector provided 17.5% of the GDP in 1985, compared with 14.5% in 1981. Following several years of declining production, output rose by 1.8% in 1984 and by a similar amount in 1985. However, the basic problems persist: the low purchasing power of the local population and the difficult financial and political situation of other Central American countries on which El Salvador depends for export markets.

Although the industrial sector plays a secondary role in the country's economy, it is the most developed in Central America. In 1992, the manufacturing sector recorded a 6.0% growth rate, accounting for 18.5% of GDP, 43% of exports, and 16% employment. This growth was fueled mainly by the apparel sector, which is the fastest growing subsector.

Other important industries are represented by food and beverages, tobacco, footwear, cement, fertilizers, and petroleum refining.

29 SCIENCE AND TECHNOLOGY

Learned societies and research institutes include organizations devoted to the study of seismology and geology, meteorology, agriculture (including the Salvadoran Institute for the Study of Coffee), and medicine. The extent to which these organizations were able to function amid the chaos of the 1980s could not be determined. Eight colleges and universities offer degrees in basic and applied sciences. In 1989, research and development expenditures totaled 291 million colones; 1,633 technicians and 142 scientists and engineers were engaged in research and development.

30 DOMESTIC TRADE

San Salvador is the chief commercial and marketing center. Marketing is comparatively simple in the predominantly rural, agrarian, and free-market economy of El Salvador.

Food is generally produced in small, scattered plots in the vicinity of urban areas and taken to market by traders (mostly women), either on their heads or by pack animals. Residents of remote rural areas usually consume most of what they produce, exchanging the remainder for other commodities. In urban areas, the business units are mainly small shops, while in rural regions, individual traders conduct their business at town marketplaces, where agricultural produce, meat, fruit, handicrafts, ceramics, and flowers are sold.

The usual business hours in the major cities are from 8 AM to noon and from 2 to 6 PM on weekdays and from 8 AM to noon on Saturday. Banking hours are from 9 to 11:30 AM and from 2:30 to 4 PM on weekdays, with a half day on Saturday.

31 FOREIGN TRADE

The main export continues to be coffee, although El Salvador has long sought to diversify its exports into other agricultural products such as cotton, as well as manufactures. Manufactured consumer and industrial goods, chemical products, machinery and transport equipment, petroleum, and various foodstuffs are the leading imports. El Salvador joined the GATT in 1991, although it is not a signatory to the GATT Subsidies Code.

Despite the strong growth in output and in export volume, export earnings rose only marginally. In 1992, exports totaled $597.5 million, an increase from $588 million in 1991. Most of the gain was from a 33% rise in exports to Central America, a result of regional trade liberalization and economic recovery. The country's primary exports were coffee, cotton, sugar, and shellfish (shrimp). Imports increased more substantially, registering a total of $1698.5 million in 1992. Major import items included consumer goods, raw materials, petroleum, food stuffs, capital goods, machinery, construction materials, and fertilizers. El Salvador's major trading partners are the US, Guatemala, Mexico, Japan, and Germany.

32 BALANCE OF PAYMENTS

El Salvador's positive trade balances in the late 1970s changed into deficits after 1980. The main reasons for this development were declines in cotton, sugar, and coffee export earnings and the virtual collapse of the CACM market. Capital inflow, principally in the form of transfers, donations, and credits from the US, has

helped cover the deficit. By 1990, the trade deficit stood at $668.2 million, up slightly from $663.8 million in 1989.

In 1992 merchandise exports totaled $588 million and imports $1,299.1 million. The merchandise trade balance was $−706.1 million. The following table summarizes El Salvador's balance of payments for 1990 and 1991 (in millions of US dollars):

	1990	1991
CURRENT ACCOUNT		
Goods, services, and income	−705.3	−839.1
Unrequited transfers	568.6	671.3
TOTALS	−136.7	−167.8
CAPITAL ACCOUNT		
Direct investment	1.7	25.3
Portfolio investment	—	—
Other long-term capital	0.2	−3.2
Other short-term capital	−12.2	−83.3
Exceptional financing	128.6	78.3
Other liabilities	—	—
Reserves	−153.5	70.0
TOTALS	−133.5	42.0
Errors and omissions	270.3	125.8
Total change in reserves	−154.0	127.4

33 BANKING AND SECURITIES

The banking system, nationalized in March 1980, consisted in 1993 of the Central Bank of El Salvador, 10 commercial and mortgage banks, four development banks, and six savings and loan associations. Two foreign banks were operating in the county as of 1993. The central bank of El Salvador, established in 1934, was nationalized in 1961. It is the sole bank of issue and the fiscal agent for the government. The commercial banks include the Mortgage Bank of El Salvador and the Commercial Bank of El Salvador. In 1991, as part of economic reforms, the government privatized six commercial banks and seven savings and loan institutions. As of December 31, 1993, the Central Reserve Bank had reserves of c8,730 million; the money supply, as measured by M2, was c23,853 million.

A stock exchange was established in San Salvador in 1964. Despite government encouragement, very few stocks are listed.

34 INSURANCE

In 1986, El Salvador had 12 insurance companies.

35 PUBLIC FINANCE

Most current public revenues come from taxes, fees, and fines. Municipal taxes and fees are subject to approval by the Ministry of the Interior. Until the early 1980s, government fiscal operations had generally shown surpluses; these enabled the government to sustain a growing volume of capital expenditures, resulting in a higher government share in total investments. The nonfinancial public sector deficit increased to 4.4% of GDP in 1991 (up from 2.5% in 1990), due to decreased coffee exports and delays in implementing tax reforms.

The following table shows actual revenues and expenditures for 1991 and 1992 in millions of colones.

	1991	1992
REVENUE AND GRANTS		
Tax revenue	4,119.2	4,915.7
Non-tax revenue	234.6	343.6
Capital revenue	16.9	0.4
Grants	512.4	243.5
TOTAL	4,783.5	5,605.8

	1991	1992
EXPENDITURES & LENDING MINUS REPAYMENTS		
General public service	532.1	761.9
Health	1,010.6	974.7
Housing and community amenities	70.8	79.8
Recreation, cultural, and religious affairs	63.8	90.6
Economic affairs and services	937.2	1,178.6
Lending minus repayments	859.4	−40.0
TOTAL	5,765.5	6,037.9
Deficit/Surplus	−982.0	−432.1

In 1990 El Salvador's total external debt stood at $2.2 billion, up 4.4% from the previous year. By 1992, external debt amounted to over $2.38 billion.

36 TAXATION

In 1985, total tax receipts accounted for 87% of current government revenues. Residents of El Salvador, whether citizens or not, are subject to progressive taxation on both domestic and foreign income ranging from 10% to 30%. Amounts received from insurance policies, interest on savings accounts, gifts, and inheritances are tax exempt. Taxes on corporate income are levied at 25% for amounts over the first $75,000, which are exempt. Other levies include a capital tax, a 10% value-added tax, and a real estate transfer tax (1.5–5% of the market value of the property). There are no local taxes.

37 CUSTOMS AND DUTIES

El Salvador's import tariff, formerly used primarily to raise revenues, has in recent years been employed to protect certain industries. As a member of the CACM, El Salvador operates under a common tariff classification system, as well as a common code and regulations. As part of President Christiani's free market economic policies, import licenses have been eliminated, and tariffs have been compressed to between 5% and 35%.

38 FOREIGN INVESTMENT

Private US investment in El Salvador totaled about $87 million in 1985, with the bulk of funds directed toward manufacturing and services. Canadian and Japanese interests are also important. Four assembly plants were operating by 1983 in a free trade zone established near San Salvador.

Presently, there are more than four operating free trade zones and a few more under development. Also, several new laws aimed at providing additional incentives for foreign investment and reactivating exports were implemented in 1991. Consequently, foreign investment in El Salvador was up to $38 million in 1992, a dramatic increase over the past few years.

39 ECONOMIC DEVELOPMENT

Prior to 1950, the government, with moderate success, followed a policy of stimulating national economic development indirectly by building roads, developing power facilities, establishing credit facilities, and extending tariff protection to some industries. The prosperity and advantageous international position of Salvadoran coffee masked the long-term need for further diversification.

During the 1950s, it became apparent that El Salvador's primary problems were the growing population and the lack of uncultivated arable land. The government therefore embarked on a program to encourage intensification of agriculture and expansion of small industry. The National Council for Economic Planning and Coordination (CONAPLAN), established in 1962, drafted a comprehensive five-year plan (1965–69) embodying general objectives proclaimed by the government in line with the aims of the US-inspired Alliance for Progress. The objectives included an annual 6% increase of the GNP; an annual 5%

decrease in illiteracy; expansion of education, health, and housing programs; extension of social security benefits to areas not yet covered; and implementation of an integrated program of agrarian reform. The stimulus for the development program was to be provided by a small group of wealthy Salvadorans, with a minimum of government participation.

In early 1973, CONAPLAN drafted a development plan for 1973–77, concentrating on production, labor, and social welfare. The basic objectives of the plan were improvement in income distribution, employment, health, nutrition, housing, and education; stimulation of the agricultural, industrial, and construction sectors; acceleration of regional development; and export diversification. The plan called for adoption of a government-sponsored investment and financing program, institutional and financial reforms, and policies to stimulate private investment. The plan set an average annual GDP growth rate of 6%. The agricultural sector was projected to grow by 4.8% per year, with priorities placed on livestock output for dairy import substitution purposes. In the industrial sector, CONAPLAN called for reorganization of work methods to increase personnel absorption, intensification of the processing of domestic raw materials, and strengthening of import substitution programs. CONAPLAN projected that the 10% annual growth rate generated by the agricultural, industrial, and construction sectors would provide substantial new job opportunities for the labor force.

The economic reforms adopted in 1980 included a land redistribution program, nationalization of coffee and sugar marketing operations, and nationalization of the banking industry. After the civil war was over, the country received sizeable influx of economic aid and private remittances. As part of the economic reform program, initiated in 1989 by newly elected President Cristiani, the country's economy has been recuperating at a fair pace. After the signing of the Chapultepec Peace Accords in January 1992, El Salvador has boosted business confidence and stimulated private investment. In 1992, GDP grew by 4.6%, an increase from 3.5% registered the previous year. Growth has been driven by high production, a reduced inflation rate of about 20%, and a stable exchange rate. Agricultural production, which constituted 25% of GDP, surged in 1992, increasing by 6.7%. Sugar, beans, and corn marked record harvests. Sugar, which is the second most important crop after coffee, brought $44.7 million in export earnings. Coffee, however, continued to suffer from low world prices.

In January 1986, a comprehensive stabilization program was announced. Among its most important measures were unification of the exchange rate at c5 = $1 (a 50% devaluation), raising of fuel prices and public transportation fares, imposition of price controls on consumer staples (food, medical supplies, and clothing), tariff hikes on nonessential imports, and a 2.5–15% rise in commercial interest rates. Because of the continuing costs from the civil war, and also from a drought the program fell short of its objectives.

US nonmilitary loans and grants to El Salvador during 1946–86 totaled $1,777.8 million; allocations rose from $58.3 million in 1980 to $182.2 million in 1982 and $322.6 million in 1986. Multilateral development assistance during the 1946–86 period amounted to $906.9 million, of which $634.2 million came from the IDB and $215.1 million from the IBRD.

In 1990, the IMF approved a stand-by agreement, which was followed by another one in January of 1992 and another one in May 1993. The World Bank extended a $75 million structural adjustment loan in 1991, and the Inter-American Development Bank has provided additional sectoral adjustment financing. It recently approved a $225 loan for the 1993-95 period.

Aided by the government's economic reforms and the IMF and World Bank supported macroeconomics programs, the country's economy will continue to prosper. The rising influx of private remittances and official transfers will boost domestic demand as well as construction activity, especially in transportation infrastructure and other public projects. The industrial sector should also benefit from this upturn. The strength of domestic demand has begun to exert pressures on domestic prices. Exports should continue to pick up in future years on the expectation of higher coffee prices. Imports will also rise, but not at the same pace of the last years. El Salvador's external deficit in goods and services will continue to be financed through private remittances and economic aid. However, economic aid is not likely to continue at levels of the last few years.

40SOCIAL DEVELOPMENT

The Salvadoran Social Security Institute was created in 1949 as an autonomous institution to provide national insurance for health, accidents, unemployment, invalidism, old age, and death. A basic social security law was passed in 1953; its program is supported by compulsory contributions from workers, employers, and the government. Most social security participants reside in urban areas.

The Ministry of Public Health and Social Welfare supervises institutions and hospitals for the infirm and aged, as well as child welfare and maternal care services. A national family planning program has been in effect since 1968, but the 1977 goal of reducing population growth rate to 2% by 1982 was not achieved. The total fertility rate in 1985–90 was 4.5. In 1989, 30,300 persons received family planning services. Abortion is permitted on narrow medical and legal grounds.

The new Family Code, passed in October 1993, removes previous provisions that discriminated against women and grants spousal rights to unmarried couples living together for at least 3 years. Women are paid lower wages than men and lack equal access to credit and property ownership.

41HEALTH

Health standards have improved considerably since 1930. Average life expectancy in 1992 was 66 years for women and men. The infant mortality rate, 139.4 per 1,000 live births in 1930 to 1934, declined to 47 by 1992. The crude death rate, 23 per 1,000 in 1930 to 1934, dropped to 7 in 1992. However, from 1979 to 1991, there were approximately 75,000 war-related deaths.

Between 1985 to 1992, only 56% of the population had access to health care services. However, immunization rates for children up to one year old were quite high between 1990 and 1992: tuberculosis (71%); diphtheria, pertussis, and tetanus (65%); polio (65%); and measles (62%).

Between 1988 and 1992, there were 64 physicians per 100,000 people, and the nurse to doctor ratio was 1.5. Between 1985 and 1990, there were 1.5 hospital beds per 1,000 inhabitants. The Ministry of Public Health and Social Welfare coordinates mobile health brigades, professional medical delegations, field offices, clinics, laboratories, and dispensaries. UNICEF, the US Institute of Inter-American Affairs, the Rockefeller Foundation, and other foreign organizations have assisted health campaigns. Total health care expenditures in 1990 were $317 million.

The principal causes of death remain gastroenteritis, influenza, malaria, measles, pneumonia, and bronchitis, caused or complicated by malnutrition, bad sanitation, and poor housing. Statistics show that from 1988 to 1991, only 47% of the population had access to safe water, and 58% had adequate sanitation. In 1990, there were 110 reported cases of tuberculosis per 100,000 people. Between the years 1985 and 1990, major causes of death were noted as communicable diseases and maternal/perinatal causes (202 per 100,000); noncommunicable diseases (385 per 100,000); and injuries (201 per 100,000). Much of the progress since the 1930s was undermined by the civil war, which overtaxed health care facilities while, in real terms, expenditures on

health care declined. The National Medical School was shut down in 1980.

42HOUSING

Inadequate housing, most critically felt in cities and towns, is endemic throughout El Salvador. More than half of all urban dwellings have earthen floors and adobe walls, and many have straw roofs. In the early 1990s, an estimated 47% of all households did not have easy access to drinking water, and 59% lacked plumbing. Housing problems have been exacerbated by the civil war, which has created hundreds of thousands of refugees. According to the latest available information for 1980–88, total housing units numbered 1,100,000, with 5.1 people per dwelling.

43EDUCATION

In 1930, 72% of Salvadorans over ten years of age were illiterate; by 1990, the figure had been reduced to about 27% (males: 23.8% and females: 30.0%). Primary education is free and compulsory, and the public school system is government controlled. Enforcement of primary-school attendance is difficult, however, and truancy is high in rural areas. Primary education lasts for nine years followed by three years of secondary education on completion of which students can obtain the "Bachillerato" diploma.

In 1991 there were 3,516 primary schools. Enrollments in 1991 were 1,000,671 primary and 94,278 secondary.

Twelve private and three public universities offer higher education. In 1990, 78,211 students were enrolled at the universities and other higher level institutions. The University of El Salvador in San Salvador, authorized in 1841 and with enrollments averaging 30,000, was a base for antigovernment agitation during the 1970s. The university was stormed and ransacked by government troops on 26 June 1980; at least 50 students and the rector were killed, and the university did not reopen for several years.

44LIBRARIES AND MUSEUMS

The National Library in San Salvador is the largest in the country, with 150,000 volumes. The library at the University of El Salvador has 91,000 volumes. In addition, there are three governmental libraries attached to the ministries of Education, Economics, and Foreign Affairs and a few small private and college libraries.

The David J. Guzmán National Museum in San Salvador, founded in 1883, is a general museum housing historical documents and pre-Columbian artifacts. The National Zoological Park in San Salvador, established in 1961, maintains a natural science museum.

45MEDIA

Ownership of domestic telephone and telegraph services has been transferred from the government to a semiautonomous agency. In 1993, there was current capacity for 152,546 telephone lines. There is an automatic telephone system in San Salvador. In 1991 there were 77 AM radio stations, including the government-owned Radio Nacional, no FM stations, and 5 commercial television stations. Radios in use increased from 398,000 in 1968 to about 2,175,000 in 1991, when television sets were estimated at 485,000.

The principal newspapers are published in the capital city, but the *Diario de Occidente*, published in Santa Ana, is the oldest daily in the country. The largest San Salvador dailies (with 1991 circulations) are *La Prensa Gráfica*, 95,000; *El Diario de Hoy*, 78,000; *El Mundo*, 40,000; *Diario Latino* 10,000; and *La Noticia*, 30,000.

46ORGANIZATIONS

Of prime importance is the Salvadoran Coffee Association, founded in 1930 to promote coffee production, distribution, and consumption; to improve quality; and to provide information and advice to growers. Its membership includes all native and foreign coffee growers in El Salvador, and it receives financial support from the government export tax on coffee. Other prominent management organizations include the National Coffee Institute, Chamber of Commerce and Industry, Salvadoran Association of Industrialists, Cattle Raisers' Association, El Salvador Sugar Cooperative, National Sugar Institute, Salvadoran Cotton Growers' Cooperative, and National Federation of Small Salvadoran Enterprises. In the early 1970s, the Salvadoran Communal Union was formed to improve peasant farming methods and to campaign for agrarian reform. In 1986, the union claimed 100,000 members.

47TOURISM, TRAVEL, AND RECREATION

A valid passport and visa are required for entry into El Salvador by all visitors except other Central American nationals, who need only a passport. Since the 1989 installation of the Christiani regime, tourism has rebounded, rising from 130,602 visitors in 1989 to 198,918 in 1991, of whom 115,230 came from Central America and 62,114 from North America. By that year, El Salvador had 3,185 hotel rooms and 6,048 beds with a 25.5% occupancy rate, and receipts from tourism totaled $157 million.

El Salvador's best-known natural wonder is Izalco, an active volcano often referred to as the Lighthouse of the Pacific because its smoke and flames are a guide to ships. Noteworthy are the cathedrals and churches of San Salvador, Santa Ana, and Sonsonate; the parks, gardens, and architecture of San Miguel; and the colonial atmosphere of San Vicente. Archaeological ruins of pre-Columbian origin are found in many parts of the country. Among the most striking are those at Tazumal, near Santa Ana, which include large pyramids and buildings with ancient carvings and inscriptions; there are more than 100 pyramid sites in El Salvador, many still unexcavated. The Pacific coast contains excellent beaches, and there is large-game fishing in the Gulf of Fonseca and in the ocean. Football (soccer) is the national sport.

48FAMOUS SALVADORANS

The national hero of El Salvador is Father José Matías Delgado (1768–1833), who raised the first call for independence. A renowned political leader was Manuel José Arce (1786–1847), who fought against the Mexican empire of Iturbide and was the first president of the United Provinces of Central America. Gerardo Barrios Espinosa (1809–65) was a liberal president during the 19th century.

Prominent Salvadoran literary figures of the 19th century were Juan José Cañas (1826–1912), a poet and diplomat and the author of the Salvadoran national anthem, and Francisco E. Galindo (1850–1900), a poet and dramatist.

Writers of note in the 20th century include Alberto Masferrer (1865–1932), an essayist and poet; Juan Ramón Uriarte (1875–1927), an essayist and educator; and Salvador Salazar Arrué (b.1899). Juan Francisco Cisneros (1823–78) is a nationally recognized painter.

Key figures in Salvadoran politics of the 1970s and 1980s are José Napoleón Duarte (b.1926) and Roberto D'Aubuisson Arrieta (b.1944?). The assassinated Roman Catholic Archbishop Oscar Arnulfo Romero y Galdames (1917–80) was well known as a defender of human rights.

49DEPENDENCIES

El Salvador has no territories or colonies.

50BIBLIOGRAPHY

Brockman, James R. *The Word Remains: A Life of Oscar Romero.* Maryknoll, N.Y.: Orbis Books, 1982.

Browning, David. *El Salvador: Landscape and Society.* Oxford: Clarendon, 1971.

Doggett, Martha. *Death Foretold: the Jesuit Murders in El Salvador.* Washington, D.C.: Georgetown University Press, 1993.

El Salvador's Decade of Terror: Human Rights Since the Assassination of Archbishop Romero. New Haven: Yale University Press, 1991.

Haggerty, Richard A. (ed.) *El Salvador, a Country Study.* 2d ed. Washington, D.C.: Library of Congress, 1990.

Haverstock, Nathan A. *El Salvador in Pictures.* Minneapolis: Lerner, 1987.

Lindo-Fuentes, Hector. *Weak Foundations: the Economy of El Salvador in the Nineteenth Century.* Berkeley: University of California Press, 1990.

Montgomery, Tommie Sue. *Revolution in El Salvador: Origins and Evolution.* Boulder, Colo.: Westview, 1982.

Parkman, Patricia. *Nonviolent Insurrection in El Salvador: the Fall of Maximiliano Hernandez Martinez.* Tucson: University of Arizona Press, 1988.

Webre, Stephen. *José Napoleón Duarte and the Christian Democratic Party in Salvadoran Politics, 1960–1972.* Baton Rouge: Louisiana State University Press, 1979.

Woodward, Ralph Lee. *El Salvador.* Santa Barbara, Calif.: Clio, 1988.

FRENCH AMERICAN DEPENDENCIES

FRENCH GUIANA

Located on the northeast coast of South America, and extending from 1°30′ to 5°30′N and from 51°4′ to 54°3′W, French Guiana (Guyane Française) has an Atlantic shoreline of about 320 km (200 mi) and a total area of some 91,000 sq km (35,000 sq mi). It is separated from Brazil by the Oyapock River in the E and the Tumuc-Humac Mountains in the S (440 km/273 mi); and from Suriname by the Maroni River (398 km/247 mi) on the W. Its length is about 400 km (250 mi) N–S, and its width is 300 km (190 mi) E–W. Several islands offshore are part of French Guiana including: the Îles du Salut (Devil's Island, Royale, and Saint-Joseph).

French Guiana consists of a small, low swampy coast called "terres basses," varying from 10 to 30 km (6–19 mi) in width, and a vast, partly unexplored interior, the "terres hautes", with grassy plateaus, equatorial forests (which cover 90% of the land area), and mountains. The mean annual temperature along the coast is 26°C (80°F) year-round. There is a rainy season from January to June; annual rainfall has a range of 350–400 cm (140–160 in). Average humidity is 85%. Endangered species in 1987 included the tundra peregrine falcon, three species of turtle (South American river, olive ridley, and leatherback), and the black caiman.

The population was estimated at 133,376 in 1993. Four-fifths of the inhabitants live in the coastal lowlands, and about 55% of the total live in Cayenne, the capital city. Most of the people are mixed white and black. In the interior are six tribes of aboriginal Indians; descendants of fugitive black slaves from Dutch Guiana (now Suriname) have settled along the rivers. Roman Catholicism is the dominant religion. The official language is French.

Amerindian tribes inhabited the region from ancient times, but their numbers probably did not exceed 25,000 on the eve of European colonization. The land now known as French Guiana was first settled by Frenchmen in 1604 and was awarded to France by the Peace of Breda in 1667. Since 1946, it has been an overseas department, sending, in 1986, two deputies and one senator to the French parliament and one representative to the French Economic and Social Council. The old penal settlements to which French prisoners were once deported have been completely liquidated. French Guiana consists of Cayenne and St.-Laurent-du-Maroni, each of which has the status of an arrondissement. The French commissioner is assisted by a popularly elected 19-member general council and a 31-member regional council.

Arable land and labor both being scarce, agriculture in French Guiana is still in a primitive state. Trade is mainly with France. The territory's exports, mainly shrimp, fish, and timber, totaled US$68.8 million in 1990; imports totaled US$435 million. Gold, and large deposits of bauxite are the chief mineral resources. The European Space Agency launches communications satellites from Kourou.

The literacy rate was 82% in 1992. Education for French Guiana's children is free and compulsory. The government has constructed 66 primary and 19 secondary schools. The Pasteur Institute, five hospitals, and other health units provide public health services. The infant mortality rate in 1993 was estimated at 16.6 per 1,000 live births.

GUADELOUPE

The French overseas department of Guadeloupe, situated among the Lesser Antilles, extends 15°52′ to 18°7′N, and 61° to 63°5′W. The length of Guadeloupe proper is 67 km (42 mi) E–W, and its width is 60 km (37 mi) N–S; its total coastline amounts to 656 km (408 mi). A narrow channel, Rivière Salée, divides Guadeloupe proper into two islands: Basse-Terre (848 sq km/327 sq mi) and Grande-Terre (585 sq km/226 sq mi). Outlying islands include Marie-Galante and La Désirade, and the Les Saintes and Petite Terre island groups, near the main islands; St. Barthélémy, about 120 km (75 mi) to the NW; and St. Martin, about 175 km (110 mi) to the NW, the northern two-thirds of it French, the southern third Dutch. Total area, including the outlying islands, is 1,780 sq km (687 sq mi). Basse-Terre is volcanic; its highest peak, La Soufrière (1,484 m/4,869 ft), erupted in the 18th and 19th centuries, and is still active. Annual rainfall ranges from 99 cm (39 in) on La Désirade to between 500 and 1,000 cm (200–400 in) on the mountains of Basse-Terre. Ferns, bamboo, mangrove, and tropical hardwoods are abundant. Endangered species include the Guadeloupe wren, green sea turtle, and leatherback turtle.

The estimated population as of 1993 was 422,114. The inhabitants are blacks or a mixture of blacks and descendants of Normans and Bretons who first settled the island in the 17th century. The people are mostly Roman Catholic. French is the official language, but a Creole dialect is widely spoken.

Guadeloupe was first settled by Arawak Indians from Venezuela about AD 200. Carib Indians, also from Venezuela, overran this agricultural and fishing community around AD 1000. Discovered by Columbus in 1493 and occupied by the French in 1635, Guadeloupe has, except for short periods during the Napoleonic wars, been French ever since. In 1648, St. Martin was shared with the Dutch. Guadeloupe became an overseas department in 1946. It is represented in the French parliament by three deputies and two senators. Local administration is similar to that of regions and departments in metropolitan France. The appointed commissioner is assisted by a 41-member general council, elected by universal suffrage, and by a newly created regional council.

There are about 1,940 km (1,200 mi) of highways, of which about 1,600 km (1,000 mi) are paved. There are no railways except for privately owned plantation lines. Marine traffic is concentrated at Pointe-à-Pitre and Basse-Terre. Steamships connect Guadeloupe with other West Indian islands, with North and South America, and with France. Air France and other airlines serve the international airport at Pointe-à-Pitre.

Sugar and bananas are the main crops. About 60% of all trade is with France. In 1988, exports yielded US$168 million; imports totaled US$1.2 billion. Sugar refining and rum distilling are the chief industries. The tourist industry accounted for receipts of US$284 million in 1991.

In 1990, elementary school enrollment totaled 59,900; secondary schools, a total of 52,300 students. Two teaching and research units—one for law and economics, the other for liberal arts and the sciences—provide higher education at the Université Antilles-Guyane in Pointe-à-Pitre. Several hundred scholarship holders study in French universities. The infant mortality rate was an estimated 9.2 per 1,000 live births in 1993, down from 17 in 1985.

MARTINIQUE

The island of Martinique is situated from 14°26′ to 14°53′N and 61°W among the Lesser Antilles in the Caribbean Sea, between the islands of Dominica and St. Lucia. It has an area of 1,110 sq km (429 sq mi), with a length of 75 km (47 mi) SE–NW and a maximum width of 34 km (21 mi) NE–SW. Its total coastline is about 350 km (220 mi). Most of the island is mountainous. The two highest peaks, Pelée (1,397 m/4,583 ft) and Carbet (1,196 m/ 3,923 ft), are volcanoes. On 8 May 1902, Mt. Pelée erupted, completely destroying the city of St. Pierre and killing 30,000 inhabitants. About 25% of the land is wooded, with both European and tropical trees represented. Average temperature is about 26°C (80°F) and average annual rainfall about 190 cm (75 in).

The 1993 estimated population was 387,656. The population, composed mostly of descendants of black Africans, Carib Indians, and Europeans, is predominantly Roman Catholic. French is the official language, but a Creole dialect is widely spoken and English is understood in tourist areas.

There are 1,864 km (1,158 mi) of roads and no railways. Steamer service connects Martinique with North and South America and France. Air France and other airlines provide air service from Lamentin Airport near Fort-de-France.

First inhabited by Carib Indians, Martinique was discovered by Columbus in 1502, and colonized by the French in 1635. Except for the periods 1762–63, 1793–1802, and 1809–15, the island has remained in French hands ever since. It is represented in the French parliament by four deputies and two senators.

Martinique's economy is agricultural. Sugarcane and bananas are the leading crops; pineapples, citrus fruit, mangoes, avocados, coffee, and cacao are also grown. Sugar refining, rum distilling, and fruit processing are the chief industries. Bananas, petroleum products, and rum are the principal exports; foodstuffs and oil are the main imports. In 1988, exports totaled US$196 million; imports totaled US$1.3 billion. Trade is mainly with France, which heavily subsidizes the budget. Tourism is important; 751,190 tourists visited Martinique in the late 1980s.

Education is compulsory through the primary and secondary levels. There is a branch of the Centre Universitaire Antilles-Guyana. The literacy rate was 93% in 1982. Martinique has 16 hospitals. The infant mortality rate in 1993 was estimated at 10.7 per 1,000 live births, down from 14 in 1985. Life expectancy was 78 years in the mid-1990s.

ST. PIERRE AND MIQUELON

The French territorial collectivity of St. Pierre and Miquelon (Territoire des Îles Saint-Pierre et Miquelon) is an archipelago in the North Atlantic Ocean, between 46°45′ and 47°10′N and 56°5′ and 56°25′W, located about 24 km (15 mi) W of Burin Peninsula on the south coast of Newfoundland. It consists of three main islands, St. Pierre, Miquelon, and Langlade—the two latter linked by a low, sandy isthmus—and several small ones. The length of the group is 43 km (27 mi) N–S, and it measures 22 km (14 mi) E–W at its widest extent. The total area is 242 sq km (93 sq mi). Although the archipelago is volcanic in origin, the highest point, Morne de la Grande Montagne, is only 393 m (1,289 ft). The temperature varies between an average daily low of –15°C (5°F) in winter and an average daily high of 22°C (72°F) in summer. The spring and autumn are very windy, and fogs are frequent throughout the year; annual precipitation averages 130 cm (51 in). Vegetation is scanty, except on Langlade, where several species of trees are found. Animal life includes seabirds, foxes, rabbits, and deer. The population in 1993 was estimated to be 6,652. Most of the people are descendants of Basque, Breton, and Norman settlers and are Roman Catholics.

The first permanent French settlement dates from 1604, and, except for several periods of British rule, the islands have remained French ever since. They became a French overseas territory in 1946, an overseas department in 1976, and a territorial collectivity in 1985. Cod fishing is the chief occupation.

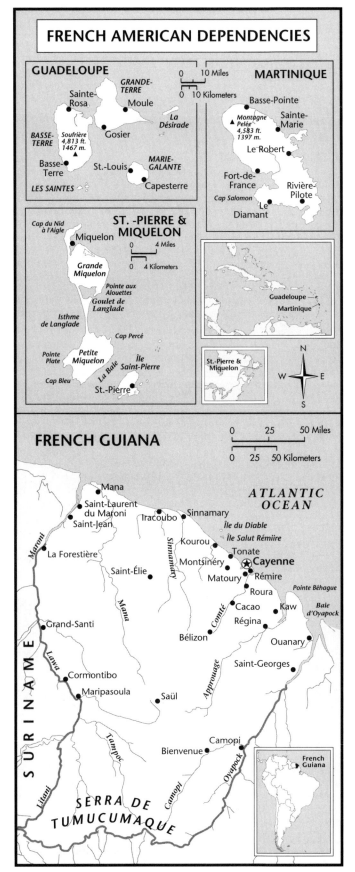

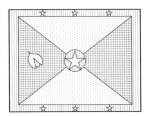

GRENADA

Grenda

CAPITAL: St. George's.

FLAG: The national flag consists of a red border surrounding a rectangle divided into two gold and two green triangles. There are seven yellow stars—three on the upper and three on the lower red border, and one large star at the apex of the four triangles—representing the six parishes and the island of Carriacou. A yellow nutmeg is represented on the hoist triangle.

ANTHEM: National Anthem beginning "Hail Grenada, land of ours, we pledge ourselves to thee."

MONETARY UNIT: The East Caribbean dollar (EC$) is a paper currency of 100 cents. There are coins of 1, 2, 5, 10, 25, and 50 cents, and 1 dollar, and notes of 5, 10, 20, and 100 East Caribbean dollars. EC$1 = US$0.3704 (or US$1 = EC$2.70).

WEIGHTS AND MEASURES: The metric system is in use.

HOLIDAYS: New Year, 1–2 January; Independence Day, 7 February; Labor Day, 1 May; Thanksgiving, 25 October; Christmas, 25 December; Boxing Day, 26 December. Movable holidays include Good Friday, Easter Monday, and Emancipation Day, 1st Monday in August.

TIME: 8 AM = noon GMT.

¹LOCATION, SIZE, AND EXTENT

Located about 160 km (100 mi) N of Trinidad and 109 km (68 mi) SSW of St. Vincent, Grenada, which includes the inhabited islands of Grenada, Carriacou, and Petite Martinique, has an area of 340 sq km (131 sq mi). Comparatively, the area occupied by Grenada is slightly less than twice the size of Washington, D.C. Grenada island extends 34 km (21 mi) NE–SW and 19 km (12 mi) SE–NW, and has a coastline of 121 km (75 mi).

Grenada's capital city, Saint George's, is located on the island's southwestern coast.

²TOPOGRAPHY

Volcanic in origin, Grenada is very hilly, with the highest peak, Mt. St. Catherine, in the Central Highlands, rising to 840 m (2,756 ft). The coastline is indented with many beaches and small bays. Several short streams cross the terrain. Lake Grand Etang is formed in the crater of a volcano at 530 m (1,740 ft) above sea level.

³CLIMATE

The tropical climate is tempered by almost constant sea breezes; the prevailing wind is from the northeast. Temperatures range from 21° to 29°C (70–84°F). Annual rainfall varies from about 150 cm (60 in) in the northern and southern coastal belts to as much as 380 cm (150 in) in the Central Highlands. There is a wet season from June to December, but rain falls periodically throughout the year.

⁴FLORA AND FAUNA

The Central Highlands support a wide variety of forest trees, and many types of tropical flowers and shrubs grow throughout the island. Characteristic wildlife includes the hummingbird, egret, dove, and wild pigeon; also to be found are armadillo, agouti, and monkeys.

⁵ENVIRONMENT

As a member of the Organization of Eastern Caribbean States (OECS) formed in 1981, Grenada shares the advantages and disadvantages of island nations in the area. Water supply is limited and, in some areas, polluted by agricultural chemicals and sewage. Forests are threatened by the expansion of farming activities and the use of wood for fuel. The nation's coasts are affected by industrial pollution which threatens the nation's tourist trade. Environmental responsibilities are vested in the Ministry of Health and Housing. As of 1987, endangered species included the Grenada hook-billed kite, tundra peregrine falcon, the green sea and hawksbill turtles, the spectacled caiman, and the Orinoco crocodile.

⁶POPULATION

The population, including dependencies, was 99,205 in 1988. Population density at that time was 280 persons per sq km (746 per sq mi). St. George's, Grenada's major city and port, had a population estimated at 7,500 in 1995 (35,742 including suburbs).

⁷MIGRATION

Grenadians have always emigrated, mainly to the UK and Canada. Emigration increased after the 1979 coup.

⁸ETHNIC GROUPS

The descendants of former African slaves, together with mixed black and white racial strains, make up about 91% of the population. The remainder consists of small groups of Asian (largely Indian) and European descent.

⁹LANGUAGES

English is the official and common language. A French African patois also is spoken.

¹⁰RELIGIONS

As of 1993, an estimated 66% of the population was Roman Catholic. Other main groups are Anglican (20%), Methodist, Seventh-day Adventist, and Presbyterian.

[11]TRANSPORTATION

In 1991, Grenada's extensive road system of 1,000 km (620 mi) included 600 km (372 mi) of paved and 300 km (186 mi) of improved roads. In 1984 there were 7,741 registered motor vehicles. The country's major port is St. George's. A new international airport at Point Salines, built largely with Cuban assistance, and scheduled for completion in 1984, was repeatedly cited by the US as posing a possible military threat to the Caribbean region. After the US-led invasion in 1983, the airport was completed with funding mainly from the US. Airline flights began in October 1984, and in 1991, the airport served 206,000 arriving and departing passengers. There are smaller airports at Pearls and on Carriacou.

[12]HISTORY

Grenada was inhabited by Arawak Indians when first discovered on 15 August 1498 by Christopher Columbus, who named it Concepción. By the 18th century, the island was known as Grenada. The origin of that name is unknown, possibly a corruption of the Spanish city of Granada. A secure harbor (at St. George's) attracted some French settlers during the 16th century. After a few failed French private ventures in 1650 and 1657, the French government annexed Grenada in 1674. The island remained under French control until 1762, when Admiral George Rodney captured it for Great Britain. The French regained Grenada in 1779, but the Versailles treaty of 1783 returned Grenada to Britain.

Sugar was Grenada's main product until the 19th century. At that time, the development of spices, especially nutmeg, coupled with the emancipation of slaves in 1834, led to a new economic base for the island. The economy flourished during the second half of the 19th century, and the cultivation of nutmeg, cloves, ginger, and cinnamon, earned Grenada the name Isle of Spice. Grenada's colonial status ended in 1958 when it joined the ill-fated Federation of the West Indies. In 1962, the federation dissolved, and in 1967, Grenada became an associated state of the UK.

On 28 February 1972, general elections resulted in the victory of Eric Matthew Gairy, who ran under the banner of the pro-independence Grenada United Labour Party (GULP). A constitutional conference was held in London during May 1973, and independence was set for the following February. Independence came on 7 February 1974, in spite of widespread strikes and demonstrations protesting Gairy's secret police force, actions that were supported by trade unions in neighboring Barbados and Trinidad and Tobago. Prime Minister Gairy ruled for five years.

On 13 March 1979, the opposition party, the New Jewel Movement, seized power, and Maurice Bishop became prime minister of the People's Revolutionary Government (PRG). Bishop suspended the constitution, jailed opposition leaders, and shut down independent newspapers. The PRG was drawn toward Cuba and its allies in the Caribbean region, as relations with the US and some of Grenada's more conservative Caribbean neighbors deteriorated.

On 19 October 1983, in the course of a power struggle within the PRG, Bishop and several followers were shot to death, and a hard-line Marxist military council, headed by Gen. Hudson Austin, took over. Six days later, thousands of US troops, accompanied by token forces from seven other Caribbean nations, invaded the island, ostensibly to protect the lives of American students there. Nearly all of the 700 Cubans then in Grenada were captured and expelled. In spite of the UN General Assembly's condemnation of the invasion, Gen. Austin was placed in detention, and the governor-general, Sir Paul Scoon, formed an interim government to prepare for elections. US combat troops were withdrawn in December 1983, but 300 support troops and 430 members of Caribbean forces remained on the island until September 1985.

Elections were held in December 1984, and Herbert Blaize and his New National Party (NNP) won 59% of the popular vote and 14 of the 15 House of Representatives seats. Prime Minister Blaize died in December 1989, and Ben Jones formed a government until the elections of 1990. Those elections elevated the National Democratic Congress (NDC) to majority status and Nicholas Brathwaite became prime minister.

[13]GOVERNMENT

The independence constitution, effective in 1974 but suspended after the 1979 coup, was reinstated after the US invasion. It provides for a governor-general appointed by the British crown and for a parliamentary government comprising independent executive, legislative, and judicial branches. Under this constitution, the bicameral legislature consists of a Senate of 13 members appointed by the governor-general, acting on the advice of the prime minister and the leader of the opposition, and a 15-seat House of Representatives. The governor-general appoints as prime minister the majority leader of the House. The cabinet of the NNP government comprises the prime minister, four senior ministers, and four ministers of state. The cabinet is the executive arm of the government and is responsible for making policy.

[14]POLITICAL PARTIES

There are six political parties in Grenada. The National Democratic Congress (NDC) is a moderate party, which took seven of 15 seats in the House of Representatives in the 1990 elections. They formed a government when one member of the opposition Grenada United Labour Party (GULP) switched allegiance.

The Grenada United Labour Party is still under the leadership of Sir Eric Gairy. Right-wing and populist in its approach, the group holds three seats in the legislature. The New National Party (NNP) is a coalition of three moderate parties, and is headed by Keith Mitchell. The NNP and a moderate group called The National Party (TNP), under Ben Jones, each hold two seats in the House.

On the left are two parties, the Maurice Bishop Patriotic Movement (MBPM) and the reconstituted New Jewel Movement (NJM). Neither is represented in the government.

[15]LOCAL GOVERNMENT

For administrative purposes, the main island is divided into six parishes and one dependency.

[16]JUDICIAL SYSTEM

The Grenada Supreme Court, in St. George's, consists of a High Court of Justice and a two-tier Court of Appeals. The Court of Magisterial Appeals hears appeals from magistrates' courts, which exercise summary jurisdiction; the Itinerant Court of Appeal hears appeals from the High Court.

On joining the Organization of Eastern Caribbean States (OECS) in 1991, the Grenada Supreme Court became the Supreme Court of Grenada and the West Indian Associated States. Under the OECS system, appeals may to taken from this court to the Privy Council in London.

[17]ARMED FORCES

The new UK- and US-trained 550-man Royal Grenada Police and Coast Guard provides internal defense. A US-trained Special Services Unit participates in the East Caribbean Defence Pact.

[18]INTERNATIONAL COOPERATION

Grenada became a member of the UN on 17 September 1974 and participates in ECLAC and all the nonregional specialized agencies except IAEA, IMO, WIPO, and WMO. Grenada joined the

OAS on 13 May 1975. It is also a member of the Commonwealth of Nations, CARICOM, G-77, and PAHO. Grenada is a signatory of the Law of the Sea and is a de facto adherent of the GATT.

[19]ECONOMY

Since 1983, Grenada's economy has been highly dependent on international trade and finance for its development. Exports of nutmeg, cocoa, spices, and bananas are the main earners of foreign exchange, with tourism close behind. Agriculture is the mainstay of the economy, with crop production utilizing about 9,000 hectares (22,240 acres) of arable land. In 1985, agriculture contributed 19.6% to GNP, 80% from crops, and 20% from livestock, fishing, and forestry. Agricultural products accounted for 93% of total exports of EC$71.4 million in 1986. Since 1984, tourism has shown signs of recovery and contributed $64 million to the economy in 1985 (6.5% of GDP). The small manufacturing sector has grown since 1984; the government emphasizes private-sector involvement and free enterprise. In 1985, manufacturing contributed 6.3% to GDP and 6–7% of exports. The public-sector investment program is financed almost entirely by aid from the US, the EC, the UK, and Canada, with technical assistance from the UNDP, the OAS, and the Commonwealth.

As a result of the signing of an agreement in 1991 between Grenada and Indonesia—the world's largest nutmeg producer—Grenada's share of the export market for the spice has decreased. Production of mace and bananas were also down in 1992, causing an overall decline in the agricultural sector. As a consequence, real economic growth fell to 0.06% in 1992 from 2.9% in 1991. Grenada continues to register a negative trade balance, but income from tourism and capital repatriation helps offset what otherwise would be foreign exchange depletion. The tourism industry continues to expand, recording 290,639 visitors to the island in 1992.

[20]INCOME

In 1992, Grenada's GNP was $210 million at current prices, or $2,310 per capita. For the period 1985–92 the average inflation rate was 5.2%, resulting in a real growth rate in per capita GNP of 4.4%.

In 1992 the GDP was $250 million in current US dollars. It is estimated that in 1991 agriculture, hunting, forestry, and fishing contributed 12% to GDP; mining and quarrying, less than 1%; manufacturing, 4%; electricity, gas, and water, 3%; construction, 8%; wholesale and retail trade, 16%; transport, storage, and communication, 12%; finance, insurance, real estate, and business services, 8%; community, social, and personal services, 2%; and other sources, 35%.

[21]LABOR

Grenada's labor force is estimated at 40,000 (1991), including a large supply of unskilled and semiskilled workers. Unemployment was estimated in 1991 at 28% of the labor force. The distribution of the employed labor force in 1991 was agriculture, 24%; construction, 8%; industry, 5%; services, 31%; and other sectors (including self-employed), 32%. The minimum hourly wage for agricultural workers in 1991 was US$0.69; for commercial workers, US$0.79; and for industrial workers, US$0.90.

Approximately 20% of the labor force is unionized. There are seven major trade unions in the country, including four civil service unions. Grenada has one of the largest per capita public sector work forces in the eastern Caribbean region. Civil servant salaries consume over 50% of government revenues and are 40% higher than all export earnings from Grenada's agricultural fishing and manufacturing sectors combined.

[22]AGRICULTURE

Numerous spices, fruits, and vegetables are grown in Grenada. The principal crops for export are nutmeg and mace, bananas,

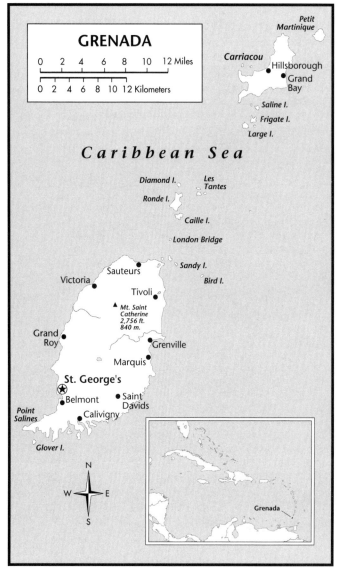

LOCATION: 12°7′N and 61°40′W. COASTLINE: 121 km (75 mi). TERRITORIAL SEA LIMIT: 12 mi.

cocoa beans, and other fresh fruits and vegetables. Production in 1992 included bananas, 11,000 tons; cocoa, 2,000 tons; and nutmeg, 1,975 tons. Banana production decreased in the 1980s due to the appearance of Moko disease. There are small scattered plots of cotton, cloves, limes, cinnamon, and coffee. Both cotton and lime oil are produced on Carriacou. Food crops consist of yams, sweet potatoes, corn, peas, and beans. Grenada is especially known for its nutmeg production, earning it the nickname "Spice Island." Prices for nutmeg, however, have been rather low in recent years due to a soft market and the collapse of the Grenada/Indonesia cartel in 1990. Late in 1991, Grenada and Indonesia (the world's largest nutmeg producer) signed a new cartel agreement which aims to constrict production, thereby driving up world prices for the spice. In 1992, exports of agricultural products amounted to US$11.5 million, or about 58% of total exports.

[23]ANIMAL HUSBANDRY

There is very little dairy farming in Grenada. Most livestock is raised by individuals for their own use. In 1992 there were an estimated 4,000 head of cattle, 23,000 sheep and goats, and 1,000 donkeys. Some 250,000 poultry were raised to supply local needs.

24FISHING
Fishing is mostly coastal. The 1991 catch was 1,990 tons, predominantly tuna and scad.

25FORESTRY
There are approximately 3,000 hectares (7,400 acres) of rain forest, about 75% of which is government-owned. Since 1957, some 320 hectares (800 acres) of forest, primarily of Honduras mahogany, blue mahoe, and teak, have been introduced. The Forestry Development Corp. was established in 1979 to develop forest resources and woodworking industries.

26MINING
There are no mining operations in Grenada except for open-face red gravel deposits.

27ENERGY AND POWER
The government and the Commonwealth Development Corp. jointly operate a private company, Grenada Electricity Services, Ltd., for the supply and distribution of electricity throughout the island. In 1991, net installed capacity of electric generating plants was about 9,000 kw; output that year totaled 52 million kwh, double the output of 1988. In 1991, a proposal was put forward to refine Venezuelan crude oil on Grenada by 1996.

28INDUSTRY
Industry is small-scale, mainly producing consumer products for local use. Local firms produce beer, oils, soap (from copra), furniture, mattresses, clothing, and a number of other items.

In 1985, the government established the Industrial Development Corporation, which by 1986 had begun construction of a 7.3-hectare (19.5-acre) industrial park near Point Salines airport. New industries that had located there by mid-1987 included data entry, protective clothing, and tire retreading.

The manufacturing sector has maintained a consistent rate of growth in recent years. In 1992, a surge within the manufacturing sector in the production of beer, cigarettes, and soft drinks was experienced.

29SCIENCE AND TECHNOLOGY
St. George's University School of Medicine has 436 students. A school of agriculture is located in Mirabeau.

30DOMESTIC TRADE
There are meat, fish, fruit, and vegetable markets, in addition to other types of small retail shops, all over the island. St. George's is the import and merchandising center. A widespread network of cooperatives exists for the distribution and sale of agricultural products.

Shopping hours are usually from 8 to 11:45 AM and 1 to 4 PM on weekdays and 8 AM to noon on Saturdays. Banks are open from 8 AM to noon, Monday through Thursday, and 8 AM to noon and 3 PM to 6 PM on Friday.

31FOREIGN TRADE
Annual trade deficits are caused by the large flow of imports, including motor vehicles and other consumer goods, fuels, and fertilizer. Virtually all exports are agricultural products, chiefly cocoa, nutmeg, and mace. The balance of trade improved in 1986, due to higher world prices for Grenada's exports. Exports in 1985 totaled EC$60.3 million, while imports reached EC$187 million.

In 1992, export of goods reached US$83.3 million. Major export items consisted of nutmeg, cocoa, bananas, mace, and textiles. Grenada exported mainly to Trinidad and Tobago, the UK, the Netherlands, and Germany.

Total merchandise imports amounted to US$128.2 million. Major import items consisted of foods, machinery and transport

equipment, manufactured goods, and fuels. Grenada imported mostly from the US, UK, Trinidad and Tobago, Canada, North Korea, and the Netherlands.

32BALANCE OF PAYMENTS
The adverse trade balance is generally offset by a flow of remittances from migrant groups abroad and by tourist expenditures. In 1992 merchandise exports totaled US$22,01 million and imports US$119.70 million. The merchandise trade balance was $–97.69 million. The following table summarizes Grenada's balance of payments for 1990 and 1991 (in millions of US dollars):

	1990	1991
CURRENT ACCOUNT		
Goods, services, and income	–60.10	–64.89
Unrequited transfers	32.11	25.78
TOTALS	–27.99	–39.10
CAPITAL ACCOUNT		
Direct investment	12.87	15.27
Portfolio investment	—	0.05
Other long-term capital	12.71	15.38
Other short-term capital	–11.09	1.22
Exceptional financing	0.01	3.65
Other liabilities	—	—
Reserves	–2.58	–2.46
TOTALS	11.92	33.12
Errors and omissions	16.07	5.99
Total change in reserves	–2.58	0.09

33BANKING AND SECURITIES
Local financial institutions include the National Commercial Bank of Grenada, the Grenada Bank of Commerce, the Grenada Development Bank, and the Grenada Cooperative Bank. Foreign banks include Barclays Bank and the Bank of Nova Scotia. There is no stock exchange.

34INSURANCE
There are a number of international firms (mainly UK, US, and Canadian) and some local interests doing business in Grenada. A full range of life and non-life insurance is available.

35PUBLIC FINANCE
Main sources of revenue are export and import duties, income tax, estate duties, and various internal rates, licenses, and taxes. The 1986 Fiscal Reform Program replaced a number of taxes, including a personal income tax, with a VAT of 20% on certain consumer goods. Current receipts and expenditures in 1992 amounted to $58.6 million and $65.7 million, respectively, with capital expenditures accounting for $19.2 million. The current deficit in 1992 was $7 million. At the end of 1992, total external public debt amounted to $71.9 million.

36TAXATION
Corporate taxes are levied on net profits at a rate of 30% on the first US$50,000 and 40% on the remainder. A debt service levy is payable on salaries over US$12,000 per year at a rate of 10%. Value-added taxes ranging from 5% on most services to 15% on locally manufactured products are also imposed.

37CUSTOMS AND DUTIES
All imports are subject to a general or preferential tariff, as well as a fixed package tax. Under the terms of a 1992 Caribbean Community agreement, Grenada is in the process of eliminating import licensing. It has adopted the Ammou External Tariff, which ranges from 5%–45%.

38 FOREIGN INVESTMENT

Since independence, the government has sought to attract foreign investment for industrial development, especially the processing of local agricultural commodities for export. Grenada offers incentives competitive with other Caribbean countries and gives high priority to foreign investment that is either 100% foreign-owned or in joint venture with nationals.

There are no free trade zones in the country, but generous investment incentive packages are available, including up to a 15-year tax holiday. There has been a considerable amount of foreign investment in the country's hotel sector, which aimed for 2,000 hotel rooms by 1994.

39 ECONOMIC DEVELOPMENT

Government policy has aimed toward sustained development of agriculture and tourism as the prime sectors of the economy, with respect to both employment and foreign exchange earnings. Import substitution has been the focal point of the agricultural development program. The PRG committed itself to nationalizing agriculture and turned the large estates that had belonged to former Prime Minister Gairy into cooperative farms.

Following the 1984 election, the Blaize government reversed the trend toward state control and embarked on an economic policy that encouraged private-sector participation and modified the fiscal system to encourage economic growth. The establishment of the Industrial Development Corporation and of the National Economic Council early in 1985 were essential components of the new policy, as was the privatizing of 18 state enterprises. The tax structure was modified in 1986 to offer incentives to the private sector.

After independence, the major aid donors were the UK and Canada. Between 1979 and 1983 considerable economic assistance came from Cuba, Libya, and the Eastern bloc. After 1983, US assistance was renewed and totaled us$59.7 million by the end of 1986. Multilateral assistance amounted to us$11.1 million in 1985–86, us$6 million of which was from the IFC and us$5 million from the IDA.

Economic growth is expected to increase slightly in 1994. Output is expected to return to 1991 levels as agricultural production expands and the tourism sector grows at a faster rate than in 1992. Additional foreign investment is expected as new tax law reforms take place. Grenada will also benefit from low interest loans under Puerto Rico's 936/Caribbean Basin Development Program after signing a Tax Information Exchange Agreement with the US.

40 SOCIAL DEVELOPMENT

A National Insurance Scheme, introduced on 4 April 1983, provides old age, disability, survivor, health, and maternity benefits to workers; it is financed by wage contributions of 4% from both employers and workers. The retirement age is 60, and pensions equal 30% of average earnings. Survivor pensions total 50% of the pension of the insured. There is also a funeral grant which amounted to EC$1,000 in 1991. A National Child Care Committee coordinates services for children. The fertility rate for 1985-90 was 4.9, and the contraceptive prevalence rate was 31%. Although Grenadian law imposes a 15-year sentence for rape, women's rights advocates claim that many cases of rape and abuse go unreported.

41 HEALTH

Grenada is divided into 10 medical districts, each headed by a medical officer. Grenada General Hospital and two other general hospitals (one on Carriacou) have a combined total of 360 beds. Other hospital facilities include a mental hospital (rebuilt after being severely damaged in the 1983 invasion) and a tuberculosis sanatorium. There are 38 health centers, which provide primary care. In 1988, there were 51 doctors and 4 dentists.

In 1992, the infant mortality rate was 29 per 1,000 live births, and life expectancy averaged 71 years for both women and men. In the same year, there were 2,200 births. From 1991 to 1992, 73% of the country's children were immunized against measles.

42 HOUSING

The Grenada Housing Authority is a government agency empowered to acquire land and construct low-income housing projects. All main population centers have been supplied with sewerage and piped-water facilities; in 1986, 77% of the population had access to potable water. As of 1981, 91% of all dwellings were detached houses, and 6% were apartments. Nearly 75% of all housing units were owner occupied, 14% were rented, and 9% were occupied rent free. The most common construction material for homes was wood (70%), followed by concrete (12%), and wood and brick combined (7%).

43 EDUCATION

Education is free and compulsory for 12 years. Primary education lasts for seven years and secondary education for five years. In 1989, there were 66 primary schools with 21,616 students and 763 teachers. Postsecondary institutions include the Technical and Vocational Institute, the Teacher Training College, and the Institute for Further Education. St. George's University Medical School, a private US institution founded in 1977, provides medical training for students from other countries, the majority from the US. Illiteracy is almost nonexistent in Grenada, with an adult literacy rate in the mid-1980s at an estimated 98%.

44 LIBRARIES AND MUSEUMS

The Grenada Public Library in St. George's has about 38,000 volumes; the St. George's University Medical School maintains a collection of over 5,000 volumes. There is also a mobile library visiting the island's schools. The Grenada National Museum, founded in 1976, is located in St. George's.

45 MEDIA

A local automatic telephone system covers the island, with connections to Carriacou. In 1993, there were approximately 20,000 telephone lines. Cable and Wireless Ltd. provides both telex and wireless telephone links overseas. Radio and television services are provided by the government-owned Radio Grenada and Grenada Television. There is 1 AM radio station, no FM stations, and 1 television station. There were some 54,000 radios and 30,000 television sets in 1991. There are five weekly newspapers, published by private companies or political parties: *The Grenadian Voice, The Informer, The Grenadian Guardian, The Grenadian Tribune,* and *The Indies Times.*

46 ORGANIZATIONS

Grenada has chapters of a number of international social and professional organizations, including a Chamber of Industry and Commerce, Lions Club, and Rotary Club. Boy Scouts and Girl Guides are available to youths.

47 TOURISM, TRAVEL, AND RECREATION

Tourism, although modest in relation to that of other Caribbean islands, was a major enterprise for Grenada before the 1979 coup and 1983 invasion. Since 1984, it has recovered rapidly, and the government is emphasizing development of the tourist infrastructure. There has also been substantial foreign investment in the hotel sector. Since 1989, American Airlines' nonstop service to Grenada has boosted tourism significantly. Stayover visitors numbered 92,497 in 1991, 49,571 from the Americas and 23,503 from Europe. There were 1,118 hotel rooms with 2,308 beds and a 65% occupancy rate. Tourism revenues were $42 million. Grenada offers the visitor a wide array of white sand beaches and excellent sailing. Two yacht harbors provide port and customs facilities. A valid passport is required for entry into Grenada, except for citizens of the UK, the US, and Canada who have two documents proving citizenship. Visas are not required by visitors from North America, the UK, France, Germany, and certain other countries.

[48]FAMOUS GRENADIANS

Theophilus Albert Marryshow (1889–1958) is known throughout the British Caribbean as "the Father of Federation." Eric Matthew Gairy (b.1922), a labor leader, became the first prime minister of independent Grenada in 1974. Maurice Bishop (1944–83) ousted Gairy in 1979 and held power as prime minister until his assassination. After the US-led invasion, the task of choosing an interim government fell to the governor-general, Sir Paul Scoon (b.1935). Herbert A. Blaize (b.1919) was elected prime minister in 1984; he was Grenada's first chief minister (1957) and its first premier (1967).

[49]DEPENDENCIES

Carriacou (34 sq km/13 sq mi), Petit Martinique, and several other islands of the Grenadines group are dependencies of Grenada.

[50]BIBLIOGRAPHY

Beck, Robert J. *The Grenada Invasion: Politics, Law, and Foreign Policy Decision Making.* Boulder, Colo.: Westview, 1993.

Brizan, George. *Grenada, Island of Conflict—From Amerindians to People's Revolution, 1498–1979.* London: Zed Books, 1984.

The Grenada Documents. London: Sherwood Press, 1987.

Heine, Jorge (ed.). *A Revolution Aborted: The Lessons of Grenada.* Pittsburgh, PA: University of Pittsburgh Press, 1990.

Lewis, Gordon K. *Grenada: The Jewel Despoiled.* Baltimore: Johns Hopkins University Press, 1987.

MacDonald, Scott B., Harald M. Sandstrom, and Paul B. Goodwin, Jr. (eds.) *The Caribbean after Grenada: Revolution, Conflict, and Democracy.* New York: Praeger, 1988.

Schoenhals, Kai P. *Grenada.* Santa Barbara, Calif.: Clio, 1990.

GUATEMALA

Republic of Guatemala
República de Guatemala

CAPITAL: Guatemala City.

FLAG: The national flag consists of a white vertical stripe between two blue vertical stripes with the coat of arms centered in the white band.

ANTHEM: *Himno Nacional,* beginning "Guatemala feliz" ("Happy Guatemala").

MONETARY UNIT: The quetzal (Q) is a paper currency of 100 centavos. There are coins of 1, 5, 10, and 25 centavos, and notes of 50 centavos and 1, 5, 10, 20, 50, and 100 quetzales. Q1 = $0.1720 (or $1 = Q5.8128). US notes are widely accepted.

WEIGHTS AND MEASURES: The metric system is the legal standard, but some imperial and old Spanish units are also used.

HOLIDAYS: New Year's Day, 1 January; Epiphany, 6 January; Labor Day, 1 May; Anniversary of the Revolution of 1871, 30 June; Independence Day, 15 September; Columbus Day, 12 October; Revolution Day, 20 October; All Saints' Day, 1 November; Christmas, 25 December. Movable religious holidays include Holy Thursday, Good Friday, and Holy Saturday.

TIME: 6 AM = noon GMT.

1 LOCATION, SIZE, AND EXTENT

Situated in Central America, Guatemala has an area of 108,890 sq km (42,043 sq mi), with a maximum length of 457 km (284 mi) NNW–SSE and a maximum width of 428 km (266 mi) ENE–WSW. Comparatively, the area occupied by Guatemala is slightly smaller than the state of Tennessee. It is bounded on the E by Belize, Amatique Bay, and the Caribbean Sea, on the SE by Honduras and El Salvador, on the S by the Pacific Ocean, and on the W and N by Mexico, with a total boundary length of 2,087 km (1,297 mi).

Guatemala has long laid claim to territory held by Belize (formerly known as British Honduras). In 1821, upon achieving independence, Guatemala considered itself the rightful inheritor of this former Spanish possession and continued to regard Belize as an administrative adjunct of Guatemala. In 1859, British rights to the area were defined in a treaty with Guatemala, but, alleging that the UK had not fulfilled its obligations, Guatemala subsequently refused to recognize the British title. In mid-1975, Guatemala demanded the cession of one-fourth of the territory of Belize as a condition for recognizing that country's sovereignty. When Belize did become independent in September 1981, Guatemala refused to recognize the new nation. In January 1983, the Guatemalan government announced that it would drop its sovereignty claim and would press instead for the cession of the southernmost fifth of Belize's territory. Guatemala's claim has been rejected not only by the UK and Belize but also by the UN General Assembly and, in November 1982, at the CARICOM heads of government conference. In mid-1986, Guatemala and the UK reestablished consular and commercial relations.

Guatemala's capital city, Guatemala City, is located in the southcentral part of the country.

2 TOPOGRAPHY

A tropical plain averaging 48 km (30 mi) in width parallels the Pacific Ocean. From it, a piedmont region rises to altitudes of from 90 to 1,370 m (300 to 4,500 ft). Above this region lies nearly two-thirds of the country, in an area stretching northwest and southwest and containing volcanic mountains, the highest of which is Mt. Tajumulco (4,211 m/13,816 ft). The larger towns and Lake Atitlán are located in basins at elevations of about 1,500 to 2,400 m (5,000 to 8,000 ft). To the north of the volcanic belt lie the continental divide and, still farther north, the Atlantic lowlands. Three deep river valleys—the Motagua, the Polochic, and the Sarstún—form the Caribbean lowlands and banana plantation area. North of it, occupying part of the peninsula of Yucatán, is the lowland forest of Petén, once the home of the Mayas. The largest lakes are Izabal, Petén Itza, and Atitlán.

Of some 30 volcanoes in Guatemala, 6 have erupted or been otherwise active in recent years. A catastrophic earthquake in February 1976 left nearly 23,000 dead, 70,000 injured, and 1 million people whose homes were partially or completely destroyed.

3 CLIMATE

Temperature varies with altitude. The average annual temperature on the coast ranges from 25° to 30°C (77° to 86°F); in the central highlands the average is 20°C (68°F), and in the higher mountains 15°C (59°F). In Guatemala City, the average January minimum is 11°C (52°F) and the maximum 23°C (73°F); the average minimum and maximum temperatures in July are, respectively, 16°C (61°F) and 26°C (79°F). The rainy season extends from May to October inland and to December along the coast, and the dry season from November (or January) to April. Because of its consistently temperate climate, Guatemala has been called the "Land of Eternal Spring."

4 FLORA AND FAUNA

Flowers of the temperate zone are found in great numbers. Of particular interest is the orchid family, which includes the white nun (monja blanca), the national flower. There is also an abundance of medicinal, industrial, and fibrous plants.

Indigenous fauna includes the armadillo, bear, coyote, deer, fox, jaguar, monkey, puma, tapir, and manatee. The national bird is the highland quetzal, the symbol of love of liberty, which reputedly dies in captivity. Lake Atitlán is the only place in the world where a rare flightless waterbird, the Atitlán (giant pied-billed) grebe, is found; this species, classified as endangered, has been protected by law since 1970. There are more than 900 other species of native birds, as well as migratory varieties. Reptiles, present in more than 204 species, include the bushmaster, fer-de-lance, water moccasin, and iguana.

⁵ENVIRONMENT

Guatemala's main environmental problems are deforestation—over 50% of the nation's forests have been destroyed since 1890—and consequent soil erosion. As of 1993, the nation gets 90% of its energy from wood. It loses 40-60,000 hectares of forest per year. The nation's water supply is also at risk due to industrial and agricultural toxins. Guatemala has 27.8 cubic miles of water with 74% used for agriculture and 17% used in farming activity. Fifty-seven percent of the people living in rural areas do not have access to pure water, while 92% of all cities have pure water. Guatemala produces 12,700 tons of industrial contaminants per year. Its cities produce 0.7 million tons of solid waste per year. The nation uses 10,000 tons of pesticides per year and 170,000 tons of fertilizer which contribute to the pollution of 63% of the nation's soil. United Nations sources show that environmental contamination is responsible for a significant number of deaths due to respiratory and digestive illnesses. Despite the establishment in 1975 of a ministerial commission charged with conserving and improving the human environment, coordination of antipollution efforts remains inadequate, and Guatemala still suffers from a lack of financial resources and well-trained personnel to implement environmental control programs.

As of 1994, 10 of Guatemala's 174 animal species and 10 in a total of 900 bird species were considered endangered. Four types of reptiles in a total of 204 and 282 plant species of 8,000 were threatened with extinction. As of 1987, endangered species in Guatemala included the giant pied-billed grebe, tundra peregrine falcon, horned guan, Eskimo curlew, California least tern, green sea turtle, hawksbill turtle, olive ridley turtle, spectacled caiman, American crocodile, and Morelet's crocodile.

⁶POPULATION

The total population at the census of 1981 was 6,054,227; it was estimated at 9,952,081 in 1994. The projected population for the year 2000 was 12,222,000, assuming a crude birthrate of 36.3 per 1,000 population, a crude death rate of 6.7, and a net natural increase of 29.6 during 1995–2000. As of 1992, 39% of the population was urban. The capital and largest city, Guatemala City, had a 1991 estimated population of 1,095,677. Other large cities (with estimated 1991 populations) are Escuintla, 63,471, and Quezaltenango, 93,439. The estimated density in 1994 was 91 persons per sq km (237 per sq mi). Most of the people live in the southern third of the country.

⁷MIGRATION

Because of persecution and civil war, Amerindian peasants began emigrating across the Mexican border in 1981; by late 1992, Mexico was harboring an estimated 49,800 Guatemalan refugees. The Guatemalan government estimated that at the end of 1992 there were 222,900 Central Americans in Guatemala without documentation. Many of these were believed to be in transit to Mexico and the US.

⁸ETHNIC GROUPS

Guatemala has a larger proportion of Amerindians in its total population than any other country in Central America. In 1983 this population was estimated at 53%. About one-half of this number were members of various communities descended from the Maya-Quiché ethnolinguistic group. Persons of mixed Amerindian and white ancestry, called mestizos, constitute about 42% of the national total. Amerindians who have become assimilated and no longer adhere to a traditional Amerindian life-style are also called ladinos, but this term is sometimes used to refer to mestizos. Blacks and mulattoes (4%) inhabit the Caribbean lowlands. The white population is estimated at about 1% of the total.

⁹LANGUAGES

Spanish is the official and commercial language. Amerindians speak some 28 dialects in five main language groups: Quiché, Mam, Pocomam, and Chol—all of the Mayan language family—and Carib.

¹⁰RELIGIONS

The constitution guarantees religious freedom. Roman Catholicism is the predominant religion (81%), with an archbishopric at Guatemala City and bishoprics at Quezaltenango, Verapaz, and Huehuetenango. In 1986, an estimated 25% of the population combined Christian beliefs with traditional Amerindian rites. About 5,000 follow tribal religions exclusively. There were about 800 Jews in Guatemala in 1990.

Protestant churches were estimated to have fewer than 500,000 adherents in 1980, but rapidly growing fundamentalist groups increased the number of Protestants to some 30% in 1991, most of whom are Pentecostals. During the civil war in the early 1980s, the government allied itself with conservative Protestant groups and accused the Catholic Church of helping the leftist guerrillas. The regime's relationship with the Church worsened when Gen. José Efraín Ríos Montt refused to listen to the Vatican's plea for clemency for six alleged subversives condemned to death by secret tribunals in March 1983 and let them be executed a few days before the visit of Pope John Paul II to Guatemala. While in Guatemala, the Pope explicitly referred to the discrimination and injustice the Amerindians had suffered.

¹¹TRANSPORTATION

In 1991, the total length of Guatemala's road system was 26,429 km (16,423 mi), of which 11% was paved. In 1992, about 260,000 motor vehicles were registered. Two international highways cross Guatemala: the 824-km (512-mi) Franklin D. Roosevelt Highway (part of the Pan American Highway system) and the Pacific Highway. Construction of a new 1,500-km (930-mi) highway network began in 1981. Guatemala Railways operates 90% of the 870 km (541 mi) of narrow-gauge railway.

Few of the rivers and lakes are important to commercial navigation. Puerto Barrios and Santo Tomás on the Caribbean coast are Guatemala's chief ports; the Pacific coast ports are Champerico and San José. In 1991, Guatemala had one registered cargo ship of 4,129 gross tons.

Aurora Airport at Guatemala City, the first air terminal in Central America, serves aircraft of all sizes, including jumbo jets. The government-owned Aviateca has a monopoly on scheduled domestic service and also flies to other Central American countries, Jamaica, Mexico, and the US.

¹²HISTORY

Three distinct stages—Mayan indigenous, Spanish colonial, and modern republican—have left their mark on the history of Guatemala. These separate ways of life persist, but are slowly merging.

Guatemala includes much of the old Mayan civilization, which may date back as early as 300 BC. The classical Mayan period lasted from about AD 300 to 900 and featured highly developed architecture, painting, sculpture, music, mathematics (including the use of zero), a 365-day calendar, roads, and extensive trade. This great pre-Columbian civilization seems to have collapsed

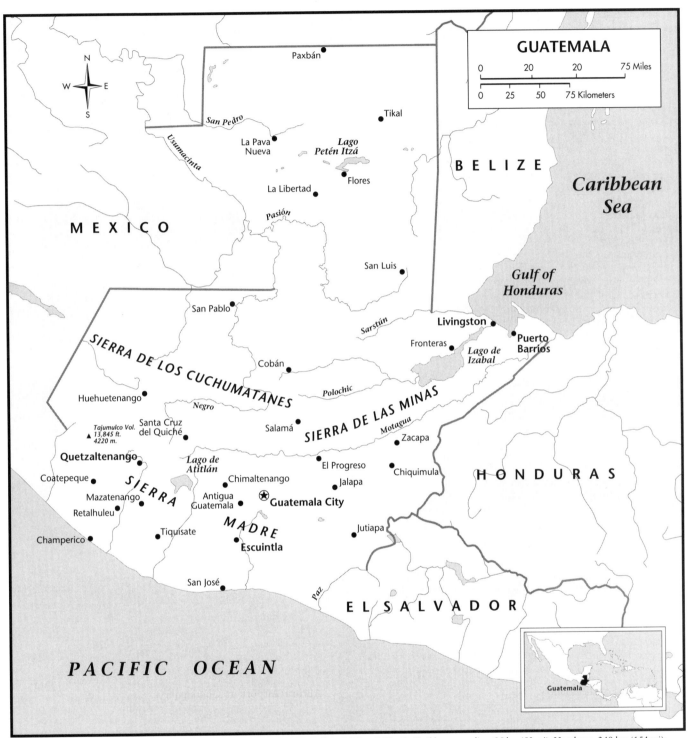

LOCATION: 13°42′ to 18°30′N; 87°30′ to 92°13′W. **BOUNDARY LENGTHS:** Belize, 261 km (162 mi); Caribbean coastline, 85 km (53 mi); Honduras, 248 km (154 mi); El Salvador, 167 km (104 mi); Pacific coastline, 245 km (152 mi); Mexico, 925 km (575 mi). **TERRITORIAL SEA LIMIT:** 12 mi.

around AD 900, and by the 12th century, the Mayas had disintegrated into a number of separate Amerindian groups. The Amerindians offered resistance to the Spanish expedition sent by Hernán Cortés from Mexico and led by Pedro de Alvarado during 1523-24, but by the end of that time, their subjugation to Spain was virtually complete.

Alvarado founded the first Guatemalan capital, Santiago de los Caballeros de Guatemala, in 1524. Because of several earthquakes, the capital was moved a number of times until it became permanently established at Guatemala City in 1776. From 1524

until 1821, Guatemala (City and Province) was the center of government for the captaincy-general of Guatemala, whose jurisdiction extended from Yucatán to Panama. Economically, this was mainly an agricultural and pastoral area in which Amerindian labor served a colonial landed aristocracy. The Roman Catholic religion and education regulated the social life of the capital. Spanish political and social institutions were added to Amerindian village life and customs, producing a hybrid culture.

In 1821, the captaincy-general won its independence from Spain. After a brief inclusion within the Mexican Empire of

Agustín de Iturbide (1822-23), Guatemala, along with present-day Costa Rica, El Salvador, Honduras, and Nicaragua, formed the United Provinces of Central America in 1824. This federation endured until 1838–39. Guatemala proclaimed its independence in 1939 under the military rule of Rafael Carrera, an illiterate dictator with imperial designs. None of his ambitions were realized, and he died in 1865.

Guatemala then fell under a number of military governments, which included three notable administrations: Justo Rufino Barrios (1871–85), the "Reformer," who was responsible for Guatemala's transition from the colonial to the modern era; Manuel Estrada Cabrera (1898–1920), whose early encouragement of reform developed later into a lust for power; and Jorge Ubico (1931—44), who continued and elaborated on the programs begun by Barrios.

Guatemalan politics changed with the election of reform candidate Juan José Arévalo Bermejo in 1945. Arévalo's popularity marked one of the first mass-based movements in Guatemalan politics. In 1951, Jacobo Árbenz Guzmán was elected. Following Arévalo's approach to land reform, Árbenz expropriated holdings of the United Fruit Co., a US firm. The US alleged Communist influence within the Árbenz government, and began mobilizing opposition against him. In the summer of 1954, Col. Carlos Castillo Armas and an army of Guatemalan exiles, backed by the CIA, invaded Guatemala from Honduras and toppled Árbenz. Castillo took over, restored expropriated properties, and ruled by decree until he was assassinated by a presidential palace guard in July 1957.

After a period of confusion, Gen. Miguel Ydígoras Fuentes became president in January 1958. His administration was essentially a military dictatorship, even though he claimed to follow democratic principles. He was particularly hard on his domestic critics, denouncing them as Communists. He was equally bombastic on the international stage, denouncing the US, quarreling with Mexico over fishing rights, and challenging the UK over Belize. He was also contemptuous of Fidel Castro, and allowed Guatemala to be a training area for the exiles in the abortive US invasion of the Bay of Pigs in April 1961.

In March 1963, Ydígoras was overthrown by Defense Minister Col. Enrique Peralta Azurdia, who declared a state of siege. For two years, Peralta ruled dictatorially, and continued to assert Guatemala's claims on Belize. In September 1965, the Peralta regime announced a new constitution and elections, and in March 1966, Dr. Julio César Méndez Montenegro was elected president. He was the first civilian president since Árbenz, and would be the last for some time. During his term, the army and right-wing counter-terrorists proceeded to kill hundreds of guerrillas, who were believed to be sponsored by Cuba, and claimed destruction of the guerrilla organization by the end of 1967. Uprooted from the countryside, the guerrillas concentrated their efforts on the capital, where, in 1968, guerrillas assassinated US Ambassador John G. Mein.

Guatemala returned to a military president, as Col. Carlos Arana Osorio was elected president in 1970. He instituted the country's first comprehensive development plan, but the plan was upset by guerrilla violence, which now engulfed the country. Ambassador Karl von Spreti of the Federal Republic of Germany (FRG) was murdered in April 1970 by leftists. Many prominent Guatemalans were killed or held for ransom. In response to the violence, Arana suspended civil liberties from November 1970 to November 1971. In 1974, Arana's candidate, Gen. Kjell Laugerud García, was confirmed by Congress as president, after an election marred by charges of fraud. Laugerud followed a centrist policy and obtained a measure of popular support. During his tenure, guerrilla violence decreased, and some political liberties were restored. The principal challenge to Laugerud's administration was the need to rebuild Guatemala after the catastrophic earthquake of February 1976.

A militant rightist, Gen. Fernando Romeo Lucas García, was elected president in 1978. As guerrilla violence continued, there was also an upsurge of activity by right-wing "death squads," which, according to unofficial Guatemalan sources, committed over 3,250 murders in 1979 and even more during 1980. In addition, hundreds of Amerindians were reportedly massacred during antiguerrilla operations. The Carter administration objected to Guatemala's deteriorating human rights record, whereupon the military charged that communist influence had reached the White House.

In January 1981, the main guerrilla groups united and escalated while the government went into crisis. The elections of March 1982 were won by Laugerud's handpicked candidate, Gen. Angel Aníbal Guevara. Three weeks later, a coup placed in power a "born-again" Protestant, Gen. José Efraín Ríos Montt. After his month-long amnesty offer to the guerrillas was rejected, he declared a state of siege in July, and the antiguerrilla campaign intensified. The government's counter-insurgency killed between 2,600 and 6,000 in 1982, and drove up to a million Guatemalans from their homes by the end of 1983. In March 1983, Ríos lifted the state of siege and announced that elections for a constituent assembly would be held in July 1984. But Ríos, who had fought off some 10 coup attempts during his administration, was overthrown in August 1983.

The new government of Brig. Gen. Oscar Humberto Mejía Victores declared that the coup was undertaken to end "abuses by religious fanatics" and pledged continued efforts to eradicate the "virus of Marxism-Leninism." Elections for a constituent assembly were held, as promised, in July 1984. In May 1985, the assembly promulgated a constitution for a new government with an elected Congress. The general elections of November 1985 were followed by a runoff election in December. The overwhelming winner was Mario Vincio Cerezo Arévalo of the Guatemalan Christian Democratic Party (DCG). He also brought a majority into Congress.

Cerezo was greeted with enthusiasm by the US, for he represented a legally-elected civilian. Political violence decreased, and Cerezo withstood two attempted coups. But Cerezo was unable to make any progress on human rights in Guatemala, and was unwilling to risk prosecution of military personnel who had been the most serious violators. As the economy worsened, political instability increased, including violence.

The elections of 11 November 1990 necessitated a runoff election, which was won by Jorge Serrano of the Movement for Solidarity and Action (Movimiento para Acción y Solidaridad—MAS). Serrano's inauguration in January 1991 marked the first transition from one elected civilian government to another in memory. Serrano promised to negotiate with insurgents and bring to justice both corrupt former officials and human rights violators.

But Serrano overplayed his hand politically. On 25 May 1993 Serrano declared a state of emergency and suspended the constitution. A week later, the military intervened and removed Serrano from office. It then restored the constitution and allowed Congress to select Serrano's successor. This unusual service of the military in defense of democracy led to the naming of Ramior de León Carpio as president on 5 June. De León, a human rights advocate, promised to bring to justice those responsible for the dismal state of human rights in Guatemala. He also proposed reductions in the military, which predictably have not been well-received by the officer corps.

13GOVERNMENT

Constitutionally, the Guatemalan government is defined as democratic and representative, and the new constitution that took effect on 14 January 1986 reaffirms that definition. Since the 1950s, however, civil disorder has often prompted the suspension of constitutional guarantees.

In theory, Guatemala is a republic. The president, who must be a native-born lay person at least 40 years old, is elected by direct vote for a five-year term and may not be reelected. The constitution calls for a popularly elected vice-president. The office of vice-president provides a guarantee of presidential succession in case of the death or disability of the chief executive. There is a five-member court of constitutionality, which officially advises the president. Its members are appointed, one each by the Supreme Court, Congress, the president, the University of San Carlos, and the bar association. The president, who has broad powers, appoints and is assisted by a cabinet of 13 ministers (as of 1994). The cabinet members traditionally resign at the end of each year so that the president may choose a new cabinet. The president, who is also commander-in-chief of the armed forces, appoints most military officers, the 22 governors, and other important public and diplomatic officials. Presidential duties include preserving public order, proposing laws, and making an annual presentation of the budget.

Each district elects at least two deputies to the unicameral National Congress by proportional representation. In districts with a population over 200,000, an additional deputy is elected to represent each additional 100,000 inhabitants or fraction exceeding 50,000. In addition, at-large representatives are elected by proportional representation from lists submitted by each political party. In 1991 there were 116 seats in Congress. Under the constitution, Congress imposes taxes, enacts the national budget, declares war and makes peace, and ratifies treaties and conventions proposed by the president. Congress elects the president of the judiciary and judges of the Supreme Court and courts of appeals. The president may veto congressional bills, but Congress may override by a two-thirds vote. All public officials must declare the amount of their incomes and property holdings before assuming their posts and after they leave office.

Citizenship is acquired at the age of 18. Voting is obligatory for literate men and women 18 years of age and older and optional for nonliterate citizens.

14POLITICAL PARTIES

Political power in Guatemala has been largely a matter of personal, rather than party, influence. Although parties have generally developed along conservative or liberal lines, political periods are commonly identified with the names of important leaders.

Under President Carlos Castillo Armas (1954–57), the Guatemalan Communist Party and other leftist parties were dissolved, and all other parties were temporarily suspended. To prevent further party proliferation, the membership necessary for party certification was raised from 10,000 to 50,000 in 1963. Only three parties were able to meet this requirement in time for the March 1966 elections: the Revolutionary Party (Partido Revolucionario—PR), a center-left party, the conservative Institutional Democracy Party (Partido Institucional Democrático—PID), formed in 1965, and the militantly anti-communist National Liberation Movement (Movimiento de Liberación Nacional—MLN).

During the 1970s, these parties remained dominant. The MLN won the presidency in 1970, and an MLD-PID coalition took the 1974 election (which was ultimately decided in Congress), defeating Gen. José Efraín Ríos Montt, representing the leftist National Opposition Front, a coalition of the several parties, including the Christian Democrats (partido de Democracía Cristiana Guatemalteca—DCG). In 1978, the PR and PID formed a center-right coalition. In the congressional voting, the MLN won 20 seats, the PID 17, the PR 14, the DCG 7, and other parties 3. In the presidential elections of March 1982, a coalition of the PR, PID, and the National Unity Front (Frente de Unidad Nacional—FUN), an extreme right-wing party formed in 1977, won a plurality of 38.9% of the vote. Congress endorsed

the PR-PID-FUN candidate, Gen. Ángel Aníbal Guevara, as president, but he was deposed in a coup later in March. All parties were suspended by the new ruler, Gen. Ríos, but political activity resumed in March 1983. In the 1980s, the Christian Democrats grew significantly, winning the 1985 presidential and congressional elections. In addition, a host of new parties entered the political arena. Many had hopeful names suggesting national reconciliation, moderation, and solutions to Guatemala's problems. Among these were the National Union of the Center (Unión del Centro Nacional—UCN), the Democratic Party for National Cooperation (Partido Democrático de Cooperación Nacional—PDCN), the Solidarity Action Movement (Movimiento para Acción y Solidaridad—MAS), the National Advancement Plan (Plan por el Adelantamiento Nacional) and the National Authentic Center (Centro Auténtico Nacional).

The left-wing guerrilla movement is represented by the Guatemalan National Revolutionary Unity (Unidad Revolucionaria Nacional Guatemalteca—URNG). Founded in 1982, these groups consists of the Guerrilla Army of the Poor (Ejército Guerrillero de los Pobres—EGP), the Committee of Peasant Unity (Comité de Unidad Campesina—CUC), the Guatemalan Workers' (Communist) Party (Partido Guatemalteco del Trabajo—PGT), the Rebel Armed Forces (Fuerzas Armadas Rebeldes—FAR), and the Organization of the People in Arms (Organización del Pueblo en Armas—ORPA). The component groups have been engaging in insurgency against the central government since the mid-1960s.

15LOCAL GOVERNMENT

Guatemala is divided into 22 departments, plus Guatemala City, each with a governor appointed by the president. Municipalities are governed by a mayor and independent municipal councils whose officials are popularly elected for two-year terms.

16JUDICIAL SYSTEM

The Constitution of 1985 establishes an independent judiciary and a human rights ombudsman. Courts of ordinary jurisdiction are the nine-member Supreme Court, 10 courts of appeals, 33 civil courts of first instance, and 10 penal courts of first instance. Judges of the Supreme Court and courts of appeals are elected by the National Congress for four-year terms; they must be native-born lay persons. Judges of first instance are appointed by the Supreme Court. Courts of private jurisdiction deal separately with questions involving labor, administrative litigation, conflicts of jurisdiction, military affairs, and other matters. An independent tribunal and office of accounts supervises financial matters of the nation, the municipalities, and state-supported institutions, such as the National University. A new criminal procedural code affording stronger due process protections took effect in July 1994.

Because military courts retain control of military personnel who commit crimes while on official business, it has been difficult to pursue actions in civil court cases involving human rights abuses by the military.

17ARMED FORCES

The national armed forces (44,600) are combined for administrative purposes, and the army provides logistical support to the navy and air force. A conscription law (30 months service) encourages recruitment. In 1993 the armed forces numbered 42,000 in the army, 1,200 in the navy, and 1,400 in the air force. The navy had 8 coastal patrol craft, and the air force had 22 aircraft and 12 armed helicopters of low readiness. The national police force of about 12,000 supplements the army in a national emergency. Defense expenditures amounted to $113 million (1990) or 1 percent of gross domestic product. More than 500,000 Guatemalan men serve in a territorial militia to combat guerrillas and other dissidents, who may number 1,500. A decree promulgated in December 1983 authorized military service for women and required the resignation of any military personnel engaged in "political militancy."

To curb human-rights abuses, the United States reduced its military assistance in the 1980s, sending no aid in 1983 and 1984. While the US ban was in effect, Israel emerged as an important supplier of arms to Guatemala, but Guatamala imported only $10 million a year in 1988–1990 and nothing in 1991.

18INTERNATIONAL COOPERATION

Guatemala is a charter member of the UN, having joined on 21 November 1945, and participates in ECLAC and all the nonregional specialized agencies except the GATT. It is also a member of G-77, IDB, OAS, and PAHO, as well as the International Coffee Organization and many other inter-American and intergovernmental organizations. Guatemala was one of the countries that formed the Union of Banana-Exporting Countries in 1974, and it was the first country to ratify the agreement establishing the CACM. In August 1987, Guatemala signed the Central American peace plan outlined by Costa Rican President Oscar Arias Sánchez.

19ECONOMY

Since the Spanish conquest, the economy of Guatemala has depended on the export of one or two agricultural products. During the colonial period, indigo and cochineal were the principal exports, but the market for them was wiped out by synthetic dyes in the 1860s. Cocoa and essential oils quickly filled the void. Coffee and bananas were introduced later, and in 1986, the chief exports in order of value were coffee, bananas, sugar, and cotton.

Since World War II, the government has encouraged light industrial production (such as tires, clothing, and pharmaceuticals). Nevertheless, in 1985, agricultural pursuits occupied 47% of the national labor force and accounted for some two-thirds of Guatemalan foreign exchange earnings. Living standards and personal income remain low, and no significant domestic market exists, except for subsistence crops. Guatemala imports finished goods, machinery, motor vehicles, foodstuffs, chemicals, and petroleum. In the absence of greater crop diversification, the economy remains at the mercy of climatic disaster and foreign market trends.

The economy boomed from 1971 through early 1974. Then, as a result of inflation (21.2% in 1973), the world energy crisis, and an annual population growth of 2.9%, the economic growth rate slowed from 7.6% for 1973 to 4.6% for 1974. During the second half of the 1970s, Guatemala's economic performance slowed further; during 1974–80, the average annual growth rate was 4.3%. By the early 1980s, the civil war, coupled with depressed world commodity prices, had led to decreases in export earnings and to foreign exchange shortages. The GDP dropped by 3.5% in 1982, the first decline in decades, and declined or was stagnant through 1986.

The annual inflation rate, which averaged 11% during 1979–81, dropped to no more than 2% in 1982. It rose thereafter, reaching 31.5% in 1985 and about 40% in the first half of 1986, and then dropped suddenly.

In the 1990s the Guatemalan economy has been growing at a healthy pace, propelled by non-traditional exports and investment. Guatemala's economy—the largest in Central America—is dominated by the private sector, which generates nearly 90% of gross domestic product. Inflation has been reduced steadily through tight fiscal and monetary policies to an average of 12% in 1993. Economic growth accelerated to an estimated 5.0% in 1993 compared with 4.6% in 1992. Agriculture and commerce continued to dominate the nation's economy. Agriculture constituted 26% of production, with commerce in second place at

24%. Agricultural production in volume rose only marginally, reflecting the dual weight of drought and depressed world commodity prices. Tremendous growth has been registered in the area of non-traditional agricultural exports in recent years. Commerce moved up 3.4%.

20INCOME

In 1992, Guatemala's GNP was $9,568 million at current prices, or $980 per capita. For the period 1985–92 the average inflation rate was 20.8%, resulting in a real growth rate in per capita GNP of 0.6%.

In 1992 the GDP was $10,434 million in current US dollars. It is estimated that in 1991 agriculture, hunting, forestry, and fishing contributed 26% to GDP; mining and quarrying, less than 1%; manufacturing, 15%; electricity, gas, and water, 3%; construction, 2%; wholesale and retail trade, 24%; transport, storage, and communication, 8%; finance, insurance, real estate, and business services, 9%; community, social, and personal services, 6%; and other sources, 7%.

21LABOR

Most of Guatemala's Amerindian population engages in subsistence agriculture and self-employed handicraft activity. In 1990, the economically active population of 2,880,000 was divided among the following sectors: agriculture, 49%; industry, 14%; services, 14.4%; trade, 12.9%; construction, 3.9%; transportation and communications, 2.5%; and other activities, 3.3%. Unemployment in 1990 was estimated at 40.7%, with an underemployment rate of 34.7%.

The trade union movement was born at the end of World War II. Directed in part by foreign Communist labor leaders and cultivated by a sympathetic administration, the labor movement grew in the next 10 years to a force claiming nearly 500 unions with 100,000 members. With the overthrow of the Árbenz government in 1954, however, the unions were dissolved. After slow reorganization, the trade unions numbered about 110 by 1974, but in 1979 and 1980, trade union activity was severely restricted. In 1992, the Labor Code of 1947 was amended to facilitate freedom of association, to strengthen the rights of working women, to increase penalties for violations of labor laws, and to enhance the role of the Labor Ministry and the courts in enforcement. There were 890 unions with 5–8% of the work force as members in 1992. These unions are independent of government and political party domination. Former President Cerezo was reputed to be an advocate of workers' rights, but in 1987, he enacted a new tax code without suggested modifications from the business community, whereupon he quickly lost the confidence of the private sector and attempted to blame retailers for the price increases caused by these policies. The labor code and social security system effectively cover less than one-fourth of the economically active population.

22AGRICULTURE

In 1991, only about 17% of the total land area of Guatemala was used for the production of annual or perennial crops, although almost two-thirds is suitable for crop or pasture use. Agriculture contributes about 25% to GDP, makes up 75% of export earnings, and employs 50% of the labor force. The principal cash crops are coffee, sugar, bananas, and cotton, followed by hemp, essential oils, and cacao. Coffee is grown on highland plantations; most of the bananas are produced along the Atlantic coastal plain. Cash crop output in 1992 included 9,788,000 tons of sugarcane, 207,000 tons of coffee, 18,000 tons of cotton, and 465,000 tons of bananas. Subsistence crop production included 1,250,000 tons of corn and 100,000 tons of dry beans, along with rice, wheat, and fruits and vegetables. Nontraditional agricultural exports have greatly increased in recent years; such products include: lychee, rambutan, melon, papaya, mango, pineapple, brocoli, okra, snow peas, celery, cauliflower, asparagus, garlic, spices and nuts, and ornamental plants.

An agrarian reform law of 1952 provided for government expropriation of unused privately owned agricultural lands, with the exception of farms of 91 hectares (225 acres) or less and those up to 273 hectares (675 acres) if two-thirds of the acreage was under cultivation. By 1954/55, 24,836 hectares (61,371 acres) had been distributed to 10,359 farmers. The law of 1952 was supplemented by an agrarian reform law of 1956, which aimed to distribute state-owned farms (fincas nacionales) to landless peasants. From 1954 to 1962, the government distributed 17,346 land titles. In 1962, the National Agrarian Improvement Institute was created to provide assistance to the new landowners and to improve their living standards. The government requires plantation owners to set aside land for the raising of subsistence crops for their tenants. An agrarian credit bank provides loans to small farmers. Some of the land farmed by Amerindians is held in common by groups of families and is never sold.

23 ANIMAL HUSBANDRY
Consumption of dairy products and meat is low, despite improvements in stock raising and dairying. The wool industry in the western highlands supplies the famed Guatemalan weavers. Hog and poultry production is ample for domestic consumption. In 1992, there were 2,097,000 head of cattle, 1,100,000 hogs, 676,000 sheep, and 114,000 horses. Guatemala exports poultry, with 10,000,000 fowl and 64,720 tons of eggs produced in 1992.

24 FISHING
Guatemalan waters are rich in fish, including shrimp, snapper, and tuna. Little has been done, however, to develop fishing into a full-fledged industry. The total catch in 1992 was 6,733 tons, nearly half of which came from inland waters.

25 FORESTRY
Despite depletion, forests are still among Guatemala's richest natural resources; they covered some 34% of the total area in 1991. The forests in the Petén region yield cabinet woods, timber, extracts, oils, gums, and dyes. Mahogany, cedar, and balsam are important export products, and chicle for chewing gum is another important commodity. In 1991, 8,049,000 cu m of roundwood were produced.

26 MINING
The principal commercial minerals are antimony, gold, iron ore, and lead. Guatemala was the 3d largest producer of antimony in Latin America, after Bolivia and Mexico. Production, however, steadily declined in the 1980s and by 1985 contributed less than 1% of GDP. Large nickel deposits in the area of Lake Izabal, with an annual production capacity of 9,000 tons in 1991, have been developed by Eximbal. A nickel plant opened in 1977 but was closed down in 1981 because of falling sales, rising fuel costs, and burdensome taxation. Gold, which was mined from the colonial period until the early 20th century, is no longer a major export item, but is still mined; in 1990, about 60 kg was produced. Reported deposits of coal, iron, gold, copper, quartz, marble, manganese, sulfur, uranium, mica, and asbestos await exploitation. Marble is exported to Mexico and other nearby countries.

27 ENERGY AND POWER
In 1991, net installed capacity was 696,000 kw; total production in that year reached 2,330 million kwh. Installations remain inadequate for the country's growing consumer needs. Guatemala City and more than 60% of the remainder of the country are supplied by a formerly U.S.-owned firm, the Electric Co. of Guatemala, which was purchased by the government in 1972 for $17 million. Most of the capacity elsewhere is provided by the government-owned Electrification Institute. An important new 300-Mw hydroelectric plant on the Chixoy River, which came on stream late in 1985, accounts for about 60% of the total electrical capacity. The remainder is produced primarily by thermal plants, with the exception of a 15,000 kw geothermal plant at Zunil. The 1988 National Electrification Plan called for another 252,000 kw of capacity from hydroelectric plants to be built on the Bobos, Grande, Samala, and Serchil Rivers.

Guatemala's principal petroleum resources lie in Alta Verapaz, Petén, Lake Izabal, and Amatique Bay, where much recent exploration has taken place. An intensive search for petroleum was carried out in 1975 near the Mexican border, and new deposits were discovered in 1981 in Alta Verapaz and the Petén basin. A pipeline between Rubelsanto and the Caribbean coastline was opened in 1981, but was attacked 17 times in 1991 by guerrillas. Production averaged 6,000 barrels per day in the 1980s, slumping to 1,352 in 1991. A combination of difficult geological conditions, guerrilla activity, poor infrastructure, and the low quality of Guatemala's oil have dissuaded most oil companies from investment. A Bahamian company, Basic Resources International, was the only company producing crude oil in 1991. Since 1983, however, oil legislation has been liberalized, and exploration activity has increased.

28 INDUSTRY
Most of the country's industrial enterprises operate on a very small scale. A small domestic market has traditionally limited Guatemala's industrial potential, although the CACM temporarily broadened the market for the country's exported manufactures. The value of Guatemala's industrial exports more than tripled between 1972 and 1978. During the 1980s, however, the industrial sector declined, partly because of the collapse of the CACM but also because of a shortage of the foreign exchange necessary to purchase basic materials.

Guatemalan factories produce beverages, candles, cement, pharmaceuticals, chemicals, cigarettes, foodstuffs, furniture, matches, molasses, rubber goods, shirts, shoes, soap, sugar, textiles, and wearing apparel. More recently established firms produce electrical machinery, refined petroleum products, metal furniture, instant coffee, pasteurized milk, plastics, plywood, aluminum, and tires. In the mid-1980s, Guatemalan firms turned to export production as internal demand contracted. The 1986 devaluation of the quetzal raised the cost of imported industrial inputs. Handmade woven and leather goods are sold to tourists and exported.

In 1993, the manufacturing sector occupied third place in the country's economy, representing 14.8% of GDP. However, Guatemalan manufacturing continues to operate well below full capacity. Local industrialists are actively engaged in marketing production output to other Central American and world markets. There is no heavy industry, although a small steel mill in Escuintla should become of great importance. Most manufacturing is devoted to light assembly and food processing operations, and is still geared largely toward the Guatemalan and Central American markets. However, efforts for diversification have continued since 1986.

The tourism industry is becoming an important sector in the country's economy, increasing year by year. Earnings from tourism increased to $275 million in 1993 from $243 million in the previous year.

29 SCIENCE AND TECHNOLOGY
In 1988, total Guatemalan expenditure on research and development amounted to Q32 million, with 925 technicians and 858 scientists and engineers engaged in research and development. The Academy of Medical, Physical, and Natural Sciences dates from 1945. In 1993, Guatemala had four scientific and technological learned societies and research institutes. The Institute of Nutrition

of Central America and Panama, administered by the Pan American Health Bureau Organization and the World Health Organization, conducts research and disseminates scientific and technical information. Four universities offer degrees in basic and applied sciences.

30DOMESTIC TRADE

Outside the capital city, markets are held on appointed days, and local fairs are held annually. Traders bring their wares on large racks atop buses or by mule, herding animals ahead of them. Prices are not fixed. Although there are some modern, tourist-oriented shops, traditional methods of commerce, with modifications, also prevail in Guatemala City. Business hours are weekdays, 8 AM to noon and 2 to 6 PM; shops also open on Saturday mornings. Normal banking hours in Guatemala City are weekdays, 9 AM to 3 PM.

31FOREIGN TRADE

Beginning in 1992, Guatemala increased sales to Western Europe and Canada, as a result of the governments efforts to diversify export production. Imports from Japan, South Korea, Mexico and Venezuela have soared also. Still, the United States remains Guatemala's largest trading partner, purchasing about 35% of its exports and supplying about 46% of its imports in 1993. Total exports amounted to $1,365 million, an increase of 4.0% from 1992. Main export items included coffee, sugar, cardamon, bananas, and other non-traditional vegetables, and textiles and apparel. Imports increased by 22.1% to a total of $2,600 million in 1993. Primary imports were machinery, electronics, petroleum products, chemicals, plastics and paper products. Guatemala's leading trading partners were the US, Mexico, Japan and Germany.

32BALANCE OF PAYMENTS

Guatemala generally finances its trade deficit through capital inflows. Capital flight caused by regional instability during the 1980s led to large deficits on the overall payments balance and brought foreign exchange reserves down from $709.6 million at the end of 1978 to $112.2 million at the end of 1982. In 1991 and 1992, with inflation lowered, private capital inflows were large enough to offset public capital outflows and the current account deficit. As a result, foreign reserves expanded in 1991, growing by $559 million. With continued soft export prices and an overvalued quetzal, however, reserves fell in 1992.

In 1992 merchandise exports totaled $1,283.7 million and imports $2,327.8 million. The merchandise trade balance was $–1,044.1 million. The following table summarizes Guatamala's balance of payments for 1991 and 1992 (in millions of US dollars):

	1991	1992
CURRENT ACCOUNT		
Goods services and income	–443.4	–1,096.4
Unrequited transfers	259.7	390.5
TOTALS	–183.7	–705.9
CAPITAL ACCOUNT		
Direct investment	90.7	94.1
Portfolio investment	71.1	11.4
Other Long-term capital	34.9	26.2
Other short-term capital	536.1	478.8
Exceptional Financing	–77.3	–6.4
Other liabilities	–1.0	—
Reserves	–554.1	20.0
TOTALS	100.4	624.1
Errors and omissions	83.3	81.8
Total change in reserves	–553.9	18.5

33BANKING AND SECURITIES

The Bank of Guatemala is the central bank and the bank of issue. The Monetary Board, an independent body, determines the monetary policy of the country. Associated with the Bank of Guatemala are four other government institutions: The National Mortgage Credit Institute (the official government mortgage bank); two development banks, the National Bank of Agricultural Development and the National Housing Bank; and the National Finance Corp., organized in 1973 to lend funds to industry, tourism, and mining to provide technical assistance.

In 1992 there were 15 domestic commercial banks and one foreign bank. The volume of private banking businesses has grown rapidly, from Q679 million in credit to the private sector in 1977 to Q8,098 million in 1993. The money supply, as measured by M2, was Q16,402.8 million in 1993.

When the Bank of Guatemala was established the Fund for the regulation of the Bond Market was created to encourage security issues. Nevertheless, Guatemalans still tend to prefer tangible investments, and there is no fully developed securities market. A stock market opened in 1987 where shares from private companies in the country and other securities are traded.

34INSURANCE

In 1986 there were 12 registered insurance companies. In addition, the National Mortgage Credit Institute, a government corporation, dealt in both life and non-life insurance policies. In 1992, life insurance in force (excluding group life) totaled Q5.6 billion.

35PUBLIC FINANCE

Fiscal policy loosened after the 1985 elections, but tax reforms in 1987 failed to generate additional income, and governmental expenditures continued to grow. By 1990, the public sector deficit was 4.7% of GDP. The Serrano administration transformed the deficit of 1990 to a slight surplus in 1991 and a virtually balanced budget in 1992. The consolidated public sector deficit amounted to 1.2% in 1991 and 1.0% in 1992. Guatemala's public sector is among Latin America's smallest, and the tax burden is one of the lightest.

The following table shows actual revenues and expenditures for 1988 and 1989 in millions of quetzalos.

	1988	1989
REVENUE AND GRANTS		
Tax revenue	1,904.20	1,853.75
Non-tax revenue	170.10	399.30
Capital revenue	0.09	0.07
Grants	216.94	174.09
TOTALS	2,291.33	2,427.21
EXPENDITURES & LENDING MINUS REPAYMENTS		
General public service	719.42	801.02
Defense	337.23	367.77
Education	457.87	541.87
Health	243.94	274.78
Social security and welfare	10.10	143.86
Housing and community amenities	22.43	72.51
Recreation, cultural, and religious affairs	99.90	—
Economic affairs and services	373.79	601.83
Adjustments	–67.03	288.20
Lending minus repayments	53.93	60.03
TOTALS	2,509.46	2,835.46
Deficit/Surplus	–218.13	–408.25

In 1987 Guatemala's total public debt stood at Q3,020.30 million, of which Q711.4 million was financed abroad. External debt as of mid-1993 stood at $1.97 billion.

36TAXATION
In 1993, the personal income tax rate ranged from 15% to 25%. Only income earned from Guatemalan sources is taxed. Capital gains are taxed at up to 34% and real estate is taxed at 25%. Corporate income is taxed at 25%, and dividends at 12.5%. An industrial development law provides tax exemptions for new industries.

Other levies include a value-added tax of 7%, a 3.5% tax on foreign exchange transactions, a 3% sale tax levied on businesses, and excise taxes on beverages, cigars, tobacco, gasoline, vehicles, and airline tickets.

37CUSTOMS AND DUTIES
Guatemala requires licenses for the importation of restricted goods, including pharmaceuticals, basic food grains, milk, coffee beans, and armaments. As a member of the CACM, Guatemala adheres to a common tariff classification system as well as a common customs code and regulations. Duties are stated as both specific and ad valorem. Import duties are generally minimal, ranging from 5–15%. There is also a value-added tax and a 3% surcharge.

38FOREIGN INVESTMENT
In 1992, total foreign direct investment amounted to $899.2 million, of which the US accounted for 75%. Other principal foreign investors include South Korea and Mexico. New foreign direct investment in 1992 totaled $99.2 million.

There is no general requirement for local participation nor any restrictions on repatriation of capital. As of 1993, the government was in the process of making the registration of foreign corporations and investments more efficiently organized. Guatemala's constitution provides the state telephone company, Guatel, with a monopoly on most telecommunication services, although Guatel does contract with US and other foreign firms to provide cellular and international services. Similarly, even though the constitution designates mineral rights as property of the state, concessions are typically granted in the form of production sharing contracts to foreign firms

39ECONOMIC DEVELOPMENT
Guatemala's economic development policy is to create rural employment through the provision of investment incentives. Such enticements include inexpensive financing by government institutions, free assistance and technical support, and preferential treatment to use government facilities and institutions for guidance.

The Law of Promotion and Development of Export Activities and Drawback ("maquiladora law") and the Law of Free Trade Zones (FTZs) provide the most significant incentives for foreign investment and apply only to the production of industrial exports outside of the Central American Common Market. Both laws allow for the suspensionn of customs duties, value-added tax, and other charges on the import of components, containers, machiners, equipment, and other materials to be used in export production. The maquiladora law grants a 10-year tax holiday on export income, as well as exemption from any export taxes. In 1992, maquiladoras employed about 15% of the domestic manufacturing work force. Industrial firms in FTZs are entitled to a 12-year income tax holiday, as well as exemption from duties, taxes, and import charges. Commercial firms receive a 5-year tax holiday.

The government's role in economic development for 1992–96 includes fiscal reform (to maintain a stable exchange rate and reduce inflation), restructuring or privatization of parastatals, and tariff and financial sector reform. In 1991, the government eliminated its arrears to the Inter-American Development Bank and froze arrears to the IBRD at $50 million. A special tax/bond levy was successful in raising Q660 million for debt arrears and priority infrastructure and social service projects.

By early 1993, the government reached agreement with the Paris Club on the terms for rescheduling its $430 million in arrears on official bilateral debt (over 80% of those arrears are with Spain). Economic assistance from the US from 1946–92 amounted to $415.0 million in loans and $1,052.9 million in grants; in 1992 alone, economic assistance loans and grants from the US came to $14.9 million and $46.9 million, respectively.

40SOCIAL DEVELOPMENT
Social security was established in 1945, and the Guatemalan Institute of Social Security (Instituto Guatemalteco de Seguridad Social) began to function three years later. At first, only accidents were covered; then maternity and hospitalization allotments for widows and orphans were provided. Finally, general medical insurance and pensions for the disabled, aged, widows, and minors were added to the system. Contributions are made by employers (3.0% of payroll in 1991), workers (1.5% of earnings), and the state (25% of cost of the benefits paid).

Guatemala began a national family planning program in 1969. For 1985-90, however, the fertility rate was still high at 5.8. The Office of Women's Affairs was created within the Ministry of Labor in 1981. After the March 1982 coup, women were appointed to several important posts, including, for the first time, the Supreme Court. Despite legal equality, women are paid significantly less than their male counterparts and are generally employed in low wage jobs. A 1991 UN study reports that they make up only one quarter of the work force.

41HEALTH
Among the chief causes of death are heart disease, intestinal parasites, bronchitis, influenza, and tuberculosis. From 1985 to 1990, major causes of death per 100,000 people were given as: communicable diseases and maternal/perinatal causes (595); noncommunicable diseases (523); and injuries (113). Malnutrition, alcoholism, and inadequate sanitation and housing also pose serious health problems. From 1988 to 1991, 62% of the population had access to safe water, and 60% had adequate sanitation. In 1990, 34% of children under 5 years of age were considered malnourished, and there were 110 reported cases of tuberculosis per 100,000 inhabitants. Guatemala does attempt to vaccinate its children, and figures from 1990 to 1992 are as follows: tuberculosis (56%); diphtheria, pertussis, and tetanus (65%); polio (69%); and measles (58%). It is estimated that the poorest half of the population gets only 60% of the minimum daily caloric requirement. From 1985 to 1992, only 34% of the population had access to health care services. Data collected from 1980 to 1993 indicates that 23% of married women (ages 15 to 49) were using contraception.

The 1993 population of Guatemala was 10.03 million. In 1992, there were 380,000 births (a rate of 38.7 per 1,000 people). In 1992, the infant mortality rate was 55 per 1,000 live births, and the overall mortality rate in 1993 was 7.7 per 1,000 people. The average life expectancy is 64 years. From 1966 to 1992, there were about 140,000 war-related deaths.

42HOUSING
Many of the nation's urban housing units and most of its rural dwellings have serious structural defects and lack electricity and potable water. In 1983, only 48% of the urban population and 28% of the rural population had sewer service. A public housing program is supervised by the National Housing Bank. The Ministry of Public Health and Welfare is charged with the improvement of rural dwellings.

As of 1981, 72% of all dwellings were detached houses, 10% were palomars (units without water and private bath facilities), another 10% were improvised, and 6% were apartments. Over 50% of all housing was owner occupied, 30% was rented, and 12% was occupied rent free.

43EDUCATION

In 1990, the adult literacy rate was estimated at 55.1% (males: 63.1% and females: 47.1%). Elementary education is free and compulsory for six years, although enforcement is lax in rural areas. In 1985, only some 73% of children aged 7–12 attended schools. In 1991 there were 9,362 primary schools with 1,249,413 pupils and 36,757 teachers. The general secondary schools had 294,907 pupils. Among Guatemala's five universities, the University of San Carlos of Guatemala, in Guatemala City, is the most important center of higher learning.

44LIBRARIES AND MUSEUMS

There are three notable libraries in Guatemala City. The National Library, with about 125,000 volumes, has collections of Guatemalan and other Central American books and newspapers. The General Archives of Central America has a 100,000-item collection of documents for the three-century colonial period, when Guatemala was the administrative center for the Central American area. The library of the Geographical and Historical Academy of Guatemala is of special value to researchers. The Central American Industrial Research Institute maintains the largest technical library in Central America. The University of San Carlos library has 205,000 volumes.

The Museum of Archaeology and Ethnology in the capital has an excellent collection of Mayan artifacts, and the Colonial Museum in Antigua Guatemala contains colonial paintings, wood carvings, iron and leather work, and sculpture.

45MEDIA

Except for a few privately controlled facilities, the government owns and operates the postal, telephone, and telegraph services. In 1993 there were 190,218 telephones, all of them automatic. The Guatemalan Telecommunications Enterprise provides international radiotelegraph and radiotelephone service. In 1992 there were 91 AM radio stations, and 25 television stations. In 1991 there were an estimated 625,000 radios and 490,000 television sets.

The leading daily newspapers published in Guatemala City (with 1991 circulations) are as follows:

	ORIENTATION	CIRCULATION
Prensa Libre (m)	Moderate liberal	72,000
El Gráfico (m)	Moderate conservative	60,000
Diario la Hora (e)	Moderate liberal	17,000
Diario de Centro América (e)	Official (government)	8,000

46ORGANIZATIONS

Artisans', consumers', service, and savings and credit cooperatives are grouped in four federations. The major employers' organizations are the General Farmers' Association, the Chamber of Commerce, the Chamber of Industry, and the National Coffee Association.

47TOURISM, TRAVEL, AND RECREATION

All visitors need a passport and may need a visa depending on nationality. Tourism has rebounded since Guatemala's return to civilian rule in 1986. In 1991, a record 513,620 foreign visitors entered the country, 406,595 from the Americas and 93,630 from Europe. There were 23,022 rooms with an 83.5 percent occupancy rate. Tourist expenditure totaled $211 million. Guatemala's main tourist attractions are Mayan ruins, such as Tikal; the numerous colonial churches in Guatemala City, Antigua Guatemala, and other towns and villages; and the colorful markets and fiestas.

48FAMOUS GUATEMALANS

The *Rusticatio mexicana,* by Rafael Landívar (1731–93), represents the height of colonial Guatemalan poetry. Outstanding figures of the romantic period were philologist Antonio José de Irisarri (1786–1868); José Batres y Montúfar (1809–44), the author of *Tradiciones de Guatemala* and many poetical works; and José Milla y Vidaurre (1822–82), a historian and novelist and the creator of the national peasant prototype, Juan Chapín. Justo Rufino Barrios (1835–85) became a national hero for his liberal, far-reaching reforms between 1871 and 1885. Enrique Gómez Carillo (1873–1927), a novelist and essayist, was perhaps better known to non-Spanish readers during his lifetime than any other Guatemalan author. Twentieth-century novelists include Rafael Arévalo Martínez (1884–1975), Carlos Wyld Ospina (1891–1956), and Flavio Herrera (1895–1968). The novelist and diplomat Miguel Ángel Asturias (1899–1974) was awarded the Nobel Prize for literature in 1967.

Mario Cardinal Casariego (b.Spain, 1909–83) became the first Central American cardinal in 1969. Among the better-known Guatemalan political personalities of the 20th century are Col. Jácobo Árbenz Guzmán (1913–71), president during 1951–54, and Gen. Miguel Ydígoras Fuentes (1896–1982), president during 1958–63.

49DEPENDENCIES

Guatemala has no territories or colonies.

50BIBLIOGRAPHY

American University. *Guatemala: A Country Study.* Washington, D.C.: Government Printing Office, 1983.

Guatemala in Pictures. Minneapolis: Lerner, 1987.

Immerman, Richard H. *The CIA in Guatemala.* Austin: University of Texas Press, 1982.

Jonas, Susanne. *The Battle for Guatemala: Rebels, Death Squads, and U.S. Power.* Boulder, Colo.: Westview Press, 1991.

Jones, Oakah L. *Guatemala in the Spanish Colonial Period.* Norman: University of Oklahoma Press, 1994.

Perera, Victor. *Unfinished Conquest: The Guatemalan Tragedy.* Berkeley: University of California Press, 1993.

Schlesinger, Stephen C., and Stephen Kinzer. *Bitter Fruit: The Untold Story of the American Coup in Guatemala.* Garden City, N.Y.: Doubleday, 1982.

Sexton, James D. *Campesino: The Diary of a Guatemalan Indian.* Tempe: University of Arizona Press, 1985.

Stoll, David. *Between Two Armies in the Ixil Towns of Guatemala.* New York: Columbia University Press, 1993.

Woodward, Ralph Lee. *Guatemala.* Santa Barbara, Calif.: Clio, 1992.

GUYANA

Cooperative Republic of Guyana

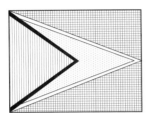

CAPITAL: Georgetown.

FLAG: A red triangle at the hoist extending to the flag's midpoint is bordered on two sides by a narrow black stripe; extending from this is a golden arrowhead pointing toward the fly and bordered on two sides by a narrow white stripe. Two green triangles make up the rest of the flag.

ANTHEM: Begins "Dear land of Guyana, of rivers and plains."

MONETARY UNIT: The Guyanese dollar (G$) of 100 cents is a paper currency tied to the US dollar. There are coins of 1, 5, 10, 25, 50, and 100 cents, and notes of 1, 5, 10, 20, and 100 Guyanese dollars. G$1 = US$0.0075 (or US$1 = G$132.80).

WEIGHTS AND MEASURES: Guyana officially converted to the metric system in 1982, but imperial weights and measures are still in general use.

HOLIDAYS: New Year's Day, 1 January; Republic Day, 23 February; Labor Day, 1 May; Caribbean Day, 26 June; Freedom Day, 7 August; Christmas, 25 December; Boxing Day, 26 December. Movable religious holidays include Good Friday, Easter Monday, Phagwah, 'Id al-'Adha, Yaou-Mun-Nabi, and Dewali.

TIME: 9 AM = noon GMT.

¹LOCATION, SIZE, AND EXTENT

Situated on the northeast coast of South America, Guyana is the third-smallest country on the continent, with an area of 214,970 sq km (83,000 sq mi), extending 807 km (501 mi) N-S and 436 km (271 mi) E-W, including disputed areas. Comparatively, the area occupied by Guyana is slightly smaller than the state of Idaho. Bounded on the N by the Atlantic Ocean, on the E by Suriname, on the S and S-W by Brazil, and on the NW by Venezuela, Guyana has a total boundary length of 2,921 km (1,815 mi).

Neither Guyana's western border with Venezuela nor its eastern border with Suriname has been resolved. Venezuela claims all territory west of the Essequibo River, an area of more than 130,000 sq km (50,000 sq mi), or over three-fifths of Guyana. Suriname claims a largely uninhabited area of 15,000 sq km (5,800 sq mi) in the southeast, between two tributaries of the Corentyne River.

Guyana's capital city, Georgetown, is located on the country's Atlantic coast.

²TOPOGRAPHY

Guyana has three main natural regions: a low-lying coastal plain, extending for about 435 km (270 mi) and ranging from 16 to 64 km (10–40 mi) in width, much of which is below high-tide level and must be protected by sea walls and drainage canals; a region of heavily forested, rolling, hilly land, about 160 km (100 mi) in width, which contains most of the mineral wealth and comprises almost five-sixths of Guyana's land area; and in the south and west, a region of mountains and savannas. There are several large rivers, including the Essequibo, Demerara, and Berbice, but few are navigable for any distance above the plains because of rapids and falls.

³CLIMATE

The climate is subtropical and rainy. The average temperature at Georgetown is 27°C (81°F); there is little seasonal variation in temperature or in humidity, which averages 80–85%. Rainfall averages 230 cm (91 in) a year along the coast, falling in two wet seasons—May to July and November to January—and 150 cm (59 in) in the southwest, where there is a single wet season, extending from April through August.

⁴FLORA AND FAUNA

The flora varies with the rainfall and soil composition. The coastal area, originally swamp and marsh with mangrove and associated vegetation, has long been cleared for farming. In inland areas of heavy rainfall there are extensive equatorial forests, with green-heart a major species; varieties of trees may number as many as 1,000. Local fauna includes locusts, moth borers, acoushi ants, bats and other small mammals. There may be more than 728 species of birds. Northwestern coastal beaches are an important breeding ground for sea turtles.

⁵ENVIRONMENT

Because at least 85% of Guyana is still wilderness, the country has so far sustained little serious environmental damage. The air is clean, but water supplies are threatened by sewage and by agricultural and industrial chemicals. One potential problem for the nation's water supply is the pollution of its wells by saltwater from the ocean. Guyana has 57.8 cubic miles of water with 99% used for farming purposes. All of the nation's city dwellers have pure water, but 29% of people living in the rural areas do not. The nation's cities produce 0.1 million tons of solid waste per year. Since 1985, the nation has experienced an increase in diseases related to water and food consumption. Kaieteur National Park is the only specifically designated conservation area. In 1994, 12 of Guyana's 198 mammal species and nine of its 728 bird species were endangered. Three types of reptiles in a total of 137 and one freshwater fish are also endangered. Sixty-eight of Guyana's 6–8,000 plant species are threatened with extinction. As of 1987, endangered species in Guyana included the tundra peregrine falcon, the black and spectacled caimans, and four species of turtle (green sea, hawksbill, olive ridley, and leatherback).

6POPULATION

Guyana's population (758,619 at the 1980 census) was estimated by the UN at 808,000 in 1992, but by the US Bureau of the Census at only 738,625 in 1994. A population of 1,883,000 is projected by the UN for the year 2000, assuming a crude birthrate of 21.9 per 1,000 population, a crude death rate of 6.5, and a net natural increase of 15.4 during 1995–2000. More than 90% of the people live on 5% of the land along the Atlantic coast; the interior is practically uninhabited. The major cities are Georgetown, the capital and chief port, with a population estimated at 195,000 in 1985; Linden, with 30,000 inhabitants; and New Amsterdam, with about 20,000.

7MIGRATION

Although specific figures are not available, there is known to have been significant outward migration in the 1980s, creating shortages of skilled workers and managers. Unofficial estimates put the number at 10,000 to 30,000 a year in the late 1980s, chiefly persons of Asian Indian extraction. Their destination was chiefly the US and Canada. In 1990 there were 123,000 Guyanese-born people in the US.

8ETHNIC GROUPS

An estimated 51% of the population is of Asian Indian descent and 43% of African descent. There are also Amerindians, Chinese, Portuguese, and other Europeans, together constituting the remaining 6%.

9LANGUAGES

English is the official language and is used in government, the schools, the press, and commerce. Also spoken are Chinese, Portuguese, Amerindian languages, and a patois used mainly by those of African descent.

10RELIGIONS

There are Christian (estimated at 50% in 1993), Hindu (33%), and Muslim (9%) communities. The major Christian denominations include Anglican, Roman Catholic (8%), Methodist, Presbyterian, and Lutheran. Public holidays include both Christmas and the Birth of the Prophet (in August each year).

11TRANSPORTATION

About 187 km (116 mi) of rail track are in service, originally built for the government-owned mining companies. The two government-owned passenger railway systems, however, have been scrapped: the Georgetown to New Amsterdam line in 1972, and the Georgetown to Parika line in 1974. Waterborne passenger and cargo service between these cities is now carried out by a government-owned transport service via the Essequibo and Berbice rivers. Georgetown is the main port, while New Amsterdam accommodates coastal and small oceangoing vessels. Springlands, on the Corentyne River, is the main port for service with Suriname. Waterfalls and rapids near the coast have prevented the development of river transportation to the interior, which contains 6,000 km (3,700 mi) of navigable waterways.

Roadways measured 7,665 km (4,763 mi) in 1991, mostly gravel or dirt. That year, Guyana had about 24,000 passenger cars, and 9,000 commercial taxis, trucks and buses. Georgetown's Timehri International Airport is served by several international carriers. Guyana Airways Corp., a government company, operates domestic and international air service.

12HISTORY

The coastline was first charted by Spanish sailors in 1499, at which time the area was inhabited by Amerindians of the Arawak, Carib, and Warrau language groups. By 1746, the Dutch had established settlements on the Essequibo, Demerara, and Berbice rivers, and had withstood French and English attempts to capture and hold the area. The English occupied the settlements in 1796 and again in 1803, and gained formal possession at the Congress of Vienna in 1815. The three main settlements were united into the Colony of British Guiana in 1831. Slavery was abolished in 1834, and many blacks settled in cooperative villages or moved into the towns. Under pressure from planters, indentured servants were brought in from India to work on the sugar plantations. As a result, most of the sugar workers still are of Asian Indian origin, while the urban population is predominantly black. This division into ethnocultural groupings later became an important factor in Guyana's politics.

The change in British imperial policy after World War II was reflected in a new constitution introduced in 1953, providing for a bicameral legislature and universal adult suffrage. Elections were held in 1953. However, the British balked after the People's Progressive Party (PPP) captured 18 of the 24 elected seats. Six months after the elections, the UK suspended the constitution, charging Communist subversion of the British Guiana government. The colony was governed on an interim basis until 1957, when general elections again were held. Again, the PPP won, with 47.9% of the votes, and Cheddi Jagan, leader of the PPP, was named chief minister.

The colony was granted full internal self-government in 1961, following four years of continued economic and social progress. In elections held under a new constitution introduced that year, the PPP won 20 of the 35 seats in the newly established Legislative Assembly. In January 1962, Jagan, who had been named prime minister, submitted an austerity program, calling for compulsory savings and a property tax. The response to this was a general strike and racial violence between Jagan's Asian Indian followers and his opponents, mainly of African descent. British troops were called upon to restore order, but the situation did not calm until July.

In the elections of December 1964, the PPP again emerged as the strongest party, but it was unable to form a government alone. As a result, the British governor called upon the leader of the People's National Congress (PNC), Forbes Burnham, to establish a government.

The following November, an independence conference held in London approved the present constitution, and on 26 May 1966, Guyana became a sovereign and independent nation. Guyana was proclaimed a cooperative republic on 23 February 1970, the 207th anniversary of a Guyanese slave revolt led by Cuffy, still a national hero. The PNC ruled as majority party between 1968 and 1992, although not without controversy.

Guyana became known to US audiences in 1978 in the wake of the Jonestown massacre. The Government of Guyana, in an attempt to colonize the nation's wilderness regions, had in 1977 allowed an American, James Warren "Jim" Jones, to establish the People's Temple commune at what became known as Jonestown, in the northwest. Many in the US had become concerned with developments in the commune, and US Representative Leo J. Ryan had gone to Jonestown to investigate. He and four other US citizens were murdered at a nearby airstrip by Jones's followers. Then, on 18 November 1978, Jones and more than 900 of his followers committed suicide by drinking poisoned punch.

Between 1980 and 1985, relations between the PNC and opposition parties deteriorated sharply, as opposition parties charged harassment and fraud. The the assassination in 1980 of Dr. Walter Rodney, a leading opposition figure, escalated the conflict. Under the administration of Forbes Burnham (1980–1985), human rights declined steadily. Burnham died in 1985 and was succeeded by First Vice-President and Prime Minister Desmond Hoyte. The new president sought to improve Guyana's relations with non-Socialist nations, particularly the US, and attempted the liberalization of the Guyanan economy.

However, by 1992 the country had grown tired of the PNC, and elected Cheddi Jagan of the PPP to the presidency. Jagan, who had been minority leader for years, received an impressive mandate with 53.4% of the vote, to 42.3% for the PNC. This translated to a solid 36 PPP seats in the National Assembly.

¹³GOVERNMENT

As of 23 February 1970, Guyana became a cooperative republic. Guyana's first president was elected by the National Assembly on 17 March 1970, and the post of governor-general was abolished. Proclamation of the cooperative republic also entailed the provision of mechanisms for the takeover of foreign enterprises.

Guyana's basic parliamentary structure dates from the constitution negotiated prior to independence in 1966. Under a new constitution approved in 1980, the unicameral National Assembly consists of 53 members elected by secret ballot under a system of proportional representation for a five-year term, plus 10 members elected by 10 regional councils, and two members elected by the National Congress of Democratic Organs. The latter, which is composed of deputies from local councils, together with the National Assembly, constitutes the Supreme Congress of the People of Guyana, which may be summoned or dissolved by the executive president. This office, created by the 1980 constitution, is filled by the leader of the majority party as both chief of state and head of government. The president appoints a cabinet including a prime minister, who also holds the title of first vice-president.

The voting age and age of majority are 18 years, and suffrage is universal. However, electoral irregularity is the rule, rather than the exception. A British-led team of observers pronounced the 1980 vote "fraudulent in every respect." Boycotts both before and after elections have also been frequent as a result of fraud charges, but the net effect of these boycotts has been to enhance the power of the majority party.

¹⁴POLITICAL PARTIES

Guyana's political parties are generally commited to socialism or some variant of it, but differ in the groups they represent and especially the ethnic groups that support them. A schism between the black and Asian Indian communities defines the major political division in the country.

In 1950, Cheddi Jagan and his wife organized the People's Progressive Party (PPP), which was anticolonial in nature, claimed to speak for the lower social classes, and cut across racial lines. Early in 1955, Forbes Burnham, who had been minister of education in Jagan's government, led a dissident PPP wing in the formation of the People's National Congress (PNC), which became the predominant political vehicle of Guyanese blacks, with Asian Indians remaining in the PPP.

Until 1992, the PNC had dominated Guyana's politics since independence. It draws its members primarily from urban blacks, and was in the majority from its first government, formed after the 1964 elections, until 1992. The PNC ideologically defines itself as socialist, but stresses the importance of a mixed economy in which the private sector is encouraged.

The PPP had been the opposition party since the 1960s, after dominating Guyanan politics in the 1950s. Appealing to Asian Indian rice farmers and sugar workers, the PPP nevertheless claims to be primarily an ideological party. Over the years, the PPP has taken an orthodox socialist position along the lines of international Communism. However, Jagan has at times called for increased foreign investment, and introduced conservative economic measures during his tenure as premier in the early 1960s. PPP opposition has been both loyal and otherwise. After the 1973 elections, the PPP boycotted the National Assembly, charging electoral fraud. In 1976 the representatives took their seats. In the 1980s the party appeared to be waning, but the 1992 elections gave a boost to this long-standing party.

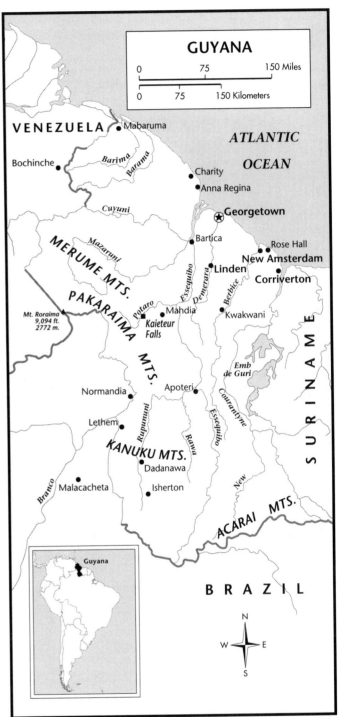

LOCATION: 1°12′ to 8°34′N; 56°29′ to 61°23′w. **BOUNDARY LENGTHS:** Atlantic coastline, 459 km (285 mi); Suriname, 600 km (372 mi); Brazil, 1,119 km (694 mi); Venezuela, 743 km (461 mi). **TERRITORIAL SEA LIMIT:** 12 mi.

Because Guyana uses a proportional representation system, small parties are accommodated within the system. One such group is the Working People's Alliance, a multi-ethnic independent party professing its own brand of Marxism. The WPA, founded in 1979, boycotted the 1980 elections on the grounds that they were bound to be rigged. In June 1980, its leader, Walter Rodney, was killed in a bomb blast. The party took one seat in the 1985 elections, and 2 seats in the 1992 elections.

The United Force was organized by Peter D'Aguilar, a wealthy brewer of Portuguese extraction in the early 1960s. Its program, called economic dynamism, was based principally on close ties with the West, encouragement of foreign enterprise, and the acquisition of foreign loans. It helped the PNC form the first non-PPP government in Guyana in 1964, but in 1968 the PNC formed a government by itself. In 1973 the UF lost the 4 seats it had won in 1968. In 1980, the UF won 2 seats, which it held until 1992, when it lost one of the two.

15LOCAL GOVERNMENT

Guyana's system of local government was restructured after independence. Guyana is divided into 10 regions, each of which is administered by a chairman and council. City and village councils administer the local communities.

16JUDICIAL SYSTEM

The Supreme Court has two divisions: the high court, which consists of the chief justice of the Supreme Court and 10 puisne justices and has both original and appellate jurisdiction; and the court of appeal (established 30 July 1966), which consists of a chancellor, the chief justice of the Supreme Court, and as many justices as the National Assembly may prescribe. The chancellor of the court of appeal is the country's chief judicial officer. Magistrates' courts exercise summary jurisdiction in lesser civil and criminal matters. The constitution of 1980 provides for an ombudsman to investigate governmental wrongdoing. English common law is followed. Although there is an ombudsman, he lacks the authority to investigate allegations of police misconduct. There is no independent body charged with responsibility for pursuing complaints of police brutality or abuse.

17ARMED FORCES

The Combined Guyana Defense Force numbered 2,000 full-time officers and troops (including a women's army corps) in 1993. Reserves and paramilitary forces numbered 2,000. The armed forces also manage a national service corps (1,500) for community development projects. Defense expenditures in 1988 were $14.4 million, subsidized by $2 million in US aid. Only the army (1,700) has modern weapons.

18INTERNATIONAL COOPERATION

Guyana became a member of the UN on 20 September 1966 and belongs to ECLAC and all the nonregional specialized agencies, except IAEA and WIPO. Guyana is also a member of the Commonwealth of Nations, CARICOM, G-77, IDB, and PAHO. It holds permanent observer status with the OAS, and is a signatory to the GATT and the Law of the Sea.

19ECONOMY

Guyana's economy is dominated by the production and processing of primary commodities, of which sugar and bauxite are the most important. Much of the country is undeveloped, with more than 90% of the population and almost all of the agriculture concentrated in the narrow coastal plain. The interior is sparsely settled, and communications are poor. The bulk of the population is engaged in agriculture, either as laborers on sugar plantations or as peasant cultivators of rice. Although sugar and rice continued to be important export earners, bauxite's share of national exports grew from 25% in 1970 to 48% in 1985. The government plays a direct role in the bauxite and sugar industries; the nation's leading bauxite-alumina producing company was nationalized in July 1971, as was a major sugar producer in 1976.

Beginning in the late 1970s, Guyana's economy suffered a severe decline, attributable both to the increasingly high costs of imported oil and petroleum products (39% of Guyana's merchandise imports in 1983) and to sagging production and prices of Guyana's exports. In 1982 there were serious shortages of basic commodities, foreign exchange reserves dwindled, and Guyana was forced to reschedule its debts. In 1985, the IMF declared Guyana ineligible for further loans because of noncompliance with fund conditions and high arrears.

However, in recent years, Guyana's economy has improved dramatically under the Economic Recovery Program (ERP) launched by the government in April 1989. This program marked a drastic reversal in government policy away from a predominantly state-controlled, socialist economy towards a more open, free market system. The government has reformed its monetary and fiscal policy establishing a free market in foreign currency, which was designed to stabilize the exchange rate and put an end to runaway inflation. The rate has remained stable at G$125 to the dollar since February 1991 and inflation has dropped from a 1989–91 annual average of 60–100% to only 14% in 1992. The government also eliminated price controls, removed import restrictions, promoted foreign investment, and divested itself of many state-owned enterprises. The program, which was designed with the assistance of International Monetary Fund (IMF) and World Bank officials, and is supported by Canada, the United Kingdom, and the United States, has proved very successful. Real GDP growth of 7.7% registered in 1992 marked the second consecutive year of strong recovery, with all of the key sectors demonstrating significant increases in production. Growth was particularly strong in the major export industries, including rice and sugar, which went up by 60% and 20% respectively. Gold exports increased by one third to $24 million. Bauxite was the only major industry to experience a decline in production of 20% from $79 million down to $56 million.

In addition, through Paris Club negotiations, debt rescheduling and debt relief agreements were reached with the United States, amounting to $422 million; the UK, amounting to $69.7 million; and Germany, amounting to $12.8 million. This left the stock of external debt, including private arrears guaranteed by the Bank of Guyana, slightly above $2 billion by the end of 1992.

20INCOME

In 1992, Guyana's GNP was $268 million at current prices, or $330 per capita. For the period 1985–92 the average inflation rate was 67.3%, resulting in a real growth rate in per capita GNP of −5.4%.

In 1992 the GDP was $375 million in current US dollars. It is estimated that in 1990 agriculture, hunting, forestry, and fishing contributed 22% to GDP; mining and quarrying, 13%; manufacturing, electricity, gas, and water, 9%; construction, 5%; wholesale and retail trade, 6%; transport, storage, and communication, 6%; finance, insurance, real estate, and business services, 5%; community, social, and personal services, 2%; and other sources, 33%.

21LABOR

The total work force in 1992 was about 250,000; unemployment reached 12.9% during the same year. The sugar industry is on the rebound, with higher wages and 6,000–7,000 new jobs added after 1990. The bauxite industry has declined in recent years; production fell by 40% in 1992 due to the deterioration of the mines at Linden. The central government employs about 18,000 people, and 46,000 more work in the public sector. Another 58,000 people work in the growing private sector. The total number of employed was registered at 121,861 in 1992. About 25% of the work force is formally organized. The public sector employs 53% of the labor force. Wages and conditions of work are increasingly regulated by government action and by collective bargaining. The Trades Union Congress is the national labor federation.

22AGRICULTURE

Agriculture, the main economic activity, provides nearly half the total value of exports and a large part of domestic food needs.

Because the narrow strip of rich, alluvial soil along the coast lies in part below the high-tide mark of the sea and rivers, and because of heavy seasonal rainfall, agricultural expansion requires heavy expenditures for flood control, drainage, and irrigation. About 2.5% of the land is used for temporary and permanent crop production.

The Guyanese economy's rebound in 1992 was primarily due to the expansion of the agricultural sector, which grew by 27.2% as a result of the successes of the sugar and rice subsectors. Increased acreage under cultivation, replanting schemes, good relations between management and labor, and favorable weather, together with improved access to foreign inputs helped to stimulate a 40% increase in sugar production. Sugar production in 1992 was estimated at 3,081,000 tons, up from the 395,000 tons produced in 1971; sugar accounted for 29% of exports in 1980 and about 36.7% in 1990. Rice production in 1992 (274,000 tons) grew by 12% in addition to a 60% increase in 1991—additional growth could have been attained were it not for infrastructural and other bottlenecks. Agricultural exports in 1992 totaled $175.7 million, up from $127 million in 1991. Other crops, grown for domestic consumption, include bananas, citrus, cassava, and yams.

23ANIMAL HUSBANDRY

Livestock in 1992 included 225,000 head of cattle, 60,000 hogs, 130,000 sheep, and 13,000,000 chickens. Other important domestic animals are goats, horses, mules, and donkeys. Extensive work is carried on to improve cattle productivity by importing breeding stock and providing artificial insemination and veterinary services.

24FISHING

Efforts are being made to increase the fish catch in order to improve the local diet and reduce imports of fish. The catch was 40,756 tons in 1991. About 3,653 tons of shrimp were harvested, mostly by a subsidiary of a US firm. The Demerara Fish Port Complex, built near Georgetown with Japanese aid, includes a fish-processing plant and office facilities.

25FORESTRY

Forests cover about 16,369,000 hectares (40,448,000 acres), but commercial exploitation has been confined to a relatively small section in the northeast. The government-operated timber plant buys lumber from private sawmills and processes it with a view to standardizing and raising the quality of timber for export. Only about 20% of the forest area is reasonably accessible for timber exploitation. Green-heart is the most important timber produced and exported. Timber production was about 175,000 cu m in 1991.

26MINING

Mineral production is second only to agriculture in importance. Guyana is the 11th largest producer of bauxite in the world, and production has recovered from the 1988–90 decline; in 1991, over 2.2 million tons were produced, up from 1.4 million tons in 1990. Gold, diamonds, and crushed stone also are mined. In 1991, gold mine output was 11,000 kg, and 18,189 metric carats of diamonds were mined. Bauxite and gold account for almost 40% of the country's exports.

Guyana's bauxite-mining industry came fully under state control on 1 January 1975, when the government nationalized Reynolds Guyana Mines Ltd., a subsidiary of the US Reynolds Metals. The government agreed to pay compensation totaling US$10 million in 13 annual installments. Guyana's largest bauxite producer, Demerara Bauxite Co. (owned by Aluminium Co. of Canada), was nationalized in July 1971 and renamed the Guyana Bauxite Co. All mineral rights are vested in the state, but the government encourages foreign investment through joint ventures or outright foreign ownership of Guyanese companies.

27ENERGY AND POWER

Net installed electric generating capacity was 114,000 kw in 1991, all of it but 2% in conventional thermal plants. Total electric energy produced was 500 million kwh. Frequent power failures have hampered production and thus impeded economic growth. The lack of reliable electricity in and around Georgetown has prompted many businesses to utilize imported small diesel-operated generators, further increasing total fuel demand.

A Canadian company found very high grade oil in the Essequibo region, also claimed by Venezuela, but further exploration was suspended when the company, unable to find the necessary financing, shut down its operations. In 1986, the government reached agreement with Venezuela on the barter of 500,000 tons of bauxite per year for 10,000 barrels of oil a day, half of which is used by the bauxite industry. In 1991, there were four oil concession holders for the Takutu Basin.

28INDUSTRY

Industrial development is a major goal of the government. Extensive surveys have been made with UK, US, and IBRD assistance, and special economic incentives, including tax holidays, are offered by the government. Credit for industrial development is made available by the government; the Small Industries Corp., created in the early 1970s, assists the development of cooperative enterprises. Industry is limited chiefly to processing bauxite, sugar, and rice for export and food and beverages for the local market. Industrial products in 1983 included wheat flour, 10,000 tons; rum, 133,000 hectoliters; and cigarettes, 408 million.

Presently, industry in general is undergoing an important transition. On 1 July 1993 the newly elected Jagan government presented it privatization policy framework paper which advertised the government's significant interest in total privatization, joint ventures, public share offers, employee and management buyouts, and leased management contracts. The government has been following a serious program of privatization of key state enterprises, such as the telephone utility—80% of which is now owned by the U.S. Virgin Islands firm Atlantic Telenetwork. The Guyana Sugar Corporation, Guysuco has recently been returned to the British firm Booker-Tate, which owned the company before it was nationalized in 1970. In this way, sugar production reached 243,000 metric tons in 1992, allowing Guyana to fulfill both its lucrative US and EC quotas for the first time since 1986. The government has also sold assets in the timber, electricity, fishing, bauxite, and rice industries.

As a result of privatization of the government-owned rice mills and the transfer of rice transactions to the common market for foreign currency, the rice industry has recovered and its production has increased to 168,000 metric tons in 1992, up from a low of just 93,000 metric tons in 1990. Also, the government expects the same result with the bauxite industry, which fell by 20% in 1992, largely due to the deterioration of the state-owned Linden Mines.

29SCIENCE AND TECHNOLOGY

The University of Guyana offers courses in the natural sciences and technical studies. The Guyana School of Agriculture Corporation was founded in 1963. The Inter-American Institute for Cooperation on Agriculture aims to stimulate and promote rural development as a means of achieving the general development and welfare of the population. The Pan-American Health Organization has an office in Georgetown.

In 1982, total research and development expenditures amounted to $2.8 million; 178 technicians and 89 scientists and engineers were engaged in research and development.

30DOMESTIC TRADE

Domestic trade is conducted largely through small retail establishments scattered throughout the settled areas, and by cooperatives. There are also traditional informal markets for the sale of agricultural products. Normal business hours are 9:30 AM to 6 PM, Monday–Friday, and 8 to 11:30 AM on Saturday. Banks are open weekdays from 8 AM to 12:30 PM and 3 to 5 PM.

31FOREIGN TRADE

Foreign trade rose steadily during the 1960s, reflecting the impact of development programs and increased exports of bauxite, which in 1968 replaced sugar as Guyana's single most important export. In the 1970s and early to mid-1980s, however, world markets for Guyana's export commodities weakened while oil import costs rose, leading to chronic trade deficits. In 1991, merchandise imports were $252 million. Fuel from Venezuela accounted for about 25% of Guyana's merchandise imports, and imports from the US, primarily capital equipment and unprocessed food, accounted for about 33%. As a result of Guyana's economic reform program, import restrictions have been removed, and import licenses are granted routinely by the Ministry of Trade, Tourism, and Industry. In the last years, seafood has accounted for over $15 million a year in foreign exchange earnings; however the forestry industry is expected to grow even more. Guyana's timber exports are expected to reach $60 million a year, surpassing the $5 million it accounted for in 1992.

32BALANCE OF PAYMENTS

Guyana generally runs a deficit on current accounts, which became increasingly severe in the 1980s. Since 1988, the governmetn has sought a policy of a free market in foreign currency and the removal of import prohibitions. By 1992, the balance of payments was recovering, assisted by a 25% increase in export earnings from 1991 (due to bountiful sugar and rice harvests). A positive capital account offset the external current account deficit of $73 million, which was further aided by foreign direct investments and $60 million official capital grants. During 1992, gross reserves increased by $34 million to total nearly $200 million, or more than four months of import costs.

33BANKING AND SECURITIES

The Bank of Guyana is the central bank. Commercial banking services are provided by the Guyana National Cooperative Bank, the country's major savings institution, by UK, Canadian, and US banks, and by one Indian bank. The money supply, as measured by M2, at the end of 1993 was G$39,783 million. At that time, demand deposits amounted to G$27,878 million, and time and savings deposits to G$27,878 million. Total domestic credit was G$120,806.

While a number of firms with publicly issued share capital are active, no large-scale securities market has developed.

34INSURANCE

In 1985, there were 11 life, 6 non-life, and 1 composite insurance company operating in Guyana. Premium revenues were divided almost equally between life insurance (51.3%) and non-life (48.7%).

35PUBLIC FINANCE

The budget follows the calendar year. Taxes finance the current account budget, with net surpluses or deficits being added to or subtracted from a general revenue balance. At the end of 1985, the IMF declared Guyana ineligible to receive further loans; most other multilateral agencies followed suit, and USAID and the Caribbean Development Bank did the same. In 1988, the government began an economic recovery program, adopting a free market. Divestment of state enterprises, elimination of price controls and subsidies, and a reform of fiscal and monetary policy have led to recent debt restructuring and forgiveness. Nevertheless, in 1991, Guyana's debt exceeded 600% of GDP and total debt service was two-thirds as large as export revenue. Since 1988, the government has reduced spending on salaries, goods and services from over 30% of GDP to less than 13%. In 1993, current expenditures and receipts totaled $100 million and $170 million, respectively. Capital expenditures amounted to $67 million in 1993; capital receipts, $6.6 million. The external debt at the end of 1992 stood at $2.1 billion.

36TAXATION

Income taxes are the major source of direct tax revenue. Personal income taxes are levied on a graduated scale up to 40%. The corporate tax rate of 35% is applied to all companies. Other taxes include property tax and consumption tax on locally manufactured goods, and stamp taxes. Local government authorities derive their revenues primarily from land, building, and service taxes. Tax evasion is a constant problem.

37CUSTOMS AND DUTIES

Customs revenues are traditionally a main source of government income, and both imports and exports are subject to charges. The common external tariffs of CARICOM are levied as well as specific duties on an ad valorem basis and, for most imports, a consumption tax. Several imported basic commodities are charged a rate below the preferential rate, while many luxury items carry rates above the general rate.

38FOREIGN INVESTMENT

Investment by foreign firms accounted for the bulk of capital formation prior to the establishment of Guyana's cooperative republic in 1970. Canadian and US capital developed the bauxite industry; between 1971 and 1975, however, the entire industry came under state control. After Guyana became a cooperative republic, the government did little to attract foreign private investment. The Hoyte government, however, began efforts to obtain foreign investments for the rehabilitation of the bauxite industry and for oil prospecting and gold mining. New legislation to simplify foreign investment procedures was being written in 1987.

In July 1992, Guyana signed a Tax Information Exchange Agreement with the US, thereby qualifying it for low-interest development and investment loans under the 936/Caribbean Basin Development Program.

The implementation of the ERP and the strong interest of the government in privatization has attracted many foreign investors and the trend is expected to continue. In addition, the government is prepared to implement arrangements designed to facilitate investors' derivation of tax benefits in their home territories as well as tax credits in Guyana. It recently signed an agreement with Canada to relieve investors from double taxation. Other investment incentives include: tax holidays, export allowances, accelerated depreciation, and special provisions for petroleum exploration and production, and for activities in gold and diamond extraction.

39ECONOMIC DEVELOPMENT

A continuing theme of Guyanese economic development policy has been the attempt, without great success, to expand agriculture and to diversify the economy. A seven-year development program (1966–72) involved public expenditure of about G$300 million. Its chief aims were to move the country's economy away from its heavy dependence on sugar, rice, and bauxite and to increase funds for scientific, vocational, and technical training and agricultural education. A key feature of the 1972–76 development plan was its emphasis on improving Guyana's health and housing standards.

A decisive change in economic orientation was marked by the proclamation on 23 February 1970 of a cooperative republic. The government embarked on a policy of cooperative socialism by nationalizing the bauxite industry, seeking a redistribution of national wealth, and fostering the establishment of cooperative enterprises for agricultural production, marketing, transportation, housing and construction, labor contracting, services, and consumer purchases. Within a decade, about 80% of the economy was in the public sector.

As economic conditions declined in the late 1970s and early 1980s, the government instituted such austerity measures as import restrictions, foreign exchange controls, cutbacks in planned government spending, and layoffs of government employees.

From 1953 through 1986, Guyana received us$115.5 million in nonmilitary loans and grants. Multilateral assistance during the same period equaled us$265.4 million, of which 42% came from the IDB and 30% from the IBRD. The 1985 declaration by the IMF that Guyana was ineligible to receive further assistance until outstanding debts with the fund had been repaid was an indication of how severe the nation's financial crisis had become. In 1983, the US had vetoed G$244.4 million in aid from US and IDB sources, and late in 1985 a barter agreement with Trinidad and Tobago was suspended because of Guyana's failure to repay outstanding loans. As a result, the Hoyte government sought a rapprochement with international lending agencies: a delegation from the IMF, the World Bank, and the IDB visited Guyana late in 1986, and early in 1987 the Guyana dollar was devalued by 56%.

In the late 1990s, primarily as a result of economic reforms, agricultural output is expected to continue growing at an stable rate. Manufacturing output is also likely to grow because of expected improvements in electricity generation and distribution and improved incentives for private investments. These factors combined should make attaining continued recovery with real growth rates in excess of 5% per year possible in the medium term.

The continuation of sound macroeconomic policies and public sector reform, together with multilateral and bilateral assistance, is crucial to sustaining these recovery efforts. The fiscal situation is expected to continue improving in the short and medium term, largely as a result of increased current revenues.

The inflation rate is likely to keep dropping, while the medium-term external position of Guyana will remain clouded by the large external debt outstanding, so that the search for debt relief and preferred lending from international donors remains essential.

40SOCIAL DEVELOPMENT

A National Insurance Scheme covering all employed persons between the ages of 16 and 65 was introduced in September 1969. In April 1971, the scheme was broadened to include self-employed persons in the same age bracket. Social welfare benefits include workers' compensation, maternity and health insurance, death benefits, disability, and old age pensions. In 1991, workers contributed 4.4% of earnings, and employers made a 6.6% payroll contribution. Retirement pensions are 40% of average weekly earnings, disability pensions are 30%, and survivor benefits are 50% of the payable old age or disability pension.

The Guyana Responsible Parenthood Association, a private family-planning agency, is affiliated with the International Planned Parenthood Federation and has been assisted by AID. The fertility rate declined from 6.64 in 1950–55 to 2.8 in 1985–90. The 1990 Equal Rights Act which was intended to end sex discrimination has proven difficult to enforce because it lacks a specific definition of discrimination. There is no legal protection against sexual harassment in the workplace or dismissal on the grounds of pregnancy.

41HEALTH

In 1990, there were 200 doctors, and 96% of the population had access to health care services. In 1993 the average life expectancy was 65 years. Infant mortality in 1992 was 49 per 1,000 live births, and the overall mortality rate in 1993 was 7.1 per 1,000 people. Public health measures had virtually eliminated malaria as a major problem. The incidence of filariasis, enteric fever, helminthiasis, nutritional deficiencies, and venereal diseases still is significant. Yellow fever remains a constant threat. In 1992, 76% of Guyana's children were vaccinated against measles.

42HOUSING

Housing is a critical problem, as is the lack of adequate water supplies and of effective waste disposal and sewage systems. Urban development plans have been prepared for Georgetown and New Amsterdam, and a number of schemes, including the construction of low-cost rental housing, have been inaugurated. Loans are made through the Guyana Cooperative Mortgage Finance Bank, founded in 1973. To spur housing development, the government established the Guyana Housing Corp. in 1974. The government provides supervision by trained personnel for those willing to build their own homes. Housing is provided by some firms for their employees. As of 1980, over 70% of all housing units were detached houses and 22% were apartments. Owners occupied 57% of all dwellings, 27% were rented, and 12% were occupied rent free. Three-fourths of all dwellings were wooden, 11% were wood and concrete, and 6% were concrete.

43EDUCATION

The adult literacy rate in 1990 was 96.4% (with males reported at 97.5% and females at 95.4%), among the highest in South America. Although educational standards are high, educational development has suffered in recent years from shortages of teachers and materials. School attendance is free and compulsory for eight years for children between the ages of 6 and 14. While primary education lasts for six years, secondary education has two phases: first five years followed by two years. In 1988, there were 414 schools at the primary level with 118,015 students enrolled.

The first students completed the one-year course at the Government Training College for Teachers in 1960. Teachers also are trained in the UK and at the University of the West Indies in Jamaica. The University of Guyana was established in 1963, and awarded its first degrees in 1967. The university has faculties in agriculture, the arts, health sciences, social sciences, education, and technology. The Kuru Kuru Cooperative College was established in 1970 to equip the Guyanese people both technically and philosophically for cooperative socialism and nation building. In 1989 at universities and all higher level institutions, there were 450 teaching staff and 4,665 students enrolled.

44LIBRARIES AND MUSEUMS

The National Library in Georgetown, with holdings of more than 195,000 volumes, also functions as a public library and has 37 branches. The Guyana Society Library (formerly the Library of the Royal Agricultural and Commercial Society) is the oldest in the country and has a collection of rare books dealing with the Amerindians of Guyana. Other important libraries include the British Council Library and the library of the US Information Agency. The University of Guyana, founded in 1963, maintains a library which has holdings of over 200,000 book and nonbook materials.

The Guyana Museum in Georgetown has a collection of flora and fauna, archaeological findings, and examples of Amerindian arts and crafts. It also has an aquarium and a zoological and botanical park.

[45]MEDIA

In 1991, Atlantic Tele-Network (ATN) purchased 80% of the Guyana Telephone and Telegraph Company. In that year there were 33,000 telephones in use. A public corporation runs the postal system. Overseas radiotelephone and cable services are provided by Cable and Wireless (W.I.), a private firm. An international telex service was inaugurated in 1967. Broadcasting is carried on by the government-owned Guyana Broadcasting Corp. There were an estimated 392,000 radios and 31,000 TV sets in use in 1991.

In 1991 there were two daily newspapers in Guyana, the state-owned *Guyana Chronicle* (circulation 14,000) and the *Mirror* (20,000), which has been reduced to irregular appearances, reportedly because of government harassment in the 1980s, temporarily becoming a Sunday paper.

[46]ORGANIZATIONS

Cooperative societies cover virtually every aspect of the economy. At the end of 1960, 588 societies were registered, with a total membership of 38,597; by October 1988 there were 1,459 societies, with a membership of some 100,000. There is a chamber of commerce in Georgetown.

[47]TOURISM, TRAVEL, AND RECREATION

Guyana's scenery varies from the flat marshy coastal plain to the savannas, plateaus, and mountains of the interior; the 226-m (740-ft) Kaieteur Falls, four times as high as Niagara, is the country's most outstanding scenic attraction. Riding, hunting, fishing, and swimming are available in the southern savanna of the Rupununi. Cricket is the national sport.

Visitors from Commonwealth countries and the US need valid passports and return or onward tickets. Visitors from most other countries need both a passport and a visa. In 1991, there were 73,000 visitor arrivals in Guyana, and tourism receipts totaled $30 million. There were 538,000 hotel rooms.

[48]FAMOUS GUYANESE

Citizens of Guyana who have established literary reputations abroad include the novelists Edgar Mittelholzer (1909–65), Edward Ricardo Braithwaite (b.1912), and the poet and novelist Jan Carew (b.1925). Linden Forbes Sampson Burnham (1923–85), former leader of the PNC, dominated Guyanese politics from 1964 until his death. Cheddi Berret Jagan, Jr. (b.1918), founder of the PPP, was chief minister from 1957 to 1961 and premier from 1961 to 1964, and has been the main opposition leader since that time, returning to office in 1992. Hugh Desmond Hoyte (b.1930) served as president from 1985 to 1992.

[49]DEPENDENCIES

Guyana has no territories or colonies.

[50]BIBLIOGRAPHY

Chambers, Frances. *Guyana*. Santa Barbara: Clio, 1989.

Daly, Vere T. *A Short History of the Guyanese People*. London: Macmillan Education, 1975.

Flanery, Michael. *The Why of Jonestown*. Pittsburgh: First Edition Books, 1980.

Guyana in Pictures. Minneapolis: Lerner, 1988.

Hintzen, Percy C. *The Costs of Regime Survival: Racial Mobilization, Elite Domination, and Control of the State in Guyana and Trinidad*. New York: Cambridge University Press, 1989.

Merrill, Tim (ed.). *Guyana and Belize: Country Studies*. 2d ed. Washington, D.C.: Library of Congress, 1993.

Singh, Chaitram. *Guyana: Politics in a Plantation Society*. Stanford, Calif.: Hoover Institution Press, Stanford University, 1988.

Spinner, Thomas J. *A Political and Social History of Guyana, 1945–1983*. Boulder, Colo.: Westview, 1984.

Williams, Brackette F. *Stains on my Name, War in my Veins: Guyana and the Politics of Cultural Struggle*. Durham: Duke University Press, 1991.

HAITI

Republic of Haiti
République d'Haïti

CAPITAL: Port-au-Prince.

FLAG: The upper half is blue, the lower half red; superimposed in the center is the national coat of arms with the motto *L'Union Fait la Force* ("Union makes strength").

ANTHEM: *La Dessalinienne (Song of Dessalines).*

MONETARY UNIT: The gourde (G) is a paper currency of 100 centimes. There are coins of 5, 10, 20, and 50 centimes and notes of 1, 2, 5, 10, 50, 100, 250, and 500 gourdes. Silver (5, 10, and 25 gourdes) and gold (20, 50, 100, 200, 1,000 gourdes) coins have also been minted. US paper currency also circulates freely throughout Haiti. G1 = $0.20 (or $1 = G5).

WEIGHTS AND MEASURES: The metric system is official for customs purposes, but French colonial units and US weights are also used.

HOLIDAYS: Independence and New Year's Day, 1 January; Forefathers Day, 2 January; Pan American Day, 14 April; Labor Day, 1 May; Flag and University Day, 18 May; National Sovereignty Day, 22 May; Assumption, 15 August; Anniversary of the Death of Dessalines, 17 October; UN Day, 24 October; All Saints' Day, 1 November; Commemoration of the Battle of Vertières and Armed Forces Day, 18 November; Discovery of Haiti, 5 December; Christmas, 25 December. Movable religious holidays include Carnival (three days before Ash Wednesday) and Good Friday.

TIME: 7 AM = noon GMT.

¹LOCATION, SIZE, AND EXTENT

Occupying the western third of the island of Hispaniola, Haiti has an area of 27,750 sq km (10,714 sq mi) including the islands of Tortuga (La Tortue), Gonâve, Les Cayemites, and Vache. Comparatively, the area occupied by Haiti is slightly larger than the state of Maryland. Extending roughly 485 km (300 mi) ENE-WSW and 385 km (240 mi) SSE-NNW, Haiti is bounded on the N by the Atlantic Ocean, on the E by the Dominican Republic, on the S by the Caribbean Sea, and on the W by the Windward Passage and the Gulf of Gonâve, with a total boundary length of 2,046 km (1,271 mi). Haiti claims Navassa Island, an uninhabited US possession about 50 km (31 mi) west of Hispaniola.

Haiti's capital city, Port-au-Prince is located on Hispaniola's west coast.

²TOPOGRAPHY

The coastline of Haiti is irregular and forms a long southern peninsula and a shorter northern one, between which lies the Gulf of Gonâve. Rising from the coastal plains to a peak height at La Selle of 2,680 m (8,793 ft) and covering two-thirds of the interior, three principal mountain ranges stretch across the country; one runs east and west along the southern peninsula, while the others stretch northwestward across the mainland. Once-fertile plains run inland between the mountains: the Plaine du Nord, extending in the northeast to the Dominican border, and the Artibonite and Cul-de-Sac plains reaching west to the Gulf of Gonâve. Of the many small rivers, the Artibonite (237 km/147 mi), which empties into the Gulf of Gonâve, and L'Estère (45 km/28 mi) are navigable for some distance.

³CLIMATE

The climate is tropical, with some variation depending on altitude. Port-au-Prince ranges in January from an average minimum of 23°C (73°F) to an average maximum of 31°C (88°F); the range in July is 25–35°C (77–95°F). The rainfall pattern is varied, with rain heavier in some of the lowlands and on the northern and eastern slopes of the mountains. Port-au-Prince receives an average annual rainfall of 137 cm (54 in). There are two rainy seasons, April–June and October–November. Haiti is subject to periodic droughts and floods, made more severe by deforestation. Hurricanes are also a menace.

⁴FLORA AND FAUNA

Tropical and semitropical plants and animals are characterized more by their variety than by their abundance. In the rain forest of the upper mountain ranges, pine and ferns as well as mahogany, cedar, rosewood, and sapin are found. Coffee, cacao, and coconut trees and native tropical fruits such as avocado, orange, lime, and mango grow wild. Many species of insects abound, but there are no large mammals or poisonous snakes. Ducks, guinea hens, and four varieties of wild pigeons rare elsewhere are plentiful, and egrets and flamingos live on the inland lakes. Reptile life includes three varieties of crocodile, numerous small lizards, and the rose boa. Tarpon, barracuda, kingfish, jack, and red snapper abound in the coastal waters.

⁵ENVIRONMENT

According to an AID report, Haiti "is suffering from a degree of environmental degradation almost without equal in the entire world." The virgin forests that once covered the entire country have now been reduced to less than 2% of the total land area. According to United Nations' sources, Haiti loses 3% of its forests every year. Deforestation has had a disastrous effect on soil fertility, because the steep hillsides on which so many Haitian farmers work are particularly susceptible to erosion. The nation loses 1.35 tons of soil per square kilometer yearly. Agricultural chemicals, such as DDT, are widely used in Haiti. These pollutants plus the use of oil with high lead content are a significant source of pollution. Not only has much of the topsoil been washed away, but the eroded slopes retain little rainfall and are vulnerable to flooding. The chief impediment to reforestation is the fact that Haiti is so intensely cultivated that allocation of land for forests means a reduction in the land available for crop growing and grazing. Foreign organizations have attempted to alleviate these problems. In 1981, an $8 million Agroforestry Outreach Project, funded primarily by the US, helped farmers plant trees throughout Haiti—over 4.5 million seedlings by 1983. The government has also agreed to set up the nation's first two national parks, with funding from the US Agency for International Development. Haiti has 2.6 cu mi of water with 68% used for farming activity. Forty-four percent of the nation's city dwellers and 65% of the rural people do not have pure water.

As of 1994, six species of mammals, one bird species, and four types of reptiles were endangered. As of 1987, endangered species in Haiti included the tundra peregrine falcon, Haitian solenodon, green sea turtle, hawksbill turtle, and American crocodile.

6POPULATION

Haiti, the most densely populated country in the Western Hemisphere, had a 1982 census population of 5,053,189. The 1992 population was estimated to be 6,764,000. The projected population for the year 2000 is 7,959,000, assuming a crude birthrate of 34.1 per 1,000 population, a crude death rate of 10.8, and a net natural increase of 23.3 for 1995–2000. The estimated population density in 1992 was 244 per sq km (631 per sq mi). An estimated 72% of the population was rural in 1990; the remaining 28% of the population was classified as urban. The district population of Port-au-Prince, the capital and largest city, was estimated as 1,143,626 in 1988. Other major cities, with estimated 1988 populations, are Jacmel, 216,600, and Les Cayes, 214,606.

7MIGRATION

Emigration from Haiti has been mainly to Cuba, other Caribbean states, Canada, and the US; illegal emigration to the US has been substantial since the 1960s. Between 1972 and 1981 more than 55,000 Haitian "boat people"—and perhaps over 100,000—arrived in Florida. During 1981–85, 43,312 Haitians were admitted legally to the US. In September 1981, the US and Haitian governments agreed to work together to halt the flow of refugees, and these efforts apparently met with success. Over the next 10 years only 28 of the 22,716 Haitians intercepted at sea were admitted to the US. Nevertheless the exodus resumed in 1994. The 1990 census found 290,000 people in the United States of Haitian origin, compared to only 90,000 in 1980.

Several thousand Haitian migratory workers travel to the Dominican Republic each year during the cane-harvest season; many more change their residences permanently. In 1986, the Haitian community in the Dominican Republic was estimated to be as many as 500,000 people.

8ETHNIC GROUPS

For the vast majority of Haiti's people, the African ethnic influence is dominant. About 95% of the inhabitants are pure black, and 5% are mulatto.

9LANGUAGES

The official languages of Haiti are French and Creole. Virtually all the people speak Creole, a mixture of early 17th-century provincial French and African tongues, with infusions of English, Spanish, and Amerindian words. Only about 20% speak French. English is used in the capital and to a lesser extent in the provincial cities, and along the Dominican border a Spanish Creole is spoken.

10RELIGIONS

The official religion is Roman Catholicism, but Haiti has freedom of worship. In 1993, affiliated Roman Catholics represented an estimated 90% of the population, and Protestants 10%. However, voodoo (vaudou), a mixture of ceremony and belief from West Africa, is still widely practiced, often in tandem with Christianity. In the mid-1980s there were an estimated 60,000 voodoo priests (houngans) in Haiti, compared with 427 Roman Catholic priests.

11TRANSPORTATION

In 1991, Haiti had 4,000 km (2,400 mi) of roads, 616 km (383 mi) of which were paved. Farm-to-market roads are few, and most produce for the local market is transported by burro or carried on foot by women. In 1980, a new highway linking Port-au-Prince with Les Cayes was opened, and several road improvement projects have been completed; overall, however, road conditions continue to deteriorate because of flooding. There were some 33,000 passenger cars and 22,000 commercial vehicles in 1992. In 1992, Haiti had a ratio of 117 inhabitants per registered vehicle, the highest amount of any country in the Western Hemisphere. Two railroad systems, the National Railroad of Haiti and the Cul-de-Sac Railroad, with a combined trackage of 301 km (187 mi), originally operated lines from Port-au-Prince to Verrettes and to Léogâne, and from Cap-Haïtien south to Bahon. By 1982, however, most of the system had become inoperative; the 40 km (25 mi) of lines that remained in 1991 were being used only for sugarcane transport.

The commercial shipping fleet consists of a few hundred small sailing vessels engaged in coastal trade and a few motorized vessels of light tonnage. The island depends chiefly on foreign shipping. During the early 1980s, the IBRD sought to stimulate intercoastal trade by building port facilities at Jérémie, Port-au-Prince, and Port-de-Paix.

Domestic air service is supplied by the privately owned Air Haiti, which connects principal cities on regular scheduled flights and also serves as an international carrier. Haiti Air Inter also provides domestic service. A jet airport at Port-au-Prince opened in 1965, and serviced 545,000 passengers in 1991.

12HISTORY

In 1492, Christopher Columbus made the European discovery of the island of Hispaniola and established a settlement near the present city of Cap-Haïtien. Within 25 years, the native Arawak Amerindians, a peace-loving, agricultural people, were virtually annihilated by the Spanish settlers. Bishop Bartolomé de las Casas, a missionary to the Amerindians, who had originally come to Hispaniola as a planter in 1502, proposed that African slaves be imported for plantation labor. Some time after 1517, a forced migration of Africans gave Haiti its black population.

About 1625, French and English privateers and buccaneers, preying on Spanish Caribbean shipping, made the small island of Tortuga their base. The French soon also established a colonial presence on nearby mainland coasts and competed with the Spaniards. In the Treaty of Ryswick (1697), Spain ceded the western third of the island (Haiti) to the French. Under French rule it became one of the wealthiest of the Caribbean communities. This prosperity, stemming from forestry and sugar-related industries, came at a heavy cost in human misery and environmental degradation.

The French Revolution in 1789 outlawed slavery in France, which inspired Haiti's nearly half million black slaves to revolt. In a series of violent uprisings, slaves killed white planters and razed estates. Although they suffered cruel reprisals, they fought on under the direction of Toussaint L'Ouverture, an ex-slave who had risen to the rank of general in the French army. By 1801 Toussaint controlled the entire island, and promulgated a

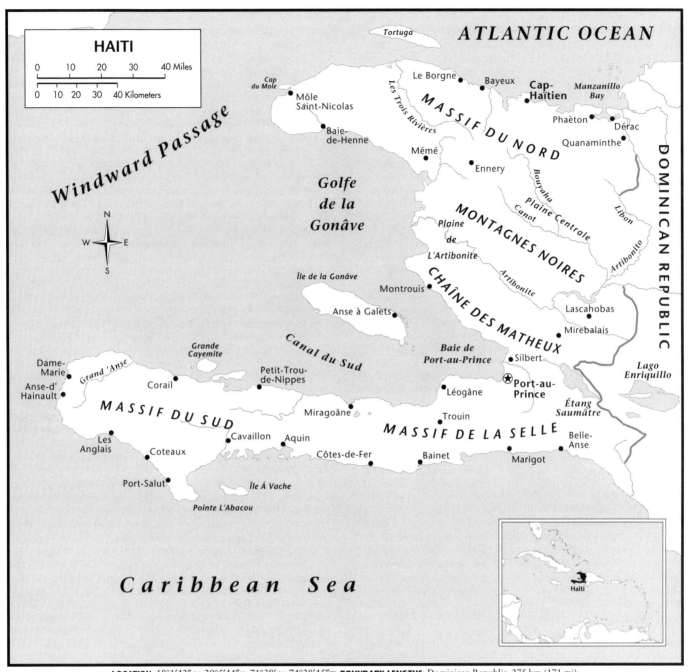

LOCATION: 18°1'42" to 20°5'44"N; 71°38' to 74°28'45"W. **BOUNDARY LENGTHS:** Dominican Republic, 275 km (171 mi); total coastline, 1,771 km (1,098 mi). **TERRITORIAL SEA LIMIT:** 12 mi.

constitution, which abolished slavery. The emperor Napoleon did not accept this move, and sent 70 warships and 25,000 men to suppress the movement. Toussaint was captured, and died in a French prison.

Jean Jacques Dessalines, another black general risen from the ranks, continued the struggle, and in 1803, the disease-decimated French army surrendered. On 1 January 1804, Dessalines proclaimed Haiti's independence. Dessalines, after assuming the title of emperor in 1804, was assassinated in 1806, and Haiti was divided into a northern monarchy and a southern republic. Under both regimes, the plantations were distributed among former slaves, and Haiti became a nation of small farmers. Haiti was reunited by Jean Pierre Boyer in 1820, and in 1822 the Haitian army conquered Santo Domingo (now the Dominican Republic). For 22 years there was one republic for the entire island. In 1844,

however, one year after Boyer was overthrown, the Dominican Republic proclaimed its independence from Haiti. In 1849, the president of Haiti, Faustin Elie Soulouque, proclaimed himself Emperor Faustin I. He was dethroned by a revolution headed by Nicholas Fabre Geffrard, who reestablished the republic and became president. In 1860, Geffrard negotiated a concordat with the Holy See that established Roman Catholicism as the national religion, although freedom of worship was retained.

A long period of political instability between 1843 and 1915, during which time Haiti had 22 dictators, culminated in the assassination of President Vilbrun Guillaume Sam and was followed by US military occupation. The occupation, which lasted 19 years, terminated in 1934 during the administration of President Sténio Vincent (1930–41), who in 1935 proclaimed a new constitution.

After World War II, another period of political instability culminated in 1950 in a coup d'etat that brought Gen. Paul Magloire to power. Magloire's economic policies led to a serious depression. In December 1956 a national sit-down strike, organized jointly by business, labor, and professional leaders, forced Magloire into exile. A period of chaos ensued, with seven governments trying to establish control.

In a September 1957 election filled with irregularities François Duvalier, a middle-class black physician known to his followers as Papa Doc, became president. He began to rule by decree in 1958, and in May 1961, he had himself elected for another six years. On 22 June 1964, Duvalier was formally elected president for life. Despite several attempted revolts, he consolidated his position, ruling largely through his security force, the Tontons Macoutes. Political opposition was ruthlessly suppressed, and thousands of suspected dissidents "disappeared." Also murdered were some 3,000 supporters of Daniel Fignolé, leader of the Peasant Workers Movement (Mouvement Ouvrier Paysan) and Duvalier's most effective opponent.

Political life under the Papa Doc regime was characterized by plots against the government and governmental counterterrorism, the latter entrusted to the security force known as the Tontons Macoutes ("bogeymen") and to other thugs known as cagoulards. Opposition leaders went into hiding or exile. The Haitian Revolutionary Movement (Mouvement Révolutionnaire Haïtien), led by Haitian exiles Luc B. Innocent and Paul G. Argelin, began operations in Colombia in February 1961.

The National Democratic Union (Union Démocratique Nationale) was founded in Puerto Rico in April 1962 by former Ambassador Pierre Rigaud, with a branch in Venezuela organized by Paul Verna and an underground movement operating in Haiti. Invasions in 1964, 1969, and 1970 met with no success. Haitian exiles in New York, Montreal, Chicago, and Washington mounted an influential anti-Duvalier campaign during the 1960s. Throughout this period, no party operated openly in Haiti except the Duvalierist Party of National Unity (Parti de l'Unité Nationale).

On 22 January 1971, Duvalier named his son Jean-Claude as his successor. Papa Doc died on 21 April 1971, and Jean-Claude, at the age of 19, became president for life the following day. The younger Duvalier sought to ease political tensions, encouraged tourism and foreign investment, and contributed to the beginnings of an economic revival. However, political arrests did not wholly cease, and there were severe economic reverses in the mid- and late 1970s.

In February 1979, elections to the National Assembly took place amid allegations of government fraud. Opposition groups were then arrested, tried, and convicted of subversion, but later released. In January and March 1982, two small exile groups tried unsuccessfully to overthrow the government by staging armed invasions. The first municipal elections of the Duvalier period were held in spring 1983. The voting resulted, for the most part, in victories for the government, partly because several opposition figures had been arrested during the campaign.

Jean-Claude proved to be an ineffectual leader, and tensions mounted as the economy stagnated after 1980. When civil disorder began to break out in the mid-1980s, the President became increasingly reclusive, occasionally attempting repression. In February 1986, following a series of demonstrations and protests, Jean-Claude and his family fled to France, and the National Governing Council (Conseil National de Gouvernement—CNG), led by Lt.-Gen. Henri Namphy, seized power.

The immediate aftermath of the CNG's takeover was euphoric. Political prisoners were released, and the dreaded Tontons Macoute, Duvalier's clandestine secret police, were disbanded. Namphy's declared purpose was to provide a transition to the inauguration of a democratically elected government. A constituent assembly, convened in October 1986, drafted a new constitution that was approved by referendum in March 1987.

Hopes for the restoration of democracy soon faded. The presidential election scheduled for November 1987 was postponed as gangs of thugs and soldiers killed at least 34 persons. The CNG attempted new elections and a new government, but those governments had no legitimacy at home or abroad. In December, 1990 a Roman Catholic priest, Jean-Bertrand Aristide, was elected with 67.5% of the votes cast.

Aristide had an ideology, a sort of egalitarian Catholic doctrine, and a political coalition of 15 parties, the National Front for Change and Democracy (FNCD), but he did not have the confidence of the military. Upset by his popularity and his foreign policy, which favored stronger hemispheric relations at the expense of US-Haitian relations, the military under General Raoul Cédras ousted him in October 1991. From exile, Aristide did not relent, and appealed to international organizations for help. The UN and OAS forged an agreement between Cédras and Aristide that was to return Aristide to the presidency in October 1993, but the military balked. Aristide promptly appealed to the Clinton administration, even as he criticized US policy, and the Clinton administration responded with sanctions against the Haitian regime in May and June of 1994. However, the impasse persisted.

In September 1994, as a last resort, the Clinton administration secured international support for a military invasion of Haiti to force Cédras from power. A US invasion force was assembled and war seemed imminent. However, at the 11th hour, Clinton sent a special delegation, headed by former US president Jimmy Carter, to negotiate a peaceful solution to the crisis. As US fighter planes were about to takeoff for Haiti, the Carter team reached an agreement with Cédras and war was adverted. American forces peacefully took control of the country, and in October 1994, restored Aristide to power.

13GOVERNMENT

Under Article 197 of the 1964 constitution, François Duvalier was appointed president for life, with the stipulation that this article be approved in a nationwide plebiscite. On 14 June, the voters were declared to have "almost unanimously" given their consent. He was granted power to dissolve the Legislative Assembly and the cabinet, and to govern by decree in case of grave conflict. A constitutional amendment in January 1971 allowed the president to choose his successor. Jean-Claude Duvalier became president for life in April 1971 and was chief of state and head of government until early 1986.

The constitution adopted in March 1987 established a president elected to a five-year term as head of state. The head of government was to be the premier, appointed by the president from the party holding the majority in both houses of the legislature, made up of a Senate and a House of Representatives. Supporters of the Duvaliers were barred from holding political office for 10 years.

Since its passage, the constitution was suspended in June 1988, and reinstated in March 1989. The leaders of the coup of October 1991 claimed to be observing the constitution, and Marc Bazin was named head of a caretaker government. But to all observers, nothing approaching a political system was present in Haiti in mid-1994.

14POLITICAL PARTIES

From the mid-19th to the mid-20th century, two major political parties, the Liberals and the Nationalists, were predominant. The Liberals, composed mainly of the wealthier and better-educated mulatto minority, advocated legislative control of government, while the Nationalists, composed mainly of the lower- and middle-class black majority, favored a strong executive. The traditional mulatto hegemony, whose wealth was inherited from the departed French colonists, was ended by Duvalier, who used the mulattoes as a scapegoat.

After Jean-Claude Duvalier became president in 1971, some political activity was allowed, but by 1982 most dissidents had again been silenced. In 1979, an opposition Haitian Christian Democratic Party (Parti Démocratique Chrétien d'Haïti—PDCH) was founded, but its leader, Sylvio Claude, was arrested in October 1980. In the 1979 legislative elections only one antigovernment candidate won a seat; he resigned in July 1981. The PDCH dropped out of the municipal election campaign in 1983 following the arrest of several party members on national security charges.

Dozens of parties emerged after the CNG ousted Jean-Claude Duvalier in February 1986. Most prominently, the National Front for Change and Democracy, which includes a number of parties including the National Cooperative Action Movement (MKN) and the National Congress of Democratic Movements (CONACOM). Other groups include the National Alliance for Democracy and Progress (ANDP), a coalition of the Movement for the Installation of Democracy in Haiti (MIDH), under current de facto Prime Minister Marc Bazin, and the National Progressive Revolutionary Party (Parti Nationaliste Progressiste Révolutionnaire—PAN-PRA), led by Serge Gilles. The Social Christian Party (Parti Social Chrétien—PSC, or PDCH-27 Juin), under Grégoire Eugène and the PDCH, now led by Joseph Douze, are also active, as well as a number of smaller parties.

15 LOCAL GOVERNMENT
As of 1986, Haiti was divided into 9 departments, 41 arrondissements, and 130 communes. Each department is headed by a number of prefects, appointed by the central government. Under the constitution, a commune is headed by an elected mayor, whose powers are strictly circumscribed. Local government is limited, and all taxes collected by the communes are paid directly into the national treasury. The first open municipal elections in 26 years took place in 1983. After the coup of 1991, local government has been inactive.

16 JUDICIAL SYSTEM
The judiciary consists of four levels: the Court of Cassation, courts of appeal, civil courts, and magistrates' courts. Judges of the Court of Cassation are appointed by the president for 10-year terms. Government prosecutors, appointed by the courts, act in both civil and criminal cases. There are also land, labor, and children's courts. Military courts function in both military and civilian cases when the constitution is suspended. The legal system is based upon the French Napoleonic Code.

Since the recent coup, the Haitian armed forces control law enforcement and public security even though the Constitution calls for separation of the police and military. Although the Constitution also calls for an independent judiciary, all judges since 1986 have been appointed and removed at the will of the government and political pressures affect the judiciary at all levels.

17 ARMED FORCES
In 1993, the Haitian armed forces consisted of about 7,400 personnel in the unified army, navy, (coast guard), air force, and police. The air corps consisted of 150 personnel and 7 aircraft, while the naval section had about 250 personnel and only small patrol boats. The Tontons Macoutes, formerly known as the National Security Volunteers, was formally disbanded in 1986 but remained a defacto mob and terrorist force under police control. Defense expenditures were budgeted at $29 million in 1991. The Haitians are armed with castoff US and European weapons.

18 INTERNATIONAL COOPERATION
Haiti is a charter member of the UN, having joined on 24 October 1945, and belongs to ECLAC and all the nonregional specialized agencies. It is also a member of IDB, OAS, PAHO, and other inter-American organizations, as well as of G-77 and the International Coffee Organization. Haiti is a signatory of the GATT and the Law of the Sea.

19 ECONOMY
Haiti is one of the world's poorest countries. The economy is basically agricultural: coffee, sugar, sisal, cotton, castor beans, cacao, plantains, and essential oils are the main products. Some cottage industries were developed in the mid-1940s, and in the late 1950s and early 1960s the mining sector, particularly bauxite and copper, grew to provide important export items. By the early 1980s, however, mining was losing its importance, and light export-oriented industry, based on cheap labor, was the main growth area.

Aside from such chronic problems as overpopulation, deforestation, and soil depletion, Haiti has suffered a series of natural and political setbacks. Hurricanes have often destroyed substantial parts of the coffee and sugar crops. In 1965, a prolonged drought caused additional crop damage and famine in large parts of the country. Another drought in 1975 crippled coffee exports and again raised the threat of famine. In 1977 and again in 1980, droughts caused widespread food shortages, and agriculture was further weakened by Hurricane David in 1979 and Hurricane Allen in 1980. (The latter destroyed more than half of the coffee crop.) The entrenchment of the Duvalier dictatorship after 1961 led to a drastic cutback in the inflow of external public funds, to the decline of tourism, and to a massive exit of technicians and entrepreneurs, who were constant targets of harassment by the militia. During 1960–70, the real GDP declined annually by 0.2%.

The accession of Jean-Claude Duvalier in 1971 improved the economy, and between 1970 and 1979, the average annual growth of the GDP was 4%. The economy took a downward turn in the early 1980s, and tourism, which had recovered during the 1970s, again suffered a severe decline.

The change of government in the mid-1980s brought no relief to the economic situation. Labor agitation prompted the withdrawal of some foreign investment, witnessed by a 15% decline in exports of assembly industries during 1985/86; that sector had been one of the few to show any dynamism in recent years. Competition from synthetics hurt jute exports, and 1985 sugar exports were hurt by a reduction of the US quota. Tourism declined still further, in part because of AIDS (which was particularly prevalent in Haiti).

The CNG has embarked on a program to eliminate the special privileges that were a feature of the Duvalier regime and to make the economy more market oriented.

On 30 September 1991 a military coup headed by General Cedras deposed the democratically elected government of President Jean-Bertrand Aristide. The Organization of American States (OAS) responded to this illegal usurpation of power by imposing economic sanction against Haiti. Former President Bush supported the OAS's efforts to restore democracy by freezing Haitian government assets and imposing a trade embargo against Haiti. A UN-mediated agreement called for President Aristide's return to power. The US subsequently suspended its trade embargo and economic sanctions were lifted on 31 August 1993. By mid-October 1993, The UN-mediated agreement collapsed and economic sanctions were reimposed immediately. President Clinton's administration continued the effort to restore democracy.

Prior to the embargo in FY 1991, the US alone supplied over 60% of Haiti's total $252 million in legal imports and purchased over 85% of its $146 million in exports.

Presently, no foreign assistance has been granted to Haiti, causing a deficit in the public sector of 5.2% of GDP in 1992.

The monetization of the deficit, along with the depreciation of the currency and consequent increase in the cost of imports, pushed Haiti's inflation rate to over 20% in FY 1992, compared with 8.4% for the preceding year.

Agriculture continues to be a predominant factor in Haiti's economy, accounting for about 66% of total employment. However, productivity has decreased considerably and agricultural products are no longer the main source of export earnings. The use of marginal areas unfit for agriculture continued to aggravate Haiti's chronic erosion problems. The trade embargo resulted in a shortage of and higher prices for spare parts, fertilizers, and agricultural tools. More than 60% of the small and medium-sized farm operations are now estimated to be highly over indebted.

20 INCOME

In 1992, Haiti's GNP was $2,479 million at current prices, or $390 per capita. For the period 1985–92 the average inflation rate was 7.7%, resulting in a real growth rate in per capita GNP of –2.9%.

In 1992 the GDP was $ 2,200 million in current US dollars. It is estimated that in 1991 agriculture, hunting, forestry, and fishing contributed 34% to GDP; mining and quarrying, less than 1%; manufacturing, 13%; electricity, gas, and water, 1%; construction, 6%; wholesale and retail trade, 16%; transport, storage, and communication, 2%; finance, insurance, real estate, and business services, 6%; community, social, and personal services, 3%; and other sources, 17%.

21 LABOR

In 1990, the labor force was estimated at 2.9 million. Agriculture employed 75%; industry and commerce, 18%; and services, 7%. The official 1990 unemployment rate was 14%, or 339,680 persons. Unofficially, however, the unemployment rate in 1991 was put at 25–50%.

Because the proportion of wage earners is relatively small, the labor movement is weak. In 1992 there were five small union federations: the Autonomous Central of Haitian Workers, the National Confederation of Haitian Teachers; the Federation of Unionized Workers; the Confederation of Haitian Workers; and the Independent General Organization of Haitian Workers. Unionization was marginal before the 1991 coup and is even more so now, with union members representing only about 1% of the work force. In industry and service organizations, the eight-hour day and the two-week paid vacation are standard. A revised pre-coup minimum wage applies to private sector workers, but is not enforced for agricultural employees. In September 1992, an estimated 30% of 10,000 enterprises paid less than minimum wage.

22 AGRICULTURE

With three-quarters of the laboring population deriving its living exclusively from the soil, farming is the mainstay of the Haitian economy. Although only about one-third of the country's land is considered suitable for cultivation because of the rugged terrain, over 32% of the land was actually being used for crop and feed production in 1991; permanent pasture land amounted to 18% of the total land area. Nevertheless, population growth outstripped agricultural growth during the 1970s, and a drought in 1985 affected the production of such important staple crops as rice, maize, and beans. Consequently, foodstuffs have had to be imported in increasing quantity.

Haitian agriculture is characterized by numerous small plots averaging slightly over 1 hectare (2.5 acres) per family, on which peasants grow most of their food crops and a few other crops for cash sale; few farms exceed 12 hectares (30 acres). Haiti employs an unusual form of farming called arboriculture. Combinations of fruit trees and various roots, particularly the manioc plant, the

Haitian bread staple, replace the grain culture of the usual subsistence-economy farming. Crops are cultivated with simple hand tools; the plow or animal power is only rarely employed, except on sugarcane plantations. Coffee is grown on humid mountain slopes, cotton on the semiarid plateaus and sea-level plains, and bananas as well as sugar on the irrigated plains, which covered about 75,000 hectares (185,300 acres) in 1991.

After two years of drought, favorable weather in 1992 helped to produce generally normal crops of sorghum and corn. Rice production fell because of fertilizer shortages and diversion of irrigation water in the Artibonite Valley to the hydroelectrical reservoir for energy production. The most important commercial product is coffee, classified as "mild." Coffee constituted 22% of Haitian exports in 1985. The trade embargo has resulted in shortages and higher prices for agricultural inputs, which has pressured farmers into selling assets or future crops at below market prices. Production in 1992 totaled 30,000 tons, as compared with the record high of 43,600 tons in 1962. Sugarcane is the second major cash crop, but production has been declining; in 1976, Haiti became a net importer of sugar. Sugarcane production in 1992 was 2,700,000 tons. Other agricultural production figures for the 1992 growing season (in thousands of tons) were bananas, 180; corn, 100; rice, 90; sorghum, 50; beans, 45; and cocoa beans, 3.

23 ANIMAL HUSBANDRY

Stock raising is generally a supplementary activity on small farms. In 1992 there were 1,300,000 head of cattle, 880,000 hogs, 1,150,000 goats, 296,000 mules and burros, 90,000 sheep, and 13,000,000 poultry. The hog population was decimated by African swine fever in 1979, and careful efforts at replacement have been unsuccessful. Poultry production has not risen enough to fill the vacuum in the rural diet. Extension work directed by the Department of Agriculture's educational center at Damien has helped to stabilize animal husbandry. Native stock has been upgraded by the introduction of hogs and cattle from abroad, particularly the zebu, which does well in the hot, dry plains. Two major stock-feeding centers operate at Port-au-Prince and Cap-Haïtien. Livestock products in 1992 included 67,000 tons of meat, 26,000 tons of goat's milk, 16,000 tons of cow's milk, and 3,300 tons of eggs.

24 FISHING

While the proximity of Haiti to the Windward Passage and the north-flowing currents off the Venezuelan coast place it in the path of major fish migrations, including tuna, marlin, bonito, and sardines, the commercial fishing industry is not developed. Reef fish, including giant grouper and rock lobster, are important as food sources because deep-sea fishing is limited. Fisheries have been successfully developed in the small ponds and in the irrigation and drainage ditches of the Artibonite Plain. Carp and tertar, a native fish, are abundant, but lack of transport and other facilities limits this important food source to local consumption. The catch was estimated at 5,150 tons in 1991.

25 FORESTRY

Originally, Haiti was endowed with abundant forest resources. Excellent stands of pine were located in the mountain rain forests of La Hotte Massif and in the Massif du Nord. (Haitian pine is high in turpentine and rosin content, making it suitable for naval stores.) Major stands of mahogany grew in the Fer à Cheval region, and small stands occurred in the island's lower mountain ranges. Tropical oak, cedar, rosewood, and taverneaux also were widespread; hardwoods included lavan (mahogany), narra, tindalo, and ipil. The intensive use of the forests for fuel, both in colonial times and in the modern era, and the clearing of woodlands for agriculture resulted in a decline of Haiti's forestland from over 2.7 million hectares (6.7 million acres) before the coming of Columbus to no more than 36,000 hectares (133,000 acres) by 1991, the majority of which was privately owned. Such deforestation has created a problem with soil erosion. Reforestation efforts have been more ambitious in design than successful in execution, although the Agroforestry Outreach Project oversaw the planting of 4.5 million seedlings between 1981 and 1983. Of the estimated 5,957,000 cu m of wood cut in 1991, almost 96% was used for fuel.

26MINING
All subsoil rights belong to the state. The Mining Law of 1976 and the creation of the Ministry of Mining and Energy Ressources in 1978 made relations between the government and private investment possible, so that mining companies may be privately owned. Haiti possesses undeveloped resources of lignite, copper, gold, silver, antimony, tin, sulfur, coal, nickel, gypsum, and porphyry. Manganese deposits are found in the Morne Macat section of the Massif du Nord; bauxite, near Miragoâne; lignite deposits in the Plateau Central; and copper deposits, both sedimentary and in veins, in the Massif du Nord. Mining is limited, and mineral exploration has generally been conducted by foreign enterprises. Production of bauxite, which peaked at 613,000 tons in 1979, and exports of which were worth $14.9 million in 1982, ceased with the 1985 closing of the Reynolds mine. Private gold mining is permitted, but the metal must be sold to the National Bank. Copper mining was suspended in 1971 because it became uneconomical. The Haitian marble industry is being developed for export possibilities; clay, limestone, gravel, salt, and stone were also produced for the domestic market.

27ENERGY AND POWER
Net installed electrical capacity totaled 153,000 kw in 1991. Output increased from 118 million kwh in 1970 to 475 million kwh in 1991. More than half of the nation's electricity was generated by petroleum-burning plants in 1991. The Péligre Dam on the Artibonite River, financed by a $30-million loan from the US Export-Import Bank, was opened in 1971, and the Gaillard Dam on the same river was opened in 1985. Power represented the government's top investment priority in the early 1970s, but a decade later the electricity supply was still erratic and inadequate. Most industrial plants have standby generators. In 1971, the government took over the foreign-owned electric company and established the Haitian Electrical Co. as a government enterprise. By 1990, however, only 10% of the population had access to electricity.

Oil exploration and drilling operations have been conducted in the Plateau Central, Gonâve Island, and the Cul-de-Sac region, but there was no commercial production as of 1991. Since the trade embargo was placed on Haiti, gasoline and other petroleum products have been regularly smuggled through the Dominican Republic to Haiti.

28INDUSTRY
Industry is primarily devoted to the processing of agricultural and forestry products. During 1970–78, the industrial sector grew by 8.3% annually; in 1979 and 1980, growth slowed to 4%; and between 1981 and 1985 growth and decline alternated annually; 1985 saw a 7% rise in overall production. Sugar refineries produced 41,300 tons of sugar in fiscal 1986. Cement production in 1986 was 246,700 tons. Other industries produce aluminum, enamelware, garments and hats, essential oils, plastics, soap, pharmaceuticals, and paint. A steel plant commenced operations in 1974, converting imported scrap into steel sections. Products of light industry in 1985/86 included 846 million cigarettes, 210,500 tons of essential oils, 127,600 tons of flour, 1,373,300 tons of detergents, 1,059,000 tons of toilet soap, and 646,700 yards of synthetic fibers. Haitian plants also assemble US-made

components to create electronic devices, toys, and leather goods. In 1986, some 140 export assembly firms employed about 40,000 people.

The handicraft industry, called petite industry, is organized on an individual basis or developed along a production-line pattern. Several handicraft factories make mahogany products, including carvings and masks. Woven sisal products are also made. Handicraft enterprises have been encouraged by five-year tax exemptions.

The trade embargo imposed in 1991 has reduced domestic and foreign demand, together with a serious shortage of energy and spare parts, leading to a virtual shutdown in the assembly industry. About 130 export firms have closed since 1990, and some 30,000 workers have been laid off since 1991. In the textile industry, about 20% of the installed production capacity was transferred to other countries. Most foreign firms, representing one third of the industry, abandoned Haiti or cut back their operations to minimum levels. Of the 25 firms operating in the electronic industry, only 6 are still present. The domestic industrial sector (cement, flour, shoes, beverages, cooking oils, and sugar) also was affected by the fall in local demand and the trade embargo. The embargo did not fully halt the supply of imported inputs but it did inflate production costs and prices and induced devaluation of the currency. As a result, domestic industrial production declined 12% in 1992.

29SCIENCE AND TECHNOLOGY
The National Council for Scientific Research, founded in 1963, coordinates scientific activities in Haiti, especially in the public health field. Four colleges and universities, including the University of Haiti, offer degrees in basic and applied sciences.

30DOMESTIC TRADE
Port-au-Prince, a free port, is the commercial center of Haiti, with Cap-Haïtien second in importance. Most Haitian products are sold in regional markets, which meet on traditionally established days, once or twice a week. Women peddlers comb the countryside for comestibles, traveling great distances to and from the markets. An IDA credit in 1983 helped in the reconstruction of the Croix-de-Bossales market in Port-au-Prince, the largest market in the country, handling about two-thirds of the food and manufactures used in the capital.

Imported goods usually are sold in small stalls (boutiques), but there are now some modern supermarkets. Specialty goods and articles for the tourist trade are offered by merchants who are generally franchised to handle specific brands. Although foreign imports, motion pictures, and soft drinks are advertised in newspapers, radio is the principal advertising medium.

Stores are generally open on weekdays from 8 AM to 5 PM in the winter. In summer, closing time is set by law at 4 PM; on Saturdays, stores close at noon. Banks are open from 9 AM to 1 PM, Monday–Friday.

31FOREIGN TRADE
Coffee has been supplanted as the main export by manufactured articles assembled in Haiti for export. Major imports include machinery, transportation equipment, electrical equipment, petroleum products, wheat, fish, textiles, paper, and pharmaceuticals. Virtually all raw materials used in the export assembly plants are imported. Instability in export prices has caused market fluctuations in foreign trade; the trade balance is generally unfavorable.

After the partial lifting of the embargo, the assembly industry production was lower than previous years. In addition, depressed world market prices negatively affected exports of coffee and cocoa. Official statistics tend to underestimate foreign trade flows. Both exports and imports have decreased since the onset of the trade embargo.

The registered trade deficit was reduced by $14.4 million in 1992 in contrast to 1991. Exports declined substantially from $162.9 million in 1991 to only $73.4 million in 1992. Similarly, imports fell dramatically from $300.4 million in 1991 to $196.5 million in 1992. This clearly reflects the consequences of the embargo on Haiti's external economy.

Up to 1991, Haiti's major export trading partners were the US, France, Japan, Italy, Belgium, and the Dominican Republic. Its major import partners included the US, Japan, France, Canada, the Netherlands Antilles, and Venezuela.

During that period, primary exports were textiles, electronics, toys and sporting goods, coffee, mangoes, sisal, essential oils, and cocoa. Primary imports included machine and manufacturing products, food and beverages, oils, and chemicals.

32 BALANCE OF PAYMENTS

Private direct investment was sufficient throughout the late 1960s and 1970s to offset a consistently sluggish export performance. As Haiti's trade position deteriorated in the late 1970s and early 1980s, international economic aid outpaced direct investment. Foreign exchange reserves stood at $25 million by the end of February 1987, up dramatically from $5 million in November 1986. Haiti, as one of the world's least-developed countries, has received funding from bilateral and multilateral lending agencies. Before the Aristide government was deposed, the balance of payments deficit was reportedly shrinking as a result of increased capital inflows. Since the coup d'etat of 30 September 1991, however, international aid (except for humanitarian aid) has been suspended.

In 1991 merchandise exports totaled $162.9 million and imports $300.4 million. The merchandise trade balance was $–137.5 million. The following table summarizes Haiti's balance of payments for 1990 and 1991 (in millions of US dollars):

	1990	1991
CURRENT ACCOUNT		
Goods, services, and income	–205.0	–261.8
Unrequited transfers	166.3	251.3
TOTALS	–38.7	–10.5
CAPITAL ACCOUNT		
Direct investment	8.2	13.6
Portfolio investment	—	—
Other long-term capital	26.6	43.1
Other short-term capital	–21.5	–26.4
Exceptional financing	3.8	1.3
Other liabilities	—	—
Reserves	–26.4	21.5
TOTALS	–9.1	53.1
Errors and omissions	47.8	–42.6
Total change in reserves	—	—

33 BANKING AND SECURITIES

The national bank of the Republic of Haiti (Banque Nationale de la République d'Haïti), the sole bank of issue and government depository, was acquired from US interests in 1934 and became the fiscal agent of Haiti in 1947. As the nation's principal commercial bank, with 12 branches in 1993, it participates in the national lottery, the national printing office and plant, the National Archives, banana development, the tobacco and sugar monopolies, the Agricultural and Industrial Development Institute, and the Agricultural Credit Bureau; it is also a depositor with the IMF and IBRD.

The first private Haitian bank, the Bank of the Haitian Union, opened in 1973. In 1993, 11 other commercial banks were in operation; including two of them Haitian, two US, one Canadian,

and one French. Commercial bank reserves were G865.9 million at the end of fiscal 1989. Money supply, as measured by M2, was G4,558.7 million as of 1989.

There is no securities exchange in Haiti. Trading in Haitian corporations that make public offerings of their bonds or equity shares is conducted on the New York over-the-counter market.

34 INSURANCE

Major world insurance companies maintain agencies or branches in Haiti, the most prominent being Sun Life of Canada, the first to enter into life insurance. The insurance classes covered are life, accident, sickness, fire, and motor.

35 PUBLIC FINANCE

In 1990, current receipts amounted to $256.32 million, while current expenditures totalled $361.06 million, resulting in a deficit of over $104 million. Capital expenditures in 1990 amounted to $62.58 million. Before the coup d'etat in 1991, the Aristide administration made a clear improvement in public finances management through several measures to improve tax collection. Following the installation of the elected government in February 1991, Haiti's aid donors indicated a willingness to increase assistance. After the putsch in September 1991, however, all international aid was suspended except for humanitarian aid.

36 TAXATION

Income tax is the most important direct tax. Corporate and personal income taxes are progressive from 9% to 30%. Since 1951, new corporations, if placed on the government's list of recommended new industries have benefited from special tax concessions, including customs duties exemption and a five-year corporate income tax exemption. In 1985, the sales tax on many consumer goods was raised from 7% to 10%.

37 CUSTOMS AND DUTIES

In February 1987, a new tariff structure replaced all remaining specific duties with ad valorem tariffs and introduced new rates of between zero and 40% except for higher rates on rice, maize, millet, flour, and gasoline. The average tariff rate was set at 20%, with rates of 10% on equipment and raw materials.

38 FOREIGN INVESTMENT

The government welcomes foreign investment, granting important concessions to new industries not competing with local production. Such enterprises are exempt from import and export duties for the life of the enterprise and enjoy a full tax exemption for the first five years of operation. Companies locating in the industrial park are entitled to tax exemption for a further three years. For companies that locate outside of the Port-au-Price metropolitan area the 100% income tax exemption is for 15 years with 15% of the income tax payable on the 16th year, reaching 100% on the 21st year. Additionally, for import and export-oriented business, there is an exemption without time limit from customs duties on imported machinery, equipment, raw materials, and accessories needed for production. Foreign capital enjoys equal status treatment with Haitian capital. The National Office for Investment Promotion is in charge of foreign investment.

Substantial foreign investment in Haiti began during World War II as a means of stimulating production of goods considered essential to the US war effort. Agricultural development was financed largely by the US Export-Import Bank and Reconstruction Finance Corp., supplemented by private foreign capital. Three major sisal plantations as well as a rubber development corporation were organized. The political atmosphere in the 1960s discouraged foreign investment, but investments rose from $1 million per year in 1969 to a $4-million average in 1970–73. During the late 1970s and early 1980s, several US companies set up their assembly operations in Haiti. As of mid-1987, the US was by far the largest investor in Haiti, with some $130 million in various enterprises. Canada, France, and the Federal Republic of Germany also had significant investments Haiti.

Presently, these countries have closed operations in Haiti. There is not incentive for foreign firms to invest in Haiti anymore as a result of political uncertainty as well as drastic economic conditions.

39ECONOMIC DEVELOPMENT

Although its annual national revenue barely covers basic operating necessities, the government supports development programs by encouraging loans and by requiring private enterprises to finance development projects. Aided by the US and various international aid organizations, the government has supported the construction of tourist facilities, public works, and irrigation and the creation of monopolies in cement, sugar marketing, tobacco, and lumbering. Government credit and development corporations for agriculture and industry have been established. The Haitian-American Society for the Development of Agriculture, founded by the government in 1941 and financed by the US Export-Import Bank, has promoted the rubber and lumber industries, the cultivation and marketing of bananas, spices, and other agricultural products, and various handicraft enterprises. The Agricultural and Industrial Development Institute was formed in 1961, financed by a loan from the IDB.

Haiti's first five-year plan, for 1971–76 was discarded. A new one, for 1976–81, focused on development of agriculture, social welfare services, human resources, and public administration; in 1977, this second plan, too, was abandoned.

During 1946–86, US nonmilitary loans and grants to Haiti totaled $539.9 million. Multilateral development assistance during 1946–86 amounted to $600.7 million, of which $284.8 million came from the IDB and $256.4 from the IDA.

With the trade embargo, the Central Bank currently is subsidizing oil consumption. Some of the problems that need to be solved included high level of unemployment, widespread poverty, risks of epidemics, inadequate infrastructure, distorted prices, a weak judiciary system, and over-staffed and dilapidated government institutions. In 1992, $85 million in humanitarian aid was directed mainly toward supplying basic medicines and food.

A particular plan that will take effect after an apparently fixed political situation includes the "Emergency Economic Recovery Program" which will require coordinated external assistance from multilateral and bilateral agencies. Its main objectives should be consolidating at democratic structure of government and reestablishing law and order as basis for economic development.

If the Emergency Economic Recovery Program fulfills all of its objectives, the economy should be stabilized in the medium term and gradual restoration of normality and growth perspectives could be anticipated, always taking into account donations from friendly governments.

40SOCIAL DEVELOPMENT

Public social insurance programs are limited, but there is a pervasive tradition of self-help and personal charity. The national lottery is the principal source of welfare funds, while an employer tax provides workers' compensation for occupational accidents to the relatively small number of industrial workers. Old age, disability, and survivor benefits are funded by 2–6% of employee wages and an equal percentage of the employer's payroll.

A social services law enacted in 1961 required every factory, enterprise of regional importance, and "foreign colony" to endow its neighborhood with social services such as dispensaries, schools, low-cost housing, and artesian wells. The law, however, contained no precise definitions of the enterprises affected, no specifications as to the extent of the services to be rendered, and

no details on control, enforcement, or penalties. A health and maternity insurance law of 1967 had not been implemented as of 1991.

The Division of Family Hygiene coordinates a limited array of maternal and child health services, including family planning. The fertility rate in 1985 was still high at 5.0, however, and only 10% of couples of childbearing age were thought to be using contraception in 1989. Abortion is allowed to save the mother's life.

Levels of violence by Haiti's military government were high in 1993 and the first half of 1994, causing thousands of refugees to flee the country, many in small boats made unsafe by overcrowding. Growing use of rape as a tactic of intimidation by the military was reported in 1994. The status of women varies, ranging from restriction to traditional occupations in rural areas, to prominent positions in the public and private sectors in urban areas.

41HEALTH

In general, sanitation facilities in Haiti are among the poorest in Latin America. Haiti lacks water in both quantity and quality, with only 39% of the population having access to safe water in 1991. City sewerage systems are inadequate, and business and residential areas often make use of septic tanks. In 1991, only 24% of Haiti's population had access to adequate sanitation.

The number of physicians increased from 412 in 1972 to 887 in 1988 and, in 1992, there were 14 doctors per 100,000 people, with a nurse to doctor ratio of 0.8. In 1988, Haiti also had 85 dentists and 691 nurses (both have declined since 1985). Half the doctors are in Port-au-Prince, and a fourth are in other principal towns, leaving a minimum of medical services for the rural population. In 1990, there were 8 hospital beds per 100,000 people.

Malaria and yaws have been combated by WHO, while other health programs have been conducted by the Rockefeller Foundation and the American Sanitary Mission. Tuberculosis has long been a serious health problem, and in 1990, there were about 333 reported cases of tuberculosis per 100,000 inhabitants. Malnutrition and gastrointestinal diseases are responsible for more than half of all deaths. Children may receive vaccinations, but the statistics are very low. In 1992, children up to one year old were vaccinated against tuberculosis (45%); diphtheria, pertussis, and tetanus (24%); polio (27%); and measles (24%). In the early 1980s, Haitians were named among the groups that had high risk factors for contracting acquired immune deficiency syndrome (AIDS), but as of late 1987, it had not been ascertained what those risk factors are.

The infant mortality rate fell from 182 per 1,000 live births in 1960 to 87 in 1992, and the general mortality rate was 11.9 per 1,000 in 1993. Death statistics for 1994 will be affected by the internal political fighting. During 1992, the average life expectancy was 56 years for women and men. The 1993 birth rate was 35.3 per 1,000 people. In 1992, there were 240,000 births, with contraception among married women (ages 15 to 49) at the rate of 10%. In 1992, still only half the population (50%) had access to health care services. Total health care expenditures for 1990 were $193 million.

42HOUSING

Although housing projects have been constructed in Port-au-Prince and in Cap-Haïtien, there is an increasing shortage of low-cost housing. Migration to the major cities has compounded the urban housing problem. Outside the capital and some other cities, housing facilities are generally primitive and almost universally without sanitation. Wooden huts are the prevalent standard for the countryside.There is as yet no official low-cost housing program. By presidential decree, the National Housing Office was established in 1966. Housing built in the 1970s in Port-au-Prince for about 18,000 people merely replaced demolished units. A new cooperative project, supervised by the National Housing Office and financed by UNDP, was initiated in 1979 in St. Martin, on the outskirts of Port-au-Prince. Housing construction is reported to have proceeded at a steady pace for the first half of the 1980s. According to the latest available information for 1980–88, total housing units numbered 890,000, with 6.1 people per dwelling.

⁴³EDUCATION
Although 80% of the students speak Creole and have only rudimentary knowledge of French, educational programs are mostly conducted in French. The Office of National Literacy and Community Action has the major responsibility for literacy programs throughout the country. The adult literacy rate was estimated at 53% in 1990 (males: 59.1% and females: 47.4%).

It was estimated that only 11% of people who had reached the age of 24 in 1982 had attended school. In 1990, Haiti had 7,306 primary schools, with 555,433 pupils and 26,208 teachers. In general secondary schools there were 184,968 students and 9,470 teachers. The sole university, the Université d'État d'Haïti (Port-au-Prince), dating from 1920, had an enrollment of 4,600 in 1986. There are also five colleges.

⁴⁴LIBRARIES AND MUSEUMS
The library of the Brothers of St.-Louis de Gonzage, the finest in Haiti, includes bound newspaper collections covering the 19th and 20th centuries and many rare works of the colonial and republican eras. The Bibliothèque Nationale contains about 22,000 volumes of Haitiana. Le Petit Séminaire, a parochial college, has an excellent library. The government has a wealth of library material dating back to colonial Saint-Domingue in the National Archives and rare papers on the Napoleonic expedition in Haiti in the famous Rochambeau Collection. Private libraries, notably the Mangones Library in Pétionville, make important contributions to Haitian scholarship.

The National Museum in Port-au-Prince dates from 1938. The Museum of the Haitian People, also in the capital, has anthropological and folklore collections, and the College of St. Pierre houses the Museum of Haitian Art, which opened in 1972.

⁴⁵MEDIA
The government owns and operates domestic telephone and telegraph communications. In 1993 there were 50,000 telephone lines, with automatic transmission in Port-au-Prince, Cap-Haïtien, Jacmel, and Les Cayes, and a new 30,000-line network was nearing completion. All America Radio and Cables, RCA Global Communications, and Western Union International provide international telephone and telegraph service. In 1993 there were 33 AM radio stations, no FM stations, and 4 TV stations, providing commercial and government broadcast services; 310,000 radios and 31,000 television sets were in use. Télé Haïti is a commercial cable station; it has also been given exclusive rights to import and sell television parts and receivers. National Television of Haiti, the official station, began broadcasting in 1985.

Freedom of the press is guaranteed by the constitution. The principal Haitian newspapers (all published in Port-au-Prince) are Le Matin, (circulation 10,500), Le Nouvelliste (6,000), L'Union (5,000), and Panorama (1,500). Le Moniteur, the official gazette, is published three times a week.

⁴⁶ORGANIZATIONS
Organizational activity in Haiti is limited. The Credit Cooperative of Les Cayes, the only cooperative of any significance, has maintained a sizable membership. Branches of the Red Cross, Rotary, Boy Scouts, and Girl Scouts function, as does the Masonic Order. The Centre d'Art, an informal artists' cooperative founded in 1944, has exhibited Haitian artists locally and internationally. There is a chamber of commerce in Port-au-Prince.

⁴⁷TOURISM, TRAVEL, AND RECREATION
For entry to Haiti, visitors generally are required to have passports. Nationals of the US, UK and its possessions, France, and Germany do not need visas. In the 1980s, tourism was adversely affected by the island's generally depressed economy and political turbulence and by the alleged link between Haitians and AIDS. In fiscal 1986, tourist arrivals totaled 117,455, down from 158,000 in fiscal 1981. Except for a brief rise in 1988, tourism remained fairly steady through 1991, when 119,000 visitors arrived, 103,000 from North America. There were 1,500 hotel rooms, and tourists spent $46 million in that year. After the ouster of President Aristide, tourism declined once again.

Port-au-Prince is a free port for a variety of luxury items. Tourist attractions include white sand beaches, numerous colonial buildings in Port-au-Prince and other cities, and the early 19th century Citadelle and Sans Souci Palace in Cap-Haïtien. Rapid divorces—granted in 24 to 48 hours—and casino gambling are among the attractions for US residents. Football (soccer) is the national sport, and cockfighting is very popular. Tourist resorts offer facilities for water sports and tennis.

⁴⁸FAMOUS HAITIANS
The national heroes of Haiti include Pierre Dominique Toussaint L'Ouverture (1743–1803), the Precursor; Jean Jacques Dessalines (1758–1806), who defeated Napoleon's army and proclaimed Haitian independence; Alexandre Sabès Pétion (1770–1818), first president of the republic established in southern Haiti; and Henri Christophe (1767–1820), king of Haiti (1811–20), who built the famous Citadelle and Sans Souci Palace. François Duvalier ("Papa Doc," 1907–71), originally trained as a physician, was elected president in 1957 and in 1964 became president for life. His son Jean-Claude Duvalier (b.1951) inherited his father's title in 1971, was ousted in 1986, and was reinstated in 1944. Leslie F. Manigat (b.1930) was president in 1988.

John James Audubon (1785–1851), an artist and ornithologist, was born in Haiti. The writers Éméric Bergeaud (1818–58), Oswald Durand (1840–1906), Philippe Thoby-Marcelin (b.1904), Jacques Roumain (1907–44), and Jean Fernand Brierre (b.1909) have won international literary recognition. Noted poets include the dramatist Pierre Faubert (1803–68), Corolian Ardouin (1812–35), Alibée Féry (1819–96), and Charles-Seguy Villavaleix (1835–1923). Haitian artists include the sculptor Edmond Laforestière (1837–1904); the primitive painter Héctor Hippolyte (1890–1948), leader of the Afro-Art Renaissance in the Caribbean; Wilson Bigaud (b.1929); and Enguérrand Gourge (b.1931). Haitian composers include Occide Jeanty (1860–1936) and Justin Elie (1883–1931); Ludovic Lamothe (1882–1953) used voodoo music in his compositions.

⁴⁹DEPENDENCIES
Haiti has no territories or colonies.

⁵⁰BIBLIOGRAPHY
Abbott, Elizabeth. Haiti: The Duvaliers and their Legacy. New York: Simon and Schuster, 1991.

DeWind, Josh. Aiding Migration: the Impact of International Development Assistance on Haiti. Boulder, Colo.: Westview Press, 1988.

Greene, Anne. The Catholic Church in Haiti: Political and Social Change. East Lansing: Michigan State University Press, 1993.

Haggerty, Richard A. (ed.) Dominican Republic and Haiti: Country Studies. 2d ed. Washington, D.C.: Library of Congress, 1991.

Laguerre, Michel S. The Military and Society in Haiti. Knoxville: University of Tennessee Press, 1993.

Plummer, Brenda Gayle. Haiti and the United States: the Psychological Moment. Athens: University of Georgia Press, 1992.

HONDURAS

Republic of Honduras
República de Honduras

CAPITAL: Tegucigalpa.

FLAG: The national flag consists of a white horizontal stripe between two blue horizontal stripes, with five blue stars on the white stripe representing the five members of the former union of Central American provinces.

ANTHEM: *Himno Nacional,* beginning "Tu bandera es un lampo de cielo" ("Thy flag is a heavenly light").

MONETARY UNIT: The lempira (L), also known as the peso, is a paper currency of 100 centavos. There are coins of 1, 2, 5, 10, 20, and 50 centavos, and notes of 1, 2, 5, 10, 20, 50, and 100 lempiras. L1 = $0.50 (or $1 = L2).

WEIGHTS AND MEASURES: The metric system is the legal standard; some old Spanish measures are still used.

HOLIDAYS: New Year's Day, 1 January; Day of the Americas, 14 April; Labor Day, 1 May; Independence Day, 15 September; Birthday of Francisco Morazán, 3 October; Columbus Day, 12 October; Army Day, 21 October; Christmas, 25 December. Movable religious holidays include Holy Thursday, Good Friday, and Holy Saturday.

TIME: 6 AM = noon GMT.

¹LOCATION, SIZE, AND EXTENT

Situated in Central America, Honduras has a total area of 112,090 sq km (43,278 sq mi), with a length of 663 km (412 mi) ENE-WSW and 317 km (197 mi) NNW-SSE. Comparatively, the area occupied by Honduras is slightly larger than the state of Tennessee. It is bounded on the N and E by the Caribbean Sea, on the S by Nicaragua and the Gulf of Fonseca, on the SW by El Salvador, and on the W by Guatemala, with a total boundary length of 2,340 km (1,454 mi). Under the terms of an arbitration award made by Alfonso XIII of Spain in 1906, Honduras received a portion of the Mosquito (Miskito) Coast, or La Mosquitia, north and west of the Coco (Segovia) River. Citing Honduras's failure to integrate the territory, Nicaragua renewed its claim to the entire Mosquito Coast in the 1950s and brought the case to the International Court of Justice (ICJ). In February 1957, Honduras created the new Department of Gracias a Dios, made up of the former Mosquitia territory. The ICJ determined in 1960 that Nicaragua was obligated to accept the 1906 arbitration ruling concerning that country's boundary with Honduras. The judges ruled, by a vote of 14–1, that once a valid arbitration award was made in an international dispute, it became effective, and remained so, despite any lapse of time in carrying it out.

The two tiny Swan Islands (Islas del Cisne), lying at 17°23′N and 83°56′W in the west Caribbean Sea some 177 km (110 mi) NNE of Patuca Point, were officially ceded by the US to Honduras on 20 November 1971. For administrative purposes, they are included under the Department of Islas de la Bahía, whose capital is Roatán on Roatán Island. The Swan Islands had been effectively held by the US, which asserted a claim in 1863 to exploit guano, and had housed a weather station and an aviation post.

The capital city of Honduras, Tegucigalpa, is located in the south central part of the country.

²TOPOGRAPHY

Honduras is mountainous, with the exception of the northern Ulúa and Aguán river valleys on the Caribbean Sea and the southern coastal area. There are four main topographic regions: the eastern lowlands and lower mountain slopes, with 20% of the land area and no more than 5% of the population; the northern coastal plains and mountain slopes, with 13% of the land and about 20% of the population; the central highlands, with 65% of the area and 70% of the population; and the Pacific lowlands and their adjacent lower mountain slopes, with 2% of the area and 5% of the population.

The width of the Caribbean coastal plain varies from practically no shore to about 120 km (75 mi), and the coastal plain of the Gulf of Fonseca is generally narrow. The highest elevations are in the northwest (almost 3,000 m/10,000 ft) and in the south (over 2,400 m/8,000 ft). Many intermontane valleys, at elevations of 910 to 1,370 m (3,000 to 4,500 ft), are settled. The old capital city, Comayagua, lies in a deep rift that cuts the country from north to south. Tegucigalpa, the modern capital, is situated in the southern highlands at about 910 m (3,000 ft). There are two large rivers in the north, the Patuca and the Ulúa. Other important features include the Choluteca, Nacaome, and Goascorán rivers in the south, Lake Yojoa in the west, and Caratasca Lagoon in the northeast.

³CLIMATE

The northern Caribbean area and the southern coastal plain have a wet, tropical climate, but the interior is drier and cooler. Temperature varies with altitude. The coastal lowlands average 31°C (88°F); from 300 to 760 m (1,000 to 2,500 ft) above sea level the average is 29°C (84°F); and above 760 m (2,500 ft) the average temperature is 23°C (73°F). There are two seasons: a rainy period, from May through October, and a dry season, from November through April. Average annual rainfall varies from over 240 cm (95 in) along the northern coast to about 84 cm (33 in) around Tegucigalpa in the south. The northwest coast is vulnerable to hurricanes, of which the most destructive, Hurricane Fifi in September 1974, claimed some 12,000 lives, caused $200 million in property damage, and devastated the banana plantations.

[4]FLORA AND FAUNA

Honduras has a rich and varied flora and fauna. Tropical trees, ferns, moss, and orchids abound, especially in the rain forest areas. Mammal life includes the anteater, armadillo, coyote, deer, fox, peccary, pocket gopher, porcupine, puma, tapir, and monkeys in several varieties. Fish and turtles are numerous in both freshwater and marine varieties. Among the reptiles are the bushmaster, coral snake, fer-de-lance, horned viper, rattlesnake, and whip snake, caiman, crocodile, and iguana. Birds include the black robin, hummingbird, macaw, nightingale, thrush, partridge, quail, quetzal, toucanet, wren, and many others.

[5]ENVIRONMENT

The major environmental problems are soil erosion and loss of soil fertility (in part because of traditional slash-and-burn cultivation) and rapid depletion of forests for lumber, firewood, and land cultivation. In 1985, deforestation had claimed 347 square miles; that figure was projected to increase dramatically in the late 1980s as a result of the construction of a charcoal-burning steel mill and a pulp and paper mill. According to the World Environment Center, Honduras could be completely deforested by the year 2000.

Enforcement of antipollution laws has been weak, and Honduras also lacks an integrated economic development and land-use policy. Rivers and streams in Honduras are threatened by pollution from mining chemicals. The nation has 24.5 cu mi of water with 91% used in farming activities. Fifteen percent of the city dwellers and 52% of people living in rural areas do not have pure water. Honduras' cities produce 0.5 million tons of solid waste per year. Air pollution results from a lack of pollution control equipment for industries and automobiles. The Secretariat of Planning, Coordination, and Budget (Secretaría de Planificación, Coordinación, y Presupuesto—SECPLAN), the Ministry of Natural Resources, and several other agencies are vested with environmental responsibilities.

As of 1994, one of Honduras' 179 species of mammals was threatened. Eleven bird species in 672 and three reptiles in a total of 161 were also threatened. Forty-three plant species of 5,000 are endangered. As of 1987, endangered species in Honduras included the tundra peregrine falcon, jaguar, three species of turtle (green sea, hawksbill, and olive ridley), and three species of crocodile (spectacled caiman, American, and Morelet's).

[6]POPULATION

The 1988 census indicated a total population of 4,248,561. In mid-1994, the population was estimated at 5,380,341, and the projected population for the year 2000 is 6,846,000, assuming a crude birthrate of 33.8 per 1,000, a crude death rate of 6.4, and a net natural increase of 27.4 during 1995–2000. Tegucigalpa, the capital and principal city, had 576,661 residents in 1988, including suburbs. San Pedro Sula, the second-largest city, had an estimated population of 300,900 in 1989; La Ceiba, 71,600; and El Progreso, 63,400. In 1988, 44% of the total population was urban. The estimated population density was 48 per sq km (124 per sq mi) in 1994, compared with 13 per sq km (33 per sq mi) in 1950. Between 1974 and 1988, the overall rate of population growth averaged 3% annually.

[7]MIGRATION

Before 1969 there was a steady flow of immigrants from El Salvador. The steps taken by the Honduran government in 1969 to curb this influx were a contributing cause of the "soccer war" with El Salvador during the same year. When the Sandinistas took over in Nicaragua in 1979, former National Guard members began to arrive in Honduras, and by 1983 there were 5,000–10,000 of them along the border. In addition, at least 25,000 Miskito Amerindians from Nicaragua and about 21,000 Salvadorans had fled to Honduras by the end of 1986. Many of them later returned, but by the end of 1992 about 100,000 citizens of Central American nations had taken refuge in Honduras.

[8]ETHNIC GROUPS

The vast majority (90–92%) of the Honduran people are mestizo, a mixture of white and Amerindian. About 5–7% of the population is Amerindian, the largest proportion being in the Copán area near the Guatemalan border. Blacks, about 2% of the population, live mostly along the north coast. Perhaps 1% of the population is white, chiefly of Spanish origin.

[9]LANGUAGES

The official language is Spanish. However, English is used widely, especially in northern Honduras. The more important Amerindian languages include Miskito, Zambo, Paya, and Xicaque.

[10]RELIGIONS

Religious freedom is guaranteed by the constitution of 1982. In 1993, an estimated 93.5% of the population was Roman Catholic. The Protestant population accounted for another 3%, and there also were small numbers of spiritists, Amerindian tribal religionists, Muslims, Buddhists, Baha'is, and Jews.

[11]TRANSPORTATION

In 1991 there were 8,950 km (5,562 mi) of highways, about 19% of which were paved. The Pan American Highway virtually bypasses Honduras, entering from El Salvador and running to the eastern Nicaraguan border. The 362-km (225-mi) Inter-Ocean Highway is the only surface connection between the Pacific and the Caribbean that includes in its path both Tegucigalpa and San Pedro Sula. In 1971, a paved highway was opened between Tegucigalpa and San Pedro Sula and west to the Guatemalan border. Tegucigalpa is served by secondary roads to the north and east, while San Pedro Sula is connected both to the important Caribbean ports of Puerto Cortés, Tela, and Trujillo and to the western Mayan shrine site of Copán. Road improvements near the Nicaraguan border have been undertaken with US military aid since 1983. In 1992 there were 140,000 motor vehicles, of which 45,000 were passenger vehicles.

Rail service exists only in the north, connecting the industrial and banana-growing northeastern coastal zone with the principal ports and cities. National Railway of Honduras, owned and operated by the government, maintains all 785 km (488 mi) of track.

Four principal ports—Puerto Cortés, Tela, La Ceiba, and Puerto Castilla—serve the country on the Caribbean side. Another Caribbean port, Roatán, is offshore, in the Bay Islands; and Puerto de Henecán, on the Pacific coast, opened in 1979, replacing Amapala as a port facility, although the latter retains a naval base. La Ceiba and Tela are primarily banana-trade ports; Puerto Castilla (completed in 1980) serves the Olancho forestry project; and Puerto Cortés and Puerto de Henecán handle general traffic. River traffic is negligible. In 1991, the Honduran merchant fleet comprised 190 vessels totaling 628,000 GRT, compared with 109 vessels totaling 339,000 GRT in 1985.

Air service is important in the transportation of passengers and cargo, and there are more than 30 landing fields in Honduras, including three international airports. Toncontín airport, about 6.4 km (4 mi) from Tegucigalpa, is served by Transportes Aéros Nacionales de Honduras/Servicio Aéreo de Honduras (TAN/SAHSA), Líneas Aéreas Costarricenses (LACSA), Challenge, and TACA airlines and the domestic carrier Lineas Aéreas Nacionales (LANSA). TAN/SAHSA flies to the US, Mexico, and other Central American countries and also provides domestic passenger service, carrying some 357,600 passengers in 1991.

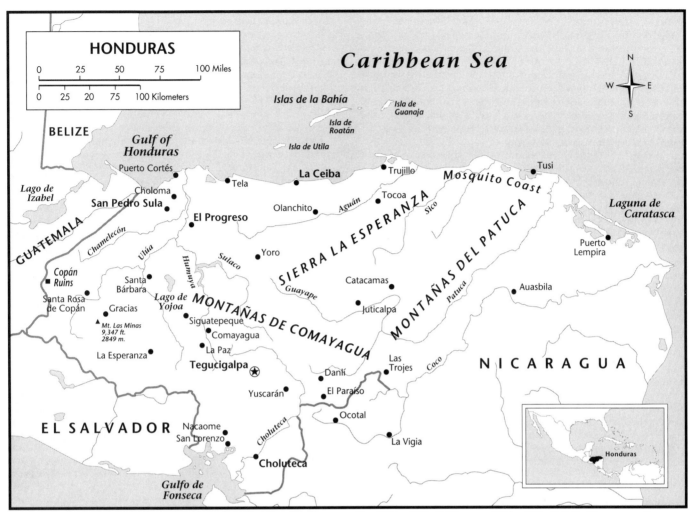

HONDURAS

| 0 | 25 | 50 | 75 | 100 Miles |
| 0 | 25 | 20 | 75 | 100 Kilometers |

LOCATION: 13° to 16°N; 83°10′ to 89°20′W. **BOUNDARY LENGTHS:** Caribbean coastline, 591 km (367 mi); Nicaragua, 922 km (573 mi);Gulf of Fonseca coastline, 74 km (46 mi); El Salvador, 335 km (208 mi); Guatemala, 248 km (154 mi). **TERRITORIAL SEA LIMIT:** 12 mi.

12HISTORY

Before the Spaniards entered the land now called Honduras, the region was inhabited by the warlike Lencas and Jicaques, Mexican Amerindian traders, and Paya hunters and fishermen. The Mayan ceremonial center at Copán in western Honduras flourished about the 8th century AD but was in ruins when Columbus reached the mainland on his fourth voyage in 1502. He named the region Honduras, meaning "depths."

Colonization began in 1524 under Gil González de Ávila. In 1536, Pedro de Alvarado, who came from Guatemala at the bidding of Hernán Cortés in Mexico, founded San Pedro Sula, and another faction founded Comayagua in 1537. After the treacherous murder by the Spaniards of an Amerindian chieftain named Lempira in 1539, his followers were subjugated. In that year, Honduras was made part of the captaincy-general of Guatemala, and for most of the period until 1821, it was divided into two provinces, Comayagua and Tegucigalpa. Some silver was produced in the mines of Tegucigalpa, but the area was otherwise ignored by the Spanish empire.

Honduras joined other provinces of Central America in declaring independence from Spain in 1821. It came under the Mexican empire of Agustín de Iturbide in 1822–23. Honduras was a member of the United Provinces of Central America from 1824 to 1838. During that time, a liberal Honduran, Francisco Morazán, became president and struggled unsuccessfully to hold the federation together. He was exiled in 1840 and assassinated in 1842.

After Honduras declared itself independent on 26 October 1838, conservatives and liberals fought for political control. From 1840 to 1876, conservative leaders held power either as presidents or as army leaders. The second half of the 19th century brought the development, by US companies, of banana growing in northern Honduras. During the administration of liberal president Marco Aurelio Soto (1876–83), there was a "golden age" in Honduran letters and education.

US corporate interests, especially the United Fruit Co. (now United Brands), and military dictators dominated Honduran political and economic life during the first half of the 20th century. Honduran politics was dominated by the conservative Gen. Tiburcio Carías Andino (1932–48). In 1948, his hand-picked successor, Juan Manuel Gálvez, took office. Gálvez proved to be more than a mere puppet, but was conservative nonetheless. When the election of 1954 produced no presidential candidate with a majority vote, he transferred the presidency to the vice-president, Julio Lozano Díaz, who governed for almost two years. After an abortive attempt to have himself elected president, Díaz was deposed in 1956 by high army officers, who set up a junta. Democratic elections were held in 1957, and José Ramón Villeda Morales of the Liberal Party was elected president.

In 1963, just before completing the final months of his six-year term, Villeda was turned out of office by a coup. The liberal government was succeeded by a conservative coalition of military, Nationalist Party, and Liberal Party leaders under an air force officer, Col. Oswaldo López Arellano. This government was legalized almost two years later by an elected constituent assembly, which adopted a new constitution and proclaimed López president in June 1965.

During López's second term, a bitter and destructive four-day war broke out in July 1969 between Honduras and El Salvador. Although the immediate cause of the war was animosity arising from a World Cup elimination-round soccer match between the two countries, the underlying causes were a long-standing border dispute and the long-term migration of some 300,000 Salvadorans in search of land, which the Honduran government made it illegal for Salvadoran immigrants to own. Salvadoran troops won the ground war, but Honduran planes controlled the air. Out of this stalemate and with the help of the OAS, a compromise ceasefire was arranged. In June 1970, the two nations accepted a seven-point peace plan, creating a "no-man's-land" demilitarized zone along their common frontier. In the fall of 1973, Honduras and El Salvador began bilateral talks to resolve their differences. Progress was slow, and it was not until October 1980 that Honduras and El Salvador signed a treaty settling the dispute.

In the 1970s, López and the military continued to dominate Honduran politics. A civilian, Ramón Ernesto Cruz Uclés, was elected president in 1971, but lasted only briefly. By 1972, Gen. López was back in power. Gen. López assumed the title of chief of state, and suspended the National Congress and all political party activities. It was later discovered that in 1974 officials in the López administration had accepted a $1.25-million bribe from United Brands in exchange for a 50% reduction in the banana tax. A Honduran investigative commission insisted on examining López's Swiss bank account, and the scandal came to be known as "Bananagate" in the US. Finally, in April 1974, López was overthrown by a group of lieutenant colonels.

This military group was something of a reformist group, seeking social reforms and the removal of the senior officer corps. Political activity continued to be banned following the coup of 1975. Meanwhile, a significant grassroots movement, the National Front of United Peasants, had come to the fore and was pressuring the successive military governments to enact a program of large-scale land redistribution.

There followed two more military governments led by Col. Juan Alberto Melgar Castro (1975–78) and Gen. Policarpo Paz García (1978–83). This period saw strong economic growth and the building of a modern infrastructure for Honduras. At the same time, there was a gradual movement toward the democratization of the system.

Elections to a constituent assembly took place in April 1980, followed by general elections in November 1981. Under a new constitution in 1982, Roberto Suazo Córdova of the Liberal Party became president. The armed forces retained broad powers, including veto power over cabinet appointments and responsibility for national security. The military continued to grow in response to domestic instability and the fighting in neighboring Nicaragua and El Salvador. By 1983, several thousand anti-Sandinista guerrillas (popularly known as "contras") in Honduras were working for the overthrow of the Sandinista government, while the Honduran army, backed by the US, was helping Salvadoran government forces in their fight against leftist guerrillas.

The Suazo government worked closely with the US on matters of domestic and foreign policy. US military presence in Honduras grew rapidly. Several joint military maneuvers took place during 1983–87, and the US CIA used Honduras during that time as a base for covert activities against the Sandinista regime. In exchange, the US sent large amounts of economic aid to Honduras. Suazo also worked closely with the Honduran military, allowing it to pursue its anti-communist agenda freely. This arrangement led to an unprecedented political stabilization in Honduras.

This stability became apparent in November 1985, when Hondurans elected José Simón Azcona Hoyo to the presidency in the first peaceful transfer of power between elected executives in half a century. Azcona was elected with only 27% of the vote, due to a peculiarity of Honduran electoral laws. Azcona attempted to distance himself from the US in foreign policy and was critical of US contra policy. He signed the Central American peace plan outlined by President Oscar Arias Sánchez of Costa Rica; however, he did not move to close down contra bases as promised.

In 1989, Rafael Leonardo Callejas of the National (conservative) Party was elected. With the Nicaraguan issue fading after the Sandinistas' electoral loss, Callejas has focused on domestic issues, applying a dose of both conservative economics and IMF austerity measures to the Honduran economy. Callejas moved to reduce the deficit and allow for a set of market adjustments, which in the short term produced a good deal of dislocation but led to higher rates of growth thereafter. Most significantly, Callejas maintained good relations with the military. In an unprecedented show of restraint, the military sat on the sidelines as voters went to the polls in November 1993.

The voters themselves showed a good deal of resentment for the Callejas reforms. The Liberal Party returned to power in the person of Carlos Roberto Reina. While it is unlikely their economic problems will be solved quickly, Honduras nevertheless had achieved a level of political stability that few could have anticipated in decades past.

13GOVERNMENT

The constitution of 1965, suspended following the 1972 coup, was superseded by a governing document adopted in November 1982. It defines Honduras as a democratic republic headed by a president who must be a native-born civilian. The president is elected by direct popular vote for a four-year term. Presidential elections are partisan, but multiple candidates often emerge from a single party. The winner of the election is the largest vote-getter from the party that gets the most total votes. Thus, in 1986, Azcona became president because the Liberal Party won more than 50% of the vote, and Azcona outpolled all other Liberal candidates. The executive branch also includes a cabinet of 12 ministers. A constitutional change approved by the legislature in November 1982 deprived the president of the title of commander-in-chief of the armed forces, transferring that responsibility to the army chief of staff.

The 1982 constitution provides for the popular election of deputies to the unicameral National Assembly, consisting of 128 deputies. The deputies, who are directly elected for four-year terms, must be natives or residents of the constituencies they represent. All men and women 18 years of age and older are eligible to vote.

14POLITICAL PARTIES

The two major parties in Honduras are the Liberal Party (Partido Liberal—PL) and the National Party (Partido Nacional—PN). Both descend from the old Liberal and Conservative Parties from the 19th century. Although generally the National Party remains more conservative in nature, the two parties are very close ideologically.

The National Party was in power from 1932 to 1954 under Carías and Gálvez. In 1965, a PN-backed constituent assembly promulgated a new constitution, designated its membership as the National Congress for a six-year term, and proclaimed Gen. Oswaldo López Arellano as president. In the 1971 elections, the PN candidate, Gen. Ramón Ernesto Cruz, received about 52% of the vote and was elected president. Their most recent success came in 1989, when Callejas became president. Currently, the National Party is split into two factions: one supports Callejas ("Callejistas") while the other supports the defeated 1993 candidate Oswaldo Ramos Soto (and are called "Oswaldistas").

The Liberals rely on their following in urban areas and among the laboring classes and have had some successes over the last half-century. In 1957, José Ramón Villeda Morales was elected to the presidency, and governed until 1963, when he was removed by a coup. The next successes came in 1981 with the election of Suazo, and then in 1985 with the election of Azcona, and in 1993 Carlos Roberto Reina.

Two minor parties occupy mildly leftist positions: the Christian Democratic Party, under Efraín Díaz, and the National Innovation and Unity Party, led by Germán Leitzelar. Neither takes more than a couple of seats in the Assembly.

15 LOCAL GOVERNMENT

Honduras is divided into 18 departments, each with a governor popularly elected for a two-year term. Departments are divided into municipalities (284 in 1986) governed by popularly elected councils. Localities with populations between 500 and 1,000 have a mayor, a legal representative, and a council member. A council member is added for each additional 1,000 residents, but the total is not to exceed 7. A special law governs the Central District of Tegucigalpa and Comayagüela.

Under the jurisdiction of the local government, municipal land is granted or lent to peasants in the district in sections known as ejidos. The ejido system is designed to aid landless peasants and has become an important function of local administration.

16 JUDICIAL SYSTEM

Judicial power is exercised by the nine-member Supreme Court (with seven substitutes) and five courts of appeal, as well as by courts of first instance and local judges. The Supreme Court appoints the judges of the courts of appeal and the courts of first instance, who, in turn, appoint local justices of the peace. The justices of the Supreme Court are elected by the National Assembly and serve for four-year terms. The Supreme Court has the power to declare laws unconstitutional.

There is a military court of first instance from which appeals can be taken to the civilian judicial system. In practice, the civilian courts are not independent. Because of underfunding and corruption, the formal resolution of legal disputes in courts is often the product of influence and political pressure.

17 ARMED FORCES

The Honduran military establishment consists of an army, an air force, a small navy, and a special security corps. Two years of compulsory peacetime military service may be required of all males between the ages of 18 and 55. Reserve status is open to men between 32 and 55 years of age (60,000 registered). The regular forces consisted of 16,800 personnel in 1993; there were 14,000 in the army, 1,800 in the air force, and 1,000 in the navy. Paramilitary police forces numbered 5,500 personnel. The defense budget in 1992 was $43.4 million or 1% of GDP.

The armed forces are equipped with US and NATO weapons that allow the army to field 3 small mixed brigades and a ground attack air force (46 combat aircraft). The armed forces are also structured for internal defense.

US military aid to Honduras increased from $3.9 million in 1980 to $81 million in 1986 and then dropped to $61 million in 1987 and $20 million in 1991. During the 1980s, the US-built Regional Military Training Center at Puerto Castilla trained anti-Sandinista rebels, contras from Nicaragua, as well as counterinsurgency forces from El Salvador. About 800 US troops maintain bases in Honduras.

18 INTERNATIONAL COOPERATION

Honduras is a charter member of the UN, having joined on 17 December 1945, and it participates in ECLAC and all the nonregional specialized agencies except IAEA and WIPO. It also belongs to G-77, IDB, OAS, PAHO, and various other inter-American and intergovernmental organizations. In 1982, Honduras joined with Costa Rica and El Salvador to set up the Central American Democratic Community to promote economic development and safeguard democracy in the region. In August 1987, Honduras and four other Central American countries signed the peace plan for Central America outlined by Costa Rican President Oscar Arias Sánchez. Honduras is a signatory to the Law of the Sea.

19 ECONOMY

Agriculture is the mainstay of the economy. About 16% of the land is arable, located mostly along the coastal plains. The principal export crops are bananas and coffee, although sugarcane and palm nuts are significant sources of foreign exchange earnings.

One of the three largest exporters of bananas in the world, Honduras is, nevertheless, by most measures, the poorest nation on the mainland of the Americas. The vast majority of banana holdings are controlled by two US companies, United Brands and Standard Fruit, and most other profitable agricultural enterprises are owned by a small number of private citizens. With its economy profoundly dependent on banana production, the country is vulnerable to crop and world market price variations. Although pastures represent the highest category of land use, production of meat and milk is low. Manufacturing remains relatively unimportant, although light manufacturing, particularly assembly operations and food processing, was conspicuous by its growth in the late 1970s.

The 1969 conflict with El Salvador was a serious blow to the Honduran economy, particularly since El Salvador was a principal Central American market for Honduran exports. Two successive years of drought and hurricane damage seriously affected banana production, retarding growth throughout the economy. Just as Honduras was recovering from these catastrophes, recording a 5% real growth in GNP in 1973, Hurricane Fifi, in September 1974, struck another blow to the economy.

Increased agricultural production sparked GDP growth of about 6.5% annually during 1976–80; inflation during the same period was 7% a year. The early 1980s brought a worsening in Honduras's economic performance, however, with the GDP growing by only 0.5% in 1981 and falling by 1.2% in 1982. The economy rebounded in 1985 and 1986, however, with GDP rising by 3% each year; public investment in development projects (largely financed by external sources), an increase in the services sector, and renewed growth in private investment were the principal reasons. Late in 1985, the Azcona administration enacted legislation to begin privatizing the economy, with 12 to 15 public companies to be sold by 1988. Inflation in 1985 and 1986 averaged 4%.

Manufacturing has exhibited modest growth due to the limited size of the local market, and to the deterioration of the Central American Common Market (CACM) during the 1980s, which resulted in a significant decline in intra-regional trade. The largest firms are found in the cement, cotton, sugar, and wood products industries, and these companies are also active exporters. Small to mid-size industrial enterprises have sprung up in textiles, household preparations, light metals, and food products.

Income per capita is one of the lowest in Latin America, although it has been improving moderately since 1990.

Shortly after taking office in January 1990, President Rafael Leonardo Callejas announced a broad-based economic reform package. The principal measures included the flotation of the currency; the restructure of the public sector; the overhaul of the tax structure; and the streamlining of import tariffs. As part of the reform program, the government cut down on price controls, liberalized interest rates, and curtailed private sector subsidies. The electoral landslide gave President Callejas a mandate to proceed with the comprehensive structural adjustment program.

Economic adjustment policies instituted during 1990 and 1991 include deregulation of restrictive pricing and marketing mechanisms, trade liberalization, lifting of interest rate ceilings, reduction of the fiscal deficit, and improved cost recovery for public utilities.

The success of the program has led to a marked improvement in the country's economic performance including an increase in the GDP growth rate from 0.1% in 1990 to 2.2% in 1991%, 4.9% in 1992, and 5.0% in 1993. Growth for 1994 was expected to drop to 4.0%. One of the reasons for the projected drop is the economic package of the new administration of President Reina, which appears to deviate from market–oriented reforms instituted by the previous administration.

20INCOME

In 1992, Honduras' GNP was $3,142 million at current prices, or $580 per capita. For the period 1985–92, the average inflation rate was 11.3%, resulting in a real growth rate in per capita GNP of 0.5%.

In 1992, the GDP was $3,284 million in current US dollars. It is estimated that in 1989 agriculture, hunting, forestry, and fishing contributed 18% to GDP; mining and quarrying, 1%; manufacturing, 14%; electricity, gas, and water, 3%; construction, 4%; wholesale and retail trade, 11%; transport, storage, and communication, 6%; finance, insurance, real estate, and business services, 14%; community, social, and personal services, 12%; and other sources, 17%.

21LABOR

In 1992, the economically active population totaled 1,728,599 (excluding the armed forces) of whom agriculture engaged 37%; services, 19.3%; manufacturing, 14.5%; commerce, 16.3%; construction, 4.2%; and other sectors, 8.7%. The unemployment rate by the end of 1991 was 15%. Underemployment amounts to about 30%.

About 20% of the labor force was unionized in 1991. The principal labor organizations in 1992 were the Confederation of Honduran Workers (CTH), the General Workers' Central (CGT), and the Unitary Confederation of Honduran Workers. The Confederation of Honduran Workers had 160,000 members in 1991.

Honduras did not have effective labor legislation until 1954. It joined the ILO in 1955 and subsequently adopted several labor codes, most notably that of 1959, which established the Ministry of Labor. The code also provided for union organization, collective bargaining, arbitration, social security, and fair labor standards, including an 8-hour day, a 44-hour week, 6-hour night shifts, and overtime, holiday, vacation, maternity, and severance pay.

22AGRICULTURE

Over 16% of the national territory is agriculturally productive; because of the uneconomical system of land use, much arable land has not been exploited. Agriculture is the primary sector of the economy, accounting for over 25% of GDP and 50% of employment in 1992. Farming methods are inefficient, and crop yields and qualities are low. The principal export crops are bananas and coffee; the major subsistence crops are corn, sorghum, beans, and rice. The average annual growth of agriculture during 1970–80 was 1.5%, with a decline of 2.8% in 1980. In 1985, the sector grew by 2.6%; in 1992, 2.1%. Crop production for 1992 included sugarcane, 3,004,000 tons; bananas, 1,086,000 tons; sorghum, 69,000 tons; beans, 46,000 tons; rice, 41,000 tons.

Since 1972, agrarian reform has been an announced priority of the national government. In January 1975, plans were made for the distribution of 600,000 hectares (1,483,000 acres) of land among 100,000 families over a five-year period. The program

was suspended in 1979 because of lack of funds and pressure from landowners; by that time, only about one-third of the goal had been met. The reform program was revived in the early 1980s, and in 1982, lands totaling 27,960 hectares (69,090 acres) were distributed to 4,000 peasant families. By 1986, however, land reform was at a virtual standstill; peasant groups, demanding immediate land distribution, staged "land invasions" and seized the offices of the National Agrarian Institute in San Pedro Sula. In 1992, agricultural production rose only 2.1%, down from the 6.1% increase of 1991. The agricultural modernization law eliminated subsidized credit to small farmers, while high commercial interest rates squeezed small farmers from the credit market.

23ANIMAL HUSBANDRY

Honduran consumption of milk and meat is traditionally low. However, pastures remain the largest category of land use in Honduras and in 1992 accounted for 25% of the total land area. Poor transportation facilities are a barrier to the development of stock raising and dairying, two potentially profitable economic activities. In 1992, the cattle population numbered 2,351,000 head; hogs, 750,000; horses and mules, 241,000; and chickens, 8,000,000. That year, 380,000 tons of milk and 31,000 tons of eggs were produced.

24FISHING

There is commercial fishing in Puerto Cortés, and other areas are served by local fishermen. A small local company operates a cannery for the domestic market on the Gulf of Fonseca. There is a commercial fishing concern on the island of Guanaja, and a large refrigeration-factory ship is engaged in freezing shrimp and lobster near Caratasca. In 1991, the total catch was 20,989 tons. In 1992, the fisheries sector grew by 14.7% due to new capital outlays in southern shrimp farms.

25FORESTRY

In 1991, about 28% of Honduras was covered by forests, including stands of longleaf pine and such valuable hardwoods as cedar, ebony, mahogany, and walnut. Total roundwood production in 1991 amounted to 6.2 million cu m, and forest products exports were valued at $29.1 million. The National Corporation for Forestry Development (Corporación Hondureña de Desarrollo Forestal), established in 1974, is charged with the overall preservation, exploitation, and exportation of Honduran forest resources. The privatization of government-owned woodlands is expected to intensify the use of forestry resources. A restriction on the export of raw wood also is causing growth in the woodworking industry for semifinished wood products.

26MINING

Lead and zinc and small amounts of cadmium, copper, silver, and gold are commercially important. The nonmetallic minerals extracted in Honduras are cement, marble, and salt. Inadequate transportation continues to hamper full development of mineral resources.

Between 1882 and 1954, the New York and Honduras Rosario Mining Co., the oldest mining company in the republic, produced over $60 million worth of gold and silver from the Rosario mines near Tegucigalpa. By the mid-1960s, these mines had become virtually nonproductive, but a mine situated west of Lake Yojoa remains active. In 1975, the government granted Industrial Alliance a lease on the Tatumbla gold and silver mines. In 1991, a study determined that the Yuscarán gold and silver deposit in El Paraíso Department would be economical for production operations. In 1992, the significant growth of construction activity boosted demand for gravel and iron, which caused the mining sector to grow by 33% that year (up from 8.3% in 1991).

In 1991, production of zinc was 38,280 tons; lead, 8,719 tons; silver, 39,359 kg; and gold, 179 kg.

[27]ENERGY AND POWER

Much of Honduras' energy is derived from the burning of wood; about 25% is provided by imported crude oil and petroleum products, supplied mainly by Mexico and Venezuela. Hydroelectricity and geothermal energy provided 10% and 3%, respectively, of Honduras' energy needs in 1991. Net installed capacity was 290,000 kw in 1991. Production in 1991 was 1,105 million kwh.

The government is playing a significant role in the development process, through the National Electric Co. The 290-Mw El Cajón hydroelectric project, financed by the IBRD, IDB, Central American Bank for Economic Integration (CABEI), OPEC, and Japan, was completed in March 1985. The $58-million Nispero hydroelectric project, financed by the IBRD and CABEI, and the rural electrification program in Aguán Valley, financed by AID at a cost of $13 million, have helped boost energy production by an average annual rate of 9% since 1981. Moreover, about three-quarters of electricity produced is now of hydroelectric origin, resulting in a reduction in the consumption of petroleum (all of which must be imported). In 1992, electricity consumption increased by 6%.

Petroleum explorations have been carried on by US companies since 1955. A Texaco refinery at Puerto Cortés began producing in 1968, and petroleum derivatives are exported to the US and Belize.

[28]INDUSTRY

Manufacturing, relatively unimportant in the national economy, has traditionally been limited to small-scale light industry supplying domestic requirements. Assembly plant operations developed in the late 1970s, especially after a free-trade zone was established at Puerto Cortés in 1975. The manufacturing sector grew by 20% in 1979 but then slackened because of low investment, shortages of raw materials, high interest rates, the decline of Central America as an export market, and regional political instability. Real industrial growth in 1984 was 2.2%, but it fell by the same rate in 1985.

San Pedro Sula is the center for matches, cigars, cigarettes, cement, meatpacking, sugar, beer and soft drinks, fats and oils, processed foods, shoes, and candles. Tegucigalpa has plants for the manufacture of plastics, furniture, candles, cotton textiles, and leather. A good-quality Panama hat is made in the departments of Copán and Santa Bárbara.

The cement industry, spurred by the needs of the El Cajón hydroelectric project, expanded greatly in the late 1970s, and Cementos de Honduras further expanded its production capacity in the early 1980s. With the completion of El Cajón, however, growth turned into decline; cement output went from 7.2 million bags in 1980 to 12.6 million in 1984 and 8.2 million in 1985. Textile production declined 11% between 1984 and 1985, wood and wood furniture by 2.5%, and nonmetallic minerals by 30%; tobacco, alcohol, and food production, however, rose by 3.4% in the same period.

The manufacturing sector has been growing over the last few years. This sector grew by 8.6% in 1992. Growth was greatest among food and tobacco products, wood and paper products, and basic metalworking and metal products. Food processing, textiles and garments, chemicals, and machinery and transport equipment constitute the major manufactured products.

The country has established a well–known apparel assembly industry. Total apparel exports to the US exceeded $200 million in 1992. A large number of firms have been interested in this market due to the lack of quotas.

[29]SCIENCE AND TECHNOLOGY

The Honduran Coffee Institute and the National Agriculture Institute are both located in Tegucigalpa. The Jose Cecilio del Valle University has engineering and computer science departments, and the National Autonomous University of Honduras has faculties of medicine, pharmacy, dentistry, and engineering. The Pan-American Agricultural School has students from 20 Latin American countries.

[30]DOMESTIC TRADE

The principal distribution centers include Puerto Cortés and San Pedro Sula; Tegucigalpa is a leading center of retail trade. In major cities, shops are comparable to those in other Central American towns. In the countryside, small markets and stores supply staple needs.

Business hours are generally 8 AM to noon and 1:30 or 2 to 5 or 6 PM on weekdays and 8 to 11 AM on Saturdays. Banks in Tegucigalpa are open from 9 AM to 3 PM, Monday through Friday.

[31]FOREIGN TRADE

Honduras exports a limited range of agricultural commodities, with bananas, coffee, and shrimp and lobster accounting for two-thirds of total exports. Important import categories are industrial raw materials and intermediate products, nondurable consumer goods, and industrial capital goods. As of 1992, the US continued to be Honduras' chief trading partner, supplying approximately 60% of its imports and purchasing over half of its exports. US exports to Honduras increased from $790 million in 1992 to $805.8 million in 1993. US imports from Honduras increased also, from $780.6 million in 1992 to $827.4 million in 1993. Other main partners in the 1990s include Japan, Mexico, and Venezuela.

Total exports went up from $816.3 million in 1992 to $881.6 million in 1993 and were expected to grow 14% to $1,005.0 million in 1994. Similarly, import growth has been significant over the last years. Total imports grew from $949.9 million in 1992 to $1,063.8 million in 1993. Imports were expected to grow at a more moderate pace (5.50%), reaching $1,122.3 million in 1994, narrowing the trade deficit by $65 million.

Honduras remains at the forefront of Central American economic integration efforts. In May 1992, Honduras signed several trade agreements with its neighbors, including a Honduran/Guatemalan Free Trade Agreement, a Honduran/Salvadoran Free Trade Agreement, and a Honduran/Salvadoran/Guatemalan Northern Tier Accord, with the intent of accelerating regional integration.

[32]BALANCE OF PAYMENTS

Since 1973, trade balances have been negative. Investment income repatriated by foreign companies in Honduras is an endemic burden on the local economy.

The Honduran authorities have generally adhered to the policies of fiscal and monetary restraint that were introduced early in 1959, following a period of exceptional strain on the country's international reserves. The fall in reserves resulted from a decline in income from the banana industry and reduced international prices for other major exports. Political instability in the region in the late 1970s to the mid-1980s, together with low commodity prices and high oil prices, had an adverse effect on the balance of payments. During 1992, import growth outpaced exports. By August 1993, the deteriorating trade balance had brought international reserves down to $79 million.

In 1992 merchandise exports totaled $842.6 million and imports $982.6 million. The merchandise trade balance was $–140 million. The following table summarizes Honduras' balance of payments for 1991 and 1992 (in millions of US dollars):

	1991	1992
CURRENT ACCOUNT		
Goods, services, and income	−387.4	−439.8
Unrequited transfers	197.3	216.1
TOTALS	−190.1	−223.7
CAPITAL ACCOUNT		
Direct investment	44.7	60.1
Portfolio investment	0.1	0.1
Other long-term capital	−123	15.7
Other short-term capital	−26.2	−89.5
Exceptional financing	184.5	212.1
Other liabilities	−1.3	−1.4
Reserves	−65.8	7.6
TOTALS	13.0	204.7
Errors and omissions	177.1	19.0
Total change in reserves	−65.5	−14.4

33BANKING AND SECURITIES

In 1950, the Central Bank of Honduras (Banco Central de Honduras), the sole bank of issue, was established to centralize national financial operations and to replace foreign currencies then in circulation. As of December 1992, capital and reserves of the Central Bank stood at L374.0 million.

Eleven commercial banks, six development banks, and three foreign banks constituted the banking sector in 1993. The Banco Atlántida, the most important commercial bank, accounts for over one-half of the total assets of private banks. US banks play a significant role in the commercial system: the Atlántida is affiliated with Chase Manhattan, and the second-largest commercial bank, the Banco de Honduras, is affiliated with Citibank of New York. As of 1992, commercial bank reserves stood at L648 million; demand deposits were L1,276 million; and time, savings, and foreign currency deposits totaled L3,151 million.

The government-controlled banks, including the National Development Bank, the National Agricultural Development Bank, and the Municipal Bank, provide credit for development projects. The National Development Bank extends agricultural and other credit—mainly to the tobacco, coffee, and livestock industries—and furnishes technical and financial assistance and other services to national economic interests. The Municipal Bank gives assistance at the local level. Honduras has no stock exchange.

34INSURANCE

The oldest insurance company in Honduras is Honduras Savings (Ahorro Hondureño), established in 1917. Five other companies deal with life insurance and other types of policies. The number and the role of foreign companies in the insurance sector have decreased because of government incentives to domestic underwriters. Total claims paid in 1985 were L35.2 million, a rise of 12% from 1983. In 1986, 48.5% of premium payments were for life and 51.5% for non-life insurance. In 1992, life insurance in force totaled L13.5 billion.

35PUBLIC FINANCE

In 1990, current revenues amounted to $1,010.9 million, while current expenditures came to $1,009.2 million. In 1990, the Callejas administration began a market-oriented reform program, which has reduced the fiscal deficit. In 1992, however, unexpectedly high public expenditures widened the fiscal deficit to 4.8% of GDP, exceeding the IMF program target level of 3%. The total external debt at the end of 1992 was $3.3 billion, with a debt service of $401 million.

36TAXATION

Honduran tax rates are progressive. The basic rate on net income begins at 12% on incomes of over L50,000 and rises to 40% on those of more than L1,000,000; a surcharge of 10% is levied on amounts over L100,000, and a surcharge of 15% is applied to incomes over L500,000. No distinction is made for tax purposes between individuals and businesses. Agricultural activities and industries classified as "basic" receive favorable depreciation rates.

The corporate tax rate is 15% on the first L100,000 of taxable income and 35% on the excess. The government also levies a general sales tax and social security taxes. In addition, there are excise taxes, mainly on beer and cigarettes, but also on imported matches, soft drinks, imported sugar, and new motor vehicles. New industries are exempted from income and production taxes and import duties for up to 10 years.

District and municipal governments obtain their revenues from taxes on amusements and livestock consumption and from permits, licenses, registrations, certifications, storage charges, transfers of real estate, and fines.

37CUSTOMS AND DUTIES

Most imports from outside the CACM are subject to customs tariffs ranging from 5%–20%, plus a 5% service fee. Duties are levied ad valorem over the cost, insurance, and freight value of goods.

In June 1992, the Central Bank of Honduras eliminated the need for most import permits and foreign exchange authorizations.

38FOREIGN INVESTMENT

Traditionally, the Honduran attitude toward foreign enterprise has been favorable. Foreign capital is treated in the same way as domestic capital; however, firms in the distribution or lumber business must have 51% Honduran ownership. Honduran economic development has been powerfully influenced by foreign investment in agriculture, industry, commerce, and other economic sectors. Since 1910, the Standard Fruit and Steamship Co. and United Brands (formerly the United Fruit Co.) have developed railroads, ports, plantations, cattle farms, lumber yards, breweries, electric power, housing, and education. All contracts, aside from commodity exports, were canceled on 15 September 1975; plans to convert banana-marketing operations into a joint venture fell through, however, and in 1976, the government instead expropriated large tracts of land from the banana producers. Mines have been developed by the New York and Honduras Rosario Mining Co. The largest corporations in Honduras in 1986 were US- and Republic of Korea (ROK)-owned.

At the beginning of the decade, the local currency devaluated continuously until July 1993, when it finally stabilized to a rate of 6.70 lempiras to the dollar. While this devaluation made Honduras exports more competitive, it also raised the cost of imports and placed a premium on access to scarce foreign exchange. Consequently, new incentives to revive the foreign exchange market had to be developed. In January 1992, the maximum tariff rate was reduced to 20% and the minimum rate to 5%, and in July 1993, a 10% surcharge on imports entering Honduras was eliminated.

In addition, a new foreign investment law was signed by then President Callejas in June 1992. The law addresses specific improvements in the investment climate, including: a reduction of government controls and intervention, guarantees of profits remittances, and access to foreign exchange for imports of goods and services required for business operations, as well as investment guarantees that are consistent with the operation of a free market economy.

39ECONOMIC DEVELOPMENT

As part of 1985/86 support fund agreements with the US AID, the Azcona administration agreed to privatize some public-sector enterprises between 1986 and 1988. Some 25% of companies under the National Investment Corp. (Corporación Nacional de Inversiones) and the National Corporation for Forestry Development were returned to the private sector. Generally, the government has emphasized the use of free-market principles in economic development.

US development loans and grants during 1946–86 totaled $995.3 million, more than 58% coming in the period since 1982. Multilateral assistance during 1946–86 amounted to $1,390.8 million, with the IDB contributing 51% and the IBRD 39%.

During the administration of President Callejas, economic policy was mostly based on neoliberal ideas. This included a move from an inward-oriented policy to an export-oriented one. In addition, privatization was deeply emphasized. During that period, GDP was characterized by consistent growth. At the same time, the country's galloping inflation fell to single digits by 1993.

The November 1993 elections gave birth to a new political era in Honduras. President Reina of the Liberal Party was expected to slow down the pace of market–oriented reforms. Continued strong growth in nontraditional exports and the prospects for moderate improvement in coffee prices should bring about a moderate improvement in the current account deficit. For this reason, the continuation of official aid and the promotion of foreign investment is expected to play an increase in role in financing Honduras external gap.

40SOCIAL DEVELOPMENT

The present Honduran Social Insurance Law covers accidents, illness, maternity, old age, occupational disease, unemployment, disability, death, and other circumstances affecting the capacity to work and maintain oneself and one's family. Effective coverage, however, has been limited mainly to maternity, sickness, and workers' compensation. Social security services are furnished and administered by the Honduran Social Security Institute and financed by contributions from employees, employers, and the government. Workers contribute 1% of their earnings toward retirement, disability, and survivor insurance, while employers pay 2% of their payroll. Old-age and disability benefits equal 40% of basic monthly earnings and an increment for contribution beyond 60 months.

The government supports family planning but, in view of the high mortality rate, considers the birthrate acceptable. The Honduran Association of Family Planning has operated since 1963. Abortion is permitted on broad health grounds. The fertility rate in 1991 was 5.1. Women represented 26.3% of the labor force in 1986. In that year, seven women were serving in the National Assembly. Cultural attitudes prevent women from full access to the educational and economic opportunities guaranteed by law. Weaknesses in the penal code prevent adequate prosecution of sex offenders.

The measure of impunity enjoyed by the civilian and military elite has resulted in continuing human-rights abuses, including killing, detentions and torture.

41HEALTH

Health conditions in Honduras are among the worst in the Western Hemisphere, and health care remains inadequate. In 1992 there were 32 physicians per 100,000 people, with a nurse-to-doctor ratio of 1.0. In 1990, there were 1.1 hospital beds per 1,000 inhabitants. The Inter-American Cooperative Public Health Service, created in 1942 under the joint sponsorship of Honduras and the US, has contributed to public health through malaria control, spraying, construction of water systems and sewage disposal plants, personnel training, and the establishment of a national tuberculosis sanatorium. US Peace Corps volunteers help train personnel for urban and rural clinics. In 1990 there were 133 reported cases of tuberculosis per 1,000 people, and 21% of children under 5 years of age were considered malnourished. In

1992, 66% of the population had access to health care services. Total health care spending in 1990 was $134 million.

Major causes of illness and death are diseases of the digestive tract, intestinal parasites, accidental homicide and suicides, influenza, pneumonia, cancer, and infant diseases. Malnutrition, impure water, poor sewage disposal, and inadequate housing are the major health problems. In 1991, about 77% of the population had access to safe water, and 61% had adequate sanitation. Immunization rates for children up to 1 year old during 1992 were as follows: tuberculosis (91%); diphtheria, pertussis, and tetanus (93%); polio (95%); and measles (89%). In 1993, the birth rate was 37.1 per 1,000 people, with about 41% of married women (ages 15 to 49) using contraception. The infant mortality rate in 1992 was 45 per 1,000 live births, and the general mortality rate in 1993 was 7.2 per 1,000 people. Life expectancy in 1992 was an average of 66 years.

42HOUSING

Many urban dwellings–and most rural dwellings– lack running water, electricity, and indoor plumbing. Estimates of the shortage in housing units grew from 173,000 in 1961 to 320,700 in 1973 and to 500,000 in 1985. As of 1981, 26% of all households had piped water, 23% had flush toilets, and 38% had electric lighting. Projects under way during the early 1980s included a $30-million low-cost housing program sponsored by the Housing Finance Corp. and a $19-million venture undertaken by the National Housing Institute.

43EDUCATION

The rate of illiteracy among adults in 1990 was estimated at 26.9% (males: 24.5% and females: 29.4%). Public education is free and compulsory for children between the ages of 7 and 15. In 1985, of primary-school-age children, 82.6% were actually in school. As of 1990 there were 7,593 primary and elementary schools, with 908,446 students and 23,872 teachers, and secondary and normal schools had 194,083 pupils and 8,507 teachers. Primary school lasts six years, followed by two levels of secondary education—three years followed by two more years.

In 1990, total university and equivalent institution enrollment was 39,324 with 3,258 teaching staff. The number of students enrolled in all other institutions of higher learning combined in 1989 was 44,849 with 3,319 teaching staff. The major university is the National Autonomous University of Honduras, founded at Tegucigalpa in 1847, with branches at San Pedro Sula and La Ceiba.

44LIBRARIES AND MUSEUMS

Although the National Archive of Honduras was established in 1880 to conserve and maintain the records of the republic, no great attention has been shown to government documents and other records in modern times. The National Archive includes land titles dating from 1580, historical documents dating from the 17th century, a newspaper collection from 1880 onward, a civil registry, and a collection of laws since 1880. The Ministry of Education has charge of the National Archive, as well as of other libraries and museums. Another noteworthy institution is the National Library in Tegucigalpa (55,000 volumes). The National University's library in Tegucigalpa contains over 200,000 volumes. The government has cleared and partially rebuilt the Mayan ruins of Copán.

45MEDIA

The government owns and operates postal, telephone, and telegraph services. Tegucigalpa and San Pedro Sula are linked by a multiplex radio relay network. The Tropical Radio Co. provides international radiotelegraph and radiotelephone service. In 1993 there were 2,300,000 telephones. During the same year, Honduras had 176 AM radio stations, 6 FM stations, and 28 television stations. There were 2,045,000 radios and 385,000 television sets.

There is no press censorship. Of four daily newspapers in the republic in 1991, two were in Tegucigalpa and two in San Pedro Sula. The country's principal newspapers (with 1991 circulation) were *La Tribuna*(50,000), and *El Tiempo (30,000)*, and *El Heraldo* (31,000), all published in Tegucigalpa, and *La Prensa*, published in San Pedro Sula.

46 ORGANIZATIONS

The Chamber of Commerce and Industries has its headquarters in Tegucigalpa; chambers of commerce also function in San Pedro Sula, La Ceiba, and other towns. Various producers' and professional associations, as well as social and cultural groups, are also active.

47 TOURISM, TRAVEL, AND RECREATION

A valid passport is needed for entry into Honduras. Visitors also need visas, except for nationals of the US, Canada, Western Europe, Japan, Australia, and New Zealand.

Honduran governments have attempted to attract foreign visitors. After a period of stagnation, tourism experienced modest growth in the late 1980s. In 1991, there were an estimated 198,000 tourist arrivals, mostly from the Americas. There were 6,739 hotel rooms and 11,946 beds, and tourists spent US$31 million. The main tourist attraction is the restoration at Copán, the second-largest city of the ancient Mayan Empire. There are also beaches on the northern coast and good fishing in Trujillo Bay and Lake Yojoa. A tourist complex in the San Pedro Sula/Tela region was completed in 1981.

48 FAMOUS HONDURANS

José Cecilio del Valle (1780–1834), a member of the French Academy of Sciences, was an intellectual, a political leader, and the author of the Central American declaration of independence. Francisco Morazán (1799–1842) was the last president of the United Provinces of Central America, which lasted from 1823 to 1839. Father José Trinidad Reyes (1797–1855) founded an institute in 1847 that became the National University. Outstanding literary figures were Marco Aurelio Soto (1846–1908), an essayist and liberal president; Ramón Rosa (1848–93), an essayist and biographer; Policarpo Bonilla (1858–1926), a politician and author of political works; Alberto Membreño (1859–1921), a philologist; Juan Ramón Molina (1875–1908), a modernist poet; Froilán Turcios (1875–1943), a novelist and writer of fantastic tales; and Rafael Heliodoro Valle (1891–1959), a historian and biographer.

49 DEPENDENCIES

Honduras has no territories or colonies.

50 BIBLIOGRAPHY

Anderson, Thomas P. *The War of the Dispossessed: Honduras and El Salvador.* Lincoln: University of Nebraska Press, 1981.

Barry, Tom. *Honduras: A Country Guide.* Albuquerque, N.M.: Inter-Hemispheric Education Resource Center, 1990.

Cline, William R. (ed.). *Economic Integration in Central America.* Washington, D.C.: Brookings Institution, 1978.

Durham, William H. *Scarcity and Survival in Central America: Ecological Origins of the Soccer War.* Stanford, Calif.: Stanford University Press, 1979.

Honduras in Pictures. Minneapolis, Minnesota: Lerner Publications Company, 1987.

Howard-Reguindin, Pamela F. *Honduras.* Santa Barbara, Calif.: Clio, 1992.

Karnes, Thomas L. *The Failure of Union: Central America, 1824–1975.* Rev. ed. Tempe: Arizona State University, Center for Latin American Studies, 1976.

MacCameron, Robert. *Bananas, Labor, and Politics in Honduras: 1954–1963.* Syracuse, N.Y.: Syracuse University Press, 1983.

MacLeod, Murdo J. *Spanish Central America: A Socioeconomic History, 1520–1720.* Berkeley: University of California Press, 1973.

Peckenham, Nancy, and Annie Street (eds.). *Honduras: Portrait of a Captive Nation.* New York: Praeger, 1985.

Ronfeldt, David F. *U.S. Involvement in Central America: Three Views from Honduras.* Santa Monica, CA: RAND, 1988.

Rosenberg, Mark B., and Philip L. Shepherd (eds.). *Honduras Confronts Its Future: Contending Perspectives on Critical Issues.* Boulder, Colo.: Lynne Rienner, 1986.

Schulz, Donald E. and Deborah S. Schulz. *The United States, Honduras, and the Crisis in Central America.* Boulder, Colo.: Westview Press, 1994.

Woodward, Ralph Lee, Jr. *Central America: A Nation Divided.* New York: Oxford University Press, 1976.

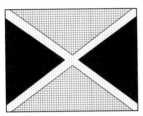

JAMAICA

CAPITAL: Kingston.

FLAG: Two diagonal yellow gold bars forming a saltire divide the flag into four triangular panels. The two side panels are black, and the top and bottom panels are green.

ANTHEM: First line, "Eternal father, bless our land..."

MONETARY UNIT: The Jamaican dollar (J$) of 100 cents was introduced on 8 September 1969. There are coins of 1, 5, 10, and 25 cents, and 1 dollar, and notes of 2, 5, 10, 20, 50, and 100 dollars. J$1 = US$0.0302 (or US$1 = J$33.100).

WEIGHTS AND MEASURES: Both metric and imperial weights and measures are used.

HOLIDAYS: New Year's Day, 1 January; Labor Day, 23 May; Independence Day, 1st Monday in August; National Heroes' Day, 3rd Monday in October; Christmas, 25 December; Boxing Day, 26 December. Movable religious holidays include Ash Wednesday, Good Friday, and Easter Monday.

TIME: 7 AM = noon GMT.

¹LOCATION, SIZE, AND EXTENT

Jamaica is an island in the Caribbean Sea situated about 160 km (90 mi) s of Cuba. It has a total area of 10,990 sq km (4,243 sq mi) and extends, at maximum, 235 km (146 mi) N–S and 82 km (51 mi) E–W. Comparatively, the area occupied by Jamaica is slightly smaller than the state of Connecticut. The total coastline is 1,022 km (634 mi).

Jamaica's capital city, Kingston, is located on the country's southeastern coast.

²TOPOGRAPHY

The greater part of Jamaica is a limestone plateau, with an average elevation of about 460 m (1,500 ft). The interior of the island is largely mountainous, and peaks of over 2,100 m (7,000 ft) are found in the Blue Mountains, which dominate the eastern part of the island; the highest point on the island is Blue Mountain Peak, at 2,256 m (7,402 ft) above sea level.

The coastal plains are largely alluvial, and the largest plains areas lie along the south coast. The island has numerous interior valleys. There are many rivers, but most are small, with rapids and falls that make navigation virtually impossible for any distance.

³CLIMATE

The climate ranges from tropical at sea level to temperate in the uplands; there is relatively little seasonal variation in temperature. The average annual temperature in the coastal lowlands is 27°C (81°F); for the Blue Mountains, 13°C (55°F).

The island has a mean annual rainfall of 198 cm (78 in), with wide variations during the year between the north and south coasts. The northeast coast and the Blue Mountains receive up to 760 cm (300 in) of rain a year in places, while some parts of the south coast receive less than 75 cm (30 in), most of it falling between May and October. The rainy seasons are May to June and September to November. The period from late August to November has occasionally been marked by destructive hurricanes.

⁴FLORA AND FAUNA

The original forest of Jamaica has been largely cut over, but in the areas of heavy rainfall along the north and northeast coasts there are stands of bamboo, ferns, ebony, mahogany, and rosewood. Cactus and similar dry-area plants are found along the south and southwest coastal area. Parts of the west and southwest consist of grassland, with scattered stands of trees.

The wild hog is one of the few native mammals, but there are many reptiles and lizards. Birds are abundant. Jamaican waters contain considerable resources of fresh- and saltwater fish. The chief varieties of saltwater fish are kingfish, jack, mackerel, whiting, bonito, and tuna; freshwater varieties include snook, jewfish, gray and black snapper, and mullet.

⁵ENVIRONMENT

Among the government agencies charged with environmental responsibilities are the Ministry of Health and Environmental Control, the Ministry of Agriculture, and the Natural Resources Conservation Authority. The major environmental problems involve water quality and waste disposal. Jamaica has 2.0 cu mi of water with 86% used for agriculture and 7% used for industrial purposes. Fifty-four percent of the people living in rural areas and 5% of the city dwellers do not have pure water. One major source of water pollution is the mining of bauxite. Another environmental problem for Jamaica is land erosion and deforestation. The nation loses 3.3% of its forests each year along with 80 tons of soil. The nation's cities produce 0.3 million tons of solid waste per year.

In the mid-1980s, sewage systems were available to only 7% of the population; coastal waters have been polluted by industrial waste, sewage, and oil spills, and groundwater has suffered contamination by red mud waste from alumina processing. Kingston has the waste disposal and vehicular pollution problems typical of a densely populated urban area.

In 1994, five of Jamaica's mammal species were endangered, and two bird species and 10 plant species are threatened with extinction. As of 1987, endangered species in Jamaica included the tundra peregrine falcon, homerus swallowtail butterfly, green sea turtle, hawksbill turtle, and American crocodile.

6POPULATION

Jamaica's 1991 population was 2,374,193, an increase of 7.6% over the 1982 census figure of 2,205,507. Population density was 216 per sq km (559 per sq mi). A population of 2,677,000 was projected for the year 2000, assuming a crude birthrate of 19.2 per 1,000 population, a crude death rate of 5.7, and a net natural increase of 13.5 during 1995–2000.

As of 1991, 587,798 people lived in the parishes of Kingston and St. Andrew, which comprise the capital city and its suburbs. Other leading cities are Spanish Town (about 92,383), Portmore (90,138), and Montego Bay (83,466). In the late 1960s, the number of Jamaicans under 15 years of age exceeded those between the ages of 15 and 59; since then, however, the median age of the population has been rising, from 16.7 in 1970 to 18.5 in 1980, 20.4 in 1985, 22.4 in 1990, and an estimated 24.3 in 1995.

7MIGRATION

Jamaica's net loss from emigration totaled 145,800 between 1891 and 1921; after a net gain of 25,800 during 1921–43, losses of 195,200 were recorded from 1943 to 1960, and 265,500 from 1960 through 1970. Until the UK's introduction of restrictions on immigration from Commonwealth countries in 1962, a large number of Jamaican workers emigrated to Great Britain. In 1964, in an effort to curb increasing migration, Jamaica passed the Foreign Nationals and Commonwealth Citizens (Employment) Act, providing Jamaicans with easier access to the island's employment market; however, domestic unemployment continued to plague Jamaica through the 1970s. During this period, Jamaica suffered from a "brain drain," losing perhaps as much as 40% of its middle class. From 1971 through 1980, 276,200 Jamaicans left the island, 142,000 for the US. In 1990 there were 343,000 Jamaican-born people in the US.

The great disparity between rural and urban income levels has contributed to the exodus of rural dwellers to the cities, where many of these migrants remain unemployed for lack of necessary skills.

8ETHNIC GROUPS

About 95% of the population is of partial or total African descent. This consists of 76% blacks, 15% mulattos, and 4% black-East Indians or black-Chinese. Other ethnic groups include East Indians (2%), Chinese (1%), and Europeans (2%). Nearly the whole population is native-born Jamaican. Black racial consciousness has been present in Jamaica at least since the beginnings of the Rastafarian sect, founded in 1930 and based on the ideas of Marcus Garvey.

9LANGUAGES

Jamaica is an English-speaking country, and British usage is followed in government and the schools. There is a recognizable patois, which is often used by Jamaican writers in novels and plays.

10RELIGIONS

There is freedom and equality of religion in Jamaica. The Church of God now claims the largest number of adherents, followed by the Baptist church. The Church of England (Anglican), formerly the dominant religion in Jamaica, now ranks third. The Roman Catholic Church is also significant. The Rastafarian movement continues to grow and is culturally influential in Jamaica and abroad. Rastas regard Africa (specifically Ethiopia) as Zion and consider their life outside Africa as a Babylonian captivity; use of marijuana, or ganja, plays an important role in the movement. The Pope visited Jamaica in August 1993.

11TRANSPORTATION

Jamaica has an extensive system of roads; in 1991 there were 18,200 km (11,300 mi) of roads, including 12,600 km (7,830 mi) of paved roads. In 1991 there were 97,500 licensed passenger cars and 18,000 commercial vehicles on the island. Motorbus service, which has greatly facilitated travel, is operated by the government-owned Jamaica Omnibus Services Co.

The standard-gauge rail system, with 370 km (230 mi) of track in 1991, is operated by the Jamaica Railway Corp. (JRC), a government enterprise formed in 1960. The main line runs diagonally across the island from Kingston to Montego Bay, and branch lines reach Port Antonio, Frankfield, and Ewarton. Most of the railway's revenue comes from freight, but there is passenger service between Kingston and Montego Bay.

Kingston, the main port, handles nearly all of the country's foreign imports but less than 10% of its exports, by weight. The remaining exports are shipped through 18 other ports, which tend to specialize in particular commodities: Montego Bay and Port Antonio in bananas and sugar, for instance, and Port Esquivel and Ocho Rios in bauxite. More than 30 shipping companies provide passenger and cargo service. The port facilities of Kingston harbor are among the most modern in the Caribbean. Jamaica itself has a small fleet, including as of 1 January 1992, three freighters and one tanker.

Air service is the major means of passenger transport between Jamaica and outside areas. Control of the two modern airports, Norman Manley International Airport (Kingston) and Sangster International Airport (Montego Bay), was assigned to the Airports Authority of Jamaica in 1974. About a dozen airlines provide scheduled international air transportation. Air Jamaica, the national airline, operates internationally in association with British Airways and British West Indian Airways. The government owns a controlling interest in Air Jamaica and has also invested in a domestic air carrier, Trans Jamaican Airlines. In 1992, Air Jamaica carried 837,200 passengers.

12HISTORY

Jamaica was discovered by Christopher Columbus in 1494 and was settled by the Spanish in the early 16th century. The Spanish used the island as a supply base and also established a few cattle ranches. The Arawak Indians, who had inhabited the island since about AD 1000, were gradually exterminated, replaced by African slaves. In 1655, the island was taken over by the English, and the Spanish were expelled five years later.

Spain formally ceded Jamaica to England in 1670 by the Treaty of Madrid. The island became a base for English privateers raiding the Spanish Main. A plantation economy was developed, and sugar, cocoa, and coffee became the basis of the island's economy. The abolition of the slave trade in 1807 and of slavery itself in 1834 upset Jamaica's plantation economy and society. The quarter million slaves were set free, and many became small farmers in the hill districts. Freed slaves were replaced by East Indians and Chinese contract workers. The economy suffered from two developments in mid-century: in 1846 the British rescinded favorable terms of trade for Jamaica, and the union blockade during the US Civil War limited commercial options for the island. Bankruptcies and abandonment of plantations followed, and dissension between the white planters and black laborers led to a crisis. An uprising by black freedmen at Morant Bay in 1865 began a struggle which necessitated the imposition of martial law. Parliament established a crown colony government in 1866, and Jamaica's new governor, Sir John Peter Grant, introduced new programs, which included development of banana cultivation, improvement of internal transportation, and reorganization of government administration. Advances in education, public health, and political representation pacified the island.

These measures did not resolve Jamaica's basic problems, stemming from wide economic and social disparities, and social unrest came to the surface whenever economic reverses beset the island. The depression of the 1930s, coupled with a blight on the banana crop, produced serious disruption and demands for political reform. A royal commission investigated the island's social and economic conditions, and recommended self-government for Jamaica. A Jamaica legislative council committee concurred, and in 1944, Jamaica had its first election. The contenders in that election were two recently formed political parties, the People's National Party (PNP), led by Norman W. Manley, and the Jamaica Labour Party (JLP), founded by Manley's cousin, Sir Alexander Bustamante.

During the 1950s, the bauxite industry and the tourist trade assumed prominent roles in the economy. The economic gains from these enterprises did little to solve Jamaica's underlying economic problems. Jamaica joined with other British Caribbean colonies in 1958 to form the Federation of the West Indies, but in a referendum in 1961 a majority of Jamaicans voted for withdrawal from the federation. The governments of the UK and Jamaica accepted the decision of the electorate, and Jamaica became an independent state on 6 August 1962, with dominion status in the Commonwealth of Nations. The PNP had supported the federation concept, so the JLP became the independence party, and Bustamante became the nation's first prime minister.

The JLP held power through the 1960s. Donald Sangster became prime minister in 1965 and was succeeded by Hugh Shearer, also of the JLP, two years later. In February 1972, the PNP regained a majority in Parliament, and the late Norman Manley's son, Michael, headed a new democratic socialist government.

Manley moved to nationalize various industries, and expanded Jamaica's programs in health and education. Manley placed price controls on a number of key products and provided consumer subsidies for others. Internationally, Manley established friendly relations with Cuba, which the US decried. Deteriorating economic conditions led to recurrent violence in Kingston and elsewhere during the mid-1970s, discouraging tourism. By 1976, Jamaica was faced with declining exports, a critical shortage of foreign exchange and investment, an unemployment rate estimated at 30–40%, and rampant currency speculation.

The PNP nevertheless increased its parliamentary majority that December in elections held during a state of emergency. Tourism suffered another blow in January 1979 with three days of rioting in Kingston, at the height of the tourist season. Meanwhile, Manley quarreled with the IMF. The IMF responded to Jamaica's request for loan guarantees by conditioning acceptance on a set of austerity measures. Manley refused to initiate many of the market-oriented measures the IMF was demanding.

Manley called for elections in the fall of 1980. The campaign was marred by somewhere between 500 and 800 deaths, and was further inflamed by PNP claims that the CIA was attempting to destabilize its government. The opposition JLP won a landslide victory, and Edward Seaga became prime minister and minister of finance. He announced a conservative economic program that brought an immediate harvest of aid from the US and the IMF. In October 1981, Jamaica broke off diplomatic relations with Cuba, and two years later it participated in the US-led invasion of Grenada.

In December 1983, Seaga called for elections, which the PNP boycotted, leaving the JLP with all 60 seats in the House of Representatives. Seaga then implemented an IMF plan of sharp austerity, pushing the economy into negative growth for two years. In May 1986, Seaga turned away from the IMF, announcing an expansionary budget. The JLP nevertheless suffered a sharp loss in July parish elections, with the PNP taking 12 of 13 municipalities. By January 1987, a new IMF agreement was in place, but their political position continued to slide.

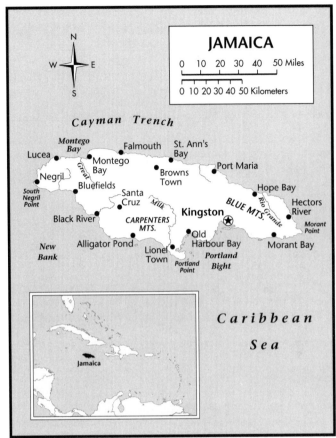

LOCATION: 17°43′ to 18°32′N; 76°11′ to 78°21′w. TOTAL COASTLINE: 1,022 km (634 mi). TERRITORIAL SEA LIMIT: 12 mi.

The 1989 elections were a good deal less tumultuous than expected. The two parties reached an agreement to control their respective partisans, and election violence was minimal. The rhetoric was also considerably less inflammatory, as the PNP's Manley ran as a more moderate candidate. Citing the deterioration of social services under Seaga, and promising to attract foreign capital, Manley was returned to the prime ministership as the PNP took a powerful 45-seat bloc in the House of Representatives. Manley reversed many of Seaga's policies, but by 1992, inflation was on the rise and the economy slowed. Unemployment hovered around 20%. Manley retired in 1992, leaving the government to Percival J. Patterson.

Patterson moved further to the right from Seaga, encouraging more market-oriented reforms. Within a year of taking office, he called for elections, in which violence erupted anew. The PNP increased its parliamentary margin to 52–8, a small consolation for a government besieged by serious political, social, and economic problems.

13GOVERNMENT

The 1962 constitution provides for a governor-general appointed by the crown, a cabinet presided over by a prime minister, and a bicameral legislature.

The Senate, the upper house, consists of 21 members appointed by the governor-general, 13 on the advice of the prime minister and 8 on the advice of the leader of the opposition. The popularly elected House of Representatives consists of 60 members (increased from 53 in 1976). The House is by far the more important of the two. The governor-general appoints both the prime minister and the leader of the opposition. The normal term of office in Parliament is five years, but elections can be called at any time. Suffrage is universal at age 18.

The cabinet consists of the prime minister and at least 11 additional ministers, appointed by the governor-general on the advice of the prime minister.

14POLITICAL PARTIES

Two political parties, the Jamaica Labour Party (JLP) and the People's National Party (PNP), dominate Jamaican politics. Their fortunes have risen and fallen dramatically over the past thirty years. Both parties have held more than three-fourths of Parliament. The JLP, founded in 1943 by Sir Alexander Bustamante, is the more conservative of the two parties. Its original political base was the Bustamante Industrial Trade Union, which Bustamante organized in 1938. The JLP held a parliamentary majority during the first 10 years of independence, and again from 1980–89 under Edward Seaga. Seaga remains opposition leader.

The PNP, founded by Norman W. Manley in 1938, holds to a moderate socialist program and from its foundation sought responsible government and independence for Jamaica. The PNP draws much of its support from the National Workers' Union, Jamaica's largest trade union. Both the JLP and PNP stand for a broad program of social reform and welfare, and economic development with the participation of foreign capital. They differ in that the PNP stands for nationalization of the means of production. The PNP formed its first government in 1972 under Michael Manley. In 1976, the PNP remained in power, increasing its majority by 10 seats in a House that had been enlarged by 7. After losing in 1980, the PNP refused to participate in the parliamentary elections called by Prime Minister Seaga for December 1983, two years ahead of schedule.

15LOCAL GOVERNMENT

Local government is patterned on that of the UK, and the unit of local government is the parish. Responsibility for local government is vested in 12 parish councils and the Kingston and St. Andrew Corporation, which represents the amalgamation of 2 parishes. Since 1947, all of the councils (called parochial boards until 1956) have been fully elective, although the members of the House of Representatives from each parish are ex-officio members of the councils. Elections are normally held every three years on the basis of universal adult suffrage.

Local government authorities are responsible for public health and sanitation, poor relief, water supply, minor roads, and markets and fire services. Revenues come largely from land taxes, supplemented by large grants from the central government.

16JUDICIAL SYSTEM

The judicial system follows British practice, with some local variations. Cases may be brought in the first instance before a lay magistrate (justice of the peace), a magistrate, or a judge in the Supreme Court, according to the seriousness of the offense or the amount of property involved. The Supreme Court also has appellate jurisdiction. Final appeal rests with the seven-member Court of Appeals, appointed on the advice of the prime minister in consultation with the leader of the opposition. The attorney general, who need not be a member of Parliament, is appointed by the governor-general on the advice of the prime minister.

A special "gun court" considers cases involving the illegal use or possession of firearms or ammunition.

The judiciary is independent but is overburdened and backlogged because of a lack of trained personnel. Recent increases in salaries, training programs for judicial personnel, and improvement in court facilities may eventually serve to improve efficiency and processing of cases.

17ARMED FORCES

The Jamaica Defense Force assumed responsibility for the defense of Jamaica following the withdrawal of British forces in 1962.

The total defense force in 1993 numbered 4,200 personnel, including 870 reserves. Designed for minimal defense missions, the armed forces have three ground force battalions, an air wing of 170 patrol and administrative aircraft, and a coast guard of five patrol boats. In 1991, Jamaica spent $20 million on defense, less than 1% of gross domestic product.

18INTERNATIONAL COOPERATION

Although Jamaica remains a member of the Commonwealth of Nations, the country's political, social, and economic ties have shifted toward participation in Latin American, Caribbean, and third-world international organizations. Jamaica was admitted to UN membership on 18 September 1962 and is a member of ECLAC and all the nonregional specialized agencies except IDA. The nation belongs to CARICOM, G-77, IDB, OAS, and PAHO, as well as the International Bauxite Association. Jamaica also participates in a number of common services with other West Indian islands, including support of the University of the West Indies; the West Indies Shipping Co., which maintains an interisland shipping service; and the West Indian Meteorological Service. Jamaica is a signatory of GATT and the Law of the Sea, and has been chosen as the site of the International Seabed Authority.

19ECONOMY

The structure of the Jamaican economy has undergone extensive changes since 1945, when it was primarily dependent on tropical agricultural products—sugar, bananas, coffee, and cocoa. The island has since become one of the world's largest producers of bauxite, though the industry suffered severely in the 1980s from high local costs and low world prices. It also has developed as a major tourist center for North Americans. Diversified light industry has also been fostered, for both the domestic market and export. Increased foreign revenues paid for improvements in transportation, electrical supplies, and other basic services.

The underlying weaknesses of Jamaica's economy, including unemployment, underemployment, and unequal distribution of income, reveal themselves when the market for Jamaica's primary products weakens, as happened during the 1930s, 1970s, and early 1980s. Under the PNP government during 1972–80, production and foreign sales of bauxite, sugar, and bananas declined; tourism dropped because of rising social unrest; investor confidence waned; and consumer prices (1975–81) increased by 325%. With the change of administrations in both Jamaica and the US during 1980–81, more than $1 billion in IMF and other credits became available. This was enough to extricate the country from its immediate payments crisis, but weak growth continued through 1986, when per capita income was 5.6% less than in 1981. The first signs of economic recovery were evident during the first quarter of 1987.

Jamaica achieved a real GDP growth of 1.5% in 1992, representing the sixth consecutive year of growth. Prior to 1991, Jamaica was the fastest growing economy in the caribbean with real GDP growth rates of 3.3% between 1987 to 1991. The liberalization of the exchange regime in late 1991 was instrumental in restoring external equilibrium in 1992. Real GDP growth was modest during 1993 because of uncertainty in the international price of bauxite/alumina, reduced consumption (due to the monetary and fiscal policies), and marginal growth in other sectors (excluding agriculture, tourism, and finance).

In recent years, Jamaica has diversified its traditional tourist market—the US and Canada—by aggressively marketing in Europe and Japan. US tourists continued to be the single largest tourist market, but it grew only 2% in 1992. In contrast, European arrivals increased by 19% and Japanese arrivals increased by 50%. This positive marketing strategy, accompanied by depreciating currency, helped Jamaica outperform other Caribbean destinations, registering 823,810 arrivals for the first half of 1993.

Agricultural production performed well during 1992, registering a growth of 12.9%. The principal crops are sugar, bananas, citrus fruits, coffee, and cocoa. Sugar is Jamaica's largest agricultural export, earning $82.5 million in 1992. Depreciation of the Jamaican dollar stimulated the production of non-traditional export crops as well.

20INCOME
In 1992, Jamaica's GNP was $3,216 million at current prices, or $1,340 per capita. For the period 1985–92, the average inflation rate was 23.9%, resulting in a real growth rate in per capita GNP of 2.9%.

In 1992, the GDP was $3,294 million in current US dollars. It is estimated that in 1991 agriculture, hunting, forestry, and fishing contributed 6% to GDP; mining and quarrying, 11%; manufacturing, 18%; electricity, gas, and water, 3%; construction, 13%; wholesale and retail trade, 23%; transport, storage, and communication, 9%; finance, insurance, real estate, and business services, 13%; community, social, and personal services, 3%; and other sources, 2%.

21LABOR
The labor force in 1991 was estimated at 1,069,000; the total unemployment rate at that time was 15.1%. Of those employed, 26.4% worked in agriculture, forestry, and fishing; 20.6% in industry; and 53% in services.

The two major trade unions are closely identified with the country's two main political parties: the National Workers' Union with the PNP and the Bustamante Industrial Trade Union with the JLP. The Trade Union Congress is a third major union. The combined membership of unions in 1991 amounted to 16.3% of those employed. There was a significant increase in strikes and industrial actions by diverse unions and professional associations in 1992. The announced layoff of 8,000 government employees, and the erosion of real wages by an annual inflation rate exceeding 100% were the principal reasons for industrial action.

Labor legislation covers such items as national insurance, employment of nationals, hours of work, minimum wages, employment of women and youths, apprenticeship, and welfare (workers' compensation and factory conditions). The industrial workweek is generally 8 hours a day for 5 days. Hours in agriculture and some of the service industries vary, but are usually longer.

22AGRICULTURE
Jamaican agriculture accounts for 5.1% of GDP, less than in most developing countries. The agricultural sector grew by 12.9% in 1992 (relative to 1991), due mainly to a 19.5% increase in domestic crop production, as well as a 5.7% increase in export crops. Attempts to offset the serious price and production problems of traditional agricultural exports by encouraging production of winter vegetables, fruits, and flowers have had limited success. Sugar, the leading export crop, is produced mainly on plantations organized around modern sugar factories that also buy cane from independent growers. Sugar has suffered badly from low world prices and cuts in US buying quotas. Raw sugar production in 1992 was 221,678 tons, down from 290,000 tons in 1978. As a result of heavy rains in early 1993, the Sugar Manufacturers' Corporation announced a lowering in estimated production levels for the year by nearly 10%. Sugar is Jamaica's largest agricultural export, earning $82.5 million in 1992. The government is in the process of privatizing all four of its sugar estates. Banana production for export grew by 1.9% to 76,723 tons in 1992. Earnings from banana exports, however, fell by about 15% to $32.3 million as a result of lower prices on world markets. Other major export crops in 1992 included cocoa, and coffee. Blue Mountain coffee, which is primarily exported to

Japan, brings in some $12 million annually in foreign exchange earnings. Jamaica also exports coconuts, pimientos, citrus fruits, ginger, tobacco, yams, peppers, and cut flowers.

The island's food needs are met only in part by domestic production, and foodstuffs are a major import item. The main food crops, grown primarily by small cultivators, are sweet potatoes and yams, rice, potatoes, manioc, tomatoes, and beans. Jamaica is a major producer of marijuana, which, however, remains illegal. The government participates in a US-funded campaign to eradicate marijuana trading.

23ANIMAL HUSBANDRY
Livestock has long been important in Jamaica's agricultural life, providing both fertilizer and protein for the local diet. Despite increases in the livestock population and in the production of meat, milk, and poultry, increased demand has resulted in continued imports of livestock products. Livestock holdings in 1992 included some 320,000 head of cattle, 440,000 goats, and 250,000 hogs. Livestock products in 1992 included 54,000 tons of poultry meat and 49,000 tons of milk.

24FISHING
The fishing industry grew during the 1980s, primarily from the focus on inland fishing. Whereas the inland catch in 1982 was 129 tons, by 1991 it had risen to 3,230 tons. Nevertheless, substantial imports have been required to meet domestic needs. The total catch in 1991 was 10,430 tons.

25FORESTRY
By the mid-1980s, only 184,000 hectares (455,000 acres) of Jamaica's original 1,000,000 hectares (2,500,000 acres) of forest remained. Roundwood production increased from 55,000 cu m in 1981 to 220,000 cu m in 1988 before falling to 180,000 cu m in 1991. The Forestry Department began a reforestation program in 1963 that was scheduled to last for 30 years; the target was to plant 1,200 hectares (3,000 acres) of timber a year. During the 1980s, reforestation averaged 1,000 hectares (2,500 acres) a year, but deforestation continued at twice that rate.

26MINING
Jamaica is the world's third-leading producer of bauxite; in 1991, production was 11,552,000 tons, or 10% of world production. Alumina production came to 3,015,000 tons in 1991, or 6% of world production. Firm bauxite and alumina demand coupled with strong prices have the government and private sector cooperating to increase production; in the 1980s, a global glut of alumina had severely hurt the Jamaican bauxite and alumina industries.

In 1991, Jamaica produced 135,844 tons of gypsum and an estimated 95,000 tons of lime. Good-quality marble is found in the Blue Mountains, and silica and clays are also exploited.

27ENERGY AND POWER
Jamaica has no coal deposits and very little hydroelectric potential. Oil and gas exploration begun in 1981 had produced no results as of 1991. Electricity is the main source of power and is almost all generated by steam from oil-burning plants. In 1991, the total amount of electricity generated by public and private sources was 2,735 million kwh. In 1977, the government granted the Jamaica Public Service Co. (JPS) a 40-year franchise as the sole supplier of public power. In 1993, however, JPS entered into a $23.2 million contract with a US. firm to build a 35,000–kw generator, whose ownership would be transferred to the private sector under the Energy Sector Privatization Program of the World Bank and the Inter-American Development Bank. Some large enterprises, such as the bauxite companies, and the sugar estates generate their own electricity.

As of 1991, imports of fuels accounted for about 19% of total imports. Jamaica has made a few ventures into alternative sources of energy, but these are still minor relative to overall demand. A large solar water-heating plant was opened in 1981, and AID has helped finance a $33.3–million power plant, also using solar energy. In 1985, Tropicana Petroleum of California and Shell invested us$23 million in an ethanol plant (mostly for export to the US) at Kingston, using sugarcane as raw material.

28 INDUSTRY

Since 1945, when Jamaican manufacturing was confined largely to processing local agricultural products and making beer, clothing, and furniture, the industrial sector has grown and diversified considerably. The island now produces a wide range of goods, including mineral fuels and lubricants, fertilizers, steel, cement, and agricultural machinery, along with footwear, textiles, paints, building materials, and processed foods. In the 1980s, the government has emphasized manufacturing for export, rather than for import substitution alone. Exports of assembled garments brought in some us$85 million in 1986, up from us$54 million in 1985.

Weak local demand, shortages of hard currency, and other problems hurt the manufacturing sector in the mid-1980s. After rising 7.2% in 1983, manufacturing output fell 2.1% in 1984 and 6.1% in 1985, recovering only mildly in 1986. In 1985, manufacturing accounted for 15.7% of real GDP. Production in 1984 included 257,000 tons of cement, 11,000 tons of steel, 3,721,000 gallons of rum and other alcoholic beverages, 26,000,000 gallons of gasoline, 31,000 tons of fertilizers, and 86,000 tons of flour.

The Jamaica Manufacturers' Association, the Jamaica Industrial Development Corp., and other groups actively promote industry, as does Jamaica National Investment Promotion, Ltd.

The manufacturing industry continued to expand through the years, as of 1993 it constituted about 20% of real GDP, employing about 11.3% of the active labor force. Export industries performed well due to the gain in competitiveness caused by the depreciation, but industries that produced mainly for the domestic market were affected negatively. In general, the industry grew by 1% in 1992 due mainly to improved performance in the apparel sector, which is the fastest-growing area of manufactured exports, accounting for 23% of Jamaica's total merchandise exports.

In addition, the processing of agricultural products, alumina, and bauxite remains responsible for much of the activity in the sector, with food, beverages, and tobacco contributing some 70% of industrial output. However, in 1992, bauxite production declined by 2.2% to 11.3 million tons over 1991. This was due mainly to loss of Soviet contracts and disruptions caused by expansion work at some of the refineries. During the first eight months of 1993, bauxite and alumina production amounted to 7.45 million tons, a decline of 3% over the corresponding period in 1992. Due to declining world-market prices for aluminum, bauxite/alumina revenues have declined disproportionally to the fall in production. Gross earnings projected for 1993 were $536 million, 4.3% lower than 1992.

29 SCIENCE AND TECHNOLOGY

Learned societies and research institutions include the Caribbean Food and Nutrition Institute, in Kingston, and the Sugar Industry Research Institute, in Mandeville. The Scientific Research Council, located in the capital and founded in 1960, coordinates research efforts in Jamaica. The University of the West Indies, with a campus in Mona, has faculties of medicine and natural sciences. An agricultural college is located in Portland. The College of Arts, Science and Technology has over 6,000 students.

30 DOMESTIC TRADE

Imports normally account for about a third of the goods distributed, and importing is in the hands of a relatively small number of firms. Competition is limited by the acquisition of import licenses, and profit margins are high. Many importers function as wholesalers and also have retail outlets. Locally produced consumer goods often are marketed through the same firms.

Retail outlets range from supermarkets and department stores, in the urban areas, to small general stores and itinerant merchants, in the rural regions. Purchases tend to be made in small quantities, and credit was extended freely until 1961, when the government instituted restrictions on retail credit and on profit margins.

Shops open weekdays between 8 and 9 AM; large stores generally close at 4:30 or 5 PM, with one early closing day each week. Food stores, drugstores, and family enterprises, however, often remain open until 9 PM or later. Banks are normally open on weekdays from 9 AM to 2 PM. Newspapers, radio, and television are the main advertising media.

31 FOREIGN TRADE

Since the discovery of bauxite deposits in the 1950s, Jamaica has become increasingly active in international trade and has gradually loosened its ties to the Commonwealth and increased its commercial contacts with North America and the Caribbean. The Jamaican government is committed to attract foreign investment in the island, and the government agency Jamaican Promotions Ltd. (Jampro) is responsible for pursuing trade and investment opportunities. Jamaica is a consumer oriented–country, that produces very little of its major necessities. The import content of manufactured goods range between 30–60%, however, Jamaica has never recorded a visible trade surplus.

In February 1991, the government implemented the new CARICOM Common External Tariff (CET), creating the first customs union in the Caribbean. This, along with other factors, such as the absence of new trade restrictions, have contributed a moderate increase in the trade balance. In 1992 and 1993, the merchandise trade balance improved, registering a decreasing deficit of $400.4 million and $313.5 million respectively. At year end 1992, the value of merchandise imports was $1,451.6 million, while merchandise exports were valued at $1,052.2 million. The US continues to be Jamaica's top trading partner, although total imports from the US declined by 5.5% to $882.7 million during 1992, 52% of Jamaica's total imports that year. Some of the major import categories include petroleum, grains, machinery and transport equipment, chemical, and poultry parts. Similarly, the US has been Jamaica's principal export market over the last two decades. In 1992, Jamaica's exports to the US increased by 13% to $386.3 million or 36.7% of total exports. Major US imports from Jamaica were bauxite and alumina ($241 million), food ($31.6 million), and garments.

32 BALANCE OF PAYMENTS

Balance-of-payments deficits in the 1960s and early 1970s were directly related to the growth of the Jamaican economy and to increased imports of capital goods and raw materials. Later in the 1970s, however, the continued deficits were symptomatic of a weakened economy, declining exports, and the flight of capital. The payments picture brightened somewhat in the first half of the 1980s (despite rising debt payments and the downturn of bauxite exports), as income from tourism and remittances from Jamaicans abroad rose, while substantial international assistance enabled Jamaica to meet its payments obligations. In 1992, a favorable balance of payments was aided by increased tourism inflows, reduced capital outflows, significant improvement in the agricultural sector, stability in the foreign exchange rate in the second half of the year, and the improved economic strength of the US, Jamaica's major trading partner.

In 1992, merchandise exports totaled $1,052.8 million and imports $1,456.7 million. The merchandise trade balance was $–403.9 million.

The following table summarizes Jamaica's balance of payments for 1991 and 1992 (in millions of US dollars):

	1991	1992
CURRENT ACCOUNT		
Goods, services, and income	−464.5	−222.8
Unrequited transfers	273.7	339.8
TOTALS	−190.8	117.0
CAPITAL ACCOUNT		
Direct investment	127.0	86.5
Portfolio investment	—	—
Other long-term capital	71.1	−82.1
Other short-term capital	30.1	185.2
Exceptional financing	−81.1	−36.7
Other liabilities	−56.2	−57.2
Reserves	85.5	−211.0
TOTALS	176.4	−115.3
Errors and omissions	14.4	−1.7
Total change in reserves	97.2	−225.9

[33]BANKING AND SECURITIES

The Bank of Jamaica, the central bank, acts as the government's banker and is authorized to act as agent for the government in the management of the public debt. It also issues and redeems currency, administers Jamaica's external reserves, oversees private banks, and influences the volume and conditions of the supply of credit. The commercial banks are required to maintain, as of 1985, a minimum liquidity of 48%.

Of the ten commercial banks in Jamaica in 1987, three were private institutions of Canadian origin. The National Commercial Bank was 100% owned by the government of Jamaica until 1986; 51% of the shares were sold that year, and the remainder were sold in 1987. The Workers Savings and Loan Bank is a government, trade union, and local private-sector joint venture.

The banks provide short-term credit, including the discounting of commercial bills of exchange, and mortgage credit. Savings accounts are available through the Workers Savings and Loan Bank. The Jamaica Development Bank, established in September 1969, is autonomous and government sponsored; the Jamaica Mortgage Bank, established in February 1972, is government owned, with j$5 million of initial authorized capital. The money supply, as measured by M2 at the end of 1991, totaled j$2,632 million in notes and coins.

Jamaica's security market merged with the stock markets in Barbados and Trinidad and Tobago in 1989. As of October 1991, 45 companies were listed on the exchange.

[34]INSURANCE

Insurance companies in Jamaica are regulated by law, through the Superintendent of Insurance of the Ministry of Finance. In 1986, there were 10 life, 20 non-life, and 5 composite insurance companies operating.

[35]PUBLIC FINANCE

In 1991/92, total revenues amounted to j$21,093 million, while recurrent expenditures came to j$18,884 million. The 1993/94 budget called for a total expenditure of j$40,370 million, an increase of 54.2% over the previous fiscal year's budget. Recurrent account expenditures rose to a larger share (70.3%) of the total budget, a reversal of the trend over the previous several years. The budget deficit of j$11 billion was financed from external debt (60%) and internal debt (40%). As a result largely of debt forgiveness, Jamaica's external debt fell by 11.4% to $3.68 billion during 1991–93. Nevertheless, debt servicing accounts for almost 40% of the fiscal budget, which limits economic expansion. Privatization of public entities has been one of the strategies used by the government to reduce the budget deficit.

[36]TAXATION

In 1993 there was a single income tax of 33.33% on all income above j$10,400. Income tax deductions include allowances for social security, retirement fund contributions, and charitable contributions. Personal allowances and deductions for medical expenses, insurance premiums, and mortgage interest are no longer applicable.

In 1987, dual taxes for agricultural and nonagricultural companies (at 35% and 45% respectively) were replaced by a single 33.33% tax on all companies except banks, which pay an additional tax of 6%. Special depreciation allowances, income tax incentives, and other benefits are still available. Other taxes include business licenses, document taxes, duties on estates, excise taxes, consumption duties, import duties, tonnage taxes, wharfage charges, retail sales taxes, fees on betting and gambling, a motor vehicle tax, and a hotel accommodations tax.

The main source of local revenue is a property tax based on the unimproved value of the land.

[37]CUSTOMS AND DUTIES

The importance of customs duties as a major source of government income is declining, and most imports are duty free. The remaining duties on imports from non-CARICOM nations range from 5% to 45%. Special rates or exemptions on dutiable imports apply to goods from members of CARICOM and signatories of the Lomé Convention.

Licenses are required for imports of certain durable and nondurable consumer goods. License applications are reviewed by the Trade Board. Most capital goods and raw materials do not require import licenses.

[38]FOREIGN INVESTMENT

Foreign investment in Jamaica has accounted for a large part of the capital formation of the post-1945 period. Until the early 1960s, new US and Canadian capital was invested heavily in the bauxite industry. Capital investment in bauxite and alumina then tapered off, but investment increased in other industries as a result of a vigorous campaign by the government. Starting in 1972, however, capital investments in the private sector fell substantially. After 1980, the JLP government of Prime Minister Seaga had some success attracting foreign capital (US$15.9 million during 1982–84), but the economic downturn of the mid-1980s again produced a decline in foreign investment. By early 1987, when 120 US companies operated in Jamaica, cumulative US investment, excluding the bauxite industry, was over US$1 billion. There are no statutory restrictions on sectors open to foreign investment, but in practice most service industries are reserved for Jamaicans.

Increased investment, particularly in the private sector, has been identified by government as an essential factor in the strategy for reviving and sustaining the economy. Government has therefore continued and initiated actions that are intended to encourage investment in a number of areas, such as those that generate foreign exchange, utilize domestic raw materials, and generate employment. The government offers a wide range of incentives, including tax holidays up to a maximum of 10 years and duty–free concessions on raw materials and capital goods for approved incentive periods. There are in existence several acts that provide major benefits for foreign investors, such as the Industrial Incentives Act, the Export Industries Encouragement Act, and The Hotel Incentives Act. Additionally, since the liberalization of exchange controls in September 1991, investors are free to repatriate without prior approval from the Bank of Jamaica. Total foreign direct investment at year end 1992 was $1.2 billion.

Of that amount, 70% or $870,000 came from the US. Other main foreign investors are the UK and Canada.

39ECONOMIC DEVELOPMENT

The election of the socialist PNP in 1972 resulted in gradual government acquisition of foreign-owned companies and nationalization of industries. At the same time, the government restricted and licensed imports and aggressively promoted foreign investment to alter the negative trade balance. These efforts failed to avert a severe decline in the Jamaican economy during the rest of the decade, and in the early 1980s the JLP government moved to reverse much of the PNP's efforts by denationalizing industries, liberalizing import regulation, and lifting wage and price controls.

From 1953 through 1986, the US extended us$903.1 million in nonmilitary loans and grants to Jamaica. Aid from multilateral agencies during the same period totaled us$1,057.4 million, of which us$693.4 million came from the IBRD and us$301 million from the IDB.

As of 1994, the short-term prospects for the Jamaican economy were mixed, with tourism expected to continue growing but mining output likely to remain flat. Jamaica's economy continues to be extremely vulnerable to external economic development, so its domestic economy must be set against the modest economy recovery in the industrialized nations, Jamaica's main trading partners.

The consolidation of the stabilization process and deepening of structural reforms are critical macroeconomic challenges. The exchange rate liberalization is likely to reduce inefficiencies in the allocation of resources and force fiscal and monetary discipline to ensure stability.

40SOCIAL DEVELOPMENT

Jamaica has pioneered in social welfare in the West Indies since 1938. Successful community development programs, which draw on local sources of leadership and use self-help methods, have long been in operation throughout the country. The Jamaica Social Welfare Commission, founded in 1937, is one of the most important voluntary agencies; it receives support from government and industry.

Government assistance is provided to those in need, and rehabilitation grants and family allowances are made. A National Insurance Scheme (NIS) came into effect in April 1966, providing benefits in the form of old age and disability health and maternity coverage, pensions, workers' compensation, widows' and widowers' pensions, and grants. The program is financed by contributions from employers, employees, and self-employed persons.

The marriage rate rose from 3.6 per 1,000 population in 1980 to 4.5 in 1984. Through the Jamaica Family Planning Board, the government has established a network of family planning clinics. By 1989, an estimated 55% of couples of childbearing age were using contraception. In 1985-90, the fertility rate was 2.7%. Abortion is available on broad health grounds. A major social concern has been high urban crime rates, especially in Kingston.

Jamaican women are guaranteed full equality under the constitution and the 1975 Employment Act, but cultural traditions, economic discrimination, and workplace sexual harassment prevent them from achieving it. Sexual assault reports increased from 1,091 in 1991 to 1,155 in 1992.

41HEALTH

The central government provides most medical services in Jamaica through the Ministry of Health. In 1985, the island was divided into 47 health districts, served by 154 health centers and dispensaries, which treat about 600,000 outpatients annually. There were 27 public hospitals in 1986, with over 5,420 beds, including 1,600 beds at Bellevue Hospital in Kingston. There were also five private hospitals with 277 beds. In 1990 there were 338 doctors and 54 dentists.

The government conducts a broad public health program, involving epidemic control, health education, industrial health protection, and campaigns against tuberculosis, venereal diseases, yaws, and malaria. In 1992, 90% of the population had access to health care services.

Public health programs have brought about a significant decrease in the death rate. The death rate was 6.2 per 1,000 people in 1993, and the infant mortality rate was 12 per 1,000 live births in 1992. Tuberculosis, hookworm, and venereal diseases remain the most prevalent diseases. Immunization programs are conducted against poliomyelitis and diphtheria, with 1992 vaccination rates for children up to one year of age as follows: tuberculosis (85%); diphtheria, pertussis, and tetanus (84%); polio (74%); and measles (63%). In 1990, 8% of Jamaica's children under 5 years old were considered malnourished.

There were 55,000 births in 1992, and life expectancy averaged 73 years for both men and women.

42HOUSING

Housing is one of the government's most pressing problems. While middle-and upper-income housing is comparable to that in neighboring areas of North America, facilities for low-income groups are poor by any standard. The problem has been aggravated by constant migration from the rural areas to the cities, causing the growth of urban slums. Most new urban housing is built of cinder block and steel on the peripheries of the cities. Rural housing is primarily built of wood and roofed with zinc sheeting. Squatter settlements surround the major cities of Jamaica. According to the latest available information for 1980-88, total housing units numbered 597,000 with 3.9 people per dwelling.

43EDUCATION

Jamaica's estimated literacy rate is 98.4% (males: 98.2% and females: 98.6%). The government devotes a large part of its budget to developing adequate education facilities. In 1990 there were 8,830 primary school teachers with 323,378 students. There were 225,240 secondary students. Education is compulsory for six years of primary education. At the secondary level, there are two stages, one of three years and one of four.

The University of the West Indies, founded in 1948 as the University College of the West Indies, achieved full university status in 1962 and serves all British Commonwealth Caribbean territories. There are faculties of arts, natural sciences, education, general studies, medicine, law, library studies, management studies, public administration, and social work at Jamaica's Mona campus; arts, natural sciences, social sciences, agriculture, engineering, international relations, and management studies at St. Augustine in Trinidad; and arts and natural sciences in Barbados. Higher technical education is provided at the College of Arts, Science, and Technology. Jamaica also has a school of agriculture, several teacher-training colleges and community colleges, and an automotive training school. At the university and higher level institutions, there were 16,018 students enrolled in 1990.

The Jamaica Movement for the Advancement of Literacy Foundation, known as JAMAL (formerly the National Literacy Board), has reached more than 100,000 students since its founding in 1972.

44LIBRARIES AND MUSEUMS

The Jamaica Library Service provides free public library programs throughout the island and assists the Ministry of Education in supplying books to primary-school libraries. The book stock of the Public Library Service totals 2,666,000 volumes, 1,473,000 in schools and 1,193,000 in parish libraries. There are 696 service points, including parish and branch libraries, book centers, and 14 bookmobiles. At the Mona campus of the University of the West Indies, there are 598,000 volumes in the main library and at two branches.

The Institute of Jamaica has a notable collection of artifacts and materials relating to the West Indies, as well as a museum and exhibition galleries focusing on Natural History, Military, and Maritime. The National Gallery of Jamaica, the African-Caribbean Institute, and Jamaica Memory Bank are also part of the Institute of Jamaica. There is a botanical garden and zoo at Hope, on the outskirts of Kingston.

45MEDIA

The Post and Telegraphs Department provides daily postal deliveries to all parts of the island and operates Jamaica's internal telegraph service. Jamaica International Telecommunications (JAMINTEL) is 51% government-owned and provides five major international services: telephone, telegraph, television, telex, and leased circuits. Telephone service is provided by the privately owned Jamaica Telephone Co. In 1993, 180,000 telephones were in use. Telephone communications between Jamaica and other countries were improved in 1963 by a 1,370-km (850-mi) submarine telephone cable linking Jamaica and Florida and in 1971 by the addition of a satellite communications system. All telephone exchanges are automatic.

Jamaica has two broadcasting companies. The privately owned Radio Jamaica Rediffusion broadcasts 24 hours a day over both AM and FM bands; it also owns an extensive wire network. The government-owned Jamaica Broadcasting Corp., with similar transmitting facilities, broadcasts FM radio and television programs. In 1992 there were approximately 1,025,000 radio sets and 320,000 television receivers.

The morning *Daily Gleaner* (circulation about 37,000 in 1986) and the evening *Daily Star* (circulation 38,000) are published by the Gleaner Co., which also publishes the *Sunday Gleaner* (81,000) and the *Weekend Star* (85,000), an overseas weekly. There are also a number of weeklies and monthlies, and in addition, several papers are published by religious groups.

46ORGANIZATIONS

The producers of the main export crops are organized into associations, and there are also organizations of small farmers. The Jamaica Agricultural Society, founded in 1895, is concerned with agricultural and rural development and works closely with the government. The cooperative movement has grown rapidly since World War II. All cooperatives must register with the government and are subject to supervision. Savings and credit groups are the most numerous, followed by marketing organizations. Consumer cooperatives have had little success.

Outside the agricultural sector, the chambers of commerce have long been the most important business groups. There is a wide range of professional and cultural organizations, including a women's federation, 4-H clubs, and an automobile association.

47TOURISM, TRAVEL, AND RECREATION

Immigration regulations require the possession of a valid passport, but visitors from the US, Canada and other Commonwealth countries, Japan and parts of Europe may stay up to six months without other valid identification. All visitors must have an onward ticket.

Jamaica is firmly established as a center for tourists, mainly from North America. Greatly expanded air facilities, linking Jamaica to the US, Canada, and Europe, were mainly responsible for the increase in tourism during the 1960s. Rising fuel costs and a weak international economy, as well as intermittent political unrest, contributed to a slowdown in the growth rate of the industry in the 1970s; between 1980 and 1986, however, the number of tourists increased by 68%. In 1991, 844,607 tourists visited the island, 544,467 of them from the US. Receipts from tourism in 1991 were estimated at $764 million. There were 17,337 hotel rooms and 35,416 beds with a 57.9% occupancy rate. Major tourist areas are the resort centers of Montego Bay and Ocho Rios. Cricket is the national sport, and excellent golf and water-sports facilities are available.

48FAMOUS JAMAICANS

Names associated with Jamaica's early history are those of Europeans or of little-known figures such as Cudjoe, chief of the Maroons, who led his people in guerrilla warfare against the English in the 18th century. George William Gordon (1820–65), hanged by the British as a traitor, was an advocate of more humane treatment for blacks. Jamaica-born Marcus Garvey (1887–1940), who went to the US in 1916, achieved fame as the founder of the ill-fated United Negro Improvement Association. In the mid-20th century, Jamaicans whose names have become known abroad have been largely political and literary figures. Sir (William) Alexander Bustamante (1894–1977), trade unionist, political leader, and former prime minister of Jamaica, and his cousin and political adversary, Norman Washington Manley (1893–1969), a Rhodes scholar and noted attorney, were leading political figures. More recently, Norman Manley's son Michael (b.1923), prime minister during 1972–80, and Edward Seaga (b.US, 1930), prime minister from 1980-89, have dominated Jamaica's political life. The novelists Roger Mais (1905–55), Vic Reid (b.1913), and John Hearne (b.1926) have built reputations in England, and the poet Claude McKay (1890–1948) played an important role in the black literary renaissance in the US. Performer and composer Robert Nesta ("Bob") Marley (1945–81) became internationally famous and was instrumental in popularizing reggae music outside Jamaica.

49DEPENDENCIES

Jamaica has no territories or colonies.

50BIBLIOGRAPHY

American University. *Area Handbook for Jamaica.* Washington, D.C.: Government Printing Office, 1976.

Bakan, Abigail B. *Ideology and Class Conflict in Jamaica: The Politics of Rebellion.* Montreal: McGill-Queen's University Press, 1990.

Bryan, Patrick E. *The Jamaican People, 1880–1902: Race, Class and Social Control.* London: Macmillan Caribbean, 1991.

Campbell, Mavis Christine. *The Maroons of Jamaica, 1655–1796: A History of Resistance, Collaboration & Betrayal.* Granby, Mass.: Bergin & Garvey, 1988.

Dawes, Hugh N. *Public Finance and Economic Development: Spotlight on Jamaica.* Lanham, Md.: University Press of America, 1982.

Floyd, Barry. *Jamaica: An Island Microcosm.* New York: St. Martin's, 1979.

Hall, Douglas. *Free Jamaica, 1838–1865: An Economic History.* New Haven, Conn.: Yale University Press, 1966.

Holt, Thomas C. *The Problem of Freedom: Race, Labor, and Politics in Jamaica and Britain, 1832–1938.* Baltimore: Johns Hopkins University Press, 1992.

Hurwitz, Samuel J. and Edith F. *Jamaica: A Historical Portrait.* New York: Praeger, 1971.

Jamaica in Pictures. Minneapolis: Lerner, 1987.

Kaufman, Michael. *Jamaica under Manley: Dilemmas of Socialism and Democracy.* London: Zed Books, 1985.

Keith, Nelson W. *The Social Origins of Democratic Socialism in Jamaica.* Philadelphia: Temple University Press, 1992.

Levi, Darrell E. *Michael Manley: The Making of a Leader.* Athens: University of Georgia Press, 1990, 1989.

Looney, Robert E. *The Jamaican Economy in the 1980s: Economic Decline and Structural Adjustment.* Boulder, Colo.: Westview Press, 1987.

Norris, Katrin. *Jamaica: The Search for an Identity.* London: Oxford University Press, 1962.

Payne, Anthony. *Politics in Jamaica.* New York, N.Y.: St. Martin's, 1988.

Stone, Carl. *State and Democracy in Jamaica.* New York: Praeger, 1986.

Waters, Anita M. *Race, Class and Political Symbols: Rastafari and Reggae in Jamaican Politics.* New Brunswick, N.J.: Transaction Books, 1985.

MEXICO

United Mexican States
Estados Unidos Mexicanos

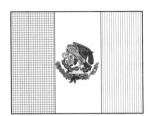

CAPITAL: Mexico City (México).

FLAG: The national flag is a tricolor of green, white, and red vertical stripes; at the center of the white stripe, in brown and green, is an eagle with a snake in its beak, perched on a cactus.

ANTHEM: *Mexicanos, al grito de guerra (Mexicans, to the Cry of War).*

MONETARY UNIT: The peso (P) is a paper currency of 100 centavos. There are coins of 1, 5, 10, 20, 50, 100, 500, 1,000 and 5,000 pesos and notes of 2,000, 5,000, 10,000, 20,000, 50,000 and 100,000 pesos. As of 1 January 1993, a new unit of currency (the new peso) was issued, worth 1,000 of the pesos that were used until 31 December 1992. P1 = $0.2976 (or $1 = P3.3598).

WEIGHTS AND MEASURES: The metric system is the legal standard, but some old Spanish units are still in use.

HOLIDAYS: New Year's Day, 1 January; Constitution Day, 5 February; Birthday of Benito Juárez, 21 March; Labor Day, 1 May; Anniversary of the Battle of Puebla (1862), 5 May; Opening of Congress and Presidential Address to the Nation, 1 September; Independence Day, 16 September; Columbus Day, 12 October; Revolution Day (1910), 20 November; Christmas, 25 December. Movable religious holidays include Holy Thursday, Good Friday, and Holy Saturday. All Souls' Day, 2 November, and Our Lady of Guadalupe Day, 12 December, are not statutory holidays but are widely celebrated.

TIME: 6 AM = noon GMT.

¹LOCATION, SIZE, AND EXTENT

Situated south of the US on the North American continent, Mexico has an area of 1,972,550 sq km (761,606 sq mi), including many uninhabited islands off the E and W coasts, which have a combined area of 5,073 sq km (1,959 sq mi). Comparatively, the area occupied by Mexico is slightly less than three times the size of the state of Texas. Mexico extends about 3,200 km (2,000 mi) SSE–NNW and 1,060 km (660 mi) ENE–WSW. Bordered on the N by the US, on the E by the Gulf of Mexico (including the Bay of Campeche), the Caribbean Sea, Belize, and Guatemala and on the S and W by the Pacific Ocean, Mexico has a total boundary length of 13,868 km (8,617 mi), including the narrow peninsula of Baja California, which parallels the W coast of the mainland.

Mexico's capital city, Mexico City, is located in the southcentral part of the country.

²TOPOGRAPHY

Mexico's dominant geographic feature is the great highland central plateau, which occupies most of the width of the country, extending from the US border to the Isthmus of Tehuantepec. It averages 900 to 1,200 m (3,000–4,000 ft) in elevation in the north and 2,100 to 2,400 m (7,000–8,000 ft) in the central part of the country. The plateau is enclosed by two high cordilleras (mountain chains), the Sierra Madre Oriental on the east and the Sierra Madre Occidental on the west, each separated from the coast by lowland plains. The ranges rise to over 3,000 m (10,000 ft), and some volcanic peaks exceed 5,000 m (16,400 ft); Pico de Orizaba, or Citlaltépetl (5,639 m/18,501 ft), Popocatépetl (5,465 m/17,930 ft), and Ixtaccíhuatl (5,230 m/17,159 ft) are the highest. The plateau falls to the low Isthmus of Tehuantepec and then rises again to Chiapas Highland to the south. The lowlands of Tabasco, Campeche, and Yucatán lie north and east of Chiapas.

There are no important inland waterways. Except for the Rio Grande (known as the Río Bravo del Norte in Mexico), which extends for about 2,100 km (1,300 m) of the boundary with the US, and the Papaloapan, an important source of waterpower, the other rivers are short; they are the Lerma, Santiago, Usumacinta (part of the boundary with Guatemala), Grijalva, Balsas, Pánuco, and the Soto la Marina. The largest lake in Mexico is Lake Chapala, in Jalisco State, covering about 1,680 sq km (650 sq mi).

³CLIMATE

The climate varies according to altitude and rainfall. The tropical and subtropical zone (tierra caliente), ranging from sea level to about 900 m (3,000 ft), consists of the coastal plains, the Yucatán Peninsula, and the lower areas of southern Mexico. These areas have a mean temperature of 25–27°C (77–81°F), with a minimum of 16°C (61°F) and a maximum of almost 49°C (120°F). The temperate zone (tierra templada), at elevations of 900 to 1,800 m (3,000–6,000 ft), has a temperate-to-warm climate and a mean temperature of 21°C (70°F). Mexico City and most other important population centers are in the cool zone (tierra fría), starting at about 1,800 m (6,000 ft), with a mean annual temperature of 17°C (63°C). The highest mountain peaks are always covered with snow.

Most of Mexico is deficient in rainfall, but two coastal belts covering about 12% of the total area—from Tampico south along the Gulf of Mexico and from the state of Colima south along the Pacific—receive an average of from 99 to 300 cm (39–118 in) per year. Annual rainfall may exceed 500 cm (200 in) in Tabasco and Chiapas, while in parts of Baja California, virtually no rain falls. Precipitation is adequate in central Mexico except at altitudes above 1,800 m (6,000 ft), while the northern states are semidesert or desert. Most of the country receives its heaviest rainfall during the summer months.

⁴FLORA AND FAUNA

Plant and animal life differs sharply with Mexico's varied climate and topography. The coastal plains are covered with a tropical rain forest, which merges into subtropical and temperate growth as the plateau is ascended. In the northern states there is a dry steppe vegetation, with desert flora over much of the area. Oaks and conifers are found in mixed forest regions along the mountain slopes. The Yucatán Peninsula has a scrubby vegetation.

Among the wild animals are the armadillo, tapir, opossum, puma, jaguar, bear, and several species of monkey, deer, and boar. Poisonous snakes and harmful insects are found, and in the coastal marshes, malarial mosquitoes pose a problem. The only remaining elephant seals in the world are on Guadalupe Island west of Baja California.

⁵ENVIRONMENT

The Secretariat for Urban Development and Ecology (SEDUE) has the principal environmental responsibility. One of Mexico's most widespread environmental problems is soil erosion; slash-and-burn agricultural practices, especially in the tropical zones, have also contributed to deforestation. Mexico loses its forest at a rate of 2,375 square miles annually due to agricultural and industrial expansion. Mexico City, located more than 2,250 m (7,400 ft) above sea level and surrounded by mountains, has chronic smog, aggravated by the presence in the metropolitan region of some 35,000 factories and more than two million motor vehicles and by open burning of garbage by slum dwellers in an attempt to dispose of the 30% of the city's refuse that is not regularly collected. Mexico contributes 2.0% of the world's total gas emissions. Transportation vehicles are responsible for 76% of the air pollution. Water pollution results from the combined impact of industrial, agricultural, and public waste. Mexico's cities produce 12.9 million tons of solid waste per year along with 164 million tons of industrial waste. Mexico has 85.8 cubic miles of water. Eighty-six percent is used in farming activity and eight percent is used for industrial activity. Fifty-one percent of the nation's rural dwellers do not have pure water. An environmental protection statute adopted in 1971 has not been widely enforced; however, SEDUE, which was created in 1982, is fostering a more coherent approach to environmental issues.

In 1992, United Nations' sources reported 242 endangered species in Mexico. In 1994, 25 of the nation's mammal species and 35 bird species are endangered. As of 1987, endangered species in Mexico included the volcano rabbit, Mexican grizzly bear (possibly extinct), Lower California pronghorn, Sonoran pronghorn, imperial woodpecker, southern bald eagle, American peregrine falcon, tundra peregrine falcon, horned guan, masked bobwhite quail, whooping crane, light-footed clapper rail, California least tern, maroon-fronted parrot, ridge-nosed rattlesnake, Bolson tortoise, five species of turtle (green sea, hawksbill, Kemp's ridley, olive ridley, and leatherback), bighead pupfish, blackfin skiffia, San Estebán Island chuckwalla, spectacled caiman, two species of crocodile (American and Morelet's), and totoaba.

⁶POPULATION

The 1990 census showed a total population of 81,249,646, more than double the 1960 census total of 34,923,129. During 1970–80, annual population growth was 3.1%, one of the highest rates in the world, but during 1985–90, the growth rate declined to 2.2%, aided by government promotion of family planning. However, the mid-1994 population estimate of 96,167,792 by the U.S. Bureau of the Census indicated a higher rate of growth than the official one. A population of 102,555,000 was projected by the UN for the year 2000, assuming a crude birth rate of 25.1 per 1,000 population, a crude death rate of 5.2, and a net natural increase of 19.9 during 1995–2000. Because of the high birth rate

(about 29 per 1,000 during 1985–90), the 1990 census showed 12.5% of the population under 5 years of age, 25.5% under 10, and 50.2% under 20. Population density in 1990 averaged 41.3 per sq km (107 per sq mi); density in 1980 ranged from 4.3 persons per sq km (11.2 per sq mi) in Baja California Sur to 5,569 persons per sq km (14,424 per sq mi) in the Federal District.

In 1990, 73% of the population was urban, and 32% of the entire population lived in fourteen urban areas (including the Federal District) of over 500,000 persons. In 1990, the metropolitan area of Mexico City had a population of 15,047,685, with 8,235,744 within the federal district. Other major cities (with their populations in 1990) were Guadalajara, 1,650,205; Monterrey, 1,069,238; Puebla, 1,057,454; León, 867,920; Ciudad Juárez, 789,522; and Tijuana, 747,381.

⁷MIGRATION

As of 1990, the US Census Bureau estimated that 13,496,000 persons of Mexican ancestry were living in the US, up from 8,740,000 in 1980. Formerly, under an agreement between the US and Mexico, there was a large annual movement of Mexican agricultural laborers (braceros) into the US. The US has since banned such border crossings, and the 1986 Immigration Act imposes stiff penalties on employers who hire illegal aliens. Nevertheless, hundreds of thousands of illegal crossings still take place annually.

In the 1970s, there was substantial internal migration to the frontier areas, especially to the northern border states, to Quintana Roo and Guerrero in the south, and to government colonization projects such as Papaloapan, Veracruz, where the construction of irrigation, flood control, and hydroelectric projects attracted many people. Most notable, however, is the migration of rural inhabitants to Mexico's already overcrowded cities, creating huge urban slums, especially in Mexico City. Guadalajara and Monterrey, although growing rapidly, are attempting to use Mexico City's experience to plan urban and industrial growth. The great disparity between rural-zone, low-income, marginal groups producing only for their own consumption and highly developed urban and industrial sectors has produced acute social problems.

Mexico has long had a liberal asylum policy; the most famous exile in Mexico was probably Leon Trotsky, a revolutionary exiled from the USSR in 1929 who lived in Mexico City from 1937 until his assassination in 1940. In the 1930s and 1940s, Spanish republican refugees in the tens of thousands settled in Mexico, as did thousands of refugees from World War II. In the 1970s and 1980s, many victims of Latin American military regimes fled to Mexico. By the end of 1993, the number of persons from Guatemala, El Salvador, and other Central American countries living in Mexico was about 400,000, not counting 45,000 Guatemalan refugees living in camps.

The largest community of US citizens living outside the US, estimated at 200,000, is found in Mexico. In all, there were officially 340,824 people of foreign birth living in Mexico in 1990.

⁸ETHNIC GROUPS

The people of Mexico are mostly mestizos, a mixture of indigenous Amerindian and Spanish heritage. There are small numbers of persons of other European heritages, and small numbers of blacks are found in Veracruz and Acapulco. An estimated 75% of the population was mestizo, 10% pure Amerindian, 15% pure white, and 1% black and other. Amerindian influence on Mexican cultural, economic, and political life is very strong.

⁹LANGUAGES

Spanish, the official language, is spoken by nearly the entire population, thus giving Mexico the world's largest Spanish-speaking community; more Mexicans speak Spanish than Spaniards. In

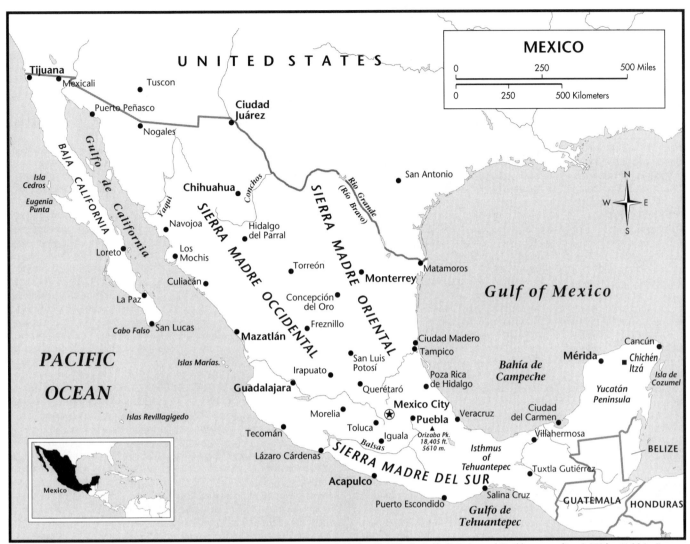

UNITED STATES

MEXICO

| 0 | 250 | 500 Miles |

| 0 | 250 | 500 Kilometers |

Tijuana
Mexicali
Tuscon
Puerto Peñasco
Nogales
Ciudad Juárez
San Antonio

Isla Cedros
Eugenia Punta

BAJA CALIFORNIA
Gulfo de California

Chihuahua
Conchos
Río Grande (Río Bravo)

Yaqui
Navojoa
Hidalgo del Parral
Los Mochis
Loreto
Torreón
Culiacán
La Paz
Concepción del Oro
Freznillo
Cabo Falso San Lucas
Mazatlán
Monterrey
Matamoros

SIERRA MADRE OCCIDENTAL
SIERRA MADRE ORIENTAL

Gulf of Mexico

Ciudad Madero
Tampico
San Luis Potosí
Poza Rica de Hidalgo
Bahía de Campeche
Mérida
Cancún
Chichén Itzá

PACIFIC

OCEAN

Islas Marías.
Islas Revillagigedo
Irapuato
Guadalajara
Querétaro
Morelia
Mexico City
Puebla
Veracruz
Ciudad del Carmen
Isla de Cozumel
Yucatán Peninsula

Mexico
Tecomán
Toluca
Iguala
Orizaba Pk. 18,405 ft. 5610 m.
Villahermosa
BELIZE

Lázaro Cárdenas
SIERRA MADRE DEL SUR
Balsas
Isthmus of Tehuantepec
Tuxtla Gutiérrez

Acapulco
Salina Cruz
GUATEMALA
HONDURAS

Puerto Escondido
Gulfo de Tehuantepec

LOCATION: 14°32′ to 32°43′N; 86°42′ to 118°22′W. **BOUNDARY LENGTHS:** US, 3,326 km (2,067 mi); Gulf of Mexico and Caribbean coastline, 2,070 km (1,286 mi); Belize, 251 km (156 mi); Guatemala, 871 km (541 mi); Pacific coastline (including Baja California), 7,339 km (4,560 mi). **TERRITORIAL SEA LIMIT:** 12 mi.

1990, about 1% of the population five years old or more spoke only indigenous Amerindian languages or dialects. In all, 7.5% spoke some Amerindian language. There are at least 31 different Amerindian language groups, the principal languages being Nahuatl, Maya, Zapotec, Otomi, and Mixtec.

¹⁰RELIGIONS

According to estimates in the early 1990s, 93.3% of Mexicans were nominally Roman Catholic, and about 5% were Protestant. About 50,000 officially followed indigenous Amerindian religions, but over 2% of the population, while professing Roman Catholicism, include strong Amerindian elements in their religion. The Jewish population was an estimated 35,000, and there were some 25,000 Buddhists and 25,000 Muslims.

Following an amendment to the constitution adopted in 1991, ecclesiastical corporations now have legal rights and can acquire property. All church buildings, including schools, however, remain national property. Priests now have political rights, and religious control over both public and private education prohibited earlier under the 1917 constitution has been restored. Roman Catholicism has increased greatly in activity since the 1940s; there have been religious processions and considerable construction of new churches in major cities. The bitter anticlericalism of the Mexican Revolution, of the 1917 constitution, and of the

administration of President Plutarco Elías Calles (1924–28) had lessened by the 1960s. In 1992, full diplomatic relations with the Vatican were established.

¹¹TRANSPORTATION

The local, state, and federal Mexican road system amounted to 210,000 km (130,500 mi) in 1991, of which 65,000 km (40,400 mi) was paved; most roads are engineered for year-round service. In 1991 there were an estimated 10,700,000 registered vehicles (as compared with 7,499,320 in 1986), including 7,400,000 passenger cars, 100,000 buses, and 3,200,000 trucks.

The major railroad system is the National Railway of Mexico. In 1991, the total route length was 20,660 km (12,800 mi), consisting of six integrated lines. A railway improvement program was carried out during 1975–79, but in the mid-1980s, the network was still in disrepair. The Mexico City subway system, totaling 120 km (75 mi) in 1987, suffers from overcrowding. The system's technology, however, is being exported to other developing countries.

Mexico's inland waterways and lakes are not important for transportation, but ocean and coastal shipping is significant. Of Mexico's 102 ocean ports, the most important are Tampico, and Veracruz, on the Gulf of Mexico; Mazatlan and Manzanillo, on the Pacific coast; and Guayamas, on the Gulf of California. These five ports together handle about 80% of total general cargo tonnage for Mexico. Most Mexican ships are operated by the government-owned Maritime Transport of Mexico. The merchant marine in 1991 had 55 vessels totaling 837,000 GRT. The oil tanker fleet, owned and operated by Mexican Petroleum (Petróleos Mexicanos—PEMEX), the government oil monopoly, included 39 tankers with a total of 704,000 GRT. In 1984, 72.4 million tons of international merchandise were loaded in Mexican ports, and 11 million tons were unloaded.

Air transportation in Mexico has developed rapidly. In 1991 there were 200 airports with permanent surface runways, 36 of them long enough to accommodate large jets. Mexican commercial aircraft performed 19,037 million passenger-km of service in 1992. The main airline company is the newly privatized Aeroméxico.

12HISTORY

The land now known as Mexico was inhabited by many of the most advanced Amerindian cultures of the ancient Americas. The Mayan civilization in the Yucatán Peninsula began about 2500 BC, flourished about AD 300–900, and then declined until its conquest by the Spanish. The Mayas had a well-developed calendar and a concept of zero; skillful in the construction of stone buildings and the carving of stone monuments, they built great cities at Chichen Itzá, Mayapán, Uxmal, and many other sites. About 1200–400 BC the Olmecs had a civilization with its center at La Venta, featuring giant carved stone heads and the first use of pyramids for worship among the Amerindians. In the early 10th century AD the Toltecs, under Ce Acatl Topiltzin, founded their capital of Tollan (now Tula) and made the Nahua culture predominant in the Valley of Mexico until the early 13th century. At that time, the Aztecs, another Nahua tribe, gained control.

The Aztec Empire, with its capital at Tenochtitlán (now Mexico City), founded in 1325, was essentially a confederation of allied and tributary communities. Skilled in architecture, engineering, mathematics, weaving, and metalworking, the Aztecs had a powerful priesthood and a complex pantheon dominated by the sun god and war god Huitzilopochtli, to whom prisoners captured from other tribes were sacrificed.

The empire was at its height in 1519, when the Spanish conquistadores, under Hernán Cortés, having set out from Cuba, landed at modern Veracruz; with superior weapons and the complicity of local chieftains, the Spaniards had conquered Mexico by 1521. First, Cortés imprisoned the emperor Montezuma II, who was wounded by stoning when he was released in an attempt to quell an uprising against the Spanish. Then Montezuma's nephew, Guatemotzin, drove the Spanish from Tenochtitlán on 30 June 1520, now called "la noche triste" ("the sad night"), during which Montezuma died, probably at the hands of the Spaniards. Eventually, Cortés returned to Tenochtitlán and defeated Guatemotzin.

The Spaniards brought Roman Catholicism to Mexico, imposed their legal and economic system on the country, and enslaved many of the inhabitants. The combination of Spanish oppression and the smallpox, influenza, and measles the conquistadores brought with them reduced the Amerindian population from an estimated 5 million in 1500 to 3.5 million a century later; not until late in the 18th century did Mexico match its pre-Columbian population. Gradually, the Spaniards extended their territory southward, to include, for a time, the captaincy-general of Guatemala, and northward as far as California, Nevada, and Colorado.

Spain ruled Mexico, as the viceroyalty of New Spain, for three centuries. Continued political abuses and Amerindian enslavement combined with the Napoleonic invasion of Spain in 1807 and consequent political uncertainty to produce a movement for independence. In 1810, a revolt against Spain was begun by a priest, Miguel Hidalgo y Costilla, and a captain, Ignacio José

Allende. Both were captured and shot by loyalists in 1811, but the revolt was continued by another priest, José María Morelos y Pavón, who proclaimed Mexico an independent republic in 1813. Morelos and his followers were defeated and he was shot in 1815, but a swing toward liberalism in Spain in 1820 altered the political picture, leading Mexico's conservative oligarchy to favor independence as a way of preserving its power. In 1821, under the leadership of a rebel, Vicente Guerrero, and a former loyalist, Agustín de Iturbide, independence was again proclaimed and this time secured. Iturbide proclaimed himself emperor in 1822 but was deposed in 1823, when a republic was established; when he returned in the following year, he was captured and shot.

In the next 25 years, there were at least 30 changes of government. Gen. Antonio López de Santa Anna, who had participated in the overthrow of Iturbide, become the dominant figure in the 1830s and 1840s and attempted to centralize the new government. Texas gained its independence from Mexico in 1836 as a result of the defeat of Santa Anna at San Jacinto; in 1845, after a period as the Republic of Texas, it joined the US. Mexico lost the subsequent war with the US (1846–48), which began over a dispute about the border of Texas; under the Treaty of Guadalupe Hidalgo, Mexico recognized the Rio Grande as the boundary of Texas and ceded half its territory (much of the present western US) in return for $45 million. In 1853, the US purchased a small portion of land from Mexico for another $10 million, which was widely regarded as further compensation for the land lost in the war.

A reform government was established in 1855 after a revolt against Santa Anna, and a new liberal constitution was adopted in 1857. Included in the reforms were laws abolishing military and clerical immunities. Article 27 prohibited corporations from holding land, and Article 123 established federal authority in matters of worship and religious discipline. Benito Juárez, the leader of the reform movement, became president in 1858. In 1861, during a period of civil strife, French troops under Emperor Napoleon III intervened in Mexico, ostensibly because Mexico had not paid its debts; in 1863, they captured Mexico City and installed Archduke Maximilian of Austria as emperor, with his Belgian wife, Carlota, as empress. After the French troops withdrew in 1866, partly because the US protested their presence and partly because Napoleon needed them in France, forces loyal to Benito Juárez and led by José de la Cruz Porfirio Díaz regained control of the country. Maximilian was executed and the republic restored in 1867; Carlota, who had returned to Europe to plead with Napoleon to protect her husband, was driven mad by his death.

Díaz twice ran for president, in 1867 and 1871, each time leading an abortive military uprising after his electoral defeat. Finally, in 1876, Díaz seized power and assumed the presidency, a position he held (except for 1880–84, when a subordinate exercised nominal power) until 1911. Under his dictatorship, Mexico modernized by opening its doors to foreign investors and managers. At the same time, the so-called Pax Porfiriana meant suppression of all dissent, by persuasion or by force, and a complete lack of concern with improving the life of the Mexican peasant; an elite corps of mounted police, the Rurales, held the rural areas in check. As the president's aging circle of associates, called Científicos (Scientists), clung tenaciously to power, resentment among the middle classes and the peasantry continued to grow.

After Díaz was once again reelected to the presidency in 1910, the Mexican Revolution erupted. This revolution, which by 1917 had claimed perhaps 1 million lives, was, on the one hand, a protest by middle-class political liberals against the stultifying Díaz regime and, on the other, a massive popular rebellion of land-hungry peasants. The interests of these two groups sometimes coincided but more often clashed, accounting for the turmoil and confusion of those years. The spark that touched off the revolution was the proclamation on 5 October 1910 of the Plan of San Luis Potosí, in which the liberal politician Francisco Indalecio Madero, who had lost the vote to Díaz, called for nullification of the election. Riots in Mexico City forced Díaz to resign and leave the country in 1911, and Madero was elected president later in that year. Meanwhile, popular revolts led by Emiliano Zapata and Pancho Villa, who refused to submit to Madero's authority, led the country into chaos. Madero, accused by the Zapatistas of not giving land to the peasants, was ousted and murdered in 1913 by Gen. Victoriano Huerta, who had conspired with the rebels. When Huerta, a corrupt dictator, was driven from power by Venustiano Carranza and Álvaro Obregón in July 1914, a full-scale civil war broke out. This phase of the revolution ended in February 1917, when a new constitution was proclaimed; this nationalistic, anti-clerical document, considered by some to be the world's first socialist constitution, embodied the principle of the one-term presidency in order to prevent the recurrence of a Díaz-type dictatorship. Article 3 established government rather than church control over schools; Article 27 provided for public ownership of land, water, and minerals; and Article 123 ensured basic labor rights. Carranza was elected president in 1917, but for the next decade Mexico was still beset by political instability and fighting between various revolutionary groups. Most of the revolutionary leaders met with violent deaths. Zapata, still regarded by many as a revolutionary hero, was assassinated in 1919, and both Carranza and Obregón (who was president during 1920–24) lost their lives in military coups.

Political stability at last came to Mexico with the formation in 1929 of an official government party that incorporated most of the social groups that had participated in the revolution; it has been known since 1945 as the Institutional Revolutionary Party (Partido Revolucionario Institucional—PRI). Although its main pillars were, at least in theory, the peasants, workers, and other popular movements, it has also been closely allied with business since the 1940s. The most outstanding political leader of the post-1929 era was Lázaro Cárdenas, president during 1934–40, who sought with some success to realize the social goals of the revolution. His reforms included massive land redistribution, establishment of labor unions with strong bargaining positions, extension of education to remote areas of the country, and in 1938, the expropriation of foreign petroleum holdings, mostly US-owned. A compensation agreement with the US was reached in 1944, when the two nations were World War II allies.

The postwar years have been marked by political stability, economic expansion, and the rise of the middle classes, but also by general neglect of the poorest segments of the population. One serious political disturbance came in 1968, the year the Summer Olympics were held in Mexico City, when the army and police clashed with students protesting political repression and human rights abuses. An economic boom during the late 1970s, brought about by huge oil export earnings, benefited a small percentage of the people, but millions of peasants continued to be only slightly better off than in 1910. Declining world oil prices in 1981 led to a severe financial crisis in 1982, a year of presidential elections. Mexico's new president, Miguel de la Madrid Hurtado, immediately introduced a series of austerity measures and promised a crackdown on corruption, which has long been a problem in Mexico. After the arrest of two government officials for misuse of public funds and fraud, the anticorruption drive appeared to languish; furthermore, public resentment of austerity increased. In 1985, when the PRI was accused of electoral malpractice in local and congressional elections, resentment boiled over in violent public protest. In October 1987, the PRI named Carlos Salinas de Gortari, a 39-year-old economist, as its candidate to succeed President de la Madrid in December 1988. In September 1993, changes in federal electoral law and practices were designed to make future elections more transparant and tamperproof.

Mexico City was devastated by a major earthquake in September 1985. The official death toll was 7,000, although unofficial estimates were as high as 20,000; in addition, 300,000 were left homeless. There was widespread protest over the fact that many of the buildings destroyed had been built in violation of construction regulations and claims that foreign emergency aid had been misappropriated by the government.

In August 1992, formal negotiations regarding the North American Free Trade Agreement were concluded, whereby Mexico would join the US and Canada in the elimination of trade barriers, the promotion of fair competition, and increased investment opportunities. NAFTA went into effect on 1 January 1994.

Early in 1994, a primarily Amerindian group calling itself the Zapatista Army of National Liberation resorted to an armed uprising, initially taking control of four municipalities in the State of Chiapas, to protest what it regarded as government failure to effectively deal with regional social and economic problems.

[13]GOVERNMENT

Mexico is a federal republic consisting of 31 states and the Federal District. Its basic political institutions are defined in the constitution of 1917.

The president, elected for a six-year term (by universal adult suffrage beginning at age 18) and forever ineligible for reelection, appoints the attorney-general and a cabinet, which may vary in number. Although the constitution established separation of powers, in practice the Mexican chief executive dominates the legislative and judicial branches. Since the president is head of state, head of government, and commander-in-chief of the armed forces, and since his party, the PRI, has enjoyed a clear majority in Congress since it was created, it has been said that the only limit placed on the power of a president of Mexico is that of time—six years in office. With the approval of Congress, the president may intervene in the states, restricting their independence; also, under congressional authorization, he has certain legislative authority, especially in the regulation and development of commerce and industry. There is no vice-president. If the president dies or is removed from office, Congress is constitutionally empowered to elect a provisional president.

The bicameral Congress, also elected by direct universal suffrage, is composed of a Senate (Cámara de Senadores), made up of 64 members (2 from each state and 2 from the Federal District), and a Chamber of Deputies (Cámara de Diputados) made up of 500 members; 1 deputy (and 1 alternate) represents each 250,000 people or fraction above 100,000, with a minimum of 2 deputies from each state, and 200 members are allocated by proportional representation from minority parties. Senators are elected for six years (half the Senate is elected every three years) and deputies for three years, and both groups are ineligible for immediate reelection. Congress meets from September through December; for the other eight months of the year there is a permanent committee consisting of 14 senators and 15 deputies. The Congress may legislate on all matters pertaining to the national government and the Federal District.

In an effort to unite various interest groups within the government party, a National Consultative Committee, composed of living ex-presidents of Mexico, was formed in 1961 by President Adolfo López Mateos (1958–64). In the final 1962 congressional session, legislation was passed to reform the 1954 electoral law. The bill was designed to give minority parties a greater chance for representation in the Chamber of Deputies by allowing them 5 seats if they received 2.5% of the total vote and 1 additional seat for each 0.5% of the vote beyond that. Under this measure, a small party could obtain up to 20 seats without winning in a single district. A new electoral reform, introduced in 1977, stipulated that the minimum number of members must be 65,000 for a party to be registered and that the party must receive 1.5% of the popular vote to have its registration confirmed. In 1986, Congress stipulated that 200 deputies would be elected by a system of proportional representation within multimember constituencies, while 300 deputies would be elected by majority vote within single-member electoral districts.

14POLITICAL PARTIES

Since 1929, the majority party and the only political group to gain national significance has been the Institutional Revolutionary Party (Partido Revolucionario Institucional—PRI), formerly called the National Revolutionary Party (Partido Nacional Revolucionario) and the Party of the Mexican Revolution (Partido de la Revolución Mexicana). The PRI includes only civilians and embraces all shades of political opinion. Three large pressure groups operate within the PRI: labor, the peasantry, and the "popular" sector (such as bureaucrats, teachers, and small business people). The PRI has faced opposition from the extreme right, represented by the Sinarquistas, and from the extreme left.

Registered opposition parties include the National Action Party (Partido de Acción Nacional—PAN), a democratic center party dating from 1939; the left-wing Popular Socialist Party (Partido Popular Socialista—PPS); the conservative Mexican Democratic Party (Partido Demócrata Mexicano—PDM); the Workers' Revolutionary Party (Partido Revolucionario de los Trabajadores—PRT); the Socialist Workers' Party (Partido Socialista de los Trabajadores—PST); and a leftist coalition, the Mexican Socialist Party (Partido Mexicano Socialista—PMS), which in July 1986 absorbed the Unified Socialist Party of Mexico (Partido Socialista Unificado de Mexico—PSUM) and the Mexican Workers' Party (Partido Mexicano de los Trabajadores—PMT). The Authentic Party of the Mexican Revolution (Partido Auténtico de la Revolución Mexicana—PARM), which was founded in 1954, generally supports the PRI.

After the 1988 elections, 6 opposition parties won 227 of the 500 seats in the Chamber of Deputies and 4 of the 64 seats in the Senate. The PAN won the governorship of Baja California Norte in 1988.

15LOCAL GOVERNMENT

Twenty-nine states of Mexico were created as administrative divisions by the constitution of 1917, which grants them those powers not expressly vested in the federal government; Mexico's two remaining territories, Baja California Sur and Quintana Roo, achieved statehood on 9 October 1974, raising the total to 31. Each state has a constitution, a governor elected for six years, and a unicameral legislature, with representatives elected by district vote in proportion to population. An ordinary session of the legislature is held annually, and extraordinary sessions may be called by the governor or the permanent committee. Bills may be introduced by legislators, by the governor, by the state supreme court, and by municipalities (a unit comparable to a US county). The president appoints the governor of the Federal District as a member of the cabinet. Many state services are supported by federal subsidies.

The principal unit of state government is the municipality. Mexico's 2,359 municipalities are governed by municipal presidents and municipal councils. State governors generally select the nominees for the municipal elections. Municipal budgets are approved by the respective state governors.

16JUDICIAL SYSTEM

Mexico's judiciary, both federal and state, has been a separate branch of government since independence. Federal courts include the Supreme Court, with 21 magistrates; 18 circuit collegiate tribunals, with 3 judges each; 9 circuit unitary tribunals, with 1 magistrate each; and 68 district courts, with 1 judge each (as of 1987). Special courts include a fiscal tribunal and boards of conciliation and arbitration.

Supreme Court magistrates are appointed for life by the president, with the approval of the Senate, and can be removed only by a guilty verdict after impeachment. The other justices are appointed for six years by the Supreme Court magistrates. The Supreme Court has both original and appellate jurisdiction in four divisions: administrative, civil, labor, and penal. Circuit courts hear appeals from the district courts.

The jury system is not commonly used in Mexico, but judicial protection is provided by the Writ of Amparo, which allows a person convicted in the court of a local judge to appeal to a federal judge. Capital punishment, except in the army for crimes against national security, was abolished by the penal code of 1 January 1930.

Although the judiciary is constitutionally independent and judges are appointed for life (unless dismissed for cause), there have been charges that judges are sometimes partial to the executive. Low pay and high caseloads contribute to a susceptibility to corruption in the judicial system. In unprecedented moves in 1993, the government issued an arrest warrant for obstructing justice and for bribery against a former Supreme Court Justice and three federal judges were dismissed for obstructing justice.

17ARMED FORCES

Total full-time strength of the armed forces was 175,000 in 1993. The army had 30,000 personnel. Regular army units included the presidential guard (a mechanized brigade group), 3 infantry brigades, 1 armored brigade, an airborne brigade, and support units. In regional garrisons, around 100 combat battalions (infantry and mechanized) and 3 artillery regiments were deployed. The navy, including naval air force and marines, had 37,000 personnel; naval vessels included 3 destroyers, and 94 patrol and coastal combatants. The air force had 8,000 personnel and 113 combat aircraft. Paramilitary forces included 14,000 rural defense militia; military reserves number 300,000. Defense expenditures in 1989 amounted to $1.5 billion or about 1 percent of gross domestic product. Military sales to Mexico from 1981 to 1991 totaled $620 million, most from the U.S.

Under the required military training program, all 18-year-old males must complete one year of part-time basic army training. In 1993 60,000 conscripts were on active duty, drawn by lottery.

18INTERNATIONAL COOPERATION

Mexico is a charter member of the UN, having joined on 7 November 1945, and participates in ECLAC and all the nonregional specialized agencies. Mexico was host to the Inter-American Conference on War and Peace in 1945, which established the Latin American basis of the UN Conference. Luis Padilla Nervo of Mexico served as president of the UN General Assembly's sixth regular session (1951–52), and Jaime Torres Bodet was first director-general of UNESCO. In May 1972, at the UNCTAD sessions in Santiago, Chile, President Luis Echeverría Álvarez proposed the Charter of Economic Rights and Duties of States, defining the relationship between developed and developing nations. The charter was approved by the UN General Assembly in 1974. The UN International Conference on Women was held in Mexico City in mid-1975. During the de la Madrid administration, Mexico joined the Contadora Group (with Colombia, Panama, and Venezuela) in an effort to bring about a negotiated settlement to the conflict in Central America. Mexico also joined the Five Continents Initiative for Peace and Disarmament, to promote a nuclear test ban and stop the spread of the arms race into outer space.

Mexico, which for many years was the only Latin American nation to recognize Fidel Castro's Cuba, has based its foreign policy on the principles of nonintervention and self-determination of peoples, which it helped incorporate into the charter of the OAS in 1948 (Articles 15–17). Mexico was a leading proponent of the

OAS resolution in July 1975 that gave member nations the right to exercise freedom of action in restoring diplomatic and commercial relations with Cuba. During President Echeverría's visit to Cuba in August 1975, the two nations signed an economic cooperation agreement.

Mexico belongs to such inter-American organizations as the IDB, PAHO, and LAIA, in addition to the OAS. As a member of G-77, Mexico has been prominent among nonaligned nations. Particularly during the presidency of José López Portillo y Pacheco (1977–82), Mexico assumed a leading role among the third-world countries; in October 1981, it hosted "North-South" talks among the leaders of 8 industrialized and 14 developing countries. Mexico is also a signatory of the Law of the Sea and a member of GATT.

19ECONOMY

Although Mexico's economy once was predominantly agricultural (with more than 55% of its economically active population employed in primary production), in terms of value, commerce and industry have long been the chief forms of production. Mexico is self-sufficient in most fruits and vegetables and in beans, rice, and sugar, and it is approaching self-sufficiency in meat and dairy products; marginal subsistence, however, is still the lot of much of Mexico's rural population.

Since 1960 there has been a gradual improvement in the social and economic integration of Mexico, but because of the rapid rate of population increase, the marginal sector is still very large. While peasant wages remained static during the 1960s, industrial wages increased more than 80%, which is part of the reason for the large-scale migration from countryside to city. Mexico City's enormous population growth has been accompanied by mass poverty.

In the early 1980s, Mexico's chief exports were petroleum, cotton, sugar, tomatoes, and coffee. A great mining nation, Mexico is the world's leading producer of silver and is well endowed with sulfur, copper, manganese, iron ore, lead, and zinc. The oil discoveries of the early 1970s led to a boom, and in the second half of the 1970s, Mexico's economy was one of the fastest growing in the world. During 1978–81, the GDP increased by 8% annually and the government embarked on an ambitious public spending program, to a great degree financed by external borrowing. About three-quarters of all export earnings were brought in by crude oil. This dependency contributed to the financial crisis of 1982, when falling world oil prices, together with rising interest rates and the overvaluation of the peso, resulted in a surging foreign debt and, by August, an acute foreign exchange shortage. The annual inflation rate, which had hovered around 30% during 1979–81, reached almost 100% in 1982. The government embarked on a stabilization program in early 1983, after reaching an agreement with the IMF that would provide $3.84 billion in credits over a three-year period. The immediate objectives were to reduce inflation and strengthen public finances and the balance of payments, laying the groundwork for a structural transformation of the economy in the medium term. A loan agreement with some 530 commercial banks, announced in February 1983, offered another $5 billion in credits to enable Mexico to meet its short-term obligations. In March 1987, Mexico signed a commercial debt rescheduling agreement that extended the repayment period and reduced the interest rate on $43.7 billion of the $104.5 billion foreign debt; the agreement also provided for $6 billion in new credits payable in 1987/88. Inflation remained high in 1987, reaching 135% by September.

In December 1987, the Pact for Stability and Economic Growth (PECE), a series of price and wage restraint agreements, was implemented. The agreements between government, labor, and the private sector combined austere fiscal and monetary restraints with price/wage controls and freer trade possibilities.

The PECE helped curb inflation to 51.6% in 1988 without incurring a recession. Gradual recovery has seen the inflation rate fall to 20% in 1991, 11.9% in 1992, and to around 10% in 1993.

Debt service restructuring, public sector reform, and privatization enacted in the early 1990s have increased the amount of savings available to finance private sector investment and have freed the fiscal resources needed for social programs. Foreign investment legislation passed in June 1991 aims to protect intellectual property.

The North American Free Trade Agreement (NAFTA), in effect as of 1 January 1994, seeks to open the domestic market to foreign trade by eliminating trade barriers between Mexico, the US, and Canada over the next 15 years.

20INCOME

In 1992, Mexico's GNP was $294,831 million at current prices, or $3,470 per capita. For the period 1985–92 the average inflation rate was 52.7%, resulting in a real growth rate in per capita GNP of 1.1%.

In 1992 the GDP was $329,011 million in current US dollars. It is estimated that in 1991 agriculture, hunting, forestry, and fishing contributed 8% to GDP; mining and quarrying, 2%; manufacturing, 22%; electricity, gas, and water, 2%; construction, 4%; wholesale and retail trade, 25%; transport, storage, and communication, 9%; finance, insurance, real estate, and business services, 13%; community, social, and personal services, 10%; and other sources, 6%.

21LABOR

Mexico's civilian labor force of 24,063,283 in 1990 was distributed as follows: government and services, 24.4%; agriculture, 25.7%; commerce, 16.1%; industry, 18.7%; building, 6.6%; transportation and communications, 4.3%; mining, 1.1%; and other sectors, 6.8%. Underemployment, Mexico's major labor problem, affected about 12% of the labor force in 1990, mainly those engaged in agriculture. According to official figures, unemployment was 13.4% in 1985; that percentage was reported as 2.7% in 1991, but was indicative of only the largest metropolitan areas. Rural unemployment was believed to be significantly higher. There is no general unemployment compensation system.

The constitution of 1917 provides that all labor disputes be submitted to boards of conciliation and arbitration consisting of an equal number of workers and employers and one government representative. Under the federal labor law, completely revised in 1970 and subsequently amended, every employee is entitled to 1 paid day of rest after every 6 days of work, 7 paid holidays, and at least 6 days of vacation after a year of employment and at least 8 days after two years. An annual bonus equal to 15 days' pay is required to be paid to all employees before Christmas, and vacation pay carries a 25% premium. Labor by women and children is limited. At least 90% of the employees of most establishments must be Mexicans. The workday is generally eight hours, and double or triple pay must be paid for overtime.

Prior to 1970, minimum wages were fixed in each municipality by commissions representing labor, management, and government. General minimum wages came into effect in 1970; by 1981, differential minimum wages were established in 89 separate sections of the country. Since January 1976, these minimums have been periodically revised. Since 1963, nearly every enterprise in Mexico has been required to distribute a percentage of annual profits to its employees. Under the revised regulations effective 15 October 1974, workers are entitled to about 10% of employers' taxable income, subject to certain exclusions and limitations.

About 25-30% of the labor force was unionized in 1991. The Confederation of Mexican Workers (CTM, founded in 1936) had about 5 million members in 1992; its secretary-general, Fidel Velázquez, played a leading role in Mexican politics from the

1940s through the 1970s. The second largest trade unions confederation is the Federation of Government Employee Unions (FSTSE), with an estimated membership between 1.5 and 1.8 million, consisting exclusively of federal workers. The Revolutionary Workers and Peasants Confederation (CROC) is considered to be the rival of the FSTSE, a rivalry encouraged by the government in order to discourage a monopoly of power by the FSTSE. The CROC had an estimated 600,000 members in 1992.

22AGRICULTURE

Agriculture's contribution to GDP fell from 15.1% in 1960 to 10.7% in 1970 and 8.4% in 1992—yet agriculture employs about 25% of the labor force. Only about 13% of Mexico's total land area is suitable for cultivation, and only 6% is cultivated with permanent crops; over 5.2 million hectares (12.8 million acres) are irrigated.

In 1960, Mexico became self-sufficient for the first time in corn; it continues to be self-sufficient in beans, rice, sugar, and most fruits and vegetables and fluctuates between being either a net importer or self-sufficient with wheat and corn. During the late 1960s, Mexico almost tripled the investment allocated for agriculture. The government continues to protect agriculture and to ensure domestic consumption through import and export duties and controls. The government supports the prices of corn, wheat, beans, and fresh eggs and then sells these and other farm products at minimal prices through retail stores operated by the National Corporation for Public Subsidies. In 1977, the government introduced a new program called the Mexican Alimentary System, seeking to foster food production through subsidized credit, cheap raw materials, guaranteed prices, and crop insurance. Because of a record harvest in 1981, the agricultural sector grew by 6% overall. The program was terminated in December 1982 because of what the government called its limited contribution to rural welfare and incomes.

In 1992, the principal crops' production totals (in tons) were as follows: sugarcane, 39,955,000; corn, 14,997,000; sorghum, 5,106,000; wheat, 3,602,000; dry beans, 804,000; soybeans, 670,000; rice, 361,000; and barley, 572,000. Principal exports are coffee, cotton, fresh fruit, sugar, tobacco, and tomatoes. In 1992, the value of agricultural exports amounted to $2.88 billion.

It has been estimated that at the time of the Mexican Revolution, only 830 landowners held 97% of the land. The principles of land reform were incorporated into Article 27 of the constitution of 1917, which provided for division of large landholdings into small farms, communally owned by villages, known as ejidos (one individual may only own 100 hectares/247 acres of irrigated land). Much of the arable land has been expropriated for the establishment of ejidos. The Agrarian Reform Law of April 1972 formally recognizes two types of landholdings, private and ejidal (in the form of lifetime land grants, which cannot be disposed of by sale or transfer). The proportion of ejidal-owned land rose from 7.5% in 1930 to 26.3% in 1960 and 47% in 1970. By 1986, 61.1% of the farm population belonged to the ejidal system, but the system yielded only about 33% of total agricultural output, and an estimated 4 million peasants remained landless. Beginning in the late 1970s, the government sought to group together ejidal holdings into larger collectives to increase production.

23ANIMAL HUSBANDRY

More than one-third of the total land area is suitable for pasture. Livestock produced in the central, southern, and southeastern states are mostly native breeds of general-purpose cattle; in the northern and northwestern states, Herefords and other improved breeds are raised for export, mainly to the US.

In 1992, the livestock population was estimated at 30.1 million head of cattle, 16.5 million hogs, 11 million goats, 6.1 million sheep, 6.1 million horses, 3.2 million donkeys, 3.2 million mules,

and 282 million chickens. Output of livestock products in 1992 included 7,204,000 tons of cows' milk, 148,000 tons of goats' milk, 1,660,000 tons of beef and veal, 932,000 tons of poultry meat, 832,000 tons of pork, and 1,160,648 tons of eggs. In 1992, the livestock subsector grew by 3.5% as a result of higher milk and poultry production. For a third successive year, milk production increased (by 8.7%), due to improved marketing systems, deregulated farm prices, and the Milk Production Development Program. Poultry farming expanded its capacity, enabling a 5.4% rise in slaughtering. Beef production, however, fell by 5.1% with beef exports plummeting 15%.

24FISHING

Mexico's principal commercial catches are shrimp, sardines, bass, pike, abalone, Spanish mackerel, and red snapper. Coastal fishing is important. The 1991 catch was 1,429,137 tons. Among the shellfish caught were shrimp, 70,580 tons; crabs, 11.248 tons; oysters, 34,604 tons; clams, 31,226 tons; and squid, 6,491 tons. The main fish caught were sardines, 467,740 tons; tuna, 116,387 tons; freshwater cichlids, 75,174; and shark, 18,300 tons.

The fishing industry is largely handled by cooperative societies, which are granted monopolies on the most valuable species of fish. Most fish processed in Mexico's canneries are consumed domestically. Ensenada, in Baja California, is Mexico's most important fisheries center. It produces most of the canned fish and virtually all of Mexico's abalone and spiny lobster exports. Mexico's first fisheries college, the Higher Institute of Marine Sciences, is located in Ensenada.

The fishing industry grew by 1.4% in 1992. Although tuna, shrimp, and sardine catches were down (1.2%, 0.5%, and 10% respectively), there were large increases in the red snapper, shark, grouper, and blue crab catches. The fishing industry had a trade surplus of $250 million in 1992.

25FORESTRY

About 41.9 million hectares (103.5 million acres) are classified as forestland. Palms are found at elevations up to 500 m (1,600 ft), while mahogany, cedar, primavera, and sapote are found from 500 to 1,000 m (1,600–3,300 ft). Stands of oak, copal, and pine grow from 1,000 to 1,500 m (3,300–4,900 ft), and conifers predominate in higher elevations.

Mexico's forestry policy is designed to protect and renew these resources, so that forests may fulfill their soil-protection functions and timber reserves may be exploited rationally and productively. Only about 30% of all forests are exploited, mostly in Chihuahua, Durango, and Michoacán. Roundwood production in 1991 was 23,617,000 cu m. In 1992, forestry activity fell by 2.4%, with production of plywood and poles especially down; only the output of sawnwood increased. Mexico's ability to supply its own wood products needs are severely restricted by the limited timber available in Mexico.

26MINING

In 1991, Mexico was the world's leading producer of silver and celestite (strontium mineral), and one of the five leading producers of mercury, antimony, white arsenic, graphite, bismuth, cadmium and fluorospar. Metallic deposits are principally in the Sierra Madre ranges. Copper, gold, and manganese are mined mainly in the northwest; lead, zinc, and silver in central Mexico; and coal and petroleum in the east. Silver and gold are often found together. Output of principal minerals in 1991 included silver (metal content of ore), 2,223,647 kg; gold, 8,937 kg; iron ore, 7,539,000 tons; sulfur, 2,094,000 tons; cadmium, 1,797 tons; zinc, 300,706 tons; barite, 203,975 tons; lead, 223,186 tons; and copper, 284,174 tons.

Mexico ranked 2d in reserves of graphite and silver, with about 15% (3.1 million tons) and 13% (37,000 tons) of total world reserves, respectively. The country was among the top five reserveholders of cadmium, with 5% of world reserves (35,000 tons); mercury, 4% (5,000 tons); and selenium, 5% (4,000 tons). As for lead reserves (3 million tons) and zinc reserves (6 million tons), Mexico ranked 6th each, with 4% each of world reserves. It also ranked 7th in molybdenum reserves (90,000 tons—1.6% of world reserves), and 8th for copper and manganese (14 million and 3.6 million tons, respectively, each accounting for less than 0.5% of world reserves). Reserves of coal, mined principally in Coahuila, were estimated at 5,985 million tons in 1991. Sulfur is found in the salt domes of the Isthmus of Tehuantepec; reserves of all forms amounted to 75 million tons in 1991. Uranium and similar deposits were discovered in the states of Chihuahua, Durango, Sonora, and Querétaro in 1958. In 1969, the country's first uranium ore smelting plant was installed at Aldama, Chihuahua, with a processing capacity of 60 tons per day. By 1990, however, Mexico was importing uranium from Germany and France for its domestic needs. An area in Baja California potentially rich in gold, silver, copper, lead, and zinc was discovered by a sensor on board the US space shuttle *Columbia* in November 1981.

Many of the mineral producing companies are being privatized, and although recent low prices for base metals (except copper) and precious metals have hurt the industry, the North American Free Trade Agreement (NAFTA) should have a significant role in attracting foreign investment into the Mexican mineral sector.

27ENERGY AND POWER

Nationalization of the electricity supply industry, begun in 1960, was completed in 1975. Net installed generating capacity as of December 1991 was 29,274,000 kw, compared with 27,338,000 kw in 1988. Public power accounted for 88% of installed capacity in 1991. The amount of electricity produced in 1991 was 126,375 million kwh, of which 19.2% was hydroelectric and 2.4% was nuclear power. The possibilities for geothermal electrical production are extensive, with over 100 thermal springs available for exploitation. Geothermal sources provided 700 Mw of electricity in 1991. Petroleum accounts for about 70% of Mexico's energy consumption.

Oil seeps were first documented in Mexico in 1543; the first oil wells were drilled in 1869. The first major discovery was the La Paz No. 1 well in the Ebano-Panuco Field in 1904. The petroleum industry was nationalized in 1938 and has since been operated by a government-owned institution, Mexican Petroleum (Petróleos Mexicanos—PEMEX). In 1961, the final compensation payment of the $18 million owed to a British concern (Mexican Eagle Oil Co.) ended Mexico's expropriation debt. The Petroleum Law of 1958 permitted PEMEX to take over all concessions held by private firms and granted it a monopoly in the petroleum industry. The company was a special target of the de la Madrid administration's crackdown on corruption in 1983. In 1991, PEMEX was the largest civilian employer in the country, with 145,000 employees. PEMEX contributes about 40% to federal tax revenues through income and sales taxes and provides from one-third to over 40% of total national export revenues. PEMEX is the world's fifth largest oil company. Private ownership of secondary petrochemical production was proposed by the government in 1991, but no further action had been taken as of June 1993.

Although the government restricts exploration and exploitation of petroleum deposits, some foreign and private exploratory drilling has been permitted. Producing fields are along the Gulf of Mexico coastline from the state of Tamaulipas to the state of Tabasco, with the richest concentrations around Tampico and in Veracruz and Tabasco. In 1974, extensive new discoveries were made in the Chiapas-Tabasco region and in Baja California. A

1975 discovery in the state of Veracruz was estimated to have reserves exceeding those of Chiapas and Tabasco combined. Estimated oil reserves as of the end of 1992 were 51.3 billion barrels, second in the Western Hemisphere after Venezuela.

Crude oil production first reached 1 million barrels per year in 1907. Mexico's production rose rapidly during the 1970s, peaking at 3,015,000 barrels per day in 1985. It stayed at about 2.8 million barrels per day through 1989 and then rose to 3,155,000 barrels per day in 1992. Mexico exports about half the oil it produces, mostly crude oil to the U.S., Spain, and the Far East. In 1992, petroleum export revenues amounted to $7.5 billion, or 30% of total exports. The majority of oil production comes from the Villahermosa District in the State of Tabasco and the offshore wells of the Bay of Campeche. Average production per well is about 220 barrels per day, except for wells in the Bay of Campeche, which produced about 16,000 barrels daily in 1989. PEMEX operated more than 3,700 wells in over 100 fields in 1989. The same two regions dominate in the production of natural gas; in 1992, Mexico produced 28.2 billion cu m of natural gas (11.1 billion cu m from offshore sources), more than any other country in Latin America. Proven reserves of natural gas were estimated at 22.7 trillion cu ft (600 billion cu m) in 1992.

28INDUSTRY

Mexico has one of the best-developed manufacturing sectors in Latin America. The largest employer (and largest industry, by revenue) is petroleum and petroleum refining, followed by iron and steel production, auto assembly, brewing, and the manufacture of paper and paper products. Between 1960 and 1971, total industrial output grew at a rate of 8.3% annually; from 1977 through 1981, the average annual growth rate was 8%. During the 1982–83 recession, industrial output declined by over 9%; it rose by over 5% in 1984, but declined by 5.6% in 1986 as a result of the steep drop in world oil prices. Since 1987, industrial growth has been positive, climbing by 2.4% in 1988 to 4.7% in 1990 before dropping to just 2.8% in 1992.

Industry accounts for about 23% of GDP and 19% of employment. The principal manufacturing industries include automobile and related parts production, steel, textiles, cement and related construction materials, chemicals and petrochemicals, paper and paper products, food processing, breweries, and glass. Manufacturing output increased by only 1.8% in 1992 (compared to 6.1% in 1990). Manufacturing sectors that grew the most in 1992 included non-metallic industries (7%), other manufacturing industries (13.1%), and the automobile industry (12%). Manufacturing has been around the re-export processing industry. These maquiladoras are usually located near the US border and owned by a foreign corporation; they contract to assemble or process imported goods brought in from the US and then re-export them duty free.

In 1992, there were some 2,042 maquiladora factories employing 494,721 workers. Baja California Norte had the most in-bond export industries in 1992, at 759; other leading states included Chihuahua, with 346; Tamaulipas, 277; Sonora, 1679; Coahuila, 166; Nuevo Leon, 83; Durango, 55; Jalisco, 45; and Yucatan, 24.

In the 1980s, the Mexican government retreated from its position of dominance over the economy, and the private sector was given an expanded role. Between 1982 and 1992, the number of state-owned enterprises was reduced from 1,155 to 223, 87 of which were in the process of being privatized. The National Financing Agency (Nafinsa), the Mexican development bank, has helped finance new industries.

29SCIENCE AND TECHNOLOGY

The National Scientific Research Academy, founded in 1884, is the principal scientific organization. Among Mexico's 39 scientific and technological learned societies and 28 scientific research institutes, the natural sciences and medicine predominate. Especially well known is the International Maize and Wheat Improvement Center, founded in Mexico City in 1964; its director, Norman Ernest Borlaug, received the Nobel Peace Prize in 1970 for his work in advancing the "green revolution."

The primary science and technology policymaking body is the National Council of Science and Technology, a decentralized public body composed of researchers, scientists, academicians, and government officials. They formulate, study, evaluate, and execute national science and technology policies. In 1989, the Consultative Council on Sciences was created to directly advise the President of Mexico on Science and Technology. In 1989, Mexico as a nation spent P1.1 billion on research and development, with 85% funded by the government.

30 DOMESTIC TRADE

Mexico City is the commercial hub of the country and is the principal distribution point for all types of commodities. Other large cities, such as Guadalajara, Monterrey, and Puebla, serve as distribution points for their respective regions. Regional marketing is dominated by the open market, with its small stalls or shops, where business is transacted on an individual bargaining basis. There are also chain stores, supermarkets, department stores (some selling by mail), and a government-operated chain of more than 2,000 discount-priced food and clothing stores.

Although most sales are for cash, the use of consumer credit is increasingly extensive, especially for automobiles, furniture, household appliances, and other expensive items. Installment sales are common. Finance companies formerly did most of the lending for installment purchases; in 1958, commercial banks began to make personal loans available to patrons desiring to purchase household goods.

Normal business hours are from 8:30 or 9 AM to 6 PM Monday through Friday, with one or two hours for lunch. Banks are open from 9 AM to 1:30 PM. Products are advertised through newspapers, radio, television, outdoor signs, and motion picture shorts and slides.

31 FOREIGN TRADE

Mexico's trade balance regularly showed a large deficit until 1982; the deficit worsened steadily through the 1970s, reaching $3.2 billion in 1979 and $4.4 billion in 1981. The surpluses of the early 1980s were mainly the result of higher exports of oil and a drastic curtailment of imports because of a foreign exchange shortage. In 1985, however, as the world oil price fell, the trade surplus dropped to $8.14 billion; it was down to $4.4 billion in 1986. Since 1989, the trade balance has been negative, increasing from a deficit of $2.6 billion in 1989 to $11.1 billion in 1991 and $20.6 billion in 1992. The surge in oil prices caused by the Iraqi invasion of Kuwait in August 1990 caused Mexico's oil export earnings to increase by 23%.

Over the years, the composition of Mexico's foreign trade has undergone some basic changes. Whereas exports formerly consisted overwhelmingly of primary agricultural products, the relative importance of manufactured exports rose during the 1960s and the early 1970s. In 1968, exports of agricultural, animal, and fishery products accounted for 50.5% of all exports, while manufactured goods accounted for 30.5%; by 1972, the shares were 43.1% and 45.6%, respectively. The trade pattern changed still further after 1974, when exports of crude oil began. In 1981, crude oil exports, valued at $13.3 billion, represented 68.7% of the total export value. In 1986, however, the lower world price brought that proportion down to 34%.

In 1992, the value of goods exported totaled $46,196 million, up 8.2% from 1991. Of 1992 exports, agricultural products accounted for 4.6% of the total value; oil, 18%; nonfuel miner-

als, 0.8%; and non-oil manufactures, 76.6%. This total includes the value of imports and exports of the in-bond *maquiladora* industry. Merchandise exports in 1992 consisted mostly of crude oil, chemicals, petrochemicals, automobiles, machinery and equipment, iron and steel products, electrical and electronic goods, textiles and clothing, coffee, oil derivatives, nonmetallic minerals, paper, and printing and publishing materials. Mexico formerly imported mainly consumer goods, but in recent years consumer goods have accounted only for just over 10% of imports (12.5% in 1992). The main bulk of imports are capital goods and industrial raw materials.

Merchandise imports in 1992 amounted to $62,129 million, up 24.3% from 1991. Some $13.937 million (22%) of those imports went to the in-bond industry for re-export and therefore did not represent a net use of foreign currency. Imports consisted primarily of machinery and transportation equipment, chemicals, electrical and electronic goods, processed foods, beverages and tobacco, iron and steel products, paper printing and publishing, textiles and clothing, oil derivatives, plastics, petrochemicals, maize, and soybeans. As of 1993, import licensing was required for importing motor vehicles, firearms and explosives, certain pharmaceuticals, certain farm commodities and equipment, shellfish, and poultry meat.

Of Mexico's exports of $22,639.4 million during the first ten months of 1991, 69.53% went to the US; 12.37% to the EC, 5.94% to Asia, 3.59% to the Latin American Integration Association, 2.22% to Canada, 1.44% to the Central American Common Market, 1.27% to the remainder of Central America, 1.2% to the rest of the Americas and the Caribbean, 0.75% to the EFTA, 0.81% to the Middle East, and the remaining 0.88% to Eastern Europe, Africa, Oceania, and elsewhere. Mexico has attempted with some success to diversify its markets; however, the US still accounted for 71.7% of imports and 81% of exports in 1992.

32 BALANCE OF PAYMENTS

Mexico's balance of payments, in deficit throughout much of the 1960s, turned favorable in the 1970s, as inflows of foreign funds rose fast enough to offset the worsening visible trade balance. In the past, receipts from tourism and border trade had contributed heavily to international receipts; in the 1970s, however, that burden was borne by long-term capital investments and long-term loans. Mexico's external position had so deteriorated by 1986 that the IMF and World Bank coordinated a financial rescue package of $12 billion and commercial banks agreed to reschedule $43.7 billion in foreign debt; an innovative feature of the agreement is that repayment was effectively tied to Mexico's oil export earnings.

The inflow of capital from abroad recently has brought about a considerable increase in the capacity to impact and a widening of the current account to 4.8% of GDP in 1991 to 6.9% of GDP in 1992. Part of this increased import capacity has caused a sizeable growth of purchases of investment goods abroad. As investments mature, however, a reversal is expected to occur and lead to a greater supply of both exportable goods and of goods that are currently imported.

In 1992 merchandise exports totaled $27,516 million and imports $48,193 million. The merchandise trade balance was $–20,677 million.

The following table summarizes Mexico's balance of payments for 1991 and 1992 (in millions of US dollars):

	1991	1992
CURRENT ACCOUNT		
Goods, services, and income	–15,971	–25,194
Unrequited transfers	2,189	2,383
TOTALS	–13,785	–22,811

	1991	1992
CAPITAL ACCOUNT		
Direct investment	4,742	5,366
Portfolio investment	9,267	14,095
Other long-term capital	6,919	723
Other short-term capital	3,410	6,232
Exceptional financing	20	—
Other liabilities	—	—
Reserves	–7,992	–1,746
TOTALS	16,366	24,670
Errors and omissions	–2,581	–1,859
Total change in reserves	–7,619	–1,935

33 BANKING AND SECURITIES

The Bank of Mexico (established 1925), in which the government owns 51% of the capital stock, is also the central bank and bank of issue. Together with the National Banking and Insurance Commission and the National Securities Commission, it supervises commercial, savings, trust, mortgage, capitalization, and investment institutions. National institutions for economic development extend agricultural and long-term industrial credit and finance and develop public works, international trade, cooperatives, and the motion picture industry; they also operate savings accounts. The National Financing Agency (founded in 1934) acts as a financing and investing corporation; it also regulates the Mexican stock market and long-term credits. As of 1993, the money supply was P349,383 million.

In September 1982, in order to stop the flight of capital, the government nationalized all 57 private banks; their combined assets were estimated at $48.7 billion. After the inauguration of President de la Madrid in December 1982, it was announced that 34% of the shares of the nationalized banks would be sold to bank workers and users and to federal, state, and municipal agencies. No single shareholder would be allowed to purchase more than 1% of the stock, and the federal government would retain a 66% controlling interest. The government had consolidated the commercial banking system into 19 financial institutions by the end of 1986. In November 1986 the government introduced a plan that would privatize 18 of Mexico's 19 state owned commercial banks. The sale of the banks began in 1987. In 1990 the government began allowing foreigners to buy up to 30% of the state's banks. By July 1992 the banking system was completely private. The only foreign bank permitted to operate within Mexico as of 1993 was Citibank; another 100 foreign banks had representatives in Mexico, however.

The National Securities Commission (founded in 1946) supervises stock transactions. The Stock Exchange of Mexico, the largest stock exchange in Latin America, was organized in its present form in 1933. It lists the stocks of the most important industrial companies, as well as a few mining stocks. Two smaller exchanges at Monterrey and Guadalajara were absorbed in 1976 by the Mexico City exchange. Trading on the exchange increased tenfold between 1976 and 1981, but dropped thereafter with the prolonged recession. It recovered to its 1979 level by 1986 and rose 124% in 1987 despite a spectacular crash in October and November of that year tied to the Wall Street's crash. The greatest part of the trading is in fixed-interest, high-yield bonds and bank deposit paper.

34 INSURANCE

Since 1935, all life, fire, marine, automobile, agriculture, accident and health, and other insurance companies have been Mexican operated. Insurance companies must be authorized by the National Banking and Insurance Commission. In 1986 there were 33 insurance companies in Mexico. In 1987 total premiums amounted to US$901 million or 1.02% of GDP, of which US$3.6 per capita were paid for life and US$7.5 for non-life insurance. Life insurance in force in 1992 totaled P538,707 billion.

35 PUBLIC FINANCE

Major sources of revenue are income taxes, a value-added tax, and public enterprise revenues. Among regular government departments, education receives the largest budget allocation, but outlays for debt service, subsidies to federal enterprises, and capital expenditures for highways, irrigation, and hydroelectric projects have exceeded regular departmental expenditures in recent years. The public-sector deficit usually increases sharply in the last year of a presidential term as the outgoing administration strives to complete its public works program.

During the 1960s, government revenues rose at a faster rate than GDP, with revenues from income taxes (including surcharges) increasing by 170% in the 1960–69 period. Budgets in the 1970s and early and mid-1980s continued to show current-account "surpluses," or minimal apparent deficits; the fact that borrowings and transfers are built into the budget structure masked the true magnitude of annual deficits. In the late 1970s and the early 1980s, real budget deficits increased substantially, reaching nearly 18% of the GDP by 1982. By slashing public spending, the government was able to bring the deficit down to 8.9% of the GDP in 1983 and 7.1% in 1984, but the collapse of the world oil price sent it up to 16.3% in 1986. By the early 1990s, however, public finances were strengthening, and a surplus (on a cash basis) of 35,054 million new pesos was recorded in 1992, equivalent to 3.4% of GDP in 1991. Non-recurrent revenues amounted to 30,123 million new pesos, mainly derived from the sale of shares of the state-owned telephone company and bank privatization. Excluding privatization revenues, the surplus in 1992 was equivalent to 1.5% of GDP in 1992, as opposed to a deficit of 1.5% in 1991. Public revenues policy in 1992 sought to widen the tax base and simplify and enforce tax administration. At the same time, public expenditures have been reoriented to provide basic infrastructure and services.

From 1988 to 1991, public expenditure in social development increased by 41.4% in real terms.

The following table shows actual revenues and expenditures for 1989 and 1990 in millions of new pesos.

	1989	1990
REVENUE AND GRANTS		
Tax revenue	77,493	88,965
Non-tax revenue	15,344	7,392
Capital revenue	4	50
Grants	—	—
TOTAL	92,841	96,407
EXPENDITURES & LENDING MINUS REPAYMENTS		
General public service	2,673	3,423
Defense	2,642	2,815
Public order and safety	524	556
Education	13,972	16,529
Health	1,810	2,279
Social security and welfare	11,183	14,776
Housing and community amenities	891	722
Recreation, cultural, and religious affairs	36	346
Economic affairs and services	11,024	15,930
Other expenditures	68,609	63,026
Adjustments	–1,490	–2,296
Lending minus repayments	716	–27,949
TOTAL	119,335	91,303
Deficit/Surplus	–26,494	5,104

The public sector borrowing requirement has fallen from 12.4% of GDP in 1988 to 4% in 1990 and 1.5% in 1991.

Public sector debt as a proportion of GDP has fallen from about 16% in 1986 to less than 2% in 1991.

³⁶TAXATION
The main sources of tax revenue in Mexico are income tax, value-added tax, and local levies on real property. The federal government also imposes excise taxes on alcohol and cigarettes, as well as production taxes on mining.

A new income tax law, effective 1 January 1987, retains the 1981 division of taxpayers into four groups: resident corporations; resident individuals; nonresident corporations and individuals taxed only on their Mexican-source income; and nonprofit organizations, which, though paying no taxes, are still required to file annual returns. Corporate income tax rates range from 17.5% to 35%, depending on the type of business. Under the new code, interest on business-related loans is no longer deductible, but most business investments henceforth may be deducted in one year instead of several. The individual tax rate ranges from 3% of the first P1,251 of taxable income to 35% on incomes in excess of P82,577.

The value-added tax (VAT), in effect since January 1980, is 10% on all types of operations (sales, services, rentals, and importation of goods and services). In 1993, there was a temporary reduced rate of 6% for medicine and food products. VAT-exempt goods and services include sales of animals, vegetables, and fruit for other than industrial use; sales of tractors, fertilizers, and pesticides; rentals of agricultural machinery; international freight; and international air passenger service.

³⁷CUSTOMS AND DUTIES
In February 1960, Mexico joined with six other Latin American nations in signing a treaty establishing LAFTA, which was superseded by LAIA on 1 January 1981.

The Mexican import tariff, primarily protective, provides for compound duties, and its rates apply equally to the products of all countries. Under the terms of a trade policy agreement with the IBRD, tariffs were lowered substantially in 1988. In general, ad valorem duties range from 0% for basic consumption goods and raw materials to 20% for consumer goods and luxury items. The average tariff in 1993 was about 11%, and most goods were subject to a 15% value-added tax.

Since mid-1985, Mexico has undertaken a major liberalization of its trade restrictions, departing from the import-substitution approach that had been followed since the 1940s. In the second half of 1985, most import licenses were abolished, and Mexico joined GATT in July 1986. Until June 1985, a license was required for 4,513 of the 8,077 items on the Import Tariff Schedule. Trade protected in this manner represented 75% of total import value. By 1991, less than 2% of all imports (14% of total import value) were subject to licensing requirements. Businesses may receive reductions of up to 100% on duties for certain industrial imports in Mexico's free trade zones, located along frontier strips and in eight other locations.

Mexico gives preferential treatment to some imports from the ten other member nations of the Latin American Integration Association and has signed a free-trade agreement with Chile. When implemented, the North American Free Trade Agreement (NAFTA) signed in 1992 and approved by all three participating countries in 1993, will gradually eliminate Mexico's tariffs on products from countries meeting NAFTA's rule of origin requirements over a 15-year period, divided into 5-year phases.

³⁸FOREIGN INVESTMENT
Federal law aims at attracting foreign investment without placing the nation "at the mercy of interests that are not those of Mexico and its citizens." Under Mexico's basic foreign investment law, adopted in 1973, capital profits and dividends may be transferred to and from Mexico. Mexico encourages the inflow of foreign funds by allowing foreigners to make bank deposits without revealing the identity of the depositors. Foreign investment in high-technology and export-oriented industries is particularly welcome. In general, foreign investment may not exceed 49% of a company's capital, nor may the foreign investor control the management of the firm. Specific investment limits include a maximum of 40% for secondary petrochemical products, 40% for automobile components, and 34% for mineral exploitation under special concessions. Sectors from which foreigners are excluded are petroleum and hydrocarbons, basic petrochemicals, radioactive minerals and nuclear energy, electricity, telegraphic communications, national highway, rail, air, and sea transportation, and coin minting.

Land and water within 100 km (62 mi) of Mexico's borders or 50 km (31 mi) of the coastline may not be foreign owned. Foreign investment is supervised by the National Foreign Investment Commission.

As of June 1992, accumulated direct foreign investment in Mexico amounted to $36 billion. Of that total, 56.2% was invested in manufacturing, 34.5% in services (including finance, insurance, transportation and communications), 7.5% in commerce (retail, restaurants and hotels), 1.4% in mining, and 0.4% in agriculture and cattle breeding. Investment trends have shifted in recent years with the service sector now the largest recipient of direct foreign investment.

The US is the leading source of direct foreign investment in Mexico, with about 61.7% of the accumulated total as of mid-1992. Other prominent sources of direct foreign investment include the UK (6.5%), Germany (5.8%), Switzerland (4.6%), Japan (4.4%), and France (4.1%).

Foreign portfolio investment (capital inflows that are used to purchase stocks, bonds, and treasury bills) in Mexico has increased dramatically; from $493 million in 1989 to $7,500 million in 1991. Portfolio investment, potentially more volatile than direct investment, has increased in Mexico because of growing confidence since debt renegotiation, increased issues of international securities, and changes in regulations to facilitate foreign purchase of most Mexican debt and equity instruments.

³⁹ECONOMIC DEVELOPMENT
Modern Mexican economic policy derives in principle from the constitution of 1917, which, in Article 27, proclaims national ownership of subsoil rights, provides for expropriation of property needed for national purposes, and provides for the breaking up of large estates and the establishment of village communal landholdings (ejidos). The property of foreign oil companies was expropriated in 1938, and production, refining, and distribution were placed under the government-controlled PEMEX. The government has also nationalized the railway and banking systems, owns most electric power plants, and partly owns some industrial establishments. Majority Mexican ownership was required in virtually all sectors until early 1984, when restrictions on foreign investment were relaxed somewhat.

The government encourages local industry by giving financial support, customs protection, and tax exemption to approved or new enterprises. The National Financing Agency has supported new industries by purchasing their stock and then reselling it to the public when the firm is established. The executive branch of government may set ceiling prices on foodstuffs, drugs, and other basic necessities, such as workers' rents.

Mexico's 1966–70 national development plan called for a total investment of $22 billion, of which $14.4 billion was to come from the private sector for expansion of petroleum and electrical projects and development of industry. In September 1968, it was reported that targets set for the program had been attained and in some instances surpassed. Major development projects in the

1970s included an attempt to increase agricultural productivity, modernization of the nation's railroads, expansion of the fishing fleet, and resettlement of some 50,000 families from the northern states to the southern states Campeche, Yucatán, and Quintana Roo.

When the exploitation of huge oil deposits began in the mid-1970s, the Mexican government embarked on an expansionist economic policy, which included an ambitious public-spending program financed to a great degree by foreign borrowing. A 17-point development program announced in 1978 created about 3 million new jobs by 1981, but it was not fully implemented because of the drop in world oil prices and the subsequent financial crisis. The crisis reached its climax in August 1982, when the government suspended all payments of foreign debt principal and had to resort to emergency credits to avoid default.

New credits from the IMF were conditional upon Mexico's acceptance of an austerity program that entailed reduction of the budget deficit from 17.9% of the GDP in 1982 to 8.5% in 1983. Other austerity measures included tax increases, increases in the prices of controlled commodities, such as bread and salt, and steps to decrease tax evasion and reduce inflation. The de la Madrid administration simultaneously pursued policies to reduce the inflated value of the peso and to generate massive trade surpluses; indicative of their effectiveness were the 1983 and 1984 surpluses, over $13 billion in each year. The government, moreover, pursued rescheduling of its foreign debt, winning agreements in 1983 ($14 billion) and 1986 ($43.7 billion). In 1985 and 1986, however, the earthquake and the fall in world oil prices undermined the recovery; export revenues plunged, and inflation soared. The 1986 rescheduling was conditional upon Mexico's agreement to increase development of the export sector and encourage efficient import-substitution policies, as well as foreign investment.

Since Mexico joined GATT in 1986, trade barriers have been eliminated and tariffs reduced. Privatization policy also has been revitalized. Whereas 33 privatizations completed in 1989 yielded $684 million, revenues from 99 privatizations in 1990 were $2.7 billion, and revenues from the 39 privatizations from 1991 to mid-1992 (excluding sales of state-owned banks) amounted to $5.1 billion. Privatizations since 1989 include: the telephone company, Telmex; Mexico's 18 commercial banks; the airlines, Aeromexico and Mexicana; two large copper mines, Cananea and Mexicana de Cobre; and two large steel companies, Sicartsa and AHMSA. Privatizations clearly have produced large one-time revenues for the government, while simultaneously reducing the government's role in the economy thus garnering savings by reducing its transfers to inefficient enterprises. Furthermore, these new profit-making private sector companies have widened the tax base.

The North American Free Trade Agreement (NAFTA), ratified in 1992 and implemented in 1994, culminated several years of trade liberalization efforts begun in 1986. NAFTA's goal is the creation of a market of 360 million consumers with $6 trillion in annual output. Tariffs on most industrial and agricultural goods will be eliminated or phased out within 15 years. NAFTA trading benefits are only given to goods produced wholly or principally in NAFTA countries. NAFTA also eliminates trade barriers and investment restrictions on participating countries autos, trucks, buses and auto parts within 10 years. NAFTA also proposes to safeguard domestic agricultural production of the dairy, egg, poultry, and sugar sectors. NAFTA also opens up foreign investment possibilities in the Mexican energy sector. NAFTA also has provisions for the textiles and services sectors, banking, investment, and intellectual property rights. Labor and environmental impacts are also addressed.

Mexico has also recently established free trade agreements with Venezuela and Colombia as a member of the Group of Three, and with several other Central American nations. Mexico also signed a free trade agreement with Chile in 1991.

40SOCIAL DEVELOPMENT

Mexico's basic social insurance program was enacted on 30 December 1942. Insurance for occupational accidents and illness is financed by an average contribution of 1.94% of payroll from the employer, varying with the risk of the job, and a subsidy of about two-thirds of the daily wage is supplied during incapacity for up to 52 weeks. Benefits are increased periodically. In the event of death, a month's wage is paid for the funeral; a widow continues to receive 40% of the wage, and children receive a smaller portion until the age of 16. Other forms of social insurance are financed by contributions of both employers and employees. Insured workers receive medical aid in addition to wage benefits, and the insured worker's family receives first-aid treatment. During pregnancy and childbirth and for a period thereafter, insured women receive obstetrical care, nursing aid, a cash subsidy, and a layette. A worker who has been 60% disabled for at least 12 months is eligible for an invalid's pension, and all residents are eligible for old age pensions at age 60.

Government employees are covered by the Security and Social Services Institute for Civil Workers. Various decentralized enterprises and government departments provide their own social security and health services. Programs for children, including a primary-school breakfast program, are overseen by the National Institute for Child Protection. The Mexican Institution for Child Welfare provides care for neglected, abandoned, or sick children.

Under a regional demographic policy introduced in 1978, the government hoped to decrease the population growth rate from 3.1% during the 1970s to 1.8% by 1988, 1.3% by 1994, and 1% by 2000. Government family-planning activities are coordinated by the Interinstitutional Council of Family Planning. The fertility rate in 1985-90 was 3.6. In 1987 an estimated 53% of couples of reproductive age used contraception. Abortion is legal only on narrow medical and legal grounds. An amendment to the 1917 constitution states that men and women are equal before the law, but the traditional concept of women as homemakers is widely accepted. However, this traditional role is beginning to change albeit slowly. Women have held top political and union leadership roles, have the right to file for separation or divorce, and the right to own property in their own name.

41HEALTH

Mexico has made slow but measurable progress in public health. The country has been in an economic crisis; over the last few years (1990 data), over 300 private hospitals have closed. Many state hospitals were closed due to inadequate budgets by the Ministry of Health. Drug prices have also been rising steeply. Mexico wanted to modernize its health centers by the end of 1994. The 1993 population was 90 million, with a birth rate of 27.9 per 1,000 people. In 1992, there were 2.5 million births, and about 53% of the married women (ages 15 to 49) were using contraception. The population is expected to increase to 104 million by the year 2000. The infant mortality rate, which was 101.7 per 1,000 live births in 1948, was reduced to 28 by 1992. The general mortality rate was 5.5 per 1,000 in 1993. Average life expectancy, meanwhile, rose from 32.4 years in 1930 to 57.6 in 1965; by 1992, average life expectancy was estimated at 70 years for men and women. Maternal mortality was 110 per 100,000 live births in 1991. Cholera, yellow fever, plague, and smallpox have been virtually eliminated, and typhus has been controlled. Permanent campaigns are waged against malaria, poliomyelitis, skin diseases, tuberculosis, leprosy, onchocercosis, and serious childhood diseases. In 1990, there were an estimated 110 cases of tuberculosis per 100,000 people, and about 14% of children under 5 years old were considered malnourished. In 1991, 76% of the population had access to safe water, and 74% had adequate sanitation. Immunization rates for children up to one year of age were as follows in 1992: tuberculosis (95%); diphtheria, pertussis, and tetanus (91%); polio (92%); and measles (91%). In 1990, the National Social Security System operated 1,639 medical units, 231 general hospitals, and 38 other health care facilities. In the same year, there were 1.3 beds per 1000 people and, in 1987, there were 60,099 beds (0.8 per 1,000). In 1988, there were 130,000 physicians and, in 1992, there were .54 doctors per 1,000 inhabitants, with a nurse to doctor ratio of 0.8. According to other data, there were 81,593 physicians in 1991. As of 1983, 33,850,264 Mexicans enjoyed free medical services and, in 1992, 72% of the population had access to health care services. Total health care expenditures were $7,648 million in 1990. Major causes of death in 1990 were noted as follows: (1) communicable diseases and maternal/perinatal causes (168 per 100,000); (2) noncommunicable diseases (490 per 100,000); and (3) injuries (102 per 100,000).

⁴²HOUSING

Mexico's public housing program dates from 1938. A 1946 law permitted credit institutions to finance low-rent projects and create a housing bank. Each of two large projects built during the 1940s and 1950s—the President Alemán Urban Center (completed in 1950) and the President Juárez Urban Center (completed in 1952), both in Mexico City—contains a school, a playground, and stores. In 1992, Mexico had 15,100,000 dwellings, of which 49.9% had piped water, 51% had flush toilets, and 87.5% had electric lighting.

Government agencies that have fostered the development of low-income housing include the Fund for Housing Operations and Bank Discounts, the National Public Works and Services Bank, the Housing Credit Guarantee and Support Fund, and the National Housing Institute. In 1974, the National Workers' Housing Fund Institute (Instituto del Fondo Nacional de la Vivienda para los Trabajadores—INFONAVIT) was created to provide housing for workers. With funds provided by employers (equal to 5% of the total salary of each worker), INFONAVIT makes direct loans to employees and provides short-term loans to finance the construction of approved multi-unit projects, which are then sold to employees covered by the program. All these efforts, however, have not come close to eradicating Mexico's housing shortage, which has been exacerbated by accelerated population growth in the 1980s. The government allocated us$1.93 billion in 1989 to build 250,000 low-cost housing units, and expected to receive an additional us$700 million from the World Bank to build more. The 1990 National Housing Plan predicted a shortage of 6.1 million homes, to be felt most severely in the outskirts of urban areas, including Mexico City, Guadalajara, Monterey, and cities in the northern states.

⁴³EDUCATION

Primary schooling is compulsory and free. Except in the Federal District, where education is administered by the federal government, schools are controlled by the states.

In 1971, children of the age for obligatory schooling (between 6 and 14) numbered 13,159,000; however, only 74% attended school. Between 1960 and 1965, for each 100 pupils enrolled in the first grade, only 23 completed the sixth, making an attrition rate of 77%. In 1984, the attrition rate was 48%. At the other end of the educational cycle, only 12,000 graduated from institutions of higher education. Approximately 97% of Mexico's youth between the ages of 16 and 25, a 1968 government report stated, were unprepared to participate in the country's development. Significant improvement had been recorded by the late 1970s, when an estimated 92% of eligible boys and 90% of eligible girls

were actually in primary school; the figures for secondary school in 1984 were 56% and 53%, respectively. In 1984, 24.5 million Mexicans were in school, 33% of the total population.

During 1965, the government established 7,000 new literacy centers, raising the total to 11,000. The literacy program helped reduce Mexico's adult illiteracy rate from 37.8% in 1960 to 12.4% in 1990 (males: 9.6% and females: 15.0%).

In 1991, Mexico had 84,606 schools with 476,616 teachers and 14,396,993 students at the primary level. In the secondary level, there were 406,998 teachers and 6,704,188 students. Nearly 818,202 of these students were in vocational schools.

Major universities include the National Autonomous University (founded in 1551), the National Polytechnic Institute, and Iberoamericana University (private), all in Mexico City, and Guadalajara University, the Autonomous University of Guadalajara, and the Autonomous University of Nuevo León. In each state there are other state and private institutions. There were 134,424 teachers and 1,310,835 students in all higher level institutions in 1990.

The government provides extracurricular education through cultural and motorized missions, community-development brigades, reading rooms, and special centers for workers' training, art education, social work, and primary education.

⁴⁴LIBRARIES AND MUSEUMS

The Mexican public libraries have somewhat limited collections, tradition being in favor of large private libraries; however there are still over 70,000,000 registered library users in the country. The National Library, now affiliated with the National University of Mexico, has about 2,000,000 volumes. Other important collections include the Library of Mexico, the Library of the Secretary of the Treasury, and the central library of the National Autonomous University in Mexico City.

The National Museum of Anthropology in Mexico City, founded in 1825, has over 600,000 anthropological, ethnological, and archaeological exhibits and a library of 300,000 volumes. Among its exhibits are the famous Aztec calendar stone and a 137-ton figure of Tlaloc, the god of rain. The National Historical Museum, attached to the National Institute of Anthropology and History, has more than 150,000 objects ranging in date from the Spanish conquest to the constitution of 1917. The National Museum of Art exhibits Mexican art from 17th century to present, while several other art museums exhibit the works of leading European artists. Many public buildings in Guadalajara and elsewhere display murals by famous Mexican painters.

⁴⁵MEDIA

National and international telegraphic service is furnished by the government-owned National Telegraph Co. The government also owns and operates the international radiotelegraph and radiotelephone facilities. A privately owned telephone company provides supplemental facilities in outlying areas. The largest utility is the government-owned Telephones of Mexico. The number of telephones in service in 1991 was 9,358,659.

The national microwave network, now complete, cost more than ₱650 million; facilities by 1968 included the central telecommunications tower in the Federal District and a land station for artificial satellite communications at Tulancingo, Hidalgo, with one of the largest antennas in the world. The network serves most of the country's larger cities.

In 1992 there were 126 television stations, including 8 cultural stations, and 868 radio stations, 45 of them cultural. In 1991, 22 million radios and over 12.7 million television sets were in use.

As of 1991, some 286 daily newspapers were published in Mexico, with the greatest number coming from Mexico City.

Leading newspapers (with their estimated average daily circulations in 1991) include the following:

MEXICO CITY	CIRCULATION
Esto (m)	400,000
El Heraldo (m)	300,000
La Prensa (m)	300,000
Ovaciones (m)	215,000
El Universal (m)	185,000
Excélsior (m)	175,000
El Nacional (m)	120,000
Novedades (m)	110,000
MONTERREY	
El Norte (m)	100,000
La Tribuna (m)	95,000
TAMPICO	
El Heraldo de Tampico (m)	95,000
El Sol de Tampico (m)	75,000
GUADALAJARA	
El Occidental (m)	85,000
Ocho Columnas (d)	81,000

Freedom of the press is guaranteed by law and to a large extent exercised in practice. However, by controlling the supply of newsprint and by providing advertising, indirect subsidies, and outright payoffs to the press, the government exerts an indirect form of press censorship. Mexico's libel and slander laws were made more restrictive in 1982.

46ORGANIZATIONS
The Mexican government supervises and promotes producer and consumer cooperatives, which are exempt from profits and dividends taxes and are given customs protection. Producer cooperatives are active in agriculture, fishing, forestry, and mining. Consumer cooperatives buy, sell, and distribute clothing, foodstuffs, and household articles.

Chambers of commerce and of industry are located in most cities, and merchants and manufacturers are required to join either or both. Headquarters for the national chamber of commerce and industry are in Mexico City. In 1993, 62 industrial chambers and 32 industrial associations represented the major economic sectors. The principal employers' organization is the Employers' Confederation of the Mexican Republic, which dates from 1929.

47TOURISM, TRAVEL, AND RECREATION
In 1991, Mexico was the world's eighth most popular tourist destination. That year 16,560,000 tourists entered Mexico, 15,673,000 of them from the United States. There were 345,159 hotel rooms and 690,318 beds with a 54.4 occupancy rate. Receipts from tourism in that year were us$4.35 billion.

Mexico's tourist attractions range from modern seaside resort areas, such as Tijuana, Acapulco, Puerto Vallarta, Punta Ixtapa/Zihuatanejo, Cozumel, and Cancún, to the Mayan ruins of Chiapas State and the Aztec monuments of the south-central regions. Mexico City, combining notable features from the Aztec, colonial, and modern periods, is itself an important tourist mecca. Veneration of the patron saints plays an important role in Mexican life, and the calendar is full of feast days (fiestas). These predominantly Roman Catholic celebrations include many ancient Amerindian rites and customs and, invariably, bands of mariachi musicians playing Mexican folk songs. Tourists from the US and Canada may enter Mexico with a tourist card stamped with a Mexican visa. All other visitors need a passport and a visa. Business travelers require non-immigrant visas and business permits.

Mexico was host in 1968 to the Summer Olympics and in 1970 and 1986 to the World Cup soccer championship; the World Cup helped tourism recover from the effects of the earthquake,

which kept many visitors away in 1985 and 1986. Mexico's most popular sports are baseball, soccer, jai-alai, swimming, and volleyball. Bullfights are a leading spectator sport; the Mexico City arena, seating 50,000 persons, is one of the largest in the world, and there are about 35 other arenas throughout the country.

48FAMOUS MEXICANS
The founder of Spanish Mexico was Hernán Cortés (1485–1547), a daring and clever Spanish conquistador. One of the great heroes in Mexican history is Guatemotzin (Cuauhtémoc, 1495?–1525), the last emperor of the Aztecs, who fought the Spanish after the death of his uncle, Montezuma II (Moctezuma or Motecuhzoma, 1480?–1520). Bartolomé de las Casas (1474–1566) and Junípero (Miguel José) Serra (1713–84) were Spanish-born missionaries who tried to improve the conditions of the Amerindians in the colonial period. Two heroes of the War of Independence were the liberal priests Miguel Hidalgo y Costilla (1753–1811) and José María Morelos y Pavón (1765–1815). The first 25 years of independence were dominated first by Manuel Félix Fernández, known as Guadalupe Victoria (1786?–1843), and then by Antonio López de Santa Anna (1794–1876). Austrian-born Maximilian (Ferdinand Maximilian Josef, 1832–67) and Belgian-born Carlota (Marie Charlotte Amélie Augustine Victoire Clémentine Léopoldine, 1840–1927) were emperor and empress from 1864 to 1867, ruling on behalf of the emperor of France. Benito Juárez (1806–72), the great leader of the liberal revolution, attempted to introduce a program of national reform. The dictator José de la Cruz Porfirio Díaz (1830–1915) dominated Mexico from 1876 to 1911. He was overthrown largely through the efforts of Francisco Indalecio Madero (1873–1913), called the Father of the Revolution. Revolutionary generals and politicians included Venustiano Carranza (1859–1920) and Álvaro Obregón (1880–1928). Two other revolutionary leaders—Doroteo Arango, known as Pancho Villa (1877?–1923), and Emiliano Zapata (1879?–1919)—achieved almost legendary status. The foremost political leader after the revolution was Lázaro Cárdenas (1895–1970). Luis Echeverría Álvarez (b.1922), who held the presidency during 1970–76, made Mexico one of the leading countries of the developing world in international forums. Miguel de la Madrid Hurtado (b.1934) became president in 1982. The diplomat Alfonso García Robles (b.1911) shared the 1982 Nobel Peace Prize for his work on behalf of disarmament.

Painters Diego Rivera (1883–1957), José Clemente Orozco (1883–1949), and David Alfaro Siqueiros (1898–1974) are renowned for their murals. The Spaniards Bernal Díaz del Castillo (1496?–1590?) and Bernardino de Sahagun (1499?–1590) wrote historical accounts of the Spanish conquest. Mexican-born Juan Ruiz de Alarcón y Mendoza (1580?–1639) became a playwright in Spain. Juana Inés de la Cruz (1651–1695), a nun, was a poet and proponent of women's rights. Outstanding novelists include José Joaquín Fernández de Lizardi (1776–1827), author of *El periquillo sarniento*; Mariano Azuela (1873–1952), author of *Los de abajo*; Martín Luis Guzmán (1887–1976), author of *El águila y la serpiente*; and Gregorio López y Fuentes (1897–1966), author of *El indio*. Well-known contemporary authors include Augustín Yáñez (1904–1980), Octavio Paz (b.1914), Juan Rulfo (1918–1986), and Carlos Fuentes (b.1928). Poets include Salvador Díaz Mirón (1853–1928), Manuel Gutiérrez Nájera (1859–95), and Amado Nervo (1870–1919). The outstanding figure in recent Mexican literary life is the diplomat, dramatist, poet, essayist, and critic Alfonso Reyes (1889–1959). Anthropologist Carlos Castaneda (b.Brazil, 1931) is widely known for his studies of mysticism among the Yaqui Amerindians. Well-known Mexican composers include Manuel Ponce (1882–1948), Silvestre Revueltas (1899–1940), Carlos Chávez (1899–1978), and Agustín Lara (1900–1970). Significant figures in the motion picture industry are the comedian Cantinflas

(Mario Moreno, b.1911), Mexican-born actor Anthony Rudolph Oaxaca Quinn (b.1916), and directors Emilio Fernández (1904–86) and Spanish-born Luis Buñuel (1900–1983).

Notable Mexican sports figures include Fernando Valenzuela, (b.1960), a pitcher for the Los Angeles Dodgers who won the Cy Young Award as a rookie.

⁴⁹DEPENDENCIES
Mexico has no territories or colonies.

⁵⁰BIBLIOGRAPHY

Aspe, Pedro, and Paul E. Sigmund. *The Political Economy of Income Distribution in Mexico.* New York: Holmes & Meier, 1984.

Bailey, John J. *Governing Mexico: The Statecraft of Crisis Management.* New York: St. Martin's Press, 1988.

Bancroft, Hubert Howe. *History of Mexico.* 8 vols. San Francisco: Bancroft, 1883–88.

Barkin, David. *Distorted Development: Mexico in the World Economy.* Boulder, Colo.: Westview Press, 1990.

Barry, Tom (ed.). *Mexico: A Country Guide.* Albuquerque, N.M.: Inter-Hemispheric Education Resource Center, 1992.

Bazant, Jan. *A Concise History of Mexico.* New York: Cambridge University Press, 1977.

Bernal, Ignacio. *The Olmec World.* Berkeley: University of California Press, 1969.

Brading, D. A. (ed.). *Caudillo and Peasant in the Mexican Revolution.* New York: Cambridge University Press, 1980.

Camp, Roderic Ai. *Entrepreneurs and Politics in Twentieth-century Mexico.* New York: Oxford University Press, 1989.

Carmichael, Elizabeth. *The Skeleton at the Feast: The Day of the Dead in Mexico.* Austin: University of Texas Press, 1992.

Cross, Harry E., and J. A. Sandos. *Across the Border: Rural Development in Mexico and Recent Migration to the United States.* Berkeley: Institute of Government Studies, University of California, 1981.

Cypher, James M. *State and Capital in Mexico: Development Policy since 1940.* Boulder, Colo.: Westview Press, 1990.

Díaz del Castillo, Bernal. *Discovery and Conquest of Mexico, 1515–1521.* New York: Farrar, Straus & Cudahy, 1956 (orig. 1632).

Grayson, George W. *The North American Free Trade Agreement.* Ithaca, N.Y.: Foreign Policy Association, 1993.

———. *The Politics of Mexican Oil.* Pittsburgh: University of Pittsburgh Press, 1981.

Hamilton, Nora. *The Limits of State Autonomy: Post-Revolutionary Mexico.* Princeton, N.J.: Princeton University Press, 1982.

Hart, John M. (John Mason). *Revolutionary Mexico: The Coming and Process of the Mexican Revolution.* Berkeley: University of California Press, 1987.

Johnston, Bruce F. (ed.). *U.S.-Mexico Relations: Agriculture and Rural Development.* Stanford, Calif.: Stanford University Press, 1987.

MacLachlan, Colin M., and Jaime E. Rodriguez. *The Forging of the Cosmic Race: A Reinterpretation of Colonial Mexico.* Berkeley: University of California Press, 1980.

Maxfield, Sylvia. *Governing Capital: International Finance and Mexican Politics.* Ithaca: Cornell University Press, 1990.

Mexican Politics in Transition. Boulder, Colo.: Westview Press, 1987.

Mexico in Pictures. Minneapolis: Lerner, 1988.

Meyer, Michael C. *The Course of Mexican History.* 4th ed. New York: Oxford University Press, 1991.

North American Free Trade Agreement: Final Text. CCH International, 1992.

Ortiz-Martínez, Guillermo. *Capital Accumulation and Economic Growth: A Financial Perspective on Mexico.* New York: Garland, 1984.

Oster, Patrick. *The Mexicans: A Personal Portrait of a People.* New York: W. Morrow, 1989.

Padgett, L. Vincent. *The Mexican Political System.* 2d ed. Boston: Houghton Mifflin, 1976.

Prescott, William Hickling. *History of the Conquest of Mexico.* New York: Harper, 1843.

Quirk, Robert E. *The Mexican Revolution and the Catholic Church, 1910–1929.* Bloomington: Indiana University Press, 1973.

Reed, John. *Insurgent Mexico.* New York: Simon & Schuster, 1969 (orig. 1916).

Ruiz, Ramón Eduardo. *The Great Rebellion: Mexico, 1905–1924.* New York: Norton, 1980.

———. *Triumphs and Tragedy: A History of the Mexican People.* New York: W.W. Norton, 1992.

Semo, Enrique. *The History of Capitalism in Mexico: Its Origins, 1521–1763.* Austin: University of Texas Press, 1993.

Shafer, Robert Jones. *Mexican Business Organizations: History and Analysis.* Syracuse, N.Y.: Syracuse University Press, 1973.

Stevens, Evelyn P. *Protest and Response in Mexico.* Cambridge: MIT Press, 1974.

Turner, Frederick C. *The Dynamic of Mexican Nationalism.* Chapel Hill: University of North Carolina Press, 1970.

Weintraub, Sidney (ed.). *U.S.-Mexican Industrial Integration: The Road to Free Trade.* Boulder, Colo.: Westview Press, 1991.

———. *Mexican Trade Policy and the North American Community.* Washington, D.C.: Center for Strategic and International Studies, 1988.

Wilkie, James W. (ed.). *Society and Economy in Mexico.* Los Angeles, Calif.: UCLA Latin American Center Publications, 1990.

NETHERLANDS AMERICAN DEPENDENCIES

NETHERLANDS ANTILLES

The five islands of the Netherlands Antilles are divided geographically into two groups: the Leeward Islands (Benedenwindse Eilanden) and the Windward Islands (Bovenwindse Eilanden). The Windward group, off the north coast of South America, comprises the islands of Curaçao, with an area of 444 sq km (171 sq mi), and Bonaire, 288 sq km (111 sq mi). Aruba, another Windward island, seceded from the Netherlands Antilles in 1986. The Leeward group, more than 800 km (500 mi) to the northeast, consists of the southern part of St. Martin (Sint Maarten), 34 sq km (13 sq mi); Saba, 13 sq km (5 sq mi); and Sint Eustatius, 21 sq km (8 sq mi). The islands total 800 sq km (309 sq mi).

All the Windward Islands have volcanic bases, partly covered with coral reefs; they are semiarid and flat, with little vegetation. The Leeward Islands are more mountainous and receive enough rainfall to enable crops and vegetation to flourish. Saba, the most fertile, is an extinct volcano with luxuriant vegetation in its crater, on its sides, and leading down to the sea. Temperatures average between 25° and 31°C (77° and 88°F); annual rainfall averages 107 cm (42 in) on the Windward Islands and 51 cm (20 in) on the Leeward Islands. In 1987, the tundra peregrine falcon and the American crocodile were endangered species. The estimated population of the Windward Islands in 1991 was 154,098, consisting of Curaçao, 142,059, and Bonaire, 16,139. The Leeward Islands had 34,619 residents, consisting of Sint Eustatius, 1,781; St. Martin, 31,722; and Saba, 1,116. Most of the inhabitants were born on the islands. Blacks predominate, but over 40 nationalities are represented, including Dutch, Surinamese, British, Latin Americans, French West Indians, Carib Amerindians, and US emigrants.

The official language is Dutch. Papiamento, a lingua franca evolved from Dutch and English with an admixture of French, Spanish, Portuguese, Arawak, and African words, is also common, principally in the Leeward Islands. English is spoken, mainly in the Windward Islands. Spanish is also widely spoken. Roman Catholicism has the most adherents on the Leeward Islands and Saba, but Protestantism is dominant on St. Martin and Sint Eustatius.

Buses, private automobiles operating as small buses on fixed routes, and taxicabs provide the only public transportation. There are no rivers or railroads. Each island has a good all-weather road system. The most important port is on Curaçao, where there is a natural harbor at Willemstad. Airline connections to nearby islands and countries are provided by Dutch Antillean Airlines (Antilliaanse Luchtvaart Maatschappij—ALM), Royal Dutch Airlines (Koninklijke Luchtvaart Maatschappij—KLM), and other international carriers.

The European discovery of the Windward Islands was made by Columbus in 1493, and that of the Leeward Islands (including Aruba) by a young Spanish nobleman, Alonso de Ojeda, who sailed with Amerigo Vespucci in 1499; hence the claim went to Spain. The Dutch fleet captured the Windward Islands in 1632 and the Leeward Islands in 1634. Peter Stuyvesant was the first governor. In 1648, St. Martin was peacefully divided between the Netherlands and France; this division still exists. During the colonial period, Curaçao was the center of the Caribbean slave trade. For a period during the Napoleonic wars (1807–15), Great Britain had control over the islands. Slavery was abolished in 1863.

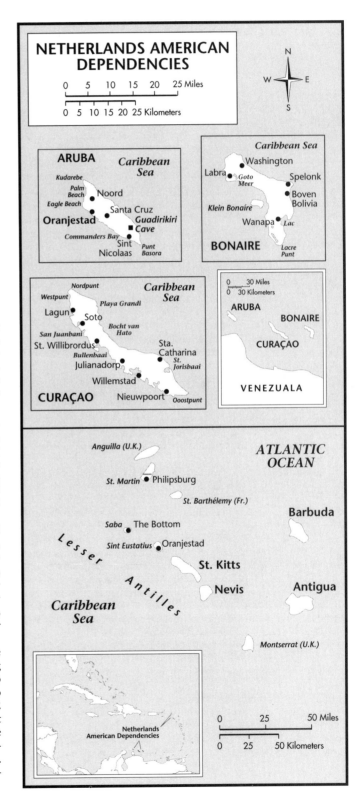

Under a 1954 statute, the Netherlands Antilles is a component of the Kingdom of the Netherlands, with autonomy in internal affairs. A governor, appointed by and representing the crown, heads the government, with a Council of Ministers as the executive body. The ministers are responsible to the Staten, a 22-member legislature (14 from Curaçao, 3 each from Bonaire and St. Martin, and 1 each from Saba and Sint Eustatius). Members are elected by general suffrage of Dutch nationals aged 18 or older. A 1951 regulation established autonomy in local affairs for each of the then-existing island communities—Aruba, Curaçao, Bonaire, and the Leeward Islands—with responsibilities divided between an elected island council, an executive council, and a lieutenant-governor. By agreements made in 1983, St. Martin, Saba, and Sint Eustatius have separate representation in the Staten, elect their own separate councils, and have their own lieutenant-governors and executive councils. Cases are tried in a court of first instance and on appeal in the Joint High Court of Justice, with justices appointed by the crown. Defense is the responsibility of the Netherlands; a naval contingent is permanently stationed in the islands, and military service is compulsory. The Netherlands Antilles is an associate member of the EEC.

The prosperity of Curaçao is inseparably linked with its oil refineries. These were built there, beginning in 1918, chiefly because of the favorable location of the islands, their good natural ports and cheap labor, and the political stability of the territory. Tankers bring crude oil from Venezuela. The economic significance of the refineries is great, not only because of their output, but also because they provide employment and stimulate other economic activities, such as shipbuilding, metal industries, shipping, air traffic, and commerce in general. The government controls the price of basic foodstuffs and participates in the setting of rates to be charged for transportation and by privately owned utilities.

The currency unit is the Netherlands Antilles guilder, or florin (NAf) of 100 cents; NAf1 = $0.5600 (or $1 = roughly NAf1.79). The GDP for 1991 was estimated at US$1.2 billion, or US$8,464 per capita. The unemployment rate (including Aruba) was 14.2% in 1992, and 60–70% of the work force was organized in labor unions. The principal agricultural products are sorghum, orange peel, aloes, groundnuts, yams, divi-divi, and some assorted vegetables. Curaçao's favorable position at the crossing of many sealanes has stimulated commerce since the earliest days of European settlement. Transit trade benefits from Curaçao's improved harbors; Willemstad is a free port, as are the islands of Saba and Sint Eustatius. In 1991, Antillean exports were valued at US$200 million and imports at US$1.6 billion. Petroleum shipments dominated the country's foreign trade, accounting for about 64% of total imports. The US accounted for 55% of 1988 exports and 12% of imports.

The Bank of the Netherlands Antilles (Bank van de Nederlandse Antillen) issues currency, holds official reserves, regulates the banking system, and acts as the central foreign exchange bank. There are 14 authorized commercial banks and several savings and loan institutions that also handle financial matters. The Netherlands Antilles Development Bank was created in 1981 to stimulate foreign investment in service industries. Tax treaties with the US have encouraged US individuals and businesses to shelter their funds in the islands. Development aid from the Netherlands totals about $60 million annually.

There are 70 nursery schools, 91 primary schools, 55 secondary schools, 1 teacher-training college, and the University Institute of the Netherlands Antilles, on Curaçao (all excluding Aruba); all schools are government-supported. The language of instruction is Dutch in the Leeward Islands, except in the International School, where classes are taught in English; English is also used in the Windward Islands. The literacy rate is more than 95%, although education is not compulsory. There are six hospitals on Curaçao and four on the other islands.

All the islands have cable, radiotelegraph, or radiotelephone connections with one another and, via the central exchange at Curaçao, with international systems. There are 9 AM, 4 FM, and 1 TV station on the islands. Broadcasts are in Papiamento, Dutch, Spanish, and English. In 1983 there were about 180,000 radios and 56,500 televisions (including Aruba). Two daily newspapers in Dutch and four in Papiamento are published in the Netherlands Antilles, along with several weekly and monthly periodicals.

Tourism is a significant source of revenue. Annually, more than 469,000 cruise passengers visit the islands; total visitors to the islands amount to over 1,000,000.

ARUBA

The island of Aruba is located off the north coast of South America, NW of Curaçao. It has an area of 193 sq km (75 sq mi). The land is basically flat and is renowned for its white-sand beaches. The temperature is almost constant at 27°C (81°F); the annual rainfall averages 60 cm (24 in). The 1993 population was 65,117. The official language is Dutch, but Papiamento, English, and Spanish are also spoken. The religion of the majority is Roman Catholic.

There are no rivers or railways. The road system connects all major cities. There are three deepwater harbors, at San Nicolas, Oranjestad, and Barcadera. Airline connections are provided by Dutch Antillean Airlines (Antilliaanse Luchtvaart Maatschappij—ALM), Royal Dutch Airlines (Koninklijke Luchtvaart Maatschappij—KLM), and other international carriers.

In March 1983, at the Hague, the governments of the Netherlands and Netherlands Antilles agreed to grant Aruba the status of a separate state. On 1 January 1986, Aruba seceded from the Netherlands Antilles, becoming a separate member of the kingdom. Full independence will be granted in 1996.

The head of government is a governor appointed by and representing the crown. A Council of Ministers, led by a prime minister, has executive power. The ministers are responsible to the Staten, a legislative body of 21 members elected by universal adult suffrage for four-year terms. Cases are tried in a court of first instance and on appeal in the Joint High Court of Justice, with justices appointed by the crown. Defense is the responsibility of the Netherlands; a naval contingent is stationed on the island, and military service is compulsory.

The two principal sources of revenue for the island are tourism and oil refining. Because of reduced worldwide demand for oil, however, Exxon closed its Lago refinery in 1985, placing a heavier economic stress on tourism. The currency unit is the Aruban florin (Af) of 100 cents. Af1 = $0.5587 (or $1 = Af1.79). The GDP for 1991 was US$900 million, or US$14,000 per capita. The unemployment rate was 3 % in 1991.

In 1991, Aruban exports were valued at US$902.4 million, and imports at US$1.36 billion. Oil refining accounted for 35% of exports. The US and the EEC were Aruba's major trading partners. There is a Central Bank of Aruba, and there are six commercial banks. The Aruba Development Bank was created in 1982 to stimulate foreign investment in service industries.

Medical care is provided free by the government; there is a modern hospital. There were 26 preprimary schools, 33 primary schools, 23 secondary schools, and 1 teacher-training college. Although education is not compulsory, the literacy rate is high (over 90%). There are 8 commercial radio stations and one television station in Aruba. There is one daily newspaper published in Dutch, one in English, and one in Papiamento. There are 72,168 telephones on the island.

Tourism is the major source of revenue. In the early 1990s 501,324 tourists visited the island. Of these visitors, 53% were from the US.

NICARAGUA

Republic of Nicaragua

República de Nicaragua

CAPITAL: Managua.

FLAG: The national flag consists of a white horizontal stripe between two stripes of cobalt blue, with the national coat of arms centered in the white band.

ANTHEM: *Salve a ti, Nicaragua (Hail to You, Nicaragua)*.

MONETARY UNIT: The córdoba (c$) is a paper currency of 100 centavos. There are coins of 5, 10, 25, and 50 centavos and 1 and 5 córdobas, and notes of 1, 2, 5, 10, 20, 50, 100, 500, 1,000, 5,000, 10,000, 20,000, 50,000, 100,000, 200,000, 500,000, 1,000,000, 5,000,000, and 10,000,000 córdobas. c$1 = us$0.014 (or us$1 = c$70).

WEIGHTS AND MEASURES: The metric system is the legal standard, but some local units are also used.

HOLIDAYS: New Year's Day, 1 January; Labor Day, 1 May; Liberation Day (Revolution of 1979), 19 July; Battle of San Jacinto, 14 September; Independence Day, 15 September; All Saints' Day, 1 November; Christmas, 25 December. Movable religious holidays include Holy Thursday and Good Friday.

TIME: 6 AM = noon GMT.

¹LOCATION, SIZE, AND EXTENT

Nicaragua, the largest of the Central American countries, has an area of 129,494 sq km (49,998 sq mi), which includes the area covered by the waters of Lake Nicaragua (about 8,000 sq km/3,100 sq mi) and Lake Managua (about 1,000 sq km/390 sq mi). Comparatively, the area occupied by Nicaragua is slightly larger than the state of New York. The country has a length of 580 km (360 mi) NE-SW and a width of 494 km (307 mi) NW-SE. Bounded on the N by Honduras, on the E by the Caribbean Sea, on the S by Costa Rica, and on the W by the Pacific Ocean, Nicaragua has a total boundary length of 2,141 km (1,330 mi).

In 1980, Nicaragua unilaterally abrogated its 1928 treaty with Colombia, confirming that nation's sovereignty over the Caribbean archipelago of San Andrés and Providencia, about 190 km (120 mi) off the Nicaraguan coast. Nicaragua also disputes the Treaty of Quita Sueño, ratified by the US Senate in July 1981, according to which Colombia received the uninhabited islands of Quita Sueño Bank, Roncador Cay, and Serrana Bank.

Nicaragua's capital city, Managua, is located in the southwestern part of the country.

²TOPOGRAPHY

The Caribbean coast, known as the Mosquito (or Miskito) Coast or Mosquitia, consists of low, flat, wet, tropical jungle, extending into pine savannas 80–160 km (50–100 mi) inland. The coastal lowland rises to a plateau covering about one-third of the total area. This plateau is broken by mountain ranges extending eastward from the main cordillera to within 64–80 km (40–50 mi) of the Caribbean coast. The mountainous central area forms a triangular wedge pointed southeast, rising at its highest to some 2,000 m (6,600 ft). The plains and lake region, in a long, narrow structural depression running northwest to southeast along the isthmus, contains a belt of volcanoes rising to 1,500 m (5,000 ft) and extending from the Gulf of Fonseca to Lake Nicaragua. In this region is located Lake Managua, at 41 m (136 ft) above sea level, which drains through the Tipitapa Channel into Lake Nicaragua, at 32 m (106 ft) above sea level, which, in turn, drains through

the San Juan River eastward into the Caribbean. Lake Nicaragua is about 160 km (100 mi) long and 72 km (45 mi) wide at the widest point, while Lake Managua is 56 km (35 mi) long by 24 km (15 mi) wide.

The principal waterways are the Coco (or Segovia) River, navigable up to 240 km (150 mi) inland from the eastern Mosquito Coast, and the San Juan, navigable to within a few miles of the Caribbean, where a series of rapids halts transportation.

Nicaragua lies in an earthquake zone. The last major earthquake, which occurred on 23 December 1972, claimed an estimated 10,000 lives and demolished a 70-square-block section of central Managua. A major volcanic eruption was that of Cerro Negro (977 m/3,205 ft), northwest of Managua, in 1968.

³CLIMATE

Except in the central highlands, the climate is warm and humid. Average humidity in Managua in June, the most humid month, is 84%; in April, the driest month, 62%. The mean temperature, varying according to altitude, is between 20° and 30°C (68° and 86°F). In Managua, monthly average temperatures range from a minimum of 23°C (73°F) and a maximum of 30°C (86°F) in January to a minimum of 26°C (79°F) and a maximum of 31°C (88°F) in July. There are two seasons: a wet season, from May to December, and a dry season, from January through April. Rainfall, however, varies according to region, and the rainy season in the eastern area may extend 9 or even 12 months. Average annual rainfall along the Mosquito Coast reaches 254–635 cm (100–250 in) as a result of the easterly trade winds blowing in from the Caribbean; the highlands also have heavy rainfall. Managua receives 114 cm (45 in), while the Pacific coast averages over 200 cm (80 in) a year.

⁴FLORA AND FAUNA

The central highlands region has extensive forests of oak and pine on the slopes, but lower valley elevations show damage from fire and agricultural activities. The largest pine savanna in the rainy tropics stands on the lowlands behind the Mosquito Coast. The wet and humid Caribbean coastal plain has an abundance of tropical forest, with wild rubber, cedar, ebony, mahogany, and rosewood attracting some exploitation.

Wildlife includes the puma, deer, monkey, armadillo, alligator, parrot, macaw, peccary, and several species of snakes (some poisonous). Lake Nicaragua contains the only freshwater sharks in the world, owing to a prehistoric geological movement that separated the lake from the Pacific Ocean, gradually changing the ocean water into fresh water.

5ENVIRONMENT

Nicaragua's major environmental problems are soil erosion, caused in part by cultivation of annual crops on steep slopes, and depletion of the upland pine forests for lumber, fuel, and human settlement. Nicaragua lost 150,000 hectares of forest land in the year 1990–1991. One contributing factor is the use of wood for fuel. Excessive or ineffective use of pesticides to control malaria, along with widespread agricultural use, has resulted in some environmental contamination. Nicaragua's cities produce 0.5 million tons of solid waste per year. The purity of the nation's water is also an environmental issue because of the prevalence of industrial pollutants in the lakes and rivers. The nation has 42.0 cu mi of water. Fifty-four percent is used for farming and 21% in industrial activity. Dumping of sewage and chemical wastes has made Lake Managua unsuitable for swimming, fishing, or drinking. Primary responsibility for resource conservation is vested in the Nicaraguan Institute of Natural Resources and Environment (Instituto Nicaragüense de Recursos Naturales y del Ambiente—IRENA), established in October 1979.

In 1994, eight of the nation's mammal species were endangered. Seven bird species and 68 plant species are threatened with extinction. Endangered species in Nicaragua included the tundra peregrine falcon, four species of turtle (green sea, hawksbill, leatherback, and olive ridley), the spectacled caiman, and the American crocodile.

6POPULATION

According to the 1971 census, the population was 1,877,952. The estimated population in 1991 was 3,999,231 and the projected population for the year 2000 was 5,169,000, assuming a crude birthrate of 36.6 per 1,000 population, a crude death rate of 6, and a net natural increase of 30.6 during 1995–2000. The average annual growth rate between 1985 and 1989 was 3.45%. An estimated 61% of the population lived in urban areas in 1991, when the population density for the nation was 33 per sq km (86 per sq mi). In 1985, the principal city and capital, Managua, had a population of 682,111. Devastated by earthquakes in 1931 and 1972, and then severely damaged during the civil war of the late 1970s, Managua was slowly being rebuilt during the early 1980s. Other major cities and their estimated 1985 populations are León, 100,982; Granada, 88,636; Masaya, 74,946; Chinandega, 67,792; Matagalpa, 36,983; Esteli, 30,653; Tipitapa, 30,078; Chichigalpa, 28,889; and Juigalpa, 25,625.

7MIGRATION

After the Sandinista takeover in 1979, thousands of Nicaraguans left the country. It was estimated in 1987 that 24,000 had fled to Honduras, 16,000 to Costa Rica, and over 200,000 to the US, chiefly to Florida. After the defeat of the Sandinistas in the 1990 elections, some 200,000 Nicaraguans returned from abroad. Some 27,800 Nicaraguans were still refugees in Costa Rica at the end of 1992. At the same time, 14,500 Central American refugees were in Nicaragua.

8ETHNIC GROUPS

The Nicaraguan population is basically mestizo, a mixture of white and Amerindian. There are no census data on racial composition, but 1980 estimates placed the mestizo component

at 69% and the white population at 14%; blacks account for 13%, and Amerindians for the remaining 4%.

Traditionally, the Atlantic littoral has been inhabited mainly by blacks from Jamaica, Belize, and various present and former British possessions in the Caribbean. The more densely populated Pacific coast highland has long been basically mestizo in composition. Most Amerindian groups in Nicaragua have been assimilated, but Miskito Amerindians, as well as Sumus, make their traditional homes on the Mosquito Coast and neighboring areas.

9LANGUAGES

Spanish is the official language and is spoken by the overwhelming majority of the population. Some Nahuatl and other Amerindian words and phrases are in common use. English is often spoken as a second language at professional levels.

10RELIGIONS

Roman Catholicism is the predominant religion, claiming the allegiance of about 89.3% of the population. The country comprises one archbishopric (with its seat at Managua) and five bishoprics (León, Granada, Bluefields, Matagalpa, and Estelí). Relations between the Sandinista government and the Roman Catholic Church deteriorated in 1982–83. There were assaults on individual churches and members of the clergy, and several antireligious incidents occurred during the visit by Pope John Paul II in March 1983. Relations began to improve in 1987, aided by the presence of a new papal nuncio, Pablo Giglio.

Protestants, especially Moravians and Anglicans, are predominant on the east coast. The Protestant community numbered about 400,000 in 1986. There are also small communities of Baha'is, Amerindian tribal religionists, spiritists (who sometimes combine elements of Christianity and African religions), Buddhists, and Jews.

11TRANSPORTATION

Main transportation arteries are concentrated in the more densely populated Pacific region. The national road network in 1991 totaled 25,000 km (15,535 mi). The Inter-American Highway from Honduras to Costa Rica was completed in 1972. The Pacific Highway begins in Granada and passes through Managua, León, and Chinandega to Corinto. In 1991 there were 70,000 registered motor vehicles. Pacific Railways of Nicaragua, government-owned, with a length of 373 km (231 mi), links Corinto to Granada.

The Naviera Nicaragüense provides regular services to Central America, the US, and Europe. The merchant fleet in 1991 consisted of two vessels (each of more than 1,000 tons), totaling 2,161 GRT. Corinto is Nicaragua's only natural harbor on the Pacific coast and the major port, handling about 60% of all waterborne trade. Other ports include Puerto Sandino and San Juan del Sur on the Pacific and Puerto Cabezas on the Atlantic coast. In 1983, Bulgaria pledged to help Nicaragua build a deepwater port at El Bluff, on the Atlantic; the port now allows ships from Europe, Africa, and the Caribbean to deliver goods to Nicaragua without passing through the Panama Canal. Under an agreement signed in 1982, the former USSR expanded fishing port facilities at San Juan del Sur. Inland waterways totaled 2,220 km (1,380 mi), including Lake Nicaragua and the San Juan River.

Air transportation is important because of limited road and railway facilities. A new state-owned airline, Aerolíneas de Nicaragua (AERONICA), provides services to El Salvador, Costa Rica, Panama, and Mexico. There is an international airport at Las Mercedes, near Managua.

12HISTORY

Nicaragua derives its name from that of the Amerindian chief Nicarao who once ruled the region. The first European contact came with Columbus in 1502. At that time the northern part of

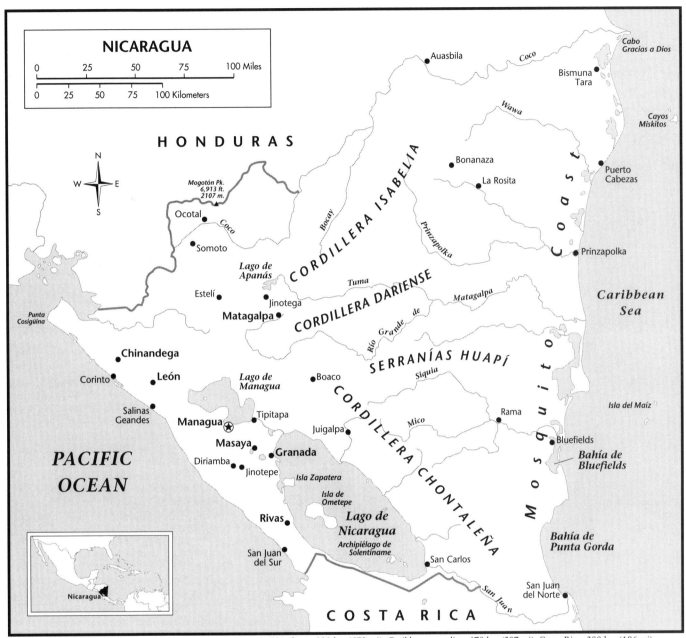

NICARAGUA

0   25   50   75   100 Miles

0   25   50   75   100 Kilometers

HONDURAS

Mogotón Pk.
6,913 ft.
2107 m.

Ocotal

Somoto

Coco

Bocay

CORDILLERA ISABELIA

Auasbila

Coco

Cabo
Gracias a Dios

Bismuna
Tara

Cayos
Miskitos

Bonanaza

La Rosita

Prinzapolka

Puerto
Cabezas

Lago de
Apanás

Estelí

Jinotega

Matagalpa

Tuma

CORDILLERA DARIENSE

Matagalpa

Río Grande de

Prinzapolka

Coast

Caribbean
Sea

Punta
Cosigüina

Chinandega

León

Corinto

Salinas
Geandes

Lago de
Managua

Boaco

SERRANÍAS HUAPÍ

Siquia

CORDILLERA CHONTALEÑA

Mico

Rama

Isla del Maíz

Mosquito

Managua

Tipitapa

Masaya

Diriamba

Granada

Jinotepe

Juigalpa

Isla Zapatera

Bluefields

Bahía de
Bluefields

PACIFIC
OCEAN

Isla de
Ometepe

Rivas

Lago de
Nicaragua

Archipiélago de
Solentiname

San Juan
del Sur

San Carlos

San Juan

San Juan
del Norte

Bahía de
Punta Gorda

Nicaragua

COSTA RICA

LOCATION: 10° to 15°N; 83° to 87°W. **BOUNDARY LENGTHS:** Honduras, 922 km (573 mi); Caribbean coastline, 478 km (297 mi); Costa Rica, 300 km (186 mi); Pacific coastline, 346 km (215 mi). **TERRITORIAL SEA LIMIT:** 200 mi.

the country was inhabited by the Sumo Amerindians, the eastern by the Miskitos, and the region around Lakes Nicaragua and Managua by agricultural tribes.

The first Spanish settlements in Nicaragua were founded by the conquistador Gil González de Ávila in 1522. The cities of Granada and León were founded in 1524 by Francisco Hernández de Córdoba. During the next 300 years—most of the colonial period—Nicaragua was ruled as part of the captaincy-general of Guatemala. On 15 September 1821, the independence of the five provinces of Central America, including Nicaragua, was proclaimed. After a brief period under the Mexican empire of Augustín de Iturbide (1822-23), Nicaragua joined the United Provinces of Central America. On 30 April 1838, Nicaragua declared its independence from the United Provinces, and a new constitution was adopted.

Nicaragua did not immediately consolidate as a nation. The Spanish had never entirely subdued Nicaragua, and the Mosquito

Coast at the time of independence was an Amerindian and British enclave, especially around the Bluefields area. Britain occupied the Mosquito Coast during the 1820s and 1830s, and maintained a significant presence thereafter. Beyond that, Nicaragua was torn apart by a bitter struggle between liberals, based in León, and conservatives, based in Granada.

Yet another factor impeding Nicaragua's development was constant foreign intervention focusing on the trade route through the country. Commodore Cornelius Vanderbilt competed with the British for control of the transisthmian traffic, a rivalry settled by the Clayton-Bulwer Treaty of 1850. In 1853, liberals led by Máximo Jérez and Francisco Castellón revolted and invited the US military adventurer William Walker to help their rebellion. Walker invaded Nicaragua in 1855, capturing Granada and suppressing Jérez, and had himself elected president in 1856. He lasted only one year, and was captured and executed in Honduras in 1860. Conservatives seized control in 1863 and ruled until 1893.

The 30-year conservative reign brought increases in coffee and banana production. Liberals successfully revolted in 1893, and José Santos Zelaya became president. Zelaya's dictatorship lasted 16 years, during which he incorporated most of the Mosquito territory into Nicaragua, developed railroads and lake transportation, enlarged the coffee plantations, and stirred up revolts among his Central American neighbors. In 1901, by the Hay-Pauncefote Treaty, Great Britain gave the US the undisputed right to build a Central American canal. Zelaya was finally deposed in 1909, after a conservative revolt.

From 1909 until 1933, the US grew in influence in Nicaragua. Conservatives immediately asked for help from Washington. The US placed an American agent in the customhouse in 1911, and US banks extended considerable credit to the bankrupt Treasury. US marines and warships arrived in 1912 in support of president Adolfo Díaz. US forces remained active in Nicaraguan politics and administered the country directly or through handpicked rulers until August 1925. During this period, the Bryan-Chamorro Treaty of 1914 allowed the US to build a canal across Nicaragua. After the marines withdrew, the liberals revolted against the US-backed conservative government of Diego Manuel Chamorro and established a government on the Mosquito Coast. The marines returned in 1926 to reimpose Díaz.

In November 1928, the marines supervised the electoral victory of the liberal José María Moncada, with whom the conservatives had made peace. The guerrilla hero Gen. Augusto César Sandino began organizing resistance to the marine occupation force in 1927, and fought the US troops to a standstill. With the inauguration of US President Franklin D. Roosevelt's "good neighbor" policy in 1933, the marines were pulled out for the last time. But the marines left a legacy, having built the Nicaraguan National Guard, headed by Anastasio ("Tacho") Somoza García.

In the following year, the liberal Juan B. Sacasa was elected to office. Also during 1934, officers of the National Guard shot Sandino after offering to negotiate a settlement with his forces. The National Guard was now unchallenged in Nicaragua, and three years later, Somoza unseated Sacasa and assumed the presidency. Somoza and his family were to rule Nicaragua directly or indirectly for the next forty-two years.

Somoza was president until 1947, making constitutional changes as necessary to prolong his term. Although he retired in 1947, he returned in 1950, and was assassinated in 1956. "Tacho's" son, Luis Somoza Debayle, was presdident of Congress, and immediately became president under the constitution. The next year, he was elected by a rather suspicious 89% of the vote.

In 1962, a law was passed prohibiting relatives within four generations from immediately succeeding Luis Somoza as president. Accordingly, in February 1963, René Schick Gutiérrez of the National Liberal Party was elected president for a four-year term. Schick died in office in August 1966 and was succeeded by his first vice-president, Lorenzo Guerrero. The presidential election of February 1967 returned the Somozas to power with an overwhelming victory for Anastasio Somoza Debayle, the younger brother of Luis.

According to Nicaraguan law, Anastasio's term in office was due to end in May 1972. But by March 1971, Somoza had worked out an agreement allowing him to stand for reelection in 1974, ruling in the interim with a three-man coalition government. Anastasio and his triumvirate drew up a new constitution, signed by the triumvirate and the cabinet on 3 April. Then, after declaring nine opposition parties illegal, Somoza easily won the September 1974 elections.

While Somoza consolidated his hold on Nicaragua, an insurgent organization, the Sandinista National Liberation Front (Frente Sandinista de Liberación Nacional—FSLN), began to agitate against his rule. At first, the group was small and confined to the foothill and mountain regions of Nicaragua. But domestic opposition to Somoza mounted, driven by the family's monopolistic and corrupt economic practices. One powerful example of the corruption was the disappearance of half the US relief aid extended to Nicaragua after a devastating 1972 earthquake. Most of the rebuilding of Managua was done by Somoza-controlled firms on Somoza's land. Throughout the 1970s, Somoza's opposition grew, and US support began to dissipate.

The FSLN used traditional unconventional tactics at first. In December, 1974, guerrillas kidnapped 13 prominent political personalities, including several members of the Somoza family. The group secured a ransom of US$1 million and the release of 14 political prisoners. Somoza responded by declaring a martial law and unleashed the National Guard. The Guard's repressive tactics created even more enemies of the Somoza regime. Repression continued throughout the 1970s, and climaxed in January 1978 with the assassination of Pedro Joaquín Chamorro, editor and publisher of the opposition newspaper La Prensa. The assassins were never found, but most felt that Somoza and the National Guard were behind the killing of this moderate leader from a prominent family.

By 1979, loss of support from the Church and the business community left Somoza without domestic allies. He had become isolated diplomatically, and after the Carter administration cut off military aid, his ability to remain in power further weakened. In May, 1979, the Sandinistas launched a final offensive. By July, Somoza had fled the country (he was assassinated on 17 September 1980 in Asunción, Paraguay). By this time, an estimated 30,000–50,000 people had died during the fighting.

Nicaragua was now ruled by a coalition Government of National Reconstruction, made up of various religious and political leaders, but dominated by the Sandinista leadership. That coalition had unraveled by mid-1980, when Alfonso Robelo and Violeta Barrios de Chamorro, widow of Pedro Chamorro, resigned from the government. Chamorro continued publishing La Prensa and preserved the paper's reputation for independence, while Robelo went into exile and supported the resistance. The Sandinistas dissolved the National Guard, and in 1982 a number of anti-Sandinista guerrilla groups (broadly referred to as the "contras") began operating from Honduras and Costa Rica. These groups consisted of former Guard members and Somoza supporters ("Somocistas") who engaged in guerrilla-style offensives, aimed at disrupting Nicaragua's agriculture and oil supplies.

The Sandinistas engaged in an ambitious program to develop Nicaragua under leftist ideals. They nationalized Somoza's land and commercial interests. They also initiated agrarian reform, and announced a series of social programs, including literacy and public health campaigns. Politically, they professed democratic ideals, but delivered only sporadically. A Statute on Rights and Guarantees was adopted, but elections were postponed. As antigovernment activity increased, the government became increasingly authoritarian. A state of emergency, proclaimed in March 1982 and extended into 1987, introduced prior censorship, particularly felt by La Prensa. Daniel Ortega emerged as the leader of the Sandinistas, and became president when elections were finally held in 1984. However, in that election, the major opposition groups withdrew from the election, making it a rather hollow victory.

In April 1981, the Reagan administration cut off aid to Nicaragua and, citing the Sandinistas' support for leftist guerrillas in El Salvador, began aiding the contras with funds channeled through the CIA. The Reagan administration sent military aid to Honduras and Costa Rica and sought increases in funding for the contras. Despite some overtures from the Sandinistas, including the expulsion of 2,200 Cuban advisors, the US continued to support the contras.

Internationally, the Sandinistas made some gains. In 1986, the World Court ruled that the US had violated international law by mining the harbors in Nicaragua. The rulings made little difference because the US refused to recognize the decision. In the US, Congress proved reluctant to fund the Nicaraguan resistance. In 1986, it was revealed that US government funds derived from covert arms sales to Iran had been secretly diverted to provide aid to the contras in violation of a US congressional ban on such aid.

On the domestic scene, the Sandinistas were less successful. Their economic policies had not produced impressive results. The inflation rate reached 33,000% in 1988 and reserves dwindled. Price controls had led to serious shortages in basic foodstuffs. Lacking any capital for investment, the situation was becoming hopeless. Attempts to pin the economic woes on the civil war fell on deaf ears as the economic situation worsened.

The Sandinistas continued to seek negotiated settlements for their internal strife. In 1986, they signed an accord with leaders of the Miskito Amerindians, granting autonomy to their region. In August 1987 Nicaragua signed the Arias peace plan for Central America. Nicaragua promised guarantees of democratic rights, and a reduction of hostilities with the contras, including a cease-fire, a reduction in the armed forces, repatriation or resettlement of refugees, and amnesty for the rebels. In exchange, the Nicaraguans were to receive guarantees of non-intervention by outside powers. Implementation was sporadic, but elections were held in 1990. The US, for its part, pledged $9 million in support of free elections, and urged all other outside donors to tie aid to the holding of elections.

The 1990 elections had a surprise winner—Violeta Chamorro. Heading a 10-party alliance called the National Opposition Union (UNO), Chamorro received 54% of the vote to Daniel Ortega's 41%. UNO also took a majority in the National Assembly. Chamorro moved to liberalize the Nicaraguan economy, but found it sluggish. Austerity measures led to dislocations and political disquiet. The US delivered miniscule amounts of economic aid, to the disappointment of hopeful Nicaraguans.

Politically, Chamorro's situation was much more tenuous. With the Sandinistas still in control of the military, Chamorro had a difficult time achieving a reduction in force. Sandinista organizations and syndicates remained, often striking against the Chamorro government. Meanwhile, the resettlement and repatriation of the contras moved slowly. Some former contras took to the field again, resuming their previous tactic of attacking civilian installations. Chamorro's own coalition, UNO, proved shaky, withdrawing support from her government in 1993 after she attempted to call for new elections. The beleaguered government persisted, but by 1994 the outlook was bleak.

13GOVERNMENT

Although constitutionally defined as a democracy, Nicaragua, between 1934 and 1979, was ruled by the Somoza family, who did not hesitate to suppress political opponents violently. The last of the constitutions promulgated during the Somoza period, effective 3 April 1974, provided for a bicameral Congress and a president elected for a six-year term, and guarantees of political rights. After the FSLN took power as the Government of National Reconstruction in July 1979, this constitution was abrogated and Congress dissolved. From July 1979 until November 1984, executive power was vested in a junta composed of five members (three members after April 1980).

The 1984 electoral reforms, created an executive branch with a president elected for a six-year term by popular vote and assisted by a vice-president and a cabinet. Legislative power is vested in a 96-member unicameral National Constituent Assembly elected under a system of proportional representation for six-year terms.

The Sandinista constitution of 1987 is still in effect. It preserves the previous institutional structure, although the National Assembly elected in 1990 had 92 seats.

14POLITICAL PARTIES

Nicaragua's traditional two parties are the National Liberal Party (Partido Liberal Nacionalista—PLN) and the Nicaraguan Conservative Party (Partido Conservador Nicaragüense—PCN). The PLN favored separation of church and state, some social legislation, no foreign interference in the political process, and limited land reform. It was supported by government employees, the National Guard, and large segments of the middle and lower classes. The PCN desired government cooperation with the Catholic Church (but also advocated freedom of religion), less government interference in private business, and a regressive tax structure. Today, the liberals are a center-right party renamed the Liberal Constitutional Party (PLC). Their power base is still in Managua, where the PLC leader, Arnoldo Alemán, is mayor. Conservatives are split into the National Conservative Party (PNC) and the Popular Conservative Alliance Party (PAPC), plus some other small parties.

When the FSLN, which was founded in 1962, came to power in July 1979, all political parties except those favoring a return to Somoza rule were permitted. Since the Somozas had all been liberals, the PLN was specifically banned.

Under the junta, Nicaragua's governing political coalition, the Patriotic Front for the Revolution (Frente Patriótico para la Revolución—FPR), formed in 1980, consisted of the FSLN, the Independent Liberal Party (Partido Liberal Independiente—PLI), the Popular Social Christian Party (Partido Popular Social Cristiano—PPSC), and the Moscow-oriented Nicaraguan Socialist Party (Partido Socialista Nicaragüense—PSN). Opposition parties included the Conservative Democratic Party (Partido Conservador Demócrata—PCD), the Nicaraguan Social Christian Party (Partido Social Cristiano Nicaragüense—PSCN), and the Social Democratic Party (Partido Social Demócrata—PSD).

The UNO coalition includes the Conservatives and the Liberals, as well as several parties formerly aligned with the Sandinistas, including the PLI and the PSD. The PLI is the party of Vice President Virgilio Godoy. Others include the Christian Democratic Union (UDC), the National Democratic Movement (MDN), the National Action Party (PAN), and the Neo-Liberal Party (PALI). Also prominent on the Nicaraguan right is the Nicaraguan Resistance Party (PRN), a legal manifestation of the contras.

15LOCAL GOVERNMENT

In July 1982, the nation's 16 departments were consolidated into 6 regions and 3 special zones, each to be administered by an official directly responsible to the central government. However, Nicaragua has returned to the old system. After the pact signed with the Miskito Indians, the South Atlantic Coast Autonomous Zone (RAAS) became a separate region, so that the number of departments is 17.

Local elections for mayoralties accompany national elections.

16JUDICIAL SYSTEM

At the head of the judicial branch is the Supreme Court in Managua whose justices are appointed by the National Assembly for six-year terms. During 1993, the Supreme Court consisted of nine judges.

The judicial system consists of both civilian and military courts. Military courts investigate, prosecute and try crimes committed by or against the police or armed forces. Therefore, the military courts have jurisdiction over citizens involved in security-related offenses. In a controversial 1993 decision, a military court exercised jurisdiction to convict a former member of the EPS (Sandinista Popular Army).

17ARMED FORCES

In 1993, the regular armed forces, a fusion of the Sandinista and Contra armies, numbered 15,000. The army had 13,000 personnel, the navy 500, and the air force 1,200 (with 16 combat aircraft). The People's Militia, founded in 1980, has been disbanded and not yet replaced. Weapons remain Russian, but foreign advisors are gone as are foreign subsidies. Nicaragua spent $70 million for defense in 1991 or 3.8% of GDP.

18 INTERNATIONAL COOPERATION

Nicaragua is a charter member of the UN, having joined on 24 October 1945, and belongs to ECLAC and all the nonregional specialized agencies except WIPO. It also participates in CACM, G-77, IDB, OAS, and PAHO, as well as the International Coffee Agreement and other inter-American and international organizations. Nicaragua is a signatory to GATT and the Law of the Sea.

The Sandinistas enjoyed considerable international support even before the July 1979 revolution. Not only Cuba, but also Panama and Venezuela backed the guerrillas, and by mid-June 1979, Brazil, Costa Rica, Ecuador, and Mexico had cut diplomatic ties with the Somoza regime. Later in that month, the OAS called for "immediate and definitive replacement" of Somoza with a broadly based democratic government. During the 1980s, relations with the US and other Western governments deteriorated, the borders with Honduras and Costa Rica grew tense, and the Sandinista government found its firmest support in Mexico and in the Communist-bloc countries. In August 1987, Nicaragua signed the Central American peace plan outlined by Costa Rican President Oscar Arias Sánchez.

19 ECONOMY

Nicaragua has long had, in effect, two economies: an export segment, producing mainly cotton, meat, coffee, and sugar; and a subsistence segment, tying a majority of both urban and rural Nicaraguans to an impoverished existence. Agriculture and forestry are still mainstays of the Nicaraguan economy, employing nearly half the labor force. During 1960–64, the GDP increased by an annual average of 8.1%, the highest rate in Latin America. Annual growth ranged from 4% to 6% during 1965–73, largely because of favorable world prices for Nicaraguan commodities. The 1972 earthquake that struck Managua caused material losses estimated at $845 million, but the agricultural sector was left largely unscathed.

The civil war of the late 1970s severely disrupted the economy. During this era, 80% of the economy was nationalized. In 1978, the GDP fell by 7.9%; in 1979, the year of the Sandinista takeover, by 25%. Massive public spending resulted in a GDP growth of 10.4% in 1980 and 7% in 1981. However, because of floods in May 1982, a weak international market for export crops, the virtual collapse of the CACM, continuing political uncertainty, mounting economic pressure by the US, and disruption by the contras, the economy suffered a GDP decline of 1.4% in 1982, and it continued to decline through 1986 (with the exception of 1983, when high world coffee prices and a bumper harvest boosted GDP by 4.6%). The average annual inflation rate during 1980–82 was 35%, but by 1984, it had risen to 50%; in 1985 it soared to 334%, and the 1986 rate was 778%. Because of shortages, rationing of soap, flour, and cooking oil was introduced in 1982.

President Violeta Chamorro instituted the first democratic government in more than 50 years when she took office in April 1990. She inherited a country with a totally controlled economy and the highest per-capita foreign debt in the world. Inflation was ascending uncontrollably and the economy was in shambles. On 31 March 1991, the Chamorro administration implemented a stringent stabilization and structural readjustment program which included: a major devaluation of the córdoba, reduction in inflation, and a plan for the re-entry of Nicaragua into the international financial market. Its main challenge was to lead Nicaragua back to a market-driven economy. President Chamorro had

to reestablish private enterprise(including the return of properties confiscated during the Sandinista era), a liberal foreign investment law, free trade zone legislation, an export promotion law, lifting of trade restrictions, and a restructuring of tax laws to encourage private sector development.

In her first two years as president, Mrs. Chamorro's government successfully stopped inflation, which amounted to only 3.9% in 1992, down from 13,490% in 1990 and 775% in 1991. It also stabilized the local currency, eliminated the fiscal deficit, and lifted many government controls. With the assistance of friendly governments, Nicaragua is renegotiating its $9 billion foreign debt and is also undertaking a privatization program that should bring additional stability to the country's economy. However, the GDP has seen no new growth under the new government; growth was expected to remain at around 2% in 1993. The unemployment rate was 54% in 1992.

Agriculture and agro-industry still represent the backbone of the economy, producing 25% of GDP, 75% percent of exports, and employing 33% of the labor force in 1992. Agricultural output has been growing at a slow pace, reflecting continuing uncertainty about property rights as well as the stricter credit terms that limit credit availability. Nonetheless, agricultural output increased 1.1% in 1992, largely due to the largest coffee harvest since 1984. Livestock operations also increased production by 5.9%. In addition, seafood production soared by 19.8%, making it the leader in export growth.

20 INCOME

In 1992, Nicaragua's GNP was $1,325 million at current prices, or $410 per capita. For the period 1985–92 the average inflation rate was 2,533.8%, resulting in a real growth rate in per capita GNP of –7.8%.

In 1992 the GDP was $1,847 million in current US dollars.

21 LABOR

In 1990, the official estimate of the total economically active population was 1,386,300, which was distributed as follows: services, 39.6%; agriculture, 30%; and industry, 16.4%. However, according to some estimates, more than 50% of the work force remains unemployed or underemployed. Officially, unemployment was at 14% in 1991.

Nicaragua became a member of the ILO in 1919, withdrew in 1938, and rejoined in 1957. The labor code, effective January 1945, was patterned on Mexican labor legislation. In establishing and protecting the rights of workers, emphasis was placed on law rather than collective bargaining. Labor unions were permitted by the Somoza regime, but the government imposed strict limits on their activities.

As of 1992, about 50% of the labor force was organized—all public and private sector workers (except military and police) have the right to form and join unions of their choice and to freely exercise the right to strike. The two major bodies are the pro-Sandinista National Workers' Front, with 7 unions, and the independent Permanent Congress of Workers, composed of 4 independent labor centrals. One major labor organization, the Christian Democratic Nicaraguan Workers' Central, remains unallied to either of the two larger umbrella groups.

22 AGRICULTURE

Nicaragua's economy is predominantly agricultural. Land under cultivation rose by 2.9% between 1974 and 1991 to 1.27 million hectares (3.1 million acres), or about 10% of the total area. Some 86,000 hectares (212,500 acres) were under irrigation in 1991. The harvest season begins in November and lasts through January; during the rest of the year, most rural laborers are unemployed. Plantings begin in May immediately before the wet season.

During the Somoza era, most of the titled land was held by large landowners (with farms of 140 hectares/346 acres or more), who owned some 60% of the land while representing only 5% of the farming population. About 36% of the farm population controlled individual holdings of less than 3.5 hectares (8.6 acres). The Sandinista government expropriated almost 1 million hectares (2.5 million acres) of land, of which over two-thirds became state farms and 280,000 hectares (692,000 acres) were turned into peasant cooperatives. An estimated 450,000 hectares (1,111,500 acres) of land were redistributed in 1985.

The main agricultural exports are coffee (21% of total exports in 1991), cotton, sugar, and bananas. Nontraditional exports are growing and include: honeydew melons, cantaloupe, sesame seed, onions, baby corn, asparagus, artichokes, and cut flowers. Sorghum, cacao, yucca, tobacco, plantains, and various other fruits and vegetables are produced on a smaller scale for the local markets. In 1992, the agricultural sector performed sluggishly, caused by uncertainty about property rights as well as stricter controls on credit availability. Cotton prices fell by 35%, causing fewer acres to be planted. In 1992, cotton accounted for only 11.9% of exports ($26 million), down from 16.1% ($44 million) in 1991. Bananas were once nearly totally decimated by Panama disease. By the late 1960s, however, production had begun a slow recovery, reaching 135,000 tons in 1992 (up from 29,000 tons in 1970). Cottonseed production has expanded from virtually zero prior to 1950 to 105,700 tons in 1985, before falling to 36,000 tons in 1992. Coffee was severely threatened by contra activities, but production of 47,000 tons in 1991 was an improvement over the 28,000 tons produced in 1990. In 1992, 2,563,000 tons of sugar were produced, largely for export. Major food crops in that year were corn, 231,000 tons; rice, 158,000 tons; sorghum, 74,000 tons; beans, 79,000 tons; and sorghum, 74,000 tons.

Agriculture was severely disrupted in 1979 and 1980 because of the revolution, but by 1981 it had recuperated. In May 1982, severe floods caused damages estimated at $180 million; the withdrawal of the Standard Fruit Co. in the following October caused losses of $400,000 per week in foreign exchange earnings. Bad weather continued to plague the sector through 1984. From 1983 to 1987, the contras sought to destabilize Nicaraguan agriculture by damaging agricultural machinery, destroying crop storage sheds, and intimidating farm workers. After eight years of steady decline, the agricultural sector grew by a modest 1–2% in 1992, mostly due to the largest coffee harvest since 1984.

23ANIMAL HUSBANDRY

Nicaragua, the second largest cattle-raising country of Central America (after Honduras), had 1,673,000 head of dairy and beef cattle in 1992. There were also 172,000 horces, 700,000 hogs, and 91,000 mules and donkeys. Total meat production in 1992 was 63,000 tons. Meat exports, perennially one of Nicaragua's most important trade commodities, were valued at us$23 million in 1981 but had fallen to us$7 million by 1987, but have rapidly increased since then to $41.6 million in 1992. In 1992, Nicaragua was given back access to the US beef products market, which caused livestock production to increase by almost 6%. Milk production in 1992 totaled 160,000 tons; eggs, 26,000 tons.

24FISHING

Commercial fishing in the lakes and rivers and along the seacoasts is limited. In 1991, the total catch amounted to 5,709 tons, almost 96% of which is from marine waters. About 80% of the marine catch comes from the Atlantic coast. Exports of shrimp and lobster expanded after the 1960s and by 1980 had reached an export value of us$25.9 million. Thereafter, however, they declined, reaching $9.3 million in 1988 before rising to $16 million in 1991. Commercial fishing is now trying to diversify its catch to include more red snapper, grouper, and flounder.

After the Sandinistas took over, the fishing industry was nationalized. After an agreement made in 1982, the former USSR expanded the fishing port at San Juan del Sur to service its tuna fleet. In late 1991, the government privatized the Atlantic seafood packaging plants, causing seafood production to rise by 19.8% in 1992.

25FORESTRY

About 28% of Nicaragua is still forested. The country has four distinct forest zones: deciduous hardwood, mountain pine, lowland pine, and evergreen hardwood. Nicaragua's largest remaining timber resources, in the evergreen hardwood zone, are largely inaccessible. Nicaragua is the southernmost area of natural North American pine lands. The most well-known cloud forest in Nicaragua is *Selva Negra (Black Forest)*, in the Matagalpa region. In 1991, roundwood production totaled 4.18 million cu m, and sawn wood production was about 222,000 cu m.

26MINING

The extent of Nicaragua's mineral resources remains largely undetermined. A government survey in 1956 uncovered commercial deposits of iron, copper, lead, and antimony, and in 1961 tungsten was reported. In the late 1970s, gold, silver, copper, lead, and zinc were being mined. Mining and quarrying contributed only 0.4% to the GDP in 1985.

During the Sandinista years, the government nationalized the mining industry. The development of gold mining was emphasized, and gold exports reached $39.9 million in 1980. In 1982, they had fallen to $15 million, and they were suspended thereafter through 1985. As of 1991, the Corporación Nicaragüenese de Minas (INMINE), a subsidiary of the government holding company, controlled most of the mineral exploration and production in the country. INMINE mines at Bonanza, El Limón, and La Libertad were producing gold and silver in 1991. Gold output in 1991 was 1,154 kg; silver, 1,543 kg.

27ENERGY AND POWER

The National Light and Power Co. (Empresa Nacional de Luz y Fuerza—ENALUF) is responsible for most of the electricity generated in Nicaragua. Production of electricity increased from 77 million kwh in 1948 to 1,043 million kwh in 1991. Nicaragua had a net capacity of 395,000 kw, of which 26% was hydroelectric. In August 1983, a geothermal electrical generating plant was opened at the foot of the Momotombo volcano; its generating capacity of 70,000 kw supplied about 17% of Nicaragua's electricity needs in 1991.

Since the US removed its economic embargo against Nicaragua in 1990, Nicaragua has been seeking to attract international exploration interests, especially along its Caribbean coast.

28INDUSTRY

Nicaraguan industry expanded during the 1970s but was severely disrupted by the civil war. In 1980, the manufacturing sector began to recuperate, and modest growth continued through 1984. In 1985, however, net output again declined, by an estimated 5%.

In the mid-1980s there were still many state enterprises, some of them created by nationalization since 1979; in 1985, the government announced plans for a mixed economy. Major industrial plants include a petroleum refinery, textile mills, tobacco and cement plants, and a number of food-processing installations. Cement production in 1983 was estimated at 298,000 tons; and wheat flour, 47,000 tons.

In 1992, the manufacturing sector contributed approximately 25% to the GDP and employed around 11% of the labor force. During that year, the industrial sector didn't experience substantial growth, due to a decline in the output of various subsectors of the manufacturing industry, which faced competition from imported non-durable consumer goods. In 1993, however, industrial production grew by about 4.6%. Manufacturing is concentrated primarily in the areas of food and tobacco processing, beverages, petroleum refining, and chemicals.

[29]SCIENCE AND TECHNOLOGY

Among Nicaragua's scientific learned societies and research institutes are the Geophysical Observatory and the Nicaraguan Society of Psychiatry and Psychology. The National Center of Agricultural Information and Documentation is part of the Ministry of Agriculture. Nicaragua has six universities and colleges offering degrees in agricultural studies and other scientific studies.

In 1987, total research and development expenditures amounted to c$989 million cordobas; 302 technicians and 725 scientists and engineers were engaged in research and development.

[30]DOMESTIC TRADE

Managua is the principal trading and distribution center, and all importers and exporters have offices there. Exporters, except those concerned with cotton, coffee, or lumber, are usually importers also. Managua has a variety of retail establishments, including department stores and numerous general stores; many small shops are in private homes. Managua also has a central market to which merchants come daily with all types of produce and domestic and imported consumer goods. Retail sales are mainly for cash.

The usual business hours are from 8 AM to noon and from 2:30 to 5:30 or 6 PM, Monday through Friday, with a half-day on Saturday. Banking hours on weekdays are from 8:30 AM to noon and from 2 to 4 PM; on Saturday, from 8:30 to 11:30 AM.

[31]FOREIGN TRADE

Nicaragua's total trade volume (in current terms) grew considerably during the 1970s because of the country's membership in the CACM and because of worldwide inflation. Following the Sandinista revolution and the virtual collapse of the CACM because of political instability in the region, Nicaraguan exports fell from $646 million in 1978 to $292 million in 1985. Imports decreased from $594 million in 1978 to $360 million in 1979 but then soared to $832 million in 1985.

By 1986, Latin American and EEC member countries, particularly the Federal Republic of Germany (FRG), accounted for the bulk of Nicaragua's trade volume, and the Communist bloc had filled the breach opened by the shutting down of US commerce. Japan is an important buyer of Nicaraguan exports. Nicaragua's exports to the US began to decline in 1982, when the Reagan administration cut Nicaragua's sugar quota by 90% in order to reduce the resources available to finance "Nicaraguan-supported subversion and extremist violence." In 1982, the US accounted for 20.7% of Nicaragua's total export value; in 1986, the proportion was 0.3%.

The Chamorro government has significantly reduced trade barriers, mainly by cutting tariffs and eliminating state monopolies. The government has issued export promotion incentives with special tax benefits for products sold outside Central America. By the Law of Free Trade Zones, Nicaragua waives all duties for imports and vehicles used in the free zones. The result has been widespread availability of US produced consumer goods in several newly established Managua supermarkets.

Nicaragua's export earnings fell by 16.4% in 1992, mainly reflecting lower prices of coffee, cotton, banana, and sugar. At the same time, the country's imports increased by 14.2% to $757 million. Petroleum imports rose by 5.2% and consumer goods imports jumped by 31.3%. At year end 1992, Nicaragua's major export-trading partners were Canada, Germany, Japan, US and CACM nations. Its major import-trading partners included: US, CACM nations, Japan, and Mexico.

[32]BALANCE OF PAYMENTS

An adverse balance of trade with Nicaragua's major trading partners is the major factor in its deficit. Incoming capital in the form of public and private loans, as well as foreign capital investment, traditionally offset amortization and interest payments abroad. In 1992, merchandise exports totaled $217.5 million and imports $735.6 million. The merchandise trade balance was $-518.1 million.

The following table summarizes Nicaragua's balance of payments for 1991 and 1992 (in millions of US dollars):

	1991	1992
CURRENT ACCOUNT		
Goods, services, and income	−849.2	−1,084.0
Unrequited transfers	844.4	388.6
TOTALS	−4.8	−695.4
CAPITAL ACCOUNT		
Direct investment	—	15.0
Portfolio investment	—	—
Other long-term capital	−518.2	−490.5
Other short-term capital	−25.4	359.6
Exceptional financing	549.8	772.1
Other liabilities	—	—
Reserves	−86.1	−1.7
TOTALS	−79.9	654.5
Errors and omissions	84.7	40.9
Total change in reserves	−82.6	−16.1

[33]BANKING AND SECURITIES

The banking system, nationalized in July 1979, is under the supervision of the comptroller general. The National Bank of Nicaragua, established in 1912, has been government-owned since 1940. In 1979, the bank was reorganized to become the National Development Bank. The Central Bank of Nicaragua (Banco Central de Nicaragua), established in 1961, is the bank of issue and also handles all foreign exchange transactions. In 1993 there were 5 other state banks and 3 foreign banks in Nicaragua. As of 1979, deposits in foreign banks were prohibited, but in May 1985, the establishment of private exchange houses was permitted. In 1990, legislation was passed that allowed for the establishment of private banks. As of 1992 there were 3 private commercial banks in Nicaragua. There are no security exchanges in Nicaragua.

[34]INSURANCE

In 1979, the Nicaraguan Institute of Insurance and Reinsurance took over all domestic insurance companies. There were four foreign insurance companies operating in Nicaragua in 1986. Life insurance in force in 1992 totaled c$0169 million.

[35]PUBLIC FINANCE

Since the mid-1960s, government spending has consistently exceeded revenues. During the Sandinista regime, detailed public finance budgets were not a priority. In 1991, budgeted revenues totaled $347 million, while expenditures amounted to $499 million. The external debt amounted to $10 billion at the end of 1991.

[36]TAXATION

In 1993, the individual income tax ranged from a base rate of 8% up to 35.5%, and the corporate income tax rate was 35.5%. Other taxes included a 10% sales tax; a luxury tax of 10% to 100%; a 2% municipal tax levied on cash sales and collections of credit sales; a stamp tax; a 1% tax on net worth; and a real estate tax at 1% of assessed value.

³⁷CUSTOMS AND DUTIES
The government's liberalized import schedule allows private sector imports for the first time in 11 years. Import licenses are easily acquired. A major tax reform in July 1990 lowered the maximum tariff rate from 61% to 20%. Duties are set on an ad valorem basis, and there is a 10% VAT and a 3% stamp tax.

³⁸FOREIGN INVESTMENT
Until the 1979 revolution, Nicaragua encouraged private investment. Virtually no restrictions were imposed on the remittance of profits or repatriation of capital. The economic and political climate for foreign investors in the 1980s was bleak, despite the claim by Nicaraguan financial authorities in 1982 that the Sandinista government was prepared to offer more favorable investment terms (including 100% foreign ownership and repatriation of profits) than the Somoza government had provided. As of 1984, direct US investment in Nicaragua had stopped completely.

As of 1994, Nicaragua had recovered some of the international credibility lost in the previous decade. Presently, under the New Foreign Investment Law, the government of Nicaragua is concentrating most of its efforts on the expansion and promotion of foreign and national investment. This law, among other things, guarantees the repatriation of invested capital and generated capital. Also, it allows for 100% foreign ownership in all areas. The Nicaraguan government is also working with the US and other first world countries to create what are known as Bilateral Investment Treaties.

³⁹ECONOMIC DEVELOPMENT
The Somoza government's 1975–79 National Reconstruction and Development Plan had as its major objectives the improvement in living conditions through increased employment, continued reconstruction of Managua, reduction in the economy's dependence on the external sector, acceleration of regional development, and strengthening of the country's role in CACM. The plan was disrupted by civil strife in the late 1970s.

After the 1979 revolution, the government nationalized banking, insurance, mining, fishing, forestry, and a number of industrial plants. An emergency economic program to reactivate the economy resulted in a 10.4% increase in the GDP in 1980. Although the government officially favored a mixed economy, in practice the private sector took second place in a development strategy that focused on public investment and control.

Between 1946 and 1979, the US provided US$269.9 million to Nicaragua in development grants and loans; aid pledged for 1980 and 1981 (before the April 1981 cutoff) totaled US$98.6 million. Within a month of the US suspension of aid, Libya had pledged US$100 million for an agricultural enterprise to be established by Libyan technicians, Cuba had offered US$64 million in aid for 1981, and the USSR and Bulgaria pledged a combined total of 30,000 tons of wheat. By 1984, 60.4% of external aid came from the Communist bloc, in contrast to 19.3% in 1979. Since 1983, an increasing proportion of aid has come in the form of credit lines and technical assistance. Between 1982 and 1985, Bulgaria and the former German Democratic Republic were estimated to have provided $200 million in credits; since 1985, the former USSR has supplied over one-half of Nicaragua's oil requirements, and Mexico continues to supply small quantities of oil despite nonpayment for past shipments by Nicaragua. Sweden and the Netherlands extended a total of $37.5 million and $78.5 million in aid, respectively, between 1981 and 1984. After 1983, the US blocked the extension of credits to Nicaragua by the IDB and IBRD. Multilateral aid to Nicaragua totaled US$736.6 million between 1946 and 1986.

In response to the macroeconomic problems that arose in 1992, a series of measures has been adopted by the Chamorro administration aimed at consolidating the stabilization process,

increasing the competitiveness of exports and establishing a base for the promotion of growth. Although there is a clear trend toward continued improvement, the country still has to resolve a number of social, economic, and political issues that are hampering its recovery. Particular problems are the property rights issue of nationalized businesses and the still–strong influence of the Sandinistas in many sectors of government.

⁴⁰SOCIAL DEVELOPMENT
Nicaragua's basic social welfare system was established during the Somoza period. A social insurance law enacted in 1956 provides for national compulsory coverage of employees against risks of maternity, sickness, employment injury, occupational disease, unemployment, old age, and death. Family allowance legislation enacted in 1982 provides benefits for children under the age of 15.

The Office of Family Welfare was established in 1976 in order to provide family planning services. After the Sandinista revolution, family planning programs were suspended; later they were integrated into a general program of women's health services. The fertility rate in 1985–90 was 5.6. Abortion is permitted only to save the woman's life. There is no official discrimination against women, who account for 25% of the labor force, and a number of women hold government positions. However, women continue to suffer de facto sex discrimination in many segments of society. They tend to hold traditionally low-paid jobs in the health, education, and textile sectors while occupying few management positions in the private sector.

⁴¹HEALTH
Slow progress in health care was made from the 1960s through the 1980s, as the crude death rate dropped from 19 per 1,000 population in 1960 to 6.8 in 1993; during 1992, the infant mortality rate was 54 per 1,000 live births, and average life expectancy was 66 years. However, malnutrition and anemia remained common, poliomyelitis and goiter were endemic, and intestinal parasitic infections (a leading cause of death) afflicted over 80% of the population. There were approximately 110 reported cases of tuberculosis per 100,000 people in 1990. Immunization rates for children up to one year old in 1992 were as follows: tuberculosis (79%); diphtheria, pertussis, and tetanus (73%); polio (86%); and measles (72%). Total health care expenditures for 1990 were $133 million. In 1992, 83% had access to health care services.

The 1993 birthrate was 40.5 per 1,000 people, with 27% of married women (ages 15 to 49) using some form of contraception. There were 163,000 births in 1992. In 1990, there were 1.8 hospital beds per 1,000 inhabitants. In 1992 there were 60 doctors per 100,000 people.

⁴²HOUSING
Both urban and rural dwellers suffer from a dire lack of adequate housing. As a result of the 1972 earthquake, approximately 53,000 residential units were destroyed or seriously damaged in the Managua area. The Sandinistas launched housing-construction and tree-planting programs, but were hampered by a shortage of hard currency to pay for the construction equipment required. As of 1971, 69% of dwellings were detached houses; 19% were rural homes called ranchos; and 9% were cuartes, or private units with some common facilities. Over 60% of all housing was owner occupied and 20% was rented.

⁴³EDUCATION
After the Sandinista takeover, a literacy campaign (with the help of 2,000 Cuban teachers) was launched in 1980. At the end of the campaign, the government claimed that the adult illiteracy rate, which was 50% in 1975, had been reduced to 13%; however, the Population Reference Bureau said that the illiteracy rate was 42% in 1985.

Primary and secondary education is free and compulsory between the ages of 6 and 13. In 1990 there were 632,882 pupils in 4,030 primary schools, with 19,022 teachers; 168,888 students in secondary schools, with 4,865 teachers; and 30,733 students in 16 institutions of higher learning, with 2,289 teachers.

The National Autonomous University of Nicaragua offers instruction in 10 faculties: medicine, law and social sciences, dentistry, chemistry, and humanities in León; and agriculture, education, economics, physical and mathematical sciences, and humanities in Managua. The Central American University, a Roman Catholic institution, opened in Managua in 1961, and the privately controlled Polytechnic University of Nicaragua, also in Managua, attained university status in 1977.

44LIBRARIES AND MUSEUMS
The National Library and the National Archives in Managua are among the largest libraries in the country. Smaller collections are found at the Municipal Library in León and at the National Autonomous University of Nicaragua. The National Museum is also located in Managua.

45MEDIA
Postal, telegraph, and telephone facilities are government-owned. Since 1990, TELCOR, the national communications company, has invested over $100 million on upgrading its facilities. Telephone service is limited to the heavily populated west coast and, except for Managua (where there is an automatic dial system), is inadequate. In 1993 there were 64,000 telephone lines, with plans underway to increase this number to 100,000 by 1994 and 230,000 by 1996. Radiotelephone circuits allow communication between the west and east coasts. In 1992 there were 45 AM radio stations, no FM stations, and 7 TV stations. Radio Católica, a Roman Catholic station, was closed by the government in 1986 but was reopened in 1987 in accordance with the democratic freedoms outlined in the Arias peace plan. The Voice of Nicaragua is the government station. In 1991, an estimated 997,000 radios and 249,000 television sets were in use. In 1991, important dailies included the Nuevo Diario, with 30,000 circulation, and the Barricada, with 38,000. Press censorship ended with the departure of the Sandinista government. La Prensa, a harsh critic of Somoza rule and of the Sandinista regime, was closed in 1986 but, in accordance with the Arias peace plan, was allowed to resume publication in 1987; its 1991 circulation was 45,000.

46ORGANIZATIONS
In 1993 there were three cooperative organizations including one for cotton growers and one for shoemakers and leather workers. There were four employer; organizaions as of that year. Of the employers' associations, the most important was the Higher Council of Private Enterprise (Consejo Superior de la Empresa Privada—COSEP).

47TOURISM, TRAVEL, AND RECREATION
Although Nicaragua has beaches on two oceans, magnificent mountain and tropical scenery, and the two largest lakes in Central America, a decade of military conflict retarded the development of the tourist industry. However, tourism has gained momentum since the advent of the Chamorro government. In 1991, 145,872 tourists arrived in Nicaragua, 118,312 from the Americas and 22,969 from Europe. Tourist revenues reached $17 million in that year. All visitors must have a valid passport. US citizens can obtain 30-day visas. Baseball is the national sport. Basketball, cockfighting, bullfighting, and water sports are also popular.

48FAMOUS NICARAGUANS
International literary fame came to Nicaragua with the publication of Azul, a collection of lyric poetry and short stories by Rubén Darío (Félix Rubén Garcia-Sarmiento, 1867–1916). Born in Metapa (renamed Ciudad Darío in his honor), Darío created a new literary style in Spanish, exemplified by "art for art's sake" and a revelry in the senses. Miguel Larreynaga (1771–1845) was an outstanding figure during the colonial period and later an ardent independence leader, teacher, jurist, and author. Santiago Arguëllo (1872–1940) was a noted poet and educator. Three modern poets are Fray Azarías Pallais (1885–1954), Alfonso Cortés (1893–1963), and Salomón de la Selva (1893–1959). Luis Abraham Delgadillo (1887–1961), a writer, educator, and musical conductor, was also Nicaragua's leading composer.

The Somoza family, which ruled Nicaragua from 1934 to 1979, included Anastasio Somoza García (1896–1956), president during 1937–47 and again during 1950–56; his oldest son, Luis Somoza Debayle (1922–67), president during 1956–63; and a younger son, Anastasio Somoza Debayle (1925–80), president during 1967–72 and again from 1974 until the 1979 revolution. The Sandinistas, who overthrew the Somoza dynasty, take their name from the nationalist Gen. Augusto César Sandino (1895–1934). José Daniel Ortega Saavedra (b.1946) emerged as the leading figure in the junta that governed Nicaragua from 1979 to 1990.

49DEPENDENCIES
Nicaragua has no territories or colonies.

50BIBLIOGRAPHY
Borge, Tomás, et al. The Sandinistas Speak: Speeches and Writings of Nicaragua's Leaders. New York: Path, 1982.
Conroy, Michael E. (ed.). Nicaragua, Profiles of the Revolutionary Public Sector. Boulder, Colo.: Westview Press, 1987.
Haverstock, Nathan A. Nicaragua in Pictures. Minneapolis: Lerner Publications Co., 1987.
Nolan, David. The Ideology of the Sandinistas and the Nicaraguan Revolution. Miami: University of Miami North-South Center, 1984.
Norsworthy, Kent. Nicaragua: A Country Guide. 2d ed. Albuquerque, N.M.: Inter-Hemispheric Education Resource Center, 1990.
Pezzullo, Lawrence. At the Fall of Somoza. Pittsburgh: University of Pittsburgh Press, 1993.
Somoza, Anastasio. Nicaragua Betrayed. Belmont, Mass.: Western Islands, 1980.
Spalding, Rose J. (ed.). The Political Economy of Revolutionary Nicaragua. Boston: Allen & Unwin, 1987.
Stanislawski, Dan. The Transformation of Nicaragua. Berkeley: University of California Press, 1983.
Walker, Thomas W. (ed.). Nicaragua in Revolution. New York: Praeger, 1982.
———. (ed.). Revolution & Counterrevolution in Nicaragua. Boulder, Colo.: Westview Press, 1991.
———. Nicaragua, the Land of Sandino. 3rd ed. Boulder, Colo.: Westview Press, 1991.
Woodward, Ralph Lee, Jr. Central America: A Nation Divided. New York: Oxford University Press, 1976.

PANAMA

Republic of Panama
República de Panamá

CAPITAL: Panama City (Panamá).

FLAG: The national flag is divided into quarters. The upper quarter next to the staff is white with a blue star; the upper outer quarter is red; the lower quarter next to the staff is blue; and the lower outer quarter is white with a red star.

ANTHEM: *Himno Nacional,* beginning "Alcanzamos por fin la victoria" ("We reach victory at last").

MONETARY UNIT: The balboa (B) of 100 centésimos is the national unit of account. Panama issues no paper money, and US notes are legal tender. Panama mints coins of 0.05, 0.10, 0.25, 0.50, 1 and 5 balboas which are interchangeable with US coins.

WEIGHTS AND MEASURES: The metric system is official, but British, US, and old Spanish units are also used.

HOLIDAYS: New Year's Day, 1 January; Martyrs' Day, 9 January; Labor Day, 1 May; National Revolution Day, 11 October; National Anthem Day, 1 November; All Souls' Day, 2 November; Independence from Colombia, 3 November; Flag Day, 4 November; Independence Day (Colón only), 5 November; First Call of Independence, 10 November; Independence from Spain, 28 November; Mother's Day and Immaculate Conception, 8 December; Christmas, 25 December. Movable religious holidays are Shrove Tuesday and Good Friday.

TIME: 7 AM = noon GMT.

1 LOCATION, SIZE, AND EXTENT

The Republic of Panama, situated on the Isthmus of Panama, has an area of 78,200 sq km (30,193 sq mi). Comparatively, the area occupied by Panama is slightly smaller than the state of South Carolina. The Canal Zone (1,432 sq km/553 sq mi), over which the US formerly exercised sovereignty, was incorporated into Panama on 1 October 1979, with the US retaining responsibility for operation of the Panama Canal and the use of land in the zone for maintenance of the canal until the year 2000.

Panama extends 406 km (252 mi) ESE-WNW and 280 km (174 mi) NNE-SSW west of the Panama Canal and 345 km (214 mi) ESE-WNW and 167 km (104 mi) nne-ssw east of the canal. Bordered on the N by the Caribbean Sea, on the E by Colombia, on the s by the Pacific Ocean, and on the w by Costa Rica, Panama has a total boundary length of 3,045 km (1,892 mi).

Panama's capital city, Panama City, is located where the Panama Canal meets the Gulf of Panama.

2 TOPOGRAPHY

Panama is a country of heavily forested hills and mountain ranges. The two principal ranges are in the eastern and western sections of the country, and a third, minor range extends southward along the Pacific coast into Colombia. The eastern Serranía del Darién parallels the Caribbean coastline, while the Serranía de Tabasará ascends westward, culminating in the Barú volcano (3,475 m/11,401 ft), formerly known as Chiriquí. Between these ranges, the land breaks into high plateaus, ridges, and valleys. The Panama Canal utilizes a gap in these ranges that runs northwest to southeast and averages only 87 m (285 ft) in altitude. Panama has more than 300 rivers, most of which flow into the Pacific, with only the Tuira River in Darién Province of any commercial importance. Both coasts of the isthmus have deep bays, but the Gulf of Panama is especially well provided with deepwater anchorages. Panama also has more than 1,600 islands, including

the Amerindian-inhabited San Blas Islands in the Caribbean (366) and the Pearls Archipelago in the Gulf of Panama (over 100). Its largest island is the penal colony Coiba, which is south of the Gulf of Chiriquí.

3 CLIMATE

Panama is tropical, but temperatures vary according to location and altitude. The annual average temperature on both coasts is 27°C (81°F), and it ranges from 10° to 19°C (50 to 66°F) at various mountain elevations. There is little seasonal change in temperature, with warm days and cool nights throughout the year. Humidity is quite high, however, averaging 80%. Rainfall averages 178 cm (70 in) in Panama City and 328 cm (129 in) in Colón. The period of lightest rainfall is from January to March.

4 FLORA AND FAUNA

Most of Panama is a thick jungle, with occasional patches of savanna or prairie. On the wet Caribbean coast, the forest is evergreen, while on the drier Pacific side the forest is semi-deciduous. Species of flowering plants exceed 2,000 and include the national flower, the Holy Ghost orchid. Mammals inhabiting the isthmus are the anteater, armadillo, bat, coati, deer, opossum, peccary, raccoon, tapir, and many varieties of monkey. Reptiles, especially alligators, are numerous along the coasts. Bird life is rich and varies according to the presence of migratory species. Fish abound, and the Pacific coast is a favored region for sport fishing.

5 ENVIRONMENT

Soil erosion and deforestation are among Panama's most significant environmental concerns. Soil erosion is occuring at a rate of 2,000 tons per year. During the late 1980s and early 1990s, Panama lost 139 square miles of its forests. Air pollution is also a problem in the urban centers due to emissions from industry and transportation. Panama contributes 0.1% of the world's total gas emissions. Water pollution directly affects 44% of the rural people who do not have pure water. Pesticides, sewage, and the oil industry cause much of the pollution. Agencies with environmental responsibilities include the Ministry of Health and Ministry of Rural Development. The Smithsonian Tropical Research Institute, in Balboa, conducts studies on the conservation of natural resources. In 1994, the nation's fish resources were threatened by water pollution. Thirteen of Panama's mammal species are endangered. Fourteen of the nation's bird species are also endangered. Of the nation's plant species, a total of 549 are threatened with extinction. Endangered species in include the red-backed squirrel monkey, tundra peregrine falcon, spectacled caiman, American crocodile, and four species of sea turtle (green sea, hawksbill, olive ridley, and leatherback).

6POPULATION

The census of 13 May 1990 showed the total population as 2,329,329, excluding residents of the former Canal Zone. The mid-1994 estimate was 2,630,791, and the UN projection for the year 2000 was 2,893,000, assuming a crude birthrate of 22.8 per 1,000 population, a crude death rate of 5.2, and a net natural increase of 17.6 during 1995–2000. Panama's estimated population density in 1990 was 30.8 per sq km (79.8 per sq mi). According to another UN estimate, 54.9% of the population was urban and 45.1% rural. Major cities and their 1990 populations were Panama City, the capital, 584,803; Colón, 140,908; and David, 102,678.

7MIGRATION

Immigration and emigration have been roughly in balance in recent years. In the 1990s, 134,953 residents entered the country and 149,028 left. There were 61,400 foreign-born persons in Panama that year, of which 13,644 were Colombians.

8ETHNIC GROUPS

The racial and cultural composition of Panama is highly diverse. According to recent estimates, some 67–70% of the inhabitants are mestizo (mixed white and Amerindian) or mulatto (mixed white and black); 14% are black (West Indians); 10–14% are white (mostly Europeans); and 5–8% are Amerindian. There is also a Chinese community of about 100,000. About 100,000 Amerindians live in isolation in eastern Panama and on the San Blas Islands.

9LANGUAGES

Spanish, the official language of Panama, is spoken by over 90% of the people, but English is a common second language, spoken by most Panamanian professionals and businesspeople. The Amerindians use their own languages.

10RELIGIONS

Although Roman Catholicism is recognized by the constitution of 1972 as the majority religion, the constitution also guarantees religious freedom, as well as separation of church and state. In 1993, an estimated 80% of the people were Catholic, 5% were Protestant, and 4.5% were Muslim. In 1985 , there were some 25,000 Baha'is, 7,400 Hindus, 5,300 Amerindian tribal religionists, 3,800 Chinese folk-religionists, 3,100 Jews, and 2,800 Buddhists.

11TRANSPORTATION

Motor vehicles transport most agricultural products. In 1991 there were 8,500 km (5,280 mi) of roads, of which 32.3% were paved. The principal highway is the National (or Central) Highway—the Panamanian section of the Pan American Highway—which runs from the Costa Rican border, via Panama City and Chepo, to the Colombian border. The 80-km (50-mi) Trans-Isthmian Highway links Colón and Panama City. Panama's rugged terrain impedes highway development, and there are few good roads in the republic's eastern sections. In 1991 there were 150,000 registered passenger cars and 46,000 trucks and buses.

Railway lines total 238 km (148 mi) of track, all government-run. The Panama Railroad parallels the canal for 77 km (48 mi) between Colón and Panama City. Other lines connect Pedregal, David, Puerto Armuelles, and Boquete and unify Bocas del Toro Province.

The Panamanian merchant marine was the world's largest in 1991, with 3,040 registered ships, totaling 46.1 million GRT. Most of the ships are foreign-owned but are registered as Panamanian because fees are low and labor laws lenient. International shipping passes almost entirely through the canal ports of Cristóbal, which serves Colón, and Balboa, which is the port for Panama City.

Panama is a crossroads for air travel within the Americas. The most widely used domestic airline is Compañía Panameña de Aviación (COPA), which also flies throughout Central America, and carried 379,400 passengers in 1991/92. Air Panama International serves passenger traffic to the US and South America, and Internacional de Aviación (INAIR) is an international passenger and cargo carrier. The main airfield is Omar Torrijos International Airport, 19 km (12 mi) east of Panama City.

The Panama Canal—built during 1904–14 by the US Army Corps of Engineers, under Col. George Washington Goethals—traverses the isthmus and is 82 km (51 mi) in length from deep water to deep water. The great technical feat involved in constructing the canal was to cut through the mountains that span the region, dam the Chagres Lake, and then design and build the three sets of double locks that raise and lower ships the 26 m (85 ft) between lake and sea levels. The first passage through the canal was completed by the S.S. Ancon on 15 August 1914. Since the US-Panama treaties went into effect on 1 October 1979, the canal has been administered by the joint Panama Canal Commission, on which the US had majority representation through the end of 1989. The canal takes ships of up to 67,000 tons; it handled cargo totaling 165.8 million tons through 14,108 transits in the 1991 fiscal year (October 1990–September 1991). An oil pipeline across the isthmus was opened in 1982 to carry Alaskan oil; its capacity is 830,000 barrels per day. The Bridge of the Americas across the canal at the Pacific entrance unites eastern and western Panama as well as the northern and southern sections of the Pan American Highway. Panama, the US, and Japan have commissioned a $20-million study to search for alternatives to the canal. The feasibility of building a new canal at sea level is to be examined; alternatively, the Panama Canal Commission has indicated its intention to increase the width of the Gaillard Cut (Corte Culebra), since larger ships are restricted to one-way daylight passage, due to the narrowness. Panama also plans to consolidate the ports of Balboa on the Pacific and Cristóbal on the Caribbean into a single container terminal system.

12HISTORY

The isthmian region was an area of economic transshipment long before Europeans explored it. It was also the converging point of several significant Amerindian cultures. Mayan, Aztec, Chibcha, and Caribs had indirect and direct contact with the area. The first European to explore Panama was the Spaniard Rodrigo de Bastidas in 1501. In 1502, Columbus claimed the region for Spain. In 1513, Vasco Nuñez de Balboa led soldiers across the isthmus and made the European discovery of the Pacific Ocean. Despite strong resistance by the Cuna Amerindians, the settlements of Nombre de Dios, San Sebastián, and, later, Portobelo were established on the Caribbean coast, while Panama City was founded on the Pacific coast. In 1567, Panama was made part of the viceroyalty of Peru. English buccaneers, notably Sir Francis Drake in the 16th century and Henry Morgan in the 17th, contested Spanish

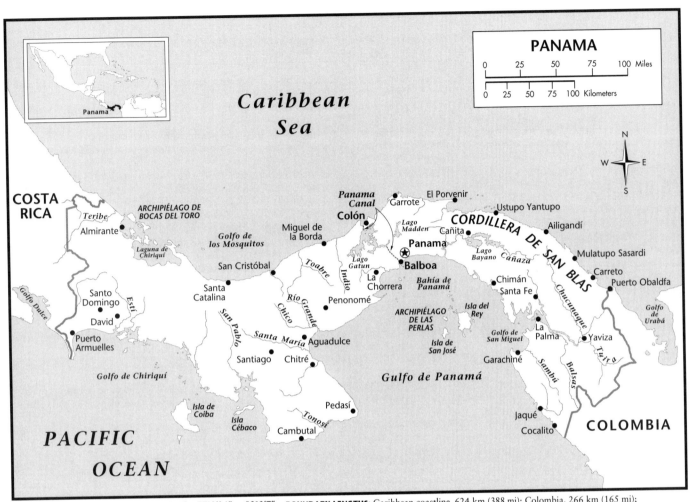

LOCATION: 7°12′9″ to 9°37′ 57″N; 77°9′24″ to 83°3′7″w. **BOUNDARY LENGTHS:** Caribbean coastline, 624 km (388 mi); Colombia, 266 km (165 mi); Pacific coastline, 1,188 km (738 mi); Costa Rica, 363 km (266 mi). **TERRITORIAL SEA LIMIT:** 200 mi.

hegemony in Panama, burning and looting its ports, including Panama City in 1671.

From the 16th until the mid-18th century, the isthmus was a strategic link in Spanish trade with the west coast of South America, especially the viceregal capital of Lima. In 1740, the isthmus was placed under the jurisdiction of the newly recreated viceroyalty of New Granada.

Panama declared its independence from Spain in 1821 and joined the Republic of Gran Colombia, a short-lived union of Colombia, Venezuela, and Ecuador, founded in 1819. In 1826, it was the seat of the Pan American Conference called by the Liberator, Simón Bolívar. When Gran Colombia was dissolved in 1829–30, Panama still remained part of Colombia. Secessionist revolts took place in 1830 and 1831, and during 1840–41.

The discovery of gold in California in 1848 brought the isthmus into prominence as a canal site linking the Atlantic and Pacific oceans. After the French failed to build one in the 1880s, they sold those rights to the US for $40 million. The US then negotiated the Hay-Herrán Treaty with Colombia in 1903. After Colombia refused to ratify the treaty, Panama seceded from Colombia and, backed by US naval forces, declared its independence on 3 November 1903. Panama then signed a canal agreement with the US and received a lump sum of $10 million and an annual rent of $250,000. The Hay-Bunau-Varilla Treaty (1903) granted the US in perpetuity an 8-km (5-mi) strip of land on either side of the canal and permitted the US to intervene to protect Panamanian independence, to defend the canal, and to maintain order in the cities of Panama and Colón and in the Canal Zone.

The US intervened to establish order in 1908—while the canal was under construction—and, after the canal had opened to traffic, in 1917 and again in 1918. In 1936, however, the US adopted a policy of nonintervention, and in 1955, the annuity was raised to $1,930,000.

During the postwar decades, the question of sovereignty over the Canal Zone was a persistent irritant in Panamanian politics. On 9 January 1964, riots broke out in the Canal Zone as Panamanians protested US neglect of a 1962 joint Panama-US flag-flying agreement. On the following day, Panama suspended relations with the US and demanded complete revision of the Canal Zone treaty. Thereafter, Panama sought sovereignty over the Canal Zone and the elimination of the concept of perpetuity on any future arrangement. Diplomatic relations were restored in April, but negotiations went slowly thereafter.

The Panamanian government turned to dictatorship in October 1968, when National Guard Brig. Gen. Omar Torrijos Herrera deposed the elected president and established a dictatorship.

Final agreement on the future of the canal and the Canal Zone came on 7 September 1977, when Gen. Torrijos and US President Jimmy Carter signed two documents at OAS headquarters in Washington, D.C. The first document, the Panama Canal Treaty, abrogated the 1903 Hay-Bunau-Varilla accord, recognized Panama's sovereignty over the Canal Zone (which ceased to exist as of 1 October 1979), and granted the US rights to operate, maintain, and manage the canal through 31 December 1999, when ownership of the canal itself would revert to Panama. Panama would receive a fixed annuity of $10 million and a subsidy of

$0.30 (to be adjusted periodically for inflation) for every ton of cargo that passed through the canal, plus up to $10 million annually from operating surpluses. The second document, the so-called Neutrality Treaty, guaranteed the neutrality of the canal for "peaceful transit by the vessels of all nations" in time of both peace and war. An additional provision added in October denied the US the right of intervention into Panamanian affairs. The treaties were ratified by plebiscite in Panama on 23 October 1977 and, after prolonged debate and extensive amendment, by the US Senate in March and April 1978. When both treaties came into force in 1979, about 60% of the former Canal Zone's total area immediately came under Panama's direct control, including 11 of 14 military bases, the Panama City-Colón railway, and ports at both ends of the canal.

The Torrijos regime was populist, with a wide appeal to the neglected lower and lower middle classes of Panama. Moreover, Torrijos established nationalist credentials by standing up to the US and demanding recognition of Panama's positions on the Canal Zone. Torrijos resigned as head of government in 1978 but continued to rule behind the scenes as National Guard commander until his death in a plane crash on 31 July 1981. Over the next few years, the National Guard, now renamed the Panama Defense Forces (PDF), came under the influence of Gen. Manuel Noriega.

On the civilian side, Aristedes Royo was elected president by the Assembly in October 1978, and was forced out of office in July 1982. His successor was the vice-president, Ricardo de la Espriella, who resigned in February 1984, just three months before scheduled presidential elections. In those elections, the economist and former World Bank official Nicolás Ardito Barletta, the military's approved candidate, won a close victory over former president Arnulfo Arias Madrid (running for the fifth time), in an election marked by voting irregularities and fraud. Barletta soon lost the confidence of the military and was forced out in September 1985. Vice-President Eric Arturo Delvalle assumed power.

By 1987, Noriega had been accused by close associates and the US of falsifying the 1984 election results, plotting the deaths of prominent opposition leaders and Gen. Torrijos, drug trafficking, giving aid to the Colombian radical group M-19 and Salvadoran rebels, and providing intelligence and restricted US technology to Cuba. Opposition forces, including the Roman Catholic Church, intensified and the government responded by banning public protest. The US Senate approved legislation cutting off aid to Panama in December 1987. In February 1988, following indictments of Noriega in US courts for drug trafficking, President Delvalle announced Noriega's dismissal. Noriega refused to step down, and the Legislative Assembly voted to remove Delvalle from office and replace him with Manuel Solís Palma, the minister of education. Delvalle went into hiding, and Panama entered a two-year period of instability and conflict.

Domestically, Noriega suffered from a lack of support. In March 1987, a general strike occurred for several weeks. Emboldened by US efforts to remove Noriega, opposition forces coalesced, even as the government became more repressive. In elections held in May 1989, opposition candidates scored overwhelming victories, forcing Noriega to annul the elections and rely on intimidation and force.

Noriega also had problems within the PDF. Dissident military leaders, with either tacit or direct US approval, attempted coups in March 1988 and in October 1989. Unable to rely on the loyalty of the PDF, Noriega created his own paramilitary force, called the "Dignity Batallions," which were nothing more than free-lance thugs called in at the dictator's whim.

Finally, the US was engaged in a series of moves calculated to bring down the Noriega regime, which eventually led to a showdown. In March 1988, President Reagan suspended preferential trade conditions and withheld canal-use payments. In April,

Reagan froze US-held Panamanian assets and suspended all private payments to Panama. Negotiations to allow Noriega to step down dissolved in May, when Noriega refused to abide by an agreement between the US and Noriega's assistants. The administration of President Bush continued pressure on Noriega, but itself came under criticism for its inability to resolve the problem. Finally, in December 1989, Noreiga played his final card, declaring war on the US and ordering attacks on US military personnel.

President Bush responded quickly, ordering the US military into Panama. The troops remained for a week, delayed when Noriega sought sanctuary in the residence of the Papal Nuncio. Noriega surrendered, and was returned to the US for trial. Immediately, the Panamanian Electoral Tribunal declared the 1989 elections valid and confirmed the results. Guillermo Endara became president, Ricardo Arias Calderón first vice-president, and Guillermo "Billy" Ford second vice-president. Legislative elections were confirmed for most Legislative Assembly seats, and in January 1991 a special election filled the remaining seats. Under President Endara, Panama made some strides toward economic recovery, but these were only impressive because the situation under Noriega had become so desperate. Politically, Endara lacked nationalist credentials, especially since he was installed by US military might. His administration was widely criticized for the continuing poor economic conditions. In May 1994, a new president, Perez Balladares was elected.

13GOVERNMENT

Under the constitution of 1972, Panama is a republic in which the president, assisted by a cabinet, exercises executive power. Reforms adopted in April 1983 changed the election of the president from an absolute majority of the National Assembly of Municipal Representatives to a direct popular vote, and a second vice-president was added. The president and the two vice-presidents must be at least 35 years of age and native Panamanians; they serve for five years and are not eligible for immediate reelection.

Legislative power is vested in the unicameral Legislative Assembly, which replaced the National Assembly of Municipal Representatives in 1984. The 67 members are elected for five-year terms by direct popular vote. Regular sessions are from 11 October to 11 November annually, and special sessions may be called by the president. Suffrage is universal for Panamanians 18 years of age or over.

The 1972 constitution conferred extraordinary decision-making powers upon the commander of the PDF, who was allowed to participate in sessions of all executive and legislative organs, to direct foreign policy, to appoint Supreme Court magistrates, and to appoint and remove ministers of state, among other responsibilities.

The PDF has now been converted into a civilian group called the Public Forces. Following a purge of PDF senior officials, the Public Forces were placed under the cabinet-level Minister of Government and Justice. The PDF budget is now public, and was cut in half in 1990.

14POLITICAL PARTIES

Personalities rather than ideological platforms tend to be the dominating force in Panamanian politics. The traditional political parties were the Liberals and the Conservatives, and their differences lay initially in the issue of church and state power. More recently, parties tended to be coalitions of the many splinter groups that had formed around local leaders. Military interventions frequently led to the banning of political parties. Such interruptions have led to an extremely splintered party system, which held together only insofar as they opposed the military regimes.

The coalition that came to power in 1990 consisted of President Endara's Arnulfista Party, led by Dr. Arnulfo Escalona, the National Liberal Republican Movement (MOLIRENA), led by second vice-president Guillermo Ford and the Christian Democratic Party PDC, led by first vice-president Ricardo Arias. Subsequently, Arias broke from the coalition, and the PDC, which holds a plurality of seats in the Legislative Assembly, is now the leader of the opposition. Also supporting the government is the Authentic Liberal Party (PLA), a splinter of the Liberal Party.

The opposition led by Arias includes the Democratic Revolutionary Party (Partido Revolucionario Democrático—PRD), founded by Gen. Torrijos, the Liberal Party (Partido Liberal—PL), Popular Nationalist Party (Partido Nacionalista Popular—PNP), the Labor Party (Partido Laborista—PALA), and the Papa Egoro Movement, led by Rubén Blades. Of these, only four hold any seats in the Legislative Assembly: the PDC (28), PRD (10), PALA (1) and PL (1).

15 LOCAL GOVERNMENT

Panama is divided into nine administrative provinces, each headed by a governor appointed by the president, and three Amerindian territories. The provinces are subdivided into 65 municipal districts, each of which is governed by a mayor and a municipal council of at least five members, including all that district's representatives in the National Assembly. There are 505 municipal subdistricts in all.

16 JUDICIAL SYSTEM

Judicial authority rests with the Supreme Court, composed of nine magistrates and nine alternates, all appointed by the president (subject to approval by the Legislative Assembly) for 10-year terms. The Supreme Court magistrates appoint judges of the superior courts who in turn appoint circuit court judges in their respective jurisdictions. There are 4 superior courts, 18 circuit courts (1 civil and 1 criminal court for each province), and at least 1 municipal court in each district.

At the local level, two types of administrative judges, "corregidores" and "night" (or "police") judges hear minor civil and criminal cases involving sentences under one year. Appointed by the municipal mayors, these judges are similar to Justices of the Peace. Their proceedings are not subject to the Code of Criminal Procedure and defendants lack procedural safeguards afforded in the regular courts.

The Constitution guarantees a right to counsel for persons charged with crimes and requires the provision of public defenders for indigent criminal defendants. Trial by jury is afforded in some circumstances.

17 ARMED FORCES

The Panamanian Defense Force disappeared with the US intervention in 1989. The new National Police Force numbers 11,000, supported by a maritime service (350 staff, 3 patrol boats) and air service (350 staff, 14 aircraft and helos). The US maintains a garrison of 10,500 in the Canal Zone. In 1990 Panama spent $75.5 million on defense or 1.5 % of GDP Panama receives about $5 million a year in US military assistance.

18 INTERNATIONAL COOPERATION

Panama is a charter member of the UN, having joined on 13 November 1945, and participates in ECLAC and all the nonregional specialized agencies. Panama belongs to G-77, IDB, OAS, and PAHO, and is signatory to the Law of the Sea but not to GATT. With Colombia, Mexico, and Venezuela, Panama in 1987 was seeking to mediate the conflicts in El Salvador and Nicaragua as a member of the so-called Contadora group, but it did not join in the signing of the peace plan proposed by Costa Rican President Oscar Arias Sánchez in that year.

19 ECONOMY

Panama's economy depends largely on its strategic location for world sea and air travel and is limited by a paucity of exportable products. Panama Canal rentals and the registry fees from Panama's large foreign-owned merchant fleet help offset a markedly unfavorable balance of trade. Panama's increase of 29.3% in shipping tolls at the time that the new Panama Canal treaties were signed on 1 October 1979 raised income from canal operations to over $400 million annually in the early 1980s. A further increase of 9.8% in 1983 helped to offset shipping losses caused by the opening in January of the trans-Panamanian oil pipeline, although 1986 earnings were only $323 million.

The Panamanian economy expanded at an annual rate of only 2% in 1974–78 but averaged over 5% in 1979–82; however, the growth rate stagnated in 1983–84 because of a slowdown in international trade. Growth in 1985 was 3.3%.

In the early 1990s, Panama rebounded from an excruciating recession brought about by a US embargo and subsequent military invasion. The US objective was the capture of General Manuel Noriega, who had installed puppet governments and was responsible for an increase in drug trafficking and money laundering. After Noriega's capture, Guillermo Endara assumed office; however, his administration was widely criticized for not fulfilling Panamanians' hope for a rapid and bountiful recovery. In May 1994, a new president, Perez Balladares, was elected.

Nonetheless, Panama's GDP grew 8% in real terms to an estimated $6 billion in 1992, making Panama's economic growth one of the most impressive in the Western Hemisphere. In 1993, the economy continued to grow, but at a slower pace. Panama's economy is based on a well-developed services sector, including the Panama Canal, banking, insurance, government, the Transisthmian oil pipeline and the Colón Free Trade Zone, accounting for 70% of total GDP. Agriculture accounts for 11% of GDP. Main primary products include bananas, sugar, shrimp, coffee, meat, dairy products, tropical fruits, rice, corn, and beans.

The US dollar is the legal tender in the nation. Panama does not print any money and, therefore, lacks an independent monetary policy.

20 INCOME

In 1992, Panama's GNP was $6,133 million at current prices, or $2,440 per capita. For the period 1985–92, the average inflation rate was 1.3%, resulting in a real growth rate in per capita GNP of –1.2%.

In 1992, the GDP was $6,001 million in current US dollars. It is estimated that in 1991 agriculture, hunting, forestry, and fishing contributed 11% to GDP; mining and quarrying, less than 1%; manufacturing, 8%; electricity, gas, and water, 4%; construction, 4%; wholesale and retail trade, 13%; transport, storage, and communication, 13%; finance, insurance, real estate, and business services, 19%; community, social, and personal services, 7%; and other sources, 20%.

21 LABOR

In 1991, the economically active population numbered 858,509. Of the employed work force of 721,800 Panamanians in 1991, government and services engaged 26.3%, commerce, 19.8%, agriculture, 22.9%, manufacturing/mining, 9.4%; construction, 3.6%; transportation and communications, 6.2%; utilities, 1.2%; business services, 4.1%; and other sectors, 6.5%. The unemployment rate rose from 11.8% in 1985 to 17% in 1990, and then declined to 15.7% in 1991.

In 1991, Panama had 177 unions with 77,527 members, including 8,595 public sector trade unionists. The province of Panama is where 71.1% of the total number of unions are found, and it has 68.7% of the country's union members. Between 1985 and 1991, organized labor lost 28,769 members. The leading unions are the Confederation of Workers of the Republic of Panama, formed in 1963, which is an affiliate of the ICFTU, and the National Center of Panamerican Workers, which is affiliated with the WFTU.

The law provides for an eight-hour day, a six-day week, minimum wages, a month's vacation with pay, maternity benefits and equal pay for women, and restrictions on the employment of minors. All employees are entitled to a one-month annual bonus in three equal installments, two of which the worker receives directly and one of which is paid into the Social Security Fund.

In March 1986, a package of laws was approved that was intended, among other things, to promote employment and productivity. Labor contracts were made more flexible by removing some guaranteed benefits for workers, particularly in small businesses. These measures were passed over labor's objections.

22AGRICULTURE

About 8.6% of the total land area was classified as arable in 1991. Farming methods are primitive, and productivity is low. The best lands are held by large owners.

Panama is self-sufficient in bananas, sugar, rice, corn, and coffee, but imports large quantities of other foods. Agricultural production grew by 3% in 1992, and by 5% in 1991. Banana exports rose by 3% in 1992 to 40.1 million boxes, with over half destined for the German market. Sugar exports fell by about 13% to $21.3 million in 1992, while coffee exports dropped by 20.4% to $10.3 million in 1992. In 1992, crop production (in tons) included raw sugar, 1,400,000; bananas, 1,110,000; rice, 165,000; corn, 95,000; and coffee, 12,000.

23ANIMAL HUSBANDRY

The Panamanian livestock industry produces sufficient meat to supply domestic demand and provides hides for export. Most cattle and hogs are tended by small herders, and dairy farming has expanded in recent years. In 1992 there were 1,400,000 head of cattle, 257,000 hogs, and 9,000,000 poultry. Milk production in 1992 totaled 127,000 tons; egg production, 12,000 tons.

24FISHING

The offshore waters of Panama abound in fish and seafood, and fisheries are a significant sector of the national economy. There is freshwater fishing in the Chiriquí River and deep-sea fishing along the Pacific and Caribbean coasts for amberjack, barracuda, bonito, corbina, dolphin, mackerel, pompano, red snapper, sailfish, sea bass, and tuna.

In 1991, the fish catch totaled 147,435 tons, as compared with 131,514 tons in 1986. In addition, 65,417 tons of anchovies were caught in 1992. Exports of shrimp and other shellfish were valued at $56.1 million in 1991; exports of fish meal were valued at $13.6 million.

25FORESTRY

Forests and woodland cover about 43% of the country's area but have been largely unexploited because of a lack of transportation facilities. Nearly all forestland is government-owned. Hardwood, particularly mahogany, is produced for export in Darién and along the Pacific coast in Veraguas. Abacá fiber, which is obtained in Bocas del Toro and is used in the making of marine cordage, is a valuable forest product. Approximately 30% of Panama's natural forests are still unused. In order to protect and preserve native forests, the National Association for the Conservation of Nature has begun a vast reforestation program. Estimated production of roundwood was 1.87 million cu m in 1991.

26MINING

Mining is limited, but there are known deposits of copper, manganese, iron, asbestos, gold, and silver. Gold mining included small-scale placer operations in Darién Province, and a gold exploration program was begun recently at Petaquilla. Commercial deposits of gold and silver were discovered in Veraguas in 1980. Salt, produced by evaporation of seawater at Aguadulce, is a major mineral product; the total salt output was an estimated 5,000 tons in 1991, down 50% from 1987.

27ENERGY AND POWER

Installation of hydroelectric stations has enabled Panama to reduce its dependence on imported oil for energy from 40% of total imports in 1975 to 15% of total imports in 1991. Panama had a net installed electric capacity of 958,000 kw in 1991; production of electric power totaled 2.9 billion kwh. All electricity in Panama is provided by the Institute of Water Power and Electrification, a government agency. About 65% of the electrical capacity is hydroelectric. A geothermal region in southwest Panama has an estimated potential of 400,000 kw. Some energy is also obtained from fuel wood and biomass residues.

After years of unsuccessful exploration, two oil strikes—one off the San Blas Islands, the other a 500-million-barrel find about 100 km (62 mi) east of Panama City—were announced in 1980. Several companies hold concessions in the Gulf of Panama, but there was no commercial production in the mid-1980s. The inauguration in 1982 of a trans-Panamanian pipeline, operated by Petroterminal de Panamá, facilitates the shipment of Alaskan crude oil to the eastern US by allowing the loading and unloading of supertankers on both sides of the isthmus; the pipeline's maximum capacity is 830,000 barrels per day. The pipeline's contribution to GDP has fallen from 6.9% in 1987 to 3.4% in 1990 and to 2.5% in 1991, reflecting reduced demand for Alaskan oil on the US east coast. Under study in the mid-1980s was a proposed $4.5 billion plant and associated pipeline for the conversion of coal to methanol.

28INDUSTRY

Limited by a small domestic market, Panamanian manufacturing contributed 8.7% of the GDP in 1985, a percentage that has been declining steadily for over a decade. Industry, generally light, consists principally of food-processing plants and firms for the production of alcoholic beverages, ceramics, tropical clothing, cigarettes, hats, furniture, shoes, soap, and edible oils. The economic readjustment program, instituted as part of the mid-1983 tightening of fiscal policy, has cut domestic demand; exports of manufactured articles, while on the increase in 1985, were inadequate to make up the difference.

Products that showed growth in 1985 were by and large in the export sector: textiles rose by 20%, shellfish products by 30%, and medications by 73%. Production of alcoholic beverages and tobacco, which have little access to foreign markets, declined by 13% and 14%, respectively. Production of petroleum derivatives, after several years of growth, fell by 1% in 1984 and 15.8% in 1985, even though exports more than tripled in the latter year. The decline was due to the sharp drop in domestic demand as a result of the conversion program to hydroelectricity, as well as reduced.

The government encourages industrialization by granting special tax concessions to new enterprises and imposing protective duties on competing foreign manufacturers. The Industrial Development Bank promotes small industries and facilitates credit on a long-term basis. A 1986 law on industrial incentives grants industrial investors a wide range of benefits, the foremost of which is tax exemptions that vary according to whether all or part of the output is earmarked for the export or domestic market.

The performance of Panama's industry has been improving in more recent years. Manufacturing, mining, utilities and construction together currently account for 19% of GDP. During the years of 1990–92, the construction industry alone was the star performer, increasing reconstruction and new-construction activities by 61.6%. A large portion of private construction was in the luxury housing sector. The government also invested large amounts in the construction of public works and low-cost housing. Panama manufactures processed food, apparel, chemical products, and construction material for the domestic market. The tourism industry has also shown important growth in recent years. Hotel occupancy went up by 6.3%, due to an increased influx of business travelers. Tourist expenditures increased by 6% to $204 million. As a result, the government is seeking to promote both foreign and domestic investment in some of the more remote areas of the country, including Coiba Island, Rio Hato, and Amador.

²⁹SCIENCE AND TECHNOLOGY

Although the shipping technology of the Panama Canal is owned and operated by the US, technicians from the US who operate the canal's facilities are to be replaced gradually by Panamanian personnel before the canal is officially turned over to Panama by the year 2000. The National Academy of Sciences of Panama (founded in 1942) advises the government on scientific matters, and the National Research Center (1976) coordinates scientific and technological research. The Smithsonian Institution has a tropical research institute in Balboa. The University of Panama has colleges of agriculture, medicine, natural sciences, dentistry, nursing, and pharmacy. Santa Maria La Antigua University has a department of technology and natural science. The Nautical School of Panama offers courses in nautical engineering. In 1986, total research and development expenditures amounted to в173,000.

³⁰DOMESTIC TRADE

Marketing and distribution are generally on a small scale, with direct merchant-to-customer sales. Many small shops in Panama City and Colón sell both native handicrafts and imported goods. Luxury items are generally untaxed in order to attract tourist sales. There are also US-style variety stores. In the rural districts, agricultural products and meat are sold at markets.

The usual business hours are 8 AM to noon and 2 to 6 or 7 PM, Monday through Saturday. Government offices are open weekdays from 8 AM to noon and from 12:30 to 4:30 PM. Banking hours in Panamanian urban centers are generally from 8:30 AM to 3 or 5 PM, Monday through Friday, although a few banks are open on Saturdays from 9 AM to 12 noon.

³¹FOREIGN TRADE

Panama's predominant export has long been fresh bananas, mainly to the US; however, sales of petroleum products have often exceeded those of bananas since 1974. The structure of imports has changed little since the 1970s, with manufactured goods, fuels and minerals, and machinery being the most important purchases from abroad; crude petroleum to supply Panama's domestic needs and growing refinery industry is the largest single import.

All things considered, Panama still offers the easiest market access in the region for most products. In addition to export opportunities available in the Colon Free Zone (where no product restrictions and a minimum of regulation apply), Panama offers a dynamic domestic market for services, franchises, processed food products, textiles, and novelties.

In 1992, the total volume of commercial transactions (import-re-exports) in the Colon Free Zone reached $9.1 billion. For the first six months of 1993, imports were $2.02 billion and re-exports were $2.34 billion.

Panama's government offers a wide variety of incentives for international and local companies to engage in international trade. Most of these incentives are tax credit-related.

At year end 1992, total merchandise exports accounted for $506 million. Exports were mainly composed of sugar, bananas, shrimp and other seafood, coffee, sugar, clothing and petroleum products. Imports recorded $1,968 million, including food products and petroleum. Main manufactured exports include foodstuffs, beverages and tobacco, 22%; textiles apparel and leather, 51%; chemicals and petroleum, carbon, rubber and plastic by-products, 17%; machinery and equipment, 3%.

As of 1992, Panama's major trading partners were as follows: the US, Central America, and Europe for exports; the US, Central America, Mexico, Japan, Venezuela, and Europe for imports.

³²BALANCE OF PAYMENTS

Panama's adverse balance of trade is largely made up by invisible foreign exchange earnings from sales of goods and services in the Colón Free Trade Zone and from the Panama Canal.

In 1992 merchandise exports totaled $5,012.2 million and imports $5,894.1 million. The merchandise trade balance was $–881.9 million.

The following table summarizes Panama's balance of payments for 1991 and 1992 (in millions of US dollars):

	1991	1992
CURRENT ACCOUNT		
Goods, services, and income	–301.4	–254.6
Unrequited transfers	108.9	111.2
TOTALS	–192.5	–143.4
CAPITAL ACCOUNT		
Direct investment	–40.3	–0.8
Portfolio investment	–0.8	–184.3
Other long-term capital	3.3	–301.1
Other short-term capital	–774.9	–347.0
Exceptional financing	621.2	576.4
Other liabilities	—	—
Reserves	–198.9	–116.4
TOTALS	–390.4	–373.2
Errors and omissions	600.9	480.6
Total change in reserves	–212.1	–111.1

³³BANKING AND SECURITIES

Since 1983, the year of the onset of Latin America's financial crisis, the Panamanian banking sector has contracted, both in number of banks and total assets. Some recovery was recorded in 1990. In 1990 there were 100 banks in Panama. Sixty of the banks in Panama have general licenses, 30 banks have international licenses, 18 foreign banks have representative offices, and two banks were government owned.The National Bank of Panama (Banco Nacional de Panamá), founded in 1904, is the principal official (but not central) bank and also transacts general banking business. Banking activities are supervised by the National Banking Commission (Comisión Bancaria Nacional). Private banks include branches of US, Japanese, Latin American, and other foreign firms. Liberalized banking regulations and use of the dollar have made Panama one of Latin America's major offshore banking centers.

The Panama City Stock Exchange, founded in 1960, has been little used by private firms, even though special tax incentives were introduced in 1969 to stimulate stock trading.

³⁴INSURANCE

There were 13 national insurance companies in Panama in 1986. Domestic companies include the General Insurance Co., the International Insurance Co., and the International Life Insurance Co. For a firm to qualify as a national insurance company, 51% of the capital must be Panamanian.

35PUBLIC FINANCE

Budget deficits are chronic, and actual expenditures consistently outrun budgeted expenditures by a substantial margin.

In 1990, however, a net budget surplus of $136.9 million was recorded, primarily due to the release of $276.4 million in formerly frozen assets by the US. Public finances remained strong in 1991, as revenues exceeded projections and more than compensated for higher wage and salary outlays. Public investment strengthened in 1991 and climbed to about $120 million (2.2% of GDP) by the end of the year.

The following table shows actual revenues and expenditures for 1990 and 1991 in billions of balboas.

	1990	1991
REVENUE AND GRANTS		
Tax revenue	982.3	1,142.9
Non-tax revenue	486.5	463.5
Capital revenue	0.7	0.4
Grants	4.9	226.6
TOTAL	1,474.3	1,833.2
EXPENDITURES & LENDING MINUS REPAYMENTS		
General public service	164.0	119.2
Defense	73.1	78.6
Public order and safety	36.7	48.2
Education	233.4	257.5
Health	281.1	348.1
Social security and welfare	277.3	336.7
Housing and community amenities	48.5	66.2
Recreation, cultural, and religious affairs	6.7	7.8
Economic affairs and services	83.5	193.8
Other expenditures	164.4	143.0
Lending minus repayments	−54.4	−81.5
TOTAL	1,314.2	1,517.5
Deficit/Surplus	160.2	315.7

At the end of 1992, total external debt amounted to $6.4 billion, with a total debt service of $872 million, for a debt service ratio of 31.7%. Since May 1990, Panama has paid all interest and principal due to the IMF, World Bank, IDB, and the International Fund for Agricultural Development.

36TAXATION

Income tax is paid to the national government by individuals and corporations on all income earned in or derived from Panama. The progressive rate rises from 4% to 30% for personal income. Corporate taxes are levied at 43.6%. Taxes in the Colón Free Trade Zone are set at 22.6%.

A real property tax ranges from 1.4% to 2.1% of assessed valuation. Property improvements are tax-exempt for the first five years. Other levies include license fees, stamp taxes, various payroll taxes, and a tax on the manufacture or importation of consumer goods.

37CUSTOMS AND DUTIES

While certain imports enter Panama duty-free, most items are subject to a single-column customs tariff from 10% to 30% ad valorem, without either conventional reductions or preferences. Import duties are by far the largest item of Panamanian indirect taxation. Tariffs for certain goods on a "strategic list" have been as high as 60% but were reduced to 40–50% as of March 1993.

Panama has a free port in the Colón Free Trade Zone, the world's second largest free-trade zone, where foreign goods enter without going through customs. Goods may be stored, assembled, processed, or repackaged for sale or shipment to another country free of duty. By 1986, this zone was the largest such facility in Latin America, with more than 1,100 foreign firms handling over $4 billion in imported goods.

38FOREIGN INVESTMENT

Private capital investment in Panama averaged 14% of the GDP during 1974–78 and, after implementation of the Panama Canal treaties in 1979, rose to 19% of GDP in 1980. About half of all capital investment has been by US companies, with Panamanian sources accounting for nearly all the remainder. Incentives apply to investments in nontraditional exports, tourism, offshore banking, and the Colón Free Trade Zone.

The Government of Panama officially promotes foreign investment and generally affords foreign investors national treatment. Panama's focus on the service sector allow countless investment opportunities in the industrial and agricultural sectors. In May 1991, The US-Panama Bilateral Investment treaty took effect, aimed at offering additional protection and alternatives to those seeking investment opportunities in industry, tourism, and an expanded range of services.

39ECONOMIC DEVELOPMENT

The national government plays a significant role in the economic life of Panama, The government owns the railroads, the national telegraph system, and credit banks. It operates all power plants, utilizes public revenues to improve transportation and communications, and has improved commerce through the Colón Free Trade Zone. Since World War II, the government has undertaken long-range planning in agriculture, industry, and housing.

Since 1968, Panama has prepared several economic development plans, but all were altered in midcourse in response to external developments; in fact, Panama's economy is more dependent on changing private investment needs than on public policy. However, the national development plan for 1968–72 was useful in its long-range programming of financial, natural, and human resources. Improved tax administration succeeded in increasing government revenues, which were earmarked for education, community development, road building, and related efforts. The plan for 1975–78 called for accelerated growth of the economy and labor force, improved national socioeconomic integration, and a greater emphasis on decentralized development. The government's economic plan for 1980–81 stressed economic recovery from the world recession and financial adjustments aimed at lowering the public-sector deficit and reducing the high growth rate of Panama's foreign debt. Since mid-1983, the government has pursued a tight fiscal policy in order to bring its finances into balance. Early in 1986, a policy designed to generate employment, create conditions for the self-sustained growth of the economy, and maintain progress in social services was announced. In February 1988, the National Bank reserves were depleted to less than $30 million, the foreign debt was the highest per capita among developing nations, and the IBRD canceled a $50-million loan for failure to meet its terms and suspended other loans because of arrears of $11 million.

As part of the Panama Canal treaties, the US pledged up to $295 million in loans, grants, and credits, beginning in 1979. Included in this aid package were Export-Import Bank support for $200 million in US exports to Panama, $75 million in AID housing assistance, and $20 million in loan guarantees from the Overseas Private Investment Corp. Overall, US nonmilitary loans and grants from 1946 through 1986 totaled $570.5 million. Multilateral assistance during the same period was $1,344.3 million, of which the IDB provided $648.2 million and the IBRD $595.8 million.

Panama was entering a period of slower growth as of 1994, as the reconstruction effort wanes and the economy was forced to rely on its traditional growth engines. A GDP of 4% was expected by year end 1994, after a very strong 8% in 1992. Inflation was very low, usually mirroring the trends in the US, because the dollar is legal tender and Panama does not have an independent monetary policy. Construction was still booming and likely stimulating other important sectors, such as the manufacturing of cement and building materials and electric power generation. Retail sales were strong, indicating a significant contribution of consumption to GDP. But, investment was decreasing amid political uncertainties and lack of action on the privatization program.

The agricultural sector had been doing poorly, due to low commodity prices. Much of Panamanian agricultural production is destined for export. For example, two-thirds of coffee production is exported. The banana industry's prospects for recovery worsened after the government decision not to sell valuable land to a subsidiary of Chiquita Brands.

Excluding re-exports from the Colón Free Zone (CFZ) the export structure of Panama will consist mainly of shrimp, meat, and other agricultural products. Prices for meat and shrimp have recovered slightly. Meat exports to Mexico have resumed, after the resolution of a dispute concerning inspections.

The unstable political situation presents the greatest risks to the outlook of Panama. The perceptions of the international community with respect to stability are very important to the growth of Panama's international services-oriented economy. In this regard, political uncertainties are a cause of concern. The CFZ, while still important due to its strategic geographical location, could lose some luster in coming years as Latin American countries proceed with trade liberalization, and establish their own ties with Asian exporters. The economy will be transformed structurally by the transfer of ownership of the Canal Zone—which is slated for late 1999—and withdrawal of the US military. The loss of tax revenues from the US military bases will represent about 6% of GDP. Organizing a management structure capable of efficiently administering the properties gradually reverting to Panamanian ownership will be a determining factor in the contribution of this infrastructure to the medium-term growth of the economy.

40 SOCIAL DEVELOPMENT
The Social Security Fund, established by the government in 1941, provides medical service and hospitalization, maternity care, pensions for disability or old age, and funeral benefits. Children are cared for through a child welfare institute, which operates under the Ministry of Labor and Social Welfare. The government has made efforts to integrate the Amerindian population of the San Blas area through land grants, basic education, and improved transportation. The social security system, extended by legislation enacted in 1975, covered more than 1 million persons in the mid-1980s.

The Panamanian Family Planning Organization, a private group established in 1966, had opened six clinics by 1969; several were later transferred to the Ministry of Health, which subsequently opened some of its own clinics. About 58% of Panamanian women of childbearing age were practicing birth control in 1985–90, at which time the fertility rate was 3.1. Abortion is prohibited.

Despite Constitutional equality, women generally do not enjoy the same opportunities as men. While Panama has a relatively high rate of female enrollment in higher education, many female graduates are still forced to take low-paying jobs. Only 5% of the country's managerial positions are occupied by women.

41 HEALTH
Public health services are directed by the Ministry of Health, whose program includes free health examinations and medical care for the needy, health education, sanitation inspection, hospital and clinic construction, and nutrition services. In 1992, 80% of the population had access to health care services.

In 1988, Panama had 4,131 physicians, 410 dentists, and 2,172 nurses (including nursing assistants). In 1990, the population per physician was 840, and there were 2.7 hospital beds per 1,000 inhabitants.

Proceeds from a national lottery support state hospitals, asylums, and public welfare. Assistance has been received from such organizations as WHO, the US Institute of Inter-American Affairs, the Pan American Sanitary Bureau, the Institute of Nutrition of Central America and Panama, and UNICEF.

During the first two decades of the 20th century, when the Panama Canal was being built, the major health threats were yellow fever, malaria, smallpox, typhoid, dysentery, and intestinal parasites. Through the efforts of Col. William Crawford Gorgas, a US military surgeon and sanitary officer, malaria was controlled and the yellow-fever mosquito was virtually eliminated. Today, the principal causes of death are cancer, heart disease, cerebrovascular disease, pneumonia and bronchopneumonia, and enteritis and diarrhea. Col. Gorgas pioneered in providing Panama City and Colón with water and sewer systems; in some areas of Panama, poor sanitation, inadequate housing, and malnutrition still constitute health hazards. In 1991, 83% of the population had access to safe water and 84% had adequate sanitation. However, in 1990, 25% of children under 5 years old were considered malnourished. Immunization rates for children up to 1 year old in 1992 were tuberculosis (98%); diphtheria, pertussis, and tetanus (82%); polio (83%); and measles (71%).

Panama's 1993 birth rate was 24.9 per 1,000 people. There were 63,000 births in 1992. The general mortality rate was 5.2 per 1,000 people in 1993, and the infant mortality rate was 18 per 1,000 live births in 1992. Maternal mortality was 60 per 100,000 live births in 1991. Average life expectancy in 1992 was 73 years.

42 HOUSING
Housing in urban areas has been a permanent problem since US construction in the Canal Zone brought a great influx of migrant laborers into Colón and Panama City. The government-established Bank of Urbanization and Rehabilitation began to build low-cost housing in 1944, and by 1950, it had built more than 1,500 units to house 8,000 people near Panama City. The Panamanian Institute of Housing estimated in 1970 that the national housing deficit of 76,000 units would increase by 7,700 units annually unless corrective measures were taken.

A 1973 housing law, designed to encourage low-income housing construction, banned evictions, froze all rents for three years, and required banks to commit half their domestic reserves to loans in support of housing construction projects. By the early 1980s, however, the shortage of low-income housing remained acute, particularly in Colón. A construction boom in the early 1980s was mainly confined to infrastructural projects and office space. According to the latest available information for 1980–88, total housing units numbered 448,000 with 4.9 people per dwelling.

43 EDUCATION
In 1990, illiteracy was estimated at 11.2% of the adult population (males, 10.6% and females, 11.7%) compared with 21% 20 years earlier. Education is free and compulsory for children aged 7 through 15. At secondary, vocational, and university levels, fees may be charged for the development of libraries and laboratories. Primary education lasts six years. In 1990 there were 2,659 primary schools with 350,931 students. At the secondary schools there were 191,251 students. Secondary education has two stages, lasting three years. The leading institution, the state-run University of Panama, was founded in Panama City in 1935. A Catholic university, Santa María la Antigua, was inaugurated in May 1965, with an initial enrollment of 233. In 1991, at all

institutions of higher learning, there were 3,308 teaching staff with 58,625 students enrolled.

44LIBRARIES AND MUSEUMS

The National Library of Panama, located in Panama City, was founded in 1892 as Biblioteca Colón and reorganized as the National Library in 1942. It is a branch of the Ministry of Education's public library system and has over 200,000 volumes. There are more than 40 other public libraries and branches. The National Archives, established in 1924, contain historical documents, books, and maps, as well as administrative papers of government agencies and a judicial section with court records. The library at the University of Panama has holdings of over 200,000 volumes.

The Instituto Panameño de Arte, founded in 1964, displays excellent collections of pre-Columbian art. Newer museums in Panama City include the Museum of Nationhood (dedicated in 1974); the Museum of Colonial Religious Art (1974); the Museum of the History of Panama (1977), which exhibits documents and objects of historical value; and the Museum of Panamanian Man (1976), with archaeological, ethnographic, and folkloric displays.

45MEDIA

The Instituto Nacional de Telecomunicaciones (INTEL) operates Panama's telephone and telegraph systems. In 1993 there were 260,000 telephones, mainly in the Panama City area. Telegraph cables link Panama to the US, to Central and South America, and to Europe. In 1992 there were 91 AM commercial radio broadcasting stations and 23 television channels; in that year there were 552,000 radios and 410,000 television sets.

Leading Panama City dailies (including the English-language *Star and Herald*), with their estimated 1991 circulations, are as follows:

La Prensa	20,000
Crítica (m)	40,000
El Siglo	36,000
La Estrella de Panamá (m)	15,000
El Panama America	32,000
The Star and Herald	25,000

46ORGANIZATIONS

The cooperative movement in Panama is limited. Producers' organizations are small, local, uncoordinated groups concerned mainly with practical education in techniques to improve production. The Chamber of Commerce, Industry, and Agriculture is in Panama City.

47TOURISM, TRAVEL, AND RECREATION

Visitors to Panama must have a valid passport and visa. Visitors who possess a valid passport can obtain a tourist visa from a Panamanian consul or buy a tourist card from a transportation company; both card and visa are valid for up to 90 days. Visas are available free of charge to US citizens.

The government encourages tourism through the Panamanian Tourist Bureau and is promoting investment in some of the most remote parts of the country, including Coiba Island, Rio Hato, and Amador. In 1991, 279,000 tourists arrived in Panama, 269,000 of them from the Americas. Revenues from tourism totaled an estimated $196 million.

Travel facilities within Panama are good; Panama City and Colón are only one hour apart by road or rail. In addition to the Panama Canal itself, tourist attractions include Panama City, beach resorts in the Pearls Archipelago and San Blas Islands, the ruins of Portobelo, and the resort of El Valle in the mountains. Water sports, tennis, golf, and horse racing are popular.

48FAMOUS PANAMANIANS

Outstanding political figures of the 19th century include Tomás Herrera (1804–54), the national hero who led the first republican movement, and Justo Arosemena (1817–96), a writer and nationalist. The international lawyer Ricardo J. Alfaro (1882–1971) and the rector of the University of Panama, Dr. Octavio Méndez Pereira (1887–1954), were well-known Panamanian nationalists. The most important political leader of the 20th century was Omar Torrijos Herrera (1929–81), who ruled Panama from 1969 until his death and successfully negotiated the Panama Canal treaties of 1979 with the US.

Important poets were Tomás del Espíritu Santo (1834–62), the nationalist Amelia Denis de Icaza (1836–1910), Federico Escobar (1861–1912), Darío Herrera (1870–1914), and Ricardo Miró (1888–1940). Panamanian-born José Benjamin Quintero (b.1924) is a noted stage director in the US. Narciso Garay (1876–1953) founded the National Conservatory of Music and served as a foreign minister. Harmodio Arias (1886–1962) was the prominent owner of the newspaper *El Panamá-America*. Leading Panamanian painters include Epifanio Garay (1849–1903), Roberto Lewis (1874–1949), Sebastián Villalaz (1879–1919), and Humberto Ivaldi (1909–47). Noteworthy among Panamanian athletes is the former world light- and welter-weight boxing champion Roberto Durán (b.1949); the former baseball star Rod (Rodney) Carew (b.1945) is also of Panamanian birth.

49DEPENDENCIES

Panama has no territories or colonies.

50BIBLIOGRAPHY

Conniff, Michael. *Black Labor on a White Canal: Panama*. Pittsburgh, Pennsylvania: University of Pittsburgh Press, 1985.

Greene, Graham. *Getting to Know the General*. New York: Simon & Schuster, 1984.

Hedrick, Basil C. and Anne K. *Historical Dictionary of Panama*. Metuchen, N.J.: Scarecrow, 1970.

Hogan, J. Michael. *The Panama Canal in American Politics: Domestic Advocacy and the Evolution of Policy*. Carbondale: Southern Illinois University Press, 1986.

LaFeber, Walter. *The Panama Canal: The Crisis in Historical Perspective*. New York: Oxford University Press, 1989.

Major, John. *Prize Possession: The United States and the Panama Canal, 1903–1979*. New York: Cambridge University Press, 1993.

McCullough, David G. *The Path Between the Seas: The Creation of the Panama Canal, 1870–1914*. New York: Simon & Schuster, 1977.

Meditz, Sandra W. and Dennis M. Hanratty (eds.). *Panama: A Country Study*. 4th ed. Washington, D.C.: Library of Congress, 1989.

Moffet, George D. *The Limits of Victory: The Ratification of the Panama Canal Treaties*. Ithaca, N.Y.: Cornell University Press, 1985.

Richard, Alfred Charles. *The Panama Canal in American National Consciousness, 1870–1990*. New York: Garland, 1990.

Torrijos Herrera, Omar. *Nuestra revolución*. Panama City: República de Panamá, 1974.

US Department of State. *The Panama Canal Treaties and Associated Agreements and Documents*. Washington, D.C.: Government Printing Office, 1979.

Watson, Bruce W. and Peter G. Tsouras (eds.). *Operation Just Cause: The U.S. Intervention in Panama*. Boulder, Colo.: Westview Press, 1991.

Zimbalist, Andrew S. *Panama at the Crossroads: Economic Development and Political Change in the Twentieth Century*. Berkeley, Calif.: University of California Press, 1991.

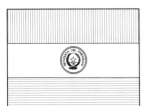

PARAGUAY

Republic of Paraguay
República del Paraguay

CAPITAL: Asunción.

FLAG: The national flag, officially adopted in 1842, is a tricolor of red, white, and blue horizontal stripes. The national coat of arms appears in the center of the white stripe on the obverse, and the Treasury seal in the same position on the reverse.

ANTHEM: *Himno Nacional,* beginning "Paraguayos, república o muerte" ("Paraguayans, republic or death").

MONETARY UNIT: The guaraní (G) is a paper currency of 100 céntimos. There are notes of 1, 5, 10, 50, 100, 500, 1,000, 5,000, and 10,000 guaraníes. G1 = $0.0005 (or $1 = G1,883.2).

WEIGHTS AND MEASURES: The metric system is the legal standard.

HOLIDAYS: New Year's Day, 1 January; San Blas Day, 3 February; National Defense Day, 1 March; Labor Day, 1 May; Independence Days, 14–15 May; Peace Day, 12 June; Founding of Asunción, 15 August; Constitution Day, 25 August; Victory Day (Battle of Boquerón), 29 September; Columbus Day, 12 October; All Saints' Day, 1 November; Our Lady of Caacupé, 8 December; Christmas, 25 December. Movable religious holidays are Holy Thursday, Good Friday, and Corpus Christi.

TIME: 8 AM = noon GMT.

¹LOCATION, SIZE, AND EXTENT

One of South America's two landlocked countries, Paraguay has a total area of 406,750 sq km (157,047 sq mi). Comparatively, the area occupied by Paraguay is slightly smaller than the state of California. The western 246,925 sq km (95,338 sq mi) of the country constitute a dry, sparsely populated region known as the Chaco, while the remaining 159,827 sq km (61,709 sq mi) lie in the more verdant east. Paraguay extends 992 km (616 mi) SSE-NNW and 491 km (305 mi) ENE-WSW. Bounded on the NE and E by Brazil, on the SE, S, and W by Argentina, and on the NW and N by Bolivia, Paraguay has a total boundary length of 3,920 km (2,436 mi).

Paraguay's capital city, Asunción, is located in the southwestern part of the country.

²TOPOGRAPHY

The eastern part of Paraguay contains luxuriant hills, meadows, and forests. The western three-fifths is a waterless, brackish prairie covered with dry grass and sparsely dotted with shadeless trees. The southward-flowing Paraguay River, the nation's most important waterway, divides the two sections; this river, which for a long time was Paraguay's principal contact with the outside world, rises in southwestern Brazil and extends for a total length of 2,549 km (1,584 mi). The Pilcomayo River, which rises in the mountains of southern Bolivia and extends about 1,600 km (1,000 mi), flows southeast, forming the southwestern border between Argentina and Paraguay, and joins the Paraguay near Asunción.

The eastern sector of Paraguay comprises the western part of the great Paraná Plateau, varying from 300 to 610 m (1,000 to 2,000 ft) in altitude. The Paraná River—called Upper (Alto) Paraná in Paraguay—flows southward from south-central Brazil through the center of the plateau, dropping in the Guairá Falls at the easternmost point in the Paraguay-Brazil frontier. Between the Guairá Falls and the confluence with the Paraguay River at

the southwestern tip of the country, the Paraná passes through a deep canyon that forms the eastern and southern frontier with Argentina.

Just west of the plateau is an area of gently rounded hills descending to the low plains that stretch westward to the Paraguay River. These hills occur in two series, one extending northwestward to the Paraguay River just north of Concepción, and the other meeting the river at Asunción. The remaining territory east of the Paraguay River is composed of lowland plain, much of it subject to annual floods.

West of the Paraguay River is the Chaco, part of the larger Gran Chaco, which includes portions of Argentina, Bolivia, and Brazil. The Gran Chaco, a vast alluvial plain composed of unconsolidated sands and clays, is crossed by the Pilcomayo and Bermejo rivers, but over much of the area there are no surface streams. The water table, however, is only a few feet below the surface, and patches of alkali frequently appear during the long dry season. In many places the groundwater is salty.

³CLIMATE

Two-thirds of Paraguay is within the temperate zone, one-third in the tropical zone. The climate varies from mild to subtropical. During the autumn and winter months (roughly April through September), temperatures generally range from 10° to 21°C (50° to 70°F); nights are occasionally colder. During the spring and summer (October through March), temperatures range from about 26° to 37°C (79° to 99°F), with extremes of 43°C (109°F) and above. Paraguay is open to dry, cold polar winds from the south and to hot, humid north winds from southwestern Brazil; sudden sharp drops in temperatures are not uncommon. Rainfall averages about 200 cm (80 in) a year along the eastern frontier with Brazil, gradually diminishing toward the west to an average of 119 cm (47 in) along the Paraguay River and 81 cm (32 in) in the Chaco. Asunción has an annual average of about 130 cm (50 in), which is moderate for its latitude. There is no definite rainy season, although violent thunderstorms sometimes occur in the summer.

4FLORA AND FAUNA

The vegetation, like the rainfall, is concentrated in the Paraná Plateau and diminishes toward the west. Tall broadleaf trees, some evergreen and some deciduous, cover eastern Paraguay, thinning out on the red sandy soils of the hilly perimeter. Scrub woodland and palm also dot the sandy plateau areas. Between the semideciduous forest and the Paraguay River, the vegetation is mostly the savanna type mixed with scattered palms. In contrast, the Chaco supports primarily deciduous scrub woodlands, luxuriant along the Paraguay River but becoming more and more xerophytic as the rainfall decreases toward the west.

The eastern forests abound in hardwoods, including indigenous varieties, such as urunday, cedron, curupay, and lapacho. Softwoods are scarce. In the northern Chaco, along the Paraguay River, there are scattered stands of quebracho and many large, spreading trees, such as the ceiba. Medicinal herbs, shrubs, and trees abound, as well as some dyewoods. Yerba maté, a holly popularly used in tea, grows wild in the northeast.

Animals found in Paraguay include the jaguar (especially numerous in the Chaco), wild boar, capybara, deer, armadillo, anteater, fox, brown wolf, carpincho, and tapir. Paraguay abounds with crocodiles along its watercourses, and the boa constrictor thrives in the west. The carnivorous piranha is common.

5ENVIRONMENT

Agencies responsible for environmental protection include the National Environmental Health Service, the Ministry of Public Health, and the Ministry of Public Works and Communications. Nearly all forests are privately owned, and little was done to develop a national forest policy until the establishment in 1973 of the National Forest Service. Paraguay's forests are currently threatened by the expansion of agriculture. By 1985, 819 square miles of forestland was lost. The absence of the trees contributes to the loss of soil through erosion. Water pollution is also a problem. Its sources include industrial pollutants and sewage. The nation has 22.6 cubic miles of reusable water. Seventy-eight percent of the water supply is used to support farming and 7% is used for industrial purposes. Forty-nine percent of the city dwellers and 91% of the rural people do not have pure water. The nation's cities contribute 0.4 million tons of solid waste per year. Some of Paraguay's cities have no facilities for waste collection. In 1994 there were five endangered mammal species. Thirty-four bird species and 15 plant species are also endangered. Endangered species include the black-fronted piping guan, glaucous macaw, black caiman, spectacled caiman, and broad-nosed caiman.

6POPULATION

According to the 1992 census, Paraguay had a population of 4,123,550, reflecting an average annual population increase of 3.6% since 1982. A population of 5,538,000 was projected for the year 2000, assuming a crude birthrate of 31 per 1,000 population, a crude death rate of 6.2, and a net natural increase of 24.8. The average density in 1992 was 10.1 persons per sq km (26.3 per sq mi). Over 98% of the population is located in the eastern two-fifths of the country; the vast western Chaco region is virtually uninhabited. The capital city, Asunción, had an estimated population of 607,000 in 1990. Other leading cities in 1983 were Puerto Presidente Stroessner (since renamed Ciudad del Este), 98,591; Pedro Juan Caballero, 41,475; Encarnación, 31,445; Pilar, 26,352; and Concepción, 25,607.

7MIGRATION

Emigration has long been a problem for Paraguay. During 1955–70, some 650,000 Paraguayans emigrated, mainly to Argentina;

more recently, the net loss from emigration has averaged an estimated 5,000–6,000 a year. Most of these emigrants live in Argentina, Uruguay, or Brazil. Much of the labor force of agricultural regions in Argentine border provinces is made up of Paraguayan nationals. The greatest exodus occurred after the 1947 civil war, but in the 1960s there were new waves of political emigration.

Immigration to Paraguay was limited to a few thousand Europeans during the 19th century. A major attempt by the Paraguayan government to encourage new settlers led to negotiations with Japan in 1959 for the immigration of 85,000 Japanese by 1990, but only about 8,000 arrived. An immigration agreement was signed with the Republic of Korea (ROK) in 1966. In 1985, the immigrant population totaled 199,500; the leading immigrant groups were Germans, Japanese, Koreans, Chinese, Brazilians, and Argentines. (It was believed, however, that the actual number of Brazilians was 300,000 to 350,000.) There were 5,417 official immigrants in 1991, of which 2,188 were Brazilians.

8ETHNIC GROUPS

According to an estimate in the late 1980s, some 95% of the population is mestizo, principally a mixture of Spanish and Guaraní Amerindian. The others are pure Amerindian (1–3%), black, or of European or Asian immigrant stock. However, another estimate puts mestizos at 76% and Europeans at 20%. During the late 1950s, Japanese settlement began in the region between Encarnación and Caazapá. Korean settlement began in 1966. Several immigrant religious groups, such as the Hutterites and Mennonites, have established colonies in Paraguay.

The Guaraní Amerindians, belonging to the Tupi-Guaraní linguistic group, had spread throughout a large area of South America east of the Andes before the Spaniards arrived. Within Paraguay, extensive intermarriage between the races resulted in almost complete assimilation. In 1986 there were about 46,700 tribal Amerindians in the backlands.

9LANGUAGES

Paraguay is a bilingual nation. Spanish, the dominant language, is taught in the schools and is spoken by about 55% of the people. However, about 90% of Paraguayans speak Guaraní, an Amerindian language that evolved from the southern dialect of the Tupi-Guaraní group. It is also the language of widely esteemed literature, drama, and popular music.

10RELIGIONS

Roman Catholicism is the official religion of Paraguay, and more than 93% of all Paraguayans are church adherents. The constitution of 1967 provides that other religions are to be guaranteed their freedom if they are not contrary to public morality and order. The president of Paraguay, who must be a Roman Catholic, appoints the archbishop and bishops in agreement with the Council of State and an ecclesiastical synod. The archbishop of Asunción is a member of the Council of State. In 1985 there were an estimated 67,000 Protestants, 22,000 members of Amerindian tribal religions, and 2,000 Buddhists. In 1990 there were 900 Jews.

11TRANSPORTATION

Inadequate transportation facilities have been a major impediment to Paraguay's development. For a long time, some 3,100 km (1,925 mi) of domestic waterways provided the chief means of transportation, with most vessels owned by Argentine interests. Hampered by the high costs and slow service of Argentine riverboats transporting cargo to and from Buenos Aires, the Paraguayan government put its own fleet of riverboats in operation. This remedy, however, did not solve the underlying problems of Paraguayan transport. Drought conditions frequently affect navigation, and while the Paraguay is open to river traffic as far as

B O L I V I A

Mayor Pablo
Lagerenza

Villazon
General
Eugenio
A. Garay

Fortín Madrejón

C H A C O

Maríscal
Estigarribia

Doctor
Pedro P. Peña

B O R E A L

Filadelfia

C H A C O   C E N T R A L

Misión
Estero

Pozo Colorado

Monte   Lindo

Fortín Gral. Delgado

Estancia Villa Rey

A R G E N T I N A

Villa Hayes

Asunción   San Lorenzo

Caacupé   Coronel Oviedo

Paraguarí

Villarrica

Caazapá

Pilar

San
Juan
Bautista

Encarnación

Corrientes

Fuerte
Olimpo

B R A Z I L

Pôrto Murtinho

Puerto
La Victoria

Apa   Bella Vista

Pedro Juan
Caballero

Ponta Porã

A M A M B A Y   M T S.

Ypané

Concepción

Jejuí-Guazú

San
Pedro

Rosario

Salto del
Guairá

Guaíra

Acaray

Santa Helena

Ciudad
del Este

Alto Paraná

Paraná

B R A Z I L

PARAGUAY

| 0 | 50 | 100 | 150 | 200 Miles |

| 0 | 50 | 100 | 150 | 200 Kilometers |

Paraguay (inset map)

Paraguay

Pilcomayo

Verde

Tueco

Pilcomayo

Bermejo

Tebicuary

Paraguay

LOCATION: 19°17′ to 27°30′s; 54°30′ to 62°28′w. **BOUNDARY LENGTHS:** Brazil, 1,339 km (832 mi); Argentina, 1,699 km (1,056 mi); Bolivia, 756 km (470 mi).

Concepción (about 290 km/180 mi north of Asunción), passage is sometimes hazardous to vessels of even medium draft. The inland waterways and the Rio de la Plata handle about 65% of Paraguay's foreign trade with Argentina, Brazil, Chile, Europe, Japan, and the US.

Asunción, the chief port, and Concepción can accommodate oceangoing vessels. In addition, Paraguay has been given free port privileges at Santos and Paranaguá, Brazil. More than 90% of Paraguay's foreign trade passes to Asunción through ports in Argentina and Uruguay. A four-year project begun in 1967 substantially improved Asunción's cargo-handling capacity.

Road construction is another critical focus of development. In 1991, highways totaled some 21,960 km (13,650 mi); of these roads, however, only 1,788 km (1,111 mi) were asphalted. Two major road projects of the 1960s were the Friendship Bridge on the Brazilian border in the Iguaçu Falls area, inaugurated in 1961, and the 770-km (480-mi) all-weather Trans-Chaco Road, which extends from Asunción to Bolivia. The Friendship Bridge permits highway travel from Asunción to the Brazilian Atlantic port of Paranaguá. A bridge over the Paraguay River, linking the western and eastern parts of the country, was inaugurated in 1978. All-weather roads connecting Asunción with Buenos Aires and Puerto Presidente Stroessner with Paranaguá have also been completed. In 1992, 80,000 passenger cars and 40,000 commercial vehicles were in use.

In 1961, the 441-km (274-mi) British-owned Paraguayan Central Railroad was sold to Paraguay for $560,000; it was subsequently renamed Ferrocarril Presidente Carlos Antonio López. There is a direct line between Asunción and Buenos Aires. Altogether there are some 970 km (603 mi) of trackage, including narrow-gauge industrial lines in the Chaco.

The modernization of Asunción's Presidente Stroessner Airport was completed in 1980. Construction of an international airport at Ciudaddel Este began in 1986. Paraguayan Air Lines (Líneas Aéreas Paraguayas—LAP) provides both domestic and international service. Three carriers provide domestic service. In 1991 there were 886 airports, 768 of which were usable; 6 had permanent runways from 1,220 to 3,700 m (4,000 to 12,140 ft).

12HISTORY

The original inhabitants of present-day Paraguay were Guaraní Amerindians of the Tupi-Guaraní language family. As many as 150,000 Amerindians may have been living in Paraguay at the time of the earliest European contacts. The first European known to have explored Paraguay was the Italian Sebastian Cabot, sailing from 1526 to 1530 in the service of Spain. The first permanent Spanish settlement, Nuestra Señora de la Asunción (Our Lady of the Assumption, present-day Asunción), was founded at the confluence of the Paraguay and Pilcomayo rivers on Assumption Day, 15 August 1537.

Paraguay's next two centuries were dominated by Jesuit missionaries, whose efforts to protect the Amerindians from Portuguese slave traders and Spanish colonists resulted in one of the most remarkable social experiments in the New World. Shortly after the founding of Asunción, missionary efforts began. The priests organized Guaraní families in mission villages (reducciones) designed as self-sufficient communes. Amerindians were taught trades, improved methods of cultivation, and the fine arts, as well as religion. Above all, they were protected from exploitation by the Spanish colonists, who sought to exploit them. As the settlements prospered and grew in number to around 30 (with over 100,000 Amerindians), the jealousy of the colonists sparked a campaign to discredit the Jesuits. Eventually, the King of Spain became convinced that the order was trying to set up a private kingdom in the New World, and in 1767, he expelled the Jesuits from the New World. Once they had left, the reducciones disappeared. As for the Spanish colony at Asunción, it dominated the area of the Río de la Plata throughout this period. However, in 1776, when Buenos Aires became the capital of the new viceroyalty of La Plata, Asunción was reduced to an outpost.

In achieving independence, Paraguay first had to fight the forces of Argentina. Buenos Aires called on Paraguay in 1810 to follow its lead in a virtual declaration of independence. Paraguay declared independence from Spain but rejected the leadership of Buenos Aires. An Argentine expedition was decisively defeated, and Paraguay completed its move toward independence by deposing the last of its royal governors in 1811.

Since then, Paraguay has been dominated by dictatorships or near-dictatorships. The first and most famous of the dictators was José Gaspar Rodríguez de Francia (known as El Supremo), originally a member of the five-man junta elected in 1811 to govern the newly independent nation. He was granted full dictatorial powers for three years in 1814 and thereafter had the term extended for life. Francia attempted to cut Paraguay off from all contact with the outside world. Commerce was suspended, foreigners were expelled, relations with the papacy were broken off, and an anticlerical campaign was begun. All criticism was stifled, and a widespread spy network was developed. However, at the same time, Francia was honest and tireless in his devotion to his personal concept of the country's welfare. Francia governed until his death in 1840. Today, he is regarded as Paraguay's "founding father."

The next dictator was Carlos Antonio López. López loosened the ties of dictatorship only slightly, but reversed Francia's paranoid isolationism. He reestablished communications with the outside world and normalized relations with the papacy. López encouraged road and railway building, improved education somewhat, and became the largest landowner and the richest man in Paraguay. He made his son Francisco Solano López commander-in-chief of the army, thereby ensuring the younger López's succession to power in 1862, when the elder López died.

During his dictatorship, Francisco Solano López provoked quarrels with Argentina, Brazil, and Uruguay, who allied and attacked Paraguay. The War of the Triple Alliance (1865–70), sometimes called the Paraguayan War, was the bloodiest in Latin American history. López, who fancied himself a Latin Napoleon, drafted virtually every male in Paraguay over the age of twelve, with no upper age limit, and insisted that his troops never surrender. The war was a disaster for Paraguay, which lost two-thirds of all its adult males, including López himself. Paraguay's population fell from about 600,000 to about 250,000. The war also cost Paraguay 55,000 square miles of territory, its economic well-being, and its pride.

For the next 50 years, Paraguay stagnated economically. The male population was replaced by an influx from Italy, Spain, Germany, and Argentina. Politically, there was a succession of leaders, alternating between the Colorado and Liberal parties. Then, a long-smoldering feud with Bolivia broke into open warfare (1932–35) after oil was discovered in the Chaco, a desolate area known as the "green hell." Although outnumbered three to one, the Paraguayans had higher morale, were brilliantly led, and were better adapted to the climate of the region. Moreover, they regarded the conflict as a national undertaking to avenge the defeat of 1870. Paraguayans conquered three-fourths of the disputed territory, most of which they retained following the peace settlement of 1938.

Although President Eusebio Ayala emerged victorious from the Chaco War, he did not last long. The war produced a set of heroes, all of whom had great ambitions. One such man, Col. Rafael Franco, took power in February 1936. In 1939, after two more coups, Gen. José Felix Estigarribía, commander-in-chief during the Chaco War, was elected president. Estigarribía was killed in an airplane crash only a year later, and Gen. Higinio Morínigo, the minister of war, was appointed president by the cabinet. Through World War II, Morínigo received large amounts of aid from the US, even though he allowed widespread Axis activity in the country. Meanwhile, he dealt harshly with domestic critics.

Morínigo retired in 1948, but was unable to find a successor. After a one-year period of instability, Federico Chávez seized control, and ruled from 1949 until 1954. In May 1954, Gen. Alfredo Stroessner, commander-in-chief of the armed forces, used his cavalry to seize power. He had himself elected president as the candidate of the Colorado Party, and then was reelected in another single-slate election in 1958, although he did permit the Liberal Party to hold its first convention in many years. With US help, he brought financial stability to an economy racked by runaway inflation, but he used terrorist methods in silencing all opposition. Exiles who invaded Paraguay simultaneously from Argentina and Brazil in December 1959 were easily routed. Six other small invasions during 1960 were also repulsed. Stroessner won a third presidential term in February 1963, despite the constitutional stipulation that a president could be reelected only once.

In August 1967, a constitutional convention approved a new governing document that not only provided for a bicameral legislature but also established the legal means for Stroessner to run for reelection. Stroessner did so in 1968, 1973, 1978, 1983, and 1988, all with only token opposition permitted. On 17 September 1980, the exiled former dictator of Nicaragua, Anastasio Somoza Debayle, who had been granted asylum by the Stroessner government, was assassinated in Asunción, and Paraguay broke off relations with Nicaragua.

During the 1980s, Stroessner relaxed his hold on Paraguay. The state of siege, which had been renewed every three months since 1959 (with a partial suspension from February 1978 to September 1980), was allowed to lapse in April 1987. Opponents of the regime gave credit to the ending of the state of siege to the US, which had kept pressure on the Stroessner administration. However, allegations of widespread human rights abuses continued to be made. In April 1987, Domingo Laíno, an opposition leader exiled in December 1982, who had tried unsuccessfully to enter the country on five earlier occasions, was allowed to return to Paraguay. Part of this liberalization may have been in response to mounting criticism from the Roman Catholic Church, whose position moved closer to that of the various dissident groups.

On 3 February 1989 Stroessner's 35-year dictatorship came to an end at the hand of Gen. Andrés Rodríguez, second in command of the Paraguayan military. Immediately after the coup, Rodríguez announced that elections would be held in May. With only three months to prepare, little opposition beyond Domingo Laíno was mounted, and Rodríguez won easily with 75.8% of the vote. There followed an immediate easing of restrictions on free speech and organization. Labor unions were recognized and opposition parties allowed to operate freely. Rodríguez promised and delivered elections in 1993. In those elections, Colorado candidate Juan Carlos Wasmosy was elected to the presidency. Paraguay had experienced an unprecedented transfer of political power through a constitution from one elected government to another. Wasmosy began to push for economic liberalization, including the sale of state-owned enterprises, but it was unclear that the military was willing to support such measures.

13 GOVERNMENT
Under the constitution of 25 August 1967, Paraguay was a republic, with substantial powers conferred on the executive. The Constituent Assembly revised the constitution on 20 June 1992, but kept most of the structure from the previous document, while limiting many of the powers Stroessner used during his administration.

The president is directly elected for a five-year term. The president is commander-in-chief of the military forces and conducts foreign relations. He appoints the 11-member cabinet, most administrators, and justices of the Supreme Court. He is advised by the Council of State, consisting of the cabinet ministers, the president of the National University, the archbishop of Asunción, the president of the Central Bank, and representatives of other sectors and the military.

The 1967 constitution provided for a bicameral legislature, consisting of the 36-member Senate and the 72-member Chamber of Deputies. By formula, two-thirds of the seats in each house are guaranteed to the party receiving the highest total vote in a general election. The remaining one-third is distributed proportionately among opposing political groups. Representatives must be at least 25 years of age and are elected for five-year terms.

Voting is by secret ballot and is compulsory for all citizens between 18–60 years of age. Women were first allowed to vote in 1963.

14 POLITICAL PARTIES
Since the end of the War of the Triple Alliance, two parties have dominated politics—the National Republican Association (Asociación Nacional Republicana), generally known as the Colorado Party, and the Liberal Party. Both parties have exemplified the uncompromising nature of Paraguayan politics and used their

position to stifle the opposition. Consequently, changes of administration have been effected principally by armed revolt.

The Colorado Party governed from its founding in 1887 until 1904, and again after 1947. Conservative and nationalistic, the Colorados split during the 1950s into two factions: the "officialist" Colorados supported the Stroessner dictatorship, while the People's Colorado Movement (Movimiento Popular Colorado—MOPOCO) styled itself a supporter of "representative democracy." Most of the MOPOCO leadership chose exile in 1959. In the 1980s the Colorados became even more divided. Three groups emerged: a "militant" pro-Stroessner faction, "traditionalists," pushing for Stroessner to step down, and a reformist "ethical" faction, which is interested in cleaning up government corruption.

The 1989 coup was engineered by a leader of the "traditionalist" faction. Wasmosy is more reformist in his approach. But even with the hefty parliamentary majority, the Colorados remain badly split and in disrepair.

The Liberal Party, like the Colorados, appeared in 1887. They seized power in 1904 and governed until 1936. Banned in 1942, the Liberals were reconstituted during the 1960s. There has never been a recognizable ideological distinction between the Liberals and Colorados, but the two parties are similar in their disunity. Liberals had, by 1982, split into three factions: the Authentic Radical Liberal Party (Partido Liberal Radical Auténtico—PLRA), the Liberal Teeté Party (Partido Liberal Teeté—PLT), and the Radical Liberal Party (Partido Liberal Radical—PLR). After 1989, the PLRA and the PLR reemerged to compete for votes, with the PLRA considerably stronger.

The Revolutionary Februarist Party (Partido Revolucionario Febrerista) consists of followers of Col. Rafael Franco's 1936 coup d'état. At the time, the Febreristas promised sweeping social changes, including the expropriation of large estates and the distribution of land among the poor. Although Franco lasted but a few months, his declarations still form the basis of party propaganda. They won a few seats in the legislature in the 1989 elections.

The Christian Democratic Party (Partido Demócrata Cristiano—PDC), founded in 1965, strongly opposed Stroessner, and refused to participate in any of his dubious electoral enterprises. In the 1989 election, they received only one legislative seat, and remain a small party.

15 LOCAL GOVERNMENT
Paraguay is divided into 19 departments, which are subdivided into districts, which, in turn, comprise municipalities (the minimum requirement for a municipality is 3,000 persons) and rural districts (partidos). A government delegate, appointed by the president, runs each department. Municipal government is exercised through a municipal board, chosen by direct election, and an executive department. In the principal cities and capitals, the executive department is headed by a mayor appointed by the minister of the interior; in other localities, the mayor is appointed by the presidents of the municipal boards. Police chiefs are appointed by the central government.

16 JUDICIAL SYSTEM
The five-judge Supreme Court exercises both original and appellate jurisdiction. There are four appellate tribunals: civil/commercial, criminal, labor, and juvenile. There are special appellate chambers for civil and commercial cases and criminal cases. Each rural district (partido) has a judge appointed by the central government to settle local disputes and to try accused persons and sentence those found guilty. Federal judges and magistrates are appointed by the executive for a term of five years coinciding with the presidential term, so that the judges of the Supreme Court and lesser tribunals are always named by the president in

power. The Council of State must approve the appointment of members of the Supreme Court and may remove them by impeachment. Justices of the peace deal with minor cases.

Although the 1992 Constitution called for selection of judges by an independent body working with the Congress and the executive, many of the active judges are holdover appointments of former dictator Alfredo Stroessner.

[17]ARMED FORCES

Paraguay's armed forces numbered 16,500 (11,000 conscripts) in 1993, about 12,500 were in the army, 3,000 in the navy (including 500 marines), and 1,000 in the air force, with 17 combat aircraft and 8 helicopters. Paraguay has compulsory military service of 18–24 months for all males between the ages of 18 and 20. Expenditures of the Ministry of Defense were $60 million in 1989, or about 1% of gross domestic product. There is a paramilitary police service of 8,000. Reservists number 45,000.

[18]INTERNATIONAL COOPERATION

Paraguay is a charter member of the UN, having joined on 24 October 1945. It participates in ECLAC and all the nonregional specialized agencies except IMO and WIPO. Paraguay also participates in G-77, IDB, LAIA, OAS, and PAHO, and has signed the Law of the Sea.

[19]ECONOMY

Landlocked Paraguay has a limited economy based predominantly on agriculture, livestock production, forestry, and the basic processing of materials. Mineral resources are negligible. In recent years, the relative importance of agriculture has declined, and the value of services has risen; however, cattle raising remains a key economic activity.

Paraguay suffered for years from runaway inflation. The IMF joined the US government in 1957 in providing for stabilization loans that enabled Paraguay to establish a free exchange system and accelerate the pace of public investment. During the 1960s, inflation ranged between 2% and 3%, but the rate increased during the following decade to 28.2% in 1979. The rate then declined to 5.1% during 1982, a recession year, but was an estimated 30% in 1986 and was projected at 40% for 1987, fueled by rapid expansion of the money supply.

Construction of the Itaipú hydroelectric project (which was finished in 1982) stimulated Paraguay's economic expansion, and the GDP grew by 11.4% annually during 1977–80. The end of the Itaipú building boom, currency devaluations in Argentina and Brazil (and thus the relative overvaluation of the guaraní), and declining international market prices for Paraguay's agricultural products led to an economic slowdown. The GDP declined by 2.5% in 1982, and in the subsequent five years, the economy went into a general decline, exacerbated by the impact of adverse weather on the agricultural sector. Gains of 3.3% in 1984 and 4% in 1985 were offset by declines of 3% and 4.3% in 1983 and 1986, respectively.

[20]INCOME

In 1992, Paraguay's GNP was $6,038 million at current prices, or $1,340 per capita. For the period 1985–92 the average inflation rate was 28.5%, resulting in a real growth rate in per capita GNP of 1.0%.

In 1992 the GDP was $6,446 million in current US dollars. It is estimated that in 1989 agriculture, hunting, forestry, and fishing contributed 30% to GDP; mining and quarrying, less than 1%; manufacturing, 17%; electricity, gas, and water, 2%; construction, 6%; wholesale and retail trade, 27%; transport, storage, and communication, 4%; finance, insurance, real estate, and business services, 2%; community, social, and personal services, 8%; and other sources, 4%.

[21]LABOR

According to 1991 estimates, the work force totaled 1,600,000, or more than one-third of the national population. In 1991, agriculture, animal husbandry, and forestry employed 44% of the work force; industry and commerce, 34%; service, 18%; and government, 4%. Unemployment in 1980 was 3%, but because of the phasing down of the Itaipú project, the rate increased to 13.9% in 1983. From 1984 through 1986, it fluctuated between 7.5% and 8.5%, but had risen back to 13% by 1991.

The 1992 constitution provides Paraguayans in both the public and private sector the freedom to form and join unions without government interference. The new constitution also protects fundamental worker rights, including the right of association. There also are provisions for antidiscrimination, employment tenure, severance pay, collective bargaining, and the right to strike.

Labor laws provide for a maximum workweek of 48 hours for day work and 42 for night work, with one day of rest. The law also provides for an annual bonus of one month's salary. The Labor Code, however, is no longer an accurate guide for analyzing work conditions, since it was superseded by the 1992 constitution. As of December 1990, the minimum wage in Asunción was $6.52 per day; the minimum wage for farm workers was equivalent to $2.90 per day.

[22]AGRICULTURE

Although land area under cultivation amounts to 5.6% of Paraguay's total land area, only 1.1% of the economically active population was engaged in agriculture in 1991. The principal areas of cultivation are in the clearings around Asunción and Encarnación. Arable land outside these regions is sparsely settled, and inhabitants there rely principally on livestock and forestry for a living. The total area under cultivation rose from 245,636 hectares (606,976 acres) in 1940–41 to about 2,235,000 hectares (5,523,000 acres) in 1991.

The two most widely cultivated crops are manioc (cassava) and corn, which, with meat, are the staples of the Paraguayan diet. Cotton, tobacco, and sugarcane are among the leading cash and export crops. A national wheat program increased production from 7,000 tons in 1965 to 23,000 tons in 1973, 55,000 tons in 1981, and 300,000 tons in 1990, eliminating the need for wheat imports. Enough beans, lentils, sweet potatoes, peanuts, coffee, and fruits are grown for home use, and slightly more than enough rice. Crops yielding edible oils are widely grown, and yerba maté is cultivated on plantations. Production of principal crops for 1992 (in tons) included sugar cane, 2,788,000; manioc, 2,200,000; soybeans, 1,315,000; corn, 466,000; cotton, 215,000; and rice, 50,000. Cotton and soybeans are the main export items; however, in 1992, heavy flooding affected cotton production (typically one-fifth of all crop production), which is based on small-scale farmers, so that overall, crop output fell by 4.1%. Lower crops and negligible growth in livestock output owing to weak demand in regional markets led to a fall in farm output for the second consecutive year. In 1992, cotton exports amounted to only $190 million (20.6% of total exports), down from $319 million (28.6%) in 1991.

[23]ANIMAL HUSBANDRY

Cattle raising forms a significant part of the country's economy, contributing 7.7% to the GDP in 1985. During the 1960s and 1970s, the meat-packing industry developed appreciably, with meat and related products constituting Paraguay's most important single export. Since the late 1970s, however, market conditions for beef exports have deteriorated, and the value of exported meat products has declined significantly. Between 1989 and 1992, meat exports dropped from $96.1 million to $50.4 million.

In 1992, livestock totaled 7,800,000 head of cattle, 2,600,000 hogs, 380,000 sheep, and 350,000 horses. Beef production was about 143,000 tons. Other livestock products in 1992 included 240,000 tons of milk and 36,000 tons of eggs.

24 FISHING
Paraguay has no appreciable fishing industry, and the consumption of fresh fish is low. The country has potential resources for fisheries, however. Dorado weighing up to 18 kg (40 lb) are caught in the Upper Paraná River, and the Paraguay River yields salmon, surubi, pacú, boga, and mandi. The catch was 13,000 tons in 1991.

25 FORESTRY
Although forest resources are immense, exploitation is limited by lack of roads and mechanized transport facilities. About half of Paraguay's total surface area consists of forest and woodland (13.3 million hectares/32.7 million acres in 1991). However, much of that lies in the western Chaco, the forest resources of which have never been exploited. Roundwood cuttings totaled 8.5 million cu m in 1991.

Exportation of logs was banned in 1973 in order to encourage the domestic lumber industry; forest products earned $27.6 million on the export market during 1992.

The chief forest products are quebracho, various cabinet and other tropical hardwoods, and oil of petitgrain. Quebracho, the source of the tannin used by the leather-tanning industry, is the wood of the greatest commercial importance. Paraguay is the world's largest producer of petitgrain oil, a perfume base distilled from the leaves and shoots of the bitter orange tree. Since wood and charcoal are the only fuels produced in Paraguay, about two-thirds of all wood cut is used for burning.

26 MINING
Paraguay's mining potential has been restricted by limited exploration, inadequate infrastructure, large fiscal and trade deficits, scarcity of foreign exchange, and limited private investment. In 1991, mining accounted for only 0.5% of the GDP. There are small deposits of iron ore, and a few mines were worked before 1865, but there had been no evidence, until recently, of any metallic mineral deposits of commercial value. Lateritic iron ore deposits along the Paraná River near Encarnación were estimated at 300 million tons with 35% iron. Manganese deposits are known to exist near the Guairá Falls. Excellent limestone, found in large quantities along the Paraguay River north of Concepción, is quarried for the cement industry. Sandstone, mica, copper, kaolin, clay, and salt have been exploited modestly, and there are known deposits of azurite, barite, gypsum, lignite, malachite, mica, peat, pyrite, pyrolusite, soapstone, and uranium. Production of mineral commodities is predominantly for domestic consumption.

Under Paraguayan law, all mineral rights belong to the government, which has sought to encourage mining development by the privatization of some state-owned companies.

27 ENERGY AND POWER
Until 1968, Paraguay relied almost entirely on thermoelectric power. In that year, however, the first turbine was inaugurated at a major hydroelectric project at the juncture of the Acaray and Monday rivers. Planned jointly with Brazil and completed by an Italian company, this first stage of the project had a capacity of 45 Mw; by 1976, the project was completed, with a capacity of 190 Mw.

In 1973, Paraguay and Brazil agreed on the joint construction of the Itaipú power plant, the world's largest hydroelectric project. The plant, which came onstream in 1984, has a generating capacity of 12,600 Mw. By the end of 1982, the total cost (shared by the two countries) exceeded $20 billion; construction of the power plant and auxiliary facilities employed a peak force of 39,000 Paraguayan workers. Itaipú produces about 70% of Paraguay's electricity. Paraguay is entitled to 50% of Itaipú's output, but can sell unused power to Brazil; Itaipú produces about 75 billion kwh of electricity per year. In 1991, Paraguay exported 90% of its domestic electrical production.

Another 1973 agreement, with Argentina, called for the joint construction on the Alto Paraná River of the Yacyretá hydroelectric plant, with an installed capacity of 4,000 Mw. The final agreement was signed in 1979, but construction was delayed because of Argentina's economic crisis and disputes over contract awards, exacerbated by reports of cost overruns and high-level corruption. In 1986 and 1987, however, construction work on the project was speeded up, due largely to the securing of 85% of the project's $6.5-billion cost. Start-up is expected by the mid-1990s. The Corpus hydroelectric project, another joint Argentine-Paraguayan venture, which would add another 6,000 Mw of generating capacity, has been postponed, due largely to budgetary constraints in Argentina.

The National Electrical Administration manages the entire power system, which was nationalized in 1947. Total power production grew from 598 million kwh in 1975 to 29,780 million kwh in 1991.

It is generally assumed that the Paraguayan Chaco contains oil, but explorations, which in the mid-1980s involved several international companies, have thus far proved fruitless.

28 INDUSTRY
Since the early 1970s, manufacturing has accounted for 16–17% of the GDP. Processing of agricultural, animal, and forestry products, mainly for export, and small-scale manufacture of consumer goods for local needs are of greatest importance. Most manufacturing is done in the Asunción area; some plants, however, are near the source of their respective raw materials. Import-substitution industries encouraged by the government include petroleum refining, foodstuffs, wood processing, and chemicals.

In 1981, work began on a cement plant (capacity 600,000 tons) at Vallemí, financed by France; it was still uncompleted by the end of 1985. A 150,000-ton Paraguayan-Brazilian steel mill at Villa Hayes was inaugurated in 1986. Food-processing plants include slaughterhouses; flour mills; sugar mills; oil mills producing cottonseed and peanut oils for domestic consumption, as well as castor, tung, coco, and palm oils for export; related industries that process the by-products of oil extraction; and mills that produce yerba maté. There are numerous sawmills. A considerable but decreasing number of hides are also produced for export. Although there is a considerable textile industry, imports still run high. Products for domestic consumption include pharmaceutical and chemical goods, finished wood and furniture, brick and tiles, cigars and cigarettes, candles, shoes, matches, soap, and small metal goods.

Cement output was 109,024 tons in 1984, but only 45,580 tons in 1985, reflecting the completion of the Itaipú hydroelectric project. Other industrial products included wheat flour, 99,500 tons; sugar, 78,135 tons; cotton cloth, 9,200 tons; carbonated beverages, 138,700 liters; and cigarettes, 41,700 packs.

29 SCIENCE AND TECHNOLOGY
In 1993 there were eight scientific and technological research institutes and learned societies in Paraguay, all of them located in Asunción. Notable among them are the Paraguayan Scientific Society, founded in 1921, and the South American Union of Engineers' Associations, established in 1935. The "Nuestra Senora de la Asuncion" Catholic University has a department of science and technology. The National University of Asuncion has faculties of medicine, dentistry, chemistry, natural sciences, and agricultural engineering. The Higher School of Philosophy, Sciences and Education was founded in 1944.

³⁰DOMESTIC TRADE

Offices of foreign concerns and most important retail establishments are in Asunción, the only significant commercial center. Most retail trade is in small shops dealing in a limited variety of goods.

Legislation in 1961 provided for governmental and private commercial credit companies to aid in the development of agricultural, livestock, and industrial activities. Consumer credit facilities have been expanding. Many of the larger Asunción stores offer installment credit.

³¹FOREIGN TRADE

Paraguay has a foreign trade typical of an agricultural country. Agricultural, animal, and forest products are exported, and foods, transportation equipment, machinery, chemicals, textiles, and other manufactured goods are imported. Except for 1976 and 1977, Paraguay has had a trade deficit every year since 1965. There were reports through the 1980s of a widespread contraband trade, stimulated by exchange differentials between Paraguay and its neighbors.

³²BALANCE OF PAYMENTS

From 1989 to 1991, because of an appreciating exchange rate, exports plummeted while imports more than doubled, creating large trade deficits in 1991 and 1992.

In 1991 merchandise exports totaled $1,117.3 million and imports $1,669.1 million. The merchandise trade balance was $–551.8 million. The following table summarizes Paraguay's balance of payments for 1990 and 1991 (in millions of US dollars):

	1990	1991
CURRENT ACCOUNT		
Goods, services, and income	−227.9	−521.0
Unrequited transfers	55.6	54.6
TOTALS	173.2	−466.4
CAPITAL ACCOUNT		
Direct investment	76.3	83.1
Portfolio investment	—	—
Other long-term capital	−142.5	−36.9
Other short-term capital	−9.6	51.8
Exceptional financing	105.0	71.9
Other liabilities	—	—
Reserves	−219.3	−298.9
TOTALS	−190.1	−129.0
Errors and omissions	362.4	595.4
Total change in reserves	−228.7	−300.7

³³BANKING AND SECURITIES

The Central Bank of Paraguay was founded in 1952 as a state-owned, autonomous agency charged with establishing the government's monetary credit and exchange policies. Recommendations in early 1961 by an economic mission of the IDB and IBRD led to the establishment of the National Development Bank to provide an effective source of medium- and long-term agricultural and industrial credits. Savings and loan institutions are regulated by the superintendent of banks. There were 7 development banks, 8 commercial banks, and 12 foreign banks in 1986. Reserves of the commercial banking sector totaled G588.2 million in December 1992.

³⁴INSURANCE

All insurance business in Paraguay is regulated by the government through the superintendent of banks. Foreign companies are permitted to operate in the country but are under stringent requirements calling for the investment of capital and reserves.

In 1986 there were 22 national insurance companies and 1 foreign company.

³⁵PUBLIC FINANCE

The public sector account balance deteriorated in the early 1990s to a deficit equivalent to 2.5% of GDP in 1992 (from a surplus of 7% of GDP in 1988). In the late 1980s, improved revenue collection and rising income from the new Itaipú power plan were largely responsible for favorable public sector account balances. In 1991/92, however, decreased tax collections (from newly-implemented tariff reforms) and large current expenditures increases (from escalating wages and pensions) caused the deficit to widen.

The following table shows actual revenues and expenditures for 1989 and 1990 in millions of guaraníes.

	1989	1990
REVENUE AND GRANTS		
Tax revenue	409,608	584,272
Non-tax revenue	113,285	207,960
Capital revenue	428	425
Grants	1,724	1,066
TOTAL	525,045	793,723
EXPENDITURES & LENDING MINUS REPAYMENTS		
General public service	71,469	131,775
Defense	57,340	79,883
Education	51,341	76,106
Health	18,498	25,981
Social security and welfare	55,801	70,822
Housing and community amenities	4,287	18,329
Recreation, cultural, and religious affairs	390	494
Economic affairs and services	61,763	76,919
Other expenditures	91,297	119,529
Lending minus repayments	14,248	2,191
TOTAL	428,349	603,064
Deficit/Surplus	96,696	190,659

Of the $1.67 billion registered external debt in 1992, approximately 40% is held by the government. The 1993 public sector long-term debt service was about $260 million.

³⁶TAXATION

The tax system is complex, with a multiplicity of taxes. Administration is marginally effective, characterized by cumbersome procedures and widespread tax evasion. There is no income tax on individual earnings from employment by others. Sole proprietorships earning $12,400 or less per year pay a flat tax of 3%.

As of 1993, corporate income tax was expected to be 30% under the new tax system. There is no excess profits tax or tax on dividends received by stockholders, except that stock dividends for nonresidents are subject to a withholding tax of 10%. A tax of 30% is levied on 90% of the value of royalties remitted abroad. Capital gains on all assets are taxed at 30%. A value-added tax is levied on goods and services, and real estate taxes are levied annually at a rate of 1%.

A 95% income tax exemption for five years is available for income sources in investments made.

³⁷CUSTOMS AND DUTIES

In general, Paraguayan customs duties have been viewed as a source of revenue and a means of conserving foreign exchange, with relatively few of the high duties being intended as protection for manufactured products. Import duties are specific, ad valorem, or both. Import pressures are such that the government has not kept a tight control on the purchase of nonessentials.

According to the Harmonized System of tariffs introduced in January 1992, customs valuation is based upon the price paid on imports plus other costs, charges, and expenses.

Paraguay has free port privileges at the Brazilian ports of Paranaguá and Santos.

³⁸FOREIGN INVESTMENT

Paraguay's economy historically has been dominated by foreign interests, in particular by those of wealthy Argentines, Britons, and Brazilians. Nevertheless, the Paraguayan government has encouraged foreign investment in recent years, as a means of developing the country. Investors willing to start or to expand needed industries receive highly favorable concessions, including freedom from many import restrictions plus special rights to retain much of the foreign exchange earned.

³⁹ECONOMIC DEVELOPMENT

To a considerable extent, Paraguay has a government-controlled economy; government agencies fix prices, control distribution, regulate production and exportation, and exercise monopolistic rights over much of the economy. In recent decades, and particularly since the IMF stabilization program went into effect in the late 1950s, some controls have been loosened. In the wake of the free exchange system have come moves to eliminate government subsidies, such as that for wheat. In agriculture there is an annual plan for acreage quotas, but the principal problem has been one of meeting the quotas rather than of the surpluses. The establishment of the National Development Bank has created a source of medium- and long-term credits favorable to agriculture and industry. Price controls and marketing quotas are particularly significant to the cattle industry. Paraguay has sought to develop closer economic ties with Brazil, the US, and Western European nations, largely to reduce the country's dependence on trade with Argentina.

Economic planning is the responsibility of the Technical Planning Secretariat for Economic and Social Development, established in 1962. The first national plan covered 1965–66; the second, 1967–68. The third plan, a medium-term, five-year program for 1969–73, was replaced by a 1972–77 development scheme calling for a 26% increase in public investment in agriculture. Regional development, also given high priority, was to be accomplished through Paraguay's utilization of its water resources in the Itaipú hydroelectric project; a parallel development program for the Alto Paraná region was retarded by delays in the Yacyretá power project. The 1977–81 development plan aimed to achieve a more equitable distribution of social resources. A plan announced in September 1986 provided for comprehensive reform in exchange rates and in investment and fiscal policies. As of mid-1987, only the part of the plan dealing with exchange rates had been implemented.

Between 1949 and 1986, Paraguay received $1,090 million in multilateral development assistance, of which $533.8 million came from the IDB and $457.6 million from the IBRD. US non-military loans and grants during the same period amounted to $203.5 million.

⁴⁰SOCIAL DEVELOPMENT

Social insurance was legally established in 1948, when the Social Security Institute was created under the Ministry of Public Health. The original law, providing for medical care and sick benefits, was modified in 1951, when health insurance became obligatory for salary and wage earners of any age who work under a written or oral contract and for apprentices not receiving wages. Legislation in 1973 authorized the institute to establish a pension plan. Self-employed persons and workers who have no contract of employment are not required to carry social insurance but may do so voluntarily by applying to the institute; coverage of foreign workers is compulsory. The program provides for free medical, surgical, and hospital care (not always available) for the worker and dependents, maternity care and cash benefits, sickness and accident benefits, retirement pensions at age 60, and funeral benefits. Unemployment insurance does not exist, but severance pay is provided. Social security contributions in 1991 included 9.5% of the gross salary withheld from the employee, 13% of payroll paid by the employer, and a government contribution equal to 1.5% of employee earnings. The government considers the rate of population growth too low, but it maintains an official family-planning program. In 1990, an estimated 48% of couples of childbearing age practiced contraception; abortion is permitted only to save the life of the mother. The fertility rate in 1985–90 was 4.6%.

Human rights abuses have reportedly diminished since the mid-1970s, but international observers have continued to cite the government's alleged reliance on secret police and informers, the widespread practice of arbitrary arrest, the Stroessner personality cult in the media, and the granting of preferred jobs and social amenities to government party members. By the end of 1986, however, torture of political prisoners had ended, due largely to US pressure. There were no political killings in 1993, although the police continued to commit human rights abuses.

Domestic violence and workplace sexual harassment remain serious problems for women. It is believed that there are many more rapes than the average of 16 per month reported to the Ministry of Health.

⁴¹HEALTH

Hospital and medical facilities are generally concentrated in Asunción and other towns. In 1990 there was 1 hospital bed for every 1,000 people. There were 2,536 doctors, 1,017 dentists, and 565 nurses in 1988. In 1992 there were 0.62 doctors per 1,000 people, with a nurse to doctor ratio of 1.7. Approximately 60% of the population has access to health care services. In 1990, total health care expenditures were $160 million.

Average life expectancy in 1992 was 67 years; the infant mortality rate averaged 28 per 1,000 live births. Overall mortality was 6.4 per 1,000 inhabitants in 1993 and maternal mortality was 300 per 100,000 live births in 1991. The principal causes of death are bacillary dysentery and other intestinal diseases, heart disease, pneumonia, and cancer. In 1990, there were approximately 166 reported cases of tuberculosis per 100,000 population, and in 1991, only 35% of the population had access to safe water and 62% had adequate sanitation. Immunization rates in 1992 for children up to one year old: tuberculosis (99%); diphtheria, pertussis, and tetanus (85%); polio (87%); and measles (86%). About 4% of children under 5 years old were considered malnourished in 1991.

The birth rate was 33 per 1,000 people for a total of 151,000 births in 1992.

⁴²HOUSING

As of the 1982 census, there were an estimated 582,700 housing units in Paraguay; 38.4% of all dwellings had electricity and 59.4% had toilets. The average number of people per household was 5.2. Between 1982 and 1988, the number of housing units rose to 755,000 with five people per dwelling. A government agency, the Paraguayan Housing and Urban Institute, was created in 1964 with an IBRD loan of $3.4 million to aid in the construction of living units for low-income families. In 1973, a National Housing Bank was established to finance low-income housing development. South Africa provided a $2-million loan in 1975 for the same purpose. Budgetary allocations to the Ministry of Housing during 1976–80 totaled G3.9 billion, or about 12% of central government current expenditures during that period.

43 EDUCATION

As of 1990, the estimated illiteracy rate was 9.9%, 11.9% for females and 7.9% for males. Elementary education is compulsory and free between the ages of 7 and 14 (ages 9 and 14 in rural areas). Primary education lasts for six years followed by secondary education in two phases—of three years each. In 1991 there were 4,649 public and private primary schools, with 720,983 students, and 169,167 students at the secondary level with 12,218 teaching staff.

The National University of Paraguay is located in Asunción, the capital city. Nuestra Señora de la Asunción Catholic University, a private institution, was founded in 1960. Total university and higher institution enrollment in 1990 was 32,884 students.

44 LIBRARIES AND MUSEUMS

Paraguay's modest cultural life is centered in Asunción, which has the nation's principal libraries and museums. The National Library and Archives, established in 1869, are located in Asunción

Asunción is the site of 11 historical, scientific, and art museums. Prominent art museums include the National Museum of Fine Arts and Antiquities; the Andrés Barbero Ethnographic Museum, devoted to Amerindian art; the Julián de la Herrería Ceramics and Fine Arts Museum; and the Museum of Modern Art of the Ministry of Education.

45 MEDIA

Since nationalization of the International Telephone Co. in 1947, a government monopoly, the National Telecommunications Administration (Administración Nacional de Telecommunicaciones—ANTELCO), has controlled the communications system. As of 1991 there were 111,514 telephones in use, most of them in Asunción. ANTELCO operates the national telegraph system; telephone, telex, and cable links have been established with most other countries. There were 47 radio stations in Paraguay in 1991, including the official stations, Radio Nacional of Asunción and Radio Encarnación, and the Catholic Radio Caritas. Paraguay also has four television stations. Radio sets in use numbered about 750,000 in 1991, television receivers 220,000. The content of programming is controlled by the government.

Although the government has proclaimed that the press is free, many opposition dailies—notably *El Enano*, a Liberal Party daily, and *ABC Color*, an independent newspaper—have been closed down, in reality government censors control the press. Newspaper readership is among the lowest in Latin America. *Patria* (circulation 30,000 in 1991), founded in 1946, is the organ of the Colorado Party. *Hoy* began publication in 1977 and in 1991 had an estimated circulation of 40,000; the circulation of *Ultima Hora*, also founded in 1977, was 40,000 in 1991.

46 ORGANIZATIONS

Several chambers of commerce promote local and international trade. Active trade associations include the Federation of Production, Industry, and Commerce, an importers' association, and various organizations of particular trades. Professional and cultural organizations also are active.

The Paraguayan Atheneum sponsors lectures, concerts, and recitals, as well as courses in foreign languages, art, and music. The Paraguayan-American Cultural Center and the Argentine-Paraguayan Institute are important binational centers. Other organizations include the Women's Center and the Youth Atheneum. Paraguay has an Academy of Language and several organizations devoted primarily to Guaraní culture, including the Academy of Guaraní Culture and the Indian Association of Paraguay.

47 TOURISM, TRAVEL, AND RECREATION

Foreign tourists entering Paraguay are required to present a valid passport; visas are not normally required for stays of less than 90 days. Visitors are urged to take precautions against tetanus, typhoid-paratyphoid, and malaria. In 1991 there were 293,794 tourist arrivals in hotels and other establishments, 34% from Argentina, 19% from Brazil, and 8% from Uruguay. Tourist receipts totaled $145 million. There were 4,766 rooms in hotels and other facilities with 10,449 beds and a 30% occupancy rate.

The monuments, museums, and parks of Asunción are the main tourist attractions. Also of interest are the Amerindian markets in and around the capital; at the famous market of Itauguá, about 30 km (18 mi) from Asunción, the makers of ñandutí lace sell their wares. Other popular tourist attractions include the world famous Iguazu Falls at Paraguay's borders with Brazil and Argentina, the San Bernardino resort, on Lake Ypacarai, and the modern boom town of Puerto Presidente Stroessner, the headquarters for construction of the Itaipú power project. Football (soccer) is Paraguay's national sport, with some 30 clubs in Asunción alone. Tennis, horse racing, boxing, basketball, and rugby football are also popular.

48 FAMOUS PARAGUAYANS

Paraguay acclaims—despite their reputations as dictators—the first three leaders of the independent nation: José Gaspar Rodríguez de Francia (El Supremo, 1761?–1840), his nephew Carlos Antonio López (1790–1862), and the latter's son Francisco Solano López (El Mariscal, 1827–70). Of nearly equal prominence is José Felix Estigarribia (1888–1940), president and Chaco War commander. Manuel Gondra (1872–1927), twice president of Paraguay, was a literary critic, educator, and diplomat. Eusebio Ayala (1875–1942), another president, was an authority on political economy and international law. Alfredo Stroessner (b.1912) was president of Paraguay from 1954 to 1989.

Leading writers include Juan Silvano Godoi (1850–1926), Manuel Domínguez (1869–1935), Pablo Max Ynsfrán (b.1894), Justo Pastor Benítez (1895–1962), former president Juan Natalicio González (1897–1966), Gabriel Casaccia (b.1907), Augusto Roa Bastos (b.1917), and Hugo Rodríguez Alcalá (b.1918). Pablo Alborno (1877–1958) and Juan Domínguez Samudio (1878–1936) were noted artists, while in music, José Asunción Flores (b.1904) is best known.

49 DEPENDENCIES

Paraguay has no territories or colonies.

50 BIBLIOGRAPHY

Hanratty, Dennis M. and Sandra W. Meditz (eds.). *Paraguay, a Country Study.* 2d ed. Washington, D.C.: Library of Congress, 1990.

Haverstock, Nathan A. *Paraguay in Pictures.* Minneapolis: Lerner Publications, 1987.

Lewis, Paul H. *Paraguay Under Stroessner.* Chapel Hill: University of North Carolina Press, 1980.

Lewis, Paul H. *Political Parties and Generations in Paraguay's Liberal Era, 1869–1940.* Chapel Hill: University of North Carolina Press, 1993.

Miranda, Carlos R. *The Stroessner Era: Authoritarian Rule in Paraguay.* Boulder, Colo.: Westview Press, 1990.

Nickson, R. Andrew. *Historical Dictionary of Paraguay.* 2d ed. Metuchen, N.J.: Scarecrow Press, 1993.

Nickson, R. Andrew. *Paraguay.* Santa Barbara, Calif.: Clio, 1987.

Roett, Riordan. *Paraguay.* Boulder, Colo.: Westview, 1986.

Warren, Harris Gaylord. *Rebirth of the Paraguayan Republic: The First Colorado Era, 1878–1904.* Pittsburgh: University of Pittsburgh Press, 1985.

Williams, John H. *The Rise and Fall of the Paraguayan Republic, 1800–70.* Latin American Monographs No. 48. Austin: University of Texas Press, 1979.

PERU

Republic of Peru
República del Perú

CAPITAL: Lima.

FLAG: The national flag consists of red, white, and red vertical stripes, with the coat of arms centered in the white band.

ANTHEM: *Himno Nacional,* beginning "Somos libres, seámoslo siempre" ("We are free; let us remain so forever").

MONETARY UNIT: The nuevo sol (ML), a paper currency of 100 céntimos, replaced the inti on 1 July 1991 at a rate of I1,000,000 = ML1, but, in practice, both currencies are circulating. There are coins of 1, 5, 10, 20, and 50 céntimos and 1 nuevo sol, and notes of 10, 20, 50, and 100 nuevos soles and 10,000, 50,000, 100,000, 500,000, 1,000,000, and 5,000,000 intis. ML1 = $0.0005 (or $1 = ML2,160).

WEIGHTS AND MEASURES: The metric system is the legal standard.

HOLIDAYS: New Year's Day, 1 January; Labor Day, 1 May; Day of the Peasant, half-day, 24 June; Day of St. Peter and St. Paul, 29 June; Independence Days, 28–29 July; Santa Rosa de Lima (patroness of Peru), 30 August; Battle of Anzamos, 8 October; All Saints' Day, 1 November; Immaculate Conception, 8 December; Christmas, 25 December. Movable holidays include Holy Thursday and Good Friday.

TIME: 7 AM = noon GMT.

¹LOCATION, SIZE, AND EXTENT

Peru is South America's third-largest country, with an area of 1,285,220 sq km (496,226 sq mi), extending about 1,290 km (800 mi) SE-NW and 560 km (350 mi) NE-SW. Comparatively, the area occupied by Peru is slightly smaller than the state of Alaska. It is bounded on the N by Ecuador and Colombia, on the E by Brazil and Bolivia, on the S by Chile, and on the W by the Pacific Ocean, with a total boundary length of 9,354 km (5,812 mi). A border dispute with Ecuador led to armed hostilities in January 1981; by 20 February, a 14-km (9-mi) demilitarized zone had been established along the disputed line.

Various offshore islands, chiefly the Chincha Islands off Pisco in southern Peru, are uninhabited, but at least 21 of these are important to the Peruvian economy and are protected by the government's guano monopoly.

Peru's capital city, Lima, is located on the Pacific coast.

²TOPOGRAPHY

Peru is divided into three contrasting topographical regions: the coast (costa), the highlands (sierra), and the eastern jungle (selva). The coastline is a narrow ribbon of desert plain from 16 to 160 km (10 to 100 mi) broad. It is scored by 50 rivers, which water some 40 oases. Only a few of these rivers, which have their source in the Andean snowbanks, reach the sea in all seasons. Although the coastal region constitutes only 12% of the national territory, it contains the ports and chief cities of Peru. Inland, the low costa rises through the steep wastes of the high costa (760–2,000 m/ 2,500–6,500 ft), then ascends abruptly to the western cordillera (Cordillera Occidental) of the Andes, which, with its ridge of towering peaks, runs parallel to the coast and forms the Peruvian continental divide. The less regular Cordillera Central and Cordillera Oriental merge in central Peru with the Cordillera Occidental. They branch off to the southeast, meeting a transverse range that becomes a crescent of peaks forming the drainage basin of the 8,300-sq-km (3,200-sq-mi) Lake Titicaca, the highest

large navigable lake in the world (about 3,800 m/12,500 ft high), which is bisected by the Peruvian-Bolivian border. Of the 10 Peruvian peaks that rise above 5,800 m (19,000 ft), Huascarán, 6,768 m (22,205 ft), is the highest.

The intermontane basins, deep-gashed canyons, and high treeless plateaus (punas) of the Andes form the sierra and constitute 27% of the country's surface. The most important rivers draining the Andes on the Atlantic watershed, such as the Marañón, Huallaga, and Ucayali, flow north or south and eventually east to form the Amazon Basin. The selva covers 61% of Peru and consists of the low selva (the Amazon rain forest) and the high selva, a steeply sloping transition zone about 100–160 km (60–100 mi) wide between the sierra and the rain forest.

³CLIMATE

Although Peru's seaboard is situated well within the tropical zone, it does not display an equatorial climate; average temperatures range from 21°C (70°F) in January to 10°C (50°F) in June at Lima, on the coast. At Cuzco, in the sierra, the range is only from 12°C (54°F) to 9°C (48°F), while at Iquitos, in the Amazon region, the temperature averages about 32°C (90°F) all year round. The cold south–north Humboldt (or Peruvian) Current cools the ocean breezes, producing a sea mist with the inshore winds on the coastal plain. Only during the winter, from May to October, does this sea mist (garúa) condense into about 5 cm (2 in) of rain.

Latitude has less effect upon the climate of the sierra than altitude. The rainy season in the Andes extends from October to April, the reverse of the coastal climate. Temperatures vary more from day to night than seasonally. The snow line ranges from 4,700 to 5,800 m (15,500 to 19,000 ft). In the eastern jungle, precipitation is heavy, from 190 to 320 cm (75 to 125 in) annually; rain falls almost continuously between October and April.

A warm Pacific west-to-east current called El Niño appears near the Peruvian coast every five or six years around Christmastime (whence its name, for the Christ child), occasionally causing

283

serious weather disturbances. During the first half of 1983, after the strongest such current ever recorded, more than 250 cm (100 in) of rain fell on Peru's northwestern plains.

⁴FLORA AND FAUNA

Peru's several climates and contrasting surface features have produced a rich diversity of flora and fauna. Where the coastal desert is not barren of life, there are sparse xerophytic shrub, cactus, and algarroba, and a few palm oases along the perennially flowing rivers from the Andes. Where the sea mist (garúa) strikes against the rising slopes between 800 and 1,400 m (2,600 and 4,600 ft), a dense belt of lomas, flowering plants, and grasses (important for grazing) grows. Perennial shrubs, candelabra cacti, and intermontane pepper trees account for much of the western slope vegetation in the higher altitudes, and forests of eucalyptus have been planted. High-altitude vegetation varies from region to region, depending on the direction and intensity of sunlight. Tola grows in profusion at 3,400 m (11,000 ft) in the southern volcanic regions; bunch puna grasses may be found at 3,700 m (12,000 ft). On the brow (ceja) of the eastern slopes, mountain tall grass and sparse sierra cactus and low shrub give way at 900 m (3,000 ft) to rain forests and subtropical vegetation. As the eastern slopes descend, glaciers are remarkably close to tropical vegetation. The 601,000 sq km (232,000 sq mi) of eastern selva, with 18 rivers and 200 tributaries, contain the dense flora of the Amazon basin. Such native plants as sarsaparilla, barbasco, cinchona, coca, ipecac, vanilla, leche caspi, and curare have become commercially important, as well as the wild rubber tree, mahogany, and other tropical woods.

For centuries, vast colonies of pelicans, gannets, and cormorants have fed on the schools of anchovies that graze the rich sea pastures of the Humboldt Current and have deposited their excrement on the islands to accumulate, undisturbed by weather, in great quantities of guano. This natural fertilizer was used by the pre-Inca peoples, who carried it on their backs to the sierra. Forgotten during the days of colonial gold greed, guano attracted the attention of scientists in 1849, when its rich nitrogen content was analyzed as 14–17%. For 40 years thereafter, Peru paid many of its bills by exporting guano to exhausted croplands of Europe. Guano has since been largely replaced in the international market by synthetic fertilizers.

The rich marine plant life off the Peruvian coast attracts a wealth of marine fauna, the most important of which are anchoveta, tuna, whale, swordfish, and marlin. Characteristic of the Andes are the great condor, ducks, and other wild fowl. The vizcacha, a mountain rodent, and the chinchilla are well known, as is the puma, or mountain lion. Peru is famous for its American members of the camel family—the llama, alpaca, huarizo, and guanaco—all typical grazing animals of the highlands. The humid forests and savannas of eastern Peru contain almost half the country's species of fauna, including parrots, monkeys, sloths, alligators, paiche fish, piranhas, and boa constrictors, all common to the Amazon Basin.

⁵ENVIRONMENT

Peru's principal environmental problems in the mid-1990s are air pollution, water pollution, soil erosion, and deforestation. Air pollution is a problem due to industrial and transportation vehicle emissions. Peru contributed 0.5% to the world's total gas emissions. Water pollution is another of Peru's environmental concerns. Its sources are industrial waste, sewage, and oil-related waste. The nation has 9.6 cu mi of water. Seventy-two percent is used to support farming and 9.0% is used for industrial activity. Thirty-two percent of the city dwellers and 76% of the rural people do not have pure water. Peru's soil is affected by erosion and pollution. A UN survey indicated that 60% of the nation's soils are subject to these problems. The nation's cities contribute 3.0

million tons of solid waste per year. Municipal and mining wastes have also degraded the quality of coastal waters, with their economically important fishery resources, and offshore oil development poses an additional threat.

The National Office for the Evaluation of Natural Resources is the principal policymaking body for resource development, while the General Department of the Environment, part of the Ministry of Health, deals with control of pollution problems; water, forest, and wildlife resources are the province of the Ministry of Agriculture. Numerous environmental protection measures have been passed, but enforcement is lax and hampered by inefficient management and scarce fiscal resources. According to a PAHO report, only 55 of 300 towns with 2,000 or more inhabitants had adequate refuse transportation and elimination systems in 1978, when the government began drawing up plans for solid-waste management. A major environmental challenge for Peru in the 1980s has been how to open the selva for agricultural development without doing irreparable harm to the ecology of the Amazon Basin.

In 1994, 29 of the nation's mammal species and 75 of its bird species are endangered. Three-hundred and sixty of its plant species are also endangered. Endangered species included the yellow-tailed woolly monkey, black spider monkey, puna rhea, tundra peregrine falcon, white-winged guan, arrau, green sea turtle, hawksbill turtle, olive ridley turtle, leatherback turtle, spectacled caiman, black caiman, Orinoco crocodile, and American crocodile.

⁶POPULATION

The population according to the 1981 census was 17,005,210. The 1994 population was estimated at 23,383,011, and the UN projects a population for the year 2000 of 26,276,000, assuming a crude birthrate of 26.7 per 1,000 population, a crude death rate of 6.8, and a net natural increase of 19.9. In 1990, some 53% of the inhabitants lived in the coastal region; the Andean sierra had 36% of the population, the eastern jungle 11%. The estimated population density in 1994 was 18.2 per sq km (47.1 per sq mi) overall. Peru's population growth rate averaged 2.1% per year between 1985 and 1995.

In 1994, an estimated 72% of the total population lived in urban areas. Metropolitan Lima, the capital, had about 6,869,209 inhabitants in 1995. Other important cities, with 1990 estimated populations, are Trujillo, 531,000; Arequipa, 634,000; and Chiclayo, 426,000. As of 1992, the birthrate was 29.3 per 1,000 and the death rate 7.7 per 1,000.

⁷MIGRATION

During the guano boom in the 1860s and 1870s, the Peruvian government imported Chinese laborers to mine the guano deposits, build railroads, and work on cotton plantations. Since then, Peru has not attracted large numbers of immigrants, although there are Japanese as well as Chinese enclaves in the coastal cities. In 1991, 377,485 Peruvians left the country and 309,136 returned. The US was the leading country of destination (38%), with Chile second. Arrivals by foreigners outnumbered departures by 7,789.

Recent governments have encouraged the movement of people into the empty areas of the eastern Andean slopes (the high selva) in order to bring the eastern provinces into the national economic mainstream. Since the 1950s, however, the main trend has been in the reverse, from the sierra to the coastal cities. Lima has received the bulk of rural migrants, and by the mid 1990s the metropolitan area of Lima supported nearly one-third of the total national population.

⁸ETHNIC GROUPS

In the early 1990s, about 45% of the inhabitants were Amerindian, 37% mestizo (of mixed Spanish and Amerindian ancestry), 15% white, and 3% black, Asian, or other.

LOCATION: 0°1′ to 18°20′s; 68°39′ to 81°19′w. **BOUNDARY LENGTHS:** Ecuador, 1,529 km (950 mi); Colombia, 1,506 km (936 mi); Brazil, 2,823 km (1,754 mi); Bolivia, 1,047 km (651 mi); Chile, 169 km (105 mi); Pacific coastline, 2,334 km (1,450 mi). **TERRITORIAL SEA LIMIT:** 200 mi.

Of the 4–7 million sierra Amerindians under Inca domination, fewer than 1 million were left when the first colonial census was taken in 1777. A failing food supply and new diseases, such as smallpox, scarlet fever, and measles, were lethal to the young. Despite continuing disease and poverty found among the Amerindians today, they have increased to more than 8 million. The main groups are the Quechua- and Aymará-speaking tribes, but there are also some other small tribes in the highlands. Peru's lowland forest Amerindians were never subjugated by Incas or by Spaniards and continue to be fishermen, hunters, and foragers. In the mid-1980s, at least 225,000 jungle Indians were grouped in 37 tribes. A 20-year plan announced in 1968 called for the full social, economic, and political integration of Peru's Amerindian population. Nevertheless, in the 1980s, sociocultural distinctions based on ethnic background were endemic to Peruvian society, with whites (especially the criollos, those of early Spanish descent) at the top of the hierarchy, mestizos and cholos (acculturated Amerindians) below them, and monolingual Quechua- or Aymará-speaking Amerindians at the bottom.

Small groups of Germans, Italians, and Swiss are important in commerce, finance, and industry. Chinese and Japanese operate small businesses, and some Japanese have been successful in agriculture.

⁹LANGUAGES

Spanish is spoken, as in all Latin America, without the use of the sound represented by *th* in *thing* characteristic of Castilian. In 1990, over 72% of the population claimed to speak only Spanish. At least 7 million Amerindians, as well as many mestizos, speak Quechua, the native tongue of the Inca peoples, the use of which was outlawed following an Amerindian revolt in 1780. A decree of 27 May 1975 granted Quechua the status of an official language, along with Spanish. Some words in modern English usage derived from Quechua are *alpaca, condor, coca, guano, Inca, llama, guanaco, vicuña, puma,* and *quinine*. Aymará is spoken by at least 700,000 people, especially in the department of Puno and around Lake Titicaca, and various other languages are spoken by tribal groups in the Amazon Basin.

¹⁰RELIGIONS

Although the population was about 90% Roman Catholic in the early 1990s, the practice of Catholicism in Peru is imbued with Amerindian elements. Roman Catholicism is the official religion, but the constitution guarantees religious freedom. Civil marriage is obligatory. Since 1929, only Roman Catholic religious instruction has been permitted in schools, state or private. Protestants numbered almost 6% in the late-1980s, and tribal religionists totaled perhaps another 1%. There were also an estimated 7,000 Buddhists and 3,300 Jews living in Peru.

¹¹TRANSPORTATION

The system of highways that was the key to the unification of the Inca Empire was not preserved by the Spanish conquerors. The lack of an adequate transportation system is still a major obstacle to economic integration and development.

Peru's railroad system, consisting of 1,884 km (1,171 mi) of track in 1991, nationalized in 1972, is subject to landslides and guerrilla attacks. The two principal railway systems, the Central and Southern railways, were built during the second half of the 19th century and were at one time owned and operated by British interests. The Central Railway, the world's highest standard-gauge railroad, connects Lima-Callao with the central sierra. The Southern Railway links Arequipa and Cuzco with the ports of Mollendo and Matarani and runs to Puno on Lake Titicaca, where steamers provide cross-lake connections with Bolivia. The Tacna-Arica Railway, totaling 62 km (39 mi) and linking Peru with Chile, is also a part of the nationalized system.

In 1991, of the estimated 56,645 km (35,200 mi) of existing roads, less than 10% were paved. The nation's highways are deteriorating, especially in the mountains, where landslides and guerrilla attacks often occur. The two primary routes are the 3,000-km (1,864-mi) north-south Pan American Highway, connecting Peru with Ecuador, Bolivia, and Chile, and the Trans-Andean Highway, which runs about 800 km (500 mi) from Callao to Pucallpa, an inland port on the Ucayali River. In the mid-1980s, most of the planned 2,500-km (1,550-mi) Jungle Edge Highway, or Carretera Marginal de la Selva, had been opened; when completed, the road system will span most of Peru along the eastern slopes of the Andes and through the selva. In 1991 there were 399,881 automobiles and 235,663 trucks and buses. About 60% of inland freight and 90% of all passengers are carried by road.

The Amazon River with its tributaries, such as the Marañón and the Ucayali, provides a network of waterways for eastern

Peru. Atlantic Ocean vessels go 3,700 km (2,300 mi) up the Amazon to Iquitos and, at high water, to Pucallpa. Peru has 11 deep-water ports, and in 1991, its merchant fleet consisted of 38 vessels over 1,000 tons, with a total GRT of 404,000. Only Peruvian ships may engage in coastal shipping. Callao, Peru's chief port, and Salaverry, Pisco, and Ilo have been expanded.

Much of Peru would be inaccessible without air transport. Faucett Airlines is the older of the two main domestic air carriers, which serve 40 airports and landing fields. An international airport serves the Lima-Callao area. The recently privatized Aeroperú, created in May 1973, provides both domestic and international services. In 1991/92, Aeroperú and Faucett flew 32.5 million km (20 million mi) and carried 2,441,200 passengers. The Peruvian Air Force also operates some commercial freight and passenger flights in jungle areas.

¹²HISTORY

Archaeological evidence indicates the Peru has been inhabited for at least 12,000 years. Perhaps as early as 6,000 years ago, the first primitive farmers appeared. Between 500 BC and AD 1000 at least five separate civilizations developed. The Paracas, on the southern coast, produced elaborately embroidered textiles. The Chavín, in the highlands, were noted for their great carved stone monoliths. The Mochica on the north coast, produced realistic pottery figures of human beings and animals. The Nazca in the south were noted for the giant figures of animals in the ground that can be seen only from the sky. The Chimú were the most developed of these groups.

The Quechua Empire, whose emperors had the title Sapa Inca, was established in the 13th century. During the next 300 years, the extraordinary empire of the Incas, with its capital at Cuzco, spread its spiritual and temporal power to northern Ecuador, middle Chile, and the Argentine plains. By means of a system of paved highways, the small Cuzco hierarchy communicated its interests to a population of 8–12 million. The intensive agriculture of scarcely tillable lands, held in common and controlled by the state, created a disciplined economy. The ayllu, a kinship group that also constituted an agrarian community, was the basic unit of the Inca Empire, economically and spiritually. The Incas were sun worshipers and embalmed their dead. Their advanced civilization used a calendar and a decimal system of counting, but never developed a wheel.

Francisco Pizarro's small band of Spaniards arrived in 1532, shortly after a civil war between the Inca half-brothers Huáscar and Atahualpa. The empire collapsed in 1533. Lima was established in 1535 and promptly became the opulent center of the Viceroyalty of Peru. It held jurisdiction over all Spanish South America except Venezuela. The Spanish imperial economy, with its huge land grants given by the crown and its tribute-collecting encomiendas, brought vast wealth and a new aristocracy to Peru. To Spain, Peru was a gold bank. Mines were exploited, and overworked Indians perished by the millions as food supplies declined.

Peru remained a Spanish stronghold into the 19th century, with little internal agitation for independence. One notable exception was the avortive revolt led by a mestizo known as Tupac Amaru II in 1780. Otherwise, Peruvian royalists helped the crown suppress uprisings in Peru and elsewhere. In the end, Peru was liberated by outsiders, José de San Martín of Argentina and Simón Bolívar of Venezuela. San Martín landed on Peruvian shores in 1820 and on 28 July 1821 proclaimed Peru's independence. The royalists were not quelled, however, until the Spaniards were defeated by forces under Bolívar at Junín and under Antonio José de Sucre at Ayacucho in 1824. The victory at Ayacucho on 9 December put an end to Spanish domination on the South American continent, although the Spanish flag did not cease to fly over Peru until 1826.

Between 1826 and 1908, Peruvian presidents ruled an unstable republic plagued by rivalries between military chieftains (caudillos) and by a rigid class system. Marshal Ramón Castilla, president from 1845 to 1851 and from 1855 to 1862, abolished Amerindian tributes and introduced progressive measures. Between the 1850s and the mid-1880s, Peru experienced an economic boom financed by sales of guano in Europe. A program of road building was implemented, and an American entrepreneur, Henry Meiggs, was hired by the government to build a railroad network in the Andes. In 1866, a Spanish attempt to regain possession of Peru was frustrated off Callao. An 1871 armistice was followed in 1879 by the formal recognition of Peruvian independence by Spain. The War of the Pacific (1879-84) followed, in which Chile vanquished the forces of Peru and Bolivia and occupied Lima from 1881 to 1883. Under the Treaty of Ancón, signed in October 1883, and subsequent agreements, Peru was forced to give up the nitrate-rich provinces of Tarapacá and Arica.

Peru entered the 20th century with a constitutional democratic government and a stable economy. This period of moderate reform came to an end in 1919, when a businessman, Augusto Leguía y Salcedo, who had served as constitutionally elected president during 1908–12, took power in a military coup and began to modernize the country along capitalistic lines. It was in opposition to Leguía's dictatorship, which had the backing of US bankers, that a Peruvian intellectual, Víctor Raúl Haya de la Torre, founded the leftist American Popular Revolutionary Alliance (APRA). In 1930, after the worldwide depression reached Peru, Leguía was overthrown by Luis M. Sánchez-Cerro, who became Peru's constitutional president in 1931 after an election which the Apristas (the followers of APRA) denounced as fraudulent. An Aprista uprising in 1932 was followed by the assassination of Sánchez-Cerro in April 1933, but the military and its conservative allies maneuvered successfully to keep APRA out of power. Manuel Prado y Ugartache served as president during World War II, a period that also brought the eruption of a border war with Ecuador in 1941. The 1942 Protocol of Rio de Janeiro, which resolved the conflict on terms favorable to Peru, was subsequently repudiated by Ecuador.

In 1945, Prado permitted free elections and legalized APRA. Haya de la Torre and the Apristas supported José Luis Bustamante y Rivera, who won the elections, and APRA (changing its name to the People's Party) received a majority in Congress. In 1948, military leaders charged the president with being too lenient with the Apristas and dividing the armed forces. A coup led by Gen. Manuel A. Odría ousted Bustamante, and APRA was again outlawed. Several hundred Apristas were jailed, while others went into exile. In January 1949, Haya de la Torre found refuge in the Colombian embassy, where he lived for the next five years. Under the rule of Odría and his military board of governors, the Peruvian economy flourished. Agriculture, industry, and education were stimulated by modernizing measures, and foreign trade prospered. Odría announced his retirement in 1956, and promoted his own candidate for the presidency. In a free election, the opposition candidate, former President Prado (tacitly supported by the outlawed APRA) returned to office.

Peru under the Prado regime was characterized by deep-rooted social unrest and political tension. Prado himself faded into the background, allowing Premier Pedro Beltrán to rule. Beltrán's economic moves stabilized Peru's financial picture, but the political problems remained. The election of 1962 was a three-way race between Haya de la Torre, Odría, back from retirement, and Fernando Belaúnde Terry, leader of the Popular Action Party (AP). Although Haya de la Torre got the most votes, he did not receive the constitutionally required one-third of the votes cast. The parties then went into negotiations, and a deal was struck giving Odría the presidency with an APRA cabinet. The military thereupon intervened, annulled the vote and suspended the newly

elected Congress. The governing junta then announced new elections for July 1963, and the same candidates ran. This time, Belaúnde received 39% of the votes cast to become president.

Belaúnde embarked on a program of agrarian reform, as well as tax incentives to promote manufacturing. However, he was caught in a crossfire between the Odriístas, who considered him a radical, and the apristas, who believed he was not doing enough. Belaúnde's AP formed a coalition with the Christian Democratic Party to control the senate, but APRA and the Odría National Union controlled the Chamber of Deputies. On top of all this, Belaúnde had to deal with two separate leftist insurgencies in Peru's highlands. As Peru approached new presidential elections, the AP began to quarrel, and opposition parties continued to sabotage Belaúnde's programs. Then a scandal concerning the granting of oil concessions to the International Petroleum Co., a subsidiary of Standard Oil of New Jersey, rocked the government. A military junta exiled Belaúnde on 3 October 1968 in a bloodless coup.

In 1969, the military government, under the presidency of Gen. Juan Velasco Alvarado, began enacting a series of social and economic reforms. This time, they did not worry about opposition, ruling instead by decree. By 1974 they had converted private landholdings into agricultural cooperatives, nationalized a number of basic industries, and had mandated profit-sharing schemes for industrial workers. A government-sponsored social mobilization agency, the National System for Support of Social Mobilization (Sistema Nacional de Apoyo a la Movilización Social—SINAMOS) was established in 1971. The military also reached out to Peru's long-neglected Amerindian population, making Tupac Amaru a national symbol, and recognizing Quechua as an official national language.

In August 1975, Velasco, whose health and political fortunes had both declined, was removed from office in a bloodless coup and replaced by Gen. Francisco Morales Bermúdez Cerruti, formerly his prime minister. The new regime moved to liberalize the system, declaring a general amnesty for post-1968 political exiles and the legalization of some previously banned publications. They subsequently announced a return to civilian government and the creation of a "fully participatory social democracy." Some state-controlled enterprises were sold, worker-participation programs were scaled down, and SINAMOS was dismantled. A Constituent Assembly was elected, and under the leadership of the perennial candidate Haya de la Torre they drew up a new constitution in 1979. New elections were held in 1980, and the AP and Belaúnde returned to power.

Belaúnde's second term was even less a success than his first. Adverse weather conditions and the world recession accompanied ill-conceived policies that led to triple-digit inflation. Austerity programs caused increased rates of unemployment, and currency problems pinched the Peruvian middle-class. Perhaps most disturbing of all, a small Maoist guerrilla group, Shining Path (Sendero Luminoso) was operating openly in the Andes, especially around Ayacucho. Despite passage of an antiterrorist law in 1981, terrorist activities intensified. During the first four months of 1983, more than 450 people lost their lives. The government's campaign against terrorism, beginning in May 1983 and continuing through 1985, resulted in the disappearance of thousands, charges of mass killings, and the granting of unlimited power to the armed forces. Meanwhile, the AP's tenuous hold on the government was slipping. The AP won only 15% of the vote in the 1983 municipal elections. By 1985, with Peru on the brink of an economic collapse, the AP received a mere 7% of the vote.

The election of 1985 was historic in two ways: it was the first peaceful transfer of power in 40 years, and it brought the first president from APRA since the party's founding in 1928. Alán García Pérez, secretary-general of APRA, won with 53% of the vote, and brought with him an APRA majority in both houses. The new president pursued populist economic policies, aimed at controlling inflation, stimulating the economy, and limiting external debt repayments. To get inflation under control, García established a strict set of price controls, dropping inflation precipitously. Salaries were then allowed to increase, which led to a dramatic surge in the production of industrial and consumer goods. García also announced that external debt service would be set at 10% of export earnings, when several times that amount would have been required to keep up with interest payments alone.

While initially successful, these programs eventually ran aground. The IMF, a constant target of García, declared Peru ineligible for any further borrowing because of the size of Peru's external debt. After its initial boom, industrial production began to sag. Food shortages became common as suppliers refused to produce with artificially low prices. By 1990, inflation had climbed to four-digit levels.

García had some success in dealing with Peru's democratic left, but the militant left was another story. By increasing the stridency of his rhetoric, especially against the US, García was able to capture leftist votes, seriously damaging the power of the United Left (Izquierda Unida—IU). However, Sendero escalated its attacks, coming down out of the mountains and striking at urban and suburban targets around Lima and Callao. In addition, the Tupac Amaru Revolutionary Movement (MRTA) merged with the Movement of the Revolutionary Left (MIR), and struck with increasing intensity. Although García had promised to get the military under control, it was soon clear that he could not function without them, and authorized a set of brutal counter-insurgent campaigns.

By 1990, Peruvians began to cast about for someone to deliver the country from its economic and social woes. Neither APRA nor the AP had any credibility. In a surprise, Alberto Fujimori, the son of Japanese immigrants, defeated conservative novelist Mario Vargas Llosa by 57% to 34%. Other candidates totaled a little over 9%. Fujimori immediately imposed a draconian set of austerity measures designed to curb inflation. These measures caused a great deal of economic dislocation, but did reduce inflation to pre-1988 levels.

Fujimori moved aggressively to combat Sendero and the MRTA-MIR. He organized and armed rural peasants to counter the increased guerrilla presence, and gave the military a broad mandate to crack down on the insurgents. The capture of Abimaél Guzmán, leader of Sendero Luminoso, was hailed as a major blow against the movement, but the violence continued. Human rights continued to deteriorate, and the military became stronger.

Domestic opposition increased as Fujimori became increasingly isolated politically. Then, in April 1992, Fujimori shut down Congress and refused to recognize any judicial decisions. The autogolpe ("self-coup") received widespread popular approval and, most significantly, the military supported Fujimori's moves. In 1992, elections were held to a Constituent Assembly charged with making constitutional reforms. Both APRA and AP refused to participate, and Fujimori's New Majority/Change 90 group took a majority of seats.

By mid-1994, with Fujimori still facing economic troubles and the military waiting in the wings, Peru's future seemed uncertain.

13GOVERNMENT
Prior to the military coup in 1968, Peru was governed under the constitution of 1933, which declared Peru to be a republic with a centralized form of government. Legislative powers were vested in a Senate and a Chamber of Deputies, of variable number. Both senators and deputies served their electoral districts for a period of six years. Under the constitution, executive power was held by the president, who, with two vice-presidents, was elected for a six-year term, with a minimum of one-third of the vote, but could

not be reelected until an intervening term had passed. Voting was obligatory for all literate Peruvian citizens aged 21 to 60.

The military leaders who seized control of the government in 1968 immediately disbanded the bicameral Congress. For the following decade, Peru was ruled by a military junta consisting of the president and the commanders of the three armed forces. The return to civilian rule began with the election of a Constituent Assembly in June 1978 and the promulgation of a new constitution on 12 July 1979. Presidential elections were held in May 1980, and Peru's first civilian government in 12 years took office in July.

Under the 1979 constitution, the president was popularly elected for a five-year term and could not be reelected to a consecutive term. The winning candidate had to win at least 50% of the vote, or face a run-off election against the second-place candidate. The National Congress consisted of a 60-member Senate and a 180-member Chamber of Deputies. All elected legislators had five-year terms. The 1979 constitution eliminated literacy as a qualification for voting, and made suffrage universal at age 18.

After the autogolpe, the constitution was suspended. The Constituent Assembly soon amended the 1979 constitution to allow a president to run for a second consecutive term. The document is still under revision.

14POLITICAL PARTIES
Throughout most of Peru's modern political history, personalities and power politics have counted for more than party platforms. There are nevertheless several parties with origins at least as far back as the 1950s.

The American Popular Revolutionary Alliance (Alianza Popular Revolucionaria Americana—APRA) was begun in 1924 by Víctor Raúl Haya de la Torre as a movement of and for Latin American workers. The five planks in its original platform were opposition to "Yankee imperialism," internationalization of the Panama Canal, industrialization, land reform, and solidarity among the world's oppressed. Controlling most of unionized labor, APRA was anti-Communist and anti-imperialist. Outlawed in 1931 and again in 1948, APRA was legalized in 1956. APRA has been historically opposed to the military, and political conditions in Peru from the 1930s until the mid-1980s have been dominated by hostility between APRA and armed forces leaders. After the death of Haya de la Torre in 1979, APRA was weakened by internal dissension. By 1985, new leadership and the failure of the Belaúnde government allowed APRA its first experience in power. APRA is currently boycotting any political arrangement under Fujimori.

The Popular Action Party (Partido de Acción Popular, or AP) was founded in 1956. Originally a reform party, it competed with APRA for the support of those favoring change in Peru. After an impressive campaign in 1956, the AP won the presidency in 1963, thanks to the military's hatred of APRA. In the 1980 presidential election, Belaúnde received 45.4% of the votes cast, compared with 27.4% for APRA candidate Armando Villanueva del Campo.

The Popular Christian Party (Partido Popular Cristiano—PPC) left the Christian Democratic Party in 1966 while the latter was allied with Belaúnde's AP government. The PPC now defines the right in Peruvian politics. The party placed third in the 1980 and 1985 elections under Luis Bedoya Reyes, former mayor of Lima. The PPC is participating in the Fujimori government, and holds 8 seats in the Constituent Assembly.

After the dissolution of Congress by the 1968 military coup, political parties continued to exist, although they were denied any role in government until the late 1970s. Ideologically, the military rulers between 1968 and 1980 reflected both strong socialist and nationalist principles.

The left has undergone a number of changes, partly as a result of military intervention, and most recently has been undermined by the activities of leftist guerrillas. The Peruvian Communist Party (Partido Comunista Peruano) was formed in 1929. Outlawed in 1948, it changed its name to the Revolutionary Labor Party (Partido Obrero Revolucionario—POR), which split in the 1980s into a number of small factions. The United Left (Izquierda Unida—IU), formed to support the candidacy of Alfonso Barrantes Lingán, took 21.3% of the 1985 ballot. Barrantes was mayor of Lima until APRA unseated him in 1986, whereupon Barrantes resigned as IU President and the coalition dissolved.

The largest active guerrilla party is Shining Path (Sendero Luminoso), a Maoist group founded in 1964. Its founder, Abimaél Guzmán, a former college professor, was captured by the government and is still imprisoned. The Sendero's strength is concentrated around Ayacucho, in the sierra southeast of Lima. Its program includes not only attacks on bridges, power lines, and urban centers but also attempts to organize highland peasants. Sendero collects tribute from peasants in exchange for protection, and encourages peasants not to sell their food crops to the cities.

A smaller group, the Tupac Amaru Revolutionary Movement (Movimiento Revolucionario Tupac Amaru—MRTA), merged with the Movement of the Revolutionary Left (MIR) to form a group that has been increasingly active. The MRTA/MIR is more urban-oriented and follows a more orthodox Marxist line than the eccentric Sendero.

15 LOCAL GOVERNMENT

In March 1987, President García promulgated a regionalization law that would replace the nation's 24 departments (and the constitutional province of Callao) with 12 regions having economic and administrative autonomy. Each region was to have an assembly of provincial mayors, directly elected members, and representatives of various institutions. However, due to inadequate funding and an uncertain political picture, these regions are not functioning, and currently exist alongside the departmental structure, which was never dismantled.

The 148 provincial subdivisions remained intact.

The 1979 constitution confirms the legal status of about 5,000 Indian communities. The first local elections since 1966 took place in November 1980 and occurr at three-year intervals.

16 JUDICIAL SYSTEM

The Peruvian legal system is based generally on the Napoleonic Code. The 1979 constitution guarantees the independence of the judiciary. Peru's highest judicial body, the 16-member Supreme Court, sits at Lima and has national jurisdiction. The 9-member Court of Constitutional Guarantees has jurisdiction in human rights cases. Superior courts, sitting in the departmental capitals, hear appeals from the provincial courts of first instance, which are divided into civil, penal, and special chambers. Judges are proposed by the National Justice Council, nominated by the president, and confirmed by the Senate; they serve permanently until age 70. Justices of the peace hear misdemeanor cases and minor civil cases.

The 1979 constitution abolished the death penalty (except for treason in time of war) and limited the jurisdiction of military tribunals; it also established the Public Ministry, including an independent attorney general, to serve as judicial ombudsman. Despite such reforms, the Peruvian judicial system still suffers from overcrowded prisons and complex trial procedures. Many accused persons (especially those accused of drug trafficking or terrorism) may spend months or even years in prison before they are brought to trial.

Although the judicial branch has never attained true independence, provisions of the 1993 constitution establish a new system for naming judges which may lead to greater judicial autonomy in the future. As of 1993, however, approximately 70% of incumbent judges and prosecutors were provisionally appointed by President Fujimori after his declaration of extraconstitutional powers in 1992. These appointees were never confirmed by Congress.

17 ARMED FORCES

Two years of military service is obligatory and universal, but only a limited number of men between the ages of 20 and 25 are drafted. After active service in the army, soldiers spend 5 years in the first or second reserve, and 20 years in the national guard. There are 188,000 army reservists.

The total strength of the armed forces in 1993 was 112,000 (including 69,000 conscripts). Army personnel numbered 75,000, navy 22,000 (including 3,000 marines), and air force 15,000, with 107 combat aircraft and 10 armed helicopters. The navy mans 9 submarines, 2 cruisers, 6 destroyers, 4 frigates, and 7 smaller combatants. Naval aviation (8 planes, 12 helos) is land based or deployed in an ASW role. About 84,000 men make up the Policia Nacional, under the direction of the Ministry of Interior and Police. Between 1981 and 1991, Peru imported arms valued at $1.8 billion with purchases from the former USSR, France, Italy, the UK and the US. In 1991, the defense establishment received approximately $430 million or 2.4% of GDP. The armed forces, advised by 50 Russians, faces 5–8,000 armed guerrillas of Sendero Luminoso and 500 terrorists of the Tupac Amaru movement.

18 INTERNATIONAL COOPERATION

Peru is a charter member of the UN, having joined on 31 October 1945, and belongs to ECLAC and all the nonregional specialized agencies. Peru also belongs to G-77, is a signatory of GATT, and cooperates with various foreign countries in cultural, scientific, trade, industrial, and health matters. A member of IDB, LAIA, OAS, and PAHO, Peru is also the seat of the five-nation Andean Group.

19 ECONOMY

Since World War II, the Peruvian economy has developed rapidly, exhibiting a rate of growth that has been among the highest in Latin America. The average annual growth of the GDP was 4.9% during 1960–70 and 3% during 1970–80, but this growth rate was tempered by the annual population growth rate of 2.8% during 1960–70 and 2.6% between 1970 and 1985. The average annual increase in per capita GDP between 1960 and 1980 was only 1.1%; between 1980 and 1985, it declined by an annual average of 3.7%. In 1985, the per capita GDP in current prices was estimated at $575, down from $1,118 in 1982. Between 1980 and 1985, the average annual inflation rate was 100%; in 1985, it was 158.3%, and although it was halved to an estimated 77.9% in 1986, the annualized rate for 1987 was projected at 100%.

The strength of Peru's economy lies in the diversity of its natural resources. Silver and gold were the prized commodities of colonial Peru. In more recent times, lead, copper, zinc, iron ore, and, since the late 1960s, petroleum have become important export earners. Fishing, including the production of fish meal, has become a major undertaking. Agriculture, which occupies about 40% of the work force, is sharply divided between two sectors: small-scale farming, producing food crops for subsistence and the domestic market, and export-oriented production. Government reforms in the early 1970s resulted in converting most large-scale private holdings into even larger cooperatives. Under the junta, major infrastructure sectors were nationalized, with a share of ownership in all other sectors being gradually turned over to workers. Protectionism and excessive public spending stifled private sector investment, leading to large balance-of-payments deficits. The last years of the junta saw the steady abandonment of the corporatist philosophy, and the civilian government installed in 1980 began to rebuild a market economy. The García government was sharply interventionist, but attempted to direct the economy by individual measures rather than a comprehensive program.

Peru has traditionally been dependent on its exports of raw materials (fish meal, cotton, copper, and sugar) to provide an exchange surplus for the import of machinery, foodstuffs, chemicals, transport vehicles, and other items. Although agricultural exports have waned in favor of mineral and fish meal exports, commercial crops have been emphasized over staple crops, and, with the population growing rapidly, per capita food production has diminished since 1950. In the late 1970s and early 1980s, Peruvian exports suffered from the low world market prices for minerals. Between 1980 and 1985, the value of exports fell by 24%; the value of imports rose by 27% until 1982 and then plunged by 49% until 1985, reflecting the recession and government curbs on imports. The 1983–85 trade surplus reflects only import controls. Foreign indebtedness rose from about $4 billion in 1975 to $13.7 billion in December 1985.

The government of President Alberto Fujimori has continued the economic restructuring program begun when he took office on 28 July 1990. The government has liberalized trade, investment, foreign exchange, labor, and land markets, as well as taken steps to eliminate public monopolies. The normalization of the democratic process gradually brought back official external support and renewed private capital inflows, which permitted the financing of the wider current account deficit and a build-up of international reserves.

In the early 1990s, the Peruvian economy came out of recession, as a result of an increase in foreign capital. Real GDP growth was 2.6% for 1991 and it contracted by 2.7% in 1992. This contraction was the result of manufacturing contracting 7.5%, mining 6.3%, and agriculture 10% because of drought. In 1993, however, the economy recovered significantly, registering a growth of 7.0% The government has continued to reduce the annual rate of inflation. Annual inflation went down from 139.2% in 1991 to 56.7% in 1992 and to 39.5% in 1993, the lowest level in 15 years. In addition, the government is pursuing its goal of privatizing all state-owned enterprises through the privatization program begun in 1991, and was poised for the privatization of major utilities and mining concerns in 1994.

Agricultural production recovered slightly in 1993, up 4.6% in the first nine months, compared with the same period of 1992. Output had dropped 5.7% in 1992. Recovery of crop production accounted for the gain in 1993, up 8.3% in January–September compared with the same period o 1992, including potatoes (up 50.5%), yellow corn (63%), and wheat (46.6%). Production of cotton and sugar continued their downward trend.

20INCOME

In 1992, Peru's GNP was $21,272 million at current prices, or $950 per capita. For the period 1985–92 the average inflation rate was 736.8%, resulting in a real growth rate in per capita GNP of –4.3%.

In 1992 the GDP was $22,100 million in current US dollars. It is estimated that in 1991 agriculture, hunting, forestry, and fishing contributed 15% to GDP; mining and quarrying, 6%; manufacturing, 58%; electricity, gas, and water, 1%; construction, 20%; wholesale and retail trade, 43%; transport, storage, and communication, 14%; finance, insurance, real estate, and business services, 41%; community, social, and personal services, 23%.

21LABOR

In 1990, the employed labor force was estimated at 5.3 million, or about one-quarter of the population. Of the total, agriculture employed 28.4%; industry, 13%; services, 29.6%; and other activities, 29%. According to official statistics, the unemployment rate was 8.3% at the end of 1990, but underemployment was estimated to be 86.4%. The labor force grew at an esti-

mated annual rate of 2.8% between 1985 and 1991.

About 5% of the total labor force was unionized in 1992; unions have played an important role in Peruvian politics in recent decades, but membership has declined as the informal labor sector grows. About 60% of the labor force worked in the informal sector in 1992. In 1990, wages equaled only 27% of the typical household's earnings in metropolitan Lima. Began in 1944 under Communist domination, the Workers' Confederation of Peru was reorganized in 1956 as the national central labor organization, now known as the Peruvian Revolutionary Workers' Center (Central de Trabajadores de la Revolución Peruana—CTRP). In 1980, this organization was incorporated into the Democratic Trade Union Front, which also includes the Communist-led General Confederation of Peruvian Workers, the National Workers' Confederation, and the APRA-affiliated General Confederation of Peruvian Workers (Confederación General de Trabajadores Peruanos—CGTP), which dates from 1944.

Legal machinery allows for conciliation or arbitration of disputes by government decree through the Ministry of Labor. An 8-hour day and a 48-hour week are the maximum in Peru, with shorter hours for women and minors. Legislation has been uneven, but the law requires in most cases that employers create healthy and safe working conditions. Dangerous and night work are regulated. The civil code prohibits labor by minors under 14. Peru's labor stability laws provide that after three months of employment a worker may be dismissed only for a "serious offense." As of 1993, labor unions existed in industries that produced 70% of Peru's GDP. Although some 15 million employee-hours were lost in 613 strikes during 1990, the number of strikes continued to fall in 1990 and 1991.

A 1970 government decree created a new entity, the industrial community, to represent workers vis-à-vis business. Companies were to share pretax profits and to contribute an additional 15% of their pretax net annual income to their respective communities. A new community law of 1977 stipulated that only 1.5% of the 15% profit participation was to be distributed to the community in cash, and the remaining 13.5% was to be issued in the form of nonvoting stock; under this system, workers were also allowed to elect one-third of the company's board of directors.

22AGRICULTURE

Only 3.7 million hectares (9.2 million acres), or 2.9% of the total land, was under cultivation in 1984. The area of available agricultural land per capita is one of the lowest in the nonindustrialized world. The major portion of the coastal farmland is devoted to the raising of export crops, while the sierra and the selva are used primarily for the production of food for domestic consumption. In various communities, the Inca system of cooperative labor and land use still remains; fields are communally planted and harvested, and the produce or the profits divided.

The Agrarian Reform Law of 1969 profoundly affected the whole of Peruvian agriculture. By 1973, most of Peru, except for the jungle on the eastern side of the Andes, was brought under the reform program. Large private landholdings were abolished. Contrary to the expectations of farm workers, however, the appropriated land was not redistributed in small individual parcels. The large estates expropriated by the government were instead reorganized into cooperatives that maintained their administrative unity and were often incorporated into still larger units, known as social-interest agricultural societies, through which they were linked on a cooperative basis. By 1980, the expropriation and redistribution of land were largely complete. Out of the 9 million hectares (22.2 million acres) of cropland and pasture originally expropriated, 8.8 million hectares (21.7 million acres) were allocated to 379,000 families. In addition, 2.9 million hectares (7.2 million acres) reverted to the state, and 1.1 million hectares (2.7 million acres) were distributed to 10,706 families in the selva. The bulk of land went to cooperatives, and only about 43,000 families received land, totaling 683,000 hectares (1,688,000 acres), in the form of private holdings. The government justified this program by arguing that the supply of arable land in inhabited zones of Peru is so small that equal distribution of the land would permit an allotment of less than 0.5 hectares (1.2 acres) per rural inhabitant; that community ownership of land accords with Peruvian traditions, especially the proto-socialist Inca heritage; and that improved equipment and technique are more easily implemented in larger enterprises. As it turned out, inadequate distribution systems and lack of technical expertise limited the productivity of the cooperatives, and by 1981, about 80% of them were operating at a loss.

Peru's agriculture is highly diversified but not well integrated. In an irrigated section of the coastal desert lowlands, more than 1,270,000 hectares (3,138,000 acres) are cultivated with cotton, sugar, rice, soybeans, pulses, fruits, tobacco, and flowers. Modern methods are widely used in this area, and as a result, output has risen at a much faster rate than population growth. The sierra, in contrast, is relatively dormant, its lands being inferior or impractical to till. The selva contributes cocoa, fruits and nuts, tea, coffee, tobacco, and forest products, but exploitation of its vast potentially productive expanses was just beginning in the early 1980s. The Agricultural Development Law of 1980 was aimed at increasing agricultural productivity through technical and financial assistance and infrastructural improvement. In early 1987, the government announced that uncultivated state-owned land on the coast and the fertile fringe of the Amazon jungle would be sold to private agroindustrial companies. The plan would undermine the 1969 agrarian reform but was expected to stimulate exports.

Staples are potatoes and corn, grown throughout Peru, but with very low yields. The leading commercial crops are rice, cotton, sugar, and barley. The principal agricultural deficiencies—wheat, livestock and meat, animal and vegetable fats, and oils—are covered by imports. Production of major crops (in thousands of tons) in 1992 included sugarcane, 7,000; potatoes, 989; rice, 827; corn, 510; coffee, 85; wheat, 73; and cotton, 34. While the sugar industry was considered extremely efficient, in the 1980s, a combination of weak financing, heavy overseas borrowing, poor pricing policy, bad weather, and outmoded equipment contributed to a serious deterioration in performance. Production fell by 34% during 1991/92 from the 6.1 million tons produced in 1989/90.

23ANIMAL HUSBANDRY

The cattle, sheep, hogs, goats, horses, and poultry brought by the Spaniards to Peru were strange to the Amerindians, whose only domestic animals were the hunting dog and the American members of the camel family—the llama, alpaca, and their hybrids—which served as carriers and for food, clothing, and fuel. Only recently domesticated, the vicuña is protected by law, and limited quantities of its fine fleece are marketed. Most Amerindians of the southern highlands are herders.

The southern Andes contain the major cattle ranges. Brown Swiss, zebu, and Holstein have been imported, and agronomists are cross-breeding stock to attain herds of greater weight or of more milk on less feed. Although 27.1 million hectares (66.9 million acres), or 21% of Peru's land area, are permanent natural pasture and meadow, areas suitable for dairy cattle are few. In 1992, the livestock population included 3.9 million head of cattle, 12 million sheep, 2.7 million hogs, 661,000 horses, 490,000 donkeys, and an estimated 62 million chickens. Livestock output in 1991 included 110,000 tons of beef, 92,000 tons of pork, 320,700 tons of poultry, 106,700 tons of eggs, and 768,000 tons of milk.

24FISHING

Commercial deep-sea fishing off of Peru's coastal belt of over 3,000 km (1,860 mi), is a major enterprise. Peruvian waters normally abound with marketable fish: bonito, mackerel, drum, sea bass, tuna, swordfish, anchoveta, herring, shad, skipjack, yellowfin, pompano, and shark. More than 50 species are caught commercially. There are over 40 fishing ports on the Peruvian coast, Paita and Callao being the most important centers.

The Peruvian fishing industry, primarily based on the export of fish meal, used in poultry feed, is among the largest in the world. Peru's fishing sector led the world during the mid-1960s, although production since then has fluctuated radically. In the 1970s, overfishing nearly lead to the disappearance of the anchovy resource. The fish meal and fish-processing industry is managed by Pescaperú, which was founded in 1973.

The key to Peru's fishing industry in any given year is the presence or absence of El Niño; this warm ocean current displaces the normally cool waters deep in the Pacific, thereby killing the microorganisms upon which other marine life depends. The recurrence of El Niño in 1982/83 caused the disappearance of anchoveta until 1985 and a sharp fall in the catch of other species. The total catch in 1991 was 6.9 million tons, including 3 million tons of anchoveta down from 3,720,173 tons in 1989. The anchoveta catch averaged 2.9 million tons from 1986 to 1991.

To suppress invasion of their rich fishing grounds by foreign powers, Peru made formal agreements with Chile and Ecuador to extend the rights to their coastal waters out to 200 nautical mi. Violations of the proclaimed sovereignty by Argentine and US fishing fleets in 1952 and 1954 gave rise to shooting incidents. Since then, US fishing boats have occasionally been seized and fined or required to purchase fishing licenses; after eight US tuna boats were taken in November 1979, the US retaliated by imposing a temporary embargo on Peruvian tuna.

25FORESTRY

About 61% of Peru's land area, or approximately 79 million hectares (195 million acres), is covered by tropical rain forests. Most of Peru's exploitable timberlands lie on the eastern slopes of the Andes and in the Amazon Basin; the arid Pacific watershed cannot support forestlands. The trees of commercial importance on the coastal plain are amarillo, hualtaco, and algarroba (cut for railway ties and for charcoal fuel). Lumber from planted eucalyptus is used locally in the sierra for ties and for props by the mining industry. Eastern Peru, however, with its abundance of rain, consists of approximately 70 million hectares (173 million acres) of forestland (more than half the country's area), most of it uncut. A precise indication of Peru's volume of standing timber has never been ascertained. The selva contains Peru's only coniferous stand, where ulcumano is logged. Cedar, mahogany, moena, tornillo, and congona (broadleaf hardwoods) are also logged. The rain forests of the Amazon lowlands contain cedar, mahogany, rubber (wild and plantation), and leche caspi (a chewing-gum base). Commercially important are tagua nuts, balata, coca, fibers, and a wide range of medicinal plants. Cultivation for illicit purposes of the coca leaf (the source of cocaine), which has long been used habitually and ritually by Andean Amerindians, was regarded as a growing problem by the Peruvian government and by the International Narcotics Control Board in the 1980s.

Lumbering is conducted chiefly in the selva, where Pucallpa and Iquitos are the main sawmill centers. Mahogany is now the principal lumber export product, sent mainly to the US and Europe; mahogany and Spanish cedar trees supply about half of Peru's lumber output, which falls far short of the nation's needs. In 1991, production of roundwood totaled 7.9 million cu m.

26MINING

Peru has long been famous for the wealth of its mines, some of which have been worked extensively for more than 300 years. Through modern techniques and equipment, a vast potential of diverse marketable minerals is gradually becoming available from previously inaccessible regions. Because many of the richest mines are found in the central Andes, often above 4,300 m (14,000 ft), their operations have been wholly dependent on the Andean Indians' adaptation to working at high altitudes.

Copper, iron, lead, and zinc, the leading metals, are mined chiefly in the central Andes, where all refining is done at the metallurgical center of La Oroya. Copper production more than tripled in 1960 with the opening of the Southern Peru Copper Corp. mine at Toquepala. The completion of a 296-km (184-mi) rail line connecting the concentrator plant at Toquepala with the port of Ilo was the last major step in the $234-million Toquepala blister-copper project. Two large mines, at Cuajone and Cerro Verde, were opened in 1977, boosting copper production by 78%. Another large mine opened at Tintaya, 247 km (153 mi) South of Cuzco, in 1985, but slumping world prices, unfavorable exchange rate policies, and labor strife kept its production and that of other mines down. In 1991, Peru's copper output was 381,991 tons, up 20% from 1990. In 1991, copper accounted for 22% of the $3,307 million export market.

Production of iron ore reached 3,593,000 tons in 1991, down from a peak of 7,928,000 tons in 1970. Exploitation of iron ore, centered in southern Peru, was exclusively for export until the steel mills at Chimbote began operations in 1958. Zinc production in 1991 stood at 627,824 tons (320,700 tons in 1960), and lead at 199,811 (compared with 156,800 in 1970). Minor industrial metals, such as tungsten, molybdenum, bismuth, and antimony, are extracted in various parts of Peru. There are huge phosphate deposits in the Sechura Desert.

From 1987 to 1991, Peru averaged as the world's second–largest producer of mine silver, producing 631 tons of the refined metal in 1991. The government intervened in 1982 to extend emergency credit to privately owned silver mines whose operations were threatened by falling prices and a market glut.

Of the 30 industrial minerals mined commercially in Peru, the important ones are salt, gypsum, marble, limestone, and anthracite. Bituminous coal deposits, in immense reserves throughout the Andes, have not been extensively developed because of difficult access.

Peru's minerals accounted for 45% of the total value of exports in 1991. Mineral deposits have traditionally been the property of the state, while development has been left largely to foreign concessions. Since the 1968 junta, however, the government has pursued a steady program of nationalization, which began in earnest in 1968 with expropriation of the Peruvian holdings of Standard Oil of New Jersey. In 1971, state mining rights were assigned to Mineroperú, a government enterprise that also represents the government's commercial interests in the mining sector. Two other state-owned mining companies are Centromín and Hierroperú. The Southern Peru Copper Corp., the single largest foreign investor in Peru, is the only remaining large-scale private company engaged in mining.

27 ENERGY AND POWER

Of the electric power produced in Peru in the 1980s, more than half was hydroelectric; the balance was thermal, mostly from diesel fuel, gasoline, and natural gas. The hydroelectric potential has been estimated at 60 million kw. Between 1960 and 1991, installed generating capacity increased from 841,000 kw to 4,137,000 kw (of which 2,396,000 kw was hydroelectric). Production increased from 2,656 million kwh to 14,828 million kwh in 1991. Consumption of electricity in 1991 averaged about 674 kwh per capita, low by Latin American standards; almost 75% of the population is without electricity.

In 1972, the government assumed control of electricity production and distribution through Electroperú, a newly created state authority. Peru's electric development plan called for the

expenditure during the 1980s of some $6 billion on new generating plants and $3 billion on distribution and transmission facilities. In 1991, the Andean Development Corporation approved a $20 million loan to Electroperú to recondition transmission lines.

In the 1970s, Peru's Upper Amazon Basin was developed into a major petroleum source; the nation's petroleum export revenues increased from $52.2 million in 1977 to $646 million in 1985. The 853-km (530-mi) trans-Andean pipeline, with a capacity of 200,000 barrels a day, was completed in 1977. Peru's petroleum fields are located in the northwestern parts of Piura and Tumbes, where oil was known in pre-Columbian days, and first produced in 1896. Exploration fell during the 1980s, causing Peru's proven oil reserves to decline to 360 million barrels in 1992, compared to the peak of 835 million barrels in 1981. Since 1969, the development of oil resources has been directed by the state-owned Petroperú. Economic reforms, improved security, and a new oil and natural gas law passed in August 1993 are now attracting increased international interest in the petroleum sector: Petroperú has leased the most important exploration blocks to foreign companies. The government expects foreign companies will be able to increase Peru's extremely low proven oil reserves.

28 INDUSTRY

Manufacturing in Peru began with the establishment of consumer goods industries, which still dominate the sector. Smelting and refining are among Peru's most important industrial enterprises. As part of a long-term industrial program through hydroelectric power development, the Chimbote steel mills began to function in 1958; by 1965, capacity reached 350,000 ingot tons. A number of foundries, cement plants, automobile assembly plants, and installations producing sulfuric acid and other industrial chemicals have also come into operation. The expansion of the fish meal industry necessitated the construction of new plants as well as the establishment of many subsidiary industries—boatyards, repair and maintenance installations, and factories for the production of tinplate and cans, paper, jute bags, and nylon fishnet. Once a major guano exporter, Peru now produces synthetic fertilizers high in nitrogen and related industrial chemicals.

During 1970–74, industrial production grew annually by about 10%, but because of protectionism and government subsidies, which tended to inhibit competition, the growth rate slowed to less than 3% in the late 1970s. Output expanded by 5.3% in 1980 and by 1.8% in 1981 but declined by 23.1% during the 1982–83 recession. Production of selected industrial goods in 1986 was as follows: crude steel, 298,801 tons; cement, 2,205,100 tons; passenger cars, 7,054; tires, 914,781; wheat flour, 874,956 tons; and raw sugar, 600,209 tons.

Aided by greater availability of electricity, Peru's manufacturing sector recuperated in 1993 after a decline of 6.2% in 1992. Manufacturing, which constitutes 22.5% of GDP, grew 9.5% in the first nine months of the year compared with the same period in 1992. Growth was concentrated in primary processing industries, which grew 16.5% in this period, as compared to the rest of the sector, which grew 6.5% In the primary product sector, fishmeal output constituted an important growth factor, increasing by 70%.

Mining and petroleum production accounted for 11% of GDP in 1992, but more than half of exports. Its output rose 9.1% in the first three quarters of 1993 compared with the same period in 1992, after falling for seven years. Mining and oil are major beneficiaries of recent economic reforms and investment laws passed since 1991. The new constitution also provides improved treatment for investors.

29 SCIENCE AND TECHNOLOGY

The Lima Academy of Exact, Physical, and Natural Sciences, in Lima, was founded in 1939. In 1993, Peru had 19 scientific and technological learned societies and 13 scientific and technological research institutes. According to the Industrial Law of 1982, enterprises may invest up to 10% of their income tax-free in research and development projects approved by the National Council of Science and Technology and carried out by the national universities. In 1984, total research and development expenditures amounted to ML159 billion soles.

30DOMESTIC TRADE

A disproportionate amount of Peru's purchasing power is concentrated in the Lima-Callao area, where selling practices increasingly follow the pattern of more commercially developed Western countries. In the highlands, where more than 60% of the population lives, retailing is done at the market level. The fiesta day, or weekly market, for the Andes Amerindian is an important social-commercial affair, where objects made at home are bartered. Barter is also the method of exchange among the forest Amerindians of the Amazon Basin. Cooperative retail outlets have been established in the large mining concerns and agricultural estates. Installment sales are increasing on vehicles, refrigerators, television sets, and agricultural and industrial equipment.

During the summer months (January to March), larger stores in Lima are open from 10:30 AM to 6:30 or 7:30 PM, Monday through Saturday; banks transact public business from 8:30 to 11:30 AM. During the remainder of the year, openings and closing are usually 30 minutes to an hour later. Advertising rates are not high. Newspapers and radio provide the best promotional coverage. Advertising by mail is practiced, but street handbills are prohibited. Aggressive sales promotion is increasingly common.

31FOREIGN TRADE

In general, Peru exports raw materials and imports capital goods and manufactures. Minerals, fish meal, and cotton are traditionally the chief exports; nontraditional exports, such as manufactured goods and processed foodstuffs, claimed under 20% of all exports.

As of 1994, the US continued to be Peru's largest trading partner, and US exports to Peru faced good prospects with the return of economic growth. In 1992, US exports to Peru rose 19%, from $840 million in 1991 to over $1 billion in 1992, constituting 25% of Peruvian imports. Exports continued to rise in 1993 and accounting for 27% of Peruvian imports in the first half. Principal exports to the US included mineral fuel oil, refined silver and jewelry, lead ore, and concentrated coffee. US imports from Peru totaled abut $739 million in 1992(about 22% of total exports), a drop from 1991. Main imports from the US include cereals, refined oil, machinery parts, chemicals, and electrical machinery.

Total Peruvian's exports in 1992 totaled $3,484 million, increasing 4.66% from $3,329 million in 1991. In 1993, exports remained nearly unchanged with a total of $3,464 million. Total imports in 1992 reached $4,051 million, up nearly 16% from $3,494 million in 1991. In 1993, imports decreased by a minimum amount, totaling $4,043 million. This caused the trade gap to narrow.

32BALANCE OF PAYMENTS

Peru's export earnings depend heavily on world market prices in metals and cotton. Peru maintained a favorable balance of trade from 1966 to 1973, but a surge in imports, coupled with a drop in fishing exports, reversed this trend. When world copper prices declined sharply, Peru's 1973 trade surplus of $113 million was transformed into a deficit of $427 million in 1974. In the following year, the current accounts deficit ballooned to $1,538 million. The trade balance began to improve in 1976 and 1977, but rising interest payments kept current accounts at a loss. An austerity program was adopted in 1978; by the end of the year, Peru had reduced the deficit on current accounts to

$192 million and had registered an overall payments surplus of $76 million, thanks in part to $383 million in public-sector loans. The surplus rose to $1,579 million in 1979 because of an extremely favorable trade performance and an additional $637-million infusion of public-sector capital. A further surplus was recorded in 1980. In 1981, however, Peru's export position was negatively affected by the worldwide recession and by lower world mineral prices, resulting in a negative trade balance, which, together with rising interest payments on the foreign debt, led once again to an overall payments deficit. During 1990, Peru experienced a trade and capital surplus, but a deficit in the current account. Exports fell, while imports grew largely due to overvalued exchange and because government subsidies promoting exports were eliminated. For the first six months of 1993 the balance of payments surplus was $410 million, up from $215 million during the same period of 1992; imports fell by 4.6% during that period, and exports remained stagnant due to a strong exchange rate.

In 1992 merchandise exports totaled $3,484 million and imports $4,051 million. The merchandise trade balance was $–567 million. The following table summarizes Peru's balance of payments for 1991 and 1992 (in millions of US dollars):

	1991	1992
CURRENT ACCOUNT		
Goods, services, and income	–1,900	–2,363
Unrequited transfers	268	283
TOTALS	–1,632	–2,080
CAPITAL ACCOUNT		
Direct investment	–7	127
Other long-term capital	–785	–494
Other short-term capital	–174	277
Exceptional financing	1,545	944
Other liabilities	—	—
Reserves	–887	–568
TOTALS	–308	286
Errors and omissions	1,940	1,794
Total change in reserves	–885	–899

33BANKING AND SECURITIES

The Central Reserve Bank, the sole bank of issue, was established in Lima in 1931 to succeed the old Reserve Bank. Also created in 1931 was the Superintendency of Banks and Insurance, an agency of the Ministry of Finance, which defines procedure and obligations of banking institutions and has control of all banks. In 1985, Peru also had 24 commercial banks (three government-owned), nine foreign commercial banks, one multinational bank, six state development banks, and five regional banks. The government-owned National Bank (Banco de la Nación) not only acts as the government's tax collector and financial agent, but also is Peru's largest commercial bank, with assets of more than $2.5 billion. Another government agency, the Caja de Ahorros, provides secured loans to low-income borrowers. The privately owned Lima Stock Exchange regulates the sale of listed securities.

34INSURANCE

Insurance companies are controlled through the Superintendency of Banks and Insurance of the Ministry of Finance. According to a law of 1952, branches controlled by foreign insurance companies may not be established in Peru, although foreign insurers may operate through Peruvian subsidiaries in which they hold only minority interest. A majority of stockholders and directors of locally incorporated stock companies must be Peruvian nationals. The National Bank has assumed exclusive control of all foreign reinsurance operations, as well as the writing of export credit insurance. Both life and non-life insurance exist. In 1986, 21 insurance companies were operating, all controlled by Peruvian interests.

35PUBLIC FINANCE

The central government publishes an annual budget representing the government's consolidated accounts (including budgetary and extrabudgetary transactions). Indirect taxes, including import and export duties, constitute the major source of government revenues. In the early 1970s, the number of state enterprises increased rapidly, which led to increased public-sector spending. As the revenues from state enterprises lagged behind expenditures, the budget deficit increased to about 10% of GDP during 1975–77, as compared to 1.7% during 1970–72. As a result of a fiscal stabilization program, the deficit was reduced to 6.5% of GDP in 1978, and to 2.5% in 1985.

In 1990, the Fujimori administration began to pursue tighter fiscal policies and attempted to avoid domestic financing of the deficit. The consolidated public sector deficit, which in 1990 was 6.5% of GDP, fell to 2.5% by 1992, despite the suspension of most foreign financing after the 5 April coup. The IMF program allowed a foreign-financed deficit of 2.9% of GPD in 1993 for increased social sector spending and investment in infrastructure. However, with lower than expected foreign financing and tax collection, the deficit that could be maintained while meeting the public sector external debt obligations was only equivalent to about 2% of GDP. By August 1993, the government had gained $452 million through the privatization of fifteen state enterprises.

In 1990 Peru's total public external debt stood at $17.3 billion (excluding interest on arrears). Including interest on arrears, the total amounts to $22 billion, or 62% of 1990 GDP and over six times the value of 1990 exports.

36TAXATION

Peru's tax system has undergone substantial revision since the 1968 coup. As of 1987 there were six main levies: income tax, excise tax, business equity tax, real estate equity tax, value-added tax, and social-security contributions. Personal income tax rates range from 6% to 37%. The corporate income tax is calculated on the basis of a tax unit (UIT), which is recalculated annually to account for the effects of inflation. Corporate income tax was 30% as of 1993. The business equity tax is set at about 2%; some enterprises, including mining and petroleum companies and those involved in development of the selva, are exempt from this tax. The value-added tax is applied to all goods and services at a single rate of 18%. The payroll tax is 6% for the employer and 3% for the employee.

37CUSTOMS AND DUTIES

Under the military government, Peru's import tariff, once moderate, was used to protect local industries from foreign competition. In addition, a number of restrictions were imposed on imports: nonessential items and goods competing with local products could not be imported, and the importation of passenger cars, most clothing articles, and cosmetics was prohibited. When the civilian government took over in 1980, it introduced a trade liberalization program. Prior permits and quantitative restrictions on imports were eliminated, and the tariff was reduced from the 155–160% range to an average of 35%. Prior to 1980, 3,344 items were subjected to import restrictions; by 1982, the number of restricted items had decreased to 125. As of 1993, import licenses were not required. Revised tariffs were 15% to 25% ad valorem, and all imports were subject to an 18% value-added tax and possible inspection.

Export processing free zones offer investors exemptions from customs duties for imports and exports.

38FOREIGN INVESTMENT

Foreign capital and industrial techniques have been integral factors in the building of a modern economy in Peru. Because domestic capital has been too limited for the large-scale development of mining and of agricultural resources, commercial and industrial laws have encouraged foreign participation.

The British were the first to gain prominence as investors in Peru, when they took over the railways in payment of debt to Peruvian bondholders in 1890. They developed oil fields and a long-distance telephone and cable service. In the mid-1950s, Swiss, German, and Canadian interests became active in the transportation and communications sectors. Private US interests ventured capital and technical aid to all sectors, especially the oil and mining industries. The fishing industry from the beginning was fostered by the US. Private companies operated tuna fleets in Peruvian waters, as well as canneries.

Between 1959 and 1961, following passage of industrial development laws and the signing of an investment guarantee treaty with the US, under which private US investors could obtain federal risk insurance against currency inconvertibility, total foreign investment almost doubled, rising from $686 million to $1,274 million. US investments continued their rapid growth rate until the 1968 coup, after which Peru's military rulers pursued a nationalist course, characterized by selective expropriation of foreign-held interests in sectors such as mining, finance, and infrastructure. In addition, a variety of strictures were placed on the uses of foreign capital, as well as on the relative proportion of foreign-to-local control. US-linked firms were the hardest struck by these measures, causing a strain in relations. In February 1974, a US-Peruvian agreement provided a compensation schedule for properties taken over during 1968–73. At the end of 1973, the US investment volume stood at $793 million, less than half the total of a decade earlier. Private investments in the 1970s continued to lag, although some new funding was being advanced in mining and petroleum. From 1977 through 1980, net direct capital investment totaled only $177 million, or less than the total for 1981 ($263 million), the first year after the restoration of civilian rule. In 1984 and 1985, after the economic slump, net direct investment was $89 million and $53 million, respectively.

After 1980, the official attitude toward foreign investment changed substantially. The 1979 constitution guarantees protection of private property, whether Peruvian or foreign, and permits foreign jurisdiction for international financial contracts. The agency responsible for foreign investment is the National Commission of Foreign Investment and Technology (Comité Nacional de Inversión Extranjera y Tecnología, or CONITE). In late 1986, the Andean Group relaxed its regulations on foreign investment; this change was expected to benefit Peru. About 1,000 foreign companies were represented in Peru in the 1980s, either directly or through subsidiaries and affiliates.

The trade and investment climate in Peru has improved significantly since President Fujimori assumed office in July 1990, although significant economic, security, and juridical problems persist.

39ECONOMIC DEVELOPMENT

After World War II, President Odría discontinued import licensing and certain price controls and enacted the Mining Code of 1950, the Petroleum Law of 1952, and the Electrical Industry Law of 1955, all with a view to reassuring sources of foreign and domestic capital of reasonable taxation and an adequate rate of earnings under liberal exploitation concessions. Given this stimulus to capital ventures, the economy expanded, and new exports, such as iron ore and coal, were developed.

After the coup of 1968, Peru's military rulers sought a profound restructuring of the country's economic life. The overall objectives were the establishment of effective state control of natural resources; strict regulation of foreign participation; creation of a manageable balance among governmental, private, and foreign sectors; and redistribution of productive sources more broadly throughout the population. Nationalization, coupled with a redistribution of ownership and management authority in major enterprises, was the cornerstone of the new policy from its incipient stages in 1968 through 1975. A five-year plan was announced in December 1968, emphasizing a reorientation from an agricultural to an industrial economy and stressing the expropriation of large estates, with redistribution of land to peasants in the sierra. In early 1969, tax and credit incentives for the formation of cooperatives and the consolidation of smaller landholding were enacted. Agrarian reform, meanwhile, was extended to the modern coastal sugar estates, which were converted into cooperatives. In nonagricultural sectors, the government began, in 1968, selective nationalization of major foreign holdings in the mining, petroleum, and infrastructure sectors. In several areas, a government presence was asserted through the creation of state-owned commercial enterprises, the most notable of which included Induperú, in industry; Mineroperú, in mining; Pescaperú, in fishing; Petroperú, in petroleum; Entelperú, in telecommunications; and COFIDE (Corporación Financiera de Desarrollo), in investment. Industrial enterprises in general were required to adopt profit-sharing and co-ownership schemes for their employees. Although strict limits were placed on foreign participation in Peruvian industry, such investments were not ruled out in principle, and in 1974, the government acted to guarantee fair settlement for US holdings expropriated during 1968–73.

On 28 July 1974, the government announced its Inca Plan (which may actually have been drawn up before the 1968 coup), a master plan that envisioned eventual transformation of all economic entities along prescribed socialist lines. Three types of enterprises were to be permitted to operate in Peru: state-owned enterprises, worker-owned collectives (industrial communities), and social-property companies (entities managed by workers but financed by the state). In late 1975, the Central Bank set up a line of credit of s2,420 million to aid the formation of social-property companies through the National Fund for Social Property. The Agrarian Bank, which had been created in July, was authorized to handle the credit requirements of the reorganized agricultural sector, as well as to ease the transformation of cooperatives and farmers' associations, many of which had existed for only a few years, into social-property entities. The order of priorities for industry placed basic industries—notably steel, nonferrous metals, chemicals, fertilizers, cement, and paper—at the high rung, followed by manufactures of capital goods, such as machine tools, and industrial research. Reinvestment of profits was stressed throughout. Special incentives were provided for industries outside the Lima-Callao area and for virtually all enterprises within the jungle region.

In the mid-1970s, the regime began to moderate the rigid price-control system instituted in its formative years. The prices of petroleum and basic consumer goods were increased, while wage increases were fixed and agricultural subsidies removed. In September 1975, the sol was devalued and financial controls were imposed to help stem inflation (reaching 40% in 1975) and to ease the trade imbalance. The Tupac Amaru Development Plan, announced in 1977, limited the structural reforms of the Inca Plan, calling for economic decentralization and encouragement of foreign investment. In the late 1970s, a number of state-controlled enterprises were sold, and worker participation was curtailed. The civilian leaders who came to power in 1980 sought to reduce government participation in the economy and to improve the efficiency of existing state enterprises. Import tariffs and export taxes were reduced, and a new investment program for 1980–85 emphasized power and irrigation projects and the construction of housing and health care facilities. These attempts to revitalize the economy were hampered by the worldwide recession and by the soft market for Peru's commodity exports. In response, the García administration reverted to an interventionist policy, imposing import controls and regulating foreign-exchange availability, as well as influencing the financial sector by threatening to nationalize the nation's banks.

Total US nonmilitary loans and grants to Peru for 1946–86 amounted to $1,195 million. During the same period, the country received $3,328 million in multilateral development aid, about 51% of it from the IBRD and 44% from the IDB.

The decade of the 1990s has attracted more investment into Peru's economy. With foreign capital flowing in recent years, Peru was poised for another year of economic recovery in 1994. Private investment was expected to play a major role in growth. Public expenditure will also rise following a financing agreement with the IMF. Privatization and the rapidly growing economy are providing the government with funds to spend on infrastructure and social programs. Private consumption will make a lesser contribution, since real income growth will remain low. Construction, fisheries, and manufacturing are expected to be the best performing sectors. Mining will benefit from a modest recovery in international metal prices. Overall, a GDP growth of 8.9% in 1994 and 7.8% in 1995 were expected. The tight monetary policy should ensure that growth will be obtained without a serious increase in inflation.

40 SOCIAL DEVELOPMENT

A modern system of social security has evolved from initial legislation provided in the 1933 constitution. Coverage and benefits were substantially broadened after the 1968 coup. Workers now receive benefits covering disability, medical attention, hospitalization, maternity, old age, retirement, and widows and orphans. The program is administered by the Ministry of Labor. Social insurance is compulsory for all employees up to the age of 60. Social security and pension funds are contributed to by employers (6% of the worker's salary in 1991) and wage earners (3%), with the balance of funds supplied by the government. Retirement pensions may be claimed at the age of 60 (men) or 55 (women) if the worker has been a contributor to the system for at least 15 years (male) or 13 years (female). Working mothers are entitled to maternity leave of 90 days at 100% pay.

Although the military government regarded the population growth as too high, it was ambivalent about family planning. An Office of Health, established by a 1977 decree, was to integrate family planning into national health services, but the same decree prohibited the establishment of organizations solely concerned with fertility control. Family planning advice is provided in government clinics, which, in 1986, began bringing information and contraception to the urban and rural population; abortion is permitted on broad health grounds. In 1991–92, 59% of couples of childbearing age used contraceptives, and the fertility rate in 1991 was 3.7.

Women are often kept from leadership roles in the public and private sectors by the force of tradition, although they are equal under the constitution.

41 HEALTH

Although Peru has made significant advances toward reducing epidemic disease, improving sanitation, and expanding medical facilities, much remains to be done. The infant mortality rate in 1992 was estimated at 46 per 1,000 live births, while the general mortality rate in 1993 was 7.6 per 1,000 people. Maternal mortality in 1991 was high at 300 per 100,000 live births. Average life expectancy in 1992 was estimated at 64 years.

Leading causes of death per 100,000 people were as follows in 1990: communicable diseases and maternal/perinatal causes (327); noncommunicable diseases (392); and injuries (53). There were about 26,000 war-related deaths in Peru from 1983 to 1992. In 1990, there were 250 reported cases of tuberculosis per 100,000 people, and 13% of children under 5 years old were considered malnourished. In 1991, still only 56% of the population had access to safe water and 57% had adequate sanitation. Immunization rates for children up to one year old in 1992 were: tuberculosis (82%); diphtheria, pertussis, and tetanus (80%); polio (81%); and measles (80%).

Health services are concentrated around metropolitan Lima. A health care plan initiated in 1981 called for the establishment of 100 health centers in rural areas and shantytowns. The central administration of all health services lies with the Ministry of Health. In 1990 there were 6,219 physicians in the public health subsector of the Ministry of Health. Health care expenditures in 1990 were $1,065 million. In 1990, there were 32,434 hospital beds at a rate of 1.5 per 1,000 people. In 1989, Peru had 21,856 physicians and, in 1988, 5,023 dentists. In 1992 there were 1.03 physicians per 1,000 people, with a nurse to doctor ratio of 0.9. In 1992, about 75% of the population had access to health care services.

The birth rate in 1993 was 29 per 1,000 people. There were 658,000 births in 1992.

42HOUSING

Successive governments since the 1950s have recognized the importance of slum clearance and public housing programs in combating disease and high mortality rates. Most housing development programs carried out by the government and by private enterprise have been in the Lima area. In Lima and other towns, several "neighborhoods" (unidades vecinales) have been completed through government efforts since the early 1960s. Each such housing complex is designed to be a self-sufficient community. A typical neighborhood is built to house 6,000 persons at moderate rentals. In rural areas, however, a conservative estimate of the housing shortage runs to a minimum of 700,000. A 1970 earthquake tremendously increased the number of homeless, displacing some 500,000 persons. The total housing stock numbered 4,906,000 in 1985.

One of the revolutionary government's early decrees gave the Housing Bank control over financing low-cost housing. Construction grew rapidly during 1970–73, to the point of depleting local cement supplies. By 1983, however, only 18% of the rural population had access to safe water; the figure for the urban population was 73%. Sewerage facilities are nearly nonexistent in rural areas. In 1981, 88% of all housing units were detached or semi-detached houses; 4% were apartments; 4% were vecindads, or semi-private dwellings; and 3% were villas. Owners occupied 73% of all dwellings, 15% were rented, and 7% were occupied by usufructus, households legally living in a dwelling belonging to another. Mud houses comprised 47% of all dwellings, 31% were constructed of bricks, 7% were clay, 7% were wood, and 5% were of stone and mud.

43EDUCATION

Education is nominally free and compulsory for children aged 6 to 15. In 1990, the adult literacy rate was estimated at 85.1% (91.5% for men and 78.7% for women). The government has been responsible for public education since 1905; free secondary education began in 1946, but with far too few public schools to meet the need. Several long-term projects have been initiated to increase literacy and raise living standards among the adults of the remote sierra and selva areas. In March 1972, new education legislation enhanced the central authority of the Ministry of Education, granting the government control over all teaching appointments in the public schools and increasing its authority over the private sector. The legislation provided for adult literacy instruction and instituted the concept of a fully staffed six-grade

"nuclear" school to serve the rural population. The 1972 law also established Quechua and Aymará as languages of instruction for non-Spanish-speaking Amerindians, especially in the lowest grades.

As of the early 1980s, the educational system consisted of three levels: nurseries and kindergartens; basic education, consisting of primary and secondary schools; and higher education (pre-university and university). In 1990 there were 27,970 primary schools, with 4,019,483 pupils and 143,025 teachers. The number of secondary school students in 1990 was 743,569 students, with 58,131 teachers. The total higher education enrollment in 1983 reached 394,000. There is a national university in virtually every major city; the oldest is the National University of San Marcos of Lima, originally founded in 1551. The National University of Engineering and the National University of Agriculture are specialized governmental institutions. The University of San Cristóbal de Huamanga in Ayacucho, founded in the 17th century, was reopened in 1960 and offers mainly technical training. Peru's hard-pressed universities can accept only a fraction of each year's applicants.

44LIBRARIES AND MUSEUMS

The National Library in Lima, with 785,615 volumes, is the largest in Peru. More than 450,000 volumes may be found in the various libraries of the University of San Marcos. The library at the National University of San Augustín at Arequipa contains over 430,000 volumes.The Library at the Pontifical Library has over 350,000 items. There are nearly 200 public libraries in Peru, the largest of them in Callao, Arequipa, and Lima.

Peru has endeavored to restore and maintain the aesthetic and historical evidences of its pre-Columbian and colonial civilizations in more than 250 public and private museums.

The City Hall in Lima contains a full record of the city's official acts since its founding. The Cathedral of Lima, with its silver-covered altars and carved stalls, contains priceless historical and religious relics. A chapel near the entrance contains the alleged remains of Francisco Pizarro. Two colonial residences, the Palacio Torre Tagle and the Quinta de Presa, have been maintained to exhibit antiques and to serve as examples of the architecture of traditional Lima. There is no law protecting old houses, however, and many have been removed to make way for new downtown buildings. Some have been privately restored, such as the headquarters of the Association of Amateur Artists, the Institute Riva Aguero, the Associated Electrical Companies, the Bullfight Museum, and the Oquendo Mansion. In the Quinta de Presa is a museum with the possessions of the actress La Perricholi, the famous favorite of the 18th-century Viceroy Amat. At Pueblo Libre in Lima are the National Museum of Archaeology and Anthropology, with exhibits of pre-Columbian civilizations, and the Museum of the Republic, a historical museum. The Rafael Larco Herrera Museum in Lima (housing the former collection of the Chiclín plantation near Trujillo) is a private museum with a vast collection of notable antiquities from the pre-Inca Chimú culture. The University of Trujillo has a museum with specimens of early Peruvian cultures. Lima's Museum of Art exhibits Peru's national art from the pre-Columbian era to the contemporary period.

45MEDIA

Peru's major telecommunications systems were developed privately by Swedish, US, and Swiss enterprises. In 1970, however, the government nationalized the Lima Telephone Co. and announced plans to take over the entire telecommunications system through its wholly owned company Entelperú. A joint Peruvian-Chilean firm operates a system in southern Peru and Arica, Chile. In 1991 there were 736,203 telephones, more than two-thirds of them in Lima. The government's Bureau of Mails and

Telecommunications operates the domestic telegraph system, using radio to reach communities not served by land lines. In 1992 there were 140 television broadcast stations and about 2.1 million sets. There were 273 radio stations, and the number of radio receivers exceeded 5.6 million.

During the period of military rule between 1968 and 1980, the press in Peru was under strict government control. The law provided severe penalties for criticizing government officials and required newspapers to publish reports from the president and cabinet ministers. In 1974, the government shut down *Caretas*, the last remaining independent political magazine, and all remaining private national newspapers with circulations of more than 20,000 were expropriated. When civilian rule returned in 1980, the press was returned to private control. However, freedom of the press has once again been restricted since the political crackdown by President Fujimori in April 1992, and journalists have been arrested by the government.

The leading Lima dailies—among them *El Comercio* (1991 circulation 126,800), *Ojo* (309,100), and *Expreso* (124,300)—are the most important newspapers and are flown daily to provincial towns. The official government organ is *El Peruano* (35,000), a daily gazette in which laws, decrees, and brief government announcements are published. Special-interest periodicals are published by learned societies, agricultural groups, and business associations.

46 ORGANIZATIONS

Although Congress had earlier authorized an organic cooperative law, nothing was done until 1941, when a number of agricultural cooperatives were created by the Ministry of Agriculture. During the early years of the military junta, cooperatives were seen as a primary device for restructuring the agricultural sector. Most expropriated commercial estates were converted to cooperative farms, while smaller holdings were consolidated into units collectively owned by farmers' associations. Subsequent reform programs, culminating in the Inca Plan of 1974, pointed toward the restructuring of Peruvian society, with the social-property company as a basic unit. However, this corporatist idea was gradually deemphasized, and when the civilian government took over in 1980, most of the reforms of the previous decade were scaled down or abandoned.

The national Society of Industries, founded in 1896, coordinates the branches of organized industry. The Office of Small Industry and the Institute for the Development of Manual Arts were established in 1946 to revive the weaving and spinning skills of Incas and to coordinate the handicraft workers in textile and other industries, such as gold and silver crafts. Numerous chambers of commerce continue to function. Of Peru's many learned societies, perhaps the most important is the Academia Peruana, affiliated with the Royal Spanish Academy of Madrid.

47 TOURISM, TRAVEL, AND RECREATION

Citizens of most Western European and Latin American countries, Canada, the US, and Japan do not need visas; for visits of up to 60 days. As part of a program to encourage foreign tourism, the government has built and manages, through its Peruvian Hotel Co., several hotels or inns and a variety of tourist services. To help increase foreign exchange earnings from tourism, Peru consolidated all government agencies dealing with tourism into an autonomous corporation in 1964. The Fund for the Promotion of Tourism was established in 1979. In the early 1980s, the government outlined new tax incentives for investment in the tourist sector.

Tourists, as well as scholars, are especially drawn to the wealth of archaeological remains on the coast and in the sierra. Chan Chan, the center of the Chimú civilization, stands in adobe ruins near Trujillo. The ruins of the 9th-century coastal city of Pachaca-

mac are just south of Lima. Inca ruins may be seen at Cuzco, Sacsahuamán (on the northern edge of Cuzco), Ollantaytambo, and Machu Picchu, as well as on Lake Titicaca islands.

The northern coastal waters are famous for big-game fishing, and the abundant resources of the sierra waters are maintained by fish culture stations. Lakes and streams have been stocked with trout throughout Peru. In Lake Titicaca, trout average 10 kg (22 lb); trout weighing as much as 21 kg (46 lb) have been caught. Other tourist attractions include beaches and water sports; several state parks offer mountain climbing, cross-country skiing, and white-water rafting. The most popular sports are football (soccer), baseball, basketball, and bullfighting. Fiestas, especially the annual celebrations of patron saints, include both Catholic and Indian rites; bands of musicians, or conjuntos, are an important part of each fiesta.

In 1991, 232,012 tourists visited Peru, 20% from the US and 30% from South America; earnings from tourism were $277 million. There were 35,142 hotel rooms with 62,081 beds.

48 FAMOUS PERUVIANS

The Inca Huayna Capac (1450?–1525) reigned from 1487 and extended the Inca Empire, encouraging public works and fine arts. On his death he left the empire to his two sons, Huáscar (1495?–1533) and Atahualpa (1500?–1533); Huáscar was executed by his half-brother, and Atahualpa, the last of the great Incas, was executed by the Spanish conquistador Francisco Pizarro (1470?–1541). Acknowledged as America's first great writer, Garcilaso de la Vega (El Inca, 1539?–1616), son of a Spanish conquistador and an Inca princess, preserved in his *Royal Commentaries of the Incas* authentic descriptions of his ancestral empire and its traditions. Manuel de Amat (fl.18th century), viceroy from 1762 to 1776, was a patron of the colonial theater and of the famous actress Micaela Villegas, known as La Perrichoti. Tupac Amaru II (José Gabriel Condorcanqui, 1742–81), partly descended from the Incas, led a revolt against Spanish rule in 1780 in which he was defeated, captured, and executed.

Dr. José Hipólito Unnúe (1758–1833), a statesman and scientist, founded Lima's medical school in 1808 and reformed Peruvian education. The hero of the War of Independence, Mariano Melgar (1792–1815), was also a poet and composer whose regional songs (yaravís) are still popular. Marshal Ramón Castilla (1797–1867) distinguished himself in two great presidential terms (1845–51, 1855–62), introducing railways and the telegraph, emancipating the slaves, abolishing Amerindian tribute, modernizing Lima, and developing the important guano industry. Ricardo Palma (1833–1919) is considered Peru's greatest literary figure; a critic, historian, and storyteller, he originated the genre called tradición, and wrote the 10-volume *Tradiciones Peruanas*. José María Valle-Riestra (1858–1925) was an important composer.

In the modern era, the work of an erudite Amerindian, Julio Tello (1880–1947), became internationally known in archaeological circles. Jorge Chávez (1887–1910) made the first solo flight across the Alps in 1910. Santos Chocano (1875–1934), César Vallejo (1895–1938), and José María Eguren (1882–1942) are considered Peru's finest modern poets. Established prose writers are Víctor Andrés Belaúnde (1883–1966), Jorge Basadre (b.1903), and Ciro Alegría (1909–67); the novelist Mario Vargas Llosa (b.1936) was the first Latin American president of PEN, an international writers' organization. Peru's best-known contemporary painter is Fernando de Szyszlo (b.1925). Other noted Peruvians include Ventura García Calderón (1866–1959), a writer; José Sabogal (1888–1956), a painter; Honorio Delgado (b.1892), a scientist; and José Carlos Mariátegui (1895–1930), a political essayist. Yma Sumac (b.1928) is an internationally known singer.

Víctor Raúl Haya de la Torre (1895–1979) originated APRA in 1924 as a Latin American workers' movement; it became Peru's most significant political force. Gen. Juan Velasco Alvarado

(1910–77), who led the military coup of 1968, ruled as president of Peru until his own ouster by a bloodless coup in 1975. Gen. Francisco Morales Bermúdez Cerruti (b.1921), president during 1975–80, prepared the country for a return to civilian rule. Fernando Belaúnde Terry (b.1913), founder of the Popular Action Party, served as president during 1963–68 and again during 1980–85. Alán García Pérez (b.1949) was elected president in 1985. In 1982, Javier Pérez de Cuéllar (b.1920) became secretary-general of the UN.

[49]DEPENDENCIES

Peru has no territories or colonies.

[50]BIBLIOGRAPHY

Alisky, Marvin. *Historical Dictionary of Peru.* Metuchen, N.J.: Scarecrow, 1979.

Becker, David G. *The New Bourgeoisie and the Limits of Dependency: Mining, Class and Power in "Revolutionary" Peru.* Princeton, N.J.: Princeton University Press, 1983.

Bingham, Hiram. *Lost City of the Incas: The Story of Machu Picchu and Its Builders.* New York: Duell, Sloan & Pearce, 1949.

Cleaves, Peter S., and Martin J. Scurrah. *Agriculture, Bureaucracy, and Military Government in Peru.* Ithaca, N.Y.: Cornell University Press, 1980.

Collier, David. *Squatters and Oligarchs: Authoritarian Rule and Policy Change in Peru.* Baltimore: Johns Hopkins University Press, 1976.

Descola, Jean. *Daily Life in Peru in the Times of the Spaniards, 1710–1820.* New York: Macmillan, 1968.

Dobyns, Henry F., and Paul L. Doughty. *Peru: A Cultural History.* New York: Oxford University Press, 1976.

Figueroa, Adolfo. *Capitalist Development and the Peasant Economy of Peru.* New York: Cambridge University Press, 1984.

Fisher, John Robert. *Peru.* Santa Barbara, Calif.: Clio Press, 1989.

FitzGerald, E. V. K. *The State and Economic Development: Peru Since 1968.* New York: Cambridge University Press, 1979.

Garcilaso de la Vega. *Royal Commentaries of the Incas and General History of Peru.* 2 vols. Austin: University of Texas Press, 1966.

Goodsell, Charles T. *American Corporations and Peruvian Politics.* Cambridge, Mass.: Harvard University Press, 1974.

Gootenberg, Paul Eliot. *Between Silver and Guano: Commercial Policy and the State in Postindependence Peru.* Princeton, N.J.: Princeton University Press, 1989.

——. *Imagining Development: Economic Ideas in Peru's "Fictitious Prosperity" of Guano, 1840–1880.* Berkeley: University of California Press, 1993.

Haas, Jonathan, Shelia Pozorski, and Thomas Pozorski (eds.). *The Origins and Development of the Andean State.* New York: Cambridge University Press, 1987.

Health Care in Peru: Resources and Policy. Boulder, Colo.: Westview Press, 1988.

Hemming, John. *The Conquest of the Incas.* New York: Harcourt Brace Jovanovich, 1970.

Hudson, Rex A. (ed.). *Peru in Pictures.* Minneapolis: Lerner, 1987.

Keatinge, Richard W. (ed.). *Peruvian Prehistory: An Overview of Pre-Inca and Inca Society.* New York: Cambridge University Press, 1988.

Klaiber, Jeffrey L. *Religion and Revolution in Peru, 1824–1976.* Notre Dame, Ind.: University of Notre Dame Press, 1977.

Lockhart, James. *Spanish Peru, 1532–1560: A Social History,* 2nd ed. Madison, Wisconsin.: University of Wisconsin Press, 1994.

Lowenthal, Abraham F. (ed.). *The Peruvian Experiment: Continuity and Change Under Military Rule.* Princeton, N.J.: Princeton University Press, 1975.

Martin, Luis. *The Kingdom of the Sun: A Short History of Peru.* New York: Scribner, 1974.

Moseley, Michael Edward. *The Incas and their Ancestors: the Archaeology of Peru.* New York: Thames and Hudson, 1992.

Palmer, David Scott. *Peru: The Authoritarian Legacy.* New York: Praeger, 1980.

Pike, Frederick B. *The Modern History of Peru.* New York: Praeger, 1967.

Prescott, William Hickling. *History of the Conquest of Peru.* Rev. ed. New York: New American Library, 1981 (orig. 1847).

The Shining Path of Peru. New York: St. Martin's Press, 1992.

Stepan, Alfred C. *The State and Society: Peru in Comparative Perspective.* Princeton, N.J.: Princeton University Press, 1978.

Stephens, Richard H. *Wealth and Power in Peru.* Metuchen, N.J.: Scarecrow, 1971.

Thorp, Rosemary. *Economic Management and Economic Development in Peru and Colombia.* Pittsburgh, Pa.: University of Pittsburgh Press, 1991.

Tullis, F. LaMond. *Lord and Peasant in Peru.* Cambridge, Mass.: Harvard University Press, 1970.

Werlich, David P. *Admiral of the Amazon: John Randolph Tucker, his Confederate Colleagues, and Peru.* Charlottesville: University Press of Virginia, 1990.

ST. KITTS AND NEVIS

Federation of Saint Kitts and Nevis

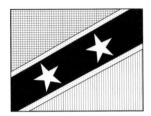

CAPITAL: Basseterre.

FLAG: Two thin diagonal yellow bands flanking a wide black diagonal band separate a green triangle at the hoist from a red triangle at the fly. On the black band are two white five-pointed stars.

ANTHEM: *National Anthem,* beginning "O land of beauty."

MONETARY UNIT: The East Caribbean dollar of 100 cents is the national currency. There are coins of 1, 2, 5, 10, and 25 cents and 1 East Caribbean dollar, and notes of 5, 10, 20, and 100 East Caribbean dollars. EC$1 = US$0.3704 (US$1 = EC$2.70).

WEIGHTS AND MEASURES: The imperial system is used.

HOLIDAYS: New Year's Day, 1 January; Labor Day, 1st Monday in May; Bank Holiday, 1st Monday in August; Independence Day, 19 September; Prince of Wales's Birthday, 14 November; Christmas, 25 December; Boxing Day, 26 December; Carnival, 30 December. Movable religious holidays include Good Friday and Whitmonday.

TIME: 8 AM = noon GMT.

¹LOCATION, SIZE, AND EXTENT

St. Kitts lies about 8 km (5 mi) SE of the Netherlands Antilles and 72 km (45 mi) NW of Antigua, in the Leeward Islands. It is 37 km (23 mi) long and 8 km (5 mi) across at its widest point, with a total area of 269 sq km (104 sq mi). Comparatively, the area occupied by St. Kitts and Nevis is slightly more than 1.5 times the size of Washington, D.C. Nevis lies about 3.2 km (2 mi) SE of St. Kitts, across a channel called the Narrows; it is 13 km (8 mi) long and 10 km (6 mi) wide, with a land area of 93 sq km (36 sq mi). Together the islands have a coastline of 135 km (84 mi).

The capital city, Basseterre, is located on St. Kitts.

²TOPOGRAPHY

St. Kitts and Nevis are of volcanic origin. In the northwest of St. Kitts is Mt. Liamuiga, the island's highest peak at 1,156 m (3,793 ft); to the south and west of Mt. Liamuiga are 210-m (700-ft) cliffs that drop straight to the sea. On the southern arm of the island lies the Great Salt Pond. Nevis's highest elevation is the central peak of Mt. Nevis, at 985 m (3,232 ft); it is usually capped in clouds. There is a black sand beach on the northwest coast.

³CLIMATE

Temperatures range from 20°C (68°F) to 29°C (84°F) all year long. Northeast tradewinds are constant. Rain usually falls between May and November, averaging 109 cm (43 in) a year. High humidity characterizes the summer months.

⁴FLORA AND FAUNA

The upper slopes of Mt. Nevis are well wooded; coconut palms, poincianas, and palmettos are profuse. Lemon trees, bougainvillea, hibiscus, and tamarind are common on both islands. There are some black-faced vervet monkeys on Monkey Hill in St. Kitts.

⁵ENVIRONMENT

Deforestation, erosion, and water pollution are among the most significant environmental problems in St. Kitts and Nevis.

Deforestation has affected the nation's wildlife population and contributed to soil erosion. The erosion of the soil produces silt which affects the living environment for marine life on the coral reefs.

Water pollution results from uncontrolled dumping of sewage into the nation's waters. Other contributing factors are pollution from cruise ships which support the nation's tourist trade. In an effort to establish a framework for the regulation of environmental issues, the government has introduced legislation. The National Conservation and Environmental Protection Act, along with the Letter Act, are aimed at monitoring the nation's most pressing environmental concerns.

⁶POPULATION

The 1980 census counted 43,309 residents, of whom St. Kitts had 33,881 and Nevis 9,428. Estimated population in 1994 was 40,980. The estimated population density overall was 157 per sq km (406 per sq mi), with the density of St. Kitts twice that of Nevis. The crude birthrate in 1990 was 23.1 per 1,000 population; the crude death rate was 10.8; and the net natural increase was 12.3.

⁷MIGRATION

There is less emigration in the current period than there was during the 1950s, largely because the economy enjoys almost full employment during the tourist and harvest seasons. During the off-season, some people migrate to other islands in search of work.

⁸ETHNIC GROUPS

The population is mainly of black African descent. As of 1985, about 5% of the population was considered to be mulatto, 3% Indo-Pakistani, and 1.5% European.

⁹LANGUAGES

English, sprinkled with local expressions, is the universal language.

[10]RELIGIONS

The Anglican Church, the largest church on the island, claims more than a third of all adherents. Other principal religious groups are the Church of God, the Methodist Church (33% in 1985), the Moravians, the Baptists, the Seventh-day Adventists, the Pilgrim Holiness Church, and the Roman Catholic Church (10.7% in 1985).

[11]TRANSPORTATION

A light, narrow-gauge railway of 58 km (36 mi) on St. Kitts is operated by the government to transport sugarcane from fields to factory and processed sugar to the pier at Rawlins Bay. In 1991 there were 300 km (186 mi) of roads on the islands, of which 42% were paved; the main roads circle each island. There are about 50 km (31 mi) of unimproved earth tracks. In 1985 there were 3,540 automobiles registered. Basseterre and Charlestown are the principal ports. A state-run motorboat service is maintained between St. Kitts and Nevis. Golden Rock International Airport is a modern facility serving Basseterre; several small airlines fly to a landing strip at Newcastle, on Nevis.

[12]HISTORY

Arawak Indians, followed by Caribs, were the earliest known inhabitants of the islands. Discovered by Columbus in 1493 and named St. Christopher, St. Kitts was the first of the British West Indies to be settled. Sir Thomas Warner established a settlement on St. Kitts in 1623, later leading colonial expeditions to Nevis in 1628 and Antigua in 1632. For a short while during this period there were French settlements at both ends of St. Kitts, and the French settlers cooperated with the British to repel a Spanish invasion. By the 1660s there were some 4,000 Europeans engaged in the sugar trade, based on a plantation system with slaves imported from Africa. The French gained control in 1664, but they lost it to the British in 1713, under the Peace of Utrecht. The French besieged the British garrison in the Brimstone Hill fortress in 1782 and once more controlled the island, but the Treaty of Versailles (1783) again returned St. Kitts to Britain. By the late 18th century, the thermal baths at Charlestown, Nevis, were attracting thousands of international tourists. Although the slaves were emancipated in 1834, many continued to work on the sugar plantations, so the sugar-based economy did not decline as rapidly as elsewhere in the West Indies.

St. Kitts, Nevis, and Anguilla (the most northerly island of the Leeward chain) were incorporated with the British Virgin Islands into a single colony in 1816. The territorial unit of St. Kitts-Nevis-Anguilla became part of the Leeward Islands Federation in 1871 and belonged to the Federation of the West Indies from 1958 to 1962. In 1967, the three islands became an associated state with full internal autonomy under a new constitution. After the Anguilla islanders rebelled in 1969, British paratroopers intervened, and Anguilla was allowed to secede in 1971, an arrangement formally recognized on 19 December 1980.

General elections were held in St. Kitts and Nevis in 1971, 1975, 1982, and 1984. A 1982 white paper on independence provoked stormy debate over the form of the constitution, spilling over into civil unrest in 1982 and 1983. Nevertheless, St. Kitts and Nevis became an independent federated state within the Commonwealth on 19 September 1983. Under the arrangement, Nevis was given its own legislature and the power to secede from the federation. Elections in June 1984 produced a clear majority for the People's Action Movement/Nevis Reformation Party coalition.

In the 21 March 1989 elections, 11 of 14 members of the National Assembly were elected, six from the People's Action Movement Party, two from Saint Kitts and Nevis Labor Party, two from the Nevis Reformation Party, and one from the Concerned Citizens Movement.

[13]GOVERNMENT

St.Kitts-Nevis is a federation of the two constituent islands. Under the constitution passed at independence in 1983, the British monarch is head of state and is represented by a governor-general, who is required to act upon the advice of the cabinet, and a deputy governor-general for Nevis.

The nation is governed under a parliamentary system, with legislative power vested in the unicameral House of Assembly, consisting of the speaker, three senators (two appointed on the advice of the prime minister and one on the advice of the leader of the opposition), and 11 elected members elected from each of 11 constituencies for up to five years. The cabinet, collectively responsible to the Assembly, consists of the prime minister (who must be able to command the support of a legislative majority), the attorney general (ex officio), and other ministers. The Nevis Island Assembly and the Nevis Island Administration (headed by the British monarch represented by the deputy governor-general) operate similarly to the federation government.

Suffrage is universal for all citizens 18 or older.

[14]POLITICAL PARTIES

There are four political parties holding seats in the House of Assembly. Although the Labour Party (also known as the Workers' League) dominated the political scene from the 1950s until 1980, it was supplanted after independence by the People's Action Movement. The PAM, led by Prime Minister Kennedy A. Simmonds, has governed ever since, It won six seats, a clear majority, in both 1984 and 1989. The Labour Party held two seats in 1984 and 1989. The Nevis Reformation Party, founded in 1970 and led by Simeon Daniel, took three seats in 1984 and 2 seats in 1989, all of which were from constituencies on Nevis. The Concerned Citizen's Movement, under Vance Amory, took one Nevisian seat in 1989.

[15]LOCAL GOVERNMENT

St. Kitts is divided into nine parishes: St. George Basseterre, St. Peter Basseterre, St. Mary Cayon, Christ Church Nichola Town, St. John Capisterre, St. Paul Capisterre, St. Anne Sandy Point, St. Thomas Middle Island, and Trinity Palmetto Point. Nevis is divided into five parishes: St. Paul Charlestown, St. John Figtree, St. George Gingerland, St. James Windward, and St. Thomas Lowland. Under the 1983 constitution, Nevis has its own legislative assembly and the right to secede under certain conditions.

[16]JUDICIAL SYSTEM

The Eastern Caribbean Supreme Court, established on St. Lucia, administers the judicial system, which is based on English common law and statutory acts of the House of Assembly. A puisne judge of the Court is responsible for St. Kitts and Nevis and presides over the Court of Summary Jurisdiction. Magistrates' courts deal with petty criminal and civil cases. The attorney general is the government's principal legal adviser.

Final appeal may be taken to the Privy Council in the UK. There are no military or political courts.

The judiciary is respected for its independence and integrity. Legal assistance is provided to indigent criminal defendants.

[17]ARMED FORCES

Antigua and Barbuda, Dominica, Grenada, St. Kitts and Nevis, St. Lucia, and St. Vincent and the Grenadines created a Regional Security System in 1985.

[18]INTERNATIONAL COOPERATION

St. Kitts and Nevis became a member of the UN on 23 September 1983 and belongs to FAO, IBRD, IMF, UNESCO, UNIDO, and WHO. It also belongs to CARICOM, OECS, the Commonwealth of Nations, G-77, OAS, and PAHO and is an adherent of GATT and a signatory of the Law of the Sea.

¹⁹ECONOMY

The economy is based on tourism and agriculture, particularly on sugar, which generated some 57% of export revenues in 1986. Sea island cotton, coconuts, sugar production and processing, and more than 40,000 visitors per year are the most important sources of jobs and earnings. The government has been making efforts to attract foreign investment, to diversify the economy industrially, to expand tourism, and to improve local food production.

The sugar industry experienced a significant decrease in production from 1986–89. As a result, the government signed a two-year management agreement with the British company Booke and Tate in August 1991. The World Bank loaned the country $1.9 million for financial support, causing sugar production to recuperate by 28% from 1990–91. As of 1994, sugar continues to be the backbone of St.Kitts and Nevis economy, constituting approximately 55% of its export revenue. Sugar production increased by 5.6% from an output of 19,499 tons in 1991 to 20,483 tons in 1992. St. Kitts and Nevis is still devoting its efforts in diversifying its economy and has had success with the development of its light manufacturing industries—mainly garments and electronics assembly, data entry, the expansion of non sugar agricultural production and tourism. Real GDP grew by about 4% in 1992.

²⁰INCOME

In 1992, St. Kitts and Nevis' GNP was $181 million at current prices, or $3,990 per capita. For the period 1985–92 the average inflation rate was 8.8%, resulting in a real growth rate in per capita GNP of 5.3%.

It is estimated that in 1989 agriculture, hunting, forestry, and fishing contributed 9% to GDP; mining and quarrying, less than 1%; manufacturing, 16%; electricity, gas, and water, 1%; construction, 12%; wholesale and retail trade, 21%; transport, storage, and communication, 14%; finance, insurance, real estate, and business services, 12%; community, social, and personal services, 4%; and other sources, 11%.

²¹LABOR

A 1991 estimate placed the labor force at 20,000, of whom unions enrolled some 6,700 workers. The St. Kitts and Nevis Trades and Labour Union, established in 1940, is associated with the Labour Party and is the only general workers' labor organization. Unemployment fluctuates during slack times between the tourist, sugar, and cotton seasons; in 1991, the unemployment rate was 13.5%. The sugar industry and civil service are the largest employers, with tourism gaining importance, especially on Nevis.

²²AGRICULTURE

Of the islands' total land area, about two-fifths is devoted to crops. The principal agricultural product of St. Kitts is sugarcane; peanuts are now the second crop. On Nevis, sea island cotton and coconuts are the major commodities. Sweet potatoes, onions, tomatoes, cabbages, carrots, and breadfruit are grown for local consumption on both islands, mostly by individual smallholders. From 1987 to 1992, agricultural products on average accounted for about 17% of total imports by value; the government has embarked on a program to substitute for food imports.

Sugar estate lands were nationalized in 1975, and the sugar factory was purchased by the government the following year. The output of raw sugar slumped between 1986 and 1989, and as a result the government entered into a management agreement with Booke and Tate of Great Britain in August 1991; a World Bank loan of $1.9 million was utilized to provide financial stability. Sugar production has since then increased by 28% from 1990 to 1991.

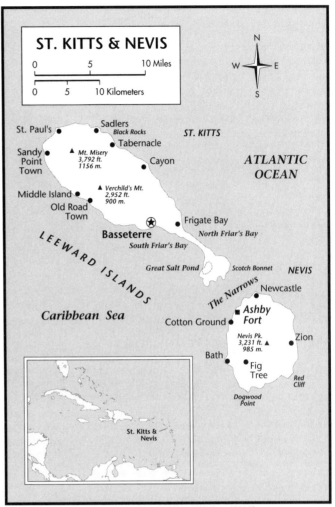

LOCATION: 17°10′ to 17°20′N; 62°24′ to 62°45′W.
TOTAL COASTLINE: 135 km (84 mi). TERRITORIAL SEA LIMIT: 12 mi.

²³ANIMAL HUSBANDRY

Pasture areas are small, covering some 2.7% of the islands. Pangola and Bermuda grasses provide the bulk of the fodder. Estimates of livestock in 1992 were cattle, 5,000 head; sheep, 15,000; pigs, 2,000; and goats, 10,000.

²⁴FISHING

Fishing is a traditional occupation that has not expanded to any great extent; the catch in 1991 was 1,750 tons (up from 1,551 tons in 1986). Some exports (primarily lobsters) are made to the Netherlands Antilles and Puerto Rico.

²⁵FORESTRY

Both islands have small stands of virgin tropical forest, with palms, poincianas, and palmettos.

²⁶MINING

Salt raking is done from time to time. Local quarrying of some materials is used to supplement the construction industry. Beach sand mining was authorized by the 1987 National Conservation and Environmental Protection Act.

²⁷ENERGY AND POWER

The government-owned electricity station supplies power to Basseterre and all rural areas. Net installed capacity was 15,000 kw in 1991, when 40 million kwh were produced, entirely from conventional sources. St. Kitts and Nevis has no fossil fuels, and all petroleum products must be imported.

28INDUSTRY

The principal manufacturing plant and largest industrial employer is the St. Kitts Sugar Manufacturing Corp., a government enterprise; it grinds and processes sugarcane for export. A brewery on St. Kitts makes beer for local consumption, and cotton is ginned and baled on Nevis. Two small electronics plants produce switches, calculators, car radios, and pocket radios. Other industries are clothing and shoe manufacturing.

The manufacturing sector experienced mixed performance in 1992, with a decrease in the production and export of electronics offset largely by an increase in the production and export of garments.

As a result of diversification and expansion, St. Kitts and Nevis has transformed small electronic plants into the largest electronics assembly industry in the Eastern Caribbean. Its apparel assembly industry has also become very successful in recent years, increasing its apparel exports to the US by 5.7% in 1992.

In addition, the tourism industry has been increasing, largely as a result of increased hotel capacity, promotional activity, and a greater emphasis on cruise ship arrivals.

29SCIENCE AND TECHNOLOGY

St. Kitts and Nevis is dependent on outside resources both for industrial technology and for advanced scientific and technical education. The government is currently developing post-secondary education; a technical school was in operation in 1987.

30DOMESTIC TRADE

General business is conducted six days a week; shops are shut on Sunday, and Thursday is early closing day. Normal banking hours are 8 AM to noon; banks are also open from 3 to 5 PM on Friday.

31FOREIGN TRADE

Exports of sugar, molasses, beer and ale, cotton, and lobsters earned an estimated us$22.5 million in 1985. Imports, mostly of foodstuffs, fuels, and manufactured goods, were estimated at us$50.4 million.

Major trading partners are the US, the UK, Canada, Japan, Trinidad and Tobago, and other CARICOM states. About two-thirds of the sugar crop is exported to the US.

The Government of St. Kitts and Nevis expanded trade incentives in recent years. Incentives include import duty exemption, hotel aid ordinance, no capital gains tax, no land tax, and no house tax among other tax credits.

Exports of goods and services have increased considerably, totaling $100.1 million in 1992. Sugar, beer and ale, lobsters, electrical equipment, and margarine and shortening are the main exports. In the same year, imports reached $139.7 million, including food, manufactured goods, machinery and transportation equipment, mineral fuels, and lubricants and related materials.

As of 1991, the major export-trading partners were the UK, the US, and CARICOM states. Its main import-trading partners were those same nations plus Japan.

32BALANCE OF PAYMENTS

Earnings from tourism and overseas remittances largely offset the trade deficit. From 1988 to 1990, the current account deficit increased from 21% of GDP to nearly 35% of GDP. In 1990/91, there was an overall balance of payments surplus, due to increased capital inflows used primarily for the construction of hotels.

In 1992, merchandise exports totaled $32.44 million and imports $99.96 million. The merchandise trade balance was $–67.51 million.

The following table summarizes St. Kitts and Nevis's balance of payments for 1991 and 1992 (in millions of US dollars):

	1991	1992
CURRENT ACCOUNT		
Goods services and income	–36.33	–25.97
Unrequited transfers	17.11	11.32
TOTALS	–19.22	–14.65
CAPITAL ACCOUNT		
Direct investment	21.44	23.87
Other long-term capital	0.85	–0.48
Other short-term capital	3.66	5.95
Reserves	–0.70	–9.57
TOTALS	25.25	19.99
Errors and omissions	–6.03	–5.12
Total change in reserves	–0.65	–9.61

33BANKING AND SECURITIES

Banking services are provided by a government savings bank and five private banks. Securities transactions on international exchanges are performed by the banks.

34INSURANCE

International, regional, and local insurance companies or agents offer life and property insurance.

35PUBLIC FINANCE

The overall public sector deficit fell from 9% of GDP in 1985 to 1.7% in 1991. Losses from the state-owned sugar corporation and recent public service wage hikes are the basis of annual deficits. Chief sources of revenue are customs and duties and the corporate income tax.

In 1992, St. Kitts and Nevis's total external debt stood at $37 million, with debt service payments of $1.3 million.

36TAXATION

There is no personal income tax for residents of St. Kitts and Nevis. Corporations are taxed at the rate of 40% of income, but the Fiscal Incentives Act offers development and tax concessions. Profit remittances are taxed at a 10% rate; there is also a land tax and a house tax.

37CUSTOMS AND DUTIES

St. Kitts and Nevis is bound by the common external tariff of CARICOM and requires an import license for certain durable and non-durable products.

38FOREIGN INVESTMENT

Like most Caribbean microstates, St. Kitts and Nevis has an investment incentives program. Joint ventures and labor-intensive industries are especially welcome. Official development assistance and resource flows from commercial sources totaled us$3.6 million in 1984.

Recent investment incentives include the following: tax holiday of up to 15 years; tax rebates of up to 5 years; and exemption from custom duties on machinery deemed to establish and update an enterprise.

39ECONOMIC DEVELOPMENT

The government has attempted to halt the decline of the sugar industry by restructuring the sector, and has encouraged agricultural diversification and the establishment of small industrial enclaves linked to the international export market. Three industrial estates have been developed, two on St. Kitts (near both the airport and the deep-water harbor) and one on Nevis. The tourist industry has received considerable government support. The Development and Finance Corp. is the principal development agency.

Some US$14.6 million in aid was received during 1970–78. Assistance from the UK totaled £3,561,000 during 1975–79; regular aid from the UK for 1983 alone was £1,000,000. During the latter year, an additional £10,000,000 was allocated by the UK for capital assistance. In 1986/87, more than US$22 million in grants and loans were committed by the US and Canada and the OAS, EEC, and other multilateral organizations.

The country's immediate plans continue to be aimed at diversifying the economy. Construction projects in the private and public sector are expected to contribute substantially to the moderate economic growth that St. Kitts and Nevis should experience in future years. In addition, the tourism industry is expected to keep growing as a result of successful promotions. A slow but sure growth in the GDP is predicted because the government is devoting efforts toward diversifying industry as well as improving the management techniques of the Sugar Manufacturing Corporation.

40SOCIAL DEVELOPMENT

Housing shortages place a continuing strain on the government's ability to establish and fund social welfare programs. Especially concerned with the problem of teen-age pregnancy, the government has established a family planning program as part of its overall health policy. The St. Kitts and Nevis Family Planning Association is an affiliate of the International Planned Parenthood Federation. The association maintains a clinic and recommends those requesting sterilization to private physicians. Effective 1978, a social security system replaced the existing provident fund as the provider of old age, disability, survivor, sickness, and maternity benefits. Coverage is compulsory for employed persons, with provisions for voluntary coverage where applicable.

The Ministry of Women's Affairs was created by the government to promote women's rights and provide counseling for abused women. A special police unit works closely with the Ministry to investigate domestic violence and rape cases.

41HEALTH

In 1990, there were 2.7 hospital beds per 1,000 people, 43 doctors, and 11 dentists. In 1988 there were 190 nurses. Hygienic education is the primary concern of the Central Board of Health. In 1992, the infant mortality rate was 34 per 1,000 live births and average life expectancy was 71 years. In 1993 there were 800 births and 99% of children were immunized against measles. An estimated 92 cases of tuberculosis per 100,000 people were reported in 1990, and the country's health care expenditures were estimated at $46.7 million that year.

42HOUSING

The government has placed emphasis on planned housing development in order to conserve agricultural lands. The Central Housing Authority began a program of low-cost home construction in 1977. The water supply, which comes from mountain springs and deep wells, is controlled by the Water Department. As of 1980, 83% of all dwellings were detached houses, 10% were apartments, and 4% were in buildings used partly for commercial purposes. Roughly 58% were owner occupied, 36% were rented, privately, 10% were occupied rent free, and 1% were rented from the government. The most common construction materials for housing were wood (50%), wood and concrete combined (25%), and concrete (20%).

43EDUCATION

The literacy rate is over 90%. Education is free and compulsory for 12 years. Elementary education lasts for seven years followed by six years of secondary education at the first level and two at the second level.

A National Committee on Education was established in 1991 to examine all aspects of the education system at the primary and secondary level. Issues under consideration were: introducing technical and vocational education and training; economic trends and the education system; administration and management of the education system; and teachers and their work conditions.

In 1991, there were 32 schools at the primary level with 350 teachers (79% were female) and 7,236 students. At the eight secondary-level schools, there were 294 teachers (56% were females) and 4,396 students enrolled. At the higher level, there were 38 teachers and 325 students enrolled. Well known for its higher education is the College of Further Education.

44LIBRARIES AND MUSEUMS

There is a public library in Charlestown. The Old Court House in Basseterre houses a museum, and there is a museum of Lord Nelson memorabilia on Nevis.

45MEDIA

The telephone system (3,805 telephones in 1991) is operated by the government, while international cable and wireless services are operated privately. ZIZ Radio and Television is owned and operated by the government; Radio Paradise broadcasts from Basseterre; and Trinity Broadcasting is based on Nevis. Radio broadcasting began in 1961, and television broadcasting in 1972. By 1991 there were some 27,000 radios and 9,000 television sets in use. There is one newspaper: the *Labour Spokesman,* founded in 1957, is published twice weekly and has a circulation of 2,000.

46ORGANIZATIONS

The St. Kitts and Nevis Chamber of Commerce has its headquarters in Basseterre, and the Nevis Cotton Growers' Association has its office in Charlestown. Many US charitable organizations have operations in the islands, including the 4-H Foundation, Planned Parenthood, Heifer Project International, Operation Crossroads Africa, and Project Concern.

47TOURISM, TRAVEL, AND RECREATION

Visitors from the US, the UK, and Canada need proof of citizenship and an onward ticket; nationals of most Commonwealth countries need passports but not visas. The chief historic attraction on St. Kitts is Brimstone Hill fortress, which towers 230 m (750 ft) above the Caribbean, took 100 years to build, and is partially restored. Beautiful beaches and the Georgian architecture of Basseterre also attract tourists. Nevis has many beaches and relic plantations and a quaint atmosphere reminiscent of the 18th century. In 1991, 83,903 stayover tourists arrived on the islands, 75,513 from the Americas. There were 1,392 hotel rooms and 2,784 beds, and tourism revenue was US$74 million. The government has dredged the main harbor on St. Kitts to accommodate cruise ships.

48FAMOUS KITTSIANS AND NEVISIANS

Sir Thomas Warner (d.1649) established the first colony on each island. US statesman Alexander Hamilton (1757–1804) was born in Charlestown.

49DEPENDENCIES

St. Kitts and Nevis has no territories or colonies.

50BIBLIOGRAPHY

Cox, Edward L. *Free Coloreds in the Slave Societies of St. Kitts and Grenada, 1763–1833.* Knoxville: University of Tennessee Press, 1984.

Hamshere, Cyril. *The British in the Caribbean.* Cambridge, Mass.: Harvard University Press, 1972.

Lowenthal, David. *West Indian Societies.* London: Oxford University Press, 1972.

Merrill, Gordon Clark. *The Historical Geography of St. Kitts and Nevis, the West Indies.* Mexico: Editorial Fournier, 1958.

Olwig, Karen Fog. *Global Culture, Island Identity: Continuity and Change in the Afro-Caribbean Community of Nevis.* Philadelphia: Harwood, 1993.

Richardson, Bonham C. *Caribbean Migrants: Environment and Human Survival on St. Kitts and Nevis.* Knoxville: University of Tennessee Press, 1983.

Sherlock, Sir Philip M. *West Indian Nations: A New History.* New York: St. Martin's, 1973.

Williams, Eric Eustace. *From Columbus to Castro: The History of the Caribbean, 1492–1969.* London: Deutsch, 1970.

ST. LUCIA

CAPITAL: Castries.

FLAG: On a blue background is a yellow triangle surmounted by a black arrowhead whose outer edges are bordered in white.

ANTHEM: *Sons and Daughters of St. Lucia.*

MONETARY UNIT: The East Caribbean dollar (EC$) of 100 cents is the national currency. There are coins of 1, 2, 5, 10, and 25 cents and 1 dollar, and notes of 5, 10, 20, and 100 East Caribbean dollars. EC$1 = US$0.3704 (or US$1 = EC$2.70).

WEIGHTS AND MEASURES: The metric system has been introduced, but imperial measures are still commonly employed.

HOLIDAYS: New Year's Day, 1 January; Carnival, 8–9 February; Independence Day, 22 February; Labor Day, 1 May; Queen's Official Birthday, 5 June; Bank Holiday, 1st Monday in August; Thanksgiving Day, 1st Monday in October; St. Lucia Day, 13 December; Christmas Day, 25 December; Boxing Day, 26 December. Movable religious holidays include Good Friday, Easter Monday, Whitmonday, and Corpus Christi.

TIME: 8 AM = noon GMT.

¹LOCATION, SIZE, AND EXTENT

The Caribbean island of St. Lucia, part of the Windward Islands group of the Lesser Antilles, is 43 km (27 mi) N–S by 23 km (14 mi) E–W and has a total area of 620 sq km (239 sq mi). Comparatively, the area occupied by St. Lucia is slightly less than 3.5 times the size of Washington, D.C. Situated between Martinique to the N and St. Vincent to the SW, St. Lucia has a total coastline of 158 km (98 mi). The capital city, Castries, is located on St. Lucia's northwest coast.

²TOPOGRAPHY

St. Lucia is a volcanic island, the younger part of which is the mountainous southern half, and the older the hilly but more nearly level northern half. The highest mountain, Mt. Gimie, rises 959 m (3,145 ft) above sea level. Better known are the two peaks on the southern coast, Grand Piton (798 m/2,619 ft) and Petit Piton (750 m/2,461 ft), which together form one of the scenic highlights of the West Indies. The lowlands and valleys of the island have fertile soil and are irrigated by many streams. The island has beautiful beaches, some with black volcanic sand.

³CLIMATE

The average yearly temperature on St. Lucia is 26°C (79°F); the warmest month is usually September, and the coolest January. The average rainfall at sea level is 231 cm (91 in) a year; on the mountain peaks, more than 380 cm (150 in). Like the rest of the West Indies, St. Lucia is vulnerable to hurricanes, which hit the Caribbean in the late summer months.

⁴FLORA AND FAUNA

Tropical sunlight, heavy rainfall, and fertile soil combine to produce an abundance of tropical flora, including hibiscus, poinciana, frangipani, orchids, jasmine, and bougainvillea. There are no large mammals on St. Lucia. Bats are common, and there are several species of small snakes. The central highlands provide nesting

places for many birds, including flycatchers, hummingbirds, pigeons, and about a hundred other species. The surrounding sea contains lobster, turtle, and conch, as well as an abundance of fish.

⁵ENVIRONMENT

Densely populated, St. Lucia has been shorn of much of its protective woodland, except for limited areas in the south-central rain forest. St. Lucia's forests are gradually being depleted by agricultural and commercial interests. The loss of forest cover contributes to the erosion of the soil, particularly in the drier, northern part of the island. The nation does not have the financial resources to develop an adequate water purification system. The population is at risk due to contamination of the water supply by agricultural chemicals and sewage. Two small areas have been set aside as nature preserves, but population pressure prevents the government from expanding them. Principal responsibility for the environment is vested in the Ministry of Agriculture's Lands, Fisheries, and Cooperatives Forestry Division and the National Trust Fund. Excessive use of herbicides and pesticides threaten the wildlife population in St. Lucia and the eastern Caribbean states in general. As of 1987, endangered species included the tundra peregrine falcon, Semper's warbler, white-breasted trembler, St. Lucia parrot, St. Lucia white-breasted thrasher, St. Lucia forest thrush, St. Lucia wren, and common iguana.

⁶POPULATION

In 1991, the population of St. Lucia was 136,041, up from about 124,000 at the time of the 1980 census. Some 51,994 persons, or 38% of the population, lived in Castries, the capital, in 1991, when the nation's population density was 221 persons per sq km (572 per sq mi).

⁷MIGRATION

Emigration has provided an escape valve for population pressure. Neighbors such as Trinidad, Guyana, and the French Caribbean islands have received the bulk of emigrants from St. Lucia, with lesser numbers going to the UK, Canada, and the US.

[8] ETHNIC GROUPS

Reliable statistics on ethnic groups are unavailable. It is estimated, however, that 90.5% of the population consists of descendants of slaves brought from Africa in the 17th and 18th centuries. Some 5.5% is mulatto and 3.2% East Indian. Approximately 0.8% of the population is of European descent.

[9] LANGUAGES

English is the official language of St. Lucia. Nearly 20% of the population cannot speak it, however. Language outreach programs are seeking to integrate these people into the mainstream of society. Almost all the islanders also speak a French patois based on a mixture of African and French grammar and a vocabulary of mostly French with some English and Spanish words.

[10] RELIGIONS

The vast majority of the population (about 79%) was Roman Catholic in 1991. There are also Anglican, Methodist, Baptist, and Seventh-day Adventist churches. The small East Indian community is divided between Hindus and Muslims.

[11] TRANSPORTATION

Direct flights to New York, Miami, Toronto, London, and Frankfurt operate out of Hewanorra International Airport, on the southern tip of the island. The smaller Vigie Airport, located near Castries, is used for flights to and from neighboring Caribbean islands. St. Lucia has two important ports: Castries, in the north, with a cargo-handling capacity of 365,000 tons per year; and Vieux Fort, at the southern tip of the island, from which ferries link St. Lucia with St. Vincent and the Grenadines.

All of the island's towns, villages, and main residential areas are linked by 760 km (472 mi) of all-purpose roads. Motor vehicles numbered 18,938 in 1992.

[12] HISTORY

Arawak and Carib Amerindians were the earliest known inhabitants of what is now St. Lucia. There is no hard evidence for the folklore that Columbus sighted St. Lucia on St. Lucy's Day in 1498, but in keeping with the tradition, 13 December is still celebrated as the date of the island's discovery. It was not settled until the mid-17th century because the Caribs defended the islands successfully for years. The French settled the islands, but the natural harbor at Castries brought English interest. The island changed hands between the British and the French no fewer than 14 times, until in 1814, the British took permanent possession. In 1838, St. Lucia came under the administration of the Windward Islands government set up by Great Britain.

Unlike other islands in the area, sugar did not monopolize commerce on St. Lucia. Instead, it was one product among many others including tobacco, ginger, and cotton. Small farms rather than large plantations continued to dominate agricultural production into the 20th century. A total of 10,328 slaves were freed when slavery was abolished in 1834. To replace the slave labor, East Indian indentured workers were brought to the island during the late 1800s.

St. Lucia has a democratic tradition which began in 1924 when a few elected positions were added to the appointed legislative council. St. Lucia became an associated state with full internal self-government in 1967 and on 22 February 1979 became an independent member of the Commonwealth.

The first three years of independence were marked by political turmoil and civil strife, as leaders of rival political parties fought bitterly. In 1982, the conservative United Workers' Party (UWP) won 14 of 17 seats in the House of Assembly. Party leader and prime minister John Compton, who had been Premier of the island from 1964 until independence, and became Prime Minister at independence, has governed ever since.

The UWP dominance was eroded in 1987, when the party won only 9 seeats. Prime Minister Compton called for new elections almost immediately, but received the same result. In 1992, the UWP increased its majority to 11 seats.

[13] GOVERNMENT

St. Lucia became independent in 1979. Under its constitution, the British monarch continues to be the titular head of government, appointing, upon recommendation of the local leaders, a governor-general to represent the crown. Executive power is effectively exercised by the prime minister and cabinet. There is a bicameral parliament consisting of a Senate with 11 members and a House of Assembly with 17 representatives. The House of Assembly has the important legislative functions. The Senate is an appointed body with little political power.

Members of the lower house are elected for a maximum period of five years. Suffrage on St. Lucia has been universal for those 18 and older since 1951, before St. Lucia achieved independence.

[14] POLITICAL PARTIES

Since 1982, the majority party has been the United Workers' Party (UWP), led by John Compton. The UWP, a coalition of the National Labour Movement and the People's Progressive Party, currently controls 11 seats in parliament. It was the party in power at the time of independence, but lost power in 1979. Although it is by reputation a more conservative party, there is in reality little ideological basis to any of the parties.

The St. Lucia Labour Party (SLP), led by Julian Hunte, is the other major party in St. Lucia. It currently holds 6 seats in the Assembly, and forms the opposition party. The Progressive Labour Party (PLP), under the leadership of George Odlum, has no representation. It is an offshoot of the SLP.

[15] LOCAL GOVERNMENT

St. Lucia is divided into 11 "quarters" for purposes of administration. Local governments are elected by popular vote.

[16] JUDICIAL SYSTEM

Both common law and statute law govern St. Lucia. The lowest court is the district or magistrate's court, above which is the Court of Summary Jurisdiction. Seated in Castries, the Eastern Caribbean Supreme Court (known as the West Indies Associated States Supreme Court upon its founding in 1967, and as the Supreme Court of Grenada and the West Indies Associated States from 1974 until 1979) has jurisdiction in St. Lucia, Anguilla, Antigua and Barbuda, the British Virgin Islands, Dominica, Montserrat, St. Kitts and Nevis, and St. Vincent and the Grenadines. It consists of the High Court, made up of a chief justice and seven puisne judges, and the Court of Appeal, made up of the chief justice and two other appellate justices. In exceptional cases, appeals may be carried to the UK Privy Council.

The constitution guarantees a public trial before an independent and impartial court. Legal counsel is afforded to indigent defendants in cases involving capital punishment.

[17] ARMED FORCES

As of 1985 there were no armed forces other than those of the police department, numbering 300. A regional defense pact, including Antigua and Barbuda, Barbados, Dominica, Grenada, Jamaica, St. Kitts and Nevis, and St. Vincent and the Grenadines, as well as St. Lucia, provides for joint coast-guard operations, military exercises, and disaster contingency plans.

¹⁸INTERNATIONAL COOPERATION

St. Lucia became a member of the UN on 12 September 1979 and belongs to ECLAC and all the nonregional specialized agencies except IAEA, ITU, and WIPO. It is also a member of the Commonwealth of Nations, as well as the Caribbean Development Bank, CARICOM, G-77, the OAS, the OECS, PAHO, and the Windward Islands Banana Growers' Association. St. Lucia is a signatory to the Law of the Sea and is a de facto adherent to GATT.

¹⁹ECONOMY

Agriculture has traditionally been the main economic activity on St. Lucia, which is the leading producer of bananas in the Windward Islands group. Tourism, with direct flights from Europe and North America, has recently become an equally important economic activity. St. Lucia's manufacturing sector grew steadily during the first half of the 1980s, with the construction of many light manufacturing and assembly plants that produce for local or export markets. The economy of St. Lucia grew at a rate of 4.1% in 1983, 5% in 1984, and 5.8% in 1985.

In 1992, real economic growth expanded appreciably by 6.6% following a deceleration in growth from 4% in 1990 to a modest 1.7% in 1991. The strong growth in the economy is credited mainly to a significant expansion in banana production, continued growth in the tourist sector and buoyant activity in the construction sector.

Agricultural output grew substantially in 1992, after registering a decline in 1991. The main contributor to this growth was the banana sector, which marked a record level of output. Banana production increased by 38.2% or 137,500 tons, of which 98.1% was exported. The uncertainty surrounding the future of the banana industry was removed to a certain degree when a new trade regime was designed by the EC Council of Agricultural Ministry. It provides continued preferential treatment to countries' banana imports. On the other hand, coconut production fell by 16.7% to 3,262.3 tons. Difficulties in obtaining labor contributed to the decline in production. Total production of nontraditional crops declined marginally by 0.9% to 4,146 tons; however, export earnings in this category increased by 17.7% to $2.4 million, owing partly to increased availability of air cargo space.

²⁰INCOME

In 1992, the GNP was $453 million at current prices, or $2,900 per capita. For the period 1985–92, the average inflation rate was 3.6%, resulting in a real growth rate in per capita GNP of 5.2%.

In 1992, the GDP was $471 million in current US dollars. It is estimated that in 1987, agriculture, hunting, forestry, and fishing contributed 16% to GDP; mining and quarrying, 0%; manufacturing, 8%; electricity, gas, and water, 4%; construction, 7%; wholesale and retail trade, 23%; transport, storage, and communication, 12%; finance, insurance, real estate, and business services, 11%; community, social, and personal services, 5%; and other sources, 14%.

²¹LABOR

In 1991, the labor force was estimated at 52,200, or about 38% of the population. Some 39% of the labor force is engaged in services (including tourism), 43% in agriculture, and 18% in manufacturing. Unemployment was unofficially estimated at 16% in 1991.

There are seven labor unions in St. Lucia, representing about 20% of the work force. The largest trade union grouping, the Industrial Solidarity Pact, includes the National Workers' Union, the St. Lucia Civil Service Association, the Prison Officers' Association, and the St. Lucia Teachers' Union.

²²AGRICULTURE

Agriculture accounted for 13.8% of GDP in 1990. The production of bananas, St. Lucia's most important crop, fluctuates as a

LOCATION: 13°53′N; 60°58′W. **TOTAL COASTLINE:** 158 km (98 mi).
TERRITORIAL SEA LIMIT: 12 mi.

result of climatic conditions and plant disease; it has gone from a low of 32,000 tons in 1975 to 160,000 tons in 1990 (48% of the Windward Islands' banana production that year) to 137,500 tons in 1992 (up 38.2% from 1991). Almost the entire production is exported. The second most important crop is coconuts, exported as oil and copra; about 25,000 tons of coconuts were produced in 1992. The production of vegetables and fruits for local consumption increased steadily since 1979, as the government sought to achieve self-sufficiency in tomatoes, onions, carrots, cabbages, and breadfruit. From 1991 to 1992, cocoa production increased by 39.5% while cocoa exports fell by 42.5%, due to substantial loss of product in processing. In 1992, the value of exported agricultural products amounted to $79.9 million, down from $85.7 million in 1990.

23 ANIMAL HUSBANDRY

Production in almost every category of animal husbandry is insufficient to satisfy local demand. There are only 12,000 head of cattle on the island, mostly grazing in the middle altitudes of the central mountain region; milk production covers only about 25% of local demand. The island has attained self-sufficiency in pork and egg production. Egg production was 54,000 tons in 1992.

24 FISHING

The establishment of the St. Lucia Fish Market Corp. in 1985, with a US$2.5-million grant from Canada, provided local fishermen with processing, storage, and marketing facilities, enabling St. Lucia to become self-sufficient in fresh fish production. In 1991, the total catch was 910 tons.

25 FORESTRY

A small timber industry processes mahogany, pine, and blue mahoe; expansion of cultivation is planned at the rate of 40 hectares (100 acres) annually. About 13% of total available land consists of forest and woodlands. Legislation is in force to protect against deforestation.

26 MINING

There is no regular commercial mining in St. Lucia. A few quarrying operations produce gravel, pumice, and sand for construction purposes.

27 ENERGY AND POWER

St. Lucia Electricity Services is responsible for the generation and supply of electricity throughout the island. In 1991, net installed capacity was 22,000 kw, electrical production 105 million kwh. St. Lucia's requirements are met through an island-wide grid serviced by two main diesel generating centers, which utilize oil imported from Venezuela and Trinidad and Tobago. The Sulfur Springs in Soufrière on the west coast, have been confirmed as a source of geothermal energy, with a potential generating capacity of 10 Mw.

28 INDUSTRY

St. Lucia's manufacturing sector is the largest and most diversified in the Windward Islands, with many light manufacturing or assembly plants producing apparel, electronic components, plastic products, and paper and cardboard boxes.

Recently, the government has devoted its efforts to the improvement of economic activity as well as development of the major export markets. As a result, the GDP of the manufacturing sector increased by 2.5% in 1992, following poor performance in the previous year. The performance of manufactures geared predominantly for export to the US was mixed. The production of the electric and assembly type products subsector increased while the production of the textile and apparel industry declined. Also, the production of refined oil increased due to an improvement in trading relations with Jamaica. With the formation of the eastern Caribbean Stated Export Development Agency (ECSEDA) in 1990, the performance of local manufacturers is expected to be significantly enhanced in the future. For this reason, several industrial sites have been established. A new deep-water port at Vieux Fort, the country's main export manufacturing center, and new facilities at the Hewanorra airport were officially opened in May 1993 at a cost of EC$94 million, about US$35 million.

Furthermore, the tourism industry has been a great contributor to the overall advancement of the economy. Total visitor arrivals increased by 9.4% in 1992. A new all-inclusive hotel, part of the Jamaican Sandals chain, opened in April 1993. Several new tourist developments are planned.

29 SCIENCE AND TECHNOLOGY

The government of St. Lucia has established a Science and Technology Division within the government's Central Planning Unit (CPU). As of 1987, three scientists were employed with the CPU. The Windward Islands Banana Growers' Association (WINBAN) maintains a research laboratory in St. Lucia serving the needs of banana growers in the region. In 1984, total expenditures on research and development amounted to $12 million; 86 technicians and 53 scientists and engineers were engaged in research and development.

30 DOMESTIC TRADE

Local produce markets are found in all the small villages and towns; they are usually most active in the early morning hours to avoid the midday heat and the afternoon tropical showers. Normal business hours are 8 AM to 12 noon and 1 to 4 PM on weekdays, except for Wednesdays, when shops close early. Banks are open from 8 AM to 12 noon on weekdays and on Fridays from 3 to 6 PM.

31 FOREIGN TRADE

The economy of St. Lucia is highly dependent on foreign trade. Agriculture is the major export earner. Foreign trade continues to be crucial to St Lucia's economy. There are two free-trade zones in St. Lucia and a variety of duty exemption and tax credits are implemented for trade inducement.

At year end 1992, exports of goods and services totaled $314.8 million, including bananas, clothing, cardboard boxes, and coconut products. Imports recorded $349.1 million, consisting mainly of manufactured goods, foodstuffs, machinery, fuels, and chemicals. St. Lucia's major export-trading partners were the UK, the US, and CARICOM. Major import-trading partners included those same nations plus Japan, and Canada.

32 BALANCE OF PAYMENTS

St. Lucia has had a negative balance of trade every year since independence; this annual deficit has been counterbalanced in part by inflows from tourism and direct investment. From 1987 to 1991, the current account deficit grew from 5% of GDP to 21% of GDP. However, the overall balance of payments improved, with reserves able to cover about four months of imports in 1991.

In 1992, merchandise exports totaled $122.78 million and imports $–275.56 million. The merchandise trade balance was $–152.78 million. The following table summarizes St. Lucia's balance of payments for 1991 and 1992 (in millions of US dollars):

	1991	1992
CURRENT ACCOUNT		
Goods services and income	–91.93	–89.78
Unrequited transfers	21.79	18.88
TOTALS	70.14	–70.90
CAPITAL ACCOUNT		
Direct investment	131.01	53.23
Portfolio investment	—	—
Other long-term capital	10.83	24.68
Other short-term capital	–0.19	4.10
Reserves	–7.74	–3.20
TOTALS	133.91	78.81
Errors and omissions	–63.77	–7.90
Total change in reserves	–4.16	–6.71

33BANKING AND SECURITIES

In early 1981, the government-owned St. Lucia National Bank and the St. Lucia Development Bank were opened. There are seven commercial banks. The money supply, as measured by M2, was EC$220.6 million at the end of 1993. St. Lucia has no securities exchange.

34INSURANCE

As of 31 December 1985, there were 27 companies registered to transact insurance business in St. Lucia, of which 3 were local companies, 17 were from other Caribbean nations, and 7 were subsidiaries of UK, Canadian, and US firms.

35PUBLIC FINANCE

In 1992, current receipts amounted to $119.41 million while current expenditures totaled $89.2 million, with capital expenditures accounting for $27.8 million. Most of the income came from customs duties, income taxes, taxes on goods and services, and taxes on international trade and transactions. In 1992, St. Lucia's total external debt stood at $95.7 million, with outstanding debt equivalent to 24.2% of GDP.

36TAXATION

The most important taxes are income tax applied on a sliding scale (from 10% to 30% in 1993); a land and house tax, also based on a sliding scale; and a corporate tax of 33.3%.

37CUSTOMS AND DUTIES

Duties on imported goods such as alcoholic beverages, motor vehicles, cigarettes, and gasoline and oil continue to be major sources of government income. Most imports except those from CARICOM nations are subject to import duties, which consist of a customs duty, a consumption tax, and a service charge. To facilitate industrial development, the government has in specific cases negotiated the elimination of both import and export duties.

38FOREIGN INVESTMENT

Firms based in Canada, the US, the UK, other EC members, Venezuela, Hong Kong, and the Republic of Korea are the principal investors in St. Lucia. Two free-trade zones operate on the island.

In recent years, the government, through the St. Lucia National Development Corporation, has set up five industrial zones in order to attract foreign investment in manufacturing and assembly type operations. More recently, a Data Entry Park was built to attract information processing operations to St. Lucia. Development incentives are available in the form of tax concessions of up to 15 years in industries prescribed as beneficial to St. Lucia, namely in the manufacturing and tourist industries.

39ECONOMIC DEVELOPMENT

Since establishing the National Development Corp. in 1971, St. Lucia has succeeded in diversifying its economy. St. Lucia has the most highly developed infrastructure of all the Windward Islands, with an international airport capable of receiving jet planes, a highway system that connects the important coastal and agricultural areas with the political and commercial centers, and a fully automated telephone system with direct dialing to most parts of the world. Development assistance from OECD countries and multilateral agencies totaled US$24 million during 1982–85.

The Caribbean Development Bank projects that there will be continued real growth in output in the major productive sectors during the 1990s. The government will continue to accelerate its agricultural diversification program.

40SOCIAL DEVELOPMENT

In 1979, the government established the National Insurance Scheme to provide all workers from age 16 to 60 with old age, disability, survivor, sickness, and maternity coverage, as well as workers' compensation. Efforts have been made to improve the status of women, especially in employment. The government, regarding the population growth rate as excessive, began a family planning program within the Ministry of Health. The St. Lucia Planned Parenthood Association is affiliated with the International Planned Parenthood Federation. In 1984, the five-year Nutrition Education Program was established to combat malnutrition. As part of increased awareness of violence against women, battering and sexual harassment are included as curriculum topics in some secondary schools. The St. Lucia Crisis Center monitors abuse and acts as an advisor and advocate for women on a number of issues. In 1993, the center was working to establish a shelter for battered women and homeless girls.

41HEALTH

There were 5 hospitals on St. Lucia with 528 beds in 1988. The main hospital, with 233 beds, is located in Castries; others are located in Vieux Fort (where a 110-bed hospital is run by the Roman Catholic Church), Soufrière, and Dennery. In addition, there are over 27 health centers scattered throughout the island. In 1990 there were an estimated 2.7 hospital beds per 1,000 people. In 1991, there were 59 doctors and 213 nurses; and in 1988 there were 9 dentists. Total health care expenditures for 1990 were estimated at $46.7 million.

Malnutrition and intestinal difficulties are the main health problems. Tuberculosis, once widespread, has been brought under control (about 92 cases per 100,000 reported in 1990). The death rate in 1985 was estimated at 5.9 per 1,000 population. The infant mortality rate in 1992 was 17 per 1,000 live births, and there were 100 deaths of children under 5 years of age that year. In 1993, 97% of children were immunized against measles. The average life expectancy is 72 years.

42HOUSING

The demand for private ownership of homes far exceeds the supply. In 1980, 83% of dwellings were detached houses and 9% were apartments. Over 50% were owner occupied, 24% were rented privately, 9% were occupied rent free, and 1% were rented from the government. The majority of housing units (74%) were built of wood, with 12% of concrete and 10% of wood and concrete.

43EDUCATION

In 1993, the literacy rate was estimated at over 72%. Education, free and compulsory for 10 years, is allocated 25% of the total annual budget. Elementary schooling lasts for seven years followed by three years of secondary education at the first stage followed by another two years. In 1988 there were 88 primary schools with 1,137 teaching staff and 33,148 students enrolled. At the general secondary level, there were 352 teachers and 6,391 students. An education complex in Castries maintains a teacher-training center, a technical school, a secretarial training center, and a branch of the University of the West Indies. The Sir Arthur Lewis Community College is to be upgraded into a full-fledged campus of the University of the West Indies.

In 1987, there were 62 teachers and 389 students enrolled in institutions of higher learning.

44LIBRARIES AND MUSEUMS

The government provides free library service. The main library with 106,000 volumes is located in Castries, and other, smaller public libraries are located in villages throughout the island.

45MEDIA

Three newspapers are published in St. Lucia. The *Voice of St. Lucia* (circulation 8,000) appears on Wednesday and Saturday;

two weeklies, the *Crusader* (4,000) and the *Star*(5,000), are published on Saturday. The telephone system is fully automatic, with an estimated 20,000 telephones in service in 1993. In the same year there were 2 AM and 2 FM radio stations and 3 TV stations. In 1991 there were 103,000 radios and 26,000 TV sets in use. Television programs consist of some local programming, videotapes, and live broadcasts originating in Barbados; television transmissions from Martinique are also received.

46ORGANIZATIONS
There is a chamber of commerce, which meets in Castries. Other organizations include the St. Lucia Hotel Association, the St. Lucia Manufacturing Association, and the St. Lucia Historical Society.

47TOURISM, TRAVEL, AND RECREATION
Tourists come by air directly from Europe, Canada, and the US, and on cruise ships sailing through the West Indies out of North American and European ports. Some 165,987 tourists visited St. Lucia in 1991, 105,338 from the Americas and 58,763 from Europe. Some 5,104 beds are available in the island's 2,750 hotel rooms. The hotel occupancy rate in 1991 was 65.9%, and tourist expenditures totaled US$172 million. Dramatic tropical scenery, beautiful beaches, and excellent water-sports facilities are St. Lucia's principal tourist attractions. Of special interest are the Piton Mountains and the Sulphur Springs (the world's only drive-in volcano). Visitors, except citizens of the US, the UK, Canada and the Windward Islands, need a valid passport and may need a visa if their government has no exemption agreement with St. Lucia.

48FAMOUS ST. LUCIANS
John G.M. Compton (b.1926), trained as a barrister and one of the founders of the United Workers' Party, has been prime minister since 1982. The writer Derek Walcott (b.1930) is best known for his epic autobiographical poem *Another Life*.

49DEPENDENCIES
St. Lucia has no territories or colonies.

50BIBLIOGRAPHY
Breen, Henry Hegart. *St. Lucia: Historical, Statistical, and Descriptive*. London: F. Cass, 1970.
Eggleston, Hazel. *Saint Lucia Diary*. Greenwich, Conn.: Devin-Adair, 1977.
Romalis, Coleman. *Barbados and St. Lucia: A Comparative Analysis of Social and Economic Development in Two British West Indian Islands*. St. Louis, 1969.

ST. VINCENT AND THE GRENADINES

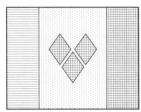

CAPITAL: Kingstown.

FLAG: Three vertical bands of blue, yellow, and green; centered on the yellow band are three green diamonds arranged in a v-pattern.

ANTHEM: *National Anthem,* beginning "St. Vincent! Land so beautiful."

MONETARY UNIT: The East Caribbean dollar (EC$) of 100 cents is the national currency. There are coins of 1, 2, 5, 10, and 25 cents and 1 dollar, and notes of 5, 10, 20, and 100 East Caribbean dollars. EC$1 = US$0.3704 (US$1 = EC$2.70).

WEIGHTS AND MEASURES: The imperial measures are used.

HOLIDAYS: New Year's Day, 1 January; Labor Day, 1 May; CARICOM Day, 5 July; Carnival, 6 July; Bank Holiday, 1st Monday in August; Independence Day, 27 October; Christmas Day, 25 December; Boxing Day, 26 December. Movable religious holidays include Good Friday, Easter Monday, and Whitmonday.

TIME: 8 AM = noon GMT.

¹LOCATION, SIZE, AND EXTENT

Located in the Windward Islands group of the Lesser Antilles, St. Vincent and the Grenadines is 34 km (21 mi) sw of St. Lucia and about 160 km (100 mi) w of Barbados. Scattered between St. Vincent and Grenada to the sw are more than 100 small islands called the Grenadines, half of which belong to St. Vincent and the other half to Grenada. The Grenadines belonging to St. Vincent include Union Island, Mayreau, Canouan, Mustique, Bequia, and many other uninhabited cays, rocks, and reefs. The land area of St. Vincent is 347 sq km (134 sq mi), and the coastline is 84 km (52 mi). Bequia, the largest of the Grenadines belonging to St. Vincent, has an area of 18 sq km (7 sq mi). The total land area of the country is 340 sq km (131 sq mi). Comparatively, the area occupied by St. Vincent and the Grenadines is slightly less than twice the size of Washington, D.C.

The capital city, Kingstown, is located on the southeast coast of the island of St. Vincent.

²TOPOGRAPHY

St. Vincent is a rugged island of volcanic formation, and the Grenadines are formed by a volcanic ridge running north–south between St. Vincent and Grenada. The highest peak on St. Vincent is Soufrière, an active volcano with an altitude of 1,234 m (4,048 ft); it has a crater lake 1.6 km (1 mi) wide. In the center of the island Richmond Peak rises to a height of 1,079 m (3,539 ft). Only 5% of the island's surface has slopes of less than 5°. The low-lying Grenadines have wide beaches and shallow bays and harbors, but most have no source of freshwater except rainfall. The highest point in the Grenadines is on Union Island, with an altitude of 308 m (1,010 ft).

³CLIMATE

The islands enjoy a pleasant tropical climate all year round, with a yearly average temperature of 26°C (79°F). The warmest month is September, with an average temperature of 27°C (81°F); the coolest is January, with an average temperature of 25°C (77°F).

The average yearly rainfall on St. Vincent is 231 cm (91 in), but in the mountainous areas the average rainfall is more than 380 cm (150 in) a year. May or June through December is the rainy season.

⁴FLORA AND FAUNA

The shallow waters of the Grenadines abound with marine life. Lobsters, conch, fish of all varieties, and turtles can be found in such areas as the Tobago Cays, which lie north of Prune (Palm) Island. Whales are frequently sighted off Petit Nevis, and large iguana can be found on some of the waterless rocks and cays.

In Kingstown, on St. Vincent, there is a famous botanical garden where the breadfruit tree was introduced to the West Indies from South Pacific islands in 1793. Some of the many birds found in St. Vincent are the Caribbean eleania, the trembler, the bananaquit, and the Antillean crested hummingbird.

⁵ENVIRONMENT

The principal recurrent threat to the environment comes from the Soufrière volcano, which on 7 May 1902 erupted violently, destroying much of northern St. Vincent and claiming 1,565 lives. After another eruption, on 13 April 1979, the volcano remained active for weeks, spewing over much of the island a pall of volcanic ash, which covered mountains, forests, and plantation fields. Forests are threatened by farming development and use of wood for commercial purposes.

Pollution from pleasure yachts and other sources has seriously affected the eastern shorelines of all the major islands of the Grenadines. In Bequia's Admiralty Bay, the pollution is so severe that swimming is dangerous. Water resources are also limited. The main contributing factors are toxic chemicals used in farming and sewage. The nation's tourist trade increases the need for more water. The nation's coast is particularly vulnerable to pollution from industrial sources.

The central highlands of St. Vincent have been set aside as a natural preservation area wherein may nest the St. Vincent parrot, the St. Vincent wren, and the St. Vincent solitaire, endangered or rare species. In the Grenadines, the hawksbill, green sea, and leatherback turtles have been declared endangered. The Tobago Cays have been proposed as a nature preserve, but aside from a few sites on Union Island there were, as of 1987, no protected areas in the Grenadines belonging to St. Vincent.

⁶POPULATION

St. Vincent and the Grenadines had, as of 1991, a population of 107,598, of whom 98,842 lived on the main island and 8,756 lived in the Grenadines belonging to St. Vincent. The population of Kingstown, the capital, was 15,670, with another 10,872 in the suburbs. The annual population growth rate averaged 2.34% a year from 1985 to 1990. Overall, the islands had a population density in 1991 of 277 per sq km (717 per sq mi).

⁷MIGRATION

Because of emigration, actual population growth has not kept pace with the natural growth. Although no reliable statistics are available, emigration is known to take place to Trinidad, Guyana, Guadeloupe, and Martinique. In the past, the UK and US also accepted substantial numbers of migrants, and Canada is still receiving arrivals from the islands.

⁸ETHNIC GROUPS

About 65% of the islanders are descendants of slaves brought from Africa. About 20% of the population is of mixed origin, and a small minority (3.5%) is of European descent. In the second half of the 19th century, 2,472 indentured laborers from Asia were brought to St. Vincent; their descendants, making up about 5.5% of the current population, are known as East Indians. About 2% of the people are Amerindians. Of the mixed group, about 1,000 persons, identified as Black Caribs, descend from the intermingling of Amerindians and Africans that occurred before European colonization.

⁹LANGUAGES

English is the official language of St. Vincent and the Grenadines. Some islanders speak a French patois, representing a mixture of African and French grammar, with a vocabulary drawing mostly upon French, along with some English and a few Spanish words. A few islanders speak French as their first language.

¹⁰RELIGIONS

The majority of the population is Protestant, about 36% Anglican (in 1985) and 40% members of other Protestant churches, but there is a significant Roman Catholic minority (10%). Members of the East Indian community profess either the Hindu or the Muslim religion.

¹¹TRANSPORTATION

St. Vincent is on the main air routes of the Caribbean, with direct flights to Trinidad and Barbados as well as the other islands to the north. The international airport is located on the southern tip of the island, near Kingstown; there is another, much smaller airport located on the east coast, north of Georgetown. In 1991, the construction industry and the infrastructure were given a minor boost with the government's announcement of an $18.5 million airport improvement and road construction and upgrading program. Small airports are located on Union, Canouan, and Mustique islands.

All of the Grenadines have excellent harbors served by a ferry service operating out of Kingstown. Until recently, the harbor at Kingstown lacked a deepwater pier, but wharf facilities were enlarged in the early 1980s, with financial support from the US. Although the main road of St. Vincent, going down the east coast and up the west coast, does not encircle the island, it does connect all the main towns with the capital. As of 1991, the islands had about 300 km (185 mi) of all-weather roads, 400 km (250 mi) of otherwise improved roads, and 300 km (185 mi) of tracks and byways. About 7,000 vehicles were registered in 1992, including 5,000 passenger cars. There is approximately one vehicle for every 16 residents.

¹²HISTORY

The Arawak Amerindians, who migrated from South America, are the earliest known inhabitants of St. Vincent and the Grenadines. Subsequently, the Caribs took control of the islands, and were there when Christopher Columbus reached St. Vincent on 22 January 1498.

St. Vincent was one of the last of the West Indies to be settled. Left to the Carib Amerindians by British and French agreement in 1660, the island continued to have a sizable Amerindian population until the first quarter of the 18th century. One of the results of this isolation from European influence was the evolution of the Black Caribs, who descend from the intermarriage of runaway or shipwrecked slaves with the Amerindians. The island was taken formally by the British in 1763, who ruled thereafter, except from 1779 to 1783 when it was in the hands of the French.

The island changed its ethnic character during the next century. When the Black Caribs and the remaining Amerindians rebelled against the British in 1795 at French instigation, most of the defeated insurgents were removed to the Bay of Honduras. Those who remained were decimated by an eruption of Soufrière in 1812. They were supplanted by African slaves, who were freed in 1834, Madeiran Portuguese, who immigrated in 1848 because of a labor shortage, and Asian indentured laborers who arrived in the latter half of the 19th century.

St. Vincent was administered as a crown colony within the Windward Islands group from 1833 until 1960, when it became a separate administrative unit linked with the Federation of the West Indies. The federation fell apart in 1962, and after lengthy discussion, St. Vincent became a self-governing state in association with the UK seven years later. On 27 October 1979, St. Vincent and the Grenadines achieved full independence as a member of the Commonwealth.

During the first months of independence, the young nation faced a rebellion on Union Island, its southernmost constituent, by a group of Rastafarians attempting to secede. The revolt was put down with military support from neighboring Barbados. In the end, 1 person was killed and 40 arrested. Otherwise, the political system has had few disruptions. The government at independence under the St. Vincent Labor Party gave way to the New Democratic Party (NDP) in 1984, with the NDP renewing its government in 1989.

¹³GOVERNMENT

When the nation became independent in 1979, it kept the then British monarch as the nominal head of government, represented by a governor-general. Executive power is in the hands of the prime minister and cabinet, who are members of the majority party in the legislature. The legislature is unicameral, a 21-seat House of Assembly. The House of Assembly consists of representatives elected from each of 15 constituencies for a maximum of five years, plus 6 senators appointed by the governor-general.

¹⁴POLITICAL PARTIES

There are two major parties and three minor parties on the islands. The majority party is the New Democratic Party (NDP), led by Prime Minister James FitzAllen Mitchell. Founded in 1975, the party has, since 1984, held a parliamentary majority. Currently, it holds all 15 elected seats in the House of Assembly.

The St. Vincent Labour Party (SVLP) was founded in 1955 and was in power at independence and governed the nation, under Robert Milton Cato, until the July 1984 elections. The SVLP held three elective seats until losing them in 1989. The current party leader is Stanley John.

A left-wing coalition, the United People's Movement, embraces the Democratic Freedom Movement, the Youyou

United Liberation Movement, and Arwee, a rural group. The Movement for National Unity, a moderate left-wing party, was founded in 1982 after the United People's Movement split. The National Reform Party, led by Joel Miguel, rounds out the list of these parties, which have never won an elected Assembly seat.

15 LOCAL GOVERNMENT
In an attempt to decentralize the government, the nation has been subdivided into eight local districts, two of which cover the Grenadines.

16 JUDICIAL SYSTEM
The islands are divided into three judicial districts, each with its own magistrate's court. Appeals may be carried to the East Caribbean Supreme Court, based in St. Lucia, and made up of the Court of Appeal and the High Court. In exceptional cases, appeals can be brought to the Judicial Committee of the UK Privy Council.

The constitution guarantees a public trial before an independent and impartial court. Legal counsel is afforded to indigent defendants in cases involving capital punishment. There are no separate security or military courts.

17 ARMED FORCES
There are no armed forces except those of the police department, a force of 489. A regional defense pact, including Antigua and Barbuda, Barbados, Dominica, Grenada, Jamaica, St. Kitts and Nevis, and St. Lucia, as well as St. Vincent and the Grenadines, provides for joint coast-guard operations, military exercises, and disaster contingency plans.

18 INTERNATIONAL COOPERATION
St. Vincent and the Grenadines became a member of the UN on 16 September 1980 and belongs to ECLAC, FAO, IBRD, ICAO, IDA, IMF, IMO, UNESCO, UPU, and WHO. As a special member of the Commonwealth of Nations, it is not represented at meetings of the heads of government. St. Vincent and the Grenadines also belongs to the Caribbean Development Bank, CARICOM, G-77, the OAS, the OECS, and the Windward Islands Banana Association. It is a signatory to the Law of the Sea and a de facto adherent to GATT.

19 ECONOMY
Agriculture is the mainstay of the economy of St. Vincent and the Grenadines, with bananas as the primary cash crop. Vegetable production for export grew significantly in the mid-1980s, as the market for arrowroot declined. Tourism is particularly important in the Grenadines, where yachting is a principal pastime. Some of the smaller cays have been wholly acquired by private interests and developed into resorts for European and North American visitors. Some industrial development has begun in St. Vincent.

In recent years, tourism and manufacturing have been expanding steadily, contributing more to the country's growth. As of 1994, St. Vincent and the Grenadines continues to rely significantly on agriculture for its economic progress—agriculture grows an average of 6.7% per year. The GDP grew by 3.6% in 1992, based primarily on increased banana production, which continues to dominate the agricultural sector.

In 1992, banana production increased by 27.2% to 82,431 tons. The Ministry of Agriculture has been implementing a crop diversification program at a cost of $100,000. Also, the Canadian Fisheries Development Project will further develop St. Vincent's fishing sub-sector by providing training for fishing improvement. This is in response to the remarkable increase in the recorded level of the size of the fish catch in last years.

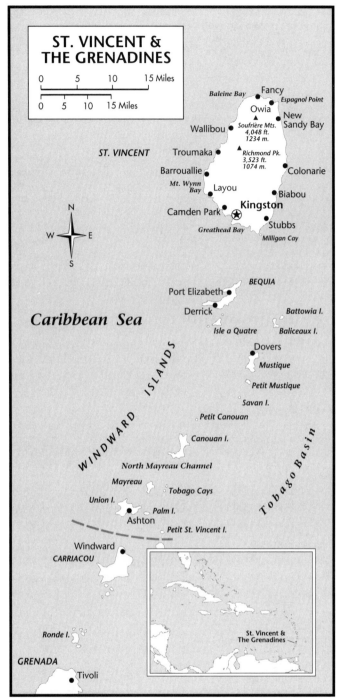

LOCATION: St. Vincent—13°6′ to 14°35′ N; 61°6′ to 61°20′ W.
TOTAL COASTLINE: 84 km (52 mi). TERRITORIAL SEA LIMIT: 12 mi.

20 INCOME
In 1992, the GNP was $217 million at current prices, or $1,990 per capita. For the period 1985–92, the average inflation rate was 4.3%, resulting in a real growth rate in per capita GNP of 4.7%. In 1992, the GDP was $171 million in current US dollars.

21 LABOR
Some 37,800 persons make up the work force. Unemployment was 30% in 1991. One of the first authentic labor unions in the West Indies was formed in St. Vincent in 1935, during the Great Depression. From this initial Workingman's Association, the labor movement in St. Vincent has developed unions for agricultural

workers, dockworkers, civil servants, and teachers. In all, the nation has some six labor unions. There is some poaching of members among competing trade unions, and the labor movement is gradually losing support. As of 1992, only slightly more than 10% of the labor force was unionized.

22AGRICULTURE

About half of St. Vincent is devoted to crop growing. Real growth in agriculture has been averaging 6.7% annually. Bananas constitute the main crop; vegetables, coconut, spices, and sugar are also important. Banana production was adversely affected by the eruption of Soufrière, which reduced exports from 30,414 tons in 1978 to 22,692 tons in 1979. Further damage was done by two hurricanes in 1979 and 1980; in the latter year, some 95% of the crop was destroyed. Production rebounded during 1981, and 83,000 tons were produced in 1982; in 1992, 76,095 tons were exported. Other crops in 1992 included coconuts, 23,000 tons; sugar cane, 2,000 tons; sweet potatoes, 2,000 tons; and plantains, 3,000 tons. Most of the agricultural products are grown on small farms; quality control is sometimes a problem, particularly in the production of bananas. There are approximately 5,000 banana growers on the island.

23ANIMAL HUSBANDRY

Rough estimates of the livestock population in 1992 include 6,000 head of cattle, 9,000 hogs, 12,000 sheep, 6,000 goats, and 100,000 poultry of all types. The island of St. Vincent does not produce enough meat, poultry, eggs, and milk to satisfy local demand.

24FISHING

At one time, St. Vincent and Bequia were the centers for a thriving whaling industry, but only six humpback whales were captured from 1982 to 1991. Since the New Kingston Fish Market opened in the late 1980s, the fish catch has rapidly increased. In 1991, the total catch amounted to 7,665 tons, up from 703 tons in 1987. Technical assistance and training to fisherman and fisheries staff is being sponsored by the Canadian Fisheries Development Project.

25FORESTRY

There is virtually no commercial forestry. Some local timber is used for residential and boat construction.

26MINING

There is no commercial mining. Some sand is extracted for local construction projects, and on some of the smaller and drier Grenadines salt is produced on a small scale for local consumption.

27ENERGY AND POWER

The electric power facilities of St. Vincent and the Grenadines are being expanded and improved to meet the growing industrial demand. Two hydroelectric plants provide 76% of the electricity generated. In 1991, total power generation amounted to 51 million kwh.

28INDUSTRY

A substantial amount of industrial activity centers on the processing of agricultural products. Because of depression in the sugar industry, sugar-processing facilities established in 1981/82 were shut down in 1985; in the same year, a beer factory began production. Nonagricultural industries include several garment factories, a furniture factory, an electronics plant, and a corrugated cardboard box plant.

The tourism industry recorded a decline in the early 1990s after steady growth of 10.5% per year between 1985 and 1990. However, the government is presently putting more emphasis

than ever in the promotion of the country as an upscale tourist destination. A new airport on Union Island was opened in April 1993. In 1992, a new airport, financed by a grant from the EC, opened on the island of Bequia.

29SCIENCE AND TECHNOLOGY

St. Vincent seeks scientific expertise to deal with problems associated with its main agricultural products, especially bananas. The computer industry in the US and Europe has opened up a new demand for paper; with proper processing, arrowroot might be able to compete in this new market, but so far the technology is lacking. Experimentation on arrowroot waste as a source of biogas has also been undertaken. A National Council for Science and Technology was created late in 1981, but as of 1993 no national policy on science and technology had yet been formulated. St. Vincent also has an Appropriate Technology Center, established in 1982.

30DOMESTIC TRADE

Kingstown is the main commercial distribution center. Local produce markets exist in all the Grenadines and in the small villages on St. Vincent. Government offices on St. Vincent are open on weekdays from 8 AM to noon and from 1 to 4:00 PM. On Saturday afternoons, most shops are closed.

31FOREIGN TRADE

St. Vincent and the Grenadines is highly dependent on foreign trade. From the agricultural sector, bananas and vegetables are the major foreign exchange earners.

Banana export earning has been rising steadily in recent years, making it the most important foreign exchange earner. At year end 1992, banana exports increased by 18.5% to 76,095 tons, accounting for 53.3% of total exports.

Exports of goods totaled US$118.6 million in 1992, including bananas, eddoes, flour, and sweet potatoes. Imports, totaling US$173.4 million were composed of food, beverages, tobacco, machinery and equipment, and manufactured goods.

32BALANCE OF PAYMENTS

The islands have had an unfavorable balance of trade since the 1950s. Income from tourism, investments, and development assistance makes up the balance. In 1991, the current account deficit was estimated at 9% of GDP, while the overall balance of payments was estimated to have remained in equilibrium.

In 1991, merchandise exports totaled US$65.70 million and imports US$110.70 million. The merchandise trade balance was US$-45.00 million. The following table summarizes St. Vincent and Grenadines' balance of payments for 1990 and 1991 (in millions of US dollars):

	1990	1991
CURRENT ACCOUNT		
Goods, services, and income	-41.5	-40.4
Unrequited transfers	33.6	31.4
TOTALS	-7.9	-9.0
CAPITAL ACCOUNT		
Direct investment	7.7	9.7
Portfolio investment	—	—
Other long-term capital	3.7	2.8
Other short-term capital	-10.8	6.6
Reserves	-4.1	3.4
TOTALS	-3.5	-22.5
Errors and omissions	11.4	-13.5
Total change in reserves	-4.1	3.4

³³BANKING AND SECURITIES

As of 1985 there were six commercial banks operating in St. Vincent and the Grenadines, of which three were Canadian, one British, and two local. Of the three government banks, one is a cooperative bank, one a savings and loan, and the third an agricultural bank. At the end of 1993, commercial banks had total assets amounting to us$30.5 million, of which ec$257.7 million consisted of loans and advances. Demand deposits were ec$62.9 million, time deposits ec$279.3 million. The government has established arrangements for offshore banking corporations, with direct connections to Swiss banking facilities. There is no securities exchange.

³⁴INSURANCE

Local insurance companies are limited in scope and importance. Representatives of insurance corporations based in the UK, Canada, and the US operate on St. Vincent.

³⁵PUBLIC FINANCE

In 1991, recurrent receipts totaled $59.2 million, and recurrent expenditures amounted to $46.7 million. Capital expenditures that year came to $36.2 million. The current account surplus reached 8% of GDP in 1991 and the overall deficit was equivalent to 2% of GDP. Most of the government's income comes from customs duties and taxes. The leading categories of expenditures are education, public works, and health. In 1992, the external debt amounted to $67.6 million.

The following table shows actual revenues and expenditures for 1989 and 1990 in millions of US dollars.

	1989	1990
REVENUE AND GRANTS		
Tax revenue	110.2	118.5
Non-tax revenue	16.4	18.5
Capital revenue	–0.3	0.2
Grants	—	—
TOTALS	126.3	121.2
EXPENDITURES & LENDING MINUS REPAYMENTS		
General public service	29.8	37.5
Defense	8.3	8.7
Education	27.4	28.1
Health	17.1	23.5
Social security and welfare	3.5	3.5
Housing and community amenities	6.6	3.7
Recreation, cultural, and religious affairs	0.6	0.4
Economic affairs and services	37.4	38.3
Other expenditures	0.1	—
TOTALS	139.8	155.4

³⁶TAXATION

The government of St. Vincent and the Grenadines levies a progressive personal income tax (ranging in 1991 from 10% to 55%), a corporate tax of 45% of net income, inheritance taxes, and a social security contribution of 2.5% of gross salary up to a maximum of us$41.75 per month.

³⁷CUSTOMS AND DUTIES

By far the most important customs revenues are from import duties. There is a duty on exported goods, but the revenue earned is relatively small. By agreement with certain private corporations, the government waives customs duties on specific items in order to stimulate industrial development. Under an October 1992 CARICOM agreement, St. Vincent and the Grenadines are in the process of eliminating import licensing. St. Vincent has adopted CARICOM's Common External Tariff, which ranges from 5% to 45%.

³⁸FOREIGN INVESTMENT

The government has encouraged foreign investment by establishing industrial estates, including both factories and homes for laborers, as well as by offering favorable tax conditions for the investor. A large multinational company handles the marketing of most banana production, and a US firm has established a children's garment factory. The government has allowed the sale of some small islands of the Grenadines, notably Mustique, owned by a Scotsman, and Prune, bought by a US family and now called Palm Island by its entrepreneur owner. Petit St. Vincent and much of Canouan are being developed by the government for tourist investments.

St. Vincent is still eager to receive foreign investment, while providing special incentives and credit tailored to the needs of the investors. St. Vincent is eligible for trade benefits under the Lomé Convention (Europe), Caribcan (Canada), and the Caribbean Basin Initiative (US).

In its current efforts to promote tourism, the government has secured a us$300 million investment in Canouan, while other negotiations are planned.

³⁹ECONOMIC DEVELOPMENT

A 1982 loan from the Caribbean Development Bank was designed to stimulate and redirect agricultural production, support the tourist industry, and contribute to the creation of the infrastructure necessary for industrial development. Between 1982 and 1984, official development assistance from OECD countries was us$17 million.

⁴⁰SOCIAL DEVELOPMENT

In 1986, legislation established a social security system, replacing the provident fund that had been in existence since 1970. Workers contribute 2.5% of earnings, while employers pay 3% of payroll distributions. Benefits are provided for old age, disability, death, sickness, and maternity. Employers fund a compulsory workers' compensation program. The worker is eligible for a pension at age 60, or earlier if incapacitated. St. Vincent has an extensive program of community development, which stimulates the formation of cooperatives and self-help programs in the rural communities. A national family planning program has been introduced as part of the government's maternal and child welfare services. A food distribution program was initiated in 1984. A new law mandating that women receive equal pay for equal work went into effect in 1990. The penalty for rape is usually 10 or more years in prison.

⁴¹HEALTH

As of 1988, Kingstown had a general hospital with 204 beds There were 3 rural hospitals, one located on the east coast of St. Vincent, the second on the west coast, and the third on Bequia. There are also hospitals for the aged and mentally ill. Approximately 34 outpatient clinics provide medical care throughout the nation. In 1990 there were an estimated 2.7 hospital beds per 1,000 inhabitants. In 1991 there were 55 doctors, 27 pharmacists, 6 dentists, and 224 nurses.

Gastrointestinal diseases continue to be a problem, although they are less so than in the past. There were an estimated 90 cases of tuberculosis per 100,000 people reported in 1990. In 1992, 100% of the children were vaccinated against measles. The death rate in 1985 was 5.9 per 1,000 population. Infant mortality in 1992 was 20 per 1,000 live births and there were 100 deaths of children under 5 years old. Life expectancy at birth in 1992 was 71 years. Estimated health care expenditures were $46.7 million in 1990.

⁴²HOUSING

Among other efforts to eliminate substandard housing conditions, the government has undertaken housing renewal projects in both rural and urban areas and has sought to provide housing for workers on industrial estates. Another government program supplies building materials at low cost to working people. As of 1980, 89% of all dwellings were detached houses and 6% were apartments. In the same year, 72% of housing units were owner occupied, 16% were rented, and 9% were occupied rent free. Dwellings were constructed primarily of wood (37%), concrete (30%), wood and concrete (17%), and stone (12%).

43 EDUCATION

Primary education, which lasts for seven years, is free but not compulsory. There are 64 primary schools. In 1990, enrollment in primary schools was 22,030 with 1,119 teaching staff. In secondary schools the same year, there were 10,719 students and 431 teachers. Secondary education at the first stage is for five years followed by two more years at the second stage. The government-assisted School for Children with Special Needs serves handicapped students.

At the postsecondary level there are a teachers' training college, affiliated with the University of the West Indies, and a technical college. Adult education classes are offered by the Ministry of Education. Vocational training is available through the Department of Public Works, and agricultural training is offered by the Ministry of Agriculture. Literacy is estimated at 85%. In 1989, students at the university and all higher-level institutions numbered 677 with 96 teaching staff.

44 LIBRARIES AND MUSEUMS

The government maintains a free public library system. The main library is in Kingstown, and there are 15 branches located throughout St. Vincent and on the larger Grenadine islands. The small National Museum in Kingstown houses ancient Indian clay pottery. The Botanical Garden in Kingstown is open to the public.

45 MEDIA

The internal telephone system of St. Vincent and the Grenadines, fully automatic, is operated by Cable and Wireless (West Indies) Ltd., which also provides telegraph, telex, and international telephone services. There were 12,000 telephones on the islands in 1983. There is one weekly newspaper, the *Vincentian News*, appearing on Friday, with a circulation of 8,000. There are 2 AM radio stations, no FM stations, and 1 television station. In 1991 there were 75,000 radios and 15,000 television sets.

46 ORGANIZATIONS

A chamber of commerce meets in Kingstown. Among the important commercial organizations is the St. Vincent Hotel Association.

47 TOURISM, TRAVEL, AND RECREATION

Tourism is oriented toward yachting, with havens located on most of the Grenadines and also at Young Island, off the southern tip of St. Vincent. Posh resorts have been created on many of the smaller Grenadines, with villas and cottages built alongside small private beaches. There are a total of 1,164 hotel rooms, including 624 classified as apartments/guest houses/cottages and villas. The number of tourist arrivals totaled 51,629 in 1991, 36,088 from the Americas and 14,013 from Europe, and tourist expenditures totaled an estimated US$52.7 million.

48 FAMOUS ST. VINCENTIANS

Robert Milton Cato (b.1915), prime minister from independence until 1984, was one of the founders of the SVLP. James FitzAllen Mitchell (b.1931), prime minister since 1984, was one of the founders of the NDP. Sir Fred Albert Phillips (b.1918) is a specialist on constitutional and international law.

49 DEPENDENCIES

St. Vincent and the Grenadines has no territories or colonies.

50 BIBLIOGRAPHY

John, Sir Rupert. *Pioneers in Nation-building in a Caribbean Mini-state*. New York: United Nations Institute for Training and Research, 1979.

Potter, Robert B. *St. Vincent and the Grenadines*. Santa Barbara, Calif.: Clio, 1992.

Shephard, Charles. *An Historical Account of the Island of Saint Vincent*. London: F. Cass, 1971.

St. Vincent and the Grenadines. Washington, D.C.: U.S. Dept. of Commerce, International Trade Administration, n.d.

Young, Virginia Heyer. *Becoming West Indian: Culture, Self, and Nation in St. Vincent*. Washington, D.C.: Smithsonian Institution Press, 1993.

SURINAME

Republic of Suriname
Republiek Suriname

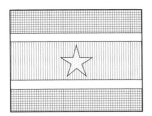

CAPITAL: Paramaribo.

FLAG: A yellow star is at the center of five stripes: a broad red band in the middle, two white bands, and a green stripe at the top and bottom.

ANTHEM: The *Surinaams Volkslied (National Anthem)* begins "God zij met ons Suriname" ("God be with our Suriname").

MONETARY UNIT: The Suriname guilder (Sf) is a paper currency of 100 cents. There are coins of 1, 5, 10, and 25 cents, and notes of 5, 10, 25, 100, and 500 guilders. Sf1 = $0.5602 (or $1 = Sf1.785).

WEIGHTS AND MEASURES: The metric system is used.

HOLIDAYS: New Year's Day, 1 January; Revolution Day, 25 February; Labor Day, 1 May; National Union Day, 1 July; Independence Day, 25 November; Christmas, 25 December; Boxing Day, 26 December. Movable religious holidays include Holi Phagwah, Good Friday, Easter Monday, and 'Id al-Fitr.

TIME: 8:30 AM = noon GMT.

¹LOCATION, SIZE, AND EXTENT

Situated on the northeast coast of South America, Suriname is the smallest independent country on the continent, with a total area of 163,270 sq km (63,039 sq mi). Comparatively, the area occupied by Suriname is slightly larger than the state of Georgia. The nation has an extension of 662 km (411 mi) NE–SW and 487 km (303 mi) SE–NW. Suriname is bordered on the N by the Atlantic Ocean, on the E by French Guiana, on the S by Brazil, and on the W by Guyana, with a total boundary length of 2,093 km (1,301 mi). Suriname also claims about 15,000 sq km (5,800 sq mi) of southeastern Guyana and some 5,000 sq km (1,900 sq mi) of southwestern French Guiana.

Suriname's capital city, Paramaribo, is located on the Atlantic coast.

²TOPOGRAPHY

Suriname is composed of thick forests, unexplored mountains, and swampy plains. Approximately 80% of the territory is classified as tropical rain forest. Several geologically old rivers, including the Maroni in the east and the Courantyne, flow northward to the Atlantic Ocean from the southern highlands near the Brazilian border; there, numerous rapids and waterfalls bar boat passage.

The coastal plain is flat and sometimes as much as 1.5 m (5 ft) below sea level, necessitating a system of sea defenses. The soils of the coastal plain are relatively fertile. A forest belt, 48–72 km (30–45 mi) wide, lies to the south, interspersed with grassy savannas. Farther south are jungle and higher ground.

³CLIMATE

The climate is tropical and moist. Daytime temperatures range from 28° to 32°C (82–90°F). At night the temperature drops as low as 21°C (70°F) because of the moderating influence of the northeast trade winds, which blow in from the sea all year. The annual rainfall in Paramaribo is about 230 cm (90 in). May to August is the main rainy season, with a lesser rainy season from November to February.

⁴FLORA AND FAUNA

Dominated by rain forest, Suriname contains many flowers but is most famous for water lilies and orchids. Tropical shrubs include hibiscus, bougainvillea, and oleander. There are over 184 species of mammals. Among the reptiles are the tortoise, iguana, caiman, and numerous snakes. Tropical birds abound, especially the white egret.

⁵ENVIRONMENT

In general, Suriname's environment and wildlife are protected from the destructive influences that threaten the majority of the world's nations. Suriname's eight nature reserves are managed by the Foundation for Nature Preservation, founded in 1969. The Suriname Wildlife Rangers Club, consisting mainly of students 15–20 years old, assists in various nature preservation activities. National responsibility for environmental matters is vested in the Ministry of Health and Environment and the Ministry of Natural Resources and Energy. The nation has 48.0 cubic miles of water. Eighty-nine percent is used for farming activities and 5% for industrial purposes. Eighteen percent of Suriname's city dwellers and 44% of all rural dwellers do not have pure water. Pollutants from the country's mining industry affect the purity of the water. Salinization of the water supply is becoming a problem for the coastal areas.

Due to the preservation of Suriname's tropical rain forest, the nation's wildlife flourishes. Eleven of the country's 184 mammal species and six of its 670 bird species are endangered. Sixty-eight types of plants are also endangered. Endangered species in Suriname include the tundra peregrine falcon, five species of turtle (South American river, green sea, hawksbill, olive ridley, and leatherback), the Caribbean manatee, and the spectacled caiman.

⁶POPULATION

According to the census of 1 July 1980, Suriname had a population of 352,041, a decrease of 7.6% from 1972, largely because of emigration to the Netherlands. The population in July 1993 was estimated at 416,321. The UN projected a population of

500,000 for the year 2000, based on a crude birthrate of 21.5 per 1,000, a crude death rate of 5.3, and a net natural increase of 16.2 during 1995–2000. Average estimated population density in 1991 was 2.5 per sq km (6.4 per sq mi); about 48% of the population was urban. Paramaribo, the capital, had a population of about 200,000 in 1990.

[7]MIGRATION

About 90,000 Surinamese resided in the Netherlands by the mid-1970s, and the number had reached some 200,000 by 1985. Emigration was about 12,000 per year in the early 1970s, but it accelerated as the date of independence approached and again after the coup of February 1980. An estimated 6,000 Surinamese fled to neighboring French Guiana by 1987, seeking refuge from a guerrilla conflict raging in the northeast. Some 1,500 were still living in camps there at the end of 1992.

[8]ETHNIC GROUPS

Suriname has one of the most cosmopolitan populations in the world. The two largest ethnic groups are the Creoles (about 35% of the population) and the Hindus (about 33%). The Bush Negroes or Bush Creoles (10%) are descended from Africans who were brought to Suriname to work as slaves on the plantations between 1650 and 1820. Other groups include the Javanese (about 16%), Chinese, and Europeans. The Amerindians (3%), Suriname's original inhabitants, include the Arawak, Carib, and Warrau groups along the riverbanks and coastal plains, and Trios, Akurios, and Wyanas along the upper reaches of the rivers.

[9]LANGUAGES

The official language is Dutch, but English is widely spoken, and the local people use a lingua franca known as Sranang-Tongo or Takki-Takki, a mixture of Dutch, African, and other languages. Hindi, Javanese, and several Chinese, Amerindian, and African languages and dialects are also spoken.

[10]RELIGIONS

The majority of the Asiatic peoples are Muslim (an estimated 19% in 1983). The Creole group is mostly Christian, about 20% Roman Catholic and 18% Protestant. Most Bush Creoles follow traditional religions, but a small proportion are either Roman Catholic or Moravian. The European sector includes small numbers of Jews, Roman Catholics, Lutherans, Moravians, and members of the Dutch Reformed Church.

[11]TRANSPORTATION

Suriname has 1,200 km (746 mi) of navigable waterways. A ferry service across the Corantijn River to Guyana began operating in 1990. There are 166 km (103 mi) of single-track railway, 86 km (53 mi) government owned and the rest industrial. Paramaribo can be reached from any town or village on the coastal plain by good all-weather roads. As of 1991, there were an estimated 8,300 km (5,158 mi) of roadways, of which 6% was paved. State-owned and private companies operate regular bus services, both local and long distance. In 1992, there were 35,000 passenger cars and 15,000 commercial vehicles. Three merchant ships were in service as of the end of 1991. Zanderij International Airport near Paramaribo can handle jet aircraft, and there are small airstrips throughout the interior. The government-owned Suriname Airways offers regularly scheduled service to the Netherlands and Curaçao.

Military operations involving the Jungle Commando and the national army have badly damaged Albina and the road connecting Moengo to the eastern border. Overall lack of proper maintenance on roads, canals, and port facilities has resulted in a degraded infrastructure and higher local transportation costs.

[12]HISTORY

Spaniards came to Suriname in the 16th century in search of gold, but did not stay when they found none. The first large-scale colonization took place under Francis, Lord Willoughby, the English governor of Barbados, who sent an expedition to Suriname in 1650 under Anthony Rowse. In 1660, the British crown granted Willoughby official rights, and it became a flourishing agricultural colony. Settlers included English colonists, African slaves, and Jewish immigrants from the Netherlands, Italy, and Brazil. In the Peace of Breda between England and the United Netherlands in 1667, Suriname became a Dutch colony.

The English held Suriname again between 1799 and 1802 and from 1804 to 1816, when the Dutch resumed control over the colony under the Treaty of Paris. With the final abolition of slavery in 1863, workers were imported from India, Java, and China. In 1954, a new Dutch statute provided for full autonomy for Suriname, except in foreign affairs and defense. A commission was set up on 5 January 1972 to prepare alternatives to the existing legal framework. In May 1974, the terms for Suriname's independence were agreed on, and Suriname became an independent country on 25 November 1975.

For five years, Suriname was a parliamentary republic under prime minister Henk Arron. On 25 February 1980, the government was overthrown in a military coup led by Désiré Bouterse. Parliament was dissolved and the constitution suspended, and in 1981 the new government declared itself a Socialist republic. Relations with the US strained as the Bouterse government moved closer to Cuba. In December 1982, as a result of the government's execution of 15 political opponents, the Netherlands and the US suspended all aid to Suriname.

The military and Bouterse continued to rule through a succession of nominally civilian governments. Still, pressure mounted for a return to genuine civilian rule. A separate challenge to the government came from a guerrilla movement under the leadership of Ronny Brunswijk. The Surinamese Liberation Army (SLA), also known as the Maroon or Bush Negro insurgency, began operating in the northeast in July 1986. It struck against various economic targets, including the Suriname Aluminium Company. The government responded with repression and the killing of civilians suspected of supporting the insurgency.

The military allowed for elections on 25 November 1987. An anti-Bouterse coalition, the Front for Democracy, won 80% of the vote and 40 of the 51 seats in the newly constituted National Assembly, but a new appointive State Council, rather than the elective National Assembly, was given law-making authority. The new president, Ramsewak Shankar lasted from 25 January 1988 until 24 December 1990. International pressure mounted, and the military soon relented, allowing for elections on 25 May 1991. Again, an anti-military coalition swept the election. The leader of the coalition, Ronald Venetiaan, was chosen president on 6 September 1991.

[13]GOVERNMENT

Between 1954 and 1975, Suriname was administered by a governor appointed by and representing the Dutch crown, with a cabinet appointed by the governor and an elected Parliament (Staten van Suriname). Under the constitution adopted by Parliament on 21 November 1975, Suriname is a republic. However, that constitution, which provided for a unicameral, 39-member Parliament directly elected for a four-year term by universal suffrage, was suspended on 15 August 1980 and Parliament was dissolved. Bouterse then ruled through a series of appointed governments, whose members represented the military, industry, trade unions, business, and political parties. In September 1987, a popular referendum approved a new constitution, which is still in effect.

The constitution provides for a unicameral 51-member National Assembly directly elected for a four-year term. The executive branch consists of the president, vice-president and prime minister, all selected by the legislature. There is also a cabinet and an appointed Council of State. The Council is a holdover from the Bouterse years. Bouterse still retains a good deal of political power within this system.

¹⁴POLITICAL PARTIES

Suriname's political parties tend to represent particular ethnic groups. The National Party of Suriname (NPS), led by President Ronald Venetiaan, draws support from the Creole population. The Progressive Reform Party (VHP), led by Jaggernath Lachmon, is East Indian, and the Indonesian Peasant's Party (KTPI), led by Willy Soemita, is more tied in name to its constituency. All three parties allied in the coalition National Front for Democracy in 1987 to defeat Bouterse's National Democratic Party. In 1991, these three parties and the Suriname Labor Party (SPA) formed the New Front (NF) and won a solid victory, gaining 30 of 51 Assembly seats, while Bouterse's NDP took 10 seats.

Another coalition formed during the 1991 elections is called Democratic Alternative '91. It includes four non-ethnic parties, representing a variety of white collar concerns. They took 9 of the remaining 11 seats in the Assembly, with the other two going to minor parties.

¹⁵LOCAL GOVERNMENT

The republic is divided into 10 districts, which include the urban district of Paramaribo. Administration is centralized, and there are no recognized municipalities.

¹⁶JUDICIAL SYSTEM

The Constitution provides the right to a fair public trial before a single judge, the right to counsel, and the right to appeal. There is a Supreme Court.

Military personnel fall under military jurisdiction and are generally not subject to civilian criminal law. Military courts follow the same procedural rules as do the civil courts with military trials held before a judge and two military personnel.

¹⁷ARMED FORCES

The Suriname National Army consists of army, air force, and naval components, with a strength of about 1,800 in 1993. No reliable fiscal estimates exist, but defense spending is below $100 million.

¹⁸INTERNATIONAL COOPERATION

Suriname was admitted to the UN on 4 December 1975 and is a member of ECLAC and all the nonregional specialized agencies except IAEA, IDA, and IFC. It also belongs to G-77, IDB, OAS, and PAHO. Suriname is a signatory of GATT and the Law of the Sea.

¹⁹ECONOMY

The bauxite industry has traditionally set the pace for Suriname's economy. Two companies, Suriname Aluminum Co. (Suralco), a wholly-owned Alcoa subsidiary, and Billiton, owned by Royal Dutch/Shell, account for about one-third of government revenue, nearly 4,000 workers, and 80% of commodity exports. In February 1987, guerrilla destruction of electricity pylons to the bauxite mines closed the industry while repairs were made.

Although agriculture is the chief means of subsistence, plantation agriculture is the weakest sector of the economy, with the notable exception of rice growing. Suriname is self-sufficient in rice, and exports large amounts; however, Suriname is a net food importer.

Next to bauxite, foreign aid is the mainstay of the country's economy. The collapse of bauxite exports in 1987 was a severe blow to the economy.

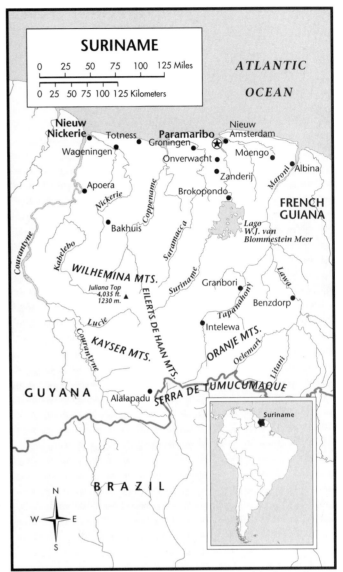

LOCATION: 2° to 6°N; 54° to 58°W. BOUNDARY LENGTHS: Atlantic coastline, 364 km (226 mi); French Guiana, 467 km (290 mi); Brazil, 593 km (369 mi); Guyana, 726 km (451 mi). TERRITORIAL SEA LIMIT: 12 mi.

Despite high expectations, the civilian government inaugurated in early 1988 proved unable to address the country's considerable economic problems and was overthrown by the military on 24 December 1990. A year later, civilian government, under the leadership of President Ronald Venetiaan, came back to power The new government inherited a nation with formidable problems. Foreign exchange reserves had reached a record low, inflation was soaring, earnings from main export sectors were falling drastically, unemployment was high, and climate for foreign investment was bad. The government began implementing of a structural adjustment program, which included the legalization of the parallel foreign exchange market, reduced government spending, privatization of key sectors of the economy, and revision of the country's investment code. In 1992, real GDP fell by 5% and average inflation accelerated to 44%, compared to 26% in 1991. Production of bauxite has not fully recovered and earnings continue to sag. Production of alumina, the current main export product and mainstay of the economy, declined by 9% in 1992. The agricultural sector is now the second-largest employer in Suriname, after the government. Agricultural exports account for 10% of export earnings. Rice is still Suriname's most important crop, grown on about half of all cultivated land. Output in 1992 increased substantially; however, it recorded a decline in 1993 mainly because of weather conditions.

²⁰INCOME

In 1992, the GNP was $1,728 million at current prices, or $3,700 per capita. For the period 1985–92 the average inflation rate was 14.7%, resulting in a real growth rate in per capita GNP of –3.2%.

It is estimated that in 1991 agriculture, hunting, forestry, and fishing contributed 11% to GDP; mining and quarrying, 3%; manufacturing, 8%; electricity, gas, and water, 5%; construction, 8%; wholesale and retail trade, 16%; transport, storage, and communication, 6%; finance, insurance, real estate, and business services, 20%; community, social, and personal services, 1%; and other sources, 22%.

²¹LABOR

Of the total working population (99,010), most are employed in the Paramaribo and Wanica regions, where about 40% are employed in services, 14% in commerce, 3% in agriculture, 9% in manufacturing, 4% in construction, 5% in transportation and communications; 2% in mining, and 13% in other sectors. Overall, agriculture accounts for about one-third of national civilian employment. About 33% of the total labor force was unemployed in 1990.

Suriname has numerous small unions, representing individual workplaces or enterprises, organized into four union federations: the General Confederation of Trade Unions, sometimes called the Moederbond (Mother Union); the Progressive Workers Organization, whose members are predominantly from the commercial and banking sectors; the Centrale 47, which includes bauxite and sugar unions; and the Central Organization for Civil Service Employees. The government destroyed the office of the Moederbond in late 1982.

²²AGRICULTURE

The chief crops are rice, sugar, plantains and bananas, citrus fruits, coffee, coconuts, and palm oil, in addition to staple food crops. With the exception of rice, the main export crop, plantation agriculture has suffered the consequences of absentee ownership. Rice production was 238,000 tons in 1991. In 1992, the rice output (aided enough by excellent weather to overcome the shortage of inputs and a deteriorating irrigation infrastructure) helped to offset a 5% drop in GDP. Sugar production dropped so substantially in the 1980s that imports were required to meet local demand. Under union pressure, the government in early 1987 agreed to a national sugar plan to improve machinery and housing, and to create employment. Production of sugar in 1992 was 45,000 tons; of bananas, 49,000 tons; of palm oil, 1,600,000 tons; of oranges, 11,000 tons; and of coconuts, 13,000 tons.

Since its creation in 1945, the Commission for the Application of Mechanized Techniques to Agriculture in Suriname has worked to reactivate several old plantations and bring new land under cultivation. The successful control of diseases and pests, introduction of water storage and irrigation schemes, and the development of new quick-growing varieties of rice have also increased total agricultural production.

²³ANIMAL HUSBANDRY

Livestock numbers are relatively small, since breeding is done primarily by small farmers who own only a few animals each. The government has tried to reduce the import of eggs, dairy products, and meat by undertaking projects to cross Dutch and local breeds of cattle and poultry. Estimated livestock numbers in 1992 included 95,000 head of cattle, 31,000 hogs, 18,000 goats and sheep, and 9,000,000 chickens.

²⁴FISHING

Fishing has become increasingly important, both on inland waterways and at sea. The chief commercial catch is shrimp, which is exported. In 1991, shrimp production was 790 tons, and the fish catch was about 4,100 tons. The Fisheries Service, founded in 1947, has worked to develop the fishing industry. Export of fish and fish products in 1992 amounted to nearly $4.5 million. Japan is the largest market for Surinamese shrimp.

²⁵FORESTRY

Approximately 95% of Suriname is covered by tropical rain forest, but existing forest resources have scarcely been touched. Initial exploitation has been confined to the more accessible strips along the riverbanks. The Suriname Forestry Service, under an FAO technical assistance program, has undertaken to survey and open up the forests for commercial use. Roundwood production was about 140,000 cu m in 1991. In August 1992, a peace agreement between the central government and insurgent groups from the interior (where timber is found) was signed. Since the fighting ended, logging has increased.

²⁶MINING

Bauxite and alumina have accounted for 80% of foreign exchange earnings in recent years. Suriname is the sixth-largest producer of bauxite and the fifth-largest producer of alumina in the world. The total quantity of bauxite mined in 1991 was some 3.2 million tons (down from 6.9 million tons in 1973). Suriname's bauxite industry has suffered in recent years from a weak market, foreign competition, and the effects of the guerrilla war, but new mines with higher grade bauxite should be replacing older depleted mines by 1995. The alumina industry, however, is threatened by the deterioration of the international alumina market.

Gold has been mined in Suriname since the second half of the 19th century. By 1991, however, mine production had fallen to 30 kg.

²⁷ENERGY AND POWER

Electricity, primarily for industry, is supplied by the 189-Mw hydroelectric power station at Afobaka, south of Brokopondo. Annual production at the plant, which was built by Alcoa, is owned and operated by the Suriname Aluminum Company (Suralco). Suriname's energy sector and bauxite industry are closely linked. Suralco also owns and operates a 47 Mw oil-fired turbine plant. Total installed generating capacity was 458 Mw in 1991; total production was 1,400 million kwh.

Gas and electricity for the city of Paramaribo are supplied by the Overseas Gas and Electric Co. Electric service throughout the rest of the country is the responsibility of the Ministry of Local Government and Decentralization. In 1990, the government-owned oil company, Staatsolie, increased production by 2.8% to 4,500 barrels per day; production by the mid-1990s was expected to expand to 6,500 barrels per day. In 1992, Staatsolie completed construction of a 60-km (37-mi) pipeline between the oil fields and distribution facilities. The Dutch government also has promised $15 million toward the construction of a new refinery. Staatsolie is also seeking joint venture partners to conduct exploration and eventual exploitation of oil reserves in western Suriname.

²⁸INDUSTRY

The major industries are mining and food processing. The bauxite industry over the years has developed into a complex of factories, workshops, power stations, laboratories, hospitals, recreational facilities, residential areas, and sports grounds. The output of refined aluminum was 28,785 tons in 1985. During the same year, Surinamese factories produced 79,480 tons of cement, 253,313 pairs of shoes, 30 million liters of soft drinks, and 14 million liters of beer.

The long term future of the mining industry depends on the companies' ability to keep production costs low and competitive and on the consolidation of peace in the country's interior. So far, the bauxite companies have reached a new agreement with the government which would give them a more favorable exchange rate. In October 1992, Golden Star negotiated an agreement with the Surinamese government for a feasibility study on possible exploitation of a large gold concession. There are also large kaolin deposits located underneath the country's bauxite which remain unexploited. Several firms have shown interest in the kaolin concessions.

29SCIENCE AND TECHNOLOGY

Research centers and scientific societies in Suriname include the Center for Agricultural Research, Geological Mining Service, and the Agricultural Experiment Station of the Ministry of Agriculture, Animal Husbandry and Fisheries, all in Paramaribo.

30DOMESTIC TRADE

There are a few supermarkets and department stores, but most urban trade is conducted in small shops. Business hours are, Monday–Saturday, 7:30 AM to 1 PM and from 4 to 6 or 7 PM. Banks are open weekdays from 6 AM to 12:30 PM, and on Saturdays 8–11 AM. Most trade in rural areas is conducted in open markets. Advertisements appear on radio and television and in the newspapers.

31FOREIGN TRADE

In 1992, exports totaled $417 million and consisted mainly of alumina, aluminum, bauxite, rice, shrimp, wood products, and bananas. Imports registered $417 million, including capital equipment, fuel and lubricating oils, cotton, flour, meat, and dairy products, raw materials and semi-manufactured goods, refined petroleum products, machinery and transport equipment, foodstuffs and consumer goods. Suriname exported mainly to Norway, Netherlands, the US, Germany, France, Brazil, and Venezuela. It imported mostly from the US, Netherlands, Trinidad and Tobago, Netherlands Antilles, Brazil, the UK, and Venezuela.

32BALANCE OF PAYMENTS

Suriname runs a persistent deficit on current accounts, which has generally been offset by a surplus on capital accounts deriving from development aid, mainly from the Netherlands. Remittances from some 200,000 Surinamese expatriates in the Netherlands are not apparent in the balance of payments because they are usually exchanged in the parallel market.

In 1989, lower profitability in the bauxite sector led to a decline in remittances which was not offset by investment inflows, resulting in the first capital account surplus in many years. Foreign exchange reserves grew by 110%; from $10 million in 1989 to $21 million in 1990. Unfortunately, these reserves were squandered by the interim government during its nine month rule and had fallen close to zero by July 1993.

In 1992, merchandise exports totaled $341.0 million and imports $–272.5 million. The merchandise trade balance was $–68.5 million.

33BANKING AND SECURITIES

The Central Bank of Suriname has acted as a bank of issue since 1 April 1957. Other banks include the Suriname People's Credit Bank, Post Office Savings Bank, Agricultural Bank, and National Development Bank.

34INSURANCE

Both Dutch and foreign insurance companies operate in Suriname.

35PUBLIC FINANCE

For years the Suriname budget operated with a deficit. However, some relief was achieved by absorbing the arrears of principal and interest on Dutch loans into the second five-year plan (1972–76), and government revenues increased following the introduction of a new bauxite levy in 1974. In 1979 and 1980 there were budget surpluses. During the 1980s, however, the military regime in power increased government intervention and participation in the economy, causing pubic employment and budget deficits to soar. The return to civilian government on 16 September 1991 enticed the Dutch government to resume its development aid program, which could annually amount to $200 million until 1996. Reforms enacted include the reduction of deficit spending, the renunciation of monetary creation as a means of financing deficits, and the deregulation of trade and business licensing systems. Adjusting for the inconsistent accounting practices of the Surinamese government, the IDB estimated that the true budget deficit in 1990 was $249 million or 16.8% of GDP (22% including arrears). According to the 1993 budget, the deficit was about 18% of GDP. The external public debt in 1991 stood at $326.4 million.

36TAXATION

Direct taxes provide only a small portion of governmental revenues, and payments have been in arrears for years. By far the greatest tax sources are the bauxite-related industries. Companies are taxed on the sum of all net profits, at progressive rates. A graduated income tax is assessed on net income.

37CUSTOMS AND DUTIES

There is a single schedule of tariffs, ranging mostly from 20% to 35% ad valorem, and most imports are subject to duty. Duty-free import of various goods is offered by the National Development Incentives program. The government has offered full and partial exemption from import duties to new businesses.

38FOREIGN INVESTMENT

Since 1973, the Suriname government has pursued a policy of joint or mixed capital ventures, offering public lands, concessionaire rights, tax write-offs, and other investment inducements. This policy was under review in the early and mid-1980s.

In an effort to attract new investment, the government has been revising the investment code, which creates an investment climate uncompetitive in comparison with that existing in most neighboring countries. Economic and business relations with the US are very important to this nation. American investment in Suriname exceeds $2 billion in replacement value, of which over 95% is accounted for by Suralco, the bauxite company. Other US firms operating in Suriname include Exxon, Texaco, IBM, and the insurance firm Alico.

In addition, Suriname is highly dependent on relations with the Dutch government. Dutch aid was suspended following the December 1990 military coup. However, the Dutch provided approximately $6 million in balance of payments support each month from October 1992 to January 1993. This aid, which is scheduled to continue for a few years, is intended for large-scale development projects involving the country's vast store of valuable natural resources, as well as for economic and social restructuring.

39ECONOMIC DEVELOPMENT

In wholesale, retail, and foreign trade, the government has been highly interventionist. Quota restrictions or outright bans on many imported items considered nonessential or in competition with local products have been announced. The government has forced price rollbacks on domestic items and imposed price controls on essential imports, resulting in some shortages. Although the government was Socialist in principle since 1981, it refrained from nationalizing Suriname's key industries, although it did increase its participation in them. The Action Program announced by the government in May 1982 called for the encouragement of small-scale industry, establishment of industrial parks, development of rural electrification and water supply projects, liberalization of land distribution, and worker participation in management of government enterprises.

In 1975, the Netherlands promised Suriname $110 million annually in grants and loans, for a period of 10–15 years. This aid program and $1.5 million in aid authorized by the US in September 1982 were suspended following the killings of prominent Surinamese in December 1982. In 1983, Brazil and Suriname reached agreement on a trade and aid package, reportedly underwritten by the US. By 1986, Suriname had signed trade agreements with several countries, among them the Netherlands.

As of 1994, Suriname is undergoing a comprehensive structural adjustment program (SAP) . This program, recommended by the EC, was designed to establish the conditions for sustained growth of output and employment with relative stability of prices, a viable balance of payments, and protection of the low-income population. However, as of the end of 1993, none of these measures had been completed. Enacting the full SAP in its proper sequence could begin to arrest the accelerating deterioration of the Surinamese economy and living conditions.

40SOCIAL DEVELOPMENT
Organized welfare programs are conducted largely by private initiative, through ethnic or religious associations. However, the government has begun to establish a social welfare system designed eventually to include a free national health service. The fertility rate in 1991 was 2.8.

The Home for Women in Crisis Situations, a private foundation, is consistently oversubscribed, indicating a national problem in the area of violence against women. The Home has doubled its capacity with private support but has not been able to obtain government funding.

41HEALTH
Tuberculosis, malaria, and syphilis, once the chief causes of death, have been controlled. In 1990, there were an estimated 92 cases of tuberculosis per 100,000 people reported. In 1992, 84% of the country's children were vaccinated against measles.

In 1993, Suriname's birth rate was 25.6 per 1,000 people. In 1992, there were 11,400 births, and average life expectancy was 70 years. The infant mortality rate was 28 per 1,000 live births in 1992, and there were 400 deaths of children under 5 years of age (a rate of 35 per 1,000). Overall mortality was 5.6 per 1,000 in 1993. In 1992, there were 1.25 doctors per 1,000 people, with a nurse to doctor ratio of 0.5. In 1990, there were an estimated 2.7 hospital beds per 1,000 people and estimated health care expenditures of $46.7 million.

42HOUSING
Housing programs are supervised by the Department of Social Affairs. As of 1 July 1980 there were 68,141 inhabited houses in Suriname and 8,208 huts; 82.3% of living quarters had electricity. Between 1988 and 1990, 82% of the urban and 94% of the rural population had access to a public water supply, while 64% of urban dwellers and 36% of rural dwellers had sanitation services.

43EDUCATION
In 1988, there were 301 elementary schools, with 2,921 teachers and 65,798 pupils, secondary schools had 34,248 pupils. Education is compulsory and free for all children aged 6 through 12.

While primary education lasts for six years, secondary education has two phases—four years followed by three years. Free primary education is offered by the government and by Roman Catholic and Protestant mission schools. The adult literacy rate in 1990 was 94.9%. The official school language is Dutch.

Higher education includes four teacher-training colleges, the Technical College, and the University of Suriname, with its Law School and a Medical Science Institute. In 1990, all higher level institutions reported 495 teaching staff with 4,319 students enrolled.

44LIBRARIES AND MUSEUMS
The main public library is at the Suriname Cultural Center in Paramaribo, with 225,000 volumes, seven branches, and two bookmobiles. The Suriname State Museum, established in 1947, is the country's only noted museum.

45MEDIA
Nearly all the towns and villages have telephone connections. There were 27,500 telephones in 1993. As of the same year there were 6 AM radio stations, 14 FM stations and 6 state-run television stations. In 1991, 274,000 radios and 56,000 television receivers were in use. There are two daily newspapers, the Dutch-language *De Ware Tijd* (circulation 30,000), and *Dagblad de West* (circulation 15,0000).

46ORGANIZATIONS
International organizations in Suriname include the YWCA, Girl Scouts, Boy Scouts, and Red Cross. The Bouterse government established the National Women's Organization and the Suriname Youth Union.

47TOURISM, TRAVEL, AND RECREATION
Most visitors need a valid passport and an onward ticket, but no visa. Tourism, a modest source of revenue during the 1970s, has fallen drastically since the 1980 coup. In 1991, there were 30,000 tourist arrivals. Hotel rooms numbered 1,550, with 3,900 beds, and tourism receipts totaled US$11 million.

48FAMOUS SURINAMESE
Lt. Col. Désiré ("Dési") Bouterse (b.1945) led the coup of February 1980.

49DEPENDENCIES
Suriname has no territories or colonies.

50BIBLIOGRAPHY
Brown, Enid. *Suriname and the Netherlands Antilles: An Annotated English-Language Bibliography.* Metuchen, N.J.: Scarecrow Press, 1992.

Chin, Henk E. *Surinam: Politics, Economics, and Society.* New York: F. Pinter, 1987.

Goslinga, Cornelis C. *A Short History of the Netherlands Antilles and Surinam.* Norwell, Mass.: Kluwer Academic Press, 1978.

Hoefte, Rosemarijn. *Suriname.* Santa Barbara, Calif.: Clio Press, 1990.

Hoogbergen, Wim S. M. *The Boni Maroon Wars in Suriname.* New York: Brill, 1990.

Price, Richard (ed.). *Maroon Societies.* New York: Anchor Press, 1973.

Sedoc-Dahlberg, Betty (ed.). *The Dutch Caribbean: Prospects for Democracy.* New York: Gordon and Breach, 1990.

Wooding, Charles J. *Evolving Culture: A Cross-cultural Study of Suriname, West Africa, and the Caribbean.* Washington, D.C.: University Press of America, 1981.

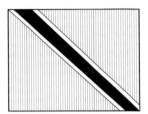

TRINIDAD AND TOBAGO

Republic of Trinidad and Tobago

CAPITAL: Port-of-Spain.

FLAG: On a red field, a black diagonal stripe with a narrow white border on either side extends from top left to bottom right.

ANTHEM: Begins, "Forged from the love of liberty, in the fires of hope and prayer."

MONETARY UNIT: The Trinidad and Tobago dollar (TT$) is a paper currency of 100 cents. There are coins of 1, 5, 10, 25, and 50 cents, and 1 dollar, and notes of 1, 5, 10, 20, and 100 dollars. TT$1 = US$0.1720 (US$1 = TT$5.8135).

WEIGHTS AND MEASURES: The metric system is official, but some imperial weights and measures are still used.

HOLIDAYS: New Year's Day, 1 January; Carnival, 14–15 February; Emancipation Day, 1st Monday in August; Independence Day, 31 August; Republic Day, 24 September; Christmas, 25 December; Boxing Day, 26 December. Movable holidays include Carnival, Good Friday, Easter Monday, Whitmonday, Corpus Christi, 'Id al-Fitr, and Dewali.

TIME: 8 AM = noon GMT.

¹LOCATION, SIZE, AND EXTENT

Situated off the northeast coast of South America at the extreme southern end of the Lesser Antilles, the islands of Trinidad and Tobago cover an area of 5,130 sq km (1,981 sq mi), with a length of 210 km (130 mi) NE–SW and a width of 93 km (58 mi) NW–SE. Comparatively, the area occupied by Trinidad and Tobago is slightly smaller than the state of Delaware. Trinidad, the main island, rectangular in shape, has an area of 4,828 sq km (1,863 sq mi), extending 143 km (89 mi) N–S and 61 km (38 mi) E–W. Cigar-shaped Tobago, 31 km (19 mi) northeast of Trinidad, has an area of 300 sq km (116 sq mi), a length of 42 km (26 mi) NE–SW, and an average width of 12 km (7.5 mi) NW–SE. Sixteen small islands are found off the coasts. The Atlantic Ocean is to the E and the Caribbean Sea to the W. Venezuela lies only 11 km (7 mi) SW across the shallow Gulf of Paria.

Trinidad and Tobago have a coastline length of 362 km (225 mi). The capital city of Trinidad and Tobago, Port-of-Spain, is located on Trinidad's Gulf of Paria coast.

²TOPOGRAPHY

Trinidad is geologically part of South America, and its topography is similar to that of the adjoining Orinoco section of Venezuela. Three hill ranges, trending east–west, cross the island roughly through the northern, central, and southern parts, respectively. The Northern Range, a continuation of the mountains of the Paria Peninsula of Venezuela, is the most extensive and rugged of the three and has peaks rising above 900 m (3,000 ft). The highest peaks on Trinidad are El Cerro del Aripo (940 m/3,084 ft) and El Tucuche (936 m/3,071 ft). Hills in the Central Range rise just over 300 m (1,000 ft). Those in the Southern Range are somewhat lower. In between these hill ranges is level or gently rolling flatland, dissected by small streams flowing from the hills. Extensive swamp areas, some of them mangrove, are found along the east, south, and west coasts. Trinidad has the world's largest natural asphalt bog, the 46-hectare (114-acre) Pitch Lake, on the southwestern coast.

Tobago is geologically part of the Lesser Antilles, and its topography, generally more irregular and rugged than Trinidad's, resembles that of Grenada, St. Vincent, and other volcanic islands to the north. A central volcanic hill core rising to over 550 m (1,800 ft) fills most of the island and reaches the sea in many places. Patches of a narrow coastal plain are scattered here and there; much of the island's limited level land is concentrated in its southwestern tip.

³CLIMATE

There is little variation in temperature conditions through the year. The mean annual temperature for the entire nation is 21°C (70°F). In Port-of-Spain the annual average is 25°C (77°F), with an average minimum of 20°C (68°F) and an average maximum of 30°C (86°F) in January; the July range is 23–31°C (73–88°F). Increasing elevation in Trinidad's Northern Range causes a corresponding decrease in temperature. Nights are generally cool.

In the northern and central hill areas and on Tobago, annual rainfall exceeds 250 cm (100 in) and probably exceeds 380 cm (150 in) in specific areas. Most hilly sections receive 200 cm (80 in) or more, while in the lowlands the average drops below 165 cm (65 in) and in certain sections below 125 cm (50 in). There is a relatively dry season from about January to May and a wet season from June to December. The dry period is not, however, a season of drought, for rain still falls every few days in most areas.

⁴FLORA AND FAUNA

The plant and animal life of Trinidad, like the geology of the island, resembles that of neighboring Venezuela. Tobago, by contrast, shows in its flora and fauna its connection with the volcanic Lesser Antillean arc. There are distinct altitudinal variations in indigenous plant life on both islands. The natural vegetation includes wild flowers, many flowering shrubs and trees, palms, giant aroids, and large broad-leaved varieties. Natural animal life includes a few species of mammals, monkeys among them, and many reptiles and birds.

⁵ENVIRONMENT

On the west coast of Trinidad is the Caroni Bird Sanctuary, famed for its marshland and mangroves, where flocks of scarlet ibis roost. Little Tobago is reputed to be the only place aside from New Guinea where the bird of paradise lives in the wild. Among environmental problems, pollution from oil spills is the most serious. Water pollution is also a problem due to mining by-products, pesticides, and fertilizers from farming activity, sewage, saltwater, and oil. The nation has 1.2 cubic miles of water with 35% used for farming and 38% used for industrial purposes. Twelve percent of the people living in rural areas do not have pure water. The nation's cities produce 0.2 million tons of solid waste per year. The land has been damaged by soil erosion due, in part, to the clearing of the land for farming. Environmental responsibility is vested in the Ministry of Energy and Natural Resources. In 1989, the Ministry of the Environment and National Service was formed to regulate the nation's treatment of its natural environment. Endangered species on Trinidad include the Trinidad piping guan, tundra peregrine falcon, loggerhead turtle, and red siskin.

⁶POPULATION

At the census of 2 May 1990, the population was 1,234,388 (Trinidad 1,184,106, Tobago 50,282). A population of 1,365,000 was projected for the year 2000, assuming a crude birthrate of 20.8 per 1,000 population, a crude death rate of 5.9, and a net natural increase of 14.9 during 1995–2000. The population density in 1990 was 241 per sq km (624 per sq mi). Port-of-Spain, the capital since 1783, had a metropolitan population of about 59,200 in 1988. The second most important town is San Fernando, with a population of 33,600. An almost continuous urban area extends from Port-of-Spain eastward to Tunapuna, westward, and northward into the Northern Range. About one-third of the population lives in Port-of-Spain or its suburbs or within 16 km (10 mi) of them. The remainder of Trinidad and virtually all of Tobago are sparsely settled. Scarborough, the main town of Tobago, has a population of approximately 4,000.

⁷MIGRATION

Lack of opportunity has encouraged the migration of numbers of people to the UK, US, and occasionally to other places abroad. In 1990 there were 119,000 people in the US who had been born in Trinidad and Tobago, up from 66,000 in 1980. This movement, however, has been counterbalanced by immigration from other islands in the Lesser Antilles, mainly from Grenada and St. Vincent, where lack of opportunity is far more critical. Some of this immigration has been legal, some not. Migration from Tobago to Trinidad is common also, and there are as many native Tobagonians living on Trinidad as on Tobago. In 1989, emigration exceeded immigration by 24,843.

⁸ETHNIC GROUPS

The population comprises Caucasians, blacks (the descendants of former slaves), East Indians (originally brought to the island as contract laborers from northern India), Syrians, and Chinese; many of the Chinese are racially and culturally intermixed. The total population in 1993 was estimated at 43% black, 40% East Indian, 14% mixed, 1% European, and 2% Chinese and other. Tobago is predominantly black.

While blacks and East Indians on Trinidad are economically interdependent, each community retains its cultural individuality: this is a life that has been called coexistence without assimilation. Intermarriage is rare, and facial and other bodily characteristics still separate the two groups, as do occupation, diet, religion, residence, agricultural landscape, sometimes dress, and often politics. Blacks are dominant in the urban areas, in the oil fields, in the poorer agricultural areas of the north, east, and southeast, and on Tobago. East Indians are dominant in the best agricultural

regions. Although outnumbered in Port-of-Spain and San Fernando, urban East Indians are apt to be economically better off than the urban blacks and are important in commerce, industry, and the professions.

⁹LANGUAGES

English is the official language; an English patois, characterized by numerous foreign words and the special pronunciations of the islands, is understood everywhere. Here and there, a French patois and Spanish are used. In rural village areas, notably in the southern part of Trinidad, East Indians, especially of the older generation, use Hindi and, less frequently, Urdu, Tamil, and Telegu.

¹⁰RELIGIONS

Complete freedom and equality are enjoyed by all religious groups. Christian churches are found on both islands; Hindu temples and Muslim mosques in the recognizable architectural styles of southern Asia are found on Trinidad. At the end of 1985, church membership was estimated at 392,000. That year the population was roughly 36.2% Roman Catholic, 23% Hindu, 13.1% Protestant, 6% Muslim, and 21.7% other or nonaffiliated.

¹¹TRANSPORTATION

In 1991 there were 8,000 km (4,970 mi) of roads of which about one-half were paved. The more densely settled sections of both islands are served by reasonably adequate roads, but large sections of Tobago either have no motorable roads or are connected by narrow, tortuous, and poorly surfaced ones. In 1992, registered motor vehicles included 120,589 passenger cars and 30,126 commercial vehicles. The Public Service Transport Corp. is responsible for road transport. Trinidad's lone remaining railway, from Port-of-Spain to San Juan, was closed down in 1968.

The largest port installation for passengers and cargo is at Port-of-Spain. Brighton is an important port for oil and asphalt loading, and there are also oil terminals at Chaguaramas, Pointe-à-Pierre, and Point Fortin. A deep-water port at Point Lisas accommodates energy-based industries at the Point Lisas industrial estate. Numerous steamship lines regularly schedule freight and passenger services from Europe and the Americas. Regularly scheduled coastal vessels connect Port-of-Spain with Scarborough. The main shipping line is the West Indies Shipping Service.

Air facilities are concentrated at Piarco International Airport, about 26 km (16 mi) southeast of Port-of-Spain. There is a secondary main airport at Crown Point, on the western tip of Tobago, and there are a dozen smaller airfields on Trinidad. Trinidad and Tobago Airways, owned by the government and formed by the merger of British West Indian Airways (BWIA) International and Trinidad and Tobago Air Services in 1980, operates domestic, regional, and international services.

¹²HISTORY

Arawak Indians inhabited what they knew as Iere—Land of the Hummingbird—before the arrival on 31 July 1498 of Christopher Columbus, who called the island La Trinidad, or "The Trinity." The Spanish took little interest in the island and did not appoint a governor until 1532. Thereafter, Spanish colonists gradually came, but skirmishes with local Indians and raids by other Europeans, including Sir Walter Raleigh, made it difficult for the Spanish to obtain a foothold there. During the early European period, the island was a supply and transshipment center for Spanish traders and fortune seekers in South America. In time, colonists established plantations and imported slave labor from West Africa. The native Indians were eventually wiped out. In 1797, a British expedition from Martinique captured Trinidad, which was ceded formally to Great Britain in 1802 by the Treaty of Amiens and became a crown colony.

During the late Spanish period and through most of the 19th century, sugar was the island's main product. The emancipation of slaves in 1834 brought severe labor shortages, and between 1845 and 1917 more than 150,000 contract workers, mostly Hindus and some Muslims from India, were brought to the island as "cheap labor" to replace the slaves. With added labor supplies and new techniques, the cocoa industry thrived, and by the late 19th century cocoa had joined sugar as a major export crop. Petroleum was discovered on south Trinidad in 1910 and since then has assumed increasing economic importance.

Tobago was also discovered by Columbus in 1498, and it too was ignored by Europeans for many years. The first colonists apparently were British from Barbados in 1616, but the local Carib Indians soon drove this group out. Other colonists followed shortly, and during the next 200 years the island changed hands many times among the Dutch, French, and British. Finally, in 1814, the British crown gained possession, which it maintained for a century and a half. Tobago was at first ruled as a separate colony, but during much of the 19th century it was administered from the Windward Islands government. It became a crown colony in 1877 and in 1888 was amalgamated with Trinidad under the colony name of Trinidad and Tobago. In 1958, the Federation of the West Indies was formed with Jamaica, Barbados, and the British Windward and Leeward Islands. Jamaica and Trinidad and Tobago withdrew in 1961, and the federation collapsed.

On 31 August 1962, Trinidad and Tobago became independent. The country retained membership in the Commonwealth as a British dominion. Eric Williams, the founder of the People's National Movement (PNM), became prime minister in 1961 and held the office till his death in 1981.

In 1976, Trinidad and Tobago declared itself a republic, and a president replaced the British monarch as chief of state. In 1980, Tobago attained a degree of self-government when it was granted its own House of Assembly. Williams's successor as majority party leader and prime minister was George Chambers, former deputy leader of the PNM. Following the victory of the National Alliance for Reconstruction (NAR) in the December 1986 elections, A.N.R. Robinson, leader of the NAR, became prime minister.

In 1991, the PNM returned to power under Prime Minister Patrick Augustus Mervyn Manning.

¹³GOVERNMENT

In 1976, Trinidad and Tobago amended its 1961 constitution. The 1976 draft preserved the bicameral legislature, but replaced the crown-appointed governor-general with a ceremonial president chosen by Parliament.

The House of Representatives is the more important of the two houses. Its 36 members are elected for five-year terms, but new elections can be called by the prime minister or by the house itself in a vote of "no confidence" in the cabinet. The party with a majority of seats in the House forms the government. The Senate consists of 31 members all appointed by the president, 16 on the advice of the prime minister, 6 on the advice of the leader of the opposition, and 9 discretionary, based on consultation with various religious, economic, and social groups.

The chief executive officer is the prime minister, who is leader of the majority party. The president holds office for five years, which is also the normal term of Parliament. Cabinet ministers are appointed primarily from the House of Representatives by the president, acting on the prime minister's recommendations.

Active participation in government by the nonwhite (black and East Indian) population began in 1925, when for the first time elected representatives were included in the otherwise appointed Legislative Council that ruled the colony. Over the years the proportion of elected members increased, and a fully electoral self-government came in December 1961. In 1976, suffrage was lowered to the age of 18.

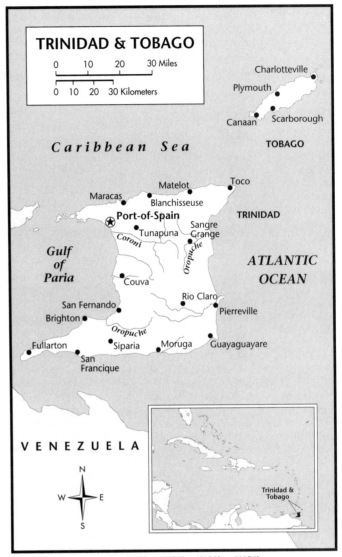

LOCATION: 10°2′ to 11°21′N; 60°30′ to 61°56′W.
TOTAL COASTLINE: 470 km (292 mi). **TERRITORIAL SEA LIMIT:** 12 mi.

¹⁴POLITICAL PARTIES

The People's National Movement (PNM), formed in 1956 by Eric Williams, has dominated politics in Trinidad and Tobago. After winning the 1961 elections, the opposition became disorganized. The PNM won all 36 seats in the House of Representatives in an election on 24 May 1972, which the opposition boycotted. The PNM monopoly was broken in June 1972, when two members of the House of Representatives broke ranks to form a splinter party.

In the 1976 elections, the first held under the republican constitution, the PNM won 24 seats and the new United Labour Front (ULF), a trade unionist party, took 10. The Democratic Action Congress (DAC) elected the two representatives from Tobago.

In the election of 1981, the PNM continued its dominance with 26 seats. The opposition ULF, the DAC, and the Tapia House Movement formed a coalition party, the Trinidad and Tobago National Alliance, and took 10 seats. The Organization for National Reconstruction (ONR), founded in 1980, took 22.3% of the vote but won no seats.

In 1986, the opposition coalition incorporated the ONR and, under the National Alliance for Reconstruction, took 33 seats to the PNM's 3 in the December elections. In 1991, the PNM returned to office, winning 21 of 36 seats. The NAR received only 2 seats, and the United National Congress, under Basdeo Panday, won 13 seats.

Party membership has been to a large extent based on race and region. The PNM is the party of the blacks. The NAR has been more powerful on the island of Tobago, where it has controlled the local assembly. Whites, Chinese, and other minorities have traditionally been anti-PNM.

[15]LOCAL GOVERNMENT

Elected county councils with certain executive powers govern the eight counties of Trinidad. The three major cities, Port-of-Spain, San Fernando, and Arima, each have a mayor and a city council. In 1980, Tobago was granted its own House of Assembly, which sits for four years and consists of 12 elected members and 3 members chosen by the majority party. In January 1987, Tobago gained full internal self-government.

[16]JUDICIAL SYSTEM

The judicial system is modeled after that of the UK, with some local variations. The Supreme Court of Judicature is made up of the High Court of Justice and the Court of Appeal. The former consists of a chief justice and 10 puisne judges. Its jurisdiction and its practices and procedures follow closely those of the High Court of Justice in England. Civil actions and proceedings are usually heard by 1 High Court judge only, but may be tried by a jury of 9 members. Criminal offenses are tried by a High Court judge with a jury of 9–12 members. The Court of Appeal consists of the chief justice and 3 other justices. There is a limited right of appeal to the Privy Council, seated in London. Minor offenses are handled by district courts, including traffic courts. The judicial system also includes the Industrial Court and Tax Appeal Board.

The judiciary is independent of the other branches and free from outside interference.

Criminal defendants are presumed innocent and are afforded the right to representation by counsel. In practice, the civil and criminal dockets are badly backlogged due to inadequate resources and inefficiency.

[17]ARMED FORCES

The Trinidad and Tobago Defense Force number an estimated 2,000 in one reinforced battalion. There are also a coastguard of 600 (10 patrol craft) and a paramilitary police force numbering 4,800. Under lend-lease agreements with the UK signed early in 1941, the US acquired several Caribbean bases, including one on Trinidad, on a 99-year lease. After local agitation over a number of years, Trinidad reclaimed the last remaining foreign base in June 1967. Two US tracking and navigation stations were closed in 1971 and 1980. Defense costs $59 million a year (1989) or around 1–2% of GDP.

[18]INTERNATIONAL COOPERATION

Trinidad and Tobago became a UN member on 18 September 1962, and it belongs to ECLAC and all the nonregional specialized agencies except IAEA, IFAD, and WIPO. Trinidad and Tobago also belongs to the Commonwealth of Nations, G-77, IDB, OAS, and PAHO and is a signatory to GATT and the Law of the Sea. The nation joined CARIFTA in 1968 and its replacement, CARICOM, in 1973; it is a member of the nonaligned movement. Oil revenues in recent years have enabled Trinidad and Tobago to contribute to regional lending agencies, such as the Caribbean Development Bank.

[19]ECONOMY

The economy of Trinidad and Tobago is oriented toward trade and tourism. The country cannot begin to feed itself and must look abroad not only for a large portion of its food, but also for the bulk of its needs in manufactured goods. Import payments are met by the export of specialized products, invisible exports, and the transshipment trade. Specialized tropical crops are produced for export, but exploitation of petroleum reserves and refining of local and imported oils and their subsequent export are the dominant factors in the country's economy. Although by far the most prosperous of Caribbean nations, the country's high degree of dependence on oil revenues has made it exceedingly vulnerable to falling oil prices in recent years. Since its petroleum reserves are likely to approach depletion during the 1990s, the government faces the necessity to diversify the economy if its standard of living is to be maintained or improved.

Trinidad and Tobago registered a mixed economic performance in the early 1990s. The country's economy was in a continuous decline from 1983 to 1990. Trinidad and Tobago experienced a real GDP growth of 2.5% in 1991; in 1992, its real growth declined again to –0.3% and to –0.9% in 1993. Among the causes for its poor economic performance are a global weakening of oil prices, an increasing debt service ($650 million in 1993), and a high public sector bill. New economic measures of trade liberalization produced a psychological shock on the domestic economy because of possible disappearance of jobs. The government is pursuing a privatization program, so far selling more than 20 state enterprises. Inflation remained at single digit levels up to 1992 (6.5%), but it soared in 1993 (11%). Nonetheless, the non-petroleum sector continued to grow during 1992, expanding by 1.4%. Agricultural production grew by approximately 8% and services by 1%. The tourism industry registered an increase of 7% in 1992 with revenues of $109 million, compared to $101 million in 1991.

[20]INCOME

In 1992, the GNP was $4,995 million at current prices, or $3,940 per capita. For the period 1985–92 the average inflation rate was 5.2%, resulting in a real growth rate in per capita GNP of –3.0%.

[21]LABOR

The economically active population in 1991 was estimated at 492,100 persons, of whom approximately 34% were engaged in services; 16.8% in commerce; 17% in construction and electricity, gas, and water supply; 15.5% in mining, quarrying, and manufacturing; 10.4% in agriculture, forestry, hunting, and fishing; 6% in transportation and communications; and 0.3% in other sectors. Unemployment averaged 13% in the early 1970s and had risen to 17% by 1973; in 1981, the rate was about 10%, after which it rose to 22% by 1989, but fell back to 17.4% in 1990.

The principal national labor federation is the Trinidad and Tobago Labour Congress; it had about 100,000 members (58,811 paid) as of 1991. Of nonaffiliated unions, the largest is the National Union of Government and Federated Workers, with a membership of about 26,500. As of 1992, about 25% of the work force was organized into 45 labor unions.

There is little protective labor legislation; work rules are subject to labor-management negotiation. Strikes are illegal in essential service industries. The normal workday is eight to nine hours, five days a week. Vacation periods vary from two to five weeks a year, depending on length of service.

[22]AGRICULTURE

About 23% of the total land area was arable in 1991, most of it on Trinidad. There are two distinct types of agricultural operations—the large estate or plantation that is managed by a specialist and employs large numbers of laborers, and the small farm cultivated by the owner (or tenant) and family. The small farms grow mainly for the home market. Crops include corn, rice, peas, beans, potatoes, other vegetables, and a wide variety of fruits. Lowland rice is grown almost entirely by Indian farmers. The large estates are interested mainly in commercial export crops, although the small farmers also grow some export crops. Agriculture accounts for only 2.5% of GDP, and grew by 5% in 1992, helping to recoup a decline of 8.4% in 1991. Increased acreage boosted rice production, and favorable weather increased the yields of the sugar and domestic crops. Coffee and cocoa production both fell because of abandonment of estates. During the 1970s and early 1980s, agriculture's traditional labor force was lured away by the booming energy sector, with foreign exchange plentiful enough to import food. By the late 1980s, however, this trend was being reversed.

The value of crops grown for the domestic market is believed to be considerably greater than that of the export crops. Sugar, the main commercial crop, is grown on a few large company-owned estates and by thousands of small farmers; modern methods allow the estates to produce about two-thirds of the sugar crop. Normally, 80% or more of the islands' production is exported. Sugar production declined from 250,000 tons in 1961 to 213,200 tons in 1971 and some 65,000 tons in 1984, the lowest output in four decades; 1992 production was 1,292,000 tons. The second major export crop, cocoa, is cultivated in the hill sections of both Trinidad and Tobago. Estates produce considerably more cocoa than smallholdings, owing to better agricultural practices and to the fact that small farmers intercrop bananas, coffee, and other crops with cocoa. Cocoa production has been in decline since 1970; the output in 1992 was 2,000 tons. Coffee is grown in much the same hill areas as cocoa, and there is about the same proportion of estate-grown to small-farmer-grown coffee. Both cocoa and coffee have been described as sick industries because of inefficiency, crop disease, and uncertain world market conditions. In 1992, coffee production was about 1,000 tons, down from 2,361 tons in 1985.

Some 90 acres (36 hectares) of ornamental flowers are being cultivated; the export value of such floriculture was about $1 million in 1993. Rice, citrus, corn, cassava, peanuts, and pigeon peas are now being grown to diversify agricultural output.

23ANIMAL HUSBANDRY

Livestock plays only an incidental role in the agricultural pattern. The water buffalo, adept at turning heavy, water-laden soils, has been brought from India by Indian farmers and is the major draft animal in rice cultivation and probably the most productive animal in the country. Cattle are kept by some small farmers, but the best stock is that on estates and government farms or in large dairies. Poor animals and poor breeding and feeding methods keep meat and milk quantity and quality low and prices high. Trinidad and Tobago relies heavily on dairy imports from Europe to satisfy domestic demand.

In 1992, the livestock population included an estimated 10,000,000 poultry, 60,000 head of cattle, 52,000 goats, 50,000 hogs, 14,000 sheep, and. 9,000 water buffalo. Animals slaughtered at abattoirs run by the government and by municipalities yielded an estimated 26,000 tons of poultry meat, and 3,000 tons of pork in 1992.

24FISHING

The fishing industry has great potential, but current production does not begin to meet local demands, and large quantities of fish must be imported. Shrimp and mackerel make up one-third of the total annual catch, with shrimp accounting for nearly half of the $2.7 million in fish and fish products exports in 1991. In 1986, the Archipelagic Waters and Exclusive Economic Zone Act defined the sovereign fishing jurisdiction around the nation, which has created new opportunities for marine fishing off the east coast of Trinidad. The harvest in 1991 was 10,283 tons, up from 3,730 tons in 1986.

25FORESTRY

Approximately half, or 219,000 hectares (541,000 acres), of the land is forested, but since this includes abandoned farmland and other cutover areas, probably less than 30% is fully constituted forest. Roughly four-fifths of the forestland is government owned or administered; however, much of the state forestland is in hill areas, inaccessible for exploitation. Several dozen small sawmills are in operation. Roundwood production in 1991 was about 72,000 cu m.

26MINING

After petroleum, natural asphalt, of which Trinidad has the largest supply in the world, is the second mineral of economic value. Southwest of San Fernando is the island's famous Pitch Lake, a 46-hectare (114-acre) deposit of oozing black asphalt; it has been mined commercially since the 19th century. The annual yield has declined from an average of 200,000 tons in the 1960s to 20,000 tons in 1991. Iron ore deposits of commercial value are reported to have been discovered in Trinidad's Northern Range. Quarrying operations on the islands involve over 1 million tons annually of limestone and cement.

27ENERGY AND POWER

Petroleum is the main source of energy and since the 1940s has been the principal industry. Despite declining oil prices, petroleum and petroleum products provided close to 40% of export earnings in 1991. However, its contribution to the GDP (including asphalt extraction and refining) decreased from about 35% in 1980 to 25% in 1991. In 1992, the petroleum sector declined by 6.1%, due to decreased drilling and exhausted wells.

Commercial petroleum operations began in 1911 and increased substantially since that time. Current mining and exploration leases cover almost 90% of the land area and the bulk of all Trinidad and Tobago's territorial waters. Natural gas reserves are estimated at 210,000 million cu m. Foreign capital has sponsored most exploration, mining, and processing. The oil fields are in the south of Trinidad and offshore in the Gulf of Paria; in 1991, roughly 75% of current production was from offshore wells.

Trinidad's oil production peaked at 245,000 barrels a day in 1977 and fell to 145,000 barrels a day in 1992. Proved crude oil reserves exceed 600 million barrels; several untapped offshore fields in the southeast were under exploration in 1992. Subsidiaries of Shell and Texaco built the two largest refineries. In August 1974, the government took over Shell (Trinidad), paying compensation of us$45 million. The company is now being operated as the Trinidad and Tobago Oil Co. Ltd. (TRINTOC). In 1985, the government acquired Texaco's onshore fields and oil refinery, which are now managed by TRINTOC. That year the government also acquired ownership of another major oil company, Trinidad-Tesoro Petroleum, which it renamed the Tesoro Petroleum Company. Natural gas production in 1992 was 5,600 million cu m, 80% of which was produced by Amoco. All natural gas produced is used domestically. About 70% is piped to ammonia, methanol, and urea plants; the state-owned steel mill; and the Trinidad and Tobago electric utility sector.

The Trinidad and Tobago Electricity Commission has been carrying out electrification throughout the islands, and generation as well as domestic, commercial, and industrial use of electric power is well ahead of other areas in the former British West Indies. More than 95% of the current total generating capacity is publicly owned and operated by the commission; the remainder is installed mainly in oil refineries and sugar factories. In 1991, total electrical energy production by public utilities was 3,525 million kwh. Net installed capacity was 1,150,000 kw, all of it in conventional thermal facilities. A submarine cable provides Tobago with electricity from Trinidad.

28INDUSTRY

Long-established industries are those processing raw materials of the farm, forest, and sea; foremost are sugar, molasses, and rum, followed by fish, lumber, fats and oils, and stock feed. Manufactured products include furniture, matches, angostura bitters, soap, and confectionery and clay products. Newer industries include concrete products, canned citrus, bottled drinks, glass, drugs, chemicals, clothing, building materials, and metal goods.

Many new industrial plants have been established under the benefits of the country's "pioneer status" provisions, which grant remission of taxes on profits during a new plant's early years, remission of import duties on production equipment and raw materials, and accelerated depreciation allowances. "Pioneer" factories are in operation producing metal products, furniture, fertilizers, cotton fabrics, and other light industrial products.

Due to a decrease in drilling activities and the maturation of existing wells, the petroleum sector declined by about 6.1% in 1992. The government has taken proper steps to increase investment and growth in the sector. It has already increased oil exploration and production by giving contracts to US companies.

In 1992, methanol production increased by 8%, but ammonia and urea output fell due to maintenance activities. The textiles, garments, and footwear subsector declined by 8%, while the food, beverages, tobacco and assembly industries registered moderate growth. The iron and steel subsector recorded an increase of 13%. Total manufacturing sector increased by 3%, continuing the trend towards a higher share of manufacturing in GDP.

29SCIENCE AND TECHNOLOGY

Among the research centers and learned societies of Trinidad and Tobago are the Commonwealth Institute of Biological Control, the Agricultural Society of Trinidad and Tobago, the Tobago District Agricultural Society, the Pharmaceutical Society of Trinidad and Tobago, and the Sugar Manufacture Association of Trinidad and Tobago. The University of the West Indies has a campus in St. Augustine with faculties of agriculture, engineering, medical sciences, and natural sciences.

30DOMESTIC TRADE

Trinidad and Tobago's wholesale trade is highly organized and highly competitive; much of it is controlled by a few managing agencies located in Port-of-Spain. These agencies are direct importers in bulk and have exclusive wholesale rights for sales in the islands, and often in other Caribbean nations and territories.

In small communities, in rural areas, on Tobago, and in less developed parts of Port-of-Spain and San Fernando, the general retail store carries a wide variety of commodities. Many of these stores are family enterprises and most are small. Business hours are generally from 8 AM to 4:30 PM, Monday through Friday, with a lunch break from noon to 1 PM. Shops are open from 8 AM to 4 PM, except on Fridays (till 6 PM) and Saturdays (till noon). Banks stay open from 9 AM to 2 PM, Monday–Thursday, and from 9 AM to 1 PM and from 3 to 5 PM each Friday.

31FOREIGN TRADE

The foreign trade of Trinidad and Tobago is very large for a country of its size, a fact attributable mainly to its petroleum processing industry, whereby crude oil is imported for processing and then reexported as gasoline, kerosene, and other petroleum products. The economy's prosperity is thus tied closely to trade (the value of imports plus exports has for many years been greater than the country's GNP). Trade, in turn, is closely linked to the price and demand structure of the world petroleum market.

Exports consist mainly of petroleum and derivatives, as well as processed agricultural products and some semi-manufactures. Imports include a wide range of manufactured items, as well as food, beverages, and some raw materials, including petroleum.

Up to 1992, the government prohibited the importation of some manufactured products without a license. The government replaced this list with supplemental tariffs, which coupled with the Caricom External Tariff (CET), reached nearly 100%. According to the government, these supplemental tariffs are to be reduced to CET levels by 1 January 1995, alleviating high prices on foreign goods.

Principal exports are petroleum and petroleum products, chemicals, and manufactured goods. Total exports at year end 1992 were $1,867 million and at year end 1993, $1,400 million. The country's major buyers were the US, Barbados, Jamaica, Canada, the UK, and Central and South America.

Imports were similarly reduced in 1993 to $1,360 million compared with $1,435.6 million in 1992. Principal imports are crude petroleum machinery, transportation equipment, manufactured goods, and fuels. Major suppliers were the US, the UK, Canada, Brazil, Venezuela, other EC countries and Jamaica.

32BALANCE OF PAYMENTS

Between 1974 and 1981, largely because of the huge increase in the value of petroleum exports, Trinidad and Tobago's payments balance was favorable. With the weakening of the market for the country's petroleum and oil refinery products in 1982, however, a deficit was recorded for the first time since the early 1970s. Foreign exchange reserves, which had reached $3.3 billion in 1981 were depleted rapidly through the 1980s, as expenditures reduced revenues (caused by lower oil prices). By 1988, foreign exchange reserves had plummeted to –$5.7 million, forcing the government to reschedule its commercial and official debt.

In 1992 merchandise exports totaled $1,661.9 million and imports $995.6 million. The merchandise trade balance was $666.3 million.

33BANKING AND SECURITIES

The Central Bank of Trinidad and Tobago (established 1964) is the central regulatory institution and the sole bank of issue. The commercial banking business is well established and is operated chiefly by Canadian, British, and American interests.

At the end of 1992, commercial banks had assets of US$99.3 million, including TT$568.6 million in claims with the Central Bank and TT$7,710 million in claims in the private sector. Liabilities included TT$1,859 million in demand deposits and TT$6,525.9 million in time and savings deposits. The money supply, as measured by M2, amounted to TT$9,221.9 million at the close of 1992.

Workers and farmers make use of the Government Savings Bank offices. There are numerous agricultural credit societies, most of which are financed by the government's Agricultural Credit Bank. Credit unions are also common. The Trinidad and Tobago Development Finance Co., jointly owned by the government and the private sector, offers medium- and long-term financing to industry.

Formerly, unincorporated and privately owned businesses were paramount; since World War II, and especially since about 1948, limited liability companies with publicly issued share capital have become increasingly important. The West Indies Stock Exchange (succeeded by West Indies Stock Brokers, Ltd.) opened a branch in Port-of-Spain in 1964. The brokerage organization became a member of the Jamaica Stock Exchange in December 1970.

34INSURANCE

Insurance firms include branches of UK and US companies and a few local companies. Their operations are highly competitive. A government-owned reinsurance company has been in operation since 1979. In 1985, there were 19 life and 21 non-life insurance companies in operation.

35PUBLIC FINANCE

After three years of deficit, the 1973 budget registered a surplus, primarily attributable to increased revenues resulting from higher petroleum prices. The budget generally remained in surplus until 1982, when a deficit of TT$2,652.4 million was registered. The budget continued in deficit, although at decreasing levels, until 1990.

In the mid-1980s, the government began an adjustment program which included currency devaluation, debt reschedulings, and the adoption of an austere budget that included public service wage reductions and decreased transfers to state enterprises.

In 1992, current receipts amounted to $1,467 million, while current expenditures totalled $1,468 million, with capital expenditures of $144 million. Public sector foreign debt reached an all-time high of $2,507.3 million in 1990, but declined to $2,366.7 million by mid-1991. The debt fell to 49.2% of GDP in 1990 from 55.3% in 1989. By the end of 1992, the total external public debt was $2.2 billion.

36TAXATION
Important sources of taxation are income taxes, a motor vehicle tax, license duties, property and building taxes, customs and excise duties (including purchase taxes), and petroleum royalties and concessions.

The individual income tax is calculated on net chargeable income, from 5% to 40%. A value-added tax of 15% is levied on many goods. To stimulate nonpetroleum industries in 1973 and 1974, purchase taxes were reduced or rescinded on locally assembled automobiles and kitchen appliances and on locally made clothing.

The basic corporate tax rate is 45%. Dividend and interest payments are taxed at 25% and royalties at 20%. The 1974 budget set forth a revised system of petroleum company taxation designed to increase corporate income taxes sixfold. Net taxable income is now calculated from tax reference prices for oil, not from realized prices. As a result of these changes, income tax revenues grew from 7.2% of GDP in 1973 to 29% in 1985.

37CUSTOMS AND DUTIES
Customs and duties are a significant source of government revenue. Most imported articles are subject to import duties, which include both specific and ad valorem levies, as well as a stamp tax, an import surcharge, a value-added tax, and excise taxes on petroleum products, tobacco, and alcoholic beverages that are sold locally.

38FOREIGN INVESTMENT
Foreign investments in Trinidad and Tobago, particularly those from British, Canadian, US, and Dutch sources, have played a major role in the development of all major manufacturing and processing industries, as well as most large agricultural enterprises. Several private banking institutions have provided development loans through the Industrial Development Corp., established by the government in 1959 to act as a liaison between investors and various government departments.

Although the government still encourages foreign investment, since the 1970s it has sought to bring about local control of all economic resources. It therefore requires foreign investment in Trinidad and Tobago to be conducted on a joint-venture basis, with majority domestic participation, most often in a 60:40 ratio. It has also acquired interests in a number of international companies and in some cases has bought out their local enterprises. It is nearly impossible for foreign investors to acquire land in Trinidad and Tobago.

The government has renewed emphasis on foreign investment in recent years with a privatization program among other incentives. By the beginning of 1991, there were already $1.2 billion of foreign investment pumping into the economy. The US is the major investor with 41.6% of total. Other major investors include the UK, Canada, and Norway.

39ECONOMIC DEVELOPMENT
To diversify the economy and raise the national standard of living, foreign capital and technical assistance are actively solicited. Major incentives are duty-free imports of equipment and raw materials, income tax holidays, accelerated depreciation allowances, unlimited carryover of losses, and repatriation of capital and profit.

The many community services upon which industry depends have been enormously improved in recent years. The government has endeavored to achieve far-reaching economic transformation through its five-year plans, beginning in 1958. The government's Special Fund for Long-Term Development has been established to make use of oil revenues in economic and social development projects.

In 1980, a 25-year development plan was announced, emphasizing industrial diversification and renewed attempts to revive agriculture and fisheries. Since 1978, the government has encouraged the development of energy-intensive industries, most of them making use of natural gas.

Monetary policy was the principal instrument used for stabilizing the economy and protecting the balance of payments and international reserves from further decline during the 1992/93 period. The broad policy of liberalization and privatization saw notable progress with respect to trade reform. Foreign investment and international trade should perform better in the following years. The external debt service burden will affect both the fiscal and external situation. Customs reforms should produce significant extra revenues and the depreciation of the exchange rate will have a positive impact on oil revenues.

While the short-term outlook is clouded, there are grounds for optimism about the medium-term prospects. The exchange rate flotation since the spring of 1993 should lower business uncertainty, reduce private capital outflows, and stimulate export industries, and may permit a gradual return to a lower interest rate environment, which should spur investment.

40SOCIAL DEVELOPMENT
All employees aged 16 through 65 are required to become members of the National Insurance System, to which employers and employees contribute two-thirds and one-third, respectively. The system provides old age, retirement, and disability pensions; maternity, sickness, and survivors' benefits; and funeral grants. A food stamp program was introduced in 1978. Public work programs are introduced in areas where seasonal unemployment in agriculture is a major problem.

Maternal and child health and family planning rank high among the government's social development priorities. The Family Planning Association of Trinidad and Tobago, affiliated with the International Planned Parenthood Federation, takes part in a family planning program established by the government in 1967. In 1985–90, an estimated 53% of married women were using contraception; the fertility rate declined from 5.03 in 1950–55 to 2.8 in 1991. Abortion is permitted on broad health grounds.

Sex discrimination is illegal. In 1993, women constituted 36% of the labor force, and many women are active in business and the professions.

41HEALTH
The general health of the population has been improving; substantial decreases have been recorded in the death rates for malaria, tuberculosis, typhoid, and syphilis. In 1990, about 9% of children under 5 years old were considered malnourished. Immunization rates for children up to 1 year old in 1992 were: diphtheria, pertussis, and tetanus (82%); polio (81%); and measles (93%).

Government health facilities include general hospitals in Port-of-Spain and San Fernando, small district hospitals, several major health centers, dental service centers, a mental hospital, and a nurses' training school. In 1991 there were 911 physicians, 562 pharmacists, and 109 dentists. In 1992, 99% of the population had access to health care services.

Improvements in sanitation have reaped impressive health benefits. Close to 75% of urban housing units are connected to sewage systems. In 1990, 97% of the population had access to safe water and 79% had adequate sanitation. As a result, reported cases of dysentery and hookworm have declined dramatically.

The 1993 birth rate was 23.3 per 1,000 people. There were 30,000 births in 1992, with an average life expectancy of 71 years. Infant mortality was 19 per 1,000 live births in 1992; maternal mortality was 110 per 100,000 live births in 1991; and general mortality was 6.3 per 1,000 people in 1993.

42HOUSING

There is an acute shortage of adequate housing, and high rents have contributed to inflation. Industrial construction has outpaced residential building in recent years. A typical rural home for a large family consists of one to three rooms plus an outside kitchen. Slums and tenements are typical of urban life. Nearly all private dwellings, urban or rural, have toilets and piped-in water. As of 1980, 76% of all housing units were detached houses, 19% were flats, and 4% were duplexes. Owners occupied 65% of all dwellings, 24% were rented, and 9% were occupied rent free.

43EDUCATION

About 98% of the population 10 years of age and older is literate. In 1990, the islands had 476 primary and intermediate schools with 7,473 teachers and 193,992 students. Secondary schools enrolled 98,741 pupils in 1988. Education is compulsory for six years. Elementary education lasts for seven years followed by secondary education, which has two phases of three years and two years, respectively.

There are four small, government-run technical colleges, a polytechnic institute, and five teachers' colleges. The University of the West Indies has a faculty of engineering, arts, and agriculture at its Trinidad campus. John F. Kennedy College, a liberal arts school outside Port-of-Spain built with a US$30-million grant from AID, has teaching facilities for about 600 students. In 1990, at the universities and equivalent institutions, 4,090 students were estimated to be enrolled with 289 teaching staff.

44LIBRARIES AND MUSEUMS

The Central Library of Trinidad and Tobago, with 442,000 volumes, and the Trinidad Public Library, with 68,000 volumes, are both in Port-of-Spain. In San Fernando, the Carnegie Free Library, with 29,790 volumes, functions as regional headquarters for rural library services to the south. The University of the West Indies library in St. Augustine has more than 296,000 volumes. The National Museum and Art Gallery is located in Port-of-Spain.

45MEDIA

Postal and internal telegraph services throughout the islands are operated by the government. As of 1991 there were 211,747 telephones. Commercial cable communication and radiotelephone services are maintained between Trinidad and all major countries of the world. Radio broadcasting is provided by the National Broadcasting Service (public) and the Trinidad Broadcasting Co., a subsidiary of Rediffusion International of London. Founded in 1962 and state-owned since 1969, the Trinidad and Tobago Television Service emphasizes school and adult-education programs. There are 2 AM and 4 FM radio stations and 5 TV stations. In 1991 there were an estimated 394,000 television sets and 615,000 radios in use.

Freedom of the press is both constitutionally guaranteed and respected in practice. There are four daily newspapers. The *Trinidad*

Guardian, a morning and Sunday paper, had an average daily circulation of 41,500 in 1991. *The Trinidad and Tobago Express*, published daily and Sunday, had a daily circulation of 46,700.

46ORGANIZATIONS

Producers of agricultural crops for export are organized into associations for solving common problems, as well as for social purposes. Among these are the Cocoa Planters Association, Cooperative Citrus Growers Association, and Sugar Manufacturers Association. Nonagricultural associations are many and varied and include the Law Society, Medical Board, Petroleum Association, Shipping Association, and Trinidad Chamber of Commerce.

47TOURISM, TRAVEL, AND RECREATION

The government offers fiscal and other incentives for the development of hotels and other tourist facilities. A deep water harbor has been commissioned in Tobago, and the newly completed Crown Point Airport is open to international traffic. As of 1991, hotels and similar establishments on the islands had a total of 2,928 rooms with a 52.1% occupancy rate. Tourist arrivals in 1991 totaled 219,836, of whom 159,735 came from the Americas and 49,292 from Europe, and tourism revenues reached US$101 million.

Outstanding tourist attractions include the mountainous areas, beaches, and reefs on both islands. Entertainment includes calypso and steel band music, both of which originated in Trinidad. Festive events include Carnival, held annually on the two days before Ash Wednesday; the Muslim festival of Hosein, which begins 10 days after the new moon in the month of Muharram; and the Hindu festival of lights, Dewali, which occurs in October or November. Cricket and football (soccer) are the most popular sports.

48FAMOUS TRINIDADIANS AND TOBAGONIANS

Eric Eustace Williams (1911–81), the main political figure of his time and the leader of Trinidad and Tobago's major political party, was instrumental in his country's achievement of independence in 1962; he was prime minister from 1961 until his death. His successor was George Michael Chambers (b.1928). A.N.R. Robinson (b.1926) became prime minister in 1986. Notable writers include Samuel Selvon (b.1923) and V.S. (Vidiadhur Surajprasad) Naipaul (b.1932).

49DEPENDENCIES

Trinidad and Tobago has no territories or colonies.

50BIBLIOGRAPHY

Bereton, Bridget. *A History of Modern Trinidad*. Portsmouth, N.H.: Heinemann Educational Books, 1982.
Hintzen, Percy C. *The Costs of Regime Survival: Racial Mobilization, Elite Domination, and Control of the State in Guyana and Trinidad*. New York: Cambridge University Press, 1989.
John, A. Meredith. *The Plantation Slaves of Trinidad, 1783–1816: A Mathematical and Demographic Enquiry*. New York: Cambridge University Press, 1988.
Magid, Alvin. *Urban Nationalism: A Study of Political Development in Trinidad*. Gainesville: University of Florida Press, 1988.
Oxaal, Iva. *Black Intellectuals and the Dilemmas of Race and Class in Trinidad*. Cambridge, Mass.: Schenkman, 1982.
Singh, Chaitram. *Multinationals, the State, and the Management of Economic Nationalism: The Case of Trinidad*. New York: Praeger, 1989.
Yelvington, Kevin A. (ed.). *Trinidad Ethnicity*. Knoxville: University of Tennessee Press, 1993.

TURKS AND CAICOS ISLANDS

CAPITAL: Grand Turk.

FLAG: The flag is a British blue ensign with the shield of the colony in the fly; the shield is yellow with a conch shell, lobster, and Turk's head cactus represented in natural colors.

ANTHEM: *God Save the Queen.*

MONETARY UNIT: The US dollar of 100 cents (US$) has been the official currency since August 1973. The Turks and Caicos crown is also in circulation.

WEIGHTS AND MEASURES: The imperial system is used.

HOLIDAYS: New Year's Day, 1 January; Commonwealth Day (May); Queen's Official Birthday (June); Emancipation Day, 1st Monday in August; Constitution Day, 30 August; Columbus Day, 2d Monday in October; Human Rights Day (October); Christmas, 25 December; Boxing Day, 26 December. Movable religious holidays include Good Friday and Easter Monday.

TIME: 7 AM = noon GMT.

¹LOCATION, SIZE, AND EXTENT

Situated in the Atlantic Ocean SE of the Bahamas, E of Cuba, and N of Hispaniola, the Turks and Caicos Islands consist of two island groups separated by the Turks Island Passage, 35 km (22 mi) across and about 2,100 m (7,000 ft) deep. The Turks group comprises two inhabited islands, Grand Turk and Salt Cay, six uninhabited cays, and numerous rocks surrounded by a roughly triangular reef bank. The Caicos group encompasses six principal islands (North Caicos, Middle Caicos, East Caicos, South Caicos, West Caicos, and Providenciales), plus numerous rocky islets, all surrounded by the Caicos Bank, a triangular shoal. The total land area of the Turks and Caicos Islands is 430 sq km (166 sq mi), with extensions of about 120 km (75 mi) E–W and 80 km (50 mi) N–S. Comparatively, the area occupied by the Turks and Caicos Islands is slightly less than 2.5 times the size of Washington, D.C. The Turks and Caicos Islands have a coastline length of 389 km (242 mi). The capital city, Grand Turk, is in the Turks Islands.

²TOPOGRAPHY

The Turks Islands are low and flat, and surrounded by reefs, sunken coral heads, and boilers. The land mass is limestone, well weathered with pockets of soil; the coastlines are indented with shallow creeks and mangrove swamps. The Caicos Bank is a triangular shoal about 93 km (58 mi) long on its northern side and 90 km (56 mi) long on its eastern and western sides. The highest elevation is only 50 m (163 ft) above sea level on Providenciales. On the north coast of Middle Caicos (which is also known as Grand Caicos) are limestone cave formations.

³CLIMATE

Days are sunny and dry and nights are cool and clear throughout the year. Temperatures on the islands range from a low of 16°C (61°F) to a high of 32°C (90°F), with the hottest period generally occurring between April and November. There are almost constant tradewinds from the east. Rainfall averages 53 cm (21 in) per year, and hurricanes are a frequent occurrence. Major hurricanes struck the islands in 1866, 1873, 1888, 1908, 1926, 1928, 1945, 1960, and 1985.

⁴FLORA AND FAUNA

The ground cover is scrubby and stunted tropical vegetation, with sea oats, mangrove, casuarina, and palmetto. There is little natural wildlife other than birds and butterflies. West Caicos island is especially noted as a sanctuary for birds, and Penniston, Gibb, and Round cays are known for their extensive varieties of butterflies. Spiny lobster, conch, clams, bonefish, snapper, grouper, and turtle are plentiful.

⁵ENVIRONMENT

Fresh water is a scarce commodity, and most islanders rely on private cisterns. Underground water is present on North and Middle Caicos, but surface water collected in ponds after rainfall becomes brackish. There have been some complaints of actual or potential ecological damage resulting from the expansion of tourist facilities on Providenciales. The government has identified the absence of environmental education as a potential problem for the preservation of the nation's natural heritage in the future. It has also introduced the National Physical Development Plan for the period from 1987 to 1997, which emphasizes the need for preservation of the nation's resources. By 1992, the government developed legislation that would create 12 national parks, 8 nature reserves, 5 sanctuaries, and 9 historic sites.

⁶POPULATION

The population in 1990 was 11,696, a 57% increase over the 1980 census figure of 7,435. Most people were living on Grand Turk (3,720), South Caicos (1,220), and North Caicos (1,305) islands. During 1991, the birthrate was 25 and the death rate 6 per 1,000 population. The 1990 population density was 27 per sq km (70 per sq mi). The major towns are Grand Turk, and Cockburn Harbour, on South Caicos.

⁷MIGRATION

Because of the islands' limited economic opportunities, it has become common for young men to emigrate to Caribbean islands (including Puerto Rico) or to the US in search of work. In the mid-1980s, many Haitian and Dominican immigrants came to

Turks and Caicos, especially to Providenciales, to work in low-wage hotel jobs unattractive to local residents.

[8]ETHNIC GROUPS
About 90% of the population is of black African descent, the remainder being of mixed, European, or North American origin.

[9]LANGUAGES
The official and universal language of the Turks and Caicos Islands is English, interspersed with a number of local colloquialisms.

[10]RELIGIONS
Most islanders are Christian; the main denominations are Baptist (25.5% in 1990), Methodist, and Anglican. Other Protestant groups and the Roman Catholic Church, which comprises 18% of the population, are also represented.

[11]TRANSPORTATION
There are about 121 km (75 mi) of roads on the islands. The main roads on Grand Turk and South Caicos are paved. Some secondary roads are surfaced with scale from the salinas (salt ponds); otherwise, the roads are merely dirt and sand tracks.

The main seaports are at Grand Turk, Cockburn Harbour on South Caicos, Providenciales, and Salt Cay. An offshore registry program with the UK enables British merchant ships to register with the Turks and Caicos Islands in order to cut crew costs while enabling the vessels to fly the Red Ensign of the UK. There are four paved runways of more than 1,220 m (4,000 ft) on Grand Turk, South Caicos, and Providenciales; in addition, there are three small unpaved landing strips among the other inhabited islands.

[12]HISTORY
Archaeological expeditions have found Arawak implements and utensils on Turks and Caicos Islands. When Juan Ponce de León arrived in 1512, Lucayan Indians had come to inhabit the islands. There is some speculation that Columbus may have made his landfall on Grand Turk or East Caicos on his first voyage of discovery in 1492. The first settlements were by Bermudians, who established solar salt pans in the 1670s. Bahamian, Bermudan, Spanish, French, and British rivalry over the prospering salt trade resulted in numerous invasions and evictions through the first half of the 18th century. In 1787, Loyalists fleeing the American Revolution established settlements and cotton and sisal plantations on several of the larger Caicos Islands. Ten years later, the islands came under the jurisdiction of the Bahamas colonial government. Slavery was abolished in 1834. In 1848, the Turks and Caicos islanders were granted a charter of separation from the Bahamas after they complained of Bahamian taxes on their salt industry.

From 1848 to 1873, the islands were largely self-governing, under the supervision of the governor of Jamaica. Following the decline of the salt industry, the islands became a Jamaican dependency until 1958, when they joined the Federation of the West Indies. When the federation dissolved and Jamaica achieved independence in 1962, Turks and Caicos became a crown colony administered by the British Colonial Office and a local council of elected and appointed members. In 1965, the governor of the Bahamas was also appointed governor of Turks and Caicos, but with the advent of Bahamian independence in 1973, a separate governor was appointed. A new constitution maintaining the status of crown colony and providing for ministerial government was introduced in September 1976. Although independence for Turks and Caicos in 1982 had been agreed upon in principle in 1979, a change in government brought a reversal in policy. The islands are still a crown colony.

The islands were shaken by scandals in the mid-1980s. In March 1985, Chief Minister Norman B. Saunders and two other ministers were arrested in Florida on drug charges and later convicted and sentenced to prison. In July 1986, a commission of inquiry found that Chief Minister Nathaniel Francis and two other ministers had been guilty of "unconstitutional behavior, political discrimination, and administrative malpractice." The governor thereupon ended ministerial government in July 1986 and, with four members of the former Executive Council, formed an Advisory Council to govern until new elections. The islands have returned to their previous form of government, and remain a dependent territory of the UK.

[13]GOVERNMENT
The constitution of 30 August 1976 established a ministerial system in which the governor, representing the crown, retains responsibility for external affairs, defense, and internal security. The governor presides over an Executive Council of seven members: the chief minister, three elected members of the Legislative Council, and three ex-officio members (the financial secretary, the chief secretary, and the attorney general). The Legislative Council has 20 seats: a speaker, elected representatives from each of 13 constituencies, 3 nominated members, and the same 3 ex-officio members from the Executive Council. In order to remain in office, the chief minister must command a majority of the Legislative Council. A number of constitutional changes have been suggested and studied, but are still under review by a special commission.

[14]POLITICAL PARTIES
The Progressive National Party (PNP) and the People's Democratic Movement (PDM) are the two major parties in the system. They differ primarily over the question of independence. The PDM had sought the rapid advance of Turks and Caicos to full internal self-government, and when the UK reportedly insisted on rapid progress toward independence, the PDM appeared to swing toward that view. The PNP, on the other hand, expressed a go-slow attitude, holding that the islands lacked the institutional and financial resources for independence.

With the PNP victory in 1980, Turks and Caicos' independence date, originally scheduled for mid-1982, was postponed indefinitely. During the 1984 elections, neither party raised the issue of independence. The PNP remains in power today. In 1991, it took 8 of the 13 elected seats in the Legislative Council.

[15]LOCAL GOVERNMENT
All local affairs are administered by the Legislative Council and Executive Council. Both of these councils sit in Grand Turk.

[16]JUDICIAL SYSTEM
Legislative Council acts, certain laws of the UK Parliament, and a few Jamaican and Bahamian statutes are the law of the land. The administration of justice is in the hands of a magistrate who sits weekly in Grand Turk and may sit in each of the other islands as necessary. The magistrate is also the registrar of deeds and of birth, marriage, death, and company records.

Appeals are heard by a nonresident member of the Eastern Caribbean Supreme Court; as of 1985, this judge was from the Bahamas. Appeals in certain cases may also be taken to the UK Judicial Committee of the Privy Council in London.

[17]ARMED FORCES
The Turks and Caicos Islands has no military force. Defense is provided by the UK, in keeping with the islands' crown colony status. The Royal Police Force is an establishment of 150 officers distributed throughout the islands.

[18]INTERNATIONAL COOPERATION
As a crown colony, Turks and Caicos does not maintain diplomatic or consular representation in other states. A UK trade commissioner based in Nassau, Bahamas, and a Canadian trade commissioner based in Kingston, Jamaica, are responsible for conducting the islands' international commerce.

¹⁹ECONOMY

Tourism and lobster fishing have replaced salt raking as the main economic activity of the islands, which are very poor. Fishing and subsistence farming are the principal occupations; underemployment and unemployment are estimated at over 40%. Important sources of income include tourism and offshore financial services. The closing of the last US military base in 1983 resulted in the loss of rental payments that accounted for 10% of government revenue. Most of the retail trade on the islands consists of imported goods.

Tourism and offshore financing remain the two major contributors to the economy. The real GDP increased an average 8.2% between 1987 and 1991. However, in 1992, economic growth was more moderate, due to a decline in the tourism and construction sectors. Tourist arrivals declined to 52,000 in 1992, a 4.5% decrease from 1991.

On the other hand, offshore financing has been increasing in popularity, earning $1.8 million in registration fees in 1990. The advantages of no direct taxation, having the dollar as the national currency, confidentiality, and the growing financial infrastructure have all contributed to the improvement of this sector of the economy.

²⁰INCOME

In 1989, the GDP was estimated at $68.5 million, or $5,000 per capita.

²¹LABOR

Most self-employed persons engage in fishing or subsistence farming. Government is the largest employer, followed by the tourist sector and financial services.

There is one labor organization, the St. George's Industrial Trade Union, with a membership of approximately 250.

²²AGRICULTURE

Of the total area of Turks and Caicos, only about 2%, or 1,000 hectares (2,500 acres), is devoted to agriculture. Crop output is small. Rainfall is low, and the cost of transporting produce from island to island is not competitive with imports.

Most agricultural production is on North Caicos, although subsistence farming occurs wherever the soil permits. Corn, beans, and other food crops are grown entirely for local or household consumption. Fruit and many other foods must be imported, with Haiti as the principal supplier.

²³ANIMAL HUSBANDRY

Cattle, hogs, and poultry are raised by householders to supplement food supplies. There is little commercial production of meat or dairy products.

²⁴FISHING

Fishing is the traditional occupation of the islanders. The harvesting of lobster and conch for export to the US and Haiti is an organized commercial operation that in 1991 yielded 210 tons of lobster and 431 tons of conch, of which a combined 313 tons were exported, valued at $3.5 million.

²⁵FORESTRY

There are no significant stands of forests.

²⁶MINING

There is no mineral wealth other than the salinas. Salt raking was the major employer and earner of export income until the 1950s; the salt was exported in coarse, rough, and fine grain for curing, tanning, and domestic uses. With the decline of foreign markets, all salt operations for export were closed down, and salt has not been taken in commercial quantities since 1974.

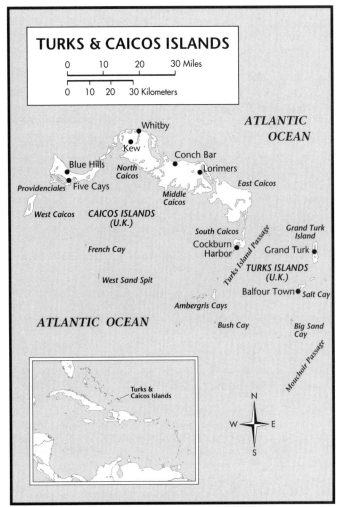

LOCATION: 21° to 22°N; 71° to 72°30′W.

²⁷ENERGY AND POWER

As of 1992, electricity generated to all the islands was provided by private companies. Electrical output in 1991 amounted to 9 million kwh, entirely from conventional thermal sources; slightly more than half the output came from public sources.

²⁸INDUSTRY

The main industry is the processing of lobster and conch for export. South Caicos and Providenciales have a combined total of four seafood processing plants.

²⁹SCIENCE AND TECHNOLOGY

The Turks and Caicos Islands are entirely dependent upon outside technological resources, and persons wishing to pursue advanced technical or scientific training must go elsewhere.

³⁰DOMESTIC TRADE

General business is of the service type, reliant upon imported goods; most businesses are located on Grand Turk. As a general rule, early closing day on the islands is Wednesday, at 1 PM.

³¹FOREIGN TRADE

The US and UK constitute the islands' principal trade partners. In 1991, the country registered $4.4 million in exports, mainly composed of lobstertails, conch, and fish meat. Imports recorded $39.8 million, including food, beverages, manufactured goods, fuel, building supplies, and tobacco.

32BALANCE OF PAYMENTS
The negative balance of trade—a consistent feature of the islands' trade pattern—had been offset by budgetary aid grants from the UK. These were ended in 1989, although British development and technical assistance funding continue. In 1991, imports exceeded exports by $35.4 million. Tourism earnings and registration fees from offshore banking, insurance, trust, and investment companies help offset the trade deficit.

33BANKING AND SECURITIES
Four commercial banks have branches on the principal islands. In addition to providing the normal financial services, the banks handle securities trading on the principal international exchanges. Turks and Caicos has been developing as an offshore financial center since 1979.

34INSURANCE
There were five life and property insurance companies as of the mid-1980s.

35PUBLIC FINANCE
In 1992/93, estimated government revenues came to $28.2 million, while expenditures totalled $23.4 million. The main sources of government revenue are customs duties, business registration fees, and philatelic sales.

36TAXATION
No taxes are levied on land, property, or income. A probate fee of 2% is levied on estates up to a maximum of $550.

37CUSTOMS AND DUTIES
Customs duties are levied on most imported items based on a single-tier customs tariff. An import license is required in most cases. Turks and Caicos Islands adhere to the customs valuation agreement of the Tokyo Round of GATT.

38FOREIGN INVESTMENT
As a means of attracting foreign capital, the Encouragement of Development Ordinance grants new investment guarantees against taxes for 10–15 years. In addition, there are no direct taxes either on income capital for individuals or companies and no exchange controls

39ECONOMIC DEVELOPMENT
The UK government has historically been the underwriter of economic development planning and investment. Tourism presents the primary economic development opportunity, but growth has so far been small. Since 1979, Turks and Caicos has experienced a flurry of offshore tax haven ventures, but as of 1987 these had little significant impact on local financial or employment conditions.

The most recent economic development has been on Providenciales, in the form of tourist facilities and retirement home construction. New hotels, resorts, and casinos were inaugurated in the islands in 1993, and the Providenciales airport was expanded at a cost of $8.5 million. In 1991, the PNP government proposed a five-year development program for economic independence called "Progress through Partnership," which aims at achieving over 15% in real growth per year for the next five years.

40SOCIAL DEVELOPMENT
The central government provides few welfare services or benefits. Churches and benevolent societies are the principal sources of charitable aid.

41HEALTH
A modern cottage hospital (30 beds) and an outpatient and dental clinic are located on Grand Turk, and there are 11 outpatient and dental clinics on South, Middle, and North Caicos, Providenciales, and Salt Cay. In 1992, there were 1.25 doctors per 1,000 people.

42HOUSING
Most old wooden shacks—vulnerable to hurricanes and frequently damaged—have been replaced by concrete block structures, especially in the towns. However, some traditional 19th-century Bermudian architecture remains. Several old churches and the library are built of limestone.

43EDUCATION
Education is free and compulsory between the ages of 4½ and 15. In 1985/86 there were 14 government primary schools and two private primary schools, with a combined total of 1,429 students, and three government secondary schools, with 815 students. Expenditure on education in 1985 was $1.4 million. There are no higher educational institutions on the islands.

44LIBRARIES AND MUSEUMS
The library in Grand Turk doubles as a museum. Its principal attraction is a display of Lucayan Indian artifacts.

45MEDIA
Telephone service operates on Grand Turk, South Caicos, and Providenciales; there were 1,450 telephones in 1993. International cable and wireless service is available through Cable and Wireless (West Indies) Ltd. There are 3 AM radio stations and several TV stations; in 1991 there were approximately 6,000 radios.

The *Turks and Caicos News* is a weekly paper published in Grand Turk. The *Turks and Caicos Current* is published every two months; the *Turks and Caicos Chronicle*, quarterly; and *The Voice*, the organ of the People's Democratic Movement, monthly.

46ORGANIZATIONS
There is a chamber of commerce on Grand Turk. Churches and benevolent societies are important centers of social life throughout the Turks and Caicos Islands.

47TOURISM, TRAVEL, AND RECREATION
A total of 54,616 tourists visited the islands in 1991, 43,390 from North America. There were 1,051 hotel rooms with 2,102 beds and a 61% occupancy rate. Revenues from tourism were US$43 million. In 1992, tourist arrivals declined slightly to 52,000, primarily due to the US recession and the demise of Pan Am, the principal airline that served the islands. Tourism is expected to rebound with the advent of service by American Airlines, and new hotel, resort and casino openings. Visitors are attracted by the beautiful beaches and by opportunities for snorkeling, diving, and sport fishing. The windmills and salinas on Salt Cay and the 19th-century architecture on Grand Turk, along with horse carriages, provide a quaint setting.

48FAMOUS PERSONS
J.A.G.S. MacCartney (1946?–1980), referred to by his countrymen as "Chief," was the first chief minister of the Turks and Caicos Islands, serving from 1976 until his death. After attending school in Jamaica, he had worked as a supply clerk and as a bartender in the Bahamas; he founded the PDM in 1975 and led it to electoral victory in 1976.

49DEPENDENCIES
Turks and Caicos has no territories or colonies.

50BIBLIOGRAPHY
Augier, F. R. *The Making of the Indies*. London: Longmans, 1965.
Boultbee, Paul G. *Turks and Caicos Islands*. Santa Barbara, Calif.: Clio Press, 1991.

UNITED KINGDOM
AMERICAN DEPENDENCIES

BERMUDA

Bermuda is a colony consisting of some 300 coral islands (20 of them inhabited), situated in the Atlantic Ocean, 933 km (580 mi) east of Cape Hatteras (US) at 32°19′N and 64°35′w. Their total area is about 54 sq km (21 sq mi). The US leases a 5.8-sq-km (2.2-sq-mi) area of land reclaimed from the sea for military purposes. The largest island, Bermuda (sometimes called Main) Island, is about 23 km (14 mi) long and has an average width of 1.6 km (1 mi). The islands are mostly flat and rocky, with luxuriant semitropical vegetation. Because Bermuda lies in the Gulf Stream, the climate is generally mild and humid, with a mean annual temperature of 21°C (70°F) and average rainfall of 147 cm (48 in). The resident civilian population in 1992 was 60,213, of which about 65% was black or mixed and 35% white, mainly of English or Portuguese descent.

Most of the 240 km (149 mi) of roads are surfaced. Public transportation is largely by bus. Hamilton, the capital, has a deepwater harbor. Kindley Field, near St. George, the former capital, is Bermuda's international airport.

The oldest British colony, the islands were uninhabited when discovered in 1503 by the Spaniard Juan de Bermúdez. Bermuda was first settled by a group of British colonists under Sir George Somers, who were wrecked there while en route to Virginia in 1609. Bermuda was acquired from a chartered company by the crown in 1684. Under the 1968 constitution, the governor, representing the sovereign, is advised by a cabinet of legislators appointed at the recommendation of the prime minister. The bicameral legislature consists of an appointed Senate of 11 members and a 40-member House of Assembly (elected by universal suffrage from 20 electoral districts). The Bermuda dollar of 100 cents is equal in value to the US dollar, which circulates freely.

Tourism is the islands' largest employer, providing about half the total national income and two-thirds of foreign exchange. In 1991 there were some 384,695 visitors, 89% of them from the US. The islands earned $454 million from tourism in 1991. Earnings from tourism have increased 5.27% a year between 1980 and 1991. Bermuda does not impose income or corporate taxes which has led to a substantial offshore financial sector. Also important to the economy are goods and services supplied to the UK and US armed forces stationed in Bermuda. Light industries produce pharmaceuticals and essences, brass electrical contacts, and cut flowers for export. Per capita GDP in 1991 was $22,400, among the highest in the world.

The chief imports are food, textiles, furniture, motor vehicles, and fuel. Commerce is principally with Canada, the UK, and the US. Visible trade balances are unfavorable, although trade in invisibles, primarily tourism and international business, more than offsets the commodity trade deficit. Exports in 1988 totaled $30 million; imports, $420 million. There is a free port at Ireland Island.

Medical services are private. One well-equipped hospital, with 233 beds, receives government support. Education is free and compulsory between the ages of 5 and 16.

In 1993 there were over 52,000 telephones on the island. The Bermuda Broadcasting Co., Ltd., runs the two commercial television stations and five of the six radio stations. There is one daily newspaper, the *Royal Gazette*.

BRITISH ANTARCTIC TERRITORY

Created on 3 March 1962 from former Falkland Islands dependencies, the British Antarctic Territory lies south of 60°s and between 20° and 80°w, and consists of the South Shetlands, 4,662 sq km (1,800 sq mi); the South Orkneys, 622 sq km (240 sq mi); and Graham Land on the Antarctic continent. The territory is governed by a UK-appointed high commissioner, who also serves as governor of the Falklands. There is no permanent habitation; the number of researchers varies from 250 to 500.

BRITISH VIRGIN ISLANDS

The British Virgin Islands consist of some 50 Caribbean islands and islets, totaling 153 sq km (59 sq mi), at about 18°25′N and 64°30′w. Until 1 July 1956, they were administered as part of the Leeward Islands. The 1992 population was 12,707, almost wholly of African descent. Road Town (population about 2,500 in 1992) on the island of Tortola is the capital. The climate is pleasantly subtropical. At least 164.5 km (99 mi) of roads are motorable. There are several airstrips on the islands.

Under the constitution of 1968, as revised in 1977, the government is headed by a British-appointed administrator, who is assisted by an Executive Council and a Legislative Council of 10 members (including 9 elected by universal adult suffrage at age 18).

The economy is interdependent with that of the US Virgin Islands, which lie to the west. The US dollar is the legal currency. Livestock raising, farming, and fishing are the principal economic activities. Light industries include distilleries for alcoholic beverages, a concrete block factory, boat building, and handicrafts. Chief imports include foodstuffs, apparel, cotton piece goods, timber, and machinery. The overwhelmingly adverse balance of trade is offset by remittances from migrant workers and an expanding tourist industry. Tourism expenditures in the British Virgin Islands in 1991 were $109 million. That year 147,030 people visited the islands. Most of the tourists were from the US mainland, Puerto Rico, and Canada. Principal trade partners include Canada, the UK, and the US. Per capita GDP was $10,479 in 1992.

The infant mortality rate—40.9 per 1,000 live births in 1982—has been reduced from 78.9 per 1,000 live births in 1960. In 1986 there was one hospital with 50 beds.

Primary education is free between the ages of 5 and 15 and compulsory up to the age of 13. In 1992 there were 11 private and 18 government primary schools and 3 public and private high schools in the British Virgin Islands. The sole newspaper, the *Island Sun*, is published weekly.

As of 1987, the Virgin Islands tree boa and Anegada ground iguana were endangered species.

CAYMAN ISLANDS

The three low-lying Cayman Islands—Grand Cayman, Little Cayman, and Cayman Brac, with a total area of 262 sq km (101 sq mi)—are situated between 79°44′ and 81°27′w and 19°15′ and 19°45′N, about 290 km (180 mi) WNW of Jamaica, of which they were formerly a dependency. Grand Cayman, flat, rockbound, and protected by coral reefs, is about 32 km (20 mi) long and 6 to 11 km (4–7 mi) broad; George Town, on Grand Cayman, is the capital and chief town. The other two islands are about 145 km (90 mi) to the NE. The 1992 population was estimated at 29,139, about 90% of whom resided on Grand Cayman. Cayman Airways is the main air carrier; the principal international airport is on Grand Cayman.

The islands were discovered in 1503 by Columbus, who named them Las Tortugas, from the turtles with which the surrounding seas abound. They were never occupied by Spaniards and were colonized from Jamaica by the British. They were a dependency of Jamaica until 1959, but severed all constitutional links with Jamaica when the latter became independent in 1962.

The 1972 constitution empowers the crown-appointed governor to make laws with the advice and consent of the Legislative Assembly. The Executive Council consists of 4 members chosen by the Assembly from among its 12 elected members, and 3 Assembly members appointed to the Council by the governor. The Legislative Assembly includes 3 ex officio members, 12 elected members, and the governor. Elections, in which all adult British residents may vote, are held every three years. Local administration is in the hands of justices of the peace and vestrymen.

The Cayman Islands dollar (CI$) is linked to the US dollar at the rate of CI$1 = US$1.20 (or US$1 = CI$0.835). Customs, duties, license and company fees, and postage and stamp taxes are the principal source of government revenue. The absence of taxes on income, capital gains, real estate, and inheritances attracts overseas investors to the region; international financial services and tourism have become principal sectors of the economy. At the end of 1990, 22,260 companies were registered in the Caymans, and 546 bank and trust companies were represented. Tourism has grown rapidly: in the early 1990s 237,000 people visited the islands annually. Although the soil is fertile and there is some farming, the agricultural sector remains small; the catching of turtles, sharks, and sponges also provides some employment. Remittances from Caymanian seamen serving on foreign ships contribute to the economy as well. The government-owned Cayman Turtle Farm, unique in the world, produces turtle meat for local consumption; exports have waned in recent years due to restrictions by the Convention on International Trade of Endangered Species of Wild Fauna and Flora.

The islands have 2 hospitals with a total 72 beds, 10 of them for extended care. In 1992, 3,199 pupils attended public elementary and secondary schools.

As of 1987, endangered species included the tundra peregrine falcon and the green sea turtle.

FALKLAND ISLANDS

The Falkland Islands (Islas Malvinas), a British crown colony in the South Atlantic, lie some 772 km (480 mi) northeast of Cape Horn, between 51° and 53°s and 57° and 62°w, and have an area of 12,173 sq km (4,700 sq mi). The two main islands, East Falkland and West Falkland, consist chiefly of hilly moorlands. The population (around 1,900 in 1992) is almost exclusively of British origin. Stanley, on East Falkland, the capital and only town, has about 1,000 inhabitants. There are no railways and few surfaced roads beyond the vicinity of Stanley. Shipping service to and from the islands is slight. There is internal air service but no international service.

The Falklands were sighted in 1592 by John Davis, an English navigator. The French founded the first colony on East Falkland, in 1764, transferring it two years later to Spain, which renamed it Soledad. The British took possession of West Falkland in 1765. Both islands were eventually abandoned. In 1820, Argentina (then the United Provinces of La Plata) colonized East Falkland. British troops occupied the islands in 1832–33, but Argentina has continued to dispute Britain's claim to the Falklands. On 2 April 1982, Argentine troops invaded the islands, precipitating a conflict with the British that cost over 1,000 lives. The UK recaptured the

islands on 14 June, and some 4,300 British soldiers remain in the Falklands following the conflict.

Under a new constitution which came into effect in October 1985, the colony is administered by an appointed governor, with an Executive Council of 5 members: 3 chosen by the 8-member Legislative Council, and 2 ex officio members, the chief executive and the financial secretary. The currency unit is the Falkland Island Pound (F£), which is equal to the pound sterling.

There is no commercial agriculture. Most households in Stanley and the outlying areas grow their own vegetables. Sheep farming, the main industry, is directed primarily to the production of wool, hides, and skins, and the manufacture of tallow. All land is divided into large sheep farms; there are some 700,000 sheep on the islands. Most commodities needed by the territory and its dependencies are imported. Exports, (mostly wool), were valued at US$14.7 million in 1990. Trade is principally with the UK. Customs is the largest source of government revenue.

All medical services are public. The islands has three physicians. The only civilian hospital was destroyed by fire in 1984; temporary facilities were used, with support from the British Military Hospital, while a replacement was constructed. Education is free and compulsory for children from 5 to 15 years of age.

Dependencies include South Georgia and the South Sandwich Islands, some 1,300 km (800 mi) E of the Falklands, with an area of 4,092 sq km (1,580 sq mi). Whaling and sealing are the main industries.

LEEWARD ISLANDS

The Leeward Islands, part of the Lesser Antilles island chain, lie east and south of Puerto Rico and north of the Windward group. Of the four territorial units that constitute the Leeward Islands, two—Antigua and Barbuda, and St. Kitts and Nevis—are independent nations covered elsewhere in this volume. The other two—Anguilla and Montserrat—retained the status of UK dependencies as of early 1988.

ANGUILLA

Anguilla, the most northerly of the Leeward chain, lies at approximately 18°N and 63°W, and has an area of 90 sq km (35 sq mi). The island is long, flat, dry, and covered with scrub; its rolling hills reach a peak elevation of 65 m (213 ft) above sea level. The average annual temperature is 27°C (81°F), with July–October being the hottest period and December–February the coolest. Rainfall averages 89 cm (35 in) a year, but there is considerable variation both from season to season and from year to year. The hurricane season, marked by occasional thunderstorms and sudden squalls, lasts from July to October.

The population was 6,963 in 1992. Most Anguillans are of African descent, with an admixture of European (especially Irish) ancestry. The population is overwhelmingly Christian: Anglicans and Methodists predominate, but there are also Seventh-Day Adventist, Baptist, Roman Catholic, and Church of God congregations. English is the official language, spoken in a distinctive island patois. Anguilla has no official capital, but The Valley serves as an administrative center. Anguilla has about 90 km (56 mi) of roads, 46 km (29 mi) of them paved. Road Bay is the main harbor, and there is daily ferry service between Blowing Point and the French-Dutch island of St. Martin (Sint Maarten), about 8 km (5 mi) away. Air service to and from Wallblake Airport is provided by the privately owned Air Anguilla and two other interisland airways.

Although discovered by Columbus in 1496, Anguilla was not settled by Europeans until 1650, when British colonists arrived from St. Kitts. From 1671, Anguilla was governed as part of the Leeward Islands, and between 1871 and 1956 the island formed (with St. Kitts and, from 1882, Nevis) part of the Leeward Islands Federation. All the Leeward Islands were consolidated into a single

UNITED KINGDOM AMERICAN DEPENDECIES

CAYMAN ISLANDS

LITTLE CAYMAN

North East Point

West End Point

CAYMAN BRAC

North Sound

Rum Point

Conch Point

Colliers

George Town

Bodden Town

GRAND CAYMAN

0 25 50 Miles

0 25 50 Kilometers

BRITISH VIRGIN ISLANDS

ANEGADA

SOMBRERO

TORTOLA *VIRGIN GORDA*

Road Town

Anguilla

St. Thomas *St. John*

U.S. VIRGIN ISLANDS

St. Martin (Neth. & Fr.)

St. Barthélemy (Fr.)

St. Croix

Saba (Neth.)

St. Eustatius (Neth.)

Barbuda

St. Kitts

ST. KITTS & NEVIS

Nevis

ANTIGUA & BARBUDA

Antigua

0 25 50 Miles

0 25 50 Kilometers

MONTSERRAT

Plymouth ✪

BERMUDA

ATLANTIC OCEAN

ST. GEORGE'S I. St. George

Murray's Anchorage

Ruth's Point

Crystal Caves

Castle Harbour

Harrington Sound

Nonsuch I.

Ireland Island N.

Grassy Bay

Ireland Island S.

Bozz I.

Tucker's Town

Flatts Village

Spanish Rock

Sue Wood Bay

Somerset

Hamilton

Great Sound

Hungry Bay

Little Sound

BERMUDA I.

0 1 2 3 Miles

0 1 2 3 Kilometers

Falkland Islands

Bermuda

Cayman Islands

U.K. Virgin Islands

N W E S

FALKLAND ISLANDS

ARGENTINA

Bahía Laura

EAST FALKLAND

Jason Is.

Pebble I.

Macbride Head

Bahía Grande

WEST FALKLAND

New I.

Beaver I.

Darwin

Stanley

Río Gallegos

Weddell I.

Falkland Sound

Cape Meredith

Sea Lion Is.

CHILE

TIERRA DEL FUEGO

Isla de los Estados

Cabo de Hornos

0 50 100 Miles

0 50 100 Kilometers

BRITISH ANTARCTIC TERRITORY

Antarctic Penninsula

SOUTH AMERICA

TRANSANTARCTIC MOUNTAINS

ANTARCTICA

Ice Shelf

territory in 1956 and, as such, were incorporated into the Federation of the West Indies two years later. With the breakup of the West Indies Federation in 1962, St. Kitts–Nevis–Anguilla reverted to colonial status. On 17 February 1967, St. Kitts–Nevis–Anguilla acquired self-government within the newly formed West Indies Associated States. After Anguilla declared its independence of the Associated States in 1969, some 300 British paratroopers temporarily took command of the island. On 10 February 1976, the UK recognized Anguilla's status as a dependency distinct from St. Kitts–Nevis, which achieved independence in 1983.

Under the Anguilla Constitution Order of 1982, the crown is represented by a governor, who presides over an appointed Executive Council and an elected House of Assembly. The Executive Council consists of the chief minister, three other ministers selected by the governor from among the members of the House of Assembly, and the attorney general and permanent secretary for finance, who serve ex officio both on the council and in the legislature. The governor also appoints two members of the House of Assembly, the remaining seven being elected to five-year terms by universal adult suffrage. In the election of March 1984, the Anguilla People's Party took two seats while the Anguilla National Alliance won four. Justice is administered by a magistrate's court, a Court of Appeal, and a High Court, whose sitting judge is provided by the Eastern Caribbean Supreme Court on St. Lucia.

The mainstays of the economy have traditionally been animal husbandry, salt production, fishing, and boat building, but tourism and offshore banking have played an increasingly important role in recent years. The principal crops are pigeon peas, corn, and sweet potatoes; sheep, goats, hogs, cattle, and poultry are raised. Salt is extracted by evaporation from two briny ponds, and live lobsters are exported to neighboring islands. The East Caribbean dollar (EC$) is used; EC$1 = US$0.3704 (or US$1 = EC$2.70). Government revenues in 1992 were budgeted at US$13.8 million and expenditures at US$15.2 million. Anguilla had 15,400 visitors in 1985. Per capita income was US$3,300 in 1992.

Education is free and compulsory between the ages of 5 and 14 years. The government maintains six primary schools and one secondary school; total school enrollment in 1984 exceeded 2,000. A 24-bed cottage hospital offers limited services. International telephone, telegraph, and telex services are available, and the government-run Radio Anguilla is on the air for more than 10 hours a day; a privately owned religious station, Caribbean Beacon Radio, operates from The Valley.

Dependencies of Anguilla include numerous offshore islets and cays, as well as Sombrero Island (5 sq km/2 sq mi), about 56 km (35 mi) to the northwest.

MONTSERRAT

Situated at 16°45′N and 62°10′W, Montserrat has an area of 103 sq km (40 sq mi). The island, which lies between Nevis and Guadeloupe, about 43 km (27 mi) southeast of Antigua, has a mountainous terrain, with two peaks rising higher than 900 m (3,000 ft). Montserrat is wholly volcanic in origin, and there are seven active volcanoes. Mean temperatures range from a minimum of 23°C (73°F) to a maximum of 31°C (88°F); June–November, the hurricane season, is the warmest time of the year, and December–March is the coolest. There is no clearly defined rainy season, although rainfall tends to be more abundant during the second half of the year; the annual average is 157 cm (62 in).

In 1992, 12,617 people lived on the island, of whom 3,500 lived in Plymouth, the capital and sole port. Most residents of Montserrat are of African ancestry. Anglicans, Methodists, Roman Catholics, and Pentecostals make up the great majority of the population. English, the official language, is spoken in an island patois. Montserrat has about 240 km (150 mi) of main surfaced roads. Blackburne Airport, about 15 km (9 mi) from the capital, has a surfaced runway of 1,030 m (3,380 ft), which opened to traffic in 1967. Montserrat Aviation Services, in cooperation with Leeward Islands Air Transport (LIAT), maintains regular flights to and from Antigua, Nevis, and St. Kitts.

Christopher Columbus, who discovered the island in November 1493, gave it the name Montserrat because its rugged terrain reminded him of the site of the Abbey of Montserrat in the Spanish highlands near Barcelona. English and Irish colonists from St. Kitts settled on the island in 1632, and the first African slaves arrived 32 years later. Throughout the 18th century, the British and French warred for possession of Montserrat, which was finally confirmed as a British possession by the Treaty of Versailles (1783). By the early 19th century, Montserrat had a plantation economy, but the abolition of slavery in 1834, the elimination of the apprentice system, the declining market for sugar, and a series of natural disasters brought the downfall of the sugar estates. In the mid-19th century, Joseph Sturge of Birmingham, England, organized a company that bought up the abandoned estates, planted them with limes (a product for which Montserrat is still famous), and sold plots of land to small farmers. From 1871 to 1956, Montserrat formed part of the Federation of the Leeward Islands, and after two years it became part of the Federation of the West Indies (1958–62). Since the breakup of the Federation, Montserrat has been separately administered, under a constitution effective 1 January 1960.

The crown is represented by an appointed governor, who presides over an Executive Council structured like that of Anguilla. There is also a Legislative Council which, like Anguilla's, includes two appointed and two ex officio members. The seven elected members of the legislature are chosen from single-member constituencies by universal adult suffrage at age 18. The legislators serve terms of up to five years. In elections held in November 1978, the People's Liberation Movement, headed by John Osborne, swept all seven elective seats; the party was returned to power in the February 1983 general election, though with a reduced majority of five seats, two having been captured by the Progressive Democratic Party. Montserrat's judicial system consists of a magistrate's court and a Court of Summary Jurisdiction; appeals are to the Eastern Caribbean Supreme Court on St. Lucia.

Tourism accounts for about one-fourth of the annual GDP; the island had some 19,244 visitors in 1991. Important crops include hot peppers (mostly for export), limes and other orchard fruits, tomatoes, and vegetables. Exports of live cattle and leather are significant, as are such light industrial exports as plastic bags, cotton garments, and electronic parts. Montserrat uses the East Caribbean dollar. In 1988, GDP per capita was US$4,500. Exports that year were US$2.3 million; imports, chiefly of industrial goods, foodstuffs, and mineral fuels, totaled US$30 million. Governmental revenues in 1991 were US$12.1 million; expenditures, US$14.3 million. Effective corporate income tax rates are 20–40%; the maximum personal income tax rate is 30%. A property tax is also levied.

The principal health facility is a 67-bed general hospital maintained by the government; provisions for social welfare include a family planning association and an old people's welfare association. Free dental care is provided by the government for all schoolchildren, elderly persons, and expectant or nursing mothers. Education is free and compulsory up to age 14. The government maintains 15 schools: 1 infant school (ages 5–7), 9 primary schools (ages 5–12), 2 all-age schools (ages 5–15), and 3 secondary schools (ages 10–19). Radio service is provided by the government-owned Radio Montserrat and by 2 commercial stations, Radio Antilles and Gem; television programs are relayed from elsewhere in the Caribbean and transmitted locally by Antilles Television and cable stations. There are two weekly newspapers, the *Montserrat Times* and the *Montserrat Reporter*.

UNITED STATES
OF AMERICA

CAPITAL: Washington, D.C. (District of Columbia).

FLAG: The flag consists of 13 alternate stripes, 7 red and 6 white; these represent the 13 original colonies. Fifty 5-pointed white stars, representing the present number of states in the Union, are placed in 9 horizontal rows alternately of 6 and 5 against a blue field in the upper left corner of the flag.

ANTHEM: *The Star-Spangled Banner.*

MONETARY UNIT: The dollar ($) of 100 cents is a paper currency with a floating rate. There are coins of 1, 5, 10, 25, and 50 cents and 1 dollar, and notes of 1, 2, 5, 10, 20, 50, and 100 dollars. Although issuance of higher notes ceased in 1969, a limited number of notes of 500, 1,000, 5,000, and 10,000 dollars remain in circulation.

WEIGHTS AND MEASURES: The imperial system is in common use; however, the use of metrics in industry is increasing, and the metric system is taught in public schools throughout the US. Common avoirdupois units in use are the avoirdupois pound of 16 oz or 453.5924277 gm; the long ton of 2,240 lb or 35,840 oz; and the short ton, more commonly used, of 2,000 lb or 32,000 oz. (Unless otherwise indicated, all measures given in tons are in short tons.) Liquid measures: 1 gallon = 231 cu in = 4 quarts = 8 pints. Dry measures: 1 bushel = 4 pecks = 32 dry quarts = 64 dry pints. Linear measures: 1 ft = 12 in; 1 statute mi = 1,760 yd = 5,280 ft. Metric equivalent: 1 m = 39.37 in.

HOLIDAYS: New Year's Day, 1 January; Birthday of Martin Luther King, Jr., 3d Monday in January; Lincoln's Birthday, 12 February (only in the northern and western states); Washington's Birthday, 3d Monday in February; Memorial or Decoration Day, last Monday in May; Independence Day, 4 July; Labor Day, 1st Monday in September; Columbus Day, 2d Monday in October; Election Day, 1st Tuesday after the 1st Monday in November; Veterans or Armistice Day, 11 November; Thanksgiving Day, 4th Thursday in November; Christmas, 25 December.

TIME: Eastern, 7 AM = noon GMT; Central, 6 AM = noon GMT; Mountain, 5 AM = noon GMT; Pacific (includes the Alaska panhandle), 4 AM = noon GMT; Yukon, 3 AM = noon GMT; Alaska and Hawaii, 2 AM = noon GMT; western Alaska, 1 AM = noon GMT.

¹LOCATION, SIZE, AND EXTENT

Located in the Western Hemisphere on the continent of North America, the US is the fourth-largest country in the world. Its total area, including Alaska and Hawaii, is 9,372,607 sq km (3,618,773 sq mi). The conterminous US extends 4,662 km (2,897 mi) ENE–WSW and 4,583 km (2,848 mi)SSE–NNW. It is bordered on the N by Canada, on the E by the Atlantic Ocean, on the S by the Gulf of Mexico and Mexico, and on the W by the Pacific Ocean, with a total boundary length of 17,563 km (10,913 mi). Alaska, the 49th state, extends 3,639 km (2,261 mi) E–W and 2,185 km (1,358 mi) N–S. It is bounded on the N by the Arctic Ocean and Beaufort Sea, on the E by Canada, on the S by the Gulf of Alaska, Pacific Ocean and Bering Sea, and on the W by the Bering Sea, Bering Strait, Chukchi Sea, and Arctic Ocean, with a total boundary length of 13,161 km (8,178 mi). The 50th state, Hawaii, consists of islands in the Pacific Ocean extending 2,536 km (1,576 mi) N–S and 2,293 km (1,425 mi) E–W, with a general coastline of 1,207 km (750 mi).

²TOPOGRAPHY

Although the northern New England coast is rocky, along the rest of the eastern seaboard the Atlantic Coastal Plain rises gradually from the shoreline. Narrow in the north, the plain widens to about 320 km (200 mi) in the south and in Georgia merges with the Gulf Coastal Plain that borders the Gulf of Mexico and extends through Mexico as far as the Yucatán. West of the Atlantic Coastal Plain is the Piedmont Plateau, bounded by the Appalachian Mountains. The Appalachians, which extend from southwest Maine into central Alabama—with special names in some areas—are old mountains, largely eroded away, with rounded contours and forested, as a rule, to the top. Few of their summits rise much above 1,100 m (3,500 ft), although the highest, Mt. Mitchell in North Carolina, reaches 2,037 m (6,684 ft).

Between the Appalachians and the Rocky Mountains, more than 1,600 km (1,000 mi) to the west, lies the vast interior plain of the US. Running south through the center of this plain and draining almost two-thirds of the area of the continental US is the Mississippi River. Waters starting from the source of the Missouri, the longest of its tributaries, travel almost 6,450 km (4,000 mi) to the Gulf of Mexico. The eastern reaches of the great interior plain are bounded on the north by the Great Lakes, which are thought to contain about half the world's total supply of fresh water. Under US jurisdiction are 57,441 sq km (22,178 sq mi) of Lake Michigan, 54,696 sq km (21,118 sq mi) of Lake Superior, 23,245 sq km (8,975 sq mi) of Lake Huron, 12,955 sq km (5,002 sq mi) of Lake Erie, and 7,855 sq km (3,033 sq mi) of Lake Ontario. The five lakes are now accessible to oceangoing vessels from the Atlantic via the St. Lawrence Seaway. The basins of the Great Lakes were formed by the glacial ice cap that moved down over large parts of North America some 25,000 years ago. The glaciers also determined the direction of flow of the Missouri River and, it is believed, were responsible for carrying soil from what is now Canada down into the central agricultural basin of the US. The great interior plain consists of two major subregions: the fertile Central Plains, extending from the Appalachian highlands to a line drawn approximately 480 km (300 mi) west of the Mississippi, broken by the Ozark Plateau; and the more arid Great Plains, extending from that line to the foothills of the Rocky Mountains. Although they appear flat, the Great Plains rise gradually from about 460 m (1,500 ft) to more than 1,500 m (5,000 ft) at their western extremity.

The Continental Divide, the Atlantic-Pacific watershed, runs along the crest of the Rocky Mountains. The Rockies and the ranges to the west are parts of the great system of young, rugged mountains, shaped like a gigantic spinal column, that runs along western North, Central, and South America from Alaska to Tierra del Fuego, Chile. In the continental US, the series of western ranges, most of them paralleling the Pacific coast, are the Sierra Nevada, the Coast Ranges, the Cascade Range, and the Tehachapi and San Bernardino mountains. Between the Rockies and the Sierra Nevada–Cascade mountain barrier to the west lies the Great Basin, a group of vast arid plateaus containing most of the desert areas of the US, in the south eroded by deep canyons. The coastal plains along the Pacific are narrow, and in many places the mountains plunge directly into the sea. The most extensive lowland near the west coast is the Great Valley of California, lying between the Sierra Nevada and the Coast Ranges. There are 71 peaks in these western ranges of the continental US that rise to an altitude of 4,267 m (14,000 ft) or more, Mt. Whitney in California at 4,418 m (14,494 ft) being the highest. The greatest rivers of the Far West are the Colorado in the south, flowing into the Gulf of California, and the Columbia in the northwest, flowing to the Pacific. Each is more than 1,900 km (1,200 mi) long; both have been intensively developed to generate electric power, and both are important sources of irrigation.

Separated from the continental US by Canadian territory, the state of Alaska occupies the extreme northwest portion of the North American continent. A series of precipitous mountain ranges separates the heavily indented Pacific coast on the south from Alaska's broad central basin, through which the Yukon River flows from Canada in the east to the Bering Sea in the west. The central basin is bounded on the north by the Brooks Range, which slopes down gradually to the Arctic Ocean. The Alaskan Peninsula and the Aleutian Islands, sweeping west far out to sea, consist of a chain of volcanoes, many still active. The state of Hawaii consists of a group of Pacific islands formed by volcanoes rising sharply from the ocean floor. The highest of these volcanoes, Mauna Loa, at 4,168 m (13,675 ft), is located on the largest of the islands, Hawaii, and is still active.

The lowest point in the US is Death Valley in California, 86 m (282 ft) below sea level. At 6,194 m (20,320 ft), Mt. McKinley in Alaska is the highest peak in North America. These topographic extremes suggest the geological instability of the Pacific Coast region. Major earthquakes destroyed San Francisco in 1906 and Anchorage, Alaska, in 1964, and the San Andreas Fault in California still causes frequent earth tremors. Washington State's Mt. St. Helens erupted in 1980, spewing volcanic ash over much of the Northwest.

³CLIMATE

The eastern continental region is well watered, with annual rainfall generally in excess of 100 cm (40 in). It includes all of the Atlantic seaboard and southeastern states and extends west to cover Indiana, southern Illinois, most of Missouri, Arkansas, Louisiana, and easternmost Texas. The eastern seaboard is affected primarily by the masses of air moving from west to east across the continent rather than by air moving in from the Atlantic. Hence its climate is basically continental rather than maritime. The midwestern and Atlantic seaboard states experience hot summers and cold winters; spring and autumn are clearly defined periods of climatic transition. Only Florida, with the Gulf of Mexico lying to its west, experiences moderate differences between summer and winter temperatures. Mean annual temperatures vary considerably between north and south: Boston, 11°c (51°F); New York City, 13°c (55°F); Charlotte, N.C., 16°c (61°F); Miami, Fla., 24°c (76°F). The Gulf and South Atlantic states are often hit by severe tropical storms originating in the Caribbean in late summer and early autumn.

The prairie lands lying to the west constitute a subhumid region. Precipitation usually exceeds evaporation by only a small amount; hence the region experiences drought more often than excessive rainfall. Dryness generally increases from east to west. The average midwinter temperature in the extreme north—Minnesota and North Dakota—is about –13°c (9°F) or less, while the average July temperature is 18°c (65°F). In the Texas prairie region to the south, January temperatures average 10–13°c (50–55°F) and July temperatures 27–29°c (80–85°F). Rainfall along the western border of the prairie region is as low as 46 cm (18 in) per year in the north and 64 cm (25 in) in the south. Precipitation is greatest in the early summer—a matter of great importance to agriculture, particularly in the growing of grain crops. In dry years, the prevailing winds may carry the topsoil eastward (particularly from the southern region) for hundreds of miles in clouds that obscure the sun.

The Great Plains constitute a semiarid climatic region. Rainfall in the southern plains averages about 50 cm (20 in) per year and in the northern plains about 25 cm (10 in), but extreme year-to-year variations are common. The tropical air masses that move northward across the plains originate on the fairly high plateaus of Mexico and contain little water vapor. Periods as long as 120 days without rain have been experienced in this region. The rains that do occur are often violent, and a third of the total annual rainfall may be recorded in a single day at certain weather stations. The contrast between summer and winter temperatures is extreme throughout the Great Plains. Maximum summer temperatures of over 43°c (110°F) have been recorded in the northern as well as in the southern plains. From the Texas panhandle north, blizzards are common in the winter, and tornadoes at other seasons. The average minimum temperature for January in Duluth, Minn., is –19°c (–3°F).

The higher reaches of the Rockies and the mountains paralleling the Pacific coast to the west are characterized by a typical alpine climate. Precipitation as a rule is heavier on the western slopes of the ranges. The great intermontane arid region of the West shows considerable climatic variation between its northern and southern portions. In New Mexico, Arizona, and southeastern California, the greatest precipitation occurs in July, August, and September, mean annual rainfall ranging from 8 cm (3 in) in Yuma, Ariz., to 76 cm (30 in) in the mountains of northern Arizona and New Mexico. Phoenix has a mean annual temperature of 22°c (71°F), rising to 33°c (92°F) in July and falling to 11°c (52°F) in January. North of the Utah-Arizona line, the summer months usually are very dry; maximum precipitation occurs in the winter and early spring. In the desert valleys west of Great Salt Lake, mean annual precipitation adds up to only 10 cm (4 in). Although the northern plateaus are generally arid, some of the mountainous areas of central Washington and Idaho receive at least 152 cm (60 in) of rain per year. Throughout the intermontane region, the uneven availability of water is the principal factor shaping the habitat.

The Pacific coast, separated by tall mountain barriers from the severe continental climate to the east, is a region of mild winters and moderately warm, dry summers. Its climate is basically maritime, the westerly winds from the Pacific Ocean moderating the

LOCATION: Conterminous US: 66°57′ to 124°44′w; 24°33′ to 49°23′n. Alaska: 130°w to 172°28′e; 51° to 71°23′n. Hawaii: 154°48′ to 178°22′w 18°55′ to 28°25′n.
BOUNDARY LENGTHS: Conterminous US: Canada, 6,416 km (3,987 mi); Atlantic Ocean, 3,330 km (2,069 mi); Gulf of Mexico coastline, 2,625 km (1,631 mi); Mexico, 3,111 km (1,933 mi); Pacific coastline, 2,081 km (1,293 mi). Alaska: Arctic Ocean coastline, 1,706 km (1,060 mi); Canada, 2,475 km (1,538 mi); Pacific coastline, including the Bering Sea and Strait and Chukchi coastlines, 8,980 km (5,580 mi). Hawaii: coastline, 1,207 km (750 mi).

extremes of both winter and summer temperatures. Los Angeles in the south has an average temperature of 13°c (56°f) in January and 21°c (69°f) in July; Seattle in the north has an average temperature of 4°c (39°f) in January and 18°c (65°f) in July. Precipitation in general increases along the coast from south to north,

extremes ranging from an annual average of 4.52 cm (1.78 in) at Death Valley in California (the lowest in the US) to more than 356 cm (140 in) in Washington's Olympic Mountains.

Climatic conditions vary considerably in the vastness of Alaska. In the fogbound Aleutians and in the coastal panhandle strip that extends southeastward along the Gulf of Alaska and includes the capital, Juneau, a relatively moderate maritime climate prevails. The interior is characterized by short, hot summers and long, bitterly cold winters, and in the region bordering the Arctic Ocean a polar climate prevails, the soil hundreds of feet below the surface remaining frozen the year round. Although snowy in winter, continental Alaska is relatively dry.

Hawaii has a remarkably mild and stable climate with only slight seasonal variations in temperature, as a result of northeast ocean winds. The mean January temperature in Honolulu is 23°c (73°F); the mean July temperature 27°c (80°F). Rainfall is moderate—about 71 cm (28 in) per year—but much greater in the mountains; Mt. Waialeale on Kauai has a mean annual rainfall of 1,168 cm (460 in), highest in the world.

The lowest temperature recorded in the US was –62°c (–79.8°F) at Prospect Creek Camp, Alaska, on 23 January 1971; the highest, 57°c (134°F) at Greenland Ranch, in Death Valley, Calif., on 10 July 1913. The record annual rainfall is 1,468 cm (578 in) at Fuu Kukui, Maui, in 1950; for a 24-hour period, 98.3 cm (38.7 in) at Yankeetown, Fla., on 5–6 September 1950; in 1 hour, 30 cm (12 in), at Holt, Mo., on 22 June 1947, and on Kauai, Hawaii, on 24–25 January 1956.

[4]FLORA AND FAUNA

As of 1994, including Alaska and Hawaii, about 29% of the US was forestland, 26% was grassland pasture, 20% was arable cropland, 2% was irrigated land, and the remaining 25% encompassed military, urban, and designated recreational and wilderness areas, among other lands.

At least 7,000 species and subspecies of indigenous US flora have been categorized. The eastern forests contain a mixture of softwoods and hardwoods that includes pine, oak, maple, spruce, beech, birch, hemlock, walnut, gum, and hickory. The central hardwood forest, which originally stretched unbroken from Cape Cod to Texas and northwest to Minnesota—still an important timber source—supports oak, hickory, ash, maple, and walnut. Pine, hickory, tupelo, pecan, gum, birch, and sycamore are found in the southern forest that stretches along the Gulf coast into the eastern half of Texas. The Pacific forest is the most spectacular of all because of its enormous redwoods and Douglas firs. In the southwest are saguaro (giant cactus), yucca, candlewood, and the Joshua tree.

The central grasslands lie in the interior of the continent, where the moisture is not sufficient to support the growth of large forests. The tall grassland or prairie (now almost entirely under cultivation) lies to the east of the 100th meridian. To the west of this line, where rainfall is frequently less than 50 cm (20 in) per year, is the short grassland. Mesquite grass covers parts of west Texas, southern New Mexico, and Arizona. Short grass may be found in the highlands of the latter two states, while tall grass covers large portions of the coastal regions of Texas and Louisiana and occurs in some parts of Mississippi, Alabama, and Florida. The Pacific grassland includes northern Idaho, the higher plateaus of eastern Washington and Oregon, and the mountain valleys of California.

The intermontane region of the Western Cordillera is for the most part covered with desert shrubs. Sagebrush predominates in the northern part of this area, creosote in the southern, and saltbrush near the Great Salt Lake and in Death Valley.

The lower slopes of the mountains running up to the coastline of Alaska are covered with coniferous forests as far north as the Seward Peninsula. The central part of the Yukon Basin is also a region of softwood forests. The rest of Alaska is heath or tundra. Hawaii has extensive forests of bamboo and ferns. Sugarcane and pineapple, although not native to the islands, now cover a large portion of the cultivated land.

Small trees and shrubs common to most of the US include hackberry, hawthorn, serviceberry, blackberry, wild cherry, dogwood, and snowberry. Wildflowers bloom in all areas, from the seldom-seen blossoms of rare desert cacti to the hardiest alpine species. Wildflowers include forget-me-not, fringed and closed gentians, jack-in-the-pulpit, black-eyed Susan, columbine, and common dandelion, along with numerous varieties of aster, orchid, lady's slipper, and wild rose.

An estimated 1,500 species and subspecies of mammals characterize the animal life of the continental US. Among the larger game animals are the white-tailed deer, moose, pronghorn antelope, bighorn sheep, mountain goat, black bear, and grizzly bear. The Alaskan brown bear often reaches a weight of 1,200–1,400 lbs. Some 25 important furbearers are common, including the muskrat, red and gray foxes, mink, raccoon, beaver, opossum, striped skunk, woodchuck, common cottontail, snowshoe hare, and various squirrels. Human encroachment has transformed the mammalian habitat over the last two centuries. The American buffalo (bison), millions of which once roamed the plains, is now found only on select reserves. Other mammals, such as the elk and gray wolf, have been restricted to much smaller ranges.

Year-round and migratory birds abound. Loons, wild ducks, and wild geese are found in lake country; terns, gulls, sandpipers, herons, and other seabirds live along the coasts. Wrens, thrushes, owls, hummingbirds, sparrows, woodpeckers, swallows, chickadees, vireos, warblers, and finches appear in profusion, along with the robin, common crow, cardinal, Baltimore oriole, eastern and western meadowlarks, and various blackbirds. Wild turkey, ruffed grouse, and ring-necked pheasant (introduced from Europe) are popular game birds.

Lakes, rivers, and streams teem with trout, bass, perch, muskellunge, carp, catfish, and pike; sea bass, cod, snapper, and flounder are abundant along the coasts, along with such shellfish as lobster, shrimp, clams, oysters, and mussels. Garter, pine, and milk snakes are found in most regions. Four poisonous snakes survive, of which the rattlesnake is the most common. Alligators appear in southern waterways, and the Gila monster makes its home in the Southwest.

During the 1960s and 1970s, numerous laws and lists designed to protect threatened and endangered flora and fauna were adopted throughout the US. Generally, each species listed as protected by the federal government is also protected by the states, but some states may list species not included on federal lists or on the lists of neighboring states. (Conversely, a species threatened throughout most of the US may be abundant in one or two states.) As of April 1993, the US Fish and Wildlife Service listed 539 endangered US species, including 292 plants, and 150 threatened species, including 64 plants. The agency listed another 493 endangered and 38 threatened foreign species by international agreement.

Threatened species, likely to become endangered if recent trends continue, include such plants as Rydberg milk-vetch, northern wild monkshood, Lee pincushion cactus, and Lloyd's Mariposa cactus. Among the endangered floral species (in imminent danger of extinction in the wild) are the Virginia round-leaf birch, San Clemente Island broom, Texas wild-rice, Furbish lousewort, Truckee barberry, Sneed pincushion cactus, spineless hedgehog cactus, Knowlton cactus, persistent trillium, dwarf bear-poppy, and small whorled pogonia.

Threatened fauna include the grizzly bear, southern sea otter, Newell's shearwater, American alligator, eastern indigo snake, bayou darter, several southwestern trout species, and Bahama and Schaus swallowtail butterflies. Among endangered fauna are the Indiana bat, key deer, black-footed ferret, northern swift fox, San Joaquin kit fox, jaguar, jaguarundi, Florida manatee, ocelot, Florida panther, Utah prairie dog, Sonoran pronghorn, Delmarva Peninsula fox squirrel, gray wolf (except in Minnesota, where it is threatened), red wolf, numerous whale species, bald eagle (endangered in most states, but only threatened in the Northwest and the Great Lakes region), Hawaii creeper, Everglade kite, brown pelican, California clapper rail, red-cockaded woodpecker, bluntnosed leopard lizard, American crocodile, desert slender salamander, Houston toad, humpback chub, several species of pupfish, 17 US species of pearly mussel, Socorro isopod, Kentucky cave shrimp, and mission blue butterfly. Several species on the federal list of endangered and threatened wildlife and plants are found only in Hawaii.

[5]ENVIRONMENT

During the 1960s and 1970s, numerous laws and lists designed to protect the environment and to preserve endangered flora and fauna were adopted throughout the US. The Council on Environmental Quality, an advisory body contained within the Executive Office of the President, was established by the National Environmental Policy Act of 1969, which mandated an assessment of environmental impact for every federally funded project. The Environmental Protection Agency (EPA), created in 1970, is an independent body with primary regulatory responsibility in the fields of air and noise pollution, water and waste management, and control of toxic substances. Other federal agencies with environmental responsibilities are the Forest Service and Soil Conservation Service within the Department of Agriculture, the Fish and Wildlife Service and the National Park Service within the Department of the Interior, the Department of Energy, and the Nuclear Regulatory Commission. In addition to the 1969 legislation, landmark federal laws protecting the environment include the Clean Air Act Amendments of 1970 and 1990, controlling automobile and electric utility emissions; the Water Pollution Act of 1972, setting clean-water criteria for fishing and swimming; and the Endangered Species Act of 1973, protecting wildlife near extinction. A measure enacted in December 1980 established a $1.6-billion "Superfund," financed largely by excise taxes on chemical companies, to clean up toxic waste dumps such as the one in the Love Canal district of Niagara Falls, N.Y. In 1992, federal revenues from major environment-related sales, funds, and taxes totaled $7.6 billion (of which Superfund comprised $1.19 billion). In 1990, pollution control costs totaled $115 billion (2.1% of GDP).

The most influential environmental lobbies include the Sierra Club (founded in 1892; 565,000 members in 1991) and its legal arm, the Sierra Club Legal Defense Fund. Large conservation groups include the National Wildlife Federation (1936; 6,200,000), the National Audubon Society (1905; 600,000), and the Nature Conservancy (1917; 550,000). Greenpeace USA (1979; 1,500,000) has gained international attention by seeking to disrupt the hunts for whales and seals.

Among the environmental movement's most notable successes have been the inauguration (and mandating in some states) of recycling programs, the banning in the US of the insecticide dichlorodiphenyltrichloroethane (DDT), the successful fight against construction of a supersonic transport (SST), and the protection of more than 40 million hectares (100 million acres) of Alaska lands (after a fruitless fight to halt construction of the trans-Alaska pipeline), and the gradual elimination of chlorofluorocarbon (CFC) production by 2000.

Outstanding problems include "acid rain," precipitation contaminated by fossil fuel wastes; inadequate facilities for solid waste disposal; the contamination of homes by radon, a radioactive gas that is produced by the decay of underground deposits of radium and which can cause cancer; runoffs of agricultural pesticides, pollutants deadly to fishing streams and very difficult to regulate; continued dumping of raw or partially treated sewage from major cities into US waterways; falling water tables in many western states; the decrease in arable land because of depletion, erosion, and urbanization; the need for reclamation of

strip-mined lands and for regulation of present and future strip mining; and the expansion of the US nuclear industry in the absence of a fully satisfactory technique for the handling and permanent disposal of radioactive wastes.

According to the United States Fish and Wildlife Service, in 1992 there were some 640 threatened and endangered species of plant and animal wildlife in the US. Endangered mammals included the red wolf, black-footed ferret, jaguar, and Sonoran pronghorn. Endangered species of rodents included the Delmarva Peninsula fox squirrel, beach mouse, salt-marsh harvest mouse, 7 species of bat (Virginia and Ozark big-eared Sanborn's and Mexican long-nosed, Hawaiian hoary, Indiana, and gray), and the Morro Ba, Fresno, Stephens', and Tipton Kangaroo rats and rice rat.

Endangered species of birds included the California condor, bald eagle, 3 species of falcon (American peregrine, tundra peregrine, and northern aplomado), Eskimo curlew, 2 species of crane (whooping and Mississippi sandhill), 3 species of warbler (Kirtland's, Bachman's, and golden-cheeked), dusky seaside sparrow, light-footed clapper rail, least tern, and San Clemente loggerhead shrike. Endangered amphibians included 4 species of salamander (Santa Cruz long-toed, Shenandoah, desert slender, and Texas blind), Houston and Wyoming toad, and 6 species of turtle (green sea, hawksbill, Kemp's ridley, Plymouth and Alabama red-bellied, and leatherback). Endangered reptiles included the American crocodile, (blunt nosed leopard and island night), and San Francisco garter snake.

Aquatic species included the shortnose sturgeon, Gila trout, 8 species of chub (humpback, Pahranagat, Yaqui, Mohave tui, Owens tui, bonytail, Virgin River, and Borax lake), Colorado River squawfish, 5 species of dace (Kendall Warm Springs, and Clover Valley, Independence Valley, Moapa and Ash Meadows speckled), Modoc sucker, cui-ui, Smoky and Scioto madtom, 7 species of pupfish (Leon Springs, Gila Desert, Ash Meadows Amargosa, Warm Springs, Owens, Devil's Hole, and Comanche Springs), Pahrump killifish, 4 species of gambusia (San Marcos, Pecos, Amistad, Big Bend, and Clear Creek), 6 species of darter (fountain, watercress, Okaloosa, boulder, Maryland, and amber), totoaba, and 32 species of mussel and pearly mussel. Also classified as endangered were 2 species of earthworm (Washington giant and Oregon giant), the Socorro isopod, San Francisco forktail damselfly, Ohio emerald dragonfly, 2 species of beetle (Kretschmarr Cave, Tooth Cave, and giant carrion), Belkin's dune tabanid fly, and 10 species of butterfly (Schaus' swallowtail, lotis, mission, El Segundo, and Palos Verde blue, Mitchell's satyr, Uncompahgre fritillary, Lange's metalmark, San Bruno elfin, and Smith's blue).

Endangered bird species in Hawaii included the Hawaiian dark-rumped petrel, Hawaiian gallinule, Hawaiian crow, 3 species of thrush (Kauai, Molokai, and puaiohi), Kauai 'o'o, Kauai nukupu'u, Kauai 'alialoa, 'akiapola'au, Maui 'akepa, Molokai creeper, Oahu creeper, palila, and 'o'u.

Endangered plants (and number of species) in the US include: aster, 32; cactus, 19; pea, 18; mustard, 14; mint, 12; mallow, 11; bellflower and pink family, 9 each; snapdragon, 8; and buckwheat, 6.

In 1990/91, state and federal expenditures for endangered species protection totaled $177 million, of which the bald eagle accounted for $24.7 million (14%; Florida Scrub Jay, $19.7 million (11%); West Indian manatee, $15.3 million (8.6%); and the northern spotted owl, $12.9 million (7.3%).

[6]POPULATION

According to census figures for 1990, the population of the US (including the 50 states and Washington, D.C.) was 248,709,873 (up from 226,542,580 in 1980), of whom 51.3% were female and 48.7% male. At the time of the first federal census, in 1790, the population of the country was 3,929,214. Between 1800 and 1850, the population almost quadrupled; between 1850 and 1900, it tripled; and between 1900 and 1950, it almost doubled. During the 1960s and 1970s, however, the growth rate slowed steadily, declining from 2.9% annually in 1960 to 2% in 1969 and to less than 1% in the 1980s. The population was estimated at 257,908,000 in mid-1993. In 1991, 7.6% of the population were under 5 years of age; 18.2% were 5 to 17; 61.6% were 18 to 64; and 12.6% were 65 or older. The median age of the population increased from 16.7 years in 1820 to 22.9 years in 1900 and to 33.1 years in 1991. The US Bureau of the Census projected a population of 275,327,023 for the year 2000.

By 1990, metropolitan areas had a total population of 197.5 million, representing a 12% increase over 1980. Suburbs have absorbed most of the shift in population distribution since 1950. In 1990 there were 8 cities with more than 1 million population: New York, 7,371,282; Los Angeles, 3,485,398; Chicago, 2,783,726; Houston, 1,630,553; Philadelphia, 1,585,577; San Diego, 1,110,549; Detroit, 1,027,974; and Dallas, 1,006,877.

[7]MIGRATION

Between 1840 and 1930, some 37 million immigrants, the overwhelming majority of them Europeans, arrived in the US. Immigration reached its peak in the first decade of the 20th century, when nearly 9 million came. Following the end of World War I, the tradition of almost unlimited immigration was abandoned, and through the National Origins Act of 1924, a quota system was established as the basis of a carefully restricted policy of immigration. Under the McCarran Act of 1952, one-sixth of 1% of the number of inhabitants from each European nation residing in the continental US as of 1920 could be admitted annually. In practice, this system favored nations of northern and western Europe, with the UK, Germany, and Ireland being the chief beneficiaries. The quota system was radically reformed in 1965, under a new law that established an annual ceiling of 170,000 for Eastern Hemisphere immigrants and 120,000 for entrants from the Western Hemisphere; in October 1978, these limits were replaced by a worldwide limit of 290,000, which was lowered to 270,000 by 1981. A major 1990 overhaul set a total annual ceiling of 700,000 (675,000 beginning in fiscal 1995), of which 480,000 would be family sponsored and 140,000 employment based.

In the 12 months ending 30 September 1992, 511,769 immigrants entered the US, 261,451 of them subject to the numerical ceiling. Some 207,822 were from Asia, 228,741 from the Americas, 62,912 from Europe, 10,137 from Africa, and 2,157 from Oceania. A direct result of the immigration law revisions has been a sharp rise in the influx of Asians (primarily Chinese, Filipinos, Indians, Japanese, and Koreans), of whom 2,738,157 entered the country during 1981–90, as compared with 153,249 during the entire decade of the 1950s.

Since 1961, the federal government has supported and financed the Cuban Refugee Program. More than 500,000 Cubans were living in southern Florida by 1980, when another 125,000 Cuban refugees arrived. Between 1975 and 1978, following the defeat of the US-backed Saigon government, several hundred thousand Vietnamese refugees came to the US. Under the Refugee Act of 1980, a ceiling for the number of admissible refugees is set annually; in fiscal 1992, 123,010 immigrants were admitted under the various refugee acts, 61,631 from the former USSR. Since Puerto Ricans are American citizens, no special authorization is required for their admission to the continental US.

Large numbers of aliens—mainly from Latin America, especially Mexico—have illegally established residence in the US after entering the country as tourists, students, or temporary visitors engaged in work or business. In November 1986, Congress passed a bill allowing illegal aliens who had lived and worked in the US since 1982 the opportunity to become permanent residents. By the end of fiscal year 1992, 2,650,000 of a potential 2,760,000 eligible for permanent residence under this bill had attained that status. In 1994 the number of illegal alien residents was estimated at 3,850,000, of which 1,600,000 were believed to be in California.

The major migratory trends within the US have been a general westward movement during the 19th century; a long-term movement from farms and other rural settlements to metropolitan areas, which showed signs of reversing in some states during the 1970s; an exodus of southern blacks to the cities of the North and Midwest, especially after World War I; a shift of whites from central cities to surrounding suburbs since World War II; and, also during the post–World War II period, a massive shift from the North and East to the Sunbelt region of the South and Southwest.

[8]ETHNIC GROUPS

The majority of the population of the US is of European origin, with the largest groups having primary ancestry traceable in 1990 to the UK (31,391,758), Germany (45,583,922), and Ireland (22,721,252); many Americans reported multiple ancestries. Major racial and national minority groups include blacks (either of US or Caribbean parentage), Chinese, Filipinos, Japanese, Mexicans, and other Spanish-speaking peoples of the Americas. Whites comprised 83.9% of the US population in 1990; blacks, 12.3%; Asians and Pacific Islanders, 3%; Native Americans (Amerindians—more commonly known as Indians, Eskimos, and Aleuts), 0.8%. Responding to a census question that cut across racial lines, 9% of Americans in 1990 described themselves as of Hispanic origin. Inequality in social and economic opportunities for ethnic minorities became a key public issue in the post–World War II period.

Some Indian societies survived warfare with land-hungry white settlers and retained their tribal cultures. Their survival, however, has been on the fringes of North American society, especially as a result of the implementation of a national policy of resettling Indian tribes on reservations. In 1890, according to the official census count, there were 248,253 Indians; in 1940, 333,909; and in 1990, 1,959,234 (including also Eskimos and Aleuts). Groups of Indians are found most numerously in the southwestern states of Oklahoma, Arizona, New Mexico, and California. The 1960s and 1970s saw successful court fights by Indians in Alaska, Maine, South Dakota, and other states to regain tribal lands or to receive cash settlements for lands taken from them in violation of treaties during the 1800s.

The black population in 1992 was estimated at 31,439,000. Some 53% of blacks still resided in the South in 1990, the region that absorbed most of the slaves brought from Africa in the 18th and 19th centuries. Two important regional migrations of blacks have taken place: (1) a "Great Migration" to the North, commencing in 1915, and (2) a small but then unprecedented westward movement beginning about 1940. Both migrations were fostered by wartime demands for labor and by postwar job opportunities in northern and western urban centers. More than three out of four black Americans live in metropolitan areas, constituting, as of 1990, 81% of the population of Gary, Ind., 66% of Washington, D.C., 67% of Atlanta, 76% of Detroit, 62% of New Orleans, 59% of Newark, N.J., and 59% of Baltimore; in New York City, which had the largest number of black residents (2,102,512), 28.7% of the population was black. Large-scale federal programs to ensure equality for blacks in voting rights, public education, employment, and housing were initiated after the historic 1954 Supreme Court ruling that barred racial segregation in public schools. By 1966, however, in the midst of growing and increasingly violent expressions of dissatisfaction by black residents of northern cities and southern rural areas, the federal Civil Rights Commission reported that integration programs were lagging. Throughout the 1960s, 1970s, and 1980s, the unemployment rate among nonwhites in the US was at least double that for whites, and school integration proceeded slowly, especially outside the South.

Included in the population of the US in 1990 were 7,226,986 persons whose lineage can be traced to Asian and Pacific nationalities, chiefly Chinese, 1,648,696; Filipino, 1,419,711; Japanese, 866,160; Indian, 786,694; Korean, 797,304; and Vietnamese, 593,213. The Chinese population is highly urbanized and concentrated particularly in cities of over 100,000 population, mostly on the West Coast and in New York City. The Japanese population has risen steadily from a level of 72,157 in 1910. Hawaii has been the most popular magnet of Japanese emigration; the Japanese population of Hawaii accounted in 1990 for 23.6% of the state's residents and 31% of the nation's total number of Japanese. Most Japanese in California were farmers until the outbreak of World War II, when they were interned and deprived of their landholdings; after the war, most entered the professions and other urban occupations.

Mexican settlement is largely in the Southwest. Spanish-speaking Puerto Ricans, who often represent an amalgam of racial strains, have largely settled in the New York metropolitan area, where they partake in considerable measure of the hardships and problems experienced by other immigrant groups in the process of settling in the US. Since 1959, many Cubans have settled in Florida and other eastern states. As of March 1990 there were 22,354,000 Hispanic Americans, of whom 60.4% were of Mexican ancestry, 12.2% Puerto Rican, and 4.7% Cuban.

⁹LANGUAGES

The primary language of the US is English, enriched by words borrowed from the languages of Indians and immigrants, predominantly European.

When European settlement began, Indians living north of Mexico spoke about 300 different languages now held to belong to 58 different language families. Only 2 such families have contributed noticeably to the American vocabulary: Algonkian in the Northeast and Aztec-Tanoan in the Southwest. From Algonkian languages, directly or sometimes through Canadian French, English has taken such words as *moose, skunk, caribou, opossum, woodchuck,* and *raccoon* for New World animals; *hickory, squash,* and *tamarack* for New World flora; and *succotash, hominy, mackinaw, moccasin, tomahawk, toboggan,* and *totem* for various cultural items. From Nahuatl, the language of the Aztecs, terms such as *tomato, mesquite, coyote, chili, tamale, chocolate,* and *ocelot* have entered English, largely by way of Spanish. A bare handful of words come from other Indian language groups, such as *tepee* from Dakota Siouan, *catalpa* from Creek, *sequoia* from Cherokee, *hogan* from Navaho, and *sockeye* from Salish, as well as *cayuse* from Chinook.

Professional dialect research, initiated in Germany in 1878 and in France in 1902, did not begin in the US until 1931, in connection with the *Linguistic Atlas of New England* (1939–43). This kind of research, requiring trained field-workers to interview representative informants in their homes, subsequently was extended to the entire Atlantic Coast, the north-central states, the upper Midwest, the Pacific Coast, the Gulf states, and Oklahoma. As of 1985, only the New England atlas, the *Linguistic Atlas of the Upper Midwest* (1973–76), and the first two fascicles of the *Linguistic Atlas of the Middle and South Atlantic States* (1980) had been published, along with three volumes based on Atlantic Coast field materials; nearing publication were atlases of the north-central states, the Gulf states, and Oklahoma. In other areas, individual dialect researchers have produced more specialized studies. The definitive work on dialect speech, the American Dialect Society's monumental *Dictionary of American Regional English,* began publication in 1985.

Dialect studies confirm that standard English is not uniform throughout the country. Major regional variations reflect patterns of colonial settlement, dialect features from England having dominated particular areas along the Atlantic Coast and then spread westward along the three main migration routes through the Appalachian system. Dialectologists recognize three main dialects—Northern, Midland, and Southern—each with subdivisions related to the effect of mountain ranges and rivers and railroads on population movement.

The Northern dialect is that of New England and its derivative settlements in New York; the northern parts of Ohio, Indiana, Illinois, and Iowa; and Michigan, Wisconsin, northeastern South Dakota, and North Dakota. A major subdivision is that of New England east of the Connecticut River, an area noted typically by the loss of /r/ after a vowel, and by the pronunciation of *can't, dance, half,* and *bath* with a vowel more like that in *father* than that in *fat.* Generally, however, Northern speech has a strong /r/ after a vowel, the same vowel in *can't* and *cat,* a conspicuous contrast between *cot* and *caught,* the /s/ sound in *greasy, creek* rhyming with *pick,* and *with* ending with the same consonant sound as at the end of *breath.*

Midland speech extends in a wide band across the US: there are two main subdivisions, North Midland and South Midland. North Midland speech extends westward from New Jersey, Delaware, and Pennsylvania into Ohio, Illinois, southern Iowa, and northern Missouri. Its speakers generally end *with* with the consonant sound that begins the word *thin,* pronounce *cot* and *caught* alike, and say *cow* and *down* as /caow/ and daown/. South Midland speech was carried by the Scotch–Irish from Pennsylvania down the Shenandoah Valley into the southern Appalachians, where it acquired many Southern speech features before it spread westward into Kentucky, Tennessee, southern Missouri, Arkansas, and northeast Texas. Its speakers are likely to say *plum peach* rather than *clingstone peach* and *snake doctor* rather than *dragonfly.*

Southern speech typically, though not always, lacks the consonant /r/ after a vowel, lengthens the first part of the diphthong in *write* so that to Northern ears it sounds almost like *rat,* and diphthongizes the vowels in *bed* and *hit* so that they sound like /beuhd/ and /hiuht/. *Horse* and *hoarse* do not sound alike, and *creek* rhymes with *meek. Corn bread* is *corn pone,* and *you-all* is standard for the plural.

In the western part of the US, migration routes so crossed and intermingled that no neat dialect boundaries can be drawn, although there are a few rather clear population pockets.

The 1990 census recorded that of 229,875,493 Americans 5 years of age or over, 198,101,862 spoke only English at home; the remaining 31,773,631 spoke a language other than English. The principal foreign languages and their speakers were as follows:

Spanish	17,310,043	Greek	387,359
French	1,920,621	Arabic	353,203
German	1,544,793	Various Native North American	331,634
Chinese	1,316,956	Russian	241,092
Italian	1,307,068	Yiddish	212,951
Tagalog	841,827	Various Scandinavian	198,261
Polish	723,161	Various Southern Slavic	170,301
Korean	625,814	Hungarian	147,621
Indic	553,882	Other Indo–European	576,196
Vietnamese	506,500	Other Slavic	270,609
Portuguese	429,440	Other West Germanic	232,075
Japanese	426,876		

The majority of Spanish speakers live in the Southwest, Florida, and eastern urban centers. Refugee immigration since the 1950s has greatly increased the number of foreign-language speakers from Latin America and Asia.

Very early English borrowed from neighboring French speakers such words as *shivaree, butte, levee,* and *prairie;* from German, *sauerkraut, smearcase,* and *cranberry;* from Dutch, *stoop, spook,* and *cookie;* and from Spanish, *tornado, corral, ranch,* and *canyon.* From various West African languages, blacks have given English *jazz, voodoo,* and *okra.*

Educational problems raised by the presence of large blocs of non-English speakers led to the passage in 1976 of the Bilingual Educational Act, enabling children to study basic courses in their first language while they learn English. A related school problem is that of black English, a Southern dialect variant that is the vernacular of many black students now in northern schools.

¹⁰RELIGIONS

US religious traditions are predominantly Judeo-Christian, and most Americans identify themselves as Protestants (of various denominations), Roman Catholics, or Jews. As of 1990, US religious bodies counted 255,173 places of worship and 137,064,509 members, or about half of the total population belong to a Judeo-Christian religious body. The largest Christian denomination is the Roman Catholic Church, with 53,385,998 members in 22,441 parishes in 1990. Immigration from Ireland, Italy, Eastern Europe, French Canada, and the Caribbean accounts for the predominance of Roman Catholicism in the Northeast, Northwest, and some parts of the Great Lakes region, while Hispanic traditions and more recent immigration from Mexico and other Latin American countries account for the historical importance of Roman Catholicism in California and throughout most of the sunbelt. More than any other US religious body, the Roman Catholic Church maintains an extensive network of parochial schools. Jewish immigrants settled first in the Northeast, where the largest Jewish population remain; in 1990, 1,843,240 Jews lived in New York out of an estimated US total of 5,982,529 adherents.

As of 1990, US Protestants groups had at least 77,695,982 adherents. Baptists predominate below the Mason-Dixon line and west to Texas. By far the nation's largest Protestant group, the Southern Baptist Convention had 18,940,682 adherents in 1990; the American Baptist Churches in the USA claimed some 1,873,731 adherents. A concentration of Methodist groups extends westward in a band from Delaware to eastern Colorado; the largest of these groups, the United Methodist Church had 11,091,032 adherents in 1990. Lutheran denominations, reflecting in part the patterns of German and Scandinavian settlement, are most highly concentrated in the north-central states, especially Minnesota and the Dakotas. Two Lutheran synods, the Lutheran church in America and the American Lutheran Church emerged in 1987. In June 1983, the two major Presbyterian churches, the northern-based United Presbyterian Church in the USA and the southern-based Presbyterian Church in the US, formally merged as the Presbyterian Church (USA), ending a division that began with the Civil War. Other Protestant denominations and their estimated adherents in 1990 were the Episcopal Church 2,445,286; Churches of Christ 1,681,013; and the United Church of Christ (Congregationalist), 1,993,459. One Christian group, the Church of Latter-day-Saints (Morman), which claimed 3,540,820 members in 1990, was organized in New York in 1830 and, since migrating westward has played a leading role in Utah's political, economic, and religious life. Notable during the 1970s and early 1980s was a rise in the fundamentalist, evangelical, and Pentecostal movements.

Several million Muslims followers of various Asian religious, a multiplicity of small Protestant groups, and a sizable number of cults also participate in US religious life.

¹¹TRANSPORTATION

The extent of the transportation industry in the US is indicated by the fact that railroads, motor vehicles, inland waterways, oil

pipelines, and domestic airways recorded a total of more than 3.5 trillion passenger-mi (5.6 trillion passenger-km) and 3.4 trillion ton-mi (5.5 trillion ton-km) of freight service in 1989. Outlays for all types of transportation products and services totaled an estimated $800 billion in 1989.

Railroads lost not only the largest share of intercity freight traffic, their chief source of revenue, but passenger traffic as well. Despite an attempt to revive passenger transport through the development of a national network (Amtrak) in the 1970s, the rail sector continued to experience heavy losses and declining revenues. In 1991/92, Amtrak had a ridership of over 38 million. In 1991 there were 14 Class I rail companies in the US operating 153,684 rail mi (247,278 km) of track. Railroads carried 1,581,871,847 tons of cargo through 23,316,176 rail carloads handled in 1991.

The most conspicuous form of transportation is the automobile, and the extent and quality of the US road-transport system are without parallel in the world. Over 190.3 million vehicles—a record number—were registered in 1992, including more than 144.2 million passenger cars and some 46.1 million trucks and buses. In 1992, there were some 4.4 million motorcycles registered as well. In 1991, auto manufacturers in the US produced 5.4 million passenger cars and 8.8 million commercial vehicles. Motor vehicle imports and exports in 1991 amounted to 4,452,922 and 962,894, respectively. Canada accounted for 36% of the imports and 66% of the exports in 1991. In 1992, 31% of the world's motor vehicles were registered in the US, down from 36% in 1985.

The US has a vast network of public roads, whose total length as of 31 December 1992 was 3,901,715 mi (6,277,859 (cm). Of that total, 80% was rural. During the 1970s, about $10 billion was spent annually on highway construction. By the late 1970s, new highway construction had slowed, and an increasing share of highway funds was allocated to the improvement of existing roads. In 1989, the Federal Highway Administration reported that 23% of the 575,000 highway bridges in the US were structurally deficient and another 19% functionally obsolete.

Major ocean ports or port areas are New York, the Delaware River areas (Philadelphia), the Chesapeake Bay area (Baltimore, Norfolk, Newport News), New Orleans, Houston, and the San Francisco Bay area. The inland port of Duluth on Lake Superior handles more freight than all but the top-ranking ocean ports. The importance of this port, along with those of Chicago and Detroit, was enhanced with the opening in 1959 of the St. Lawrence Seaway. Waterborne freight consists primarily of bulk commodities such as petroleum and its products, coal and coke, iron ore and steel, sand, gravel and stone, grains, and lumber. The US merchant marine industry has been decreasing gradually since the 1950s. In 1991, the US had the 10th-largest registered merchant shipping fleet in the world (by GRT), with 394 privately-owned vessels of more than 1,00 gross registered tons. The total US merchant fleet, including government-owned vessels, was 619 ships (with a total of 15,466,000 GRT) in 1991.

In 1983, the US had 96 certified air carriers, more than double the number in 1978, when the Airline Deregulation Act was passed. Revenue passengers carried by the airlines in 1940 totaled 2.7 million; by 1992, the figure was 473 million. The US in 1992 had 17,846 airports, of which 5,545 were public. US international carriers performed some 214,628 million passenger-km (133,370 million passenger-mi) of service in 1992, along with 8,166 million freight ton-km (5,074 million freight ton-mi). By the end of the year, the US had 692,095 active pilots, of whom 148,365 were on commercial routes. An estimated 198,475 general aviation aircraft flew a total of 30,055,000 hours in 1991.

¹²HISTORY

The first Americans—distant ancestors of the American Indians—probably crossed the Bering Strait from Asia at least 12,000 years ago. By the time Christopher Columbus came to the New World in 1492 there were probably no more than 2 million Native Americans living in the land that was to become the US.

Following exploration of the American coasts by English, Portuguese, Spanish, Dutch, and French sea captains from the late 15th century onward, European settlements sprang up in the latter part of the 16th century. The Spanish established the first permanent settlement at St. Augustine in the future state of Florida in 1565, and another in New Mexico in 1599. During the early 17th century, the English founded Jamestown in Virginia Colony (1607) and Plymouth Colony in present-day Massachusetts (1620). The Dutch established settlements at Ft. Orange (now Albany, N.Y.) in 1624, New Amsterdam (now New York City) in 1626, and at Bergen (now part of Jersey City, N.J.) in 1660; they conquered New Sweden—the Swedish colony in Delaware and New Jersey—in 1655. Nine years later, however, the English seized this New Netherland Colony and subsequently monopolized settlement of the East Coast except for Florida, where Spanish rule prevailed until 1821. In the Southwest, California, Arizona, New Mexico, and Texas also were part of the Spanish empire until the 19th century. Meanwhile, in the Great Lakes area south of present-day Canada, France set up a few trading posts and settlements but never established effective control; New Orleans was one of the few areas of the US where France pursued an active colonial policy.

From the founding of Jamestown to the outbreak of the American Revolution more than 150 years later, the British government administered its American colonies within the context of mercantilism: the colonies existed primarily for the economic benefit of the empire. Great Britain valued its American colonies especially for their tobacco, lumber, indigo, rice, furs, fish, grain, and naval stores, relying particularly in the southern colonies on black slave labor.

The colonies enjoyed a large measure of internal self-government until the end of the French and Indian War (1745–63), which resulted in the loss of French Canada to the British. To prevent further troubles with the Indians, the British government in 1763 prohibited the American colonists from settling beyond the Appalachian Mountains. Heavy debts forced London to decree that the colonists should assume the costs of their own defense, and the British government enacted a series of revenue measures to provide funds for that purpose. But soon, the colonists began to insist that they could be taxed "only with their consent," and the struggle grew to become one of local versus imperial authority.

Widening cultural and intellectual differences also served to divide the colonies and the mother country. Life on the edge of the civilized world had brought about changes in the colonists' attitudes and outlook, emphasizing their remoteness from English life. In view of the long tradition of virtual self-government in the colonies, strict enforcement of imperial regulations and British efforts to curtail the power of colonial legislatures presaged inevitable conflict between the colonies and the mother country. When citizens of Massachusetts, protesting the tax on tea, dumped a shipload of tea belonging to the East India Company into Boston harbor in 1773, the British felt compelled to act in defense of their authority as well as in defense of private property. Punitive measures—referred to as the Intolerable Acts by the colonists— struck at the foundations of self-government.

In response, the First Continental Congress, composed of delegates from 12 of the 13 colonies—Georgia was not represented— met in Philadelphia in September 1774, and proposed a general boycott of English goods, together with the organizing of a militia. British troops marched to Concord, Mass., on 19 April 1775 and destroyed the supplies that the colonists had assembled there. American "minutemen" assembled on the nearby Lexington green and fired "the shot heard round the world," although no one knows who actually fired the first shot that morning. The

British soldiers withdrew and fought their way back to Boston.

Voices in favor of conciliation were raised in the Second Continental Congress that assembled in Philadelphia on 10 May 1775, this time including Georgia; but with news of the Restraining Act (30 March 1775), which denied the colonies the right to trade with countries outside the British Empire, all hopes for peace vanished. George Washington was appointed commander in chief of the new American army, and on 4 July 1776, the 13 American colonies adopted the Declaration of Independence, justifying the right of revolution by the theory of natural rights.

British and American forces met in their first organized encounter near Boston on 17 June 1775. Numerous battles up and down the coast followed. The British seized and held the principal cities but were unable to inflict a decisive defeat on Washington's troops. The entry of France into the war on the American side eventually tipped the balance. On 19 October 1781, the British commander, Cornwallis, cut off from reinforcements by the French fleet on one side and besieged by French and American forces on the other, surrendered his army at Yorktown, Va. American independence was acknowledged by the British in a treaty of peace signed in Paris on 3 September 1783.

The first constitution uniting the 13 original states—the Articles of Confederation—reflected all the suspicions that Americans entertained about a strong central government. Congress was denied power to raise taxes or regulate commerce, and many of the powers it was authorized to exercise required the approval of a minimum of nine states. Dissatisfaction with the Articles of Confederation was aggravated by the hardships of a postwar depression, and in 1787—the same year that Congress passed the Northwest Ordinance, providing for the organization of new territories and states on the frontier—a convention assembled in Philadelphia to revise the articles. The convention adopted an altogether new constitution, the present Constitution of the United States, which greatly increased the powers of the central government at the expense of the states. This document was ratified by the states with the understanding that it would be amended to include a bill of rights guaranteeing certain fundamental freedoms. These freedoms—including the rights of free speech, press, and assembly, freedom from unreasonable search and seizure, and the right to a speedy and public trial by an impartial jury—are assured by the first 10 amendments to the constitution, adopted on 5 December 1791; the constitution did however recognize slavery, and did not provide for universal suffrage. On 30 April 1789 George Washington was inaugurated as the first president of the US.

During Washington's administration, the credit of the new nation was bolstered by acts providing for a revenue tariff and an excise tax; opposition to the excise on whiskey sparked the Whiskey Rebellion, suppressed on Washington's orders in 1794. Alexander Hamilton's proposals for funding the domestic and foreign debt and permitting the national government to assume the debts of the states were also implemented. Hamilton, the secretary of the treasury, also created the first national bank, and was the founder of the Federalist Party. Opposition to the bank as well as to the rest of the Hamiltonian program, which tended to favor northeastern commercial and business interests, led to the formation of an anti-Federalist party, the Democratic-Republicans, led by Thomas Jefferson.

The Federalist Party, to which Washington belonged, regarded the French Revolution as a threat to security and property; the Democratic-Republicans, while condemning the violence of the revolutionists, hailed the overthrow of the French monarchy as a blow to tyranny. The split of the nation's leadership into rival camps was the first manifestation of the two-party system, which has since been the dominant characteristic of the US political scene. (Jefferson's party should not be confused with the modern Republican Party, formed in 1854.)

The 1800 election brought the defeat of Federalist President John Adams, Washington's successor, by Jefferson; a key factor in Adam's loss was the unpopularity of the Alien and Sedition Acts (1798), Federalist-sponsored measures that had abridged certain freedoms guaranteed in the Bill of Rights. In 1803, Jefferson achieved the purchase from France of the Louisiana Territory, including all the present territory of the US west of the Mississippi drained by that river and its tributaries; exploration and mapping of the new territory, notably through the expeditions of Meriwether Lewis and William Clark, began almost immediately. Under Chief Justice John Marshall, the US Supreme Court, in the landmark case of *Marbury v. Madison,* established the principle of federal supremacy in conflicts with the states and enunciated the doctrine of judicial review.

During Jefferson's second term in office, the US became involved in a protracted struggle between Britain and Napoleonic France. Seizures of US ships and the impressment of US seamen by the British navy led the administration to pass the Embargo Act of 1807, under which no US ships were to be put out to sea. After the act was repealed in 1809, ship seizures and impressment of seamen by the British continued, and were the ostensible reasons for the declaration of war on Britain in 1812 during the administration of James Madison. An underlying cause of the War of 1812, however, was land-hungry westerners' coveting of southern Canada as potential US territory.

The war was largely a standoff. A few surprising US naval victories countered British successes on land. The Treaty of Ghent (24 December 1814), which ended the war, made no mention of impressment and provided for no territorial changes. The occasion for further maritime conflict with Britain, however, disappeared with the defeat of Napoleon in 1815.

Now the nation became occupied primarily with domestic problems and westward expansion. Because the US had been cut off from its normal sources of manufactured goods in Great Britain during the war, textiles and other industries developed and prospered in New England. To protect these infant industries, Congress adopted a high-tariff policy in 1816.

Three events of the late 1810s and the 1820s were of considerable importance for the future of the country. The federal government in 1817 began a policy of forcibly resettling the Indians, already decimated by war and disease, in what later became known as Indian Territory (now Oklahoma); those Indians not forced to move were restricted to reservations. The Missouri Compromise (1820) was an attempt to find a nationally acceptable solution to the volatile dispute over the extension of black slavery to new territories. It provided for admission of Missouri into the Union as a slave state but banned slavery in territories to the west that lay north of 36°30′. As a result of the establishment of independent Latin American republics and threats by France and Spain to reestablish colonial rule, President James Monroe in 1923 asserted that the Western Hemisphere was closed to further colonization by European powers. The Monroe Doctrine declared that any effort by such powers to recover territories whose independence the US had recognized would be regarded as an unfriendly act.

From the 1820s to the outbreak of the Civil War, the growth of manufacturing continued, mainly in the North, and was accelerated by inventions and technological advances. Farming expanded with westward migration. The South discovered that its future lay in the cultivation of cotton. The cotton gin, invented by Eli Whitney in 1793, greatly simplified the problems of production; the growth of the textile industry in New England and Great Britain assured a firm market for cotton. Hence, during the first half of the 19th century, the South remained a fundamentally agrarian society based increasingly on a one-crop economy. Large numbers of field hands were required for cotton cultivation, and black slavery became solidly entrenched in the southern economy.

The construction of roads and canals paralleled the country's growth and economic expansion. The successful completion of the Erie Canal (1825), linking the Great Lakes with the Atlantic, ushered in a canal-building boom. Railroad building began in earnest in the 1830s, and by 1840, about 3,300 mi (5,300 km) of track had been laid. The development of the telegraph a few years later gave the nation the beginnings of a modern telecommunications network. As a result of the establishment of the factory system, a laboring class appeared in the North by the 1830s, bringing with it the earliest unionization efforts.

Western states admitted into the Union following the War of 1812 provided for free white male suffrage without property qualifications and helped spark a democratic revolution. As eastern states began to broaden the franchise, mass appeal became an important requisite for political candidates. The election to the presidency in 1928 of Andrew Jackson, a military hero and Indian fighter from Tennessee, was no doubt a result of this widening of the democratic process. By this time, the US consisted of 24 states and had a population of nearly 13 million.

The relentless westward thrust of the US population ultimately involved the US in foreign conflict. In 1836, US settlers in Texas revolted against Mexican rule and established an independent republic. Texas was admitted to the Union as a state in 1845, and relations between Mexico and the US steadily worsened. A dispute arose over the southern boundary of Texas, and a Mexican attack on a US patrol in May 1846 gave President James K. Polk a pretext to declare war. After a rapid advance, US forces captured Mexico City, and on 2 February 1848, Mexico formally gave up the unequal fight by signing the Treaty of Guadalupe Hidalgo, providing for the cession of California and the territory of New Mexico to the US. With the Gadsden Purchase of 1853, the US acquired from Mexico for $10 million large strips of land forming the balance of southern Arizona and New Mexico. A dispute with Britain over the Oregon Territory was settled in 1846 by a treaty that established the 49th parallel as the boundary with Canada. Thenceforth the US was to be a Pacific as well as an Atlantic power.

Westward expansion exacerbated the issue of slavery in the territories. By 1840, abolition of slavery constituted a fundamental aspect of a movement for moral reform, which also encompassed woman's rights, universal education, alleviation of working class hardships, and temperance. In 1849, a year after the discovery of gold had precipitated a rush of new settlers to California, that territory (whose constitution prohibited slavery) demanded admission to the Union. A compromise engineered in Congress by Senator Henry Clay in 1850 provided for California's admission as a free state in return for various concessions to the South. But enmities dividing North and South could not be silenced. The issue of slavery in the territories came to a head with the Kansas-Nebraska Act of 1854, which repealed the Missouri Compromise and left the question of slavery in those territories to be decided by the settlers themselves. The ensuing conflicts in Kansas between northern and southern settlers earned the territory the name "bleeding Kansas." in 1860, the Democratic Party, split along northern and southern lines, offered two presidential candidates. The new Republican Party, organized in 1854 and opposed to the expansion of slavery, nominated Abraham Lincoln. Owing to the defection in Democratic ranks, Lincoln was able to carry the election in the electoral college, although he did not obtain a majority of the popular vote. To ardent supporters of slavery, Lincoln's election provided a reason for immediate secession. Between December 1860 and February 1861, the seven states of the Deep South—South Carolina, Mississippi, Florida, Alabama, Georgia, Louisiana, and Texas—withdrew from the Union and formed a separate government, known as the Confederate States of America, under the presidency of Jefferson Davis. The secessionists soon began to confiscate federal property in the South. On 12 April 1861, the Confederates opened fire on Ft. Sumter in the harbor of Charleston, S.C., and thus precipitated the US Civil War. Following the outbreak of hostilities, Arkansas, North Carolina, Virginia, and Tennessee joined the Confederacy.

For the next four years, war raged between the Confederate and Union forces, largely in southern territories. An estimated 360,000 men in the Union forces died of various causes, including 110,000 killed in battle. Confederate dead were estimated at 250,000, including 94,000 killed in battle. The North, with great superiority in manpower and resources, finally prevailed. A Confederate invasion of the North was repulsed at the battle of Gettysburg, Pa. in July 1863; a Union army took Atlanta in September 1864; and Confederate forces evacuated Richmond, the Confederate capital, in early April 1865. With much of the South in Union hands, Confederate Gen. Robert E. Lee surrendered to Gen. Ulysses S. Grant at Appomattox Courthouse in Virginia on 9 April.

The outcome of the war brought great changes in US life. Lincoln's Emancipation Proclamation of 1863 was the initial step in freeing some 4 million black slaves; their liberation was completed soon after the war's end by amendments to the Constitution. Lincoln's plan for the reconstruction of the rebellious states was compassionate, but only five days after Lee's surrender, Lincoln was assassinated by John Wilkes Booth as part of a conspiracy in which US Secretary of State William H. Seward was seriously wounded.

During the Reconstruction era (1865–77), the defeated South was governed by Union Army commanders, and the resultant bitterness of southerners toward northern Republican rule, which enfranchised blacks, persisted for years afterward. Vice President Andrew Johnson, who succeeded Lincoln as president, tried to carry out Lincoln's conciliatory policies but was opposed by radical Republican leaders in Congress, who demanded harsher treatment of the South. On the pretext that he had failed to carry out an act of Congress, the House of Representatives voted to impeach Johnson in 1868, but the Senate failed by one vote to convict him and remove him from office. It was during Johnson's presidency that Secretary of State Seward negotiated the purchase of Alaska (which attained statehood in 1959) from Russia for $7.2 million.

The efforts of southern whites to regain political control of their states led to the formation of terrorist organizations like the Ku Klux Klan, which employed violence to prevent blacks from voting. By the end of the Reconstruction era, whites had reestablished their political domination over blacks in the southern states and had begun to enforce patterns of segregation in education and social organization that were to last for nearly a century.

In many southern states, the decades following the Civil War were ones of economic devastation, in which rural whites as well as blacks were reduced to sharecropper status. Outside the South, however, a great period of economic expansion began. Transcontinental railroads were constructed, corporate enterprise spurted ahead, and the remaining western frontier lands were rapidly occupied and settled. The age of big business tycoons dawned. As heavy manufacturing developed, Pittsburgh, Chicago, and New York emerged as the nation's great industrial centers. The Knights of Labor, founded in 1869, engaged in numerous strikes, and violent conflicts between strikers and strikebreakers were common. The American Federation of Labor, founded in 1886, established a nationwide system of craft unionism that remained dominant for many decades. During this period, too, the woman's rights movement organized actively to secure the vote (although woman's suffrage was not enacted nationally until 1920), and groups outraged by the depletion of forests and wildlife in the West pressed for the conservation of natural resources.

During the latter half of the 19th century, the acceleration of westward expansion made room for millions of immigrants from Europe. The country's population grew to more than 76 million by 1900. As homesteaders, prospectors, and other settlers tamed the frontier, the federal government forced Indians west of the Mississippi to cede vast tracts of land to the whites, precipitating a series of wars with various tribes. By 1890, only 250,000 Indians remained in the US, virtually all of them residing on reservations.

The 1890s marked the closing of the US frontier for settlement and the beginning of US overseas expansion. By 1892, Hawaiian sugar planters of US origin had become strong enough to bring about the downfall of the native queen and to establish a republic, which in 1898, at its own request, was annexed as a territory by the US. The sympathies of the US with the Cuban nationalists who were battling for independence from Spain were aroused by a lurid press and by expansionist elements. A series of events climaxed by the sinking of the USS *Maine* in Havana harbor finally forced a reluctant President William McKinley to declare war on Spain on 25 April 1898. US forces overwhelmed those of Spain in Cuba, and as a result of the Spanish-American War, the US added to its territories the Philippines, Guam, and Puerto Rico. A newly independent Cuba was drawn into the US orbit as a virtual protectorate through the 1950s. Many eminent citizens saw these new departures into imperialism as a betrayal of the time-honored US doctrine of government by the consent of the governed.

With the marked expansion of big business came increasing protests against the oppressive policies of large corporations and their dominant role in the public life of the nation. A demand emerged for strict control of monopolistic business practice through the enforcement of antitrust laws. Two US presidents, Theodore Roosevelt (1901–9), a Republican and Woodrow Wilson (1913–21), a Democrat, approved of the general movement for reform, which came to be called progressivism. Roosevelt developed a considerable reputation as a trustbuster, while Wilson's program, known as the New Freedom, called for reform of tariffs, business procedures, and banking. During Roosevelt's first term, the US leased the Panama Canal Zone and started construction of a 42-mi (68-km) canal, completed in 1914.

US involvement in World War I marked the country's active emergence as one of the great powers of the world. When war broke out in 1914 between Germany, Austria-Hungary, and Turkey on one side and Britain, France, and Russia on the other, sentiment in the US was strongly opposed to participation in the conflict, although a large segment of the American people sympathized with the British and the French. While both sides violated US maritime rights on the high seas, the Germans, enmeshed in a British blockade, resorted to unrestricted submarine warfare. On 6 April 1917, congress declared war on Germany. Through a national draft of all able-bodied men between the ages of 18 and 45, some 4 million US soldiers were trained, of whom more than 2 million were sent overseas to France. By late 1917, when US troops began to take part in the fighting on the western front, the European armies were approaching exhaustion, and US intervention may well have been decisive in ensuring the eventual victory of the Allies. In a series of great battles in which US soldiers took an increasingly major part, the German forces were rolled back in the west, and in the autumn of 1918 were compelled to sue for peace. Fighting ended with the armistice of 11 November 1918. President Wilson played an active role in drawing up the 1919 Versailles peace treaty, which embodied his dream of establishing a League of Nations to preserve the peace, but the isolationist bloc in the Senate was able to prevent US ratification of the treaty.

In the 1920s, the US had little enthusiasm left for crusades, either for democracy abroad or for reform at home; a rare instance of idealism in action was the Kellogg-Briand Pact (1928), an antiwar accord negotiated on behalf of the US by Secretary of State Frank B. Kellogg. In general, however, the philosophy of the Republican administrations from 1921 to 1933 was expressed in the aphorism "The business of America is business," and the 1920s saw a great business boom. The years 1923–24 also witnessed the unraveling of the Teapot Dome scandal: the revelation that President Warren G. Harding's secretary of the interior, Albert B. Fall, had secretly leased federal oil reserves in California and Wyoming to private oil companies in return for gifts and loans.

The great stock market crash of October 1929 ushered in the most serious and most prolonged economic depression the country had ever known. By 1933, an estimated 12 million men and women were out of work; personal savings were wiped out on a vast scale through a disastrous series of corporate bankruptcies and bank failures. Relief for the unemployed was left to private charities and local governments, which were incapable of handling the enormous task.

The inauguration of the successful Democratic presidential candidate, Franklin D. Roosevelt, in March 1933 ushered in a new era of US history, in which the federal government was to assume a much more prominent role in the nation's economic affairs. Proposing to give the country a "New Deal," Roosevelt accepted national responsibility for alleviating the hardships of unemployment; relief measures were instituted, work projects were established, the deficit spending was accepted in preference to ignoring public distress. The federal Social Security program was inaugurated, as were various measures designed to stimulate and develop the economy through federal intervention. Unions were strengthened through the National Labor Relations Act, which established the right of employees' organizations to bargain collectively with employers. Union membership increased rapidly, and the dominance of the American Federation of Labor was challenged by the newly formed Congress of Industrial Organizations, which organized workers along industrial lines.

The depression of the 1930s was worldwide, and certain nations attempted to counter economic stagnation by building large military establishments and embarking on foreign adventures. Following German, Italian, and Japanese aggression, World War II broke out in Europe during September 1939. In 1940, Roosevelt, disregarding a tradition dating back to Washington that no president should serve more than two terms, ran again for reelection. He easily defeated his Republican opponent, Wendell Willkie, who, along with Roosevelt, advocated increased rearmament and all possible aid to victims of aggression. The US was brought actively into the war by the Japanese attack on the Pearl Harbor naval base in Hawaii on 7 December 1941. The forces of Germany, Italy, and Japan were now arrayed over a vast theater of war against those of the US and the British Commonwealth; in Europe, Germany was locked in a bloody struggle with the Soviet Union. US forces waged war across the vast expanses of the Pacific, in Africa, in Asia, and in Europe. Italy surrendered in 1943; Germany was successfully invaded in 1944 and conquered in May 1945; and after the US dropped the world's first atomic bombs on Hiroshima and Nagasaki, the Japanese capitulated in August. The Philippines became an independent republic soon after the war, but the US retained most of its other Pacific possessions, with Hawaii becoming the 50th state in 1959.

Roosevelt, who had been elected to a fourth term in 1944, died in April 1945 and was succeeded by Harry S Truman, his vice president. Under the Truman administration, the US became an active member of the new world organization, the United Nations. The Truman administration embarked on large-scale programs of military aid and economic support to check the expansion of communism. Aid to Greece and Turkey in 1948 and the Marshall Plan, a program designed to accelerate the economic recovery of Western Europe, were outstanding features of US postwar foreign policy. The North Atlantic Treaty (1949) established a defensive alliance among a number of West European nations and the US. Truman's Point Four program gave technical

and scientific aid to developing nations. When, following the North Korean attack on South Korea on 25 June 1950, the UN Security Council resolved that members of the UN should proceed to the aid of South Korea. US naval, air, and ground forces were immediately dispatched by President Truman. An undeclared war ensued, which eventually was brought to a halt by an armistice signed on 27 June 1953.

In 1952, Dwight D. Eisenhower, supreme commander of Allied forces in Europe during World War II, was elected president on the Republican ticket, thereby bringing to an end 20 years of Democratic presidential leadership. In foreign affairs, the Eisenhower administration continued the Truman policy of containing the USSR and threatened "massive retaliation" in the event of Soviet aggression, thus heightening the Cold War between the world's two great nuclear powers. Although Republican domestic policies were more conservative than those of the Democrats, the Eisenhower administration extended certain major social and economic programs of the Roosevelt and Truman administrations, notably Social Security and public housing. The early years of the Eisenhower administration were marked by agitation (arising in 1950) over charges of Communist and other allegedly subversive activities in the US—a phenomenon known as McCarthyism, after Republican Senator Joseph R. McCarthy of Wisconsin, who aroused much controversy with unsubstantiated allegations that Communists had penetrated the US government, especially the Army and the Department of State. Even those who personally opposed McCarthy lent their support to the imposition of loyalty oaths and the blacklisting of persons with left-wing backgrounds.

A major event of the Eisenhower years was the US Supreme Court's decision in *Brown v. Board of Education of Topeka* (1954) outlawing segregation of whites and blacks in public schools. In the aftermath of this ruling, desegregation proceeded slowly and painfully. In the early 1960s, sit-ins, "freedom rides," and similar expressions of nonviolent resistance by blacks and their sympathizers led to a lessening of segregation practices in public facilities. Under Chief Justice Earl Warren, the high court in 1962 mandated the reapportionment of state and federal legislative districts according to a "one person, one vote" formula. It also broadly extended the rights of defendants in criminal trials to include the provision of a defense lawyer at public expense for an accused person unable to afford one, and established the duty of police to advise an accused person of his or her legal rights immediately upon arrest.

In the early 1960s, during the administration of Eisenhower's Democratic successor, John F. Kennedy, the Cold War heated up as Cuba, under the regime of Fidel Castro, aligned itself with the Soviet Union. Attempts by anti-Communist Cuban exiles to invade their homeland in the spring of 1961 failed despite US aid. In October 1962, President Kennedy successfully forced a showdown with the Soviet Union over Cuba in demanding the withdrawal of Soviet-supplied "offensive weapons"—missiles—from the nearby island. On 22 November 1963, President Kennedy was assassinated while riding in a motorcade through Dallas, Texas; hours later, Vice President Lyndon B. Johnson was inaugurated president. In the November 1964 elections, Johnson overwhelmingly defeated his Republican opponent, Barry M. Goldwater, and embarked on a vigorous program of social legislation unprecedented since Roosevelt's New Deal. His "Great Society" program sought to ensure black Americans' rights in voting and public housing, to give the underprivileged job training, and to provide persons 65 and over with hospitalization and other medical benefits (Medicare). Measures ensuring equal opportunity for minority groups may have contributed to the growth of the woman's rights movement in the late 1960s. This same period also saw the growth of a powerful environmental protection movement.

US military and economic aid to anti-Communist forces in Viet-Nam, which had its beginnings during the Truman administration (while Viet-Nam was still part of French Indochina) and was increased gradually by presidents Eisenhower and Kennedy, escalated in 1965. In that year, President Johnson sent US combat troops to South Viet-Nam and ordered US bombing raids on North Viet-Nam, after Congress (in the Gulf of Tonkin Resolution of 1964) had given him practically carte blanche authority to wage war in that region. By the end of 1968, American forces in Viet-Nam numbered 536,100 men, but US military might was unable to defeat the Vietnamese guerrillas, and the American people were badly split over continuing the undeclared (and, some thought, ill-advised or even immoral) war, with its high price in casualties and materiel. Reacting to widespread dissatisfaction with his Viet-Nam policies, Johnson withdrew in March 1968 from the upcoming presidential race, and in November, Republican Richard M. Nixon, who had been the vice president under Eisenhower, was elected president. Thus, the Johnson years—which had begun with the new hopes of a Great Society but had soured with a rising tide of racial violence in US cities and the assassinations of civil rights leader Martin Luther King, Jr., and US Senator Robert F. Kennedy, among others—drew to a close.

President Nixon gradually withdrew US ground troops from Viet-Nam but expanded aerial bombardment throughout Indochina, and the increasingly unpopular and costly war continued for four more years before a cease-fire—negotiated by Nixon's national security adviser, Henry Kissinger—was finally signed on 27 January 1973 and the last US soldiers were withdrawn. The most protracted conflict in American history had resulted in 46,163 US combat deaths and 303,654 wounded soldiers, and had cost the US government $112 billion in military allocations. Two years later, the South Vietnamese army collapsed, and the North Vietnamese Communist regime united the country.

In 1972, during the last year of his first administration, Nixon initiated the normalization of relations—ruptured in 1949—with the People's Republic of China and signed a strategic arms limitation agreement with the Soviet Union as part of a Nixon-Kissinger policy of pursuing détente with both major Communist powers. (Earlier, in July 1969, American technology had achieved a national triumph by landing the first astronaut on the moon.) The Nixon administration sought to muster a "silent majority" in support of its Indochina policies and its conservative social outlook in domestic affairs. The most momentous domestic development, however, was the Watergate scandal, which began on 17 June 1972 with the arrest of five men associated with Nixon's reelection campaign, during a break-in at Democratic Party headquarters in the Watergate office building in Washington, D.C. Although Nixon was reelected in 1972, subsequent disclosures by the press and by a Senate investigating committee revealed a complex pattern of political "dirty tricks" and illegal domestic surveillance throughout his first term. The president's apparent attempts to obstruct justice by helping his aides cover up the scandal were confirmed by tape recordings (made by Nixon himself) of his private conversations, which the Supreme Court ordered him to release for use as evidence in criminal proceedings. The House voted to begin impeachment proceedings, and in late July 1974, its Judiciary Committee approved three articles of impeachment. On 9 August, Nixon became the first president to resign the office. The following year, Nixon's top aides and former attorney general, John N. Mitchell, were convicted of obstruction and were subsequently sentenced to prison.

Nixon's successor was Gerald R. Ford, who in October 1973 had been appointed to succeed Vice President Spiro T. Agnew when Agnew resigned following his plea of *nolo contendere* to charges that he had evaded paying income tax on moneys he had received from contractors while governor of Maryland. Less than a month after taking office, President Ford granted a full pardon

to Nixon for any crimes he may have committed as president. In August 1974, Ford nominated Nelson A. Rockefeller as vice president (he was not confirmed until December), thus giving the country the first instance of a nonelected president and an appointed vice president serving simultaneously. Ford's pardon of Nixon, as well as continued inflation and unemployment, probably contributed to his narrow defeat by a Georgia Democrat, Jimmy Carter, in 1976.

President Carter's forthright championing of human rights—though consistent with the Helsinki accords, the "final act" of the Conference on Security and Cooperation in Europe, signed by the US and 34 other nations in July 1974—contributed to strained relations with the USSR and with some US allies. During 1978–79, the president concluded and secured Senate passage of treaties ending US sovereignty over the Panama Canal Zone. His major accomplishment in foreign affairs, however, was his role in mediating a peace agreement between Israel and Egypt, signed at the camp David, Md., retreat in September 1978. Domestically, the Carter administration initiated a national energy program to reduce US dependence on foreign oil by cutting gasoline and oil consumption and by encouraging the development of alternative energy resources. But the continuing decline of the economy because of double-digit inflation and high unemployment caused his popularity to wane, and confusing shifts in economic policy (coupled with a lack of clear goals in foreign affairs) characterized his administration during 1979 and 1980; a prolonged quarrel with Iran over more than 50 US hostages seized in Tehran on 4 November 1979 contributed to public doubts about his presidency. Exactly a year after the hostages were taken, former California Governor Ronald Reagan defeated Carter in an election that saw the Republican Party score major gains throughout the US. The hostages were released on 20 January 1981, the day of Reagan's inauguration.

Reagan, who survived a chest wound from an assassination attempt in Washington, D.C. in 1981, used his popularity to push through significant policy changes. He succeeded in enacting income tax cuts of 25 percent, reducing the maximum tax rate on unearned income from 70 percent to 50 percent, and accelerating depreciation allowances for businesses. At the same time, he more than doubled the military budget, in constant 1985 dollars, between 1980 and 1989. Vowing to reduce domestic spending, Reagan cut benefits for the working poor, reduced allocations for food stamps and Aid to Families With Dependent Children by 13 percent, and decreased grants for the education of disadvantaged children. He slashed the budget of the Environmental Protection Agency and instituted a flat rate reimbursement system for the treatment of Medicare patients with particular illnesses, replacing a more flexible arrangement in which hospitals had been reimbursed for "reasonable charges."

Reagan's appointment of Sandra Day O'Connor as the first woman justice of the Supreme Court was widely praised and won unanimous confirmation from the Senate. However, some of his other high-level choices were extremely controversial—none more so than that of his secretary of the interior, James G. Watt, who finally resigned on October 1983. To direct foreign affairs, Reagan named Alexander M. Haig, Jr., former NATO supreme commander for Europe, to the post of secretary of state; Haig, who clashed frequently with other administration officials, resigned in June 1982 and was replace by George P. Shultz. In framing his foreign and defense policy, Reagan insisted on a military buildup as a precondition for arms-control talks with the USSR. His administration sent money and advisers to help the government of El Salvador in its war against leftist rebels, and US advisers were also sent to Honduras, reportedly to aid groups of Nicaraguans trying to overthrow the Sandinista government in their country. Troops were also dispatched to Lebanon in September 1982, as part of a multinational peacekeeping force in Beirut, and to Grenada in October 1983 to oust a leftist government there.

Reelected in 1984, President Reagan embarked on his second term with a legislative agenda that included reduction of federal budget deficits (which had mounted rapidly during his first term in office), further cuts in domestic spending, and reform of the federal tax code. In military affairs, Reagan persuaded Congress to fund on a modest scale his Strategic Defense Initiative, commonly known as Star Wars, a highly complex and extremely costly space-based antimissile system. In 1987, the downing of an aircraft carrying arms to Nicaragua led to the disclosure that a group of National Security Council members had secretly diverted $48 million that the federal government had received in payment from Iran for American arms to rebel forces in Nicaragua. The disclosure prompted the resignation of two of the leaders of the group, Vice Admiral John Poindexter and Lieutenant Colonel Oliver North, as well as investigations by House and Senate committees and a special prosecutor, Lawrence Walsh. The congressional investigations found no conclusive evidence that Reagan had authorized or known of the diversion. Yet they noted that because Reagan had approved of the sale of arms to Iran and had encouraged his staff to assist Nicaraguan rebels despite the prohibition of such assistance by Congress, "the President created or at least tolerated an environment where those who did know of the diversion believed with certainty that they were carrying out the President's policies."

Reagan was succeeded in 1988 by his vice president, George Bush. Benefiting from a prolonged economic expansion, Bush handily defeated Michael Dukakis, Governor of Massachusetts and a liberal Democrat. On domestic issues, Bush sought to maintain policies introduced by the Reagan administration. His few legislative initiatives included the passage of legislation establishing strict regulations of air pollution, providing subsidies for child care, and protecting the rights of the disabled. Abroad, Bush showed more confidence and energy. While he responded cautiously to revolutions in Eastern Europe and the Soviet Union, he used his personal relationships with foreign leaders to bring about comprehensive peace talks between Israel and its Arab neighbors, to encourage a peaceful unification of Germany, and to negotiate broad and substantial arms cuts with the Russians. Bush reacted to Iraq's invasion of Kuwait in 1990 by sending 400,000 soldiers to form the basis of a multinational coalition which he assembled and which destroyed Iraq's main force within seven months.

One of the biggest crises that the Bush administration encountered was the collapse of the savings and loan industry in the late eighties. Thrift institutions were required by law to pay low interest rates for deposits and long-term loans. The creation of money market funds for the small investor in the eighties which paid higher rates of return than savings accounts prompted depositors to withdraw their money from banks and invest it in the higher yielding mutual funds. To finance the withdrawals, banks began selling assets at a loss. Congress and state legislatures tried to help banks compete by passing legislation permitting interest rates paid on deposits to fluctuate, allowing thrift institutions to enter new markets such as commercial loans, lowering minimum capital requirements and allowing thrifts to postpone reporting losses from bad loans and to overestimate the value of their assets. At the same time, Congress increased deposit insurance from $40,000 to $100,000 per account. The deregulation of the savings and loan industry, combined with the increase in federal insurance, encouraged many desperate savings institutions to invest in high risk real estate ventures—ventures for which no state supervision or regulation existed. When the majority of such ventures predictably failed, the federal government found itself compelled by law to rescue the thrifts, costing taxpayers over $100 billion.

In his bid for reelection in 1992, Bush faced not only Democratic nominee Bill Clinton, Governor of Arkansas, but also third party candidate Ross Perot, a Dallas billionaire who made his fortune in the computer industry. In contrast to Bush's first run for the presidency, when the nation had enjoyed an unusually long period of economic expansion, the economy in 1992 was just beginning to recover from a recession. Although data released the following year indicated that a healthy rebound had already begun in 1992, the public perceived the economy during election year as weak. Clinton took advantage of this perception in his campaign, focusing on the financial concerns of what he called "the forgotten middle class." He also took a more centrist position on many issues than more traditional Democrats, promising fiscal responsibility and economic growth. Clinton defeated Bush, winning 43 percent of the vote to Bush's 38 percent. Perot garnered 18 percent of the vote.

Almost as soon as he assumed office, Clinton came under pressure from various liberal groups who had supported his candidacy to establish their objectives as official policy. In particular, having endorsed gay rights during his campaign, Clinton faced demands in the first few weeks of his tenure to remove the ban on homosexuals in the military. Ultimately he reached a compromise with the top commanders of the Army, Navy, Air Force, Marines and Coast Guard who opposed lifting the ban. Colloquially termed "Don't ask, don't tell," the policy stipulated that gay members of the armed forces must refrain from homosexual behavior on or off base and that commanders, in turn, could not request information about sexual preference without clear evidence of homosexual behavior.

Clinton's major achievements since becoming President have included passage of a budget designed to raise revenue and thereby lower the deficit, which had ballooned during the Reagan and Bush years, as well as revitalize the economy and reduce spending. The budget, which narrowly passed, increased taxes on individuals in the highest income bracket and on energy while lowering the capital gains tax for small businesses and reducing spending on defense and health care. Clinton also persuaded Congress to approve the North American Free Trade Agreement, which removed or reduced tariffs on most goods moving across the borders of the United States, Canada, and Mexico. Although supporters and critics agreed that the treaty would create or eliminate relatively few jobs—two hundred thousand—the accord prompted heated debate. Labor strenuously opposed the agreement, seeing it as accelerating the flight of factory jobs to countries with low labor costs such as Mexico, the third largest trading partner of the US. Business, on the other hand, lobbied heavily for the treaty, arguing that it would create new markets for American goods and insisting that competition from Mexico would benefit the American economy.

Perhaps the most controversial initiative of the Clinton administration has been its health care plan. Crafted by a task force headed by Clinton's wife, Hillary, the administration's health care bill would provide all Americans with insurance for a basic package of services. The component of the plan provoking the greatest debate would require employers to pay most of the cost of a standard package for their employees, although a company's expenditure on the health care of its workers would be limited to a certain percentage of the payroll and small businesses would be eligible for subsidies. The plan also aroused the opposition of some physicians in its proposal to limit federal reimbursement of medical services and of the drug industry in its provision to cap the price of drugs.

[13]FEDERAL GOVERNMENT

The Constitution of the United States, signed in 1787, is the nation's governing document. In the first 10 amendments to the Constitution, ratified in 1791 and known as the Bill of Rights, the federal government is denied the power to infringe on rights generally regarded as fundamental to the civil liberties of the people. These amendments prohibit the establishment of a state religion and the abridgment of freedom of speech, press, and the right to assemble. They protect all persons against unreasonable searches and seizures, guarantee trial by jury, and prohibit excessive bail and cruel and unusual punishments. No person may be required to testify against himself, nor may he be deprived of life, liberty, or property without due process of law. The 13th Amendment (1865) banned slavery; the 15th (1870) protected the freed slaves' right to vote; and the 19th (1920) guaranteed the franchise to women. In all, there have been 26 amendments, the last of which, in 1971, reduced the voting age to 18. The Equal Rights Amendment (ERA), approved by Congress in 1972, would have mandated equality between the sexes; only 35 of the required 38 states had ratified the ERA by the time the ratification deadline expired on 30 June 1982.

The US has a federal form of government, with the distribution of powers between the federal government and the states constitutionally defined. The legislative powers of the federal government are vested in Congress, which consists of the House of Representatives and the Senate. There are 435 members of the House of Representatives. Each state is allotted a number of representatives in proportion to its population as determined by the decennial census. Representatives are elected for two-year terms in every even-numbered year. A representative must be at least 25 years old, must be a resident of the state represented, and must have been a citizen of the US for at least seven years. The Senate consists of two senators from each state, elected for six-year terms. Senators must be at least 30 years old, must be residents of the states from which they are elected, and must have been citizens of the US for at least nine years. One-third of the Senate is elected in every even-numbered year.

Congress legislates on matters of taxation, borrowing, regulation of international and interstate commerce, formulation of rules of naturalization, bankruptcy, coinage, weights and measures, post offices and post roads, courts inferior to the Supreme Court, provision for the armed forces, among many other matters. A broad interpretation of the "necessary and proper" clause of the Constitution has widened considerably the scope of congressional legislation based on the enumerated powers.

A bill that is passed by both houses of Congress in the same form is submitted to the president, who may sign it or veto it. If the president chooses to veto the bill, it is returned to the house in which it originated with the reasons for the veto. The bill may become law despite the president's veto if it is passed again by a two-thirds vote in both houses. A bill becomes law without the president's signature if retained for 10 days while Congress is in session. After Congress adjourns, if the president does not sign a bill within 10 days, an automatic veto ensues.

The president must be "a natural born citizen" at least 35 years old, and must have been a resident of the US for 14 years. Under the 22nd Amendment to the Constitution, adopted in 1951, a president may not be elected more than twice. Each state is allotted a number of electors based on its combined total of US senators and representatives, and, technically, it is these electors who, constituted as the electoral college, cast their vote for president, with all of the state's electoral votes customarily going to the candidate who won the largest share of the popular vote of the state (the District of Columbia also has three electors, making a total of 538 votes). Thus, the candidate who wins the greatest share of the popular vote throughout the US may, in rare cases, fail to win a majority of the electoral vote. If no candidate gains a majority in the electoral college, the choice passes to the House of Representatives.

The vice president, elected at the same time and on the same ballot as the president, serves as ex officio president of the Senate.

The vice president assumes the power and duties of the presidency on the president's removal from office or as a result of the president's death, resignation, or inability to perform his duties. In the case of a vacancy in the vice-presidency, the president nominates a successor, who must be approved by a majority in both houses of Congress. The Congress has the power to determine the line of presidential succession in case of the death or disability of both the president and vice president.

Under the Constitution, the president is enjoined to "take care that the laws be faithfully executed." In reality, the president has a considerable amount of leeway in determining to what extent a law is or is not enforced. Congress's only recourse is impeachment, to which it has resorted only twice, in proceedings against presidents Andrew Johnson and Richard Nixon. Both the president and the vice president are removable from office after impeachment by the House and conviction at a Senate trial for "treason, bribery, or other high crimes and misdemeanors." The president has the power to grant reprieves and pardons for offenses against the US except in cases of impeachment.

The President nominates and "by and with the advice and consent of the Senate" appoints ambassadors, public ministers, consuls, and all federal judges, including the justices of the Supreme Court. As commander in chief, the president is ultimately responsible for the disposition of the land, naval, and air forces, but the power to declare war belongs to Congress. The president conducts foreign relations and makes treaties with the advice and consent of the Senate. No treaty is binding unless it wins the approval of two-thirds of the Senate. The president's independence is also limited by the House of Representatives, where all money bills originate.

The president also appoints as his cabinet, subject to Senate confirmation, the secretaries who head the departments of the executive branch. As of 1985, the executive branch included the following cabinet departments: Agriculture (created in 1862), Commerce (1913), Defense (1947), Education (1980), Energy (1977), Health and Human Services (1980), Housing and Urban Development (1965), Interior (1849), Justice (1870), Labor (1913), State (1789), Transportation (1966), and Treasury (1789).The Department of Defense—headquartered in the Pentagon, the world's largest office building—also administers the various branches of the military: Air Force, Army, Navy, defense agencies, and joint-service schools. The Department of Justice administers the Federal Bureau of Investigation, which originated in 1908; the Central Intelligence Agency (1947) is under the aegis of the Executive office. Among the several hundred quasi-independent agencies are the Federal Reserve System (1913), serving as the nation's central bank, and the major regulatory bodies, notably the Environmental Protection Agency (1970), Federal Communications Commission (1934), Federal Power Commission (1920), Federal Trade Commission (1914), and Interstate Commerce Commission (1887).

Regulations for voting are determined by the individual states for federal as well as for local offices, and requirements vary from state to state. In the past, various southern states used literacy tests, poll taxes, "grandfather" clauses, and other methods to disfranchise black voters, but Supreme Court decisions and congressional measures, including the Voting Rights Act of 1965, more than doubled the number of black registrants in Deep South states between 1964 and 1992. In 1960, only 29.1% of the black voting-age population was registered to vote; by 1992, that percentage had risen to 64%.

14POLITICAL PARTIES

Two major parties, Democratic and Republican, have dominated national, state, and local politics since 1860. These parties are made up of clusters of small autonomous local groups primarily concerned with local politics and the election of local candidates to office. Within each party, such groups frequently differ drastically in policies and beliefs on many issues, but once every four years, they successfully bury their differences and rally around a candidate for the presidency. Minority parties have been formed at various periods in US political history, but most have generally allied with one of the two major parties, and none has achieved sustained national prominence. The most successful minority party in recent decades—that of Texas billionaire Ross Perot in 1992—was little more than a protest vote. Various extreme groups on the right and left, including a small US Communist Party, have had little political significance on a national scale; in 1980, the Libertarian Party became the first minor party since 1916 to appear on the ballot in all 50 states. Independent candidates have won state and local office, but no candidate has won the presidency without major party backing.

Traditionally, the Republican Party is more solicitous of business interests and gets greater support from business than does the Democratic Party. A majority of blue-collar workers, by contrast, have generally supported the Democratic Party, which favors more lenient labor laws, particularly as they affect labor unions; the Republican Party often (though not always) supports legislation that restricts the power of labor unions. Republicans favor the enhancement of the private sector of the economy, while Democrats generally urge the cause of greater government participation and regulatory authority, especially at the federal level.

Within both parties there are sharp differences on a great many issues; for example, northeastern Democrats in the past almost uniformly favored strong federal civil rights legislation, which was anathema to the Deep South; eastern Republicans in foreign policy are internationalist-minded, while midwesterners of the same party constituted from 1910 through 1940 the hard core of isolationist sentiment in the country. More recently, "conservative" headings have been adopted by members of both parties who emphasize decentralized government power, strengthened private enterprise, and a strong US military posture overseas, while the designation "liberal" has been applied to those favoring an increased federal government role in economic and social affairs, disengagement from foreign military commitments, and the intensive pursuit of nuclear-arms reduction.

President Nixon's resignation and the accompanying scandal surrounding the Republican Party hierarchy had a telling, if predictable, effect on party morale, as indicated by Republican losses in the 1974 and 1976 elections. The latent consequences of the Viet-Nam and Watergate years appeared to take their toll on both parties, however, in growing apathy toward politics and mistrust of politicians among the electorate. As of 1992, Democrats enjoyed a large advantage over Republicans in voter registration, held both houses of congress, had a majority of state governorships, and controlled most state legislative bodies. The centers of Democratic strength were the South and the major cities; most of the solidly Republican states were west of the Mississippi. Ronald Reagan's successful 1980 presidential bid cut into traditional Democratic strongholds throughout the US, as Republicans won control of the US Senate and eroded state and local Democratic majorities. On the strength of an economic recovery, President Reagan won reelection in November 1984, carrying 49 of 50 states (with a combined total of 525 electoral votes) and 58.8% of the popular vote; the Republicans retained control of the Senate, but the Democrats held on to the House. Benefiting from a six-year expansion of the economy, Republican George Bush won 54% of the vote in 1988. As Reagan had, Bush successfully penetrated traditionally Democratic regions. He carried every state in the South as well as the industrial states of the north. Bush's approval rating reached a high of 91% in March of 1992 after he decisively defeated Saddam Hussein in the Gulf War. By July of 1992, however, that rating had plummeted to 25%, in part because Bush appeared to be disengaged from domestic issues,

particularly the 1991 recession. Bill Clinton, governor of Arkansas and twenty years younger than Bush, presented himself as a "New Democrat. " He took more moderate positions than New Deal Democrats, including calling for a middle class tax cut, welfare reform, national service, and such traditionally Republican goals as safe streets.

The presidential race took on an unpredictable dimension with the entrance of Independent Ross Perot, a Texas billionaire. Perot attacked the budget deficit and called for shared sacrifice. He proved to be somewhat erratic, however. He withdrew from the race in July and then re-entered in October. Clinton won the race with 43% of the vote. Bush received 38% and Perot captured 18%. The Republicans gained nine seats in the House of Representatives while the Democrats picked up one seat in the Senate. As of 1993, 25 state legislatures were under Democratic control, 16 legislatures were split, and eight legislatures were under Republican control.

The 1984 election marked a turning point for women in national politics. Geraldine A. Ferraro, a Democrat, became the first female vice-presidential nominee of a major US political party; no woman has ever captured a major-party presidential nomination. As of 1994, seven women served in the U.S. Senate, 22 women held seats in the U.S. House of Representatives, and four women occupied state governorships.

The candidacy of Jesse L. Jackson in the 1984 presidential election, the first black ever to win a plurality in a statewide presidential preference primary, likewise marked the emergence of black Americans as a political force, especially within the Democratic Party. As of 1993, the US had 8,015 black elected officials, including the mayors of some of the nation's largest cities. One black woman, Carol Moseley Braun, won election to the Senate, becoming the first black senator. There were 38 blacks in the House. The House also had 17 Hispanics as of 1993.

[15]LOCAL GOVERNMENT

Governmental units within each state comprise counties, municipalities, and such special districts as those for water, sanitation, highways, parks. and recreation. There are more than 3,000 counties in the US; more than 19,000 municipalities, including cities, villages, towns, and boroughs; nearly 15,000 school districts; and at least 28,000 special districts. Additional townships, authorities, commissions, and boards make up the rest of the more than 82,000 local governmental units.

The states are autonomous within their own spheres of government, and their autonomy is defined in broad terms by the 10th Amendment to the US Constitution, which reserves to the states such powers as are not granted to the federal government and not denied to the states. The states may not, among other restrictions, issue paper money, conduct foreign relations, impair the obligations of contracts, or establish a government that is not republican in form. Subsequent amendments to the Constitution and many Supreme Court decisions added to the restrictions placed on the states. The 13th Amendment prohibited the states from legalizing the ownership of one person by another (slavery); the 14th Amendment deprived the states of their power to determine qualifications for citizenship; the 15th Amendment prohibited the states from denying the right to vote because of race, color, or previous condition of servitude; and the 19th, from denying the vote to women.

Since the Civil War, the functions of the state have expanded. Local business—that is, business not involved in foreign or interstate commerce—is regulated by the state. The states create subordinate governmental bodies such as counties, cities, towns, villages, and boroughs, whose charters they either issue or, where home rule is permitted, approve. States regulate employment of children and women in industry, and enact safety laws to prevent industrial accidents. Unemployment insurance is a state function, as are education, public health, highway construction and safety, operation of a state highway patrol, and various kinds of personal relief. The state and local governments still are primarily responsible for providing public assistance, despite the large part the federal government plays in financing welfare.

Each state is headed by an elected governor. State legislatures are bicameral except Nebraska's, which has been unicameral since 1934. Generally, the upper house is called the senate, and the lower house the house of representatives or the assembly. Bills must be passed by both houses, and the governor has a suspensive veto, which usually may be overridden by a two-thirds vote.

The number, population, and geographic extent of the more than 3,000 counties in the US—including the analogous units called boroughs in Alaska and parishes in Louisiana—show no uniformity from state to state. The county is the most conspicuous unit of rural local government and has a variety of powers, including location and repair of highways, county poor relief, determination of voting precincts and of polling places, and organization of school and road districts. City governments, usually headed by a mayor or city manager, have the power to levy taxes; to borrow; to pass, amend, and repeal local ordinances; and to grant franchises for public service corporations. Township government through an annual town meeting is an important New England tradition.

During the late 1960s and 1970s and again in the late 1980s and early 1990s, a number of large cities began to suffer severe fiscal crises brought on by a combination of factors. Loss of tax revenues stemmed from the migration of middle-class residents to the suburbs and the flight of many small and large firms seeking to avoid the usually higher costs of doing business in urban areas. Low-income groups, many of them unskilled blacks and hispanic migrants, came to constitute large segments of city populations, placing added burdens on locally funded welfare, medical, housing, and other services without providing the commensurate tax base for additional revenues.

16JUDICIAL SYSTEM

The Supreme Court, established by the US Constitution, is the nation's highest judicial body, consisting of the chief justice of the US and eight associate justices. All justices are appointed by the president with the advice and consent of the Senate. Appointments are for life "during good behavior," otherwise terminating only by resignation or impeachment and conviction.

The original jurisdiction of the Supreme Court is relatively narrow; as an appellate court, it is open to appeal from decisions of federal district courts, circuit courts of appeals, and the highest courts in the states, although it may dismiss an appeal if it sees fit to do so. The Supreme Court, by means of a writ of certiorari, may call up a case from a district court for review. Regardless of how cases reach it, the Court enforces a kind of unity on the decisions of the lower courts. It also exercises the power of judicial review, determining the constitutionality of state laws, state constitutions, congressional statutes, and federal regulations, but only when these are specifically challenged.

The Constitution empowers Congress to establish all federal courts inferior to the Supreme Court. On the lowest level and handling the greatest proportion of federal cases are the district courts—numbering 94 in 1986, including one each in Puerto Rico, Guam, the Virgin Islands, the Northern Mariana Islands, and the District of Columbia—where all offenses against the laws of the US are tried. Civil actions that involve cases arising under treaties and laws of the US and under the Constitution, where the amount in dispute is greater than $5,000, also fall within the jurisdiction of the district courts. District courts have no appellate jurisdiction; their decisions may be carried to the courts of appeals, organized into 13 circuits. These courts also hear appeals from decisions made by administrative commissions. For most

cases, this is usually the last stage of appeal, except where the court rules that a statute of a state conflicts with the Constitution of the US, with federal law, or with a treaty. Special federal courts include the Court of Claims, Court of Customs and Patent Appeals, and Tax Court.

State courts operate independently of the federal judiciary. Most states adhere to a court system that begins on the lowest level with a justice of the peace and includes courts of general trial jurisdiction, appellate courts, and, at the apex of the system, a state supreme court. The court of trial jurisdiction, sometimes called the county or superior court, has both original and appellate jurisdiction; all criminal cases (except those of a petty kind) and some civil cases are tried in this court. The state's highest court, like the Supreme Court of the US, interprets the constitution and the laws of the state.

The grand jury is a body of from 13 to 24 persons that brings indictments against individuals suspected of having violated the law. Initially, evidence is presented to it by either a justice of the peace or a prosecuting county or district attorney. The trial or petit jury of 12 persons is used in trials of common law, both criminal and civil, except where the right to a jury trial is waived by consent of all parties at law. It judges the facts of the case, while the court is concerned exclusively with questions of law.

17ARMED FORCES

The armed forces of the United States of America in 1993 numbered 1.9 million on active duty and 1.7 million in the Ready Reserve, a category of participation that allows regular training with pay and extended active duty periods for training. The Standby and Retired Reserve includes about 200,000 experienced officers and NCOs who can be recalled in a national emergency. Membership in all of these forces is voluntary and has been since 1973 when conscription expired in the death throes of the Viet Nam war. The active duty force includes 212,600 women who serve in all grades and all occupational specialties except direct ground combat units and some aviation billets.

From plans drawn in 1989–90 the armed forces are reducing their personnel numbers and force structure by about one-third because of the diminished threat of a nuclear war with the former Soviet Union or a major conflict in central Europe. Despite the interlude of the Gulf War, 1990–91, the force reductions will continue in the 1990s, which in turn will force some restructuring of the active duty forces. Their emphasis will be on rapid deployment to deter or fight major regional conflicts much like the Gulf War, in Korea, elsewhere in the Middle East, or Latin America (e.g. Cuba). The essence of the conventional force debate is whether the US can or should maintain forces to fight two regional conflicts simultaneously.

For the purposes of administration, personnel management, logistics, and training, the traditional four military services in the Department of Defense remain central to strategic planning. The U.S. Army numbers 674,800 (77,700 women) soldiers on active duty, divided roughly between seven heavy (armored or mechanized) divisions, six light (infantry airborne, airmobile) divisions, and five independent brigades as well as two armored cavalry regiments, 7 aviation brigades, and 17 air defense battalions. Army special operations missions go to 5 Special Forces groups, an airborne ranger regiment, an aviation group, and a psychological warfare group with civil affairs and communications support units. The Army has 15,629 main battle tanks, 5,371 infantry fighting vehicles, almost 20,000 other tracked vehicles, almost 6,000 towed or self-propelled artillery, 433 aircraft, and 8,000 armed and transport helicopters. Also in the process of reorganization, the Army National Guard (443,150; 32,200 women) will emphasize the preparation of combat units up to division size for major regional conflicts while the Army Reserve (680,900; 121,000 women) will prepare individuals to fill active units or provide combat support or service support/technical/medical units upon mobilization. In addition, the National Guard retains a residual state role in suppressing civil disturbances and providing disaster relief.

The U.S. Navy (546,650; 55,100 women) has shifted from its role in nuclear strategic deterrence and control of sea routes to Europe and Asia to the projection of naval power from the sea. Naval task forces normally combine three combat elements: air, surface, and subsurface. The Navy mans 87 nuclear-powered attack submarines with one configured for special operations; most of these boats can launch cruise missiles at land targets.

Naval aviation is centered on 12 carriers (6 nuclear-powered) and 11 carrier aircraft wings. Including its armed ASW helicopters and armed long-range ASW patrol aircraft—as well as a large fleet of communications and support aircraft—the Navy controls 1,735 aircraft and 421 armed helicopters. Naval aviation reserves provide 5 more wings for carrier deployment. The surface force is awesome: 38 cruisers (21 with advanced anti-air suites), 45 destroyers, 83 frigates, 35 amphibious ships, 24 mine warfare ships, and 30 patrol and coastal combatants. More ships are kept in ready reserve or are manned by surface line reserve units. The fleet support force numbers 162 specialized ships for global logistics that are not base-dependent.

The Marine Corps, a separate naval service, is organized into three active divisions and three aircraft wings of the Fleet Marine Force, which also include 3 Force Service Support groups. The Marine Corps (193,000; 8,800 women) emphasizes amphibious landings, but trains for a wide-range of contingency employments. It can draw upon some 43,000 reserve Marines in a fourth division and aircraft wing as well as 60,000 individual ready reservists. The Marine Corps has 400 main battle tanks, 1,500 other mobile warfare tracked and wheeled combat vehicles, and about 1,000 artillery pieces, which makes it more formidable than most armies. It flies 410 aircraft and almost 500 helicopters as well.

The U.S. Air Force (499,300; 71,000 women) provides 3,500 aircraft, all but about 300 now dedicated to non-strategic roles in support of forward deployed ground and naval forces. The Air Force stresses the missions of air superiority and interdiction with complementary operations in electronic warfare and reconnaissance, but it also mans a transport fleet of around 1,200 aircraft. Air Force personnel manage the US radar and satellite early-warning and intelligence effort. The Air Force Reserve and Air National Guard (roughly 200,000 active reserves) provide a wide range of flying and support units, and their flying squadrons have demonstrated exceptional readiness and combat skills on contingency missions. Air Force reserves, for example, provide the backbone of the air refueling and transport fleets.

The armed forces are deployed in functional unified or specified commands for actual missions. The Strategic Command controls the strategic nuclear deterrence forces: 1000 ICMBs (2,450 warheads), 25 Navy fleet ballistic missile submarines with 504 missiles (6,000 warheads) and 270 operational long-range bombers (2,400 missiles and bombs). These forces are undergoing reduction to conform with the START arms limitation treaty of 1991, as amended in 1992, so that by the year 2003 the US will have only 500 ICMBs, 1,728 SLBMs, and 95 nuclear-armed bombers. Strategic Command is complemented by Space Command/North American Air Defense Command. The conventional forces are assigned to a mix of geographic and functional commands: Atlantic Command, European Command, Central Command, Southern Command and Pacific Command as well as Transportation Command and Special Operations Command. The Army also maintains a Forces Command for ground forces in strategic reserve in the US. Major operational units are deployed to Germany, Korea, and Japan as part of collective security alliances. About one-third of active duty personnel are assigned to

overseas billets (1–3 years) or serve in air, naval, and ground units that serve short tours on a rotational basis. The US has one battalion in Egypt and observes with four other peacekeeping missions.

Patterns of defense spending reflect the movement away from Cold War assumptions and confrontation with the former Soviet Union and the People's Republic of China. During the 1980s when defense spending hovered around $300 billion a year and increased roughly 30 percent over the decade, defense spending absorbed roughly 6 percent of the gross domestic spending, 25 percent of federal spending, and 16 percent of net public spending. In the early 1990s when the defense budget slipped back to the $250–$260 billion level, the respective percentages were 4.5, 18, and 11, the lowest levels of support for defense since the Korean War (1950). US military assistance abroad shows similar trends. From 1981 to 1991 the US sold $118 billion in arms abroad and provided some outright grants, military training, and other support services, most in dollar value to its NATOs allies, Sau'di Arabia, Israel, South Korea, and Japan. This spending is also declining.

[18]INTERNATIONAL COOPERATION

The US, whose failure to join the League of Nations was a major cause of the failure of that body, is a charter member of the UN, having joined on 24 October 1945. The US participates in ECE, ECLAC, ESCAP, and all the nonregional specialized agencies except UNESCO, from which it withdrew at the end of 1984, charging the agency with mismanagement and with bias against Western nations; there are also policy differences between the US and several other UN agencies. The US contributes about 25% of the total funds required for the upkeep of the UN, far more than does any other nation. In the mid-1980s, the US participated in more than 70 intergovernmental organizations, including the Asian Development Bank, OECD, the IMF and IBRD (World Bank), and international councils and commissions on various industries. The US also participates actively in the Permanent Court of Arbitration. Hemispheric agencies include the Inter-American Committee on the Alliance for Progress, IDB, OAS, IADB, and PAHO.

NATO is the principal military alliance to which the US belongs. The ANZUS alliance is a mutual defense pact between Australia, New Zealand, and the US; in 1986, following New Zealand's decision to ban US nuclear-armed or nuclear-powered ships from its ports, the US renounced its ANZUS treaty security commitments to New Zealand. The nation is a signatory of GATT but refused to sign the Law of the Sea because of unwillingness to relinquish rights over seabed mining; in keeping with international practice, however, the US does maintain a 200-mi coastal economic zone. In 1986, the US approved the 1948 UN convention against genocide.

[19]ECONOMY

In variety and quantity, the natural resources of the US probably exceed those of any other nation, with the possible exception of the former Soviet Union. The US is among the world's leading exporters of coal, wheat, corn, and soybeans. However, because of its vast economic growth, the US depends increasingly on foreign sources for a long list of raw materials. The extent of US dependence on oil imports was dramatically demonstrated during the 1973 Arab oil embargo, when serious fuel shortages developed in many sections of the country.

By the middle of the 20th century, the US was a leading consumer of nearly every important industrial raw material. The industry of the US produced about 40% of the world's total output of goods, despite the fact that the country's population comprised about 6% of the world total and its land area about 7% of the earth's surface.

In recent decades US production has continued to expand, though at a slower rate than that of most other industrialized nations. While the value of US exports of manufactured goods increased from $12.7 billion to $132.4 billion between 1960 and 1983, the US share of all world industrial exports decreased from 25.3% to 19.4%.

The US gross national product (GNP) more than tripled between 1979 and 1983 to $3.3 trillion in current dollars; measured in constant 1972 dollars, the increase was 41.3%. Although in absolute terms the US far exceeds every other nation in the size of its GNP, the US GNP per capita—$14,093 in 1983—was surpassed by Switzerland among the industrialized nations and by several oil-producing countries of the Middle East. According to preliminary data, the GNP rose by 6.8% in 1984, the best growth performance since 1951.

Inflation is an ever-present factor in the US economy, although the US inflation rate tends to be lower than that of the majority of industrialized countries: for the period 1970–78, for example, consumer prices increased by an annual average of 6.7%, less than in every other Western country except Austria, Luxembourg, Switzerland, and West Germany, and well below the price increase in Japan. Thus, the double-digit inflation of 1979–81 came as a rude shock to most Americans, and economists and politicians vied with each other in blaming international oil price rises, federal monetary policies, and US government spending for the problem.

The following table shows the erosion of the purchasing power of the dollar between 1950 and 1992:

	PRODUCER PRICES (1982=$1.00)	CONSUMER PRICES (1982–84=$1.00)
1950	3.546	4.151
1955	3.279	3.732
1960	2.994	3.373
1965	2.933	3.166
1970	2.545	2.574
1975	1.718	1.859
1980	1.136	1.215
1985	0.955	0.928
1990	0.839	0.766
1992	0.812	0.713

Consumer prices rose 6.2% in 1982 and 3.0% in 1993.

The US unemployment rate was 6.4% in 1993. Pockets of concentrated unemployment in the central cities, especially among young nonwhites, constitute one of the nation's most serious social and economic problems.

The following table shows major components of the GDP for 1970, 1980, and 1991 (in billions of dollars):

	1970	1980	1991
Agriculture, forestry, fisheries	29.8	66.7	108.6
Mining	18.7	112.6	91.8
Construction	51.1	128.7	223.4
Manufacturing	253.1	588.3	1026.2
Transportation	88.0	242.2	506.0
Wholesale and retail trade	168.8	436.6	907.2
Finance, insurance, real estate	146.3	418.4	1,039.7
Services	120.5	377.0	1,089.8
Government and government enterprises	134.2	324.2	720.6
TOTALS	1,010.7	2,708.0	5,722.9

Industrial activity within the US has been expanding southward and westward for much of the 20th century, most rapidly since World War II. Louisiana, Oklahoma, and especially Texas are centers of industrial expansion based on petroleum refining; aerospace and other high technology industries are the basis of the new wealth of Texas and California, the nation's leading manufacturing state. The industrial heartland of the US is the east–north–central region, comprising Ohio, Indiana, Illinois, Michigan, and Wisconsin, with steelmaking and automobile manufacturing among the leading industries. The Middle Atlantic states (New Jersey, New York, and Pennsylvania) and the Northeast are also highly industrialized; but of the major industrial states in these two regions, Massachusetts has taken the lead in reorienting itself toward such high technology industries as electronics and information processing.

During Reagan's first term from 1980–84, the nation endured two years of severe recession followed by two years of robust recovery. The inflation rate was brought down, and millions of new jobs were created. The economic boom of the early and mid-eighties, however, coincided with a number of alarming developments. Federal budget deficits, caused by huge increases in the military budget and by rising costs of entitlement programs such as medicaid and medicare, averaged more than $150 billion annually. By 1992, the total deficit reached $290 billion, or $1,150 for every American. As a percentage of GDP, the federal debt rose from 26.8 percent in 1980 to 51.1 percent in 1992. Private debt also ballooned. From 1982 to 1988, corporate borrowing rose dramatically, and household borrowing grew twice as fast as personal income.

The eighties also witnessed a crisis in the banking industry. High inflation and interest rates forced banks to increase the interest on short term deposits, for which their revenues from fixed interest, long term mortgage rates proved increasingly inadequate. The debt crisis in developing countries created problem loans which severely reduced the value of commercial bank portfolios in the 1980s. Desperate to augment their assets and encouraged by the deregulation of the banking industry which loosened restrictions on investments and reporting procedures, banks invested in speculative real estate ventures. When the real estate boom of the early eighties collapsed in the latter part of the decade, thousands of banks became insolvent and a credit crunch ensued.

Structural changes in the economy during the eighties included a continuation in the decline in manufacturing's percentage of GDP, from 23.8 percent in 1980 to 18.4 percent in 1992. Productivity declined or grew slowly, dropping 0.9 percent in 1980 and increasing 0.8 percent in 1985, 1.2 percent in 1991, and 3.1 percent in 1992. The stagnation of productivity, in turn, slowed the growth of real compensation per hour and real median family income. The eighties also witnessed a widening disparity between the affluent and the poor. The share of the nation's income received by the richest 5 percent of American families rose from 18.6 percent in 1977 to 24.5 percent in 1990, while the share of the poorest 20 percent fell from 5.7 percent to 4.3 percent. The lowest wages, such as the minimum wage, fell in real terms while while the top wages, including those of chief executive officers, rose.

Externally, the nation's trade position deteriorated in the 1980s. The low rate of savings created a need for investment from international sources which was readily filled. A high level of foreign investment combined with an uncompetitive US dollar to create a trade deficit. By 1992, the U.S. posted a deficit on current accounts of $520 billion.

In 1990, the American economy plunged into a recession. Factors contributing to the slump included rising oil prices following Iraq's invasion of Kuwait, a sharp increase in interest rates, and declining availability of credit. Output fell 1.6 percent and 1.7 million jobs were cut. Unemployment rose from 5.2 percent in 1989 to 7.5 percent in 1991. Recovery has come slowly. From 1989 to 1990, real GDP grew 1.5 percent per year, and the civilian unemployment rate has remained above 6 percent since November 1990.

²⁰INCOME

Total earned income increased by 4.9% from 1991 to 1992, rising from $3.1 trillion to $3.3 trillion. Nonfarm personal income grew by 5.1% during that time, from $4.8 trillion to $5.1 trillion. The GDP amounted to $5,951 billion in 1992 ($23,400 per capita), for a real growth rate of 2.1%. The inflation rate in 1992 was 3%. According to the ILO, general consumer prices for goods (excluding shelter) rose by 65.6% between 1980 and 1992.

The median household disposable income in 1992 was $33,769. New Jersey, Alaska, Connecticut, Hawaii, and New Hampshire had the highest median household disposable incomes (above $41,300). Among the lowest (under $27,200) were those of Mississippi, West Virginia, Arkansas, Oklahoma, and Kentucky. Some 36.9 million persons from nearly 8 million families lived below the US federal poverty level in 1992. Approximately 14.5% of all US residents live in poverty; the proportion has steadily risen since 1973 (with ebbs in the late 1970s and late 1980s), but is still far below the 1959 level of 22.4%. By race, 33% of blacks in the US were considered by the federal government as impoverished, comprising 29% of the total. Nearly 12% of all whites lived in poverty, but made up 66% of all US residents living in poverty. Among persons of Hispanic origin (who may be of any race), 29.3% lived in poverty, representing 18% of the nation's total impoverished population.

²¹LABOR

About 117,600,000 persons constituted the country's civilian and military labor force in 1992. During 1992, a recession year, the unemployment rate reached 7.4%. A federal survey in March 1991 revealed the following nonfarm employment patterns:

	TOTAL ESTABLISHMENTS	EMPLOYEES	ANNUAL PAYROLL (1,000)
Agricultural srvices:			
Forestry, and fisheries	91,286	543,652	9,086,006
Mining	30,445	716,859	26,230,900
Construction	577,792	4,671,221	122,713,468
Manufacturing, of which:	373,999	13,383,368	544,846,450
Food and kindred products	(20,256)	(1,471,535)	(35,452,450)
Textile mill products	(6,233)	(621,591)	(12,288,813)
Lumber and wood products	(34,458)	(635,024)	(13,017,102)
Paper and allied products	(6,364)	(622,656)	(19,617,953)
Chemical and allied products	(12,224)	(865,888)	(31,979,226)
Petroleum and coal products	(2,235)	(114,872)	(4,881,000)
Leather and leather products	(1,990)	(108,028)	(1,841,719)
Primary metals	(6,867)	(688,280)	(21,819,871)
Electric and public utilities	(16,810)	(1,481,703)	(43,971,427)
Transportation and public utilities	224,855	5,584,484	168,896,216
Wholesale trade	478,456	6,218,875	182,766,852
Retail trade	1,547,316	19,600,024	247,010,947
Finance, insurance, and real estate	577,140	6,860,177	200,255,750
Services	2,141,727	29,575,248	639,368,055
Other	137,634	147,635	3,900,151
TOTALS	6,200,650	92,301,543	2,145,074,795

As of 1991, federal, state, and local governments employed 18,554,000 persons. Agriculture engaged 3,115,372 Americans during 1992.

Earnings of workers vary considerably with type of work and section of country. The national average wage was $11.45 per hour for industrial workers in 1992. The average workweek for nonfarm employees in 1992 was 39.6 hours (35 hours in 1983), ranging from 41.1 hours in manufacturing to 28.8 hours in retail trade.

There were 39 national labor unions with over 100,000, the largest being the National Educational Association with 2,000,000 members. In 1990, 16.1% of the nonagricultural work force belonged to labor unions, down from 18.8% in 1984. The most important federation of organized workers in the US is the American Federation of Labor–Congress of Industrial Organizations (AFL–CIO), whose affiliated unions had 14,100,000 members in 1992. The major independent industrial and labor unions and their estimated 1993 memberships are the International Brotherhood of Teamsters, 1,700,000, and the United Automobile Workers, 1,197,000. Most of the other unaffiliated unions are confined to a single establishment or locality. US labor unions exercise economic and political influence not only through the power of strikes and slowdowns but also through the human and financial resources they allocate to political campaigns (usually on behalf of Democratic candidates) and through the selective investment of multibillion-dollar pension funds.

The National Labor Relations Act of 1935 (The Wagner Act), the basic labor law of the US, was considerably modified by the Labor-Management Relations Act of 1947 (the Taft-Hartley Act) and the Labor-Management Reporting and Disclosure Act of 1959 (the Landrum-Griffin Act). Closed-shop agreements, which require employers to hire only union members, are banned. The union shop agreement, however, is permitted; it allows the hiring of nonunion members on the condition that they join the union within a given period of time.

In the mid-1990s, 19 states had right-to-work laws, forbidding the imposition of union membership as a condition of employment. Under the Taft-Hartley Act, the president of the US may postpone a strike for 90 days in the national interest. The act of 1959 requires all labor organizations to file constitutions, bylaws, and detailed financial reports with the secretary of labor, and stipulates methods of union elections. The National Labor Relations Board seeks to remedy or prevent unfair labor practices and supervises union elections, while the Equal Employment Opportunity Commission seeks to prevent discrimination in hiring, firing, and apprenticeship programs.

The number of work stoppages and of workers involved reached a peak in the late 1960s and early 1970s, declining steadily thereafter. In 1992 there were 35 major stoppages involving 364,000 workers, the lowest number at least since World War II; a major stoppage was defined as one involving 1,000 workers or more for a minimum of one day or shift.

²²AGRICULTURE

In 1992, the US produced a huge share of the world's agricultural commodities, including soybeans, 52%; corn for grain (maize), 46%; cotton, 17%; oats, 13%; wheat, 12%; and tobacco, 10%. Much US farm produce is exported; among other products, 60% or more of the soybeans and corn that cross international boundaries originates in the US. Agricultural exports reached an all-time high of $48.2 billion in 1992, accounting for 10.8% of all merchandise exports.

The gross farm income of $189.5 billion in 1991 included $167.3 billion in farm marketings (crops, 48%; livestock and livestock products, 52%); $8.2 billion in government payments; $0.6 billion in the value of products consumed on the farm; and $5.3 billion in the rental value of dwellings. Average net cash income per farm was $21,285.

About 15% of the total US land area was actively used for crops in 1991; another 26% was grassland pasture. Between 1930 and 1991, the number of farms in the US declined from 6,546,000 to an estimated 2,095,740. The total amount of farmland increased from 399 million hectares (986 million acres) in 1930 to 479 million hectares (1.18 billion acres) in 1959 but declined to 397 million hectares (980 million acres) in 1991. From 1930 to 1991, the size of the average farm tripled from 61 to 189 hectares (from 151 to 468 acres), a result of the consolidation effected by large-scale mechanized production. The farm population, which comprised 35% of the total US population in

1910, declined to 25% during the Great Depression of the 1930s and dwindled to 1.8% by 1991.

A remarkable increase in the application of machinery to farms took place during and after World War II. Tractors, trucks, milking machines, grain combines, corn pickers, and pickup bailers became virtual necessities in farming. In 1920 there was less than one tractor in use for every 400 hectares (1,000 acres) of cropland harvested; by 1991 there were 15. Two other elements essential to US farm productivity are chemical fertilizers and irrigation. In 1991, $13.7 billion allocated to fertilizers and lime represented more than 9% of farm operating expenses.

Substantial quantities of corn, the most valuable crop produced in the US, are grown in almost every state; its yield and price are important factors in the economies of the regions where it is grown. The following table reports figures for production and value of selected US crops in 1992:

	PRODUCTION (1,000 METRIC TONS)
Corn for grain	240,774
Soybeans	59,780
Wheat	66,920
Oats	4,276
Cotton	9,211
Tobacco	764
Sorghum	22,455
Potatoes	18,671

In 1991, of the total cash receipts of $80,547 million, feed crops accounted for 23.6%; oil-bearing crops, 15.6%; vegetables, 14% fruits and tree nuts, 12.3%; food grains, 8.5%; cotton, 6.9%; tobacco, 3.6%; and other crops (e.g., sugar crops, mushrooms, grass seeds, floriculture and ornamentals), 15.5%.

23ANIMAL HUSBANDRY
The livestock population at the end of 1992 included an estimated 99.5 million head of cattle (approximately 10.2% were milk cows), 57.7 million hogs, 10.7 million sheep and lambs, 1.4 billion chickens, and 90 million turkeys (35% of the world's total). Meat production in 1992 amounted to 30,876,000 metric tons, of which poultry accounted for 12,026,000 tons; beef, 10,607,000 tons; pork, 7,817,000 tons; and mutton and lamb, 158,000 tons. Some 4,175,700 metric tons of eggs were produced in 1992. Milk production totaled 68,966,000 metric tons in that year, with Wisconsin, California, and New York together accounting for much of the total. Wisconsin, Minnesota, and California account for more than half of all US butter production, which totaled 632,000 metric tons in 1992; in that year, the US was the world's largest producer of cheese, with more than 3.3 million metric tons (23% of the world's total). The US produced an estimated 17% of the world's meat supply in 1992.

24FISHING
The US, which ranked sixth in the world in commercial fish landings in 1991, nevertheless imports far more fish and fishery products than it exports.

The 1991 commercial catch was 5.5 million metric tons (4.3 million metric tons excluding the weight of mollusk shells) and had a value of $3.3 billion. Food fish make up 74% of the catch, and nonfood fish, processed for fertilizer and oil, 26%. About 4,600 firms in the US were engaged in processing fishery products in 1991, with an average employment of nearly 73,000. The commercial fishing fleet comprised some 78,000 vessels and boats in 1992. Commercial landings by US fishermen at ports in the 50 states were a record 4.4 million metric tons valued at $3.7 billion in 1992. That year, commercial landings by US fishermen at ports outside the 50 states (or transferred to internal water processing vessels) amounted to an additional 271,000 metric tons, valued

at $205.5 million. Domestic aquaculture yielded 273,883 metric tons in 1991 (81% finfish, 19% shellfish), with a value of $634.5 million.

Alaska pollock, with landings of 1.3 million metric tons (3.0 billion lb), was the most important species in quantity and 4th in value for 1992, accounting for 31% of the commercial fishery landings in the US. Other species' rankings for quantity (and value) in 1992 included: salmon, 3d (1st); flounder, 4th (7th); crab, 5th (3d); cod, 6th (5th); and shrimp, 2d (7th). Menhaden was the second most important species in quantity, but was low in value. Per capita finfish and shellfish consumption (edible meat basis) was 6.7 kb (14.8 lb) in 1992.

Pollution is a problem of increasing concern to the US fishing industry; dumping of raw sewage, industrial wastes, spillage from oil tankers, and blowouts of offshore wells are the main threats to the fishing grounds. Overfishing is also a threat to the viability of the industry in some areas, especially Alaska, which accounted for 58% of the 1992 commercial landing.

25FORESTRY
US forest and woodlands covered about 286.8 million hectares (708.8 million acres) in 1991. Major forest regions include the eastern, central hardwood, southern, Rocky Mountain, and Pacific coast areas. The National Forest System accounts for approximately 27% of the nation's forestland. Large private lumber companies control extensive tracts of land in Maine, Oregon, and several other states.

Domestic production of roundwood during 1991 amounted to 495.8 million cu m (17,507 million cu ft), of which softwoods accounted for roughly 80%. The US, the world's 2d-leading producer of newsprint, attained an output of 6.2 million metric tons in 1991. To satisfy its needs, the US imported 6.8 million metric tons of newsprint in 1991, primarily from Canada, the major world producer. Other forest products in 1991 included 58.9 million metric tons of wood pulp, 72.7 million metric tons of paper and paperboard (excluding newsprint), 30.4 million cu m (1,073 million cu ft) of wood-based panels, 6.8 million cu m (240 million cu ft) of particleboard, and 17.3 million cu m (611 million cu ft) of plywood. Rising petroleum prices in the late 1970s sparked a revival in the use of wood as home heating fuel, especially in the Northeast. Fuelwood and charcoal production amounted to 85.9 million cu m in 1991.

Throughout the 19th century, the federal government distributed forestlands lavishly as a means of subsidizing railroads and education. By the turn of the century, the realization that the forests were not inexhaustible led to the growth of a vigorous conservation movement, which was given increased impetus during the 1930s and again in the late 1960s. Federal timberlands are no longer open for private acquisition, although the lands can be leased for timber cutting and for grazing. In recent decades, the states also have moved in the direction of retaining forestlands and adding to their holdings when possible. As of 30 September 1992, the US Forest Service managed 77.2 million hectares (191 million acres) of forest, including 13.8 million hectares (34 million acres) of designated wilderness.

26MINING
Rich in a variety of mineral resources, the US is a world leader in the production of many important mineral commodities, such as aluminum, cement, copper, pig iron, lead, molybdenum, phosphates, potash, salt, sulfur, uranium, and zinc. In 1991, mining in the US employed some 698,100 persons at over 33,600 establishments. Personal income from mineral production in 1991 amounted to $32.1 billion, of which fuels accounted for $25.6 billion. The total US nonfuel mineral production value in 1992 was $31,174 million. The leading mineral-producing states are Texas, Louisiana, Oklahoma, and New Mexico, which are

important for petroleum and natural gas, and Kentucky, West Virginia, and Pennsylvania, important for coal. Iron ore supports the nation's most basic nonagricultural industry, iron and steel manufacture. Domestic demand for iron ore is satisfied by imports from Canada and Venezuela. The major domestic sources are in the Lake Superior area: Minnesota and Michigan lead all other states in iron ore yields. In 1991, the US ranked 1st in the mining of salt, phosphate, and elemental sulfur; 2d in lead, gold, and silver; 3d in uranium and nitrogen in ammonia; 4th in potash and cement; and 6th in iron ore.

The following table shows volume for selected mineral production in 1991 (excluding fossil fuels):

	VOLUME
Cement*	69,853,000 tons
Iron ore	56,596,000 tons
Phosphate	48,096,000 tons
Salt*	35,943,000 tons
Nitrogen in ammonia	12,692,000 tons
Sulfur, elemental	10,816,000 tons
Potash	1,749,000 tons
Copper	1,631,000 tons
Zinc	547,000 tons
Lead	447,000 tons
Uranium	3,585 tons
Silver	1,848 tons
Gold	289,885 kg

*Includes Puerto Rico

27 ENERGY AND POWER

The US, with about 4.7% of the world's population, consumed 24.1% of the world's energy in 1991. The US accounted for about 26% of the world's electricity generation in 1991. In 1993, the US produced 24% of the world's coal, 25% of its natural gas, but only 2.7% of its crude oil. Coal supplied about 55% of the energy used by electric utilities in 1992; nuclear sources, 22.7%; natural gas, 9.7%; waterpower, 8.4%; petroleum, 3.2%; and geotherman, wood, waste, wind, photovoltaic, and solar thermal energy, 1%. Increased use of natural gas was the most spectacular development in the commercial marketing of fuel after World War II; between 1950 and 1971, its share of total US energy production doubled from 20% to 40%, although by 1992 the proportion had declined to 27.4%. Coal accounted for 32.3% of US energy production (Btu basis) that year; crude oil, 22.8%; liquefied natural gas, 3.5%; nuclear electric power, 10%; hydroelectricity, 3.8%; geothermal and other energy, 0.2%.

Mineral fuel production in the US dates to the early 1800s; by 1810–19 US coal production was about 230,000 tons, and fuelwood production equivalent to 26 million tons of coal. In 1854, a refinery opened in Brooklyn, New York, to process shale and coal to kerosene. In 1859, the US recorded its first oil production, and in 1870, John D. Rockefeller established the Standard Oil Company in Cleveland, Ohio, which controlled 90% of the Western oil market from 1880 to 1910, and was split into 33 independent companies in 1911 under the Sherman Antitrust Act. In 1957, the nuclear portion of domestic electricity generation accounted for less than 0.05% of the nation's total; by 1992 it accounted for 22.1%.

Proved recoverable reserves of crude oil totaled an estimated 31.2 billion barrels at the end of 1993 (3.1% of the world's total). Reserves of natural gas were about 165 trillion cu ft (4.7 trillion cu m) in 1993, equivalent to about 3.3% of the world's proved reserves. Recoverable coal reserves amounted to 240,560 million tons at the end of 1993 (47% anthracite and bituminous), equivalent to 23.1% of the world's total. Mineral fuel production in 1992 included an estimated 607 million tons of hard coal (2d after China) and 295 million tons of lignite (1st); 508.2 billion cu m

(17.9 trillion cu ft) of natural gas (2d after Russia), and 416.6 million tons of crude oil (2d after Saudi Arabia). The 1973 Arab oil embargo and subsequent fuel shortages prompted a host of governmental measures aimed at increasing development of oil and gas resources, including an easing of restrictions on oil drilling on the continental shelf.

In 1991, public utilities and private industrial plants generated 2.8 trillion kwh of electricity, of which steam and internal combustion plants produced over 1.9 trillion and hydroelectric plants 275 billion. Installed generating capacity at electric utilities totaled 694.8 million kw in 1992. In 1992, nuclear-powered generators in operation had an estimated total capability of 99 million kw and generated 618.8 billion kwh of electricity (up from 51.8 million kw and 251.1 billion kwh in 1980). Because of cost and safety problems, the future of the nuclear power industry in the US was in doubt in the late 1980s. The number of operable nuclear power units totaled 109 in 1992, an increase of only nine units since 1986.

During the 1980s, some attention was focused on the development of solar power, synthetic fuels, geothermal resources, and other energy technologies. Such energy conservation measures as mandatory automobile fuel-efficiency standards and tax incentives for home insulation were promoted by the federal government, which also decontrolled oil and gas prices in the expectation that a rise in domestic costs to world-market levels would provide a powerful economic incentive for consumers to conserve fuel. Since the mid-1980s, net petroleum imports have risen from 4.29 million barrels per day in 1985 to 6.89 million barrels daily in 1992, which is still below the record level of 8.56 million barrels per day set in 1977.

28 INDUSTRY

Although the US remains one of the world's preeminent industrial powers, manufacturing no longer plays as dominant a role in the economy as it once did.

Between 1979 and 1991, manufacturing employment fell from 20.9 million to 18.4 million, or from 21.8% to 16.2% of national employment. In 1988, manufacturing's gross product amounted to nearly $949 billion. Throughout the 1960s, manufacturing accounted for about 29% of total national income; by 1987, the proportion was down to about 19%. In 1989, manufacturing accounted for about 22.7% of GDP.

Leading industrial centers are the metropolitan areas of Chicago, Los Angeles, New York, Detroit, and Philadelphia. The Midwest leads all other regions by virtue of its huge concentration of heavy industry, including the manufacturing of automobiles, trucks, and other vehicles.

In 1991, personal income derived from manufacturing earnings in durable goods amounted to nearly $399.3 billion; for nondurable goods, $657 billion. Leading manufacturing industries of durable goods in 1989 (as a percentage of all manufacturing) included: non-electrical machinery (18.9%), electric and electronic equipment (9.8%), motor vehicles and equipment (5%), and other transportation equipment (6.9%). The principal manufacturing industries of nondurable goods in 1989 and their contribution to the output of the manufacturing sector were: chemicals and allied products (8.1%), food and kindred products (7.6%), printing and publishing (4.9%), and petroleum and coal products (4.8%).

Large corporations are dominant especially in sectors such as steel, automobiles, pharmaceuticals, aircraft, petroleum refining, computers, soaps and detergents, tires, and communications equipment. The 500 largest firms, as ranked by *Fortune* magazine in 1992, had over $2.3 trillion in sales and assets of over $2.5 trillion, but registered just $10.5 million in net profits. Fourteen companies had profits exceeding $1 billion in 1992, led by Philip Morris ($4.94 billion), Exxon ($4.77 billion), and General Electric ($4.73 billion). Companies with the largest amount of sales in 1992 were General Motors, $132.8 billion; Exxon, $103.5 billion; Ford, $100.8 billion; International Business Machines (IBM), $64.1 billion; General Electric, $62.2 billion; Mobil, $57.4 billion; Philip Morris, $50.2 billion; DuPont, $37.6 billion; and Chevron, $37.5 billion. Although General Motors led in sales, it also led in losses, at nearly $23.5 billion. Ford and IBM reported losses of $7.3 and $4.9 billion, respectively, in 1992. The growth of multinational activities of US corporations has been rapid in recent decades, with capital expenditures by US-owned foreign affiliates peaking at $42.4 billion in 1980 before declining to $37.7 billion in 1983.

The history of US industry has been marked by the introduction of increasingly sophisticated technology in the manufacturing process. Advances in chemistry and electronics have revolutionized many industries through new products and methods: examples include the impact of plastics on petrochemicals, the use of lasers and electronic sensors as measuring and controlling devices, and the application of microprocessors to computing machines, home entertainment products, and a variety of other industries. Science has vastly expanded the number of metals available for industrial purposes, notably such light metals as aluminum, magnesium, and titanium. Integrated machines now perform a complex number of successive operations that formerly were done on the assembly line at separate stations. Those industries have prospered that have been best able to make use of the new technology, and the economies of some states—in particular California and Massachusetts—are largely based on it. On the other hand, certain industries—especially clothing and steelmaking—have suffered from outmoded facilities that (coupled with high US labor costs) force the price of their products above the world market level. In 1991, the US was the world's 2d-leading steel producer (after Japan) at 79.7 million metric tons, but also the world's 2d-largest steel importer (after Germany) at 14.3 million metric tons. Employment in the steel industry has fallen from 521,000 in 1974 to 190,000 in 1992. Automobile manufacturing was another ailing industry in the 1980s: passenger car production fell from 7,098,910 in 1987 to 5,438,579 in 1991, while commercial vehicle production declined from 3,825,776 in 1987 to 3,371,942 in 1991. Moreover, included among the cars assembled domestically were an increasing number of vehicles produced by US subsidiaries of foreign firms and containing a substantial proportion of foreign-made parts.

29 SCIENCE AND TECHNOLOGY

In 1993, an estimated $180 billion was spent on research and development. Since 1980, industry has matched the federal government as a source of research and development funding; the proportions in 1993 were 46% from industry and 54% from the federal government.

In 1994 NASA's budget was $14.7 billion. In 1960 NASA spent only $1.1 billion. Launching of the space shuttle orbiter "Columbia" began in 1981; a fleet of four reusable shuttles, which would replace all other launch vehicles was planned. However, the January 1986 "Challenger" disaster, in which seven crew members died, cast doubt on the program. The three remaining shuttles were grounded, and the shuttles were redesigned for increased safety. A new shuttle, "Endeavor" was built to take "Challenger's" place. Reagan, following the "Challenger's" disaster, banned the shuttle from commercial use for nine years. The shuttles were launched back into space beginning with the shuttle "Atlantic's" launch into space in September 1988.

An estimated 949,200 scientists participated in research and development and teaching in 1988. In 1990, 15,243 engineering, mathematics and science doctorates were awarded. Some 96,514 patents were granted in 1991 to individuals, US corporations and foreigners.

30 DOMESTIC TRADE

Retail sales in the US exceeded $1.8 million ($7,200 per capita) in 1991, with California, Florida, New York, and Texas leading in volume. The growth of great chains of retail stores, particularly in the form of the supermarket, was one of the most conspicuous developments in retail trade following the end of World War II. Nearly 100,000 single-unit grocery stores went out of business between 1948 and 1958; the independent grocer's share of the food market dropped from 50% to 30% of the total in the same period. In 1987, there were just 137,584 grocery retailers in the US.

Multiunit chain stores account for over one-third of the total retail trade, but in certain kinds of retail business (variety and department store trade) the chain is the dominant mode of business organization. With the great suburban expansion of the 1960s emerged the planned shopping center, usually designed by a single development organization and intended to provide different kinds of stores in order to meet all the shopping needs of the particular area. Franchised businesses now account for about 40% of all retail sales in the US, and are projected to exceed 50% of retail sales by 2000.

Installment credit is a major support for consumer purchases in the US. Most US families own and use credit cards. Credit cards accounted for 35% of the $385 billion in consumer loans made to individuals by insured commercial banks in 1992.

The US advertising industry is the world's most highly developed. Particularly with the expansion of television audiences, spending for advertising has increased almost annually to successive record levels. Advertising expenditures in 1992 nearly reached an estimated $131.74 billion, up from $66.58 billion in 1982 and $11.96 billion in 1960. From 1990 to 1991, advertising expenditures decreased by $2.24 billion, due to recession. Of the 1992 total, about 58% was spent in national media and 42% in local media. Newspaper advertising accounted for 23.4%, television advertising 58%, and radio advertising 6.6%. Direct mail advertising (chiefly letters, booklets, catalogs, and handbills) accounted for 19.3%, magazines 5.3%, and other media 23%.

31 FOREIGN TRADE

In the realm of foreign commerce, the US led the world in value of exports and imports in 1992. Exports of domestic merchandise, raw materials, agricultural and industrial products, and military goods amounted in 1992 to nearly $416 billion, up from $394 billion in 1990. General imports for 1992 were valued at $535.5 billion, a record high, leaving a deficit of more than $119 billion; in 1987, the net merchandise trade deficit reached $159.6 billion. Except for mineral fuels and lubricants, the US merchandise trade balance worsened in all major product groups from 1982 to 1986. One import category that proliferated between the late 1970s and the mid-1980s was telecommunications apparatus—mainly television sets and video cassette recorders from Japan—which increased some ninefold between 1975 and 1985. Road vehicles accounted for nearly 14% of imports in 1992. One rapidly growing export category was computers, which rose from $1.2 billion in 1970 to $30.9 billion in 1992; grain exports rose from $2.6 billion to a peak of $19.5 billion in 1981, but declined to $11.47 billion in 1992.

Leading merchandise export by value in 1992 (in billions) included: transportation equipment, $38.7; road vehicles, $37.9; electric machinery and parts, $37.4; office and automatic data processing machines, $31; miscellaneous manufactured articles, $23.3; industrial machinery and parts, $18.9; and power-generating machinery, $18.5.

Leading merchandise imports by value in 1992 (in billions) included: road vehicles, $74.5; electric machinery and parts, $39.7; office and automatic data processing machines, $36.4;

apparel and clothing, $31.2; miscellaneous manufactured articles, $28.5; telecommunications and sound-reproduction equipment, $25.8; and power-generating machinery, $15.9.

Principal trading partners in 1992 (and the respective US merchandise exports and imports in millions of dollars) were:

	EXPORTS	IMPORTS
Canada	90,423	100,724
Japan	46,856	96,858
Mexico	40,469	35,588
UK	22,410	22,000
Germany	20,348	28,750
France	14,579	14,658
Taiwan	14,470	24,599
South Korea	13,859	16,657
Netherlands	13,427	5,729
Belgium–Luxembourg	9,964	4,697
Singapore	9,526	11,316
Hong Kong	9,027	9,800
Australia	8,731	3,666
Italy	8,573	12,205
China	7,453	25,702
Brazil	5,741	7,615
Venezuela	5,316	8,180

32BALANCE OF PAYMENTS

Since 1950, the US has generally recorded deficits in its overall payments with the rest of the world, despite the fact that it had an unbroken record of annual surpluses up to 1970 on current-account goods, services, and remittances transactions. The balance of trade, in the red since 1975, reached a record deficit of $166 billion in 1986. In 1992, exports totaled $442.3 billion, while imports stood at $544.1 billion

The nation's stock of gold declined from a value of $22.9 billion at the start of 1958 to $10.5 billion as of 31 July 1971, only two weeks before President Nixon announced that the US would no longer exchange dollars for gold. On 12 February 1973, pressures on the US dollar compelled the government to announce a 10% devaluation against nearly all of the world's major currencies. US international gold reserves thereupon rose from $14.4 billion in 1973 to $15.9 billion at the end of 1974; as of 1992, US gold reserves stood at $1,022 million, up from $514 million in 1990. Total US international reserves in 1991 were $159.22 billion, comprising $45.93 billion in foreign exchange, $11.24 billion in SDRs, $92.56 in monetary gold, and $9.49 billion as an IMF reserve position.

The following table shows the US balance of payments for 1991 and 1992 (in billions of dollars):

	1991	1992
CURRENT ACCOUNTS		
Trade, net	−73.81	−96.14
Other goods, services, and income, net	58.83	62.72
Transfers, net	6.50	−32.88
TOTALS	−8.48	−66.30
CAPITAL ACCOUNTS		
Direct investment, net	−5.16	−32.42
Portfolio investment	8.60	14.19
Other long-term capital	12.56	0.62
Other short-term capital	−14.22	54.19
TOTALS	1.78	36.58
Errors and omissions	−15.08	−12.34
Liabilities constituting foreign authorities' reserves	16.02	38.14
Net change in reserves	5.76	3.92

Between 1985 and the end of 1987, the value of the dollar declined by about 50% against the Japanese yen as well as other major currencies; this decline in value, which accelerated in 1987, was part of a Reagan administration strategy to make foreign imports more expensive to US consumers and US exports more competitive in world markets. From 1988 to 1990, the value of the dollar fell by over 12% against the yen, but rose by almost 14% from 1990 to 1992.

33BANKING AND SECURITIES

The Federal Reserve Act of 1913 provided the US with a central banking system. The Federal Reserve System dominates US banking, is a strong influence in the affairs of commercial banks, and exercises virtually unlimited control over the money supply.

Each of the 12 federal reserve districts contains a federal reserve bank. A board of nine directors presides over each reserve bank. Six are elected by the member banks in the district: of this group, three may be bankers; the other three represent business, industry, or agriculture. The Board of Governors of the Federal Reserve System (usually known as the Federal Reserve Board) appoints the remaining three, who may not be officers, directors, stockholders, or employees of any bank and who are presumed therefore to represent the public.

The Federal Reserve Board regulates the money supply and the amount of credit available to the public by asserting its power to alter the rediscount rate, by buying and selling securities in the open market, by setting margin requirements for securities purchases by altering reserve requirements of member banks in the system, and by resorting to a specific number of selective controls at its disposal. The Federal Reserve Board's role in regulating the money supply is held by economists of the monetarist school to be the single most important factor in determining the nation's inflation rate.

Member banks increase their reserves or cash holdings by rediscounting commercial notes at the federal reserve bank at a rate of interest ultimately determined by the Board of Governors. A change in the rediscount rate, therefore, directly affects the capacity of the member banks to accommodate their customers with loans. Similarly, the purchase or sale of securities in the open market, as determined by the Federal Open Market Committee, is another device whereby the amount of credit available to the public is expanded or contracted. The same effect is achieved in some measure by the power of the Board of Governors to raise or lower the reserves that member banks must keep against demand deposits. Credit tightening by federal authorities in early 1980 pushed the prime rate—the rate that commercial banks charge their most creditworthy customers—above 20% for the first time since the financial panics of 1837 and 1839, when rates reached 36%. As federal monetary policies eased, the prime rate dropped below 12% in late 1984 and has remained low.

Combined assets of all commercial banks as of 31 December 1992 exceeded $5.5 trillion. Net loans and leases amounted to nearly $2 trillion; aggregate domestic deposits were $2,698,954 million dollars; deposits with banks in foreign countries, foreign governments, and official institutions were $101,170 million.

The nation's leading commercial bank-holding companies on the basis of total assets as of 1990 were Citicorp—the largest commercial bank in the US, with (in thousands) $112,586,000 in deposits, followed by Bank America, $77,027,000 in deposits.

Under the provisions of the Banking Act of 1935, all members of the Federal Reserve System (and other banks that wish to do so) participate in a plan of deposit insurance (up to $100,000 for each individual account as of 1994) administered by the Federal Deposit Insurance Corporation (FDIC). As of 1992, the deposits of 14,146 commercial banks were insured by the FDIC. All national banks, of which there were 4,013 in 1990, are regulated members of the Federal Reserve System; most of the 361 mutual savings banks are registered by the FDIC. State banks, of which there were 1,021 in 1990, are also regulated by the FDIC.

Savings and loan associations are insured by the Federal Savings and Loan Insurance Corporation (FSLIC). Individual accounts were insured up to a limit of $100,000. Savings and loans failed at an alarming rate in the 1980s. In 1989 the government signed legislation that created the Resolution Trust Corporation. The RTC's job is to handle the savings and loans bailout, expected to cost taxpayers $345 billion through 2029. Assets of all FSLIC-insured institutions totaled $814.6 billion in 1983, of which $516.4 billion consisted of mortgage loans and contracts. The 10,962 federally chartered credit unions had 26.8 million members and combined assets of $54.5 billion in 1983; during the same year there were 8,200 state-chartered credit unions with 21.8 million members and $43.8 billion in assets.

When the New York Stock Exchange opened in 1817 its trading volume was 100 shares a day. In 1992, 51.4 billion shares with a value of $1.7 trillion were traded on the NYSE. The two other major stock markets in the US are AMEX and NASDAQ. The total shares traded on these markets were 103.4 billion in 1992. The 51.4 billion shares traded on the NYSE represented 49.7% of the total shares traded. AMEX's 3.6 billion shares represented 3.5%; NASDAQ's 48.4 billion shares represented 46.8% of the total.

As of 1990 51 million people, 21.1% of the population, owned stock or mutual funds; up from 13.5% in 1980. Nearly 30 million people owned stocks listed on the NYSE and 26 million owned mutual funds, up 130% since 1985.

[34]INSURANCE

At the end of 1991, 2,105 companies were dispensing ordinary life, group, industrial, and other kinds of life insurance policies. The overwhelming majority of US families have some life insurance with a legal reserve company, the Veterans Administration, or fraternal, assessment, burial, or savings bank organizations. As of 31 December 1992, life insurance policies in force totaled $10.4 trillion. Payments to policyholders totaled $218.6 billion.

Hundreds of varieties of insurance may be purchased. Besides life, the more important include accident, fire, hospital and medical expense, group accident and health, automobile liability, automobile damage, workers' compensation, ocean marine, and inland marine. Americans buy more life and health insurance than any other group except Canadians and Japanese. Of the world's ten largest insurers in 1991, three—Prudential Insurance, Metropolitan Life, and Aetna Life and Casualty—were American. During the 1970s, many states enacted a "no fault" form of automobile insurance, under which damages may be awarded automatically, without recourse to a lawsuit.

Life insurance company assets totaled $1,551 billion on 31 December 1991. The property and liability insurance industry had about $601.4 billion in assets in 1991. Premiums written totaled $223 billion, including private automotive coverage, $82.8 billion; accident and health coverage, $5.1 billion; workers' compensation coverage, $31.3 billion; homeowners' insurance, $19.3 billion; commercial multiperil policies, $17 billion; and fire insurance, $7.2 billion.

[35]PUBLIC FINANCE

Under the Budget and Accounting Act of 1921, the president is responsible for preparing the federal government budget. In fact, the budget is prepared by the Office of Management and Budget (established in 1970), based on requests from the heads of all federal departments and agencies and advice from the Board of Governors of the Federal Reserve System, the Council of Economic Advisers, and the Treasury Department. The president submits a budget message to Congress in January. Under the Congressional Budget Act of 1974, the Congress establishes, by concurrent reso-

lution, targets for overall expenditures and broad functional categories, as well as targets for revenues, the budget deficit, and the public debt. The Congressional Budget Office monitors the actions of Congress on individual appropriations bills with reference to those targets. The president exercises fiscal control over executive agencies, which issue periodic reports subject to presidential perusal. Congress exercises control through the comptroller general, head of the General Accounting Office, who sees to it that all funds have been spent and accounted for according to legislative intent. The fiscal year runs from 1 October to 30 September.

The public debt, subject to a statutory debt limit, rose from $43 billion in 1939/40 to more than $3.3 trillion in 1993. In fiscal year 1991/92, the federal deficit reached $290 million, a record high. President Clinton introduced a taxing and spending plan to reduce the rate of growth of the federal deficit when he began his term in 1993. The Clinton Administration calculated the package of tax increases and spending would cut the deficit by $500 billion over a four year period. The following table shows actual revenues and expenditures of the federal government for fiscal years 1991 and 1992 in billions of dollars.

	1991	1992
REVENUE AND GRANTS		
Tax revenue	1,027.00	1,052.92
Non-tax revenue	89.55	91.73
Capital revenues	0.22	0.15
Grants	43.15	11.91
TOTALS	1,159.92	1,156.71
expenditures and lending minus repayments		
General public services	105.46	115.23
Defense	308.86	297.37
Public order and safety	13.82	16.11
Education	24.72	25.33
Health	196.54	231.11
Social security	373.49	411.32
Housing and community	36.55	38.40
Amenities	—	—
Recreations, cultural and religious affairs	3.72	3.08
Economic affairs and services	144.07	87.93
Other expenditures	220.40	217.83
Adjustments	1.42	1.43
Lending minus repayments	3.38	0.82
TOTALS	1,433.43	1,445.96
Deficit/Surplus	−272.51	−289.25

[36]TAXATION

Measured as a proportion of the GDP, the total US tax burden is less than that in most industrialized countries. Federal, state, and local taxes are levied in a variety of forms. The greatest source of revenue for the federal government is the personal income tax, which is paid by citizens and resident aliens on their worldwide income.

The Tax Reform Act of 1986, which took full effect in 1988, reduced 14 graduated tax brackets, ranging from 11% to 50%, to two brackets, 15% and 28%, in taxing individual income. In 1993, the 15% rate applied to the first $22,100 in taxable income of a single person, $18,450 of a married person filing separately, $36,900 of a married couple filing jointly, and $29,600 of a head of household. There was also a maximum rate of 31% for income over a certain amount, ranging from $44,575 for married taxpayers filing separately to $88,150 for joint returns. The 1986 law also eliminated the capital gains exclusion, reduced other deductions, and curtailed the number of value of tax shelters. The top corporate tax rate was reduced from 46% to 34%, and the investment tax credit was eliminated. Excise taxes are levied on certain motor vehicles, personal air transportation, some motor fuels (excluding gasohol), alcoholic beverages, tobacco products, tires and tubes, telephone charges, and gifts and estates.

37CUSTOMS AND DUTIES

Customs receipts amounted to an estimated $19.2 billion in 1992/93. Under the Trade Agreements Extension Act of 1951, the president is required to inform the US International Trade Commission (known until 1974 as the US Tariff Commission) of contemplated concessions in the tariff schedules. The commission then determines what the "peril point" is; that is, it informs the president how far the tariff may be lowered without injuring a domestic producer, or it indicates the amount of increase necessary to enable a domestic producer to avoid injury by foreign competition. Similarly, the act provides an "escape clause, "—in effect, a method for rescinding a tariff concession granted on a specific commodity if the effect of the concession, once granted, has caused or threatens to cause "serious injury" to a domestic producer. The Trade Expansion Act of 1962 grants the president the power to negotiate tariff reductions of up to 50% under the terms of GATT.

In 1974, Congress authorized the president to reduce tariffs still further, especially on goods from developing countries. As the cost of imported oil rose in the mid-1970s, however, Congress became increasingly concerned with reducing the trade imbalance by discouraging "dumping" of foreign goods on the US market. The International Trade Commission is required to impose a special duty on foreign goods offered for sale at what the commission determines is less than fair market value.

Most products are dutiable under most-favored-nation (MFN) rates or general duty rates. The import tariff schedules contain over 10,000 classifications, most of which are subject to interpretation. Besides duties, the US imposes a 17% "user fee" on all imports. Excise taxes and harsher maintenance fees are also imposed on certain imports. Under the terms of the North American Free Trade Agreement (NAFTA), which was approved by Congress in 1993, tariffs on goods qualifying as North American under the rules of origin will be phased out over a 15-year period.

38FOREIGN INVESTMENT

From the end of World War II through 1952, US government transfers of capital abroad represented an annual average of about $5,470 million, or 88.3% of the overall national average, while private investments averaged roughly $730 million a year, or about 11.7%. Portfolio investment represented less than $150 million a year, or only 2.5% of the annual aggregate.

After 1952, however, direct private investment began to increase and portfolio investment rose markedly. In the late 1950s, new private direct investment was increasing yearly by $2 billion or more, while private portfolio investment and official US government loans were climbing by a minimum annual amount of $1 billion each. During the 1960–73 period, the value of US-held assets abroad increased by nearly 12% annually; from the mid-1970s through the early 1980s, it rose most years by at least 15%. Direct investments abroad had a market value of $802 billion in 1991. Much of US foreign investment was in the EC, Canada, and Latin America.

Direct foreign investments in the US have risen rapidly, from $6.9 billion in 1960 to $27.7 billion in 1975 and an estimated $182.95 billion at the end of 1985. As of 31 December 1991, foreign direct investment in the US was valued at an estimated $487 billion; total foreign assets were over $2.3 trillion. About one-third of the investment volume is in manufacturing. The most important sources are the UK, the Netherlands, Japan, and Canada; European countries account for two-thirds of direct foreign investments.

39ECONOMIC DEVELOPMENT

By the end of the 19th century, regulation rather than subsidy had become the characteristic form of government intervention in US economic life. The abuses of the railroads with respect to rates and services gave rise to the Interstate Commerce Commission in 1887, which was subsequently strengthened by numerous acts that now stringently regulate all aspects of US railroad operations.

The growth of large-scale corporate enterprises, capable of exercising monopolistic or near-monopolistic control of given segments of the economy, resulted in federal legislation designed to control trusts. The Sherman Antitrust Act of 1890, reinforced by the Clayton Act of 1914 and subsequent acts, established the federal government as regulator of large-scale business. This tradition of government intervention in the economy was reinforced during the Great Depression of the 1930s, when the Securities and Exchange Commission and the National Labor Relations Board were established. The expansion of regulatory programs accelerated during the 1960s and early 1970s with the creation of the federal Environmental Protection Agency, Equal Employment Opportunity Commission, Occupational Safety and Health Administration, and Consumer Product Safety Commission, among other bodies. Subsidy programs were not entirely abandoned, however. Federal price supports and production subsidies have made the government a major force in stabilizing US agriculture. Moreover, the federal government has stepped in to arrange for guaranteed loans for two large private firms—Lockheed in 1971 and Chrysler in 1980—where thousands of jobs would have been lost in the event of bankruptcy.

As the 1980s began, there was a general consensus that, at least in some areas, government regulation was contributing to inefficiency and higher prices. Thus, the Carter administration moved to deregulate the airline, trucking, and communications industries; subsequently, the Reagan administration relaxed government regulation of bank savings accounts and automobile manufacture as it decontrolled oil and gas prices. The Reagan administration also sought to slow the growth of social-welfare spending and attempted, with only partial success, to transfer control over certain federal social programs to the states and to reduce or eliminate some programs entirely.

Some areas of federal involvement, however, seem safely entrenched. Old age and survivors' insurance, unemployment insurance, and other aspects of the Social Security program have been accepted areas of governmental responsibility since the 1930s. Federal responsibility has also been extended to insurance of bank deposits, to mortgage insurance, and to regulation of stock transactions. The government fulfills a supervisory and regulatory role in labor-management relations. Labor and management customarily disagree on what the role should be, but neither side advocates total removal of government from this field.

From the end of World War II until the end of 1952, US government transfers of capital abroad represented an annual average of about $5,470 million, or 88.3% of the overall national average, while private investments averaged roughly $730 million a year, or about 11.7%. Portfolio investment represented less than $150 million a year, or only 2.5% of the annual aggregate.

After 1952, however, direct private investment began to increase, and portfolio investment rose markedly. In the late 1950s, new private direct investment was increasing yearly by $2 billion or more, while private portfolio investment and official US government loans were climbing by a minimum annual amount of $1 billion each. During the 1960–73 period, the value of US-held assets abroad increased by nearly 12% annually; from the mid–1970s, they rose most years by at least 15%, reaching $887.4 billion in 1983. Through 1980, direct private investments abroad represented the largest share of US overseas assets; since 1981, however, foreign lending by US banks has come to dominate US investment holdings. The rapid rise of US bank lending abroad gave the US a vital economic as well as political stake in the financial stability of developing nations, especially in Latin America, since a serious default by a principal debtor nation would threaten all creditor institutions.

The US share of global wealth creation fell from over 40 percent in 1970 to slightly more than a third in the early nineties. This shift indicated not that the competitiveness of US industries had declined but rather that faster technological, transport and information transfer had enabled other countries to catch up, thus limiting US economy's global market share. In 1994, US productivity and costs were comparable to those of its major competitors in both manufacturing and services industries.

During the 1980s, the Reagan and Bush administrations sought to implement a variety of "supply side" reforms—to limit government intrusion in markets by reducing taxation and regulation. These measures were intended to promote saving and investment and thus raise productivity and long term growth. Contrary to the government's projections, however, the personal savings ratio continued to fall, from 7.9 percent in 1980 to 4 percent in 1989 and 1993. Gross private investment's share of US GDP declined from 17.3 percent in 1980 to 12.9 percent in 1992. The decline in savings relative to investment created a market receptive to foreign savings which transformed the US from a large creditor to a large debtor and increased its vulnerability to economic fluctuations in other countries. Whereas the US had a current account credit of 14.5 percent of GDP in 1980, in the early nineties it had a current account deficit of 2 percent of GDP. The trade deficit in 1992 was largely in goods and investment. In 1992, the US traded goods deficit consisted of a $49 billion deficit with Asia excluding Japan, $50 billion with Japan itself, and $10 billion with Canada. The US had a surplus of $7 billion with eastern and western Europe and a $6 billion surplus with Latin America.

In 1993, Congress approved the North American Free Trade Agreement, which extended the Free Trade Agreement between Canada and the United States to include Mexico. NAFTA, by eliminating tariffs and other trade barriers, created a free trade zone with a combined market size of $6.5 trillion and 370 million consumers. The effect on employment was uncertain—estimates varied from a loss of 150,000 jobs over the next ten years to a net gain of 200,000. Labor intensive goods-producing industries, such as apparel and textiles, were expected to suffer, while it was predicted that capital goods industries would benefit. It was anticipated that US automakers would benefit in the short run by taking advantage of the low wages in Mexico and that US grain farmers and the US banking, financial, and telecommunications sectors would gain enormous new markets.

40 SOCIAL DEVELOPMENT

Social welfare programs in the US depend on both the federal government and the state governments for resources and administration. Old age, survivors', disability, and the Medicare (health) programs are administered by the federal government; unemployment insurance, dependent child care, and a variety of other public assistance programs are state administered, although the federal government contributes to all of them through grants to the states. State public expenditures for all social welfare programs—including income maintenance, health, education, and welfare services—reached $156.3 billion in 1992, representing 21.2% of all state outlays. Federal expenditures on social welfare rose during the first four years of the Reagan presidency, but his administration did succeed in slowing the growth rate from 15% in 1980 to 13.7% in 1981 and 6.3% in 1982. Eligibility requirements for many social programs were tightened, operating budgets were slashed, and some programs—notably public-sector employment under the Comprehensive Employment and Training Act (CETA)—were eliminated entirely.

In 1990, 13,285,000 Americans received $18.5 billion under the aid to families with dependent children (AFDC) program. Attempts to get clients "off the welfare rolls and onto the payrolls" have foundered on the reality that many of them are not easily employable, especially in a tight job market. Moreover, the job training and day care facilities necessary to allow some poor people to find and hold jobs are often no less expensive than direct public assistance.

The Food and Nutrition Service of the US Department of Agriculture oversees several food assistance programs. In 1991, eligible Americans took part in the food stamp program, at a cost to the federal government of $17,388,720. Eligible pupils participated in the school lunch program, at a federal cost of $3.3 billion. During the same year, the federal government also expended money for school breakfasts, on nutrition programs for the elderly, and in commodity aid for the needs. The present Social Security program differs greatly from that created by the Social Security Act of 1935, which provided that retirement benefits be paid to retired workers aged 65 or older. Since 1939, Congress has attached a series of amendments to the program, including provisions for workers who retire at age 62, for widows, for dependent children under 18 years of age, and for children who are disabled prior to age 18. Disabled workers between 50 and 65 years of age are also entitled to monthly benefits. Other measures increased the number of years a person may work; among these reforms was a 1977 law banning mandatory retirement in private industry before age 70. The actuarial basis for the Social Security system has also changed. In 1935 there were about nine US wage earners for each American aged 65 or more; by the mid–1990s, however, the ratio was closer to three to one.

In 1940, the first year benefits were payable, $35 million was paid out. By 1983, Social Security benefits totaled $268.1 billion, paid to more than 40.6 million beneficiaries. The average monthly benefit for a retired worker with no dependents in 1960 was $74; in 1983, the average benefit was $629.30. Under legislation enacted in the early 1970s, increases in monthly benefits are pegged to the inflation rate, as expressed through the Consumer Price Index. Employers, employees, and the self-employed are legally required to make contributions to the Social Security fund. Wage and salary earners pay Social Security taxes under the Federal Insurance Contributions Act (FICA). As the amount of benefits and the number of beneficiaries have increased, so has the maximum FICA payment, which for 1984 was $2,533, based on a rate of 6.7% on earnings up to $37,800, plus another 0.3% supplied from general revenues; the 1960 maximum was $144. In 1990, Social Security Taxes leveled off at 7.65% on earnings up to $51,300. Among workers with many dependents, the Social Security tax deduction can now exceed the federal income tax deduction.

In January 1974, the Social Security Administration assumed responsibility for assisting the aged, blind, and disabled under the Supplemental Security Income program. In 1991, some 5,118,470 disabled Americans—1,464,684 aged, 3,569,237 disabled, and 84,549 blind—received over $18 billion. Medicare, another program administered under the Social Security Act, provides hospital insurance and voluntary medical insurance for persons 65 and over, with reduced benefits available at age 62. Medicare hospital insurance covered some 29,866,000 Americans in 1990. Medicaid is a program that helps the needy meet the costs of medical, hospital, and nursing-home care.

The laws governing unemployment compensation originate in the states. Therefore, the benefits provided vary from state to state in duration (generally from 26 to 39 weeks) and amount (ranging from $111 to $222 weekly in 1991); the national average was about $170 per week. Workers' Compensation payments in 1992 amounted to $34 billion; federal and state outlays for vocational rehabilitation exceeded $1.2 billion in 1983.

Private philanthropy plays a major role in the support of relief and health services. The private sector plays an especially important role in pension management.

The federal agency ACTION, established in 1971, coordinates several US social service agencies. Chief among these is Volunteers in Service to America—VISTA—which was created in 1964 to marshal human resources against economic and environmental problems. ACTION also administers activities for the young and the aged, including the Retired Senior Volunteer Program and the Foster Grandparent Program, enlisting elderly persons to work with children who have special physical, mental, or emotional needs.

41HEALTH

The US health care system is among the most advanced in the world. In 1993, health expenditures were to have reached a projected $903.3 billion, equivalent to 14.4% of GDP. Escalating health care costs resulted in several proposals for a national health care program in the 1970s, early 1980s, and early 1990s. During 1991, the US Census Bureau estimated that 35.4 million US residents (14% of the population) were without any form of health insurance. At the end of 1991, over 40 such proposals had been aired by politicians, medical and business groups, labor unions, and academicians. Most reform measures rely either on market-oriented approaches designed to widen insurance coverage through tax subsidies, on a federally controlled single-payer plan, or on mandatory employer payments for insurance coverage.

Life expectancy for someone born in 1992 was a US record high 75.7 years. Males could expect to live 72.3 years, females 79.0 years. By race and gender, white females had the highest US average expected lifespan, with 79.7 years; next came black females, 73.9; white males, 73.2; and black males, 65.5. The female and male life expectancies for all other races in 1992 were 75.6 and 67.8 years, respectively. Infant mortality has fallen from 3,830 per 100,000 live births in 1945 to 848.7 in 1992. In 1990, 1,429,577 legal abortions were obtained in the US, for a rate of 24 per 1,000 women ages 15–44 and a ratio of 345 per 1,000 live births.

The overall death rate is comparable to that of most nations—8.5 per 1,000 population in 1992. Leading causes of death (number, rate per 100,000 population, and percent of total deaths) in 1992 were: heart disease, 720,480, 282.5 (33.1%); cancer, 521,090, 204.3 (23.9%); cerebrovascular diseases, 143,640, 56.3 (6.6%); chronic obstructive pulmonary diseases, 91,440, 35.8 (4.2%); accidents and adverse effects, 86,310, 33.8 (4.0%); pneumonia and influenza, 76,120, 29.8 (3.5%); diabetes mellitus, 50,180, 19.7 (2.3%); human immunodeficiency virus (HIV) infection, 33,590, 13.2 (1.5%); suicide, 29,760, 11.7 (1.4%); and homicide and legal intervention, 26,570, 10.4 (1.2%). Other leading causes of death in 1992 included chronic liver disease and cirrhosis, nephritis, septicemia, arteriosclerosis, and from certain conditions originating in the perinatal period. Cigarette smoking has been linked to heart and lung disease; about 20% of all deaths in the US were attributed to cigarette smoking in 1990. First identified in 1981, HIV infection (resulting in acquired immune deficiency syndrome—AIDS) has risen; by 1992, there were just under 250,000 AIDS cases reported in the US.

Medical facilities in the US included 6,634 hospitals in 1991 (down 4.3% from 1981), with 1,202,000 beds (down 11.8%). There were 653,062 physicians as of 1 January 1992, for a ratio of 255 per 100,000 population (or 209 active physicians in patient care per 100,000 population). Of the total number of active classified physicians, the largest areas of activity were internal medicine, 14.9%; family practice, 8.8%; pediatrics, 6.9%; and psychiatry, 6.1%. As of March 1992, there were 2,239,816 registered nurses (82.7% employed in nursing), for a ratio of 726 employed nurses per 100,000 population. Dentists numbered 156,738 in 1993.

Per capita health care expenditures have risen from $247 in 1967 to about $3,380 in 1993. National health care spending is projected to rise to $1.6 trillion by 2000 ($5,712 per capita) comprising over 16% of GDP. Hospital care represented 38% of national health care spending in 1991, even though hospitals' proportion of total health care expenditures has fallen slightly since 1980. Hospital costs amounted to over $225 billion in 1991, equivalent to about $5,360 per admission or $752.10 per inpatient day. In fiscal 1992, services provided for 30.9 million Medicaid recipients required $90.8 billion in federal expenditures, while 35.6 million Medicare enrollees received benefit payments of $129.1 billion. Medicare payments have lagged behind escalating hospital costs; in 1991, over 60% of hospitals lost money on Medicare patients. In addition, the Federal Hospital Insurance Trust Fund, which pays for Medicare's hospital expenditures, is projected to be exhausted by 2002. Meanwhile, the elderly population in the US is projected to increase to 13% of the total population by 2000, and to 18% by 2020, thus exacerbating the conundrum of health care finance.

42HOUSING

The housing resources of the US far exceed those of any other country, with 102,264,000 housing units as of April 1990, 91,946,000 of which were occupied. In 1991, 57.2% of all US homes were occupied by their owners, 31.9% were rented, 8.3% were vacant, and 2.6% were occupied seasonally. The majority of rental tenants are found in the large metropolitan areas. As of 1990, an estimated 94 million year-round households possessed and used electrical appliances. Of these, 90.3% had color television sets; 71.7% had washing machines; 64.7% had clothes dryers; 42.7% had electric dishwashers; 29.1% had room air conditioners; and 36.6% had central air conditioning. As of 1980, about 22% of all housing units in the US had been built before 1940; 8% during 1940–49; 13% during 1950–59; 15% from 1960 to 1969; 23% during 1970–79; and 19% since 1980. The median construction year was 1964.

Construction of housing following World War II set a record-breaking pace; 1986 was the 38th successive year during which construction of more than one million housing units was begun. In that year, 1,810,000 new units were started, the highest number in eight years. After 1986, however, housing starts dropped for five years in succession, hitting 1,014,000 in 1991 and then climbing to 1,200,000 in 1992. The great bulk of the housing stock, as well as of homes produced, consists of one-family houses. Perhaps the most significant change in the housing scene has been the shift to the suburbs made possible by the widespread ownership of automobiles.

43EDUCATION

Education is compulsory in all states and a responsibility of each state and the local government. However, federal funds are available to help meet special needs at primary, secondary, or higher levels. Generally, formal schooling begins at the age of six and continues up to age 17. But, each state specifies the age and circumstances for compulsory attendance.

"Regular" schools which educate a person towards a diploma or degree include both public and private schools. Public schools are controlled and supported by the local authorities as well as state or federal governmental agencies. Private schools are controlled and supported by religious or private organizations. Elementary schooling is from grade 1 to grade 8. High schools cover grades 9 through 12. Colleges include junior or community colleges, offering two-year Associate Degrees; regular four-year colleges and universities; and graduate or professional schools. The school year begins in September and ends in June.

The enrollment rate of three- to five-year olds in preprimary schools was 37% in 1970. It went up to 56% in 1991. Persons 25 years old and over completing college education was 11% in 1970 and 21% in 1992.

The total number of children enrolled in public schools in 1990 was 52 million. Private schools had 8 million the same year. It is projected that by the year 2003, public schools will have 61 million students and private schools 9.5 million students.

In 1990, elementary schools had 27 million students in public schools and 4 million in private schools. Secondary schools had 14 million public and 1 million private school students. Colleges had 10.7 million students in public and 2.97 million in private colleges. In 1991–92, there were 3,500 two-year and four-year colleges and universities. Governmental expenditure on education (state and federal) was $280,713 million in 1988–89. The literacy rate is estimated to be 98% (males 97% and females 98%).

⁴⁴LIBRARIES AND MUSEUMS

Of the 32,414 libraries in the US in 1993, 9,050 were public, with 6,215 branches; 4,619 were academic; 1,871 were government; 1,925 wre medical; and a number were religious, military, legal, and specialized independent collections.

The foremost library in the country is the Library of Congress, with holdings of more than 80 million items (including more than 20 million books and pamphlets) in the mid–1990s. Other great libraries are the public libraries in New York, Philadelphia, Boston, and Baltimore, and the John Crerar and Newberry libraries in Chicago. Noted special collections are those of the Pierpont Morgan Library in New York; the Huntington Library in San Marino, Calif.; the Folger Shakespeare Library in Washington, D.C.; the Hoover Library at Stanford University; and the rare book divisions of Harvard, Yale, Indiana, Texas, and Virginia universities. Among the leading university libraries, as judged by the extent of their holdings in 1993, are those of Harvard, Yale, Illinois (Urbana-Champaign), Michigan (Ann Arbor), California (Berkeley), Columbia, Stanford, Cornell, California (Los Angeles), Chicago, Wisconsin (Madison), and Washington (Seattle)— each having more than 4 million bound volumes.

There are about 5,000 nonprofit museums in the US. The most numerous type is the historic building, followed in descending order by college and university museums, museums of science, public museums of history, and public museums of art. Eminent US museums include the American Museum of Natural History, the Metropolitan Museum of Art, and the Museum of Modern Art, all in New York City; the National Gallery of Art and the Smithsonian Institution in Washington, D.C.; the Boston Museum of Fine Arts; the Art Institute of Chicago and the Chicago Museum of Natural History; the Franklin Institute and Philadelphia Museum of Art, both in Philadelphia; and the M. H. de Young Memorial Museum in San Francisco.

⁴⁵MEDIA

All major electric communications systems are privately owned but regulated by the Federal Communications Commission. The US uses wire and radio services for communications more extensively than any other country in the world. In 1990 97.8% of all US households had telephones.

Radio serves a variety of purposes other than broadcasting. It is widely used by ships and aircraft for safety; it has become an important tool in the movement of buses, trucks, and taxicabs. Forest conservators, fire departments, and the police operate with radio as a necessary aid; it is used in logging operations, surveying, construction work, and dispatching of repair crews. In 1993, broadcasting stations on the air comprised 11,420 radio stations (both AM and FM) and, as of 1 May 1993 1,541 television stations. In 1993 98% of all US households owned at least one TV set. The expanding cable television industry, with 11,385 cable

systems as of 1993, served 57.6 million subscribers.

The Post Office Department of the US was replaced on 1 July 1971 by the US Postal Service, a financially autonomous federal agency. In 1992 160 billion pieces of mail passed through 39,613 post offices, branches and substations with about 711,369 employees. In addition to mail delivery, the Postal Service provides registered, certified, insured, express and COD mail service, issues money orders, and operates a postal savings system. Since the 1970s, numerous privately owned overnight mail and package delivery services have been established.

In 1992 there were 1,570 daily newspapers in the US, with a combined circulation of 60 million plus. Circulation has hovered around the 60 million mark for the past 15 years or so. Twenty large newspaper chains account for 57% of the total daily circulation. The following newspapers, all in the English language, reported average daily circulation of 500,000 or more in 1991:

Wall Street Journal	1,795,448
USA Today	1,418,477
Los Angeles Times	1,177,253
New York Times	1,110,562
Washington Post	791,289
Long Island/New York Newsday	762,639
New York Daily News	759,068
Chicago Tribune	723,178
Detroit Free Press	598,418
San Francisco Chronicle	553,433
Chicago Sun-Times	531,462
Boston Globe	504,675
Philadelphia Inquirer	503,603

Modern Maturity, published by AARP, had a circulation 22,879,886 in 1992. The two general circulation magazines that appealed to the largest audiences were *Reader's Digest*, 16,258,476, and *TV Guide*, 14,498,341. *Time* and *Newsweek* were the leading newsmagazines, with a weekly circulation of 4,203,991 and 3,240,131 respectively.

The US book-publishing industry consists of the major book companies (mainly in the New York metro area), nonprofit university presses distributed throughout the US, and numerous small publishing firms. In 1992 44,528 book titles were published in the US.

⁴⁶ORGANIZATIONS

The 1987 Census of Service Industries counted 67,997 membership organizations, embracing 40,415 civic, social, and fraternal associations, 12,299 business associations, 5,610 professional associations, and 9,673 other bodies.

A number of industrial and commercial organizations exercise considerable influence on economic policy. The National Association of Manufacturers and the US Chamber of Commerce, with numerous local branches, are the two central bodies of business and commerce. Various industries have their own associations, concerned with cooperative research and questions of policy alike.

Practically every profession in the US is represented by one or more professional organizations. Among the most powerful of these are the American Medical Association, comprising regional, state, and local medical societies; the American Bar Association, also comprising state and local associations; the American Hospital Association; and the National Education Association.

Many private organizations are dedicated to programs of political and social action. Prominent in this realm are the National Association for the Advancement of Colored People (NAACP), the Urban League, the American Civil Liberties Union (ACLU), Common Cause, and the Anti-Defamation League. The League of Women Voters, which provides the public with nonpartisan information about candidates and election issues, sponsored televised debates between the major presidential candidates in 1976, 1980, 1984, 1988, and 1992. The National Organization for Women, and the National Rifle Association have each mounted nationwide lobbying campaigns on issues affecting their members. During 1981–82, political action committees (PACs), affiliated with corporations, labor unions, and other groups, gave $83 million to candidates for the House and Senate—an increase of 50% over 1979/80. Probably the best-financed and most influential PAC by the mid–1980s was the National Conservative Political Action Committee (NCPAC), which provided $9.8 million to President Reagan's reelection campaign in 1984. In 1988 there were 4,268 PAC's that disbursed $364.2 million to candidates for the House and Senate and other elected offices.

The great privately endowed philanthropic foundations and trusts play an important part in encouraging the development of education, art, science, and social progress in the US. In the early 1980s there were nearly 32,401 foundations, with combined assets exceeding $142 billion. Foundations, donations, and charitable bequests in 1983 provided a total of $81 billion in grants. Prominent foundations include the Carnegie Corporation and the Carnegie Endowment for International Peace, the Ford Foundation, the Guggenheim Foundation, the Mayo Association for the Advancement of Medical Research and Education, and the Rockefeller Foundation. Private philanthropy was responsible for the establishment of many of the nation's most eminent libraries, concert halls, museums, and university and medical facilities; private bequests were also responsible for the establishment of the Pulitzer Prizes. Merit awards offered by industry and professional groups include the "Oscars" of the Academy of Motion Picture Arts and Sciences, the "Emmys" of the National Academy of Television Arts and Sciences, and the "Grammys" of the National Academy of Recording Arts and Sciences.

Funds for a variety of community health and welfare services are funneled through United Way campaigns, which raises annually $2 billion. The American Red Cross has over 3,000 chapters, which paid for services and activities ranging from disaster relief to blood donor programs. Private organizations supported by contributions from the general public lead the fight against specific diseases.

The Boy Scouts of America, the Girl Scouts of the USA, rural 4–H Clubs, and the Young Men's and the Young Women's Christian Associations are among the organizations devoted to recreation, sports, camping, and education.

The largest religious organization in the US is the National Council of the Churches of Christ in the USA, which embraces 32 Protestant and Orthodox denominations, whose adherents total more than 42 million. Many organizations, such as the American Philosophical Society, the American Association for the Advancement of Science, and the National Geographic Society, are dedicated to the enlargement of various branches of human knowledge. National, state, and local historical societies abound, and there are numerous educational, sports, and hobbyist groups.

The larger veterans' organizations are the American Legion, the Veterans of Foreign Wars of the US, the Catholic War Veterans, and the Jewish War Veterans. Fraternal organizations, in addition to such international organizations as the Masons, include indigenous groups such as the Benevolent and Protective Order of Elks, the Loyal Order of Moose, and the Woodmen of the World. Many, such as the Ancient Order of Hibernians in America, commemorate the national origin of their members. One of the largest fraternal organizations is the Roman Catholic Knights of Columbus.

47TOURISM, TRAVEL, AND RECREATION

Foreign visitors to the US numbered approximately 42,723,000 in 1991. Of these visitors, 44% came from Canada, 18% from Mexico, 18% from Europe, and 12% from Eastern Asia and the Pacific. In 1991, travelers to the US from all foreign countries spent $45.5 billion, and another $18.8 billion for fares on US carriers to and from their home countries. There were 3,080,000 hotel rooms with 5,544,000 beds. With a few exceptions, such as Canadians entering from the Western Hemisphere, all visitors to the US are required to have passports and visas.

A total of $39.4 billion was spent by US tourists abroad on 41,835,000 trips.

With the advent of a new, more environmentally conscious political administration in 1992, ecotourism is expected to grow. Among the most striking scenic attractions in the US are the Grand Canyon in Arizona; Carlsbad Caverns in New Mexico; Yosemite National Park in California; Yellowstone National Park in Idaho, Montana, and Wyoming; Niagara Falls, partly in New York; and the Everglades in Florida. The US had a total of 49 national parks as of August 1987. Popular coastal resorts include those of Florida, California, and Cape Code in Massachusetts. Historical attractions include the Liberty Bell and Constitution Hall in Philadelphia; the Statue of Liberty in New York City; the White House, the Capitol, and the monuments to Washington, Jefferson, and Lincoln in the District of Columbia; the Williamsburg historical restoration in Virginia; various Revolutionary and Civil War battlefields and monuments in the East and South; the Alamo in San Antonio; and Mt. Rushmore in South Dakota. Among many other popular tourist attractions are the movie and television studios in Los Angeles; the cable cars in San Francisco; casino gambling in Las Vegas and in Atlantic City, N.J.; thoroughbred horse racing in Kentucky; the Grand Ole Opry in Nashville, Tenn.; the many jazz clubs of New Orleans; and such amusement parks as Disneyland (Anaheim, Calif.) and Walt Disney World (near Orlando, Fla.). For abundance and diversity of entertainment—theater, movies, music, dance, and sports—New York City has few rivals. In April, 1993, Amtrak began the country's first regularly scheduled transcontinental passenger service, from Los Angeles to Miami.

Americans' recreational activities range from the major spectator sports—professional baseball, football, basketball, ice hockey, and soccer, horse racing, and collegiate football and basketball—to home gardening. Participant sports are a favorite form of recreation, including jogging, aerobics, tennis, and golf. Skiing is a popular recreation in New England and the western mountain ranges, while sailing, power boating, rafting, and canoeing are popular water sports. In 1994, the United States hosted the World Cup Soccer Championship.

48FAMOUS AMERICANS

Printer, publisher, inventor, scientist, statesman, and diplomat, Benjamin Franklin (1706–90) was America's outstanding figure of the colonial period. George Washington (1732–99), leader of the colonial army in the American Revolution, became first president of the US and is known as the "father of his country." Chief author of the Declaration of Independence, founder of the US political party system, and third president was Thomas Jefferson (1743–1826). His leading political opponents were John Adams (1735–1826), second president, and Alexander Hamilton (b.West Indies, 1755–1804), first secretary of the treasury, who secured the new nation's credit. James Madison (1751–1836), a leading figure in drawing up the US Constitution, served as fourth president. John Quincy Adams (1767–1848), sixth president, was an outstanding diplomat and secretary of state.

Andrew Jackson (1767–1845), seventh president, was an ardent champion of the common people and opponent of vested interests. Outstanding senators during the Jackson era were John Caldwell Calhoun (1782–1850), spokesman of the southern planter aristocracy and leading exponent of the supremacy of states' rights over federal powers; Henry Clay (1777–1852), the great compromiser, who sought to reconcile the conflicting views of the North and the South; and Daniel Webster (1782–1852), statesman and orator, who championed the preservation of the Union against sectional interests and division. Abraham Lincoln (1809–65) led the US through its most difficult period, the Civil War, in the course of which he issued the Emancipation Proclamation. Jefferson Davis (1808–89) served as the only president of the short-lived Confederacy. Stephen Grover Cleveland (1837–1908), a conservative reformer, was the strongest president in the latter part of the 19th century. Among the foremost presidents of the 20th century have been Nobel Peace Prize winner Theodore Roosevelt (1858–1919); Woodrow Wilson (1856–1924), who led the nation during World War I and helped establish the League of Nations; and Franklin Delano Roosevelt (1882–1945), elected to four terms spanning the Great Depression and World War II. The presidents during the 1961–88 period have been John Fitzgerald Kennedy (1917–63), Lyndon Baines Johnson (1908–73), Richard Milhous Nixon (1913–94), Gerald Rudolph Ford (Leslie Lynch King, Jr., b.1913), Jimmy Carter (James Earl Carter, Jr., b.1924), and Ronald Wilson Reagan (b.1911).

Of the outstanding US military leaders, four were produced by the Civil War: Union generals Ulysses Simpson Grant (1822–85), who later served as the eighteenth president, and William Tecumseh Sherman (1820–91); and Confederate generals Robert Edward Lee (1807–70) and Thomas Jonathan "Stonewall" Jackson (1824–63). George Catlett Marshall (1880–1959), army chief of staff during World War II, in his later capacity as secretary of state under President Harry S Truman (1884–1972), formulated the Marshall Plan, which did much to revitalize Western Europe. Douglas MacArthur (1880–1964) commanded the US forces in Asia during World War II, oversaw the postwar occupation and reorganization of Japan, and directed UN forces in the first year of the Korean conflict. Dwight D. Eisenhower (1890–1969) served as supreme Allied commander during World War II, later becoming the thirty-fourth president.

John Marshall (1755–1835), chief justice of the US from 1801 to 1835, established the power of the Supreme Court through the principle of judicial review. Other important chief justices were Edward Douglass White (1845–1921), former president William Howard Taft (1857–1930), and Earl Warren (1891–1974), whose tenure as chief justice from 1953 to 1969 saw important decisions on desegregation, reapportionment, and civil liberties. The justice who enjoyed the longest tenure on the court was William O. Douglas (1898–1980), who served from 1939 to 1975; other prominent associate justices were Oliver Wendell Holmes (1841–1935), Louis Dembitz Brandeis (1856–1941), and Hugo Lafayette Black (1886–1971).

Indian chiefs renowned for their resistance to white encroachment were Pontiac (1729?–69), Black Hawk (1767–1838), Tecumseh (1768–1813), Osceola (1804?–38), Cochise (1812?–74), Geronimo (1829?–1909), Sitting Bull (1831?–90), Chief Joseph (1840?–1904), and Crazy Horse (1849?–77). Other significant Indian chiefs were Hiawatha (fl. 1500), Squanto (d.1622), and Sequoya (1770?–1843). Historical figures who have become part of American folklore include pioneer Daniel Boone (1734–1820); silversmith, engraver, and patriot Paul Revere (1735–1818); frontiersman David "Davy" Crockett (1786–1836); scout and Indian agent Christopher "Kit" Carson (1809–68); James Butler "Wild Bill" Hickok (1837–76); William Frederick "Buffalo Bill" Cody (1846–1917); and the outlaws Jesse Woodson James (1847–82) and Billy the Kid (William H. Bonney, 1859–81).

Inventors and Scientists

Outstanding inventors were Robert Fulton (1765–1815), who developed the steamboat; Eli Whitney (1765–1825), inventor of the cotton gin and mass production techniques; Samuel Finley

Breese Morse (1791–1872), who invented the telegraph; and Elias Howe (1819–67), who invented the sewing machine. Alexander Graham Bell (b.Scotland, 1847–1922) gave the world the telephone. Thomas Alva Edison (1847–1931) was responsible for hundreds of inventions, among them the long-burning incandescent electric lamp, the phonograph, automatic telegraph devices, a motion picture camera and projector, the microphone, and the mimeograph. Lee De Forest (1873–1961), the "father of the radio," developed the vacuum tube and many other inventions. Vladimir Kosma Zworykin (b.Russia, 1889–1982) was principally responsible for the invention of television. Two brothers, Wilbur Wright (1867–1912) and Orville Wright (1871–1948), designed, built, and flew the first successful motor-powered airplane. Amelia Earhart (1898–1937) and Charles Lindbergh (1902–74) were aviation pioneers. Pioneers in the space program include John Glenn (b.1921), the first US astronaut to orbit the earth, and Neil Armstrong (b.1930), the first man to set foot on the moon.

Benjamin Thompson, Count Rumford (1753–1814), developed devices for measuring light and heat, and the physicist Joseph Henry (1797–1878) did important work in magnetism and electricity. Outstanding botanists and naturalists were John Bartram (1699–1777); his son William Bartram (1739–1832); Louis Agassiz (b.Switzerland, 1807–73); Asa Gray (1810–88); Luther Burbank (1849–1926), developer of a vast number of new and improved varieties of fruits, vegetables, and flowers; and George Washington Carver (1864–1943), known especially for his work on industrial applications for peanuts. John James Audubon (1785–1851) won fame as an ornithologist and artist.

Distinguished physical scientists include Samuel Pierpont Langley (1834–1906), astronomer and aviation pioneer; Josiah Willard Gibbs (1839–1903), mathematical physicist, whose work laid the basis for physical chemistry; Henry Augustus Rowland (1848–1901), who did important research in magnetism and optics; and Albert Abraham Michelson (b.Germany, 1852–1931), who measured the speed of light and became the first of a long line of US Nobel Prize winners. The chemists Gilbert Newton Lewis (1875–1946) and Irving Langmuir (1881–1957) developed a theory of atomic structure.

The theory of relativity was conceived by Albert Einstein (b.Germany, 1879–1955), generally considered the greatest mind in the physical sciences since Newton. Percy Williams Bridgman (1882–1961) was the father of operationalism and studied the effect of high pressures on materials. Arthur Holly Compton (1892–1962) made discoveries in the field of X rays and cosmic rays. The physical chemist Harold Clayton Urey (1893–1981) discovered heavy hydrogen. Isidor Isaac Rabi (b.Austria, 1898–1988), nuclear physicist, did important work in magnetism, quantum mechanics, and radiation. Enrico Fermi (b.Italy, 1901–54) created the first nuclear chain reaction, in Chicago in 1942, and contributed to the development of the atomic and hydrogen bombs. Also prominent in the splitting of the atom were Leo Szilard (b.Hungary, 1898–1964), J. Robert Oppenheimer (1904–67), and Edward Teller (b.Hungary, 1908). Ernest Orlando Lawrence (1901–58) developed the cyclotron. Carl David Anderson (1905–91) discovered the positron. Mathematician Norbert Wiener (1894–1964) developed the science of cybernetics.

Outstanding figures in the biological sciences include Theobald Smith (1859–1934), who developed immunization theory and practical immunization techniques for animals; the geneticist Thomas Hunt Morgan (1866–1945), who discovered the heredity functions of chromosomes; and neurosurgeon Harvey William Cushing (1869–1939). Selman Abraham Waksman (b.Russia, 1888–1973), a microbiologist specializing in antibiotics, was co-discoverer of streptomycin. Edwin Joseph Cohn (1892–1953) is noted for his work in the protein fractionalization of blood, particularly the isolation of serum albumin. Philip

Showalter Hench (1896–1965) isolated and synthesized cortisone. Wendell Meredith Stanley (1904–71) was the first to isolate and crystallize a virus. Jonas Edward Salk (b.1914) developed an effective killed-virus poliomyelitis vaccine, and Albert Bruce Sabin (1906–93) contributed oral, attenuated live-virus polio vaccines.

Adolf Meyer (b.Switzerland, 1866–1950) developed the concepts of mental hygiene and dementia praecox and the theory of psychobiology; Harry Stack Sullivan (1892–1949) created the interpersonal theory of psychiatry. Social psychologist George Herbert Mead (1863–1931) and behaviorist Burrhus Frederic Skinner (b.1904) have been influential in the 20th century.

A pioneer in psychology who was also an influential philosopher was William James (1842–1910). Other leading US philosophers are Charles Sanders Peirce (1839–1914); Josiah Royce (1855–1916); John Dewey (1859–1952), also famous for his theories of education; George Santayana (b.Spain, 1863–1952); Rudolf Carnap (b.Germany, 1891–1970); and Willard Van Orman Quine (b.1908). Educators of note include Horace Mann (1796–1859), Henry Barnard (1811–1900), and Charles William Eliot (1834–1926). Noah Webster (1758–1843) was the outstanding US lexicographer, and Melvil Dewey (1851–1931) was a leader in the development of library science. Thorstein Bunde Veblen (1857–1929) wrote books that have strongly influenced economic and social thinking. Also important in the social sciences have been sociologists Talcott Parsons (1902–79) and William Graham Sumner (1840–1910) and anthropologist Margaret Mead (1901–78).

Social Reformers

Social reformers of note include Dorothea Lynde Dix (1802–87), who led movements for the reform of prisons and insane asylums; William Lloyd Garrison (1805–79) and Frederick Douglass (Frederick Augustus Washington Bailey, 1817–95), prominent abolitionists; Elizabeth Cady Stanton (1815–1902) and Susan Brownell Anthony (1820–1906), leaders in the women's suffrage movement; Clara Barton (1821–1912), founder of the American Red Cross; economist Henry George (1839–97), advocate of the single-tax theory; Eugene Victor Debs (1855–1926), labor leader and an outstanding organizer of the Socialist movement in the US; Jane Addams (1860–1935), who pioneered in settlement house work; Robert Marion La Follette (1855–1925), a leader for progressive political reform in Wisconsin and in the US Senate; Margaret Higgins Sanger (1883–1966), pioneer in birth control; Norman Thomas (1884–1968), Socialist Party leader;and Martin Luther King, Jr. (1929–68), a central figure in the black civil rights movement and winner of the Nobel Peace Prize in 1964.

Religious leaders include Roger Williams (1603–83), an early advocate of religious tolerance in the US; Jonathan Edwards (1703–58), New England preacher and theologian; Elizabeth Ann Seton (1774–1821), the first American canonized in the Roman Catholic Church; William Ellery Channing (1780–1842), a founder of American Unitarianism; Joseph Smith (1805–44), founder of the Church of Jesus Christ of Latter-day Saints (Mormon) and his chief associate, Brigham Young (1801–77); and Mary Baker Eddy (1821–1910), founder of the Christian Science Church. Paul Tillich (b.Germany, 1886–1965) and Reinhold Niebuhr (1892–1971) were outstanding Protestant theologians of international influence.

Famous US businessmen include Éleuthere Irénée du Pont de Nemours (b.France, 1771–1834), John Jacob Astor (Johann Jakob Ashdour, b.Germany, 1763–1848), Cornelius Vanderbilt (1794–1877), Andrew Carnegie (b.Scotland, 1835–1919), John Pierpont Morgan (1837–1913), John Davison Rockefeller (1839–1937), Andrew William Mellon (1855–1937), Henry Ford (1863–1947), and Thomas John Watson (1874–1956).

Literary Figures

The first US author to be widely read outside the US was Washington Irving (1783–1859). James Fenimore Cooper (1789–1851) was the first popular US novelist. Three noted historians were William Hickling Prescott (1796–1859), John Lothrop Motley (1814–77), and Francis Parkman (1823–93). The writings of two men of Concord, Mass.—Ralph Waldo Emerson (1803–82) and Henry David Thoreau (1817–62)—influenced philosophers, political leaders, and ordinary men and women in many parts of the world. The novels and short stories of Nathaniel Hawthorne (1804–64) explore New England's Puritan heritage. Herman Melville (1819–91) wrote the powerful novel *Moby-Dick*, a symbolic work about a whale hunt that has become an American classic. Mark Twain (Samuel Langhorne Clemens, 1835–1910) is the best-known US humorist. Other leading novelists of the later 19th and early 20th centuries were William Dean Howells (1837–1920), Henry James (1843–1916), Edith Wharton (1862–1937), Stephen Crane (1871–1900), Theodore Dreiser (1871–1945), Willa Cather (1873–1947), and Sinclair Lewis (1885–1951), first US winner of the Nobel Prize for literature (1930). Later Nobel Prize–winning US novelists include Pearl Sydenstricker Buck (1892–1973), in 1938; William Faulkner (1897–1962), in 1949; Ernest Hemingway (1899–1961), in 1954; John Steinbeck (1902–68), in 1962; Saul Bellow (b.Canada, 1915), in 1976; and Isaac Bashevis Singer (b.Poland, 1904–91), in 1978. Among other noteworthy writers are James Thurber (1894–1961), Francis Scott Key Fitzgerald (1896–1940), Thomas Wolfe (1900–1938), Richard Wright (1908–60), Eudora Welty (b.1909), John Cheever (1912–82), Norman Mailer (b.1923), James Baldwin (1924–87), John Updike (b.1932), and Philip Roth (b.1933).

Noted US poets include Henry Wadsworth Longfellow (1807–82), Edgar Allan Poe (1809–49), Walt Whitman (1819–92), Emily Dickinson (1830–86), Edwin Arlington Robinson (1869–1935), Robert Frost (1874–1963), Wallace Stevens (1879–1955), William Carlos Williams (1883–1963), Marianne Moore (1887–1972), Edward Estlin Cummings (1894–1962), Hart Crane (1899–1932), and Langston Hughes (1902–67). Ezra Pound (1885–1972) and Nobel laureate Thomas Stearns Eliot (1888–1965) lived and worked abroad for most of their careers. Wystan Hugh Auden (b.England, 1907–73), who became an American citizen in 1946, published poetry and criticism. Elizabeth Bishop (1911–79), Robert Lowell (1917–77), Allen Ginsberg (b.1926), and Sylvia Plath (1932–63) are among the best-known poets since World War II. Robert Penn Warren (1905–89) won the Pulitzer Prize for both fiction and poetry and became the first US poet laureate. Carl Sandburg (1878–1967) was a noted poet, historian, novelist, and folklorist. The foremost US dramatists are Eugene (Gladstone) O'Neill (1888–1953), who won the Nobel Prize for literature in 1936; Tennessee Williams (Thomas Lanier Williams, 1911–83); Arthur Miller (b.1915); and Edward Albee (b.1928). Neil Simon (b.1927) is among the nation's most popular playwrights and screenwriters.

Artists

Two renowned painters of the early period were John Singleton Copley (1738–1815) and Gilbert Stuart (1755–1828). Outstanding 19th-century painters were James Abbott McNeill Whistler (1834–1903), Winslow Homer (1836–1910), Thomas Eakins (1844–1916), Mary Cassatt (1845–1926), Albert Pinkham Ryder (1847–1917), John Singer Sargent (b.Italy, 1856–1925), and Frederic Remington (1861–1909). More recently, Edward Hopper (1882–1967), Georgia O'Keeffe (1887–1986), Thomas Hart Benton (1889–1975), Charles Burchfield (1893–1967), Norman Rockwell (1894–1978), Ben Shahn (1898–1969), Mark Rothko (b.Russia, 1903–70), Jackson Pollock (1912–56), Andrew Wyeth (b.1917), Robert Rauschenberg (b.1925), and Jasper Johns (b.1930) have achieved international recognition.

Sculptors of note include Augustus Saint-Gaudens (1848–1907), Gaston Lachaise (1882–1935), Jo Davidson (1883–1952), Daniel Chester French (1850–1931), Alexander Calder (1898–1976), Louise Nevelson (b.Russia, 1899–1988), and Isamu Noguchi (1904–88). Henry Hobson Richardson (1838–86), Louis Henry Sullivan (1856–1924), Frank Lloyd Wright (1869–1959), Louis I. Kahn (b.Estonia, 1901–74), and Eero Saarinen (1910–61) were outstanding architects. Contemporary architects of note include Richard Buckminster Fuller (1895–1983), Edward Durrell Stone (1902–78), Philip Cortelyou Johnson (b.1906), and Ieoh Ming Pei (b.China, 1917). The US has produced many fine photographers, notably Mathew B. Brady (1823?–96), Alfred Stieglitz (1864–1946), Edward Steichen (1879–1973), Edward Weston (1886–1958), Ansel Adams (1902–84), and Margaret Bourke-White (1904–71).

Entertainment Figures

Outstanding figures in the motion picture industry are D. W. (David Lewelyn Wark) Griffith (1875–1948), Sir Charles Spencer "Charlie" Chaplin (b.England, 1889–1978), Walter Elias "Walt" Disney (1906–66), and George Orson Welles (1915–85). John Ford (1895–1973), Howard Winchester Hawks (1896–1977), Frank Capra (b.Italy, 1897), Sir Alfred Hitchcock (b.England, 1899–1980), and John Huston (1906–87) were influential motion picture directors; Mel Brooks (Kaminsky, b.1926), George Lucas (b.1944), and Steven Spielberg (b.1947) have achieved remarkable popular success. Woody Allen (Allen Konigsberg, b.1935) has written, directed, and starred in comedies on stage and screen. World-famous American actors and actresses include the Barrymores, Ethel (1879–1959) and her brothers Lionel (1878–1954) and John (1882–1942); Humphrey Bogart (1899–1957); James Cagney (1899–1986); Spencer Tracy (1900–1967); Helen Hayes Brown (b.1900); Clark Gable (1901–60); Joan Crawford (Lucille Fay LeSueur, 1904–77); Cary Grant (Alexander Archibald Leach, b.England, 1904–86); Greta Garbo (Greta Louisa Gustafsson, b.Sweden, 1905); Henry Fonda (1905–82) and his daughter, Jane (b.1937); John Wayne (Marion Michael Morrison, 1907–79); Bette (Ruth Elizabeth) Davis (1908–89); Katharine Hepburn (b.1909); Judy Garland (Frances Gumm, 1922–69); Marlon Brando (b.1924); Marilyn Monroe (Norma Jean Mortenson, 1926–62); and Dustin Hoffman (b.1937). Among other great entertainers are W. C. Fields (William Claude Dukenfield, 1880–1946), Al Jolson (Asa Yoelson, b.Russia, 1886–1950), Jack Benny (Benjamin Kubelsky, 1894–1974), Fred Astaire (Fred Austerlitz, 1899–1987), Bob (Leslie Townes) Hope (b.England, 1903), Bing (Harry Lillis) Crosby (1904–78), Frank (Francis Albert) Sinatra (b.1915), Elvis Aaron Presley (1935–77), and Barbra (Barbara Joan) Streisand (b.1942). The first great US "showman" was Phineas Taylor Barnum (1810–91).

Composers and Musicians

The foremost composers are Edward MacDowell (1861–1908), Charles Ives (1874–1954), Ernest Bloch (b.Switzerland, 1880–1959), Virgil Thomson (b.1896), Roger Sessions (1896–1985), Roy Harris (1898–1979), Aaron Copland (1900–90), Elliott Carter (b.1908), Samuel Barber (1910–81), John Cage (1912–92), and Leonard Bernstein (1918–90). George Rochberg (b.1918), George Crumb (b.1929), Steve Reich (b.1936), and Philip Glass (b.1937) have won more recent followings. The songs of Stephen Collins Foster (1826–64) have achieved folk-song status. Leading composers of popular music are John Philip Sousa (1854–1932), George Michael Cohan (1878–1942), Jerome Kern (1885–1945), Irving Berlin (Israel Baline, b.Russia, 1888–1989), Cole Porter (1893–1964), George Gershwin (1898–1937), Richard Rodgers (1902–79), Woody Guthrie (1912–67), Stephen Joshua Sondheim (b.1930), Paul Simon (b.1941), and

Bob Dylan (Robert Zimmerman, b.1941). Preeminent in the blues traditions are Leadbelly (Huddie Ledbetter, 1888–1949), Bessie Smith (1898?–1937), and Muddy Waters (McKinley Morganfield, 1915–83). Leading jazz figures include the composers Scott Joplin (1868–1917), James Hubert "Eubie" Blake (1883–1983), Edward Kennedy "Duke" Ellington (1899–1974), and William "Count" Basie (1904–84), and performers Louis Armstrong (1900–1971), Billie Holiday (Eleanora Fagan, 1915–59), John Birks "Dizzy" Gillespie (b.1917), Charlie "Bird" Parker (1920–55), John Coltrane (1926–67), and Miles Davis (b.1926).

Many foreign-born musicians have enjoyed personal and professional freedom in the US; principal among them were pianists Artur Schnabel (b.Austria, 1882–1951), Arthur Rubinstein (b.Poland, 1887–1982), Rudolf Serkin (b.Bohemia, 1903), Vladimir Horowitz (b.Russia, 1904–89), and violinists Jascha Heifetz (b.Russia, 1901–87) and Isaac Stern (b.USSR, 1920). Among distinguished instrumentalists born in the US are Benny Goodman (1909–86), a classical as well as jazz clarinetist, and concert pianist Van Cliburn (Harvey Lavan, Jr., b.1934). Singers Paul Robeson (1898–1976), Marian Anderson (1897–1993), Maria Callas (Maria Kalogeropoulos, 1923–77), Leontyne Price (b.1927), and Beverly Sills (Belle Silverman, b.1929) have achieved international acclaim. Isadora Duncan (1878–1927) was one of the first US dancers to win fame abroad. Martha Graham (b.1893) pioneered in modern dance. George Balanchine (b.Russia, 1904–83), Agnes De Mille (b.1905), Jerome Robbins (b.1918), Paul Taylor (b.1930), and Twyla Tharp (b.1941) are leading choreographers; Martha Graham (1893–1991) pioneered in modern dance.

Sports Figures

Among the many noteworthy sports stars are baseball's Tyrus Raymond "Ty" Cobb (1886–1961) and George Herman "Babe" Ruth (1895–1948); football's Samuel Adrian "Sammy" Baugh (b.1914), Jim Brown (b.1936), Francis A. "Fran" Tarkenton (b.1940), and Orenthal James Simpson (b.1947); and golf's Robert Tyre "Bobby" Jones (1902–71) and Mildred "Babe" Didrikson Zaharias (1914–56). William Tatum "Bill" Tilden (1893–1953) and Billie Jean (Moffitt) King (b.1943) have starred in tennis; Joe Louis (Joseph Louis Barrow, 1914–81) and Muhammad Ali (Cassius Marcellus Clay, b.1942) in boxing; William Felton "Bill" Russell (b.1934) and Wilton Norman "Wilt" Chamberlain (b.1936) in basketball; Mark Spitz (b.1950) in swimming; Eric Heiden (b.1958) in speed skating; and Jesse Owens (1913–80) in track and field.

⁴⁹DEPENDENCIES

As of January 1988, US dependencies, in addition to those listed below, included American Samoa, Guam, Midway, Wake Island, and the Northern Mariana Islands; see the *Asia* volume. Sovereignty over the Panama Canal Zone was transferred to Panama on 1 October 1979; the canal itself will not revert to effective Panamanian control until 31 December 1999.

Navassa

Navassa, a 5-sq-km (2-sq-mi) island between Jamaica and Haiti, was claimed by the US under the Guano Act of 1856. The island, located at 18°24′N and 75°1′W, is uninhabited except for a lighthouse station under the administration of the coast guard.

Puerto Rico

Puerto Rico—total area 8,897 sq km (3,435 sq mi)—is the smallest and most easterly of the Greater Antilles, which screen the Caribbean Sea from the Atlantic proper. It lies between 17°51′ and 18°31′N and 65°13′ and 67°56′W, being separated from the Dominican Republic on the island of Hispaniola to the w by the Mona Passage, 121 km (75 mi) wide, and from the Virgin Islands on the E by Vieques Sound and the Virgin Passage. Roughly rectangular, the main island of Puerto Rico extends 179 km (111 mi) E–W and 58 km (36 mi) N–S. It is crossed from east to west by mountain ranges, the most prominent being the Cordillera Central, rising to nearly 1,338 m (4,390 ft). The coastal plain is about 24 km (15 mi) wide at its broadest point, and approximately one-third of the island's land is arable. About 50 short rivers flow rapidly to the sea. Islands off the coast include Mona and Desecheo to the W and Vieques and Culebra to the E. The mildly tropical climate is moderated by the surrounding sea, and seasonal variations are slight. The prevailing winds are the northeast trades. In San Juan on the northern coast, mean temperatures range from 24°C (75°F) for January to 27°C (81°F) for July. Mean annual rainfall varies from 74 cm (29 in) on the south coast to 150 cm (59 in) in San Juan and may total more than 380 cm (150 in) on the northern mountain slopes in the interior. Tropical fruits and other vegetation abound. As of 1991, endangered species on the island included the Puerto Rican plain pigeon, Puerto Rican parrot, Puerto Rican boa, giant anole, and hawksbill, leatherback, olive ridley, and green sea turtles.

The official census figure for 1990 was 3,522,039; the 1991 estimate was 3,551,000. In 1990, San Juan, the capital, had a population of 415,000, with a metropolitan area of more than 1 million. The population has more than doubled since 1930, despite extensive migration to the US mainland. Improved economic conditions on the island and diminishing opportunities in the US had slowed the trend by 1970; net out-migration was 493,000 during 1950–60, 200,000 during 1960–70, and 65,000 during 1970–80. However, out-migration increased to 157,300 during 1980–85. About three million Puerto Ricans reside in the US. Thousands commute annually between Puerto Rico and the US.

Many Puerto Ricans are of mixed black and Spanish ancestry; nearly all of the Amerindian inhabitants were exterminated in the 16th century. Spanish is the official language, but many Puerto Ricans also speak English, which is required as a second language in the schools. The Roman Catholic religion is predominant, but evangelical Protestant sects also have wide followings.

San Juan is the busiest commercial air center in the Caribbean, and there is excellent air service to New York, Miami, other points in the Caribbean, and Latin America. More than 40 steamship companies provide overseas freight and passenger service; San Juan, Ponce, and Mayagüez are the principal ports. In 1993 there were 13,762 km (8,546 mi) of paved highway; trucks carry the bulk of overland freight.

Archaeological finds indicate that at least three Amerindian cultures settled on the island now known as Puerto Rico, long before its European discovery by Christopher Columbus on 19 November 1493. The first group, belonging to the Archaic Culture, are believed to have come from Florida. Having no knowledge of agriculture or pottery, they relied on the products of the sea; their remains have been found mostly in caves. The second group, the Igneri, came from northern South America. Descended from South American Arawak stock, the Igneri brought agriculture and pottery to the island; their remains are found mostly in the coastal areas. The third culture, the Taíno, also of Arawak origin, combined fishing with agriculture. A peaceful, sedentary tribe, the Taíno were adept at stonework and lived in many parts of the island; to these Amerindians, the island was known as Borinquén.

Columbus, accompanied by a young nobleman named Juan Ponce de León, landed at the western end of the island—which he called San Juan Bautista (St. John the Baptist)—and claimed it for Spain. Not until colonization was well under way would the island acquire the name Puerto Rico ("rich port"), with the name San Juan Bautista applied to the capital city. The first settlers arrived on 12 August 1508, under the able leadership of Ponce de

León, who sought to transplant and adapt Spanish civilization to Puerto Rico's tropical habitat. The small contingent of Spaniards compelled the Taíno, numbering perhaps 30,000, to mine for gold; the rigors of forced labor and the losses from rebellion reduced the Taíno population to about 4,000 by 1514, by which time the mines were nearly depleted. With the introduction of slaves from Africa, sugarcane growing became the leading economic activity.

Puerto Rico was briefly held by the English in 1598, and San Juan was besieged by the Dutch in 1625; otherwise, Spanish rule continued until the latter part of the 19th century. The island was captured by US forces during the Spanish-American War, and under the Treaty of Paris (December 1898) Puerto Rico was ceded outright to the US. It remained under direct military rule until 1900, when the US Congress established an administration with a governor and an executive council, appointed by the US president, and a popularly elected House of Delegates. In 1917, Puerto Ricans were granted US citizenship.

In 1947, Congress provided for popular election of the governor, and in 1948, Luis Muñoz Marín was elected to that office. A congressional act of 1950, affirmed by popular vote in the island on 4 June 1951, granted Puerto Rico the right to draft its own constitution. The constitution was ratified by popular referendum on 3 March 1952. Puerto Rico's new status as a free commonwealth voluntarily associated with the US became effective on 25 July. The commonwealth status was upheld in a plebiscite in 1967, with 60.5% voting for continuation of the commonwealth and 38.9% for Puerto Rican statehood. In 1993 the plebiscite vote drew nearly 1.7 million voters or 73.6 percent of those eligible. The voters choose to keep the commonwealth status 48.4% to 46.2% for statehood, and 4.4% for independence.

The Commonwealth of Puerto Rico enjoys almost complete internal autonomy. The chief executive is the governor, elected by popular vote to a four-year term. The legislature consists of a 27-member Senate and 51-member House of Representatives elected by popular vote to four-year terms. The Supreme Court and lower courts are tied in with the US federal judiciary, and appeals from Puerto Rican courts may be carried as far as the US Supreme Court.

The Popular Democratic Party (PDP) was the dominant political party until 1968, when Luis A. Ferré, a New Progressive Party (NPP) candidate, who had supported the statehood position in the 1967 plebiscite, won the governorship. The NPP also won control of the House, while the PDP retained the Senate. The PDP returned to power in 1972 but lost to the NPP in 1976 and again, by a very narrow margin, in 1980; in 1984, it took roughly two-to-one majorities in both houses. The PDP has remained in control of the government in each election after 1984. The party is pro-commonwealth. There is a small but vocal independence movement, divided into two wings: the moderates, favoring social democracy, and the radicals, supporting close ties with the Fidel Castro regime in Cuba.

For more than 400 years, the island's economy was based almost exclusively on sugar. Since 1947, agriculture has been diversified, and a thriving manufacturing industry has been established; since 1956 there has been increasing emphasis on hotel building to encourage the expansion of the tourist industry. By 1990/91, the gross commonwealth domestic product reached $30,000 million, up from $15,794 million in 1985/86. In 1955, income from manufacturing was nearly equal to income from agriculture; in 1990/91 manufacturing generated $12 billion, or 40% of domestic output. By 1985/86, the gross commonwealth product reached $15,794 million, up from $7,129 million in 1975. In 1955, income from manufacturing was nearly equal to income from agriculture, but by 1970, it was roughly five times that of agriculture; in 1984/85, net income from manufacturing was estimated at $6,886 million, while net income from agriculture was only $418 million. The leading industrial products were pharmaceuticals, clothing, electrical machinery, rum, and processed foods. Sugar processing, once the dominant industry, now plays a lesser role. In 1952, there were only 82 labor-intensive plants on the island. By 1990 there were 2,000 plants—most capital intensive in Puerto Rico.

US taxes do not apply in Puerto Rico, since the commonwealth is not represented in Congress. New or expanding manufacturing and hotel enterprises are granted exemptions of varying lengths and degrees from income taxes and municipal levies. In 1940, when annual income per capita was $118, agricultural workers made as little as 6 cents an hour, and the illiteracy rate was 70%. By 1990/91, however, personal income per capita was $6,330, and the illiteracy rate was about 10.9%. Nevertheless, inequalities of income continue to plague Puerto Rico, and some 60% of the population was estimated to be living in poverty in 1985. Unemployment was 15.2% in 1991.

In 1991, Puerto Rico's exports totaled $21.3 million, imports totaled $15.9 million. In the 1990 season, an estimated 3,500,000 tourists visited Puerto Rico, spending nearly $1.3 billion.

Education was allotted 20% of total government expenditures in 1989/90, and health and welfare were allotted another 26%. In 1989/90, 651,225 pupils were enrolled in public day schools and 145,768 in accredited private schools. Attendance at the University of Puerto Rico reached nearly 55,626 in 1989/90. Other institutions of higher learning are the Catholic University in Ponce and the Inter-American University in San Germán.

In 1985, 854,000 telephones were in operation; as of 1990, 100 radio stations (50 AM, 63 FM) and 9 television stations were broadcasting regularly. The three largest Spanish-language daily newspapers, all from San Juan, are *El Vocero de Puerto Rico* (259,129 daily circulation in 1992), *El Nuevo Día* (228,685), and *El Mundo* (97,236).

Virgin Islands of the United States

The Virgin Islands of the United States lie about 64 km (40 mi) n of Puerto Rico and 1,600 km (1,000 mi) sse of Miami, between 17°40′ and 18°25′n and 64°34′ and 65°3′n. The island group extends 82 km (51 mi) n–s and 80 km (50 mi) e–w with a total area of at least 353 sq km (136 sq mi). Only 3 of the more than 50 islands and cays are of significant size: St. Croix, 218 sq km (84 sq mi) in area; St. Thomas, 83 sq km (32 sq mi); and St. John, 52 sq km (20 sq mi). The territorial capital, Charlotte Amalie, on St. Thomas, has one of the finest harbors in the Caribbean.

St. Croix is relatively flat, with a terrain suitable for sugarcane cultivation. St. Thomas is mountainous and little cultivated, but it has many snug harbors. St. John, also mountainous, has fine beaches and lush vegetation; about two-thirds of St. John's area has been declared a national park. The subtropical climate, with temperatures ranging from 21° to 32°c (70–90°f) and an average temperature of 25°c (77°f), is moderated by northeast trade winds. Rainfall, the main source of fresh water, varies widely, and severe droughts are frequent. The average yearly rainfall is 114 cm (45 in), mostly during the summer months.

The population of the US Virgin Islands was 96,569 at the time of the 1980 census, about triple the 1960 census total of 32,099, and was estimated at 98,942 in 1992. St. Croix has two principal towns: Christiansted and Frederiksted. Economic development has brought an influx of new residents, mainly from Puerto Rico, other Caribbean islands, and the US mainland. Most of the permanent inhabitants are descendants of slaves who were brought from Africa in the early days of Danish rule, and about 80% of the population is black. English is the official and most widely spoken language.

Some of the oldest religious congregations in the Western Hemisphere are located in the Virgin Islands. A Jewish synagogue

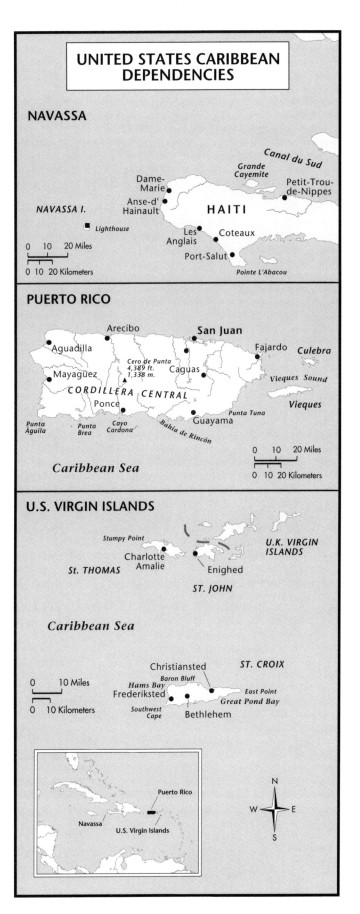

there is the second oldest in the New World, and the Lutheran Congregation of St. Thomas, founded in 1666, is one of the three oldest congregations in the US. As of 1993, Baptists made up an estimated 42% of the population, Roman Catholics 34%, and Episcopalians 17%.

In 1993 there were 856 km (531.6 mi) of roads in the US Virgin Islands. Cargo-shipping services operate from Baltimore, Jacksonville, and Miami via Puerto Rico. In addition, weekly shipping service is available from Miami. In 1991 arrivals of tourists who spent one night or more in a hotel or similar establishment numbered 374,090. Both St. Croix and St. Thomas have airports, with St. Croix's facility handling the larger number of jet flights from the continental US and Europe.

Excavations at St. Croix in the 1970s uncovered evidence of a civilization perhaps as ancient as AD 100. Christopher Columbus, who reached the islands in 1493, named them for the martyred virgin St. Ursula. At this time, St. Croix was inhabited by Carib Indians, who were eventually driven from the island by Spanish soldiers in 1555. During the 17th century, the archipelago was divided into two territorial units, one controlled by the British, the other (now the US Virgin Islands) controlled by Denmark. The separate history of the latter unit began with the settlement of St. Thomas by the Danish West India Company in 1672. St. John was claimed by the company in 1683, and St. Croix was purchased from France in 1733. The holdings of the company were taken over as a Danish crown colony in 1754. Sugarcane, cultivated by slave labor, was the backbone of the islands' prosperity in the 18th and early 19th centuries. After brutally suppressing several slave revolts. Denmark abolished slavery in the colony in 1848. A long period of economic decline followed, until Denmark sold the islands to the US in 1917 for $25 million. Congress granted US citizenship to the Virgin Islanders in 1927. In 1931, administration of the islands was transferred from the Department of the Navy to the Department of the Interior, and the first civilian governor was appointed. In the late 1970s, the Virgin Islands government began to consider ways to expand self-rule. A UN delegation in 1977 found little interest in independence, however, and a locally drafted constitution was voted down by the electorate in 1979.

The chief executive of the Virgin Islands is the territorial governor, elected by direct popular vote (prior to 1970, territorial governors were appointed by the US president). Constitutionally, the US Congress has plenary authority to legislate for the territory. Enactment of the Revised Organic Act of the Virgin Islands on 22 July 1954 vested local legislative power—subject to veto by the governor—in a unicameral legislature. Since 1972, the islands have sent one nonvoting representative to the US House of Representatives. Courts are under the US federal judiciary; the two federal district court judges are appointed by the US president. Territorial court judges, who preside over misdemeanor and traffic cases, are appointed by the governor and confirmed by the legislature. The district court has appellate jurisdiction over the territorial court.

Tourism has supplanted agriculture as the islands' principal economic activity. The number of tourists rose dramatically throughout the late 1960s and early 1970s, from 448,165 in 1964 to 1,116,127 in 1972/73, but has stagnated since that time. Today, tourism accounts for more than 70% of GDP and 70% of the employment. Rum remains an important manufacture, with petroleum refining (on St. Croix) a major addition in the late 1960s. Economic development is promoted by the US-government-owned Virgin Islands Corp. In 1987 the national product per capita was $11,000. The unemployment rate was 3.7% in 1992. Exports for 1990 totaled $2.8 billion while imports cost $3.3 billion. The island's primary export is refined petroleum products. Raw crude oil constitutes the Virgin's Island's most expensive import. Between 1970 and 1989 economic aid from the

nation's of the world totaled $42 million. In 1985, per capita income was $7,780, and the unemployment rate was 5.8%. Exports in 1985 totaled $3,357.1 million, of which petroleum products constituted 95%; imports of crude petroleum for refining accounted for 82% of the Virgin Islands' $3,740.6-million import bill. The total operating budget of the Virgin Islands government in 1985 was $263.3 million, of which taxes on personal and corporate income provided 50%.

The territorial Department of Health provides hospital and medical services, public health services, and veterinary medicine. Education is compulsory. The College of the Virgin Islands is the territory's first institution of higher learning. There are 58,931 telephones and 98,000 radios in use on the islands. In 1988 the Virgin Islands had 12 radio stations (4 AM, 8 FM) and 4 television stations. In 1983, the islands had eight radio stations and three television stations. There were 52,314 telephones in 1985.

50 BIBLIOGRAPHY

Ahlstrom, Sydney E. *A Religious History of the American People.* New Haven: Yale University Press, 1972.

Barone, Michael. *The Almanac of American Politics.* Washington, D.C.: Hational Journal, 1992.

Becker, Carl Lotus. *The Declaration of Independence: A Study in the History of Political Ideas.* New York: Knopg, 1960.

Bennett, Lerone. *Before the Mayflower: A History of Black America.* 6th ed. New York: Penguin, 1993.

Brown, Richard Maxwell. *No Duty to Retreat: Violence and Values in American History and Society.* New York: Oxford University Press, 1991.

Commager, Henry Steele (ed.). *Documents of American History.* Englewood Cliffs, N.J.: Prentice-Hall, 1988.

Glassborow, Jilly, Gillian Freeman (eds.). *Atlas of the United States.* New York: Macmillan, 1986.

Hart, James David (ed.). *Oxford Companion to American Literature.* New York: Oxford University Press, 1983.

Harvard Encyclopedia of American Ethnic Groups. Cambridge: Harvard University Press, 1980.

Josephy, Alvin M., Jr. *Now that the Buffalo's Gone: A Study of Todya's American Indians.* Norman, Okla.: Univ. of Oklahoma Press, 1984.

Kammen, Michale (ed.). *The Origins of the American Constitution: A Documentary History.* New York: Penguin, 1986.

———. *People of Paradox: An Inquiry Concerning the Origins of American Civilization.* New York: Oxford University Press, 1980.

Martis, Kenneth C. *The Historical Atlas of Political Parties in the United States Congress 1789–1989.* New York: Macmillan, 1989.

McNickle, D'Arcy. *Native American Tribalism: Indian Survivals and Renewals.* New York: Oxford University Press, 1993.

Mencken, Henry Louis. *The American Language.* New York: Knopf, 1963.

Morison, Samuel Eliot. *The Oxford History of the American People.* New York: New American Library, 1972.

Nevins, Allan. *Ordeal of the Union.* New York: Collier Books, 1992.

Reed, Carroll E. *Dialects of American English.* Amherst: University of Massachusetts Press, 1977.

Tocqueville, Alexis de. *Democracy in America.* New York: A. Knopf, 1994.

United States Government Manual. Washington, D.C.: US Government Printing Office, 1935-date.

US Bureau of the Census. *Historical Statistics of the United States, Colonial Times to 1970.* Washington, D.C.: US Government Printing Office, 1879-date.

Who's Who in America: A Biographical Dictionary of Notable Living Men and Women. Chicago: Marquis, 1899—.

URUGUAY

Oriental Republic of Uruguay
República Oriental del Uruguay

CAPITAL: Montevideo.

FLAG: The flag, approved in 1830, has four azure blue horizontal stripes on a white background; on a white canton is a golden sun with 16 rays (sometimes 20 rays are shown), alternately straight and wavy. This "Sun of May" symbolizes Uruguay's independence.

ANTHEM: *Himno Nacional,* which begins "Orientales, la patria o la tumba" ("Easterners [Uruguayans], our country or death").

MONETARY UNIT: The Uruguayan peso (UP), of 100 centésimos replaced the new peso in 1993 at the rate of UP1=1,000 new pesos. There are coins of 10, 20, and 50 centésimos and 1, 2, 5, and 10 new pesos, and notes of 1, 5, 10, 20, and 50 Uruguayan pesos. UP1 = $0.2133 ($1 = UP4.6880).

WEIGHTS AND MEASURES: The metric system is the legal standard, but some traditional measures are also used.

HOLIDAYS: New Year's Day, 1 January; Epiphany, 6 January; Landing of the 33, 19 April; Labor Day, 1 May; Battle of Las Piedras, 18 May; Birthday of Artigas, 19 June; Constitution Day, 18 July; Independence Day, 25 August; Columbus Day, 12 October; All Souls' Day, 2 November; Blessing of the Waters, 8 December; Christmas Day, 25 December.

TIME: 9 AM = noon GMT.

¹LOCATION, SIZE, AND EXTENT

The second-smallest South American country, Uruguay is situated in the southeastern part of the continent. It has an area of 176,220 sq km (68,039 sq mi), extending 555 km (345 mi) NNW–SSE and 504 km (313 mi) ENE–WSW. Comparatively, the area occupied by Uruguay is slightly smaller than the state of Washington. Bounded on the N and NE by Brazil, on the SE and S by the Atlantic Ocean, and on the W by Argentina, Uruguay has a total boundary length of 2,063 km (1,279 mi). The Uruguay River and the Río de la Plata separate Uruguay from Argentina. The Cuareim and Yaguarón rivers and the Laguna Merín separate it from Brazil.

Uruguay's capital city, Montevideo, is located in the southern part of the country on the Atlantic coast.

²TOPOGRAPHY

The general character of the land is undulating hills, with a few forest areas along the banks of the numerous streams. Southern Uruguay consists mostly of rolling plains and is an eastward extension of the Argentine pampas. The Atlantic coastline is fringed with tidal lakes and sand dunes. Low, unbroken stretches of level land line the banks of the two border rivers, the Uruguay and the Plata. The northern section is broken by occasional ridges and low ranges (cuchillas), alternating with broad valleys, and is a southern extension of Brazil. The highest point in the country, Mirador Nacional, near Montevideo, is 501 m (1,644 ft) above sea level. The most noteworthy feature of the northwest landscape is the Cuchilla de Haedo. The Cuchilla Grande runs northeastward from the southern region to the Brazilian border. The Negro, which rises in Brazil, crosses Uruguayan territory and flows into the Uruguay River, which separates Uruguay from Argentina.

³CLIMATE

The climate is temperate; the average temperature in June, the coolest month, is about 15°C (59°F), and the average for January, the warmest month, is 25°C (77°F). The weather is transitional between the weather of the humid Argentine pampas and that of southern Brazil. Rainfall is evenly distributed throughout the year; it averages about 109 cm (43 in), varying from 97 cm (38 in) in Montevideo to nearly 127 cm (50 in) farther north. There are from 120 to 180 sunny days a year. Frost is virtually unknown.

⁴FLORA AND FAUNA

Uruguay is primarily a grass-growing land, and the vegetation is essentially a continuation of the Argentine pampas. Forest areas are relatively small. The most useful hardwoods are algarobo, guayabo, quebracho, and urunday; other hardwoods include arazá, coronilla, espinillo, lapacho, lignum vitae, and nandubay. The acacia, alder, aloe, eucalyptus (imported from Australia), ombú, poplar, and willow are common softwoods. Palms are indigenous to the valleys. Rosemary, myrtle, scarlet-flowered ceibo, and mimosa are common. Most of the valleys are covered with aromatic shrubs, and the rolling hills with white and scarlet verbena.

Large animals have virtually disappeared from the eastern regions. The carpincho (water hog), fox, deer, nutria, otter, and small armadillo roam the northern foothills. On the pampas are the hornero (ovenbird), quail, partridge, and crow. The avestruz (a small ostrich similar to the Argentine rhea), swan, and royal duck are found at lagoons. Fish include pompano, salmon, and corvina. The principal reptiles are cross vipers and tortoises. Seals are found on Lobos Island, near Punta del Este.

⁵ENVIRONMENT

Air and water pollution are environmental concerns in Uruguay. Air pollution is a problem particularly in the larger population centers. Its primary sources are industry and an energy plant in Brazil. Water pollution from mining and industrial sources threatens the nation's water supply. Uruguay has 14.2 cu mi of water with 91% used for farming activity and 3% for industrial purposes. Ninety-five percent of the nation's rural people do not have pure water. The nation is also subject to other environmental problems due to drought, flooding, and fires. Erosion of the soil affects the nation's agricultural productivity. The nation's cities produce 0.5 million tons of solid waste per year. Forty percent of the urban population does not have adequate waste disposal facilities. Government agencies with environmental responsibilities include the Division of Environmental Health, within the Ministry of Public Health; the Ministry of Agriculture; and the Interior Ministry. As of 1994, 5 of Uruguay's mammal species and 11 of its bird species are endangered. Fourteen types of plants are threatened with extinction. Endangered species included the tundra peregrine falcon, glaucous macaw, two species of turtle (green sea and leatherback), and two species of crocodile (spectacled caiman and broad-nosed caiman).

6POPULATION

According to the 1985 census, the population of Uruguay was 2,955,241. It was estimated at 3,094,000 in 1990 and 3,179,000 in 1994. A population of 3,274,000 was projected for the year 2000, assuming a crude birthrate of 16.8 per 1,000 population and a crude death rate of 10.4 during 1995–2000. Uruguay's average annual population increase (0.56% during 1985–90) was the lowest in Latin America. In 1985, 44% of the population lived in Montevideo; in 1990, 88.8% of the estimated population was urban. The estimated population density in 1994 was 18 per sq km (46.7 per sq mi).

7MIGRATION

The basic sources of immigration to Uruguay were Spain and Italy. English, French, German, Slavic, and Swiss immigrants also settled in various departments (provinces). In the 1930s, restrictions were placed on immigration, and the importation of seasonal farm workers was stopped. There were 103,002 foreign-born people in 1985. Immigration came to 1,471 in 1990. Substantial emigration by Uruguayans for political or economic reasons occurred during the mid-1970s and early 1980s. Official figures suggest that about 180,000 Uruguayans left between 1963 and 1975 and 150,000 between 1975 and 1985. Most of them were young and, on average, better educated than the total population. Argentina was the main destination.

8ETHNIC GROUPS

The inhabitants of Uruguay are primarily (about 88%) white and of European origin, mostly Spanish and Italian; a small percentage is descended from Portuguese, English, and other Europeans. Mestizos (those of mixed white and Amerindian lineage) represent 8% of the population, and mulattoes and blacks about 4%. The indigenous Charrúa Amerindians were virtually wiped out early in the colonial era.

9LANGUAGES

Spanish is the official language. Uruguayan Spanish, like Argentine Spanish, has been somewhat modified by the Italians who migrated in large numbers to both countries. In general, the language of Uruguay is softer than that of Castile, and some words are different from those commonly used in Spain. The gauchos have influenced the language, particularly in words dealing with their way of life.

10RELIGIONS

Since 1919, church and state have been separated, and the constitution, as revised in 1966, guarantees religious freedom. Uruguay is the only Latin American nation that approaches religious pluralism. Contrasting with the 90% formal Roman Catholic affiliation in other Latin American countries, 78% of Uruguayans identified themselves as Roman Catholics in 1993, but many claimed no church membership. Only 3% of the population was Protestant, and there were some 24,000 Jews in 1990.

11TRANSPORTATION

After World War II, the Uruguayan government purchased the British-owned railroads and nationalized the entire system. The railroads are run by the State Railway Administration. Four main lines connect the western and northern areas with Montevideo. In 1991 there were 3,005 km (1,867 mi) of standard-gauge, government-owned track.

Highways have surpassed railroads as the principal means of conveyance of passengers and freight. In 1991 there were 49,900 km (31,007 mi) of roads, of which 13.4% were paved. The Investment and Economic Development Commission's 10-year plan (1965–74) provided about $87 million for highway construction and improvement. A five-year plan for transport and public works, covering the years 1983–87 and partly financed by the IBRD and IDB, provided for construction of 10,000 km (6,200 mi) of new roads. In 1986, the IDB approved a loan of $36 million to help finance a highway development project. Two sections of highway (Routes 1 and 5) in addition to a main artery funneling traffic into Montevideo were scheduled for improvement. In 1992, there were 350,000 motor vehicles, two-thirds of which were passenger cars.

Montevideo is the major Uruguayan port. Colonia and Nueva Palmira are free ports. There are some 1,600 km (994 mi) of inland waterways, of which the most important are the Plata and the Uruguay; the latter has a depth of 4.3 m (14 ft) as far as Paysandú. There were four vessels in the merchant fleet in 1991, with GRT of 59,000.

Carrasco, an airport 19 km (12 mi) from the center of Montevideo, is used by most international carriers between Europe, Brazil, and Argentina and serviced 821,000 passengers in 1991. A new airport was opened at Artigas in 1973. Frequent air service links Buenos Aires with Montevideo. The state-owned Primeras Líneas Uruguayas de Navegación Aérea (PLUNA) offers service to the principal departmental capitals as well as international flights.

12HISTORY

During the 16th century, only a few Spanish expeditions landed on the Banda Oriental, or east bank of the Uruguay River. Most of them were driven off by the native Charrúa Amerindians. Jesuit and Franciscan missionaries landed in 1624, and formed permanent settlements. By 1680, Portuguese from Brazil had founded Colonia do Sacramento as a rival to Buenos Aires, on the opposite bank of the estuary. Thereafter the area was a focal point for Spanish-Portuguese rivalry.

Montevideo was founded in 1726, and Uruguay became part of the viceroyalty of La Plata, which the Spaniards established in Buenos Aires in 1776. During the Napoleonic Wars, the British invaded the region of La Plata and captured Buenos Aires and Montevideo (1806–7), but they were forced out in 1807. After Buenos Aires refused to give Uruguay autonomy, the Uruguayan national hero, José Gervasio Artigas, declared Uruguay independent in 1815. A year later, Brazilians attacked Montevideo from the north, but Artigas led a revolutionary movement against them. The struggle continued from 1816 to 1820, when the Portuguese captured Montevideo and Artigas had to flee to Paraguay. Uruguay was annexed to Brazil in 1821 and was known as the Cisplatine Province.

On 25 August 1825, Juan Antonio Lavalleja, at the head of a group of patriots called the "treinta y tres orientales" ("33 easterners"), issued a declaration of independence. After a three-year fight, a peace treaty signed on 28 August 1828 guaranteed

Uruguay's independence. Disappointed in his hopes for the presidency, Lavalleja launched a series of rebellions. During this period of political turmoil and civil war, the two political parties around which Uruguayan history has traditionally revolved, the Colorados (reds) and the Blancos (whites), were founded. Uruguay's first President, Gen. José Fructuoso Rivera, an ally of Artigas, founded the Colorados. The second president, Brig. Gen. Manuel Oribe, a friend of Lavalleja, founded the Blancos.

The 19th century was largely a struggle between the two factions. Some measure of national unity was achieved in the 1860s. In 1865, Uruguay allied with Brazil and Argentina to defeat Paraguay in the Paraguayan War (1865–70), also known as the War of the Triple Alliance and the Triguarantine War. However, it was not until the election of José Batlle y Ordóñez as president in 1903 that Uruguay matured as a nation.

The Batlle administrations (1903–7, 1911–15) marked the period of greatest progress. A distinguished statesman, Batlle initiated the social welfare system codified in the Uruguayan constitution. From then on, Uruguay's social programs, funded primarily by earnings of beef and wool in foreign markets, gave Uruguay the sobriquet "Switzerland of South America."

After World War II, the Colorados ruled, except for an eight-year period from 1958–66. It was during the administration of President Jorge Pacheco Areco (1967–72) that Uruguay entered a political and social crisis. As wool declined in world markets, export earnings no longer kept pace with the need for greater social expenditures. Political instability resulted, most dramatically in the emergence of Uruguay's National Liberation Movement, popularly known as the Tupamaros. This well-organized urban guerrilla movement mounted a campaign of kidnaping, assassination and bank robbery while espousing Marxist and nationalist ideals.

In November 1971, Colorado candidate Juan María Bordaberry Arocena was elected president, and the Colorados retained control of the Senate and Chamber of Deputies. After Bordaberry assumed office in March 1972, the Tupamaros ended a temporary truce and returned to the offensive. Their activities, coupled with the worsening economic situation, exacerbated Uruguay's political uncertainty. Gradually, the military assumed a greater role in government, and by 1973 was in control of the system. By the end of 1973, the Tupamaros had been crushed.

Military officers were named in 1974 to head all state-run enterprises, including the Central Bank. The 1966 constitution was suspended, and political activity was banned. Military leaders ousted Bordaberry from the presidency in 1976, because of his refusal to restore constitutional rule, and they named a new president, Aparicio Méndez Manfredini. Uruguay was denounced by the OAS and other international organizations for human rights violations; in 1979, Amnesty International estimated the number of political prisoners at 6,000. In mid-1981, the military government began to negotiate with leaders of the traditional parties, and in September 1981, a "transitional" president, Georgio Álvarez Armellino, was installed.

Intraparty elections took place in November 1982, followed by legislative and presidential voting in November 1984. The moderate government of Colorado candidate Julio María Sanguinetti Cairolo took office in March 1985. Lacking a majority in Congress, Sanguinetti worked closely with other political leaders to reach a consensus on major issues.

One of the first acts of the new government was to release all political prisoners. Another was to declare amnesty for former military and police leaders accused of human rights violations. In an attempt to reconcile warring factions, the government initiated a "social dialogue" with employers and union leaders to reduce social tension. However, slow progress on the economic front led to the 1989 election of the Blanco candidate, Luis Alberto Lacalle.

URUGUAY

0   40   80   120 Miles

0   40   80   120 Kilometers

ARGENTINA

BRAZIL

Embalse del Salto Grande

Artigas
Cuareim

Chajari

Rivera

Arapey Grande

Salto

Daymán

Tacuarembó

Aceguá

CUCHILLA DE HAEDO

SANTA ANA HILLS

Tacuarí

Queguay Grande

Achar

Negro

Melo

Paysandú

Laguna Baygorria

Laguna Rincón del Bonete

Laguna Palmar

Yaguarón

Treinta y Tres

Laguna Merín

Durazno

Mercedes

Yi

CUCHILLA GRANDE

Cebollatí

San Salvador

Trinidad

San José

Florida

Santa Lucía

Laguna Negra

Carmelo

Mt. Catedral 1,683 ft. 513 m.

Colonia

San José

Canelones

Minas

Rocha

Las Piedras

Pando

San Carlos

Montevideo

Maldonado

Río de la Plata

ATLANTIC

OCEAN

Uruguay

LOCATION: 30°06′ to 35°02′s; 53°05′ to 59°29′w. **BOUNDARY LENGTHS:** Brazil, 1,003 km (623 mi); Atlantic coastline, 565 km (351 mi); Argentina, 495 km (308 mi). **TERRITORIAL SEA LIMIT:** 200 mi.

Lacalle engaged in an ambitious attempt to liberalize the Uruguayan economy. He emphasized deficit reduction, reforms in education, labor, and the civil service, as well as the privatization of state enterprises. However, these plans were dealt a serious blow in 1993, when a plebiscite failed to ratify a set of proposals for liberalization. After almost a century, Batlle's system is still alive, and Uruguay remains resistant to change.

13GOVERNMENT

The constitution of 1830 underwent numerous revisions, notably in 1917, 1934, 1952, and 1966. This constitution provided for a republican government, divided into three branches: executive, legislative, and judicial. From 1951 to 1966, the executive consisted of a colegiado, or council, of nine ministers, six from the majority party and three from the minority. In the 1966 elections, however, the electorate reinstated the positions of president and a vice-president, popularly elected for a five-year term, together with a council of ministers.

According to the constitutional revision of 1966, the Congress (or General Assembly) consisted of the Senate and the Chamber of Deputies. The Senate had 30 popularly elected senators, plus the vice-president of the republic as the thirty-first voting member. The Chamber of Deputies had 99 deputies, popularly elected by departments (provinces). The right to vote was extended to all citizens 18 years of age or older, with female suffrage in local elections as early as 1919 and in national elections in 1934.

From June 1973, when President Bordaberry dissolved the Assembly and suspended the constitution, until March 1985, Uruguay was ruled by executive decree, subject to veto by the military, with legislative functions carried out by the 25-member Council of State, appointed by the executive. A new constitution, providing for the permanent participation of the armed forces in government by means of a National Security Council, was drafted by the Council of State but rejected by 57.2% of the voters in a referendum on 30 November 1980. In March 1985, democracy was restored under President Sanguinetti; in July, the government set up a National Constituent Assembly to devise constitutional reforms that would be submitted to the electorate for ratification.

¹⁴POLITICAL PARTIES

Uruguay has Latin America's oldest two-party system. The Colorados (reds) and Blancos (whites), formed during the conflicts of the 1830s and 1840s, persisted into the 1990s. The Colorados are traditional Latin American liberals, representing urban business interest, and favoring limitation on the power of the Catholic Church. The Blancos (officially called the National Party) are conservatives, defenders of large landowners and the Church.

For more than 90 years, until the 1958 elections, the executive power was controlled by the Colorados. Under such leaders as Batlle, the party promulgated a progressive program advocating public education, advanced labor laws, government ownership of public utilities, and separation of church and state. After eight years of Blanco government, the Colorado party regained power in the 1966 election.

The results of the November 1971 balloting were so close that the final tabulation took more than two months to ascertain; the Colorados won 36.3% of the vote, the Blancos 35.7%, and the Broad Front (Frente Amplio, a left-wing coalition that included the Tupamaros) 16.6%. These three groups, plus the Christian Democratic Party (Partido Democrático Cristiano—PDC), formed in 1962 from the former Catholic Civic Union, made up the Uruguayan party system at the time of the military takeover.

Political activities were suspended following the constitutional crisis of June 1973, and in December 1973, the Communist and Socialist parties were outlawed. In June 1980, the military began to liberalize, as they permitted political meetings of nonleftist groups. In November 1982, they allowed for intraparty elections in each of three parties: the Colorados, the National Party (Blancos), and the small Civic Union (an outgrowth of the Christian Democrats). In the voting, party candidates who had campaigned against the military's proposed constitution in 1980 took more than 60% of the vote.

Neither Blanco leader Wilson Ferreira Aldunate nor Broad Front leader Líber Seregni Mosquera was allowed to participate in the elections, but both retained their party posts. In the November 1984 elections, Colorado candidate Julio María Sanguinetti Cairolo won the presidency with 38.6% of the vote. The Colorados also won pluralities in the Chamber of Deputies and the Senate. Aldunate and Seregni frequently consulted with President Sanguinetti, and previously outlawed parties were legalized. In 1985, the PDC and FIDEL belonged to the Broad Front, and the National Liberation Movement (Movimiento de Liberación Nacional—MLN), also known as the Tupamaros, reconstituted years after their destruction in 1973, announced its intention to give up violence and join the Broad Front as a legal party.

In 1989, Blanco candidate Lacalle took 37% of the vote. Jorge Batlle, of the Colorado party, unable to capitalize on name recognition, received 29%, while Liber Seregni Mosquera of the Broad Front received 20%. The Blancos also carried a plurality in each house of the legislature, followed respectively by the Colorados, the Broad Front and the "New Space" Coalition, which consists of the PDC and the Civic Union.

¹⁵LOCAL GOVERNMENT

Uruguay territorially is divided into 19 departments (provinces). Under the 1966 constitutional revision, each department had a unicameral legislature, but all 19 legislatures were dissolved by President Bordaberry on 28 June 1973. Following the installation of the democratic government in 1985, the departments returned to their pre-1973 status of limited autonomy under the central government.

¹⁶JUDICIAL SYSTEM

Most of the nation's legal system was suspended in 1973, but in 1981, the military government restored the independence of the civilian judiciary. In that year, a Supreme Judicial Council was empowered to name Supreme Court justices and supervise the judiciary. Below the Supreme Court are appellate courts and lower civil and criminal courts, justices of the peace, electoral and administrative courts, and an accounts court. A parallel military court system operates under its own procedure. When the Supreme Court hears cases involving the military, two military justices join the Court. Civilians are tried in the military court only in time of war or insurrection. The judiciary is structurally independent of the executive and this separation of powers is respected in practice.

¹⁷ARMED FORCES

The armed forces numbered 24,700 in 1993. The active army consists of volunteers between the ages of 18 and 45 who contract for one or two years of service. The army numbered 17,200, organized in 4 regional divisions of 10 brigades; the navy (including naval air force and infantry) numbered 4,500 with 3 frigates and 8 patrol and coastal combatants. The air force had 3,000 men and 37 combat aircraft. Since 1985, the military has redirected its attention to more traditional concerns of national defense and to reducing its size from 32,000 to 24,700. Defense spending was $168 million (1988) or 2.2% of GDP.

¹⁸INTERNATIONAL COOPERATION

Uruguay is a charter member of the UN, having joined on 18 December 1945, and belongs to ECLAC and all the nonregional specialized agencies except IDA. It is also a member of the G-77, IDB, OAS, and PAHO. In 1927, the Inter-American Child Institute was established in Montevideo as a center for social action, study, and information in the Americas. Under the direction of the OAS, it has given particular attention to health and nutrition. In 1961, Montevideo became the seat of the Latin American Free Trade Association, which was superseded by LAIA in 1981. Uruguay is also a signatory of GATT, the Law of the Sea, and the Treaty for the Prohibition of Nuclear Weapons in Latin America (1967).

Uruguay and Argentina signed an agreement in 1973 guaranteeing free passage in perpetuity on the Río de la Plata. In 1985, the presidents of the two countries signed an agreement "to promote the economic and social integration of the two countries."

¹⁹ECONOMY

Uruguay's economy is based on the production and processing of agricultural commodities. In 1986, exports of wool and meat accounted for 39% of the total goods sold outside the country.

Owing partly to the need for industrial products generated by World War II and the Korean War and also to government efforts to attain a fair degree of economic self-sufficiency and to improve foreign trade, the value of Uruguay's industrial output doubled between 1936 and 1960. Between 1960 and 1970, the industrial growth rate leveled off, as the limitations of an import-substitution strategy became apparent. Encouragement of nontraditional manufactured exports led to industrial production increases averaging 6% annually during 1973–79. The lack of natural resources obliges Uruguay to import most raw materials needed by its industries.

Since 1982 the economic situation has been characterized by an uninterrupted fall in real output, low levels of investment, high unemployment, mounting inflation, a severe imbalance in public finance, and a massive accumulation of arrears in private-sector debt to the domestic banking system, which in turn has caused a potentially critical situation in the country's financial institutions. Moreover, export earnings declined persistently between 1982 and 1985. The GDP (discounting inflation) rose by 0.7% for the entire year (after falling by 9.8%, 4.7%, and 1.8% in 1982, 1983, and 1984, respectively). Inflation dipped slightly in 1986, registering 74% as compared with 83% in 1985. The new government, in mid-1985, was able to negotiate agreements with the IMF and creditor banks that produced a standby credit from the fund, a renegotiation of debt with foreign banks, and an economic-financial program with the IMF intended to reduce the public-sector deficit, inflation, and the money supply and to remedy the balance-of-payments disequilibrium. Some measure of success was achieved by 1987.

However, strong economic growth was most significant in 1992. Real GDP grew by 7.4% as a result of a successful policy reform. There was a strong external demand and substantial capital inflows. Tourism, in particular, contributed to the economy. In 1993, real GDP grew by only 1.5%. Inactivity in oil refining over most of the year resulted in a decrease in growth. The production of cattle and crops, mainly wheat and barley, increased substantially, causing agricultural total output to recover rapidly. Brazil's demand for Uruguayan rice rose, causing rice output to report gains as well. Also, Uruguay experienced an expansion in food processing and textiles. Manufacturing industries experienced a decrease in 1993 stemming from the fall in Argentinean demand. Construction was the most dynamic sector. Inflation in 1993 was 52.8%, slightly lower than in 1992 (58.9%).

[20]INCOME

In 1992, the GNP was $10,444 million at current prices, or $3,340 per capita. For the period 1985–92, the average inflation rate was 78.6%, resulting in a real growth rate in per capita GNP of 2.9%.

In 1992, the GDP was $11,405 million in current US dollars. It is estimated that in 1991, agriculture, hunting, forestry, and fishing contributed 10% to GDP; mining and quarrying, less than 1%; manufacturing, 25%; electricity, gas, and water, 3%; construction, 4%; wholesale and retail trade, 12%; transport, storage, and communication, 6%; finance, insurance, real estate, and business services, 26%; community, social, and personal services, 10%; and other sources 4%.

[21]LABOR

About 1,380,000 Uruguayans, or an estimated 44% of the population, were in the civilian labor force in 1992. In that year, manufacturing and construction accounted for 28% of the labor force; agriculture, 5%; and services, 67%. The unemployment rate rose from 8.94% in 1991 to 9.03% in 1992.

The 8-hour day and 48-hour week were instituted in 1915. Civil servants and employees of state-run businesses and other public services have a 30-hour week. The law provides for one day of rest after every six days of work and grants holidays with pay, plus an annual vacation bonus.

In 1943, industrial wage boards with seven members (three for the government, two for the employers, and two for the employees) were established to fix minimum wages and settle wage disputes. Because the wage boards (consejos de salarios) were slow in reaching decisions, the Uruguayan labor force tended to use the strike as a first resort rather than as a last one, to force the initiation of negotiations. Since December 1968, wages and prices have been controlled by the Price and Wage Commission. Unemployment and dismissal compensation, old age and liability pensions, workers' accident compensation, and family allowances are provided by law.

In July 1973, the National Workers' Convention, which claimed 400,000 members, was declared illegal. Laws enacted in August 1974 restricted trade-union membership to "free and nonpolitical" trade unions. Political activity by union officials was banned, as were strikes in the public sector, health, and commerce. Labor conditions returned to pre-1973 conditions with the 1985 government changeover. The sudden release of years of frustration triggered some 250 strikes in the ensuing year. As of March 1993, seven general strikes had occurred within the previous three years, for a ratio of one hour of general strike for every 114 worked. Uruguay's sole labor confederation, the Inter-Union Workers Assembly-National Federation of Workers (PIT-CNT) had a membership of 155,000 at the end of 1992, down over 21% from the previous year.

[22]AGRICULTURE

Uruguay has a primarily agricultural and pastoral economy, but the importance of these sectors has been declining. Agriculture and animal husbandry together contributed 10.8% of the GDP in 1992, compared with 13.6% in 1985, and engaged about 4.5% of the economically active population in 1991. About 77% of Uruguay's land area is devoted to stock raising and 7.4% to the cultivation of crops. In pasturage, large farms predominate, with farms of more than 1,000 hectares (2,500 acres) accounting for two-thirds of all farmland. Crops are grown mainly on small farms of less than 100 hectares (250 acres).

The principal crops (in thousands of tons) are listed below:

	1990	1991	1992
Sugarcane	506	509	430
Rice	347	493	622
Wheat	539	201	265
Sugar beets	208	156	160
Potatoes	174	196	155
Sorghum	59	90	126
Barley	133	138	214
Corn	112	124	82
Oats	43	51	26
Sunflowers	45	57	41
Linseed	1	6	4

[23]ANIMAL HUSBANDRY

Livestock is the basis of the economy. As of 1992, Uruguay surpassed Argentina and now has more sheep than any other South American country. The production costs of stock raising are low, and the quality of the product is generally high. Hereford, Shorthorn, and Aberdeen Angus breeds account for 90% of all beef cattle, with Hereford the most numerous. Corriedale represents about 70% of the sheep stock, followed by Ideal (11%).

Uruguay is especially suited to the raising of sheep and cattle. In 1992 there were 9.5 million head of cattle, up from 8.7 million in 1990. In 1992, slaughtered livestock amounted to 360,000 tons (compared with 341,000 tons in 1981); about 82% was used for domestic consumption and 18% for export. Sheep (25.7 million in 1992) are raised mainly for wool. Other livestock in

1992 included 475,000 horses and 215,000 hogs. Milk production in 1992 reached 1,100,000 tons. Exports of beef and dairy products amounted to $135.3 million and $60.7 million, respectively, in 1991. Wool exports totaled $430.7 million that year. Leather exports in 1990 were valued at over $116.2 million, contributing 7% to total exports.

The Ministry of Agriculture and Fisheries is responsible for stock raising and breeding, control of animal diseases, and improvement of existing grassland and arable resources. The National Meat Board acts as consultant to the government. The meat-packing industry, taken over by the government in 1958, has been restored to the private sector. The government encourages local production through a system of special licenses or customs documents for imported meat and livestock. Furthermore, imports of bull semen and embryos are numerically restricted and must comply with animal health requirements.

24FISHING

Fishing underwent rapid growth in the 1970s. The government-promoted fishing industry made an average annual catch of about 6,000 tons in the 1960s; the catch increased to 20,600 tons in 1972 and, despite temporary setbacks, to 143,170 tons by 1991. Fish exports in 1991 amounted to $101.6 million.

There are three fishing zones on the southern coast: the low zone, from Colonia to Piriápolis; the middle zone, from Piriápolis to Punta del Este, which is considered one of the finest fishing areas in the world; and the high zone, from Punta del Este to the Brazilian border. Important sea fish are corvina negra (a kind of bass), mullet, sole, anchovy, mackerel, whiting, and shark. The finest freshwater fish is the dorado, a type of salmon.

25FORESTRY

Uruguay has some 669,000 hectares (1,653,000 acres) of forestland. About 8,200 hectares (20,300 acres) were reforested annually from 1981 to 1986. The principal species cultivated are eucalyptus and pine; domestic woods are used primarily for windbreaks, fence posts, and firewood. Lumber suitable for building and construction is imported. Roundwood removals totaled 3,829,000 cu m in 1991.

26MINING

Mineral resources are limited and undeveloped. There are deposits of manganese, iron, lead, and copper, and in the Rivera Department, gold is mined in small quantities. More important are stones, such as the marble quarried for local use, limestone, granite, quartz, gypsum, and dolomite. Sand, common stone, and talcum are exported. Agates, opals, and onyx are found in Salto and Artigas. All products of the subsoil belong to the state. In 1991, Uruguay exported clays, gravel, limestone, precious stones, and sands valued at $12 million.

27ENERGY AND POWER

Uruguay's power is provided by hydroelectric and diesel-generating plants. Supplying electricity for light, power, and traction has been a state monopoly since 1897. Uruguay's installed capacity was 1,795 Mw in 1991; total output was 7,017 million kwh, of which 87% came from hydroelectric sources. In the mid-1970s, the government imposed a mandatory program to curb power consumption because of rising fuel import costs.

The first hydroelectric plant, at the site of the Rincón del Bonete Dam on the Río Negro, was completed in 1949 and has a capacity of 128 Mw. The dam created the Lago Artificial del Río Negro, the largest artificial lake in South America. In 1960, a generating plant with a capacity of 108 Mw was completed at Rincón de Baygorría, also on the Río Negro. Several large hydroelectric projects were under way in the early 1980s. The Salto Grande Dam on the Río Uruguay, built with Argentina and having a potential capacity of

1,890 Mw, began producing electricity in 1979; as a result, Uruguay became a net exporter of electricity to Argentina in the mid-1980s. In 1991, Uruguay exported 26% of its generated electricity. The 300-Mw Palmar hydroelectric power station, financed and built by Brazil, started production in 1981 and was completed by the mid-1980s. In the mid-1980s there were also plans for a 150-Mw plant at Isla González on the Río Negro and another 150-Mw plant on the Río Cebollatí in eastern Uruguay.

By a bilateral agreement signed with the US, Uruguay is entitled to receive atomic equipment and to lease nuclear fuels. Negotiations on contracts with four US oil companies for offshore oil exploration were concluded by the Uruguayan government in February 1975, but no commercially exploitable resources were found. In 1991, the state-owned refinery at Montevideo processed 2.9 million barrels of distillate fuel oil, 2.6 million barrels of residual fuel oil, and 1.9 million barrels of gasoline. Consumption included 30,000 barrels per day of crude petroleum in 1992, when fuel and energy accounted for 11.7% ($241 million) of total imports.

28INDUSTRY

Although foreign trade depends mainly on agricultural production, the production of industrial goods for domestic consumption is increasing, primarily in the fields of textiles, rubber, glass, paper, electronics, chemicals, cement, light metallurgical manufactures, ceramics, and beverages. World War II spurred the industrial growth of Uruguay, and now local industry supplies most of the manufactured products used. Most industry is concentrated in and around Montevideo. In 1985, manufacturing accounted for 27.9% of GDP. The structure of industrial output in 1985, by value, was as follows: food, 29.4%; chemicals and petroleum derivatives, 28.5%; textiles, 9.4%; beverages, 4.9%; electrical appliances, 2.9%; tobacco, 2.6%; nonmetallic minerals, 2.5%; electrical appliances, 1.8%; and other sectors, 18%.

The transport equipment sector was affected by the trade liberalization measures of the automobile assembly industry in 1992. Construction, which was performing negatively in previous years, recovered as a result of strong investment, especially in private residential and commercial building.

29SCIENCE AND TECHNOLOGY

The UNESCO Regional Office for Science and Technology in Latin America and the Caribbean is located in Montevideo; its activities include teaching engineering, hydrology, earth sciences, marine sciences, and ecology. Learned societies include the Pediatrics Society, the Association of Uruguayan Engineers, and the Chemical and Pharmaceutical Association of Uruguay. The University of the Republic has faculties of agronomy, sciences, engineering, medicine, dentistry, chemistry, and veterinary medicine. The Institute of Higher Studies offers courses in biological climatology and mathematics. In 1987, 2,093 scientists and engineers were engaged in research and development.

30DOMESTIC TRADE

Although the Uruguayan population is small, it has a relatively high purchasing power. The reasonably good transportation system allows easy shipment of agricultural products to the capital. Chambers of commerce and other trade associations play an active role in interpreting local market demands. Normal business hours are 8:30 or 9 AM to 12 noon and 2 to 6:30 or 7 PM, Monday through Friday, and 9 AM to 12:30 PM, Saturday. Banks open on weekdays from 1 to 5 PM in Montevideo but from 7:30 to 11:30 AM in the interior.

31FOREIGN TRADE

Uruguay traditionally relies on foreign sales of wool, hides, and meat and meat products for its export revenues, which have been increasing over the past several years. In 1992, exports kept rising, amounting to $1,702 million, compared with $1,604.7 million in 1991. Imports followed the same trend, totaling $1,537 million in 1991 and $1,941.2 million in 1992. The trade deficit in 1993 reached $679 million with a 3.4% fall in exports. Traditional exports decreased by 15.9% as a consequence of the fall in the activity of the livestock sector, whereas non-traditional exports had an increase of 2.1%. Imports mostly grew in the consumer goods sector and, to a lesser extent, in the capital assets sector. Over the period there was a greater concentration of commercial interchange in the region (Argentina and Brazil), which accounts for 62% of the total deficit.

32BALANCE OF PAYMENTS

Uruguay derives its foreign exchange largely from raw material exports. Deficits are common, owing to fluctuating world markets for Uruguay's agricultural exports and high dependency on imports for raw materials and fuels. Traditionally, multilateral assistance, income from tourism, and inflow of capital from other Latin American countries have tended to offset the negative trade picture. The balance of payments in the early 1980s was affected by the continued weakening of international prices of Uruguay's exports. In 1991, the value of merchandise exports fell by 5.2% while merchandise imports increased 21%. The merchandise trade account consequently fell from a surplus of $351 million in 1990 to a deficit of $31.8 million in 1991. In 1992, the merchandise trade balance deteriorated further; imports increased by 26% due to increased domestic demand, reduced tariffs, and currency appreciation, so that the trade deficit exceeded 2% of GDP.

In 1992 merchandise exports totaled $1,702.5 million and imports $1,941.2 million. The merchandise trade balance was $-238.7 million. The following table summarizes Uruguay's balance of payments for 1991 and 1992 (in millions of US dollars):

	1991	1992
CURRENT ACCOUNT		
Goods, services, and income	2.4	–235.6
Unrequited transfers	40.1	28.6
TOTALS	42.5	207.0
CAPITAL ACCOUNT		
Direct investment	—	—
Portfolio investment	109.4	229.1
Other long-term capital	–292.6	126.9
Other short-term capital	–246.0	–253.9
Exceptional financing	75.0	50.7
Other liabilities	–2.2	—
Reserves	–145.9	–1.9
TOTALS	502.3	150.9
Errors and omissions	459.8	56.1
Total change in reserves	–59.9	–05.8

33BANKING AND SECURITIES

The Bank of the Republic, established in 1896, is a state bank with a government-appointed director. It operates as both a public and a private bank. It is the financial agent of the government; it also acts as an autonomous agency and, as a commercial bank, makes loans and receives deposits. It participates in determining financial policies and the allocation of foreign exchange for imports. One of its main functions is to provide rural credit.

The 1966 constitutional revision created a Central Bank, which is responsible for currency circulation, thus permitting the Bank of the Republic to concentrate on public and private credit. The third state bank is the Mortgage Bank. In 1985 there were also 12 principal commercial banks and 11 foreign banks.

Currency in circulation amounted to UP22 million at the end of 1993. At the end of 1993, claims on the central government totaled UP5,815.7 million. At that time, time and savings, and foreign currency deposits amounted to UP15,641.5 billion in the private banking system. A policy of regular minidevaluations was introduced in mid-1975; currency stability was established in the late 1970s, but in November 1982, the peso, regarded as overvalued, was allowed to float freely. Tax-free offshore banking was authorized in August 1982.

The Montevideo Stock Exchange is small but highly active. In the 1960s, prices steadily declined, owing to the growing shortage of investment money, the large outstanding debts of Uruguayan banks, and a shift of private investment to government bonds. In the 1970s, however, securities transactions followed the general inflationary trend.

34INSURANCE

The State Insurance Bank, launched early in 1912 by a bond issue, was granted a monopoly on all insurance in 1919; however, private insurance companies were allowed to continue issuing life, fire, and marine insurance if they had been in business prior to creation of the state bank. In 1990, premiums totaled $US41.8 or 2.1% of the GDP. An October 1993 law demonopolized the insurance industry.

35PUBLIC FINANCE

Under inflationary pressures, budgetary expenditures have generally exceeded revenues, although the picture improved in the late 1970s, mainly as a result of higher taxes and customs receipts. The budget deficit in 1975 represented 25% of total expenditures; this ratio fell to 9% in 1978 and 0.5% in 1981, then rose to 36.4% in 1982 and declined to 16% in 1985.

When civilian rule was restored in 1985, the IMF sponsored a new standby adjustment program, which rescheduled $2.1 billion of medium- and long-term commercial bank debt. Recovery stagnated from 1988 to 1990, due to labor disputes, drought, and economic instabilities in Argentina and Brazil. The public sector deficit thus increased to 7.6% of GDP in 1989. Public sector deficit reduction reforms enacted in 1990 reduced the deficit to 3.6% of GDP in 1990, and to 1.1% in 1991.

The following table shows actual revenues and expenditures for 1990 and 1991 (in millions of pesos):

	1990	1991
REVENUE AND GRANTS		
Tax revenue	2,461	5,510
Non-tax revenue	132	238
Capital revenue	–5	20
Grants	—	—
TOTAL	2,598	5,768
EXPENDITURES & LENDING MINUS REPAYMENTS		
General public service	175	467
Defense	233	363
Public order and safety	135	249
Education	187	376
Health	114	275
Social security and welfare	1,274	2,993
Housing and community amenities	2	9
Recreation, cultural, and religious affairs	14	21
Economic affairs and services	222	425
Other expenditures	706	552
Adjustments	—	—
Lending minus repayments	23	34
TOTAL	2,562	5,584
Deficit/Surplus	36	184

In 1991, Uruguay's total public debt stood at UP5,560 million, of which UP4,384 million was financed abroad.

36TAXATION

Personal income and inheritance taxes were repealed in 1974; the only individual direct taxes are a levy on farm profits and a net worth tax. The most important indirect tax is the value-added tax, which consists of a basic rate of 21%. Real property is taxed at a 1.25–2.5% rate in Montevideo and at a 2% rate in rural areas. The tax rate on industrial and commercial income in 1993 was 30%.

37CUSTOMS AND DUTIES

Uruguay's traditionally protectionist policies were relaxed as a result of the economic reforms of 1974. Average tariff rates fell from 139% in 1976 to 36% in 1982. As of August 1992, the Customs Unified Rate ranged from 10% on all raw materials to 24% on finished goods. Almost all goods can be imported into Uruguay without restrictions or licenses.

Uruguay has four free trade zones.

38FOREIGN INVESTMENT

The Foreign Investment Law of 28 March 1974 closed certain industries to foreign investors: public water and drainage services; railroads; alcohol and petroleum refineries; electric, telephone, local telegraph, mail, and port services; insurance; and issuance of mortgage bonds. Private investment in other sectors is generally welcome. If a foreign company is considered to be of national benefit, it is entitled to credit assistance and tax exemptions.

39ECONOMIC DEVELOPMENT

Monopolies are permitted in the fields of banking, postal services, water supply, ports, railway and bus transport, light and power, telephone and cable services, fuels, alcohol, seal fishing, meat packing, milk cooperatives, lotteries, and betting. Industrial and commercial activities of the state must be organized as "autonomous entities." Other public services may be organized as autonomous entities, as decentralized services, or as divisions under a ministry. Exceptions to this policy are state-operated postal and telephone services, customs houses, port administration, and public health.

The Committee on Investment and Economic Development, established in 1960, published a 10-year plan for 1965–75 for production, investment, and consumption. The plan, stressing industrial development and external financing, was superseded in April 1973 by the National Development Plan for 1973–77, prepared by the Planning and Budget Office. This plan projected an annual growth rate of 3.8% in real GDP, or 2.5% per capita; increases in exports and imports were projected at 10.1% and 14.9%, respectively. Domestic investment was to rise at an annual rate of 15.1%—a projection quite remote from the actual average annual increase of 3.2% realized during 1966–72.

After the severe slump of the early 1980s, the decisive actions by the new government injected life into the economy. In particular, restructuring the heavy domestic debt burden of the industrial and agricultural sectors increased confidence in the economy. Early indications were that domestic investment rose sharply in 1986 from the low level of 8.4% of GDP recorded in 1985. Some industries, moreover, are running at full capacity.

Between 1946 and 1986, Uruguay received nonmilitary loans and grants totaling $176.3 million from the US. Assistance from multilateral sources amounted to $851.9 million, of which $551.4 million came from the IBRD and $238.4 million from the IDB. Aid from all sources dropped sharply during the five years that followed the 1973 military takeover.

President Lacalle's administration continued the fiscal adjustment program in 1990 to reduce the budget deficit. A state enterprises reforms law passed in September 1991 permits partial privatization of certain state-owned enterprises. Progress in privatization will generate an immediate inflow of cash from investors which can be used to reduce the deficit. Uruguay has become an important trade partner and provider of services to those countries. The flip side of this relationship is Uruguay's dependency on and vulnerability to economic developments in its neighboring countries. A recovery in prices for Uruguay's main exports is currently underway. This should help contain the current account deficit to about the present levels.

40SOCIAL DEVELOPMENT

Uruguay has frequently been referred to as South America's first welfare state. The social reform movement began under the leadership of José Batlle y Ordóñez in the early 1900s. Social legislation now provides for a day of rest in every week (plus Saturday afternoon), holidays with pay, minimum wages, annual cash and vacation bonuses, family allowances, compensation for unemployment or dismissal, workers' accident compensation, retirement pensions for rural and domestic workers, old age and disability pensions, and special consideration for working women and minors. The state also provides care for children and mothers, as well as for the blind, deaf, and mute. Free medical attention is available to the poor, as are low-cost living quarters for workers.

The government's social security system is divided into six main funds: civil service and teachers, industrial and commercial, rural workers and domestic servants, family allowances, banking, and the professions. Employers withhold 10–13% of each employee's gross earnings and contribute 15–20% of payroll to the appropriate fund.

In the early 1980s, the government indicated that fertility and population growth rates were too low. However, they continued to fall: the fertility rate dropped from 2.6 in 1980–85 to 2.4 in 1985–90. In 1991, 22,400 persons accepted family planning services. Officials sought to encourage the return of those who had left Uruguay during the mid-1970s; the election of a civilian government and the growing sense of confidence among Uruguayans that the difficult problems facing the country can be dealt with make the success of this effort more likely.

In 1993, women made up about one third of the work force but tended to be concentrated in lower paying jobs. Nevertheless, many attend the national university and pursue professional careers.

41HEALTH

The government traditionally has placed great emphasis on preventive medicine and on the sociological approach to public health problems. The US Institute of Inter-American Affairs and the Uruguayan Ministry of Public Health created the Inter-American Cooperative Public Health Service, which built four health centers and clinics. For the region, life expectancy is high (72 years in 1992); infant mortality is low (20 per 1,000 live births in 1992); and the ratios of doctors and beds to the population are exceptionally good. In 1992 there were 2.90 doctors per 1,000 population (about 9,100 doctors), with a nurse to doctor ratio of 0.2; in 1990 there were 4.6 hospital beds per 1,000 people (about 15,000 beds).

The Commission for the Fight against Tuberculosis is under the jurisdiction of the Ministry of Health; as a result of its efforts, tuberculosis is almost unknown in Uruguay today. In 1990, there were only 15 per 100,000 cases of tuberculosis reported. The commission also deals with the social and economic effects of various diseases. The major causes of death are heart diseases, cancer, and digestive disorders. Degenerative diseases rank higher as a cause of death in Uruguay than in most other Latin American countries. The general mortality rate was 10.3 per 1,000 people in 1993, and maternal mortality was 36 per 100,000 live births in 1991. In 1990, about 9% of children under 5 years old were considered malnourished.

In 1991, 75% of the population had access to safe water and 61% had adequate sanitation. In 1992, 82% had access to health care services. Immunization rates in 1992 for children up to 1 year old were: tuberculosis (99%); diphtheria, pertussis, and tetanus (93%); polio (93%); and measles (93%). Total health care expenditures in 1990 were $383 million.

42 HOUSING
The housing situation is more favorable in Uruguay than in most Latin American countries. Housing construction, dominated by the public sector, is financed in large part by the Mortgage Bank. The National Institute of Low-Cost Housing builds low-cost dwellings for low-income workers and pensioners. In 1985, 98% of all housing units were made of durable materials including stone masonry, wood, zinc, or concrete. Owners occupied 56% of all dwellings, 23% were rented, 17% were occupied by usufructus (households legally inhabiting someone else's living quarters) and 1% were occupied by members of housing cooperatives. Of all housing units, 92% had private toilet facilities and 74% had water piped indoors.

43 EDUCATION
Adult illiteracy in 1990 was 3.8% (males, 3.4% and females, 4.1%), among the lowest in Latin America. Education in elementary, secondary, and technical schools and at the University of the Republic in Montevideo is free. Elementary education, which lasts six years, is compulsory. Secondary education is in two stages of three years followed by three more.

In 1991 there were 2,413 primary schools, with 15,747 teachers and 340,789 students. There were 276,482 students in secondary and technical schools. Enrollment at all institutions of higher learning was 73,660, out of which universities and equivalent institutions had 62,587 students and others had 11,073 students enrolled in 1991.

44 LIBRARIES AND MUSEUMS
The National Library, founded in 1816, contains over 900,000 volumes, primarily modern but including a sizable historical collection. Other major collections are found in the special libraries of the Council of Secondary, Basic, and Higher Education (105,000 volumes), and the Museum of Natural History (150,000), and at the Pedagogic Library (118,000).

The National Historical Museum exhibits artifacts of local Amerindian cultures and of Uruguayan historical development. The National Museum of Fine Arts has paintings by prominent Uruguayan artists, including Juan Manuel Blanes and Joaquín Torres García.

45 MEDIA
The state owns the telegraph and telephone services. In 1991 there were 528,674 telephones, most of them in the metropolitan Montevideo area. Uruguay in 1991 had 110 radio stations and 33 television stations; color television broadcasting was introduced in 1981. The number of radio receivers was about 1,880,000 in 1991; the number of television sets was 720,000.

The first newspaper in the Banda Oriental was the *Southern Star*, published by the British in 1807 during their brief occupation of Montevideo. *El Día*, founded by José Batlle y Ordóñez in 1886, helped lay the foundation for the social reforms of the first two decades of the 20th century. Under the military regime established in 1973, periodicals were forbidden to report on internal security matters. Censorship was imposed, and more than 30 newspapers and periodicals were closed down; many newspapers were harassed with suspensions of one day or more. In November

1974, *Marcha*, a weekly read throughout Latin America and Uruguay's last remaining opposition periodical, was shut down. Press censorship was relaxed in the late 1970s, but the military government still reserved the right to detain and question editors and reporters and to suspend publication for four or more issues. Press freedom was restored in 1985.

In 1991, Uruguay had 33 daily newspapers, with a combined circulation of more than 694,000. There were 6 dailies in Montevideo, with their affiliations and circulations (in 1991) as follows:

	AFFILIATION	CIRCULATION
El Diario (e)	Conservative	170,000
El País (m)	Conservative	65,000
La Mañana (m)	NA	40,000
La Republica	NA	25,000
Últimas Noticias	NA	12,000
El Observador Economico	NA	10,000

46 ORGANIZATIONS
The two most important organizations of livestock farmers are the Rural Association and the Rural Federation, both founded by large landowners. Other employers' organizations include the Importers' and Wholesalers' Association and the Uruguayan Exporters' Union. The Chamber of Industries is a powerful organization, with a representative on the Export and Import Control Commission.

47 TOURISM, TRAVEL, AND RECREATION
Tourism, one of Uruguay's major enterprises, enjoys government support. The state owns many hotels along the coast, especially in the area of Punta del Este, 145 km (90 mi) east of Montevideo and one of the more sophisticated resorts in South America. Montevideo has been promoted as the "city of roses" because of its many parks and gardens.

A visitor must have a valid passport; for most visitors, no visa is required. The maximum stay is three months, plus an extension for another three months. In 1991, 1,509,962 tourists visited Uruguay, 68% from Argentina and 9% from Brazil. Tourism receipts totaled US$333 million.

The most popular sport in Uruguay is soccer; there is an intense rivalry between supporters of the two major teams, the Peñarol and the Nacional. Other popular sports include basketball, cycling, tennis, pelota, golf, and water sports. Uruguayan soccer teams won the World Cup in 1930 and 1950; the first World Cup competition was hosted by Uruguay in 1930.

48 FAMOUS URUGUAYANS
The national hero of Uruguay is José Gervasio Artigas (1764–1850), who led the fight for independence against Brazil and Portugal. Juan Antonio Lavalleja (1786?–1853) directed the uprising that established Uruguay's independence in 1828. The nation's first two presidents were Gen. José Fructuoso Rivera (1790?–1854) and Brig. Gen. Manuel Oribe (1796?–1857), the founders of the Colorados and Blancos, respectively. One of Uruguay's greatest citizens was José Batlle y Ordóñez (1856–1929), who served twice as president of the country. José Pedro Varela (1845–79) was Uruguay's chief educational reformer.

One of the most respected defenders of Latin America's cultural tradition was José Enrique Rodó (1872–1917), whose *Ariel* and *Motivos de Proteo* fostered the idea of the superiority of Latin American culture. Juan Zorrilla de San Martín (1855–1933) was a 19th-century romantic poet whose finest work, *Tabaré*, describes Uruguay at the time of the Spanish conquest. Eduardo Acevedo Díaz (1851–1924) won fame as the writer of a gaucho novel, *Soledad* (1894). Other significant novelists are Carlos Reyles (1868–1938) and Javier de Viana (1872–1925). Horacio Quiroga (1878–1937) is regarded as one of Latin America's foremost short-story writers. The poets Julio Herrera y Reissig (1875–1910) and Juana de Ibarbourou (1895–1979) have attained a devoted audience beyond the borders of Uruguay. Emir Rodríguez Monegal (b.1921) is a leading writer and literary scholar.

The painter Juan Manuel Blanes (1830–1901) is best known for his *Episode of the Yellow Fever.* Pedro Figari (1861–1938) painted vivid scenes of early 19th-century Uruguay. Joaquín Torres García (1874–1949) founded his painting style on the principles of universalism and constructivism. Eduardo Fabini (1883–1951) is Uruguay's best-known composer. Francisco Curt Lange (b.Germany, 1903), Latin America's foremost musicologist, founded various inter-American institutions and publications for the promotion of music of the Americas.

[49]DEPENDENCIES

Uruguay has no territories or colonies.

[50]BIBLIOGRAPHY

Finch, Martin H. J. *A Political Economy of Uruguay Since 1870.* New York: St. Martin's, 1982.

Finch, Martin. *Uruguay.* Santa Barbara, Calif.: Clio, 1989.

Gillespie, Charles. *Negotiating Democracy: Politicians and Generals in Uruguay.* New York: Cambridge University Press, 1991.

Gonzalez, Luis E. *Political Parties and Redemocratization in Uruguay.* Washington, D.C.: Woodrow Wilson International Center for Scholars, 1985.

Hudson, Rex A. and Sandra W. Meditz (eds.). *Uruguay, a Country Study.* 2d ed. Washington, D.C.: Dept. of the Army, 1992.

Sosnowski, Saul and Louise B. Popkin (eds.). *Repression, Exile, and Democracy: Uruguayan Culture.* Durham, N.C.: Duke University Press, 1993.

Weinstein, Martin. *Uruguay, Democracy at the Crossroads.* Boulder, Colo.: Westview Press, 1988.

VENEZUELA

Republic of Venezuela
República de Venezuela

CAPITAL: Caracas.

FLAG: The national flag, adopted in 1930, is a tricolor of yellow, blue, and red horizontal stripes. An arc of seven white stars on the blue stripe represents the seven original states.

ANTHEM: *Himno Nacional,* beginning "Gloria al bravo pueblo" ("Glory to the brave people").

MONETARY UNIT: The bolívar (B) is a paper currency of 100 céntimos. There are coins of 5, 25, and 50 céntimos and 1, 2, and 5 bolívars, and notes of 5, 10, 20, 50, 100, 500, and 1,000 bolívars. B1 = $0.0087 (or $1 = B114.721).

WEIGHTS AND MEASURES: The metric system is the legal standard.

HOLIDAYS: New Year's Day, 1 January; Declaration of Independence and Day of the Indian, 19 April; Labor Day, 1 May; Army Day and Anniversary of the Battle of Carabobo, 24 June; Independence Day, 5 July; Bolívar's Birthday, 24 July; Civil Servants Day, 4 September; Columbus Day, 12 October; Christmas, 25 December; New Year's Eve, 31 December. Movable holidays are Carnival (Monday and Tuesday before Ash Wednesday), Holy Thursday, Good Friday, and Holy Saturday. Numerous other bank holidays and local festivals are observed.

TIME: 8 AM = noon GMT.

¹LOCATION, SIZE, AND EXTENT

Venezuela, located on the northern coast of South America, covers an area of 912,050 sq km (352,144 sq mi), extending 1,487 km (924 mi) WNW–ESE and 1,175 km (730 mi) NNE–SSW. Comparatively, the area occupied by Venezuela is slightly more than twice the size of the state of California. It is bordered on the N by the Caribbean Sea and the Atlantic Ocean, on the E by Guyana, on the S by Brazil, and on the W by Colombia, with a total boundary length of 7,609 km (4,718 mi). There are 72 offshore islands.

Venezuela claims more than 130,000 sq km (50,000 sq mi) of Guyanese territory west of the Essequibo River, constituting over three-fifths of Guyana. In 1985, UN mediation was requested by both countries. Conflicting maritime claims with Colombia in the Gulf of Venezuela remain unresolved, despite negotiations since 1970. Demarcation of the land boundary between the two nations began in February 1982. In August 1987, a Colombian naval vessel entered the disputed area in an apparent attempt to pressure Venezuela to be more flexible.

Venezuela's capital city, Caracas, is located in the northern part of the country on the Caribbean Sea coast.

²TOPOGRAPHY

Venezuela has four principal geographical divisions. In the north emerges a low extension of the Andes chain; to the west lies the hot basin of Lake Maracaibo; to the southeast spread the great plains (llanos) and forests; and south of the Orinoco River lie the unoccupied and largely unexplored Guiana Highlands, accounting for about half the country's total area. The Orinoco, which is more than 2,900 km (1,800 mi) long and has over 70 mouths and a delta of nearly 23,300 sq km (9,000 sq mi), drains four-fifths of Venezuela. There are more than 1,000 other rivers. About 90% of the nation's population is concentrated between the northeastern plateau of the Andes, on which is located the capital, Caracas, and another Andean extension to the west along the Venezuela-

Colombia border, covering approximately one-fourth of the total national area.

Outstanding geographical features include Angel Falls (979 m/3,212 ft high) in the Guiana Highlands of southeastern Venezuela, the highest waterfall in the world; the navigable Lake Maracaibo in the west, which is about 80 km (50 mi) wide and 210 km (130 mi) long and is accessible to ocean shipping; and Pico Bolívar in the Sierra Nevada de Mérida, the highest peak in Venezuela (5,007 m/16,427 ft).

³CLIMATE

Although Venezuela lies entirely within the torrid zone, generally there are four climatic regions, based mainly on altitude: tropical, up to 760 m (2,500 ft) in elevation; subtropical, from 760 to 1,830 m (2,500 to 6,000 ft); temperate, from 1,830 to 2,740 m (6,000 to 9,000 ft); and cold, above 2,740 m (9,000 ft). In the tropical region, including the cities of Maracaibo and Ciudad Bolívar, mean annual temperatures range from 24° to 35°C (75° to 95°F). In the subtropical region, where Caracas is situated, the means range from 10° to 25°C (50° to 77°F). In January, in Caracas, the average minimum and maximum temperatures are 15°C (59°F) and 26°C (79°F), respectively; the range in July is 17–26°C (63–79°F). During the wet season (May to October), the llanos and forest areas are swampy, green, and lush. Upon the advent of the dry season, the same areas become dry, brown, and parched. There is perpetual snow on several peaks of the Cordillera de Mérida.

⁴FLORA AND FAUNA

The natural vegetation of the tropical zone varies from the rain forest regions of the lower Maracaibo Basin to the grasslands of the llanos. In the areas of insufficient rainfall are found xerophytic plants, as well as mimosa. The subtropical zone, tierra templada, was originally almost covered by a luxuriant forest and is now the nation's principal agricultural region. In the temperate region, only a small portion of the total land area, wild vegetation is sparse and scrubby. In the páramo region, from about 2,740 to 4,880 m (9,000 to 16,000 ft) in elevation, vegetation becomes even thinner and barely affords an existence for the few sheep and cattle raised by the local Amerindian population. Above 3,050 m (10,000 ft), the only vegetation seen is the espeletia, similar to the century plant, which grows to a height of 1.8–2.1 m (6–7 ft).

The wild animals of Venezuela are abundant because of their relative isolation from human disturbance. The forests are populated with tapirs, sloths, anteaters, and a variety of monkeys. In the mountains are puma, margay, vampire bats, and deer. Semi-wild horses, donkeys, and cattle are found in the plains. The forests are rich in tropical birds such as the cacique, crested coquette, heron, umbrella bird, manakin, cock-of-the-rock, parrot, macaw, and aigrette. Aquatic fowl include the pelican, heron, flamingo, and a muscovy duck weighing up to 9.1 kg (20 lb). More than 32 species of eagles are found in Venezuela. There are numerous reptiles, including the rattlesnake, coral snake, bushmaster, anaconda, and boa. Crocodiles are found in the lowland rivers. Fish, shellfish, tortoises, and sand tortoises are also plentiful.

5ENVIRONMENT

Pressing environmental problems—such as urbanization, the unrestricted use of pesticides, the pollution of waterways with untreated industrial wastes, increasing air pollution and gasoline consumption by automobiles, and the uncontrolled exploitation of soil and forest resources—led the government in 1976 to establish the Ministry of Environment and Renewable Natural Resources. The basic legislative instrument is the Organic Law of the Environment; other laws govern the protection of soils, forests, and water supplies, regulate public sanitation, and seek to prevent the contamination of waterways by oil. Incentives are offered to industry to avoid environmental damage. Air pollution results from the activity of power plants, industrial emissions, and transportation vehicles. Venezuela contributes .5% of the world's total gas emissions. Water pollution is the result of contaminants from the oil industry, mining activity, and industrial chemicals. The nation's cities also contribute 3.6 million tons of solid waste per year. Venezuela has 205.4 cu mi of water. Forty-six percent is used for farming activity and 11% for industrial purposes. In the previous decade, deforestation threatened the country's forestlands at a rate of 1,000 square miles per year. In the Andes area, Venezuela loses up to 300 tons of soil per hectare due to land erosion by rivers. Measures designed to prevent forest depletion include suspension of logging permits and a large-scale afforestation program. The shrinking oil market of recent years has reduced funding for the ministry; its budget was cut by nearly 35% between 1984 and 1986. Staffing consequently has been reduced, to the detriment of the ministry's reputation as a model for the third world.

In 1994, 19 of the nation's mammal species and 34 bird species were endangered. Also 106 plant types were also endangered. As of 1987, endangered species in Venezuela included the tundra peregrine falcon, red siskin, giant otter, five species of turtle (green sea, hawksbill, olive ridley, leatherback, and South American river), and three species of crocodile (spectacled caiman, American, and Orinoco).

6POPULATION

The last census, taken in 1990, placed the total population of Venezuela at 18,105,265, 17% above the 1981 census figure. The annual population growth rate between 1981 and 1990 was 2.5%. A population of 23,622,000 was projected for the year 2000, assuming a crude birthrate of 23.9 per 1,000 population, a crude death rate of 5.4, and a net natural increase of 18.5 during 1995–2000. Population density in 1990 was 20.1 per sq km (52.1 per sq mi); in the relatively populous northwest, the density is

more than twice the national average. Nevertheless, Venezuela remains one of the least densely populated countries in the Western Hemisphere. The Amazon River area (nearly 20% of the total) averaged only 0.3 population per sq km (0.8 per sq mi) in 1990. In 1990, 84.1% of the population was urban. The population of the largest cities in 1990 was Caracas, 1,822,465; Maracaibo, 1,249,670; Valencia, 903,621; Maracay, 354,196; and Barquisimeto, 625,450.

7MIGRATION

For a time, Venezuela encouraged large-scale immigration in the hope that the newcomers would help increase the nation's food production. Although the yearly totals of foreigners entering Venezuela were high, a large portion of these immigrants remained only briefly. In the decades immediately before and after World War II, nearly 500,000 Europeans—mostly from Italy, Spain, and Portugal—came to Venezuela. By 1990, however, only 5.7% of the resident population was of foreign birth. In 1989 there were 18,893 immigrants and 9,643 emigrants. Along most of the land border, population is sparse, and there is little movement across frontiers. An estimated 300,000 illegal immigrants, most of them Colombians, were living in Venezuela in 1985. Internal migration in the 1980s was chiefly from the federal district to adjoining areas and eastward from the state of Zulia, in the far northwest.

8ETHNIC GROUPS

The original inhabitants of Venezuela were Amerindians, predominantly Caribs and Arawaks. The bulk (about 68%) of the present population is mestizo (mixed race); an estimated 21% is unmixed white, 8–10% black, and 2% Amerindian.

9LANGUAGES

The official language is Spanish. It is fairly standardized throughout the country among the educated population, but there are marked regional variations.

10RELIGIONS

Venezuelans are constitutionally guaranteed freedom of religion provided that a faith is not contrary to public order or to good customs. The 1961 constitution also stipulates that no citizen may refuse to obey the law on religious grounds. In 1993, an estimated 92.1% of the population was Roman Catholic. In 1987, there were about 350,000 Protestants, 200,000 tribal religionists, 122,000 Afro-American spiritists, and 20,000 Jews.

11TRANSPORTATION

The most important mode of domestic cargo and passenger transport is shipping over the country's more than 16,000 km (9,900 mi) of navigable inland waterways, which include 7,100 km (4,410 mi) navigable to oceangoing vessels. A large percentage of Venezuelan tonnage is carried by ships of the government-owned Venezuelan Navigation Co. In 1991, the merchant fleet had 62 vessels of over 1,000 gross tons, for a total GRT of 824,000. Shallow-draught ships are able to reach Colombian river ports in the wet season. Shallow-draught river steamers are the principal means of transportation from the eastern llanos to Puerto Ordaz, which, thanks to constant dredging, is also reached by deep-draught oceangoing vessels. Dredging operations also are maintained in Lake Maracaibo to allow the entry of oceangoing tankers. The government has invested substantially in the port of La Guaira, hoping to make it one of the most modern in the Caribbean area. Puerto Cabello handles the most cargo, and Maracaibo is the main port for oil shipments.

Highway and railroad construction is both costly and dangerous because of the rough mountainous terrain in the areas of dense population. Nevertheless, the government has undertaken massive highway construction projects throughout the country.

LOCATION: 0°35′ to 12°11′N; 60°10′ to 73°25′w. **BOUNDARY LENGTHS:** Total coastline, 2,816 km (1,750 mi); Guyana, 743 km (462 mi); Brazil, 2,000 km (1,243 mi); Colombia, 2,050 km (1,274 mi). **TERRITORIAL SEA LIMIT:** 12 mi.

Major ventures include the completion of the Caracas-La Guaira Autopista, which links the capital with its airport at Maiquetía and its seaport at La Guaira, and a section of the Pan-American Highway connecting Carora with the Colombian border. The General Rafael Urdañeta Bridge crosses the narrow neck of water connecting Lake Maracaibo with the Gulf of Venezuela and provides a direct surface link between Maracaibo and the east. By 1991, Venezuela had 22,780 km (14,155 mi) of paved highway, and 55,005 km (34,180 mi) of gravel and dirt roads. In 1992 there were 1,532,572 passenger cars and 449,135 commercial vehicles in Venezuela.

Venezuela's two railroads carry mostly freight. Rail transportation is concentrated in the northern states of Lara, Miranda, Carabobo, Aragua, and Yaracuy, with branches connecting the principal seaports with the important cities of the central highlands. There were 542 km (337 mi) of track in 1991, 363 km (226 mi) of which were owned and operated by the government.

Much of the equipment is antiquated, and the linking of lines is difficult because of the different gauges in use. The government planned in the early 1980s to build a 3,900-km (2,420-mi) railroad network by the end of the decade; however, the financial crisis that began in 1983 has scaled the program down to 2,000 km (1,200 mi) over 20 years. The first 7.2-km (4.5-mi) section of a government-financed metro line in Caracas was opened in 1983.

Cities and towns of the remote regions are linked principally by air transportation. Venezuela has three main airlines, the government-owned Aerovías Venezolanas S.A. (AVENSA), Línea Aeropostal Venezolana (LAV), and Venezolana Internacional de Aviación, S.A. (VIASA); VIASA, an overseas service, is jointly run by AVENSA and LAV. A new airline, Aeronaves del Centro, began domestic flights in 1980. Of the 61 commercial airports in the mid-1980s, 7 were international. The government has expanded Simón Bolívar International Airport at Maiquetía, near Caracas, to accommodate heavy jet traffic.

¹²HISTORY

No more than 400,000 Amerindians were living in the land now known as Venezuela when Christopher Columbus landed at the mouth of the Orinoco in August 1498, on his third voyage of discovery. The nation received its name, meaning "Little Venice," from Alonso de Ojeda, who sailed into the Gulf of Venezuela in August 1499 and was reminded of the Italian city by the native huts built on stilts over the water.

The first Europeans to settle Venezuela were Germans. Holy Roman Emperor Charles V granted the Welsers, a German banking firm, the right to colonize and develop Venezuela in exchange for the cancellation of a debt. Lasting a little less than 20 years, the administration of the Welsers was characterized by extensive exploration and organization of the territory but also by brutality toward the native population. In 1546, the grant was rescinded, and Venezuela was returned to the Spanish crown.

Under the Spanish, Eastern Venezuela was governed under the audiencia (region under a royal court) of Santo Domingo, and the western and southern regions became a captaincy-general under the viceroyalty of Peru. Settlement of the colony was hampered by constant wars with the Amerindians, which did not stop until after a smallpox epidemic in 1580. Meanwhile, the province was carved into encomiendas (hereditary grants), which were given to the conquistadores as rewards. By the end of the century, however, the encomienda system was abandoned, and existing grants were declared illegal. The cabildos, or town councils, won more authority, and a national consciousness began to develop. In 1717, the western and southern provinces were incorporated into the viceroyalty of New Granada, and in 1783, the area of present-day Venezuela became a captaincy-general of Caracas.

The war for independence against Spain began in 1810. Francisco de Miranda ("El Precursor"), a military adventurer, was named leader of the Congress of Cabildos, which declared the independence of Venezuela on 5 July 1811. Royalist factions rallied to overthrow the new republic, which was weakened when an earthquake destroyed revolutionary strongholds and left royalist centers virtually untouched. Miranda was captured and sent to die in a dungeon in Cádiz, but Simón Bolívar, a native of Caracas who had served under Miranda, was able to flee to Colombia. He then led an army across the Andes into Venezuela, declaring "War to the death and no quarter to Spaniards." In August 1812, he entered Caracas and assumed the title of Liberator ("El Libertador"). He was defeated and forced to flee to Jamaica, as the royalists again took control of the capital. In December 1816, Bolívar, after landing in eastern Venezuela, established his headquarters in Angostura, now Ciudad Bolívar. He was aided by Gen. José Antonio Páez. The Congress of Angostura convened in October 1818 and in February 1819 elected Bolívar president of the Venezuelan republic. In the spring, he crossed the Andes, and in July, he entered Tunja, Colombia, with about 3,000 men and, after defeating the Spaniards at Boyacá, entered Bogotá. Under Bolívar's leadership, Gran Colombia (Greater Colombia) was formed from Colombia, Ecuador, and Venezuela, with Bolívar as its president and military dictator. The end of the Venezuelan war of independence came with Bolívar's victory at Carabobo in June 1821.

In 1830, Venezuela seceded from Gran Colombia. A period of civil wars lasted from 1846 to 1870, when the caudillo Antonio Guzmán Blanco assumed power. Guzmán was overthrown in 1888, and a few years later, Joaquín Crespo, a former puppet president of Guzmán, seized power. The next dictator, Cipriano Castro (1899–1908), was a colorful, if controversial, figure. A drunkard and a libertine, Castro also put Venezuela deeply into debt. When Castro refused to repay its outstanding loans, Germany, Great Britain, and Italy sent gunboats to blockade the Venezuelan coast. After mediation by the US and a decision favorable to the European creditors by the Permanent Court of Arbitration

at The Hague, Venezuela met its obligations by 1907. The next year, Castro sought medical care in Paris, leaving Venezuela in the hands of Juan Vicente Gómez. Gómez (1908–35) promptly seized power and ruled as dictator until his death. Although uneducated and practically illiterate, Gómez had a mind for business and proved a capable administrator. During his dictatorship, agriculture was developed and oil was discovered, making Venezuela one of the richest countries in Latin America. Oil concessions attracted US, British, and Dutch companies, initiating an era of oil wealth that continues today.

After the death of Gómez, Venezuela began to move toward democracy. During two military governments, opposition parties were permitted, allowing the Democratic Action (Acción Democrática—AD) to organize. In 1945, the military scheduled an election, but many feared a fraudulent outcome. The AD deposed the military and a junta named AD leader Rómulo Betancourt provisional president. Betancourt set elections for 1947, and conducted the first free election for president in Venezuelan history. The AD candidate Rómulo Gallegos, a distinguished novelist, was elected overwhelmingly, but the military intervened in November 1948. A bloodless army coup replaced Gallegos with a military junta, which ruled for four years. In December 1952, during the scheduled presidential election Marcos Pérez Jiménez seized power and became an absolute dictator.

Venezuela took its last steps toward full democracy after January 1958, when a popular revolt with military backing drove Pérez Jiménez from power. An interim government consisting of a military junta held elections in December 1958, and Betancourt was chosen president. Venezuela has had fair and free elections ever since.

The Betancourt government (1959–64) instituted modest programs for fiscal and agrarian reform, school construction, the elimination of illiteracy, and diversification of the economy. However, a depression beginning in 1960 trammeled these efforts and aggravated dissatisfaction with the regime. Betancourt was challenged by political instability coming from several fronts. The military, looking for a chance to return to power, engaged in several attempts to overthrow the government. Betancourt also opposed Fidel Castro, and allied with the US against him. Castroites in Venezuela responded with a guerrilla campaign under the FALN (Armed Forces for National Liberation). Betancourt charged the Castro government with attempting to subvert his government by supporting the FALN. In February 1964, the OAS formally charged Cuba with an act of aggression against Venezuela as a result of the discovery of an arms shipment to guerrillas in November 1963.

The AD was reelected in December 1964, when Raúl Leoni won the presidency over five other candidates. In 1966, supporters of Pérez Jiménez staged an unsuccessful military uprising. In the same year, in a drive against continued Castro-supported terrorism, President Leoni suspended constitutional guarantees and empowered the police to make arrests without warrants, to hold suspects without bail for an indefinite period, and to enter the quarters of suspected terrorists without judicial permission.

In 1968, Venezuela passed another test of democracy by transferring power peacefully from AD to the opposition Social Christian Party. The victor, Rafael Caldera Rodríguez, governed along the same lines as his AD predecessors, maintaining a set of social programs and benefitting from increasing oil revenues. At this point, Venezuela's future seemed assured, and public expenditures increased. The AD returned in 1973 with the victory of Carlos Andrés Pérez. By 1976, Pérez had brought about complete nationalization of the oil and iron industries. In the December 1978 elections, Luis Herrera Campíns, leader of the Social Christian Party, won the presidency. The next year, Venezuela received its first rude awakening, when the oil market dropped, thus threatening the foundations of the Venezuelan economic and political systems. There was soon a financial crisis, as Venezuela struggled to make payments on its overextended debt. The crisis culminated in the devaluation of the national currency, the bolívar, which dropped to one-third of its previous value against the dollar. Venezuelan consumers responded angrily, and the early 1980s were years of unrest. In the elections of December 1983, the AD returned to power behind presidential candidate Jaime Lusinchi.

Lusinchi's tenure was marred by scandal and trouble in the midst of the world petroleum crisis. While the economy floundered through the 1980s, the government maintained public confidence by stressing a "social pact" with guarantees of housing, education, and public health. Some progress has been made in boosting non-oil exports, particularly in agriculture and mining, and the government has promoted import substitution, particularly in food and manufacturing.

The 1988 elections brought back Carlos Andrés Pérez, who had been elected president 15 years earlier. Pérez immediately imposed austerity measures, removing government subsidies on a number of consumer goods, including gasoline. Prices rose and Caracas was rocked by rioting on a scale not seen since the uprising of 1958. The military was called in to quell the disturbances, but when the trouble finally died down, thousands had been killed or injured. The situation continued to deteriorate, and in 1992 Venezuela was shocked by two military coup attempts. Venezuelan democracy was in jeopardy.

Pérez did not reverse this trend when a major scandal broke, with allegations of embezzlement and theft in office. Pérez was suspended from office, and Ramón José Velásquez was named interim president until the regularly scheduled elections in December 1993.

In that election, Venezuelans chose Rafael Caldera, who ran under a coalition of four parties. The election of Caldera, who had been president during the brighter years of 1968–73, demonstrated the level of impasse in the Venezuelan system. Unable to produce new leadership, former presidents were being returned to office. Even though Venezuela remained one of the wealthiest countries in Latin America, instability was rapidly increasing.

13 GOVERNMENT

Venezuela is a republic governed under the constitution of 23 January 1961, which replaced the constitution of 1953. The constitution stresses social, economic, and political rights.

The president is elected by direct popular vote for a five-year term. He cannot run for reelection until 10 years after the completion of a term. The president must be a native citizen, at least 30 years of age, and a "layman." Presidential duties include the selection and removal of cabinet ministers and all other administrative officers and employees of the national government, as well as the appointment of state governors. The president is commander-in-chief of the armed forces, directs foreign affairs, and may make and ratify international treaties, conventions, and agreements. The president may declare a state of emergency and order the suspension of constitutional guarantees, but must seek the authorization of Congress within 10 days. He has the right to introduce bills and defend them before Congress. He may veto legislation, but a two-thirds majority of Congress can override it. There is no vice-president, and if the president cannot complete a term of office, the Congress, meeting in joint session, must select a new president by secret vote.

The Congress is bicameral, with a 47-member Senate and 200-seat Chamber of Deputies (as of 1987). The Senate consists of two members elected from each of the states and the Federal District by direct vote. Additional senators are chosen to represent minorities. Also, former presidents who were popularly elected and served more than one-half their terms are eligible for ex officio membership. Senators must be native-born and at least 30

years of age. Seats in the Chamber of Deputies are apportioned among the states, territories, and the Federal District, with a minimum of two deputies for each state and one deputy for each territory. Elected by direct suffrage, the deputies must be native-born and at least 21 years of age. Both houses are elected concurrently with the president for five-year terms. Special sessions of Congress may be called by the president. Joint sessions are held to break deadlocks or disagreements and to authorize the declaration of a state of emergency or the suspension of constitutional guarantees. A bill may be introduced in either house but must be passed by both in order to become law.

There is compulsory adult suffrage at age 18, except for members of the armed forces and criminals. Foreigners may also be enfranchised by an act of Congress. Women won the right to vote in 1947.

14 POLITICAL PARTIES

Before the formation of the existing political parties, political conflict in Venezuela was confined to the traditionally Latin American centralist-federalist debate, with few actual differences in governments. Since the late 1950s, however, a stable party system has evolved. Each political group has its own ballot with its own distinctive color and symbol, so that illiteracy is no barrier to political participation. Constitutional provisions barring the military from political involvement further ensure the stability of the party system and the continuity of elected civilian leadership.

Since 1958, the dominant force in Venezuelan politics has been the Democratic Action Party (Acción Democrática—AD). It grew out of the socialist movement, which unified under the National Democratic Party (PDN). That party splintered, with a Moscow-oriented group forming the Communist Party of Venezuela (PCV), and the nationalist and democratic-socialist faction creating the AD.

The left has been fragmented throughout modern Venezuelan politics. After the split between AD and the PCV, the advent of Fidel Castro in Cuba caused further fragmentation. A group of AD members left the party to form the Movement of the Revolutionary Left (Movimiento de Izquierda Revolucionaria—MIR). The Armed Liberation Forces (Fuerzas Armadas de Liberación Nacional—FALN) took to the field and attempted a guerrilla uprising. The PCV remained loyal to Moscow and at times battled the FALN openly. All of these movements were denounced by the AD, and in 1962 the MIR and PCV were barred from political activity. The FALN, which never bothered with legal political action, was subdued, and with it hopes of a Castroite takeover died. The most recent development from the left has been the emergence of the Movement for Socialism (Movimiento al Socialismo—MAS), which took 10% of the vote for the Chamber of Deputies in 1988. MAS is an attempt to rejuvenate the left by a faction of the old PCV.

The right has been characterized by small parties without much chance of electoral success alone. Some have been able to form coalitions with larger parties to achieve some success within the system. One such party, the Democratic Republican Union (Unión Republicana Democrática—URD), has been a governing partner with the AD during the Leoni government (1963-1968). COPEI is a Christian Democratic party, with the center-right implications of that movement. It has succeeded as an opposition party to the AD, occasionally taking advantage of splits in the AD's governing coalition or within the AD itself.

In 1947, the AD won the first free elections ever held in Venezuela. The PCV also fielded a candidate. A third group was the Committee for Free Elections (Comité de Organización Política Electoral Independiente—COPEI), also known as the Social Christian Party (Partido Social-Cristiano). These three parties were outlawed during the dictatorship of Pérez Jiménez, and carried on their activities clandestinely. In December 1958, after

Pérez Jiménez had been driven from power, free elections were held. The presidential victor, Rómulo Betancourt, formed a coalition government of the AD, COPEI, and URD.

The AD and COPEI reached several agreements over the years to cooperate with each other and to exclude the more leftist parties from the Venezuelan system. After the COPEI victory in 1968, Venezuela became a more competitive two-party system, with AD and COPEI competing for power. Agreements between AD and COPEI in 1970 and 1973 called for cooperation in appointive posts, so that the competition has been controlled. AD and COPEI have dominated the system since, although the recent election of Caldera as the candidate of a four-party coalition suggests a movement away from the two-party arrangement.

15LOCAL GOVERNMENT

Venezuela is divided into 21 states, 1 federal territory (Amazonas), the Federal District, and 72 offshore islands. All of the islands, except for the Isla de Margarita and adjacent islands (which together constitute a state), are classified as federal dependencies. As of 1985, the states were subdivided into 156 districts and 613 municipalities.

State government is weak, the state governor being the agent of the federal government directed to enforce national laws and decrees, as well as state legislation. Governors (including the governor of the Federal District) are appointed by the president and may be removed by him. According to the constitution the states are autonomous and are guaranteed the right to regulate their own affairs. They are given all powers not reserved to the nation or the municipalities. In reality, these powers are very few, since control of elections, education, health, agriculture, and labor is delegated to the national government. The state legislatures are unicameral.

Municipalities are autonomous in the election of their officials, in all matters within their competence, and in action regarding the collection, creation, and expenditure of their revenues. City councils vary from 5 to 22 members according to population. The council is popularly elected and, in turn, selects the mayor. According to the constitution, elections for state and municipal offices may not be held more often than once every two years or less often than once every five years.

16JUDICIAL SYSTEM

The Supreme Court of Justice consists of nine judges and nine alternates, elected by joint session of Congress for a period of nine years; a third of the judges are newly elected every three years. The Supreme Court, which organizes and directs the other courts and tribunals of the republic, is the nation's highest tribunal, and there is no appeal of its decisions. It can declare a law or any part of a law or any regulation or act of the president unconstitutional. The Court decides in cases of conflict of laws and tries cases of litigation between different governmental units. It determines whether there are grounds for the trial of the president, a member of Congress, members of the Court itself, or certain executive officials.

The lower branches of the judiciary include courts of appeal and direct courts, whose judges and magistrates are appointed by the Supreme Court. Each state has its own supreme court, superior court, district courts, and municipal courts. The territories have civil and military judges. The jury system is not used. There is no capital punishment; the maximum prison sentence is 30 years.

Although there are some procedural due process safeguards in criminal trials, there is no presumption of innocence. Public defenders are appointed for persons unable to afford a defense attorney. The justice system in general is inefficient, badly backlogged, and susceptible to corruption.

17ARMED FORCES

Venezuela's armed forces are professional and well equipped. In principle, all male Venezuelans 18 years of age are required to serve two years (or 30 months in the Navy) and then to remain in the reserve until the age of 45. In practice, however, many males, including workers in essential occupations, students, and heads of households, are exempted from service.

Total armed strength in 1993 was 75,000, including 23,000 volunteers in the Fuerzas Armados de Cooperacion, an internal security force. The army had 34,000 regulars, including 6 infantry divisions and 12 specialized brigades and 2,000 airmen. The navy had 11,000 men, including 6,000 marines; naval strength included 2 submarines and 6 frigates. The air force had 7,000 personnel, 102 combat aircraft, and 30 armed helicopters.

Defense expenditures in 1991 were $1.95 billion, or 4% of the GDP. During 1981–91, Venezuela imported armaments worth $2 billion, half from the US.

18INTERNATIONAL COOPERATION

Venezuela is a charter member of the UN, having joined on 15 November 1945, and participates in ECLAC and all the nonregional specialized agencies except the IDA. The nation is also a member of G-77, IDB, LAIA, OAS, OPEC, and PAHO. On 31 December 1973, Venezuela entered the Andean Pact. In June 1974, Caracas hosted the third UN Conference on the Law of the Sea; as of late 1987, however, it had declined to endorse the Law of the Sea treaty. Venezuela was instrumental in the creation of the 25-member Latin American Economic System in October 1975.

Venezuela reportedly played an important role in supporting the Sandinistas in their successful effort to win power in Nicaragua. After the Sandinista takeover, Venezuela sought to reduce tensions between Nicaragua on one side and Honduras and the US on the other.

19ECONOMY

During the colonial era and until the development of petroleum resources, the exportation of coffee and cocoa and the raising of cattle and goats provided the main support for the economy. However, agriculture now accounts for only about 7.5% of the GDP. For over 40 years the economy has been completely dominated by the petroleum industry; in the mid-1980s, oil exports accounted for 90% of all export value, and in 1986 petroleum accounted for about 16% of the GDP. The Venezuelan economy is therefore greatly influenced by petroleum market conditions and by OPEC pricing policies. The second most important mineral product is iron, and Venezuela's mineral wealth is augmented almost daily by discoveries of additional reserves. Industrial development is fostered by government policy.

The average annual GDP growth during 1970–80 was 5%, with a peak of 7.2% during 1974–77. In the late 1970s, the economy began to stagnate, and the GDP growth rate declined from 3.2% in 1978 to zero in 1979, with a negative rate of 1.5% recorded in 1980. Weakening world oil prices contributed to further economic stagnation in the 1980s, when Venezuela had difficulty meeting its obligations on short-term loans accumulated by state-owned enterprises. Soft markets for oil, iron ore, and aluminum aggravated Venezuela's financial problems in 1982; the GDP declined by 5.6% in 1983, by 1.4% in 1984, and by 0.4% in 1985. Improvement was shown in 1986 with a 5.2% growth. Inflation, however, increased from 7.6% in 1985 to 11.6% in 1986.

After severe adjustments experienced during 1989 and 1990, the main economic indicators improved considerably in 1991 and 1992. Real GDP was 10.4% and 7.3% respectively. Growth was led by expansion of the petroleum sector due mainly to the Persian Gulf War. Growth was also a result of the liberalization of the economy, including a privatization program initiated in 1990 and restructuring of the public sector enterprises. However, growth slowed down and real GDP contracted by 1.0% in 1993. Economic growth was estimated to decrease further by 0.4% by the end of 1994. Inflation remains stubbornly high. Prices increased 40.8% in 1990, a sharp decline from the 84.3% recorded in 1989. Inflation dropped in 1991 to 31%, but in 1992 and 1993 it climbed again, recording 32% and 38% respectively. The most dynamic sectors of the economy were trade, transportation, communications, manufacturing, and most significantly, construction; meanwhile the agricultural sector was weaker.

Agricultural output has decreased by nearly 8% since 1989. Thus Venezuela remains a net agricultural importer.

20 INCOME

In 1992, Venezuela's GNP was $58,901 million at current prices, or $2,900 per capita. For the period 1985–92 the average inflation rate was 35.8%, resulting in a real growth rate in per capita GNP of 1.1%.

In 1992, the GDP was $61,137 million in current US dollars. It is estimated that in 1991 agriculture, hunting, forestry, and fishing contributed 6% to GDP; mining and quarrying, 19%; manufacturing, 20%; electricity, gas, and water, 2%; construction, 5%; wholesale and retail trade, 18%; transport, storage, and communication, 6%; finance, insurance, real estate, and business services, 12%; community, social, and personal services, 4%; and other sources, 8%.

21 LABOR

Venezuela's economically active population in 1991 was 7,417,929. The distribution of employment among major economic sectors was as follows: services, 33.5%; trade, restaurants, and hotels, 20.9%; manufacturing, 16.3%; agriculture, 11.3%; construction, 9.2%; transportation and communications, 5.7%; mining and quarrying, 1%; and other sectors, 2.1%. The unemployment rate was 9.5% at the end of 1991. From 1985 to 1991, the labor force size increased annually by 3.1%.

The Venezuelan labor movement, for all practical purposes, had its inception in 1928 with the formation of the Syndicalist Labor Federation of Venezuela. The movement, stunted by the Gómez dictatorship, grew rapidly after his death. After the election of Betancourt, the Confederation of Venezuelan Workers (Confederación de Trabajadores de Venezuela—CTV), which had been founded between 1945 and 1948 but outlawed by Pérez Jiménez, was reconstituted. Another major labor federation, the National Movement of Workers for Liberation (Movimiento Nacional de Trabajadores para la Liberación—MONTRAL), was formed in 1974. In 1992, about 25% of the labor force was unionized.

There are fiscal controls over union funds. A strike may not be called before a conciliation attempt is made. Voluntary arbitration is encouraged, but arbitration may be ordered by the Ministry of Labor; awards are binding on all parties for a period of not less than six months. If no agreement is reached, a strike may be called 120 hours after the government labor inspector has been notified.

The comprehensive Labor Code enacted in 1990 extends to all public sector and private sector employees (except members of the armed forces) the right to form and join unions of their choosing. The Code also lowered the maximum workweek to 44 hours; the constitution provides for a minimum wage, the right to strike, and the right of labor to organize into unions. Labor laws include provisions for an eight-hour day, a paid vacation of at least 15 workdays a year, compulsory profit sharing, severance pay, death and disability payments, medical services, and social security.

22 AGRICULTURE

In 1992, agriculture accounted for 5.5% of the GDP and grew by 2.3% over 1991; it also engaged 11.3% of the economically active population in 1991. Although the Venezuelan economy grew by 7.3% in 1992, growth in the agricultural sector remained weak compared to construction and manufacturing.

Venezuela does not have the rich soil of many other Latin American countries. In 1992, 3,905,000 hectares (9,649,000 acres), or 4.4% of the total land area, were used for temporary or permanent crops. The most highly developed agricultural region is the basin of Lake Valencia, west of Caracas and inland from Puerto Cabello. The principal crop of this area is coffee. Before oil came to dominate the economy, coffee accounted for 40–60% of all income from exports.

The main field crops are sugarcane, rice, corn, and sorghum, and the chief fruits are bananas, plantains, oranges, coconuts, and mangoes. The most important agricultural items for industrial use are cotton, tobacco, and sisal. Two varieties of tobacco grow in Venezuela, black and Virginia blond; the latter is used for the most part to make certain popular brands of US cigarettes under license. Sisal is grown and widely used to make cordage and bags for sacking grains and coffee. Thin strings of the fiber are also employed in hammocks, household bags, doormats, hats, and sandals. Agricultural production in 1992 (in thousands of tons) included sugarcane, 6,700; bananas, 1,215; corn, 904; rice, 595; sorghum, 528; plantains, 510; oranges, 440; potatoes, 215; cotton, 24; tobacco, 15; and sisal, 9.

Under an agrarian reform law of 1960, three kinds of land are subject to expropriation by the government: uncultivated lands; lands worked indirectly through renters, sharecroppers, and other intermediaries; and lands suitable for cultivation that are being devoted to livestock raising. Compensation is paid for expropriated lands. Between 1960 and 1980, 8,467,000 hectares (20,922,295 acres) of land were distributed to 155,200 farming families who had never previously owned property. However, the land reform was adversely affected by mass migration of rural people to the cities.

23 ANIMAL HUSBANDRY

Since colonial days, cattle raising has been the dominant livestock industry in Venezuela. Chiefly criollos, or Spanish longhorns, the cattle are raised on the unfenced ranges of the llanos. Crossbreeding with shorthorns has been going on since the last half of the 19th century and with zebu since 1915. The government has made considerable progress in eradicating tick and other infestations and hoof and mouth disease. It buys breeding stock from the US, and finances programs to improve cattle production, processing, and distribution. The government also offers a subsidy for pasteurized milk, thereby helping to expand and improve the dairy industry.

Venezuela's livestock population in 1992 included 1,727,000 hogs, 1,530,000 goats, and 525,000 sheep and significant herds of cattle. Egg production in 1992 was 122,265 tons, and the milk output fell to 1,485,000 tons. Beef production increased from 147,000 tons in 1963 to 361,000 tons in 1992. In the same year, pork production was 102,000 tons; goat meat, 7,000 tons; mutton, 2,000 tons; and poultry, 415,000 tons.

24 FISHING

With its 2,816 km (1,750 mi) of open coast, Venezuela has vast fishing potential. Fish and fish products currently play a relatively minor role in Venezuela's international trade, but fish are extremely important domestically. Venezuela has the highest per capita fish consumption in Latin America, about three times that of the US. The principal fishing areas are La Guaira, the Paraguaná peninsula, and the Cariaco–Margarita–Carúpano area. The total catch in 1991 was 352,835 tons, up from 284,235 tons in 1986.

The fish-canning industry, begun in the 1940s, has had difficulty finding a market, since there has long been a preference for imported canned fish of higher quality. In recent years, however, exports to the US, the Netherlands Antilles, and other countries have increased. Tariff protection and improvements in quality have helped the industry.

25 FORESTRY

Partly because of the inaccessibility of forest areas, Venezuela's high-quality wood is very much underdeveloped. It is also misused by small farmers, who clear land for farming by burning trees without replacing them. The greatest concentration of forests lies south of the Orinoco, but the areas currently utilized are in the states of Portuguesa and Barinas. Cedar and mahogany are the principal trees cut; rubber, dividivi, mangrove bark, tonka beans, oil-bearing palm nuts, and medicinal plants are also produced. Roundwood output was about 1.3 million cu m in 1991. Virtually all sectors of the forest products industry rely on domestic output to meet domestic demand.

26 MINING

In 1991, production of mineral commoities was valued at $441 million; the principal commodities were aluminum, cement, diamonds, ferroalloys, gold, and iron ore. The most important metal mined in Venezuela is iron. Iron ore production peaked at 26.4 million tons in 1974 before declining to 9.4 million tons in 1983, mainly because of the worldwide recession and the consequent lessening of the demand for steel. Production subsequently rose steadily, reaching an estimated 21.2 million tons in 1991. Average 1987–91 annual production amounted to 19,338,000 tons, for a relative ranking of tenth highest in the world.

Iron mining was developed mainly by the Orinoco Mining Co., a subsidiary of US Steel, and by Iron Mines of Venezuela, a subsidiary of Bethlehem Steel. The industry was nationalized as of 1 January 1975, when a state-owned regional development corporation assumed control. Late in 1975, a new state enterprise, the Orinoco Ferrominerals Co., was established to run the nationalized concerns.

Gold, the first metal found in Venezuela, reached its production peak in 1890. Gold was exported until 1950, but thereafter it became necessary to import small amounts. Production has risen sharply from 570 kg in 1981 to 7,700 kg in 1990, but fell by 43% to 4,215 kg in 1991. Since about 1930, both industrial and gem diamonds have been mined in the Gran Sabana region in the southern part of the state of Bolívar. Diamond production in 1974 had an extraordinary increase of 468,200 carats over 1973, to a total of 1,249,000 carats, but output has since fallen to an estimated 213,557 carats in 1991. Deposits of high-grade bauxite, estimated at 300 million tons in 1991, were discovered at Los Pijiguaos, in the northern part of the state of Bolívar; at 1,992,448 tons, bauxite production reached a new high in 1991.

Other minerals extracted are sulfur, gypsum, limestone, salt (produced as a government monopoly), granite, clay, and phosphate rock. Minerals known to exist but not yet exploited are manganese (with deposits estimated at several million tons), mercury, nickel, magnesite, cobalt, mica, cyanite, and radioactive materials.

27 ENERGY AND POWER

With vast petroleum deposits, extensive waterways, and an abundance of natural gas, Venezuela possesses great electric power potential. In 1986, Venezuela ranked fourth in Latin America in electric power capacity and production and first on a per capita basis. In 1991, net installed capacity was 17,726 Mw; total electrical output in 1991 was an estimated 57,150 million kwh. Until 1975, the electric power companies were privately owned, with the exception of the government-run Maracay Electric. In March

1975, President Pérez announced that all electrical generating companies would be taken over by the state by the end of the year and merged into a single enterprise.

In 1964, a regional development agency, the Corporación Venezolana de Guayana, began construction of one of the world's largest hydroelectric power plants. Constructed across the Caroní River, which flows out of the Guiana Highlands, the Guri Dam began operations in November 1968, generating 525,000 kw of power. The capacity of the Guri project increased to 14.5 million kw in 1982.

In 1992, Venezuela was the world's sixth-largest oil producer. Together with Mexico and the North Sea area, Venezuela is among the most important producers of oil for the Western world outside of the Middle East, with a net exportable capacity of about 2 million barrels of oil per day (including 1.3 million barrels per day to the US). In 1992, proven oil reserves were estimated at 63 billion barrels, with an estimated 200 billion barrels of heavy oil in the Orinoco tar belt. While 73% of Venezuela's 1992 crude production was light- or medium-weight, almost 72% of its reserves are heavy or extra-heavy. Venezuela will soon need more advanced refineries to process heavy crude into more marketable products.

Petroleum production in Venezuela began in 1917, when 120,000 barrels were produced. Until the early 1960s, Venezuela was the world's third-largest producer of oil after the US and the USSR. Output rose to a peak daily volume of 3,700,000 barrels a day in 1970, but in 1974, President Pérez began a conservation program, resulting in cutbacks to 2,400,000 barrels a day by mid-1975. The 1986 collapse of oil prices eventually forced OPEC producers to set production quotas, which cut Venezuela's output to 1.5 million barrels per day for the first half of 1987. At the end of 1993, OPEC's crude oil production quota for Venezuela was 2,359,000 barrels per day. Production has risen from 1,902,000 barrels per day in 1988 to an estimated 2,500,000 barrels per day in 1993.

The output of natural gas was an estimated 25,700 million cu m in 1992. Proven reserves totaled 3.6 trillion cu m in 1992. The long-awaited Cristobal Colon liquefied natural gas (LNG) export project received congressional approval in August 1993. The $5.6-billion multinational project plans to eventually export up to 6 million tons of LNG annually, supplied by four offshore Caribbean fields utilizing 77 km (48 mi) of pipelines.

Venezuela's present production of crude oil is confined mostly to the western part of the country. The eastern shore and floor of Lake Maracaibo produce about 50% of the total oil extracted. A plan to develop the Orinoco tar belt was reactivated in the late 1980s after having been postponed in 1982 because of petroleum price cuts. Production in the initial phases is to be directed to the thermoelectric sector as a substitute for coal and conventional oil. Early in 1991, however, British Petroleum announced that it would not participate in the development of the Orinoco heavy oil belt.

On 29 August 1975, President Pérez signed into law the Oil Industry Nationalization Act, under which all concessions to private oil companies were rescinded as of 31 December 1975. A state holding company, Petroleum of Venezuela (Petróleos de Venezuela—Petrovén), was established in September 1975 with an initial capitalization of $465 million. Petrovén obtained a 50-year renewable monopoly over Venezuelan petroleum production, beginning 1 January 1976. Initial compensation of B1.9 billion was paid in December 1975; total compensation was to exceed B5 billion, of which B3.9 billion was to be paid in tax-free five-year bonds, at 6% interest, redeemable only in Venezuelan oil. Within a few years, Petrovén's 16 original operating companies were combined into four: Lagovén, the largest subsidiary, formerly the Creole Petroleum Corp., an Exxon affiliate; Maravén, formerly a Royal Dutch/Shell subsidiary; Menevén, formerly Mene Grande, a Gulf subsidiary; and Corpovén, including the former Mobil Oil subsidiary. In 1986, the state holding company (now referred to as PDVSA) was reorganized. Menevén was absorbed into Corpovén, and the new operating company ceded its operations in Zulia and Falcón states to Maraven and Lagovén. PDVSA's reputation has developed into what may be considered the best-managed state oil company in the world; the company has major refinery, pipeline, and service station networks in Europe, the US, and elsewhere in the Caribbean. PDVSA invested $276 million for oil exploration in 1993, an increase of 73% over 1992. Without investment, production of oil would decline by 20% annually, due to the necessity for maintenance of aging oil fields.

28INDUSTRY

Since the economy was largely dependent on oil, Venezuela neglected other domestic industries for decades in favor of importing goods to satisfy Venezuelan consumer needs. Especially since the inauguration of the fifth national plan (1975–80), however, industrial diversification has become a high priority. The government has encouraged industrial development through protective tariffs and tax exemptions for reinvestment.

Although much of Venezuela's petroleum is exported in crude form, petroleum refining is a major industry. Petroleum products amounted to 327.7 million barrels in 1985, up from 317 million in 1975. The steel industry produced 3 million tons of steel in 1985. By 1983, Venezuela had the largest aluminum-smelting capacity in Latin America, with a total of 400,000 tons a year. Bauxite imported from Brazil, Suriname, and Jamaica is processed in plants owned by the three state companies, Interalumina, Venalum, and Alcasa. Production in 1986 was 2.7 million tons. When the bauxite deposits at Los Pijiguaos are fully developed, the Venezuelan aluminum industry will no longer need to import bauxite. Other industries include shipbuilding, automobile production (149,902 vehicles in 1986), and fertilizer manufacture.

The government is involved in a massive decentralization project to relocate industries in the peripheral cities and the interior. Valencia, the capital of Carabobo State, is a major new industrial center. A second major industrial development scheme has made Ciudad Guayana the hub of a vast industrial area with a 160-km (100-mi) radius. Puerto Ordaz was founded on the basis of the Cerro Bolívar iron ore deposits. West of Puerto Ordaz are the Matanzas steel mill, with a yearly capacity of 750,000 tons, and the Interalumina bauxite-processing installation.

The manufacturing sector suffered from a recession in 1989, but was recovering at a steady rate during the next four years. It grew at more than 10% in 1992. Sales of cement, major appliances, and motor vehicles rose while others, such as aluminum, fell in 1993. At the beginning of 1994, the world aluminum market was back on track, with total production increasing by 1.8% in the first quarter of the year. However, the automobile industry decreased by 28.1% in the same period as demand contracted substantially.

In 1992, Venezuela's total oil production decreased slightly. However, PDVSA's current plans call for an increase in petroleum production capacity from 2.833 million barrels per day to 4 million barrels per day by the year 2002. The government is presently allowing foreign companies to produce Venezuelan oil, which will benefit the industry.

29SCIENCE AND TECHNOLOGY

The main research organization is the Venezuelan Institute of Scientific Investigations, founded in 1959. The National Council of Scientific and Technological Investigations and the State Ministry for Science and Technology direct and coordinate research activities. Among the principal research institutes, academies, and learned societies are the National Academy of Medicine, the Academy of Sciences, the Venezuelan Association for the Advancement of Science, the Institute of Experimental Medicine, and the Venezuelan Institute of Petroleum Technology. In 1986, research and development expenditures amounted to B4,447.7 million. In 1983, 2,692 technicians and 4,568 scientists and engineers were engaged in research and development.

30DOMESTIC TRADE

The three primary distribution centers for import trade are the Caracas area, the Maracaibo area, and an area that centers on Ciudad Bolívar and serves the vast inland llanos region and the Guiana Highlands. In Caracas are located the main offices of the national and foreign banks, many of the important industries, and the largest commercial houses. Wholesale importers and distributors in Caracas cover the entire country by means of branch offices, stores, and traveling salespeople. The most common and widespread form of retail selling used to be door-to-door, but the number of large shopping centers and supermarkets has risen in recent years.

Stores are usually open from 9 AM to 1 PM and from 3 to 7 PM, Monday through Saturday. Normal banking hours are from 8:30 to 11:30 AM and from 2 to 4:30 PM on weekdays only.

31FOREIGN TRADE

Since 1950, Venezuelan foreign trade has gone through three basic phases. The first, between 1950 and 1957, was marked by a boom in oil exports. The second, from the late 1950s through the 1960s, was characterized by a decline in world petroleum prices and a general drop in investment; during this period, exports of iron ore became increasingly important. The third phase, beginning in the early 1970s and especially since 1973, has featured a fivefold rise in petroleum prices and a staggering increase in the value of Venezuela's petroleum exports, despite declining output and the 1986 plunge in world oil prices.

After several years of negotiations, the Group of Three (Colombia, Mexico, and Venezuela) signed a free-trade agreement in Cartagena. The agreement goes into effect on 1 January 1995 and commits the countries to lifting most trade restrictions over a 10-year period. Imports have been increasing in recent years. Total imports amounted to $13.0 billion in 1993, compared with $12.4 billion in 1992. However, the first quarter of 1994 recorded a decrease, reflecting a weaker domestic economy. Venezuela's principal imports were composed of machinery and transport equipment, chemicals, construction material, and agricultural products. The country's major suppliers included the US (47.7%), Germany(6.1%), Italy(4.7%), Brazil(4.1%) and other countries (29.6%).

Total merchandise exports have remained almost unchanged. Exports totaled $14.4 billion in 1993, mainly composed of petroleum and derivatives, aluminum, steel, iron ore, coal, gold, coffee, and cocoa.

32BALANCE OF PAYMENTS

Venezuela has enjoyed an enviable balance-of-payments position for many years. Although the country was forced to import goods to satisfy the demand for many industrial, construction, and household items, its income from exports has more than equaled its expenditures for imports.

Between 1957 and 1961, gold reserves declined, despite the favorable balance of trade, chiefly because of capital flight attributable to political instability and the debt of $1.4 billion left by deposed dictator Marcos Pérez Jiménez. After 1962, foreign exchange reserves rose markedly, reaching $8,937 million at the end of 1985.

Venezuela has experienced foreign exchange problems throughout the 1980s, largely because of lower world oil prices.

Venezuela's balance of payments position deteriorated from its strong performance of 1991. In 1992, the merchandise trade surplus fell to $1.6 billion, a drop of 66% from 1991, primarily due to the economic recovery program and a decline in exports. Political uncertainty triggered a net short-term capital outflow of $155 million in 1992; however, high interest rates later reversed that outflow.

In 1992, merchandise exports totaled $13,955 million and imports $12,266 million. The merchandise trade balance was $1,689 million.

The following table summarizes Venezuela's balance of payments for 1991 and 1992 (in millions of US dollars):

	1991	1992
CURRENT ACCOUNT		
Goods, services, and income	2,085	–3,009
Unrequited transfers	–349	–356
TOTALS	1,736	–3,365
CAPITAL ACCOUNT		
Direct investment	1,769	545
Portfolio investment	192	61
Other long-term capital	234	2,046
Other short-term capital	–654	–523
Exceptional financing	553	309
Other liabilities	–90	298
Reserves	–2,224	1,031
TOTALS	–220	3,767
Errors and omissions	–1,516	–402
Total change in reserves	–2,259	1,032

33 BANKING AND SECURITIES

The Central Bank of Venezuela (Banco Central de Venezuela) is the fiscal agent of the government, responsible for fixing the rediscount rates, holding the country's gold and foreign exchange reserves, making collections and payments on behalf of the Treasury, and buying foreign exchange acquired from the oil companies and from exporters and reselling it to the government or to commercial banks. It also maintains the accounts of the National Coffee Fund, cooperates with government departments and other institutions in the work of special commissions, and is the sole note-issuing agency.

The state banking system consists of the Central Bank, the Industrial Bank of Venezuela, the Workers' Bank, seven regional and development banks controlled by the Venezuelan Development Corp., and the Agricultural Bank. In the private sector there are commercial banks, investment banks, mortgage banks, and savings and loan associations. The country's first mortgage bank, the Mortgage Bank of Urban Credit, initiated operations in Caracas in 1958. The Bank Law of 30 December 1970 requires that all banks be at least 80% Venezuelan-owned.

Total reserves minus gold reserves of the Central Bank totaled B9,216 million at the end of 1993; as of 1993, the Central Bank also held B1,341.1 billion in foreign assets. Demand deposits in commercial banks were B256.5 billion at the end of 1993. Time, savings, and foreign currency deposits in commercial banks were B1,093.9 billion at the end of 1993. The money supply, as measured by M2, stood at B1,504.4 billion in 1993.

Until 1972, Caracas was the only city in Latin America with two stock exchanges: the Commercial Exchange of Caracas (founded in 1947), controlled by the Caracas Chamber of Commerce, and the Commercial Exchange of Miranda State (founded in July 1958). The two exchanges were merged under the terms of the 1972 Capital Markets Law, which regulates the trading of securities and the activities of brokerage houses. There is also an exchange in Valencia. The National Securities Commission,

established in 1973, oversees public securities transactions. Not all the securities of Venezuelan corporations are listed on the exchanges, and new securities are constantly added. Investments are encouraged by the tax structure.

34 INSURANCE

All insurance companies operating in Venezuela are under the direction of the Ministry of Development. Venezuelan nationals must own 80% of the capital of foreign insurance companies. The insurance industry is an important source of investment capital.

In 1984 there were 57 insurance companies. Per capita premiums in 1990 totaled US$43.5 million or 1.9% of the GDP.

35 PUBLIC FINANCE

Petroleum provided about 67% of total revenues in 1992, while personal and corporate income taxes represented only 29% of total current revenues that year. Although revenues increased greatly in the 1980s, expenditures grew at an even faster rate, mainly because of heavy spending to combat the nation's pressing social and industrial problems. The expansion of the national budget was reflected in a fourfold increase in the government bureaucracy, from about 300,000 to 1,200,000 employees during the 1973–83 period. Falling oil prices in the mid-1980s caused a severe deterioration in the deficit.

Budget surpluses were recorded in 1990 and 1991 (0.4% and 0.8% of GDP, respectively), due neither to fiscal restraint nor enhanced tax collections, but largely to privatization revenues. In 1992, however, the government recorded a deficit equivalent to 6.1% of GDP. The Treasury, in conjunction with the Central Bank (its financial agent) sold B40 billion in short-term instruments to finance its temporary cash imbalances. A fiscal deficit was widely anticipated for 1993 (even though central government expenditures were cut by 15% from 1992) due to a revenue shortfall.

36 TAXATION

Almost all forms of taxation are the responsibility of the federal government; it apportions federal tax revenue to the states, which, in turn, allocate a portion to the municipalities. Cities may levy taxes on such items as water and other municipal services and the exploitation of community lands.

Venezuela first imposed an income tax in 1943; the present tax structure dates from 1966, as subsequently amended. The basic personal income tax is progressive and ranges from 10% to 30%. However, different rates apply to income from special sources, including oil and other minerals and proceeds from gambling or lotteries. Corporate income tax ranges from 20% to 30%. Oil income is taxed at a flat rate of 67.7%, and mining income at 60%. Exemptions are available for income from new industries or from agriculture, livestock, reforestation, and fishing.

Municipal license fees range from 0.25% to 10% of gross revenue. There are also inheritance and gift taxes, a real estate tax, an entertainment and advertising tax (in the Federal District), and minor excise taxes on liquor, tobacco, cigarettes, and petroleum products.

37 CUSTOMS AND DUTIES

Duties on imports are calculated ad valorem, according to government classifications. At the government's discretion, export duties may also be levied. Transit duties are required on certain goods, including hides, cocoa, coffee, cotton, and plumes. The government may also increase duties on items coming in from certain countries; in addition, it may establish import quotas or subject imports to licensing to protect domestic industry.

In 1967, Venezuela joined LAFTA, which became LAIA in 1981. The ad valorem duty system was adopted in 1973 to harmonize Venezuela's tariff structure with that of the other members of the Andean Pact. In 1979, the government began to turn away from protectionism by lowering tariff barriers, and a maximum rate of 100% ad valorem was established. In November 1982, however, strict import controls were imposed, affecting about 1,200 items. This measure, which was an open violation of Andean Pact rules, came at a time when Venezuela was approaching a foreign exchange crisis and when other Andean Pact members were also imposing import restrictions. The measure has since been relaxed, but the government restricts imports by manipulating the availability of foreign exchange. Maximum duty levels in 1993 were at 20% of the cost, insurance, and freight value of the goods.

Venezuela has three free trade zones: the Isla de Margarita, the Paraguaná Peninsula Industrial Zone, and the Port of La Guaira.

38FOREIGN INVESTMENT

Before the 1970s, over 97% of the total foreign investment in Venezuela was made by firms representing the US and the UK. In 1972, total foreign investment in Venezuela was $5.5 billion, of which 85% was in oil. Of the total, 68% came from US sources, 9% from the Netherlands, and 7% from the UK. Until the mid-1970s, Venezuela imposed few restrictions on foreign investment. Beginning in 1974, however, new foreign investors were required to obtain advance authorization from the Superintendency of Foreign Investment (Superintendencia de Inversiones Extranjeras—SIEX). After the nationalization of petroleum companies on 1 January 1976, the total foreign investment in Venezuela declined sharply; by the end of 1984, registered US direct investment was $1.7 billion, or approximately 55% of all foreign holdings in Venezuela. As of 1985, the sectors closed to foreign capital were public services, domestic trade, professional services, extractive industries, and banking, finance, and insurance. Tax credits are available for investment in the manufacture of industrial goods and in transportation, construction, and agriculture.

The economic reforms started in 1989 have stripped away many barriers to trade and investment. Foreign investors have almost the same rights as national investors. Foreigners may now buy shares in national or mixed companies, and repatriation of dividends and their reinvestment of profits is also unrestricted. For the first time, the government has begun to take the necessary steps to open up production activities of the petroleum sector to foreign participation on a contract basis.

New foreign investment continues to flow into the country. New investment totaled $812.4 million in 1992. Accumulated foreign direct investment totaled $5.9 billion at year end. The United States accounted for 54% of total foreign investment with $3.2 billion. Other principal foreign investors included the Netherlands and the United Kingdom.

39ECONOMIC DEVELOPMENT

In recent decades, government influence over the economy has increased. The iron and petroleum industries were nationalized in 1974–75, and the electrical generating industry is also a state enterprise. The government operates the salt and match industries; sets the prices of pharmaceuticals, petroleum products, milk, meat, and other consumer goods and services; and controls rents. The terms of Andean Pact membership also obligate the government to keep a tight rein on foreign trade and foreign investment.

Development policy from the 1950s through the late 1970s stressed import substitution, industrialization, and foreign investment. A new policy, inaugurated in 1979 and formulated in detail in the sixth national development plan (1981–85), was intended to eliminate price controls and reduce protectionism. The government also sought to reduce Venezuela's dependence on oil by industrial diversification, to pay more attention to agriculture, and to devote greater resources to social development, particularly

housing, education, public services, and health. The economic crisis of the early 1980s led to a partial abandonment of the new policy. An economic adjustment program aimed at decreasing inflation, limiting imports, and cutting government spending was announced in April 1983, and further austerity measures were imposed in February 1984. In late 1986, the Lusinchi government announced a three-year plan to stimulate the economy through government spending.

Between 1962 and 1986, Venezuela received $186.6 million in nonmilitary loans and grants from the US; development assistance from multilateral agencies totaled $1,277.9 million during 1953–86. Between 1974 and 1981, the nation extended development aid totaling $6.5 billion to other Western Hemisphere countries.

Venezuela's economic reform program, initiated in 1989, is transforming the country from a traditionally state-dominated, oil driven economy, toward a more market-oriented, diversified, and export-oriented economy. However, the financial crisis persists and the new administration of President Rafael Caldera has assumed direct control over the banking system. Much of the liquidity created by the support for the financial sector has been soaked up by the Central Bank, but a more permanent solution needs to be found for this new internal debt. A fiscal package of sorts is now in place and the government has presented a budget for 1995. The government has confirmed its commitment to selling off state enterprises. The bolivar was in free-fall before the government announced exchanged controls. Annual consumer inflation rose above 60% in June 1994, but government pressure and excess liquidity has kept interest rates down. The recovery of world oil markets and growth in non-traditional exports, combined with falling imports, will keep the trade balance in surplus in 1994–95.

Venezuela's Extended Fund Facility (EFF) with the IMF was exhausted in March 1993. The country drew about $2.5 billion under the $5.0 billion program. Government officials were discussing a continuation of the program.

40SOCIAL DEVELOPMENT

The social security system derives from a law effective 1 January 1967. The system covers medical care, maternity benefits, incapacity and invalidity, retirement and survivors' pensions, burial costs, and a marriage bonus. The employer is assessed 9–11% of each employee's salary, with the worker paying 4%.

In July 1982, a reform of the Venezuelan civil code extended women's rights, but women are still underrepresented in political and economic life. The government considers the fertility rate (3.5 in 1985–90) satisfactory but has been expanding family-planning programs. By 1985, an estimated 49% of married women were using contraception. Abortion is permitted only to save the woman's life. Women comprise roughly half the student body of most universities, and have advanced in many professions.

41HEALTH

Despite strenuous government efforts in the field of public health, Venezuela lacks a sufficient number of physicians for its booming population. There were 531,000 births in 1992, with a birthrate of 26.2 per 1,000 population in 1993. In 1992 there were 1.55 physicians per 1,000 inhabitants, with a nurse to doctor ratio of 0.5.

Great strides have been made in improving public health conditions. The death rate was reduced from 21.4 per 1,000 population in 1915 to 10.8 in 1950 and 5.4 in 1993. The infant mortality rate, 50.2 per 1,000 live births in 1971, fell to 20 in 1992. Life expectancy rose to an average of 70 years in 1992. Leading causes of death per 100,000 people in 1990 were: communicable diseases and maternal/perinatal causes (151); noncommunicable diseases (449); and injuries (110). Venezuela is virtually free of malaria, typhoid, and yellow fever. To maintain this status, the Department of Health and Social Welfare continues its drainage and mosquito control programs. It also builds aqueducts and sewers in towns of fewer than 5,000 persons. In 1991, 89% of the population had access to safe water, and 92% had adequate sanitation. In 1990 there were 44 cases of tuberculosis per 100,000; about 5% of children under 5 years old were considered malnourished. Immunization rates for children up to 1 year old in 1992 were: tuberculosis (82%); diphtheria, pertussis, and tetanus (66%); polio (72%); and measles (61%). Total health care expenditures were $1,747 million in 1990.

42 HOUSING

In 1928, the Workers' Bank was founded as a public housing agency of the federal government. Between 1959 and mid-1966, 23,881 low-cost units (houses and apartments) were constructed by the Workers' Bank throughout the country. By spending 75% of housing funds in rural areas and 25% in the cities, reversing the earlier ratio, the government hoped to cut down the exodus of peasants to Caracas and to maintain the nation's essential agricultural labor force. A housing program was initiated in 1958 to build low-cost rural homes to replace the existing mud and thatch huts with homes built of cement and earthen blocks with asbestos roofs. The average cost of each home came to an estimated B4,450, payable in 20 years at a rate of B20 monthly. By 1969, the government had built 104,598 cheap and comfortable homes for lower-income groups, 57,675 in cities and 46,923 in rural areas.

Construction of low-cost housing units during the early 1970s proceeded at a rate of about 100,000 a year. During 1977–81, the public sector built 167,325 housing units, and the private sector 71,922. In 1981, the government introduced new low-interest housing loans, but that policy did not prevent a housing slump that persisted from 1982 through 1986 as a result of the general recession; housing units built by the public sector in 1986 totaled 91,666. The total number of housing units in 1992 was 3,384,000.

In 1992, 63% of all housing units were detached or semi-detached houses, 17% were apartments, and 16% were improvised living quarters. Owners occupied 75% of all dwellings and 18% were rented.

43 EDUCATION

Venezuela has made considerable progress in education in recent years. The rate of literacy in 1990 was 88.1% (86.7% for men and 89.6% for women). An extensive literacy campaign has been conducted by the Venezuelan business community since 1980. Public education from kindergarten through university is free, and education is compulsory for children aged 7 through 13. Approximately 20% of the national budget is assigned to education.

Preprimary schools are being established throughout the country by the government. After nine years of elementary school, children undergo two to three years of secondary school which comes in two stages: the first is designed to provide a general education in the sciences and the humanities; the second prepares students for the university and offers specialization in philosophy and literature, physical science and mathematics, or biological science. In 1990–1991, there were 4,052,947 students enrolled in elementary schools; 281,419 in secondary schools; and 550,030 in colleges and universities. Technical and vocational schools provide instruction in industry and commerce, the trades, nursing, and social welfare.

There are 17 universities, both national and private, including the University of Venezuela (founded in 1725), Los Andes University (1785), Simón Bolívar University (1970), and the Open University (1977). Leading private institutions were the Andrés Bello Catholic University (1953), Santa María University (1953), and the Metropolitan University of Caracas (1970). Over 74 institutes

of higher learning, colleges, and polytechnic institutes exist where students pursue at least 180 different fields or professions.

44 LIBRARIES AND MUSEUMS

Venezuela has over 100 public and private libraries. The largest is the National Library, which was founded in 1833 and has over 2 million volumes. In 1958, the family of Pedro Manuel Arcaya donated his library of 100,000 rare volumes to the National Library. Both the National Library and National Archives are in Caracas, as are many of the largest libraries in the country. Other libraries include the Library of the Venezuelan Institute for Scientific Research (500,000 volumes), the National Academy of History (120,000), the Central University of Venezuela (280,000), Los Andes University in Mérida (250,000), the Andrés Bello Catholic University in Caracas (112,000), and public libraries and reading rooms in most state capitals.

Bolívar Museum, founded in 1911, has about 1,500 exhibits dealing with the life of Simon Bolivar and his fellow-workers in the independence movement. The National Pantheon, located in the restored Church of the Holy Trinity (dating from 1744), contains the ashes of Bolívar and the remains of other national heroes. The Birthplace of Bolívar (Bolívar's house) is also a national museum. The Fine Arts Museum, founded in 1938, contains paintings and sculpture by Venezuelan and foreign artists. The Natural Science Museum contains about 30,000 scientific exhibits. Other notable institutions include the Phelps Ornithological Collection, containing exhibits of thousands of Venezuelan birds, and the Museum of Colonial Art. All the preceding museums are located in Caracas. In Ciudad Bolívar is the Talavera Museum, with pre-Columbian and colonial exhibits; in Maracaibo are the Natural Science Museum and Urdañeta Museum of Military History; and in Trujillo is the Cristóbal Mendoza Museum.

45 MEDIA

Venezuela is covered by a network of telephone, telegraph, and radiotelephone services and is also served by international cable and radiotelephone systems. In 1991, the government sold 40% of the state-owned CANTV to a consortium led by GTE. In the same year there were 1,749,325 telephones in use. There were 204 radio stations in 1991 and 63 television stations, and an estimated 8,820,000 radios and 3,200,000 television sets were in use.

During the Gómez dictatorship, there was no freedom of the press whatsoever; suppression of newspapers was again practiced under the dictatorship of Marcos Pérez Jiménez. With the overthrow of his regime in January 1958, complete press freedom was proclaimed. Restrictions were imposed again during the second Betancourt administration. Since the late 1960s, however, the press has been generally open and free.

Leading Venezuelan newspapers (all published in Caracas, except as noted), with their 1991 circulations, are as follows:

Diario Ultimas Noticias	285,000
Meridiano	200,000
El Nacional	155,000
El Mundo	150,000
Diario 2001	130,000
El Universal	125,000
Diariode Caracas	103,000
Panorama (Maracaibo)	100,000

46 ORGANIZATIONS

Organization membership for the vast majority of Venezuelans is confined to labor unions and professional societies. Industrial, commercial, and agricultural management associations—including the National Association of Merchants and Industrialists, the Industrial Chamber of Caracas, the National Confederation of Associations of Agricultural Producers, and the National

Association of Metallurgical Industries and Mining—are represented nationally by the Federation of Chambers, which is influential in shaping foreign and domestic trade policy.

⁴⁷TOURISM, TRAVEL, AND RECREATION

Since the early 1970s, Venezuela has sought foreign investors for the construction, rehabilitation, and management of top-ranking hotels. As Venezuela's economic restructuring progresses, the tourism sector is expected to undergo privatization. The number of foreign visitors in 1991 was an estimated 598,328. Of these, 335,016 came from the Americas and 254,885 from Europe. There were 68,063 hotel rooms and 142,601 beds with a 69.6% occupancy rate. Tourism receipts totaled US$365 million.

A valid passport and visa issued by a Venezuelan consulate are required for entry into the country. Tourist attractions include Angel Falls, the many remembrances of Bolívar, numerous beach resorts, and the duty-free shopping and superb water sports facilities of the Isla de Margarita.

The most popular sports are baseball, football (soccer), bullfighting, cockfighting, horse racing, and water-related activities. Cultural life in the national capital offers, among other attractions, the Ballet Nuevo Mundo de Caracas.

⁴⁸FAMOUS VENEZUELANS

No participant in the drama of Venezuelan history is as well known, both nationally and internationally, as the great Liberator of the South American revolution, Simón Bolívar (1783–1830). Renowned for his military genius and his ability to lead and to inspire, Bolívar also wrote brilliantly and prophetically on government and international politics. Among the greatest Venezuelan literary figures was Andrés Bello (1781–1864), outstanding in journalism, history, law, literary criticism, philology, and philosophy. Francisco de Miranda (1750–1816) is credited with bringing the first printing press to Venezuela and publishing the first newspaper, *La Gaceta de Caracas*; he was also an adventurer who, for a three-month period in 1812, held dictatorial powers in Venezuela. For most of the period since independence was achieved, a series of dictators held power, including Antonio Guzmán Blanco (1829–99), Cipriano Castro (1858?–1924), Juan Vicente Gómez (1857?–1935), and Marcos Pérez Jiménez (b.1914). Rómulo Betancourt (1908–81) was a leading political leader in the decades after World War II.

Simón Rodríguez (1771–1854), called the "American Rousseau," was the leading liberal scholar of the prerevolutionary period. Other writers of note were Rafael María Baralt (1810–60) and Juan Vicente González (1811–66), known as the father of Venezuelan national literature. Fermín Toro (1807–74) introduced the Indianist movement into Venezuelan romantic poetry. José Antonio Calcaño (1827–97) was referred to as "the Nightingale" for the flowing beauty of his verse. A noted poet and educator was Cecilio Acosta (1831–81), who was also Venezuela's first corresponding member of the Spanish Royal Academy and the author of the Venezuelan penal code. Juan Antonio Pérez Bonalde (1846–92) is considered Venezuela's greatest romantic novelist of the 19th century. The outstanding writers of the later 19th century were Manuel Díaz Rodríguez (1868–1927) and Manuel Fombona Palacio (1857–1903). The most famous contemporary writer, Rómulo Gallegos (1884–1969), gained world renown almost overnight with the publication of his *Doña Bárbara* in 1929. Other prominent contemporary writers include the poet, novelist, and playwright Mariano Picón Salas (1901–65); the novelist Ramón Díaz Sánchez (1903–68); the economist, historian, and novelist Arturo Uslar-Pietri (b.1906); and the poet Juan Liscano (b.1915).

Father Pedro Palacios y Sojo (fl.18th cent.) founded the Chacao Conservatory. Among his students were José Cayetano Carreño (1774–1836), José Angel Lamas (1775–1814), and Juan José Lan-

daeta (1780–1812). Teresa Carreño (1853–1917) won world fame as a concert pianist. Leading contemporary composers are Vicente Emilio Sojo (b.1887), Juan Bautista Plaza (1898–1965), Juan Lecuna (1898–1954), and María Luisa Escobar (b.1903).

The first painter of note in Venezuela was Juan Llovera (1785–1840). His son Pedro Llovera (fl.19th cent.) became the teacher of the first generation of 19th-century Venezuelan painters. Important Venezuelan painters of the 19th century were Martín Tovar y Tovar (1828–1902) and Arturo Michelena (1863–98). The best-known 20th-century painter is Tito Salas (b.1889).

The outstanding pioneer of Venezuelan science was José María Vargas (1786–1854). The first university courses in mathematics and physics were introduced by Vargas's contemporary Juan Manuel Cajigal (1802–56). Modern scientists include Gaspar Marcano (1850–1910), famous for his investigations of blood; Luis Razetti (1862–1932), biologist and physician; and Arnoldo Gabaldón (b.1909), director of an antimalaria campaign from 1945 to 1948.

⁴⁹DEPENDENCIES

The small, sparsely inhabited Caribbean islands off the Venezuelan coast are federal dependencies, with a total area of 120 sq km (46 sq mi). The chief economic activity in the region is fishing and pearl diving.

Venezuela has two territories, Delta Amacuro and Amazonas. Delta Amacuro, with an area of 40,200 sq km (15,521 sq mi), is located in the northeastern corner of Venezuela, bordering Guyana, and had an estimated population of 92,610 as of 1993. The capital city of Tucupita is situated on Caño Mánamo, one of the principal channels of the Orinoco Delta. It is a commercial center for the petroleum-producing area to the west and is also a loading port for cocoa exports. The remote Amazonas territory, located in the southern corner of Venezuela and bordered by Colombia and Brazil, is larger in area, with 175,750 sq km (67,857 sq mi). Its population was estimated at 80,000 in 1993, and its capital, Puerto Ayacucho, is scarcely more than a trading post.

⁵⁰BIBLIOGRAPHY

Baloyra, Enrique, and John Martz. *Political Attitudes in Venezuela*. Austin: University of Texas Press, 1979.

Betancourt, Rómulo. *Venezuela, Oil, and Politics*. Boston: Houghton Mifflin, 1979.

Blank, David E. *Venezuela: Politics in a Petroleum Republic*. New York: Praeger, 1984.

Bond, Robert D. (ed.). *Contemporary Venezuela and Its Role in International Affairs*. New York: New York University Press, 1977.

Boue, Juan Carlos. *Venezuela: The Political Economy of Oil*. New York: Oxford University Press, 1993.

Fox, Geoffrey. *The Land and People of Venezuela*. New York: Harper Collins, 1991.

Gall, Norman. *Oil and Democracy in Venezuela*. Hanover, N.H.: American Universities Field Staff, 1973.

Gilmore, Robert L. *Caudillism and Militarism in Venezuela, 1810–1910*. Athens: Ohio University Press, 1964.

Haggerty, Richard A. (ed.). *Venezuela: A Country Study*. 4th ed. Washington, D.C.: Library of Congress, 1993.

Hellinger, Daniel. *Venezuela: Tarnished Democracy*. Boulder, Colo.: Westview Press, 1991.

Levine, Daniel H. *Conflict and Political Change in Venezuela*. Princeton, N.J.: Princeton University Press, 1973.

Lombardi, John V. *Venezuela: The Search for Order, the Dream of Progress*. New York: Oxford University Press, 1982.

———. *Venezuelan History: A Comprehensive Bibliography*. Boston: G. K. Hall, 1977.

Naim, Moises. *Paper Tigers and Minotaurs: The Politics of Venezuela's Economic Reforms*. Washington, D.C.: Carnegie

Endowment for International Peace, 1993.

Oropeza, Luis. *Venezuela's Tutelary Pluralism: A Critical Approach to Venezuelan Democracy.* Cambridge, Mass.: Harvard University Press, 1984.

Petras, James F., *et al. The Nationalization of Venezuelan Oil.* New York: Praeger, 1977.

Rudolph, Donna K. and G. A. *Historical Dictionary of Venezuela.* Metuchen, N.J.: Scarecrow, 1971.

Salazar-Carrillo, J. *Oil in the Economic Development of Venezuela.* Stanford, Calif.: Stanford University Press, 1976.

Silva Michelena, José A. *The Illusion of Democracy in Dependent Nations: The Politics of Change in Venezuela,* Vol. 3. Cambridge, Mass.: MIT Press, 1971.

Smith-Perera, Roberto. *Energy and the Economy in Venezuela.* Cambridge, Mass.: Harvard University, 1988.

Tugwell, Franklin. *The Politics of Oil in Venezuela.* Stanford, Calif.: Stanford University Press, 1975.

Venezuela in Pictures. Minneapolis: Lerner, 1987.

Waddell, D. A. G. (David Alan Gilmour). *Venezuela.* Oxford, England, and Santa Barbara, Calif.: Clio Press, 1990.

INDEX TO COUNTRIES AND TERRITORIES

This alphabetical list includes countries and dependencies (colonies, protectorates, and other territories) described in the encyclopedia. Countries and territories described in their own articles are followed by the continental volume (printed in *italics*) in which each appears, along with the volume number and first page of the article. For example, Argentina, which begins on page 7 of *Americas* (Volume 3), is listed this way: Argentina—*Americas* 3:7. Dependencies are listed here with the title of the volume in which they are treated, followed by the name of the article in which they are dealt with. In a few cases, an alternative name for the same place is given in parentheses at the end of the entry. The name of the volume *Asia and Oceania* is abbreviated in this list to *Asia*.

Adélie Land—*Asia:* French Pacific Dependencies: French Southern and Antarctic Territories
Afars and the Issas, Territory of the—*Africa:* Djibouti
Afghanistan—*Asia* 4:1
Albania—*Europe* 5:1
Algeria—*Africa* 2:1
American Samoa—*Asia:* US Pacific Dependencies
Andaman Islands—*Asia:* India
Andorra—*Europe* 5:11
Angola—*Africa* 2:13
Anguilla—*Americas:* UK American Dependencies: Leeward Islands
Antarctica—*United Nations:* Polar Regions
Antigua and Barbuda—*Americas* 3:1
Arctic—*United Nations:* Polar Regions
Argentina—*Americas* 3:7
Armenia—*Europe* 5:17
Aruba—*Americas:* Netherlands American Dependencies: Aruba
Ashmore and Cartier Islands—*Asia:* Australia
Australia—*Asia* 4:11
Austria—*Europe* 5:23
Azerbaijan—*Asia* 4:27
Azores—*Europe:* Portugal

Bahamas—*Americas* 3:25
Bahrain—*Asia* 4:35
Bangladesh—*Asia* 4:41
Barbados—*Americas* 3:31
Basutoland—*Africa:* Lesotho
Bechuanaland—*Africa:* Botswana
Belarus—*Europe* 5:37
Belau—*Asia:* Palau
Belgium—*Europe* 5:43
Belize—*Americas* 3:37
Benin—*Africa* 2:21
Bermuda—*Americas:* UK American Dependencies
Bhutan—*Asia* 4:51
Bolivia—*Americas* 3:43
Bonin Islands—*Asia:* Japan (Ogasawara Islands)
Borneo, North—*Asia:* Malaysia
Bosnia and Herzegovina—*Europe* 5:55
Botswana—*Africa* 2:31

Bouvet Island—*Europe:* Norway
Brazil—*Americas* 3:55
British Antarctic Territory—*Americas:* UK American Dependencies
British Guiana—*Americas:* Guyana
British Honduras—*Americas:* Belize
British Indian Ocean Territory—*Africa:* UK African Dependencies
British Virgin Islands—*Americas:* UK American Dependencies
Brunei Darussalam—*Asia* 4:57
Bulgaria—*Europe* 5:71
Burkina Faso—*Africa* 2:39
Burma—*Asia:* Myanmar
Burundi—*Africa* 2:47

Caicos Islands—*Americas:* Turks and Caicos Islands
Cambodia—*Asia* 4:63
Cameroon—*Africa* 2:55
Canada—*Americas* 3:73
Canary Islands—*Europe:* Spain
Cape Verde—*Africa* 2:65
Caroline Islands—*Asia:* Federated States of Micronesia; Palau
Carriacou—*Americas:* Grenada
Cayman Islands—*Americas:* UK American Dependencies
Central African Republic—*Africa* 2:71
Ceuta—*Europe:* Spain
Ceylon—*Asia:* Sri Lanka
Chad—*Africa* 2:79
Chile—*Americas* 3:93
Chilean Antarctic Territory—*Americas:* Chile
China—*Asia* 4:79
Christmas Island (Indian Ocean)—*Asia:* Australia
Christmas Island (Pacific Ocean)—*Asia:* Kiribati
Cocos Islands—*Americas:* Costa Rica
Cocos (Keeling) Islands—*Asia:* Australia
Colombia—*Americas* 3:107
Columbus, Archipelago of—*Americas:* Ecuador (Galapagos Islands)
Comoros—*Africa* 2:87
Congo—*Africa* 2:93
Cook Islands—*Asia:* New Zealand
Coral Sea Islands—*Asia:* Australia
Corn Islands—*Americas:* Nicaragua

Costa Rica—*Americas* 3:121
Côte d'Ivoire—*Africa* 2:101
Croatia—*Europe* 5:83
Cuba—*Americas* 3:131
Curaçao—*Americas:* Netherlands American Dependencies:
  Netherlands Antilles
Cyprus—*Asia* 4:103
Czech Republic—*Europe* 5:99
Czechoslovakia—*Europe:* Czech Republic; Slovakia

Denmark—*Europe* 5:109
Diego Garcia—*Africa:* UK African Dependencies:
  British Indian Ocean Territory
Diego Ramirez Island—*Americas:* Chile
Djibouti—*Africa* 2:111
Dominica—*Americas* 3:141
Dominican Republic—*Americas* 3:147
Dubai—*Asia:* United Arab Emirates
Dutch Guiana—*Americas:* Suriname
Easter Island—*Americas:* Chile
East Germany—*Europe:* German Democratic Republic
Ecuador—*Americas* 3:157
Egypt—*Africa* 2:117
El Salvador—*Americas* 3:169
England—*Europe:* UK
Equatorial Guinea—*Africa* 2:131
Eritrea—*Africa* 2:137
Estonia—*Europe* 5:123
Ethiopia—*Africa* 2:143

Falkland Islands—*Americas:* UK American Dependencies
  (Malvinas)
Faroe Islands—*Europe:* Denmark
Federated States of Micronesia—*Asia* 4:113
Fiji—*Asia* 4:119
Finland—*Europe* 5:129
Formosa—*Asia:* Taiwan
France—*Europe* 5:141
French African Dependencies—*Africa* 2:153
French American Dependencies—*Americas* 3:179
French Guiana—*Americas:* French American Dependencies
French Pacific Dependencies—*Asia* 4:125
French Polynesia—*Asia:* French Pacific Dependencies
French Somaliland—*Africa:* Djibouti
French Southern and Antarctic Lands—*Asia:* French
  Pacific Dependencies
Fujairah—*Asia:* United Arab Emirates

Gabon—*Africa* 2:155
Galapagos Islands—*Americas:* Ecuador
Gambia—*Africa* 2:165
Georgia—*Europe* 5:161
German Democratic Republic (GDR)—*Europe:* Germany
Germany—*Europe* 5:167
Germany, East—*Europe:* Germany
Germany, Federal Republic of (FRG)—*Europe:* Germany
Germany, West—*Europe:* Germany
Ghana—*Africa* 2:171
Gibraltar—*Europe:* UK
Gilbert Islands—*Asia:* Kiribati
Graham Land—*Americas:* UK American Dependencies:
  British Antarctic Territory
Great Britain—*Europe:* UK
Greece—*Europe* 5:183
Greenland—*Europe:* Denmark
Grenada—*Americas* 3:181
Grenadines—*Americas:* Grenada; St. Vincent and the Grenadines

Guadeloupe—*Americas:* French American Dependencies
Guam—*Asia:* US Pacific Dependencies
Guatemala—*Americas* 3:187
Guiana, British—*Americas:* Guyana
Guiana, Dutch—*Americas:* Suriname
Guiana, French—*Americas:* French American Dependencies
Guinea—*Africa* 2:181
Guinea-Bissau—*Africa* 2:191
Guinea, Portuguese—*Africa:* Guinea-Bissau
Guyana—*Americas* 3:197

Haiti—*Americas* 3:205
Heard and McDonald Islands—*Asia:* Australia
Honduras—*Americas* 3:215
Honduras, British—*Americas:* Belize
Hong Kong—*Asia:* UK Asian and Pacific Dependencies
Howland, Baker, and Jarvis Islands—*Asia:* US Pacific
  Dependencies
Hungary—*Europe* 5:195

Iceland—*Europe* 5:207
Ifni—*Africa:* Morocco
India—*Asia:* 4:127
Indochina—*Asia:* Cambodia; Laos; Viet Nam
Indonesia—*Asia* 4:147
Inner Mongolia—*Asia:* China
Iran—*Asia* 4:165
Iraq—*Asia* 4:179
Ireland—*Europe* 5:217
Ireland, Northern—*Europe:* UK
Irian Jaya—*Asia:* Indonesia
Israel—*Asia* 4:189
Italy—*Europe* 5:229
Ivory Coast—*Africa:* Côte D'Ivoire

Jamaica—*Americas* 3:225
Jammu and Kashmir—*Asia:* India; Pakistan
Jan Mayen Island—*Europe:* Norway
Japan—*Asia* 4:203
Johnston Atoll—*Asia:* US Pacific Dependencies
Jordan—*Asia* 4:221
Juan Fernandez Island—*Americas:* Chile

Kampuchea—*Asia:* Cambodia
Kashmir—*Asia:* India; Pakistan
Kazakhstan—*Asia* 4:231
Kazan Islands—*Asia:* Japan (Volcano Islands)
Kenya—*Africa* 2:197
Khmer Republic—*Asia:* Cambodia
Kiribati—*Asia* 4:239
Korea, Democratic People's Republic of (DPRK)—*Asia* 4:245
Korea, North—*Asia:* Korea, Democratic People's Republic of
Korea, Republic of (ROK)—*Asia* 4:255
Korea, South—*Asia:* Korea, Republic of
Kuwait—*Asia* 4:265
Kyrgzstan—*Asia* 4:273

Laccadive, Minicoy, and Amindivi Islands—*Asia:* India:
  Lakshadweep
Lakshadweep—*Asia:* India
Lao People's Democratic Republic—*Asia* 4:281
Laos—*Asia:* Lao People's Democratic Republic
Latvia—*Europe* 5:243
Lebanon—*Asia* 4:291
Leeward Islands—*Americas:* UK American Dependencies;
  Antigua and Barbuda; St. Kitts and Nevis
Lesotho—*Africa* 2:209

Liberia—*Africa* 2:217
Libya—*Africa* 2:225
Liechtenstein—*Europe* 5:249
Line Islands—*Asia:* Kiribati
Lithuania—*Europe* 5:255
Luxembourg—*Europe* 5:261

Macau—*Asia:* Portuguese Asian Dependency
Macedonia—*Europe* 5: 269
Macquarie Island—*Asia:* Australia
Madagascar—*Africa* 2:235
Madeira—*Europe:* Portugal
Malagasy Republic—*Africa:* Madagascar
Malawi—*Africa* 2:245
Malaya—*Asia:* Malaysia
Malaysia—*Asia* 4:301
Malden and Starbuck Islands—*Asia:* Kiribati
Maldive Islands—*Asia:* Maldives
Maldives—*Asia* 4:315
Mali—*Africa* 2:253
Malta—*Europe* 5:277
Malvinas—*Americas:* UK American Dependencies
  (Falkland Islands)
Mariana Islands—*Asia:* US Pacific Dependencies
Marquesas Islands—*Asia:* French Pacific Dependencies:
  French Polynesia
Marshall Islands—*Asia* 4:321
Martinique—*Americas:* French American Dependencies
Matsu Islands—*Asia:* Taiwan
Mauritania—*African* 2:263
Mauritius—*Africa* 2:271
Mayotte—*Africa:* French African Dependencies
Melilla—*Europe:* Spain
Mexico—*Americas* 3:1235
Micronesia, Federated States of—*Asia:* Federated States
  of Micronesia
Midway—*Asia:* US Pacific Dependencies
Moldova—*Europe* 5:283
Monaco—*Europe* 5:287
Mongolia—*Asia* 4:327
Montserrat—*Americas:* UK American Dependencies:
  Leeward Islands
Morocco—*Africa* 2:277
Mozambique—*Africa* 2:289
Muscat and Oman—*Asia:* Oman
Myanmar—*Asia* 4:335

Namibia—*Africa* 2:297
Nauru—*Asia* 4:349
Navassa—*Americas:* US
Nepal—*Asia* 4:355
Netherlands—*Europe* 5:293
Netherlands American Dependencies—*Americas* 3:251
Netherlands Antilles—*Americas:* Netherlands American
  Dependencies
Nevis—*Americas:* St. Kitts and Nevis
New Caledonia—*Asia:* French Pacific Dependencies
New Guinea—*Asia:* Papua New Guinea
New Hebrides—*Asia:* Vanuatu
New Zealand—*Asia* 4:365
Nicaragua—*Americas* 3:253
Nicobar Islands—*Asia:* India
Niger—*Africa* 2:305
Nigeria—*Africa* 2:313
Niue—*Asia:* New Zealand
Norfolk Island—*Asia:* Australia
North Borneo—*Asia:* Malaysia

Northern Ireland—*Europe:* UK
Northern Mariana Islands—*Asia:* US Pacific Dependencies
Northern Rhodesia—*Africa:* Zambia
North Korea—*Asia:* Korea, Democratic People's Republic of
North Viet-Nam—*Asia:* Viet Nam
Northwest Territories—*Americas:* Canada
Norway—*Europe* 5:307
Nosy Boraha and Nosy Be—*Africa:* Madagascar
Nyasaland—*Africa:* Malawi

Ocean Island—*Asia:* Kiribati (Banaba)
Ogasawara Islands—*Asia:* Japan (Bonin Islands)
Okinawa—*Asia:* Japan
Oman—*Asia* 4:379
Outer Mongolia—*Asia:* Mongolia

Pacific Islands, Trust Territory of the—*Asia:* Federated States of
  Micronesia; Marshall Islands; Palau; US Pacific Dependencies
Pakistan—*Asia* 4:385
Pakistan, East—*Asia:* Bangladesh
Palau—*Asia* 4:399
Palmyra Atoll—*Asia:* US Pacific Dependencies
Panama—*Americas* 3:263
Papua New Guinea—*Asia* 4:403
Paracel Islands—*Asia:* China (Xisha Islands)
Paraguay—*Americas* 3:2273
Peru—*Americas* 3:283
Peter I Island—*Europe:* Norway
Petit Martinique—*Americas:* Grenada
Philippines—*Asia* 4:411
Phoenix Islands—*Asia:* Kiribati
Pitcairn Island—*Asia:* UK Asian and Pacific Dependencies
Poland—*Europe* 5:321
Polar Regions—*United Nations* 1:213
Portugal—*Europe* 5:335
Portuguese Asian Dependency—*Asia* 4:425
Portuguese Timor—*Asia:* Indonesia
Puerto Rico—*Americas:* US

Qatar—*Asia* 4:427
Queen Maud Land—*Europe:* Norway
Quemoy Islands—*Asia:* Taiwan

Ras al-Khaimah—*Asia:* United Arab Emirates
Réunion—*Africa:* French African Dependencies
Rhodesia—*Africa:* Zimbabwe
Río Muni—*Africa:* Equatorial Guinea
Romania—*Europe* 5:345
Ross Dependency—*Asia:* New Zealand
Ruanda-Urundi—*Africa:* Burundi; Rwanda
Russia—*Europe* 5:357
Rwanda—*Africa* 2:325
Ryukyu Islands—*Asia:* Japan

Sabah—*Asia:* Malaysia
St. Christopher—*Americas:* St. Kitts and Nevis
St. Christopher and Nevis—*Americas:* St. Kitts and Nevis
St. Helena—*Africa:* UK African Dependencies
St. Kitts—*Americas:* St. Kitts and Nevis
St. Kitts and Nevis—*Americas* 3:299
St. Lucia—*Americas* 3:305
St. Pierre and Miquelon—*Americas:* French American
  Dependencies
St. Vincent and the Grenadines—*Americas* 3:311
Sala y Gómez Island—*Americas:* Chile
Samoa, American—*Asia:* US Pacific Dependencies
Samoa, Western—*Asia:* Western Samoa

San Ambrosio Island—*Americas:* Chile
San Andrés and Providentia—*Americas:* Colombia
San Felix Island—*Americas:* Chile
San Marino—*Europe* 5:367
São Tomé and Príncipe—*Africa* 2:333
Sarawak—*Asia:* Malaysia
Sa'udi Arabia—*Asia* 4:433
Scotland—*Europe:* UK
Senegal—*Africa* 2:339
Serbia and Montenegro—*Europe* 5:371
Seychelles—*Africa* 2:349
Sharjah—*Asia:* United Arab Emirates
Sierra Leone—*Africa* 2:355
Sikkim—*Asia:* India
Singapore—*Asia* 4:443
Slovakia—*Europe* 5:383
Slovenia—*Europe* 5:391
Society Islands—*Asia:* French Pacific Dependencies:
  French Polynesia
Solomon Islands—*Asia* 4:455
Somalia—*Africa* 2:365
Somaliland, French—*Africa:* Djibouti
South Africa—*Africa* 2:375
South Arabia, Federation of—*Asia:* Yemen
Southern Rhodesia—*Africa:* Zimbabwe
Southern Yemen—*Asia:* Yemen
South Georgia—*Americas:* UK American Dependencies:
  Falkland Islands
South Korea—*Asia:* Korea, Republic of
South Viet-Nam—*Asia:* Viet Nam
South West Africa—*Africa:* Namibia
Spain—*Europe* 5:403
Spanish Guinea—*Africa:* Equatorial Guinea
Spanish Sahara—*Africa:* Morocco
Spratly Islands—*Asia:* Viet-Nam
Sri Lanka—*Asia* 4:461
Sudan—*Africa* 2:391
Suriname—*Americas* 3:317
Svalbard—*Europe:* Norway
Swan Islands—*Americas:* US
Swaziland—*Africa* 2:401
Sweden—*Europe* 5:415
Switzerland—*Europe* 5:429
Syria—*Asia* 4:473

Tahiti—*Asia:* French Pacific Dependencies: French Polynesia
Taiwan—*Asia* 4:483
Tajikistan—*Asia* 4:495
Tanganyika—*Africa:* Tanzania
Tanzania—*Africa* 2:407
Thailand—*Asia* 4:501
Tibet—*Asia:* China
Timor, East—*Asia:* Indonesia
Tobago—*Americas:* Trinidad and Tobago
Togo—*Africa* 2:417
Tokelau Islands—*Asia:* New Zealand
Tonga—*Asia* 4:515
Transkei—*Africa:* South African Homelands
Trinidad and Tobago—*Americas* 3:323
Tristan da Cunha—*Africa:* UK African Dependencies:
  St. Helena
Trust Territory of the Pacific Islands—*Asia:* Federated
States of Micronesia; Marshall Islands; Palau; US Pacific
  Dependencies
Tuamotu Islands—*Asia:* French Asian Dependencies:
  French Polynesia
Tunisia—*Africa* 2:425
Turkey—*Asia* 4:521
Turkmenistan—*Asia* 4:535
Turks and Caicos Islands—*Americas* 3:331
Tuvalu—*Asia* 4:541

Uganda—*Africa* 2:435
Ukraine—*Europe* 5:439
Umm al-Qaiwain—*Asia:* United Arab Emirates
Union of Soviet Socialist Republics (USSR)—*Asia:* Azerbaijan;
  Kazakhstan; Kyrgzstan; Tajikistan; Turkmenistan; Uzbekistan;
  Europe: Armenia; Belarus; Estonia; Georgia; Latvia; Lithuania;
  Moldova; Russia; Ukraine
United Arab Emirates (UAE)—*Asia* 4:545
United Arab Republic—*Africa:* Egypt
United Kingdom (UK)—*Europe* 5:447
United Kingdom African Dependencies—*Africa* 2:443
United Kingdom American Dependencies—*Americas* 3:335
United Kingdom Asian and Pacific Dependencies—*Asia* 4:553
United States of America (US)—*Americas* 3:359
United States Pacific Dependencies—*Asia* 4:555
Upper Volta—*Africa:* Burkina Faso
Uruguay—*Americas* 3:375
Uzbekistan—*Asia* 4:559

Vanuatu—*Asia* 4:565
Vatican—*Europe* 5:469
Venda—*Africa:* South African Homelands
Venezuela—*Americas* 3:385
Viet Nam—*Asia* 4:571
Viet-Nam, North—*Asia:* Viet Nam
Viet-Nam, South—*Asia:* Viet Nam
Virgin Islands, British—*Americas:* UK American Dependencies
Virgin Islands of the US—*Americas:* US
Volcano Islands—*Asia:* Japan (Kazan Islands)

Wake Island—*Asia:* US Pacific Dependencies
Wales—*Europe:* UK
Wallis and Futuna—*Asia:* French Asian Dependencies
Western Sahara—*Africa:* Morocco
Western Samoa—*Asia* 4:587
West Germany—*Europe:* Germany, Federal Republic of
West Irian—*Asia:* Indonesia
Windward Islands—*Americas:* Dominica; St. Lucia; St. Vincent
  and the Grenadines

Xisha Islands—*Asia:* China (Paracel Islands)

Yemen, People's Democratic Republic of (PDRY)—*Asia:* Yemen
Yemen, Republic of—*Asia* 4:593
Yemen Arab Republic (YAR)—*Asia:* Yemen
Yugoslavia—*Europe:* Bosnia and Herzegovina; Croatia;
  Macedonia; Slovenia
Yukon Territory—*Americas:* Canada

Zaire—*Africa* 2:445
Zambia—*Africa* 2:457
Zimbabwe—*Africa* 2:467

ISBN 0-8103-9881-8

90000

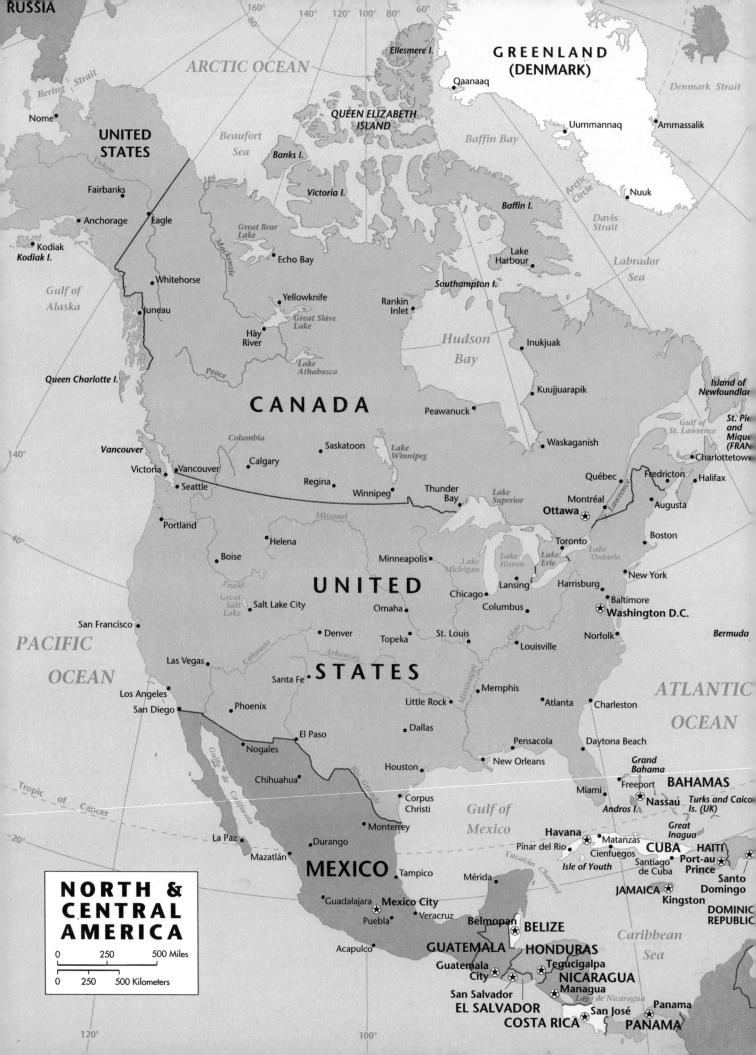

RUSSIA

ARCTIC OCEAN

GREENLAND
(DENMARK)

Bering Strait

Nome

UNITED
STATES

Fairbanks

Anchorage
Eagle

Kodiak
Kodiak I.

Whitehorse

*Gulf of
Alaska*

Juneau

Queen Charlotte I.

*Beaufort
Sea*

Ellesmere I.

Qaanaaq

QUEEN ELIZABETH
ISLAND

Uummannaq

Ammassalik

Denmark Strait

Banks I.

Victoria I.

Baffin Bay

Arctic Circle

Nuuk

Echo Bay

Mackenzie

Yellowknife

*Great Bear
Lake*

Håy
River

*Great Slave
Lake*

*Lake
Athabasca*

Peace

Columbia

Saskatoon

Calgary

Regina

Baffin I.

Lake
Harbour

*Davis
Strait*

*Labrador
Sea*

Rankin
Inlet

Southampton I.

*Hudson
Bay*

Inukjuak

Kuujjuarapik

CANADA

Peawanuck

Waskaganish

*Island of
Newfoundlan*

Victoria
Vancouver

Vancouver

Seattle

Portland

Boise

Helena

*Lake
Winnipeg*

Winnipeg

Thunder
Bay

*Lake
Superior*

Missouri

Minneapolis

*Lake
Michigan*

*Lake
Huron*

Lansing

Chicago

Québec

Montréal

St. Lawrence

Ottawa

Toronto

*Lake
Erie*

*Lake
Ontario*

*Gulf of
St. Lawrence*

*St. Pie
and
Mique
(FRAN*

Charlottetow

Fredricton

Halifax

Augusta

Boston

New York

PACIFIC

OCEAN

San Francisco

Las Vegas

Los Angeles

San Diego

Snake

*Great
Salt
Lake*

Salt Lake City

UNITED

Denver

Colorado

Phoenix

Nogales

Chihuahua

Gulfo de California

La Paz

Mazatlán

Durango

Omaha

Topeka

Santa Fe

STATES

El Paso

Rio Grande

Corpus
Christi

Monterrey

MEXICO

Guadalajara

Puebla

Acapulco

Mexico City

Veracruz

St. Louis

Arkansas

Little Rock

Dallas

Houston

Columbus

Ohio

Louisville

Memphis

Mississippi

Atlanta

Pensacola

New Orleans

Harrisburg

Baltimore

Washington D.C.

Norfolk

Charleston

Daytona Beach

*Gulf of
Mexico*

Tampico

Mérida

Yucatán Channel

Miami

Andros I.

Havana

Pinar del Rio

Isle of Youth

Belmopan

BELIZE

GUATEMALA

HONDURAS

Guatemala
City

San Salvador

EL SALVADOR

Tegucigalpa

NICARAGUA

Managua

Lago de Nicaragua

San José

COSTA RICA

PANAMA

Bermuda

ATLANTIC

OCEAN

*Grand
Bahama*

Freeport

BAHAMAS

Nassau

*Turks and Caico
Is. (UK)*

*Great
Inagua*

Matanzas

Cienfuegos

CUBA

HAITI

Santiago
de Cuba

Port-au-
Prince

JAMAICA

Kingston

Santo
Domingo

DOMINICA
REPUBLIC

*Caribbean
Sea*

Panama

PANAMA

Tropic of Cancer

160°   140°   120°   100°   80°   60°

80°

60°

40°

140°

40°

20°

20°

120°

100°

NORTH &
CENTRAL
AMERICA

| 0 | 250 | 500 Miles |

| 0 | 250 | 500 Kilometers |